ENCYCLOPEDIA OF PHYSICS

EDITED BY

S. FLÜGGE

VOLUME XV

LOW TEMPERATURE PHYSICS II

WITH 318 FIGURES

Springer-Verlag Berlin Heidelberg GmbH
1956

HANDBUCH DER PHYSIK

HERAUSGEGEBEN VON

S. FLÜGGE

BAND XV

KÄLTEPHYSIK II

MIT 318 FIGUREN

Springer-Verlag Berlin Heidelberg GmbH
1956

ISBN 978-3-662-38851-8   ISBN 978-3-662-39773-2 (eBook)
DOI 10.1007/978-3-662-39773-2

Softcover reprint of the hardcover 1st edition 1956

Contents.

Low Temperature Magnetism.

By

J. VAN DEN HANDEL.

With 32 Figures.

This article is intended as an introduction to the following contributions of this volume. Therefore it does not give a complete treatment of magnetism, but only deals with those subjects which are important at low temperatures and for whose study low temperatures are necessary. Diamagnetism, ferromagnetism and the magnetic properties of metals will not be considered [1].

I. Introduction.

1. Historical remarks. The temperature dependence of the magnetic susceptibilities in general gives information on the existence and the magnitude of free magnetic moments, μ, as was studied in the first phase of magnetic research. It also gives at low temperatures in many cases more detailed information on the surroundings of the magnetic ions as their interaction with the neighbours influences the structure of the lowest energy levels. Especially after the second world war, this subject has been intensively studied and special methods were discovered and developed in order to get more knowledge of these details.

Starting with a few remarks on the first phase, we must primarily mention the fundamental work by P. CURIE, who found that for many salts the dependence of the magnetic susceptibility, χ — that is the magnetization, σ, devided by the magnetic fieldstrength, H — on the temperature, T, can be represented by a law of the form

$$\chi = \frac{C}{T} \text{ (CURIE's law)}$$

where C is a constant depending on the salt used. A theoretical explanation was given by P. LANGEVIN according to which the increasing order of the magnetic dipoles oriented by the magnetic field gives rise to an increase of χ for decreasing temperature. This theory was extended by P. WEISS who assumed an interaction between the dipoles, the effect of which he represented by an extra magnetic field proportional to the existing magnetization. In this way he was able to explain the existence of ferromagnetism below a special temperature, the CURIE point, T_C. From his formula it is seen that for temperatures higher than T_C the ferromagnetic substance has a paramagnetic behaviour. Here the susceptibilities can be represented by the "CURIE-WEISS law",

$$\chi = \frac{C}{T - \Theta},$$

where $\Theta = T_C$. As it was found that this law was obeyed by many paramagnetic substances as well, it was thought that here too a similar explanation could be used, the only difference being that the interactions were much weaker. In

---

[1] Detailed articles on magnetism will be found in vol. XVIII of this Encyclopedia.

this theory the value of C proves to be the same as in the case where no inter-actions exist. It is

$$C = \frac{N \mu^2}{3 k T},$$

where N is the number of magnetic dipoles in the amount of substance being considered and k is Boltzmann's constant. For the following C will be taken per gram-atom, thus N is Avogadro's number. The susceptibility per gram-atom will be called χ_A, whereas χ and $\varkappa$ indicate the susceptibility per gram and per cm³ respectively. From the temperature dependence of χ_A the value of μ can be found. The measurements at low temperatures have proved to be especially useful as more accuracy was attainable because of the higher values of χ_A.

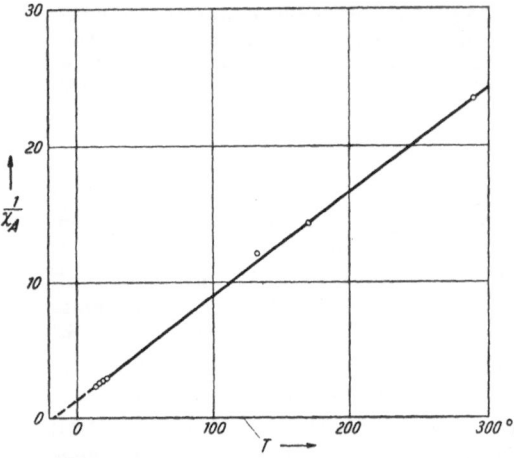

Fig. 1. $1/\chi_A$ as a function of T for Dy₂O₃ which follows a Curie-Weiss law.

The results of the measurements of the magnetic moments were in good agreement with those of the calculations following the rules given by Hund in the case of the ions of the rare earths (with the exception of Sm⁺⁺⁺ and Eu⁺⁺⁺), but a complete failure of these rules was found for the ions of the elements of the iron group (the ions from Ti⁺⁺⁺ to Cu⁺⁺)[1]. In this group a much better agreement was obtained with the calculations based on the assumption made by D.M. Bose and by E.C. Stoner that only the spins of the electrons of the incomplete shell, and not their orbital motions, contribute to the magnetic moments. Especially in the first half of this group, this assumption found good confirmation.

In most cases a Curie-Weiss law was followed. The values of $1/\chi$ vs T then gave rise to a straight line when plotted against one another (see Fig. 1). At lower temperatures, however, (below the temperatures of liquid nitrogen for many salts) deviations from this law were found, called by Kamerlingh Onnes "cryo-magnetic anomalies" [1] (comp. Fig. 2).

It was not before more material at very low temperatures was available that a new phase in the study of magnetism began. To start with, there was the suggestion of Becquerel [2] and of Miss Brunetti [3] that the deviations from the behaviour of free magnetic dipoles originated in the influence of the inhomo-geneous electric fields of the surrounding ions on the magnetic ion. This idea was worked out by Bethe [4] in a general way. He showed that the influence of these fields could result in a partial or complete removal of the degeneracy of the energy levels of the free magnetic ions. As an extension of this develop-ment, Kramers [5] showed that for ions with an odd number of electrons in the incomplete shell giving rise to the magnetic properties, the inhomogeneous electric fields can not completely remove the degeneracy. The levels must in that case be at least double (Kramers degeneracy). Only a magnetic field can remove this last degeneracy. When an even number of electrons is responsible for the magnetic moment, the levels can be single so that no degeneracy is left.

---

[1] Though Pd, Pt and U also give magnetic ions they will not be considered here as their cases are more complicated and practically no low temperature work has been done with them.

In many cases the energetic distances of these splittings are of the order of kT at room temperature or smaller. It is evident that it is then useful to extend the measurements to temperatures which are much lower than those which are characteristic for the splitting.

About 1930 VAN VLECK [6], [7] gave a systematic development of the theory of magnetism using the new quantum mechanics. He and his students worked out on the basis of this theory many special cases, and so did KRAMERS and his students, especially in connection with the experimental work of BECQUEREL and coworkers in the KAMERLINGH ONNES Laboratory in Leiden. After the second world war much work in this field was done by PRYCE and his coworkers in Oxford in connection with the experimental work in the Clarendon Laboratory.

2. Brief review of the newer research methods. In the mean time methods other than direct susceptibility measurements were introduced for the study of the magnetic properties. They had to do with the phenomenon of magnetization itself, or with the study of the splittings of the energy levels of the ions whose magnetic properties were studied. The magnetization was studied by means of the *paramagnetic relaxation*, especially by GORTER and coworkers. This phenomenon originates in the fact that the process of magnetization is not instantaneous. As a result there exist phase differences between magnetization and fieldstrength in alternating fields and these cause losses, whereas, on the other hand, the susceptibility[1] depends on the frequency. These two phenomena are called the paramagnetic absorption and dispersion. Their study gives among other things information on the interaction of the magnetic ions with their surroundings and on the energy transfer from the magnetic dipoles to the crystalline lattice and viceversa. This transfer becomes less effective when the temperature decreases and therefore this effect can best be studied at low temperatures.

The splitting of the lowest levels of the ions can be found very well by means of the *paramagnetic resonance*. When a substance with a separation δ between two low lying levels is placed in a magnetic field with a frequency $\nu = \delta/h$, energy can often be absorbed from this alternating field. The experimental methods, developed since the war with the aid of radartechniques, also opened possibilities to study these separations, which are in many cases of the order of about 1 cm^{-1}, calling for frequencies of about 3×10^{10} sec^{-1}, corresponding to a wavelength of 1 cm. This "spectroscopy" at centimeter wavelengths has already given much information that was unobtainable from the susceptibility measurements. Many laboratories are working in this field. Only the work of BLEANEY and coworkers in Oxford will be referred to in this introduction. As the levels become, in general, narrower at low temperatures, this method gives more results at low than at high temperatures. The shape of the absorption curves, when the absorption is plotted as a function of frequency, gives information about the interactions of the magnetic ions with each other, sometimes through intermediary ions.

A field of research that has also many connections with the others is that of *adiabatic demagnetization*. The splittings of the lowest levels are especially important here because they determine the specific heat. This field will be treated separately in the following article by DE KLERK.

In the following sections, we shall treat the various methods in more detail and some results will be given. Also, some words will be said about antiferromagnetism, as it seems that many paramagnetic salts become antiferromagnetic at low temperatures.

---

[1] A correct definition of the susceptibility will be given in Chapter IV.

II. Effects of magnetic and of electric fields on the energy levels of the magnetic ions.

3. Short review of Van Vleck's theory [6], [7]. In a free ion, the lowest energy level, characterized by a value j for the total angular momentum, a value which can be calculated using the rules derived by Hund from the analysis of the spectra, is $(2j+1)$ times degenerate. In a magnetic field, this degeneracy is removed and a group of $(2j+1)$ equidistant levels is found. Their mutual separation is $g\,\mu_B H$ where $\mu_B = \dfrac{e}{2mc}\dfrac{h}{2\pi}$, the Bohr magneton, and g is Landé's splitting factor. When $g=2$ and $H=10000$ Oe, the separation is $0.9\ \mathrm{cm^{-1}}$. These levels are characterized by m_j, which can have the values $j, j-1, \ldots, -j$, corresponding to values $m_j\,g\,\mu_B$ of the components μ_H of the magnetic moment in the direction of H. The total magnetization per gram-mol, σ, is therefore:

$$\sigma = N\mu_B g\,\frac{\sum\limits_{m_j} m_j\, e^{m_j g \mu_B H/kT}}{\sum\limits_{m_j} e^{m_j g \mu_B H/kT}} = N\mu_B g j\, B_j\!\left(\frac{g j \mu_B H}{kT}\right). \qquad (3.1)$$

B_j is called a Brillouin function. It can also be written as follows:

$$B_j(y) = \frac{2j+1}{2j}\,\mathrm{Cot}\,\frac{(2j+1)\,y}{2j} - \frac{1}{2j}\,\mathrm{Cot}\,\frac{y}{2j}.$$

In reality one is not concerned with free ions, but usually with ions placed in crystal lattices.

In the paramagnetic case, where exchange interactions can be neglected, the influence of the lattice is threefold: 1. The magnetic field acting on one dipole is not constant but is continuously varying because of magnetic interactions with the neighbours, which are changing their orientation and which precess about the magnetic field in which they are placed. It may be remarked here that even when no external field is applied, the field for one ion is not always zero and an r.m.s. field H_i can be introduced. 2. The average field in which the magnetic ions are placed is usually not equal to the external field. The difference depends on the shape of the sample used (because of the demagnetizing field) and on the crystal structure (the average magnetic field at the position of the dipoles depends on the relative positions and mean orientation of the neighbours). 3. The other ions, including the non magnetic ones and especially the water-molecules, set up an inhomogeneous electric field at the position of the magnetic dipole being considered.

The results of these effects are: 1. A broadening of the energy levels as a result of the fluctuations mentioned. Two reasons can be given. The first is due to the fact that the magnetic field at the position of the magnetic ions varies from position to position in the lattice at any one time; the second is found in the transitions induced by the time variations of the field. 2. A displacement of the energy levels in a magnetic field. This effect is accounted for by applying a correction, $\varepsilon\sigma$, proportional to the magnetization, σ, to the value of the magnetic field (cf. Sect. 7). 3. A partial or complete removal of the degeneracy. In the next section this splitting will be considered somewhat more closely. For the moment we accept the fact of its existence. We consider now, with Van Vleck, the case, where this splitting is such, that the levels which are present have energy separations which are either small or large compared with kT. The large distances will be characterized by the quantum number n. In that case Van Vleck has derived, for values of H/T for which no saturation effects occur, the following results which we shall write down here without any derivation. (A more extended treatment of VAN VLECK's theory is found in Vol. XVIII.)

$$\sigma = N \left(\frac{\mu^2}{3kT} + \alpha \right) H$$

with

$$N\alpha = \frac{2}{3} N \sum_{n'(\neq n)} \frac{|\mu^0(n\,n')|^2}{h\nu(n'\,n)}. \tag{3.2}$$

Diamagnetism is neglected here, as it is in general unimportant at low temperatures. $\mu^0(n\,n')$ are the high frequency elements of the magnetic moment matrix, the index ^{0} indicating that this matrix is taken when no external field is applied. μ is the value of the permanent magnetic moment. Its matrix is formed from that of μ^0 by dropping the high-frequency elements. For the rare earth elements the intervals between the levels of the lowest multiplet are in general between 10^3 and 10^4 cm^{-1}. Only for Sm and Eu are the distances between the lowest two multiplet levels smaller than 10^3 cm^{-1} (for Eu even smaller than 300 cm^{-1}). Therefore we are sure that, except for Sm and Eu, only the lowest multiplet level is occupied at room temperature and below. In the case of the elements of the iron group the situation is different. For the free ions the splitting of the lowest multiplet is narrower, but in crystals (and it is only the crystalline state that is dealt with here) the influence of the crystalline field causes a decoupling of the orbital and spin magnetism, so that one may consider first only the orbital moment and then treat the influence of the spin as a perturbation. The distance of the lowest orbital state to the next one is several times 10^4 cm^{-1} so that at temperatures below room temperature only the lowest orbital state is occupied. This state is split by the crystalline field and the result is, in general, that only a single, thus non magnetic, lowest level is occupied at temperatures in the region of liquid air or lower. In that case only the effect of the spin remains to a first approximation. This spin level is again split by a combined effect of spin orbit and crystalline field interactions.

As long as the crystalline field splitting of the lowest level of the rare earth or iron group ion is small compared to kT, Eq. (3.2) can be used. In the rare earth group the value of μ^2 is then equal to $g^2 j(j+1) \mu_B^2$. When, however, the temperature is so low that kT becomes smaller than the mutual distances of the sublevels, each of these has to be considered separately and the effect of each, weighted with its own population, has to be taken.

One starts, then, with:

$$\chi_A = -\frac{N}{H} \frac{\sum \frac{\partial W}{\partial H} e^{-W/kT}}{\sum e^{-W/kT}}. \tag{3.3}$$

W is the energy of a state and $\partial W/\partial H$ the average value of the component in the direction of H of its magnetic moment.

With $W = W_0 + W_1 H + W_2 H^2 + \cdots$ (3.3) can be transformed into

$$\chi_A = N \frac{\sum \left(\frac{W_1^2}{kT} - 2W_2 \right) e^{-W/kT}}{\sum e^{-W/kT}}. \tag{3.4}$$

In order to be able to calculate this expression, it is necessary to know something more about the splitting pattern. As will be seen in Sect. 4, this exhibits in many cases a group of double levels. When the distances of the second, third, ... doublet from the lowest are $\delta_2, \delta_3, \ldots$ one can write

$$\sigma = N \frac{(\mu_1 + \nu_1 H) e^{\mu_1 H/kT} + (-\mu_1 + \nu_1 H) e^{-\mu_1 H/kT} + [(\mu_2 + \nu_2 H) e^{\mu_2 H/kT} + (-\mu_2 + \nu_2 H) e^{-\mu_2 H/kT}] e^{-\delta_2/kT} + \cdots}{e^{\mu_1 H/kT} + e^{-\mu_1 H/kT} + [e^{\mu_2 H/kT} + e^{-\mu_2 H/kT}] e^{-\delta_2/kT} + \cdots}. \tag{3.5}$$

The μ_i are the components of the magnetic moment in the direction of H connected with the different states. When, instead of a double level, a single one occurs, the part between square brackets becomes $\nu_i H$.

When kT is much smaller than δ_2 the higher doublets have no longer any direct influence and the expression reduces to

$$\sigma = N \left(\mu_1 \operatorname{Tan} \frac{\mu_1 H}{kT} + \nu_1 H \right). \tag{3.6}$$

Moreover, as long as $\mu_1 H/kT \ll 1$ one can write as well

$$\sigma = N \left(\frac{\mu_1^2 H}{kT} + \nu_1 H \right). \tag{3.7}$$

For higher temperatures Eq. (3.2) is valid. The transition from this equation to (3.7) gives rise to the cryomagnetic anomalies.

It may be remarked here that, because of the law of the "spectroscopic stability", a quantum mechanical sum rule (see ref. [7]), Eq. (3.2) is valid in a rather wide temperature region, independently of the exact splitting-pattern as long as the splittings are smaller than kT.

4. Splitting of levels in a crystalline electric field. Bethe [4] studied the influence of electric fields of different symmetry on the energy levels of magnetic ions.

The state of a magnetic ion can be found from the Schrödinger equation $\mathscr{H}\psi = E\psi$ where $\mathscr{H}$ is the Hamiltonian. For a free ion the states may be degenerate, but when the ion is placed in a crystalline field this degeneracy is in general reduced. The way in which it is reduced depends on the symmetry of this field. When the ion is rotated over a special angle (e.g. $\pi/2$ round a quaternary axis, $\pi/3$ round a hexagonal axis) or reflected with respect to a plane, etc., the resulting state of the system must be identical with the original state. This property must be found also in the properties of the eigenfunctions of the Schrödinger equation. The solutions of these equations form groups and it is possible by means of group theoretical methods to deduce some particulars on the multiplicity of the solutions in the crystalline field, without the exact knowledge of the form of the potential function describing the field and of its magnitude. In this way it is possible, for example, to derive that a state for which $j = \frac{5}{2}$, thus having sixfold degeneracy for the free ion, is split in a crystalline field with cubic symmetry into a double and a fourfold level. Which of the two has the higher energy, and what the energy separation of the two is, cannot be determined without knowing more details of the function V.

Table 1 gives a review of the splittings of levels with different values of L (or J) in fields of different symmetry.

5. Some examples. In order to illustrate the preceding sections, some examples will now be given of experimental data which can be explained using the theory outlined above.

A first qualitative confirmation, especially with regard to Kramers' theorem of the persistence of twofold degeneracy in the case of odd numbers of electrons, was obtained by Gorter and de Haas [8]. Their results for the susceptibilities of the octohydrated sulfates of Pr and Nd are represented in Fig. 2. The lowest sublevel of the Nd^{+++} ion, having three electrons in the $4f$ shell, will be at least double in the crystalline electric field; that of the Pr^{+++} ion with two $4f$ electrons may be single and therefore non-magnetic. Fig. 2 shows that this is just what happens. Penney and Schlapp [9] were able to give an excellent theoretical

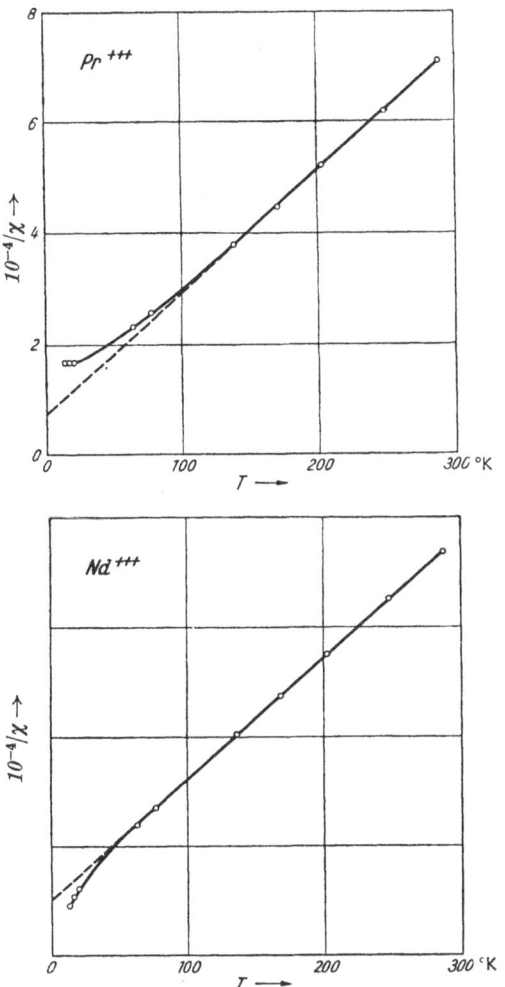

Fig. 2. $1/\chi$ as a function of T for Pr sulfate 8 aq and Nd sulfate 8 aq. The curves are drawn after the calculations of PENNEY and SCHLAPP.

Table 1. Splitting of energy levels in crystalline fields of various symmetry.

L, S or J	0	1	2	3	4	5	6
Degeneracy of the free ion	1	3	5	7	9	11	13
Splitting in:							
Cubic field		3	2+3	1+3+3	1+2+3+3	2+3+3+3	1+1+2+3+3+3
Hexagonal field		1+2	1+2+2	1+1+1+2+2	1+1+1+2+2+2	1+1+1+4×2	5×1+4×2
Tetragonal field		1+2	1+1+1+2	1+1+1+2+2	5×1+2+2	5×1+2+2+2	7×1+2+2+2
Rhombic field		1+1+1	5×1	7×1	9×1	11×1	13×1

S or J	1/2	3/2	5/2	7/2	9/2	11/2	13/2
Degeneracy of the free ion	2	4	6	8	10	12	14
Splitting in:							
Cubic field	2	4	2+4	2+2+4	2+4+4	2+2+4+4	2+2+2+4+4
Hexagonal field } Tetragonal field } Rhombic field	2	2+2	3×2	4×2	5×2	6×2	7×2

In this table the numbers indicate the multiplicity of the sublevels. E. g. a level with $J = 3$ is split in a hexagonal field into three single and two double levels.

representation of these results assuming a field with cubic symmetry. It must be admitted, however, that probably just as good agreements could have been obtained with other fields; and as a matter of fact the existing fields must have a different symmetry, as was shown, among others, by the experiments on paramagnetic resonance. Several other salts were also treated and in these cases one started with a cubic field and applied trigonal or hexagonal corrections. As ELLIOTT and STEVENS [10], [12] pointed out, this starting point is in many cases not correct. The field in the rare earth ethylsulfates (on which much experimental work is done because of the fact that they give hexagonal crystals) is predominantly trigonal, as was also shown by Ketelaar's X-ray measurements [11]. Starting with a trigonal field Stevens and Elliott [10], [12] obtained a good agreement with the results of the resonance experiments (see Sect. 16) of Bleaney et al. and a rather good agreement with some susceptibility measurements.

Some examples will now be given of good qualitative and quantitative confirmation of the theoretical predictions for salts of elements belonging to the iron group. They may be preceded by some general remarks.

It has already been mentioned that the ions of these elements have only spin magnetism (Bose-Stoner hypothesis). The explanation for this behaviour is to be found in the splitting of the orbital levels in the crystalline field. As this is larger than the multiplet splitting[1], or, in other words, as the energetic influence of the crystalline field on the electron orbits is larger than the spin-orbit coupling, one can to a first approximation consider only the orbits and treat the influence of the spins as a perturbation. The spin orbit interaction energy, $\lambda \boldsymbol{L} \cdot \boldsymbol{S}$, comes in a second approximation. When the lowest orbital level is single, it is non-magnetic with respect to the orbits, and only spin magnetism remains. A further approximation gives the influence of higher levels which is smaller if they are at greater energetic distances. The "quenching" of the orbital moments is in general less complete in the second than in the first half of the iron group, where for most ions a better agreement with the Bose-Stoner values for the magnetic moments is found. The principal reason is the much greater value of λ, the spin orbit coupling constant, in the second half of the group.

As is seen from Table 2, states with n, $n-5$, $n+5$ and $10-n$ electrons in the $3d$ shell have the same value of L, and thus the same orbital degeneracy.

Table 2.

Number of $3d$ electrons	1	2	3	4	5	6	7	8	9
Ion	Ti^{+++}	Ti^{++}	V^{++}	Cr^{++}	Mn^{++}	Fe^{++}	Co^{++}	Ni^{++}	Cu^{++}
	V^{++++}	V^{+++}	Cr^{+++}	Mn^{+++}	Fe^{+++}				
Electronic state of the free ion . . .	$^2D_{\frac{3}{2}}$	$^3F_2$	$^4F_{\frac{3}{2}}$	$^5D_0$	$^6S_{\frac{5}{2}}$	$^5D_4$	$^4F_{\frac{9}{2}}$	$^3F_4$	$^2D_{\frac{5}{2}}$

In a cubic field their splittings have the same character. All D terms are split into a double and a triple state; all F terms are split into a singlet and two triplets. But the order may be different in the different cases. Van Vleck has shown [13] that in similar crystalline fields the states with n and $(5+n)$ electrons have the same order, whereas in those with $(5-n)$ and $(10-n)$ electrons, this order is reversed.

Moreover, Van Vleck has shown that the order depends also on the type of crystalline field. In the expression $V_c = D(x^4 + y^4 + z^4)$ for the cubic part of the crystalline potential, the sign of D is decisive in this respect. As Gorter [14] has pointed out, this sign is positive when the magnetic ion is surrounded by six negative (e.g. oxygen) ions, forming an octahedron, whereas in the case of four negative ions forming a tetrahedron, this sign is negative. For a negative value of D the order of the sublevels proves to be reversed with respect to fields with a positive value of D. Let us start with an ion of which much is known. From

---

[1] But smaller than the spin-spin or orbit-orbit coupling.

X-ray measurements it can be seen that Cr^{+++} in tuttonsalts and in many other salts is surrounded by six oxygen ions in an octahedral arrangement. Thus D is positive. On the other hand it is known from susceptibility measurements that CURIE's law is followed down to very low temperatures and that the magneton number agrees nicely with the BOSE-STONER value. This can be understood when the single level is assumed to be the lowest (see Fig. 3). As the distance to the first triplet level is of the order of 10^4 cm^{-1}, this level will not be occupied and has only some influence on the susceptibility by means of the high frequency elements of the magnetic moment, giving a contribution to the temperature independent term [c.f. Eq. (3.2)], and through the LS coupling. As the lowest orbital level is a singlet, it is non-magnetic and only the spin, which can now be introduced, contributes to the magnetic moment.

Similar remarks can be made for the Ni^{++} ion. Again CURIE's law is followed very well and, just as in the Cr^{+++} salts, here too the anisotropy is very small (only some 2 or 3%).

For Co^{++}, on the other hand, an ion with 7 electrons in the $3d$ shell, the pattern of Fig. 3 must be turned upside down. Indeed, there are great differences with Cr^{+++} and Ni^{++}. In a cubic field the lowest term is a triplet. This is split into three singlets when the field potential contains a rhombic part. Though here again a singlet level lies lowest, still, the distance to the next highest level is too small to give a simple magnetic behaviour. The splitting of the triplet in most cobalt salts is comparable to kT at room temperature, so that at not too low temperatures a finite population is found in the higher state. Moreover the low frequency non-diagonal elements of the magnetic moment matrix which have $h\nu$ in the denominator, also give an important contribution to the magnetization. In going to lower temperatures, the populations vary, and so does the value of μ_{eff} which is defined by means of the expression $\chi_A = \dfrac{N\mu_{\text{eff}}^2}{3kT}$. Assuming values for the constants of the rhombic field, it is possible to calculate the values of μ_{eff} in different directions of a crystal. This was done by PENNEY and SCHLAPP [15] who made their calculations with the following expression for the potential of the electric field:

$$V = A(x^2 - z^2) + D(x^4 + y^4 + z^4). \qquad (5.1)$$

Fig. 3. Splitting of the seven fold lowest orbital level of Cr^{+++}, having $L = 3$, in the cubic field of the alums. The two groups of three lines indicate two groups of three coinciding levels.

These authors carried out numerical calculations with a large and a small value of A (200 and 40 cm^{-1}), whereas they took for D 1200 cm^{-1}, in agreement with the value found for nickel salts in similar fields. At higher temperatures no choice between the two values of A was possible in the case of cobalt ammonium-sulphate. The variation of μ_{eff} with T is too small to determine whether the deviations depend on the magnitude of the rhombic contribution or on the fact that the expression (5.1) is only a rather rough approximation. But, from the measurements at low temperatures by A. BOSE [16] (down to 80° K) and JACKSON [17] (down to 1.5° K), both on the same salt, it was possible to conclude that the rhombic contribution could certainly not be represented by an expression with a large value of A.

The fact that a triplet is lowest causes, as was already indicated, a great deviation from the BOSE-STONER value for the magnetic moment and also a large anisotropy [13]. As an example some of BOSE's data [16] for $CoSO_4(NH_4)_2SO_4$ ·

$6H_2O$, the cobalt ammonium tuttonsalt, are given in Table 3. The BOSE-STONER value for the (effective) moment is 3.87.

Table 3.

	296° K	84.7° K	(in Bohr magnetons)	296° K	84.7° K
χ_{A_1}	0.116	0.425	μ_{eff_1}	5.25	5.39
χ_{A_2}	0.084	0.191	μ_{eff_2}	4.47	3.61
χ_{A_3}	0.099	0.283	μ_{eff_3}	4.86	4.39

The fourth element with an F-term is V^{+++} (or Ti^{++}). Here, again, the triplet is lowest. Calculations of SIEGERT [18] gave values for χ which are in good agreement with measurements on a powdered sample of vanadium-ammonium-alum by VAN DEN HANDEL [18], as is shown in Fig. 4. The tendency of χ towards a temperature independent behaviour at very low temperatures, which is evident from the figure, is caused by the splitting due to the spin orbit interaction of the threefold spinlevel. The splitting in this case proves to be about 4.6 cm^{-1} (6.9° K).

As GORTER stated, the sign of D changes, when the magnetic ion is surrounded by four instead of six negative ions and the splitting pattern is inverted. Not many cases are known of a four coordination. It exists in the blue salts Cs_2CoCl_4 and Cs_3CoCl_5. The nearest neighbours of the Co^{++} ions are four chlorine atoms placed at the corners of a tetrahedron around the Co^{++} ion. The principal values of χ for Cs_2CoCl_4 at 296.8 and 83.8° K are given in Table 4, deduced from BOSE's paper [16].

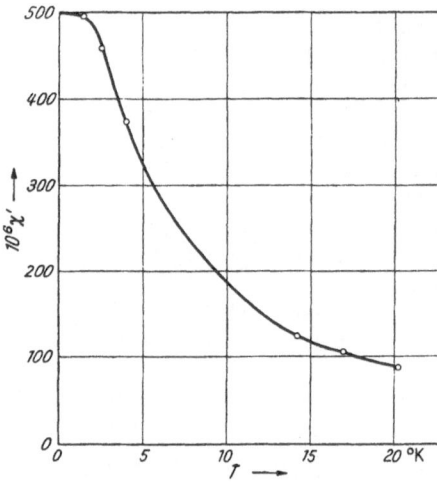

Fig. 4. χ', the susceptibility per gram, corrected for diamagnetism, *versus* T for $V_2(NH_4)_2(SO_4)_4 \cdot 24H_2O$. The curve gives the calculated values, the points indicate measured values.

Table 4.

	296.8° K	83.8° K		296.8° K	83.8° K
χ_{A_1}	0.085	0.293	μ_{eff_1}	4.51	4.42
χ_{A_2}	0.084	0.286	μ_{eff_2}	4.48	4.37
χ_{A_3}	0.081	0.275	μ_{eff_3}	4.41	4.29

The average value of the effective moment for cobalt-ammonium-sulfate is 4.87 μ_B at room temperature and 4.52 μ_B at 85° K; for Cs_2CoCl_4 the corresponding values are: 4.46 μ_B and 4.36 μ_B. Also from these values a difference in behaviour is evident. The moment found for the second salt is closer to the BOSE-STONER value. At the same time the small variation of this moment with T indicates smaller deviations from a CURIE law.

To conclude this section, some brief remarks will be made on paramagnetic molecules. Whereas most of the paramagnetic substances are ionic compounds of elements of the transition groups, there are also some paramagnetic molecules. The best known of these are NO, O_2 and the free radicals. In Sect. 18 a few remarks will be made about the latter.

NO is normally in a $^2\Pi$ state. The spectroscopically determined distance between $^2\Pi_{\frac{1}{2}}$ and $^2\Pi_{\frac{3}{2}}$, of which the first has the lower energy, is 120.9 cm^{-1}. As this distance is comparable with kT at room temperature, CURIE's law is not followed when the temperature is lowered. Theoretical predictions of VAN VLECK [7] on the temperature dependence were in good agreement with the experimental results of several research groups, of which WIERSMA, DE HAAS and CAPEL [19] worked at the lowest temperature (112.8° K) where only low pressures could be used.

The O_2 molecule is normally in a $^3\Sigma$ state. As the splitting of this state is very small, the gas should follow CURIE's law when the temperature is lowered. The experiments [20] show small deviations from this law, probably caused by a temperature independent term. Measurements at higher densities show a decrease in molecular susceptibility, attributed by WIERSMA and GORTER [21] to a formation of some O_4. The research on the susceptibility of solid oxygen, carried out by KANDA, HASEDA and OTSUBO [22] down to 1.6° K, shows sudden changes of χ at the transition points (23.7 and 43.7° K) which were already known from the specific heat measurements. In the α-phase (below 23.7° K) as well as in the β-phase (between the two transition points), χ decreases with T. This may be due to some antiferromagnetic order; the forming of O_4 at high densities may, however, also play a role here.

III. Older research methods.

a) Measurements of the paramagnetic susceptibilities.

6. Apparatus. For the measurement of paramagnetic susceptibilities, in general, the FARADAY or GOUY methods are used. Here the force is measured when the sample is brought into an inhomogeneous magnetic field. The force on a small sample with a magnetic moment, $d\sigma$, is grad $(\boldsymbol{H} \cdot d\sigma)$. When the gradient of the field is in the z-direction and $d\sigma$ can be represented by $\chi H\, dm$, the force is $\chi H \frac{dH}{dz} dm$. In the FARADAY method the samples are chosen so small that one can neglect the error due to taking the average value of $H\frac{dH}{dz}$ for the whole sample. In the GOUY method a rod (e.g. a tube filled with a powder) is used. When ϱ is the mass per cm. of the rod the force is $\int \chi H \frac{dH}{dz} \varrho\, dz = \frac{1}{2} \varrho \chi (H_0^2 - H_1^2)$. When H_0 is the fieldstrength in the centre of the field, it can be measured with great accuracy. Often the rod is chosen so long that H_1^2 can be neglected, or that in any case the accuracy with which H_1 has to be measured is not very great. As a rule the accuracy of the method is limited by the homogeneity of ϱ along the rod. The FARADAY method must be used when χ is not independent of H as in the case when the substance is ferromagnetic or antiferromagnetic and when saturation effects become important. For the low temperature measurements the balances used in measuring the forces must be constructed in such a way that they can be placed and handled in the closed space containing the cooling liquid. Even when the sample is not in direct contact with the liquid, still, the balance cannot be brought into contact with the air because of condensation effects.

A short description will now be given of some of these balances.

a) Sometimes a normal balance (e.g. a chain balance), placed under a glass cover, is used.

b) In the KAMERLINGH ONNES Laboratory in Leiden a special balance of similar type is used (Fig. 5). The force is compensated by a set of coils. In a fixed coil (Fig. 5a) two other coils can move. They are wound in opposite senses; one is attracted, the other one is repulsed by the fixed coil. In this way the influence of the stray field of the magnet is almost completely neutralized. Only the second derivative of the field enters into the resulting force but this correction can practically always be neglected.

c) Sucksmith constructed a balance [23] in which the restoring force is given by the elasticity of a ring, r, carrying the sample, s, (Fig. 6). Small mirrors fastened to the ring on the places of strongest deformation reflect a light beam and thus give it a deviation. The

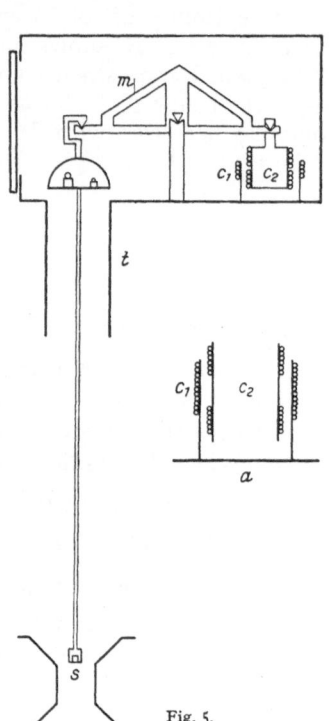

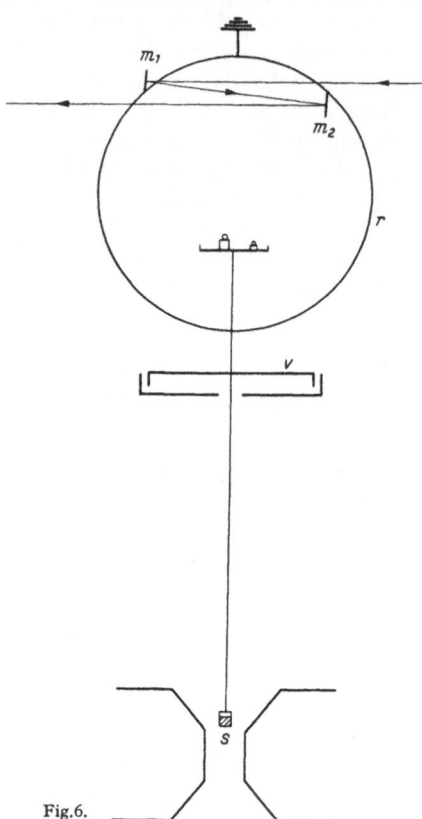

Fig. 5. Magnetic balance for low temperature measurements. The dewar vessels can be fastened to the tube t; s is the sample, m is a mirror, c_1 and c_2 are the compensating coils, separately drawn in a.

Fig. 6. Principle of the magnetic balance after Sucksmith. r: ring; m_1, m_2: mirrors; s: sample; v: damping vane.

force is not compensated and it is measured by reading the displacement of an index on a scale. The accuracy is not quite as good as that of the balances quoted under a) and b).

d) Torsion balances of a construction as shown in Fig. 7 are also used [24]. Here the force on the sample is compensated. Very high accuracies can be obtained by this method.

e) For small values of the magnetic field the so called "couple balance" was designed by Schultz [25] for his study of the magnetic properties of anhydrous salts. This instrument, comparable with a magnetometer, proved to be very satisfactory.

Torsion balances of the type of the Curie balance are generally used at higher temperatures only and will not be treated here.

In order to give an impression of the accuracy which is necessary, the following example is given: When different values of H and thus of $\dfrac{dH}{dz}$ are used, a normal region for $H \dfrac{dH}{dz}$ is from a few times 10^6 Oe²/cm to 50 or 100×10^6 Oe²/cm. Samples having weigths of the order of magnitude of 50 mg are normal. At low temperatures, susceptibilities of a few times 10^{-3} per gram are possible in exceptional cases. But most measurements are as a rule extended up to room temperature and here they are often only of the order of magnitude of several times 10^{-6}. It must, therefore, be possible to measure forces of the order of a few mg with an accuracy of at least 1%.

Another method for measuring susceptibilities, a method that can only be used at low temperatures where the susceptibilities are not too small, is the induction method. The value of a self-inductance or a mutual inductance depends on the permeability of a substance in the coil. For small samples the filling

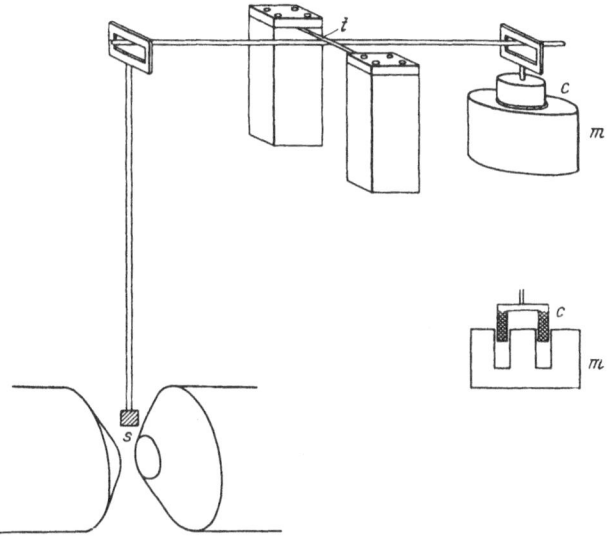

Fig. 7. Principle of a torsion balance. t: torsion fibre or band; s: sample; c: coil; m: fixed permanent magnet. Lower right a cut through this compensation set (for which Mc GUIRE and LANE [24] use a loudspeaker coil and magnet) is shown.

factor is small, sometimes being as low as 1%. Even when the volume susceptibility is about 10^{-3}, so that the permeability is about 1.01, the relative change of the self-inductance or mutual inductance, caused by the presence of the paramagnetic sample, is only about 10^{-4} and this has to be measured with an accuracy of at least 1%.

In the researches on adiabatic demagnetization the induction methods have been greatly developed and now they are generally used at temperatures below 1° K. A description is given in the article on adiabatic demagnetization in this volume (p. 71). It was also used by KANDA et al [22] for the measurement of the susceptibility of solid oxygen at low temperature. W. E. HENRY [26] describes an apparatus which he calls a magnetic moment differential fluxmeter. A set of two coils is placed in a BITTER type solenoidal magnet (with which magnetic fields up to over 5×10^4 Oe were reached) with its axis vertical. The sample can be displaced from one coil into the other one. The change of the flux is proportional to the magnetic moment of the sample.

Anisotropies can be determined by measuring the maximum torque on a crystal in a homogeneous magnetic field, as was done by JACKSON [17], by FEREDAY and WIERSMA [27] and by KRISHNAN and collaborators [28].

7. Corrections. At low temperatures the corrections for diamagnetism of the salt itself and of the carrier become less important than at higher temperatures, though in many cases they are not negligible. On the other hand the influence of the demagnetizing field and of the field of the surrounding ions may become important. For a spherical crystal with a cubic lattice these two are equal and of opposite sign so that they compensate each other. But, in general, this equality does not occur and the resultant might be of the same order of magnitude as each. Let us again give an example: For a sphere the demagnetizing field is $\frac{4}{3}\pi\sigma = \frac{4}{3}\pi\varkappa H$. When $\varkappa$, the volume susceptibility, is about 3×10^{-3}, $\frac{4}{3}\pi\varkappa$ is about 10^{-2}, so that a correction of about 1% to the applied field is necessary because of this effect. For a substance like $Gd_2(SO_4)_3 \cdot 8H_2O$ for which the susceptibility per cm.^{3} at $1°$ K is 0.064 for low fieldstrengths, $\frac{4}{3}\pi\varkappa$ is about 0.27. These corrections are especially important below $1°$ K. It then becomes also necessary to calculate the effect for a powder instead of for one single crystal. Calculations were carried out by de Klerk [29] and are treated in the following article by this author.

b) The influence of magnetic and electric fields on the spectra.

8. A direct method of checking the results of the theory developed by Bethe is to study the spectra of salts containing magnetic ions. In general, this has to be done at low temperatures as the thermal agitation broadens the spectral lines. Moreover the absorption lines starting from levels higher than the fundamental one disappear at low temperatures, so that a discrimination between the levels is possible.

Though rather much work has been done on this subject, it is still very difficult to get enough information in this way as long as the work is restricted to the visible and ultraviolet spectral regions. The identification of the lines is often difficult and, in many cases, the lines are not resolved and continuous absorption is found in the near ultraviolet region. Still, some general confirmations can be obtained and a few examples will be indicated here.

About 1937 Gobrecht [30] studied the absorption and emission spectra of rare earth ions at liquid air temperatures. Of his many results attention is drawn to the fluorescence spectrum of terbium-sulfate showing seven line groups, corresponding with the seven levels $J = 0, 1, \ldots, 6$ of the lowest multiplet ($L = 3$, $S = 3$). The splitting of the separate levels in the crystalline field was also found. From this splitting it was evident that the field in the octohydrated sulfates cannot be cubic. Hund's rule, saying that the multiplets in the second half of the rare earth group are inverted with respect to those in the first half, was also demonstrated. For Sm^{+++} and Eu^{+++}, where the lowest multiplet levels are rather close together, the temperature dependence of the intensities of some absorption lines indicated the variation in population of the levels. It was also possible to derive a value for the screening constant from these spectra.

Much work in the absorption spectra of crystals containing rare earth ions at low temperatures was carried out by Freed and Spedding [31]. The latter's result with neodymium-sulfate gave him the possibility to make a choice between the values for the susceptibilities which were obtained by several groups of physicists and which did not agree with each other. Accepting crystalline fields of the same strength as those which were supposed to exist in Pr sulfate by Penney and Schlapp [9], Spedding finds good agreement between the measured and calculated positions of the energy levels in Er and Dy sulfate [32]. Freed and Harwell [31] have, among other things, carried out a very thorough investigation of the absorption spectrum of Sm ethylsulfate down to the temperature region of liquid hydrogen.

Finally some words may be said on the results obtained by BECQUEREL, who carried out an elaborate study of the influence of the magnetic field and the temperature on the spectra of rare earth ions in crystals [33]. Most of this work was done on the absorption spectra of the natural crystals xenotime (containing Er and Gd phosphate) and tysonite (containing CeF_3). Here, many details of the general picture are confirmed.

In a magnetic field many of the absorption lines prove to be double, indicating that the asymmetry of the crystalline field is enough to leave only the KRAMERS degeneracy. When the light travels through the crystal parallel to the axis and parallel to the magnetic field (longitudinal case) the absorption lines are split into two lines having a different sense of circular polarization. The experiment shows that the right hand vibration may be displaced to the higher as well as to the lower frequencies and that their relative distance may be much larger than the classical LORENTZ value (for erbium ions in xenotime 8.6 times this value was found).

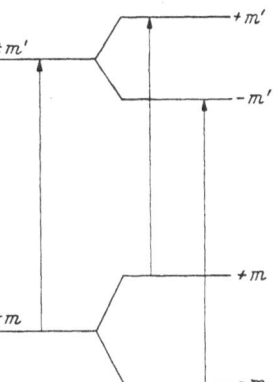

Fig. 8. Splitting of an absorptionline of a paramagnetic crystal caused by the splitting of two double energy levels. $m' = m \pm 1$.

Supposing that only double degenerate levels occur in the crystalline field, Fig. 8 shows what happens in a longitudinal magnetic field. When $\Delta m = +1$ or -1 the circular vibration has the same or the opposite direction than the current. When $(m'g' - mg)$, giving the difference in the displacement of the absorption lines compared with the original line, has the same sign as $(m' - m)$, the effect has the ordinary classical sense, which means that the absorption of circularly polarized light of the same direction as the current is displaced towards the high frequencies. Since opposite signs of $(m'g' - mg)$ and $(m' - m)$ also occur, it is evident that there occur as well such splittings in which the right hand as such in which the left hand component is displaced to higher frequencies.

At low temperatures there is a general tendency for one of the components to increase in intensity at the cost of the other one. This is the spectral evidence of paramagnetic saturation.

The centre of the two lines into which an absorption line is split in the magnetic field does not, in general, coincide with the position of the original line (for $H = 0$). This fact shows the existence of higher order terms in the energy of a state as a function of H (the quadratic ZEEMAN effect).

Also the effect of the direction of the magnetic field relative to that of the crystalline electric field was studied. For the results, as well as for many other details, the reader is referred to the original literature.

c) The FARADAY effect.

9. Origin of the effect and comparison with the magnetization. In 1845 FARADAY found that the plane of polarization of a linearly polarized light beam, passing through a substance in the direction of an applied magnetic field, is rotated. Later on, it became evident that this rotation was associated with the magnetization and in 1928 J. BECQUEREL and W. J. DE HAAS [34] showed that this effect could be separated into two effects, the first of which is independent of temperature and was called the diamagnetic rotation—though it must be stated that this name was not chosen very well, as the effect is not connected with diamagnetism—, whereas the second, the paramagnetic effect, depends on temperature in the same way as does the paramagnetic susceptibility. An explanation of these effects (Cotton, Dorfmann, Ladenburg [35]) is represented in Fig. 9. The splitting of the absorption lines in a magnetic field (see a) causes the diamagnetic effect, in producing a difference, Δn, in the refractive index for the two oppositely circularly polarized beams. The rotation, ϱ, is connected with Δn by the relation $\varrho = \dfrac{d}{\lambda_0} \Delta n \cdot \pi$, where d is the path of the beam inside

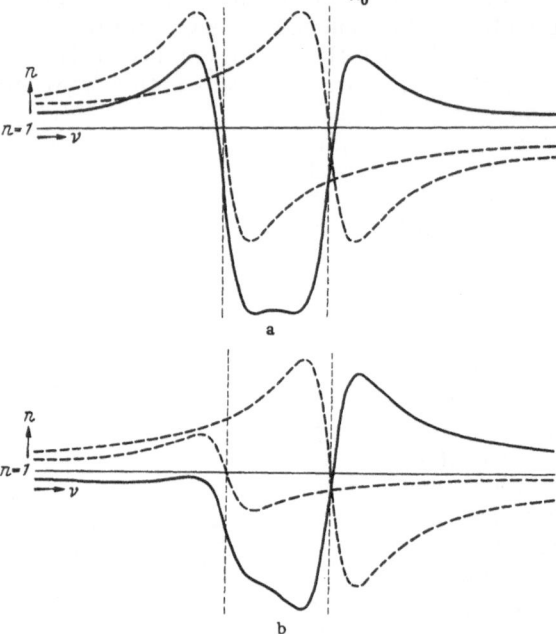

the substance and λ_0 is the wavelength in vacuum of the light used. This effect, shown by all transparent substances, has a positive sign, which means that the rotation takes place in the same direction as the current in the coil which produces the magnetic field. In order to illustrate the magnitude of this effect, it may be stated that in water at 20°C the rotation is 2.18 degrees per mm, for $H = 10000$ Oe.

At low temperatures the populations of the levels change as was already discussed and this causes the paramagnetic effect as is demonstrated in Fig. 9 b, where, just as in Fig. 9 a, the effect of only one absorption line, wich gives a doublet in the longitudinal field, is drawn. It is evident that this rotation is directly connected with paramagnetism. In general, it has a negative sign, though, as was

Fig. 9 a and b. Origin of the diamagnetic (a) and paramagnetic (b) rotation of the plane of polarization. The vertical broken lines indicate the positions if the two Zeeman components of the considered absorption line in the magnetic field. The full lines indicate Δn, the difference of the two refractive indices, each of which is indicated by a broken line.

seen in the preceding section, a positive sign could also be expected. This is actually found in Ni salts.

Supposing that, if all ions were in one of the two states the rotation of the plane of polarization were ϱ_∞, and $-\varrho_\infty$ if they were in the other state, the real rotation would be

$$\varrho = \frac{\varrho_\infty \, e^{\mu H/kT} - \varrho_\infty \, e^{-\mu H/kT}}{e^{\mu H/kT} + e^{-\mu H/kT}} = \varrho_\infty \, \mathrm{Tan} \, \frac{\mu H}{kT}.$$

In this simple case the rotation and the magnetization which can be represented by $\sigma = N\mu \, \mathrm{Tan} \, \mu H/kT$ are proportional with a constant of proportionality $N\mu/\varrho_\infty$. Van Vleck and Hebb [36] showed that, under special conditions, the proportionality is conserved even if, at higher temperatures, other levels with other magnetic moments are also playing a role and if, furthermore, temperature independent terms originating in the non-diagonal elements of the magnetic moment matrix are taken into consideration in the case of the magnetization as well as in that of the rotation. The conditions mentioned above are fulfilled in the ions of the rare earths, with the exception of Sm^{+++} and Eu^{+++}, and in the S state ions of the iron group. As an example of this proportionality, Fig. 10 gives the $1/\chi$ *vs* T curves for crystals of the ethyl-sulfates of Pr, Nd and Dy, measured in the direction of the hexagonal axis. The curves are drawn between the points representing the magnetizations. From the rotations the factor $N\mu/\varrho_\infty$ can be calculated The values of the magnetizations obtained from the rotations by multiplication with this factor are also indicated. The agreement is seen to be satisfactory.

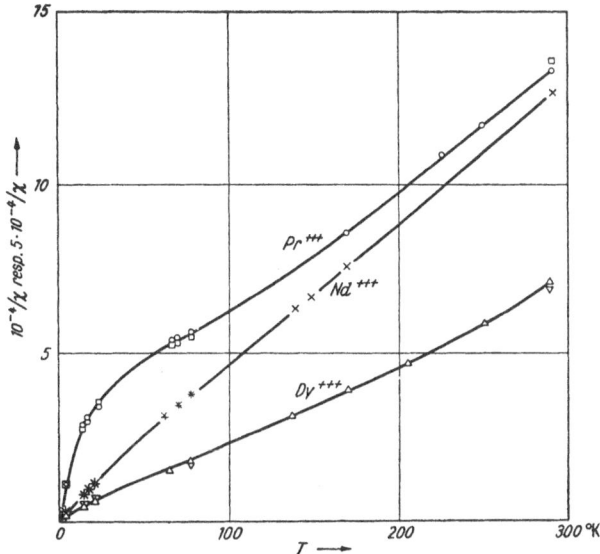

Great advantages of these measurements are their accuracy when ϱ_∞ is not too small and the speed with which many measurements can be done. The great accuracy is seen in Fig. 11 representing the results for dysprosium ethyl-sulfate.

As ϱ is in practice proportional to H as long as no saturation occurs, the ratio ϱ/H is generally given. When taken per cm, this is called VERDET's constant.

Fig. 10. Reciprocal of the susceptibility in the direction of the hexagonal axis as a function of T for Pr-, Nd-, and Dy ethylsulfate. For the Dy salt the scale is five fold enlarged. $\odot$, $\times$, $\triangle$ indicate the susceptibility measurements; $\square$, $+$, $\triangledown$ are calculated from the rotations.

10. Some results. A systematic research was carried out for the ethyl-sulfates of Ce, Pr, Nd, Sm, Gd, Dy Er [*37*]. For most of these salts direct magnetic measurements were also carried out, confirming the proportionality property [*38*]. The value of ϱ_∞ per mm. thickness of the crystal, which depends on the active absorption bands, varies from 0° (Sm) and 1.8° (Gd) to 426° (Ce). In all cases with the exception of Gd the measurements at the lowest temperatures can be represented by a formula of the type

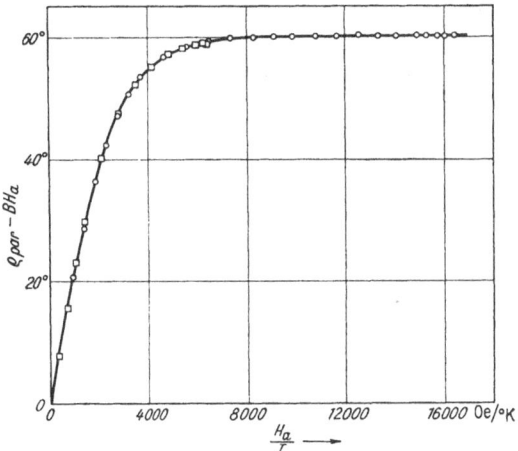

$$\varrho = A \operatorname{Tan} \frac{\mu H}{kT} + BH, \qquad (10.1)$$

indicating that the lowest level is a (KRAMERS) doublet. The distance δ

Fig. 11. Saturation curve for Dy ethyl-sulfate (calculated from the paramagnetic rotations). The Tan-curve represents the theoretical values of the paramagnetic rotation, ϱ_{par}, minus the temperature-independent contribution.

to the next higher level is different for the ions of different elements; e.g. in the case of the Ce salt it is only small (after measurements of the paramagnetic resonance only 6.6 cm⁻¹). As a result not all values of ϱ in the temperature region between 1.5° K and 4.2° K can be represented by the Tan-function. In the case of Pr⁺⁺⁺ δ is about 26 cm⁻¹ and deviations from (10.1) are already found in the liquid hydrogen temperature region. For Er all values between 1.5° K and 20.4° K can be represented by the same simple law of the form (10.1) indicating that, here, δ is much greater.

Some data on the magneto-rotations of the ethyl-sulfates at low temperatures are given in Table 5. They are supplemented by some results of direct susceptibility measurements.

Table 5.

	Element						
	Ce	Pr	Na	Sm	Gd⁴	Dy	Er
Lowest level of the free ion . . .	$^2F_{\frac{5}{2}}$	$^3H_4$	$^4I_{\frac{9}{2}}$	$^6H_{\frac{5}{2}}$	$^2S_{\frac{7}{2}}$	$^6H_{15/2}$	$^4I_{15/2}$
$-A^1$ in degrees	426³	59.75	113.85	0	1.823	60.15	12.58
$-10^6 B$	—	17.0	13.5	—	0.00	3.02	0.00
μ/μ_B for the free ion	2.56	3.62	3.68	0.84	7.94	10.6	9.6
$\mu_{\parallel}/\mu_B{}^2$ at He temperatures . . .	1.786	0.758	1.764	$0.29_8$⁵	7.94	5.66	6.02
$\mu_{\perp}/\mu_B{}^2$ at He temperatures . .	—	0.0	1.034	$0.30_2$⁵	7.94	0.00	—

[1] For Ce and Nd the green mercury line was used, for the others the yellow mercury line.

[2] $\mu_{\parallel}$ is the moment parallel to the hexagonal axis, $\mu_{\perp}$ is the moment perpendicular to this axis.

[3] Only for temperatures below 1.5° K.

[4] For Gd, where a Brillouin function should be used instead of (10.1), A indicates the saturation value of the paramagnetic rotation.

[5] After the resonance measurements of Boyle and Scovil (see [12]).

The method is restricted to monoaxial crystals (glasses, which were sometimes studied, are not very useful because the surroundings of the magnetic ions are not well defined). In other crystals the double refraction prevents reliable rotation measurements. Only weak double refractions can be accepted. The linearly polarized light is then transformed into an elliptically polarized beam. When the ellipticity is not too great the rotation of the large axis can be measured and, by a method indicated by Poincaré [39], the rotation can be calculated when the double refraction of the crystal is known.

A description of a method to determine the rotatory power in a direction normal to the axis, for monoaxial crystals with a small double refraction, is given by Becquerel [40], who used it for tysonite. For the ratio of the Verdet constants in the direction of the optical axis, V_A, and in the direction of a diagonal binary axis perpendicular to the first one, V_N, the values of Table 6 were found.

Table 6.

T	293	77.4	10.32	14.23	1.7 °K
V_N/V_A	0.83	0.58	0.33	0.30	0.21

For the corresponding values of the magnetic moment he found

$$\mu_A = 0.687\,\mu_B \qquad \mu_N = 0.572\,\mu_B.$$

The same method was used by Lévy and the author [41] for $NiSO_4 \cdot 6H_2O$.

Of the measurements on salts of elements of the iron group attention is drawn to those on $NiSiF_6 \cdot 6H_2O$, as they suggested to Opechowski and Becquerel [42] a theoretical treatment of the magnetization of this salt. The lowest level, with $S = 1$, is split into a double and a single level with a relative distance δ, for which 0.301 cm^{-1} was found from the results at helium temperatures. Here as in many other cases it was possible to compare the theoretical and experimental values because of the accuracy of these measurements and their great number. Later paramagnetic resonance measurements of Penrose and Stevens [43] gave a value for δ which varied with T (probably due to thermal dilatation of the crystal). At 195° K a value of 0.35 cm⁻¹ was found, and at $T = 20°$ K and 14° K only 0.12 cm⁻¹. OLLOM and VAN VLECK [44] tried to solve this discrepancy by introducing some isotropic and anisotropic exchange interaction into the calculations. After redoing the calculations with the new value of δ and an extra term of the exchange interaction, BECQUEREL [45] came to the conclusion that the agreement between theory and experiments was not completely satisfactory so that, perhaps, some, for the moment unknown, influence has to be taken into account.

Another nickel salt for which the rotation measurements were performed which gave rise to rather extended theoretical considerations was $NiSO_4 \cdot 6 H_2O - \alpha$. Measurements and discussions exist [46] on this salt which possesses not only a paramagnetic rotatory power but also a natural rotatory power, the magnitude of which is a function of H as was shown by LÉVY [47].

The study of some iron carbonates has led to the discovery of the meta-magnetism [48], which is now considered as a type of antiferromagnetism on which more will be said in Chapter VI.

IV. Paramagnetic relaxation.

11. Introduction. So far, some results have been presented of the study of paramagnetic crystals in constant magnetic fields. When the substances are placed in alternating fields special phenomena are found. In 1932 the effects were already discussed by WALLER [49] and in 1936 GORTER [50], unaware of WALLER's article, proved their existence.

If the magnetization would follow the magnetic fieldstrength instantaneously, there would not be any difference between the magnetizations in constant or alternating fields. But there are several reasons for the existence of a time lag between the field and the magnetization, and thus for the existence of a phase difference between the two in the case of alternating fields. As a result, there is in this case an energy dissipation and therefore an *absorption*. Because of this phase difference, showing that it is difficult for the magnetization, σ, to follow the fieldstrength, H, there is a dependence of σ on the frequency ν: When ν increases, σ (or χ) decreases. So there is also a *dispersion*.

12. Spin-Lattice relaxation. One of the theoretical descriptions is due to CASIMIR and DU PRÉ [51] and though it has only a limited applicability, an outline of it will be given here.

In this theory it is assumed that the system of magnetic dipoles is in internal thermal equilibrium. This system is called the spin system as the substances which studied contained Gd-ions or ions belonging to the iron group. The Gd-ions are in an S-state whereas in the lowest state of the other ions the orbital contribution is largely quenched, so that the moment is practically only due to the spin. The heat developed in this spin system has to be given to the lattice and vice versa and this transition is difficult. Associated with it is a *relaxation time*, τ, which depends on the specific heat of the spin system at constant field-strength, and a conductivity factor, α, for the transition of heat from the spin system to the lattice. With these assumptions, thermodynamical reasoning gives the result

$$\left.\begin{aligned} \frac{\chi'}{\chi_0} &= \frac{F}{1 + \nu^2 \varrho^2} + 1 - F = \frac{F}{1 + \omega^2 \tau^2} + 1 - F, \\ \frac{\chi''}{\chi_0} &= \frac{F \nu \varrho}{1 + \nu^2 \varrho^2} \qquad\qquad = \frac{F \omega \tau}{1 + \omega^2 \tau^2}. \end{aligned}\right\} \qquad (12.1)$$

χ' and χ'' are the real and the negative imaginary parts of the complex suceptibility which can be introduced as follows: If the magnetic field is

$$H = H_0 + H_1 e^{i\omega t}, \tag{12.2}$$

the magnetization, which varies with the same frequency, will be

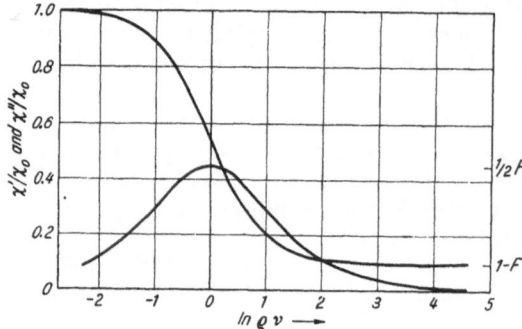

Fig. 12. χ'/χ_0 and χ''/χ_0 as functions of ln $\varrho\nu$ following eq. (12.1). For F the value 0.9 is chosen.

$$\sigma = \sigma_0 + \sigma_1 e^{i\omega t} \tag{12.3}$$

(where σ_1 can have a complex value because of the phase difference between H and σ). The static part of the susceptibility, then, is defined by

$$\frac{\sigma_0}{H_0} = \chi_0. \tag{12.4}$$

The high frequency susceptibility is

$$\frac{\sigma_1}{H_1} = \chi' - i\chi'' = |\chi| e^{i\varphi}. \tag{12.5}$$

Therefore $\chi' = \left|\dfrac{\sigma_1}{H_1}\right| \cos \varphi$ and $\chi'' = \left|\dfrac{\sigma_1}{H_1}\right| \sin \varphi$.

F and ϱ are constants, the second one being called the *relaxation constant*. It is related to the relaxation time τ by the relation

$$\varrho = 2\pi\tau. \tag{12.6}$$

If c_H is the specific heat in a constant field one finds that

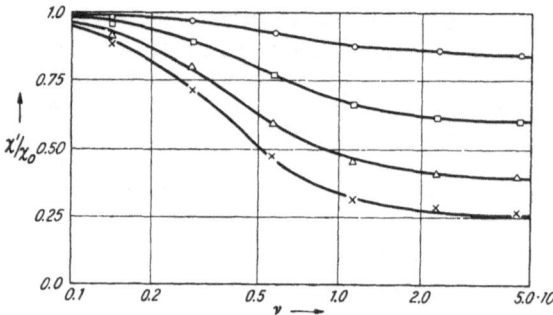

Fig. 13. χ'/χ_0 vs ln ν for Gd sulfate $8\,aq$ after measurements of Broer. $T = 77°$ K. $\odot$: $H = 800$ Oe; $\square$: $H = 1600$ Oe; $\triangle$: $H = 2400$ Oe; $\times$: $H = 3200$ Oe.

$$\varrho = \frac{2\pi c_H}{\alpha}. \tag{12.7}$$

In Fig. 12 χ'/χ_0 and χ''/χ_0 are represented as functions of ln $\varrho\nu$. As is seen, the top of the χ''/χ_0 curve as well as the steepest slope in the χ'/χ_0 curve, the dispersion curve, are both found for ln $\varrho\nu = 0$.

The constant F is given by

$$F = \frac{c_H - c_\sigma}{c_H}, \tag{12.8}$$

or, if Curie's law is satisfied to a good approximation, by

$$F = \frac{CH^2}{b + CH^2}. \tag{12.9}$$

Here, c_σ is the specific heat for a constant magnetization. C is Curie's constant, b is the constant in the expression for the specific heat $\left(c_\sigma = \dfrac{b}{T^2}\right.$; contributions to b are given by the splitting of the lowest level in the crystalline electric field, by the magnetic interaction of the magnetic dipoles and by the exchange interactions).

Fig. 13 to 15 give some results of the measurements on several salts. In Fig. 13 the values of χ'/χ_0 are represented as a function of the frequency for different fieldstrengths and for $T = 77°$ K for $Gd_2(SO_4)_3 \cdot 8H_2O$ [52]. When the frequency is increased the susceptibility decreases from χ_0 to $\chi_{ad} = (1 - F)\,\chi_0$. The dependence of χ_{ad} on H, as is demanded by the expression for F, is evident from the figure. From these dispersion curves it is possible to calculate the values of ϱ which depend on H. The specific heat and the heat conductivity factor, α, are strongly dependent on T. This is also the case with ϱ. Whereas at liquid air temperatures values of about 10^{-7} sec are found for the relaxation time, one finds

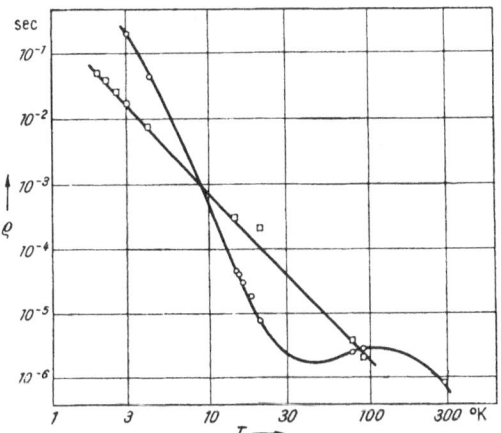

Fig. 14. The relaxation constant as a function of the temperature in a constant magnetic field. $\odot$: $Gd_2(SO_4)_3 \cdot 8H_2O$ at 3200 Oe; $\square$: $CrK(SO_4)_2 \cdot 12H_2O$ at 4000 Oe.

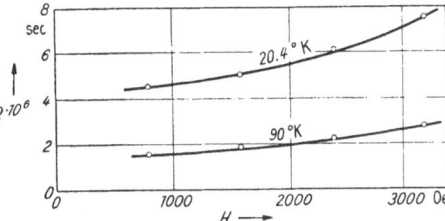

Fig. 15. The relaxation constant as a function of the magnetic fieldstrength for $Gd_2(SO_4)_3 \cdot 8H_2O$ at 20.4° K and 90° K.

values of the order of 10^{-2} sec in the temperature region of liquid helium. In some cases still larger values were found. For example van der Marel [53] found a value of about 0.7 sec for $CuK_2(SO_4)_2 \cdot 6H_2O$ at about 2° K and 3000 Oe. For a constant fieldstrength the values of ϱ as a function of T are represented in Fig. 14 for two cases. For a constant temperature its dependence on H is shown in Fig. 15.

Another representation of the quantities displayed in Fig. 12 is shown in Fig. 16, where χ'/χ_0 is plotted as a function of χ''/χ_0 for chromic alum. (In Fig. 17 the same data, obtained by du Pré, are plotted as functions of $\ln \nu$.) From (12.10) one can derive the relation:

$$\left\{\frac{\chi'}{\chi_0} - (1 - F) - \frac{1}{2}F\right\}^2 + \left\{\frac{\chi''}{\chi_0}\right\}^2 = \frac{1}{4}F^2. \qquad (12.10)$$

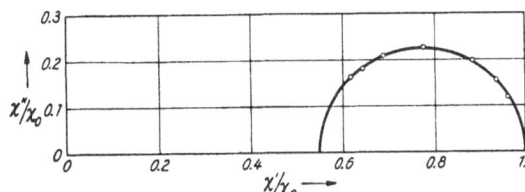

Fig. 16. χ''/χ_0 vs χ'/χ_0 for chromic potassium alum. $T = 2.06°$ K; $H = 785$ Oe.

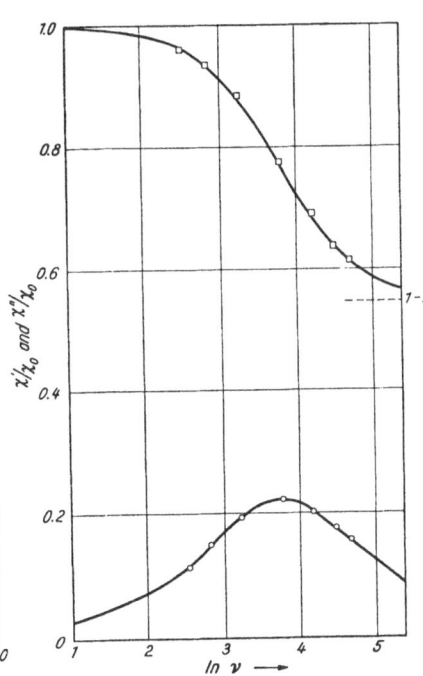

Fig. 17. χ''/χ_0 and χ'/χ_0 as functions of $\ln \nu$ for chromic alum. F is found to be 0.45 in this case.

Thus the values χ'/χ_0, χ''/χ_0 are found on a semi-circle. In many cases, especially at low temperatures, only a part of a circle was obtained with the centre at some distance from the χ''/χ_0 axis. Perhaps such a behaviour is connected with the fact that the description which was given above, and which led to only one relaxation time, is too simple. In 1941 Van Vleck [54] drew attention to the fact that it is probably not the energy transition from the spin system to the lattice, but the heat transfer from the low frequency lattice vibrations to the other lattice vibrations of the crystal, that is the bottle neck for the heat transfer from the spin system to the bath. Gorter, van der Marel, and Bölger [55] examined this situation somewhat more closely and concluded that the experimental facts are represented satisfactorily if one takes into account two heat transitions, viz. from the spin system to the system of the low frequency lattice vibrations, the bandwidth of which depends on the temperature, and from these low frequency vibrations to the other lattice vibrations, which are in a good heat contact with the bath. It depends on circumstances as to which of the two is the bottleneck. The authors conclude that in most of the investigations carried out in the temperature region of liquid helium the relaxation phenomena are determined by the second transition. Deviations from the representation with only one relaxation constant, as they were represented in Fig. 12 and in Fig. 16, can probably be explained in this way. The relaxation constants which are originally found are then something like an average of a continuous group of relaxation constants which may be introduced in this new conception.

From the measurements one can not only derive information about the transition of energy from one system to another one, but also about the specific heats. When ϱ (for the moment we persist in accepting only one relaxation constant) and F are known, c_H can be calculated. This value is of great importance for the measurements at temperatures below $1°$ K, obtained by the method of adiabatic demagnetization. It is therefore useful to compare these results with those of the demagnetization experiments and with those obtained from measurements on the paramagnetic resonance. This will not be done here, but the reader is referred to the article on adiabatic demagnetization for more details.

13. Spin-spin relaxation. If the frequency of the alternating field is increased a second relaxation region is found. The relaxation time connected with these losses is of the order of 10^{-9} sec, and is independent of temperature. Its origin can be described in the following way: In a magnetic field a degenerate lowest level of a magnetic ion is in general split and the populations of the sublevels depend on H and T. An alternating field of frequency ν will cause a stimulated emission and absorption of photons $h\nu$ which bring the system from one state to another with an energy difference $\Delta E = h\nu$. Though the transition probabilities are equal, still there is a net absorption caused by the difference in population of the levels. As this absorbed energy is quickly enough transported to the lattice, it will not cause an appreciable temperature rise of the spin system. Also without an external magnetic field there exists an absorption originated by the magnetic interaction of the dipoles. This interaction can be described as a magnetic field acting on the dipoles. This field is changing continuously because of the thermal agitation so that its average value is zero. The root-mean-square, H_i, however, will not disappear. It can be shown that

$$H_i^2 = 2\mu^2 \sum_j r_{ij}^{-6} = 2 g^2 \mu_B^2 j (j+1) \cdot \sum_j r_{ij}^{-6}. \tag{13.1}$$

The spin system as a whole shows a very great number of energy levels which are all broadened as a result of the continuously changing internal field. In this

energy band transitions can take place and because of the fact that the density
of these levels decreases when the distance from the lowest level increases, here
again a net absorption results. In connection with this absorption a relaxation
constant, ϱ', which is valid as long as the frequency is much smaller than $1/\varrho'$,
can be defined by means of the relation

$$\varrho' = \frac{A}{16\pi^2 \chi_{ad} \nu^2}, \tag{13.2}$$

in which A is the absorption coefficient. BROER has calculated the following
value for ϱ' when no external field is present [56].

$$\varrho' = \sqrt{\frac{\pi}{2}} \frac{h}{2\mu_B H_i}. \tag{13.3}$$

The spin relaxation can also be described as follows: The intensity of a transition between states indicated by p and q is proportional to M_{pq}, the transition
probability. Because of the finite values of the M_{pq}, it takes a finite time to restore
equilibrium after it has been disturbed and there will be a phase difference between
the magnetization and the field strength in an alternating field. This gives rise
to an absorption which can formally be described with a relaxation constant.
As in many cases the frequencies used in the experiments were smaller than $1/\varrho'$,
the influence on χ''/χ_0 could be described with sufficient accuracy by a term proportional to ϱ', so that the combined influence can be represented by the expression:

$$\frac{\chi''}{\chi_0} = F \frac{\nu \varrho}{1 + \nu^2 \varrho^2} + (1 - F) \nu \varrho'. \tag{13.4}$$

The values of ϱ' which were found for chromic alum, iron alum and gadolinium
sulfate 8 aq are 1.4×10^{-9}, 0.9×10^{-9} and 0.3×10^{-9} [57]. They are not very
dependent on the strength of a parallel magnetic field, but usually decrease rapidly
when a perpendicular field is applied.

14. Third relaxation. In 1948 DE VRIJER and GORTER [58] discovered a new
relaxation phenomenon. It was found in several chromic alums in magnetic

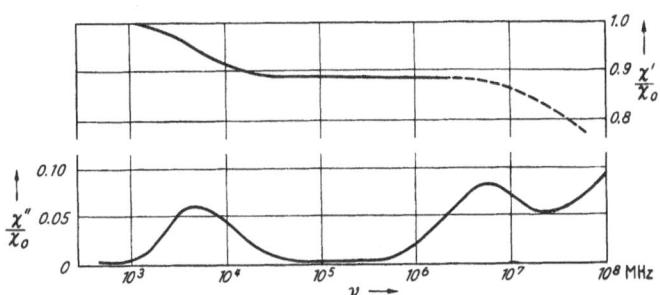

Fig. 18. χ'/χ_0 and χ''/χ_0 for chromic potassium alum as functions of ν at $T = 20.4°$ K and $H = 320$ Oe in a large frequency
interval, showing the lattice- and the third relaxation regions and the beginning of the spin relaxation region.

fields smaller than 600 Oe, at frequencies of the order of 10^7 sec^{-1}. The relaxation constant proved to be independent of the temperature used. At higher
temperatures this effect overlaps the spin lattice relaxation, discussed in Sect. 12,
but in the temperature region of liquid hydrogen and helium the two relaxations
are widely separated as a result of the shift of the lattice relaxation towards much
smaller values of ν. There might be some connection between this third relaxation
and certain anomalies of ϱ'.

In Fig. 18 a provisional sketch of the absorption and dispersion curves for chromic alum in a large frequency region are given.

15. Experimental methods. Two methods were developed for the measurement of χ' and χ''. In the first one they were measured separately, in the other one both parts of the complex susceptibility were determined simultaneously.

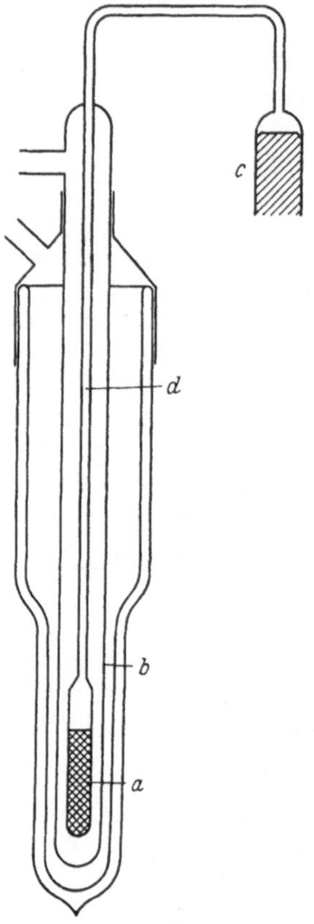

In the first group of measurements the influence of the salt on the frequency of a tuned circuit was determined by means of a heterodyne beat method. Descriptions are given by Broer and Schering [59] and by de Vrijer [57]. In this way the value of χ' could be found. χ'' was measured as follows [60]. In an alternating field with amplitude H_0, the heat developed is $W = \pi \nu \chi'' m H^2$ per second (m is the mass of the sample). If the vessel contains some gas in addition to the magnetic substance, it can be used simultaneously as a gas-thermometer and as a calorimeter. Thus, if the heat capacities are known, the relation between the change in pressure of the gas per second and the value of χ'' can be calculated. In Fig. 19 a schematic picture of the apparatus is given.

In the second method the values of χ' and χ'' can be measured with a Hartshorn mutual induction bridge [61] constructed for a large frequency region (the Leiden bridge covers a region from 3 to 1200 cycles). As a similar bridge is described in the article on adiabatic demagnetization (p. 75), the reader is referred to that article. The difference is that the bridge for the demagnetization experiments is only used for one frequency.

Fig. 19. Apparatus for the measurement of χ''. The container a with the salt is connected through a capillary tube d with the manometer c and can be isolated thermally by pumping off the tube b.

V. Paramagnetic resonance.

16. Discussion of the effect. Though a more complete treatment of this subject is given elsewhere (Vol. XVIII), some words will be said on it, because in using this method, very important results have been obtained, especially for the low temperature magnetic research work.

In Sect. 5 it was already stated, that it was not possible to obtain an exact knowledge of the crystalline electric field from the susceptibility measurements. One could only assume a field with a special symmetry. The expression for the potential V contains some constants. With such an expression the splitting of the lowest ionic energy level could be calculated as well as the magnetic moments associated with the sublevels. Using this picture the susceptibility can be calculated as a function of T. This function still contains the constants from the expression for V. These are now chosen in such a way that the calculated susceptibilities agree as well as possible with the values found experimentally. Even if this fit is rather good it is not certain that a potential with a different symmetry would not have given as good or even better agreement.

Another difficulty is that one is not at all sure that the principal axes for the magnetic properties coincide with the crystallographic axes. The situation may even be different for different ions. As an example one can consider an ion in a cubic field. Such a situation is always more or less distorted for values of $J > \frac{1}{2}$ (JAHN-TELLER effect [62]). There might be a displacement along one of the four trigonal axes. As these axes are equivalent there is no preference for any one of them and, as a result, the crystal contains four groups of magnetic ions which will show a different behaviour in a magnetic field. For the time being, this last complication will be neglected.

It is now very important, in order to obtain more information, to attack the problem in another way and to try to perform a direct measurement of the separation of the sublevels in the crystalline field. The paramagnetic resonance method can be used for this purpose. By means of this method, which was for the first time successfully applied by ZAVOISKY and HALLIDAY [63] in 1945, one can carry out spectroscopy at cm wavelengths and the relative positions of the sublevels can be found with good accuracy. The ambiguity of the expression for V is now much smaller and a much better agreement can be obtained between theory and experiment.

The principle of the method is as follows: When an ion is in an energy state at a distance ΔE from another state and when transitions between these two states are allowed, they can be stimulated by an alternating magnetic field of frequency $\nu = \dfrac{\Delta E}{h}$. In many cases these energy differences arise completely or partially due to an applied magnetic field. In the first case $\Delta E = g\mu_B H \Delta m$ where g is LANDÉ's splitting factor and Δm is the difference in magnetic quantum number. Because of the selection rule for m, $\Delta m = \pm 1$, the resonance condition is:

$$\Delta \nu = \frac{g\,\mu_B\,H}{h} \ . \tag{16.1}$$

The transitions can be stimulated by an alternating magnetic field of this frequency, applied at right angles to H. As ν or H is varied a large absorption is found when (16.1) is satisfied.

Sometimes, hyperfine structures are found in these spectral lines [64]. They were predicted by GORTER [65] and discovered by PENROSE [66] and find their origin in the nuclear magnetic moments. In order to discover these hyperfine structures, which produce splittings of the order of magnitude of 100 Oe (when the frequency is kept constant), it is necessary to reduce the width of the spectral lines which is normally of the order of a few thousand oersteds at room temperature. This line width is caused by the sum of several effects. One of them is the paramagnetic relaxation which was discussed in the preceding chapter. This relaxation (which was studied for non-resonant frequencies) also has several origins. The spin-lattice relaxation depends strongly on the temperature as was stated in Sect. 13 and, as the contribution to the width of the absorption line, because of this spin-lattice interaction is proportional to $1/\tau$, the broadening is greatly reduced at low temperatures where τ may become of the order of 10^{-2} sec This dependence on the relaxation time can be considered as a consequence of the fact that τ is proportional to the lifetime of a state and therefore inversely proportional to its width. At low temperatures this contribution to the broadening of the spectral lines is very much reduced and, at temperatures in the liquid helium region, it is negligible compared with the temperature independent contribution of the spin-spin interactions. These cause a variable magnetic field superposed on the external field and on the other hand a shortening of the lifetime of the spin states as they induce transitions. This contribution can be reduced by using diluted crystals, in which the dipoles are at considerable distances from each other. In many cases the ion, for which the hyperfine structure was studied, was introduced as an "impurity" into a non-magnetic salt. In such a manner a halfwidth of the order of 6 Oe can be obtained, arising from the magnetic moments of the protons of the water of crystallization. Bleaney and coworkers [67] indicated that a further reduction in line-width is possible by replacing these protons by deuterons, as the magnetic moment of D is about $\frac{1}{3}$ of that of H. In this way a half-width of about 2 Oe was obtained. The only possibility for a further reduction of the line-width is to use crystals without water of crystallization. This was done by Bowers [68].

17. Experimental methods. From the relation $\nu = -g\beta H/h$ it is seen, that for $g=2$ the use of a field of 10^4 Oe requires a frequency of about $3 \cdot 10^{10}$ corresponding to a wavelength of about 1 cm. Though resonance could also be obtained in weaker fields, e.g. in a field of 10 Oe with a wavelength of 10 m, the wavelengths between 1 and 10 cm are preferred, as for longer waves the line-width would become comparable with or greater than the frequency used (as long as one is not working at low temperatures). Another advantage of the use of higher frequencies is the fact that the intensities of the lines are greater than at lower frequencies (they are

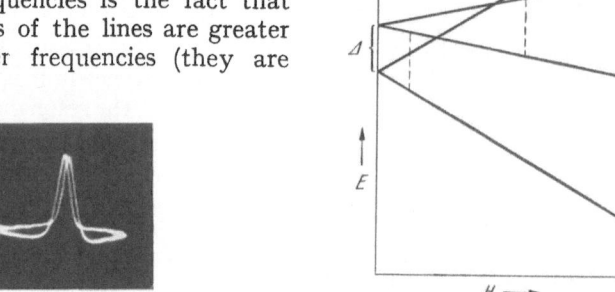

Fig. 20. Oscillogram for diphenylpicrylhydrazyl (a free radical) at 3.2 cm. The resonance field is obtained when the intersection near the top of the two absorption curves is in the centre of the screen.

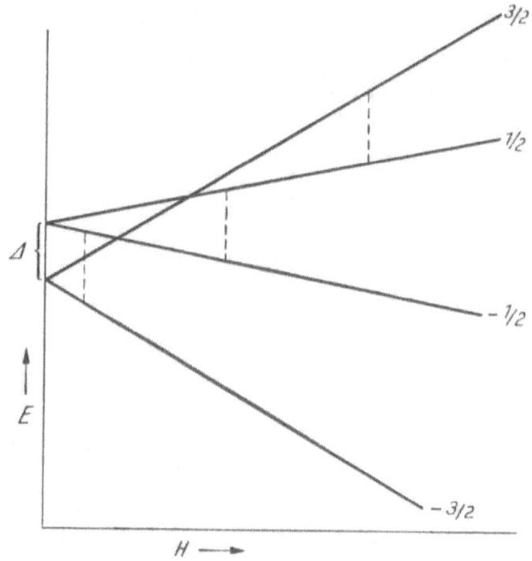

Fig. 21. Energy splitting of a degenerate level with $j = \frac{3}{2}$ as a function of H. $\varDelta$ is the splitting in the crystalline field. The dotted lines correspond to possible transitions $(\varDelta j = 1)$ with equal energy difference.

about proportional to ν). In the temperature region of liquid helium only this last argument counts, though here the intensity of the oscillating magnetic field must sometimes be reduced in order to prevent saturation effects from becoming important. These effects will appear if the relaxation time is long, which means that it takes a rather long time to restore temperature equilibrium after a change in the populations of the levels caused by a stimulated transition.

Whereas at higher temperatures the lines may be rather wide and therefore the constancy of the magnetic field in time and space and its accurate measurement are not so very important, the situation may be different at temperatures in the liquid helium region. If possible the magnetic field should be constant within about one oersted over the sample. For very accurate measurements of the magnetic field the proton resonance is often used. In order to get the homogeneity required the pole faces have to be given a special shape. Sometimes just a ring shim on the border of the front of the poles is sufficient.

If ZEEMAN effects are measured the frequency is, in general, kept constant. Together with a constant field of about the required intensity a low frequency alternating field having an amplitude of about 100 Oe is applied, so that the field sweeps through the absorption region. In order to display the absorption line on an oscilloscope screen, it is convenient to apply the voltage giving the magnetic field sweep to the horizontal oscilloscope plates. When, as is sometimes done, a phase difference is given to the applied voltage with respect to the magnetic field modulation, the two absorption curves corresponding to sweeping back and forth through the resonance are displaced with respect to each other, thus allowing for a more accurate adjustment of the resonance field (see Fig. 20).

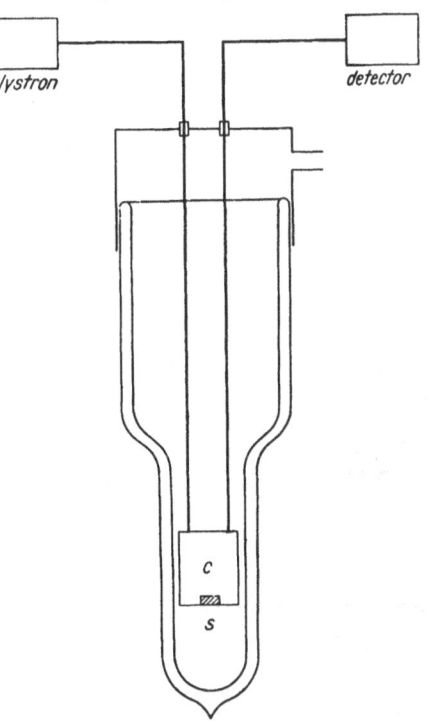

Fig. 22. Schematic drawing of a cryostat for resonance measurements (BLEANEY, UBBINK).
c: cavity; s: sample.

If a splitting of the fundamental level of a magnetic ion in a crystalline electric field is measured it is very improbable that the high frequency generator has exactly the proper frequency. In that case a magnetic field is applied in order to cause, in addition to the electric field splitting, also a magnetic splitting. In Fig. 21 a simple example is given as an illustration. Here, it is assumed that the lowest level is a spin level with $S = \frac{3}{2}$, as is found e.g. in the Cr^{+++} ion. In a cubic field this level is split into two double levels which can, in their turn, be split in a magnetic field. If this field is applied in the direction of a trigonal axis the splitting takes place in the indicated way. Transitions in a high-frequency field perpendicular to the constant field H are possible for $\Delta m = \pm 1$. As is seen in the figure, resonance can be observed at constant frequency for three values of H. From these field-strengths the values of the original splitting, Δ, and of g can be calculated if the cubic symmetry is beyond doubt.

In most cases a reflex klystron is used as a generator. The generated wave is conducted through a coaxial cable or a wave guide to a resonant cavity which is placed between the pole pieces of an electromagnet. For longer wavelengths the part in the cryostat consists in general of a coaxial line as a wave guide would take up too much space and would give too great a heat conduction to the cooling bath. For shorter waves a waveguide is used. A second coaxial cable or waveguide connects the cavity to a detector which measures the relative intensities of the output signal, from which the absorption in the substance can be calculated. A low temperature cryostat for the measurement of the paramagnetic resonance is represented schematically in Fig. 22.

The cavity must occasionally be evacuated or filled with helium gas in order to prevent the condensation of air or other gases at low temperatures.

The substances, sometimes powders and sometimes single crystals, are positioned such that the high frequency magnetic field has a great intensity and the electric field is weak. This last condition is given because in this way the dielectric losses are reduced.

18. Some results. Only few results will be quoted here. For the others the reader is referred to Vol. XVIII. It will be evident, that the knowledge of the splitting of the lowest level enables one to calculate its contribution to the specific heat.

It was already stated that the resonance experiments had revealed a gradual change in the relative distance, δ, of the two levels in which the lowest energy level with $J = S = 1$ of the Ni^{++} ions in the fluo-silicate is split by the crystalline field. Penrose and Stevens [43] found a decrease of δ from 0.35 cm^{-1} at 195° K to 0.12 cm^{-1} in the temperature region of liquid hydrogen. It was attributed to a thermal contraction of the crystal. This possibility should be kept in mind in calculating susceptibilities.

An interesting group of substances is formed by the free radicals. These organic combinations which contain a free valency and, as a consequence, an unpaired electron spin have, so far as they have been studied magnetically, susceptibilities which can be attributed to one spin [69]. Only at very low temperatures the radicals which were measured down to the helium region showed a small deviation from a Curie-Weiss law, probably caused by an antiferromagnetic exchange interaction. A much greater influence of this interaction is found when studying the paramagnetic resonance. The line width that should be of the order of 100 Oe if only the dipole-dipole interaction is considered is in reality of the order of 10 Oe, in some cases even much smaller [70]. This effect was attributed by Van Vleck and Gorter [71] to the exchange interaction and is known as "exchange narrowing". So far only one of these free radicals, Würster's blue perchlorate, has shown antiferromagnetic properties and these were even found at rather high temperatures, in the neighbourhood of 190° K, as was demonstrated by Pake and Townsend.

VI. Antiferromagnetism.

19. Review of some antiferromagnetic manifestations and their theoretical explanation. When measuring the susceptibilities of paramagnetic salts at low temperatures, Théodoridès and later Woltjer and Kamerlingh Onnes found anomalies at hydrogen temperatures for the anhydrous salts Cr_2O_3, $CrCl_3$, $CoCl_2$ and $NiCl_2$ [72]. The susceptibilities depend on the magnetic fieldstrength and are for all values of H smaller than would correspond to the extrapolated values at higher temperatures, where a Curie-Weiss law is followed. Afterwards, these measurements were extended by Woltjer and Wiersma [73], Schultz [74], Becquerel and van den Handel [48], Bizette and Tsai [75], Stout and Griffel [76] etc. In the mean time Schubnikow and coworkers had measured anomalies in the specific heats [77] of these salts and had found hysteresis effects [78]. These results gave the impression that these salts had become ferromagnetic. In Fig. 23 the behaviour of $CoCl_2$ is given as an example. It should, however, be stated that, whereas the susceptibility of $CoCl_2$ increases when H increases, other salts may show a different behaviour (e.g. a decrease for $CrCl_3$, a maximum for $FeCl_2$). The temperature of the maximum in the specific heat, and that below which anomalies in χ are found, are very near each other and also near the value of Θ in the equation $\chi(T - \Theta) = C$, valid at higher temperatures (for $CoCl_2$, $\Theta = 20°$; the maximum in the specific heat was found at 24.9° K).

There are, however, important differences with ferromagnetism e.g.: even for high values of H there is no approach to saturation and the values of χ are not greater than for paramagnetic substances.

NÉEL had interpreted a group of effects, which were first observed in certain metals, in terms of the existence of what he called "antiferromagnetism" [79]. This interpretation later proved to be applicable to the behaviour of a number of salts, probably including the anhydrous salts mentioned above.

Only an outline of this theory, which was afterwards extended by VAN VLECK [80], will be given here. The impression seems justified that anti-ferromagnetism is a rather general property of matter and that many paramagnetic substances become antiferro-magnetic when the temperature is low enough. The origin of this state is found in an exchange interaction between the dipoles, giving rise to a preference for antiparallel orientation[1]. Below a certain transition temperature, T_N, the NÉEL point, comparable with the CURIE point for ferromagnetic substances, a spontaneous order sets in. In simple cases the magnetic dipoles form two groups, situated in two sublattices (which can be realised in several ways depending on the crystalline structure). In each of the two groups a parallel orientation exists, with increasing degree of order when the temperature decreases, just as in the ferromagnetic state and approaching a saturation value for $T = 0$. In the absence of an external magnetic field the equal magnetizations, σ_1 and σ_2, of the two sublattices have an antiparallel orientation so that the total magnetization is zero. A nice confirmation of

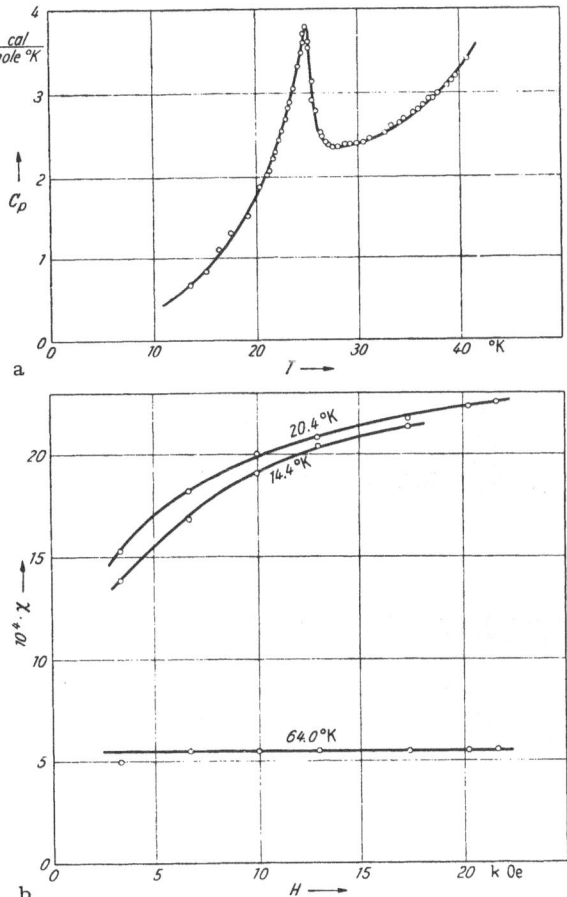

Fig. 23 a and b. Antiferromagnetic behaviour of $CoCl_2$. (a) The curve represents the specific heats, measured by TRAPEZNIKOWA, SCHUBNIKOW and MILJUTIN; (b) the curves show the susceptibilities, measured by WOLTJER.

this hypothesis is given by SHULL and collaborators [81] who used the method of neutron diffraction. For several antiferromagnetic substances the pattern for $T < T_N$ could be interpreted with a lattice constant twice as large as that which existed for $T > T_N$. In a weak magnetic field in the direction of σ_1 and σ_2, so weak that the interaction energy of the dipole with its surroundings is much greater than the magnetic energy of the dipole in this field, no magnetization takes place when $T = 0$. For a temperature between zero and the NÉEL point a small magnetization would be found. In a field perpendicular to σ_1 and σ_2 these magneti-zations are "bent" a little bit in the direction of H. Here a temperature

---

[1] In special cases it seems that magnetic interactions can already give rise to antiferro-magetism, as was pointed out by Miss O'BRIEN for the chrome alums [104].

independent susceptibility is found. In a magnetic field there is always a tendency for the spontaneous magnetizations to have an orientation perpendicular to the field as this causes a lowering of the free energy. This tendency will be counteracted if there exists an anisotropic binding of the magnetizations to the crystal lattice, as is geneially the case. Then a preferred direction can be found. If H is applied along this direction, $\chi = 0$ for $T = 0$, and is only relatively small for $T \neq 0$ but below T_N and for fieldstrengths below a critical value, the threshold value H_{thr} (see Fig. 24). Above H_{thr} the state with σ_1 and σ_2 perpendicular to H will have a lower free energy (or energy for $T = 0$) than the parallel state. The decrease of the magnetic energy is then greater than the increase of the energy because of the anisotropy. A flopping over from the parallel to the normal position occurs and now the susceptibility $\chi_{||}$ is almost equal to $\chi_\perp$, in a field $H_\perp$, perpendicular to the a-direction. In general a small difference between $\chi_{||}$ and $\chi_\perp$ will exist because of the anisotropy of the lattice.

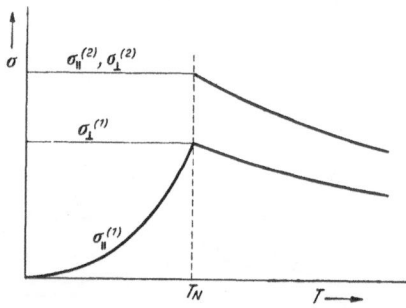

Fig. 24. The magnetizations in the preferred direction ($\sigma_{||}$) and perpendicular to it ($\sigma_\perp$) as functions of T for a constant fieldstrength in an antiferromagnetic single crystal after the theory of Néel and Van Vleck, when there is isotropy above T_N. $\sigma_{||}^{(1)}$ and $\sigma_\perp^{(1)}$ are found when $H < H_{\text{thr}}$; $\sigma_{||}^{(2)}$ and $\sigma_\perp^{(2)}$ are found when $H > H_{\text{thr}}$.

In the theories of Néel and Van Vleck, as well as in the extension of Gorter and Haantjes [82] and others [83], the influence of the exchange interaction is described in a ormal way by molecular fields which are fntroduced in the same way as in the Weiss itheory of ferromagnetism. If the sublattices are characterized by means of the indices 1 and 2 the field in which an ion of lattice 1 or 2 are placed is:

$$\left.\begin{aligned} H_1 &= H_{\text{ext}} - \alpha\,\sigma_2, \\ H_2 &= H_{\text{ext}} - \alpha\,\sigma_1. \end{aligned}\right\} \tag{19.1}$$

If not only the effect of nearest neighbours is taken into account as was done in (19.1) one obtains instead:

$$\left.\begin{aligned} H_1 &= H_{\text{ext}} - \alpha\,\sigma_2 - \beta\,\sigma_1, \\ H_2 &= H_{\text{ext}} - \alpha\,\sigma_1 - \beta\,\sigma_2, \end{aligned}\right\} \tag{19.2}$$

β can be positive or negative depending on the question of whether the ions in the same sublattice have a ferromagnetic or an antiferromagnetic influence on each other. If an anisotropy is included α and β have a tensor character.

Whereas the treatment of Néel, Gorter and Haantjes, is only valid at the absolute zero, Nagamiya [84] and Yosida [85] considered the case of $T \neq 0$ but restricted themselves to small values of $H (H < H_{\text{thr}})$.

Other theoretical considerations start from the order-disorder theory (e.g. Li[86], Kasteleyn and van Kranendonk [87], Brooks and Domb [88]) or use the spin wave theory (e.g. Kramers and Heller [89], Hulthén [90], Anderson [91], Nakamura [92], Tessman [93], Kubo [94], van Kranendonk and Van Vleck [95]. A review of the literature is found in the articles of Newell and Montroll [96] of Nagamiya, Yosida and Kubo [97] and of Poulis and Gorter [98].

20. Discussion of a special case. As one salt presenting a not too complicated case of antiferromagnetism was rather intensively studied, this will be taken as an example. This salt is $CuCl_2 \cdot 2H_2O$ which has orthorhombic crystal structure. In Leiden measurements were done on the magnetization in static and alternating magnetic fields [99], on the specific heat [100], the electronic resonance [101] and the proton resonance [102].

Some of the results of the magnetization measurements in static fields are collected in Fig. 25 and 27. In Fig. 25 the magnetization is represented as a

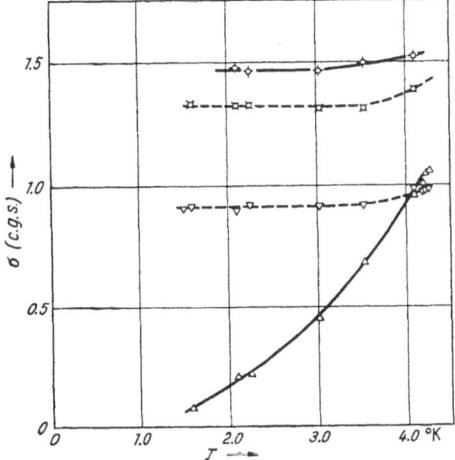

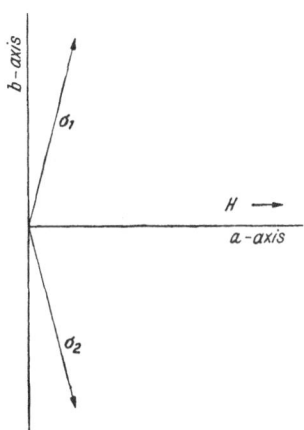

Fig. 25. $\sigma_{||}$ and $\sigma_{\perp}$ vs T for a single crystal of $CuCl_2 \cdot 2H_2O$. $\triangle$ and $\triangledown$ for $H_1 < H_{thr}$; $\oplus$ and $\boxtimes$ for $H_2 > H_{thr}$; —— $||$ axis, – – – $\perp$ axis.

Fig. 26. Directions of the magnetizations of the sublattices when H is parallel to the preferred direction and $> H_{thr}$.

function of temperature for two values of H, one lower and one higher than H_{thr}, parallel $(H_{||})$ and perpendicular $(H_{\perp})$ to the preferred direction. This is, in $CuCl_2 \cdot 2H_2O$, the a-direction. Of the directions perpendicular to the a axis, magnetization parallel to the b axis leads to a lower free energy than along the c axis. For $H_{||} > H_{thr}$ the spontaneous antiparallel magnetizations σ_1 and σ_2 will therefore be parallel to the b axis, aside from a small deviation caused by the magnetic field (Fig. 26). At first sight there seems to be good agreement between Fig. 24 and 25. There are, however, several serious discrepancies. In the first place, the value of the magnetization at the NÉEL point, which point was determined by the resonance- and specific heat measurements and found to be 4.3° K, is much lower than the value extrapolated from the suscepti- bility measurements at higher temperatures, which can be re- presented by $\chi(T + 5) = C$.

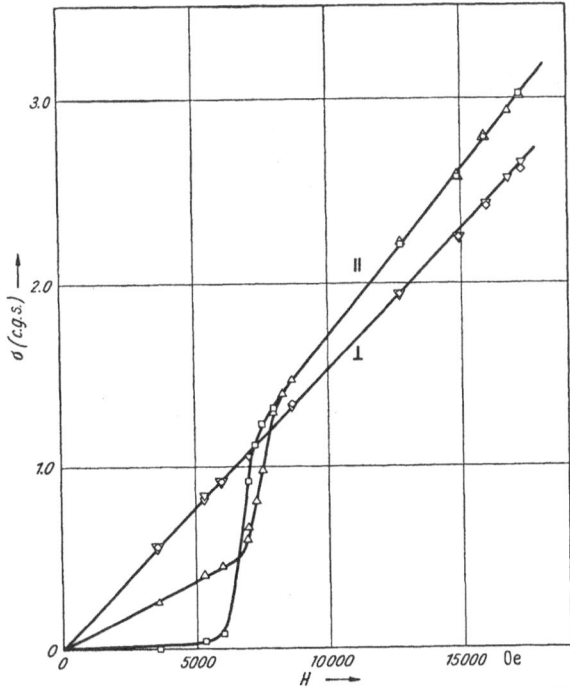

Fig. 27. $\sigma_{||}$ and $\sigma_{\perp}$ vs H for a single crystal of $CuCl_2 \cdot 2H_2O$ at 1.59° K ($\square$ and $\diamondsuit$) and 3° K ($\triangle$ and $\triangledown$). The curves for $\sigma_{\perp}$ coincide for all values of H, those for $\sigma_{||}$ only when $H > H_{thr}$.

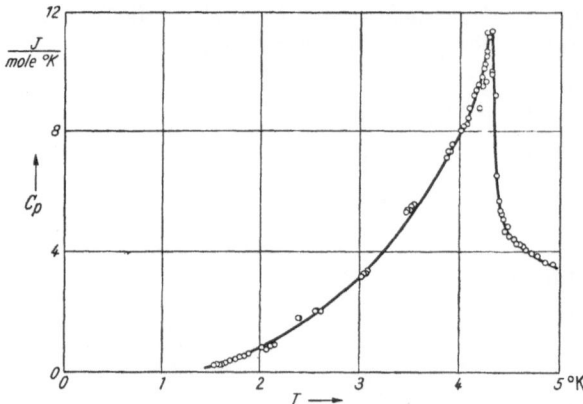

Fig. 28. Specific heat of CuCl₂ · 2H₂O as a function of T.

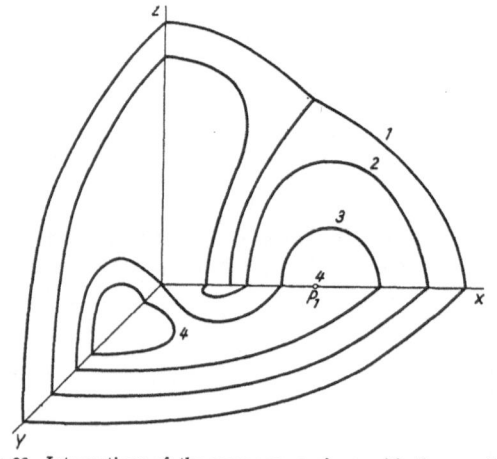

Fig. 29. Intersections of the resonance surfaces with the coordinate planes for different values of the frequency. 1 corresponds to $\omega = 0$.

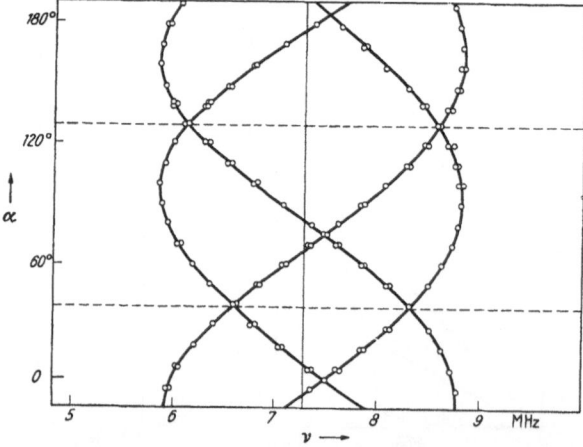

Fig. 30. Rotation diagram for CuCl₂ · 2H₂O. The positions of the proton resonance lines are plotted as functions of the direction of the magnetic field in the ab plane. At 7.26 MHz the position of the proton resonance line in water is indicated. The directions of the a- and b-axes are found at 125° resp. 35°. $T = 4.13°$ K; $H = 1705$ Oe.

In the second place, there is an anomalous behaviour near the Néel point. When the temperature is increased the magnetization does not decrease immediately when the Néel point is passed. Whereas the first fact is also found for other salts, the second one seems not to be a general property. In Fig. 27, representing the magnetization as a function of the magnetic field-strength for directions parallel to the a and b directions and for two different temperatures, the threshold value of H manifests itself in a sudden rise of the magnetization. Above H_{thr} $\sigma_{||}$ and $\sigma_{\perp}$ do not coincide because of an anisotropy in μ and α.

The specific heat of the salt, which was measured by Friedberg is represented in Fig. 28. It is seen that when T has increased to values larger than T_N the specific heat has not immediately its normal value. There is an appreciable change in entropy above T_N. For CuCl₂ · 2H₂O, about $\frac{1}{3}$ of the orientational entropy is released above the Néel point. This indicates a persistence of a short range order when T_N is passed, a fact that was also suggested by the magnetization measurements. Similar results were also obtained for other salts.

Using the theory of Gorter and Haantjes, extended by Ubbink [101] with regard to the antiferromagnetic resonance, the precession frequencies of the system of magnetic ions can be calculated in the external field if the antiferromagnetic interaction is represented by a molecular field. Or, what amounts to the same thing, for a certain frequency of a high frequency field, the static external magnetic field at which the resonance will occur can be calculated. Zero frequency is obtained for those values of H where the antiferromagnetic order is transformed into a state of saturated paramagnetism. The two sublattices now have parallel magnetizations. For both, the molecular field has a direction opposite to the external field. The two fields have the same magnitude here. A frequency zero is also obtained on a critical hyperbola in the ac plane of the crystal. When the vector $\boldsymbol{H}$ surpasses this curve, a flopping over of the magnetizations σ_1 and σ_2 occurs. The intersection with the a axis is found for $H = H_{\text{thr}}$. On this hyperbola the directions of the dipoles are indeterminate. The calculations of UBBINK have led him to a representation of the resonance surfaces for different values of the

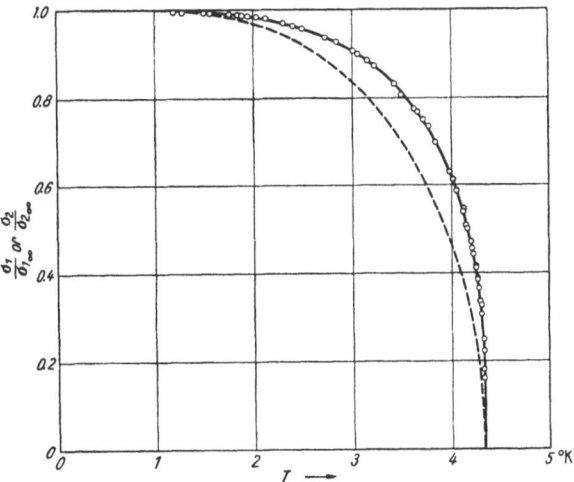

Fig. 31. —o— spontaneous magnetizations σ_1 or σ_2 divided by their maximum value ($\frac{1}{2} N \mu_B$ per mole) as a function of T after the measurements of POULIS. - - - - - calculated values after the molecular field theory when for T_N the experimental value is taken.

frequency. The intersections of a group of these surfaces with the coordinate planes for not too high frequencies is represented in Fig. 29. Along the axes are plotted the values of $\mu_i H_i$ instead of the values of the components of $\boldsymbol{H}$, so that the anisotropy of g is included. It is seen that, for a constant frequency, resonances may occur for different values of the magnetic field strength. The measurements of the antiferromagnetic resonance were in general in reasonably good agreement with this representation.

The protons of the water molecules were used as indicators of the magnetic field in the neighbourhood of the Cu^{++} ions. As two groups of Cu^{++} ions exist causing in general a strengthening or a weakening of the external magnetic field, H_e, the proton resonance lines may be displaced in a constant field to a higher or a lower frequency with respect to that which would be found for free protons. For a given value of T and of H_e the amount of the displacement depends on the angle between the magnetic field and the a-axis. The fact that the pattern obtained by plotting the resonance frequency as a function of this angle is, to a high approximation, symmetric with respect to the position of the free proton line, indicates the existence of an antiferromagnetic order. As an example such a pattern is reproduced in Fig. 30 for which $H_e < H_{\text{thr}}$. From the maximum value of the displacement, found when H_e is parallel to the a-axis, the values of σ_1 and σ_2 can be deduced. Poulis finds a dependence on T as is represented in Fig. 31. Until now no satisfactory agreement with theory is obtained in this respect.

The results obtained with this salt caused Gorter to develop a phase diagram for antiferromagnetic substances [103]. From the σ-H diagrams at different temperatures the transition lines between different phases in the H-T plane may be drawn. In the simple case of $CuCl_2 \cdot 2H_2O$ the diagram has the form indicated in Fig. 32. The transition from the antiferromagnetic to the paramagnetic state is a second order transition (no change in the magnetization); that which corresponds to the flopping over of the submagnetizations is a first order transition. In [103] these diagrams are discussed under various conditions, sometimes leading to more complicated behaviour than in the example given.

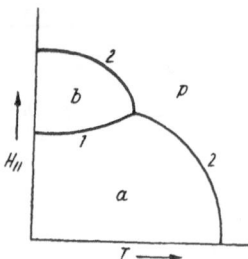

Fig. 32. Type of phase diagram of an antiferromagnetic substance when the magnetic field is in the preferred direction a indicates the antiferromagnetic region where σ_1 and σ_2 are parallel and antiparallel to $H_{||}$, b is the region where they are oriented as in Fig. 26. These two regions are separated by the threshold curve. In p the substance is in the paramagnetic state. The numbers 1 and 2 indicate the order of the transitions.

Bibliography.

[1] Rapp. 3me conseil de Physique Solvay, Brussels 1921. — Commun. Phys. Lab. Univ. Leiden Suppl., No. 44a, I, § 5 (1921).
[2] Becquerel, J.: Proc. Kon. Nederl. Akad. Wetensch. **32**, 749 (1929). — Commun. Phys. Lab. Univ. Leiden Suppl. No. 68a.
[3] Brunetti, R.: Rend. Accad. naz. Lincei (6) **7**, 238 (1928); **9**, 754 (1929).
[4] Bethe, H.: Ann. Phys. (5) **3**, 133 (1929). — Z. Physik **60**, 218 (1930).
[5] Kramers, H. A.: Proc. Kon. Nederl. Akad. Wetensch. **33**, 959 (1930).
[6] Van Vleck, J. H.: Phys. Rev. **29**, 727 (1927); **31**, 587 (1928); **35**, 467 (1931).
[7] Van Vleck, J. H.: Electric and Magnetic Susceptibilities. Oxford: Clarendon Press 1932.
[8] Gorter, C. J., and W. J. de Haas: Proc. Kon. Nederl. Akad. Wetensch. **34**, 1243 (1931). — Commun. Kamerlingh Onnes Lab. Univ. Leiden No. 218b.
[9] Penney, W. G., and R. Schlapp: Phys. Rev. **41**, 194 (1932).
[10] Elliott, R. J., and K. W. H. Stevens: Proc. Phys. Soc. Lond. A **64**, 205 (1951). — Proc. Roy. Soc. Lond., Ser. A **215**, 437 (1952); **219**, 387 (1953).
[11] Ketelaar, J. A. A.: Physica, 's-Grav. **8**, 619 (1937).
[12] Bleaney, B., and K. W. H. Stevens: Rep. Progr. Phys. **16**, 108 (1953).
[13] Van Vleck, J. H.: Phys. Rev. **41**, 208 (1932).
[14] Gorter, C. J.: Phys. Rev. **42**, 437 (1932).
[15] Penney, W. G., and R. Schlapp: Phys. Rev. **42**, 666 (1932).
[16] Bose, A.: Indian J. Phys. **22**, 195, 276 (1948).
[17] Jackson, L. C.: Phil. Trans. Roy. Soc. Lond., Ser. A **224**, 1 (1923). — Commun. Phys. Lab. Univ. Leiden, No. 163 (1923). — Thesis Univ. Leiden 1923.
[18] Siegert, A.: Physica, 's-Grav. **3**, 85 (1936); **4**. 138 (1937). — Handel, J. van den, u. A. Siegert: Physica, 's-Grav. **4**, 871 (1937). — Commun. Kamerlingh Onnes Lab. Univ. Leiden No. 249e.
[19] Wiersma, E. C., W. J. de Haas and W. H. Capel: Proc. Kon. Nederl. Akad. Wetensch. **33**, 1119 (1930). — Commun. Phys. Lab. Univ. Leiden No. 212b.
[20] Wiersma, E. C., W. J. de Haas and W. H. Capel: Proc. Kon. Nederl. Akad. Wetensch. **34**, 494 (1931). — Commun. Phys. Lab. Univ. Leiden No. 215b. — Woltjer, H. R., C. W. Coppoolse and E. C. Wiersma: Proc. Kon. Nederl. Akad. Wetensch. **32**, 1329 (1929) — Commun. Phys. Lab. Univ. Leiden No. 201d.

[21] WIERSMA, E. C., and C. J. GORTER: Physica, 's-Grav. 12, 316 (1932). — Commun. Kamerlingh Onnes Lab. Univ. Leiden Suppl. No. 73a.
[22] KANDA, E., T. HASEDA and A. ÔTSUBO: Science Rep. RITU, A 7, 1 (1955).
[23] SUCKSMITH, W.: Phil. Mag. 8, 158 (1929). — A theoretical treatment of the deformations of the ring is given by: SWINDLEHURST, E: Proc. Leeds Phil. and Lit. Soc. 5, 224 (1949).
[24] HUTCHISON, T. S., and J. REEKIE: J. Scient. Instr. 23, 209 (1946). — McGUIRE, T. R., and C. T. LANE: Rev. Sci. Instrum. 20, 489 (1949).
[25] SCHULTZ, B. H.: Physica, 's-Grav. 6, 137 (1939). — Commun. Kamerlingh Onnes Lab. Univ. Leiden No. 253d. — Thesis Univ. Leiden 1940.
[26] HENRY, W. E.: Proc. N.B.S. Semicentennial symp. on Low Temp. Phys., p. 237, 1951. Phys. Rev. 87, 1133 (1952). Rev. Mod. Phys. 25, 163 (1953).
[27] FEREDAY, R. A., and E. C.WIERSMA: Physica, 's-Grav. 2, 575 (1935). — Commun. Kamerlingh Onnes Lab. Univ. Leiden No. 237a.
[28] KRISHNAN, K. S., and S. BANERJEE: Phil. Trans. Roy. Soc. Lond. 234, 265 (1935). — KRISHNAN, K. S., A. MOOKHERJI and A. BOSE: Phil. Trans. Roy. Soc. Lond. 238, 125 (1939).
[29] KLERK, D. DE: Thesis Univ. Leiden 1948.
[30] GOBRECHT, H.: Ann. Phys. 28, 673 (1937).
[31] FREED, S., and F. H. SPEDDING: Phys. Rev. 34, 945 (1929). — FREED, S.: Phys. Rev. 38, 2122 (1931). — Commun. Kamerlingh Onnes Lab. Univ. Leiden, No. 222a. — FREED, S., and J. G. HARWELL: Proc. Kon. Nederl. Akad. Wetensch. 35, 979 (1932). — Commun. Kamerlingh Onnes Lab. Univ. Leiden No. 222b. — SPEDDING, F. H.: Phys. Rev. 50, 574 (1936). — SPEDDING, F. H., H. F. HAMLIN and G. C. NUTTING: J. Chem. Phys. 5, 191 (1937).
[32] SPEDDING, F. H.: J. Chem. Phys. 5, 160 (1937). — Phys. Rev. 50, 574 (1936).
[33] BECQUEREL, J.: Commun. Phys. Lab. Univ. Leiden No. 103 (1908); Suppl. No. 20 (1909); No. 177 (1925). — Proc. Kon. Nederl. Akad. Wetensch. 32, 749 (1929). — Commun. Phys. Lab. Univ. Leiden Suppl. No. 68a.
[34] BECQUEREL, J., and W.J. DE HAAS: Commun. Phys. Lab. Univ. Leiden No. 193a (1928).
[35] COTTON, A.: Le phénomène de Zeeman. Coll. Scientia 1899. — DORFMANN, J.: Z. Physik 17, 98 (1923). — LADENBURG, R.: Z. Physik 46, 168 (1928).
[36] VAN VLECK, J. H., and M. H. HEBB: Phys. Rev. 46, 17 (1934).
[37] BECQUEREL, J.: Physica, 's-Grav. 3, 705 (1936). — Commun. Kamerlingh Onnes Lab. Univ. Leiden No. 243d. — BECQUEREL, J., and J. VAN DEN HANDEL: Physica, 's-Grav. 3, 1133 (1936); 4, 345, 543 (1937); 5, 753, 857 (1938); 8, 711 (1940). — Commun. Kamerlingh Onnes Lab. Univ. Leiden No. 244a, b, c, d, e, 259d.
[38] JACKSON, L. C.: C. R. Acad. Sci. Paris 177, 154 (1923). — Commun. Phys. Lab. Univ. Leiden No. 168a. — HAAS, W. J. DE, J. VAN DEN HANDEL and C. J. GORTER: Phys. Rev. 43, 81 (1933). — Commun. Kamerlingh Onnes Lab. Univ. Leiden No. 228b. — HANDEL, J. VAN DEN: Physica, 's-Grav. 8, 513 (1941). — Commun. Kamerlingh Onnes Lab. Univ. Leiden No. 263a. — HANDEL, J. VAN DEN, and J.C. HUPSE: Physica, 's-Grav. 9, 225 (1942). — Commun. Kamerlingh Onnes Lab. Univ. Leiden No. 263b. — HANDEL, J. VAN DEN: Thesis Univ. Leiden 1940.
[39] POINCARÉ, H.: Théorie mathématique de la lumière, Vol. II, Chap. XII. 1892. See also: BECQUEREL, J.: Commun. Phys. Lab. Univ. Leiden No. 191c (1920).
[40] BECQUEREL, J.: Commun. Phys. Lab. Univ. Leiden No. 191c (1928). — Z. Physik 52, 342 (1928). — J. Phys. Radium 9, 337 (1928). — BECQUEREL, J.: Commun. Phys. Lab. Univ. Leiden No. 211a (1930). — BECQUEREL, J., and W. J. DE HAAS: Commun. Phys. Lab. Univ. Leiden No. 211b, c (1930).
[41] LÉVY, M., and J.VAN DEN HANDEL: Physica 17, 737 (1951). — Commun. Kamerlingh Onnes Lab. Univ. Leiden No. 286c. — LÉVY, M.: Thesis Univ. Paris 1949.
[42] BECQUEREL, J., et W. OPECHOWSKI: Physica, 's-Grav. 6, 1039 (1939).
[43] PENROSE, R. P., and K. W. H. STEVENS: Proc. Phys. Soc. Lond. A 63, 29 (1949).
[44] OLLOM, J. F., and J. H. VAN VLECK: Physica 17, 205 (1951).
[45] BECQUEREL, J.: Physica 18, 183 (1952).
[46] BECQUEREL, J., J.VAN DEN HANDEL et H.A. KRAMERS: Physica 17, 717 (1951). — Commun. Kamerlingh Onnes Lab. Univ. Leiden No. 286b.
[47] LÉVY, M.: Thesis Paris 1949. — Ann. de Phys. 5, 153, 310 (1950).
[48] BECQUEREL, J., et J. VAN DEN HANDEL: J. Phys. Radium 10, 10 (1939). — Commun. Kamerlingh Onnes Lab. Univ. Leiden No. 255b.
[49] WALLER, I.: Z. Physik 79, 370 (1932).
[50] GORTER, C. J.: Physica, 's-Grav. 3, 503 (1936).
[51] CASIMIR, H. B. G., and F. K. DU PRÉ: Physica, 's-Grav. 5, 507 (1938). — Commun. Kamerlingh Onnes Lab. Univ. Leiden Suppl. No. 85a.
[52] BROER, L. J. F., and C. J. GORTER: Physica, 's-Grav. 8, 621 (1943).

[53] Marel, L. C. v. d.: Conf. low temp. phys. Paris 1955. Physica 1956 or 1957. — Thesis Univ. Leiden to appear in 1957.
[54] Van Vleck, J. H.: Phys. Rev. 59, 724, 730 (1941).
[55] Gorter, C. J., L. C. v. d. Marel and B. Bölger: Physica 21, 103 (1955). — Commun. Kamerlingh Onnes Lab. Univ. Leiden Suppl. No. 109c.
[56] Broer, L. J. F.: Thesis Univ. Amsterdam 1945.
[57] Vrijer, F. W. de: Thesis Univ. Leiden 1951.
[58] Vrijer, F. W. de, and C. J. Gorter: Physica, 's-Grav. 14, 617 (1949).
[59] Broer, L. J. F., and D. C. Schering: Physica, 's-Grav. 10, 631 (1943).
[60] Gorter, C. J.: Physica, 's-Grav. 3, 503 (1936). — Commun. Kamerlingh Onnes Lab. Univ. Leiden No. 241e. — Brons, F., and C. J. Gorter: Physica, 's-Grav. 5, 999 (1938), see also ref. 57.
[61] Haas, W. J. de, and F. K. du Pré: Physica, 's-Grav. 6, 705 (1939). — Commun. Kamerlingh Onnes Lab. Univ. Leiden No. 258a.
[62] Jahn, H. A., and E. Teller: Proc. Roy. Soc. Lond., Ser. A 161, 220 (1937). — Jahn, H. A.: Proc. Roy. Soc. Lond., Ser. A 164, 117 (1937). — Van Vleck, J. H.: J. Chem. Phys. 7, 72 (1939).
[63] Zavoisky, E.: J. Phys. USSR. 9, 211 (1945); 10, 197 (1946). — Cummerow, R. L., and D. Halliday: Phys. Rev. 70, 433 (1946).
[64] Bleaney, B.: Physica 17, 175 (1951).
[65] Gorter, C. J.: Physica, 's-Grav. 14, 504 (1948). — Commun. Kamerlingh Onnes Lab. Univ. Leiden Suppl. No. 97d.
[66] Penrose, R. P.: Nature, Lond. 163, 992 (1949). — Commun. Kamerlingh Onnes Lab. Univ. Leiden No. 278f..
[67] Bleaney, B., K. D. Bowers and D. J. E. Ingram: Proc. Phys. Soc. Lond. A 64, 758 (1951).
[68] Bowers, K. D.: Proc. Phys. Soc. Lond. A 65, 860 (1952).
[69] Müller, E., u. I. Müller-Rodloff: Ann. Chem. 521, 81 (1936). — Handel, J. van den: Physica 18, 921 (1952). — Commun. Kamerlingh Onnes Lab. Univ. Leiden No. 291b. — Gerritsen, H. J., R. Okkes, H. M. Gijsman and J. van den Handel: Physica 20, 13 (1954). — Commun. Kamerlingh Onnes Lab. Univ. Leiden No. 294c. — Holden, A. N., C. Kittel, F. R. Merritt and W. A. Yager: Phys. Rev. 75, 1614 (1949); 77, 147 (1950).
[70] Holden, A. N., W. A. Yager and F. R. Merritt: J. Chem. Phys. 19, 1319 (1951). — Townes, C. H., and J. Turkevitch: Phys. Rev. 77, 148 (1950).
[71] Gorter, C. J., and J. H. Van Vleck: Phys. Rev. 72, 1126 (1947). — Commun. Kamerlingh Onnes Lab. Univ. Leiden Suppl. No. 97a. — Van Vleck, J. H.: Phys. Rev. 74, 1168 (1948).
[72] Théodoridès, Ph.: J. Phys. (VI) 3, 1 (1922). — Woltjer, H. R.: Commun. Phys. Lab. Univ. Leiden No. 173b. — Woltjer, H. R., and H. Kamerlingh Onnes: Commun. Phys. Lab. Univ. Leiden No. 173c (1925).
[73] Woltjer, H. R., and E. C. Wiersma: Proc., Kon. Nederl. Akad. Wetensch. 32, 735 (1929). — Commun. Phys. Lab. Univ. Leiden No. 201a.
[74] Schultz, B. H.: Thesis Univ. Leiden 1940. — Physica, 's-Grav. 7, 413 (1940). — Commun. Kamerlingh Onnes Lab. Univ. Leiden No. 259b. — Haas, W. J. de, et B. H. Schultz: J. Phys. Radium 10, 7 (1939). — Commun. Kamerlingh Onnes Lab. Univ. Leiden No 255a. Haas, W. J. de, B. H. Schultz and Miss J. Koolhaas: Physica, 's-Grav. 7, 57 (1940). — Commun. Kamerlingh Onnes Lab. Univ. Leiden No. 259a.
[75] Bizette, H., et B. Tsai: C. R. Acad. Sci. Paris 207, 449 (1938); 209, 205 (1939); 212, 119 (1941).
[76] Stout, J. W., and M. Griffel: Phys. Rev. 76, 144 (1949).
[77] Trapeznikowa, O. N., u. L. W. Schubnikow: Phys. Z. Sowjet. 7, 66, 255 (1935). — Trapeznikowa, O. N., L. Schubnikow u. G. A. Miljutin: Phys. Z. Sowjet. 9, 237 (1936). — Miljutin, G. A., and S. S. Shalyt: C. R. Acad. URSS. 24, 680 (1939).
[78] Schubnikow, L. W., u. S. S. Shalyt: Phys. Z. Sowjet. 11, 566 (1937). — Shalyt, S. S.: C. R. Acad. URSS. 20, 657 (1938).
[79] Néel, L.: Ann. Phys. (11) 5, 232 (1936); (12) 3, 137 (1948).
[80] Van Vleck, J. H.: J. Chem. Phys. 9, 85 (1941).
[81] Shull, C. G., and J. S. Smart: Phys. Rev. 76, 1256 (1949). — Shull, C. G., W. A. Strauser and E. O. Wollan: Phys. Rev. 83, 333 (1951). — Erickson, R. A.: Phys. Rev. 90, 779 (1953).
[82] Gorter, C. J., and J. Haantjes: Physica 18, 285 (1952). — Commun. Kamerlingh Onnes Lab. Univ. Leiden Suppl. No. 104b.
[83] See for literature e.g. Peski-Tinbergen, T. van, and C. J. Gorter: Physica 20, 592 (1954). — Commun. Kamerlingh Onnes Lab. Univ. Leiden Suppl. No. 109a, or Nagamiya, T., K. Yosida and R. Kubo: Adv. Physics 4, 1 (1955).

[*84*] NAGAMIYA, T.: Progr. Theor. Phys. **6**, 342 (1951).
[*85*] YOSIDA, K.: Progr. Theor. Phys. **6**, 691 (1951).
[*86*] YIN-YUAN LI: Phys. Rev. **80**, 457 (1950); **84**, 721 (1951).
[*87*] KASTELEYN, P. W., and J. VAN KRANENDONK: Physica, 's-Grav. **22** (1956).
[*88*] BROOKS, J. E., and C. DOMB: Proc. Roy. Soc. Lond., Ser. A **207**, 343 (1951).
[*89*] HELLER, G. u. H. A. KRAMERS: Proc. Kon. Nederl. Akad. Wetensch. **37**, 378 (1934).
[*90*] HULTHÉN, L.: Proc. Kon. Nederl. Akad. Wetensch. **39**, 190 (1936).
[*91*] ANDERSON, P. W.: Phys. Rev. **86**, 694 (1952).
[*92*] NAKAMURA, T.: Progr. Theor. Phys. **7**, 539 (1952).
[*93*] TESSMAN, J. R.: Phys. Rev. **88**, 1132 (1952).
[*94*] KUBO, R.: Phys. Rev. **87**, 568 (1952). — Rev. Mod. Phys. **25**, 344 (1953).
[*95*] KRANENDONK, J. VAN, and J. H. VAN VLECK: Rev. Mod. Phys.
[*96*] NEWELL, G. F., and E. W. MONTROLL: Rev. Mod. Phys. **25**, 353 (1953).
[*97*] NAGAMIYA, T., K. YOSIDA and R. KUBO: Adv. Physics **4**, 1 (1955).
[*98*] POULIS, N. J., and C. J. GORTER: Progress in low temp. phys., p. 245. Amsterdam: North-Holland Publ. Comp. 1955.
[*99*] HANDEL, J. VAN DEN, H. M. GIJSMAN and N. J. POULIS: Physica **18**, 862 (1952). — Commun. Kamerlingh Onnes Lab. Univ. Leiden No. 290c. — MAREL, L. C. V. D., J. V. D. BROEK, J. D. WASSCHER and C. J. GORTER: Physica **21**, 685 (1955). — Commun. Kamerlingh Onnes Lab. Univ. Leiden No. 300d.
[*100*] FRIEDBERG, S. A.: Physica **18**, 714 (1952). — Commun. Kamerlingh Onnes Lab. Univ. Leiden No. 289d.
[*101*] UBBINK, J.: Proc. int. conf. low temp. phys., p. 163. Oxford 1951. — Thesis Univ. Leiden 1953. — Physica **19**, 9, 919 (1953). — Commun. Kamerlingh Onnes Lab. Univ. Leiden Suppl. No. 105b, c. — UBBINK, J. B., J. A. POULIS, H. J. GERRITSEN and C. J. GORTER: Physica **19**, 928 (1953). — Commun. Kamerlingh Onnes Lab. Univ. Leiden No. 293a.
[*102*] POULIS, N. J.: Thesis Univ. Leiden 1952. — POULIS, N. J., and G. E. G. HARDEMAN: Physica **18**, 201, 315, 429 (1952); **19**, 391 (1953); **20**, 719 (1954). — Commun. Kamerlingh Onnes Lab. Univ. Leiden No. 287a, 288b, c, 291d, 294a.
[*103*] GORTER, C. J., and TINEKE VAN PESKI-TINBERGEN: Physica **22**, 273 (1956). — Commun. Kamerlingh Onnes Lab. Univ. Leiden Suppl. No. 110b. — GORTER, C. J.: Conf. low temp. phys., Paris 1955.
[*104*] O'BRIEN, MARY C. M.: Bull. Amer. Phys. Soc., Ser. II **1**, 290 (1956).

General References.

VAN VLECK, J. H.: Electric and Magnetic Susceptibilities. Oxford: Clarendon Press 1932.
STONER, E. C.: Magnetism and Matter. London: Methuen 1934.
BATES, L. F.: Modern Magnetism. Cambridge University Press 1939.
Congrès sur le magnétisme. Strasbourg 1939.
Washington Conference on Magnetism, 1952, published in Rev. Mod. Phys. **25**, 1 (1953).

On paramagnetic relaxation.

GORTER, C. J.: Paramagnetic Relaxation. Amsterdam: Elsevier Publishing Co. 1947.
COOKE, A. H.: Rep. Progr. Phys. **13**, 276 (1950).

On paramagnetic resonance.

GORDY, W.: Rev. Mod. Phys. **20**, 668 (1948).
GORDY, W., W. V. SMITH and R. F. TRAMBARULO: Microwave Spectroscopy. New York: John Wiley & Sons; London: Chapman & Hall 1953.
BLEANEY, B., and K. W. H. STEVENS: Rep. Progr. Phys. **14**, 108 (1953).

On antiferromagnetism.

Colloque International de ferromagnétisme et d'antiferromagnétisme de Grenoble, 1950, publié dans le J. Phys. Radium **12**, 149 (1951).
NAGAMIYA, T., K. YOSIDA and R. KUBO: Adv. Physics **4**, 1 (1955).
POULIS, N. J., and C. J. GORTER: Progress in low temp. phys., p. 245. Amsterdam: North-Holland Publ. Comp. 1955.

Adiabatic Demagnetization.

By

D. DE KLERK.

With 122 Figures.

A. Fundamental considerations.

I. Introduction.

1. The basic principles of refrigeration. The concept of "low temperature" has been different at different times. In the days when air was first liquefied by CAILLETET and PICTET temperatures of 90 to 50° K were considered as extremely low. At the present time, however, since liquefiers are commercially available, which give the possibility for relatively inexperienced people to liquefy helium in reasonable quantities, few cryogenics physicists would consider the temperature of liquid hydrogen (20 to 14° K) as being "low". The liquid helium range extends from 4.2° K to roughly 1° K (see Sect. 2) and the aim of this article is the discussion of the region of still lower temperatures which, at the present time, is considered as "very low".

The requirements which a thermodynamic system must fulfill in order to be suitable to obtain temperatures below that of its surroundings were discussed very precisely by SIMON[1,2]. A low temperature is characterized not only by a low energy (small thermal motion of the particles), but also by a low entropy (small degree of disorder in the system). For any refrigeration process a working substance is needed of which the entropy depends both on the temperature and on an externally variable parameter. The cooling is carried out in two steps. First the parameter is varied isothermally in the direction in which the entropy is decreased. During this step, mechanical work is put into the system and heat is removed from it. Suppose the decrease of entropy is ΔS and takes place at a temperature T. If the variation of the parameter is performed reversibly (which is favourable for highest efficiency of the process) the amount of heat ΔQ removed from the substance is equal to

$$\Delta Q = T \Delta S. \tag{1.1}$$

It should be clear that during this stage of the process only the part of the entropy is diminished which is determined by the parameter, the part due to the temperature is unaffected. During the second stage the parameter is varied in the opposite direction, but now the process is performed adiabatically. In this step the entropy of the whole system is constant, but part of the entropy due to the temperature is shifted to that given by the parameter. This is accompanied by a fall of the temperature of the system.

The principle of the method can be nicely demonstrated with the help of a (T, S)-diagram as shown in Fig. 1. Here a, b, c and d are curves of constant parameter. For each curve the entropy increases with increasing temperature.

---

[1] F. E. SIMON: Science Museum Handbook, book **3**, 58 (1937).
[2] F. E. SIMON: Physica, 's-Grav. **16**, 753 (1950).

Suppose the process is started at T_i; if the parameter is varied from the value a to c the entropy decreases from S_0 to S_1 and the amount of heat removed from the substance is $(S_0 - S_1) T_i$. After this the parameter is reduced adiabatically to the value a again and the temperature T_f is obtained.

It is clear that the actual drop in temperature depends widely on the shapes of the curves of Fig. 1. According to NERNST's law the system must have zero entropy at the absolute zero of temperature for every value of the parameter. This means that the curves of Fig. 1 converge at low temperatures and finally coincide at absolute zero as shown in Fig. 2. This implies the unattainability of the absolute zero for any process of refrigeration.

The background of NERNST's statement is that at sufficiently low temperatures the disorder in the system is removed by the interaction forces between the ele-

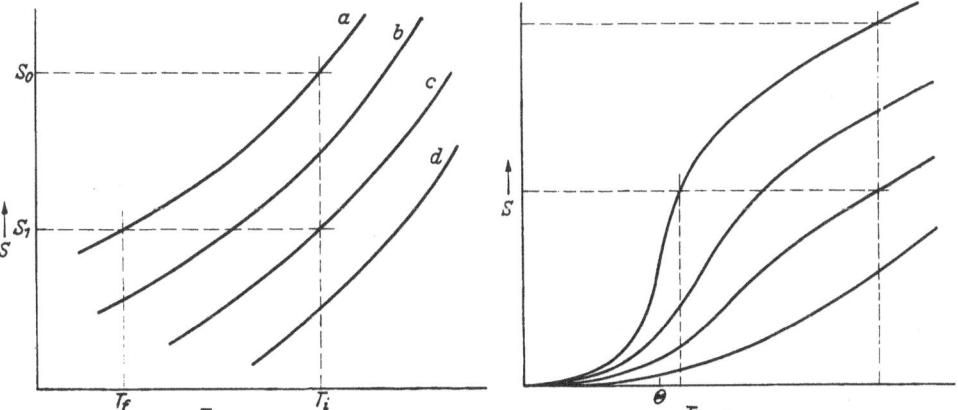

Fig. 1. Entropy versus temperature diagram.   Fig. 2. Entropy versus temperature diagram near absolute zero.

mentary particles. This takes place in a region where the interaction energy E is comparable to the thermal energy kT. Hence a characteristic temperature Θ may be introduced, of the order of E/k, where the system enters a new ordered phase or state. Here a steep fall occurs in the higher curves of Fig. 2 and the specific heat at constant parameter (equal to $T \, \partial S/\partial T$) shows a pronounced maximum. [In the case of a first order transition we have a discontinuity in the (S, T)-curves and a latent heat.] At temperatures well below Θ the entropy depends very little on the parameter and here the working substance looses its effectiveness as a coolant.

The occurence of the specific heat maximum in the region near Θ must be considered as an advantage from a technical point of view. A heat leak occurs as soon as the temperature of the working substance falls below that of the surroundings. Though precautions may be taken to keep it as small as possible it can never be avoided completely, and the larger the specific heat the smaller is the effect of the heat leak on the temperature of the substance. Moreover, if some other material must be cooled down for investigations, the cooling procedure is the more effective the larger is the specific heat of the working substance at the low temperature. So it is clear that, if full advantage should be taken of the working substance, it must be applied in the region of its Θ, hence at the lowest temperatures that can be reached with it. If investigations must be made in different regions of low temperature, substances with different values of Θ are needed.

2. The process of adiabatic demagnetization. The considerations of Sect. 1 apply very well in the case of a gas. In this case the parameter is the pressure, Θ is the boiling point, and the specific heat hump corresponds to the heat of vaporization. The entropy decreases with increasing pressure so that a cooling effect is obtained from isothermal compression followed by adiabatic expansion.

For many years the penetration of a new region of low temperatures has been identical to the liquefaction of a new gas with a lower boiling point than the foregoing one. The last gas in this sequence was helium, liquefied in 1908 by KAMERLINGH ONNES.

The lowest temperature that can be reached with liquid helium by reducing the vapour pressure depends on the capacity of the pumping aggregate and the thermal insulation of the dewar. KAMERLINGH ONNES himself, in 1922, reached a temperature of $0.83°$ K[1]. KEESOM, in 1932, using a set of powerful oil diffusion pumps, obtained $0.71°$ K[2]. (These are the figures as they were originally given by the authors themselves, according to the present vapour pressure data[3] they should be 0.81 and $0.726°$ K). In 1939 it was shown by BLAISSE, COOKE and HULL[4,5] that temperatures of the order of $0.7°$ K can also be reached with diffusion pumps of moderate size if a constriction is applied in the dewar; but it was clear that this was about the limit that could be reached by the method. It should be stressed, however, that the here given figures have the characteristics of records; with the ordinary rotatory pumps used in most laboratories temperatures of about 1.1 or $1.0°$ K can be reached under normal conditions. In this article temperatures of the order of $1°$ K will be quoted as the *"lowest helium temperatures"*.

Below $1°$ K no gas is available for liquefaction any more, so that a different process is needed to enter this region. The first proposal for a new method was published in 1926, independently by DEBYE[6] and GIAUQUE[7]. It was not before 1933, however, that the first experimental results were reported, almost simultaneously from Leiden[8], Berkeley[9] and Oxford[10]. This method is now generally known as the *process of "adiabatic demagnetization"* or *"magnetic cooling"*.

DEBYE and GIAUQUE pointed out that some paramagnetic salts fulfill the requirements of Sect. 1 very nicely. If the magnetic ions in the lattice are fairly far apart ("diluted") so that their interaction energies are very small as compared to the thermal energy at $1°$ K, the spatial orientation is still random at that temperature and the entropy is considerable. In a magnetic field of such a strength, that the potential energy of the magnetic ions is of the same order of magnitude as their thermal energy, big part of the ions is oriented parallel to the field and the entropy is noticeably lower. Hence, if a suitable salt is magnetized isothermally (in heat contact with a cryostat of liquid helium) and then demagnetized adiabatically (the heat contact with the helium being broken), the temperature of the salt falls well below the temperature of the liquid helium.

---

[1] H. KAMERLINGH ONNES: Commun. Kamerlingh Onnes Lab. Leiden, No. 159; Trans. Faraday Soc. **18**, No. 53 (1922).

[2] W. H. KEESOM: Leiden Commun. No. 219a; Proc. Roy. Soc. Amst. **35**, 136 (1932).

[3] H. VAN DIJK and D. SHOENBERG: Nature, Lond. **164**, 151 (1949).

[4] B. S. BLAISSE, A. H. COOKE and R. A. HULL: Physica, 's-Grav. **6**, 231 (1939).

[5] A. H. COOKE and R. A. HULL: Nature, Lond. **143**, 799 (1939).

[6] P. DEBYE: Ann. Phys. **81**, 1154 (1926).

[7] W. F. GIAUQUE: J. Amer. Chem. Soc. **49**, 1864, 1870 (1927).

[8] W. J. DE HAAS, E. C. WIERSMA and H. A. KRAMERS: Leiden Commun. No. 229a; Physica, 's-Grav. **1**, 1 (1933/34).

[9] W. F. GIAUQUE and D. P. MCDOUGALL: Phys. Rev. **43**, 768 (1933).

[10] N. KURTI and F. SIMON: Nature, Lond. **133**, 907 (1934).

The parameter in this process is the magnetic field, the characteristic temperature Θ is the CURIE or NÉEL temperature of the salt.

The technique of adiabatic demagnetization has been in use now for over twenty years. Before 1940 investigations were only made in Leiden, Berkeley, Oxford and Cambridge. After the war, when the number of low temperature laboratories increased enormously, larger or smaller demagnetization installations were set up in many places. In the first experiment in Leiden a temperature of 0.27° K was reached. At present a temperature of a few hundredths of a degree absolute can be made without exceptional difficulties, and even temperatures of the order of a thousandth of a degree have been produced.

A completely new region of temperatures has been opened for investigation, but since the technique in the demagnetization region is widely different from that at higher temperatures new experimental problems were encountered. The use of a cryostat filled with a liquefied gas has many advantages: The thermal contact between the liquid and an immersed object is good; the temperature is reasonably homogeneous and the homogeneity can even be improved by stirring; the temperature can be set to a desired value and kept constant there by adjusting the pressure at which the liquid is boiling; a heat leak causes an evaporation of the liquid at constant temperature without influencing the temperature itself; and the vapour pressure of the liquid provides a useful secondary thermometer which may be calibrated against the gas thermometer. All these advantages are lost when a paramagnetic salt is used as a coolant. In this case a heat leak causes a rise of temperature, and since the heat conductivity of a paramagnetic salt is very bad at the lower temperatures (see Sect. 19) a heat leak may spoil the homogeneity of the temperature noticeably. For the same reason the thermal equilibrium between the salt and an object under investigation becomes doubtful at the lower temperatures. Since no suitable gas is available in the demagnetization region the determination of thermodynamic temperatures becomes a problem by itself.

In spite of these difficulties the process of adiabatic demagnetization has given rise to a large number of new investigations. The most obvious experiments are those concerning the magnetic properties of the paramagnetic salts themselves and the determination of the absolute temperatures reached with them; but also other materials have been cooled with a salt in order to make measurements with them. In recent years properties of liquid helium have been investigated, several new superconductors have been detected and the electric and thermal conductivities of metals have been measured.

3. Energy levels of paramagnetic salts. It was stated in Sect. 2 that a paramagnetic salt is suitable for the demagnetization process if the interaction energies between the magnetic ions are small as compared with the thermal energy at 1° K, and if the potential energy of the ions in a magnetic field that can be obtained by technical means is of the same order as the thermal energy or even larger. This is equivalent to the statement that the distances between the energy levels of the salt in zero field must be small as compared with kT, whereas the separation in the field should be at least of the same order of magnitude as kT.

We shall consider this in some more detail.

If the energy levels of a magnetic ion are $E_1, E_2, \ldots E_n$ then the partition function for a system of N ions (N being AVOGADRO's number) is given by:

$$Z = \left(\sum_n e^{-E_n/kT} \right)^N. \tag{3.1}$$

The free energy obeys:

$$F = -kT \ln Z, \tag{3.2}$$

and the entropy and magnetic moment:

$$S = -\left(\frac{\partial F}{\partial T}\right)_H, \tag{3.3}$$

$$M = -\left(\frac{\partial F}{\partial H}\right)_T. \tag{3.4}$$

Suppose the angular momentum of the paramagnetic ions in the ground state is $\hbar\sqrt{J(J+1)}$, where J is the inner quantum number and $\hbar$ PLANCK's constant divided by 2π. The ground level, in the absence of a magnetic field, is $(2J+1)$-fold degenerate, hence, if the higher levels can be considered as unoccupied, the partition function obeys:

$$Z = (2J+1)^N, \tag{3.5}$$

and hence:

$$S = R\ln(2J+1), \tag{3.6}$$

$$M = 0. \tag{3.7}$$

These formulae cannot hold rigorously for two reasons. First the entropy due to the lattice vibrations has been neglected, but in most practical cases below $1°$ K this gives only rise to a small correction (see Sect. 38), though it sets a lower limit to the temperatures that can be reached. More important is, however, that the $(2J+1)$-fold degeneracy is not complete since in that case the entropy should be $R\ln(2J+1)$ down to absolute zero, which is contrary to NERNST's law. The interaction forces in the crystal, quoted in Sect. 1 cause a small level splitting or broadening which modifies the expression for the partition function. If the temperature is so high that kT is large as compared with the splitting, the sublevels are nearly equally populated and formulae (3.5) and (3.6) are still approximately valid. This is not true, however, for the region where kT and the level splitting are of the same order of magnitude, hence in the region of the characteristic temperature Θ mentioned in Sect. 1. Here the populations of the levels depend widely on temperature and so does the entropy.

Let us consider now the influence of a magnetic field.

In the temperature region where the influence of the interaction forces can be neglected, the level separation due to a magnetic field can be considered as proportional to the field, the distance between two subsequent sublevels being $g\mu_B H$. Here g ist the splitting factor, μ_B the BOHR magneton and H the field acting on the ions. Now the partition function obeys:

$$Z = \left(\sum_{m=-J}^{+J} e^{m\,g\mu_B H/kT}\right)^N \tag{3.8}$$

and expressions for the entropy and magnetic moment can be derived with the help of (3.3) and (3.4). The exact formulae will be discussed in Sect. 29. For small fields $(g\mu_B H \ll kT)$ we may develop:

$$Z = (2J+1)^N\left\{1 + \frac{1}{2}\frac{C}{R}\frac{H^2}{T^2}\right\}, \tag{3.9}$$

$$S = R\ln(2J+1) - \frac{1}{2}C\frac{H^2}{T^2}, \tag{3.10}$$

$$M = C\frac{H}{T}, \tag{3.11}$$

where C is an expression in J, g, μ_B and k, see Sect. 29. Formula (3.11) is well known as CURIE's law. It is obeyed by many paramagnetics for not too large values of H/T, hence in the case that the level separation due to the field is so small that all the sublevels are still approximately equally populated.

In the region of temperatures where the interactions cannot be neglected these considerations do not apply any more. In strong fields ($g\mu_B H$ much larger than the level splitting due to the interactions) the distances between the sublevels are still approximately equal to $g\mu_B H$ so that (3.8) is more or less valid, but in small fields this is certainly not true. Large deviations from (3.9) are found and also CURIE's law is no more valid.

It is also clear what happens when after the isothermal magnetization the field is removed adiabatically. As long as the distances between the energy levels are equal to $g\mu_B H$ the partition function is a function of H/T only and so are S and M. Hence, if S is constant M is constant, the temperature decreases proportionally to the field and the distribution of the ions over the levels is unaffected. In low fields, however, where the interaction forces become of the same order of magnitude as the field strength, the level distances are no longer proportional to the field. The ions are redistributed over the levels in order to keep the entropy constant, the magnetic moment decreases and the temperature approaches a value determined by the level scheme in zero field. The weaker the interaction forces acting on the ions are, the smaller is the level splitting and the lower is the final temperature.

If a relaxation time of finite length should be involved in the level transitions the entropy could not be constant. In this case the demagnetization is no more a reversible (quasi-static) process and the final temperature is higher than in the case of a strictly isentropic demagnetization. At present, however, we shall assume that a demagnetization is purely reversible.

4. Suitability of salts for the demagnetization process. With the help of the above considerations we can discuss the suitability requirements of a paramagnetic salt for the demagnetization process in some more detail. First, the energy $g\mu_B H$ in a field of about 10000 oersteds should be at least of the order of kT at $1°$ K. Then, the level splittings and broadenings due to the interaction forces must be small as compared with kT at $1°$ K and higher levels must be so high that their influence on the partition function can be neglected.

The first condition is fulfilled by many of the ions of the iron group and the rare earths. It is not fulfilled for elements with a nuclear spin, since nuclear magnetic moments are about a factor 1000 smaller than electronic moments so that a nuclear demagnetization can only be successful if either magnetic fields of at least a million oersteds are used or starting temperatures of the order of $0.01°$ K.

The interaction forces in the salt crystal have different origins: magnetic coupling between the ions (either dipole or exchange interaction); STARK effects caused by electric fields due to usually non-magnetic surrounding atoms; or hyperfine structure. The details will be discussed in Chapt. C, but some general remarks can be made already here.

The smaller the interactions, the lower is the characteristic temperature Θ and the lower are the temperatures that can be reached with the salt. The ideal case in this respect is a salt with small magnetic coupling and no STARK splitting and hyperfine structure at all. (It should be kept in mind, however, that if investigations must be performed at moderately low temperatures it may be advantageous to use a salt with a somewhat higher Θ, hence with stronger interactions, see Sect. 1.) Since CURIE's law is obeyed only well above Θ a qualitative and preliminary criterium for the suitability of a salt is the validity of this law in the liquid helium region.

Magnetic interactions are the weaker the larger the distances between the magnetic ions in the lattice are. Paramagnetic alums and tutton salts are very suitable in this respect. Here the level broadening is usually of the order of a few hundredths of a cm^{-1} so that the fall of the entropy takes place at a temperature of a few hundredths of a degree absolute (one cm^{-1} corresponds to 1.438° K). The interaction forces can still be decreased by "diluting" the crystal, i.e. replacing part of the magnetic ions by equivalent non-magnetic ions.

The influence of Stark splittings on the orbital momentum is often rather large, but the spin is only influenced through the spin-orbit interaction and this second order effect is small. Hence, magnetic ions with a vanishing orbital momentum may be suitable. It is possible, however, if a salt shows orbital magnetism, that the higher orbital levels, due to the crystalline Stark effect, are so high that they are unoccupied at 1° K. This effect is named "quenching of the higher levels" and the result is that the ions show only spin magnetism as well. Salts in which this effect occurs may also be used for the demagnetization process. A limitation on the choice of salts is still imposed by a theorem of Kramers[1]. It states that for an ion with an odd number of electrons (even number of spin levels, $2S+1$) the degeneracy cannot be removed completely by an electric field, a two-fold degeneracy must remain at least which can only be removed by a magnetic field. If the ion has an even number of electrons (odd number of spin levels) the electric field removes the degeneracy completely[2]. Since it is obvious that a singlet level cannot exhibit paramagnetism, only ions with an odd number of electrons can be used for adiabatic demagnetization experiments. Ions with zero orbital momentum are Gd^{+++}, Fe^{+++} and Mn^{++}; ions in which the orbital magnetism is quenched by the crystalline Stark effect are Ti^{+++}, Cr^{+++}, Co^{++} and Cu^{++}. The second order splitting of the spin levels in these salts is of the order of a few tenths of a cm^{-1} so that its influence becomes perceptible at a few tenths of a degree absolute.

The splitting due to hyperfine structure (interaction with the magnetic moment of the nucleus or with its electric quadrupole moment) is usually of a smaller order of magnitude than that of the Stark effect; it does not spoil the suitability of a salt for the adiabatic demagnetization process, but it sets a lower limit to the temperatures that can be reached.

From the discussion given here it is clear that the exact knowledge of the level schemes of paramagnetic salts is of predominant importance. Approximate data can be obtained from paramagnetic relaxation investigations[3-5] and from the demagnetization experiments themselves. The microwave technique[6-8], developed after the war, gives the possibility to measure the level distances for diluted salts in magnetic fields. The extrapolation to zero field, however, may involve some problems, and usually the level scheme of a diluted salt differs somewhat from that of the concentrated one.

[1] H. A. Kramers: Proc. Acad. Sci. Amst. 33, 959 (1930).
[2] H. A. Jahn and E. Teller: Proc. Roy. Soc. Lond., Ser. A 161, 220 (1937).
[3] C. J. Gorter: Paramagnetic relaxation. Amsterdam 1947.
[4] L. C. van der Marel: Kolloid-Z. 134, 32 (1953).
[5] A. H. Cooke: Rep. Progr. Physics 13, 276 (1950).
[6] D. M. S. Bagguley, B. Bleaney, J. H. E. Griffiths, R. P. Penrose and B. I. Blumpton: Proc. Phys. Soc. Lond. 61, 542 (1948).
[7] B. Bleaney and K. H. W. Stevens: Rep. Progr. Physics 16, 108 (1953).
[8] K. D. Bowers and J. Owen: Rep. Progr. Physics 18, 304 (1955).

5. The (T, S)-diagram of a paramagnetic salt. The (T, S)-diagram of a salt suitable for the demagnetization process is shown in Fig. 3. The upper curve represents the entropy in zero magnetic field. In the neighbourhood of T_0, which is of the order of 1° K, the entropy per mole is equal to $R \ln (2J+1)$. At higher temperatures the lattice vibrations become important, they give a rise in entropy which, according to DEBYE's formula, is proportional to T^3. At T_0 this contribution is usually still small. Below T_0 the entropy first depends little on temperature, but a decrease occurs in the neighbourhood of the characteristic temperature Θ. Here the specific heat shows a maximum.

Lines of constant magnetic field are also shown in Fig. 3. Below Θ the entropy depends very little on the field strength; here the interaction forces give an

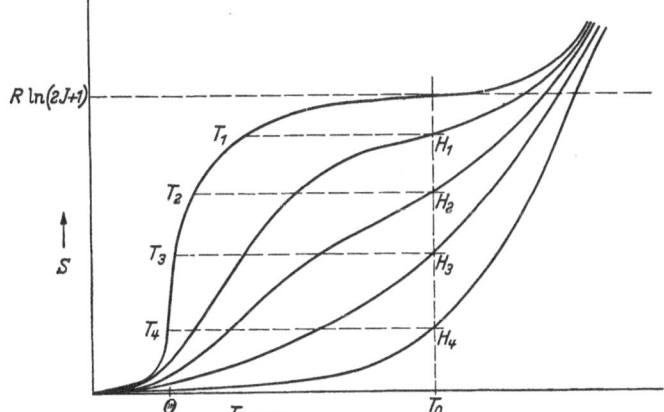

Fig. 3. Entropy versus temperature diagram of a paramagnetic salt.

appreciable amount of order in the crystal. Well above T_0 the decrease due to a field is also small; here the aligning influence of a field is mostly disturbed by the thermal motion. In between is the region where the entropy depends strongly on the field and here the demagnetization method is most effective.

If different interactions occur simultaneously in a salt, giving rise to level splittings of different orders of magnitude, the zero field curve has a more complicated character. Suppose a salt has a fourfold degenerate ground level which is split by the crystalline STARK effect into two twofold levels 0.2 cm^{-1} apart, and suppose each of them shows a broadening due to magnetic interaction of 0.01 cm^{-1}. Then the entropy decreases from $R \ln 4$ to $R \ln 2$ in a region near 0.3° K, and from $R \ln 2$ to zero near $T = 0.015$° K. In this case we have two characteristic temperatures, Θ_1 and Θ_2. The specific heat shows two maxima and the entropy has another horizontal part between them. If the entropy decrease due to a magnetic field is smaller than $R \ln 2$ we obtain temperatures of the order of Θ_1, if the decrease of entropy is between $R \ln 2$ and $R \ln 4$ temperatures near Θ_2 are reached. CURIE's law, which is valid at high temperatures, breaks down near Θ_1, but a new CURIE law with a different value for C [Eq. (3.11)] may occur in the region between Θ_1 and Θ_2.

If two or more interactions of the same order of magnitude occur in a salt the effects may overlap partly and the entropy diagram shows a rather unsurveyable pattern.

6. Other methods to obtain temperatures below 1° K. Before we discuss the demagnetization method on a quantitative basis it should be pointed out that it is not necessarily the only process by which the region below $1°$ K can be penetrated. Every process fulfilling the requirements given in Sect. 1 could be applied. Until now two more methods have been proposed and, though they are in general less suitable than the demagnetization method, we shall describe them briefly.

First, there is the *adiabatic magnetization of a superconductor*[1, 2]. The entropy of a superconducting metal below its transition point is lower in the superconducting state than in the normal state. Hence, if a magnetic field is applied isothermally the entropy increases suddenly when the field passes through the critical value. If it is applied adiabatically the temperature falls to the point where the entropy in the normal state is the same as it was in the superconducting state at the initial temperature.

The course of the process is completely different from the adiabatic demagnetization of a paramagnetic salt. In the latter, the temperature decreases gradually as the field is diminished and the final temperature depends both on the starting temperature and the field. In the case of a superconductor the temperature is unaffected as long as the field is smaller than the critical field, then it drops suddenly and after that it is constant again. The final temperature depends only on the initial temperature.

In some respects the method is advantageous. Reasonably low temperatures can be obtained with fields smaller than 1000 oersteds. Starting from $1°$ K, with tantalum a temperature of $0.07°$ K is reached, the transition field being 905 oersteds. Further the problem of thermal equilibrium at the lowest temperatures encountered in the work with paramagnetic salts (see Sect. 2) is less serious in the case of a metal.

For some experiments, however, it may be undesirable that the low temperature can only be produced in a magnetic field of the order of a thousand oersteds. Further the specific heat of a metal at the lowest temperatures is much smaller than that of a paramagnetic salt, hence a small heat leak causes a much faster rise in temperature. (The specific heat hump near Θ occurring in a paramagnetic salt is replaced here by a latent heat at the transition temperature.) MENDELSSOHN[3] remarked, however, that since the specific heat of a metal increases proportionally to T a superconductor may have some advantages over a paramagnetic salt at the higher temperatures, e.g. between 0.3 and $1°$ K.

The second method makes use of the *mechanocaloric effect in liquid helium II*. If liquid helium II flows through narrow tubes or slits (of the order of 10^{-3} mm diameter) the resistance to the normal component is very large, but the superfluid component passes through easily, so that a separation of the phases is achieved. Since the superfluid component seems to have zero entropy the temperature of the liquid flowing out of the tubes is reduced with a simultaneous rise of the temperature at the entrance of the tubes. The phenomenon was first observed experimentally by DAUNT and MENDELSSOHN[4]. KAPITZA concluded[5] that appreciable drops in temperature might be obtained from a process based on it, for instance forcing liquid helium through a tube highly packed with fine grains of powder. A detailed analysis given by SIMON[6], however, demonstrated that, although the actual lowering of the temperature may be important, the

[1] K. MENDELSSOHN and J. R. MOORE: Nature, Lond. **133**, 413 (1934).
[2] K. MENDELSSOHN, J. G. DAUNT and R. B. PONTIUS: Actes du VIIᵉ Congrès Intern. du Froid, The Hague, vol. 1, p. 445 (1936).
[3] K. MENDELSSOHN: Nature, Lond. **169**, 366 (1952).
[4] J. G. DAUNT and K. MENDELSSOHN: Nature, Lond. **143**, 719 (1939).
[5] P. KAPITZA: J. Phys. USSR. **5**, 59 (1941).
[6] F. E. SIMON: Physica, 's-Grav. **16**, 753 (1950).

method is not very suitable for most practical purposes. The reason is that by this method practically all the entropy is removed at once, so that a temperature far below the characteristic temperature (i.e., the lambda point) is obtained where the specific heat is very small again. Even if a liter could be cooled to 0.01° K the smallest heat leak that has been accomplished until now should raise the temperature by a factor two within a fraction of a second.

II. Thermodynamics of the demagnetization process.

7. Quantities of field and susceptibility. Before we discuss the adiabatic demagnetization from a thermodynamic point of view we shall introduce a few definitions of field quantities.

If a paramagnetic substance is placed in a magnetic field it shows a magnetic moment which, in general, can be calculated with the help of the formulae of Sect. 3. In this section, the quantity H was introduced as "the field acting on the magnetic ions" and this formulation requires some further explanation.

Suppose, a solenoid produces a field of a certain value in empty space. We fill up part of the volume inside the solenoid with a substance containing magnetic ions. Then the field inside the medium, defined as the field in a long narrow cavity parallel to the lines of force, is different from the field of the empty coil. Moreover the interaction forces between the magnetic ions, which also have an aligning influence, may be described in terms of a fictive magnetic field, sometimes called the WEISS field. The sum of the field inside the medium and this fictive WEISS field may be denoted as the "local" field. In this article we represent the field of the solenoid by H_{ext}, the field inside the medium by H_{int} and the local field by H_{loc}.

Now we may introduce three susceptibilities:

$$\chi_{ext} = M/H_{ext}, \tag{7.1}$$

$$\chi_{int} = M/H_{int}, \tag{7.2}$$

$$\chi_{loc} = M/H_{loc}. \tag{7.3}$$

The permeability is defined as:

$$\mu = 1 + 4\pi\chi_{int}. \tag{7.4}$$

The relation between H_{ext} and H_{int} can be derived from the MAXWELL equations but, generally speaking, this is not a simple problem since a homogeneous H_{ext} does not necessarily entail a homogeneous H_{int}, even if the medium is homogeneous. The exception is the case that the medium has the shape of an ellipsoid[1] and then the relation is given by:

$$H_{int} = H_{ext} - \varepsilon M/V. \tag{7.5}$$

M/V is the "intensity of magnetization", it is the magnetic moment per unit volume. ε is the demagnetization coefficient of the ellipsoid which depends on the axial ratio. For the case of a prolate spheroid of an isotropic substance (the case that has been investigated most often in adiabatic demagnetization work) we have, if the field is parallel to the axis of revolution:

$$\varepsilon_l = 4\pi \frac{1-e^2}{e^2}\left[\frac{1}{2e}\ln\frac{1+e}{1-e} - 1\right], \tag{7.6}$$

---

[1] J. C. MAXWELL: A treatise on electricity and magnetism, ed. 3, vol. 2, p. 69. Oxford: Clarendon Press 1904.

and if the field is perpendicular to this axis:

$$\varepsilon_p = 4\pi \left[\frac{1}{2e^2} - \frac{1-e^2}{4e^3} \ln \frac{1+e}{1-e} \right]. \tag{7.7}$$

Here

$$e = \sqrt{1 - \frac{a^2}{c^2}} \tag{7.8}$$

and c is the axis of revolution. (7.6) and (7.7) fulfill the relation:

$$2\varepsilon_p + \varepsilon_l = 4\pi. \tag{7.9}$$

For a sphere $\varepsilon_p = \varepsilon_l = \frac{4}{3}\pi$ and for a cylinder $\varepsilon_l = 0$, $\varepsilon_p = 2\pi$.

From (7.5) we may derive the relation between χ_{ext} and χ_{int}:

$$\chi_{\text{int}} = \frac{\chi_{\text{ext}}}{1 - \varepsilon \chi_{\text{ext}}/V}. \tag{7.10}$$

The difference between H_{int} and H_{loc} at the position of a certain ion is due to the magnetic dipole and exchange interactions with the neighbouring ions. Since the evaluation of their influence is a many particle problem it is difficult to give exact formulae, but some approximations valid at not too low temperatures may be derived.

The simplest method is the application of the LORENTZ theory for dielectrics[1]. If all the dipoles of a substance are equal and parallel they give a total contribution to the field at the position of one ion proportional to the intensity of magnetization, and the proportionality factor depends on the crystal structure. In the case of a cubic lattice we have:

$$H_{\text{loc}} = H_{\text{int}} + \frac{4}{3}\pi M/V, \tag{7.11}$$

which, according to (7.4) may also be written:

$$H_{\text{loc}} = \frac{\mu+2}{3} H_{\text{int}} = H_{\text{int}} \left(1 + \frac{\mu-1}{3}\right). \tag{7.12}$$

From (7.12) and (7.5) it follows for the case of an ellipsoid:

$$H_{\text{loc}} = H_{\text{ext}} + \left(\frac{4}{3}\pi - \varepsilon\right) M/V. \tag{7.13}$$

For the proof of (7.11) we refer to the original publication. For the susceptibilities we may derive:

$$\chi_{\text{loc}} = \frac{\chi_{\text{int}}}{1 + \frac{4}{3}\pi \chi_{\text{int}}/V}, \tag{7.14}$$

$$\chi_{\text{loc}} = \frac{\chi_{\text{ext}}}{1 + \left(\frac{4}{3}\pi - \varepsilon\right) \chi_{\text{ext}}/V}. \tag{7.15}$$

In the case of a spherical sample (7.13) and (7.15) are simplified to:

$$H_{\text{loc}} = H_{\text{ext}}, \tag{7.16}$$

$$\chi_{\text{loc}} = \chi_{\text{ext}}. \tag{7.17}$$

The assumption that all the dipoles in the medium are equal and parallel may be justified in the case of a dielectric (polarization of atoms), it is not for a paramagnetic (orientation of ions). ONSAGER pointed out[2] that the average field at the position of an ion (both in space and time) is the field as calculated

---

[1] H. A. LORENTZ: The theory of electrons, p. 138, 306. Leipzig 1909.
[2] L. ONSAGER: J. Amer. Chem. Soc. **58**, 1486 (1936).

from (7.12), but this is not the field exerting the aligning force on the ion. The ion itself has a polarizing influence on the medium surrounding it and this is responsible for part of the field at the position of the ion. This contribution, called by Böttcher[1] the "reaction field" changes its direction with the dipole (assuming that the medium around the dipole is isotropic) and, therefore, does not lead to an orientation of the ion (though it does lead to a term in the energy). The problem is now to calculate the field at the position of one ion in the lattice in the case that the ion itself is missing. This is a difficult problem and in order to obtain an approximate solution Onsager replaced the paramagnetic medium by a continuum with permeability μ and with a spherical cavity of the size of the missing ion. In this case we may derive from the Maxwell equations:

$$H_{\text{loc}} = \frac{3\mu}{2\mu + 1} H_{\text{int}} = H_{\text{int}} \left(1 + \frac{\mu - 1}{2\mu + 1}\right), \qquad (7.18)$$

which is equivalent to:

$$H_{\text{loc}} = H_{\text{int}} \frac{H_{\text{int}} + 4\pi M/V}{H_{\text{int}} + \frac{8}{3}\pi M/V}. \qquad (7.19)$$

For the susceptibilities it follows:

$$\chi_{\text{loc}} = \chi_{\text{int}} \frac{1 + \frac{8}{3}\pi\chi_{\text{int}}/V}{1 + 4\pi\chi_{\text{int}}/V}, \qquad (7.20)$$

and it is not difficult to derive the relations between H_{loc} and H_{ext} and between χ_{loc} and χ_{ext}. For high temperatures (μ close to unity) the relations (7.13) and (7.19) become identical so that here the Lorentz and Onsager approximations give the same results.

Van Vleck[2] tried to calculate H_{loc} from the actual magnetic interactions between the ions. Since an exact solution was impossible he developed the partition function in terms of $1/T$ and evaluated the contributions of the first few terms. At high temperatures his results are equivalent to those of Lorentz and Onsager, at lower temperatures he finds values in between, but at the lowest temperatures, unfortunately, his series development converges so slowly that many more terms should be required in order to obtain satisfactory results. We shall discuss van Vleck's method in more detail in Sect. 32. For the relation between H_{loc} and H_{int} he finds:

$$H_{\text{int}} = H_{\text{loc}} - \frac{4}{3}\pi M/V + \frac{12\eta M^2/V^2}{H_{\text{loc}}} + \cdots \qquad (7.21)$$

where η depends on the crystal structure and the ionic moment, see Sect. 32.

We come to the conclusion that the relation between H_{ext} and H_{int} for the case of a homogeneous and isotropic ellipsoid does not offer basic problems, but the situation about H_{loc} is less satisfactory. At the lower temperatures the models introduced by Lorentz and Onsager are too schematic to give reliable results, the series development of van Vleck is fundamentally correct but converges too slowly.

Many investigations in the demagnetization region have been performed with homogeneous samples (single crystals or compressed pills of approximately crystalline density), but also investigations have been made with loosely packed powders of roughly two thirds of the crystalline density. This brings up the problem of the determination of H_{int} inside the grains of a powder. A general solution proves to be impossible, but simplified models have been discussed by several authors.

[1] C. J. F. Böttcher: Physica, 's-Grav. 9, 937 (1942).
[2] J. H. van Vleck: J. Chem. Phys. 5, 320 (1937).

DE KLERK, in his thesis[1], divided the problem into two stages: (a) one grain of salt composed of paramagnetic ions (the "microscopic" stage); (b) the sample as a whole consisting similarly of a large number of grains (the "macroscopic" stage). In both stages an H_{ext}, H_{int} and H_{loc} could be defined in such a way that H_{loc} in stage (b) was equal to H_{ext} in stage (a). The relation between H_{ext} and H_{int} in the microscopic stage depends on the shapes of the individual grains; in the macroscopic stage it depends on the shape of the sample as a whole and on the filling factor (defined as the fraction of solid material in the powder). The relation between H_{int} and H_{loc} for each stage can be derived with either the LORENTZ, ONSAGER or VAN VLECK approximation.

The method was applied by DE KLERK to the case of an ellipsoidal sample consisting of spherical grains. Some of his formulae had been given earlier by other authors[2,3], but the conditions under which each formula is valid follows more precisely from his derivation. In the case that the LORENTZ approximation is applied to both stages (a) and (b) we obtain:

$$H_{int} = H_{ext} - \varepsilon f M/V - \tfrac{4}{3}\pi (1 - f)\, M/V, \tag{7.22}$$

$$H_{loc} = H_{ext} + (\tfrac{4}{3}\pi - \varepsilon)\, f M/V, \tag{7.23}$$

where H_{ext} refers to the macroscopic stage (the magnetic field in the absence of the sample), and H_{int} and H_{loc} refer to the microscopic stage (H_{int} being the field in a long narrow cavity parallel to the lines of force inside one grain of the salt, and H_{loc} the field acting on a magnetic ion). Further ε is the demagnetization coefficient of the sample, f is the filling factor (defined above) and M/V is the intensity of magnetization for the salt of crystalline density. For the formulae applying the ONSAGER approximation to either the macroscopic or the microscopic stage, or to both, we refer to the original publication.

BREIT[2], and POLDER and VAN SANTEN[4] considered also some cases of ellipsoids containing non-spherical grains.

8. Thermodynamics of magnetization. The laws of thermodynamics applied to a substance in a magnetic field may be expressed in two ways:

$$T\,dS = dU + M\,dH, \tag{8.1}$$

$$T\,dS = dU' - H\,dM, \tag{8.2}$$

where

$$U' = U + MH. \tag{8.3}$$

All terms concerning non-magnetic work (like $p\,dv$) have been omitted but they can be neglected in most magnetic investigations and certainly in the region below $1°$ K.

It has been a subject of thorough discussion whether (8.1) or (8.2) is the "correct" relation, hence whether U or U' is the true internal energy of the sample. For a survey, we refer to GARRETT's recent monograph[5]. The present author's point of view is that the answer to this question is arbitrary. The term MH, the difference between U and U', is the energy due to the simultaneous presence of the salt and the magnet and it is only a matter of taste whether it should be included into the energy of the salt or of the magnet[6]. Moreover

[1] D. DE KLERK: Thesis, Leiden 1948, p. 16—20.
[2] G. BREIT: Leiden Commun. Suppl. No. 46; Proc. Kon. Acad. Amst. **25**, 293 (1922).
[3] C. J. F. BÖTTCHER: Rec. Trav. chim. Pays-Bas **64**, 47 (1945).
[4] D. POLDER and J. H. VAN SANTEN: Physica, 's-Grav. **12**, 257 (1946).
[5] C. G. B. GARRETT: Magnetic cooling, p. 17. Cambridge (Mass.) 1954.
[6] H. B. G. CASIMIR: Magnetism and very low temperatures, p. 22. Cambridge 1940.

both equations lead to identical results if any relation between parameters of state is derived from them. If we name U the energy then U' is the magnetic analog of the enthalpy and vice versa.

Another problem is the exact meaning of H in these equations. In Sect. 7 we introduced three field quantities: H_{ext}, H_{int} and H_{loc}, and it is not a priori evident that all of them can be inserted at choice in (8.1) and (8.2). Let us suppose, as an example, that the equations are correct for H_{ext}, so that we may write for (8.1):

$$T\,dS = dU + M\,d H_{ext}, \tag{8.4}$$

then, according to (7.5) we have:

$$T\,dS = d(U + \tfrac{1}{2}\varepsilon M^2/V) + M\,d H_{int}, \tag{8.5}$$

so that H_{int} may also be used for H in (8.1), but then a different expression results for the internal energy. This is not surprising; for instance it is obvious that the specific heat at constant H_{ext} is different from the specific heat at constant H_{int}.

The case of H_{loc} is more complicated. If we apply the LORENTZ approximation then (8.4), according to (7.11), may be written:

$$T\,dS = d\left(U + \tfrac{1}{2}(\varepsilon - \tfrac{4}{3}\pi) M^2/V\right) + M\,d H_{loc} \tag{8.6}$$

which results, again, in a different expression for the energy. Difficulties occur if $H_{loc} - H_{int}$ is no more a function of M alone. Let us suppose that we may write:

$$H_{int} = H_{loc} + f(M, H_{loc}). \tag{8.7}$$

We choose H_{loc} and S as independent variables, then (8.5) may be written:

$$T\,dS = dU + M\,dH_{loc} + M\left(\frac{\partial f}{\partial H_{loc}}\right)_S dH_{loc} + M\left(\frac{\partial f}{\partial S}\right)_{H_{loc}} dS. \tag{8.8}$$

If we introduce a function $\varphi(M, H_{loc})$ in such a way that:

$$\left(\frac{\partial \varphi}{\partial H_{loc}}\right)_S = M\left(\frac{\partial f}{\partial H_{loc}}\right)_S \tag{8.9}$$

we have:

$$\left(T + \left(\frac{\partial \varphi}{\partial S}\right)_{H_{loc}} - M\left(\frac{\partial f}{\partial S}\right)_{H_{loc}}\right) dS = d(U + \varphi) + M\,dH_{loc}. \tag{8.10}$$

Now if it is possible to choose $\varphi(M, H_{loc})$ in such a way that beside (8.9) we have:

$$\left(\frac{\partial \varphi}{\partial S}\right)_{H_{loc}} = M\left(\frac{\partial f}{\partial S}\right)_{H_{loc}} \tag{8.11}$$

no troubles occur, but if $f(M, H_{loc})$ is such that this is impossible the introduction of H_{loc} entails not only a variation of the expression for the internal energy, but also a modification of the lefthand member of the equation.

The condition that (8.9) and (8.11) can be fulfilled simultaneously is:

$$\left(\frac{\partial M}{\partial S}\right)_{H_{loc}} \left(\frac{\partial f}{\partial H_{loc}}\right)_S = \left(\frac{\partial M}{\partial H_{loc}}\right)_S \left(\frac{\partial f}{\partial S}\right)_{H_{loc}}. \tag{8.12}$$

Now we have:

$$\left(\frac{\partial f}{\partial H_{loc}}\right)_S = \left(\frac{\partial f}{\partial H_{loc}}\right)_M + \left(\frac{\partial f}{\partial M}\right)_{H_{loc}} \left(\frac{\partial M}{\partial H_{loc}}\right)_S, \tag{8.13}$$

$$\left(\frac{\partial f}{\partial S}\right)_{H_{loc}} = \left(\frac{\partial f}{\partial M}\right)_{H_{loc}} \left(\frac{\partial M}{\partial S}\right)_{H_{loc}}, \tag{8.14}$$

and substitution in (8.12) gives:

$$\left(\frac{\partial M}{\partial S}\right)_{H_{\text{loc}}}\left(\frac{\partial f}{\partial H_{\text{loc}}}\right)_M = 0. \tag{8.15}$$

Since $(\partial M/\partial S)_{H_{\text{loc}}}$ is not zero in general the condition is that $f(M, H_{\text{loc}})$ in (8.7) is only a function of M.

This condition is fulfilled in the LORENTZ approximation, but not in the case of the ONSAGER and VAN VLECK relations (7.19) and (7.21). Hence at the higher temperatures, where the LORENTZ formula gives correct results, we may insert H_{ext}, H_{int} and H_{loc} at will. At the lower temperatures, though the ONSAGER and VAN VLECK approximations are not really satisfactory, it is probable that the correct relation between H_{int} and H_{loc} contains terms both in M and H. Here complications may arise due to the two extra terms in (8.10), and the application of even apparently simple thermodynamic relations may lead to wrong results at the lowest temperatures.

Until now the discussion of the meaning of U, U' and H in (8.1) and (8.2) has been restricted to pure thermodynamics. Some remarks, however, can be made also from a statistical point of view. In Sect. 3 expressions were given for Z, F, S and M [Eqs. (3.1 to (3.4)]. If they are combined with (8.1) and with:

$$U = F - T\left(\frac{\partial F}{\partial T}\right)_H, \tag{8.16}$$

$$F = U - TS, \tag{8.17}$$

then we have a complete and consistent set of equations. In this case U, and not U', must be considered as the internal energy. This, however, is not a proof that U is the "correct" expression for the energy. If we introduce:

$$Z' = Ze^{-\frac{MH}{kT}}, \tag{8.18}$$

$$F' = -kT\ln Z + MH, \tag{8.19}$$

$$S = -\left(\frac{\partial F'}{\partial T}\right)_M, \tag{8.20}$$

$$H = \left(\frac{\partial F'}{\partial M}\right)_T, \tag{8.21}$$

$$U' = F' - T\left(\frac{\partial F'}{\partial T}\right)_M, \tag{8.22}$$

$$F' = U' - TS, \tag{8.23}$$

together with (8.2) then, again, we have a consistent set of equations, but now U' may be considered as the internal energy.

H, in Sect. 3, was introduced as "the field acting on the magnetic ions", hence as H_{loc}. In this case, the H occurring in (3.3), (3.4) and (8.16) is also H_{loc}. But, again, this is not a proof that H_{loc} is the "correct" field in (8.1) and (8.2). Suppose the difference between H_{loc} and H_{int} (or H_{ext}) is proportional to M, e.g.

$$H_{\text{loc}} = H_{\text{int}} + \alpha M. \tag{8.24}$$

In this case we may add a factor $\exp(-\frac{1}{2}\alpha M^2/kT)$ to the partition function and then (8.1) and (8.2) become valid for H_{int}. Difficulties occur only at the lowest temperatures where $H_{\text{loc}} - H_{\text{int}}$ becomes a function of both M and H.

9. Thermodynamics of adiabatic demagnetization. In this section we derive some relations from (8.1) and (8.2) without going further into the question which is the exact meaning of H in each case.

Applying the condition that S is a total differential, Eqs. (8.1) and (8.2) may be written:

$$T \, dS = c_H \, dT + T \left(\frac{\partial M}{\partial T} \right)_H dH, \qquad (9.1)$$

$$T \, dS = c_M \, dT - T \left(\frac{\partial H}{\partial T} \right)_M dM, \qquad (9.2)$$

where c_H and c_M are the specific heats at constant field and constant magnetic moment:

$$c_H = \left(\frac{\partial U}{\partial T} \right)_H = T \left(\frac{\partial S}{\partial T} \right)_H, \qquad (9.3)$$

$$c_M = \left(\frac{\partial U'}{\partial T} \right)_M = T \left(\frac{\partial S}{\partial T} \right)_M. \qquad (9.4)$$

From Eqs. (9.1) and (9.2) it follows also:

$$c_H = c_M - T \left(\frac{\partial M}{\partial T} \right)_H \left(\frac{\partial H}{\partial T} \right)_M \qquad (9.5)$$

which, in the case that CURIE's law [see Eq. (3.11)] is valid, is equivalent to:

$$c_H = c_M + \frac{CH^2}{T^2}. \qquad (9.6)$$

From Eqs. (9.1) and (9.3) we have:

$$\left(\frac{\partial T}{\partial H} \right)_S = - \left(\frac{\partial M}{\partial S} \right)_H. \qquad (9.7)$$

The proof that adiabatic demagnetization produces a cooling effect follows from Eq. (9.1). For a process at constant entropy, it gives:

$$c_H \, dT = - T \left(\frac{\partial M}{\partial T} \right)_H dH. \qquad (9.8)$$

Since in ordinary paramagnetism $(\partial M / \partial T)_H$ is negative, a negative dH entails a negative dT. In some special cases, e.g. antiferromagnetism, $(\partial M / \partial T)_H$ may become positive and then demagnetization causes a heating effect.

The course of temperature with the field along an isentropic follows from the integration of Eq. (9.7):

$$T - T_0 = - \int_{H_0}^{H} \left(\frac{\partial M}{\partial S} \right)_H dH. \qquad (9.9)$$

The course of the magnetic moment on an isentropic is somewhat more complicated. Suppose we have a salt at a temperature T in zero field. In the case that a field H is switched on isothermally, we obtain a magnetic moment $M(H, T)$; if the field is switched on adiabatically the moment is $M(H, T')$, where T' is the temperature in the field. For small fields we may write:

$$M(H, T') - M(H, T) = \left(\frac{\partial M}{\partial T} \right)_H (T' - T), \qquad (9.10)$$

which, according to Eq. (9.9) is equal to:

$$M(H, T') - M(H, T) = - \left(\frac{\partial M}{\partial T} \right)_H \int_{0}^{H} \left(\frac{\partial M}{\partial S} \right)_H dH. \qquad (9.11)$$

The slopes of the isothermal and adiabatic magnetization curves for given values of H and T are related as:

$$\left(\frac{\partial M(H, T)}{\partial H} \right)_S - \left(\frac{\partial M(H, T)}{\partial H} \right)_T = \left(\frac{\partial M}{\partial T} \right)_H \left(\frac{\partial T}{\partial H} \right)_S, \qquad (9.12)$$

or, according to Eq. (9.7):

$$\left(\frac{\partial M(H,T)}{\partial H}\right)_S - \left(\frac{\partial M(H,T)}{\partial H}\right)_T = -\left(\frac{\partial M}{\partial T}\right)_H \left(\frac{\partial M}{\partial S}\right)_H. \tag{9.13}$$

Both factors of the last term are zero for $H=0$, so that the isothermal and the adiabatic magnetization curves have equal slopes in zero field (this may be not true in the case that a remanent magnetic moment occurs). Now suppose we have an isothermal and an adiabatic magnetization curve starting from the same zero field temperature. For the relation between the slopes in a field H we must add a term which, like in Eq. (9.10), accounts for the temperature difference. For small fields it may be written as $(\partial^2 M/\partial T\,\partial H)\,(T'-T)$, so that, if Eq. (9.9) is taken into account, Eq. (9.13) is replaced by:

$$\left(\frac{\partial M(H,T')}{\partial H}\right)_S - \left(\frac{\partial M(H,T)}{\partial H}\right)_T = -\left(\frac{\partial M}{\partial T}\right)_H \left(\frac{\partial M}{\partial S}\right)_H - \left(\frac{\partial^2 M}{\partial T\,\partial H}\right)\int_0^H \left(\frac{\partial M}{\partial S}\right)_H dH. \tag{9.14}$$

The decrease of entropy during the isothermal magnetization at the initial temperature T_i may be derived from Eq. (9.2):

$$S_0 - S = \int_0^{M(H,T_i)} \left(\frac{\partial H}{\partial T}\right)_M dM, \tag{9.15}$$

where S_0 is the zero field entropy. Now Eq. (9.15) must be equal to the difference in entropy at zero field between the initial and final temperatures of the demagnetization, so that we have:

$$\int_0^{M(H,T_i)} \left(\frac{\partial H}{\partial T}\right)_M dM = \int_{T_f}^{T_i} \frac{c_0}{T} dT, \tag{9.16}$$

where c_0 is the zero field specific heat of the salt.

III. Absolute temperature determination.

10. Thermometry in general. The fundamental definition of absolute temperature is the one introduced by KELVIN, already more than a hundred years ago[1, 2]. It may be based on a reversible CARNOT cycle. Let us suppose that the heat absorbed isothermally at the higher temperature (T_1) is ΔQ_1, and that the heat delivered isothermally at the lower temperature (T_2) is ΔQ_2. Then, if ΔS is the difference in entropy of the two isentropics, we have the relations:

$$\frac{\Delta Q_1}{T_1} = \frac{\Delta Q_2}{T_2} = \Delta S. \tag{10.1}$$

These relations are independent of the working substance, the only requirement being reversibility of the cycle. If we know one of the temperatures, or the entropy difference, we can base the determination of the other temperature on this relation.

For the practical execution of thermometry it is not really necessary to carry out CARNOT cycles in which experimental errors are often prohibitively large. Temperature is introduced in the second law of thermodynamics as an integrating denominator and it can be proved that this is identical to the KELVIN

[1] W. THOMSON: Phil. Mag. **33**, 313 (1848).
[2] J. P. JOULE and W. THOMSON: Phil. Trans. **144**, 321 (1854).

temperature. Hence if a relation is derived from the second law of thermo-dynamics between the temperature and other properties of state, this relation can be used to establish the temperature scale as well[1,2].

The best known example of the above is the ideal gas. In fact, for most physicists the "absolute temperature scale" is not the KELVIN scale but the ideal gas scale. The gas thermometer can be used from very high temperatures down to the liquid helium region. But no gases are available in the demagnetization region so that for thermometry here we must either look for some other application of the second law or go back to KELVIN's definition itself.

Standard practice in thermometry at higher temperatures is that the direct measurements are made with a "secondary thermometer". This is a substance with an easily measurable property which depends strongly and in a single-valued way on temperature. The thermometer is calibrated empirically against the absolute temperature scale. For the measurements it is brought into thermal contact with the substance under investigation. Different thermometers are needed for different regions of temperature; at room temperature, for instance, we may use the mercury thermometer or the platinum resistance thermometer.

Application of a separate secondary thermometer is impracticable in the region below $1°$ K since thermal equilibrium is difficult to achieve at the lower temperatures (see Sect. 2). The problem is solved most easily if a temperature dependent property of the salt itself is used (then the salt is its own secondary thermometer). Such a property is called a "thermometric parameter". The consequence is, however, that the calibration of the parameter against the thermodynamic scale must be repeated for not only each new salt under investigation, but also for different samples of the same salt, since the results are not always identical. It even happens that the data obtained with the same sample during different helium runs are somewhat different.

11. Thermometric parameters. It follows from (9.8) that the stronger M depends on temperature the larger is the cooling effect of the demagnetization process. Hence, if a paramagnetic salt is suitable for adiabatic demagnetization, its magnetic quantities make useful thermometric parameters. In fact, no other quantities have been used up to the present time.

The susceptibility M/H has been used as a thermometric parameter for many years. If CURIE's law is valid it is inversely proportional to the temperature and this is the reason why also the quantity C/χ [C being the CURIE constant of the salt, see Eq. (3.11)] is in use as a parameter. It is named the "magnetic temperature" and denoted by T^*:

$$T^* = \frac{C}{\chi} = \frac{CH}{M}. \tag{11.1}$$

If CURIE's law is valid T^* is equal to T; marked deviations from CURIE's law are found at the lower temperatures (see Sect. 3) and here T and T^* may widely diverge.

In Sect. 7 we introduced three quantities for both H and χ and each of them might be inserted into Eq. (11.1). Originally T^* was defined with the help of the external field:

$$T^* = \frac{C}{\chi_{\text{ext}}}, \tag{11.2}$$

but the difficulty was encountered that then the T^* values measured for different samples of the same salt were different if the axial ratios were not the same.

---

[1] P. S. EPSTEIN: Textbook of Thermodynamics, p. 74, New York 1949, fifth printing.
[2] F. E. SIMON: Sci. Progr. **133**, 31 (1939).

For this reason KURTI and SIMON[1] introduced a new quantity, denoted by $T^{(*)}$ and defined as:

$$T^{(*)} = \frac{C}{\chi_{\text{loc}}}. \tag{11.3}$$

If we apply the LORENTZ approximation for H_{loc} then Eq. (7.15) leads to:

$$T^{(*)}_{\text{Lor.}} = T^* + \varDelta, \tag{11.4}$$

where

$$\varDelta = (\tfrac{4}{3}\pi - \varepsilon)\, C/V. \tag{11.5}$$

The symbol $T^{(*)}$ was chosen because it is the T^* value which is measured in the case of a spherical sample, see Eq. (7.17).

If we apply the ONSAGER and VAN VLECK relations for H_{loc} different expressions result for the connection between $T^{(*)}$ and T^*. According to Eqs. (7.20) and (7.21) they may be expressed by:

$$T^{(*)}_{\text{Ons.}} = \frac{(T^{(*)}_{\text{Lor.}} - \tfrac{4}{3}\pi\, C/V)\,(T^{(*)}_{\text{Lor.}} + \tfrac{8}{3}\pi\, C/V)}{T^{(*)}_{\text{Lor.}} + \tfrac{4}{3}\pi\, C/V}, \tag{11.6}$$

$$T^{(*)}_{\text{v. Vl.}} = T^{(*)}_{\text{Lor.}} \left\{ 1 - 12\eta \left(\frac{C/V}{T^{(*)}_{\text{Lor.}}} \right)^2 + \cdots \right\}. \tag{11.7}$$

In the case of a powdered sample, as was pointed out in Sect. 7, we have a microscopic and a macroscopic stage, and in each of them we may insert either the LORENTZ, ONSAGER, or VAN VLECK approximations. The relation for $T^{(*)}$ is different for each of the cases. Some formulae have been given by DE KLERK[2].

In some publications the difference between T^* and $T^{(*)}$ is made very rigorously. In the present article, we shall follow the system that is used in most of the papers on the subject; we shall always give the quantity as reduced to spherical shape and, if no doubt can arise, denote it simply by T^*.

At the very lowest temperatures, T^* and χ cease to be suitable thermometric parameters for reasons that will be explained later (see e.g. Sect. 58). Here they must be replaced by other quantities, for instance the heat absorption coefficient from an alternating magnetic field, or the remanent magnetic moment of the salt.

12. Temperature determinations based on KELVIN's definition. For the determination of temperatures with the help of (10.1) we must measure the variation of entropy when a well-known amount of heat is supplied to the salt.

The entropy determination is not a problem in these investigations. The decrease in entropy during the isothermal magnetization can be calculated from (9.15). For this it is necessary to know how M depends on H and T at the initial temperature. We shall assume that this relation, the magnetic equation of state, is known. It can often be represented by a BRILLOUIN function (see Sect. 29), sometimes with an experimental correction. Eq. (9.15) represents also the difference in entropy between the states at the initial and final temperatures in zero field since the demagnetization is an isentropic process. Now a number of demagnetizations can be performed from different initial fields; each time the thermometric parameter is measured as a function of time after the demagnetization and its value is extrapolated back to the time of the demagnetization. This provides us with an experimental relation between the parameter and the entropy so that afterwards variations in entropy can be derived from the measured variations in the parameter.

[1] N. KURTI and F. SIMON: Phil. Mag. **26**, 849 (1938).
[2] D. DE KLERK: Thesis, Leiden 1948, p. 18—20.

The main problem in these experiments is finding a method of heat supply that is strictly homogeneous over the sample. This is an absolute necessity at the lower temperatures where the heat conductivity of the salt is very poor (see Sects. 2 and 19). A metal or carbon electric heater[1, 2] or an induction heater[3] (see Sect. 47) can be used down to 0.2° K but below this temperature they are unsatisfactory. Two other methods are available: irradiation with gamma rays[4] and heat absorption from an alternating magnetic field[5]. Both methods are in use and in some cases the agreement is not too good (see Sect. 58).

The gamma ray method can be used throughout the whole region below 1° K. The absorption coefficient of a paramagnetic salt for gamma rays is small, hence the penetration depth is large. Thus for not too thick samples the absorption is rather homogeneous. This can still be improved by a suitable arrangement of the gamma sources around the sample. If sufficient care is given to this point no afterperiod in the heating experiment is found. It is difficult to estimate the heat absorption in absolute units, but the standard procedure is to derive this from a separate experiment in a region where the temperature scale and the specific heat of the salt are known.

The method has been criticized on different occasions. PLATZMAN[6] remarked that it is possible that at low temperatures a large part of the absorbed energy is not converted into heat, but stored in the crystal. KURTI and SIMON[7] pointed out, however, that this does not imply that the method gives uncorrect results. The transfer of radiation energy into thermal energy can be described with the help of a relaxation time. If this relaxation time is either very short or very long as compared with the times involved in the experiment (e.g. shorter than a second or longer than a day) no error can be caused. If it is of the order of some minutes (comparable with a heating period) considerable after-effects in the heating must be found. If it is of the order of some hours differences must be found between experiments with a "virgin" specimen and one that has been irradiated during the preceding hour or so. None of these effects were found so far. If the gamma ray absorption or the relaxation time is very different at the low temperature and at the higher temperature, where the absorption from the gamma ray source is checked, similar effects as the above should be noticed. If the change is very sharp with temperature an actual heat evolution should be found at the transition temperature. The absence of such effects makes it plausible that the results obtained with the method are sufficiently reliable.

The method of heat absorption from an alternating magnetic field has the advantage that no troubles occur involving the homogeneity of the heat supply over the sample. Since in most cases the heat absorption is the stronger the lower the temperature, small inhomogeneities will even be decreased automatically.

The method has two main disadvantages. First, it can only be applied if relaxation or hysteresis effects occur in the salt. This restricts the practical application to the determination of only the lowest temperatures that can be reached with each individual salt. Secondly, it is sometimes difficult to discriminate between the heat absorption in the salt and the a.c. losses in the bridge

[1] P. H. KEESOM: Thesis, Leiden 1948.

[2] W. F. GIAUQUE, J. W. STOUT and C. W. CLARK: J. Amer. Chem. Soc. 60, 1053 (1938).

[3] W. F. GIAUQUE and J. W. STOUT: J. Amer. Chem. Soc. 60, 388 (1938).

[4] N. KURTI and F. SIMON: Phil. Mag. 26, 840 (1938).

[5] H. B. G. CASIMIR, W. J. DE HAAS and D. DE KLERK: Leiden Commun. No. 256 b; Physica, 's-Grav. 6, 255 (1939).

[6] R. L. PLATZMAN: Phil. Mag. 44, 497 (1953).

[7] N. KURTI and F. SIMON: Phil. Mag. 44, 501 (1953).

network. Corrections can be applied, but if the heat absorption in the salt is small the accuracy may become insufficient.

The practical execution of the method is very simple: the real and imaginary parts of the a.c. susceptibility of the salt are determined simultaneously with a bridge method (see Sect. 26). The real part is used as a thermometric parameter [it is related to T^* according to (11.1)] and the heat supply per second follows from the phase angle. Hence both the entropy and heating data are derived from the same measurement. It requires some experience to obtain a reasonable number of bridge compensations in a short time if both components vary rapidly with time.

13. The theoretical method. A method that can be considered as a somewhat different application of Kelvin's formula consists in replacing the caloric measurements by theoretical considerations.

Suppose it is possible from the geometry of the lattice of the paramagnetic salt and from the interactions in the lattice to evaluate the exact expression of the partition function. Then, with the help of (3.3) and (3.4), we may derive M and S as functions of H and T. From these we have relations for the case of zero field which can be expressed:

and:

$$\chi = \chi(T) \quad \text{or} \quad T^* = T^*(T), \tag{13.1}$$

$$S = S(T) \quad \text{or} \quad c_0 = c_0(T). \tag{13.2}$$

Elimination of T gives:

$$S = S(T^*) \tag{13.3}$$

and this can be checked directly since it must be identical to the experimental relation between the decrease of entropy and T^* (the thermometric parameter) as was described at the beginning of Sect. 12. [A small correction must be applied in (13.2) and (13.3) for the specific heat of the lattice which is proportional to T^3.] If the curves coincide over a long range of temperatures it is plausible that the partition function is correct, and also the relations derived from it, especially the $T^*(T)$ relation.

Satisfactory results can be expected from this method in a region where reliable theoretical relations can be given[1, 2], hence at the higher temperatures where the deviations from Curie's law are still relatively small.

Also this method has been criticized on different occasions. The main objection is that it is too indirect. Until now, however, it is the method with which the highest precision has been obtained.

14. Temperature determinations based on applications of the second law. It was stated in Sect. 10 that any relation between T and other parameters of state derived from the second law of thermodynamics is as fundamental for absolute temperature determinations as Kelvin's definition itself. Such a relation is, for instance, Eq. (9.9).

If M is measured on a number of isentropics as a function of H we can calculate $(\partial M/\partial S)_H$ as a function of H and S. According to (9.9), integration of this quantity along an isentropic gives the temperature difference between any two points of the isentropic.

The most obvious application of this method, proposed by Giauque[3, 4], is that the integral be extended from the initial field of the demagnetization to

---

[1] J. H. van Vleck: J. Chem. Phys. **5**, 320 (1937).

[2] M. H. Hebb and E. M. Purcell: J. Chem. Phys. **5**, 338 (1937).

[3] W. F. Giauque and D. P. MacDougall: Phys. Rev. **47**, 885 (1935).

[4] W. F. Giauque: Phys. Rev. **92**, 1339 (1953).

zero field. This gives, immediately, the difference between the initial and final temperatures. Unfortunately this process is unsuitable for the lower temperatures, since a small relative uncertainty in the initial temperature may make the precision of the final temperature unsatisfactory. This objection does not apply in the case of a method based on KELVIN's definition (Sect. 12) where ratios of temperatures are determined rather than differences, see (10.1). The many graphical differentiations and integrations involved in the calculations are other sources of inaccuracy.

Still the method has some interesting applications. If the course of an isentropic can be predicted on a theoretical basis it can be checked by experiment. Further, if the temperature in zero field can be derived from one of the other methods the variation of temperature in moderate fields can be determined. This may be of some importance, for instance in the case of investigations on other substances where a magnetic field is required (e.g. superconductors).

A somewhat indirect method of absolute temperature determination, first proposed by GARRETT[1,2] is based on (9.11)and (9.14). If we restrict ourselves to relatively small fields the magnetization curve may be developed:

$$M = \chi(T)\,H + \psi(T)\,H^3 + \cdots \tag{14.1}$$

where $\psi(T)$ is due to saturation effects. In this case Eqs. (9.11) and (9.14) may be written as:

$$\frac{M}{H} = \chi\left\{1 - \left[\frac{1}{2}\,\varXi - \frac{\psi}{\chi}\right]H^2 + \cdots\right\} \tag{14.2}$$

and:

$$\left(\frac{\partial M}{\partial H}\right)_S = \chi\left\{1 - 3\left[\frac{1}{2}\,\varXi - \frac{\psi}{\chi}\right]H^2 + \cdots\right\}, \tag{14.3}$$

with:

$$\varXi = \frac{1}{\chi}\left(\frac{\partial \chi}{\partial S}\right)_{H=0}\left(\frac{\partial \chi}{\partial T}\right)_{H=0}. \tag{14.4}$$

It should be noted that here the meaning of χ is different from the definition given earlier in this article (see Sect. 7). In the present formulae, according to Eq. (14.1), it is only the zero field susceptibility:

$$\chi = \left(\frac{M}{H}\right)_{H=0} = \left(\frac{\partial M}{\partial H}\right)_{H=0}. \tag{14.5}$$

Now χ and either M/H or $(\partial M/\partial H)_S$ can be derived from the measurement of adiabatic magnetization curves, hence, if ψ/χ is small or can be accounted for, we may calculate $\varXi$ as a function of the entropy. With the help of experimental values of χ and $(\partial \chi/\partial S)_{H=0}$ we can derive $(\partial \chi/\partial T)_{H=0}$. From the relation between $(\partial \chi/\partial T)_{H=0}$ and χ we can integrate T at any point starting from a known temperature.

A simple case occurs when the salt obeys a CURIE-WEISS law:

$$\chi = \frac{C}{T - \Theta} = \frac{C}{T*}, \tag{14.6}$$

and the specific heat is proportional to $1/T^2$ (see Sect. 32), so that:

$$S = S_0 - \frac{1}{2}\,\frac{A}{T^2}. \tag{14.7}$$

---

[1] C. G. B. GARRETT: Cérémonies LANGEVIN-PERRIN, p. 43. Paris 1948.
[2] C. G. B. GARRETT: Proc. Roy. Soc. Lond., Ser. A **203**, 375 (1950).

Then we have:

$$\Xi = \frac{C}{A}\,\frac{T^3}{(T-\Theta)^3} = \frac{C}{A}\,\frac{(T^*+\Theta)^3}{T^{*3}} \tag{14.8}$$

so that, if $\sqrt[3]{\Xi}$ is plotted against $1/T^*$ a straight line is found from which the values of C/A and Θ can be derived.

The main difficulty of the method is the evaluation of ψ/χ. In most cases it must be derived from a theoretical expression for the magnetization curve. For this reason satisfactory results can only be expected at relatively high temperatures.

B. Experimental methods.

I. Introduction.

15. General description. Adiabatic demagnetization experiments require an apparatus with which a paramagnetic salt can first be magnetized isothermally at the lowest temperature that can be reached with liquid helium and then demagnetized adiabatically. Equipment is needed for the temperature determination, and for the cooling of other materials together with the salt so that investigations can be carried out with these materials.

Some of the typical difficulties encountered in the investigations below $1°$ K were already mentioned in Sect. 2. The heat capacity of a piece of paramagnetic salt of reasonable dimensions (e.g. 25 cm³) is much smaller than that of a cryostat filled with liquid helium. Hence much more care must be given to the thermal insulation. At the higher temperatures a heat leak causes a steady rise in the temperature of the sample, but at the lower temperatures, where the thermal conductivity of the paramagnetic salts becomes very poor, considerable inhomogeneities in the temperature may arise within a short time. It may happen, for instance, that the heating from the lowest temperatures to $1°$ K takes many hours, but that the time available for the actual experiments, due to the inhomogeneous heat leak, is only some minutes. For the same reason, special precautions must be taken for good thermal contact between the salt and a substance cooled with it for investigations, especially if the temperature of the substance is derived from the thermometric parameter of the salt (see Sect. 11).

The salt is mounted in a sample tube inside the liquid helium vessel. During the isothermal magnetization heat contact is accomplished, usually by admitting some helium gas in the sample tube. Thermal insulation is achieved before the demagnetization by pumping the gas away.

The cryostat is mounted between the poles of an electromagnet or along the axis of a high power solenoid. Since it is important that the starting temperature of the demagnetization is as low as possible, the helium is evaporated under reduced pressure. This is done by means of a high-capacity vacuum pump, operating through a large diameter pumping line.

The temperature measurement depends, as was pointed out in Sect. 11, on the measurement of some magnetic quantity of the salt. It is usually determined with the help of an induction bridge method. Since the bridge settings may be influenced by the presence of the magnet, it is desirable that, after the demagnetization, the magnet and the cryostat are separated. This is certainly true in the case of an iron magnet. If the magnet is relatively small, one can have the cryostat in a fixed position and mount the magnet on rails or on an elevating mechanism. In the case of a very heavy magnet this is impossible and the cryostat must be movable. This entails either flexible pumping lines or a setup in which the pumps move with the cryostat. GIAUQUE pointed out[1] that in the case of an iron-free coil magnet the separation is not strictly necessary. But still then, eddy currents may occur both in the metal tubing of the coil and in the winding itself. Since these eddy currents depend on the susceptibility of the salt a satisfactory correction for their influence on the bridge settings is difficult, especially if an a.c. method is used. Though it is possible to adopt special precautions in the construction of the bridge coils[2, 3], many investigators prefer separation also in the case of a coil magnet.

In the present chapter we restrict ourselves to the description of the apparatus needed for the production and measurement of the low temperature itself, such as cryostats with their pumping installations, methods for the thermal insulation of the sample, construction of magnets, and bridge networks. Auxiliary equipment, for instance devices for thermal contact between a salt and other substances and apparatus needed for investigations with these substances will be described in Chapt. F, see also Sect. 50.

For details on the Leiden demagnetization apparatus we refer to the theses of DE KLERK[4] and STEENLAND[5] and to some of the Leiden communications[6-8]; a description of the Oxford setup was given by KURTI and SIMON[9] and by HULL[10]; a detailed paper on the installation at the Bureau of Standards was recently published by DE KLERK and HUDSON[11]; data on the Berkeley setup may be found in articles by GIAUQUE and his co-operators[12-16].

II. Demagnetization Cryostats.

16. The dewars and vacuum pumps. Liquid helium dewars may be made either of glass or of metal. A glass demagnetization cryostat, typical for the KAMERLINGH ONNES Laboratory at Leiden is shown in Fig. 4. It consists of two coaxial dewars. The internal dewar contains the helium, the external one is filled with liquid hydrogen or nitrogen for the protection of the helium. Brass rings are waxed to the tops of the vessels and fit tightly into metal caps, so that the cryostat is in a rigid and well defined position in the magnet. The dewars are made with a narrow lower portion (the "tail") so that, although they contain a large quantity of refrigerant, they fit into a magnet with a relatively small pole gap.

The cap of the helium dewar is connected by a wide tube P to the vacuum pump for the vapour pressure. A manometer is connected to another tube M,

---

[1] W. F. GIAUQUE: Phys. Rev. **92**, 1339 (1953).

[2] H. B. G. CASIMIR, D. BIJL and F. K. DU PRÉ: Leiden Commun. No. 262a; Physica, 's-Grav. **8**, 449 (1941).

[3] W. F. GIAUQUE, J. J. FRITZ and D. N. LYON: J. Amer. Chem. Soc. **71**, 1657 (1949).

[4] D. DE KLERK: Thesis, Leiden 1948, p. 31—45.

[5] M. J. STEENLAND: Thesis, Leiden 1952, p. 10—27.

[6] W. J. DE HAAS and E. C. WIERSMA: Leiden Commun. No. 236a; Physica, 's-Grav. **2**, 81 (1935).

[7] W. J. DE HAAS and E. C. WIERSMA: Leiden Commun. No. 236b; Physica, 's-Grav. **2**, 335 (1935).

[8] H. B. G. CASIMIR, W. J. DE HAAS and D. DE KLERK: Leiden Commun. No. 256a; Physica, 's-Grav. **6**, 241 (1939).

[9] N. KURTI and F. SIMON: Proc. Roy. Soc. Lond., Ser. A **149**, 152 (1935).

[10] R. A. HULL: Phys. Soc. Conf. Report, p. 72. Cambridge 1947.

[11] D. DE KLERK and R. P. HUDSON: J. Res. Nat. Bur. Stand. **53**, 173 (1954).

[12] W. F. GIAUQUE and D. P. MACDOUGALL: J. Amer. Chem. Soc. **57**, 1175 (1935).

[13] D. P. MACDOUGALL and W. F. GIAUQUE: J. Amer. Chem. Soc. **58**, 1032 (1936).

[14] W. F. GIAUQUE, J. J. FRITZ and D. N. LYON: J. Amer. Chem. Soc. **71**, 1657 (1949).

[15] J. J. FRITZ and W. F. GIAUQUE: J. Amer. Chem. Soc. **71**, 2168 (1949).

[16] T. H. GEBALLE and W. F. GIAUQUE: J. Amer. Chem. Soc. **74**, 3513 (1952).

since it is standard practice to derive the temperature of the liquid helium from
its vapour pressure. Tube HV connects the sample tube to the high vacuum
pump and its manometer system. Valve V_3 is used to admit some helium gas
from the cryostat into the sample tube if isothermal conditions are required.
Care should be taken, however, that not too much gas is admitted in order to
prevent the helium from con-
densation. Further tube S is
available for the accomodation
of the transfer syphon of the
liquefier, and the inlet I for
electrical leads entering the
cryostat.

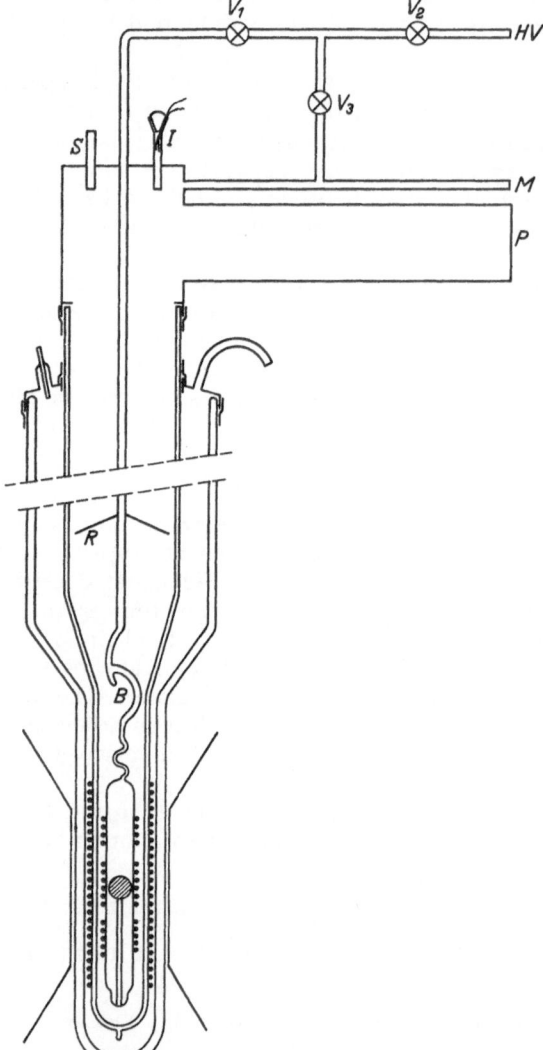

Fig. 4. Glass demagnetization cryostat as used in Leiden and the
National Bureau of Standards.

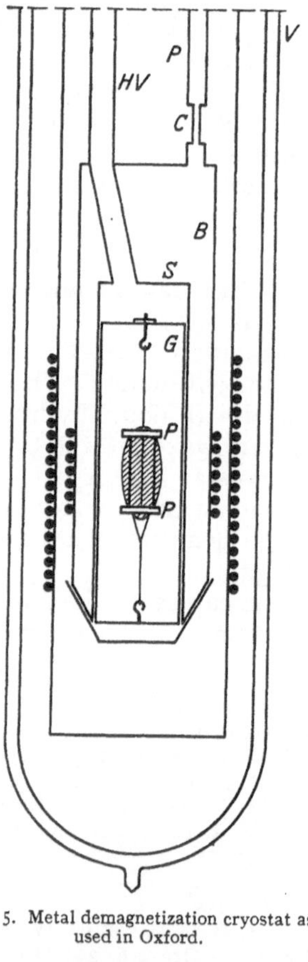

Fig. 5. Metal demagnetization cryostat as
used in Oxford.

Since pyrex dewars are somewhat transmissible for helium gas they must
be repumped after a few runs. In some laboratories the helium dewar is pumped
continuously during a run.

The "tail" part of a metal cryostat, typical for the CLARENDON Laboratory,
at Oxford, is shown in Fig. 5. The demagnetization cryostat is built together
with a SIMON expansion liquefier, not shown in Fig. 5. The glass dewar V contains

liquid hydrogen. It surrounds both the liquefier and the helium vessel B. In this setup the helium is completely surrounded by low temperature equipment (B does not have a cap of room temperature, the tubes P and HV are soldered to the reservoir of the liquefier). Due to this construction, runs of reasonable length can be made with surprisingly small amounts of liquid. P is the pumping line for the vapour pressure pump. S is the sample tube. It is closed at the bottom by a greased ground joint. The ground joint avoids the presence of solder which might become superconductive, but the sample tube does not withstand any over-pressure of gas, unless the grease is frozen, and only a small over-pressure then. The ground joint is reliably gas-tight, but if a small amount of helium condenses in S as He II, it may leak through the joint and spoil the vacuum of the cryostat.

As was stated in Sect. 15, a low starting temperature for the demagnetization is important. Hence a low evaporation of the helium and a powerful pumping installation are essential. The heat flow to the liquid helium can be decreased drastically by surrounding the container completely with low temperature equipment (as in Fig. 5). If this is impossible it is advisable to insert some radiation screens in the dewar (R in Fig. 4). In some laboratories it is standard practice to make the upper part of the helium dewar single walled and to keep the level in the outer dewar above the ring seal. This gives a noticeable decrease of the heat flow along the inner wall of the dewar.

If the top of the helium vessel is at a temperature below the lambda point it is advantageous to have a constriction in the pumping line (C in Fig. 5). It decreases the creep velocity of the film and reduces the rate of its evaporation, resulting in a lower ultimate pressure (see Sect. 2). The room temperature part of the pumping line, however, should be as short and wide as possible. A big rotary pump is usually applied (with a capacity of the order of a hundred cubic meters per hour), but a noticeable improvement may be obtained, if a high capacity oil ejector ("booster") pump is inserted in the system.

The high vacuum pump for removing the exchange gas from the sample tube is usually an oil or mercury-diffusion pump with a capacity of at least ten liters per second.

After the demagnetization the cryostat and the magnet are usually separated, see Sect. 15. In the case of a heavy magnet the cryostat must be movable. In the Leiden setup this problem was solved with the help of a large tripod bearing two flat ground joints. One is part of the 6″ pumping line of the helium vapour pressure pump, the other is inserted in the high vacuum line for the sample tube (tubes P and HV of Fig. 4). The cryostat is suspended from the two pumping lines, they can rotate around the ground joints. At the Bureau of Standards in Washington the cryostat is mounted on the far end of a wooden framework that pivots on a vertical steel pillar. The framework carries the entire high vacuum installation and a 3″ copper pumping line which is connected to the vapour pressure pump by a reinforced flexible rubber hose.

17. Samples and sample tubes. The paramagnetic salt is usually in the shape of a sphere or a prolate spheroid for reasons pointed out in Sect. 7. The salt can be used as a single crystal ground in the desired shape, as a powder compressed to approximately the crystalline density, or as small crystals loosely packed in a container. It depends on the particular experiment which of the methods is the more useful. If the magnetic properties of the salt itself are investigated, a single crystal may be advisable; if liquid helium must be cooled with the salt small crystals should be preferred (see Sect. 68); in the case that a heat contact between a salt and a metal must be brought about, a compressed pill is better (see Sect. 70). The salt should be protected against deterioration between sub-sequent helium runs. Covering a single crystal or a compressed pill with a thin layer of celluloid or glyptal is not absolutely effective. It is better to keep the sample at liquid nitrogen temperature between runs. It is not advisable to eva-cuate the sample tube at room temperature.

The sample tube may be made of glass, metal or plastic. A metal tube has the advantage that it can easily be dismantled if low melting solder is used. A glass container, however, cannot influence the magnetic measurements with an a.c. measuring bridge through eddy currents or ferromagnetic impurity. Moreover, the evacuation of exchange gas is markedly more rapid in the case of a glass container. It was shown by GIAUQUE, GEBALLE, LYON and FRITZ[1] that it is possible to make sample tubes of methyl methacrylate plastic (plexiglas). The material can be sealed together and electric leads can be applied through it. It is not vacuum-tight at room temperature, but it is at low temperatures, even when immersed in liquid helium II. The pumping line connecting the tube to room temperature must be made of metal, but it proves to be possible to make metal to plexiglas connections vacuum tight.

A typical Leiden sample tube, made of soft glass, is shown in Fig. 6. Soft glass is preferred over pyrex in order to decrease the possibility of overheating the salt and hence damaging it, during assembly of the tube. Some paramagnetic salts lose their water of crystallization even at 25° C.

The tube is well silvered or painted black with a material not cracking at low temperature. This is important since even little holes in the silvering or painting may cause a noticeable heat leak due to radiation. For the same reason some bends and radiation traps are inserted in the pumping line of the high vacuum pump (at B in Fig. 4); they are silvered or painted black as well.

The salt is mounted in the sample tube on a thin walled glass pedestal. In the case of a single crystal ground in the shape of a sphere or ellipsoid it is placed in a glass "egg cup", if powdered salt is used it is enclosed in a glass container. The heat leak along the glass pedestal may be decreased by an extra-thin mid-section. It is reduced drastically by inserting a piece of paramagnetic salt as shown in Fig. 6. Precautions must be taken, however, that this piece of salt does not influence the magnetic measurements with the sample itself.

Assembly and dismantling of the sample tube is made easier by making a ground joint in the sample tube. If it is carefully made and the proper grease is applied, it is vacuum tight even when immersed in liquid helium II.

A typical metal sample tube is visible in Fig. 5. It is made of cupronickel in order to decrease eddy currents. The sample is suspended between taut fibres of artificial silk, attached to a light metal cage (G in Fig. 5). The cage is made of german-silver tubing, 0.1 mm thick, and has close fitting brass caps. Most of the metal is cut away from the tube and the caps in order to make the cage springy, so that it just fits inside the sample tube when it is pushed inside. The connection of the fibres to the sample is made with two thin perspex collars (P in Fig. 5). The top collar is fastened to the bottom fibre and vice versa. It is advantageous to incorporate a light helical spring at the end of one of the fibres, particularly when the apparatus is subject to vibration. The heat leak along the suspension fibres may be decreased, similar to the method of Fig. 6, by inserting pieces of paramagnetic salt in the fibres.

---

[1] W. F. GIAUQUE, T. H. GEBALLE, D. N. LYON and J. J. FRITZ: Rev. Sci. Instrum. **23**, 169 (1952).

Fig. 7 shows two sample tubes as used in Berkeley, one made of pyrex and one of plexiglass. They are somewhat similar to the Leiden design, but the sample is suspended from the top. The chamber E can be filled with liquid helium. The chamber is closed at its top by a thin diaphragm, pierced by a small hole. The starting temperature of the demagnetization can be reduced below the temperature of the cryostat by pumping the helium in E, but the presence of an unknown amount of helium in thermal contact with the sample may introduce some difficulties, especially in the case of caloric measurements. The electrical leads shown at B are connected to a carbon thermometer heater glued to the outside of the sample, see Sect. 74. In the plastic apparatus the carbon thermometer is located on the wall I, and the sample space J is filled with some gas for thermal contact.

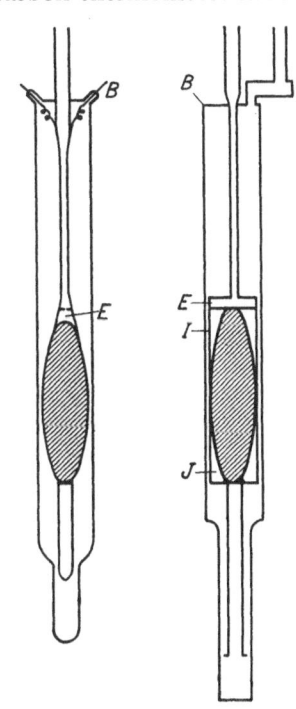

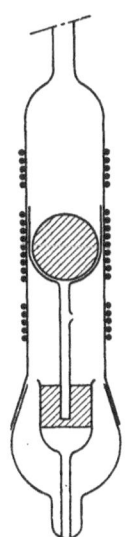
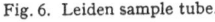

Fig. 6. Leiden sample tube.

Fig. 7. Berkeley sample tubes. The lefthand one is made of pyrex, the righthand one of plexiglass.

It is possible to make and break thermal contacts with the help of a super-conductor. For the contact between a paramagnetic salt and a helium bath this method has only been used by STEELE and HEIN[1] so far. The method will be discussed in Sect. 79.

18. Data on the thermal insulation of paramagnetic samples. The importance of good thermal insulation after adiabatic demagnetization was pointed out in Sect. 15. Two methods of suspension of a sample were described in Sect. 17, but very little research has been performed on the amount of heat leaking to the salt for each way of suspension. From experience we know that the heat leak to a sample mounted on a pedestal in a glass tube is about 30 ergs per minute, provided no holes or cracks occur in the silvering or painting of the tube. Insertion of a piece of paramagnetic salt in the pedestal as shown in Fig. 6, decreases the heat leak, under favourable circumstances, to a few ergs per minute.

---

[1] M. C. STEELE and R. A. HEIN: Phys. Rev. **87**, 908 (1952).

COOKE and HULL[1] made some investigations in the Clarendon Laboratory. If a cylinder of paramagnetic salt was placed loosely in a metal container from which it was separated by glass distance pieces they found a heat leak of about 25 ergs per minute. If the sample was suspended between two taut fibres as shown in Fig. 5 the heat leak was reduced to 5 ergs per minute. A special trick was applied in these experiments to improve the vacuum in the sample tube, the so-called "baking out process". It consists of raising the temperature of the helium bath to 1.5° K before pumping out the exchange gas and lowering it again just before the demagnetization. The exact starting temperature of the demagnetization is unknown in this case, and for some investigations this may be a disadvantage.

In their next experiment, COOKE and HULL inserted extra pieces of paramagnetic salt in the suspension fibres as described in Sect. 17. The result was that the heat flow to the central sample was reduced to one erg per minute, even in the case that the temperature of the central sample was somewhat higher than that of the other two pieces of salt. This result suggests that most of the heat leak in the case of only one sample is due to conduction along the fibres, but that the residual heat leak of one erg is due to helium gas not removed from the tube. The most probable explanation is that there is a layer of helium on the inside of the sample tube with a non-negligible vapour pressure causing a continuous transfer of heat to the sample, independent of the temperature of the sample itself.

Some experiments were made in which the vacuum was not good. The only difference in most of the results was that the top sample warmed faster, but the raise in temperature of the central sample was unaltered. The explanation is that now, on demagnetization, the gas is removed by condensation on the cold samples instead of by pumping, but the subsequent behaviour is not affected. Only the top sample warms faster since a large amount of helium gas falls down the pumping tube and condenses as a thick film on this sample and on the upper fibre. Only in case that the final temperature of the demagnetization was 0.5° K or higher, the heat leak to all three samples was markedly higher if the vacuum was not good. This, however, is not surprising since the vapour pressure of liquid helium at 0.5° K is still non-negligible.

The conclusion following from the experiments of COOKE and HULL is that for experiments below 0.4° K once the bulk of the helium gas is removed further extensive pumping (e.g. baking out) is of no advantage, even if good insulation is desired. The essential point is the use of a thin suspension for good insulation, and the use of extra pieces of salt in the suspension for still better insulation. On the other hand, in experiments above 0.4° K a good vacuum is of high importance and here the baking out procedure is of great help.

More recent experiments, also made at Oxford, showed[2] that in some cases remarkably large heat leaks may occur which are in distinct disagreement with the data of COOKE and HULL. The explanation is that an important source of heat influx may be the vibration of the salt in the sample tube. It should be realized that dropping a body of one gram over a distance of one centimeter produces already a thousand ergs of mechanical energy and this is more than sufficient for calorimetric investigations below 1° K. The vibrations are much more dangerous in the case of fibre suspension than if the sample is mounted on a glass pedestal. Precautions should be taken that the vibrations of the rotary pump for the vapour pressure of the helium are kept well away from the

[1] A. H. COOKE and R. A. HULL: Proc. Roy. Soc. Lond., Ser. A 181, 83 (1942).
[2] R. A. HULL, K. R. WILKINSON and J. WILKS: Proc. Phys. Soc. Lond. A 64, 379 (1951).

cryostat; it was found that even the boiling of the mercury in a relatively small diffusion pump may cause a noticeable rise in the temperature of the sample[1]. Sometimes, if fibre suspension is applied, increasing the number of fibres may even decrease the heat leak[2].

19. Thermal equilibrium in a demagnetized sample. The presence of a heat leak as described in Sect. 18 may result into an inhomogeneous warming up of the sample. Though the consequences are very important for adiabatic demagnetization investigations in general, there are only few quantitative data available for a discussion of the problem.

The thermal conductivity of chromium potassium alum below 1° K was investigated by KURTI, ROLLIN and SIMON[3] and by GARRETT[4]. A long single crystal was demagnetized in such a way that the ends were cooled to different temperatures and the approach to temperature equilibrium was derived from measurements of T^* (see Sect. 11) made at both ends of the sample. KURTI, ROLLIN and SIMON made the demagnetizations from an inhomogeneous field, GARRETT demagnetized from a homogeneous field and set up the temperature gradient separately with a set of coils producing a linearly varying field. It was found that the heat conductivity is a steep function of temperature; above 0.3° K the equilibrium time was too short to be measured, below 0.14° K it was too long. Between 0.14° K and 0.3° K the heat conductivity was found to be proportional to T^3.

Let us suppose that the behaviour of the heat conductivity is more or less the same for all the salts used for adiabatic demagnetization. It is clear that a temperature difference in the region well below 0.1° K, once set up, remains practically unaltered in the course of time. In the case of a constant heat flow, the effect accumulates and the temperature differences become larger and larger. Since the measurement of a thermometric parameter gives an average over the sample, the inhomogeneities make the results unreliable some time after the demagnetizations. We quote two well-known examples.

1. For most paramagnetic salts the susceptibility shows a maximum as a function of temperature (see Sect. 28). Suppose the salt is demagnetized to a temperature somewhat below this maximum. Then a homogeneous heat supply (e.g. with gamma radiation or in an alternating magnetic field) produces an increase of the susceptibility. In the case of an inhomogeneous heat leak, however, the bulk of the salt remains at the low temperature and a small fraction is heated to a much higher temperature well above the maximum of the susceptibility; in this case a decrease of the suceptibility is measured[5].

2. Suppose we have two thermometric parameters and their courses with temperature are widely different. The relation between the two can be derived performing a large number of demagnetizations from different fields and measuring both parameters simultaneously immediately after each demagnetization. The curve obtained in this way may be completely different from the result found by making only one demagnetization from the highest field and measuring both quantities as functions of time during the warming up.

---

[1] J. DARBY, J. HATTON, B. V. ROLLIN, E. F. W. SEYMOUR and H. S. SILSBEE: Proc. Phys. Soc. Lond. A **64**, 861 (1951).

[2] S. BERNSTEIN, L. D. ROBERTS, C. P. STANFORD, J. W. T. DABBS and T. E. STEPHENSON: Phys. Rev. **94**, 1243 (1954).

[3] N. KURTI, B. V. ROLLIN and F. SIMON: Physica, 's-Grav. **3**, 266 (1936).

[4] C. G. B. GARRETT: Phil. Mag. **41**, 621 (1950).

[5] See for instance: J. M. DANIELS and N. KURTI: Proc. Roy. Soc., Lond., Ser. A **221**, 243 (1954).

From the comparison of such measurements it is possible to obtain a qualitative estimation of the inhomogeneity of the temperature of a sample, but it is difficult to obtain quantitative data. A well-known procedure is to consider the salt, some time after the demagnetization, as to be divided in two parts: a fraction $(1 - \alpha)$ still at the original low temperature, and a much warmer fraction α. The temperature of the warm part is assumed to be constant, the only consequence of the heat leak being a shift of the border between the two regions leading to an increase of α. This picture was first proposed by COOKE and HULL[1]. It was further developed by DE KLERK, STEENLAND and GORTER, who neglected the susceptibility of the warm part of the sample[2], and by DANIELS and KURTI, who made also an estimation of the susceptibility of the warm part[3]. For the details we refer to the original papers. It follows, for instance, that in the case of a 25 gram sample of chromium potassium alum at the lowest temperatures a heat leak of 30 ergs per minute gives rise to a fraction α of three per cent within ten minutes. By then the best time for investigations is over.

From the considerations given here it follows that, in order to keep the influence of the heat leak on the results of the measurements small, it is advisable 1. to use big samples (assuming that the heat leak is independent of the size of the sample), 2. to keep the heat leak small, and 3. to make the measurements in as short a time as possible after the demagnetization. There are, however, a few possibilities to improve the homogeneity of the temperature some time after the demagnetization. If the salt is placed in an alternating magnetic field most heat is usually developed in its coldest parts (see Sect. 12) so that the temperature differences are decreased. This is done at the cost of an increase of the average temperature, and in case a fraction is at such a high temperature that the a.c. heating is negligible it is impossible to homogenize the temperature completely. A better method is to magnetize the sample for a short time adiabatically to a temperature of about $0.5°$ K where the heat conductivity is much better, at least in the case of a single crystal.

III. Magnets.

20. Introduction. Two types of magnets are in use for adiabatic demagnetization experiments: iron core electromagnets and iron free coil magnets. The fields easily obtained in an iron magnet are limited by the saturation to about 20 kilo-oersteds. If much higher fields are required (up to 100000 oersteds) the contribution of iron is relatively small and high power iron free solenoids are used.

The energy consumption, in the case of an iron magnet, is reasonably small, of the order of 25 kilowatts. It may be obtained from storage batteries or from a motor-generator set. The latter method has the advantage that the complete control of the magnet current, and the action of the protection devices (against, for instance, failure of the cooling supply, overload, surge on current break, etc.) can take place through the exciter field. A coil magnet takes much larger energies, up to several megawatts. It may be energized from a generator or from a set of large mercury rectifiers, the power being supplied by the mains' electric plant. Since the power consumption is an appreciable part of the energy of a city the experiments must usually be performed during the night. (In the case

---

[1] A. H. COOKE and R. A. HULL: Proc. Roy. Soc. Lond., Ser. A **181**, 83 (1942).
[2] D. DE KLERK, M. J. STEENLAND and C. J. GORTER: Leiden Commun. No. 282a; Physica, 's-Grav. **16**, 571 (1950).
[3] J. M. DANIELS and N. KURTI: Proc. Roy. Soc. Lond., Ser. A **221**, 243 (1954).

of a generator a diesel motor might also be used but the operation is rather troublesome in many laboratories.)

In the case of an iron magnet the magnet itself is the expensive piece of equipment; a suitable power supply is available in most laboratories and the cooling can be performed from the mains' water supply. In the case of a coil magnet the coil can be constructed in the workshop of the laboratory, but now the power supply and the cooling installation are the complicated parts of the setup.

Several automatic control circuits have been· developed for the elimination of slow drifts in a magnet current[1, 2], but in the case of adiabatic demagnetization experiments manual adjustment proves to be satisfactory. A more serious source of trouble may be a fast ripple in the current (e.g. the commutator ripple of the generator). If the ripple in the field is large it causes relaxation heating in the paramagnetic salt during the evacuation of the exchange gas such that, by the time that the field is removed, the starting temperature of the demagnetization is much higher than that of the liquid helium bath. A coil magnet has a small self-inductance (some milli-henries), so that the ripple in the field is almost proportional to the ripple in the voltage. Big chokes and condenser batteries must usually be applied to suppress it. For an iron magnet, however, the situation is more favourable. For frequencies above a few hundred cycles per second the solid mass of iron in the magnet becomes ineffective so that its choking effect is small; but, on the other hand, the iron is ineffective as well in the contribution of the ripple current to the magnetic field and this is the reason why in most iron magnets no special precautions are needed to suppress the ripple current. In the case of the magnet of the Bureau of Standards[3] the voltage ripple was one per cent, but the ripple in the field was less than one part in 2×10^5.

21. Iron magnets. The standard type iron magnet as used in most laboratories is shown schematically in Fig. 8. It was developed by WEISS[4], as long ago as 1907. It consists of a U-shaped yoke, Y, made of carbon steel which is magnetically very soft. The pole pieces AA' and BB' are cylinders of the same material, the pole tips A and B are truncated cones made of cobalt steel which has a very high saturation magnetization.

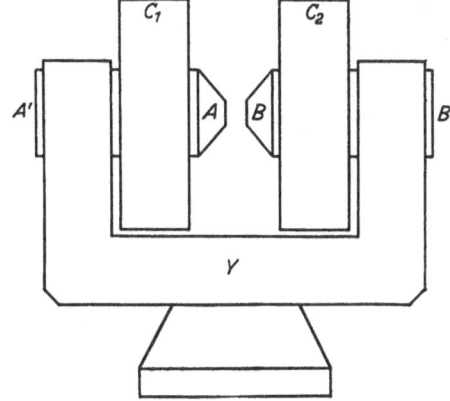

Fig. 8. WEISS magnet.

WEISS pointed out that, if the magnetization of the poles is homogeneous, the contribution of the pole tips to the field is maximum if the half angle of the cones, ϑ, is $54° 44'$ ($\tan \vartheta = \sqrt{2}$). Since the poles are never completely saturated the angle must be somewhat larger in practice, e.g. $60°$. The condition for most homogenous field, however, is: $\tan \vartheta = \sqrt{\frac{2}{3}}$, $\vartheta = 39° 14'$.

Flat coils of large diameter (C_1 and C_2) are used for the excitation of the magnet, they are located close to the pole tips. In this way the demagnetizing

[1] M. E. PACKARD: Rev. Sci. Instrum. **19**, 435 (1948).

[2] H. S. SOMMERS, P. R. WEISS and W. HALPERN: Rev. Sci. Instrum. **20**, 244 (1949); **22**, 612 (1951).

[3] D. DE KLERK and R. P. HUDSON: J. Res. Nat. Bur. Stand. **53**, 173 (1954).

[4] P. WEISS: J. de Phys. **6**, 353 (1907).

action is checked as much as possible and the direct contribution of the coils to the field is maximum.

The cooling is performed with water from the mains' supply or by circulating oil through the magnet. In some designs the windings themselves are hollow tubes and the cooling liquid passes through them. If large energies are dissipated the coils are made of bare copper strip with distance pieces and the liquid is pumped through the slits.

By increasing the dimensions of a magnet and the amount of iron we do not increase the maximum field very much beyond a certain value (determined by the saturation value of the B of the pole tips), but we do increase the volume of the pole gap over which this field may be maintained. The largest magnet of the WEISS type regularly used for adiabatic demagnetization work is the one of the KAMERLINGH ONNES Laboratory[1,2]. It consists of 12 tons of iron. The diameter of the pole pieces is 40 cm, that of the pole faces is 10 cm. Using 80 kilowatts it produces a field of 24 kilo-oersteds in a pole gap of 6 cm.

The design of the WEISS magnet has been modified and improved by several investigators. The Bellevue magnet[3,4] has a symmetrical yoke and the pole pieces (AA' and BB' of Fig. 8) are shaped conically; their diameter varies from 121 to 75 cm and the diameter of the pole faces is 6 cm. The magnet's weight is 120 tons, and taking 93 kilowatts, it produces a field of 36 kilo-oersteds in a pole gap of 5 cm. In a magnet disigned by BITTER and REED[5] the yoke is axially symmetrical around the poles. The poles have a conical shape and the windings are located much closer to the field space than in the original WEISS magnet. It consists of two tons of iron and it produces the same field as the Leiden magnet at an energy consumption of only 20 kilowatts.

22. Coil magnets. The field in the centre of an iron free coil magnet[6,7] can be expressed by:

$$H = G \sqrt{\frac{Wf}{\varrho \, r_i}}, \tag{22.1}$$

where W is the power dissipated in the magnet, ϱ the resistivity of the coil material, r_i the inner radius of the coil and f the "filling factor", i.e. the volume occupied by the coil metal divided by the total volume of the winding space. G is a dimensionless factor depending on the shape of the winding space and on the distribution of the current density in it.

Values for G were calculated for several models of magnets. BITTER showed[6] that the maximum value that can be reached for any coil is: $G = 0.272$ (H being expressed in oersteds, W in watts, ϱ in ohm cm and r_i in cm). This coil, however, extends to infinity in all directions. For a finite coil with a rectangular winding space and uniform current density the maximum value for G is 0.179. In this case $r_e = 3r_i$, $l = 4r_i$ (see Fig. 9). Values for G between 0.18 and 0.21 can be reached if the current density near the centre is larger than at the outside of the coil.

In general the homogeneity of the field is the smaller the higher is the value of G and in practice the value of G is adapted to the homogeneity requirements of the experiment.

[1] W. J. DE HAAS: Physica, Nederl. Tijdschr. Natuurk. **12**, 113 (1932).
[2] G. HÄDER: Siemens-Z. **8**, 3 (1930).
[3] A. COTTON: C. R. Acad. Sci. Paris **187**, 77 (1928).
[4] A. COTTON and G. DUPOUY: C. R. Acad. Sci. Paris **190**, 544 (1930).
[5] F. BITTER and F. E. REED: Rev. Sci. Instrum. **22**, 171 (1951).
[6] F. BITTER: Rev. Sci. Instrum. **7**, 482 (1936).
[7] F. BITTER: Rev. Sci. Instrum. **10**, 373 (1939).

For a given value of f the field is independent of the geometrical composition of the coil inside the winding space. The actual number of turns and the cross-section of the conductors is entirely determined by the impedance of the power supply to which the magnet should be adapted. In the case of low impedance (high current and low voltage) few turns of thick metal should be used. In the case of high impedance (low current and high voltage) many turns of thin material are needed. High impedance coils are made of square wire or flat strip wound into layers or "pancakes"[1]. A nice system for low impedance coils was developed by BITTER. The turns of his magnets consist of flat copper discs separated by thin insulating sheets and joined together at their edges. In this type of coil the current density is higher near the axis than at the exterior, resulting into a higher value for G (see above). For the details of the construction we refer to the original papers[2, 3].

If the power is dissipated at a low voltage the cooling may be achieved with the help of water. Distilled water should be preferred over mains' water in order to prevent the magnet from corrosion. In the case of a high voltage coil some non-inflammable organic fluid should be used. A low viscosity and a large specific heat are advantageous. (The necessity of cooling is one reason why the filling factor f cannot be chosen too close to unity.) The liquid is pumped through

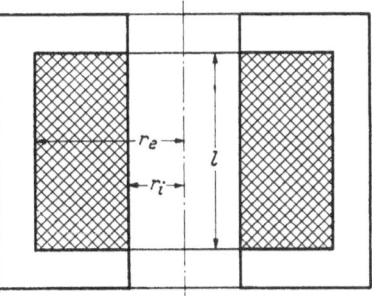

Fig. 9. Cross-section of a coil magnet covered with iron.

the magnet and a heat exchanger at a high speed, up to several hundred cubic meters per hour. The heat is carried away from the heat exchanger with ordinary tap water, or even river water. The flow of liquid through the magnet and the heat exchanger should be well turbulent, and the heat contact with the coil metal must be as good as possible. For this reason bare copper is used, the turns being kept apart by distance pieces or a wrapping of suitable thread.

The field of a coil may be increased somewhat by covering the outside with iron[4], as shown in Fig. 9. Since the contribution is the less significant the higher the field, iron is only applied in smaller coil magnets[5]. In very strong coil magnets[6-8] it is omitted.

IV. Bridge methods.

23. Introduction. It was pointed out in Sects. 10 and 11 that the temperature determination in the region below 1° K is based on the measurement of a magnetic quantity of the salt, a so-called "thermometric parameter". Suitable parameters are, for instance, the susceptibility [or the magnetic temperature which is directly related to the susceptibility, see Eq. (11.1)], the heat absorption coefficient from an alternating magnetic field, and the remanent magnetic moment.

[1] J. M. DANIELS: Proc. Phys. Soc. Lond. B **63**, 1028 (1950).
[2] F. BITTER: Rev. Sci. Instrum. **7**, 482 (1936).
[3] F. BITTER: Rev. Sci. Instrum. **10**, 373 (1939).
[4] F. BITTER: Rev. Sci. Instrum. **7**, 479 (1936).
[5] E. MENDOZA: Cérémonies LANGEVIN-PERRIN, p. 54, Paris 1948.
[6] B. TSAI: Phys. Soc. Cambr. Conf. Report, p. 89, 1947.
[7] J. M. DANIELS: Proc. Phys. Soc. Lond. B **63**, 1028 (1950).
[8] F. GAUME: C. R. Acad. Sci. Paris **223**, 719 (1946).

In the first experiments of DE HAAS, WIERSMA and KRAMERS the susceptibility was measured with a balance method[1]. The balance was mounted in a box connected to the cap of the cryostat. The salt was suspended from one arm by means of a long quartz rod; it was placed in the inhomogeneous part of the field and the susceptibility was derived from the force acting on the salt. At the present time this method is only used in BRISTOL[2], most other laboratories use induction bridge methods.

In an induction method we observe the voltage in a coil when the field in it is varied. The flux through the coil may be represented by:

$$\varphi = \alpha H + \beta M. \tag{23.1}$$

Here H is the field in the absence of the salt and M is the magnetic moment. α and β are geometrical factors. The variation of the flux on variation of the field is:

$$\Delta\varphi = \alpha \Delta H + \beta \Delta M = \left(\alpha + \beta \frac{\Delta M}{\Delta H}\right) \Delta H. \tag{23.2}$$

$\Delta\varphi$ is proportional to the voltage induced in the coil. If H is produced by the coil itself, $(\alpha + \beta \Delta M/\Delta H)$ is proportional to its self-inductance coefficient; if H is produced by a separate coil, $(\alpha + \beta \Delta M/\Delta H)$ is proportional to the mutual inductance. In both cases α is the contribution of the empty coil and $\beta \Delta M/\Delta H$ is due to the salt. If only small fields are applied so that CURIE's law is valid $\Delta M/\Delta H$ is equal to the susceptibility.

From the voltage in the coil we can derive the self-inductance or mutual inductance apart from a constant unknown factor. This, however, is not serious. The susceptibility of the salt in the liquid helium region and its dependence on temperature are usually known. By measuring $\Delta\varphi/\Delta H$ as a function of temperature in the liquid helium region we can determine α and β empirically.

The measurements may be performed in two ways: ballistically or with the help of a.c. In the first case the field is varied suddenly and the voltage in the coil is derived from the ballistic deflection of a galvanometer. If a small field is reversed the quantity determined in the experiment is the susceptibility. If a large field is switched on in steps the magnetization curve is found by adding up the deflections. At the lower temperatures hysteresis effects are often found in paramagnetic salts. In this case the shape of the loop and the value of the remanent moment may be measured by switching the field on and off in steps in both directions.

In the case of a.c. measurements, an alternating magnetic field is applied and the a.c. voltage in the coil is measured. If the a.c. field is small the quantity derived from the experiment is the susceptibility again. At the lower temperatures relaxation effects occur in most paramagnetic salts. They result into a phase shift between the field and the magnetic moment. The susceptibility, then, can be divided into two components, one is denoted by χ' and is in phase with the field, the other one is denoted by χ'' and it is in quadrature with the field. In this case the susceptibility (often referred to as the "dynamic susceptibility") may be represented by a complex quantity:

$$\chi = \chi' - i \chi''. \tag{23.3}$$

Both components can be determined with the a.c. bridge (see Sect. 26). If χ'' is small, χ' is equal to the static susceptibility as measured ballistically; if χ'' is of an order of magnitude comparable with χ' this is no longer true.

[1] W. J. DE HAAS, E. C. WIERSMA and H. A. KRAMERS: Leiden Commun. No. 229a; Physica, 's-Grav. 1, 1 (1933/34).

[2] E. MENDOZA and J. G. THOMAS: Phil. Mag. 42, 291 (1951).

If the amplitude of the alternating magnetic field is h_0 the heat absorption from the magnetic field per second [represented by $\int H\,dM$, see Eq. (8.2)] is equal to:

$$\tfrac{1}{2}h_0^2\,\omega\,\chi'' \tag{23.4}$$

where ω is the frequency of the alternating field multiplied by 2π. It follows that χ'' is the heat absorption coefficient referred to at the beginning of this section.

24. Alternating current and ballistic bridges. In most adiabatic demagnetization work, the a.c. method must be preferred to the ballistic one. A higher precision can be reached and more measurements can be obtained per unit time. A disadvantage is, however, that the whole apparatus inside the cryostat must be made of insulating material, since all metal parts give rise to eddy currents which influence the bridge settings, especially the χ'' values (see Sect. 26).

At the higher temperatures where no relaxation or hysteresis effects occur χ'' is practically zero. Here one can obtain about twenty bridge settings per minute. At the lower temperatures, where both χ' and χ'' are significant, the bridge is balanced with two components which are both functions of temperature and hence, during a heating curve, of time. In this case balancing the bridge requires some experience, but still several settings may be obtained per minute.

The a.c. method gives valuable information on both χ' and χ'' but at the lowest temperatures the heat absorption may give rise to a too rapid heating of the sample. In this case the ballistic method must be preferred which, moreover, gives information on the remanent magnetic moment and on the shape of the hysteresis loop.

The rate at which ballistic measurements can be taken depends on the vibration period of the galvanometer. If a telescope and scale are used the vibration period cannot be reduced below about six seconds without seriously impairing the reliability of the readings. In this case one can take about six measurements per minute. This number can be increased noticeably by using a faster galvanometer and recording the deflections photographically. Still it is not advisable to make the vibration period of the galvanometer too short. If it is of an order of magnitude comparable with a relaxation time of the salt, double deflections are found; the galvanometer, although critically damped, moves very fast in one direction and then in the opposite direction[1, 2]. The interpretation of these deflections proves to be complicated and it is preferable to use a somewhat slower galvanometer. In the investigations with chromium potassium alum[2] it was found that a galvanometer with a vibration period of 1.5 sec is about the fastest that can be used in practice.

As was indicated above, metal parts in the cryostat are less dangerous for ballistic measurements than in the case of an a.c. bridge. Still, under certain circumstances, the eddy currents may give rise to double deflections similar to those occurring in the case of a relaxation effect in the salt. It was shown by GIAUQUE and his coworkers[3, 4], however, that they can be eliminated if the proper precautions are taken in the construction of the bridge coils, see Sect. 25.

For descriptions of a large number of bridge networks we refer to HAGUE's book[5]. In the present article we restrict ourselves to a short discussion of the

[1] D. DE KLERK: Proc. N. B. S. Semicentennial Symposium on Low Temperature Physics, 1951, p. 211.

[2] J. A. BEUN, M. J. STEENLAND, D. DE KLERK and C. J. GORTER: Leiden Commun. No. 300a; Physica, 's-Grav. **21**, 651 (1955).

[3] W. F. GIAUQUE and J. W. STOUT: J. Amer. Chem. Soc. **61**, 1384 (1939).

[4] W. F. GIAUQUE, J. J. FRITZ and D. N. LYON: J. Amer. Chem. Soc. **71**, 1657 (1949).

[5] B. HAGUE: Alternating current bridge methods. London 1946.

a.c. Anderson bridge for self-inductance measurements[1, 2], and of the ballistic and a.c. Hartshorn bridge[3-9] for mutual inductance measurements, since these are the only bridges which are in use for adiabatic demagnetization investigations.

25. The ballistic Hartshorn bridge. The principle of a ballistic Hartshorn bridge is shown in Fig. 10. M_1 is a set of coils surrounding the sample (see for instance Fig. 5). It is connected in series with another set of coils, M_2, and the connections are made in such a way that the secondary voltages in M_1 and M_2, due to a variation in the primary current, have opposite signs. The coefficient of mutual inductance of M_2 can be varied, for instance by varying the number of turns on the secondary coil.

In the Berkeley laboratory, it is standard practice to set M_2 to such a value that the bridge is exactly balanced. The galvanometer is used as a zero detector and the values of the susceptibility are derived from the settings of M_2. This method gives the highest precision, but it has the disadvantage that the double deflections due to eddy currents, quoted in Sect. 24, are the more pronounced the better the bridge is balanced. In Leiden and in the Bureau of Standards, the bridge is only approximately balanced and the values of the susceptibility (and the remanence or the shape of the hysteresis loop, see Sect. 23) are derived from the settings of M_2 and the residual deflections of G. Moreover, this method works faster.

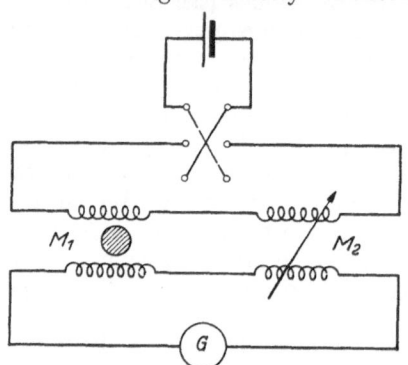

Fig. 10. Ballistic Hartshorn bridge.

The stability of the bridge is improved by mounting big part of M_2 together with M_1 in the cryostat. In Leiden, the secondary coil in the cryostat is wound in three sections (see Fig. 4 and 6). The central section surrounds the salt specimen; it consists of several hundreds of turns, the number being the larger the weaker is the paramagnetism of the salt. The remaining two sections are wound above and below the central section in the opposite direction, each containing half the number of turns. With this arrangement the mutual inductance in the absence of the salt [the term α in Eq. (23.2)] is zero (apart from a small correction due to inhomogeneity of the primary field) and the voltage over the secondary coils is practically proportional to the susceptibility of the salt.

In Leiden and in the Bureau of Standards, the variable part of M_2 is a separate set of coils mounted outside the cryostat. It consists of two decades, of approximately 30 and 300 microhenries per unit, and it is constructed in such a way that, if a number of secondary turns is switched into the circuit, an equivalent resistance is removed from it. In this way the ballistic sensitivity of the

[1] W. F. Giauque and D. P. MacDougall: J. Amer. Chem. Soc. **57**, 1175 (1935).
[2] H. van Dijk: Proceedings of the Third Symposium on Temperature, p. 199, Washington 1954.
[3] D. de Klerk: Thesis, Leiden 1948, p. 36.
[4] M. J. Steenland: Thesis, Leiden, 1952, p. 12.
[5] D. Bijl: Thesis, Leiden, 1950, p. 89.
[6] D. de Klerk and R. P. Hudson: J. Res. Nat. Bur. Stand. **53**, 173 (1954).
[7] R. A. Erickson, L. D. Roberts and J. W. T. Dabbs: Rev. Sci. Instrum. **25**, 1178 (1954).
[8] W. F. Giauque and J. W. Stout: J. Amer. Chem. Soc. **61**, 1384 (1939).
[9] W. F. Giauque, J. J. Fritz and D. N. Lyon: J. Amer. Chem. Soc. **71**, 1657 (1949).

galvanometer is independent of the setting of M_2. Since the variable part of M_2 is calibrated in absolute units of mutual inductance, it can also be used for the calibration of the sensitivity of the galvanometer (by switching the primary coil of the cryostat out of the circuit).

In Berkeley, the variable part of M_2 is mounted inside the cryostat somewhat removed from the salt. This has the advantage that the resistance of the secondary circuit is low and that the stability of the bridge is better. Moreover, the influence of the eddy currents in the metal tubing of the big magnet coil surrounding the cryostat is automatically balanced. A large number of wires passes into the cryostat, connecting each of the bridge coils to a switch, but this does not give rise to a too large heat flow into the liquid helium. For calibration purposes a separate set of coils was constructed, used at room temperature. For the compensation of the eddy currents due to the magnet these coils are placed in an instrument duplicating the metal parts of the magnet in sufficient detail.

26. The alternating current HARTSHORN bridge. The a. c. HARTSHORN bridge is shown schematically in Fig. 11. The power supply SG is usually an audio-frequency signal generator. Frequencies between 20 and several hundred cycles are used. For the stability of the readings it is advisable to use only frequencies which are not too close to the harmonics of the main's power supply. The detector G (a vibration galvanometer, oscilloscope or headphones, usually preceded by an amplifier) is used as a zero-instrument so that M_2 must be continuously

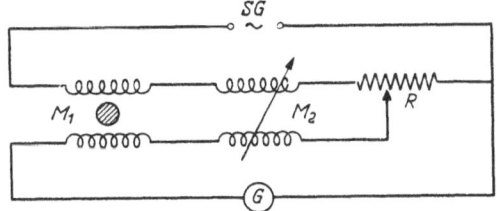

Fig. 11. Alternating current HARTSHORN bridge.

variable. In Leiden and in the Bureau of Standards, M_2 consists of four decades and a continuous variometer. Mutual inductances up to three millihenries can be measured with a precision better than a hundredth of a microhenry. In these bridges the secondary coils are made of tenfold stranded wire. This has the advantage that the ten coils of one decade are identical to a high precision (a few parts in 10^4), but the disadvantage is that the capacitive coupling between the coils is not negligible, so that the bridge can only be used for low frequencies (up to about 500 cycles). This system was abandoned in a bridge recently constructed in Oak Ridge. In this bridge all the coils were wound separately, the layers of the different coils being kept well apart by polysterene sheet. Small trimming coils were needed for each of the individual coils in order to adjust the exact ratios, but the bridge can be used up to 16 kilocycles without difficulties.

If the susceptibility of the salt is represented by Eq. (23.3) M_1 fulfills the relation:

$$M_1 = M_0 + \beta (\chi' - i \chi'').$$ (26.1)

Here M_0 is the mutual inductance of the empty coils (which is approximately balanced if the coils in the cryostat are arranged in three sections as described in Sect. 25) and β is a geometrical factor. It is obvious that only $M_0 + \beta \chi'$ can be balanced by adjusting M_2, since they both give rise to voltages in quadrature with the primary current. The voltage due to $\beta \chi''$ is in phase with the primary current and hence it can be balanced by setting the potentiometer R. It is found that the equilibrium conditions of the bridge are:

$$M_2 = M_0 + \beta \chi'$$ (26.2)

and
$$R = \beta \omega \chi''$$ (26.3)

so that G reaches its zero position only if both M_1 and R are adjusted to the proper values.

If eddy currents occur in metal parts of the cryostat, the magnet and the pumping lines, or if capacitive or inductive leaks occur in the bridge, the deflection of G can still be compensated by adjusting M_2 and R, but now the interpretation of the bridge settings is more difficult. Corrections must be applied, both in χ' and χ'' (those in χ'' usually being the largest), which may be determined from measurements where M_1 is replaced by a variable mutual inductance which is free of a. c. losses.

In Leiden, the a.c. and ballistic HARTSHORN bridges are built together into one network. The alteration of one bridge into the other can be made in a few seconds with the help of some switches.

27. The ANDERSON bridge. The principle of the a. c. ANDERSON bridge is given in Fig. 12. The equilibrium conditions are:

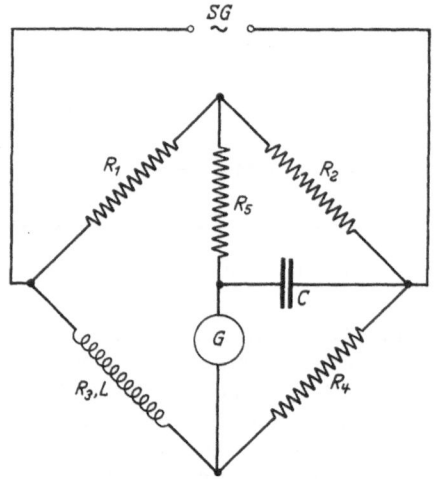

Fig. 12. ANDERSON bridge.

$$R_1 R_4 = R_2 R_3,$$ (27.1)

$$L = C R_4 \left[R_5 \left(\frac{R_1}{R_2} + 1 \right) + R_1 \right].$$ (27.2)

If L, due to the presence of the salt, is represented by:

$$L = L_0 + \beta (\chi' - i\chi''),$$ (27.3)

where β is a geometrical factor, the equilibrium conditions become:

$$R_1 R_4 = R_2 (R_3 + \beta \omega \chi''),$$ (27.4)

$$L_0 + \beta \chi' = C R_4 \left[R_5 \left(\frac{R_1}{R_2} + 1 \right) + R_1 \right].$$ (27.5)

The bridge can be balanced by adjusting R_4 and R_5 to the proper values. The consequences of eddy currents and a.c. leaks in the bridge are similar to those in the case of a HARTSHORN bridge.

In the bridge used in Berkeley L consists of two equal coils each surrounding one half of the sample. If they are connected in series they give an approximately homogeneous field over the sample. If measurements are made in an external field (produced by the big coil magnet, see Sect. 25) the coils are connected oppositely so that they are insensitive to fluctuations in the field of the magnet. In this case the measuring field is very inhomogeneous, but since it is only a small fraction of the external field the magnetization of the sample is not much influenced. The loss in sensitivity is only 15%. For measurements in zero external field the coils must be connected in the ordinary way.

C. Magnetic investigations at relatively high temperatures.
I. Theoretical considerations.

28. Introduction. The most striking phenomenon in the magnetic behaviour of the salts used for the demagnetization process is the occurrence of a maximum in the susceptibility. Below this maximum is the region where the relaxation and hysteresis effects occur which were quoted in Sect. 23. Here the phenomena are very similar to those of ferromagnetism and antiferromagnetism at higher temperatures. In the region above the maximum such phenomena are not encountered. Here the behaviour is paramagnetic.

Experiments on the magnetic and caloric behaviour of several salts have been performed in both regions. In general the agreement between different investigations is better at the higher temperatures than below the maximum. Theoretical considerations were also given for some salts. For the region below the maximum the theories have a more qualitative character than for the paramagnetic region.

In the present chapter we restrict ourselves to the discussion of theoretical and experimental results in the region above the maximum. Chapter D will be devoted to the lower temperatures.

The problem of low temperature paramagnetism is the computation of the energy levels of the salt under the joint influence of the external field and the crystalline interactions. These interactions are, as was stated in Sect. 4, the STARK splitting due to the electric field of the non-magnetic atoms surrounding the paramagnetic ion, the hyperfine splitting due to the magnetic and electric interactions with the nucleus, and the magnetic and exchange interactions with neighbouring magnetic ions.

The influence of all these contributions to the HAMILTONian may be computed and the eigen values may be derived from a succession of perturbation calculations, in which the largest term is taken into account first and the smallest one last. After this the partition function can be set up and the thermodynamic quantities may be found from the relations given in Sect. 3.

In the present article we shall not go through all the details of these calculations. In the next section we derive the magnetic moment and entropy of a salt neglecting the crystalline interactions. In Sects. 30, 31 and 32 we give a short discussion of each of the interactions; after this we discuss the results obtained with the individual salts.

29. Normal paramagnetism. The influence of crystalline interactions may be negelected, as was stated in Sect. 4, under the conditions that the zero field distance between the low lying levels is small as compared with kT, and that the higher levels are so far away that they can be considered as unoccupied. In this case the lower levels in zero field may be considered as degenerate; in a field they are separated and the distances between the sublevels are equal to $g\mu_B H$ (see Sect. 3). The partition function is represented by Eq. (3.8) and according to Eqs. (3.2), (3.3) and (3.4) we have for the entropy S and magnetic moment M:

$$
\left.
\begin{aligned}
\frac{S}{R} &= \ln \sum_{-J}^{+J} e^{-mg\alpha} + \frac{\displaystyle\sum_{-J}^{+J} mg\alpha\, e^{-mg\alpha}}{\displaystyle\sum_{-J}^{+J} e^{-mg\alpha}} \\
&= \ln \operatorname{Sin} \tfrac{1}{2}(2J+1)g\alpha - \ln \operatorname{Sin} \tfrac{1}{2} g\alpha - \tfrac{1}{2}(2J+1)g\alpha \operatorname{Cot} \tfrac{1}{2}(2J+1)g\alpha - \\
&\qquad\qquad - \tfrac{1}{2} g\alpha \operatorname{Cot} \tfrac{1}{2} g\alpha .
\end{aligned}
\right\} \quad (29.1)
$$

$$
\left.
\begin{aligned}
\frac{M}{R} &= \frac{\mu_B}{k} \frac{\displaystyle\sum_{-J}^{+J} mg\, e^{-mg\alpha}}{\displaystyle\sum_{-J}^{+J} e^{-mg\alpha}} \\
&= \frac{\mu_B}{k} \left\{ \tfrac{1}{2}(2J+1)g \operatorname{Cot} \tfrac{1}{2}(2J+1)g\alpha - \tfrac{1}{2} g \operatorname{Cot} \tfrac{1}{2} g\alpha \right\} .
\end{aligned}
\right\} \quad (29.2)
$$

Here R is the gas constant, g is the splitting factor, and α satisfies the relation:

$$\alpha = \frac{\mu_B H}{kT}, \tag{29.3}$$

where H is the local field as defined in Sect. 7. The other quantities have the same meanings as in Sect. 3. (Sin and Cot are the hyperbolic functions.) If we introduce:

$$B(J) = \tfrac{1}{2}(2J+1)\, g \operatorname{Cot} \tfrac{1}{2}(2J+1)\, g\alpha - \tfrac{1}{2} g \operatorname{Cot} \tfrac{1}{2} g\alpha, \tag{29.4}$$

we have:

$$\frac{M}{R} = \frac{\mu_B}{k}\, B(J), \tag{29.5}$$

$$\frac{S}{R} = -\alpha B(J) + \int B(J)\, d\alpha. \tag{29.6}$$

$B(J)$ is called a Brillouin function.

With these formulae we are in the case, described in Sect. 3, that both S and M are functions of H/T only. During an adiabatic demagnetization S is constant, hence M is constant and T falls proportionally to H. It is self-evident under these circumstances, that $(\partial M/\partial H)_S$ (sometimes called the "adiabatic susceptibility") is zero for any value of H and T; and that the specific heat at constant magnetic moment c_M, equal to $T(\partial S/\partial T)_M$, is also zero (as well as the zero field specific heat c_0).

For small values of H we may develop Eqs. (29.1) and (29.2) into:

$$\frac{S}{R} = \ln(2J+1) - \frac{1}{6} J(J+1) \frac{g^2 \mu_B^2}{k^2} \frac{H^2}{T^2}, \tag{29.7}$$

$$\frac{M}{R} = \frac{1}{3} J(J+1) \frac{g^2 \mu_B^2}{k^2} \frac{H}{T}. \tag{29.8}$$

Eq. (29.8) is Curie's law [see Eq. (3.11)] with:

$$\frac{C}{R} = \frac{1}{3} J(J+1) \frac{g^2 \mu_B^2}{k^2}. \tag{29.9}$$

Obviously, the conditions given here cannot be satisfied down to zero field. Deviations occur due to the crystalline interactions whose influence on the partition function will be discussed in more detail in the next few sections. For temperatures where kT is large as compared with the level separations it follows that the influence on the magnetic moment is small, but the specific heat at constant magnetic moment instead of being zero, satisfies the relation:

$$\frac{c_M}{R} = \frac{A}{T^2}, \tag{29.10}$$

where $k\sqrt{A}$ is of the order of the level splitting. In this case M is no more constant during an adiabatic demagnetization and hence the adiabatic susceptibility is not zero. Assuming the validity of Curie's law and of Eq. (29.10), the course of temperature with field on an isentropic follows from Eqs. (9.1) and (9.6):

$$\frac{T}{T_0} = \sqrt{1 + \frac{C}{A} H^2}, \tag{29.11}$$

where T_0 is the zero field temperature. For strong fields $(H \gg \sqrt{A/C})$ it follows that H/T is constant as was to be expected, for small fields the rise in temperature is proportional to H^2. It is obvious that the demagnetization process cannot be successful for a salt with $\sqrt{A/C}$ of the order of 10^4 oersteds or higher. As a matter of fact the value of A/C has often been used as a criterion for the suitability of a salt for the demagnetization process.

Some remarks should be made on the values of J and g. In the case of free magnetic ions (paramagnetic gas) J may be derived from HUND's stability rules and g follows from the LANDÉ formula:

$$g = 1 + \frac{J(J+1) + S(S+1) - L(L+1)}{2J(J+1)}. \tag{29.12}$$

The effective J and g for an ion in a crystal may be widely different. In case that the orbital levels are quenched (see Sect. 4) we have $L=0$, $J=S$ and $g=2$, but in practice the susceptibility may even become anisotropic so that g must be replaced by a tensor. This may be due, for instance, to inclomplete quenching of the orbital levels, see Sect. 4 (in this case J may become smaller than S) or to a combined action of the electric field and the spin-orbit coupling.

30. The crystalline STARK effect. In most paramagnetic salts used for the adiabatic demagnetization process the magnetic ion is surrounded by six molecules of water of crystallization, or oxygen atoms, giving an electric field of approximately cubic symmetry at the ion. As was pointed out in Sect. 4, this field may remove the degeneracy of the lowest orbital level to such an extent that only the lowest sublevel is occupied at liquid helium temperatures, the splitting being of the order of 10^4 cm^{-1}. The lowest orbital sublevel, referred to here as the ground state, exhibits a spin degeneracy which cannot be removed by the direct influence of the electric field. A small indirect effect, however, is possible[1] due to the spin-orbit coupling, giving rise to a splitting of the order of a few tenths of a cm^{-1}. If the spin degeneracy of the ground state is even (the only case considered here, see Sect. 4) it cannot be removed completely by an electric field, but, according to KRAMERS' theorem[2], an even degeneracy is left for each sublevel. This can only be removed by a magnetic field.

It is obvious now that no STARK splitting can occur if the ground state of an ion has only two-fold spin degeneracy. A four-fold level cannot be split by an electric field if it has exactly cubic symmetry[1], but usually the octahedron formed by the water of crystallization is slightly distorted and a trigonal or tetragonal component occurs in the field. This component can split a fourfold level. In the case of a sixfold level a field of cubic symmetry may cause a splitting[1], but still then a trigonal component may give a contribution which is of the same order of magnitude as the splitting due to the cubic field, even if the trigonal field is much weaker[3].

For the discussion of the STARK effect as far as adiabatic demagnetization is involved we can restrict ourselves to the following aspects: (a) the influence on the zero field entropy and specific heat, (b) the influence on the susceptibility in small fields (hence on the course of T^* with T), (c) the influence on the magnetization curve and the entropy of the salt in strong fields at the initial temperature [hence the modification of Eqs. (29.1) and (29.2)].

The zero field specific heat can easily be calculated if the level scheme is known for zero field. Suppose the overall splitting is $k\delta$ and we have n_1 KRAMERS doublets at an energy $\alpha_1 k\delta$, n_2 doublets at $\alpha_2 k\delta$, etc., where $\alpha_1, \alpha_2, \ldots$ are

---

[1] J. H. VAN VLECK and W. G. PENNEY: Phil. Mag. **17**, 961 (1934).
[2] H. A. KRAMERS: Proc. Acad. Sci. Amst. **33**, 959 (1930).
[3] A. ABRAGAM and M. H. L. PRYCE: Proc. Roy. Soc. Lond., Ser. A **205**, 135 (1951).

numbers varying from zero to one, then the partition function for a gram ion of paramagnetic salt is given by:

$$Z_s = \left(\sum_r 2n_r\, e^{-\alpha_r\, \delta/T} \right)^N. \tag{30.1}$$

The entropy and specific heat can be derived from Eq. (30.1) with the help of Eqs. (3.2), (3. 4) and (9.4). The exact formulae will be discussed later for the ndividual salts. For relatively high temperatures we may develop:

$$Z_s = \left(\sum 2n_r - \frac{\delta}{T} \sum 2n_r \alpha_r + \frac{1}{2} \frac{\delta^2}{T^2} \sum 2n_r \alpha_r^2 + \cdots \right)^N, \tag{30.2}$$

$$\frac{S}{R} = \ln \sum 2n_r - \frac{1}{2} \frac{\delta^2}{T^2} \left\{ \left(\frac{\sum 2n_r \alpha_r^2}{\sum 2n_r} \right) - \left(\frac{\sum 2n_r \alpha_r}{\sum 2n_r} \right)^2 \right\}, \tag{30.3}$$

$$\frac{c_0}{R} = \frac{\delta^2}{T^2} \left\{ \left(\frac{\sum 2n_r \alpha_r^2}{\sum 2n_r} \right) - \left(\frac{\sum 2n_r \alpha_r}{\sum 2n_r} \right)^2 \right\} \tag{30.4}$$

in agreement with Eq. (29.10).

The zero field level scheme must be derived from the electric field pattern at the position of the ion. It turns out, however, that the configuration of the non-magnetic atoms around the ion (deviation from cubic symmetry) must be known with a high precision and it proves to be impossible to obtain sufficient information from the X-ray diffraction pattern. A more or less inverse procedure could be followed. The level distances may be derived from paramagnetic resonance experiments and the empirical data are inserted in formula (30.1). This procedure, however, leads often to problems as described at the end of Sect. 4. For this reason the constants in Eq. (30.1) are often adapted empirically to the results of the demagnetization experiments.

The calculation of the susceptibility in weak fields is somewhat more complicated. For this purpose we must know the variation of the energy levels with an applied field. This is far from linear and, moreover, depends on the orientation of the field with respect to the crystalline axes. As an example, Figs. 13 and 14 show the level patterns for the chromium alums for fields in the directions of the cubic axis and the trigonal axis. If there are several ions in the unit cell of the crystal with different symmetry axes (e.g. in the case of trigonal symmetry) the results for a given field direction must be averaged over the ions.

The deviations from Curie's law in weak fields due to the Stark splitting may be expressed by:

$$\chi = \gamma(T) \frac{C}{T}, \tag{30.5}$$

or:

$$T = \gamma(T)\, T^* \tag{30.6}$$

where χ is defined as χ_{loc} of Sect. 7 and T^* stands for $T^{(*)}$. In these relations $\gamma(T)$ may depend on the orientation of the magnetic field and approaches to the value one for high temperatures. In order to calculate $\gamma(T)$ we have to set up the secular determinant and to find its roots[1,2,3]. Results for individual salts will be given later. It should be pointed out that for relatively high temperatures $\gamma(T)$ can be developed into a series in $1/T$, where the coefficient of the $1/T$-term may be represented as a Curie-Weiss Θ. It may be shown[4] that the

[1] J. H. van Vleck: The theory of electric and magnetic susceptibilities. Oxford 1932.
[2] W. G. Penney and R. Schlapp: Phys. Rev. 41, 194 (1932).
[3] R. Schlapp and W. G. Penney: Phys. Rev. 42, 666 (1932).
[4] J. H. van Vleck and W. G. Penney: Phil. Mag. 17, 961 (1934).

value for Θ averaged over all the possible field directions in the crystal (hence the Θ measured for a powdered sample) is zero.

The influence of the STARK splitting on the magnetic moment and the entropy of a salt at the initial temperature of the demagnetization can be calculated from Eqs. (3.3) and (3.4) if the course of the energy levels with the field is known. The latter may be derived from theoretical considerations[1] or from paramagnetic resonance experiments (see Sect. 4). The effect has been neglected in the calculations of most investigators. Only for the case of the chromium alums the corrections to Eqs. (29.1) and (29.2) were computed by HUDSON[2] and by DANIELS and KURTI[3]. At 1° K the influence is small for fields over a few thousands oersteds, as can be seen from Fig. 13. For the numerical values of the corrections we refer to the original papers.

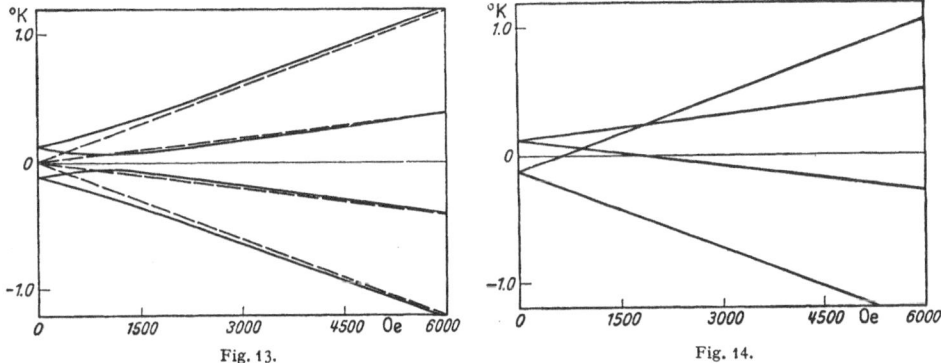

Fig. 13.   Fig. 14.

Fig. 13. Energy levels of the Cr^{+++}-ion in the alums for a field parallel to the cubic axis. The dotted lines are the levels if no STARK splitting should occur. The energy values are divided by the BOLTZMANN constant, so that they are expressed in "equivalent degrees KELVIN".

Fig. 14. Energy levels of the Cr^{+++}-ion in the alums for a field parallel to the trigonal axis.

31. Hyperfine structure. If the nucleus of the paramagnetic ion has a spin I a small splitting of the ground level occurs which is referred to as hyperfine splitting. It consists of two contributions: (a) magnetic interaction between the magnetic moments of the nucleus and the electrons, (b) electric interaction between the electric quadrupole moment of the nucleus and the gradient of the electric field at the nucleus produced by the electrons. The first contribution is usually of the order of 10^{-2} cm^{-1}, the second one is smaller.

For the discussion of the hyperfine structure we are interested in the same aspects as in the case of the STARK splitting (Sect. 30): the zero field specific heat, the susceptibility, and the course of the magnetization and the entropy with field at the initial temperature.

Since the hyperfine splitting is much smaller than the STARK splitting it is obvious that the influence on the magnetization curve and the entropy at the high temperature can be neglected. For the same reason, in the region above the maximum of the susceptibility (to which we restrict ourselves in this chapter, see Sect. 28), we have only to consider the term in the specific heat proportional to $1/T^2$.

[1] L. J. F. BROER: Physica, 's-Grav. **9**, 547 (1942).
[2] R. P. HUDSON: Phys. Rev. **88**, 570 (1952).
[3] J. M. DANIELS and N. KURTI: Proc. Roy. Soc. Lond., Ser. A **221**, 243 (1954).

In the case that we have a symmetry axis the interaction energy can usually [1] be expressed by:

$$W = A\,S_z I_z + B\,(S_x I_x + S_y I_y) + P\left(I_z^2 - \tfrac{1}{3} I(I+1)\right). \tag{31.1}$$

The z-axis is the axis of symmetry; A, B and P are constants which can be determined from paramagnetic resonance experiments (like the g-values of Sect. 30). The first two terms are due to the magnetic interaction with the nucleus, the last one is the contribution of the electric interaction with the nuclear quadrupole moment. Inserting Eq. (31.1) into the partition function we may derive [2]:

$$\frac{c}{R} = \frac{1}{(3kT)^2}\left\{(A^2 + 2B^2)\,S(S+1)\,J(J+1) + \tfrac{1}{5} P^2 I(I+1)\,(2I-1)\,(2I+3)\right\}. \tag{31.2}$$

BLEANEY showed [3] that in the susceptibility the hyperfine structure gives rise only to terms proportional to $1/T^3$ and higher for any crystal direction. Hence the CURIE constant is not affected and no WEISS Θ is introduced (see Sect. 30). If we expand:

$$\chi = \frac{C}{T}\left(1 - \frac{Q}{k^2 T^2} + \cdots\right), \tag{31.3}$$

the contribution to Q/k^2 due to the magnetic interaction, is of the order of 10^{-4}, the contribution from the electric interaction with the nuclear quadrupole moment is still smaller.

32. Magnetic dipole and exchange interactions. STARK splitting and hyperfine structure are fundamentally single particle problems. Exact solutions of the quantum mechanical equations can essentially be given though, due to our limited knowledge of the crystallographic data (e.g. the exact location of the water molecules), some parameters must be adapted empirically.

For the interactions between the magnetic ions the situation is more or less reversed. The relative positions of the ions in the lattice are known with sufficient precision, but now the problem is a statistical one and a rigorous solution is usually impossible. We don't have an ion with a degenerate ground state which is split into a small number of levels (e.g. $2J+1$), but we must consider the crystal as a whole. The ground state consists of an energy band with a large number of levels, each representing a possible state of the crystal.

VAN VLECK [4] solved the problem by expanding the partition function into a power series in $1/T$ of which he could evaluate the first few terms. By this method he obtained an approximate solution, valid at the higher temperatures, which did not contain any experimental parameters. In the region, however, where kT is comparable with the band width, his series expansion converges too slowly and no satisfactory solution is obtained.

Two kinds of interactions between the ions must be taken into account: magnetic dipole interaction and electrostatic exchange effects. The widths of the energy bands are of the order of 10^{-2} cm^{-1}. Since exchange energies decrease rapidly with the distance between the ions, it was expected originally that, for the fairly dilute salts used for the demagnetization process, the influence of the exchange interaction should be much smaller than that of the magnetic dipole interaction. This, however, proved to be not true. It was shown by DE KLERK [5] that salts exist for which the exchange is even larger than the dipole

[1] A. ABRAGAM and M. H. L. PRYCE: Proc. Roy. Soc. Lond., Ser. A **205**, 185 (1951).
[2] B. BLEANEY: Phys. Rev. **78**, 214 (1950).
[3] B. BLEANEY: Phil. Mag. **42**, 441 (1951).
[4] J. H. VAN VLECK: J. Chem. Phys. **5**, 320 (1937).
[5] D. DE KLERK: Leiden Commun. No. 270c; Physica, 's-Grav. **12**, 513 (1946).

interactions. KRAMERS[1] suggested the possibility of an indirect exchange (the so-called "super exchange") originated by means of excited states of intermediate diamagnetic atoms; it is rather probable now that all the exchange effects found in the dilute salts considered here, are due to this mechanism.

The interaction energy between two magnetic ions may be represented by:

$$w_{ij} = r_{ij}^{-3}[(1 + v_{ij})\,\mu_i \cdot \mu_j - 3\,(r_{ij}^{-2})\,(\mu_i \cdot r_{ij})\,(\mu_j \cdot r_{ij})].\tag{32.1}$$

Here v_{ij} is due to the exchange interaction. For isotropic salts we have $\mu_i = g\mu_B J_i$, where J_i is identical with J of Sect. 29.

At high temperatures the specific heat is proportional to $1/T^2$, as in the cases of STARK splitting and hyperfine structure. VAN VLECK derived from Eq. (32.1) that this term obeys:

$$\frac{c}{R} = \frac{1}{6}\,Q\,\frac{\tau^2}{T^2},\tag{32.2}$$

where τ is a characteristic temperature:

$$\tau = g^2\,\mu_B^2\,\frac{N}{V}\,J(J + 1)/k.\tag{32.3}$$

N/V is the number of magnetic ions per unit volume, so that τ is three times the CURIE constant per cm³ [see Eq. (29.9)]. Q is a geometrical factor depending on the crystal structure:

$$Q = \left(\frac{N}{V}\right)^{-2} \sum_{j \neq i} r_{ij}^{-6}\,(2 + v_{ij}^2).\tag{32.4}$$

If the exchange term is neglected we have for a face centered cubic structure (the alums): $Q = 14.4$, and for the Tutton salts: $Q = 17.6$.

Relations for the influence of the ionic interactions on the susceptibility were first given by LORENTZ and ONSAGER. Their calculations were based on classical models, the formulae were already discussed in Sect. 7. VAN VLECK's series expansion method can be applied if a term $-J_{z_i}g\mu_B H$ is added to the energy of the ions. If H_{ext} is chosen for H, the value of the susceptibility depends on the shape of the sample as was to be expected, see Sect. 7. In this case the result is:

$$\chi_{ext} = \frac{C}{T - \left[\left(\frac{4}{3}\pi - \varepsilon\right)\frac{C}{V} - \Theta\right] + \frac{4}{3}\eta\,\frac{\tau^2}{T} + \cdots},\tag{32.5}$$

with:

$$\eta = \frac{1}{12}\left[1 + \frac{3}{8J(J+1)}\right]Q.\tag{32.6}$$

Θ is due to the exchange interaction. The relation (32.5) is equivalent to Eqs. (7.21) and (11.7) if the exchange is neglected.

The case that the exchange interaction gives the only contribution to the specific heat was discussed by OPECHOWSKI[2]. We assume that v_{ij} of Eq. (32.1) is independent of the orientations of the ions relative to r_{ij}, and that its value decreases so rapidly with distance that only the exchange between nearest neighbouring ions needs be considered. If the number of neighbours of an ion is z, the specific heat at high temperatures obeys:

$$\frac{c}{R} = z\,\frac{r_{ij}^{-6}\,v_{ij}^2\,g^4\,\mu_B^4\,S^2(S + 1)^2}{6k^2\,T^2};\tag{32.7}$$

Θ in Eq. (32.5) is equal to:

$$\Theta = -z\,\frac{r_{ij}^{-3}\,v_{ij}\,g^2\,\mu_B^2\,S(S + 1)}{3k},\tag{32.8}$$

---

[1] H. A. KRAMERS: Physica, 's-Grav. **1**, 182 (1934).
[2] W. OPECHOWSKI: Physica, 's-Grav. **4**, 181 (1937).

so that we have a relation between the specific heat constant and the WEISS constant:

$$\frac{c\,T^2}{R} = \frac{3}{2}\,\frac{\Theta^2}{z}\,.$$

(32.9)

Since STARK splitting, hyperfine structure, magnetic dipole interaction and exchange effects all contribute to the specific heat with terms proportional to $1/T^2$, it is difficult to distinguish between them. One possibility is to make experiments with diluted salts (see Sect. 4) with different concentrations of the magnetic ions[1,2]. The STARK and hyperfine splittings are only little influenced by dilution[3], but the ionic interactions are. Further it is possible to derive the STARK and hyperfine structure contributions from paramagnetic resonance experiments. The magnetic dipole interaction may be computed from VAN VLECK's relation (32.2) and if these contributions are subtracted from the experimental value of the specific heat one may consider the rest remaining as the exchange contribution. Now it should be possible to check this contribution with the help of formula (32.9), but unfortunately several salts were found for which this relation is not fulfilled[1,4], the WEISS constant being much too small to account for the exchange part of the specific heat. The explanation is probably that the assumption of isotropic exchange (v_{ij} being independent of the orientations of the ions) is too simple, especially in the case of KRAMERS' super-exchange[5]. It is even possible that exchange effects with different signs occur between different kinds of neighbours.

Some remarks should be made on the region of temperatures where the specific heat of the salt cannot be represented any more by a term in $1/T^2$ only. The difficulty is that in the region where VAN VLECK's series expansion ceases to converge rapidly it is not very valuable to calculate a few more terms since the whole series must be taken into account. Further, if the level broadening due to the magnetic interaction is not really small as compared with the STARK splitting the two specific heat humps overlap partly and the whole problem becomes exceedingly complicated, even if hyperfine splitting and exchange interaction are neglected entirely. It may happen, for instance, that the total specific heat is smaller than the value due to the STARK splitting alone. The problem has been discussed by VAN VLECK[6]. The partition function may be expressed by:

$$Z = Z_s\left(1 + \Omega\cdot\frac{\tau^2}{T^2} + \cdots\right)$$

(32.10)

where Z_s is the partition function due to the STARK effect alone [Eq. (30.1)]. The factor $(1 + \Omega\tau^2/T^2)$ gives rise to a term in the specific heat:

$$\frac{c}{R} = \tau^2\,T\,\frac{d^2}{d\,T^2}\left(\frac{\Omega}{T}\right),$$

(32.11)

and to an entropy:

$$\frac{S}{R} = \ln 2 + \tau^2\,\frac{d}{d\,T}\left(\frac{\Omega}{T}\right).$$

(32.12)

HEBB and PURCELL[7] have calculated the function Ω for several salts. The results will be given in the sections dealing with the individual salts.

[1] R. J. BENZIE and A. H. COOKE: Nature, Lond. **164**, 837 (1949).
[2] R. J. BENZIE and A. H. COOKE: Proc. Phys. Soc. Lond. A **63**, 210, 213 (1950).
[3] D. DE KLERK and D. POLDER: Leiden Commun. No. 262d; Physica, 's-Grav. **8**, 508 (1941).
[4] C. G. B. GARRETT: Proc. Roy. Soc. Lond., Ser. A **203**, 392 (1950).
[5] W. OPECHOWSKI: Physica, 's-Grav. **14**, 237 (1948).
[6] J. H. VAN VLECK: J. Chem. Phys. **5**, 320 (1937).
[7] M. H. HEBB and E. M. PURCELL: J. Chem. Phys. **5**, 338 (1937).

II. Results obtained with individual salts.

33. The chromium alums in general. In the alums the trivalent chromium ions are located on a face-centered cubic lattice, each ion being surrounded by an octahedron of six water molecules. There are four non-equivalent ions in the unit cell. The water octahedrons are somewhat distorted, giving rise to a trigonal component in the electric field which, for each of the ions, is parallel to one of the body diagonals of the cell.

It was found from crystallographic investigations[1,2] that there are three different types of alum structures, referred to as α, β and γ. The differences in the dimensions in the unit cell do not exceed one part in 300 for the different structures and they are apparently due to differences in the sizes of the monovalent ions. Of the chromium alums, the potassium, ammonium and rubidium alums have the α structure, the methylamine and caesium alums have the β structure, and the sodium alum has the γ structure[3]. In the α type, the water octahedron is more distorted than in the β structure.

The free chromium ion is in a ${}^4F$ state, but due to the complete quenching of the orbital levels (see Sects. 30 and 4) the effective state in the alums is ${}^4S$. The fourfold ground level is split by the trigonal component of the electric field into two KRAMERS doublets a distance $k\delta$ apart. Since δ is of the order of $0.25°$ K (see below) the magnetic moment and entropy at $1°$ K can be represented with the BRILLOUIN function with $J = S = \frac{3}{2}$ and $g = 2$ [see Eqs. (29.1) and (29.2)]. For the magnetic moment this has been verified by experiment[4,5]. A small correction to the entropy due to the splitting was calculated by HUDSON[6] and by DANIELS and KURTI[7].

The influence of the STARK splitting on the specific heat and the entropy below $1°$ K can be expressed [see Eq. (30.1)] by:

$$Z_s = 2(1 + e^{-\delta/T}), \tag{33.1}$$

$$\frac{c}{R} = \frac{\delta^2}{T^2}\frac{e^{-\delta/T}}{(1 + e^{-\delta/T})^2} = \frac{\delta^2}{4T^2}\left(1 - \frac{\delta^2}{4T^2} + \cdots\right), \tag{33.2}$$

and

$$\frac{S}{R} = \ln 2(1 + e^{-\delta/T}) + \frac{\delta}{T}\frac{e^{-\delta/T}}{1 + e^{-\delta/T}} = \ln 4 - \frac{\delta^2}{8T^2}\left(1 - \frac{\delta^2}{8T^2} + \cdots\right). \tag{33.3}$$

The influence of the STARK splitting on the susceptibility can be given in terms of the function $\gamma(T)$, see Eqs. (30.5) and (30.6). In the case of a powdered sample (average over all the possible orientations of the field in the crystal) we have[8]:

$$\gamma(T) = \frac{\left(3 + 4\dfrac{T}{\delta}\right) + \left(3 - 4\dfrac{T}{\delta}\right)e^{-\delta/T}}{5(1 + e^{-\delta/T})}. \tag{33.4}$$

Since the only isotope with a nuclear spin, Cr^{53}, has an abundance of 9.4% and a spin value of $3/2$, the influence of hyperfine structure on the specific heat can

[1] H. LIPSON: Proc. Roy. Soc. Lond., Ser. A **151**, 347 (1935).

[2] H. LIPSON and C. A. BEEVERS: Proc. Roy. Soc. Lond., Ser. A **148**, 664 (1935).

[3] D. M. S. BAGGULEY and J. H. E. GRIFFITHS: Proc. Roy. Soc. Lond., Ser. A **204**, 188 (1950).

[4] C. J. GORTER, W. J. DE HAAS and J. VAN DEN HANDEL: Leiden Commun. No. 222d; Proc. Kon. Acad. Amst. **36**, 158 (1933).

[5] W. E. HENRY: Phys. Rev. **88**, 559 (1952).

[6] R. P. HUDSON: Phys. Rev. **88**, 570 (1952).

[7] J. M. DANIELS and N. KURTI: Proc. Roy. Soc. Lond., Ser. A **221**, 243 (1954).

[8] M. H. HEBB and E. M. PURCELL: J. Chem. Phys. **5**, 338 (1937).

be neglected[1,2]. Also exchange effects are small in chromium alums. The contribution of the magnetic interaction to the specific heat may be derived from Eq. (32.11) with[3]:

$$\Omega = \frac{\frac{1}{12} Q}{(1 + e^{-\delta/T})^2} \times$$
$$\times \left\{ \left(\frac{3}{50} + \frac{223}{150} \frac{T}{\delta} \right) + \left(\frac{88}{75} + \frac{8}{15} \frac{T}{\delta} \right) e^{-\delta/T} + \left(\frac{49}{150} - \frac{143}{150} \frac{T}{\delta} \right) e^{-2\delta/T} \right\} \tag{33.5}$$

with $Q = 14.4$ (see Sect. 32). For high temperatures this leads to:

$$\frac{c}{R} = 2.40 \frac{\tau^2}{T^2} \tag{33.6}$$

and

$$\frac{S}{R} = \ln 2 - 1.20 \frac{\tau^2}{T^2}. \tag{33.7}$$

Consequently, the coefficient of the $1/T^2$ term in the specific heat [see Eq. (29.10)] obeys:

$$A = \tfrac{1}{4} \delta^2 + 2.40 \, \tau^2. \tag{33.8}$$

34. Chromium methylamine alum. $Cr(NH_3CH_3)(SO_4)_2 \cdot 12H_2O$. Molecular weight: 492.4, density: 1.66.

This is a salt for which the theoretical relations discussed in the foregoing sections are nicely fulfilled. The first experiments below 1° K were performed by DE KLERK and HUDSON[4]. They used a powdered sample of spherical shape. Investigations with single crystals ground into spheres were performed by HUDSON and McLANE[5] and by BEUN, STEENLAND, DE KLERK and GORTER[6]. In these experiments the initial field of the demagnetization was applied parallel to a cubic axis so that its influence on each of the four ions in the unit cell was the same (see Sect. 33). The coils of the mutual inductance bridge were in the direction of another cubic axis (see Fig. 4). GARDNER and KURTI[7] used a compressed pill of powdered salt, 95% of the crystalline density. It was turned on the lathe into an ellipsoid with axial ratio four to one.

Demagnetizations were performed from a number of fields starting from well known temperatures and T^* was measured immediately after each demagnetization. The decrease of entropy during an isothermal magnetization is derived from Eq. (29.1) with $J = 3/2$ and $g = 2$, applying the correction for the STARK splitting as calculated by HUDSON and by DANIELS and KURTI, see Sect. 33.

The results should fit in (see Sect. 13) with the sum of the entropy differences, between the initial and final temperatures, due to the STARK splitting [Eq.(33.3)], the magnetic interaction [Eqs. (32.13) and (33.5)] and the lattice entropy (Sect 13). In order to check this, the measured T^*-values must be reduced to the corresponding T-values with the help of Eq. (33.4). In order to carry out the computations, the quantities δ and τ must be known. τ, according to Eq.(32.3) is equal to 0.0189° K and δ may be either derived from paramagnetic resonance experiments or it may be adapted directly to the experimental results.

[1] B. BLEANEY and K. D. BOWERS: Proc. Phys. Lond. A **64**, 1135 (1951).
[2] K. D. BOWERS: Proc. Phys. Soc. Lond. A **65**, 860 (1952).
[3] M. H. HEBB and E. M. PURCELL: J. Chem. Phys. **5**, 338 (1937).
[4] D. DE KLERK and R. P. HUDSON: Phys. Rev. **91**, 278 (1953).
[5] R. P. HUDSON and C. K. McLANE: Phys. Rev. **95**, 932 (1954).
[6] J. A. BEUN, M. J. STEENLAND, D. DE KLERK and C. J. GORTER: Leiden Commun. No. 301a; Physica, 's-Grav. **21**, 767 (1955).
[7] W. E. GARDNER and N. KURTI: Proc. Phys. Soc. Lond. A **223**, 542 (1954).

The lattice specific heats of several alums were measured recently by Kapadnis at Leiden (unpublished). His value for aluminum potassium alum was $c/R = 4.03 \times 10^{-4} T^3$, for aluminum ammonium alum: $c/R = 3.82 \times 10^{-4} T^3$, for chromium methylamine alum: $c/R = 4.70 \times 10^{-4} T^3$, for chromium potassium alum: $c/R = 4.95 \times 10^{-4} T^3$.

It was found from all the experiments quoted above that if the value of δ is adapted at one temperature (for instance at 0.5° K), the experimental and theoretical curves coincide nicely down to 0.1° K. This is only true if the Onsager or the van Vleck approximation [see Eqs. (11.6) and (11.7)] is used for

Table 1. *Chromium methylamine alum* (de Klerk and Hudson).

H and T_i are the initial field and temperature, S/R is calculated from Eq. (29.1) with the corrections given in the text. T^*_{Lor} is the magnetic temperature measured in the experiment (for a spherical sample), from this T^*_{Ons} and $T^*_{V. VI.}$ were calculated with Eqs. (11.6) and (11.7) (they are not given in the table) and T_{Lor}, T_{Ons} and $T_{V. VI.}$ were derived from these, applying Eq. (33.4) with $\delta = 0.275°$ K. T_{theor} was derived from S/R with the help of Eqs. (33.3), (32.13) and (33.5).

H oersteds	T_i ° K	S/R	T^*_{Lor}	T_{Lor} ° K	T_{Ons} ° K	$T_{V. VI}$ ° K	T_{theor} ° K
1363	1.166	1.3644	0.674	0.670	0.669	0.670	0.663
1451	1.164	1.3623	0.629	0.625	0.624	0.624	0.632
1451	1.158	1.3621	0.650	0.646	0.645	0.646	0.630
1717	1.165	1.3555	0.549	0.544	0.543	0.543	0.555
2005	1.164	1.3470	0.505	0.500	0.499	0.499	0.493
2293	1.157	1.3366	0.439	0.433	0.431	0.432	0.437
2597	1.160	1.3252	0.398	0.392	0.390	0.391	0.391
2866	1.164	1.3142	0.367	0.360	0.358	0.359	0.356
3533	1.161	1.2823	0.302	0.294	0.292	0.293	0.291
4367	1.165	1.2374	0.245	0.235	0.232	0.234	0.237
4936	1.163	1.2019	0.223	0.212	0.209	0.211	0.209
5794	1.187	1.1532	0.193	0.181	0.178	0.179	0.179
6730	1.196	1.0934	0.167	0.153	0.149	0.151	0.152
7610	1.163	1.0175	0.149	0.134	0.130	0.132	0.127
8805	1.175	0.9386	0.128	0.112	0.107	0.109	0.106
10265	1.210	0.8587	0.113	0.0960	0.0906	0.0928	0.0870
11705	1.189	0.7548	0.0936	0.0757	0.0692	0.0718	0.0636
11705	1.156	0.7345	0.0899	0.0720	0.0652	0.0678	0.0579
13075	1.214	0.6895	0.0782	0.0603	0.0532	0.0559	0.0426
13075	1.164	0.6581	0.0662	0.0491	0.0416	0.0442	0.0307
13610	1.199	0.6505	0.0638	0.0469	0.0394	0.0419	0.0280

the influence of the magnetic interaction on T^*. If the Lorentz formula is applied, differences are already found below 0.3° K. Small systematical differences between the values of de Klerk and Hudson and those of Beun, Steenland, de Klerk and Gorter, could be eliminated by applying the Onsager formula[1] for the calculation of H_{int} in a powder, see Sect. 7.

Some of the results of Hudson and de Klerk are shown in Table 1 and in Fig. 15. The deviations below 0.1° K may be due to the fact that here van Vleck's series expansion is no more satisfactory for the evaluation of the magnetic interaction.

The δ values given by the different authors are nicely in agreement. De Klerk and Hudson give: $\delta = 0.275°$ K; Hudson and McLane: $\delta = 0.270°$ K and 0.267° K (for two different samples); Gardner and Kurti: $\delta = 0.27°$ K; Beun, Steenland, de Klerk and Gorter: $\delta = 0.275°$ K. The first crystal

---

[1] C. J. F. Böttcher: Rec. Trav. chim. Pays-Bas **64**, 47 (1945).

used by HUDSON and McLANE at the National Bureau of Standards was of the same origin as the one used by BEUN, STEENLAND, DE KLERK and GORTER at Leiden.

The specific heat measurements of KAPADNIS above 1° K give: $\delta = 0.273°$ K.

Paramagnetic resonance measurements were performed by BLEANEY[1]. From his results he computed: $\delta = 0.245°$ K. In these experiments, however, the splitting is not measured directly. The level separation is measured in a field and the zero field splitting is calculated with the help of theoretical assumptions on the shift of the levels with the field. In BLEANEY's experiments the field was parallel to the body diagonal of the cube and for the calculations it was assumed that the splitting is due to a trigonal field with symmetry about this axis. This is the case for the rubidium and caesium alums, but unpublished measurements by BAKER[2] showed that, for the case of the methylamine alum, this assumption

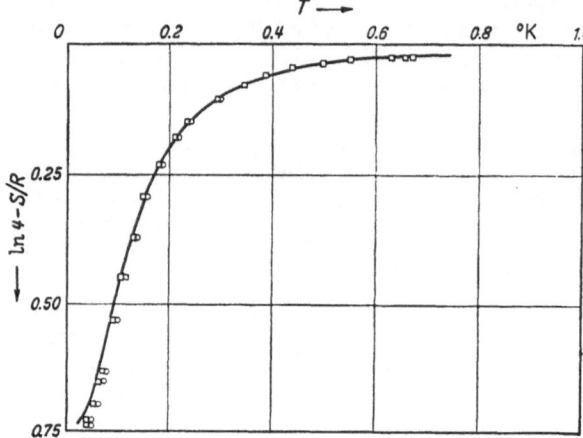

Fig. 15. Entropy versus temperature diagram for a powdered sample of chromium methylamine alum (according to DE KLERK and HUDSON). O LORENTZ approximation. □ ONSAGER approximation. The theoretical curve is calculated with $\delta = 0.275°$ K.

is not correct. Measurements on the dielectric constant by GRIFFITHS and POWELL[3] showed that a crystalline transition takes place at 160° K and recent measurements of BLEANEY (unpublished) indicated that below this temperature

Table 2. *Chromium methylamine alum* (GARDNER and KURTI).

S/R, $T_{Lor.}$, $T_{Ons.}$, $T_{V. VI.}$ and $T_{theor.}$ have the same meaning as in table 1; $\delta = 0.27$. $T^*_{Lor.}$ is the measured magnetic temperature reduced to spherical shape ($T^{(*)}$) with Eq. (11.4). c^* is $dQ/d\,T^*_{Lor.}$ and c is the specific heat dQ/dT. T is derived from the caloric measurements applying Eq. (34.1).

S/R	$T^*_{Lor.}$	c^*/R	c/R	T °K	$T_{Lor.}$ °K	$T_{Ons.}$ °K	$T_{V. VI.}$ °K	$T_{theor.}$ °K
1.325	0.404	0.117	0.115	0.396	0.398	0.394	0.397	0.382
1.300	0.336	0.166	0.162	0.326	0.329	0.325	0.327	0.318
1.250	0.266	0.250	0.239	0.254	0.257	0.253	0.255	0.245
1.200	0.224	0.339	0.321	0.210	0.215	0.207	0.211	0.202
1.150	0.196	0.410	0.381	0.180	0.185	0.177	0.181	0.174
1.100	0.175	0.466	0.425	0.157	0.163	0.154	0.158	0.152
1.050	0.157	0.503	0.449	0.137	0.144	0.134	0.139	0.134
1.000	0.143	0.518	0.452	0.122	0.129	0.118	0.124	0.119
0.950	0.132	0.518	0.431	0.109	0.117	0.107	0.111	0.106
0.900	0.121	0.503	0.386	0.096	0.105	0.095	0.099	0.094
0.850	0.111	0.455	0.232	0.077	0.095	0.083	0.087	0.083
0.800	0.102	0.366	0.202	0.065	0.085	0.073	0.077	0.072
0.750	0.092	0.250	0.160	0.048	0.074	0.061	0.065	0.060
0.700	0.083	0.174	0.174	0.039	0.065	0.052	0.056	0.046
0.650	0.071	0.138	0.373	0.032	0.053	0.040	0.043	0.029
0.600	0.061	0.240	0.647	0.028	0.045	0.030	0.030	—

[1] B. BLEANEY: Proc. Roy. Soc. Lond., Ser. A **204**, 203 (1950).
[2] W. E. GARDNER and N. KURTI: Proc. Roy. Soc. Lond., Ser. A **223**, 542 (1954).
[3] J. H. E. GRIFFITHS and J. A. POWELL: Proc. Phys. Soc. Lond. A **65**, 289 (1952).

the symmetry is lower than cubic. The investigations have not yet been finished, but it is well possible, under these conditions, that a correct interpretation of the paramagnetic resonance data leads to a higher splitting parameter. An eventual residual difference from the value derived from the demagnetization and relaxation experiments might be accounted for by a small exchange effect; preferably anisotropic exchange since otherwise a noticeable CURIE-WEISS Θ might occur (see Sect. 32) and this has not been found in the experiment.

Calorimetric determinations of absolute temperatures (Sect. 12) were made by GARDNER and KURTI using gamma ray heat supply. Two 250 millicurie radium sources could be rotated round the axis of the cryostat, or a cylindrical source with radioactive silver wires placed on a circle of 6 cm diameter could be used. The intensity of the latter source could be varied up to 60 millicuries by altering the number of wires. From the measured variation of T^* the quantity c^*, equal to dQ/dT^*, could be derived. The absolute temperature, according to Eq. (10.1) obeys:

$$T = \frac{dQ}{dS} = \frac{c^*}{dS/dT^*}. \tag{34.1}$$

Some smoothed values are shown in Table 2. Above 0.1° K they are in good agreement with the magnetic measurements quoted above.

35. Chromium potassium alum. $CrK(SO_4)_2 12H_2O$. Molecular weight: 499.4, density: 1.83.

This is probably the salt which, below 1° K, has been most thoroughly investigated. The first experiments were made by DE HAAS and WIERSMA[1].

Magnetic measurements (see Sect. 13) at relatively high temperatures were carried out by the Leiden group[2-5] and by AMBLER and HUDSON[6]. Caloric measurements were performed by BLEANEY[7,8] and by KEESOM[9].

The salt is very similar to the methylamine alum, but in general the agreement between the experiments and the theoretical formulae (Sect. 33) is somewhat less satisfactory. The magnetic investigations of CASIMIR, DE HAAS and DE KLERK are shown in Table 3 and Fig. 16. If the splitting

Table 3. *Chromium potassium alum* (CASIMIR, DE HAAS and DE KLERK).

H, T_i, S/R, T_{Lor}^*, T_{Lor} and T have the same meaning as in Table 1, $\delta = 0.27°$ K.

H oersteds	T_i °K	S/R	T_{Lor}^*	T_{Lor} °K	T_{theor} °K
823	1.184	1.3739	0.877	0.875	0.874
1022	1.177	1.3709	0.784	0.782	0.784
1209	1.174	1.3674	0.701	0.699	0.702
1645	1.158	1.3567	0.570	0.566	0.563
1905	1.157	1.3494	0.508	0.502	0.502
2183	1.155	1.3402	0.453	0.448	0.448
2762	1.152	1.3176	0.365	0.359	0.360
3572	1.149	1.2778	0.288	0.280	0.280
4152	1.153	1.2483	0.251	0.242	0.242
5805	1.148	1.1383	0.178	0.166	0.157
8120	1.142	0.9683	0.124	0.108	0.111
10310	1.143	0.8153	0.095	0.077	0.076
12060	1.142	0.7032	0.078	0.060	0.047

[1] W. J. DE HAAS and E. C. WIERSMA: Leiden Commun. No. 236a; Physica, 's-Grav. **2**, 81 (1935).

[2] H. B. G. CASIMIR, W. J. DE HAAS and D. DE KLERK: Leiden Commun. No. 256c; Physica, 's-Grav. **6**, 365 (1939).

[3] H. B. G. CASIMIR, D. DE KLERK and D. POLDER: Leiden Commun. No. 261a; Physica, 's-Grav. **7**, 737 (1940).

[4] D. DE KLERK, M. J. STEENLAND and C. J. GORTER: Leiden Commun. No. 278c; Physica, 's-Grav. **15**, 649 (1949).

[5] J. A. BEUN, M. J. STEENLAND, D. DE KLERK and C. J. GORTER: Leiden Commun. No. 300a; Physica, 's-Grav. **21**, 651 (1955).

[6] E. AMBLER and R. P. HUDSON: Phys. Rev. **95**, 1143 (1954).

[7] A. H. COOKE: Proc. Phys. Soc., Lond. A **62**, 269 (1949).

[8] B. BLEANEY: Proc. Roy. Soc. Lond., Ser. A **204**, 216 (1950).

[9] P. H. KEESOM: Thesis, Leiden 1948.

parameter is adapted at the higher temperatures, as was done in the case of the methylamine alum (Sect. 34), the agreement between the theoretical and experimental curves is good down to 0.2° K. Small deviations, however, begin already at this temperature. They were found systematically for all the samples which have been investigated, and they are always quantitatively the same. Between 0.2 and 0.1° K the experimental temperatures are somewhat lower than the theoretical ones; below 0.1° K they are higher. Only T_{Lor} is given in Fig. 16 and Table 3. There was no point in calculating T_{Ons} and $T_{\mathrm{V. VI}}$; they are still lower than T_{Lor} so that discrepancies in the region between 0.2° K and 0.1° K should become even larger.

The caloric investigations, performed by BLEANEY with gamma ray heating (Sect. 12), are represented in Table 4. BLEANEY found that between 0.3° K and 0.09° K the quantity $H\,T^{*}_{\mathrm{Lor}}/T_i$ was fairly constant, and this could be used in order to simplify the calculation of absolute tempe-

Table 4. *Chromium potassium alum* (BLEANEY).

S/R, $T^{*}_{\mathrm{Lor.}}$, c/R and T have the same meaning as in Table 2.

S/R	T^{*}_{Lor}	T °K	c/R
1.378	1.000	1.000	0.0160
1.364	0.604	0.600	0.042
1.352	0.485	0.480	0.064
1.339	0.406	0.400	0.089
1.329	0.368	0.360	0.108
1.314	0.330	0.320	0.130
1.295	0.291	0.280	0.163
1.266	0.252	0.240	0.206
1.228	0.215	0.200	0.266
1.199	0.195	0.180	0.296
1.160	0.174	0.160	0.325
1.115	0.156	0.140	0.350
1.057	0.138	0.120	0.374
0.982	0.121	0.100	0.391
0.883	0.103	0.080	0.40
0.770	0.086	0.060	0.39
0.704	0.079	0.050	0.36
0.666	0.075	0.045	0.33

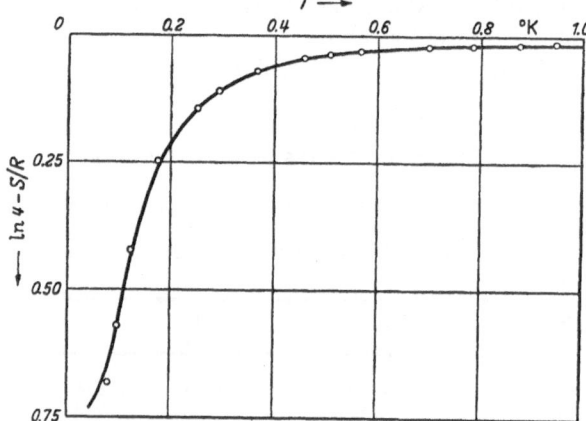

Fig. 16. Entropy versus temperature diagram for a powdered sample of chromium potassium alum (according to CASIMIR, DE HAAS and DE KLERK). ○ LORENTZ approximation. The theoretical curve is calculated with $\delta = 0.270$° K.

ratures from Eq. (34.1). It followed that the experimental temperature values down to 0.05° K were in much better agreement with the ONSAGER and VAN VLECK approximations than with the LORENTZ formula, see Fig. 17. The absolute specific heat data, derived from dQ/dT^{*} with the help of the empirical T^{*} versus T relation are given in Table 4 and in Fig. 18. Curve C is the theoretical STARK specific heat for a value of δ adapted at the higher temperatures (see below). The discrepancy between the theoretical and the experimental curves below 0.2° K proves to be in agreement with the deviation between the theoretical and experimental data obtained from the Leiden magnetic measurements shown in Fig. 16.

P. H. KEESOM made specific heat measurements using a heating wire and a phosphorbronze thermometer. The results are in good agreement with the gamma ray measurements of BLEANEY down to 0.3° K. Below this temperature, KEESOM's values are much lower. This, however, is the region where the thermal equilibrium inside the sample becomes already somewhat doubtful and the results may be explained by assuming a good heat contact between the thermometer and the heating wire, but not with the bulk of the salt. Since the thermo-meter and the heating wire were wound simultaneously on the sample, alternately a turn of each, this possibility is not excluded.

The splitting parameters given by different authors show also larger differences than in the case of the methylamine alum. Assuming $\tau = 0.0204°$ K

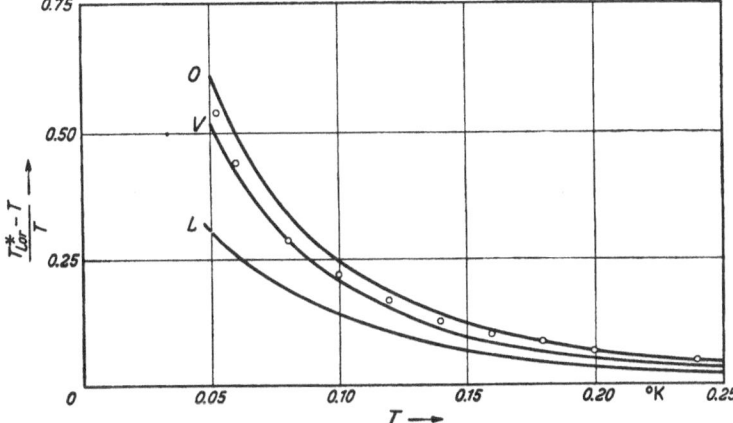

Fig. 17. Relation between magnetic and absolute temperatures for a compressed pill of chromium potassium alum (according to BLEANEY). Curve L: LORENTZ's formula. Curve O: ONSAGER's formula. Curve V: VAN VLECK's formula. The curves are calculated for $\delta = 0.240°$ K. ○ Experimental values.

[according to Eq. (32.3)], CASIMIR, DE HAAS and DE KLERK found: 0.270° K for a powdered sample; CASIMIR, DE KLERK and POLDER gave, for a single crystal of ellipsoidal shape: $\delta = 0.263°$ K; DE KLERK, STEENLAND and GORTER, for a spherical single crystal: $\delta = 0.251°$ K; BEUN, STEENLAND, DE KLERK and GORTER: $\delta = 0.240°$K and $\delta = 0.250°$ K for two different samples, both spherical single crystals. AMBLER and HUDSON gave: $\delta = 0.250°$ K; BLEANEY's value is: $\delta = 0.240°$ K; KEESOM's: $\delta = 0.285°$ K.

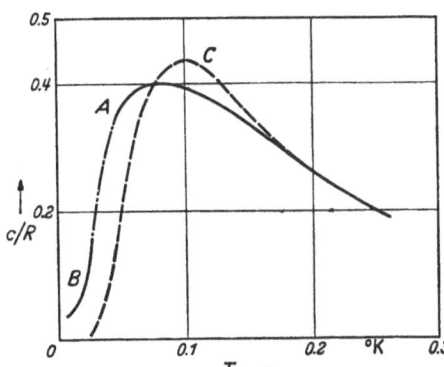

Fig. 18. Specific heat of chromium potassium alum (according to BLEANEY). Curve A: Experimental curve of BLEANEY. Curve B: Experimental curve of DE KLERK, STEENLAND and GORTER. Curve C: Theoretical STARK specific heat for $\delta = 0.245°$ K.

All the Leiden investigations, except those by KEESOM, were made by the same method in the same apparatus. The differences are well beyond the experimental error and we are sure that they are due to the samples themselves. Maybe it is the method of preparation, but, as BIJL[1] suggested, there is another possibility. The salt shows a transition in its crystalline structure below 160° K[2], like chromium methylamine alum (see Sect. 34), only more gradual. It is well possible that the rate at which the salt is cooled through the transition region influences somewhat the magnetic properties at low temperatures. HUDSON and McLANE[3] reported a very pronounced dependence on the cooling rate in the case of chromium methylamine alum at still lower temperatures, see Sect. 57.

[1] D. BIJL: Leiden Commun. No. 276b; Physica, 's-Grav. 14, 684 (1949).
[2] B. BLEANEY: Proc. Roy. Soc. Lond., Ser. A 204, 203 (1950).
[3] R. P. HUDSON and C. K. McLANE: Phys. Rev. 95, 932 (1954).

Caloric experiments at liquid helium temperatures by KAPADNIS (unpublished) gave: $\delta = 0.247°$ K. Paramagnetic relaxation investigations in the same region by CASIMIR, BIJL and DU PRÉ[1] yielded $\delta = 0.260°$ K; KRAMERS, BIJL and GORTER[2] found $\delta = 0.251°$ K. Relaxation measurements in the liquid nitrogen region give, on the average, slightly lower values. GORTER, DIJKSTRA and VAN PAEMEL[3] found $\delta = 0.243°$ K; STARR[4]: $\delta = 0.231°$ K; BROER[5]: $\delta = 0.243°$ K.

Paramagnetic resonance measurements revealed an interesting phenomenon. At room-temperature[6] the value of the splitting was found to be $0.172°$ K; at $193°$ K[7] it was $0.079°$ K, but at lower temperatures BLEANEY[7] found two splittings. In the nitrogen region they were $0.374°$ K and $0.15°$ K, at hydrogen temperatures the values are $0.388°$ K and $0.15°$ K.

This peculiar behaviour of the level splitting must be related to the gradual transition in the crystalline structure mentioned above. It seems that below the transition there are two different kinds of ions in the lattice with different splittings. The intensities of the absorption lines suggest that both kinds are equally abundant.

The consequence of the two groups of ions with different level splittings must be a specific heat composed of two peaks of the shape of C in Fig. 18, hence a curve with a broader and lower maximum. This is in qualitative agreement with BLEANEY's result but, unfortunately, it proves to be impossible to explain his curve quantitatively. The high temperature tail (corresponding to an adapted δ of $0.24°$ K) may be accounted for if 15% of the ions have the splitting of $0.388°$ K and 85% have the $0.22°$ K splitting, but this is in distinct disagreement with the conclusion from the spectral intensities quoted above. The specific heat curve below $0.2°$ K cannot be explained by any distribution of the splittings between different percentages of ions. Even the assumption of a third splitting of a value outside the range of BLEANEY's microwave apparatus, could not provide agreement between the theoretical curve and the experiment.

Another remarkable fact is the following. In Sect. 5 it was pointed out that if the STARK splitting and the magnetic interaction are of different orders of magnitude the entropy versus temperature curve may have a horizontal part between τ and δ. Now the S versus T^* curve of chromium potassium alum shows that such a horizontal part exists (see for instance Fig. 19), but it is not located at $S = R \ln 2$, as should be expected, but at a much lower entropy. This, however, is in qualitative agreement with BLEANEY's specific heat data since it is obvious from Fig. 18 that curve A corresponds to a larger entropy content than curve C.

A large hyperfine splitting might account for the behaviour of the salt, but the only chromium isotope with a nuclear magnetic moment, Cr^{53}, has an abundance of only 9.4% and a small spin value, see Sect. 33. The occurrence of a large exchange interaction is unlikely as well since the shapes of the resonance lines at room temperature[6] are exactly what one would expect from the magnetic dipole interaction.

[1] H. B. G. CASIMIR, D. BIJL and F. K. DU PRÉ: Leiden Commun. No. 262a; Physica, 's-Grav. **8**, 449 (1941).

[2] H. C. KRAMERS, D. BIJL and C. J. GORTER: Leiden Commun. No. 280a; Physica, 's-Grav. **16**, 65 (1950).

[3] C. J. GORTER, L. J. DIJKSTRA and O. VAN PAEMEL: Physica, 's-Grav. **9**, 673 (1942).

[4] C. STARR: Phys. Rev. **60**, 241 (1941).

[5] L. J. F. BROER: Physica, 's-Grav. **13**, 352 (1947).

[6] D. M. S. BAGGULEY and J. H. E. GRIFFITHS: Proc. Roy. Soc. Lond., Ser. A **204**, 188 (1950).

[7] B. BLEANEY: Proc. Roy. Soc. Lond., Ser. A **204**, 203 (1950).

Some years ago, the chromium potassium alum was considered as a suitable salt for absolute thermometry in the region between 1° K and 0.1° K. Since that time, however, the unexplained properties described in the present section were discovered and the general point of view is now[1] that chromium methylamine alum is better for the purpose. For this salt (see Sect. 34) the experimental results are in agreement with theory down to 0.1° K and only one level splitting has been found from paramagnetic resonance investigations.

36. Diluted chromium alums. In Sect. 4 it was pointed out that the magnetic dipole and exchange interactions (hence the value of τ, see Sect. 32) can be reduced by increasing the distances between the magnetic ions. For this reason some investigations have been performed with "diluted" chromium alums, in which part of the magnetic ions was replaced by equivalent non-magnetic ions, viz. aluminum.

The first experiments were made by DE HAAS and WIERSMA[2]. They used a concentration of one chromium ion in 14.4 aluminum ions. It was found that the T^* values, obtained by demagnetizing from strong fields, were substantially lower than in the case of the normal chromium alum. If, however, moderate fields were applied there was not much difference between the two salts.

The experiments were repeated in more detail by DE KLERK and POLDER[3].

Table 5. *Diluted chromium potassium alum 1:13*
(DE KLERK and POLDER).

H, T_i, S/R, T_{Lor}^*, T_{Lor} and T have the same meaning as in Table 1; $\delta = 0.30$ °K.

H oersteds	T_i °K	S/R	T_{Lor}^*	T_{Lor} °K	T_{theor} °K
1 650	1.159	1.370	0.706	0.704	0.830
1 925	1.145	1.361	0.597	0.592	0.654
2 520	1.142	1.339	0.495	0.489	0.479
3 030	1.181	1.323	0.417	0.411	0.409
3 060	1.174	1.321	0.406	0.400	0.399
4 160	1.174	1.265	0.299	0.289	0.286
4 200	1.149	1.256	0.287	0.277	0.278
4 960	1.162	1.214	0.242	0.230	0.238
4 980	1.170	1.215	0.242	0.230	0.234
6 140	1.171	1.137	0.194	0.180	0.183
7 290	1.162	1.053	0.161	0.145	0.148
8 210	1.149	0.980	0.130	0.112	0.125
8 270	1.180	0.992	0.127	0.109	0.129
9 400	1.173	0.912	0.109	0.090	0.107
10 340	1.178	0.849	0.091	0.070	0.091
11 500	1.176	0.771	0.060	0.043	0.072
12 550	1.184	0.714	0.037	0.025	
14 100	1.176	0.622	0.023	0.014	
16 300	1.173	0.515	0.016	0.010	
17 200	1.162	0.468	0.014	0.0087	
18 650	1.162	0.412	0.012	0.0076	

They used a powdered sample in the shape of an ellipsoid, containing one chromium ion in 13 aluminum ions. Results are given in Table 5. For the calculation of T_{theor} it was assumed that the magnetic interaction could be entirely neglected ($\tau = 0$). The difficulty in computing the entropy was the correction for the lattice specific heat. Since this contribution was effectively fourteen times larger than in the case of the undiluted chromium alum it was not justified to derive its value from the experiments with the undiluted salt (see Sect. 34). The simplest solution was to adapt both the STARK splitting and the lattice specific heat in such a way that the best agreement was reached between T_{theor} and T_{Lor}. This was obtained for a lattice specific heat obeying $c/R = 0.0217\,T^3$, which is essentially higher than fourteen times the value for the undiluted chromium potassium alum (see Sect. 34).

---

[1] D. DE KLERK: Proceedings of the Third Symposium on Temperature, Washington 1954, p. 251.

[2] W. J. DE HAAS and E. C. WIERSMA: Leiden Commun. No. 236b; Physica, 's-Grav. **2**, 305 (1935).

[3] D. DE KLERK and D. POLDER: Leiden Commun. No. 262d; Physica, 's-Grav. **8**, 508 (1941).

The best value for δ proved to be 0.30° K but the agreement between T_{theor} and T_{Lor} is somewhat less satisfactory than in the case of the undiluted chromium alum. BIJL[1] made paramagnetic relaxation experiments in the liquid helium region using a sample of the same origin as that of DE KLERK and POLDER. His value for the STARK splitting was 0.281° K. This is somewhat lower, but it should be realized that for a diluted sample the susceptibility values are small so that the precision of the investigations is limited.

The δ-value for the diluted chromium alum is higher than for the undiluted salt, so that its specific heat per gram-ion chromium is larger. This is not un-

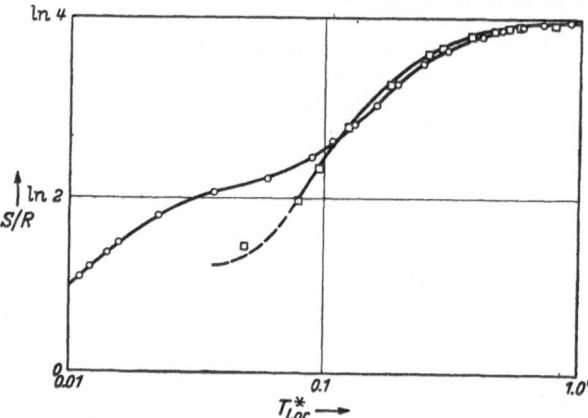

reasonable. The STARK splitting is due to the distorsion of the octahedron of water molecules surrounding the ion (see Sect. 30), and it is plausible that this distorsion can be somewhat larger for a mixed crystal than for a normal salt. This may also account for the fact that the agreement between T_{Lor} and T_{theor} of Table 5 is not so good as in the case of the undiluted material. The distorsion of the cubes in a mixed crystal may be not the same for all the ions of the lattice and hence the splittings for individual ions may diverge noticeably. The consequence

Fig. 19. Entropy versus magnetic temperature diagram for diluted and normal chromium potassium alum. ○ Diluted chromium alum (according to DE KLERK and POLDER). ☐ Normal chromium alum (according to CASIMIR, DE HAAS and DE KLERK). The dotted part of the curve was constructed with the help of the data of DE KLERK, STEENLAND and GORTER.

may be that effectively the diluted salt is even more complicated in its behaviour than the normal chromium potassium alum with its two splittings, see Sect. 35.

Due to the large value of the splitting parameter demagnetizations from relatively small fields (for which the entropy decrease is smaller than $R\ln 2$) give rise to somewhat higher final temperatures than the undiluted salt This is clear from Fig. 19 where T_{Lor}^* is plotted against the entropy for both the diluted and the undiluted chromium potassium alum. In this region of entropies the application of a diluted salt does not provide any advantage over the undiluted material. With strong fields, however, for which the entropy decrease is larger than $R\ln 2$, temperatures are reached where the only significant contribution to the specific heat is the magnetic interaction. Here the final temperatures for a diluted salt are noticeably lower than for the normal salt.

It should be noticed from Fig. 19 that for the diluted chromium potassium alum the plateau in the entropy curve (see Sect. 35) occurs at the correct value of the entropy, viz. $R\ln 2$.

37. Chromium nitrate. $Cr(NO_3)_3\, 9H_2O$. Molecular weight: 400.2, density: 2.41.

Some experiments with chromium nitrate were reported by CASIMIR, DE KLERK and POLDER[2]. A powdered sample of ellipsoidal shape was used, the filling factor being 0.514. The results are shown in Table 6.

---

[1] D. BIJL: Leiden Commun. No. 262c; Physica, 's-Grav. **8**, 497 (1941).

[2] H. B. G. CASIMIR, D. DE KLERK and D. POLDER: Leiden Commun. No. 261a; Physica, 's-Grav. **7**, 737 (1940).

The value of τ [see Eq. (32.3)] for this salt is $0.0336°$ K. The quantity Q of Eq. (32.4) depends on the crystalline structure. Since this is unknown the value for the chromium alums was taken, so that Eqs. (33.2), (33.3), (33.4), (33.5), (33.8) and (32.13) could be applied. A small mistake in Q does not influence the value of the splitting parameter δ very much.

CASIMIR, DE KLERK and POLDER reported a splitting parameter of $\delta = 0.283°$ K. This was adapted at the higher temperatures (about $0.6°$ K) and only T_{Lor} was considered. We rediscussed the data taking also T_{Ons} and $T_{V.VI}$ into account. The results are also shown in Table 6. It was found that between 0.4 and $0.05°$ K

Table 6. *Chromium nitrate* (CASIMIR, DE KLERK and POLDER).

H, T_i, S/R, T_{Lor}^*, T_{Lor}, T_{Ons}, $T_{V.VI}$ and T_{theor} have the same meaning as in Table 1; $\delta = 0.275°$ K.

H oersteds	T_i °K	S/R	T_{Lor}^*	T_{Lor} °K	T_{Ons} °K	$T_{V.VI}$ °K	T_{theor} °K
850	1.173	1.372	0.904	0.901	0.900	0.900	0.870
1110	1.193	1.369	0.804	0.801	0.800	0.800	0.782
1380	1.193	1.364	0.704	0.700	0.698	0.699	0.678
1370	1.181	1.363	0.704	0.700	0.698	0.699	0.674
1636	1.162	1.356	0.598	0.594	0.592	0.593	0.585
2160	1.183	1.342	0.495	0.490	0.488	0.489	0.479
2760	1.159	1.317	0.388	0.382	0.379	0.381	0.378
3020	1.188	1.310	0.371	0.364	0.361	0.362	0.360
3546	1.158	1.280	0.310	0.302	0.299	0.300	0.299
4090	1.175	1.254	0.276	0.267	0.263	0.265	0.263
5460	1.166	1.167	0.207	0.196	0.191	0.193	0.192
5450	1.164	1.168	0.205	0.194	0.189	0.191	0.192
6460	1.157	1.097	0.174	0.161	0.155	0.158	0.158
7520	1.152	1.018	0.151	0.137	0.130	0.133	0.131
8730	1.147	0.927	0.128	0.112	0.104	0.108	0.105
10440	1.146	0.806	0.104	0.086	0.077	0.081	0.077
11490	1.145	0.738	0.089	0.071	0.061	0.065	0.061
12630	1.145	0.668	0.077	0.059	0.048	0.053	0.047
13380	1.146	0.627	0.060	0.044	0.030	0.035	0.039
14160	1.147	0.584	0.048	0.033	0.017	0.022	0.033

the agreement between T_{theor} and T_{Ons} and T_{Lor} was good if a δ of $0.275°$ K was chosen, but then noticeable deviations were found at the highest temperatures. If the value of CASIMIR, DE KLERK and POLDER is applied, the agreement above $0.5°$ K is better, but in this case the results are less satisfactory in the neighbourhood of $0.2°$ K.

The only other value of the splitting parameter was reported by TEUNISSEN[1], derived from paramagnetic relaxation measurements in the liquid nitrogen region. He found $\delta = 0.296°$ K. This is higher than our value, but it is quite possible that, as in the case of the chromium potassium alum, the value depends somewhat on temperature.

It was found that a flatter part occurs in the entropy curve at $S = R\ln 2$, that is at the correct value. Still the salt is less suitable as a standard substance for absolute thermometry than the chromium methylamine alum. This is partly due to the larger uncertainty in the splitting parameter quoted above, and partly to the fact that the salt is chemically less stable.

38. Iron ammonium alum. $Fe(NH_4)(SO_4)_2 \cdot 12H_2O$. Molecular weight: 482.2; density: 1.70.

---

[1] P. TEUNISSEN: Thesis, Groningen, 1939.

Iron ammonium alum is one of the salts that have been in use since the very beginning of adiabatic demagnetization work, both in Leiden and in Oxford[1,2,3]. It was applied by several investigators for cooling down other materials below 1° K, see Chapt. E. The number of investigations concerning the magnetic and caloric properties below 1° K, however, is much smaller than in the case of the chromium alums, so that our knowledge of its behaviour is less complete. Another disadvantage of the salt is that it is chemically less stable.

No crystallographic investigations have been made with the salt, but probably it has an α structure (see Sect. 33). The iron ion is in a ^6S-state, so that orbital magnetism is absent. Since the STARK splitting of the sixfold spin level is of the order of a few tenths of a degree the magnetic moment and entropy at liquid helium temperatures can be described using the BRILLOUIN function with $J = 5/2$, $g = 2$, see Eqs. (29.1) and (29.2); the magnetic moment has been confirmed by experiment[4].

Theoretical relations for the behaviour below 1° K have first been given by HEBB and PURCELL[5] on the assumption that the STARK splitting is due to an electric field of cubic symmetry. This may split the ground level into a twofold and a fourfold degenerate level[6]. Assuming that the twofold level is the lower one they derived [see Eqs. (30.1), (3.2), (3.3), (9.4) and (30.5)]:

$$Z_s = 2(1 + 2e^{-\delta/T}), \tag{38.1}$$

$$\frac{c}{R} = 2 \frac{\delta^2}{T^2} \frac{e^{-\delta/T}}{(1 + 2e^{-\delta/T})^2} = \frac{2}{9} \frac{\delta^2}{T^2}\left(1 + \frac{1}{3}\frac{\delta^2}{T^2} + \cdots\right), \tag{38.2}$$

$$\frac{S}{R} = \ln 2(1 + 2e^{-\delta/T}) + 2\frac{\delta}{T}\frac{e^{-\delta/T}}{1 + 2e^{-\delta/T}} = \ln 6 - \frac{1}{9}\frac{\delta^2}{T^2}\left(1 + \frac{1}{6}\frac{\delta^2}{T^2} + \cdots\right), \tag{38.3}$$

$$\gamma(T) = \frac{\left(5 + 32\frac{T}{\delta}\right) + \left(26 - 32\frac{T}{\delta}\right)e^{-\delta/T}}{21(1 + 2e^{-\delta/T})}. \tag{38.4}$$

The only isotope with a nuclear spin, Fe^{57}, has an abundance of only 2.2% so that hyperfine splitting needs not be taken into account. Neglecting exchange coupling HEBB and PURCELL derived for the magnetic interaction [see Eq. (32.11)]:

$$\left.\begin{aligned}\Omega = {} & \frac{1.20}{(1 + 2e^{-\delta/T})^2} \times \\ & \times \left\{\left(\frac{25}{441} + \frac{64}{49}\frac{T}{\delta}\right) + \left(\frac{772}{441} + \frac{64}{21}\frac{T}{\delta}\right)e^{-\delta/T} + \left(\frac{676}{441} - \frac{640}{147}\frac{T}{\delta}\right)e^{-2\delta/T}\right\},\end{aligned}\right\} \tag{38.5}$$

so that, for high temperatures, the specific heat and entropy due to the magnetic interaction obey the relations

$$\frac{c}{R} = 2.40\frac{\tau^2}{T^2}, \tag{38.6}$$

and

$$\frac{S}{R} = \ln 2 - 1.20\frac{\tau^2}{T^2} \tag{38.7}$$

[1] W. J. DE HAAS and E. C. WIERSMA: Leiden Commun. No. 236b; Physica, 's-Grav. 2, 335 (1935).
[2] N. KURTI and F. SIMON: Nature, Lond. 135, 31 (1935).
[3] N. KURTI and F. SIMON: Proc. Roy. Soc. Lond., Ser. A 149, 152 (1935).
[4] W. E. HENRY: Phys. Rev. 88, 559 (1952).
[5] M. H. HEBB and E. M. PURCELL: J. Chem. Phys. 5, 338 (1937).
[6] J. H. VAN VLECK and W. G. PENNEY: Phil. Mag. 17, 961 (1934).

where $\tau = 0.0472°$ K, according to Eq. (32.3), so that the coefficient of the $1/T^2$ term in the specific heat is equal to:

$$A = \frac{2}{9}\,\delta^2 + 2.40\,\tau^2. \tag{38.8}$$

It was found from paramagnetic resonance experiments[1,2,3], however, that the assumption of a STARK field of cubic symmetry cannot be correct. The value of the level splitting derived from measurements in the 1,0,0-direction is widely different from the value obtained from adiabatic demagnetization and paramagnetic relaxation experiments; and the absorption pattern in the 1,1,0-direction is more complicated than can be explained with a field of cubic symmetry. For this reason, MEYER[4] made new calculations on the assumption of the simultaneous presence of cubic and trigonal fields. In this case, the ground level is split into three KRAMERS doublets.

MEYER assumed a potential of the electric field:

$$U = A \sum_i U_{\text{cub}}^{(i)} + C \sum_i U_{\text{trig}}^{(i)}, \tag{38.9}$$

and found for the zero field energy levels:

$$E_{1,\,2} = \tfrac{1}{2}\left(- a + c \pm \sqrt{9a^2 + 6ac + 81c^2}\right), \tag{38.10}$$

$$E_3 = a - c, \tag{38.11}$$

where a and c are proportional to A and C of Eq. (38.9). Now the STARK specific heat at high temperatures obeys:

$$\frac{c}{R} = \frac{2\,(a^2 + 7\,c^2)}{T^2}, \tag{38.12}$$

so that, according to Eq. (38.2), if the trigonal field is neglected ($c = 0$) we have: $\delta = 3\,a$ and the splitting parameter derived empirically from Eq. (38.8) obeys:

$$\delta = 3\,a\,\sqrt{1 + 7c^2/a^2}. \tag{38.13}$$

The complete expression for the STARK specific heat becomes:

$$\frac{c}{R} = \frac{1}{T^2}\,\frac{(E_1 - E_2)^2\,e^{-(E_1+E_2)/T} + (E_1 - E_3)^2\,e^{-(E_1+E_3)/T} + (E_2 - E_3)^2\,e^{-(E_2+E_3)/T}}{e^{-(E_1+E_2)/T} + e^{-(E_1+E_3)/T} + e^{-(E_2+E_3)/T}}, \tag{38.14}$$

and the function $\gamma(T)$, according to BLEANEY and TRENAM[5] (under the assumptions of MEYER) is equal to:

$$\gamma(T) = \frac{\{a_1 - T(b_2 + b_3)\}e^{-E_1/T} + \{a_2 + T(b_3 - b_1)\}e^{-E_2/T} + \{9 + T(b_1 + b_2)\}e^{-E_3/T}}{35\,(e^{-E_1/T} + e^{-E_2/T} + e^{-E_3/T})}, \tag{38.15}$$

where:

$$a_1 = (25\,p^4 - 10\,p^2 q^2 + 19\,q^4), \tag{38.16}$$

$$a_2 = (19\,p^4 - 10\,p^2 q^2 + 25\,q^4), \tag{38.17}$$

$$b_1 = (32\,p^2 + 20\,q^2)/(E_2 - E_3), \tag{38.18}$$

$$b_2 = (20\,p^2 + 32\,q^2)/(E_1 - E_3), \tag{38.19}$$

$$b_3 = (108\,p^2 q^2)/(E_1 - E_2). \tag{38.20}$$

[1] R. T. WEIDNER, P. R. WEISS, C. A. WHITMER and D. R. BLOSSER: Phys. Rev. 76, 1727 (1949).

[2] D. BIJL: Thesis, Leiden 1950, p. 144.

[3] J. UBBINK, J. A. POULIS and C. J. GORTER: Leiden Commun. No. 283b; Physica, 's-Grav. 17, 213 (1951).

[4] P. H. E. MEYER: Leiden Commun. Suppl. No. 103e; Physica, 's-Grav. 17, 899 (1951).

[5] B. BLEANEY and R. S. TRENAM: Proc. Roy. Soc. Lond., Ser. A 223, 1 (1954).

Further p and q satisfy the relations $p^2 + q^2 = 1$; $q/p = \tan\frac{1}{2}\alpha$ with:

$$\tan\alpha = \frac{4a\sqrt{5}}{27c}. \qquad (38.21)$$

Not very much work has been done to verify the theoretical relations for iron ammonium alum experimentally. Old measurements of KURTI and SIMON[1] were discussed by HEBB and PURCELL[2], but the agreement was not satisfactory.

CASIMIR, DE HAAS and DE KLERK[3] made some experiments with a sample of rather impure material. They restricted themselves to the region where CURIE's law is still valid $(\gamma(T) = 1)$,

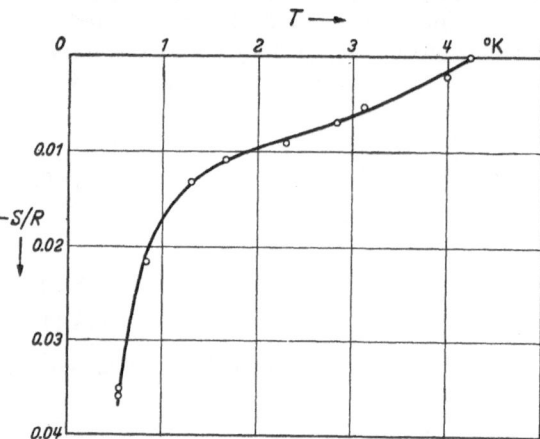

Fig. 20. Entropy versus temperature diagram for an impure sample of iron ammonium alum (according to CASIMIR, DE HAAS and DE KLERK). O Experimental points. The theoretical curve is calculated with $S/R = -\frac{1}{2}A/T^2 + \frac{1}{3}BT^3$, see text. The zero point of the entropy is arbitrary.

and where the STARK specific heat is proportional to $1/T^2$. This is the case for temperatures above $0.5°$ K. Demagnetizations were performed from the boiling point of liquid helium. Here the lattice specific heat still gives an important contribution to the entropy of the salt, so that it could be derived from the experiments with reasonable accuracy. The results of the measurements are shown in Fig. 20; they are in good agreement with the relation:

$$\frac{S}{R} = -\frac{1}{2}\frac{A}{T^2} + \frac{1}{3}BT^3, \qquad (38.22)$$

with $A = 0.0165 \ (° K)^2$ and $B = 0.000363 \ (° K)^{-3}$.

Some demagnetizations were performed starting from solid hydrogen temperatures. It followed that the specific heat between $4.2°$ K and $9.0°$ K is larger than predicted from Eq. (39.22) with the value of B given here. This was corroborated by experiments of KURTI, LAÎNÉ and SIMON[4]. DUYCKAERTS[5] measured the specific heat in the region between $2°$ K and $20°$ K. He found $B = 0.00046 \ (° K)^{-3}$, but an anomaly was superimposed on the DEBYE specific heat with a maximum near $14°$ K. The origin of this anomaly could not be explained in a satisfactory way.

Recently KAPADNIS made specific heat measurements in the liquid helium

Table 7. *Iron ammonium alum* (CASIMIR, DE HAAS and DE KLERK).

H, T_i, S/R, T_{Lor} and T_{theor} have the same meaning as in Table 1; $\delta = 0.183°$ K.

H oersteds	T_i ° K	S/R	T_{Lor} ° K	T_{theor} ° K
615	1.164	1.7847	0.76	0.756
825	1.164	1.7791	0.62	0.614
1055	1.172	1.7715	0.51	0.513
1260	1.173	1.7630	0.44	0.441
2180	1.162	1.7074	0.269	0.268

region. He found $B = 0.000424 \ (° K)^{-3}$ with small deviations at the highest temperatures. The results have not yet been published.

[1] N. KURTI and F. SIMON: Proc. Roy. Soc. Lond., Ser. A **149**, 152 (1935).
[2] M. H. HEBB and E. M. PURCELL: J. Chem. Phys. **5**, 338 (1937).
[3] H. B. G. CASIMIR, W. J. DE HAAS and D. DE KLERK: Leiden Commun. No. 256a; Physica, 's-Grav. **6**, 241 (1939).
[4] N. KURTI, P. LAÎNÉ and F. SIMON: C. R. Acad. Sci. Paris **208**, 173 (1939).
[5] G. DUYCKAERTS: Bull. Soc. Roy. Sci. Liège **6**, 193 (1945).

De Haas, Casimir and de Klerk repeated their experiments with a sample of higher purity, a powdered ellipsoid with filling factor 0.66 being used. The demagnetizations were now performed from 1.16° K and the results are given in Table 7. The coefficient A of the specific heat, Eq. (39.22), for this sample

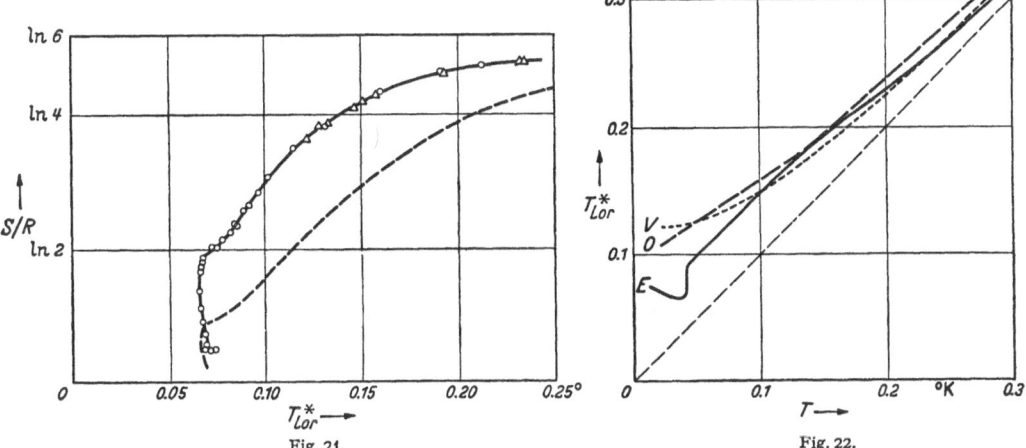

Fig. 21.

Fig. 22.

Fig. 21. Entropy versus magnetic temperature diagram for a compressed pill of iron ammonium alum (according to Kurti and Simon). The dotted line represents the heat supply integrated from the lowest temperature obtained in the experiment. This curve is given in arbitrary units.

Fig. 22. Relation between magnetic and absolute temperatures for a compressed ellipsoid of iron ammonium alum (according to Kurti and Simon). Curve O: Onsager's formula. Curve V: van Vleck's formula. Curve E: Experimental values from caloric measurements. Curves O and V are calculated for a cubic electric field with $\delta = 0.20°$ K.

is 0.0128 $(°\,K)^2$. Caloric measurements of van Dijk and Keesom[1] were not in very good agreement with these results, but recent calorimetric experiments by Kapadnis gave: $A = 0.0135\ (°\,K)^2$.

More recent experiments, performed by Kurti and Simon were reported by Cooke[2]. He published an S versus T_{Lor}^* plot, see Fig. 21. A more or less horizontal part in this curve at $S = R\ln 2$ gives a qualitative confirmation of the supposition that the lowest Stark level is a doublet. Caloric experiments with gamma radiation provided data on the absolute temperatures; they were published as a T_{Lor}^* versus T-curve, see Fig. 22. Above 0.1° K the results were in reasonable agreement with the Onsager and van Vleck approximations on the assumption of only a cubic Stark field [Eqs. (38.3) and (38.4)].

The data as published by Cooke were converted by Mendoza, in his "Demagnetizer's Vademecum"[3], into a table of entropies versus absolute temperatures. The data reduced to the units of the present article are given in Table 8.

Table 8. *Iron ammonium alum* (Kurti and Simon).

H, T_i, and S/R have the same meaning as in Table 1; T is derived from caloric experiments; $\delta = 0.20°$ K.

H/T_i oersteds $(°\,K)^{-1}$	S/R	T °K
0	1.792	1.000
1000	1.766	0.465
1500	1.735	0.330
2000	1.693	0.237
2500	1.643	0.172
3000	1.585	0.137
3500	1.523	0.113
4000	1.459	0.0993
5000	1.325	0.0812
6000	1.195	0.0687
7000	1.074	0.0563
8000	0.963	0.0470

[1] H. van Dijk and W. H. Keesom: Leiden Commun. No. 260b; Physica, 's-Grav. 7, 970 (1940).

[2] A. H. Cooke: Proc. Phys. Soc. Lond. A **62**, 269 (1949).

[3] E. Mendoza: „The Demagnetizer's Vademecum", unpublished, no date, probably 1952.

The splitting parameter δ has been calculated by several investigators on the assumption of only a cubic field, hence with the help of Eq. (38.8). The results of Casimir, de Haas and de Klerk with the pure sample (see above) give $\delta = 0.183°$ K, the experiments of Kurti and Simon $\delta = 0.20°$ K. The calorimetric data of Kapadnis at liquid helium temperatures yield $\delta = 0.192°$ K. Paramagnetic relaxation measurements of du Pré[1] in the liquid helium region gave $\delta = 0.187°$ K; Kramers, Bijl and Gorter[2] found at the same temperatures $\delta = 0.186°$ K; and Benzie and Cooke[3] $\delta = 0.20°$ K. Dijkstra, Gorter and Volger[4] found, from relaxation experiments in the liquid nitrogen region, $\delta = 0.200°$ K; Broer[5]: $\delta = 0.200°$ K; Starr[6]: $\delta = 0.193°$ K.

The paramagnetic resonance experiments in the 1,0,0-direction, however, gave much lower values. Bijl[7] found $\delta = 0.052°$ K; Weidner, Weiss, Whitmer and Blosser[8] $\delta = 0.046°$ K; Ubbink, Poulis and Gorter[9] $\delta = 0.052°$ K. Meyer's analysis of the spectrum in the 1,1,0-direction gave $a = 0.0184°$ K, $c = 0.023°$ K, so that the apparent splitting parameter, according to Eq. (38.13), is equal to $0.193°$ K, in excellent agreement with the data obtained from the demagnetization, paramagnetic relaxation and calorimetric experiments.

In view of this result it should be interesting to discuss the results of Kurti and Simon of Table 8 with the formulae derived for the case of simultaneously a cubic and a trigonal field, Eqs. (38.14) and (38.15).

39. Titanium cesium alum. $TiCs(SO_4)_2 \cdot 12 H_2O$. Molecular weight: 589, density about 2. This salt was used in the early days of demagnetization work, both by de Haas and Wiersma[10,11] and by Kurti and Simon[12]. De Haas and Wiersma showed that rather low T^*-values (viz. $0.0055°$) could be obtained with it.

At first sight the salt is very attractive from a theoretical point of view. The ground level of the free titanium ion is in a ^2D state. It is split by an electric field of cubic symmetry into an orbital doublet and a triplet, the triplet being the lower one. In the alum the triplet is further split into three spin doublets which, according to Kramers' theorem (see Sect. 30), are not affected by electric fields any more.

If only the lowest Kramers doublet is occupied at liquid helium temperature, the magnetic moment and entropy follow the Brillouin function with $J = S = \frac{1}{2}$, $g = 2$ [see Eqs. (29.1) and (29.2)]. The Stark specific heat is zero and $\gamma(T) = 1$ for all temperatures. The specific heat due to the magnetic dipole interaction, according to Hebb and Purcell[12] is equal to:

$$\frac{c}{R} = 2.40 \left(\frac{\tau}{T}\right)^2 \left(1 - \frac{29}{16}\frac{\tau}{T} - \frac{45}{8}\left(\frac{\tau}{T}\right)^2 + \cdots\right), \qquad (39.1)$$

[1] F. K. du Pré: Leiden Commun. No. 258c; Physica, 's-Grav. **7**, 79 (1940).

[2] H. C. Kramers, D. Bijl and C. J. Gorter: Leiden Commun. No. 280a; Physica, 's-Grav. **16**, 65 (1950).

[3] R. J. Benzie and A. H. Cooke: Proc. Phys. Soc. Lond. A **63**, 213 (1950).

[4] L. J. Dijkstra, C. J. Gorter and J. Volger: Physica, 's-Grav. **10**, 337 (1943).

[5] L. J. F. Broer: Physica, 's-Grav. **13**, 353 (1947).

[6] C. Starr: Phys. Rev. **60**, 241 (1941).

[7] D. Bijl: Thesis Leiden 1950, p. 144.

[8] R. T. Weidner, P. R. Weiss, C. A. Whitmer and D. R. Blosser: Phys. Rev. **76**, 1727 (1949).

[9] J. Ubbink, J. A. Poulis and C. J. Gorter: Leiden Commun. No. 283b; Physica, 's-Grav. **17**, 213 (1951).

[10] W. J. de Haas and E. C. Wiersma: Leiden Commun. No. 236c; Physica, 's-Grav. **2**, 438 (1935).

[11] W. J. de Haas and E. C. Wiersma: Leiden Commun. No. 241c; Physica, 's-Grav. **3**, 491 (1936).

[12] M. H. Hebb and E. M. Purcell: J. Chem. Phys. **5**, 338 (1937).

with $\tau = 0.0038°$ K, see Eq. (32.3). It follows that, if hyperfine splitting and exchange interaction can be neglected, the magnetic and caloric behaviour of the salt can be predicted quantitatively. It was shown by HEBB and PURCELL, however, that the measurements of KURTI and SIMON are not in agreement with the theoretical formulae.

Measurements of magnetization curves by VAN DEN HANDEL[1] indicated that the separations between the three lowest doublets are only of the order of 100 cm^{-1} so that the susceptibility in the liquid helium region is noticeably influenced by the higher doublets. BENZIE and COOKE[2] found for the splitting factor: $g = 1.12$; BOGLE and COOKE (unpublished) derived from the paramagnetic resonance experiments: $g_{||} = 1.25$ and $g_{\perp} = 1.14$.

The high temperature contribution of the specific heat follows from the paramagnetic relaxation experiments of BENZIE and COOKE. They found $cT^2/R = 3.9 \times 10^{-5}$, whereas the magnetic dipole interaction, Eq. (39.1), accounts only for 0.3×10^{-5}. Investigations with a diluted sample indicated that hyperfine splitting (due to the isotopes Ti^{47} and Ti^{49} which, together, amount to 13% abundance) gives rise to a contribution of 0.4×10^{-5}, so that by far the largest part of the specific heat must be due to exchange interaction.

The conclusion from the magnetic investigations described here is, that the titanium cesium alum has very little of the ideal properties assumed in the early investigations. Moreover the salt is chemically very unstable, oxidizing readily in air. For these reasons the interest in the substance has been lost in later years, at least for adiabatic demagnetization work.

40. The Tutton salts in general. The Tutton salts have the general formula $M''M_2'(XO_4)_2 \, 6H_2O$, where M'' is a divalent ion and M' a monovalent one. We restrict ourselves to the sulphates in which M'' is a magnetic ion with an odd number of electron spins (of which only Mn, Cu and Co have been investigated at low temperatures).

The crystalline structure was investigated by HOFMANN[3]. The unit cell is monoclinic and, though there are small deviations between different salts, the ratios of the dimensions of the cell are never very much different from $a:b:c = 1.47:2:1$, the length of the c-axis being roughly 6.20 Å. The b-axis is perpendicular to the a, c-plane and the angle between the a and c-axes is about 106°. There are two non-equivalent magnetic ions in the unit cell, denoted by A and B. Ion A is located at $(0, 0, 0)$ and ion B at $(\frac{1}{2}, \frac{1}{2}, 0)$.

Each ion is surrounded by six molecules of water of crystallization, four of them forming a square; they are removed about 2 Å from the ion. The other two are located on the line perpendicular to the square at a somewhat larger distance from the ion, e.g. 2.3 Å. From this it is plausible that the crystalline STARK effect can be described with the superposition of a cubic and a tetragonal electric field.

The symmetry axes (T_1 and T_2) of the tetragonal fields of the two ions are parallel to the same plane through the b-axis. The angle between this plane and the c-axis is denoted by ψ. In the plane the tetragonal axes make equal angles, α, with the a, c-plane as shown in Fig. 23. The angles ψ and α may be rather different for different Tutton salts. It is obvious now that the principal axes of magnetization of the crystal, denoted by K_1, K_2 and K_3, are the following: K_1 is the intersection of the T_1, T_2-plane with the a, c-plane, K_2 is in the a, c-plane perpendicular to K_1, and K_3 coincides with the b-axis.

[1] J. VAN DEN HANDEL: Thesis, Leiden 1940.
[2] R. J. BENZIE and A. H. COOKE: Proc. Roy. Soc. Lond., Ser. A **209**, 269 (1951).
[3] W. HOFMANN: Z. Kristallogr. **78**, 279 (1931).

The principal values of the splitting parameter g are the same for both ions, so that we can give $g_{\parallel}$ along the appropriate T-axis and $g_{\perp}$ at right angles to this axis. From these we may calculate the effective g-value for any direction in the crystal from:

$$g_{\text{eff}} = \sqrt{l_1^2 g_{\parallel}^2 + l_2^2 g_{\perp}^2 + l_3^2 g_{\perp}^2}, \tag{40.1}$$

where l_1, l_2 and l_3 are the direction cosines with respect to the tetragonal axis and two lines perpendicular to it [1].

No complete theoretical description is available of the behaviour of the Tutton salts below 1° K which accounts for (a) the shape of the Stark specific heat peak and the $\gamma(T)$-function (Sect. 30) under the joint influence of the cubic and tetragonal fields; (b) the $\Omega(T)$-function for the magnetic dipole and exchange

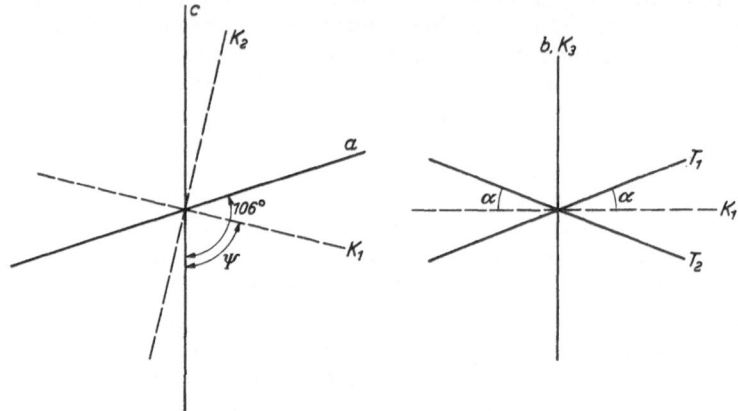

Fig. 23. Structure of the Tutton salts.

interactions (Sect. 32); and (c) the hyperfine structure. It should be noticed that in the Tutton salts hyperfine splitting and exchange interaction play a much more important role than in the alums. Some calculations have been made [2,3,4], however, concerning the $1/T^2$-term in the specific heat and the occurrence of a Curie-Weiss Θ due to the magnetic interactions. We shall refer to them in the discussions of the individual salts.

41. Manganese ammonium sulphate. $Mn(NH_4)_2(SO_4)_2\,6H_2O$. Molecular weight: 391; density: 1.83.

The most striking phenomenon encountered with this salt is the high temperature of the Curie point. It is located somewhat above 0.1° K and can be obtained with fields of the order of 6000 oersteds. At this temperature, however, the specific heat increases rapidly, so that the temperatures obtained with the very highest fields are not much lower, see Sect. 61.

The manganese ion is in a ^6S state so that the orbital magnetism can be neglected entirely. Resonance experiments [5] show that g is 2 and isotropic with a high degree of precision, and since no Curie-Weiss constant occurs, the magnetic behaviour in the liquid helium region can be represented by the Brillouin

[1] A. Abragam and M. H. L. Pryce: Proc. Roy. Soc. Lond., Ser. A **205**, 135 (1951).
[2] J. M. Daniels: Proc. Phys. Soc. Lond. A **66**, 673 (1953).
[3] D. Polder: Leiden Commun. Suppl. No. 92b; Physica, 's-Grav. **9**, 709 (1942).
[4] C. G. B. Garrett: Proc. Roy. Soc. Lond., Ser. A **203**, 392 (1950).
[5] B. Bleaney and D. J. E. Ingram: Proc. Roy. Soc. Lond., Ser. A **205**, 336 (1951).

function with $S = \frac{5}{2}$, see Eqs. (29.1) and (29.2). This was corroborated by experiment[1,2].

The most complete demagnetization experiments at the higher temperatures are those of COOKE and HULL, published by COOKE[3]. As in the case of iron ammonium alum he gave a T^* versus S curve, derived from a number of demagnetizations, and a T^* versus T diagram, obtained from gamma ray heating experiments. These two diagrams are shown in Figs. 24 and 25. MENDOZA[4] derived a table from them giving the absolute temperature as a function of the entropy; it is given in Table 9.

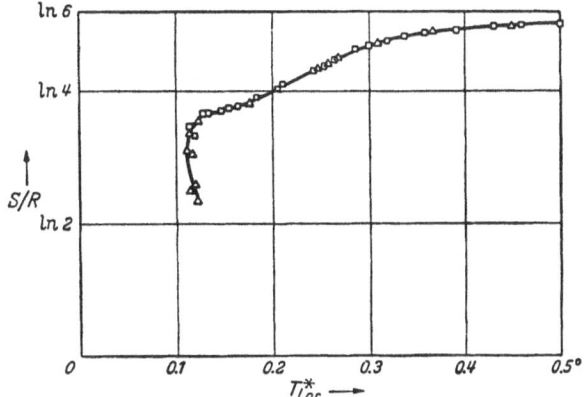

Fig. 24. Entropy versus magnetic temperature diagram for a compressed pill of manganese ammonium sulphate (according to COOKE and HULL).

Table 9. *Manganese ammonium sulphate* (COOKE and HULL).

H, T_i and S/R have the same meaning as in Table 1; T is derived from caloric experiments; $\delta = 0.33°$ K.

H/T_i oersteds $(° K)^{-1}$	S/R	T $°K$
0	1.792	1.000
2000	1.693	0.360
3000	1.585	0.250
3500	1.523	0.215
4000	1.459	0.188
4500	1.392	0.165
5000	1.325	0.145

Measurements of STEENLAND, VAN DER MAREL, DE KLERK and GORTER[5] were mainly restricted to the lower temperatures; they will be discussed in Chapt. D.

The experiments of COOKE and HULL were discussed under the assumption of only a cubic electric field. In this case the sixfold spin level is split into a

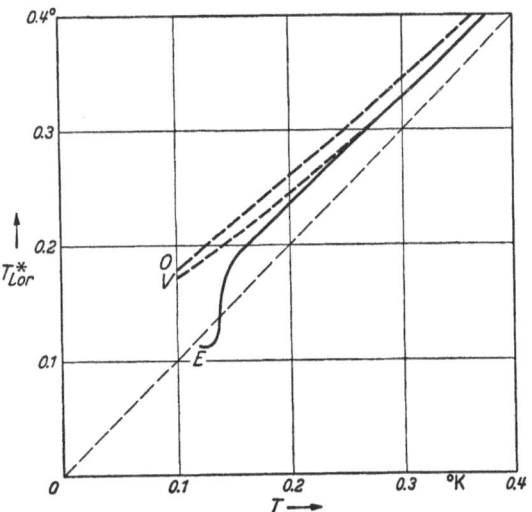

Fig. 25. Relation between magnetic and absolute temperatures for a compressed ellipsoid of manganese ammonium sulphate (according to COOKE and HULL). Curve O: ONSAGER's formula. Curve V: VAN VLECK's formula. Curve E: Experimental values from caloric measurements. Curves O and V are calculated for a cubic electric field with $\delta = 0.33°$ K.

twofold and a fourfold degenerate level, as was pointed out in Sect. 38, and the formulae derived by HEBB and PURCELL for the case of iron ammonium alum can be applied. Reasonable agreement with the VAN VLECK formula was

[1] L. C. JACKSON and H. KAMERLINGH ONNES: Proc. Roy. Soc. Lond., Ser. A **104**, 671 (1923).
[2] R. A. ERICKSON and L. D. ROBERTS: Phys. Rev. **93**, 957 (1954).
[3] A. H. COOKE: Proc. Phys. Soc. Lond. A **62**, 269 (1949).
[4] E. MENDOZA: "The Demagnetizer's Vademecum." Unpublished, no date, probably 1952.
[5] M. J. STEENLAND, L. C. VAN DER MAREL, D. DE KLERK and C. J. GORTER: Leiden Commun. No. 279c; Physica, 's-Grav. **15**, 906 (1949).

obtained for temperatures above $0.25°$ K with a splitting parameter $\delta = 0.33°$ K, see Fig. 25. The flat part in the S versus T^* curve of Fig. 24 at $S = R \ln 4$ might indicate that for this salt the fourfold level is the lower one.

A discussion on the assumption of simultaneous cubic and tetragonal fields has never been given. There is not very much point in this, however, since paramagnetic resonance measurements by Bleaney and Ingram[1] showed that a hyperfine splitting occurs of the same order of magnitude as the Stark splitting (the only isotope, Mn^{55}, has a nuclear spin $\frac{5}{2}$). Due to the joint actions of the Stark splitting and the nuclear interaction, the ground level is split into six singlets and fifteen doublets spread over $0.408°$ K. They give rise, together, to a high temperature contribution in the specific heat $cT^2/R = 0.0154$ and this is in reasonable agreement with the value found by Benzie, Cooke and Whitley[2] from paramagnetic relaxation experiments. For a sample of infinite dilution they found $cT^2/R = 0.017$.

The specific heat of the salt of normal concentration, according to the demagnetization measurements of Cooke and Hull, is $cT^2/R = 0.034$. The relaxation measurements of Benzie, Cooke and Whitley gave 0.032. Benzie and Cooke[3] found, in the liquid helium region: 0.033; Bijl[4], between $4°$ K and $20°$ K: 0.034; and Broer[5], at liquid nitrogen temperatures and above: 0.034. The contribution for the magnetic dipole interaction, calculated from Eq. (32.2) with $Q = 17.6$ and $\tau = 0.062$ [according to Eq. (32.3)], is $cT^2/R = 0.011$. Hence, if we take 0.033 as an average for the total specific heat, and 0.016 for the contribution of the Stark splitting and nuclear interaction, we have still 0.006 left for exchange interaction.

42. Copper potassium sulphate. $CuK_2(SO_4)_2 6H_2O$. Molecular weight: 403, density: 2.22. The free copper ion is in a ^2D state. This state is split by the cubic component of the electric field into an orbital doublet and a triplet. The tetragonal component and the spin-orbit interaction give a further splitting into five Kramers doublets of which only the lowest one is populated.

Measurements by Reekie[6] showed that the susceptibility for a powder is larger than predicted by the Brillouin function with $S = \frac{1}{2}$, $g = 2$. Miss Hupse[7] showed that the susceptibility is anisotropic, and according to Polder[8] this is due to the proximity of the higher levels. Paramagnetic resonance experiments with a diluted salt[9] gave $g = 2.45$ parallel to the tetragonal axis, and $g = 2.14$ perpendicular to it.

The absence of Stark specific heat was demonstrated by demagnetization experiments of Ashmead[10]. Measurements of Casimir, de Klerk and Polder[11] showed that the specific heat at the higher temperatures was not proportional to $1/T^2$. [Only T^* was measured, but if no Stark splitting occours $\gamma(T) = 1$ and $T^* = T$ in first approximation, see Sect. 30.]

[1] B. Bleaney and D. J. E. Ingram: Proc. Roy. Soc. Lond., Ser. A **205**, 336 (1951).
[2] C. J. Gorter et al.: Progress in Low Temperature Physics, p. 283. Amsterdam 1955.
[3] R. J. Benzie and A. H. Cooke: Proc. Phys. Soc. Lond. A **63**, 213 (1950).
[4] D. Bijl: Leiden Commun. No. 280b; Physica, 's-Grav. **16**, 269 (1950).
[5] L. J. F. Broer: Physica, 's-Grav. **13**, 353 (1947).
[6] J. Reekie: Proc. Roy. Soc. Lond., Ser. A **173**, 367 (1939).
[7] J. C. Hupse: Leiden Commun. No. 265b; Physica, 's-Grav. **9**, 633 (1942).
[8] D. Polder: Leiden Commun. Suppl. No. 92b; Physica, 's-Grav. **9**, 709 (1942).
[9] B. Bleaney, K. D. Bowers and D. J. E. Ingram: Proc. Phys. Soc. Lond. A **64**, 758 (1951).
[10] J. Ashmead: Nature, Lond. **143**, 853 (1939).
[11] H. B. G. Casimir, D. de Klerk and D. Polder: Leiden Commun. No. 261a; Physica, 's-Grav. **7**, 737 (1940).

The explanation was given by DE KLERK[1]. In his experiment a powdered sphere of copper potassium sulphate was demagnetized in thermal equilibrium with chromium potassium alum (see Sect. 69). A comparison of the susceptibilities of the two salts in the region down to $0.3°$ K revealed that a CURIE-WEISS law is obeyed, hence:

$$T = \alpha\, T^* + \Theta, \qquad (42.1)$$

where α is an experimental factor close to unity. It is due to the fact that in the liquid helium region Θ is too small to be observed, so that it is neglected in the χ versus T calibration (see Sect. 23). The value for Θ found by DE KLERK was $0.052°$ K, corroborated by KRAMERS, WASSCHER and GORTER[2] and by unpublished measurements of ROBERTS and DABBS. Table 10 gives the T^*-values of CASIMIR, DE KLERK and POLDER with the T-values calculated from Eq. (42.1) using DE KLERK's values of α and Θ.

Further investigations were made by GARRETT[3]. He measured the susceptibility in zero external field and in a field parallel to the small a.c. measuring field. From the values obtained for $(\partial M/\partial H)_S$ he could derive the $\varXi$ parameter as a function of temperature, see Eqs. (14.1), (14.3) and (14.4). The correction term ψ/χ was calculated from the theoretical expression for the magnetization curve at high temperatures. Some of his values are collected in Table 11. If $\sqrt[3]{\varXi}$ is plotted against $1/T^*$ a straight line is found down to $T^* = 0.025°$, and this confirms the validity of the CURIE-WEISS law, see Eq. (14.8). (A deviation is also found at the highest temperatures, but this is due to the influence of

Table 10. *Copper potassium sulphate* (CASIMIR, DE KLERK and POLDER).

H, T_i, S/R and T_{Lor}^* have the same meaning as in Table 1, T was derived from T_{Lor}^* with $\Theta = 0.052°$ K, see text.

H oersteds	T_i °K	S/R	T_{Lor}^*	T °K
601	1.143	0.6924	0.554	0.585
857	1.155	0.6916	0.424	0.460
1118	1.160	0.6906	0.318	0.358
1115	1.136	0.6905	0.308	0.348
1384	1.160	0.6893	0.258	0.300
1384	1.160	0.6893	0.255	0.297
1383	1.142	0.6892	0.227	0.270
2188	1.154	0.6835	0.139	0.186
2192	1.132	0.6831	0.141	0.188
2774	1.150	0.6777	0.105	0.153
3570	1.157	0.6682	0.0772	0.127
3570	1.150	0.6678	0.0766	0.126
4130	1.150	0.6596	0.0623	0.112
4900	1.153	0.6468	0.0493	0.0997
5570	1.132	0.6320	0.0401	0.0908
5580	1.160	0.6345	0.0404	0.0910
5540	1.157	0.6350	0.0408	0.0914
6095	1.153	0.6232	0.0357	0.0866
6540	1.179	0.6166	0.0321	0.0831
7200	1.157	0.5995	0.0272	0.0784

the lattice specific heat.) GARRETT's Θ proved to be smaller than that of DE KLERK and of the other investigators given above. He found $\Theta = 0.034°$ K; and since the same value was obtained from paramagnetic relaxation experiments by BENZIE and COOKE[4], and from unpublished demagnetization experiments by ASHMEAD, it seems that further investigations are required on this point.

If the corrections of Eq. (42.1) are applied it is found that the specific heat obeys a $1/T^2$ law down to $0.2°$ K. DE KLERK found: $cT^2/R = 6.8 \times 10^{-4}$; GARRETT: $cT^2/R = 6.0 \times 10^{-4}$. Paramagnetic relaxation experiments by BIJL[5] in the liquid helium region gave: $cT^2/R = 5.4 \times 10^{-4}$; BENZIE and COOKE[4] found, at the same temperatures: $cT^2/R = 6.0 \times 10^{-4}$; and BROER and KEMPERMAN[6], at liquid nitrogen

[1] D. DE KLERK: Leiden Commun. No. 270c; Physica, 's-Grav. **12**, 513 (1946).
[2] H. C. KRAMERS, J. D. WASSCHER and C. J. GORTER: Leiden Commun. No. 288c; Physica, 's-Grav. **18**, 329 (1952).
[3] C. G. B. GARRETT: Proc. Roy. Soc. Lond. Ser. A **203**, 375 (1950).
[4] R. J. BENZIE and A. H. COOKE: Proc. Phys. Soc. Lond. A **63**, 213 (1950).
[5] D. BIJL: Leiden Commun. No. 280b; Physica, 's-Grav. **16**, 269 (1950).
[6] L. J. F. BROER and J. KEMPERMAN: Physica, 's-Grav. **13**, 465 (1947).

temperatures: $cT^2/R = 6.5 \times 10^{-4}$. Direct calorimetric determinations by RAYNE[1] (see Sect. 75) gave: $cT^2/R = 5.8 \times 10^{-4}$.

The contribution of the magnetic dipole interaction to the specific heat [according to Eq. (32.2) with $\tau = 0.00677$ and $Q = 17.6$] is $cT^2/R = 1.35 \times 10^{-4}$. The hyperfine splitting contribution (both copper isotopes Cu[63] and Cu[65] have

Table 11. *Copper potassium sulphate* (GARRETT).

T^*_{Lor} and T have the same meaning as in Table 1; H is the external field, applied parallel to the measuring field (see text), Ξ and ψ/χ are defined in Sect. 14.

T^*_{Lor}	H oersteds	$\Xi - \dfrac{2\psi}{\chi}$ (oersteds)$^{-2}$	$\dfrac{2\psi}{\chi}$ (oersteds)$^{-2}$	Ξ (oersteds)$^{-2}$	T °K
0.999	43.0	3.27×10^{-6}	0.00×10^{-6}	3.27×10^{-6}	1.019
0.713	43.0	5.63	-0.01	5.62	0.737
0.578	36.0	7.9	0.0	7.9	0.603
0.401	35.8	11.1	0.0	11.1	0.429
0.172	25.5	14.9	-0.1	14.8	0.203
0.110	26.0	19.9	-0.2	19.7	0.141
0.0789	19.0	26.3	-0.3	26.0	0.111
0.0646	19.2	31.1	-0.4	30.7	0.097
0.0319	11.5	89.0	-0.9	88.1	0.064
0.0234	3.39	133.1	-1.1	132.0	
0.0189	2.37	179	-1	178	

a nuclear spin $I = \frac{3}{2}$) was determined by BENZIE and COOKE[2] from paramagnetic relaxation experiments on successively diluted samples. Extrapolation to zero concentration gave $cT^2/R = 1.1 \times 10^{-4}$. Demagnetization experiments with a diluted sample[3] suggested: $cT^2/R = 1.3 \times 10^{-4}$ and paramagnetic resonance[4] gave: $cT^2/R = 1.4 \times 10^{-4}$.

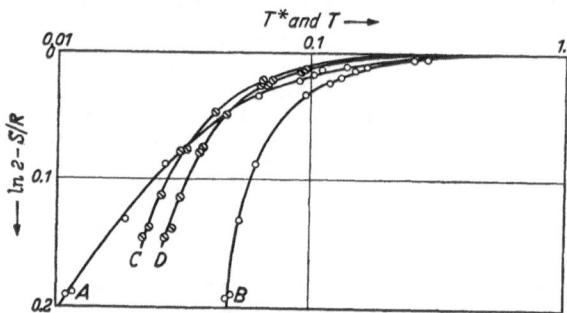

Fig. 26. Entropy versus magnetic and absolute temperature curves for normal and diluted copper potassium sulphates (according to GARRETT). Curve A: S versus T* for normal salt. Curve B: S versus T for normal salt. Curve C: S versus T* for diluted salt. Curve D: S versus T for diluted salt.

If we take $cT^2/R = 6.0 \times 10^{-4}$ as an average value of the total specific heat and subtract the magnetic dipole and hyperfine splitting contributions we have left 3.6×10^{-4} for exchange interactions. If we assume that the exchange is purely isotropic, OPECHOWSKI's relation (32.9) leads to a number of eleven neighbouring ions between which exchange occurs if DE KLERK's Θ is assumed, and five for GARRETT's Θ. In the case of anisotropic exchange these numbers become higher.

Investigations below 1°K with a diluted copper potassium sulphate, in which 86.8% of the copper ions were replaced by magnesium, were made by GARRETT[5]. The most remarkable result was that at the lower temperatures T^* for the normal salt was lower than for the diluted sample, though the absolute temperatures

[1] J. RAYNE: Phys. Rev. 95, 1428 (1954).
[2] R. J. BENZIE and A. H. COOKE: Nature, Lond. 164, 837 (1949).
[3] C. G. B. GARRETT: Proc. Roy. Soc. Lond., Ser. A 203, 392 (1950).
[4] A. ABRAGAM and M. H. L. PRYCE: Proc. Roy. Soc. Lond., Ser. A 206, 164 (1951).
[5] C. G. B. GARRETT: Proc. Roy. Soc. Lond., Ser. A 203, 375 (1950).

were lower for the diluted salt as should be expected. The results for both salts are shown in Fig. 26. For the diluted salt GARRETT found $\Theta = 0.0048°$ K and $cT^2/R = 1.98 \times 10^{-4}$.

43. Copper sulphate. $CuSO_4, 5H_2O$. Molecular weight: 249.6; density: 2.284. Crystallographic investigations by BEEVERS and LIPSON[1] showed that there are two ions in the unit cell. Each of them is surrounded by six oxygen atoms. Four of these belong to water molecules, they are arranged in a square around the copper ion; the other two belong to SO_4-groups, they are located on the line perpendicular to the square at a slightly greater distance. The electric field acting on the copper ions then is of cubic symmetry with a tetragonal component. Due to this field the higher orbital levels are unoccupied, see Sect. 42. The angle between the tetragonal axes of the two copper ions in the unit cell is about 80° according to the crystallographic investigations, but paramagnetic resonance experiments[2] suggest that it is nearly 90°.

REEKIE[3] measured the susceptibility with a powder, and BENZIE and COOKE[4] with a single crystal. A CURIE-WEISS law was found with $\Theta = -0.6°$ K and an anisotropic CURIE constant. Paramagnetic resonance experiments by BAGGULEY and GRIFFITHS[2] gave for the splitting factor in the direction of the tetragonal axis: $g_{||} = 2.07$, and perpendicular to it: $g_{\perp} = 2.26$.

The first demagnetization experiments were made by ASHMEAD[5]. He measured the specific heat with gamma ray irradiation. Beside the rise at the lowest temperatures, similar to that of copper potassium sulphate, a maximum was found near 1° K which could not be accounted for theoretically. The presence of this maximum was confirmed by DUYCKAERTS[6], who made specific heat measurements in the liquid helium region.

More recently investigations were made by GEBALLE and GIAUQUE[7] in the region down to 0.25° K. It was found again that the salt does not obey a CURIE law, but it follows from the results that a CURIE-WEISS law is satisfied between 1 and 8° K with the Θ value of BENZIE and COOKE. Magnetic measurements were made in fields up to 8500 oersteds, see Sect. 52. The zero field temperatures reached in the experiments are given in Table 12. An auxiliary carbon thermometer was attached to the sample for caloric measurements.

The specific heat data are shown in Fig. 27. At the higher temperatures they are in good agreement with the results of DUYCKAERTS. The maximum occurs at 1.37° K, but a small extra peak was found at 0.75° K. There is a slight indication that this peak occurs also in ASHMEAD's curve. At the lowest temperatures the specific heat rises again, as was found in the results of ASHMEAD.

It turned out that the general course of the specific heat can be represented by formula (33.2), hence the specific heat curve for a salt with two levels; except that the values are a factor two too low, see the dotted line in Fig. 27. Consequently the entropy decrease is only $\frac{1}{2} R \ln 2$. An explanation suggested by GEBALLE and GIAUQUE is that the two ions in the unit cell give rise to the existence of two systems of ions with different environments. One of them might behave more or less like the ions in copper potassium sulphate, whereas the

[1] C. A. BEEVERS and H. LIPSON: Proc. Roy. Soc. Lond., Ser. A **146**, 570 (1934).
[2] D. M. S. BAGGULEY and J. H. E. GRIFFITHS: Proc. Roy. Soc. Lond., Ser. A **201**, 366 (1950).
[3] J. REEKIE: Proc. Roy. Soc. Lond., Ser. A **173**, 367 (1939).
[4] R. J. BENZIE and A. H. COOKE: Proc. Phys. Soc. Lond. A **64**, 124 (1951).
[5] J. ASHMEAD: Nature, Lond. **143**, 853 (1939).
[6] G. DUYCKAERTS: Bull. Soc. roy. Sci. Liège **10**, 284 (1941).
[7] T. H. GEBALLE and W. F. GIAUQUE: J. Amer. Chem. Soc. **74**, 3513 (1952).

other should have a level splitting of the order of 2 cm.$^{-1}$. The origin of this level splitting, however, is completely obscure. It follows from the paramagnetic resonance experiments that exchange interaction plays an important role in copper sulphate. The exchange integral between two like ions is different from that between unlike ions, but a quantitative explanation on this basis seems to be difficult[1].

No indication of relaxation or hysteresis heating in alternating magnetic fields has been found down to the lowest temperatures that were obtained.

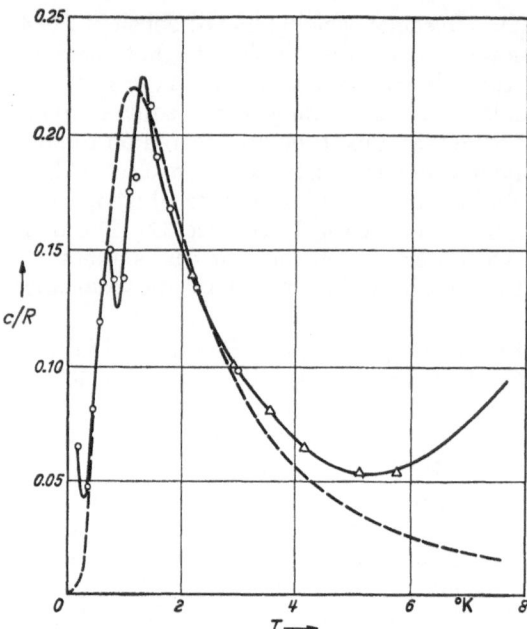

Fig. 27. Specific heat of copper sulphate (according to Geballe and Giauque). ○ Geballe and Giauque. △ Duyckaerts. The dotted line is the theoretical curve, see text.

Table 12. *Copper sulphate* (Geballe and Giauque).

T is derived from caloric measurements.

$\dfrac{S_T - S_{1°K}}{R}$	T °K
+ 0.0126	1.090
− 0.0218	0.849
− 0.0509	0.700
− 0.0781	0.564
− 0.113	0.286
− 0.128	0.222

44. Cobalt ammonium sulphate. $Co(NH_4)_2(SO_4)_2 6H_2O$. Molecular weight: 395.2; density: 1.902. Cobalt ammonium sulphate is a Tutton salt. At first sight, it might be expected to behave rather analogously to the chromium salts, the cobalt ion having three holes in the $3d$-shell instead of three electrons. This simple picture, however, proves to be wrong. The lowest state is not a quartet but a doublet, so that application of the Brillouin function with $S = \frac{3}{2}$ does not lead to correct results.

The ^4F ground state of the free cobalt ion is split by the cubic part of the electric field into a doublet and a triplet, of which the latter is lower. The triplet is split by the joint action of the tetragonal component and the spin-orbit coupling into three Kramers doublets, roughly 10^2 cm^{-1} apart. The influence of the higher doublets cannot be neglected, and down to hydrogen temperatures no Curie law is found. A Curie law is obeyed, however, at liquid helium temperatures, but the susceptibilities are highly anisotropic. Paramagnetic resonance experiments[2] gave in the direction of the tetragonal axis: $g_{||} = 6.45$, and perpendicular to it: $g_\perp = 3.06$, the angle α of Sect. 40 being $33°$.

The Curie constant in any direction of the crystal can be calculated from Eq. (29.9) with (40.1), inserting $J = \frac{1}{2}$ and the above data for $g_{||}$ and $g_\perp$. In Table 13 the values of $\Sigma g^2 l^2$ of Eq. (40.1) and the Curie constants are given for the K_1, K_2 and K_3 axes. The results are in good agreement with the experimental data.

[1] C. G. B. Garrett: Magnetic cooling, p. 68. Cambridge (Mass.) 1954.
[2] B. Bleaney and D. J. E. Ingram: Proc. Roy. Soc. Lond., Ser. A **208**, 143 (1951).

GARRETT[1] found for the ratio of the susceptibilities in the three directions: $1:0.291:0.603$. MALAKER[2] found for the CURIE constant in the K_2-direction: $C/R = 1.03 \times 10^{-8}$.

Table 13. *Anisotropy data for cobalt ammonium sulphate, according to* GARRETT.

The first three columns have been recalculated, using more recent paramagnetic resonance data of BLEANEY and INGRAM.

axis	$\Sigma g^2 l^2$	C/R	ratio	Θ	down to	$c T^2/R$
K_1	32.1	3.62×10^{-8}	1	$-0.005°$ K	$0.5°$ K	42×10^{-4}
K_2	9.36	1.06	0.292	-0.017	0.2	42.0
K_3	18.9	2.13	0.589	$+0.050$	0.25	43.6

Experiments below $1°$ K were performed both by GARRETT and by MALAKER. GARRETT made investigations in the directions of the K_1, K_2 and K_3 axes; MALAKER restricted himself to the K_2 axis, but he investigated also two diluted samples. The decrease of entropy in a magnetic field can be derived from Eq. (29.1) with (40.1). It follows from the values of $\Sigma g^2 l^2$ given above that a field applied parallel to the K_1 axis is much more effective than in the other directions.

GARRETT's values for T^* as a function of entropy for the three axes are given in Fig. 28. Also the absolute temperatures are shown. They were obtained from the measurement of the Ξ-parameter in an external field, see Sect. 14.

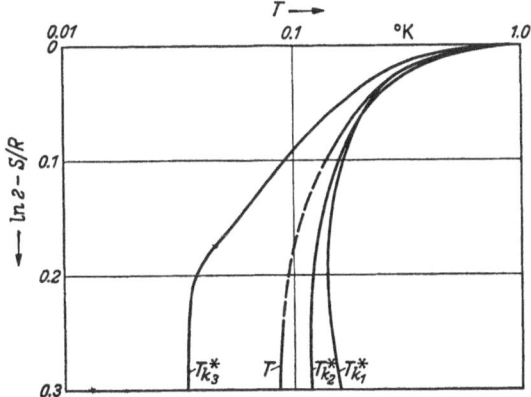

Fig. 28. Entropy versus temperature diagram for a single crystal of cobalt ammonium sulphate (according to GARRETT). T^* values are given for the three principal axes of magnetization. Absolute temperatures for $\ln 2 - S/R < 0.1$ are obtained from measurements of the Ξ-parameter. The values for $\ln 2 - S/R > 0.24$ are derived from caloric measurements with a.c. heating.

Both, GARRETT and MALAKER made measurements of the Ξ-parameter. MALAKER found for the K_2-axis: $\Xi = 2.43 \times 10^{-6}$ oersteds^{-2}, independent of temperature down to $0.125°$ K and this indicates that the salt follows a CURIE law, see Sect. 14. GARRETT, however, found a dependence on temperature, and since $\sqrt{\Xi}$ was a linear function of $1/T^*$ (see Fig. 29) his conclusion was that the salt follows a CURIE-WEISS law. It seems that further investigations on this point are required. GARRETT's Θ-values for the three orientations are given in Table 13, together with the temperatures at which the CURIE-WEISS laws break down. The ψ/χ correction for the saturation of the magnetic moment (see Sect. 14) was calculated from the expression for the magnetization curve at higher temperatures[3].

The coefficients of the $1/T^2$-term in the specific heat derived from GARRETT's measurements are given in Table 13. The results for the three orientations are in good mutual agreement. A plot of S versus $1/T^2$ gave slightly higher

[1] C. G. B. GARRETT: Proc. Roy. Soc. Lond., Ser. A **206**, 242 (1951).
[2] S. F. MALAKER: Phys. Rev. **84**, 133 (1951).
[3] Notice that in GARRETT's original paper the ψ/χ values of tables 1 and 3 have the wrong signs, and that in his Fig. 4 K_1 and K_3 are reversed.

values, viz. 43, 43.4 and 43.0×10^{-4} for the three orientations, but this difference is not significant. We may take $cT^2/R = 43 \times 10^{-4}$ as an average. MALAKER's value is 42.5×10^{-4}.

For the two diluted samples, investigated by MALAKER, the concentrations of the cobalt ions were 0.436 and 0.174. The specific heat coefficients were 27.45 and 20.61. A plot of cT^2/R against concentration showed that cT^2/R decreases linearly with decreasing concentration. Extrapolation to zero concentration gives $cT^2/R = 16.1 \times 10^{-4}$. This result is in good agreement with the value for the hyperfine splitting contribution obtained from paramagnetic resonance investigations. BLEANEY and INGRAM found $cT^2/R = 16.6 \times 10^{-4}$ (the only isotope, Co^{59}, has a nuclear spin $I = \frac{7}{2}$).

The contribution of magnetic dipole interaction to the specific heat of the undiluted salt must be calculated from Eqs. (32.2) and (32.3) with $Q = 17.6$.

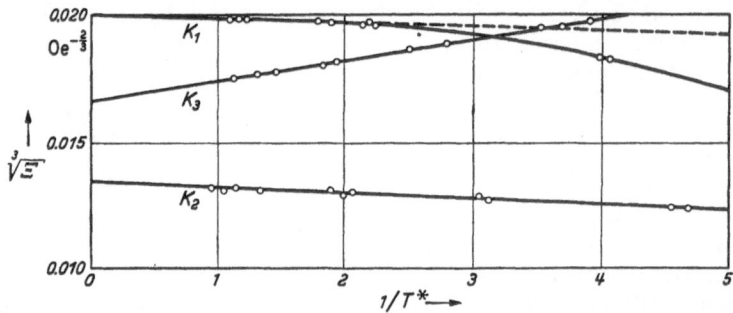

Fig. 29. Dependence of Ξ on T^* for a single crystal of cobalt ammonium sulphate for the three principal axes of magnetization (according to GARRETT). The CURIE-WEISS law is fulfilled as long as $\sqrt[3]{\Xi}$ is a linear function of $1/T^*$.

Since g is strongly anisotropic it seems plausible to replace g^2 by the average over the three principal axes, hence $\overline{g^2} = 20.1$. Under this assumption we obtain $cT^2/R = 21.8 \times 10^{-4}$. If we subtract the hyperfine splitting and dipole interactions from the total specific heat it is found that only 5×10^{-4} is left for exchange interaction.

45. Cobalt sulphate. $CoSO_4, 7H_2O$. Molecular weight: 281.1; density: 1.948. Experiments below $1°K$ were performed by FRITZ and GIAUQUE[1]. They used a powdered sample of ellipsoidal shape. The entropy calculations were based on the assumption that in this salt the cobalt ion behaves like the trivalent chromium ion (see Sect. 44), hence, that the lowest state is a spin quartet. In this case a BRILLOUIN function can be applied with $J = \frac{3}{2}$ and an isotropic splitting factor $g = 2$.

Paramagnetic resonance experiments by BLEANEY and INGRAM[2] showed that g is strongly anisotropic, and that the magnetic behaviour is, in general, more analogous to that of cobalt ammonium sulphate than to that of the chromium alums. For this reason it seems better to assume $J = \frac{1}{2}$ with rather high g-values. On this basis, however, it is difficult to give a quantitative interpretation of the investigations on a powdered sample with grains of random orientation.

FRITZ and GIAUQUE found that in the liquid helium region CURIE's law is not completely fulfilled, in such a way that χT decreases somewhat with decreasing temperature. Therefore, it is not well possible to define a T^*-scale. Below $1° K$ the decrease of χT becomes much steeper, but it should be kept in mind

---

[1] J. J. FRITZ and W. F. GIAUQUE: J. Amer. Chem. Soc. **71**, 2168 (1949).
[2] B. BLEANEY and D. J. E. INGRAM: Proc. Roy. Soc. Lond., Ser. A **208**, 143 (1951).

that all the absolute temperature determinations below 1° K are uncertain. They were determined from caloric measurements with a carbon thermometer-heater and the entropy variations were computed on the assumption of a fourfold degenerate level as described above. Table 14 gives some susceptibility values together with the initial fields and temperatures from which they were obtained. The value of χT at 1.145° K is 2.045 e.m.u. per mole, at 4.224° K it is 2.146 e.m.u. per mole.

Below 1° K an appreciable increase in the specific heat was found. BLEANEY and INGRAM suggested that it is due to hyperfine splitting or magnetic interactions; probably the latter, since cobalt sulphate is a relatively concentrated salt.

No indication was found of relaxation or hysteresis phenomena down to the lowest temperatures obtained in the experiments (probably of the order of 0.1° K).

Table 14. *Cobalt sulphate* (FRITZ and GIAUQUE).

H and T_i have the same meaning as in Table 1; χ is the molar susceptibility obtained after demagnetization.

H oersteds	T_i °K	χ e.m.u./mole
2400	2.119	3.26
6000	2.683	6.55
8768	2.569	11.40
8752	2.238	14.14
692	1.156	3.31
2384	1.163	6.58

46. Gadolinium sulphate. $Gd_2(SO_4)_3 \, 8H_2O$. Gram-ionic weight: 373.0; density: 3.010. The free gadolinium ion is in an ^8S state, so that the orbital magnetism needs not be taken into account. The eightfold degenerate spin level is split by a cubic field into two doublets with a quartet in between[1], the spacings being in the ratio 3:5. A field of lower symmetry may give a further splitting of the quartet. If these STARK splittings are relatively small as compared with 1° K the magnetic moment and entropy may be described by the BRILLOUIN function with $J = \frac{7}{2}$ and $g = 2$, Eqs. (29.1) and (29.2), and the validity of these formulae has been confirmed by experiment[2,3].

A theoretical discussion was given by HEBB and PURCELL[4] on the assumption of a cubic electric field only. According to the above the contribution of the STARK splitting to the partition function [see Eq. (30.1)] is either:

$$Z_s = 2(1 + 2e^{-3\delta/8T} + e^{-\delta/T}) \tag{46.1}$$

or

$$Z_s = 2(1 + 2e^{-5\delta/8T} + e^{-\delta/T}), \tag{46.2}$$

δ being the overall splitting. Since slightly better agreement is obtained with the first formula, we discuss this one only. The entropy and specific heat [see Eqs. (3.2), (3.3) and (9.4)] are:

$$\frac{S}{R} = \ln 2(1 + 2e^{-3\delta/8T} + e^{-\delta/T}) + \frac{\delta}{T}\,\frac{\frac{3}{4}e^{-3\delta/8T} + e^{-\delta/T}}{1 + 2e^{-3\delta/8T} + e^{-\delta/T}}$$
$$= \ln 8 - \frac{33}{512}\frac{\delta^2}{T^2} + \cdots \tag{46.3}$$

and

$$\frac{c}{R} = \frac{9}{32}\frac{\delta^2}{T^2}e^{-3\delta/8T}\frac{1 + \frac{32}{9}e^{-5\delta/8T} + \frac{25}{9}e^{-\delta/T}}{(1 + 2e^{-3\delta/8T} + e^{-\delta/T})^2}$$
$$= \frac{33}{256}\frac{\delta^2}{T^2}\left(1 - \frac{2}{11}\frac{\delta}{T} - 0.0608\frac{\delta^2}{T^2} + \cdots\right). \tag{46.4}$$

---

[1] J. H. VAN VLECK and W. G. PENNEY: Phil. Mag. **17**, 961 (1934).
[2] H. R. WOLTJER and H. KAMERLINGH ONNES: Leiden Commun. No. 167c; Proc. Roy. Acad. Amst. **32**, 772 (1932).
[3] W. E. HENRY: Phys. Rev. **88**, 559 (1952).
[4] M. H. HEBB and E. M. PURCELL: J. Chem. Phys. **5**, 338 (1937).

The dependence of susceptibility on temperature may be expressed [see Eqs. (30.5) and (30.6)] by:

$$\gamma(T) = \frac{\left(\frac{7}{27} + \frac{320}{81}\frac{T}{\delta}\right) + \left(\frac{130}{189} - \frac{6016}{2835}\frac{T}{\delta}\right)e^{-3\delta/8\,T} + \left(\frac{3}{7} - \frac{64}{35}\frac{T}{\delta}\right)e^{-\delta/T}}{1 + 2e^{-3\delta/8\,T} + e^{-\delta/T}}. \quad (46.5)$$

There are some odd isotopes of gadolinium, but the hyperfine splitting seems to be very small[1]. Neglecting exchange coupling, HEBB and PURCELL derived for the magnetic interaction [see Eq. (32.11)]:

$$
\left.
\begin{aligned}
\varOmega(T) = \frac{\frac{1}{12}Q}{(1 + 2e^{-3\delta/8\,T} + e^{-\delta/T})^2} &\left\{\left(\frac{49}{729} + \frac{2560}{729}\frac{T}{\delta}\right) + \left(\frac{1060}{729} + \frac{5696}{945}\frac{T}{\delta}\right)e^{-3\delta/8\,T} + \right. \\
&+ \left(\frac{16900}{35721} + \frac{520192}{178605}\frac{T}{\delta}\right)e^{-3\delta/4\,T} + \left(\frac{2}{9} - \frac{4096}{945}\frac{T}{\delta}\right)e^{-\delta/T} + \\
&+ \left.\left(\frac{548}{441} - \frac{5696}{945}\frac{T}{\delta}\right)e^{-11\delta/8\,T} + \left(\frac{9}{49} - \frac{512}{245}\frac{T}{\delta}\right)e^{-2\delta/T}\right\}
\end{aligned}
\right\} \quad (46.6)
$$

so that for high temperatures, the entropy and specific heat due to magnetic interactions become:

$$\frac{S}{R} = \ln 2 - \frac{1}{12}Q\frac{\tau^2}{T^2} \quad (46.7)$$

and

$$\frac{c}{R} = \frac{1}{6}Q\frac{\tau^2}{T^2}. \quad (46.8)$$

Table 15. *Gadolinium sulphate* (GIAUQUE and MacDOUGALL).

H, T_i, S/R and T^{*}_{Lor} have the same meaning as in Table 1.

Table 16. *Gadolinium sulphate (according to* VAN DIJK).

In this Table T^* is the value as derived from the experiment (hence not $T(^*)$), c^* is dQ/dT^*, c is dQ/dT and T is derived from the caloric measurements, see text. S is integrated from c/T.

H oersteds	T_i °K	$\dfrac{S_T - S_{1^\circ\mathrm{K}}}{R}$	T^{*}_{Lor}	T^*	c^*/R	T °K	c/R	S/R
1030	1.760	+0.051	1.434	1.500	0.121	1.500	0.121	2.016
1520	1.722	+0.032	1.215	1.400	0.135	1.400	0.134	2.007
1650	1.732	+0.022	1.148	1.302	0.151	1.300	0.148	1.996
1700	1.715	+0.022	1.162	1.205	0.173	1.200	0.166	1.984
1960	1.720	+0.010	1.019	1.110	0.199	1.100	0.188	1.968
2250	1.740	−0.005	0.990	1.016	0.230	1.000	0.216	1.949
2510	1.740	−0.023	0.882	0.922	0.270	0.900	0.252	1.923
2630	1.722	−0.033	0.843	0.829⁵	0.326	0.800	0.302	1.893
2450	1.495	−0.059	0.764	0.736	0.399	0.700	0.369	1.848
2480	1.497	−0.061	0.762	0.643	0.494	0.600	0.456	1.782
3050	1.708	−0.068	0.737	0.551	0.622	0.500	0.574	1.692
3250	1.715	−0.084	0.713	0.504⁵	0.706	0.450	0.650	1.626
3820	1.745	−0.128	0.625	0.458	0.803	0.400	0.745	1.546
4490	1.735	−0.195	0.543	0.411⁵	0.909	0.350	0.861	1.435
5210	1.720	−0.274	0.475	0.362⁵	0.997	0.300	1.01	1.294
6040	1.700	−0.368	0.410	0.342	1.011	0.280	1.06	1.223
6170	1.690	−0.386	0.402	0.320	1.007	0.260	1.11	1.143
7800	1.718	−0.544	0.327	0.309	0.992	0.250	1.13	1.098
7840	1.700	−0.557	0.322	0.297	0.962	0.240	1.13	1.052
8000	1.710	−0.568	0.319	0.285	0.924	0.230	1.15	0.997
7660	1.499	−0.650	0.287	0.272	0.884	0.220	1.18	0.947
				0.250	0.807	0.213	2.69	—

[1] B. BLEANEY, R. J. ELLIOTT, H. E. D. SCOVIL and R. S. TRENAM: Phil. Mag. **42**, 1062 (1951).

Hence, the coefficient of the $1/T^2$ term in the specific heat is equal to:

$$A = \frac{33}{256} \delta^2 + \frac{1}{6} Q \tau^2. \tag{46.9}$$

According to Eq. (32.3), we have $\tau = 0.189°$ K, and Q depends on the crystalline structure, which is unknown. VAN DIJK and AUER[1] applied the value for a simple cubic lattice, viz. 16.8; HEBB and PURCELL used 17.9 without further justification.

Experiments below $1°$ K were performed by GIAUQUE and MACDOUGALL[2], and by VAN DIJK[3]. Samples of cylindrical shape were used in both investigations, so that the evaluations of internal fields are somewhat doubtful. The data obtained by GIAUQUE and MAC-DOUGALL are shown in Table 15. They were discussed by HEBB and PURCELL with the help of the formulae given above. Fig. 30 shows the results. The circles are the T^*_{Lor}-values of Table 15. The dots give the absolute temperatures calculated with the ONSAGER formula (11.6), applying Eq. (46.5). The curve represents the relation between S and T calculated from Eqs. (46.3), (32.13) and (46.6); the value of the splitting parameter giving best agreement between the dots and the curve is $\delta = 1.4°$ K.

We may compare GIAUQUE and MACDOUGALL's δ with that obtained from specific heat investigations at higher temperatures, applying Eq. (46.9). It should be kept in mind, however, that, due to the high value of δ, the quantity $c T^2/R$

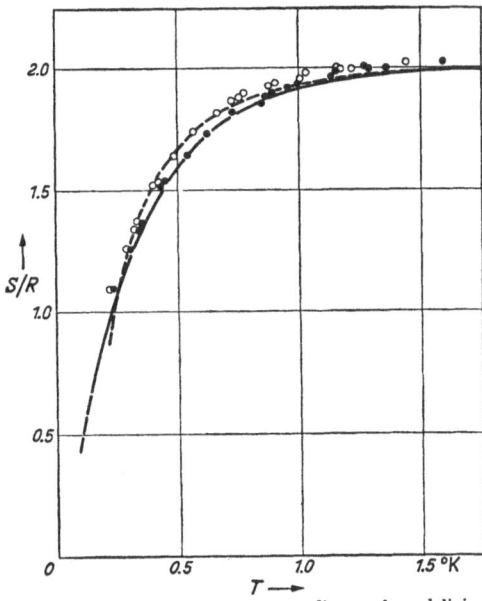

Fig. 30. Entropy versus temperature diagram for gadolinium sulphate (according to GIAUQUE and MACDOUGALL). The fully-drawn curve is the theoretical relation for $\delta = 1.4°$ K. O T^* LORENTZ. ● T ONSAGER. The dotted line is the S versus T curve of VAN DIJK.

in the liquid helium region is no longer a constant. This was confirmed by paramagnetic relaxation experiments of BENZIE and COOKE[4] and by caloric measurements of VAN DIJK and AUER[1], but it was overlooked in the calculations of some authors so that they arrived at too low a value of δ[5,6]. BENZIE and COOKE found $\delta = 1.35°$ K, and VAN DIJK and AUER: $\delta = 1.346°$ K. Paramagnetic relaxation experiments by BROER and GORTER[7] at liquid nitrogen temperatures, and by DE VRIJER, VOLGER and GORTER[8] at room temperature gave $\delta = 1.4°$ K.

[1] H. VAN DIJK and W. U. AUER: Leiden Commun. No. 267b; Physica, 's-Grav. 9, 785 (1942).

[2] W. F. GIAUQUE and D. P. MACDOUGALL: J. Amer. Chem. Soc. 57, 1175 (1935).

[3] H. VAN DIJK: Leiden Commun. No. 270a; Physica, 's-Grav. 12, 371 (1946).

[4] R. J. BENZIE and A. H. COOKE: Proc. Phys. Soc. Lond. A 63, 213 (1950).

[5] W. J. DE HAAS and F. K. DU PRÉ: Leiden Commun. No. 258a; Physica, 's-Grav. 6, 705 (1939).

[6] H. B. G. CASIMIR: Magnetism and very low temperatures, p. 79. Cambridge 1940.

[7] L. J. F. BROER and C. J. GORTER: Physica, 's-Grav. 10, 621 (1943).

[8] W. F. DE VRIJER, J. VOLGER and C. J. GORTER: Physica, 's-Grav. 11, 412 (1946).

It may be of some interest that VAN DIJK and AUER found for the lattice specific heat $c/R = 14.6 \times 10^{-5} T^3$; CLARK and KEESOM[1] gave the same value.

VAN DIJK[2] made caloric investigations with the help of a constantan heating wire and a phosphorbronze thermometer glued to the surface of the sample.

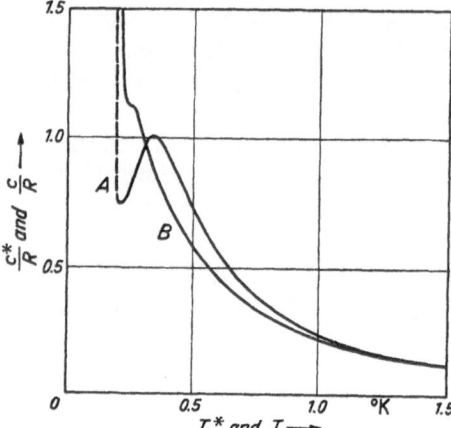

Fig. 31. Specific heat of gadolinium sulphate (according to VAN DIJK). Curve A: c^*/R versus T^*. Curve B: c/R versus T.

Experiments were first carried out in zero field below $1°$ K, and then in magnetic fields at higher temperatures. Absolute temperatures below $1°$ K may be derived from these data applying the first two terms of Eq. (10.1), which may be written:

$$\frac{T}{T_H} = \frac{c^*}{c_H} \frac{dT^*}{dT_H}. \qquad (46.10)$$

Here T_H and c_H refer to the experiments in the magnetic field at the high temperature, and T, T^* and c^* to the low temperature. The T^* (T_H) relation was determined separately from a large number of magnetizations and demagnetizations.

The resulting c^* versus T^*, and c versus T curves are given in Fig. 31 and Table 16. It is remarkable that the maximum and minimum of the c^* curve have practically vanished in c. This is due to the course of T^* with T, which is separately shown in Fig. 32. It was found that this curve is in better agreement with the ONSAGER and VAN VLECK relations than with the LORENTZ formula (see Sect. 7).

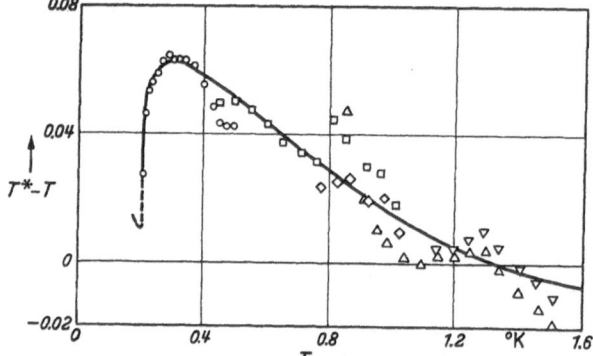

Fig. 32. Relation between magnetic and absolute temperature for gadolinium sulphate (according to VAN DIJK).

Table 16 also gives the entropy values as calculated from the specific heat data. It follows from Fig. 30 that there are some systematical differences from the results of GIAUQUE and MacDOUGALL.

47. Other gadolinium salts. Experiments were made by GIAUQUE and his coworkers on gadolinium compounds in which the gadolinium ions are further apart than in the sulphate. In general, it was found that the specific heat per gram-ion is the smaller, and the magnetic temperature obtained from a given initial field is the lower, the less the salt is concentrated.

Results obtained by MacDOUGALL and GIAUQUE[3] for gadolinium nitrobenzene sulphonate, $Gd(C_6H_4NO_2SO_3)_3 \cdot 7H_2O$, are shown in Table 17. They were discussed by HEBB and PURCELL[4] applying the formulae of Sect. 46. Since the salt is appreciably more diluted than gadolinium sulphate, the influence of the

[1] C. W. CLARK and W. H. KEESOM: Leiden Commun. No. 240a; Physica, 's-Grav. 2, 1075 (1935).
[2] H. VAN DIJK: Leiden Commun. No. 270a; Physica, s-Grav. 12, 371 (1946).
[3] D. P. MacDOUGALL and W. F. GIAUQUE: J. Amer. Chem. Soc. 58, 1032 (1936).
[4] M. H. HEBB and E. M. PURCELL: J. Chem. Phys. 5, 338 (1937).

magnetic interaction is much smaller, so that it could be entirely neglected in the calculations. Fig. 33 shows the theoretical entropy curve for the STARK splitting only, with $\delta = 1.4°$ K, hence the same value as for gadolinium sulphate, see Sect. 46. The circles represent the experimental T_{Lor}^* values and the dots are the absolute temperatures, also calculated with the LORENTZ approximation. There was no point in evaluating T_{Ons} and $T_{v. vl}$ since down to the lowest temperatures they are practically identical to T_{Lor}. This is due to the low value of the CURIE constant per cm³, see Sect. 11. Satisfactory agreement was found between the dots and the theoretical curve.

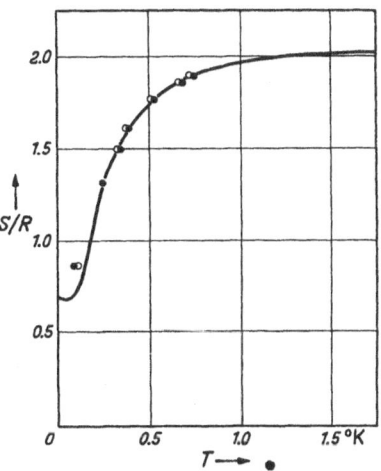

MACDOUGALL and GIAUQUE[1] made some investigations with gadolinium anthraquinone sulphonate, $Gd(C_{14}H_7SO_5)6$ or $7H_2O$. The results are shown in Table 18. The dilution is somewhat higher than for the nitrobenzene sulphonate and the magnetic temperatures are slightly lower. Afterwards the water of crystallization was removed from the sample by means of long evacuation with a diffusion pump and the experiments were repeated. Table 19 shows that now the final magnetic temperatures were even higher than in the case of gadolinium sulphate.

More detailed investigations were made with gadolinium phospho-molybdate,

$$Gd(PMo_{12}O_{40})30H_2O,$$

first by GIAUQUE and MACDOUGALL[2], later by GIAUQUE, STOUT, EGAN and CLARK[3]. This salt is still more diluted than the foregoing ones, and, moreover, it has a cubic structure[4].

Fig. 33. Entropy versus temperature diagram for gadolinium nitrobenzene sulfonate (according to MACDOUGALL and GIAUQUE). The curve is the theoretical STARK entropy for $\delta = 1.4°$ K. ○ T^* LORENTZ. ● T ONSAGER.

CURIE's law is satisfied in the liquid helium region. Entropy values were calculated using the BRILLOUIN function with $S = \frac{7}{2}$, $g = 2$, see Eq. (29.1). Some data are collected in Table 20.

Table 17. *Gadolinium nitrobenzene sulphonate* (MACDOUGALL and GIAUQUE).

H, T_i, S/R and T_{Lor}^* have the same meaning as in Table 1.

H oersteds	T_i °K	$\dfrac{S_T - S_{1°K}}{R}$	T_{Lor}^*
2840	1.56	-0.076	0.694
3215	1.56	-0.115	0.642
4050	1.56	-0.203	0.504
5630	1.62	-0.352	0.358
6610	1.62	-0.461	0.315
8040	1.56	-0.651	0.2435
8090	0.940	-1.194	0.0978

Table 18. *Gadolinium anthraquinone sulphonate* (MACDOUGALL and GIAUQUE).

H, T_i, S/R and T_{Lor}^* have the same meaning as in Table 1.

H oersteds	T_i °K	$\dfrac{S_T - S_{1°K}}{R}$	T_{Lor}^*
1830	1.62	$+0.024$	1.12
2936	1.64	-0.052	0.801
3836	1.62	-0.146	0.568
3650	1.60	-0.177	0.513
5615	1.52	-0.389	0.328
6480	1.52	-0.492	0.273
8280	1.62	-0.602	0.224
8000	1.54	-0.660	0.207

[1] D. P. MACDOUGALL and W. F. GIAUQUE: J. Amer. Chem. Soc. **58**, 1032 (1936).
[2] W. F. GIAUQUE and D. P. MACDOUGALL: J. Amer. Chem. Soc. **60**, 376 (1938).
[3] W. F. GIAUQUE, J. W. STOUT, C. J. EGAN and C. W. CLARK: J. Amer. Chem. Soc. **63**, 405 (1941).
[4] J. L. HOARD: Z. Kristallogr. A **84**, 217 (1933).

Table 19. *Anhydrous gadolinium anthraquinone sulphonate* (MacDOUGALL and GIAUQUE).

H, T_i, S/R and T_{Lor}^* have the same meaning as in Table 1.

H oersteds	T_i °K	$\dfrac{S_T - S_{1\,°K}}{R}$	T_{Lor}^*
2200	1.56	+0.010	1.03
2810	1.56	−0.040	0.878
3830	1.56	−0.143	0.692
5730	1.56	−0.358	0.476
6485	1.54	−0.448	0.423
8100	1.56	−0.619	0.340
8060	1.54	−0.621	0.339
8180	1.56	−0.629	0.337

Table 20. *Gadolinium phospho-molybdate* (MacDOUGALL and GIAUQUE).

H, T_i, S/R and T_{Lor}^* have the same meaning as in Table 1.

H oersteds	T_i °K	$\dfrac{S_T - S_{1\,°K}}{R}$	T_{Lor}^*
1220	1.410	−0.006	0.923
1630	1.422	−0.029	0.782
2285	1.426	−0.081	0.576
2840	1.430	−0.136	0.459
3820	1.429	−0.249	0.3290
5720	1.428	−0.486	0.1950
6580	1.423	−0.595	0.1625
8220	1.400	−0.796	0.1165

Caloric measurements were made by GIAUQUE and MacDOUGALL with the help of an induction heater. This is a ring of an alloy with temperature-independent resistance in which heat can be developed with the help of an alternating magnetic field. It has the advantage that no supply wires are needed, but the amount of heat is only known apart from a geometrical factor which must be derived from separate experiments in the liquid helium region. A correction must be applied for the permeability of the salt present in the proximity of the heater. A loop with a diameter of 2.2 cm was made of 0.08 mm. wire. It consisted of gold with 0.1% silver. The heat was generated with a 60 cycle field of 175 or 350 oersteds r.m.s.

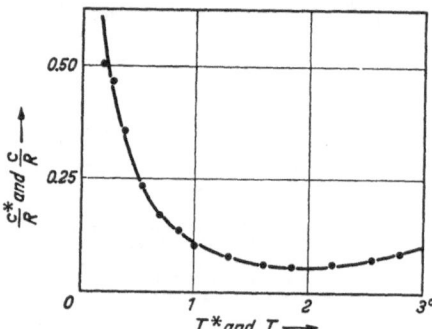

Fig. 34. Specific heat of gadolinium phospho-molybdate (according to GIAUQUE, STOUT, EGAN and CLARK). The curve represents c^* versus T^* as derived from the entropy data of table 20. The points give c versus T, calculated from the caloric experiments.

The experiments were repeated with a higher precision by GIAUQUE, STOUT, EGAN and CLARK using a carbon thermometer heater. The results are shown in Fig. 34. The curve represents the c^* data as calculated from the S versus T^* curve of Table 20. The points are the c versus T values, obtained from the caloric measurements applying:

$$T = \frac{dQ/dT^*}{dS/dT^*} . \tag{47.1}$$

see Sect. 10. Since the points coincide with the curve within the accuracy of the experiments, it was concluded that down to 0.2° K T is equal to T^*.

It should be emphasized that the fact that the T^* values reached from a given initial field are the lower, the more the salt is diluted, is not in agreement with the HEBB and PURCELL theory. For the nitrobenzene sulphonate the ions are already so far apart that the influence of the magnetic interaction can be neglected. If the STARK splitting is the same for all gadolinium compounds (as is suggested by the results obtained with the sulphate and the nitrobenzene sulphonate) the conclusion should be that further dilution cannot influence the final T^* any more. Maybe there is a small deviation from the cubic symmetry which is different for different compounds; maybe there is a hyperfine structure or some exchange.

48. Cerium magnesium nitrate. $Ce_2Mg_3(NO_3)_{12} \, 24H_2O$. Gram-ionic weight: 714.8; density: 2.0. The salt crystallizes in the trigonal system. The lattice structure is unknown, but the positions of the cerium ions were investigated by POWELL (unpublished). They are located in a simple rhomboëdral lattice with a unit cell of side 8.51 Å, and interaxial angle of 79.5°. The long diagonal of the rhomboëdra coincides with the trigonal axis of the crystal. There is only one ion in the unit cell.

The free cerium ion is in a ^2F state. It is split by the crystalline field into a number of KRAMERS doublets. The ground state is strongly anisotropic. Paramagnetic resonance experiments by COOKE, DUFFUS and WOLF[1] gave for the directions of the trigonal axis and perpendicular to it: $g_{||} = 0.25$, $g_\perp = 1.84$. No hyperfine structure was found, the stable cerium isotopes having no nuclear moments.

The results of the resonance experiments were confirmed by direct susceptibility measurements, performed by the same authors with a SUCKSMITH balance. Perpendicular to the symmetry axis they found:

$$\chi = \frac{C}{T} + \alpha \qquad (48.1)$$

with $C/R = 0.36 \times 10^{-8}$, leading to $g_\perp = 1.8$. Within the experimental accuracy this is in agreement with the value from the resonance measurements. α/R is 0.28×10^{-9}, it is due to the proximity of the higher KRAMERS doublets. The susceptibility in the direction of the symmetry axis was too small to be measured with reasonable precision. BECQUEREL, DE HAAS and VAN DEN HANDEL[2], who made measurements of the FARADAY effect in the direction of the symmetry axis, came to the same conclusion.

Paramagnetic relaxation experiments, also performed by COOKE, DUFFUS and WOLF, gave for the direction perpendicular to the trigonal axis: $cT^2/C = 1970$ which, for $g = 1.84$, leads to $cT^2/R = 7.5 \times 10^{-6}$. The value is somewhat higher than that obtained from adiabatic demagnetization experiments (see below), viz. 6.4×10^{-6}. The difference is probably due to the accuracy with which the measurements could be made. The specific heat of cerium magnesium nitrate is much smaller than that of any other salt quoted in the foregoing sections. The value for copper potassium sulphate, for instance, is 85 times larger, that for chromium potassium alum 2300 times. Down to 0.5° K the lattice specific heat is even larger than the magnetic contribution.

The theoretical specific heat, for magnetic dipole interaction only, was calculated by DANIELS[3]. Neglecting the contribution of $g_{||}$ he found $cT^2/R = 6.6 \times 10^{-6}$. Since there are no STARK splitting and hyperfine structure contributions (see above) the conclusion is that practically the whole specific heat is accounted for by magnetic dipole coupling, very little being left for the possibility of exchange interaction.

Demagnetization experiments were performed by DANIELS and ROBINSON[4]. An ellipsoid was used of axial ratio 6:1. It was cut from a single crystal with its long axis perpendicular to the crystallographic trigonal axis. Both the magnetizing and the measuring field were applied parallel to the long axis.

---

[1] A. H. COOKE, H. J. DUFFUS and W. P. WOLF: Phil. Mag. **44**, 623 (1953).
[2] J. BECQUEREL, W. J. DE HAAS and J. VAN DEN HANDEL: Leiden Commun. No. 218a; Proc. Kon. Acad. Wetensch. Amst. **34**, 1231 (1931).
[3] J. M. DANIELS: Proc. Phys. Soc. Lond. A **66**, 673 (1953).
[4] J. M. DANIELS and F. N. H. ROBINSON: Phil. Mag. **44**, 630 (1953).

Fig. 35 shows some of the results of the entropy determinations. It was found that from $S = R\ln 2$ down to $S = 0.43R$ (between $T^* = 1°$ and $T^* = 0.0035°$) the decrease of entropy is proportional to $1/T^{*2}$ satisfying the relation:

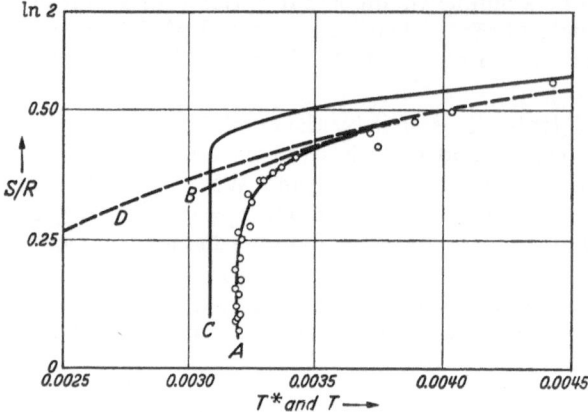

$$\left.\begin{array}{l} \ln 2 - S/R \\ \quad = 3.2 \times 10^{-6}/T^{*2}, \end{array}\right\} \quad (48.2)$$

leading to the high temperature specific heat value quoted above. Between $S = 0.28R$ and $S = 0.1R$ the magnetic temperature was constant, within the accuracy of the measurements, at $0.0032°$.

Absolute temperatures were calculated from the relation (34.1). Since above $0.01°$ both c^* and the entropy decrease proportionally to $1/T^{*2}$ it follows that $T = T^*$.

Fig. 35. Entropy versus temperature diagram for cerium magnesium nitrate (according to DANIELS and ROBINSON.) Curve A: Experimental S versus T^*-data. Curve B: Theoretical $1/T^2$-curve. Curve C: S versus absolute T, calculated by DANIELS and ROBINSON. Curve D: S versus absolute T, present calculations.

Below $S = 0.3R$ the quantity T^* is no longer suitable as a thermometric parameter. In this region the method was modified in the following way. Demagnetizations were performed from well-known entropy values and in each case the total amount of heat was determined, needed for

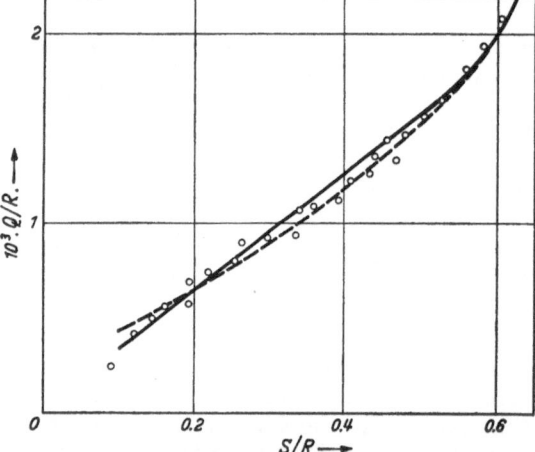

heating the sample to a temperature that could be measured with the T^* thermometer. From these data a diagram was made of the integrated heat supply (with an arbitrary zero point) as a function of the entropy. The absolute temperature is equal to the slope of the curve.

The results are shown in Fig. 36. From the curve drawn by DANIELS and ROBINSON it followed that between $S = 0.1R$ and $S = 0.4R$ the absolute temperature is constant at $0.00308°$ K. Above $0.006°$ K the difference between T and T^* is zero within the experimental accuracy. Some results are shown in Fig. 35 and in Table 21, column three.

Fig. 36. Heat content versus entropy diagram for cerium magnesium nitrate (according to DANIELS and ROBINSON). The absolute temperatures were derived from the slope of the full-drawn curve. We made new calculations, based on the dotted line.

In some respects the results are slightly unsatisfactory. First $c^* T^{*2}/R$ is constant over a wider range of temperatures than $c T^2/R$ (see Fig. 35); then, between $S = 0.28R$ and $S = 0.4R$, T is independent of the entropy but T^* is not. For this reason, we reconsidered the data of DANIELS and ROBINSON at the lowest temperatures. Neglecting the lowest point of Fig. 36, we drew a new curve through the experimental points (the dotted line). The temperatures derived from graphical differentiation are given (after being somewhat smoothed) in the last column of Table 21 and in Fig. 35, curve D. They look more reasonable than the values calculated by DANIELS and ROBINSON, but the main conclusion is probably that the precision of the method at the lower temperatures is unsufficient.

Apart from the difficulties at the lowest temperatures cerium magnesium nitrate has a number of attractive properties. From a theoretical point of view, it is the only substance that we have at the present time for which the magnetic properties depend entirely, or nearly entirely, on the magnetic dipole interaction and more detailed investigations at the lowest temperatures should be of considerable interest. (No remanence was found in a preliminary experiment made at Leiden.) From an experimental point of view it is important that very low temperatures can be obtained with reasonable field values; and that down to rather low temperatures T is equal to T^*. Moreover, due to the very high anisotropy, it is possible, after a demagnetization, to switch on a field in the direction of the trigonal axis without influencing the temperature very much. It should be realized, however, that the salt is not practicable for investigations in which a powdered sample or a compressed pill is required (for instance, if good thermal contact must be made with other materials under investigation). In this case considerable temperature differences may arise between the individual crystals which, at the lower temperatures, are not balanced in any reasonable length of time, see Sect. 19.

Table 21. *Cerium magnesium nitrate* (DANIELS and ROBINSON).

S/R and T^*_{Lor} have the same meaning as in Table 1. T_{DR} are the absolute temperatures as calculated from the caloric measurements by DANIELS and ROBINSON. T_{Rec} are the present recalculations from the same experimental data.

S/R	T^*_{Lor}	T_{DR} °K	T_{Rec} °K
0.600	0.00586	0.00586	0.00586
0.590	556	546	548
0.580	532	512	523
0.570	509	479	500
0.560	490	450	480
0.550	472	425	463
0.525	435	376	430
0.500	409	343	400
0.475	388	323	379
0.450	366	312	356
0.425	349	308	337
0.400	338	308	321
0.350	328	308	290
0.300	320	308	265
0.250	320	308	246
0.200	320	308	234
0.150	320	308	225
0.100	320	308	—

49. Cerium ethyl sulphate. $Ce(C_2H_5SO_4)_3\,9H_2O$. Molecular weight: 677.7; density unknown. The salt has hexagonal symmetry and the magnetic properties are symmetrical around the hexagonal axis. There is one ion in the unit cell.

The ^2F ground state is split into three KRAMERS doublets of which the highest lies at about 190° K. The lower two, however, are only 6.6° K apart, so that CURIE's law is not obeyed in the liquid helium region[1,2]. BOGLE, COOKE and WHITLEY[3] found for the g-values of the lowest $(J=\pm\tfrac{5}{2})$ doublet: $g_{||}=3.80$, $g_\perp=0.2$; and for the next higher $(J=\pm\tfrac{1}{2})$ one: $g_{||}=1.0$, $g_\perp=2.25$. Below 1° K only the lowest doublet is occupied.

The first demagnetization experiments were performed by DE HAAS and WIERSMA[4]. They found a steep rise in the specific heat near $T^*=0.1°$.

[1] R. A. FEREDAY and E. C. WIERSMA: Leiden Commun. No. 237a; Physica, 's-Grav. 2, 575 (1935).

[2] J. BECQUEREL, W. J. DE HAAS and J. VAN DEN HANDEL: Leiden Commun. No. 244e; Physica, 's-Grav. 5, 857 (1938).

[3] G. S. BOGLE, A. H. COOKE and S. WHITLEY: Proc. Phys. Soc. Lond. A 64, 931 (1951).

[4] W. J. DE HAAS and E. C. WIERSMA: Leiden Commun. No. 236b; Physica, 's-Grav. 2, 335 (1935).

Unpublished measurements of Cooke, Whitley and Wolf showed[1] that the specific heat at the higher temperatures obeys $c\,T^2/R = 11 \times 10^{-4}$. This is much larger than can be accounted for by magnetic dipole interaction. Daniels[2] calculated for this contribution: $c\,T^2/R = 1.9 \times 10^{-4}$. Finkelstein and Mencher[3] suggested the possibility that part of the specific heat is due to quadrupole-quadrupole coupling between the electronic charge distributions of different atoms.

III. The influence of magnetic fields.

50. Technical details. The magnetization curve of a substance describes the relation between its magnetic moment and the applied field. In general, magnetization curves are measured at constant temperature; in the region below 1° K however, it is easier to work at constant entropy.

An obvious method to determine a magnetic moment is to switch the field on or off suddenly and to measure the voltage over a coil surrounding the sample with the help of a ballistic galvanometer. The deflection, apart from a contribution due to the empty coils which can be accounted for, is proportional to the magnetic moment. This method was used by de Haas and Wiersma in their early demagnetization work at Leiden.

It is better to measure the voltage when a small variation is applied to the field. The quantity measured in this case is $(\partial M/\partial H)_S$ and the magnetization curve can be found by integration. Since now the voltage induced in the coil is much smaller, a more sensitive galvanometer can be used so that a higher precision is obtained in the final results.

The experiments can be carried out using an induction bridge as described in Sect. 25. The axis of the bridge coils must be parallel to that of the field solenoid and the variation of the field is achieved by reversing a small current in the primary coil of the bridge. The difficulty in these experiments is the coupling of the bridge coils with the windings of the magnet and its metal parts. It is nearly impossible, for instance, to make the measurements with an alternating current bridge. In this case, both χ' and χ'' (see Sect. 23) are noticeably altered by the presence of the magnet. But even in the case of ballistic measurements serious difficulties are encountered. The galvanometer shows double deflections as described in Sect. 24, and the interpretation of the results is difficult.

The problem was solved by Giauque, Fritz and Lyon[4] by mounting all the compensation coils of the mutual inductance bridge inside the cryostat so that they were influenced by the magnet in the same way as the sample coils, see Sect. 25. It is best to have the secondary coils of the bridge balanced with respect to the primary field and the magnet simultaneously. In this case, small fluctuations in the magnet current do not influence the galvanometer. This, however, involves some difficulties in the design of the coils and the solution of Giauque, Fritz and Lyon was, not to use the primary coil at all, but to produce the small field variation by shunting a suitable resistance across the magnet. The primary bridge coils are used for zero field measurements and calibration purposes only.

A different method was proposed by Casimir[5]. The magnetic field is applied perpendicular to the coils of the mutual inductance bridge. The field is produced

---

[1] C. J. Gorter et al: Progress in Low Temperature Physics, p. 239. Amsterdam 1955.
[2] J. M. Daniels: Proc. Phys. Soc. Lond. **66**, 673 (1953).
[3] R. Finkelstein and A. Mencher: J. Chem. Phys. **21**, 472 (1953).
[4] W. F. Giauque, J. J. Fritz and D. N. Lyon: J. Amer. Chem. Soc. **71**, 1657 (1949).
[5] H. B. G. Casimir: Magnetism and very low temperatures, p. 20. Cambridge 1940.

by an iron free HELMHOLTZ coil and the coupling between the mutual inductance bridge and the field coils is negligible; it is possible to make the measurements with alternating current.

The method is explained in Fig. 37. If the measuring field h is small as compared with the applied field H, so that it does not influence the degree of saturation of the salt's magnetic moment, we have:

$$\frac{m}{h_{\text{int}}} = \frac{M}{H_{\text{int}}}, \tag{50.1}$$

where m and M are the components of the magnetic moment in the directions of h and H. For the case of an ellipsoidal sample, Eq. (50.1) is equivalent to:

$$\frac{M}{H_{\text{ext}}} = \frac{m/h_{\text{ext}}}{1 + (\varepsilon_1 - \varepsilon_2)\left(\frac{m}{V}/h_{\text{ext}}\right)} \tag{50.2}$$

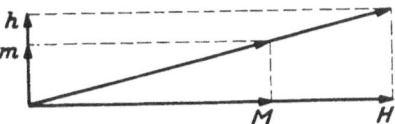

where ε_1 and ε_2 are the demagnetization coefficients of the ellipsoid in the H and h directions, see Sect. 7. Since m/h_{ext} follows from the bridge settings, M can be computed from the H_{ext} values. It should be emphasized that Eqs. (50.1) and (50.2) are only valid in the case of an isotropic sample.

Fig. 37. Measurement of magnetization curve in transverse fields.

For obvious reasons $(\partial M/\partial H)_S$ is often referred to as $\chi_{\parallel}$, whereas m/h_{ext} is denoted by $\chi_{\perp}$.

A method developed by HENRY[1] for the measurement of magnetic moments is to move the sample in a homogeneous field from one coil into another one, wound in the opposite direction. HENRY made very accurate experiments with this method in the liquid helium region, but it does not seem to be very suitable for work below 1° K. Another possibility is to move the coil up and down around the cryostat. In this case the accuracy is limited, due to the small fraction of the volume of the coil occupied by the salt, but the method has been applied occasionally in order to investigate the occurrence of remanence in a very direct way.

51. The alums. Measurements of magnetization curves at constant entropy in fields up to 500 oersteds were made at Leiden for both chromium methylamine alum[2] and chromium potassium alum[3]. Spherical single crystals were investigated; the applied field and the measuring field were mutually perpendicular (see Sect. 50), each being parallel to a cubic axis.

The values of the susceptibility M/H for both salts are given in Tables 22 and 23. From these data the quantity $(\partial M/\partial S)_H$ may be calculated as a function of H and S and then, applying Eq. (9.9), the temperature variation with field. The results are given in the same tables and in Figs. 38 and 39.

In small fields the variation of T is proportional to H^2, obeying:

$$\varDelta T = -\frac{1}{2}\left(\frac{\partial \chi}{\partial S}\right)_{H=0} H^2. \tag{51.1}$$

For larger fields the curves become more linear; they can be easily extrapolated to the initial fields and temperatures of the demagnetizations. At the higher

[1] W. E. HENRY: Phys. Rev. **88**, 559 (1952).
[2] J. A. BEUN, M. J. STEENLAND, D. DE KLERK and C. J. GORTER: Leiden Commun. No. 301a; Physica, 's-Grav. **21**, 767 (1955).
[3] J. A. BEUN, M. J. STEENLAND, D. DE KLERK and C. J. GORTER: Leiden Commun. No. 300a; Physica, 's-Grav. **21**, 651 (1955).

Table 22. *Chromium methylamine alum* (BEUN, STEENLAND, DE KLERK and GORTER).

Variation of susceptibility and temperature with magnetic field for a spherical single crystal. The applied field was perpendicular to the measuring field. The first number in each block gives the quantity M/H, the second one is the temperature difference from the zero field temperature.

S/R	$T_{H=0}$ °K	M/H (e.m.u. per mole) and ΔT (millidegrees absolute)								
		$H=0$	$H=40$	$H=80$	$H=130$	$H=170$	$H=260$	$H=340$	$H=430$	$H=540$
1.358	0.590	3.12	3.11	3.10	3.08	3.07	2.99	2.91	2.81	2.68
		0	0.57	2.27	5.98	10.20	22.32	38.61	61.74	96.62
1.228	0.241	7.28	7.26	7.23	7.17	7.11	6.92	6.70	6.43	6.08
		0	0.26	1.02	3.25	4.51	10.31	17.19	26.66	40.47
1.065	0.141	11.35	11.32	11.26	11.14	11.02	10.62	10.18	9.66	9.08
		0	0.21	0.86	2.21	3.78	8.66	14.28	21.89	32.50
0.790	0.072	18.24	18.11	17.87	17.46	17.02	15.84	14.69	13.65	12.43
		0	0.15	1.04	2.82	4.67	9.73	14.85	20.70	27.57
0.700	—	21.62	21.40	20.90	20.04	19.29	17.46	15.96	14.51	13.09
		0	0.55	2.07	4.93	7.55	13.94	19.41	25.35	31.41

Table 23. *Chromium potassium alum* (BEUN, STEENLAND, DE KLERK and GORTER).

Variation of susceptibility and temperature with magnetic field for a spherical single crystal. The applied field was perpendicular to the measuring field. The first number in each block gives the quantity M/H, the second one is the temperature difference from the zero field temperature.

S/R	$T_{H=0}$ °K	M/H (e.m.u. per mole) and ΔT (millidegrees absolute)								
		$H=0$	$H=44.7$	$H=89.4$	$H=126$	$H=173$	$H=251$	$H=339$	$H=426$	$H=500$
1.355	0.550	3.77	3.77	3.76	3.74	3.72	3.65	3.54	3.40	3.27
		0	0.80	2.90	5.82	10.88	22.56	40.14	61.54	83.04
1.323	0.379	5.22	5.21	5.18	5.13	5.05	4.90	4.74	4.60	4.46
		0	0.54	2.08	4.16	7.64	15.64	27.56	42.38	55.66
1.280	0.284	6.88	6.87	6.83	6.79	6.72	6.57	6.33	6.09	5.90
		0	0.38	1.54	2.94	5.82	11.72	20.72	31.92	43.76
1.166	0.185	10.21	10.18	10.12	10.03	9.88	9.59	9.20	8.78	8.36
		0	0.35	1.40	2.71	5.09	10.38	18.25	27.65	36.82
1.049	0.136	13.84	13.80	13.67	13.51	13.24	12.72	12.06	11.46	10.92
		0	0.42	1.52	2.85	5.28	10.52	17.80	26.30	33.91
0.905	0.089	18.36	18.26	18.01	17.65	17.13	16.08	14.97	13.99	13.09
		0	0.48	1.61	3.26	5.62	10.82	17.78	25.54	32.35
0.808	0.076	21.82	21.68	21.27	20.78	20.03	18.61	17.08	15.64	14.54
		0	0.43	1.54	2.97	5.54	10.85	17.93	25.42	31.85
0.662	—	26.54	26.27	25.59	24.79	23.62	21.62	19.58	17.88	16.51
		0	0.39	1.42	2.62	4.37	8.18	13.32	19.12	24.28
0.587	—	28.82	28.44	27.37	26.33	25.01	22.92	20.81	18.86	17.28
		0	0.31	1.13	2.13	3.74	7.21	12.02	17.39	21.95

entropies, the values of $(\partial\chi/\partial S)_{H=0}$ are rather large (plots of S versus χ will be discussed later, see Sects. 57 and 58); here the variation of T with field is large. For the case of potassium alum a region occurs between $S=1.3R$ and $S=0.7R$, where χ in zero field is practically a linear function of S. Here the variation of temperature with field is nearly independent of entropy as can be seen from Table 23. Such a region does not occur for the methylamine alum (compare Figs. 44 and 51). For the latter salt, the ΔT versus H curves show a downward curvature in fields of the order of 300 oersteds at entropies below $0.8R$. This phenomenon was not found for the potassium alum down to $S = 0.5R$. It is related with the proximity of the susceptibility maximum (see Sect. 28) which occurs for the methylamine alum at a much higher entropy than for the potassium alum, see Sect. 35.

If we plot M/H versus H^2 the parameter Ξ can be calculated from the initial slope, applying Eq. (14.2). From this, we may calculate the course of temperature with entropy for zero field as was pointed out in Sect. 14. The difficulty in the case of chromium alums is, however, that the correction ψ/χ is not very small

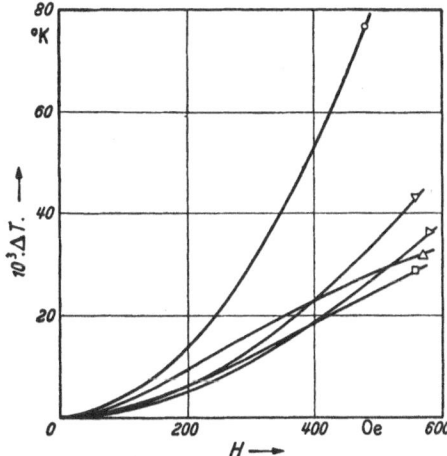

Fig. 38. Variation of temperature with magnetic field on adiabatic magnetization curves for a spherical single crystal of chromium methylamine alum (according to BEUN, STEENLAND, DE KLERK and GORTER).

O $S/R = 1.358$, $T_{H=0} = 0.590$.
▽ $S/R = 1.228$, $T_{H=0} = 0.241$.
▷ $S/R = 1.065$, $T_{H=0} = 0.141$.
□ $S/R = 0.790$, $T_{H=0} = 0.072$.
△ $S/R = 0.700$.

Fig. 39. Variation of temperature with field on adiabatic magnetization curves for a spherical single crystal of chromium potassium alum (according to BEUN, STEENLAND, DE KLERK and GORTER).

O $S/R = 1.355$, $T_{H=0} = 0.550$.
▽ $S/R = 1.323$, $T_{H=0} = 0.379$.
▷ $S/R = 1.26$, $T_{H=0} = 0.284$.
□ $S/R = 0.90$, $T_{H=0} = 0.089$.
△ $S/R = 0.60$.
◇ $S/R = 0.50$.

as compared with Ξ. If we derive ψ/χ from the BRILLOUIN function with $g = 2$, $J = 3/2$ [see Eq. (29.2)] we obtain:

$$\frac{\psi}{\chi} = -\frac{17}{15}\left(\frac{\mu_B}{k}\right)^2\frac{1}{T^2}\bullet \tag{51.2}$$

In the case of HUDSON's formula (see Sect. 33) this must be multiplied by a power series in $(\delta/T)^2$ which converges very slowly below $0.2°$ K.

BEUN, STEENLAND, DE KLERK and GORTER made some calculations for the case of chromium potassium alum. Down to $0.2°$ K there was no systematical difference between the dT^*/dT values, calculated from the Ξ data and those obtained from the HEBB and PURCELL formula (33.4). Due to the two graphical differentiations involved in the calculations, however, the scatter in the points obtained from the Ξ data was rather bad. For the results we refer to the original paper[1].

KURTI[2] published measurements of magnetization curves for iron ammonium alum. They were performed with the applied field parallel to the measuring

[1] J. A. BEUN, M. J. STEENLAND, D. DE KLERK and C. J. GORTER: Leiden Commun. No. 300a; Physica, 's-Grav. **21**, 651 (1955).
[2] N. KURTI: J. de Phys. **12**, 281 (1951).

field (see Sect. 50). The investigations were made up to 75 oersteds. Only the results obtained at the lowest entropies (below $S = 0.6\,R$) were discussed in some detail. They are described in Sect. 66.

52. Other salts. GEBALLE and GIAUQUE[1] made measurements of magnetization curves on copper sulphate. The applied field was parallel to the measuring field. The results are represented in Table 24. $(\partial M/\partial H)_S$ is the quantity measured in the experiment (see Sect. 50) and the M values were obtained by integration along the isentropics.

Table 24. *Copper sulphate* (GEBALLE and GIAUQUE).

Measurements with an ellipsoidal single crystal in magnetic fields. The first number in each block gives $(\partial M/\partial H)_S$ in e.m.u. per mole; the second number is the magnetic moment, also in e.m.u. per mole; the third number gives the internal energy, $(U_T - U_{1°K})/R$, in °K; the fourth number is the thermodynamic temperature T in °K.

$\dfrac{S_T - S_{10\,K}}{R}$	$\left(\dfrac{\partial M}{\partial H}\right)_S$, M, $\dfrac{U_T - U_{10\,K}}{R}$ and T				
	$H=0$	$H=1000$	$H=2400$	$H=6200$	$H=8500$
$+0.0126$	0.280	0.274	0.248	0.174	0.144
	0	278	643	1448	1811
	$+0.0132$	$+0.0116$	$+0.0038$	-0.0452	-0.091
	1.090	1.098	1.137	1.306	1.430
-0.0218	0.342	0.324	0.280	0.175	0.144
	0	334	737	1611	1972
	-0.0204	-0.0223	-0.0312	-0.087	-0.136
	0.849	0.862	0.918	1.138	1.259
-0.0509	0.418	0.383	0.304	0.175	0.144
	0	407	885	1759	2119
	-0.0422	-0.0447	-0.055	-0.118	-0.172
	0.700	0.718	0.769	1.016	1.140
-0.0781	0.554	0.449	0.305	0.170	0.140
	0	512	1030	1881	2232
	-0.059	-0.063	-0.076	-0.144	-0.201
	0.564	0.595	0.664	0.897	1.038
-0.113	0.930	0.529	0.280	0.140	0.124
	0	754	1273	2012	2326
	-0.075	-0.080	-0.098	-0.174	-0.235
	0.286	0.366	0.560	0.777	0.902
-0.128	1.332	0.540	0.265	0.140	0.124
	0	870	1384	2112	2425
	-0.079	-0.085	-0.106	-0.186	-0.249
	0.222	0.292	0.497	0.737	0.852

The procedure for the calculation of the absolute temperatures was the following. The differences in internal energy U (see Sect. 21) of the several isentropics at $H=0$ were derived from the caloric measurements described in Sect. 43. The variation of U along each isentropic could be calculated from the relation [cf. Eq. (8.1)]:

$$\Delta U = - \int_0^H M\, dH . \tag{52.1}$$

In this way the U values of Table 24 were calculated with an arbitrary zero point.

---

[1] T. H. GEBALLE and W. F. GIAUQUE: J. Amer. Chem. Soc. **74**, 3513 (1952).

Subsequently, the entropies of the first column were calculated; not from the magnetization curve at the initial temperature (the behaviour of the salt does not obey a BRILLOUIN function), but from the relation:

$$\Delta S = \int \frac{1}{T} dU, \quad (52.2)$$

which is valid at constant magnetic field [cf. Eq. (8.1)]. It was applied to the U values at the field of 8500 oersteds, where the absolute temperatures could still be measured with the help of a carbon thermometer (see Sect. 43). Finally, all the temperatures of Table 24 could be calculated applying:

$$T = \left(\frac{\partial U}{\partial S}\right)_H. \quad (52.3)$$

Table 25. *Cobalt sulphate* (FRITZ and GIAUQUE).

Adiabatic magnetization curves for a powdered ellipsoid. $(\partial M/\partial H)_S$ is given in e.m.u. per mole. H_i and T_i are the initial field and temperature of the demagnetization, H is the applied field.

$\frac{H_i}{T_i}=1280$		$\frac{H_i}{T_i}=1990$		$\frac{H_i}{T_i}=3200$	
H	$\left(\frac{\partial M}{\partial H}\right)_s$	H	$\left(\frac{\partial M}{\partial H}\right)_s$	H	$\left(\frac{\partial M}{\partial H}\right)_s$
0	2.92	0	6.21	0	11.47
800	0.81	800	1.40	776	2.17
1600	0.21	1600	0.365	1185	1.15
2400	0.18	2400	0.152	2373	0.231
6000	0.13	6000	0.036	5390	0.028

Measurements on cobaltous sulphate were made by FRITZ and GIAUQUE[1]. A powdered sample was used (see Sect. 45) and the applied field was parallel to the measuring field. Values of $(\partial M/\partial H)_S$ on three isentropics are given in Table 25. The zero field susceptibilities can be compared with those of Table 14.

Table 26. *Gadolinium phospho-molybdate* (GIAUQUE and MACDOUGALL, and GIAUQUE, STOUT EGAN and CLARK).

Adiabatic magnetization curves. The first number in each block gives the quantity $(\partial M/\partial H)_S$ in e.m.u. per mole; the second number is the magnetic moment in e.m.u. per mole, the third number is the temperature difference from the zero-field temperature.

$T_{H=0}$		$H=0$	$H=50$	$H=100$	$H=250$	$H=500$	$H=1000$	$H=2000$	$H=4000$	$H=8000$
					$(\partial M/\partial H)_S$, M and ΔT (millidegrees)					
0.798	$H_i=1633$	9.81	9.78	9.72	8.84	7.03	3.95	1.42	0.48	0.38
	$T_i=1.429$	0	490	978	2375	4380	7020	9400	10980	12460
		0		0.94	3.8	19	80	289	—	—
0.664	$H_i=2040$	11.79	11.76	11.65	10.76	8.09	4.44	1.52	0.42	0.31
	$T_i=1.419$	0	589	1174	2880	5240	8240	10860	12410	13550
		0		0.94	3.8	20	82	265	738	—
0.487	$H_i=2760$	16.06	15.95	15.45	13.83	10.10	5.25	1.73	0.34	0.19
	$T_i=1.430$	0	801	1587	3800	6820	10370	13460	15120	15860
		0		0.95	3.9	21	78	234	620	—
0.339	$H_i=3750$	23.09	22.95	22.20	18.67	12.60	6.07	1.99	0.34	0.08
	$T_i=1.423$	0	1152	2282	5312	9222	13460	17010	18860	19390
		0		0.96	3.9	21	72	205	504	—
0.202	$H_i=5660$	38.30	37.90	35.10	25.84	15.00	6.52	2.32	0.46	0.02
	$T_i=1.451$	0	1909	3744	8320	13280	18160	22080	24340	24890
		0		1.04	4.1	21	57	169	393	—
0.172	$H_i=6420$	45.60	44.00	40.00	28.25	15.42	6.55	2.38	0.53	0.02
	$T_i=1.444$	0	2247	4350	9480	14720	19680	23640	26100	26690
		0		1.14	5.0	23	57	144	343	—
0.126	$H_i=8000$	62.20	60.00	51.60	31.20	15.72	6.52	2.11	0.49	0.10
	$T_i=1.433$	0	3081	5881	11940	17470	22500	26250	28330	29250
		0		1.42	5.6	24	58	132	299	—

[1] J. J. FRITZ and W. F. GIAUQUE: J. Amer. Chem. Soc. 71, 2168 (1949).

We shall not give the absolute temperature values as calculated by FRITZ and GIAUQUE, since they were based on wrong assumptions concerning the entropy of the salt, see Sect. 45.

Gadolinium phospho-molybdate was investigated by GIAUQUE and MAC-DOUGALL[1], and later by GIAUQUE, STOUT, EGAN and CLARK[2]. The results are given in Table 26. The M values were calculated from the measured $(\partial M/\partial H)_S$ data. The zero field temperatures were derived from caloric experiments (see Sect. 47), the temperatures in fields from the magnetic data.

D. Magnetic investigations at the lowest temperatures.

I. Cooperative effects.

53. Introduction. As was stated in Sect. 28 most paramagnetic salts used for the demagnetization process show a maximum in the susceptibility; below this maximum the magnetic properties undergo radical alterations. The present chapter deals with the phenomena occurring in this region.

In general the situation is much more complicated and less surveyable than at the higher temperatures. A theoretical interpretation of the phenomena is far from complete. Moreover the quantitative results obtained in different experiments may be different by an order of magnitude and it even happens that the data found with the same sample in the same apparatus are noticeably different on subsequent helium days.

The general course of the results may be described in the following way.

Near and below the temperature of the maximum, the susceptibility as measured with a ballistic inductance bridge (denoted by χ, whereas χ' and χ'' are the real and imaginary components of the dynamic susceptibility, see Sect. 23) depends on the value of the measuring field. Measurements with an a.c. bridge show that χ'' becomes important in this region. It has already a noticeable value somewhat above the temperature of the maximum of χ, but a steep rise takes place near this maximum. Here χ' is markedly lower than the ballistic susceptibility.

The value of χ' decreases with increasing frequency, whereas χ'' increases with increasing frequency. This behaviour suggests the occurrence of a relaxation effect. Usually, however, it is impossible to describe the curves quantitatively with one relaxation time and the experimental values are not in agreement with the extrapolation of the spin-spin or spin-lattice relaxations found in the liquid helium region.

The a.c. experiments suggest relaxation times of the order of 10^{-3} sec., but beside these there are also much longer times involved, which influence the ballistic measurements, giving rise to double deflections as described in Sect. 24.

For several salts χ'' increases with decreasing temperature, but in a few cases χ'' shows also a maximum at a temperature somewhat below that of the maximum of χ'. Both χ' and χ'' depend on the value of the measuring field. In most cases χ'' is very small as compared with χ', only a few percents at the lowest temperatures. For manganese ammonium sulphate, however, χ''/χ' reaches an appreciable value, viz. 0.4.

If the susceptibility is defined as M/H_{ext} (see Sect. 7) it is obvious that it depends on the shape of the sample. In some cases, however, the results obtained with ellipsoids of different excentricities do not give the same values for M/H_{int}.

---

[1] W. F. GIAUQUE and D. P. MACDOUGALL: J. Amer. Chem. Soc. **60**, 376 (1938).
[2] W. F. GIAUQUE, J. W. STOUT, C. J. EGAN and C. W. CLARK: J. Amer. Chem. Soc. **63**, 405 (1941).

The susceptibilities are strongly influenced by external magnetic fields, but the curves for different salts are widely different. In the case of chromium potassium alum, for instance, at the lowest temperatures, the susceptibility, as measured for a sphere, decreases to about half its value in a field smaller than 50 oersteds. In the case of chromium methylamine alum, however, after a slight decrease, a steep increase is found, followed by a pronounced maximum.

If lines of constant magnetic field are drawn in a susceptibility versus entropy diagram the low field curves show a maximum in χ. The maximum moves to lower entropies with increasing field strength, and for quite moderate fields it vanishes below the region accessible to the experiments. (In the case of the chromium alums it was found that a second maximum occurs at higher fields, but the origin of this maximum is completely obscure and in the present section we leave it out of the discussion.)

Below the locus of the maxima the quantity $(\partial M/\partial S)_H$ is positive so that, according to Eq. (9.7), the temperature on an isentropic magnetization curve decreases with increasing field. On the locus of the maxima, T passes through a minimum. If the field is increased to the initial value of the demagnetization, the temperature must go up to the initial temperature.

It can easily be shown that, if T on an isentropic shows a minimum, S on an isothermal exhibits a maximum. In a T versus H plot, lines of constant entropy and magnetic moment both show minima.

In general it is found that the temperature drop on an isentropic in low fields is small, so that effectively there is not much difference between the isothermal and the adiabatic magnetization curves in this region.

Hysteresis effects have been found at the lower temperatures in several salts. The shape of a hysteresis loop can be measured by switching a field on and off in a number of steps in both directions observing the ballistic galvanometer deflections, see Sect. 23. If we are only interested in the remanent magnetic moment (e.g. as a thermometric parameter, see Sect. 11) a very simple loop of only four deflections is sufficient. During such experiments it was found that the value of the remanent moment may depend somewhat on the number of steps in which the loop is passed through.

The general tendency is that the remanent moment increases with increasing maximum field of the loop and with decreasing temperature. There are, however, exceptions from this rule. In the case of chromium potassium alum the remanent moment passes through a maximum when the field is increased; in the case of chromium methylamine alum the remanent moment shows a maximum with falling temperature.

If the hysteresis loops are plotted as a function of H_{ext} they are very narrow and the remanent moments are exceedingly small. If, however, the loops are plotted against H_{int} (see Sect. 7) they show a more familiar shape. The coërcitive fields are very small under all circumstances.

The hysteresis effects start at a well defined temperature T_c, which is slightly higher than the temperature of the susceptibility maximum. If a magnetic field is applied parallel or perpendicular to the small measuring field, the hysteresis phenomena decrease rapidly and they vanish already in fields of some tens of oersteds.

A transition curve where the hysteresis effects disappear can be constructed in a T versus H diagram. This curve does not coincide with the locus of the maxima of the susceptibility mentioned above, the region of the hysteresis effects being somewhat wider. The two curves, however, are rather close together and the difference is neglected in most theoretical work.

The occurrence of a χ'' in the hysteresis region must be partly due to the hysteresis effects themselves. A proof that also relaxation plays still an important role follows from the fact that χ'' is not frequency independent, as it should be in the case of pure hysteresis losses. But also the relaxation effects decrease rapidly outside the hysteresis region. This follows both from the smallness of χ'' and from the absence of double deflections in ballistic measurements somewhat outside the region (see Sect. 24).

The specific heat of all paramagnetic salts shows a steep increase in the neighbourhood of the susceptibility maximum.

In the following sections we give first a survey of theoretical work dealing with the phenomena described here; after this we discuss in some detail the experimental results obtained with various salts. We will present the data in the form of graphs and not in tables (contrary to what we did in Chap. C), since we believe that at the present time most of the data are of qualitative interest only.

54. The theoretical problem. The phenomena mentioned in the foregoing section must be due to interactions between the magnetic ions giving rise to cooperative effects. The only way to come to a satisfactory theoretical description is to give a very rigorous discussion of both the magnetic dipole and exchange interactions. In general, the dipole interaction presents more difficulties than the exchange since the forces are of long range.

No complete theoretical picture is available at the present time, but there are two methods to come to approximate solutions. One is to derive formulae valid at high temperatures and to extrapolate them to the lower region. This is the method of Sect. 32. It is hardly probable, however, that the formulae obtained in this way will keep their validity below T_c (see Sect. 53). The other method starts from the other end. It consists of finding the configuration of the magnetic ions with the lowest free energy at absolute zero (as a function of an applied magnetic field) and then introduce the influence of the temperature as a perturbation. It is possible, however, that formulae obtained in this way are only valid at temperatures low as compared with T_c, hence for each salt at temperatures essentially below the region that can be obtained with it by the magnetic cooling method (see Sect. 1).

Let us first investigate whether the formulae discussed in Sect. 32 lead to the occurrence of a transition temperature. The LORENTZ formula [Eq. (7.15)] predicts a CURIE point:

$$T_c = (\tfrac{4}{3}\pi - \varepsilon)\, C/V, \tag{54.1}$$

so that the value depends on the axial ratio of the sample, and no cooperative effects can occur in the case of a sphere. This is in distinct disagreement with experimental evidence. ONSAGER's formula [Eq. (7.20)] does not predict a CURIE point at all. VAN VLECK's relation (32.5), if the term with η and all higher terms are neglected, gives cooperative effects for a sphere in the case of non-vanishing exchange interaction only. Since, however, it must be expected that T_c is of the same order of magnitude as τ, so that the denominator of Eq. (32.5) converges only very slowly or not at all, it is not justified to derive conclusions from the first terms of VAN VLECK's formula alone.

It follows that the theories of Sect. 32 do not give satisfactory results for the region near and below T_c. Before going into the details of other interaction theories we give a discussion on the basis of the WEISS molecular field.

55. The molecular field theory of antiferromagnetism. WEISS[1,2] gave a phenomenological theory of the interactions in a magnetic substance, as long ago as 1907. It was based on the assumption that the interactions can be accounted for with the help of a virtual magnetic field at the positions of the magnetic ions, proportional to the magnetization of the substance per unit volume (see Sect. 7).

Though, at first sight, this assumption is rather crude it has yielded some remarkably successful results. If only the magnetic dipole interaction is taken into account the theory leads in first approximation to the LORENTZ formula of Sect. 7. HEISENBERG[3] showed that, if the WEISS field is a consequence of the exchange interaction between neighbouring ions, it leads to an explanation of ferromagnetism as it occurs at normal temperatures.

NÉEL[4-6], and later BITTER[7] and VAN VLECK[8] investigated the consequence of accepting a negative ratio between the WEISS field and the magnetization. The system of magnetic ions was split into two sublattices, A and B, in such a way that each ion of A is surrounded by ions of B only, and vice versa. It was assumed that the ions of each sublattice experience a WEISS field proportional to the magnetization of the other sublattice, but in the opposite sense. Under these conditions it is found that a transition temperature T_c occurs, above which the salt is essentially paramagnetic, following a CURIE-WEISS law:

$$\chi = \frac{C}{T - \Theta}, \tag{55.1}$$

but below this temperature the sublattices have spontaneous magnetizations in opposite directions. This configuration of the magnetic ions is denoted as "antiferromagnetic".

The direction of spontaneous magnetization of the two sublattices is determined by the crystalline anisotropy. The course of susceptibility with temperature depends on the orientation of the measuring field. If the field is parallel to the direction of the spontaneous magnetization, the susceptibility below the transition point decreases to zero with falling temperature, so that χ shows a maximum at T_c. If the field is perpendicular to the spontaneous magnetization, the susceptibility below T_c is independent of temperature. In the case of a powdered sample the susceptibility at absolute zero is two thirds of that at T_c.

The influence of large magnetic fields was studied by GARRETT[9], NAGAMIYA[10], YOSIDA[11], GORTER and HAANTJES[12], and Mrs. VAN PESKI and GORTER[13]. For a field perpendicular to the direction of spontaneous magnetization the susceptibility is independent of the field strength. If it is applied parallel to the spontaneous magnetization, the susceptibility increases with increasing field strength,

[1] P. WEISS: J. de Phys. **4**, 661 (1907).
[2] P. WEISS: Ann. Phys., Paris **17**, 97 (1932).
[3] W. HEISENBERG: Phys. Z. **49**, 619 (1928).
[4] L. NÉEL: Ann. Phys., Paris **18**, 5 (1932).
[5] L. NÉEL: Ann. Phys., Paris **5**, 232 (1936).
[6] L. NÉEL: Ann. Phys., Paris **3**, 137 (1948).
[7] F. BITTER: Phys. Rev. **54**, 79 (1938).
[8] J. H. VAN VLECK: J. Chem. Phys. **9**, 85 (1941).
[9] C. G. B. GARRETT: J. Chem. Phys. **19**, 1154 (1951).
[10] T. NAGAMIYA: Progr. Theor. Phys. **6**, 342 (1951).
[11] K. YOSIDA: Progr. Theor. Phys. **6**, 691 (1951).
[12] C. J. GORTER and J. HAANTJES: Leiden Commun. Suppl. No. 104b; Physica, 's-Grav. **18**, 285 (1952).
[13] T. VAN PESKI-TINBERGEN and C. J. GORTER: Leiden Commun. Suppl. No. 109a; Physica, 's-Grav. **20**, 592 (1954).

but an anomaly is found at a field H_c, where the magnetizations of the two sublattices orient themselves perpendicular to the field. If the field is increased further, another anomaly is encountered at a field H_g, where a transition takes place from the antiparallel configuration perpendicular to the field to a parallel orientation in the direction of the field.

The relation between H_g and T plotted in the H, T-plane is called the "critical field curve". The shape is somewhat similar to the transition curve of a superconductor, see Fig. 40. GARRETT derived for the relation between T_c (the intersection with the T-axis) and H_g^0 (the intersection with the H-axis):

$$kT_c = \mu H_g^0, \tag{55.2}$$

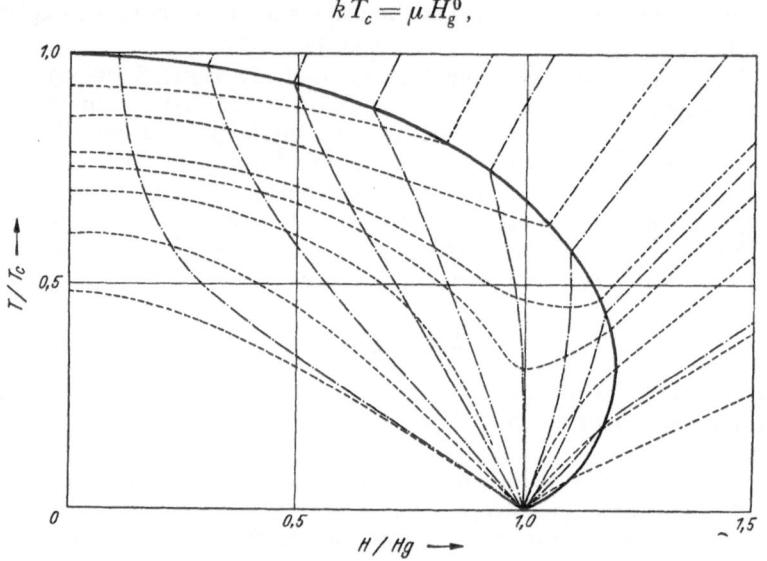

Fig. 40. Temperature versus magnetic field diagram for an antiferromagnetic crystal (according to GARRETT). —·—·— Lines of constant magnetic moment. ——— Lines of constant entropy.

where μ is the magnetic moment of the dipoles. The more rigorous, strictly threedimensional, theory of GORTER and HAANTJES leads to:

$$kT_c = 2\mu H_g^0. \tag{55.3}$$

This was derived for a lattice of rhombic symmetry.

The case of a crystal with cubic symmetry was discussed by Mrs. VAN PESKI and GORTER. Only the behaviour at absolute zero was discussed, and the influence of a field parallel to a cubic axis. It was assumed that the preferred orientation for the spontaneous magnetization of the two sublattices is parallel to a cubic axis.

If a field is applied parallel to one of the cubic axes, the sublattices orient themselves immediately parallel to one of the other cubic axes, so that the first transition field, H_c, does not occur, only H_g being found. Under these conditions the choice of the cubic axis for the zero field orientation depends on the direction in which a field has been applied previously, and in practical cases this may lead to hysteresis effects in small fields.

Let us suppose that the field is applied in the x-direction and that the zero field orientation of the sublattices is parallel to the y-axis. Then the susceptibility as measured in the x-direction is the quantity $\chi_\parallel$ introduced in Sect. 50.

For $\chi_\perp$ there are two possibilities; it can be measured parallel to the zero field orientation of the sublattices or perpendicular to it. These two possibilities may be indicated by $\chi_\perp (y)$ and $\chi_\perp (z)$. The difference vanishes for fields larger than H_g where both sublattices are parallel to the x-axis.

Three expressions could be introduced into the HAMILTONian for the cubic symmetry. They lead to three solutions for the χ versus H diagram at absolute

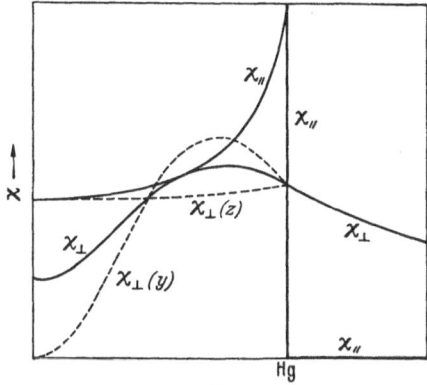

Fig. 41. Susceptibilities $\chi_{||}$ and $\chi_\perp$ for an antiferromagnetic crystal with cubic symmetry at absolute zero (according to Mrs. VAN PESKI and GORTER). The field is applied parallel to a cubic axis. For further description see text.

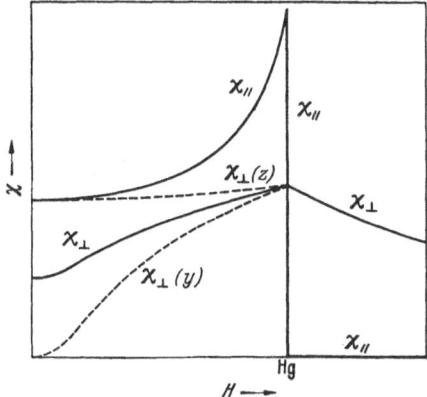

Fig. 42. Susceptibilities $\chi_{||}$ and $\chi_\perp$ for an antiferromagnetic crystal with cubic symmetry at absolute zero (according to Mrs. VAN PESKI and GORTER). The field is applied parallel to a cubic axis. For further description see text.

zero, represented in Figs. 41, 42 and 43. The $\chi_\perp$ curve for fields below H_g is the average of $\chi_\perp (y)$ and $\chi_\perp (z)$. This is what is probably measured in an actual experiment.

The case of Fig. 41 represents a preference of the two spin systems for the cubic axes independently of one another. Fig. 42 is concerned with an anisotropic interaction between the two sublattices. The curve for $\chi_{||}$ is the same as in Fig. 41, but $\chi_\perp$ is essentially different. The solution of Fig. 43 is somewhat of a mixture of the two foregoing cases. H_g proves to be larger, $\chi_{||}$ is constant below H_g and zero above it. The curves for $\chi_\perp$ are somewhat similar to those of Fig. 42.

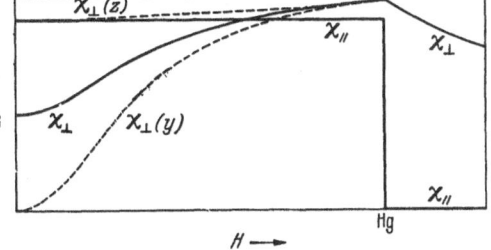

Fig. 43. Susceptibilities $\chi_{||}$ and $\chi_\perp$ for an antiferromagnetic crystal with cubic symmetry at absolute zero (according to Mrs. VAN PESKI and GORTER). The field is applied parallel to a cubic axis. For further description see text.

ANDERSON[1,2] discussed the possibility that an antiferromagnetic crystal should consist of several pairs of antiparallel sublattices of different orientations, taking into account next-to-nearest-neighbour interaction. In this case it is no more true that the susceptibility of a powder at absolute zero is two thirds of that at the transition temperature.

If only two antiparallel sublattices occur with interactions between ions of different sublattices alone the Θ derived from measurements in the paramagnetic region [Eq. (55.1)] is related to T_c by:

$$\Theta = -T_c. \qquad (55.4)$$

[1] P. W. ANDERSON: Phys. Rev. **79**, 350 (1950).
[2] P. W. ANDERSON: Phys. Rev. **79**, 705 (1950).

If also interactions between the ions of one sublattice are introduced this is no more true and Θ may even become positive[1].

The above considerations indicate that almost any susceptibility curve can be explained if the proper theoretical assumptions are made. Though this is a somewhat unsatisfactory state of affairs, it may be in agreement with the fact that fairly large differences are found in the results obtained with different salts.

NÉEL[2] introduced the hypothesis that antiferromagnetics may be split into domains, like the WEISS domains of a ferromagnetic. Each domain consists of two antiparallel sublattices, the orientations of neighbouring domains being different. Since the antiferromagnetic domains have a very small net magnetic moment, their shapes are much more irregular than in the case of ferromagnetic domains and there is very little direct correlation between the orientations in neighbouring domains. If a field is applied, the domains tend to orient themselves perpendicular to the field and, apart from this, the walls may undergo a reversible or irreversible shift.

A direct experimental proof of the occurrence of antiferromagnetic domains is difficult. The complication is encountered, in the theoretical interpretation of the measurements, that one has to distinguish between the magnetic properties of the sample as a whole and those of each domain, separately.

It is very difficult to give a satisfactory explanation of the hysteresis effects in an antiferromagnetic. It is well possible, since the remanent moments are very small, that they are only secondary effects caused, for instance, by impurities in the crystal. If a domain structure occurs, however, hysteresis may also be due to irreversible phenomena in the walls[3].

56. Interaction theories. The first effort to find the state of lowest energy of a magnetic crystal at abolute zero was made by SAUER[4]. He considered a cube of 125 ions and calculated the field at the central lattice point, for various possible arrangements of the ions.

LUTTINGER and TISZA[5] made calculations for cubic lattices with magnetic dipole interaction, also at absolute zero. The free energy of the crystals could be computed for various configurations of the dipoles. The danger of the method is apparently that the configuration with the really lowest energy may be overlooked in the calculations.

For a face-centred cubic lattice (the case of the alums) the energetically most favourable configuration proved to be one with dipoles aligned parallel in chains, neighbouring chains being antiparallel. For a spheroid of excentricity larger than 6:1, however, the free energy is lower for the parallel orientation. This is due to the contribution of the demagnetizing energy of the spheroid.

SAUER and TEMPERLEY[6] considered the influence of non-zero temperature with the help of the BRAGG-WILLIAMS approximation, hence assuming long-range order. As in the case of the molecular field theories (see Sect. 55) the lattice was split into two sublattices of antiparallel orientations. Parameters r_1 and r_2 were introduced giving the fractions of dipoles with wrong orientation in each sublattice. For any temperature equilibrium, values of r_1 and r_2 can be calculated as a function of the applied magnetic field by minimizing the free energy of the crystal.

---

[1] J. S. SMART: Phys. Rev. **86**, 968 (1952).

[2] L. NÉEL: Proc. Conf. Theor. Phys. 1953 Tokyo, p. 703.

[3] J. A. BEUN, M. J. STEENLAND, D. DE KLERK and C. J. GORTER: Leiden Commun. No. 301a; Physica, 's-Grav. **21**, 767 (1955).

[4] J. SAUER: Phys. Rev. **57**, 142 (1940).

[5] J. M. LUTTINGER and L. TISZA: Phys. Rev. **70**, 954 (1946); **72**, 257 (1947).

[6] J. SAUER and H. N. V. TEMPERLEY: Proc. Roy. Soc. Lond., Ser. A **176**, 203 (1940).

It is found that a critical field curve occurs, like in molecular field theories, inside which the antiparallel configuration is stable. For the case of a spherical sample it follows;

$$kT_c = 2\mu H_g^0, \tag{56.1}$$

as was also derived by GORTER and HAANTJES, see Eq. (55.3). The transition on the critical field curve for temperatures below $2T_c/3$ is first order with a latent heat and a discontinuity in M. This has not been confirmed by the experiments. Variation of the shape of the sample leaves T_c unaltered, but the transition curve moves toward the T-axis with increasing excentricity. For an axial ratio of about 6:1 the region of antiferromagnetic order vanishes and the behaviour of the salt becomes ferromagnetic.

ZIMAN[1] applied the BETHE method (hence assuming short range order) to the case of magnetic dipole interaction. His results were rather similar to those given above. The ratio $kT_c/\mu H_g^0$ is of the order of unity and depends on the number of nearest neighbouring ions.

KITTEL[2] demonstrated the possibility that the magnetic dipole interaction may lead to a domain structure. This, however, is not the antiferromagnetic domain structure as proposed by NÉEL (see Sect. 55), but a configuration of antiparallel domains of the order of 10^{-4} cm., the orientation being parallel inside each domain.

The occurrence of such domains gives rise to appreciable complications. The transition from the antiferromagnetic to the ferromagnetic order in a spheroid of axial ratio approximately 6:1 (see above) is ruled out; and in the interpretation of experimental results we have to make a sharp difference between the "technical" magnetization curve of the sample as a whole and the behaviour of the individual domains.

Conclusions about the question whether this domain structure occurs or not might be derived from measurements of the BARKHAUSEN effect.

At the end of this section, attention should be drawn to the possibility that the spin wave method might provide a suitable way of approach to the solution of the problem. The method has been worked out successfully for exchange ferromagnetism[3-5] and antiferromagnetism[6-11], but, as far as we know, it has not been applied to the problem of magnetic dipole interaction. Since, however, the results obtained from it are most reliable if the number of excitation waves is small, the possibility exists that the formulae derived for a given salt are only valid at temperatures below the region that can be reached by demagnetization of the salt itself, see Sect. 54.

II. Results obtained with individual salts.

57. Chromium methylamine alum. The relation between entropy and susceptibility as determined by BEUN, STEENLAND, DE KLERK and GORTER[12] is shown in Fig. 44. The measurements were performed with single crystals of spherical

---

[1] J. M. ZIMAN: Proc. Phys. Soc. Lond. A **64**, 1108 (1951).

[2] C. KITTEL: Phys. Rev. **82**, 965 (1951).

[3] F. BLOCH: Z. Physik **61**, 206 (1930).

[4] F. BLOCH: Z. Physik **74**, 295 (1931).

[5] T. HOLSTEIN and H. PRIMAKOFF: Phys. Rev. **58**, 1098 (1940).

[6] L. HULTHÈN: Proc. Kon. Acad. Wetensch. Amst. **39**, 190 (1936).

[7] G. HELLER and H. A. KRAMERS: Proc. Kon. Acad. Wetensch. Amst. **37**, 378 (1934).

[8] J. M. ZIMAN: Proc. Phys. Soc. Lond. A **65**, 540 (1952).

[9] J. M. ZIMAN: Proc. Phys. Soc. Lond. A **65**, 548 (1952).

[10] J. M. ZIMAN: Proc. Phys. Soc. Lond. A **66**, 89 (1953).

[11] R. KUBO: Phys. Rev. **87**, 568 (1952).

[12] J. A. BEUN, M. J. STEENLAND, D. DE KLERK and C. J. GORTER: Leiden Commun. No. 301a; Physica, 's-Grav. **21**, 767 (1955).

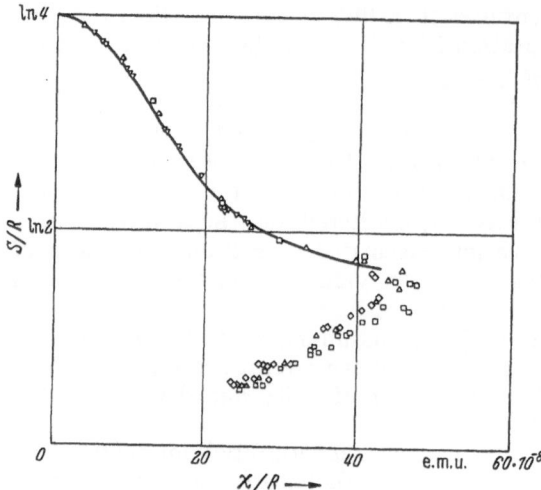

Fig. 44. Entropy versus susceptibility diagram for two spherical single crystals of chromium methylamine alum (according to BEUN, STEENLAND, DE KLERK and GORTER).

△ χ' for first sphere, $\nu = 225$ c/sec.
□ χ for first sphere, measuring field 1.10 oersteds, free period of ballistic galvanometer 1.3 sec.
▽ χ' for second sphere, $\nu = 225$ c/sec.
◇ χ for second sphere, measuring field 1.10 oersteds, free period of ballistic galvanometer 1.3 sec.

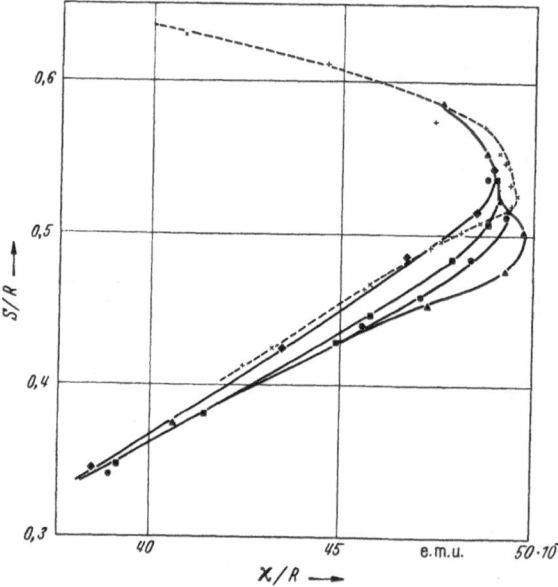

Fig. 45. Entropy versus susceptibility diagram for a spherical single crystal of chromium methylamine alum in the region of the susceptibility maximum (according to HUDSON and McLANE).

+ χ', measuring field 0.46 oersteds, $\nu = 210$ c/sec.
× χ', measuring field 0.46 oersteds, $\nu = 150$ c/sec.
▲ χ, measuring field 1.72 oersteds, free period of ballistic galvanometer 5.6 sec.
● χ, measuring field 3.43 oersteds, free period 5.6 sec.
■ χ, measuring field 6.86 oersteds, free period 5.6 sec.
◆ χ, measuring field 10.92 oersteds, free period 5.6 sec.

shape, mounted in such a way that one cubic axis was parallel to the small measuring field. Two samples were investigated, and systematical differences up to several percents were found in the susceptibilities.

The results at the highest entropies are those discussed in Sect. 34. A sudden rise in susceptibility occurs at $S = R \ln 2$, followed by a maximum near $S = 0.5 R$. Below this maximum the a.c. susceptibility χ' is noticeably smaller than the ballistic susceptibility χ. It was found that χ' depends only very little on the frequency and the amplitude of the measuring field.

The susceptibility in the vicinity of the maximum as measured by HUDSON and McLANE[1] is shown in Fig. 45. These authors also investigated two spherical single crystals and found differences of several percents in the susceptibilities. (Only the data of one sample are represented in Fig. 45.) Taking this into account, the agreement between the Leiden and Washington results is not bad.

HUDSON and McLANE found that the susceptibility values near the maximum are largely influenced by the rate of precooling the crystal from room temperature to the liquid nitrogen region. This must be connected with the transition in the crystalline structure mentioned in Sect. 34. By cooling very slowly (several hours) it was possible to obtain reproducible results.

There is an indication that χ' has a double maximum, one peak occurring at $S = 0.541 R$, the other at $0.562 R$. Also the ballistic susceptibility shows a double maximum, but only for weak measuring fields. The lower maxi-

[1] R. P. HUDSON and C. K. McLANE: Phys. Rev. **95**, 932 (1954).

mum decreases with increasing field and vanishes already at about ten oersteds, see Fig. 45. It is also the Leiden experience that only near the maximum χ depends on the measuring field.

The ballistic measurements performed at Oxford with an ellipsoidal sample 4:1 of compressed powder lead to a different result. GARDNER and KURTI[1] found that χ is constant from $S = 0.50\,R$ down to $S = 0.36\,R$, the lowest entropy reached in the experiments.

The imaginary part of the a.c. susceptibility, χ'', as measured by HUDSON and MCLANE is shown in Fig. 46. The χ'' values are quite small over the whole region with the exception of a sharp peak near the maximum of χ and χ'. Similar results were obtained in Leiden. It is not absolutely sure that the field dependence as shown in Fig. 46 is real. HUDSON and MCLANE even suggested the possibility that χ'' might show a double maximum like χ'.

A striking difference between the Leiden and Washington experiments is found in the frequency dependence of χ''. In Leiden it was observed that χ'' increases with increasing frequency. From this it was concluded that the relaxation times occurring in the salt are very short. The Washington people found that χ'' is nearly independent of frequency and this leads to long relaxation times below the maximum. In accordance with these conclusions hardly any double deflections (see Sect. 24) were found in the Leiden ballistic

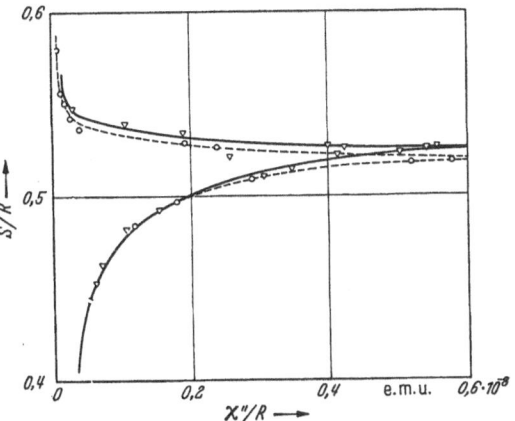

Fig. 46. Entropy versus imaginary part of a.c. susceptibility diagram for a spherical single crystal of chromium methylamine alum (according to HUDSON and MCLANE). ▽ Measuring field 0.30 oersteds. ○ Measuring field 0.45 oersteds.

measurements (only occasionally close to the maximum) but very pronounced time effects were found in the Washington experiments.

The supposition seems to be justified that the relaxation times of the Leiden and Washington samples were really strongly divergent. This is the more remarkable, since one of the Leiden samples was of the same origin as one of the Washington samples. Maybe, again, this is connected to the rate of cooling to liquid air temperature.

Due to the very steep course of χ'' with entropy near the maximum the variation with time during a heating period is very fast. Consequently, it is difficult to extrapolate χ'' to the time of the demagnetization so that, for absolute temperature determinations, χ'' is impracticable as a thermometric parameter (see Sect. 11).

The remanent magnetic moment as a function of the measuring field and the entropy, according to BEUN, STEENLAND, DE KLERK and GORTER, is shown in Fig. 47. The results are in reasonable agreement with those of HUDSON and MCLANE. In Leiden the starting point of the hysteresis was found to be $S = 0.54\,R$. HUDSON and MCLANE gave $S = 0.53\,R$, GARDNER and KURTI $S = 0.50\,R$. The remanent moments are considerably smaller than those found for chromium potassium alum (see Sect. 58).

[1] W. E. GARDNER and N. KURTI: Proc. Roy. Soc. Lond., Ser. A **223**, 542 (1954).

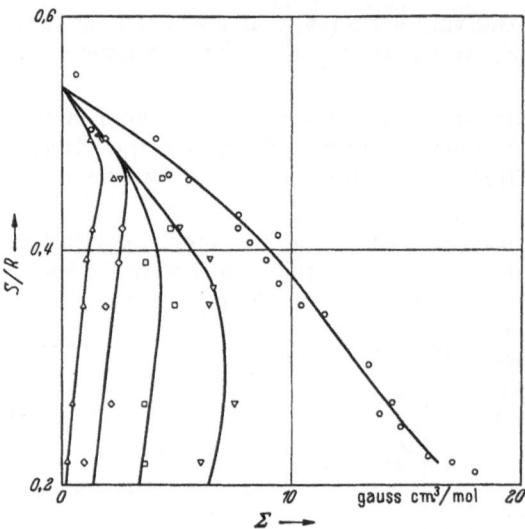

Fig. 47. Entropy versus remanent magnetic moment diagram for a spherical single crystal of chromium methylamine alum (according to BEUN, STEENLAND, DE KLERK and GORTER). Free period of ballistic galvanometer 1.3 sec.

△ Measuring field 1.10 oersteds.
◇ Measuring field 2.20 oersteds.
□ Measuring field 4.39 oersteds.
▽ Measuring field 8.78 oersteds.
○ Measuring field 21.95 oersteds.

It appears that the remanent moment, for a given field H_{ext}, shows a maximum as a function of entropy. Until now, chromium methylamine alum is the only salt showing this behaviour. A consequence is that the remanent moment is unfeasible as a thermometric parameter.

The magnetic method of absolute temperature determination is difficult for this salt. χ, χ', χ'' and Σ all are unsatisfactory as thermometric parameters below the susceptibility maximum (see above). χ'' is rather small (even at its maximum it is much smaller than for chromium potassium alum) so that it is difficult to distinguish between the heat absorption from the alternating field and the a.c. losses in the bridge (see Sect. 12). Moreover, the rapid variation of χ'' with time during a heating period is

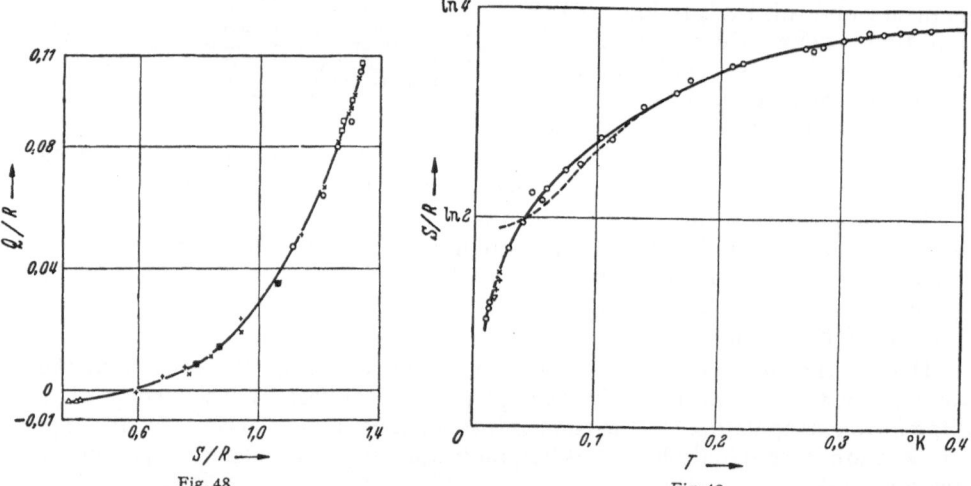

Fig. 48.   Fig. 49.

Fig. 48. Heat content versus entropy diagram for chromium methylamine alum (according to GARDNER and KURTI). The zero point of Q/R is arbitrary, see text. Different symbols refer to different helium runs. The triangles represent points for which Q/R was determined by heating into the region above the maximum.

Fig. 49. Entropy versus absolute temperature diagram for chromium methylamine alum (according to GARDNER and KURTI).

○ T^* thermometer.   + Σ thermometer.   △ Hysteresis heating.
× The CURIE point.   —— Experimental curve.   — — — Theoretical curve for $\delta = 0.27°$ K.

another source of inaccuracy. After many tedious experiments HUDSON and McLANE came to the conclusion that the absolute temperature of the CURIE point (defined as the starting point of the remanent moment) lies most probably in the region between 0.015 and 0.020°K.

The best absolute temperature determinations are those by GARDNER and KURTI. They used gamma radiation for heat supply. Above the maximum χ (or T^*) was the parameter. The quantity $c^* = dQ/dT^*$ was calculated and from this the total heat content $Q = \int c^* \, dT^*$ could be computed as a function of entropy (with an arbitrary zero). The results are given in Fig. 48.

Below the susceptibility maximum the method was modified as described in Sect. 48. After the demagnetization the total energy was determined, necessary to heat the sample to a temperature well above the susceptibility maximum, where χ could be used as a parameter. The data obtained with this method are also shown in Fig. 48 (the triangles).

The absolute temperature for each value of the entropy is equal to the slope of the curve of Fig. 48. The results are plotted in Fig. 49, together with some additional measurements made with the remanence thermometer and by applying hysteresis heat. The temperature of the CURIE point was found to be 0.020° K.

Recent Leiden measurements with a.c. heating gave $T = 0.020°$ K for the CURIE point and $T = 0.002°$ K for the lowest temperature ($S = 0.26\,R$). Experiments with gamma ray heating, however, gave definitely higher values for the lowest temperature.

58. Chromium potassium alum. Measurements below the susceptibility maximum of this salt were performed in Leiden, Oxford and Washington. The results are usually somewhat different for different samples. Fig. 50 gives susceptibility values obtained in Leiden[1] with four samples. They all were spherical single crystals mounted with one cubic axis parallel to the field of the magnet and one parallel to the

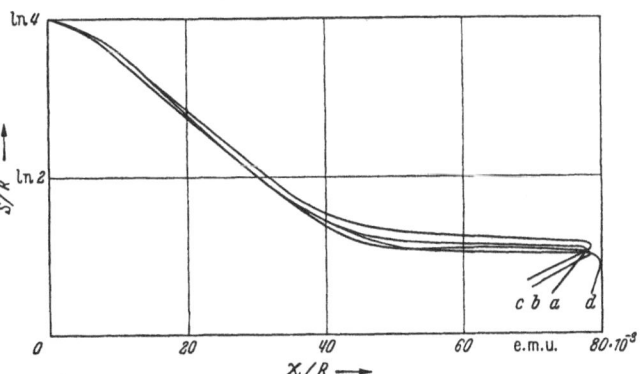

Fig. 50. Entropy versus susceptibility diagram for four spherical single crystals of chromium potassium alum (according to BEUN, STEENLAND, DE KLERK and GORTER). $S > 0.40\,R$: a.c. measurements. $S < 0.40\,R$: ballistic measurements with galvanometers of various free periods. Curve a: samples 1 and 2, measuring field 1.08 oersteds, free period 7 sec. Curve b: sample 3, measuring field 1.08 oersteds, free period 0.2 sec. Curve c: sample 4, July 1951, measuring field 0.33 oersteds, free period 1.3 sec. Curve d: sample 4, January 1953, measuring field 1.08 oersteds, free period 1.3 sec.

measuring field. At the higher temperatures, where χ and χ' are equal, the χ' values have been plotted since they can be measured with a higher precision. At the lower temperatures we have plotted χ.

The differences between the four curves are partly due to the different measuring techniques (see the subscript of the figure). It is sure, however, that also part of the differences is caused by the properties of the samples themselves; for instance, the differences at the higher temperatures give rise to the different values of the splitting parameter quoted in Sect. 35.

The results obtained with one of the samples[2,3] are shown in more detail in Fig. 51. Both χ and χ' show a sudden rise just before the maximum. The

[1] J. A. BEUN, M. J. STEENLAND, D. DE KLERK and C. J. GORTER: Leiden Commun. No. 300a; Physica, 's-Grav. **21**, 651 (1955).

[2] D. DE KLERK, M. J. STEENLAND and C. J. GORTER: Leiden Commun. No. 278c; Physica, 's-Grav. **15**, 649 (1949).

[3] M. J. STEENLAND: Thesis, Leiden 1952.

rise is steeper than in the case of chromium methylamine alum (compare Figs. 44 and 51), and it takes place at a smaller entropy value, well below $R \ln 2$.

The susceptibility maximum is noticeably higher than in the case of the methylamine alum; to such an extent that, in the maximum, the correction for the demagnetizing field [see Eq. (7.5)] is very large, χ_{int} being at least a factor twenty higher than χ_{ext}. This is distinctly not true in the case of the methylamine alum. For the latter salt χ_{int} at the maximum is still of the same order of magnitude as C/T_{max}; for the potassium alum χ_{int} at the maximum is much larger than C/T_{max}.

Below the maximum, χ is smaller than χ', as in the case of the methylamine alum, but the difference is more pronounced. The value of χ' is very little influenced by the frequency of the measuring field, but χ depends on the free period of the ballistic galvanometer. This was demonstrated by making measurements with different galvanometers alternately in the same heating period. Double deflections were found (see Sect. 24) in the neighbourhood of the maximum, especially if a galvanometer with a short free period was used (e.g. 0.2 sec.).

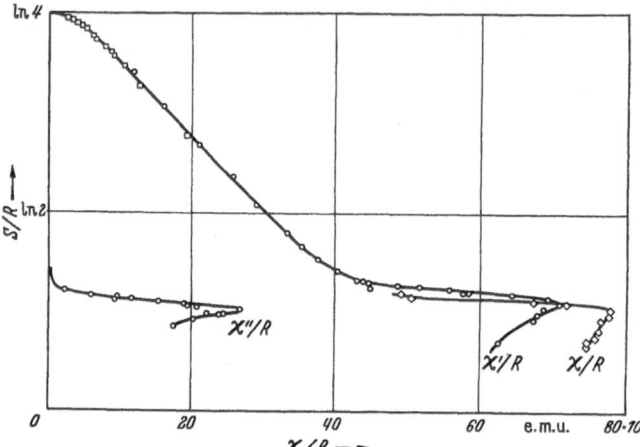

Fig. 51. Entropy versus susceptibility diagram for a spherical single crystal of chromium potassium alum (according to data of DE KLERK, STEENLAND and GORTER). χ''/R is plotted on a tenfold magnified scale. χ/R: measuring field 1.08 oersteds, free period of ballistic galvanometer 7 sec. χ'/R and χ''/R: amplitude of measuring field 0.183 oersteds, $\nu = 225$ c/sec.

These phenomena indicate the occurrence of relaxation effects and this is confirmed by the behaviour of χ''.

Some values of χ'' are shown in Fig. 51[1]. It is already perceptible above the maximum of χ', where no hysteresis effects were found. χ'' depends on the frequency of the measuring field. Experiments with several frequencies during the same heating period gave a proportionality to $\nu^{1.7 \pm 0.15}$.

The χ'' values of Fig. 51 show a maximum at an entropy somewhat below the maximum of χ and χ'. The values are noticeably larger than those obtained for the methylamine alum, and the maximum is less pronounced (see Sect. 57). Still χ'' is much smaller than χ', the ratio χ''/χ' never exceeding 0.03 (it should be noticed that χ'' in Fig. 51 is plotted on a tenfold magnified scale).

Though the behaviour of both χ and χ'' indicates the occurrence of relaxation effects, it proves to be impossible to describe all the phenomena with one relaxation time. The time derived from the ballistic experiments is of the order of 10^{-2} sec., whereas the a.c. measurements suggest times shorter than 10^{-3} sec.

Measurements by AMBLER and HUDSON[2] with a spherical single crystal gave slightly higher values for χ and χ' near the maximum than the Leiden experi-

---

[1] D. DE KLERK, M. J. STEENLAND and C. J. GORTER: Leiden Commun. No. 278c; Physica, 's-Grav. 15, 649 (1949).
[2] E. AMBLER and R. P. HUDSON: Phys. Rev. 95, 1143 (1954).

ments. DANIELS and KURTI[1], who used an ellipsoid of compressed powder with axial ratio 6:1 found somewhat lower values for χ; they made only ballistic measurements.

The main qualitative difference between the Leiden results and those of AMBLER and HUDSON is that the maximum of χ found by the latter authors is less sharp; the maximum of χ' has about the same shape in both experiments. AMBLER and HUDSON found no double deflections with a ballistic galvanometer of 5.6 sec. Moreover χ'' did not show a maximum; it approached a constant value at about the entropy of the Leiden maximum, but at lower entropies it increased again.

Some data on the remanent magnetic moment, Σ, obtained by STEENLAND, DE KLERK and GORTER[2,3] are given in Fig. 52.

The CURIE point, defined as the point where a remanent moment appears first, was found at $S = 0.40\,R$, slightly higher than the entropy of the susceptibility maximum. This is in good agreement with the value given by AMBLER and HUDSON ($S = 0.42\,R$), and with old measurements of KURTI, LAÎNÉ and SIMON[4] ($S = 0.44\,R$).

The values of Σ found for this salt are appreciably larger than those for the methylamine alum (see Sect. 57), but still they are only a few percents of the moment in the field of 1.08 oersted which, itself, is about one percent of the saturation moment.

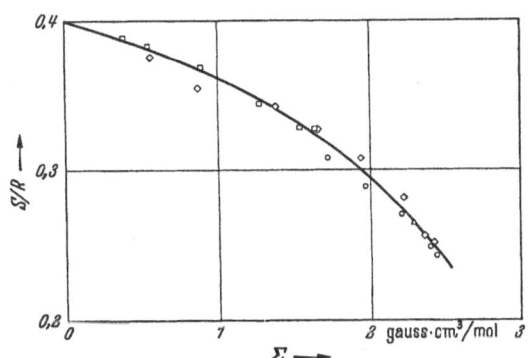

Fig. 52. Entropy versus remanent magnetic moment diagram for a spherical single crystal of chromium potassium alum (according to STEENLAND). Measuring field 1.08 oersteds, free period of ballistic galvanometer 7 sec. Different symbols refer to different helium runs.

The field dependence of the remanent moment was also investigated by STEENLAND, DE KLERK and GORTER. For a given entropy Σ increases first with the measuring field, then it decreases. The maximum moves to higher field values with decreasing entropy.

The values of the remanent moment are not too well reproducible. For one sample it was found that Σ increased by about 20% in the course of a year; the data obtained with different samples may be different by even a factor of three. This, together with the smallness of Σ, may indicate that the remanent moment is entirely due to spurious effects, such as impurities in the crystal or lattice defects.

Complete hysteresis loops were measured with maximum fields of 4.32 and 12.95 oersteds. Each loop was described in 24 steps. Some data on the loops of 4.32 oersteds are collected in Fig. 53. If the loops are plotted against H_{ext} they become very long and narrow (see the lower righthand block of Fig. 53). In ferromagnetism, however, it is standard practice to "shear" the loops, i.e. plot them as a function of H_{int}, see Eq. (7.5). The sheared loops have a more familiar shape than the unsheared ones, as can be seen from Fig. 53. The difficulty in these experiments is, however, that the term $\varepsilon M/V$ in Eq. (7.5) is of the

[1] J. M. DANIELS and M. KURTI: Proc. Roy. Soc. Lond., Ser. A **221**, 243 (1954).

[2] M. J. STEENLAND, D. DE KLERK and C. J. GORTER: Leiden Commun. No. 278d; Physica, 's-Grav. **15**, 711 (1949).

[3] M. J. STEENLAND: Thesis, Leiden 1952.

[4] N. KURTI, P. LAÎNÉ and F. SIMON: C. R. Acad. Sci., Paris **204**, 675 (1937).

same order of magnitude as H_{ext}, so that small inaccuracies in the density of the sample, or small deviations from the spherical shape, may introduce appreciable errors into the shapes of the sheared loops.

It was found that the remanent moment calculated from the loops of 12.95 oersteds was always somewhat larger than the directly measured remanent moment. This, again, is probably due to relaxation effects. Such differences were not found in the loops of 4.32 oersteds.

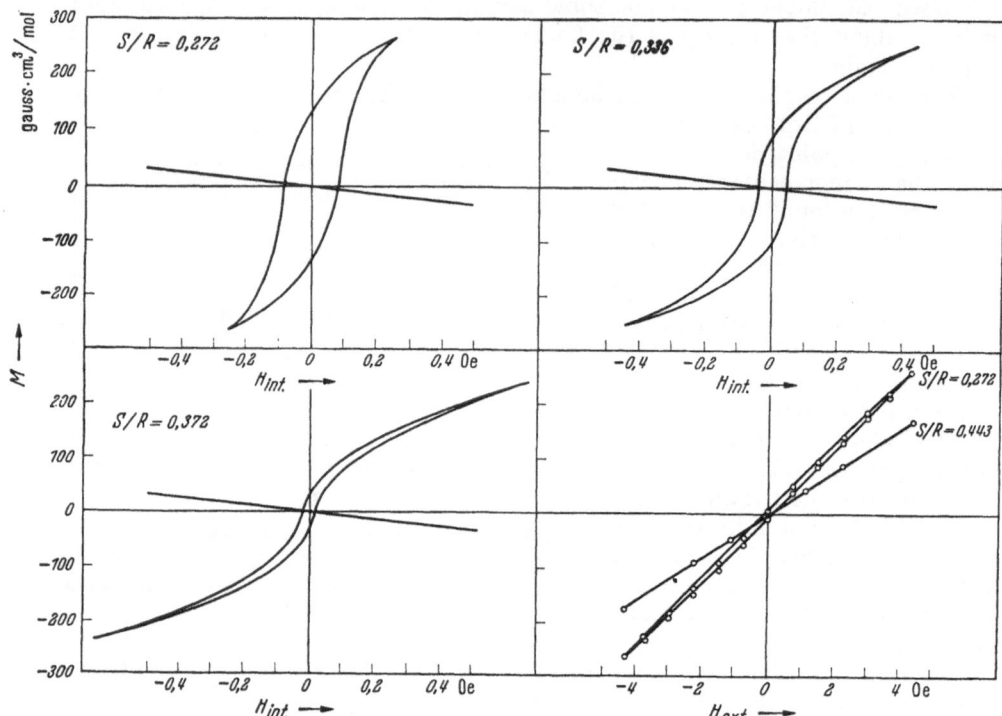

Fig. 53. Hysteresis loops for a spherical single crystal of chromium potassium alum (according to STEENLAND, DE KLERK and GORTER). Maximum field of the loops: $H_{ext} = 4.30$ oersteds. The unsheared loop of the lower righthand block of the figure is the same as the sheared loop of the higher lefthand block. The lower righthand block shows also a magnetization curve slightly above the CURIE point.

It should be noticed that all the hysteresis loops were measured under adiabatic conditions. It was found from measurements in external magnetic fields, however, that the temperature variations in fields of the order of ten oersteds are very small (see Sect. 65) so that, for further discussion of the results, the difference between adiabatic and isothermal loops can probably be neglected.

Several methods of absolute temperature determination have been applied for this salt, but until recently the results were rather unsatisfactory.

Early Leiden experiments[1] were made with a.c. heating, χ'' being noticeably larger than in the case of the methylamine alum (see above). χ' (or T^*) was the thermometric parameter above the susceptibility maximum. At and below the maximum it was no longer very suitable, since its variation with entropy is too slow and it is no more a single valued function of temperature. χ'' increases strongly in the region of the maximum of χ' and it can be used as a parameter

[1] D. DE KLERK, M. J. STEENLAND and C. J. GORTER: Leiden Commun. No. 278c; Physica, 's-Grav. **15**, 649 (1949).

there. A third parameter was needed below the maximum of χ'' and Σ could be applied for this purpose.

Furthermore some measurements were made with a different method of heat supply[1]. A number of hysteresis loops were described at such a speed that relaxation heating could be neglected (e.g. one loop per second for a few minutes), the area of the loop being determined in a separate experiment during the same helium-run. Σ, again, was used as a thermometric parameter.

The results of these experiments are represented by curve KSG of Fig. 54. They are in striking disagreement with gamma ray experiments performed at Oxford. In these investigations T^* was used as a parameter throughout the whole region of temperatures. Results obtained by DANIELS and KURTI[2] are represented

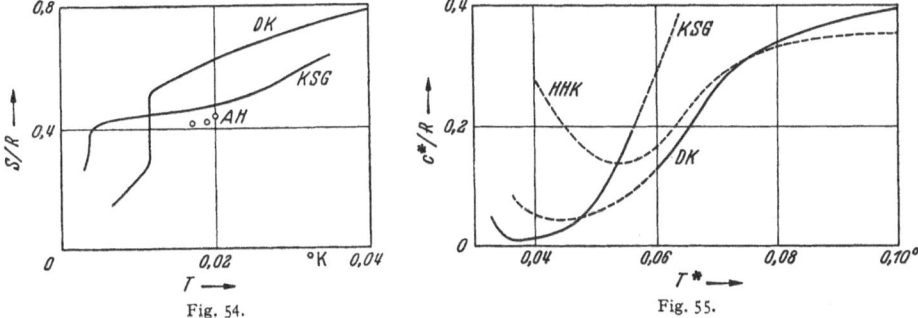

Fig. 54.   Fig. 55.

Fig. 54. Entropy versus absolute temperature diagram for chromium potassium alum. *KSG*: Curve obtained by DE KLERK, STEENLAND and GORTER. *DK*: Curve obtained by DANIELS and KURTI. *AH*: Points obtained by AMBLER and HUDSON.

Fig. 55. c^* versus T^* diagram for chromium potassium alum. *KSG*: curve obtained by DE KLERK, STEENLAND and GORTER. *DK*: curve obtained by DANIELS and KURTI. *HHK*: curve obtained by HUDSON, HUNT and KURTI.

by curve *DK* of Fig. 54. Three points given by AMBLER and HUDSON[3] are also shown in Fig. 54. They applied a.c. heating and T^* was the parameter.

The fact that the discrepancies between the results on different samples are largely due to the caloric measurements is demonstrated in Fig. 55. Here c^* (defined as dQ/dT^*) is plotted against T^*. Two of the curves represent the Leiden and Oxford data quoted above, the third curve shows older Oxford results, obtained by HUDSON, HUNT and KURTI.

Recently new experiments were performed in Leiden by BEUN, STEENLAND, DE KLERK and GORTER, in which both a.c. and gamma ray heating were applied; they have not yet been published. A new thermometric parameter was introduced, viz. the susceptibility in a longitudinal field of 13 oersteds. It was found (see Sect. 65) that this quantity does not show a maximum as a function of the entropy, hence it is suitable throughout the whole region of temperatures. The results were in satisfactory agreement with curve *DK* of Fig. 54.

An explanation of the discrepancies is rather difficult. The possibility is not a priori excluded that the thermal properties of different samples of chromium potassium alum are really widely different. This assumption, however, is unsatisfactory and, as DANIELS and KURTI remarked, it is more likely that the region of the STARK splitting should be influenced than the region of the magnetic interactions.

[1] M. J. STEENLAND, D. DE KLERK and C. J. GORTER: Leiden Commun. No. 278d; Physica, 's-Grav. **15**, 711 (1949).
[2] J. M. DANIELS and N. KURTI: Proc. Roy. Soc. Lond., Ser. A **221**, 243 (1954).
[3] E. AMBLER and R. P. HUDSON: Phys. Rev. **95**, 1143 (1954).

Recently a new analysis was made of the old Leiden measurements quoted above. It revealed that it is possible, within the experimental accuracy, to shift the χ versus S curve in such a way that the high temperature part of the S versus T diagram reaches agreement with the recent results. The reason is that the almost horizontal part of the χ versus S curve is involved (see Fig. 51). The quantity entering into the calculations is $d\chi/dS$ and a small variation in the curve may influence the slope appreciably.

The deviations at the lowest temperatures might be due to a stray heat influx for which the proper allowance has not been made. A difficulty encountered with this correction is the following. As was pointed out in Sect. 19, it is often supposed that, some time after the demagnetization, the sample may be described as a cold core, still at the original low temperature, surrounded by a much warmer shell. If, during the heating period, the shape of the cold core does not remain similar to that of the sample as a whole, the demagnetizing field may become widely different and this may influence the observed susceptibilities noticeably. It is difficult to make an estimate of this effect on a quantitative basis, but it is well possible that the shapes of cold cores of different samples change in different ways with time, especially when the methods of suspension are not the same. Maybe this accounts for the discrepancies as observed in the experiments.

At the present time it seems plausible that the best relation between entropy and temperature for chromium potassium alum is curve DK of Fig. 54.

59. Diluted chromium potassium alum. DE KLERK, STEENLAND and GORTER[1] carried out some experiments at the lowest temperatures obtainable with a mixed crystal of chromium potassium alum and aluminum potassium alum. The sample consisted of a glass sphere filled with small crystals containing 21.3 aluminum ions for each chromium ion.

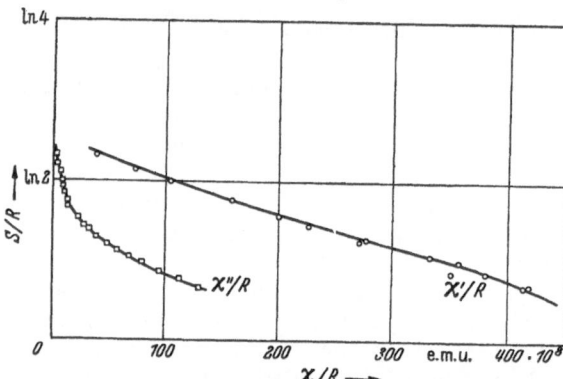

Fig. 56. Entropy versus susceptibility diagram for a powdered sample of spherical shape of diluted chromium potassium alum 1:23.1 (according to DE KLERK, STEENLAND and GORTER). χ''/R is plotted on a tenfold magnified scale.

The χ' and χ'' values, obtained at a frequency of 225 c/sec., are shown in Fig. 56. The results proved to be independent of the amplitude of the measuring field between 0.183 and 1.83 oersted. The molar values of χ' and χ'' were much larger than in the case of the undiluted salt (cf. Fig. 51) but the ratio χ''/χ' is of the same order of magnitude, its value being 0.028 at the lowest temperature. No maxima were found in χ' and χ'', but they may occur at still lower temperatures.

Since the specific heat is very small a normal heat leak causes already an appreciable rise in temperature after demagnetization. Therefore it was difficult to make reliable ballistic measurements. No conclusive information could be obtained about the occurrence of hysteresis effects, but it seems improbable that they should appear when no maximum is found in χ'.

[1] D. DE KLERK, M. J. STEENLAND and C. J. GORTER: Leiden Commun. No. 282a; Physica, 's-Grav. **16**, 571 (1950).

Absolute temperature determinations were made with the help of a.c. heating. Due to the absence of a maximum in χ', the quantity T^* could be used as a thermometric parameter throughout the whole region. As a consequence of the smallness of the specific heat, special attention had to be given to the correction for the residual heating rate. Curve II of Fig. 57 gives the relation between S and T if this correction is applied on the basis of the model of Sect. 19; curve I gives the results if the correction is entirely neglected. The correct values are probably between the two curves. The lowest temperature is then $0.0014°$ K $\pm 10\%$.

The T^* values are also shown in Fig. 57, together with the T^* values obtained by DE KLERK and POLDER with a sample 1:13, see Sect. 36.

60. Iron ammonium alum. The first measurements with this salt at the lowest temperatures were performed by KURTI, LAÎNÉ, ROLLIN and SIMON[1-3] in the Bellevue Laboratory. It was the first time that a susceptibility maximum was discovered in an alum. Remanent magnetic moments were found starting at a slightly higher temperature than the susceptibility maximum, and also hysteresis loops were measured.

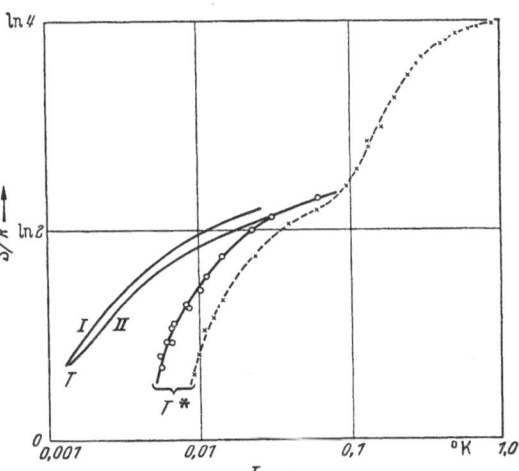

Fig. 57. Entropy versus temperature diagram for diluted chromium potassium alum 1:23.1 (according to DE KLERK, STEENLAND and GORTER). Curve I: S/R versus T, if no correction is applied for the residual heating. Curve II: S/R versus T, if a correction is applied, see text. ○ S/R versus T^*. × S/R versus T^* for a sample 1:13, see Sect. 36.

More recently experiments were made in Oxford and Leiden The Leiden susceptibility data[4,5] are shown in Fig. 58. The measurements were performed with a glass sphere filled with small crystals, and with a solid ellipsoid with axial ratio 2:1 ground from a big piece of salt which was no single crystal.

Above $S = R \ln 2$, χ' was equal to χ, below $R \ln 2$ it was smaller. Maxima occurred both in χ and χ', but not in χ''. The maxima were less sharp than in the cases of the chromium alums. The maximum in χ was at a somewhat lower entropy than that of χ'. The χ''/χ' values were of the same order of magnitude as for chromium potassium alum, not exceeding 0.046 at the lowest temperatures.

Below the susceptibility maximum, it was found that χ decreases with increasing measuring field, but χ' and χ'' increase with the amplitude of the field, see Fig. 58. Experiments with 225 and 525 c/sec. alternately in one heating period showed that χ' decreases slightly with increasing frequency. Near $S = R \ln 2$, the quantity χ'' depends stronger on frequency than at the lowest temperatures.

χ and χ' for the ellipsoid were about a factor two larger than for the sphere, but further the shapes of the curves were very similar for both samples. The differences must be due to the difference in the demagnetizing fields. It was found, however, that it is impossible to reduce the susceptibility values of the

[1] N. KURTI, P. LAÎNÉ, B. V. ROLLIN and F. SIMON: C. R. Acad. Sci., Paris 202, 1576 (1936).
[2] N. KURTI, P. LAÎNÉ and F. SIMON: C. R. Acad. Sci., Paris 204, 675 (1937).
[3] N. KURTI, P. LAÎNÉ and F. SIMON: C. R. Acad. Sci., Paris 204, 754 (1937).
[4] M. J. STEENLAND, D. DE KLERK, M. L. POTTERS and C. J. GORTER: Leiden Commun. No. 284b; Physica, 's-Grav. 17, 149 (1951).
[5] M. J. STEENLAND: Thesis, Leiden 1952.

ellipsoid exactly to those of the spherical sample, the values calculated from the ellipsoid being somewhat smaller than those measured with the sphere. The

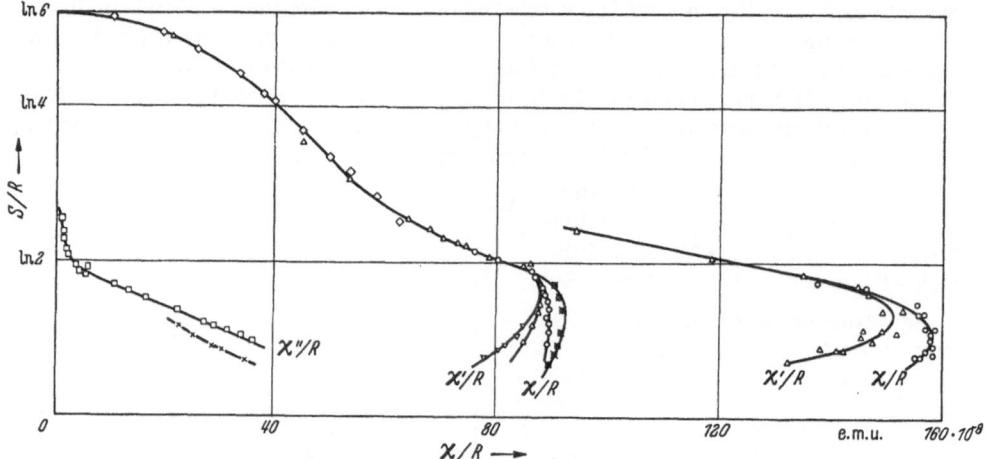

Fig. 58. Entropy versus susceptibility diagram for iron ammonium alum (according to STEENLAND, DE KLERK, POTTERS and GORTER). The two righthand curves refer to the solid ellipsoid with axial ratio 2:1, the other curves refer to the powdered sphere. χ''/R is plotted on a tenfold magnified scale.

○ χ/R, measuring field 4.31 oersteds.
⊗ χ/R, measuring field 1.08 oersteds.
△ χ'/R, measuring field 0.610 oersteds; $\nu = 225$ c/sec.
▽ χ'/R, measuring field 0.182 oersteds; $\nu = 225$ c/sec.

□ χ''/R, measuring field 0.610 oersteds; $\nu = 225$ c/sec.
× χ''/R, measuring field 0.182 oersteds; $\nu = 225$ c/sec.
◇ χ'/R, measurements of 1939, see Sect. 38, $\nu = 50$ c/sec.

reason is probably that the demagnetization coefficient of a sphere filled with small crystals of arbitrary shape cannot be described by the formulae of Sect. 7.

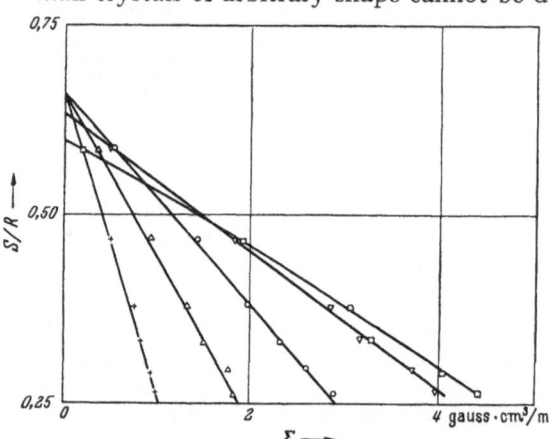

Fig. 59. Entropy versus remanent magnetic moment diagram for a powdered sphere of iron ammonium alum (according to STEENLAND, DE KLERK, POTTERS and GORTER).

+ measuring field 1.08 oersteds.
△ measuring field 2.16 oersteds.
○ measuring field 4.31 oersteds.
▽ measuring field 8.63 oersteds.
□ measuring field 12.94 oersteds.

The differences could be eliminated by introducing a Δ of $0.08 \times 4\pi/3 \times C/V$, see Sect. 11.

Taking these corrections into account, it follows that the Leiden susceptibility values are in satisfactory agreement with the Oxford data of KURTI and SIMON, published by COOKE[1]. The experiments were made with a compressed ellipsoid of axial ratio 3:1. The corresponding T^* values were already given in Fig. 21 of the present article, see also Fig. 62.

The values of the remanent magnetic moment are noticeably different during subsequent helium runs (up to about 20%), but the entropy of the CURIE point seems to be very little influenced.

Some values for the Leiden powdered sphere, all obtained during the same run, are shown in Fig. 59. The remanent moment for a given measuring field increases approximately linearly with decreasing entropy. For weak fields the entropy

[1] A. H. COOKE: Proc. Phys. Soc. Lond. A **62**, 269 (1949).

of the starting point of Σ is independent of the field strength. For stronger fields, however, this entropy is somewhat lower, and small negative Σ values were found above the zero point. Since the occurrence of ferrimagnetism[1] seems improbable we are inclined to ascribe these effects to relaxation phenomena.

If we define the CURIE point as the point where a remanent moment starts in weak fields, it occurs, according to Fig. 59, at $S = 0.66R$, slightly above the maxima of χ and χ'. The value is in good agreement with that given by KURTI[2], viz. $S = 0.65R$. There is no indication that different values for the CURIE point are found for samples of different ex-centricities, as might be expected on the basis of LORENTZ's theory of magnetic interactions, see Sect. 54. The same conclusion was reached by ASHMEAD (unpublished)[3] who investigated poly-crystalline spheroids of various excen-tricities. The remanent moment for a given field increased with increasing excentricity, but the CURIE tempera-ture and the value of its entropy were not influenced.

If we plot the remanent moment as a function of the measuring field at constant entropy, it follows from Fig. 59 that at the lowest entropies Σ increases with increasing field, where-as above $S = 0.50R$ Σ passes through a maximum. This was confirmed by experiments of KURTI.

A hysteresis loop, measured by KURTI at about $S = 0.40R$ with a maxi-mum field of 5 oersteds, is shown in Fig. 60. Both the unsheared and the sheared loop (see Sect. 58) are plotted. The remanent magnetic moment in the unsheared loop is much larger than

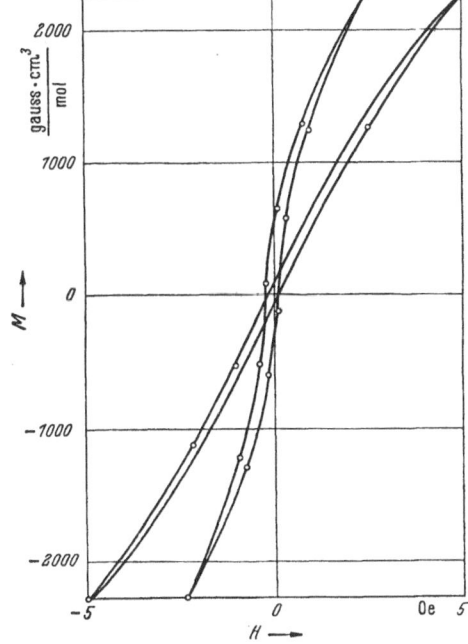

Fig. 60. Hysteresis loop for a very prolate sample of iron ammonium alum (8:1) at $T = 0.03°$ K ($= 0.7\,T_c$) (accord-ing to KURTI). Maximum field: $H_{ext} = 5$ oersteds. Curve 1: unsheared loop. Curve 2: sheared loop.

the values of Fig. 59, and the difference between the sheared and the unsheared loops is relatively small (compare, for instance, with the data on chromium potassium alum of Fig. 53). This is due to the fact that the sample was much more prolate than any one of the Leiden samples, viz. 8:1.

Four sheared loops, determined in Leiden with the ellipsoidal sample are plotted in Fig. 61. One of the loops has a slightly negative slope at the centre, and this is probably due to the fact, pointed out already in Sect. 58, that the term $\varepsilon M/V$ for an ellipsoid 2:1 is of the same order as H_{ext}, see Eq. (7.5), so that a small inaccuracy in ε may influence the values of H_{int} largely. The loops obtained with the powdered sphere, if sheared, showed still more pronounced negative slopes. If the empirical correction of $0.08 \times 4\pi/3$, quoted above, was applied to the demagnetizing field the effect was greatly decreased. Probably KURTI's loop of Fig. 60 is the only one with reliable values of H_{int}, due to the large excentricity of the sample.

[1] L. NÉEL: Proc. Conf. Theor. Phys. 1953, Tokyo, p. 703.

[2] N. KURTI: J. de Phys. 12, 281 (1951).

[3] C. G. B. GARRETT: Magnetic Cooling, p. 72. Cambridge (Mass.) 1954.

Absolute temperature determinations were made in Leiden with a.c. and hysteresis heating, and in Oxford with gamma ray heating.

The Leiden measurements could not be extended above the CURIE point, since here χ'' was too small. Between $S = 0.7R$ and $0.4R$ the χ'' was used as a thermometric parameter, below $0.6R$ the Σ could also be applied. The results are shown in Fig. 62, together with the T^* data. The precision was not too good, but it follows that there are no systematical differences between the results obtained with different parameters, nor are there between the results for the sphere and the ellipsoid. The CURIE point obeys: $S = 0.64R$, $T = 0.030°$ K.

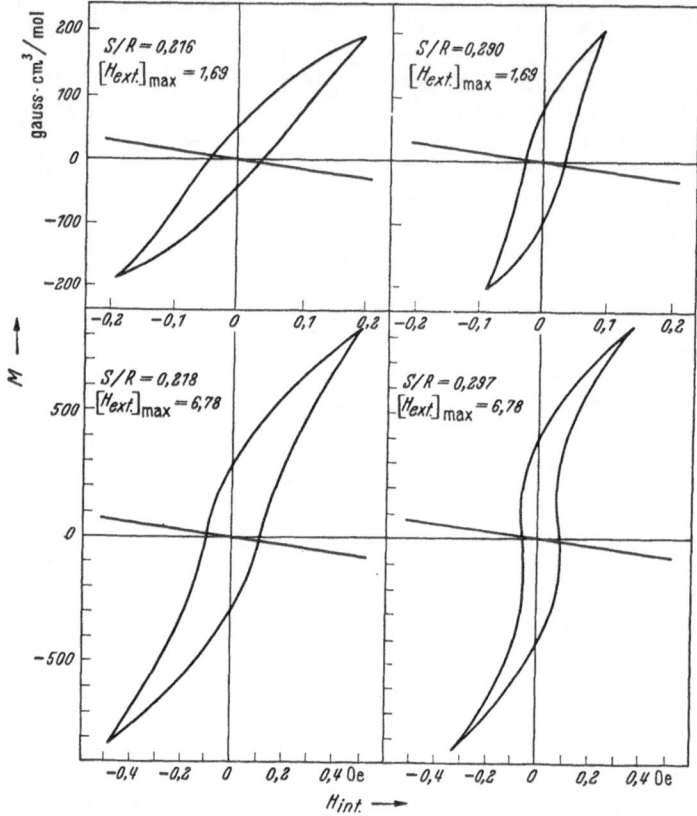

Fig. 61. Sheared hysteresis loops for an ellipsoid of iron ammonium alum with axial ratio 2:1 (according to STEENLAND, DE KLERK, POTTERS and GORTER).

The Oxford gamma ray data[1] are also shown in Fig. 62. The T^* values are the ones of Fig. 21. They are in reasonable agreement with the Leiden results, as was stated above. The absolute temperatures, however, calculated by MENDOZA[2] from Figs. 21 and 22, are noticeably higher than the Leiden results. The value for the CURIE point is: $S = 0.65R$, $T = 0.043°$ K.

61. Manganese ammonium sulphate. This is the first salt with which indications were found for the occurrence of a CURIE point below $1°$ K[3,4].

[1] A. H. COOKE: Proc. Phys. Soc. Lond. A **62**, 269 (1949).
[2] E. MENDOZA: ,,The Demagnetizer's Vademecum.'' Unpublished, no date, probably 1952.
[3] N. KURTI and F. SIMON: Proc. Roy. Soc. Lond., Ser. A **149**, 152 (1935).
[4] N. KURTI, P. LAÎNÉ and F. SIMON: C. R. Acad. Sci., Paris **204**, 675 (1937).

The susceptibility data given by STEENLAND[1] are shown in Fig. 63. Two samples were investigated; a powdered sphere[2] and a powdered ellipsoid of axial ratio 3:1. The experimental values obtained during subsequent helium runs showed usually somewhat larger discrepancies than those obtained with other salts.

The susceptibility maximum is very pronounced; it is localized at a rather high entropy value, viz. $S = 1.28R$, see Sect. 41. Below the maximum χ' increases

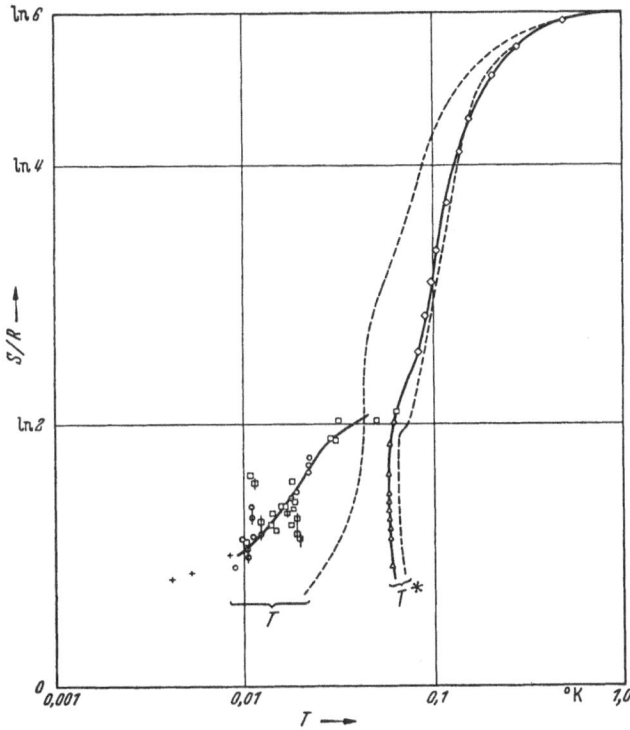

Fig. 62. Entropy versus magnetic and absolute temperature diagram for iron ammonium alum (according to STEENLAND DE KLERK, POTTERS and GORTER). The dotted lines represent the Oxford data, the full drawn curves give the Leiden results.

◇ T^*, data of 1939, see Sect. 38.   Spheroid 2:1: ⊡ T, a.c. heating and χ'' parameter.
Powdered sphere: △ T^*, a.c. data of Fig. 58   ⊕ T, a.c. heating and Σ parameter.
        at 0.610 oersteds and $\nu = 225$ c/sec.   + T, hyst. heating and Σ parameter.
        □ T, a.c. heating and χ'' parameter.
        ○ T, a.c. heating and Σ parameter.

with increasing measuring field, and it depends strongly on the frequency. The ballistic susceptibility χ is also noticeably larger than χ'. Unfortunately, in the case of the sphere only one ballistic point was measured.

The most striking phenomenon is that χ'' is much larger than for any one of the other salts that have been investigated (it should be emphasized that in Fig. 63 χ'' is not plotted on a tenfold scale, as was done in Figs. 51, 56, 58 and 65). At the lowest temperature ($S = 0.243\,R$) χ''/χ' reaches a value 0.333 at $\nu = 225$ c/sec. and 0.425 at $\nu = 525$ c/sec. Since the remanent moments found for this salt are very small (see below), these phenomena must be ascribed to relaxation effects.

The ballistic susceptibilities obtained with the ellipsoid can be reduced to those for the sphere with the help of the correction for the demagnetizing field

[1] M. J. STEENLAND: Thesis, Leiden 1952.
[2] M. J. STEENLAND, L. C. VAN DER MAREL, D. DE KLERK and C. J. GORTER: Leiden Commun. No. 279c; Physica, 's-Grav. 15, 906 (1949).

(see Sect. 7). They are also in satisfactory agreement with the values of COOKE and HULL, published by COOKE[1]. These data were already given in Fig. 24 of the present article, compare also Fig. 64.

If χ' for the ellipsoid is reduced to spherical shape the values near the susceptibility maximum lead still to satisfactory results, but discrepancies are found at the lower temperatures, the differences between the measured χ' values for the two samples being too small. The explanation is that the relaxation phenomena quoted above give a phase shift between M and H_{ext}, and hence between H_{ext} and the demagnetizing field, so that the calculation of H_{int} becomes more complicated than usually. STEENLAND, assuming that the field in the grains can be described by BREIT's formula (see Sect. 7), calculated that the phase lag between H_{int} and H_{ext} is about 10°. This leads to relaxation times of a few times 10^{-4} sec., but for a given entropy the time is not independent of the frequency. Apparently the relaxation phenomena cannot be described with one single relaxation time.

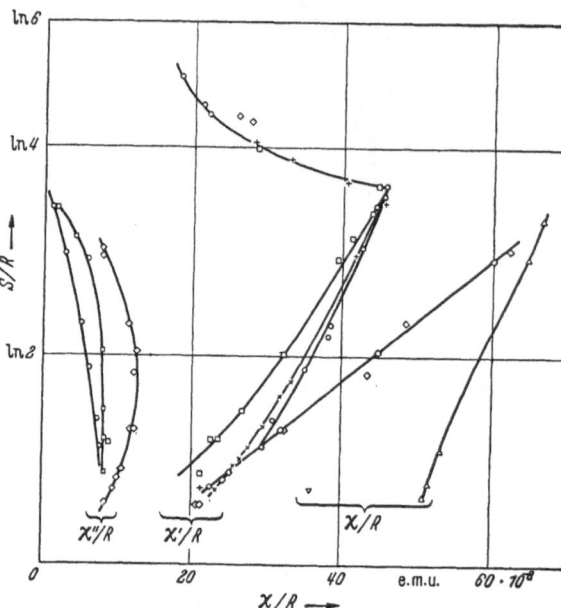

Fig.63. Entropy versus susceptibility diagram for manganese ammonium-sulphate (according to STEENLAND).
Powdered sphere:
  $\triangledown$ χ/R; measuring field 1.08 oersteds.
  $+$ χ'/R and χ''/R; measuring field 0.183 oersteds, $\nu = 225$ c/sec.
  $\times$ χ'/R and χ''/R; measuring field 0.610 oersteds, $\nu = 225$ c/sec.
  $\bigcirc$ χ'/R and χ''/R; measuring field 0.183 oersteds, $\nu = 225$ c/sec.
  $\square$ χ'/R and χ''/R; measuring field 1.22 oersteds, $\nu = 525$ c/sec.
Powdered spheroid 3:1:
  $\triangle$ χ/R; measuring field 3.38 oersteds.
  $\diamond$ χ'/R and χ''/R; measuring field 0.143 oersteds, $\nu = 225$ c/sec.

The remanent magnetic moments found in Leiden are much smaller than those for the alums, see Sects. 57, 58 and 60. In the case of the sphere, at $S = 0.249R$, a value $\Sigma = (0.10 \pm 0.05)$ gauss cm.^{3}/mol was found in a field of 1.08 oersted. The value for the ellipsoid, under the same conditions, is about 0.3 gauss cm.^{3}/mol. KURTI[2], investigating a compressed ellipsoid of axial ratio 6:1, found a much larger remanent moment, viz. 23.5 gauss cm.^{3}/mol. Maybe this is due to the different way of preparation of the sample, or the rate of precooling, see Sect. 57.

Absolute temperature determinations were made in Leiden with a.c. heating and in Oxford with gamma rays.

The a.c. experiments were performed below the susceptibility maximum, since χ'' was too small above it. The only useful thermometric parameter was χ''. At the lowest temperatures, however, χ'' became more and more constant (see Fig. 63), so that no good measurements could be made there. The experiments with $\nu = 525$ c/sec. could be made down to $S = R \ln 2$, those with $\nu = 225$ were extended to a somewhat lower entropy. Due to the bad thermal conductivity

[1] A. H. COOKE: Proc. Phys. Soc. Lond. 62, 269 (1949).
[2] N. KURTI: Proc. Int. Conf. Low. Temp. M.I.T. 1949, p. 59.

of the sample the experiments had to be finished within two minutes after each demagnetization.

The results are shown in Fig. 64. The precision is poor, but it is clear that the curve is very steep, so that the specific heat is large. This is probably due to the hyperfine splitting of the Mn ion, see. Sect. 41.

The data of COOKE and HULL are also shown in Fig. 64. They were obtained with an ellipsoidal sample of axial ratio 4:1. The T^* values are in reasonable agreement with the Leiden data, as was stated above. The absolute temperatures suggest also a big specific heat, but the values are noticeably higher than those obtained in Leiden, to such an extent that for the Leiden data T is lower than T^*, whereas in the Oxford results below the susceptibility maximum T is higher than the corresponding T^*.

If we identify the CURIE point with the susceptibility maximum the Leiden value is $S = 1.28R$, $T = 0.12°$ K and the Oxford value[1] $S = 1.27R$, $T = 0.15°$ K.

62. Copper potassium sulphate. Experiments at the lowest temperatures were performed in Leiden with a glass sphere filled with small crystals[2,3,4].

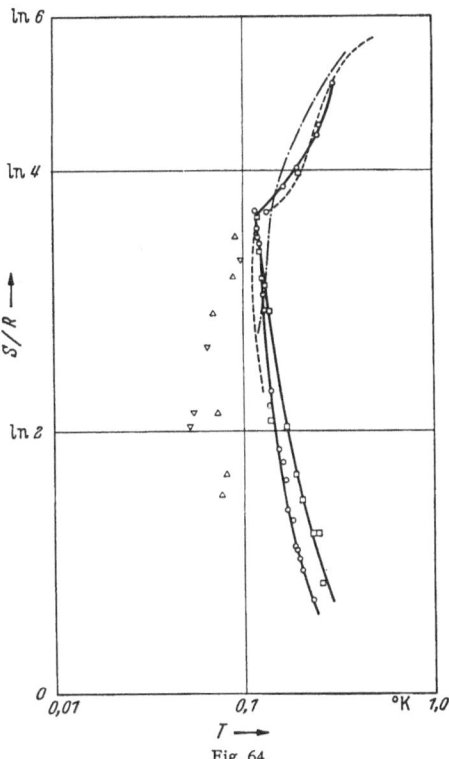

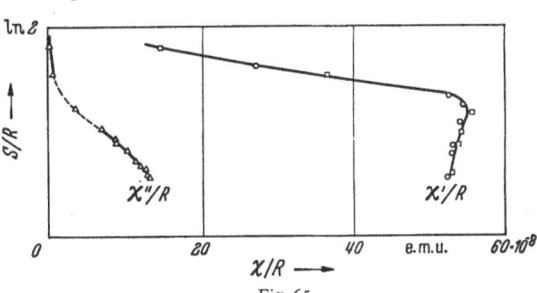

Fig. 64.

Fig. 65.

Fig. 64. Entropy versus magnetic and absolute temperature diagram for manganese ammonium sulphate (according to STEENLAND, VAN DER MAREL, DE KLERK and GORTER).

⊙ T^*, Leiden data for $\nu = 225$ c/sec.   △ T, Leiden data for $\nu = 225$ c/sec.
□ T^*, Leiden data for $\nu = 525$ c/sec.   ▽ T, Leiden data for $\nu = 525$ c/sec.
— — — T^*, Oxford ballistic data.   — · — · — T, Oxford data.

Fig. 65. Entropy versus susceptibility diagram for a powdered sphere of copper potassium sulphate (according to STEENLAND, DE KLERK, BEUN and GORTER). χ''/R is plotted on a tenfold magnified scale.

□ χ'/R, measuring field 0.609 oersteds, $\nu = 225$ c/sec.
○ χ'/R, measuring field 6.09 oersteds, $\nu = 225$ c/sec.
△ χ''/R, measuring field 0.609 oersteds, $\nu = 225$ c/sec.

The susceptibility data are shown in Fig. 65. The measurements were carried out at $\nu = 225$ c/sec. The susceptibility maximum was found at $S = 0.42R$. Below, χ' decreases only little with falling entropy and it depends slightly on the

[1] N. KURTI: J. de Phys. **12**, 281 (1951).
[2] M. J. STEENLAND, L. C. VAN DER MAREL, D. DE KLERK and C. J. GORTER: Leiden Commun. No. 279c; Physica, 's-Grav. **15**, 906 (1949).
[3] M. J. STEENLAND, D. DE KLERK, J. A. BEUN and C. J. GORTER: Leiden Commun. No. 284d; Physica, 's-Grav. **17**, 161 (1951).
[4] M. J. STEENLAND: Thesis, Leiden 1952.

measuring field. It was found that the ballistic susceptibilities (not shown in Fig. 65) are about one percent larger than the χ' values.

The quantity χ'' is very small above the maximum of χ'. Below this maximum χ'' increases with falling entropy, but it does not show a maximum. The phase angle is small, the highest value found for χ''/χ' is 0.0242, at $S = 0.19R$.

Remanent magnetic moments are very small. Σ, in a field of 1.08 oersted, was found to be (0.03 ± 0.01) gauss cm.^{3}/mol at $S = 0.20R$.

Absolute temperature determinations were performed with the help of a.c. heating. χ'' was the only useful thermometric parameter. The results are shown in Fig. 66. The T^* values are in good agreement with the experiments of CASIMIR, DE KLERK and POLDER, see Sect. 42. T and T^* intersect at $0.01°K$, and the absolute temperatures can be extrapolated without difficulties to the high temperature values, derived by GARRETT from the Ξ parameter, see Sect. 42.

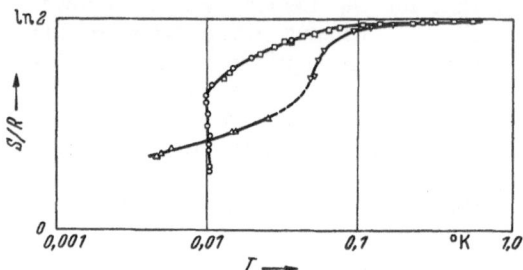

Fig. 66. Entropy versus magnetic and absolute temperature diagram for copper potassium sulphate (according to STEENLAND, BEUN, DE KLERK and GORTER).

○ T^*, data of STEENLAND, BEUN, DE KLERK and GORTER.
□ T^*, data of CASIMIR, DE KLERK and POLDER.
△ T, data of STEENLAND, BEUN, DE KLERK and GORTER.
▽ T, data of GARRETT.

The absolute temperature curve as drawn in Fig. 66 suggests the occurrence of a maximum in the specific heat near the maximum of χ'.

It was pointed out in Sect. 42 that the susceptibility of copper potassium sulphate is noticeably anisotropic. If also χ'' should exhibit such an anisotropy, the heat absorption in different grains of the powder might be widely different. In that case temperature differences may arise which, due to the bad heat conductivity, are not equalized in a reasonable time. If this effect occurs, appreciable mistakes may originate in the absolute temperature values of Fig. 66.

63. Cobalt ammonium sulphate. Measurements at the lowest temperatures were performed by GARRETT[1] with a spherical single crystal.

The susceptibilities along the three principal magnetic axes (see Sect. 44) are shown in Fig. 67. The experiments were performed at $\nu = 40$ c/sec. The sequence of the three curves is not the same as that of the T^* curves of Fig. 28. This is due to the fact that the CURIE constants in the directions of the different magnetic axes are widely different, see Table 13.

The χ'' values are only given for the K_3 axis. For the other orientations χ'' is so small that the points should almost coincide with the S/R axis. The quantity χ''/χ' along the K_3 axis reaches the value 0.224 at $S = 0.055R$; this is rather high as compared with other salts. Only manganese ammonium sulphate shows a larger value, see Sect. 61. χ''/χ' is much smaller for the other orientations, viz. 0.0038 at $S = 0.017R$ for the K_1 axis, and 0.022 at $S = 0.158R$ for the K_2 axis. These values are more like those obtained with the alums.

Remanent magnetism and hysteresis loops have not been investigated.

Absolute temperature determinations were performed with the help of a.c. heating, χ'' being the thermometric parameter. Experiments could only be made below the CURIE point and with the alternating field parallel to the K_3 axis. The results are shown in Fig. 68, they can be extrapolated without difficulties to the data derived from the parameter Ξ, see Sect. 44. The steep part of the

---

[1] C. G. B. GARRETT: Proc. Roy. Soc. Lond., Ser. A **206**, 242 (1951).

curve is due to the occurrence of the CURIE point. It is localized at $T = 0.084°$ K, and the corresponding T^* values in the directions of the K_1, K_2 and K_3 axes are $0.137°$, $0.113°$ and $0.0342°$.

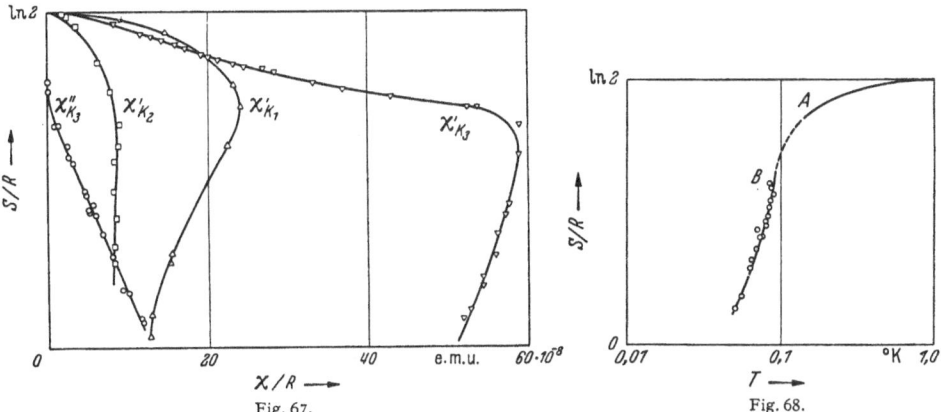

Fig. 67. Entropy versus susceptibility diagram for a spherical single crystal of cobalt ammonium sulphate (according to GARRETT).

$\triangle$ χ'/R for the K_1 axis. $\square$ χ'/R for the K_2 axis. ∇ χ'/R for the K_3 axis. $\bigcirc$ χ''/R for the K_3 axis.

Fig. 68. Entropy versus absolute temperature diagram for cobalt ammonium sulphate (according to GARRETT). Curve A: derived from the Ξ parameter, see Sect. 44. Curve B: derived from experiments with a.c. heating.

III. The influence of magnetic fields.

64. Chromium methylamine alum. The influence of a magnetic field, as was pointed out in Sect. 50, may be investigated with the applied field either parallel or perpendicular to the small measuring field. Since most salts show marked anisotropies below their CURIE points, Eqs. (50.1) and (50.2) lose their validity, so that magnetic moments can only be derived from the measurements in longitudinal fields. Still the investigations in transverse fields are of some importance, since they may provide information on the anisotropies. For several salts both kinds of experiments have been performed.

The course of the susceptibility with a transverse field for chromium methylamine alum is shown in Fig. 69. The experiments were performed

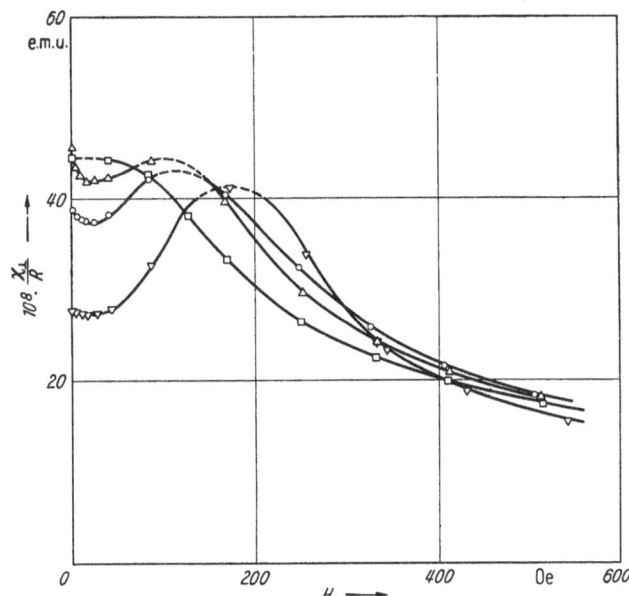

Fig. 69. Susceptibility $\chi_\perp$ as a function of a transverse field for a spherical single crystal of chromium methylamine alum (according to BEUN, STEENLAND, DE KLERK and GORTER). Both the measuring field and the applied field parallel to cubic axes

$\square$ $S = 0.537\,R$, $\triangle$ $S = 0.455\,R$, $\bigcirc$ $S = 0.366\,R$, ∇ $S = 0.200\,R$.

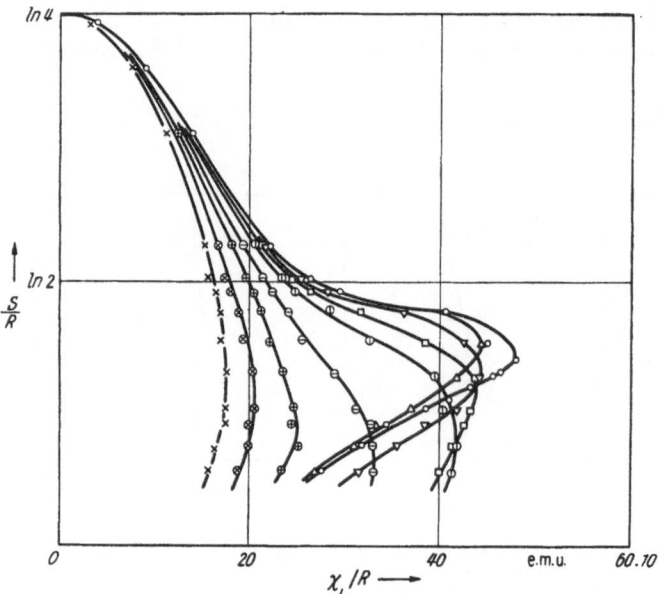

Fig. 70. Entropy versus susceptibility diagram in transverse fields ($\chi_\perp$) for a spherical single crystal of chromium methylamine alum (according to BEUN, STEENLAND, DE KLERK and GORTER). Lines of constant field strength.

$S > 0.79\,R$　a.c. measurements.　　$S < 0.79\,R$　ballistic measurements.
　　○ 0 oersteds.　□ 130 oersteds.　⊕ 340 oersteds.
　　△ 20 oersteds.　⊕ 170 oersteds.　⊠ 430 oersteds.
　　▽ 80 oersteds.　⊖ 260 oersteds.　× 540 oersteds.

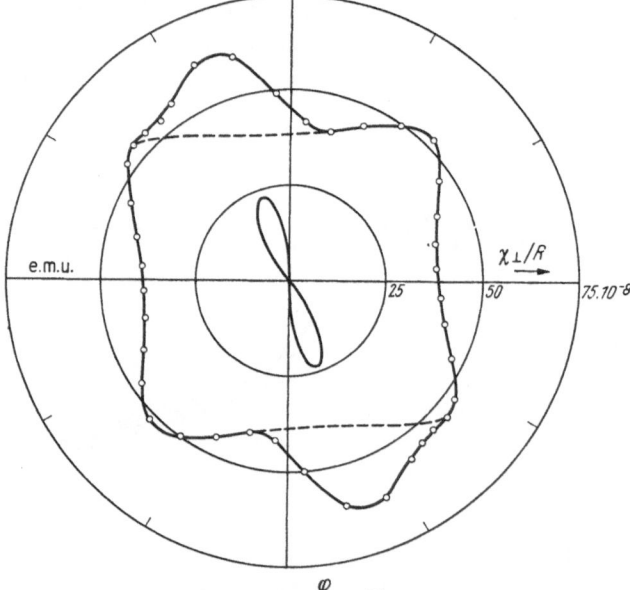

Fig. 71. Polar diagram of susceptibility $\chi_\perp$ versus orientation of the transverse field for a spherical single crystal of chromium methylamine alum (according to BEUN, STEENLAND, DE KLERK and GORTER). The straight lines are the directions of the cubic axes. ○ Experimental points, $S = 0.20\,R$, $H_\perp = 170$ oersteds. The experimental curve may be split into a diagram with quaternary symmetry and an excess curve with twofold symmetry.

by BEUN, STEENLAND, DE KLERK and GORTER[1] with a spherical single crystal mounted in such a way that both the applied field and the measuring field were parallel to cubic axes. The investigations were carried out ballistically with a measuring field of 1.08 oersted and with a galvanometer with a free period of 1.3 sec.

In small fields, up to about 20 oersteds, the susceptibility shows a slight decrease. After this a marked increase is found, followed by a maximum. Finally the susceptibility falls again to a rather low value. A survey of the results is given in the (S, χ)-diagram of Fig. 70.

Some measurements performed with a.c. showed that the course of χ' with field is very similar to that of χ. The χ'' showed a pronounced maximum at the same field value as χ'; there it was about a factor two larger than in zero field. The remanent moment decreases very steeply with the applied field. At $S = 0.36\,R$, in a field of ten oersteds, it is already smaller than one tenth of its initial value.

The occurrence of strong anisotropies was demonstrated by rotating the applied field around the axis of the measuring field. One of the polar diagrams ob-

[1] J. A. BEUN, M. J. STEENLAND, D. DE KLERK and C. J. GORTER: Leiden Commun. No. 301a; Physica, 's-Grav. 21, 767 (1955).

tained in this way is shown in Fig. 71. The most surprising result is that the symmetry is not quaternary, as might be expected around a cubic axis, but binary. It is possible, however, to split the curve formally into two parts, one with the expected fourfold symmetry (with its sides parallel to the cubic axes), and an excess curve with binary symmetry. The orientation of the excess curve proved to be different during subsequent helium runs (compare Figs. 71 and 72). Maybe this is due to processes during the precooling to liquid air temperature, see Sect. 57.

For a given field the anisotropy is the more pronounced the lower the entropy is, see Fig.72. The anisotropy is small for relatively small fields (see the curve for 42.5 oersteds in Fig. 73), but at higher fields it becomes more pronounced and increases relatively with increasing field.

One polar diagram was measured above the Curie point (at $S = 0.70R$ and in a field of 130 oersteds); no anisotropy was observed there.

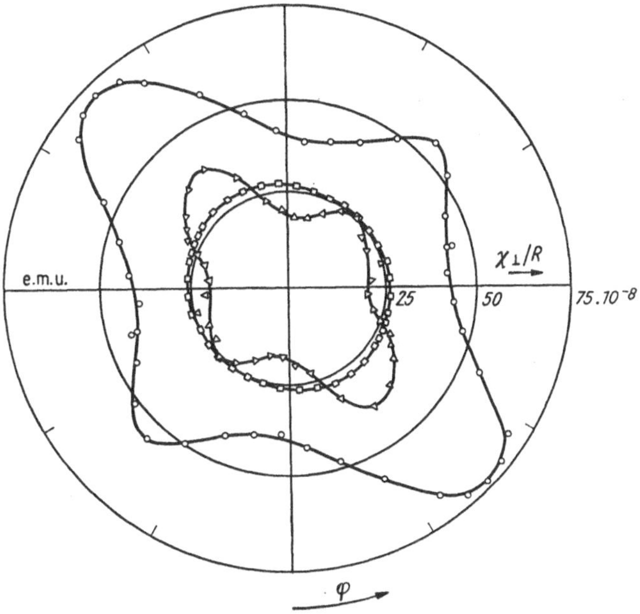

Also Hudson and McLane[1] made investigations in perpendicular fields, with the applied field parallel to a cubic axis. The experiments were carried out with

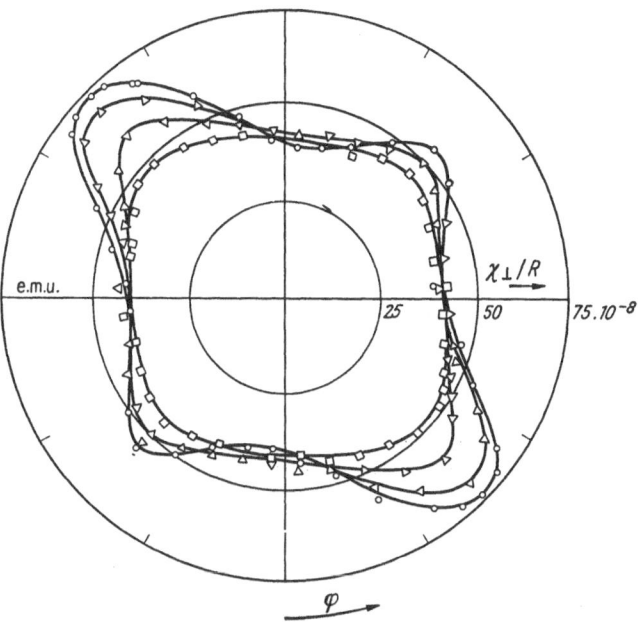

Fig. 72. Polar diagram of susceptibility $\chi_\perp$ versus orientation of the transverse field for a spherical single crystal of chromium methylamine alum (according to BEUN, STEENLAND, DE KLERK and GORTER). The straight lines are the directions of the cubic axes. Curves for $H_\perp = 170$ oersteds at various entropy values.

$\bigcirc\ S = 0.20\,R$, $\quad\triangle\ S = 0.26\,R$, $\quad\triangledown\ S = 0.33\,R$, $\quad\square\ S = 0.44\,R$.

Fig. 73. Polar diagram of susceptibility $\chi_\perp$ versus orientation of the transverse field for a spherical single crystal of chromium methylamine alum (according to BEUN, STEENLAND, DE KLERK and GORTER). The straight lines are the directions of the cubic axes. Curves at $S = 0.20\,R$ for various values of the field.

$\square\ H_\perp = 42.5$ oersteds, $\quad\bigcirc\ H_\perp = 170$ oersteds, $\quad\triangle\ H_\perp = 425$ oersteds.

[1] R. P. Hudson and C. K. McLane: Phys. Rev. 95, 932 (1954).

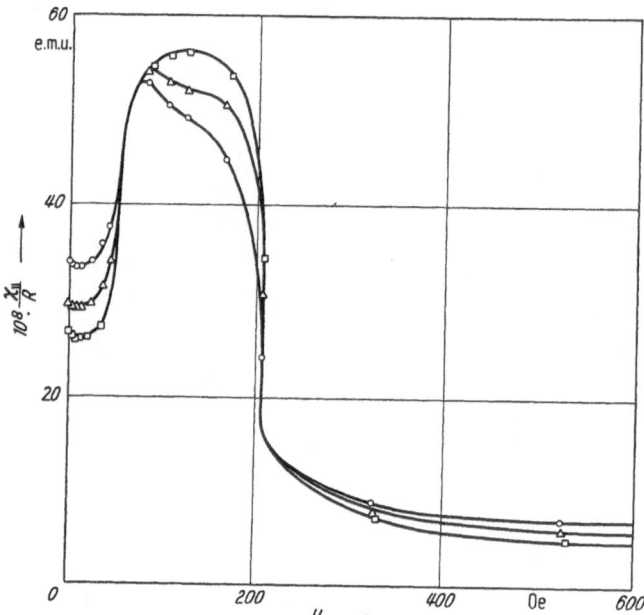

Fig. 74. Susceptibility $\chi_{||}$ as a function of a longitudinal field for a spherical single crystal of chromium methylamine alum (according to Beun, Steenland, de Klerk and Gorter). The fields were applied parallel to a cubic axis.

○ $S = 0.318\,R$,   △ $S = 0.253\,R$,   □ $S = 0.200\,R$.

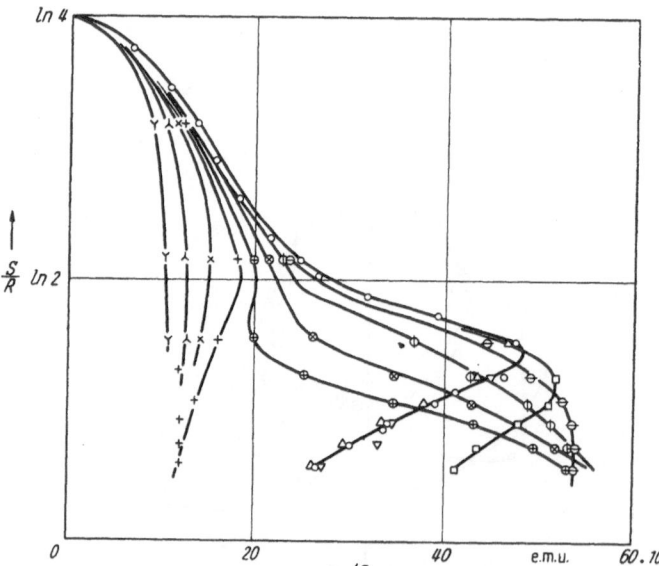

Fig. 75. Entropy versus susceptibility diagram in longitudinal fields ($\chi_{||}$) for a spherical single crystal of chromium methylamine alum (according to Beun, Steenland, de Klerk and Gorter). Lines of constant field strength. Ballistic measurements.

○ 0 oersteds.	⊖ 71 oersteds.	+ 230 oersteds.
△ 9 oersteds.	⦶ 106 oersteds.	× 318 oersteds.
▽ 27 oersteds.	⊠ 141 oersteds.	⋏ 424 oersteds.
□ 53 oersteds.	⊕ 177 oersteds.	Υ 531 oersteds.

a.c. at 210 c/sec. The susceptibility curves were similar to those of Fig. 69, but the height of the maximum increased with falling entropy, to such an extent that, in the χ versus S diagram, the curve for 180 oersteds exhibited a higher maximum than the zero field curve, contrary to the Leiden results of Fig. 70. A possible explanation is that in the sample of Hudson and McLane the excess curve with twofold symmetry happened to be parallel to the cubic axis.

Further it was found in Hudson and McLane's experiments that the small decrease of the susceptibility in fields below 20 oersteds did not occur at the lowest entropies, below $S = 0.48\,R$.

Susceptibility curves obtained in Leiden from experiments in longitudinal fields are shown in Fig. 74. The applied field was parallel to a cubic axis. The measurements were performed ballistically; the small measuring field was 1.08 oersted, and the free period of the galvanometer was 1.3 sec.

Qualitatively the results are similar to those in transverse fields (compare Fig. 69), but the maxima are much higher and the rise and fall of the curves are almost vertical. They take place at 60 and 210 oersteds.

An indication is found for a double maximum, but it disappears again at the lowest temperatures.

AMBLER and HUDSON[1] made measurements in longitudinal fields, also by the ballistic method, the measuring field being 1.72 oersted, the free period of the galvanometer 5.6 sec. The susceptibility versus field curves were very similar to the Leiden results, though the agreement was not quantitative. Indications for a double maximum were found below $S = 0.45 R$.

The Leiden entropy versus susceptibility diagram is shown in Fig. 75. Here, contrary to Fig. 70, the susceptibilities in moderate fields reach values noticeably higher than the maximum of the zero field curve.

Since the susceptibility in a longitudinal field is equal to $(\partial M/\partial H)_S$ (see Sect. 50) the magnetization curves can be obtained by integration. Some Leiden data are given in Fig. 76 (the figure contains also a few curves above the CURIE point). Due to the increase of the susceptibility in small fields the curves well below the CURIE point start with a concave part. The sharp fall in the susceptibility at 210 oersteds results in a sharp bend of the magnetization curve (very pronounced in curve E). Above this bend the course of M with H is convex. At the bend, the magnetic moment

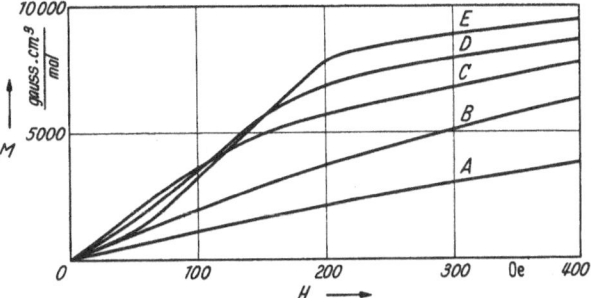

Fig. 76. Magnetization curves for a spherical single crystal of chromium methylamine alum (according to BEUN, STEENLAND, DE KLERK and GORTER). Fields applied parallel to a cubic axis. The value of the saturation magnetization is 16 700 gauss cm.³/mol.

Curve A: $S = 1.111 R$, Curve B: $S = 0.738 R$, Curve C: $S = 0.529 R$, Curve D: $S = 0.372 R$, Curve E: $S = 0.200 R$.

is about one half of the saturation moment and the internal field in the sample was calculated to be 98 oersteds.

Also AMBLER and HUDSON calculated the magnetization curves from their measurements in longitudinal fields. They compared them with the curves computed from the experiments of HUDSON and McLANE in transverse fields on the assumption that anisotropy is absent [hence applying Eq. (50.1)]. Large deviations were found in strong fields, but reasonable agreement was obtained below about 100 oersteds. From this the conclusion was derived that the anisotropy in low fields is small and this is in agreement with the shape of the curve for 42.5 oersteds in Fig. 73.

The magnetic moment data may also be plotted in an M versus S diagram with lines of constant magnetic field. The values of AMBLER and HUDSON are shown in Fig. 77. Up to 120 oersteds the curves show a maximum and the locus of the maxima is indicated by the dotted line. Inside the region bounded by the dotted line $(\partial M/\partial S)_H$ is positive so that, according to Eq. (9.9) application of a magnetic field gives a decrease in temperature. Outside this region the magnetocaloric effect has the normal sign. Some of the Leiden data on the variation of temperature with the applied field are collected in Fig. 78.

An explanation of the susceptibility curves of Fig. 74 may be given on the basis of the occurrence of antiferromagnetic domains as introduced by NÉEL[2], see Sect. 55. It seems[3] that each curve may be derived into four intervals:

[1] E. AMBLER and R. P. HUDSON: Phys. Rev. **96**, 907 (1954).

[2] L. NÉEL: Proc. Conf. Theor. Phys., Tokyo 1953, p. 703.

[3] J. A. BEUN, M. J. STEENLAND, D. DE KLERK and C. J. GORTER: Leiden Commun. No. 301a; Physica, 's-Grav. **21**, 767 (1955).

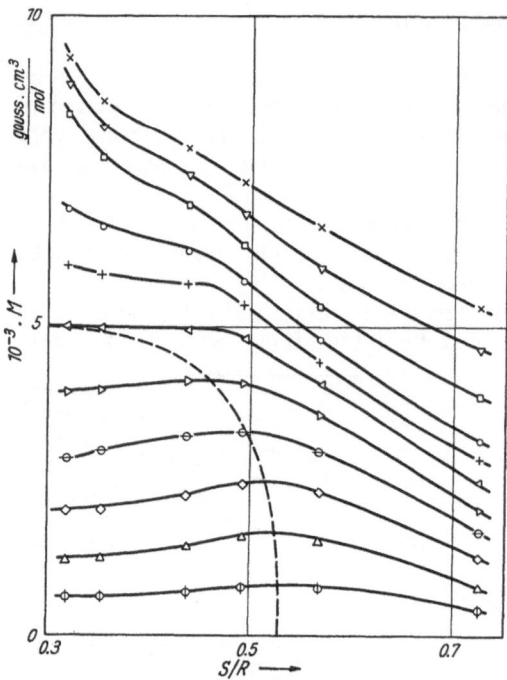

Fig. 77. Magnetic moment versus entropy diagram for a spherical single crystal of chromium methylamine alum (according to AMBLER and HUDSON). The magnetic fields were applied parallel to a cubic axis. The dotted line is the locus of the maxima.

  ⊕ 20 oersteds.   ▷ 100 oersteds.   □ 200 oersteds.
  △ 40 oersteds.   ◁ 120 oersteds.   ▽ 250 oersteds.
  ◇ 60 oersteds.   + 140 oersteds.   × 300 oersteds.
  ⊖ 80 oersteds.   ○ 160 oersteds.

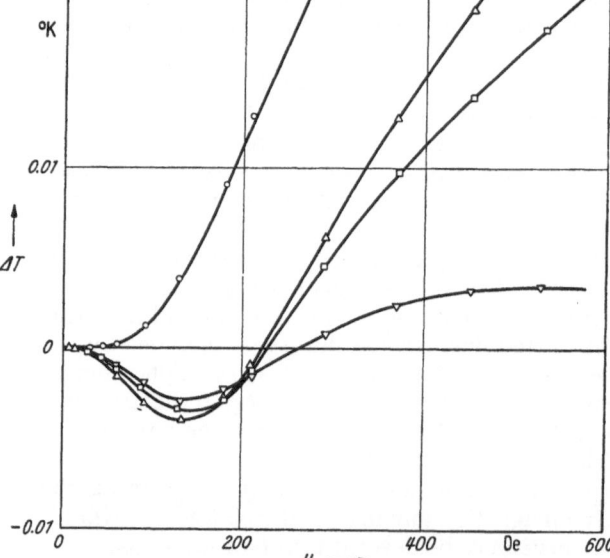

Fig. 78. Variation of temperature with field on adiabatic magnetization curves for a spherical single crystal of chromium methylamine alum (according to BEUN, STEENLAND, DE KLERK and GORTER).

  ○ $S = 0.500 R$,   △ $S = 0.400 R$,   □ $S = 0.300 R$,   ▽ $S = 0.200 R$.

(a) The region up to about 10 oersteds. Here the susceptibility shows a modest but steep fall with increasing field while the weak hysteresis phenomena are rapidly reduced.

(b) The region between about 10 and 60 oersteds. Here the susceptibility is approximately constant though, as a matter of fact, there is a flat minimum between the fall below 10 oersteds and the spectacular rise at about 60 oersteds.

(c) The interval between 60 and 210 oersteds. At about 60 oersteds the susceptibility rises to a level well above the zero field value and a decrease occurs near 210 oersteds. In between, it is more or less constant. In this region a marked crystalline anisotropy shows up in $\chi_\perp$ with a binary superposed on a quaternary symmetry.

(d) The region above 210 oersteds. Here $\chi_\parallel$ has a low value while $\chi_\perp$ decreases smoothly. The crystalline anisotropy persists.

The border between the regions a and b is not very marked. In the picture of the antiferromagnetic domains the small decrease in the susceptibility and the occurrence of hysteresis phenomena in region a may be ascribed to the magnetization in 180° walls between domains. This magnetization is easily oriented into the direction of the external field. The contribution to the magnetic moment due to this effect was found to be at most 0.03% of the saturation.

The approximately constant susceptibility which persists in region b may be due to a combination of three causes. (I) The spins in domains with orientations roughly perpendicular to the field are bent over into the direction of the field. This effect leads to a susceptibility independent of H and T (see Sect.55). (II) Spins in domains oriented almost parallel to the field flop over into the field direction. This gives a susceptibility increasing rapidly with T and slightly with H. (III) Walls between domains whose orientations make arbitrary angles with the field undergo a reversible shift. As NÉEL pointed out, the MAXWELL pressure acting on the walls is proportional to the square of the field, so that this effect entails a susceptibility proportional to H.

The larger contribution to the rapid rise of the susceptibility at about 60 oersteds is probably due to a switching over of domains oriented parallel to the field to an orientation perpendicular to it. Here we have the field H_c quoted in Sect. 55. It is difficult to predict a value for H_c; it depends widely on the anisotropy. Mrs. VAN PESKI and GORTER showed that $H_c = 0$ for a cubic crystal, if the preferred orientation of the ions is parallel to a cubic axis (see Sect. 55). Nothing is known about the latter point and, since the symmetry of the salt proves to be lower than cubic, it is not surprising that H_c has a finite value.

In the interval c we have apparently to do with antiferromagnetic directions perpendicular to the field which are gradually bent over towards the direction of the field (similar to effect I in region b).

At about 210 oersteds we reach the field where the antiparallel magnetizations of the sublattices have been bent into the direction of the field. Here antiferromagnetism goes over into paramagnetism (in a transition of the second order).

In region d paramagnetic saturation should be complete at absolute zero, with $\chi_{\parallel} = 0$ and $\chi_{\perp}$ equal to the saturation moment divided by H. At the lowest temperature reached in the experiments the magnetic moment at 210 oersteds is about half the saturation; $\chi_{\parallel}$ is small, and $\chi_{\perp}$ is approximately proportional to H^{-1}.

Lines of constant magnetic moment in the (M, S) diagram of Fig. 77 are equivalent [according to Eq. (9.7)] to the lines of constant entropy in the (T, H) diagram as given by GARRETT in Fig. 40. The reason that the (M, S) diagram was preferred by AMBLER and HUDSON is that the uncertainty in the absolute temperatures is appreciably larger than that of the entropies, see Sect. 57.

The locus of the maxima in Fig. 77 was identified by AMBLER and HUDSON with the critical field curve of Sect. 55, hence with the borderline between the antiferromagnetic and paramagnetic regions. (The same assumption had been made earlier, for the case of cobalt ammonium sulphate, by GARRETT[1, 2], see Sect. 67). In the considerations given above, however, the borderline between the two regions is the locus of the sharp fall in the $\chi_{\parallel}$ versus H curves of Fig. 74. There is no reason why this should coincide with the maxima of the constant field curves of Fig. 77. In fact, at the lower entropies, it is localized well above the dotted line of Fig. 77.

The dotted line of Fig. 77, as was pointed out before, is the locus where the magnetocaloric effect reverses its sign. KURTI remarked[3] that a decrease of temperature on an isentropic magnetization curve follows quite naturally from the antiferromagnetic arrangement. It is equivalent with a rise of entropy on an isothermal magnetization curve, and this may be caused by the fact that the antiferromagnetic order, in the first instance, is spoilt by the application of a field. KURTI's argument, however, does not necessarily entail that the transition from the antiferromagnetic to the paramagnetic region must coincide

[1] C. G. B. GARRETT: Proc. Roy. Soc. Lond., Ser. A **206**, 242 (1951).
[2] C. G. B. GARRETT: Magnetic Cooling, p. 75. Cambridge (Mass.) 1954.
[3] N. KURTI: J. de Phys. **12**, 281 (1951).

with the point where the magnetocaloric effect reverses its sign. Also in Fig. 40 the critical field curve does not coincide with the locus of the minima of the lines of constant entropy.

Let us identify the field of 210 oersteds with the value of H_g at absolute zero. The quantity $kT_c/\mu H_g^0$, assuming $T_c = 0.020°$ K (see Sect. 57) and $\mu = \mu_B$, is then equal to 1.44. The values predicted from theory vary between 1 and 2, see Sects. 55 and 56.

A remarkable fact is that anisotropies of importance do not occur in relatively weak fields. Apparently, the anisotropy is not caused by the antiferromagnetic behaviour. The discovery of anisotropies with a symmetry lower than cubic came quite unexpectedly. The fact that the orientation of the binary component may be different during different helium runs, gives the impression that some secondary cause determines its direction, maybe a deviation from the spherical shape, or strains in the crystal. Probably it is related with the result of BLEANEY, who observed a lower than cubic symmetry in his paramagnetic resonance experiments at and below liquid air temperatures (see Sect. 34). It would be desirable to obtain data on the $\chi_{\parallel}$ in other directions than the cubic axis.

65. Chromium potassium alum. The course of susceptibility with an applied field in the case of chromium potassium alum is quite different from that for chromium methylamine alum. The $\chi_{\perp}$ versus S diagram as given by BEUN,

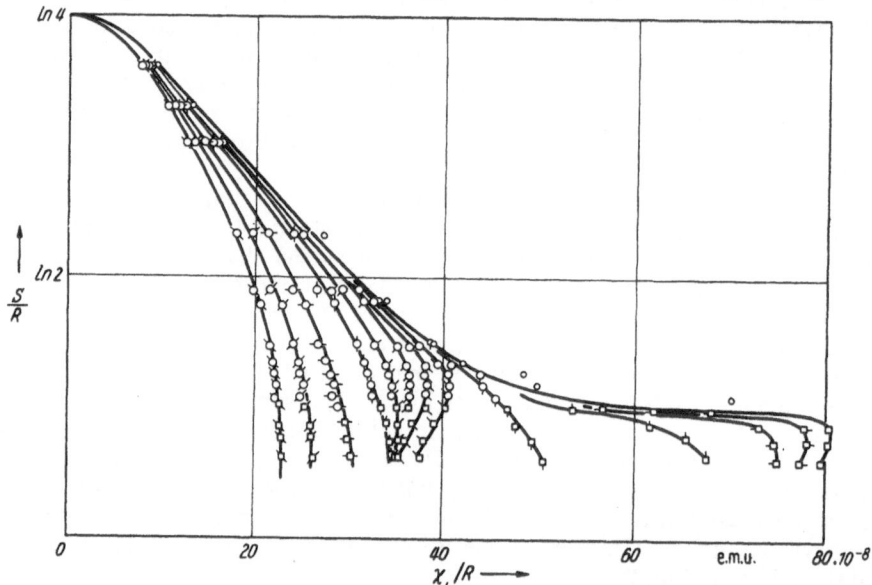

Fig. 79. Entropy versus susceptibility diagram in transverse fields ($\chi_{\perp}$) for a spherical single crystal of chromium potassium alum (according to BEUN, STEENLAND, DE KLERK and GORTER). Lines of constant field strength.

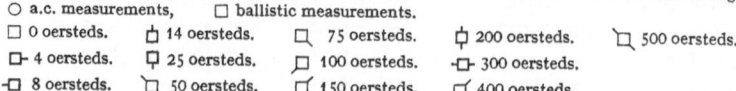

○ a.c. measurements,   □ ballistic measurements.

□ 0 oersteds.	⊡ 14 oersteds.	⊡ 75 oersteds.	⊡ 200 oersteds.	⊡ 500 oersteds.
⊡ 4 oersteds.	⊡ 25 oersteds.	⊡ 100 oersteds.	⊡ 300 oersteds.	
⊡ 8 oersteds.	⊡ 50 oersteds.	⊡ 150 oersteds.	⊡ 400 oersteds.	

STEENLAND, DE KLERK and GORTER[1] is shown in Fig. 79. It was obtained with one of the four spherical single crystals of Fig. 50 (curve d), the applied field and the measuring field both being parallel to cubic axes.

[1] J. A. BEUN, M. J. STEENLAND, D. DE KLERK and C. J. GORTER: Leiden Commun No. 300a; Physica, 's-Grav. **21**, 651 (1955).

The susceptibility maximum is shifted rapidly to lower entropies with increasing field strength. It disappears already at 8 oersteds, but a new maximum is found in the region between 50 and 200 oersteds.

The most remarkable result is the steep decrease of $\chi_\perp$ in small fields. At about 50 oersteds it has reached already half its zero field value, the further decrease in fields up to 500 oersteds being small. Due to the fact that in general $\chi_\perp$ decreases monotonously with increasing field, the curves of Fig. 79 do not intersect, contrary to those of Fig. 70. Only at the lowest entropy, the curves between 75 and 200 oersteds come together, giving rise to a very weak maximum in the $\chi_\perp$ versus field curve. This can be seen from Fig. 80 which shows both $\chi_\perp$ and $\chi_{\parallel}$ at $S = 0.20\,R$, the lowest entropy reached in the experiments.

Anisotropies in $\chi_\perp$ occur below the CURIE point. The results obtained at the lowest entropy[1] ($S = 0.20\,R$) are shown in Fig. 81. The anisotropies are much smaller than in the case of the chromium methylamine alum. In a polar diagram, like Fig. 73, they should hardly be visible. It follows that the symmetry in strong fields is quaternary, in weak fields it is binary. In between (from about 50 to 250 oersteds) a transition region occurs where the curves are very complicated and unsurveyable. At present, it is impossible to give an interpretation of these phenomena.

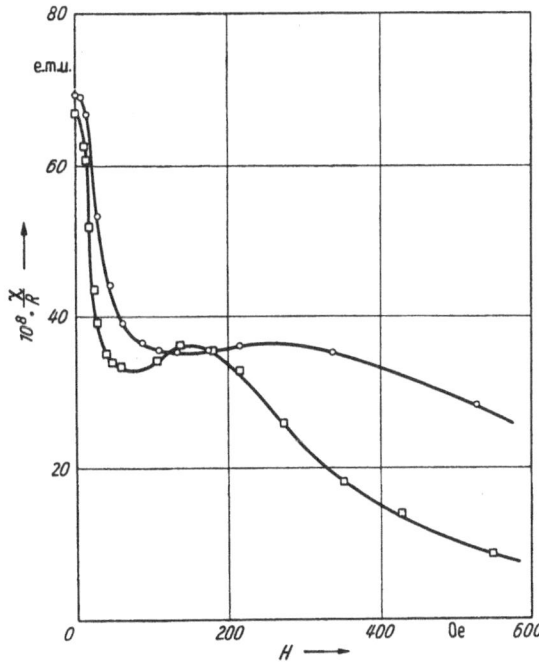

Fig. 80. Susceptibilities $\chi_\perp$ and $\chi_{\parallel}$ at $S = 0.20\,R$ for a spherical single crystal of chromium potassium alum (according to BEUN, STEENLAND and DE KLERK). $\bigcirc \chi_\perp$. $\square \chi_{\parallel}$.

Due to the anisotropies in $\chi_\perp$ the values of the magnetic moment and of the variation of temperature with an applied field can only be derived from measurements of $\chi_{\parallel}$, see Sect. 64.

The $\chi_{\parallel}$ versus entropy diagram, obtained recently by BEUN, STEENLAND, DE KLERK and GORTER[2] is shown in Fig. 82. The experiments were made with a new spherical single crystal, not one of Fig. 50. The fields were applied parallel to a cubic axis. The course of $\chi_{\parallel}$ in general is fairly analogous to that of $\chi_\perp$. The susceptibility decreases strongly in weak fields, and the lines intersect only at the lowest entropies giving rise to the small maximum in the $\chi_{\parallel}$ curve of Fig. 80.

Due to the high susceptibility values in small fields the magnetization curves (M versus H, obtained by integration of $\chi_{\parallel}$) start at the lowest entropies with a rather steep part. In quite small fields, however, the slope is already much smaller. The curves are convex over the whole region and one half of the saturation moment is reached, at the lowest entropy ($S = 0.20\,R$), in a field of about

[1] J. A. BEUN, M. J. STEENLAND and D. DE KLERK: Report Low Temperature Conference, Paris 1955, to be published.
[2] J. A. BEUN, M. J. STEENLAND, D. DE KLERK and C. J. GORTER: To be published.

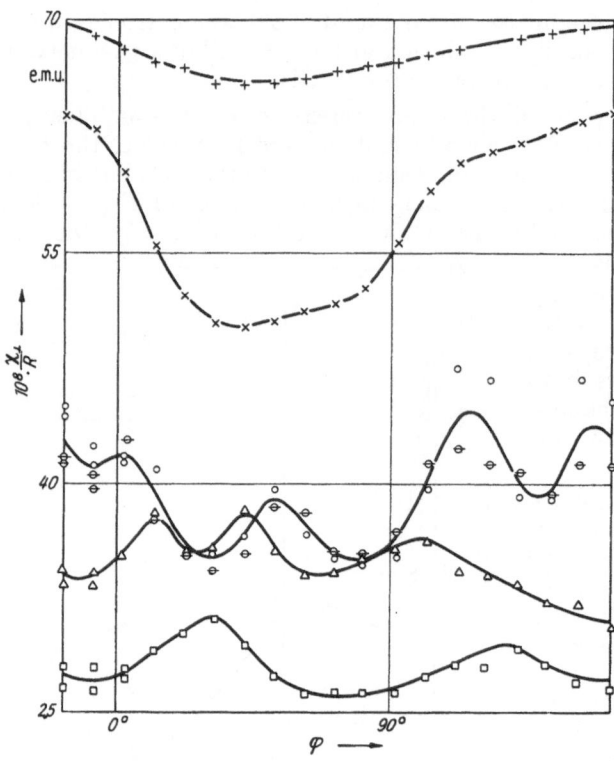

Fig. 81. Susceptibility $\chi_\perp$ as a function of the orientation of the applied field $H_\perp$ for a spherical single crystal of chromium potassium alum at $S = 0.20\,R$ (according to BEUN, STEENLAND and DE KLERK). The directions 0° and 90° are those of the cubic axes.

+   12.74 oersteds.
×   25.48 oersteds.
○   76.45 oersteds.
⊖   138.6 oersteds.
△   254.8 oersteds.
□   403.5 oersteds.

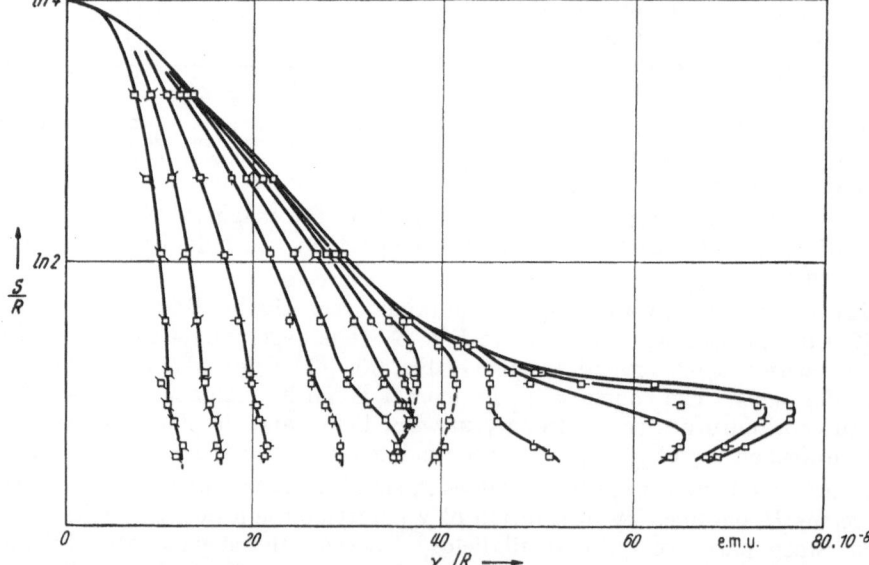

Fig. 82. Entropy versus susceptibility diagram in longitudinal fields ($\chi_{\parallel}$) for a spherical single crystal of chromium potassium alum (according to BEUN, STEENLAND, DE KLERK and GORTER). Lines of constant field strength.

□ 0 oersteds.	⛀ 14 oersteds.	▢ 75 oersteds.	⛏ 200 oersteds.	⛝ 500 oersteds.
▢ 4 oersteds.	⛝ 25 oersteds.	▢ 100 oersteds.	⊟ 300 oersteds.	
⊣ 8 oersteds.	⛝ 50 oersteds.	▱ 150 oersteds.	◹ 400 oersteds.	

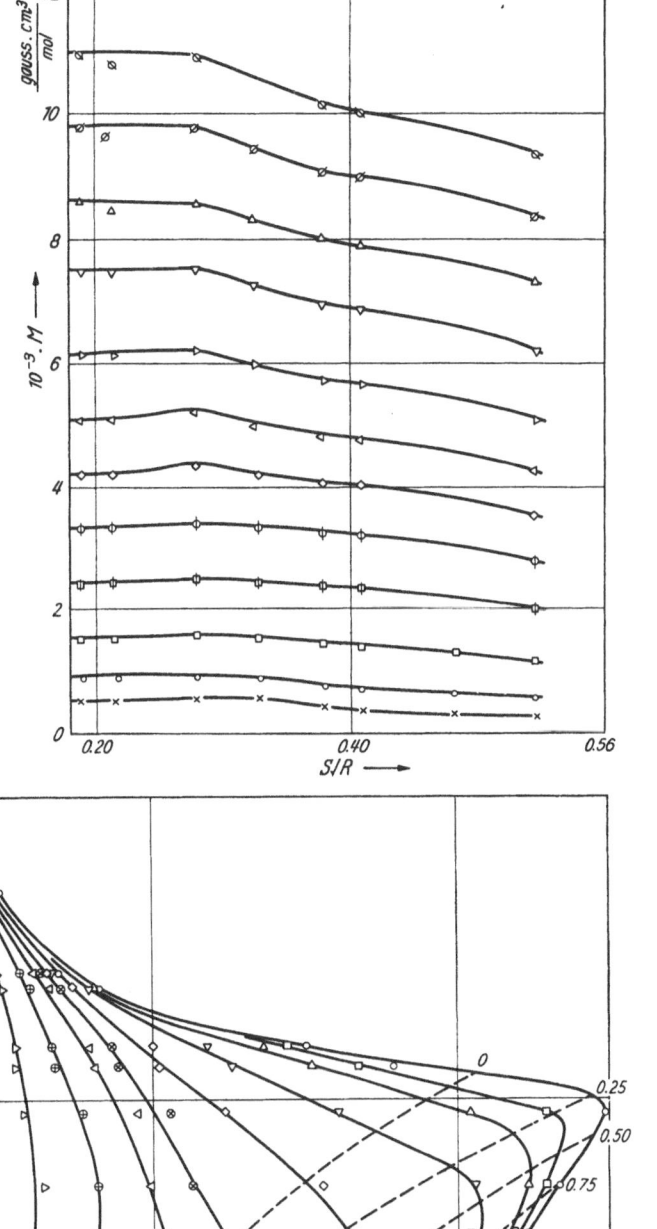

Fig. 83. Magnetic moment versus entropy diagram for a spherical single crystal of chromium potassium alum (according to BEUN, STEENLAND, DE KLERK and GORTER). The magnetic fields were applied parallel to a cubic axis.

×	10 oersteds.
O	20 oersteds.
□	40 oersteds.
⊞	70 oersteds.
⊕	100 oersteds.
◇	130 oersteds.
◁	160 oersteds.
▷	200 oersteds.
▽	260 oersteds.
△	320 oersteds.
⌀	400 oersteds.
⌀	500 oersteds.

Fig. 84. Entropy versus susceptibility diagram in transverse fields ($\chi_\perp$) for a spherical single crystal of chromium potassium alum (according to BEUN, STEENLAND, DE KLERK and GORTER). Curves of constant field strength (full lines) and of constant remanent magnetic moment (broken lines). For the latter the value of the remanent moment is given at the righthand side in gauss cm.³/mol.

O 0 oersteds.	□ 2 oersteds.	△ 4 oersteds.	▽ 8 oersteds.	◇ 14 oersteds.	⊗ 20 oersteds.	◁ 25 oersteds.
⊕ 35 oersteds.	▷ 50 oersteds.	× 75 oersteds.	+ 100 oersteds.	⊞ 125 oersteds.	⊟ 160 oersteds.	

300 oersteds. Sharp bends, like in the case of the methylamine alum (see Fig. 76), do not occur. If the magnetization curves are plotted against the internal field they start almost vertically[1].

The M versus S diagram is shown in Fig. 83. Up to 300 oersteds the curves show a maximum at an entropy which is not very much influenced by the value of the field strength. Due to this, the course of temperature with field on an isentropic is markedly different for entropies just above and below this maximum. Above, T increases gradually with increasing field, but below, a noticeable decrease is found in fields up to about 300 oersteds. The quantitative results are widely divergent for different samples.

The remanent magnetic moment, in fields of a few oersteds, is almost independent of the field strength. Sometimes it increases by a few percents[2]. In quite small fields, however, it decreases sharply and, at the lowest entropy, it disappears already in a field of 30 oersteds (the value is the same for $H_\perp$ and $H_{||}$). Fig. 84 gives part of a $\chi_\perp$ versus S diagram (corresponding to curve c of Fig. 50) with lines of constant remanent moment (the dotted lines, the value of Σ being given for each line in gauss cm.[3] per mole). It appears that the locus of the susceptibility maxima is inside the region where hysteresis occurs. Also the relaxation effects giving rise to double deflections of the ballistic galvanometer (see Sect. 24) were only found inside the hysteresis region.

The magnetic properties of chromium potassium alum and chromium methylamine alum seem to be strongly divergent. A closer consideration of Figs. 69, 74 and 80 may indicate, however, that there is some qualitative agreement. The slight decrease in weak fields for the methylamine alum is very pronounced in the case of potassium alum, but the spectacular maximum of methylamine alum is very small for potassium alum and occurs only at the very lowest entropies. An examination of Figs. 79 and 82, however, might suggest that this maximum may become more pronounced at still lower entropies (it should be realized that the entropy $0.2 R$ is more below the CURIE point of methylamine alum than below that of potassium alum).

In the picture of antiferromagnetic domains as given in Sect. 64 the behaviour of the potassium alum might be explained assuming that for this salt the domains are smaller and the walls between the domains are thicker than in the case of the methylamine alum.

It is suggested by Fig. 82 that in a field of about 13 oersteds $\chi_{||}$ does not show a maximum as a function of the entropy. This entails that $\chi_{||}$ in a field of 13 oersteds provides a very suitable thermometric parameter. If the field is kept well constant, it can be measured with a higher precision than χ'' or Σ. The recent Leiden absolute temperature measurements quoted at the end of Sect. 58 were made with the help of this parameter. It follows from Fig. 79 that also $\chi_\perp$ in a field of about 25 oersteds could be used.

66. Iron ammonium alum. Measurements in longitudinal fields were performed by KURTI[3] with an ellipsoid of axial ratio 8:1. The $\chi_{||}$ versus H curve for $S = 0.47 R$ (corresponding to a zero field temperature of $0.03°$ K, see Sect. 60) is shown in Fig. 85, curve A. The susceptibility decreases gradually with increasing field strength. No maximum as in the case of chromium methylamine alum

---

[1] C. J. GORTER et al: Progress in Low Temperature Physics, p. 307. Amsterdam 1955.
[2] M. J. STEENLAND: Thesis, Leiden 1952.
[3] N. KURTI: J. de Phys. **12**, 281 (1951).

is found, and no steep fall occurs as in the case of chromium potassium alum. This result was confirmed qualitatively in Leiden[1, 2] with a powdered sphere.

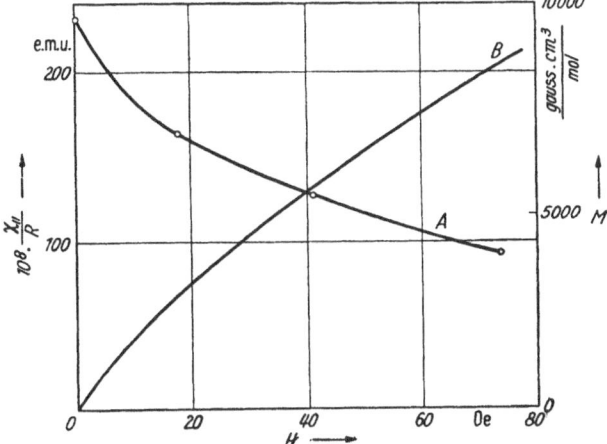

Fig. 85. Susceptibility $\chi_{||}$ (curve A) and magnetic moment M (curve B) as functions of a longitudinal field for a compressed spheroid 8:1 of iron ammonium alum at $S = 0.47R$ (according to KURTI). The value of the saturation magnetization is 27 800 gauss cm.³/mol.

The magnetization curve (M versus H_{ext}), integrated from $\chi_{||}$, is represented by curve B of Fig. 85. It is convex over the whole region and reaches 30%

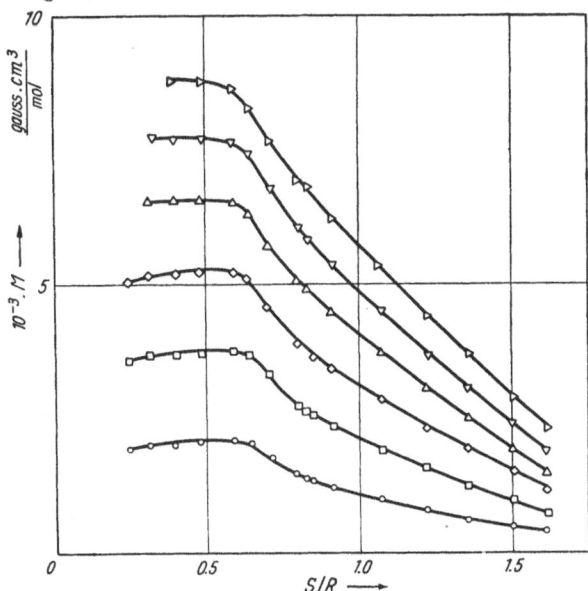

Fig. 86. Magnetic moment versus entropy diagram for a compressed spheroid 8:1 of iron ammonium alum (according to KURTI).
○ 12.8 oersteds.   □ 25.6 oersteds.   ◇ 38.4 oersteds.   △ 51.2 oersteds.   ▽ 64.0 oersteds.   ▷ 76.8 oersteds.

of the saturation moment in a field of 80 oersteds. If M is plotted as a function of H_{int}, it starts almost vertically as in the case of chromium potassium alum, see Sect. 65.

[1] M. J. STEENLAND, D. DE KLERK, M. L. POTTERS and C. J. GORTER: Leiden Commun. No. 284 b; Physica, 's-Grav. **17**, 149 (1951).
[2] M. J. STEENLAND: Thesis, Leiden 1952, p. 58.

The M versus S diagram with lines of constant H_{ext} is shown in Fig. 86. The curves show maxima in fields up to 50 oersteds, similar to the results obtained with the chromium alums, see Sects. 64 and 65. The variations of temperature obtained by integration [according to Eq. (9.9)] is shown in Fig. 87. Below $S = 0.60\,R$ the curves show a minimum. The minima become deeper and shift to larger fields with decreasing entropy. It is obvious, however, that at absolute zero the decrease of temperature on an isentropic must become zero again. An indication for this behaviour was found in the case of chromium methylamine alum where the decrease of temperature at the lowest entropy is smaller than at $S = 0.40\,R$, see Fig. 78.

In general, due to the smallness of ΔT in relatively low fields, there is little difference between the isentropic and the isothermal magnetization curves for the ammonium iron alum.

It was found in the Leiden experiments that the remanent magnetic moment decreases very steeply for small field values, much steeper than in the case of chromium potassium alum, see Sect. 65. The field where the remanent moment disappears, however, is somewhat larger than in the case of chromium potassium alum, viz. 50 oersteds at $S = 0.27\,R$. It follows that for the determination of remanent moments in zero field a good earth field compensation is required.

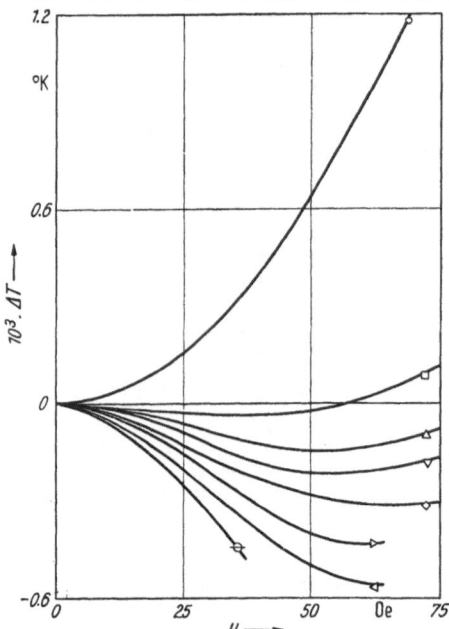

Fig. 87. Variation of temperature with field on adiabatic magnetization curves for a compressed spheroid 8:1 of iron ammonium alum (according to KURTI).
○ $S = 0.60\,R$.   ▽ $S = 0.45\,R$.   ◁ $S = 0.30\,R$.
□ $S = 0.55\,R$.   ◇ $S = 0.40\,R$.   ⊖ $S = 0.25\,R$.
△ $S = 0.50\,R$.   ▷ $S = 0.35\,R$.

67. Cobalt ammonium sulphate. Susceptibility measurements in longitudinal fields were made by GARRETT[1] with a spherical single crystal. The results for the K_1 axis (the axis of easiest magnetization at high temperatures, see Sect. 44) are shown in Fig. 88. For low entropies the curves show a pronounced maximum, but the rise and fall are not so very steep as in the case of the chromium

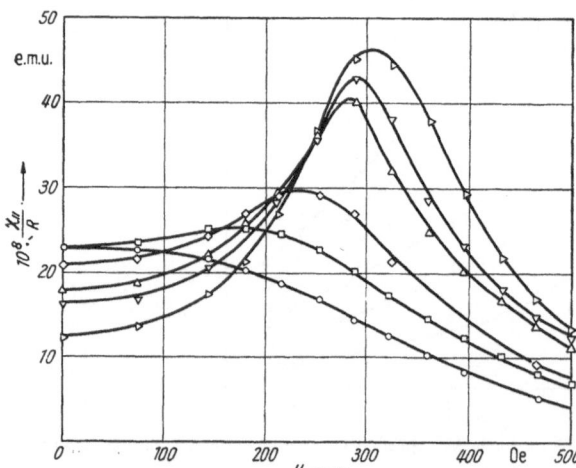

Fig. 88. Susceptibility $\chi_{\|}$ as a function of a longitudinal field for a spherical single crystal of cobalt ammonium sulphate (according to GARRETT). The fields were applied parallel to the K_1 axis.
○ $S = 0.543\,R$.   ◇ $S = 0.381\,R$.   ▽ $S = 0.223\,R$.
□ $S = 0.441\,R$.   △ $S = 0.289\,R$.   ▷ $S = 0.043\,R$.

methylamine alum (cf. Fig. 74). We do not know whether a small decrease of $\chi_{\|}$ occurs in weak fields, since no experiments were made in fields of the order of 10 oersteds.

[1] C. G. B. GARRETT: Proc. Roy. Soc. Lond., Ser. A **206**, 242 (1951).

No maxima were found in the K_2 and K_3 directions, nor were there in investigations with a powdered sample.

The values of the magnetic moment could be integrated from the data of Fig. 88; from the results the variation of temperature with field on the isentropics was evaluated, see for instance Sect. 64. Since the zero field temperatures were known from caloric experiments with a.c. heating (see Sect. 63) a T versus H diagram could be composed with lines of constant S and M. It is given in Fig. 89; the largest values of M at the right hand side being about two thirds of the saturation moment.

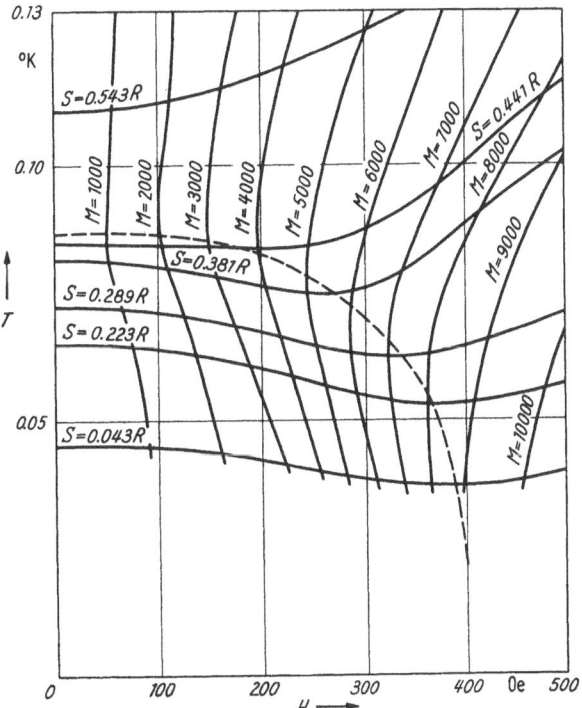

The T versus H diagram, as pointed out in Sect. 64, is equivalent to the M versus S diagrams given in Figs. 77, 83 and 86 for other salts. The dotted line of Fig. 89 is the locus of the minima of the two sets of curves [they must coincide for both sets according to Eq. (9.7)]. The locus proves to be not very much different from that of the maxima of the susceptibility curves of Fig. 88.

GARRETT supposed[1] that the dotted line of Fig. 89 is the critical field curve of Sect. 55, hence the borderline between the paramagnetic and antiferromagnetic regions. It was pointed out in Sect. 64 that this is not necessarily correct. In the case of chromium methylamine alum, this borderline was identified by BEUN, STEENLAND, DE KLERK and

Fig. 89. Temperature versus magnetic field strength diagram with lines of constant entropy and magnetic moment for a spherical single crystal of cobalt ammonium sulphate (according to GARRETT). The fields were applied parallel to the K_1 axis. The dotted line is the locus of the minima. The values of S and M are given with the curves, the latter in gauss cm.³/mol.

GORTER with the locus of the steep fall in the $\chi_\parallel$ versus H curves of Fig. 74. The absence of a steep fall in the curves of Fig. 88 is probably due to the fact that the crystalline symmetry of the cobalt ammonium sulphate is completely different from that of the alums. The Tutton salts have two ions in the unit cell, each with a tetragonal symmetry axis, see Sect. 40. No theoretical picture is available for the interactions between the magnetic ions in such a crystal.

E. Other investigations below 1° K.

I. Heat transfer and thermal equilibrium.

68. The achievement of thermal contact. The aim of demagnetization work is not only to investigate the magnetic, caloric and thermometric properties of paramagnetic salts, but also to cool down other materials with a salt in order

---

[1] C. G. B. GARRETT: Magnetic Cooling, p. 75 Cambridge (Mass.) 1954.

to make investigations on them. In the latter kind of experiments the salt is the thermostat, often also the thermometer, and special techniques had to be developed for achieving good thermal contact between the salt and the substance under investigation. Since heat transfer takes place through the thermal vibrations of the lattice it must be expected that the problem becomes the more serious the lower the temperature is.

Suppose the substance under investigation (for instance a metal wire, the resistance of which is to be measured) is connected to the paramagnetic salt by means of a "transfer medium" (e.g. liquid helium or some kind of glue) then the thermal equilibrium is accomplished in the following steps:

1. the heat transfer from the spin system to the lattice of the salt.
2. the establishment of thermal equilibrium in the salt itself.
3. the heat transfer from the lattice of the salt to the transfer medium.
4. the heat conduction of the transfer medium.
5. the heat transfer from this medium to the substance under investigation.
6. the establishment of equilibrium in the substance under investigation.

Experimental values for the times involved in each of these processes are hard to obtain; usually the experiment yields only the sum of several of them (for instance of 3, 4 and 5). Little is also known about each step from theory.

The equilibrium time between the spin system and the lattice of a paramagnetic salt should be closely related to the spin-lattice relaxation time as determined from paramagnetic relaxation experiments. This relaxation time has been determined in two different ways; (1) with the help of the paramagnetic saturation method, and (2) by placing the salt in a field $H = H_c + h e^{i\omega t}$, where ω is an audiofrequency. In the first set of experiments, performed with diluted paramagnetic alums, Eschenfelder and Weidner[1] found relaxation times of the order of 10^{-3} sec. in the liquid helium region; they were proportional to T^{-1}. The second method[2,3] gives relaxation times of the order of 10^{-2} sec. and the temperature dependence varies between T^{-2} and T^{-5}. An explanation of this discrepancy was suggested by Gorter, van der Marel and Bölger[4]. The long relaxation times of the second method are due to the heat transfer of a small band of the system of lattice oscillations, excited by energy transition in the spin system, to the helium bath. Consequently, the times found in the first method are the ones of interest in the present considerations.

It should be noticed that in the relaxation experiments quoted here the salt was in direct thermal contact with the helium of the cryostat, making the lattice specific heat effectively infinite. Casimir[5] pointed out that, if the salt is thermally insulated, the relaxation time is smaller by a factor $c_L/(c_L + c_H)$, where c_L is the specific heat of the lattice. Since this factor is approximately proportional to T^5 it follows that down to very low temperatures the lattice must follow the temperature of the spin system in a negligible time.

The thermal conductivity of chromium potassium alum, as measured by Kurti, Rollin and Simon, and by Garrett, was already discussed in Sect. 19. It decreases steeply with falling temperature; below 0.14° K the equilibrium time becomes too long to be measured. Investigations with iron ammonium alum, also by Kurti, Rollin and Simon, gave very similar results.

[1] A. H. Eschenfelder and R. T. Weidner: Phys. Rev. 92, 869 (1953).
[2] H. C. Kramers, D. Bijl and C. J. Gorter: Leiden Commun. No. 280a; Physica, 's-Grav. 16, 65 (1950).
[3] D. Bijl: Leiden Commun. No. 280b; Physica, 's-Grav. 16, 269 (1950).
[4] C. J. Gorter, L. C. van der Marel and B. Bölger: Leiden Commun. Suppl. No. 109c; Physica, 's-Grav. 21, 103 (1955).
[5] H. B. G. Casimir: Leiden Commun. Suppl. No. 85c; Physica, 's-Grav. 6, 156 (1939).

Experiments on the thermal conductivities of other materials will be discussed in Sect. 78. In the case of a non-superconducting metal the heat conductivity is reasonably good down to very low temperatures. Since it is mainly due to the free electrons[1] it is proportional to T. For metals in the superconducting state, however, the heat conductivity is much smaller.

Liquid helium has a very good thermal conductivity at $1°$ K. It follows from the experiments, however, (see Sect. 70) that it decreases strongly with falling temperature. At about $0.1°$ K or $0.2°$ K it is of the same order of magnitude as that of He I. Here, under normal experimental conditions, thermal equilibrium can still be reached in a short time through a thin layer of liquid, not through a long narrow capillary.

Very little is known about the heat conductivities of adhesives and glues. Probably they are not too good, but if a really thin layer is applied the equilibrium time may be reasonably short.

Residual helium gas in a sample may act as a transfer medium as long as the pressure is well above 10^{-6} mm. Extrapolation of the vapour pressure curve suggests that this may be true for temperatures above $0.4°$ K.

Data on heat transfer from one medium to another are hard to obtain. It is difficult to separate them from the heat conductivities of the media themselves. MENDOZA[2] could explain his heat conduction and superconductivity results by introducing an empirical coefficient of heat transfer between a salt and a metal proportional to T^2, hence

$$dQ = \beta A T^2 dT \tag{68.1}$$

where Q is the heat flow per second and A the area of contact. β was of the order of 300 ergs sec.$^{-1}$ cm.$^{-2}$ degree^{-3}. GOODMAN[3], in later experiments, found $\beta = 4 \times 10^4$ ergs sec.$^{-1}$ cm.$^{-2}$ degree^{-3}. It is not surprising that this coefficient is widely different for different experimental conditions, depending, for instance, on the tightness of the contact between salt and metal. At $0.2°$ K GOODMAN's value leads to a surface layer conductivity of 1.6×10^3 ergs sec.$^{-1}$ cm.$^{-2}$ degree^{-1}, whereas the thermal conductivity of chromium potassium alum at this temperature is 4×10^3 ergs sec.$^{-1}$ cm.$^{-1}$ degree^{-1}, that of copper is 4×10^6 ergs sec.$^{-1}$ cm.$^{-1}$ degree^{-1} and that of liquid helium 10^5 ergs sec.$^{-1}$ cm.$^{-1}$ degree^{-1}.

It follows from the above data that the best transfer media are non-superconducting metals and liquid helium; but it also follows that the main sources of troubles at the lower temperatures are the large resistance in the contact layer between two media and the small heat conductivities of the salts themselves. The heat transfer between two media may be improved by achieving an intimate contact over a large area. The consequence of the bad heat conductivity of the salts is that, even if a piece of salt is in good thermal contact with a transfer medium, only the outer layer is active as a coolant.

In some cases this is not too serious. If the heat capacity of the substance under investigation is much smaller than that of the salt a reasonably low temperature is still reached. But if the specific heat of the substance is large, or if appreciable amounts of heat are developed in it (e.g. in the case of experiments on electric or thermal conductivity) a noticeable difference from the temperature of the bulk salt may occur. In this case it is impossible to derive the temperature of the substance from a thermometric parameter of the salt.

An improvement is obtained by powdering the salt and embedding it in a transfer medium with a good thermal conductivity. This can be easily done

---

[1] C. J. GORTER et al.: Progress in Low Temperature Physics, p. 187. Amsterdam 1951.

[2] E. MENDOZA: Cérémonies LANGEVIN-PERRIN, p. 61. Paris 1948.

[3] B. GOODMAN: Proc. Phys. Soc. Lond. A **66**, 217 (1953).

with the help of liquid helium. In the case of a metal the best solution is to have thin sheets or wires not too far apart in the salt powder. Good heat transfer is achieved by compressing the sample hydraulically, usually after addition of a binding agent, see Sect. 70.

69. Liquid helium as a transfer medium. Liquid helium, as was pointed out in Sect. 68, is a feasible transfer medium. The main problem of an apparatus containing liquid helium is the film creeping out of the sample tube, which may cause a heat leak to the bath.

The first solution was given by KURTI, ROLLIN and SIMON[1]. A thick walled metal capsule is partly filled with powdered salt. Helium gas of about 120 atmos-

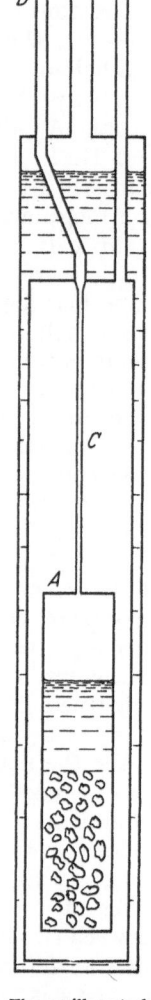

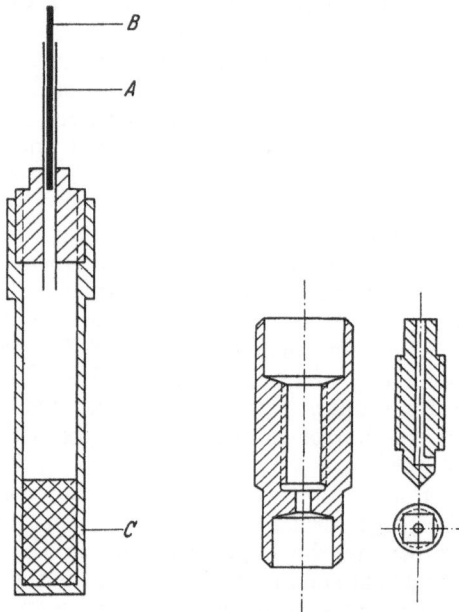

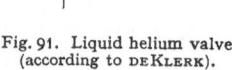

Fig. 90. Liquid helium capsule (according to HULL, WILKINSON and WILKS).

Fig. 91. Liquid helium valve (according to DE KLERK).

Fig. 92. The capillary technique (according to HUDSON, HUNT and KURTI).

pheres is admitted and the capsule is sealed off. At low temperature the helium is condensed and covers the salt completely. The substance under investigation may be soldered to the outer wall of the capsule.

A capsule is shown in Fig. 90. It consists of an alloy with low electric conductivity, e.g. cupro nickel or phosphor bronze, in order to keep heating by eddy currents as small as possible. After introduction of the salt the screwed plug is soldered in while the other end of the capsule is kept cool, so that the salt is not deteriorated by the heat. Helium is admitted through the capillary A. In the bore of this capillary is a wire of soft solder B. The gas is sealed in the

[1] N. KURTI, B. V. ROLLIN and F. SIMON: Physica, 's-Grav. **3**, 266 (1936).

capsule by hammering the capillary flat and then applying heat so that the solder runs[1]. Finally the capillary is cut above the solder seal.

Good results have been obtained with these capsules, but sometimes they leak and under certain circumstances it may be undesirable to have large amounts of metal in the sample. It is practically impossible, for instance, to make susceptibility measurements with a.c.

A solution in which the high pressure filling of helium at room temperature was avoided was given by DE KLERK[2]. A valve was constructed as shown in Fig. 91. The seat consisted of chrome-iron, both ends being sealed to glass tubes. The plug was made of steel. After filling the cryostat the appropriate amount of helium gas was condensed into the sample and then the valve was closed by means of a long metal rod which could be lifted afterwards. The measuring coils for the mutual inductance bridge were wound in such a way that the field was zero at the position of the valve.

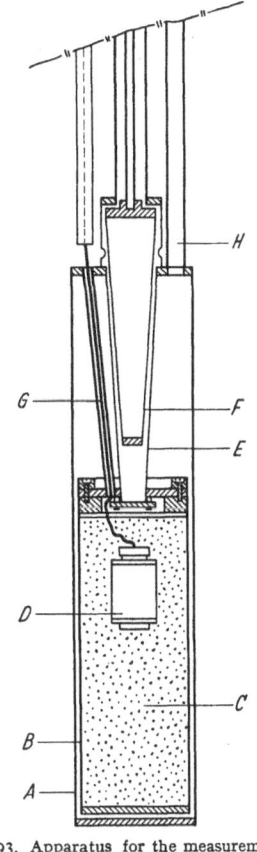

Fig. 93. Apparatus for the measurement of sound velocity in liquid helium (according to CHASE and HERLIN).

The difficulty with these valves is that the use of grease is impossible. The conical end of the plug must be so well centered in the seat that the helium film of about 3.5×10^{-6} cm. thickness does not creep through. This is a very high demand and one is never sure whether a valve that has worked satisfactorily during one run will be good during the next one. Under the best circumstances the heating up time from about 0.05 to $1°$ K was roughly two hours.

If a container with liquid helium II is suspended from a tube the film flow through the tube is roughly proportional to its circumference and the main source of heat leak is recondenzation into the container[3]. According to these considerations HUDSON, HUNT and KURTI[4] constructed an apparatus in which the valve was replaced by a long narrow capillary as shown in Fig. 92. The capillary was made of german silver, it was 7 cm. long and had an internal diameter of 0.2 mm. The upper end was connected to a diffusion pump in order to prevent the film evaporating above the capillary from recondenzation into the sample.

With this simple arrangement the heat leak proved to be of the same order of magnitude as in the case of a valve, and for this reason the valve technique has been abandoned in recent years. Glass capillaries of about the same dimensions were used in the Leiden experiments on the specific heat of liquid helium and the propagation of second sound below $1°$ K, see Sect. 70.

[1] R. A. HULL, K. R. WILKINSON and J. WILKS: Proc. Phys. Soc. Lond. A 64, 379 (1951).
[2] D. DE KLERK: Leiden Commun. No. 270c; Physica, 's-Grav. 12, 513 (1946).
[3] B. V. ROLLIN and F. SIMON: Physica, 's-Grav. 6, 219 (1939).
[4] R. P. HUDSON, B. HUNT and N. KURTI: Proc. Phys. Soc. Lond. A 62, 392 (1949).

An interesting apparatus which may be considered as a combination of the valve and the capillary techniques was recently described by Chase and Herlin[1]; it was originally designed by Ashmead. It is represented in Fig. 93. A is the sample tube and B the salt container. The space in between is evacuated through the pumping line H. E and F are thin walled stainless steel cones, machined and lapped to fit together as closely as possible. If the inner cone is lifted helium flows from the bath into B and the liquid between the cones provides good thermal contact. If F is seated the heat flow is reduced to a very low value, and the thermal insulation is sufficient to keep the experimental chamber cold for more than an hour. No exchange gas is needed in this apparatus, so that pumping the vacuum space when the field is on is eliminated. This reduces the magnetization time appreciably.

The apparatus was used for measurements of the velocity of sound in liquid helium (see Sect. 71). The necessary equipment was mounted in the experimental chamber D. The supply wires were brought through the vacuum space by means of the stainless steel tubes G which were filled with vaseline. The vaseline freezes at low temperatures and prevents the helium from flowing into the sample.

70. Heat transfer between solids. Thermal equilibrium between a salt and a metal at the lower temperatures is inadequate if the metal is glued to the surface of the salt sample. The occurrence of large inhomogeneities in the salt's temperature is clearly demonstrated in Fig. 94. This represents a heating curve at 0.35° K obtained by van Dijk[2] during his specific heat measurements on gadolinium sulphate (see Sect. 46). A phosphorbronze thermometer and a heating wire were wound on the sample together, alter-

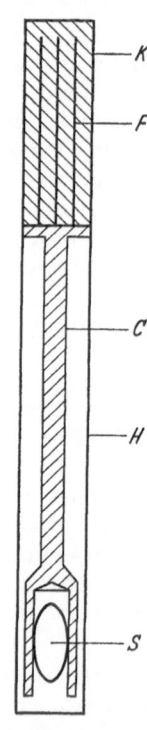

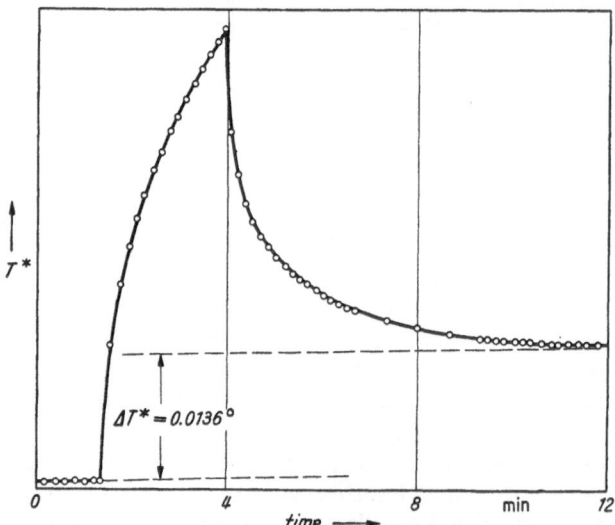

Fig. 94. Local overheating during specific heat measurements (according to van Dijk).

Fig. 95. Apparatus with vane technique (after Goodman and Mendoza).

nately a turn of each. The course of the temperature as derived from the phosphor bronze thermometer shows a local overheating which is only slowly equalized.

[1] C. E. Chase and M. A. Herlin: Phys. Rev. **97**, 1447 (1955).
[2] H. van Dijk: Leiden Commun. No. 270a; Physica, 's-Grav. **12**, 371 (1946).

A good solution for a metal-to-salt contact, as was indicated in Sect. 68, was first given by Mendoza[1,2], see Fig. 95. Thin copper sheets or vanes F were soldered to a copper rod C. The spaces between the sheets were well filled with powdered salt mixed with a binding agent and then the sample was compressed hydraulically under a pressure of 2000 atmospheres.

The total contact area of the vanes in Mendoza's apparatus was 30 cm.[2]. The binding agent was a solution of a plastic cement in acetone. The latter proves to evaporate fastly and completely from the compressed sample. In Fig. 95, S is the substance under investigation, a superconducting ellipsoid gripped in a cup at the lower end of the copper rod. H is a cylindrical shield of copper foil in good thermal contact with the salt K, protecting S from stray heat. Vertical slots were cut in H and in the cup holding S in order to reduce eddy currents. It was found that thermal equilibrium between the salt and S is reached five minutes after demagnetizing the salt to 0.1° K.

The basic idea of Mendoza was developed and modified by several investigators. The general trend in recent years has been to enlarge the area of contact between the salt and the metal. Darby, Hatton, Rollin, Seymour and Silsbee[3] in their two-stage demagnetization experiments (see Sect. 80), replaced the vanes by six copper wires of 0.2 mm. diameter. In later experiments this number was appreciably increased. In a recent Leiden experiment initiated by Wheatly (unpublished) 500 wires were soldered to copper frames. They were embedded in a mixture of equal quantities of chromium potassium alum and silver chloride and the whole sample was compressed to 2000 atmospheres. The total contact area was 100 cm.[2]. Since silver chloride is very plastic it provides an intimate contact between the wires and the salt. Thermal equilibrium between two pills of this type, connected by a copper rod, was reached in about three quarters of an hour at 0.06° K.

In recent experiments in Oxford[4] the sample was not compressed at all. Vanes with a large total area (larger than in the above Leiden experiment), were used and powdered salt was inserted between them by shaking the apparatus vigorously. Glycerin was used as a binding agent. Thermal equilibrium between such a sample of chromium potassium alum and a similar one of cerium magnesium nitrate was obtained at 0.025° K within an hour.

Dabbs, Roberts and Bernstein[5] made an experiment on the polarization of indium nuclei. Twenty sheets of the metal were soldered to silver wires of 12 cm. length and iron ammonium alum was crystallized around the other ends of the wires. The whole unit was mounted on rigid insulators in a silver cage which was cooled by another sample of iron alum. The salts were cooled magnetically to 0.035° K and the temperature reached with the indium as derived from the fractional change in the transmitted neutron intensity was 0.043° K.

Heer, Barnes and Daunt[6], in their magnetic refrigerator (see Sect. 81), used a finned copper shaft as shown in Fig. 96 surrounded by a brass cylinder. Iron ammonium alum, mixed with small pieces of copper wire and with silicone vacuum stopcock grease, was compressed inside this unit under 200 atmospheres. Satisfactory contact was reached down to at least 0.1° K.

[1] E. Mendoza: Cérémonies Langevin-Perrin, p. 53. Paris 1948.
[2] B. B. Goodman and E. Mendoza: Phil. Mag. **42**, 594 (1951).
[3] J. Darby, J. Hatton, B. V. Rollin, E. F. W. Seymour and H. B. Silsbee: Proc. Phys. Soc. Lond. A **64**, 861 (1951).
[4] F. N. H. Robinson: Thesis, Oxford 1954.
[5] J. W. T. Dabbs, L. D. Roberts and S. Bernstein: Phys. Rev. **98**, 1522 (1955).
[6] C. V. Heer, C. B. Barnes and J. G. Daunt: Rev. Sci. Instrum. **25**, 1088 (1954).

STEELE and HEIN[1, 2], in their experiments on carbon thermometers and on the superconductivity of titanium, pressed chromium potassium alum without a binding agent around a single copper fin B mounted on a brass base C as shown in Fig. 97. The copper specimen holder E was screwed tightly to C. The thermal contact was satisfactory the first time the apparatus was brought to liquid helium temperature. It was found, however, that it deteriorated to some extent once the apparatus was allowed to warm up to room temperature. This was ascribed to the salt's breaking away from the copper fin, possibly due to the difference in thermal expansion. If a new salt pill was used for each helium run the results were quite reproducible.

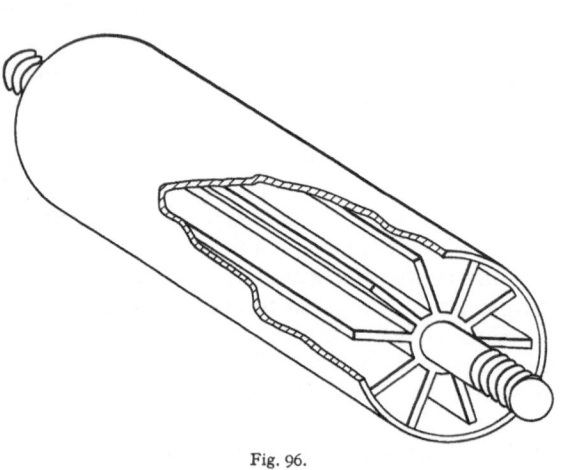

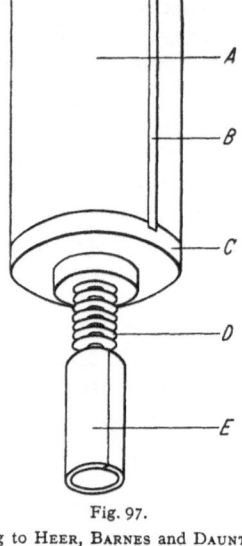

Fig. 96.   Fig. 97.

Fig. 96. Construction of the working unit of the magnetic refrigerator (according to HEER, BARNES and DAUNT).

Fig. 97. Apparatus for the investigation of carbon composition thermometers and superconductors (according to CLEMENT QUINNELL, STEELE, HEIN and DOLECEK).

A strongly simplified vane technique has been in use for many years in investigations on the occurrence of superconductivity. It was introduced by KURTI and SIMON[3] in 1935. Small grains of the metal are mixed with powdered salt and compressed to a solid pill. After demagnetization the susceptibility of the pill is followed during the warming up. The disappearance of superconductivity is evidenced by a fairly sudden discontinuity in the susceptibility curve, since a superconductor behaves like a completely diamagnetic substance with volume susceptibility $-1/4\pi$, whereas the susceptibility of a normal metal can be neglected. The transition curve may be derived by observing heating curves in magnetic fields.

The method has some obvious disadvantages:

1. In pressing the pill the metal is subjected to considerable stresses.

2. If particles of large size are used the thermal contact may become bad owing to the difference in thermal expansion. This is particularly dangerous in investigations on the threshold curve because of the caloric effects occurring during the transition (see Sect. 6).

[1] M. C. STEELE and R. A. HEIN: Phys. Rev. 92, 243 (1953).

[2] J. R. CLEMENT, E. H. QUINNELL, M. C. STEELE, R. A. HEIN and R. L. DOLECEK: Rev. Sci. Instrum. 24, 545 (1953).

[3] N. KURTI and F. SIMON: Proc. Roy. Soc. Lond., Ser. A 151, 610 (1935).

3. Application of a magnetic field influences the temperature of the salt somewhat.

4. The magnetic field inside the pill is different from that outside, the difference is unknown if the superconducting particles are of irregular shape and distributed at random.

These disadvantages are avoided by using an apparatus like Fig. 95. On the other hand the method described here is very simple. It is very useful if one is mainly interested in the question whether a substance becomes superconductive or not.

Finally we want to mention an interesting apparatus developed by Cooke[1], in which the susceptibilities of two salts can be compared. A single crystal sphere of cerium magnesium nitrate is covered by a spherical shell of a different salt, the latter being powdered and mixed with some grease. Cerium magnesium nitrate, as was stated in Sect. 48, is highly anisotropic, to such an extent that parallel to the symmetry axis the susceptibility is almost zero. Moreover it has a very small specific heat and Curie's law is obeyed down to a few thousandths of a degree. If, after demagnetization, thermal equilibrium is reached, a susceptibility measurement in the direction of easy magnetization of the cerium magnesium nitrate gives the sum of the two susceptibilities, whereas a measurement perpendicular to it gives only the susceptibility of the shell.

In the following sections, we give a survey of non-magnetic investigations that have been performed in the demagnetization region. We restrict ourselves to the experimental details as far as they are of interest for work below 1° K, and to a short description of the results. For the theoretical discussions of the results we refer to other chapters of the Encyclopedia.

II. Experimental results.

71. Investigations on liquid He⁴. Preliminary experiments on the heat conductivity of liquid helium were performed by Kurti and Simon[2] and by De Klerk[3]. Kurti and Simon used a twin capsule connected by a capillary of 18 mm. length and 0.5 mm. diameter. De Klerk applied two glass spheres filled with powdered salt connected by a glass tube 10.5 cm. long and 3.2 mm. internal diameter; the apparatus was closed by a helium valve, see Fig. 91. In both experiments a temperature difference was set up between the two salt pills, and the heat conductivity of the liquid was derived from the course of the temperatures of the two samples with time.

It was found that the heat conductivity decreases rapidly with falling temperature. At about 0.2° K it was of the same order of magnitude as that of He I.

More recent experiments were made by Fairbank and Wilks[4], also with a capsule technique. Complete experimental details have not yet been published, but the measurements were made with a german silver tube of 0.29 mm. internal diameter. Heat was supplied at one end and the heat conductivity was derived from the course of temperature of two thermometers soldered to the tube. Since the heating of the whole sample to 1° K took several hours, it was possible to obtain good equilibrium conditions for each measurement.

The results are shown in Fig. 98. A pronounced break occurs in the curve between 0.6 and 0.7° K. Below, the heat flow is normal, i.e. proportional to the

---

[1] A. H. Cooke: Houston Low Temperature Conference programme, 1953, p. 26.

[2] N. Kurti and F. Simon: Nature, Lond. **142**, 207 (1938).

[3] D. de Klerk: Leiden Commun. No. 270c; Physica, 's-Grav. **12**, 513 (1946).

[4] H. A. Fairbank and J. Wilks: Phys. Rev. **95**, 277 (1954).

temperature gradient. Above, this is probably no more true. Here the curve is very steep; it can be extrapolated to the values obtained above 1° K.

The first investigations on the specific heat of liquid helium below 1° K were performed by PICKARD and SIMON[1] and by KEESOM and WESTMYZE[2]. The latter authors found a proportionality to T^6 down to 0.6° K, whereas PICKARD and SIMON obtained a T^3-law below 0.8° K.

More recent investigations were made by HULL, WILKINSON and WILKS[3] with the help of a capsule filled with iron ammonium alum. A manganin heater was connected to the outside of the capsule and the temperature values were derived from the susceptibility of the salt. Between 1.4 and 0.6° K a proportionality to $T^{6.2}$ was found, below 0.6° K the specific heat of the liquid helium proved to be so small as compared with that of iron ammonium alum that it could not be measured with a reasonable precision.

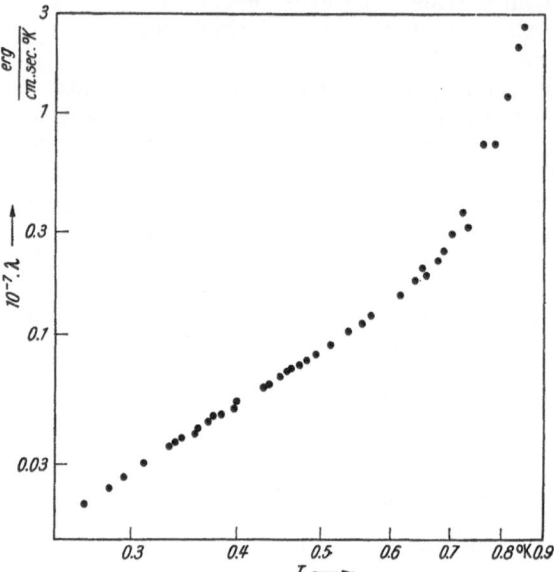

Fig. 98. Thermal conductivity of liquid helium as a function of temperature (according to FAIRBANK and WILKS).

This difficulty was avoided in experiments of KRAMERS, WASSCHER and GORTER[4] by replacing the iron ammonium alum by copper potassium sulphate. The latter salt has a specific heat which is roughly one twentieth of that of iron ammonium alum and the measurements could be extended with a reasonable precision to 0.25° K. A consequence of the smaller specific heat is a smaller cooling capacity, but it proved to be possible, by choosing the correct helium to salt ratio, to obtain a sufficiently low temperature with a reasonable field value. The authors used 14 grams of copper potassium sulphate and 1.8 grams of helium and a temperature of 0.1° K was reached with a field of 12500 oersteds.

A difficulty encountered with the copper potassium sulphate is that the experimental values for the CURIE-WEISS Θ and the coefficient of the $1/T^2$-term in the specific heat were noticeably different in different investigations, see Sect. 42. This introduces some uncertainty into the interpretation of the measurements at the lower temperatures. The salt used by KRAMERS, WASSCHER and GORTER was of the same origin as that investigated by DE KLERK, and the calculations performed with his constants gave the most satisfactory results.

A glass apparatus with a capillary was used, see Sect. 69. Heat was supplied with a carbon resistor and the temperature was obtained from the susceptibility of the salt.

[1] G. PICKARD and F. SIMON: Abstracts of papers communicated to the Royal Society of London 1939, p. 521.
[2] W. H. KEESOM and W. K. WESTMYZE: Physica, s-Grav. **8**, 1044 (1941).
[3] R. A. HULL, K. R. WILKINSON and J. WILKS: Proc. Phys. Soc. Lond. A **64**, 379 (1951).
[4] H. C. KRAMERS, J. D. WASSCHER and C. J. GORTER: Leiden Commun. No. 288c; Physica, 's-Grav. **18**, 329 (1952).

The results are shown in Fig. 99. A rather sharp bend occurs between 0.6 and 0.7° K, like in the heat conductivity curve of Fig. 98. Below it is the region where only the phonons make a contribution to the specific heat. Here the slope of Fig. 99 gives a proportionality to T^3. If we substract this contribution from the values above 0.7° K no unique power of T is found for the excess curve. LANDAU derived for the roton contribution to the specific heat:

$$c_r = B f(T) e^{-\Delta/kT},\qquad\qquad(71.1)$$

where $f(T)$ is a slowly varying function of temperature and Δ is the energy gap between the lowest energy levels of rotons and phonons. Since both B and Δ are unknown it is difficult to estimate whether the results above 0.7° K are in agreement with this formula or not.

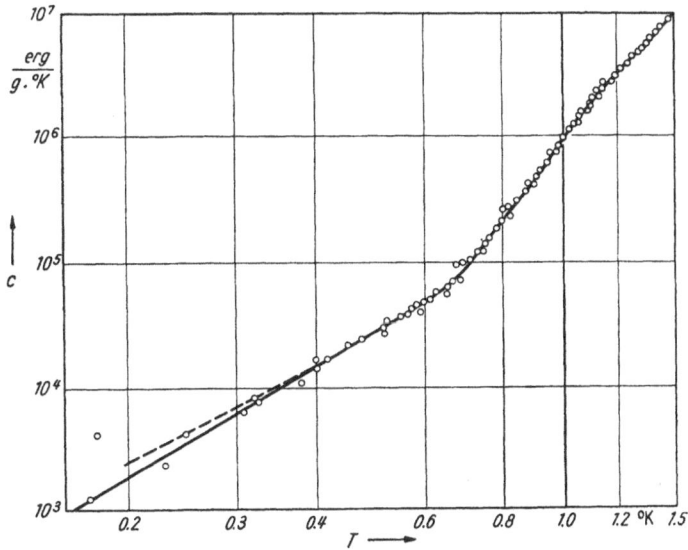

Fig. 99. Specific heat of liquid helium (according to KRAMERS, WASSCHER and GORTER). The points and the full line were obtained with DE KLERK's data on copper potassium sulphate, the broken curve with those of GARRETT.

Experiments on the propagation of sound in liquid helium were made by CHASE and HERLIN[1] using the apparatus of Fig. 93. The sound velocity as measured in cavities of 3.94 and 1.96 cm. length depended only little on temperature. At 0.1° K it was (240 ± 5) m/sec., whereas the value extrapolated from measurements above 1° K was (239 ± 2) m/sec.

The attenuation is represented in Fig. 100. A pronounced double maximum occurs near 0.9° K. This is in agreement with KHALATNIKOV's prediction that the attenuation is determined by two relaxation times, one due to phonon-phonon interaction, the other to phonon-roton interaction. The attenuation falls smoothly to zero at absolute zero. Above 0.3° K it is proportional to $T^{2.8}$, below 0.3° K it is somewhat steeper.

The first experiments on the propagation of heat waves in liquid helium ("second sound") below 1° K were performed by PELLAM and SCOTT[2] and by ATKINS and OSBORNE[3]. Though in both experiments the thermal insulation was very poor, so that no good equilibrium was reached between the helium

[1] C. E. CHASE and M. A. HERLIN: Phys. Rev. 97, 1447 (1955).

[2] J. R. PELLAM and R. B. SCOTT: Phys. Rev. 76, 869 (1949).

[3] K. R. ATKINS and D. V. OSBORNE: Phil. Mag. 41, 1078 (1950).

and the salt, it was demonstrated that the velocity increases rapidly below 1° K and that the pulses are noticeably broadened at the lower temperatures.

More recent experiments by de Klerk, Hudson and Pellam[1] and by Kramers, Mrs. van Peski, Wiebes, van den Burg and Gorter[2] showed that the theoretical speed limit at absolute zero as predicted by Landau:

$$v_2 = v_1/\sqrt{3}, \tag{71.2}$$

(where v_2 is the velocity of second sound and v_1 that of ordinary sound) is exceeded in such a way that v_2 seems to become equal to v_1. A glass apparatus

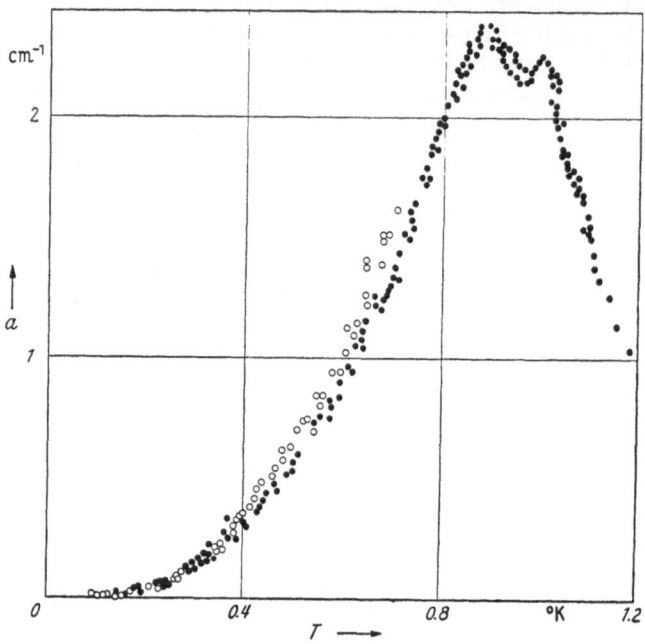

Fig. 100. Attenuation of sound in liquid helium at 12.1 Mc/sec. (according to Chase and Herlin). ○ Length of cavity 3.94 cm., ● length of cavity 1.96 cm.

with a capillary technique was used in both experiments. De Klerk, Hudson and Pellam used an ellipsoidal sample of chromium potassium alum, below which the second sound cavity was mounted. Kramers et al. had the cavity embedded in the salt. This has the advantage of a better thermal contact between salt and helium, but the disadvantage is that the demagnetizing field of the salt is less precisely known.

In both experiments the transmitter and receiver consisted of carbon resistor sheets. De Klerk, Hudson and Pellam used square wave modulated a.c. pulses of 22.5×10^3 c/sec. carrier wave frequency. Per second, 88 pulses were generated with a duration of 80 to 100 microseconds each. Kramers et al., in order to decrease the heat input, used single pulses of 20 microseconds. The receiver, in both experiments, was connected to an oscilloscope. Due to a small pick up, both the transmitted and the received pulse appeared on the screen. The second sound velocity could be derived from the time delay. The data were recorded photographically.

[1] D. de Klerk, R. P. Hudson and J. R. Pellam: Phys. Rev. 93, 28 (1954).
[2] H. C. Kramers, T. van Peski-Tinbergen, J. Wiebes, F. A. W. van den Burg and C. J. Gorter: Leiden Commun. No. 296b; Physica, 's-Grav. 20, 743 (1954).

De Klerk, Hudson and Pellam found that above 0.5° K the second sound velocity shows a tendency towards leveling off at about $v_1/\sqrt{3}$. At lower temperatures, however, it increases again (see Fig. 101). The received pulse is very narrow near 1° K. It spreads out in the region between 0.8 and 0.5° K, but below 0.5° K the width changes little with temperature. For pulses of very low energy the spreading was less than for large energies.

Kramers, Mrs. van Peski, Wiebes, van den Burg and Gorter performed experiments with cavities of different lengths, mainly 1.60, 3.05 and 6.25 cm. A detailed investigation was made of the pulse shape.

Above 0.9° K the pulses are well-bunched, below this temperature they begin to spread out and the velocities measured in the longer cavity are somewhat lower than those found in the shorter ones. Below 0.7° K

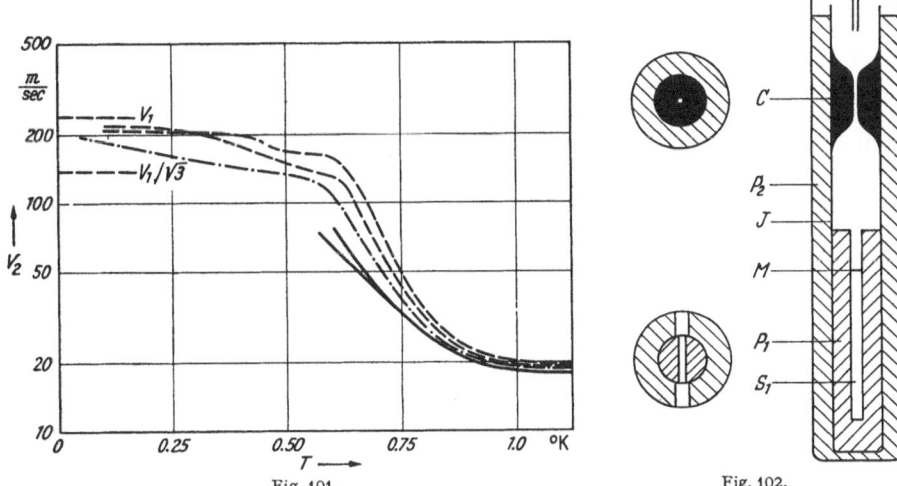

Fig. 101.   Fig. 102.

Fig. 101. Second sound velocity in liquid helium (according to Kramers, van Peski-Tinbergen, Wiebes, van den Burg and Gorter). – – – Velocity of start of signal; upper curve: 1.60 cm. cavity; lower curve: 6.25 cm. cavity. –·–·– Velocity of start of signal according to De Klerk, Hudson and Pellam; 5.1 cm. cavity. ——— Velocity of second sound (see text); upper branch: 1.60 cm. cavity; lower branch: 6.25 cm. cavity.

Fig. 102. Apparatus for the measurement of liquid helium film creep (according to Ambler and Kurti).

the pulses become very wide, a large tail effect developing rapidly. The leading slope of the pulse shows a minimum at 0.56° K for the longer cavities only. A sharp edge develops at still lower temperatures at the start of the signal, mainly for the longer tubes. By then the velocity of the starting point becomes independent of temperature.

At the higher temperatures, where the pulses are well-bunched, a very short transmitted puls (δ-function) develops into a received pulse of approximately Gaussian shape. The authors pointed out that, in this case, the second sound velocity follows from the position of the maximum rather than from that of the start. The results are shown in Fig. 101, together with those of De Klerk, Hudson and Pellam.

An explanation for the low temperature behaviour (below 0.5° K) follows from the assumption that here the phonon free path length becomes of the order of the second sound wavelength or the dimensions of the cavity. In this case there is no sense in speaking about second sound at all. The sharp edge of the received pulse may be due to phonons going the direct way with the velocity v_1. The value derived from all three tubes (if a retardation is introduced of 8 microseconds, probably caused by the KAPITZA thermal resistances at the surfaces of heater and thermometer) is (236 ± 4) m/sec., in good agreement with the value of CHASE and HERLIN quoted above. The large broadening of the pulse is due to phonons arriving after many diffuse collisions with the wall.

MAYPER and HERLIN[1] made second sound measurements with liquid helium under pressure. The apparatus was very similar to that of DE KLERK, HUDSON and PELLAM, and the results at normal pressure were not too much different. The heating time after demagnetization was only three or four minutes.

It followed that increasing the pressure leads to a decrease of v_2 at the higher temperatures, but to an increase at lower temperatures. The latter result is in agreement with the fact that the number of phonons at a given temperature diminishes with increasing pressure.

Measurements on film creep below $1°$ K were made by AMBLER and KURTI[2]. A very refined apparatus was built in which one could actually see the helium level below $1°$ K. It is shown in Fig. 102. A compressed cylinder of manganese ammonium sulphate P_1 was split in half longitudinally by a slit S_1. It was mounted in a glass beaker I with a constriction C. The beaker was surrounded by a ring P_2, also made of manganese ammonium sulphate, with a slit in its lower half in line with S_1. In order to be sure that the constriction was at the same temperature as the salt, the ring surrounded I completely, and was stuck to it by means of a cold setting plastic. Helium could be condensed into I through the fine capillary L. A further pill (not shown in Fig. 102) was attached to L and acted as a thermal shield. Narrow slits were applied in the silvering of the cryostat and the vacuum jacket so that the helium level M could be observed with the help of a small mercury lamp with filters transmitting only the green light. If care was taken that no light fell directly on the salt, and if the slit was only illuminated during the actual time of observation (a few seconds) the total heating-up time was about one hour.

The creep rate values above $1°$ K were somewhat higher than those usually observed for glass, and differences up to 15% were found between different runs. This was probably due to slight contamination of the creep surface. For the purpose of comparison the results of each run were multiplied by a factor so as to give identical transfer rates at $1.2°$ K. The data obtained in this way are shown in Fig. 103. After the flat portion between 1.5 and $0.8°$ K a further increase was found at lower temperatures.

LESENSKY and BOORSE[3] made some experiments on creep over pure and contaminated copper down to $0.75°$ K. The variation of the helium level was derived from the variation in the capacity of a condenser, caused by the dielectric constant of the helium. Also in their results an increase of creep rate was found below $1°$ K.

Measurements of the fountain effect were made by BOTS and GORTER[4, 5]. The apparatus is shown in Fig. 104. The glass capillary A was connected by tube B, filled with jeweler's rouge (Fe_2O_3), to the glass vessel E containing chromium potassium alum. The temperature of the helium in A was determined with the phosphorbronze thermometer F, that of the helium in E followed from

[1] V. MAYPER and M. A. HERLIN: Phys. Rev. 89, 523 (1953).
[2] E. AMBLER and N. KURTI: Phil. Mag. 43, 260 (1952).
[3] L. LESENSKY and H. A. BOORSE: Phys. Rev. 87, 1135 (1952).
[4] G. J. C. BOTS and C. J. GORTER: Phys. Rev. 90, 1117 (1953).
[5] G. J. C. BOTS: Report of the Paris Low Temperature Conference, 1955, to be published.

the susceptibility of the salt. After demagnetization the integrated fountain effect between the temperatures of E and A could be determined by observing the liquid level in A.

The results of the measurements down to 0.8° K were in good agreement with H. London's formula:

$$\frac{dp}{dT} = \varrho\,S \qquad (71.3)$$

where ϱ is the density of the liquid and S its entropy. At lower temperatures, if S was derived

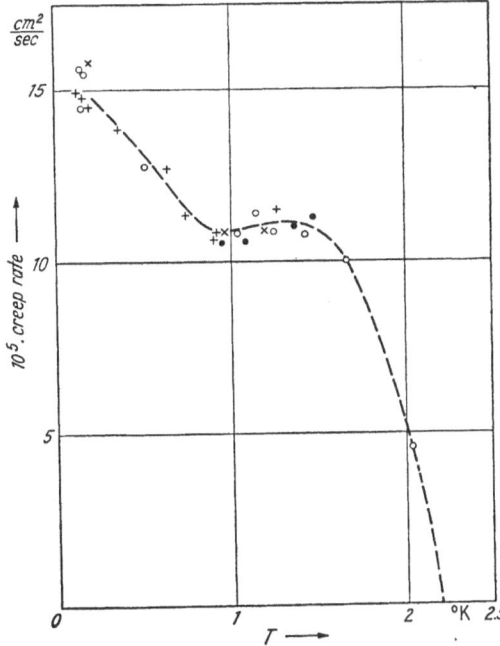

Fig. 103. Creep velocity of liquid helium film over glass (according to Ambler and Kurti). Different symbols refer to different helium runs.

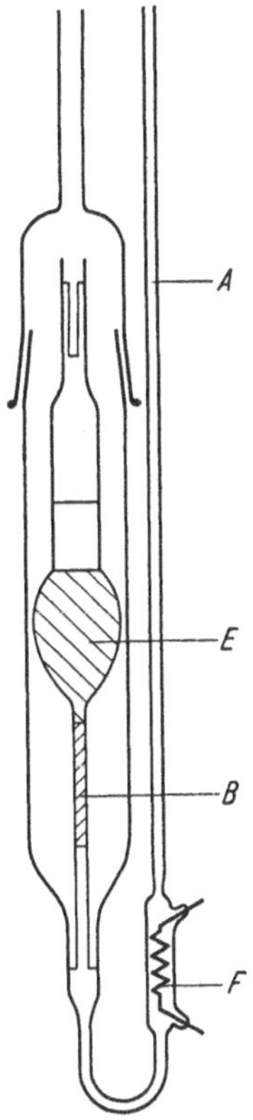

Fig. 104. Apparatus for the measurement of the fountain effect (according to Bots).

from the specific heat data of Kramers, Wasscher and Gorter, the experimental values were somewhat too small.

72. Solid helium. Webb, Wilkinson and Wilks[1] made some investigations on solid helium.

The specific heat was measured with the help of the apparatus of Fig. 105. A is the phosphorbronze calorimeter, suspended from nylon threads. A manganin heating wire was varnished to the outside. The calorimeter was filled with a mixture of iron ammonium alum and aluminum alum 1:4. This ratio was chosen in such a way that the iron alum provided sufficient cooling capacity and thermometric sensitivity without masking the specific heat of the helium at the lower temperatures. The diamagnetic aluminum alum served to distribute the iron alum uniformly over the calorimeter without introducing an appreciable heat capacity. About one half of the volume was occupied by the salt mixture. Helium could be admitted to the space between the grains through the capillary

---

[1] F. J. Webb, K. R. Wilkinson and J. Wilks: Proc. Roy. Soc. Lond., Ser. A **214**, 546 (1952).

C (0.13 mm. inner diameter and 9 cm. length). The calorimeter was filled to a certain pressure, the temperature of the cryostat being 0.05° K above the corresponding melting temperature. Then the cryostat was cooled rapidly causing the capillary to block so that the helium in the calorimeter solidified at constant volume.

The experiments were extended to about 0.5° K. It was found that the specific heat decreased with falling temperature, the value per gram being the lower,

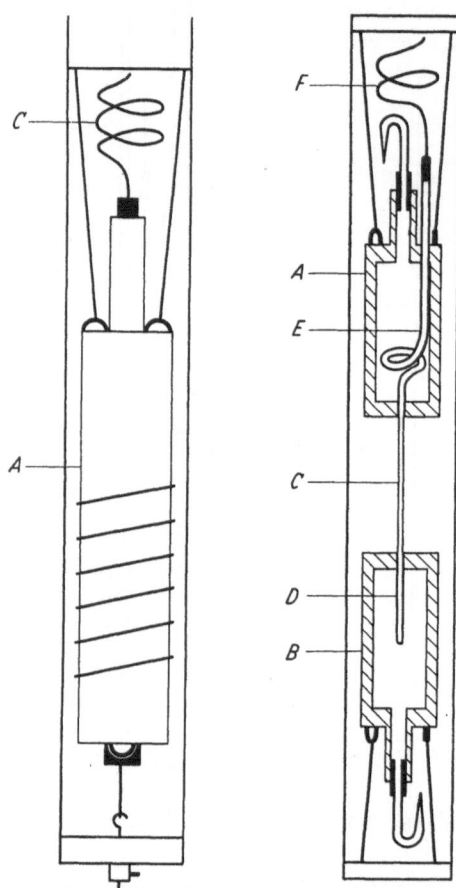

the larger was the density. If the results were expressed as values of a DEBYE Θ it appeared that, between 1.5 and 0.8° K, Θ increased with falling temperature. A slight decrease of Θ was found below 0.8°K, the interpretation of the latter result being not very clear.

The heat conductivity of solid helium was investigated by the same authors with the apparatus of Fig. 106. The helium was solidified in the german silver capillary C of 2 cm. length and 0.5 mm. inner diameter. The capsules A and B, filled with iron ammonium alum and liquid helium, served as thermometers. In order to improve the heat contact, the capillary C was hard soldered to copper capillaries D and E which protruded into the capsules. The helium was solidified into D, C and E through F applying the same technique as in the specific heat experiment. A heater was wound on each of the capsules. After demagnetization a temperature difference was set up between the capsules and the thermal conductivity of the helium in C was

Fig. 105. Apparatus for the measurement of the specific heat of solid helium (according to WEBB, WILKINSON and WILKS).

Fig. 106. Apparatus for the measurement of the heat conductivity of solid helium (according to WEBB, WILKINSON and WILKS).

derived from the course of the temperature difference with time and from the heat capacities of the capsules.

A maximum was found in the heat conductivity slightly below 1° K. This is due to the fact that the mean free path of the elastic waves reaches the order of magnitude of the diameter of the capillary. A proportionality to $T^{2.3}$ was found at lower temperatures, whereas CASIMIR's theory predicts a third power law[1].

At the lowest temperatures (about 0.25° K) the heat conductivity was of the same order of magnitude as that of liquid helium (see Fig. 98), but at 1° K the conductivity of the liquid is much better. The authors suggested that solid helium might be a suitable contact medium below 1° K, the advantages being

[1] H. B. G. CASIMIR: Leiden Commun. Suppl. No. 85 b; Physica, 's-Grav. **5**, 495 (1938)

that the heat leak through the filling capillary is small due to the absence of a creeping film, and that the contact with the salt, the walls and eventual further substances under investigation is very tight so that the KAPITZA contact resistance (see Sect. 71) may be well lower than in the case of liquid helium. It should be pointed out, however, that the method can only be applied if no objections exist against the use of a metal apparatus.

73. Experiments on He³ and on mixtures of He³ and He⁴. The first experiments on the specific heat of He³ were reported by DE VRIES and DAUNT[1]. The calorimeter contained 13 mm.³ of 96% purity. The connection between the calorimeter and the paramagnetic salt was made with the help of a superconducting thermal switch (see Sect. 79) so that, after the demagnetization, the thermal contact could be broken. The temperatures were measured with the help of a carbon resistance thermometer (see Sect. 74) calibrated against the susceptibility of the salt.

Experiments were made down to 0.57° K. The specific heat decreased with falling temperature, but the value became rather constant at the lowest temperatures so that it could not be easily extrapolated to absolute zero.

The data were in reasonable agreement with those obtained by ROBERTS and SYDORIAK[2]. The latter authors made experiments with He³ of 99.9% purity. It was enclosed in a vacuum jacketed ¾ cm.³ copper sphere, filled with iron ammonium alum of 55% packing factor. Temperatures down to 0.54° K were reached by pumping the He³ (hence not by adiabatic demagnetization). The specific heat followed from the measurement of warming rates (due to the heat leak) when the calorimeter contained different amounts of He³. The iron ammonium alum served only as a thermometer; its susceptibility was calibrated against the vapour pressure of the He³.

Also OSBORNE, ABRAHAM and WEINSTOCK[3] made specific heat measurements. They used a calorimeter consisting of a copper vessel of 1.87 cm.³ connected to an external He³ filling line. Four copper vanes were hard-soldered to the vessel and the whole unit was mounted inside a copper container, the space in between being filled with iron ammonium alum of 0.7 times crystalline density. One atmosphere of He⁴ gas was admitted to the container at room temperature in order to improve the thermal equilibrium. A strip carbon resistor was fastened to one of the vanes serving both as a heater and as a thermometer. It was calibrated against the susceptibility of the salt.

The first experiments of ABRAHAM, OSBORNE and WEINSTOCK were made in the region down to 0.42° K; they were in good agreement with the other results, quoted above. Quite recently the investigations were extended[4] to 0.23° K. A survey of the data is shown in Fig. 107.

All the specific heat experiments demonstrated clearly that liquid He³ does not show the behaviour of an ideal FERMI-DIRAC gas. The specific heat of such a gas with a degeneration temperature of 4.98° K (calculated in accordance with the density and atomic mass of He³) is represented by curve C of Fig. 107.

Entropy differences can be calculated from the specific heat values, and by combining them with the vapour pressure data it follows that the entropy is equal to $R \ln 2$ at about 0.5° K. At this temperature the orientation of the spins of the He³ nuclei must be still almost random. At 0.23° K, however, the entropy is well below $R \ln 2$ so that here the spins must be appreciably ordered. This

[1] G. DE VRIES and J. G. DAUNT: Phys. Rev. **93**, 631 (1954).
[2] T. R. ROBERTS and S. G. SYDORIAK: Phys. Rev. **93**, 1418 (1954).
[3] D. W. OSBORNE, B. M. ABRAHAM and B. WEINSTOCK: Phys. Rev. **94**, 202 (1954).
[4] B. M. ABRAHAM, D. W. OSBORNE and B. WEINSTOCK: Phys. Rev. **94**, 551 (1955).

is in agreement with the susceptibility measurements of FAIRBANK, ARD and WALTERS described below.

ABRAHAM, OSBORNE and WEINSTOCK pointed out that an almost linear part may be subtracted from the specific heat (curve D of Fig. 107) which, in its shape, is rather similar to the specific heat of He I above 2.5° K. The remaining part (curve E) gives then the spin specific heat. The entropy calculated from curve E amounts to $R \ln 2$ and the fraction of the spins which are not aligned antiparallel may be computed from the course of this entropy with temperature.

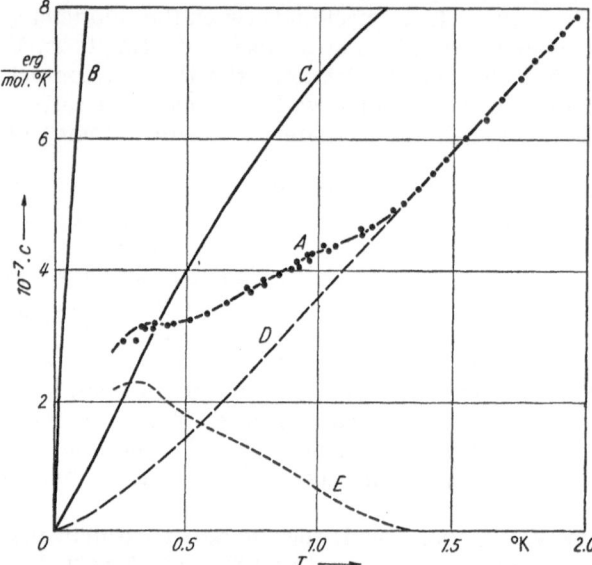

Fig. 107. Specific heat of liquid He³ (according to ABRAHAM, OSBORNE and WEINSTOCK). Curve A: Experimental results. Curve B: Ideal FERMI-DIRAC gas with degeneracy temperature 0.45° K. Curve C: Ideal FERMI-DIRAC gas with degeneracy temperature 4.98° K. Curve D: Estimated non-spin specific heat. Curve E: Estimated spin specific heat.

The values obtained in this way are in satisfactory agreement with the directly measured susceptibility data.

The nuclear susceptibility was measured by FAIRBANK, ARD and WALTERS[1]. He³ of 99% purity was enclosed in a cavity which was in thermal contact with a copper rod. Chromium potassium alum was pressed around this rod at a distance of six inches from the cavity. Temperature measurements were made with the help of a carbon resistor, calibrated against the susceptibility of the salt. An independent check on the temperature was obtained by placing a salt containing protons in direct contact with the He³ and measuring the strength of the proton magnetic resonance signal. The susceptibility of the He³ followed from the amplitude of the nuclear magnetic resonance at 30 megacycles per second in a field of 10000 oersteds.

The quantity χT was almost constant above 1° K. Large deviations should be expected well above 1° K for an ideal FERMI-DIRAC gas with a degeneration temperature of 4.98° K (see above). A noticeable decrease of χT was found below 1° K, the results being in agreement with a degeneration temperature of 0.45° K. This degeneration temperature, however, is in distinct disagreement with the specific heat data (see curve B of Fig. 107) so that it seems better, at present, to hold to the picture of ABRAHAM, OSBORNE and WEINSTOCK in which the specific heat is split into a nuclear contribution and a non-nuclear part.

The melting pressure was determined by WEINSTOCK, ABRAHAM and OSBORNE[2]. The blocked capillary technique was applied in an apparatus rather similar to that used by the same authors for specific heat measurements (see above). Four copper vanes were soldered to the capillary; it was surrounded by iron ammonium alum and enclosed in a copper container. One atmosphere of helium gas was admitted to the container at liquid nitrogen temperature in order to improve the thermal contact.

[1] W. M. FAIRBANK, W. B. ARD and G. K. WALTERS: Phys. Rev. 95, 566 (1954).
[2] B. WEINSTOCK, B. M. ABRAHAM and D. W. OSBORNE: Phys. Rev. 85, 158 (1952).

The results are shown in Fig. 108. Down to about 0.5° K the data obey the relation:

$$p \text{ (atmospheres)} = 26.8 + 13.1\, T^2, \qquad (73.1)$$

but below this temperature the melting pressure approaches rapidly to a constant value. Three explanations are possible for this behaviour: (1) The entropy difference between the liquid and the solid becomes zero at about 0.5° K due to a phase transition in the liquid. (2) The heat contact between the salt and the He³ breaks down at this temperature. Since in this experiment the heat contact

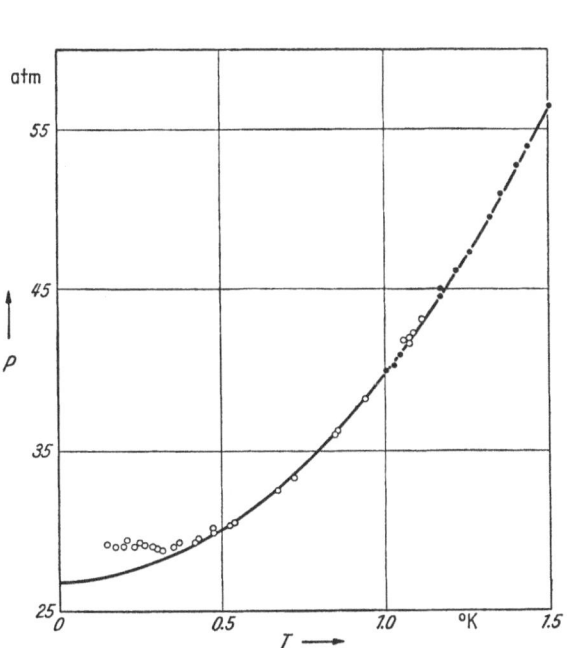

Fig. 108. Melting pressure of He³ (according to WEINSTOCK, ABRAHAM and OSBORNE). ○ Experiments with the capillary in the salt. ● Experiments with the capillary in the bath.

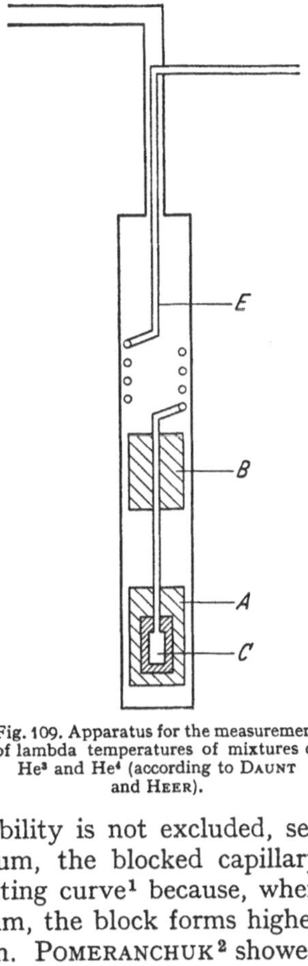

Fig. 109. Apparatus for the measurement of lambda temperatures of mixtures of He³ and He⁴ (according to DAUNT and HEER).

was mainly brought about by helium gas, this possibility is not excluded, see Sect. 68. (3) If the melting pressure shows a minimum, the blocked capillary method continues to record the minimum of the melting curve[1] because, when the temperature is reduced below that of the minimum, the block forms higher up in the capillary at the temperature of the minimum. POMERANCHUK[2] showed that a minimum occurs in the melting curve if the alinement of the nuclear spins in the solid state takes place at a much lower temperature than in the liquid.

Obviously more investigations are needed in order to decide between these three possibilities. It is clear from Fig. 108, however, that the melting pressure is positive at absolute zero so that there the liquid phase is the stable one, like in the case of He⁴.

Lambda temperatures of mixtures of He³ and He⁴ were measured by DAUNT and HEER[3]. The apparatus is shown schematically in Fig. 109. A pill of chromium

[1] C. J. GORTER et al.: Progress in Low Temperature Physics, p. 86. Amsterdam 1955.
[2] I. POMERANCHUK: J. exp. theor. Phys. (USSR.) 20, 1919 (1950).
[3] J. G. DAUNT and C. V. HEER: Phys. Rev. 79, 46 (1950).

potassium alum A was compressed around a 6 mm.[3] copper reservoir C. The latter was partly filled with the mixture under investigation through the narrow steel capillary E. A second pill B, pressed around the capillary higher up, acted as a thermal barrier. A and B were demagnetized. As long as C was below the lambda point of the mixture the heating was fast, due to film creep and recondenzation through the capillary. The lambda point was marked by a sudden decrease in slope of the heating curve.

The results are given in Fig. 110, together with those obtained by ABRAHAM, WEINSTOCK and OSBORNE above 1° K. Several theoretical relations have been given for the dependence of the lambda point on the He[3] concentration. Some of them are given in Fig. 110. Unfortunately the spread of the experimental points is so large that it is difficult to decide which of the theoretical curves is in agreement with the measurements. The supposition seems justified, however, that T_λ for pure He[3] is zero, in other words, that no superfluidity occurs in pure He[3].

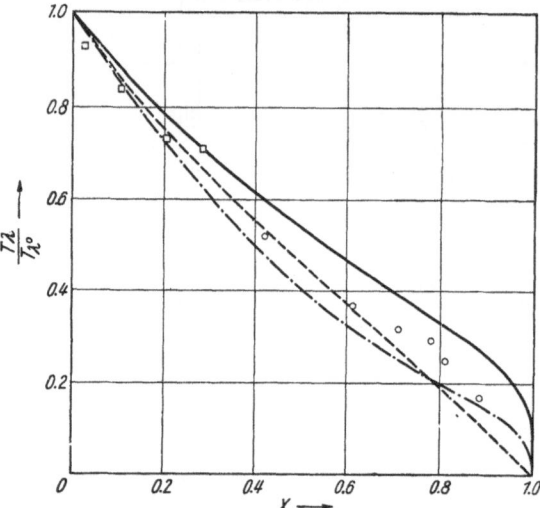

Fig. 110. Variation of the lambda temperature T_λ with the concentration X of He[3]. T_λ^0 is the lambda point for pure He[4]. ○ Experimental points of DAUNT and HEER. □ Experimental points of ABRAHAM, WEINSTOCK and OSBORNE. ——— Theoretical curve of DE BOER. —·—·— Theoretical curve of DE BOER and GORTER. ——— Theoretical curve of MIKURA.

Measurements of the second sound velocity of mixtures of He[3] and He[4] were made by KING and FAIRBANK[1]. The cavity was constructed similar to those used for pure He[4] (described in Sect. 71), but the dimensions were much smaller. The distance between the two carbon sheet resistors (transmitter and receiver) was 8.54 mm. and the diameter of the cavity was 2.8 mm. The cavity was filled through a stainless steel capillary of 0.15 mm. inner diameter. The single pulse method was used with transmitted pulses of 20 to 60 microseconds.

The cavity was screwed to a copper rod. Circular copper discs were fastened perpendicular to it. Chromium potassium alum was inserted between the discs and the whole unit was compressed to 2500 atmospheres. The temperature measurements were made with a carbon resistance thermometer, calibrated against the susceptibility of the salt.

Investigations were made on mixtures with He[3] concentrations varying from 0.017 to 4.30%. Above 1° K the second sound velocity increased with increasing concentration but a maximum of about 35 m/sec. was found just below 1° K. At lower temperatures v_2 decreased somewhat with increasing concentration. Practically no pulse broadening was found below 0.6° K.

In pure He[4] the density of the normal component becomes exceedingly small below 0.6° K and the phonon free path increases strongly. This results into a high velocity of pulse propagation, up to v_1, and a signal broadening (see Sect. 71). In a mixture the He[3] atoms play the predominant role at the lowest temperatures

[1] J. C. KING and H. A. FAIRBANK: Phys. Rev. 93, 21 (1954).

so that the increase of velocity and the pulse broadening do not occur. POMERAN-CHUK derived for diluted mixtures at the lowest temperatures, assuming that the He³ moves with the normal phase:

$$v_2^2 = \frac{5\,k\,T}{3\,\mu}, \tag{73.2}$$

where μ is the effective mass of the He³ atom in the liquid. It appears that the results of KING and FAIRBANK are in good agreement with this formula if one assumes that μ varies slightly in a linear way with temperature. The value decreases from 3.6 times the mass of the free He³ atom at 1.8° K to 2.0 times this mass at 0.2° K.

74. Phosphorbronze and carbon resistance thermometers. In the early days of demagnetization work some experimenters cooled a phosphorbronze wire with a paramagnetic salt, mainly for the purpose of investigating the possibility of resistance thermometry in the region below 1° K.

VAN DIJK, KEESOM and STELLER [1] found that down to 0.25° K the resistance decreased about linearly with the T^* of gadolinium sulphate. Heat contact was achieved with the help of some liquid helium.

ALLEN and SHIRE [2] cooled a phosphorbronze wire to $T^* = 0.025°$ with the help of iron ammonium alum. A capsule technique was used in most of their experiments, but in some cases the salt was compressed inside a german silver tube and the wire was glued to the outside. Over the whole region the resistance decreased with falling temperature. Deviations from linearity may have been due to the differences between T^* and T, but also to somewhat insufficient heat contact at the lower temperatures.

Phosphorbronze thermometers have the disadvantage that they are strongly dependent on the measuring current and on magnetic fields. For this reason they have been mostly abandoned in demagnetization work of recent years, and replaced by carbon resistors. The latter are much better in this respect, but precautions must be taken that the whole thermometer is in good thermal equilibrium with the substance under investigation.

GIAUQUE, STOUT and CLARK [3], in their early experiments, made use of thin layers of carbon ink on glass. Between 290 and 1.63° K the resistance increased from 57 to 780000 ohms. Since this sensitivity was almost too good new experiments were made with lamp black. A thin sheet of lens paper (0.004 cm.) was applied to the outside of the glass salt container with ethyl alcohol and, when still wet, painted with lamp black in alcohol. A coating of collodion was applied afterwards.

Excess paper was cut away until a U-shaped strip remained of such a width that the resistance had the desired order of magnitude. After drying it was surprisingly stable. Between 1 and 0.129° K the resistance increased from 44×10^3 to 59×10^3 ohms, the measuring current being 4×10^{-7} amp. The values of the resistance were well reproducible during one helium run, differences up to several percents were found between subsequent runs, but the ratio of the resistances at 4.2 and 290°K varied only a few tenths of a percent. In a magnetic field the resistance increased proportionally to H^2, but the effect was very small; a field of 8200 oersteds at 1.5°K entailed a rise of a quarter of a percent. In later

---

[1] H. VAN DIJK, W. H. KEESOM and J. P. STELLER: Leiden Commun. No. 252g; Physica, 's-Grav. **5**, 625 (1938).

[2] J. F. ALLEN and E. S. SHIRE: Nature, Lond. **139**, 878 (1937).

[3] W. F. GIAUQUE, J. W. STOUT and C. W. CLARK: J. Amer. Chem. Soc. **60**, 1053 (1938).

experiments, FRITZ and GIAUQUE[1] found that the quantity $(\Delta R/R)(T/H)^2$ was a slowly varying function of temperature. Between 20 and 4° K it was practically constant, at lower temperatures it decreased gradually to zero.

GEBALLE, LYON, WHELAN and GIAUQUE[2] made a nice investigation on the influence of the particle size on the sensitivity of a carbon thermometer. Several thermometers were constructed as described above from commercial carbon samples of various average grain sizes. The relative increase of the resistance with falling temperature for fine carbon particles was noticeably larger than for larger ones. The quantity $R_{4.2°K}/R_{295°K}$ amounted to 1.245 for a sample with average particle size of 12×10^{-6} cm., for a sample with grains of 2×10^{-6} cm. it was equal to 825. Any value in between could be obtained by choosing the proper particle size. Fig. 111 shows the sensitivity, defined as $(dR/dT)/R$, as a function of temperature for different grain sizes.

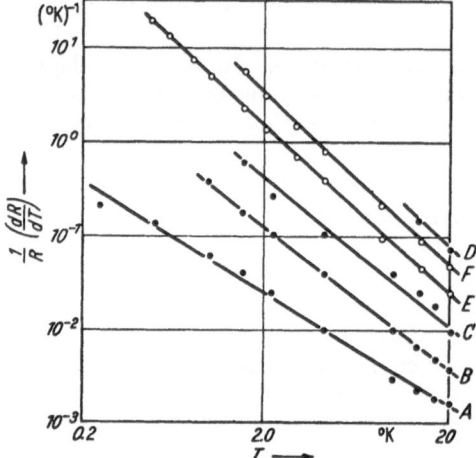

The influence of a magnetic field could be described with the formula of FRITZ and GIAUQUE, quoted above. The effect was somewhat more pronounced for the smaller particles than for the larger ones, but it was small under all circumstances.

Adsorption of small amounts of helium gas increased the resistance somewhat, but a small decrease was found if the saturated vapour pressure was approached. The variations in the resistance involved were equivalent to temperature differences of a few millidegrees.

Fig. 111. Sensitivity data for carbon thermometers (according to CLEMENT, QUINNELL, STEELE, HEIN and DOLECEK). ● Results of GEBALLE, LYON, WHELAN and GIAUQUE. Curve A: Average particle size 12×10^{-6} cm. Curve B: Average particle size 8×10^{-6} cm. Curve C: Average particle size 5×10^{-6} cm. Curve D: Average particle size 2×10^{-6} cm. ○ Results of CLEMENT, QUINNELL, STEELE, HEIN and DOLECEK. Curve E: Nominal room temperature resistance 10 ohms. Curve F: Nominal room temperature resistance 270 ohms.

The carbon thermometers described above are very satisfactory for low temperature work, the only disadvantage being that they must be recalibrated for each new helium run. For this reason CLEMENT and QUINNELL[3] investigated some commercial radio resistors. They came to the conclusion that the one watt resistors manufactured by the ALLEN-BRADLEY Company gave satisfactory results. Resistors between nominal values of 10 and 270 ohms at room temperature were investigated. In order to improve the heat transfer to the thermometer, the insulation was ground off and the carbon was covered with a thin layer of glyptal. The course of the resistance with temperature could be expressed by the empirical formula:

$$\log R + \frac{K}{\log R} = A + \frac{B}{T}. \tag{74.1}$$

BROWN, ZEMANSKY and BOORSE[4] investigated a 0.5 watt resistor, also of ALLEN-BRADLEY. It was cooled seven times to liquid helium temperature and the

[1] J. J. FRITZ and W. F. GIAUQUE: J. Amer. Chem. Soc. 71, 2168 (1949).
[2] T. H. GEBALLE, D. N. LYON, J. M. WHELAN and W. F. GIAUQUE: Rev. Sci. Instrum. 23, 489 (1952).
[3] J. R. CLEMENT and E. H. QUINNELL: Rev. Sci. Instrum. 23, 213 (1952).
[4] A. BROWN, M. W. ZEMANSKY and H. A. BOORSE: Phys. Rev. 84, 1050 (1951).

calibration curves of different runs coincided in general within $0.002°$ K. The resistance was found to obey:

$$\log R = A + \frac{B}{T} + \frac{C}{T^2} - K\,T^2. \tag{74.2}$$

DE NOBEL[1], in his thesis, proposed a formula:

$$\log \frac{R}{R_0} = \frac{a}{T} + \frac{b}{\sqrt{T}}. \tag{74.3}$$

All these formulae, with the proper choice of constants, result into almost coinciding curves. It should be noticed that Eq. (74.1) is dimensionally uncorrect.

It was also found by CLEMENT and QUINNELL that application of a magnetic field gives a small increase in resistance, proportional to H^2; and that the sensitivity $(dR/dT)/R$ becomes larger with increasing nominal value.

Most of the investigations on carbon resistors discussed here were made above $1°$ K. CLEMENT, QUINNELL, STEELE, HEIN and DOLECEK[2] investigated two ALLEN-BRADLEY resistors of nominally 2.7 and 10 ohms, in the region below $1°$ K. The apparatus of Fig. 97 was used, the thermometer was cemented into the cylindrical holder E for good heat contact with the salt. The course of the resistance with temperature could be expressed by formula (74.1) down to $0.3°$ K. The sensitivity $(dR/dT)/R$ is shown in Fig. 111. The curves fit in well with those obtained by GIAUQUE and his coworkers.

The general conclusion from these investigations is that the ALLEN-BRADLEY resistors show a better reproducibility between runs, whereas GIAUQUE's carbon films permit a larger variety of sensitivities.

HOWLING, DARNELL and MENDOZA[3] investigated an Erie "ceramicon" radio resistor. It was connected to the salt with the vane technique, see Sect. 70. Between 1 and $0.1°$ K the resistance increased by about 50%. This is much less than in the case of an ALLEN-BRADLEY resistor, but still sufficient for many purposes of thermometry. The influence of a magnetic field on the resistance was very small.

The authors proposed to use a carbon resistor for absolute temperature determination below $1°$ K. The resistance of the carbon in this case plays the role of the thermometric parameter: it is measured as a function of the entropy of the salt and its variation is determined on supplying heat (see Sect. 12). The difficulty is that at the lower temperatures the equilibrium time between the salt and the resistor becomes long, see Sect. 70. The authors reported a time of seven minutes at $0.1°$ K and of 14 min. at $0.05°$ K. This restricts the application of the method to the region above $0.1°$ K.

75. Specific heats of metals.
RAYNE[4] made investigations on the specific heats of some metals. The cylindrical sample was connected by a copper wire to a cylindrical salt pill, the thermal contact with the latter being achieved with the vane technique (see Sect. 70). A manganin heater was glued to the metal by means of glyptal. The supply wires consisted of tinned manganin; at helium temperatures the tin becomes superconductive, providing high electric conductivity but stil good thermal insulation.

The salt pill, in most experiments, consisted of copper potassium sulphate. The T^* values were measured ballistically and GARRETT's Θ was used for reduction

[1] J. DE NOBEL: Thesis, Leiden 1954, stelling III.
[2] J. R. CLEMENT, E. H. QUINNELL, M. C. STEELE, R. A. HEIN and R. L. DOLECEK: Rev. Sci. Instrum. **24**, 545 (1953).
[3] D. H. HOWLING, F. J. DARNELL and E. MENDOZA: Phys. Rev. **93**, 1416 (1954).
[4] J. RAYNE: Phys. Rev. **95**, 1428 (1954).

to absolute temperatures (see Sect. 42). The measuring field was 30 oersteds. The ratio of the amounts of salt and metal was chosen in such a way that at about 0.5° K they had equal heat capacities.

The specific heat of the copper potassium sulphate was measured in a separate experiment. It was found to obey, down to 0.1° K, $cT^2/R = (5.8 \pm 0.2) \times 10^{-4}$ (°K)^2, in good agreement with other determinations, see Sect. 40. Marked deviations occurred above 0.65° K. They were too large to be accounted for by the lattice specific heat. Probably they were due to a helium film evaporating from the surface of the salt pill. A thin cylindrical lucite case was then glued around the salt with cold setting araldite, but the effect was not eliminated completely. Similar desorption effects were found in the measurements with the metal samples, making the results above 0.65° K unreliable.

Since the electron specific heat of a metal is expected to be proportional to T the heat capacity of salt and metal together should obey:

$$c_{\text{tot}} = \frac{A}{T^2} + \gamma T \tag{75.1}$$

(the lattice specific heats can be neglected below 1° K). This is equivalent to:

$$c_{\text{tot}} T^2 = A + \gamma T^3, \tag{75.2}$$

so that if $c_{\text{tot}} T^2$ is plotted versus T^3 a straight line should be found the intersept of which on the vertical axis determines the specific heat of the salt, the slope giving the electron specific heat of the metal.

The experiments could be extended down to 0.15° K. No after-periods were observed, except at the lower temperatures for metals with bad heat conductivities like tungsten. This suggests that good thermal contact with the salt existed down to the lowest temperatures.

Straight lines in the $c_{\text{tot}} T^2$ versus T^3 diagrams were found for copper, silver, platinum, palladium, tungsten and molybdenum (except in the region above 0.65° K, but this was due to the helium adsorption quoted above). The values of γ for most of these metals were in reasonable agreement with those obtained by other investigators at higher temperatures. The values for tungsten, given in the literature are widely divergent, so that it is difficult to make a comparison. RAYNE's value, however, seems to be not unreasonable. The value for palladium is about 20% lower than that given by PICKARD and SIMON, but this may be due to a difference in the purities of the samples.

Remarkable results were found for sodium. In the first place, it was impossible to cool it with copper potassium sulphate below 0.85° K. For this reason new experiments were made using iron ammonium alum. A temperature of 0.3° K was reached then. A pronounced maximum in the specific heat was found at 0.87° K. Due to the uncertainty caused by the helium film (see above) the exact shape of the peak is not very well known. It appears that the entropy content of the peak as calculated from the specific heat curve is a factor ten smaller than that derived from the final temperature reached after demagnetization.

A possible explanation is that a transition in the lattice structure occurs of the martensitic type. The hysteresis involved with such a transition might account for the difference in entropy on cooling and on warming.

SAMOILOV[1] measured the specific heat of cadmium in both the superconducting and the normal states. The heat flow scheme of his calorimeter is shown

---

[1] B. N. SAMOILOV: Dokladi Akad. Nauk, SSSR. **86**, 281 (1952).

in Fig. 112. The cylinder under investigation, C, was connected to the iron ammonium alum pill, A, by a copper wire W_1. The thermal contact with the salt pill was achieved with the vane technique, see Sect. 70. The dimensions of the copper wire were chosen in such a way (0.1 mm. diameter and 30 cm. length) that, after the demagnetization, it took about an hour for the cadmium sample to cool to the temperature of the iron ammonium alum. The advantage was that, if heating periods cf about ten seconds were applied in the specific heat determinations, the heat flow to the salt was negligible so that the heat capacity of the cadmium alone was measured.

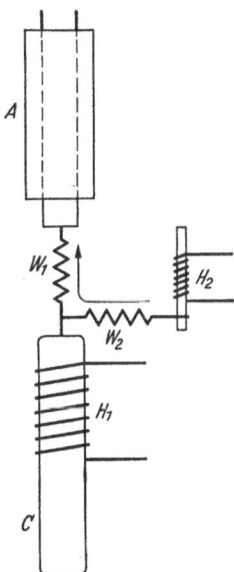

Fig. 112. Heat flow scheme of Sa-moilov's calorimeter for specific heat measurements on cadmium.

Difficulties were encountered in the region above 0.5° K since here the specific heat of the salt becomes rather small and helium desorption from the salt begins, so that the temperature rise due to the heat leak increases appreciably. This was obviated in the following way. The cadmium was connected by a second copper wire W_2, similar to W_1, to a heater H_2. The salt was kept in the neighbourhood of 0.1° K and the temperature of the cadmium was kept at a higher equilibrium temperature by generating a constant heat flow through W_2 and W_1.

The calorimetric experiments were performed with the heater H_1, glued in a helical groove on the surface of the cadmium sample. The temperature was measured with a phosphorbronze thermometer mounted in a small capsule filled with He II. This whole unit was mounted inside a cavity in the lower end of the metal, thermal equilibrium being achieved with a copper wire. The thermometer was calibrated against the susceptibility of the iron ammonium alum, the latter being determined ballistically. The supply wires of the thermometer and the heaters were made of tinned constantan. Copper vanes, inserted in the wires, were embedded in the salt in order to decrease the heat leak to the cadmium.

The vacuum of the calorimeter was improved with an "adsorption pump", a space filled with activated charcoal at helium temperature, connected by a short tube to the vacuum space. Before admitting exchange gas the tube could be closed off with the help of a valve.

It was found that the specific heat in the normal state obeys:

$$c = 7.11 \times 10^3 \, T + 1.942 \times 10^{10} \, (T/300)^3 \, \text{erg/mol degree}. \qquad (75.3)$$

The accuracy reached for the superconducting state was smaller, due to the small region in which the measurements could be made (the normal transition point is about 0.55° K, see Sect. 77) and to irregularities in the phosphorbronze thermometer below 0.4° K. It appeared that the specific heat was roughly proportional to T^3, but the possibility of a small linear term was not excluded, see Sect. 77.

76. Electric conductivity of non-superconducting metals. For most non-superconducting metals the resistance falls off with decreasing temperature to a constant value, the "residual resistance".

In 1934 DE HAAS, DE BOER and VAN DEN BERG[1] found that a minimum occurs in the case of gold and since that time resistance minima have been discovered for several more metals. It is of high importance to know the course of the resistance when the temperature approaches absolute zero and for this reason several investigations have been performed on metallic conductivity below 1° K.

The first investigations on gold were made by DE HAAS, CASIMIR and VAN DEN BERG[2]. The wire was enclosed, together with a phosphorbronze wire acting as a thermometer, in a glass tube filled at room temperature with one atmosphere of helium. This tube was placed in the salt container into which about the same helium pressure was admitted. It follows that the thermal equilibrium between the salt and the gold specimen depended almost entirely on the conductivity through the gas. A smooth curve was found down to 0.4° K; below this temperature the points scattered badly. VAN DER LEEDEN[3] noticed that the results above 0.4° K could be described with the help of a term proportional to $T^{-\frac{1}{2}}$.

New measurements were made by MENDOZA and THOMAS[4] with an apparatus in which the vane technique was applied. A hollow cylindrical copper former was connected to one of the vanes embedded in the salt. The former had a spiral groove on its outside holding the wire under investigation; it was insulated from the wire with a thin coating of bakelite varnish. The ends of the wire were connected by short current and potential leads to the vanes and the other ends of the vanes were soldered to thin tin-covered constantan wires. They passed, along the outside of the salt, to platinum glass seals leading into the helium bath. The whole unit was suspended by a nylon thread from a quartz rod which was attached to a SUCKSMITH balance, the latter being used for the T^* determination of the salt (see Sect. 23). It was found that the gold resistance increased with falling temperature, but much steeper than proportional to $T^{-\frac{1}{2}}$. No simple law was obeyed.

CROFT, FAULKNER, HATTON and SEYMOUR[5] made experiments, first in the region down to 0.2° K, then also down to 0.0065° K. In the first experiment, the heat contact between the salt and the gold wire was made with liquid helium in an apparatus using the capillary technique (see Sect. 69). Gadolinium sulphate was used as a coolant. The susceptibility was measured ballistically and the reduction from T^* to T was carried out with the help of VAN DIJK's data (Sect. 46). The supply wires were brought into the metal salt container through holes sealed with Araldite, a thermo-setting plastic. The results between 0.25 and 2° K obeyed the relation:

$$\frac{R}{R_0} = a \log \frac{\Theta}{T}. \tag{76.1}$$

Small deviations found below 0.25° K were probably due to uncertainties in the T^* to T relation.

A double demagnetization technique was used in the second experiment, see Sect. 80. The first stage sample consisted of iron ammonium alum, the second was made of diluted chromium potassium alum (see Sects. 36 and 59) with a chromium concentration of 5%. The gold wire with its current and potential leads was compressed inside the second stage. Higher up, the four supply wires were

---

[1] W. J. DE HAAS, J. DE BOER and G. J. VAN DEN BERG: Leiden Commun. No. 233b; Physica, 's-Grav. 1, 1115 (1933/34).

[2] W. J. DE HAAS, H. B. G. CASIMIR and G. J. VAN DEN BERG: Leiden Commun. No. 251c; Physica, 's-Grav. 5, 225 (1938).

[3] P. VAN DER LEEDEN: Thesis, Leiden 1940.

[4] E. MENDOZA and J. G. THOMAS: Phil. Mag. 42, 291 (1951).

[5] A. J. CROFT, E. A. FAULKNER, J. HATTON and E. F. W. SEYMOUR: Phil. Mag. 44, 289 (1953).

compressed in the iron ammonium alum sample. Further details on the apparatus and on the two stage demagnetization technique are given in Sect. 80. The results obtained with this apparatus are shown in Fig. 113. They can also be represented approximately by formula (76.1).

Several other metals were investigated by MENDOZA and THOMAS[1,2]. Silver samples showed minima like gold. The resistance of copper was constant in the helium region, but below 1° K it increased somewhat. Magnesium specimens, though they had been cut from the same piece of wire, showed minima located anywhere between 0.7 and 25° K. No minimum occurred in aluminum. Molybdenum showed a minimum, but below 0.1° K the resistance became constant again. A very slight mini-mum was detected in cobalt, none was found in tungsten.

Several explanations have been given for the occurrence of resistance minima; for instance scattering of electrons by uncompletely filled d-shells of impurity atoms[3], limitation of the electron mean free path due to the dimensions of the sample, maybe even due to internal boundaries[4], a rearrangement of the lattice causing a gap at the top of the electron distribution which results into semiconductor behaviour[5]. MacDonald[6] made some experiments on alloys above 1° K and none of the existing theories could account for his results.

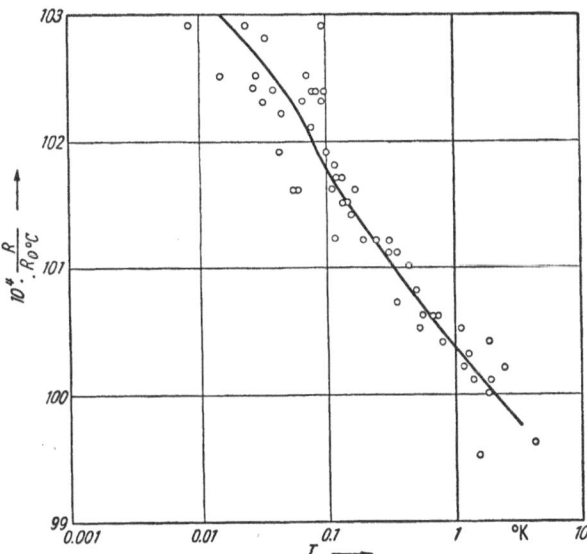

Fig. 113. Resistance of gold (according to CROFT, FAULKNER, HATTON and SEYMOUR).

77. Electric and magnetic properties of superconductors. Two methods are in use for investigations on superconductivity, see Sect. 70. In the first one, small grains of the metal are mixed with the salt and pressed together into a solid pill. The susceptibility of the pill consists of the paramagnetic contribution of the salt and the diamagnetic one of the superconductor. The transition curve can be derived from the observation of the heating curves of the pill for various values of the magnetic field. Typical heating curves, according to DAUNT and HEER[7], are shown in Fig. 114. Point a is the transition temperture. The occurrence of a region of excess positive susceptibility of the superconductor in the presence of a magnetic field (between b and a) was explained by STEELE[8]. It is due to the fact that, in the intermediate state, the negative magnetic moment decreases with time because the fraction of the volume in the superconducting

[1] E. MENDOZA and J. G. THOMAS: Phil. Mag. 42, 291 (1951).
[2] J. G. THOMAS and E. MENDOZA: Phil. Mag. 43, 900 (1952).
[3] A. N. GERRITSEN and J. KORRINGA: Phys. Rev. 84, 604 (1951).
[4] D. K. C. MACDONALD: Phil. Mag. 42, 756 (1951).
[5] J. C. SLATER: Phys. Rev. 84, 179 (1951).
[6] D. K. C. MACDONALD: Phys. Rev. 88, 148 (1952).
[7] J. G. DAUNT and C. V. HEER: Phys. Rev. 76, 1324 (1949).
[8] M. C. STEELE: Phys. Rev. 87, 1137 (1952).

state decreases. This gives rise to a positive contribution in $(\partial M/\partial H)_S$. In some cases (for instance titanium) the region of apparent paramagnetism was missing in the heating curve and this was ascribed by STEELE and HEIN[1] to non-ideality of the superconductor.

The second method is based on MENDOZA's vane technique (see Fig. 95). This method has the advantage that a field can be applied to the metal without influencing the temperature of the salt. For further comparison of the two methods we refer to Sect. 70.

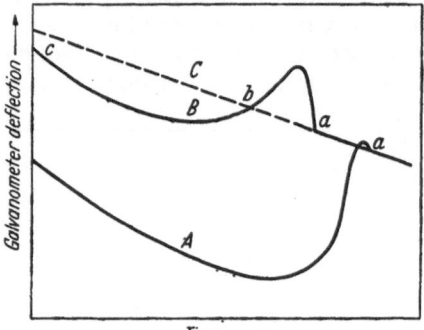

Fig. 114. Change of susceptibility with time for a mixed pill of paramagnetic salt and superconducting metal (according to DAUNT and HEER). Curve A: Heating curve in zero field. Curve B: Heating curve in a constant magnetic field. Curve C: Heating curve of the salt alone. a is the transition point.

If the vane technique is applied the transition curve is usually derived from the mutual inductance of two coils surrounding the metal[1,2]. An interesting modification, in which only a very small amount of metal is needed, was described by SAMOILOV[3]. The principle of the method is shown in Fig. 115. The sample (R, L), a wire of 10 mm. length and 0.4 mm. diameter was connected by the vane technique to the paramagnetic salt. It was connected as a shunt between the coils a and b, which were parts of the mutual inductances M_1 and M_2. If an alternating current passes through coil A an a.c. voltage is generated into B of which the amplitude depends on the impedance of (R, L). A steep increase is found in the voltage over B when (R, L) passes through its transition temperature.

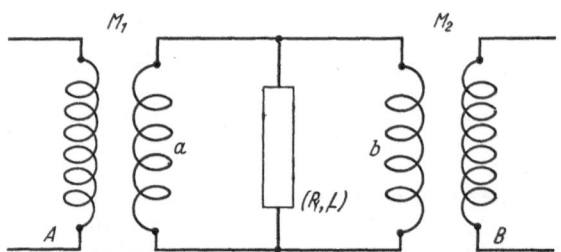

Fig. 115. Measuring circuit for the determination of the transition curve of a superconductor (according to SAMOILOV).

Coils a and b were mounted inside the sample space, they consisted of ten turns of lead wire each, a resistance of 10^{-7} ohm being inserted in both of them in order to suppress persistent currents in a short time. Coils A and B were mounted in the liquid helium. No supply wires into the sample space were needed in this setup and, if the numbers of turns on A and B were well adapted, a variation in the voltage over (R, L) of 10^{-9} volts resulted into a variation of at least 10^{-5} volts over B.

The first metal for which a transition temperature was discovered with the help of the demagnetization technique was cadmium. The first experiments were performed by KURTI and SIMON[4,5] in 1934 and 1935 with a compressed mixture of salt and metal. The transition temperature in zero field, T_c, was 0.54° K and the slope of the transition curve at T_c was 100 oersteds per degree.

[1] M. C. STEELE and R. A. HEIN: Phys. Rev. **92**, 243 (1953).
[2] B. B. GOODMAN and E. MENDOZA: Phil. Mag. **42**, 594 (1951).
[3] B. N. SAMOILOV: Doklady Akad. Nauk, SSSR. **81**, 791 (1951).
[4] N. KURTI and F. SIMON: Nature, Lond. **133**, 907 (1934).
[5] N. KURTI and F. SIMON: Proc. Roy. Soc. Lond., Ser. A **151**, 610 (1935).

Smith and Daunt [1] found by the same method: $T_c = 0.602°$ K. The transition curve could be represented by a parabola:

$$H = H_0\left[1 - \left(\frac{T}{T_c}\right)^2\right]. \qquad (77.1)$$

H_0, the critical field at absolute zero was calculated to be 33.8 oersteds, and the slope of the transition curve at T_c to be 112 oersteds per degree. If Eq. (77.1) is valid the specific heat of the electrons in the normal state, assuming a reversible transition, obeys:

$$c_{el} = \gamma\, T = \frac{V}{8\pi}\left(\frac{\partial H}{\partial T}\right)_{T=T_c} \cdot T, \qquad (77.2)$$

where V is the atomic volume of the metal. The value of γ following from the experiments of Smith and Daunt is 6.44×10^3 erg/mol. degree[2].

Steele and Hein [2], applying the same method, investigated small cadmium grains of spherical shape. The transition point was $0.65°$ K, the slope of the transition curve was the higher the smaller were the particles. If this phenomenon was related to the penetration depth it followed that λ_0 is of the order of 10^{-4} cm. This is rather high as compared to other superconductors, but in general λ_0 increases with decreasing T_c.

Goodman and Mendoza [3] made investigations on cadmium with the vane technique. The transition temperature was derived from two coils surrounding the metal as mentioned above. They found a parabolic transition curve with $T_c = 0.560°$ K, $H_0 = 28.8$ oersteds, the slope at T_c was 103 oersteds per degree and γ was 5.35×10^3 erg/mol. degree[2].

Samoilov [4], applying the method described above, found $T_c = 0.547°$ K, $H_0 = 28.4$ oersteds, the slope at T_c was 104 oersteds per degree and $\gamma = 5.56 \times 10^3$ erg/mol. degree[2]. The value for γ was 20% lower than that derived from Samoilov's specific heat measurements (see Sect. 75), the difference being well beyond the accuracy of the experiments. Samoilov suggested that the difference might be due to a small linear term in the specific heat of the superconducting state; Clement [5] showed that the discrepancy may also be solved by assuming a small term proportional to T^3 in the magnetic transition curve.

Titanium is a "hard" superconductor. It is difficult to obtain in a strain-free state and annealing is rather ineffective. Results obtained by different investigators diverge widely. Old investigations by Meissner et al.[6,7] and by de Haas and van Alphen [8] gave transition temperatures between 1.1 and 1.8° K. Shoenberg [9] did not find superconductivity down to 1° K; he ascribed the earlier results to impurities.

The first investigations on titanium below 1° K were performed by Daunt and Heer [10] with a compressed pill of salt and metal. T_c was $0.527°$ K and the slope of the transition curve at T_c was 470 oersteds per degree. The same material was reinvestigated by Smith and Daunt [11]. After annealing for $2\frac{1}{2}$ hours at 800° C the hardness had decreased only very little. The transition point was then

[1] T. S. Smith and J. G. Daunt: Phys. Rev. 88, 1172 (1952).
[2] M. C. Steele and R. A. Hein: Phys. Rev. 87, 908 (1952).
[3] B. B. Goodman and E. Mendoza: Phil. Mag. 42, 594 (1951).
[4] B. N. Samoilov: Doklady Akad. Nauk, SSSR. 81, 791 (1951).
[5] J. R. Clement: Phys. Rev. 92, 1578 (1953).
[6] W. Meissner: Z. Physik 60, 181 (1930).
[7] W. Meissner, H. Franz and H. Westerhoff: Ann. Physik 13, 555 (1932).
[8] W. J. de Haas and P. M. van Alphen: Leiden Commun. No. 212e; Proc. Roy. Acad. Amst. 34, 70 (1931).
[9] D. Shoenberg: Proc. Cambridge Phil. Soc. 36, 84 (1940).
[10] J. G. Daunt and C. V. Heer: Phys. Rev. 76, 715 (1949).
[11] T. S. Smith and J. G. Daunt: Phys. Rev. 88, 1172 (1952).

0.558° K and the slope of the transition curve 450 oersteds per degree. A new, relatively soft, sample was investigated by Smith, Gager and Daunt[1]. After annealing in vacuum for three hours at 670° C the hardness was only 67 dph numbers. The sample consisted of a crystal bar pressed inside the chromium alum pill. The transition point was 0.387° K, the initial slope 89.5 oersteds per degree. These values are appreciably lower than those obtained for the harder samples. The transition curve was approximately parabolic with $H_0 = 20$ oersteds and $\gamma = 4.60 \times 10^3$ erg/mol. degree^2.

Two samples of titanium were investigated by Steele and Hein[2]; the first one consisted of small pieces of wire compressed in a salt pill, the other was a crystal bar on which experiments were made in an apparatus very similar to that of Fig. 97. A second holder was connected to the screw for accommodating a carbon thermometer which served for checking the heat contact with the salt. The samples were not annealed, the hardness was not too much different from that of the material used by Smith, Gager and Daunt. For the first sample the transition curve was a parabola with $T_c = 0.37°$ K, $H_0 = 86$ oersteds and a slope at T_c of 465 oersteds per degree. No parabola was obeyed in the second experiment. T_c was 0.49° K and the slope at T_c was 400 oersteds per degree. The region of apparent paramagnetism below T_c (see above) was missing in these experiments. For this reason the samples were considered as not very ideal superconductors and no γ values were calculated from the results. The data are in reasonable agreement with those for the unannealed sample of Daunt and Heer, but no agreement is found with the values of Smith, Gager and Daunt.

Several more metals have been investigated below 1° K. Measurements with zirconium and hafnium were made by Kurti and Simon[3,4] and by Smith and Daunt[5]. Both are hard superconductors. Kurti and Simon found for zirconium: $T_c = 0.70°$ K, the initial slope of the transition curve being 400 oersteds per degree. Smith and Daunt found that the influence of annealing is rather large. Before annealing the results were: $T_c = 0.565°$ K, initial slope: 335 oersteds per degree. After annealing the transition curve had acquired parabolic shape with $T_c = 0.546°$ K, initial slope 171 oersteds per degree, $H_0 = 46.6$ oersteds and $\gamma = 16.4 \times 10^3$ erg/mol. degree^2. In the case of hafnium, Kurti and Simon found $T_c = 0.35°$ K. Smith and Daunt did not find superconductivity before annealing; thereafter T_c was 0.347° K and the initial slope was 230 oersteds per degree. An estimate showed that only part of the volume of the metal had become superconductive.

Zinc and aluminum were investigated by Daunt and Heer[6] and by Goodman and Mendoza[7]. The transition curves, as found by Daunt and Heer, were not very exact parabolae. Their data for zinc are: $T_c = 0.95°$ K, initial slope 98 oersteds per degree, $\gamma = 5.68 \times 10^3$ erg/mol. degree^2; for aluminum: $T_c = 1.17°$ K, initial slope 136 oersteds per degree, $\gamma = 10.8 \times 10^3$ erg/mol. degree^2. Goodman and Mendoza found transition curves of parabolic shape with, for zinc: $T_c = 0.905°$ K, $H_0 = 52.5$ oersteds, initial slope 116 oersteds per degree and $\gamma = 4.85 \times 10^3$ erg/mol. degree^2; for aluminum: $T_c = 1.197°$ K, $H_0 = 106.0$ oersteds, initial slope 177 oersteds per degree and $\gamma = 12.3 \times 10^3$ erg/mol. degree^2. Also gallium was investigated by Goodman and Mendoza[7]. The transition

[1] T. S. Smith, W. B. Gager and J. G. Daunt: Phys. Rev. 89, 654 (1953).
[2] M. C. Steele and R. A. Hein: Phys. Rev. 92, 243 (1953).
[3] N. Kurti and F. Simon: Nature, Lond. 135, 31 (1935).
[4] N. Kurti and F. Simon: Proc. Roy. Soc. Lond., Ser. A 151, 610 (1935).
[5] T. S. Smith and J. G. Daunt: Phys Rev. 88, 1172 (1952).
[6] J. G. Daunt and C. V. Heer: Phys. Rev. 76, 1324 (1949).
[7] B. B. Goodman and E. Mendoza: Phil. Mag. 42, 594 (1951).

curve was a parabola with $T_c = 1.103°$ K, $H_0 = 50.3$ oersteds, initial slope 91.2 oersteds per degree and $\gamma = 3.80 \times 10^3$ erg/mol. degree^2.

GOODMAN and SHOENBERG[1] made some measurements with uranium. The results were strongly divergent for different samples. T_c varied from 0.75° K to 1.3° K. Marked differences occurred also in the slopes of the transition curves. In general they were rather steep, of the order of 2000 oersteds per degree. Uranium was also investigated by ALEKSEYEVSKY and MIGUNOV[2], the transition point being 1.3° K. GOODMAN[3] made experiments with osmium and ruthenium. Both showed parabolic transition curves with, for the osmium: $T_c = 0.71°$ K, $H_0 = 65$ oersteds, and for the ruthenium: $T_c = 0.47°$ K, $H_0 = 46$ oersteds.

Finally we quote the metals that did not become superconductive, down to the temperatures indicated. Gold (0.05° K), copper (0.05° K), bismuth (0.05° K), magnesium (0.05° K) and germanium (0.05° K) investigated by KURTI and SIMON[4]; silicon (0.073° K), chromium (0.082° K), antimony (0.152° K), tungsten (0.070° K), beryllium (0.064° K) and rhodium (0.086° K), investigated by ALEKSEYEVSKY and MIGUNOV[2]; lithium (0.08° K), sodium (0.09° K), potassium (0.08° K), barium (0.15° K), yttrium (0.10° K), cerium (0.25° K), praseodymium (0.25° K), neodymium (0.25° K), manganese (0.15° K), palladium (0.10° K), iridium (0.10° K) and platinum (0.10° K), investigated by GOODMAN[3]; cobalt (0.06° K), molybdenum (0.05° K) and silver (0.05° K) investigated by THOMAS and MENDOZA[5].

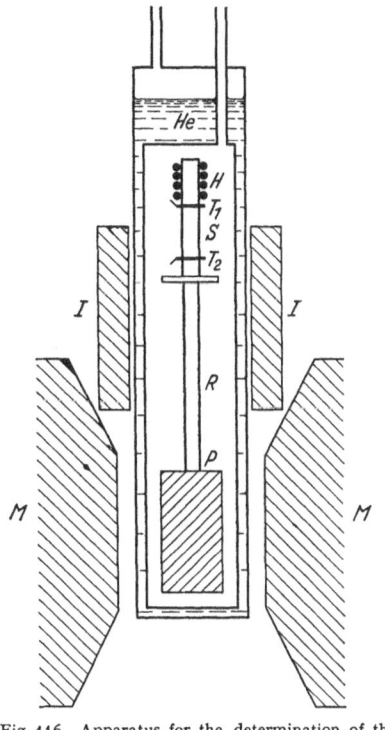

Fig. 116. Apparatus for the determination of the heat conductivity of superconductors (according to MENDELSSOHN).

78. Thermal conductivity of metals. Two techniques are in use for the determination of thermal conductivities of metals below 1° K. In the first one, the rod or wire under investigation is connected between two salt pills which are demagnetized to somewhat different temperatures. The heat conductivity follows from the course of the temperatures of the pills with time. The connections between the rods and the salt pills are made by copper inserts, the heat contact with the salt being achieved with the vane technique, or sometimes by high pressure molding. A correction must be applied for the thermal resistance of the contact layer between salt and metal on the basis of Eq. (68.1). Still systematical errors may be made due to inhomogeneities in the temperatures of the salt pills.

The second method is illustrated in Fig. 116. The metal specimen S is connected by the copper rod R to the salt pill P, which serves only as a cooling agent and a heat sink. Energy is supplied with the heater H and temperature measurements are made with the help of the thermometers T_1 and T_2. They consist of

[1] B. B. GOODMAN and D. SHOENBERG: Nature, Lond. **165**, 441 (1950).
[2] N. ALEKSEYEVSKY and L. MIGUNOV: J. Phys. USSR. **11**, 95 (1947).
[3] B. B. GOODMAN: Nature, Lond. **167**, 111 (1951).
[4] N. KURTI and F. SIMON: Proc. Roy. Soc. Lond., Ser. A **151**, 610 (1935).
[5] D. SHOENBERG: Superconductivity. Cambridge: Univ. Press 1952.

narrow rings of carbon black forming conducting bridges between the specimen and the leads. They are calibrated against the susceptibility of the salt when no heat is supplied. Under normal experimental conditions the heat leak along the supply wires of H, T_1 and T_2 and the energy developed in the thermometers are small as compared to the heat supplied in H.

A difficulty was encountered when superconducting metals were investigated. The big field used for the adiabatic demagnetization renders the specimen temporarily non-superconductive and sometimes it happens that, after the demagnetization, a residual magnetic flux is trapped in the metal. Since the heat conductivity in the normal state is much better than in the superconducting state this gives rise to spurious effects. For this reason the long copper rod R was inserted and the iron shield I was applied.

The only investigation on a non-superconductor is that on copper by NICOL and TSENG[1]. The measurements were performed with the first method. It was found that the heat conductivity was proportional to the temperature as should be expected in the case of electron conductivity limited only by impurity scattering.

The heat conductivity K_s of a superconductor, if the temperature is so low that practically all the electrons are in the superconducting state, is essentially the lattice conductivity, proportional to T^3. The heat conductivity of the normal state, K_n, is the electron conductivity, proportional to T. Since K_s and K_n are equal at T_c the quantity K_s/K_n becomes very small at the lowest temperatures. Due to this, it is usually impossible to choose the dimensions of the specimen under investigation in such a way that both K_s and K_n can be measured with satisfactory precision. Standard practice is to apply a sample for which K_s can be determined experimentally, the value of K_n being extrapolated linearly from the data obtained in the liquid helium region.

The method with the specimen between two salt pills was used by HEER and DAUNT[2] for tin and tantalum, by GOODMAN[3] for tin. In the results of HEER and DAUNT the value of K_s/K_n for tin was about 1/40 at 0.65° K, for tantalum it was somewhat less than 1/60 at 0.55° K. For both metals K_s/K_n was approximately equal to $(T/T_c)^2$. Also GOODMAN found a steep decrease of K_s with falling temperature, the slope decreased with increasing impurity of the sample.

The second method was used by OLSEN and RENTON[4] for lead, and by MENDELSSOHN and RENTON[5] for tin, indium, thallium, columbium, tantalum and aluminum. Niobium was investigated by MENDELSSOHN[6]. Single crystals of lead, tin and indium and a polycrystalline sample of niobium gave results for K_s proportional to T^3 at the lower temperatures. The coefficients were compatible with those predicted from theory for the lattice conductivity[6]. Deviations from the T^3 law began at 0.9° K, for lead, at 0.55° K for tin, and at 0.7° K for indium. The excess conductivity was probably due to the beginning of electron conduction. The precision of the measurements was not sufficient to decide on the dependence of the electron conduction on T. A polycrystalline sample of thallium gave an exponential rise of K_s with temperature. Polycrystalline samples of tin, columbium, tantalum and aluminum gave rather irregular and unsurveyable results, they were ascribed to frozen-in magnetic flux as described above.

[1] J. NICOL and T. P. TSENG: Phys. Rev. 92, 1062 (1953).
[2] C. V. HEER and J. G. DAUNT: Phys. Rev. 76, 854 (1949).
[3] B. B. GOODMAN: Proc. Phys. Soc. Lond., A 66, 217 (1953).
[4] J. L. OLSEN and C. A. RENTON: Phil Mag. 43, 946 (1952).
[5] K. MENDELSSOHN and C. A. RENTON: Phil. Mag. 44, 776 (1953).
[6] C. J. GORTER et al.: Progress in Low Temperature Physics, p. 194. Amsterdam 1955

Some measurements were made by the same authors on the thermal conductivity in the intermediate state. A minimum was found in some cases, but the interpretation is not very clear.

III. The thermal valve and its applications.

79. Thermal valves. Some years after the first demagnetization experiments the desirability was suggested[1] of a "thermal valve" or "thermal switch". This is an instrument which provides the possibility to break, after the demagnetization, the heat contact between the paramagnetic salt and a substance cooled with it. The importance of such an instrument is obvious for specific heat measurements below 1° K and for more-stage demagnetizations (see Sect. 80). It might also be used instead of exchange gas thus reducing the time of magnetization.

The first effort to construct a thermal switch was made by MENDOZA[2]. Two salt pills were connected to copper rods with the vane technique, the rods being connected to each other with the help of copper sheet, 4 mm. wide and 0.02 mm. thick. The heat contact could be broken by lifting the upper pill somewhat, so that the copper sheet was torn up. The disadvantage of this switch is obviously that it can be used only once.

A switch proposed by KURTI[3,4] is based on the fact that, at the lower temperatures, the thermal conductivity of liquid helium becomes rather bad (see Sect. 71). Heat transfer can be achieved through a thin layer, not through a narrow tube. KURTI's idea was to connect the salt and the other substance by a tube filled with liquid helium in which a copper rod can be moved. If the rod is in such a position that it passes through both substances, the heat transfer is good. If the rod is lifted so that it is removed several centimeters from one of the samples the contact is broken. Since the rod must be operated from outside the cryostat, the film creep may give rise to undesirable heat leak. Also moving parts in an apparatus below 1° K may introduce some difficulties.

Another possibility was suggested by WILKINSON and WILKS[5]. At not too low temperatures the heat conductivity of solid helium is much worse than that of the liquid (see Sect. 72). If the pressure of the apparatus is increased until the helium solidifies, the heat contact is broken, when the helium is allowed to melt again the thermal equilibrium is restored.

The only thermal switch which has yielded successful experimental results makes use of a superconductor. A metal in the superconducting state, at temperatures well below T_c, has a much lower thermal conductivity than in the normal state (see Sect. 78). A switch based on this principle can be operated by applying and removing a magnetic field. A metal with a transition temperature well above 1° K must be applied, and care should be taken that no appreciable magnetic flux is trapped in the superconducting state.

This switch was proposed independently by several authors[6,7,8]. Its practical applicability was first demonstrated by HEER and DAUNT[7]. A chromium potas-

---

[1] F. SIMON: Réunion d'études sur le magnétisme, Strasbourg vol. III, p. 1. 1939.

[2] E. MENDOZA: Cérémonies LANGEVIN-PERRIN, p. 67. Paris 1948.

[3] N. KURTI: Cérémonies LANGEVIN-PERRIN, p. 34. Paris 1948.

[4] F. SIMON, N. KURTI, J. F. ALLEN and K. MENDELSSOHN: Low temperature physics, four lectures, p. 59. London 1952.

[5] K. R. WILKINSON and J. WILKS: Proc. Phys. Soc. Lond. A **64**, 89 (1951).

[6] C. J. GORTER: Cérémonies LANGEVIN-PERRIN, p. 76. Paris 1948.

[7] C. V. HEER and J. G. DAUNT: Phys. Rev. **76**, 854 (1949).

[8] K. MENDELSSOHN and J. L. OLSEN: Proc. Phys. Soc. Lond., A **63**, 2 (1950).

sium alum pill was connected by a tantalum wire, 56 cm. long and 0.017 cm. in diameter, to the liquid helium bath. The heating curve is shown in Fig. 117. Before and after the magnetic field was applied the heat leak was about 7 ergs per minute. The heat gain during the two minutes that the field was on, was 1.5×10^4 ergs.

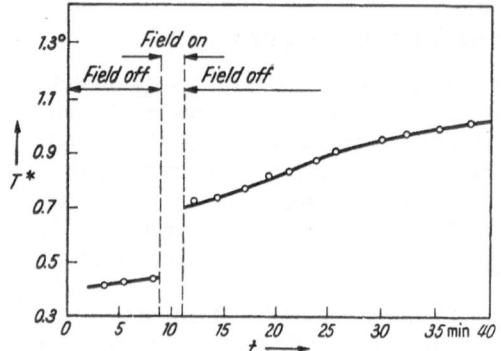

Fig. 117. Efficiency of a superconducting thermal switch (according to HEER and DAUNT).

80. Applications of the thermal valve; cascade demagnetization. A superconducting thermal switch, replacing exchange gas, was applied in the experiments of STEELE and HEIN[1] on the superconductivity of small spherical cadmium particles, quoted in Sect. 77.

DE VRIES and DAUNT[2] used a thermal switch for the determination of the specific heat of He³. After the demagnetization the heat contact between the salt and the helium container was broken and the specific heat of the liquid was determined separately, see Sect. 73.

An important application of the thermal valve is the cascade or more-stage demagnetization. It is explained in Fig. 118. A and B are paramagnetic salt pills, connected by the thermal valve V. The first stage A is demagnetized in the ordinary way while V is in the open (non-superconducting) state. Then B is magnetized, V being still open so that the heat of magnetization is carried off to A. Subsequently V is closed and B is demagnetized. Since the starting temperature of the second demagnetization is well below that of the first one, appreciably lower temperatures may be reached than by ordinary one-stage demagnetization.

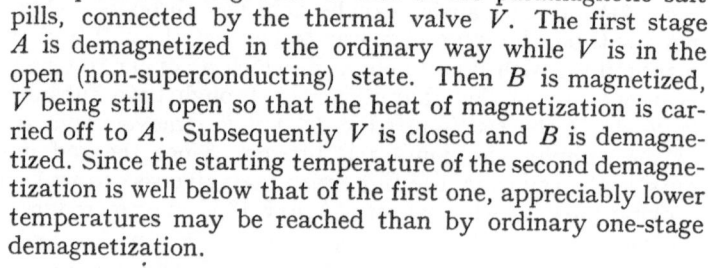

The limit imposed to the starting temperature of the second demagnetization is that, in the open position of the thermal valve, equilibrium between the two samples must be reached in a short time (e.g. some minutes). The bottleneck of this equilibrium proves to be the heat transfer between the valve and the salt pills. With the techniques described in Sect. 70 this temperature cannot be too far below 0.1° K.

The ratio of the masses of the two salt pills must be such that the temperature of A is not raised too much by the heat of magnetization of B. This can be estimated from the course with temperature of the specific heat of A in zero field and that of B in the field. MENDOZA showed[3] that, if both pills are made of chromium potassium alum, the ratio must be of the order of 1:20 if the field is about 10000 oersteds.

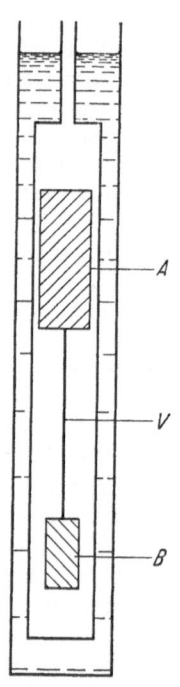

Fig. 118. Schematic design of a two-stage demagnetization apparatus.

It is obvious that the application of the same salt in both stages is not the most effective method if one wants

[1] M. C. STEELE and R. A. HEIN: Phys. Rev. **87**, 908 (1952).
[2] G. DE VRIES and J. G. DAUNT: Phys. Rev. **93**, 631 (1954).
[3] E. MENDOZA: Cérémonies LANGEVIN-PERRIN, p. 68. Paris 1948.

to reach the lowest possible temperature. In fact, the requirements for the stages A and B are widely different. The temperature reached with A needs not be exceptionally low, not below the limit quoted above, but a large amount of entropy must be removed from it, at least considerably more than the entropy of magnetization of B. For B a salt is required which, when magnetized at 0.1° K to almost its saturation moment, reaches a very low final temperature. For this purpose we need a salt with very weak magnetic and crystalline interactions, hence, referring to Sect. 4, a salt with a very low Θ.

Two stage demagnetization experiments were performed by DARBY, HATTON and ROLLIN[1] and by DARBY, HATTON, ROLLIN, SEYMOUR and SILSBEE[2]. The apparatus is shown in Fig. 119. Stage A consists of a pill of iron ammonium alum, 57 mm. long and 16 mm. in diameter. In the first experiments, B consisted of copper potassium sulphate, diluted with 90% of zinc potassium sulphate. Since, however, this salt has appreciable nuclear and exchange specific heat (see Sect. 42) it was replaced later by diluted chromium potassium alum 1:20, a pill 23 mm. long and 9 mm. in diameter.

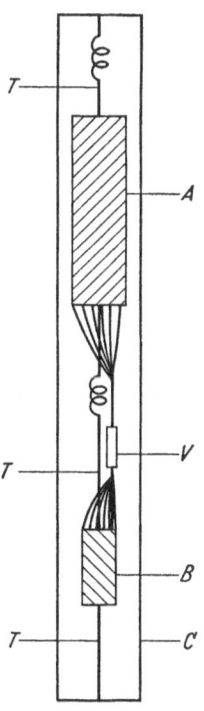

Six copper wires were compressed in each pill, they were connected to the thermal valve V, the latter consisting of a lead wire 3 cm. long and 0.3 mm. in diameter. The two salt pills were suspended in a metal cage C with the help of cotton threads T, maintained in tension by springs. The distance between the two pills was such that the upper one had no disturbing influence on the measurement of the susceptibility of the lower one.

The course of an experiment was the following: First both stages were magnetized in the field of an electromagnet. After removing the exchange gas, the magnet was slowly lowered so that the upper stage was demagnetized while the lower stage remained in the field. This operation was carried out very slowly so that the temperature difference between the two stages was always very small, making the process nearly quasistatic, and thus reducing the entropy transfer to the upper stage as much as possible. When the lowering of the magnet was completed, the residual field at the upper stage was compensated with a small permanent magnet and after waiting about 30 minutes for equilibrium the magnetic field was reduced to zero. As soon as the stray field on the lead wire fell below 800 oersteds, the wire became superconducting breaking effectively the thermal contact between the stages.

Fig. 119. Cascade demagnetization apparatus (according to DARBY, HATTON, ROLLIN, SEYMOUR and SILSBEE).

When the process was carried out with a field of 4200 oersteds, the T^* reached with the diluted chromium alum was 0.009°, corresponding (according to the measurements of DE KLERK, STEENLAND and GORTER, see Sect. 59) to an absolute temperature of 0.003° K. For a single demagnetization starting from 1.05° K a field of 15 200 oersteds would be required. The heating rate after the demagnetization, if care was taken that mechanical vibrations of the apparatus were eliminated (see Sect. 18), corresponded to a leak of about one erg per minute. The temperature remained below 0.01° K for about three minutes.

[1] J. DARBY, J. HATTON and B. V. ROLLIN: Proc. Phys. Soc. Lond. A **63**, 1179 (1950).

[2] J. DARBY, J. HATTON, B. V. ROLLIN, E. F. W. SEYMOUR and H. B. SILSBEE: Proc. Phys. Soc. Lond. A **64**, 861 (1951).

With a starting field of 9000 oersteds the lowest T^* was 0.0045°. The absolute temperature, following from the extrapolation of the results of DE KLERK, STEENLAND and GORTER, is approximately 10^{-3} °K. The temperature remained below 0.01° K for forty minutes.

Finally it should be mentioned that, in the experiments of CROFT, FAULKNER, HATTON and SEYMOUR[1] on the electric resistivity of gold (see Sect. 76), use was made of a cascade demagnetization technique for obtaining the lowest temperatures; the gold wire was embedded in the second stage. The current and voltage leads passed through both stages and a piece of lead wire was inserted in each of them between the stages. The four of them acted as the thermal switch together.

81. The magnetic refrigerator. A disadvantage of the adiabatic demagnetization technique described thus far is that, due to the inevitable heat leak, experiments are always carried out with steadily rising temperature. Though precautions can be taken to keep the influence of the leak small (see Sect. 18) it may be advantageous, under certain circumstances, to make experiments under truly isothermal conditions. This may be achieved with the help of a cyclically operating refrigerating machine, as was pointed out by DAUNT and HEER[2]. A schematic layout is shown in Fig. 120.

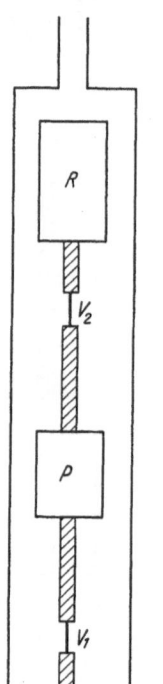

Fig. 120. Schematic diagram of the magnetic refrigerator.

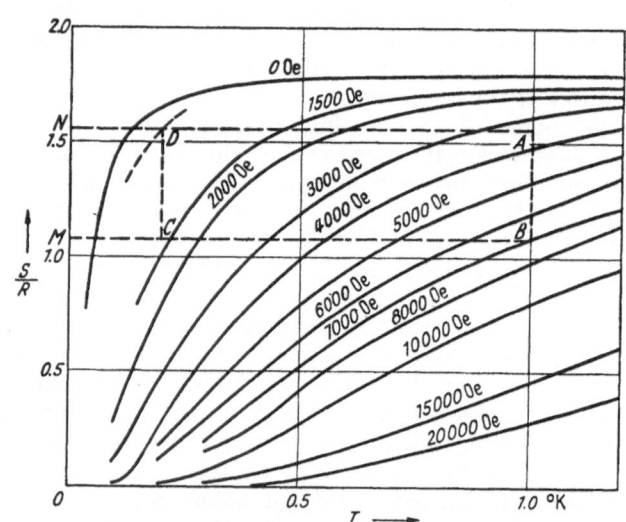

Fig. 121. Entropy versus temperature diagram for iron ammonium alum (according to HEER, BARNES and DAUNT).

The paramagnetic salt P is the working substance. The heat contact with the liquid helium bath B can be made and broken with the thermal valve V_1. R is the reservoir to be cooled; it is connected to P by the valve V_2.

The cycle of operation, in its simplest form, is the following. First, V_1 is opened and P is magnetized. After the heat of magnetization has been carried off to the bath V_1 is closed, P is demagnetized and V_2 is opened. As soon as

[1] A. J. CROFT, E. A. FAULKNER, J. HATTON and E. F. W. SEYMOUR: Phil. Mag. **44**, 289 (1953).

[2] J. G. DAUNT and C. V. HEER: Phys. Rev. **76**, 985 (1949).

thermal equilibrium between R and P is obtained V_2 is closed, P is magnetized and V_1 is opened again. This cycle can be repeated as often as desired. The temperature of R falls gradually until an equilibrium is reached where the heat extraction per cycle is equal to the amount of heat leaking in during a cycle.

The performance of the refrigerator is noticeably increased if a true CARNOT cycle is carried out. This is illustrated in Fig. 121. (The zero field curve of this figure is in agreement with the data of Sect. 38, the constant field curves were calculated from the zero field data with the help of the BRILLOUIN function, see Sect. 29. The latter is probably not completely justified, but the figure may still give an impression of the performance of the apparatus). The demagnetization of P is carried out adiabatically from B to C, further isothermally from C to D. This is achieved by opening V_2 at the point C and then decreasing the field at such a speed that the heat flow from R to P is compensated by the action of the field. Subsequently, P is magnetized adiabatically until the temperature of the bath is reached (point A) and the further magnetization to B is carried out isothermally with V_1 open. The energy extraction from R per cycle is equal to the area $CDMN$. A salt must be chosen for P for which this quantity is large.

It is obvious that, when the refrigerator has reached its equilibrium state, the temperature of R is not exactly constant, a ripple occurring of the frequency of the cycle of P. The amplitude is the smaller the larger is the heat capacity of R, but a large heat capacity increases the time necessary to attain the final equilibrium. Temperatures have been obtained down to 0.2° K, and it is clear that the application of a refrigerator as described here is only useful in the case of an exceptionally large heat leak, or when large amounts of heat are developed in the experiment itself. A possibility suggested by HEER, BARNES and DAUNT[1], for instance, is the establishment of a visible bath of liquid helium down to 0.2° K.

A refrigerator in which the thermal contacts were made and broken mechanically was constructed by COLLINS and ZIMMERMAN[2]. Both P and R were sealed brass cylinders filled with iron ammonium alum with a small amount of helium gas. It was found that the raising and lowering of P between its two contact surfaces gave rise to an appreciable heat development due to mechanical vibrations and this set a lower limit to the temperatures reached with the refrigerator. The starting temperature was 1.13° K, the magnetic field applied to P was 1850 oersteds and the final temperature of R was 0.73° K.

HEER, BARNES and DAUNT[3] developed a refrigerator (see Fig. 122) in which the thermal valves consisted of lead ribbon. The working unit P was made of 15 grams of iron ammonium alum compressed around a finned copper shaft as described in Sect. 70, see Fig. 96. The reservoir R was constructed in a very similar way with the help of chromium potassium alum. Experiments could be performed with different values of the heat capacity of the reservoir simply by having R in a constant magnetic field. Copper rods were inserted between the salt pills and the valves in order to keep the different magnetic fields well separated.

The magnetic field applied to P was produced with the help of an iron-covered solenoid magnet, M_2, giving fields up to 8000 oersteds. It was cooled by circulating oil between the layers. Similar magnets of smaller size, M_1 and M_3, were

[1] C. V. HEER, C. B. BARNES and J. G. DAUNT: Phys. Rev. **91**, 412 (1953).
[2] S. C. COLLINS and F. J. ZIMMERMAN: Phys. Rev. **90**, 991 (1953).
[3] C. V. HEER, C. B. BARNES and J. G. DAUNT: Rev. Sci. Instrum. **25**, 1088 (1954).

used for the operation of the valves. They gave a field of 1000 oersteds with a current less than 7 amp. The control of the three magnets was all done automatically by a system of clocks, relays, rheostats run with motors, etc.

There are several reasons why the cycle of a refrigerator deviates in practice from a true CARNOT cycle. It proves to be impossible to maintain exact isothermal and adiabatic conditions. Difficulties are especially encountered on the iso-

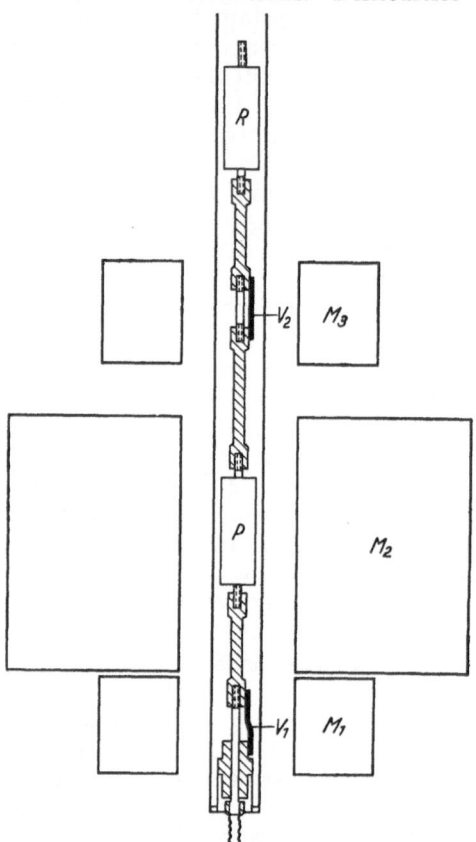

Fig. 122. Magnetic refrigerator (according to HEER, BARNES and DAUNT).

thermal CD. The heat conductivity of V_2 and the specific heat of P are both marked functions of temperature. It is very difficult to adjust the field in such a way that the heat flow from R to P is nicely balanced.

Heat leaks occur through the suspensions and through residual gas. Eddy currents in the metal parts of the salt pills, due to the variations in the magnetic fields, are still more serious sources of heat development. It is essential that all the alterations in magnetic fields are applied smoothly. Also eddy currents due to ripples in the magnets and to the a.c. measuring field should be kept as small as possible.

The switches V_1 and V_2 are not perfect. They transmit some heat in the superconducting state, so that a heat flow takes place through both valves on each of the isothermals. Further they are no ideal short circuits when non-superconductive. This has the consequence that, on AB, the temperature of P is somewhat higher than that of the bath; on CD it is somewhat lower than that of R, making the cycle essentially non-reversible.

Some of the deviations from ideality quoted here can be treated numerically, some cannot. An extensive discussion was given by HEER, BARNES and DAUNT. They lead to an appreciably smaller performance of the refrigerator than that calculated from the CARNOT cycle.

A complete cycle of the refrigerator of HEER, BARNES and DAUNT was passed through in two minutes, 43 sec. being used for the path AB and 49 sec. for CD. The initial field was 7000 oersteds. The field at C was always 3000 oersteds, that at D was decreased, in the course of the cooling process from 1800 to 300 oersteds. If the chromium potassium alum pill R was kept in a constant field of 3000 oersteds (see above) a temperature of 0.3° K was attained in about 40 minutes, 0.2° K was reached in an hour and a half. The heat extraction rate at 0.55° K was 4.2×10^4 ergs per cycle, at 0.26° K it was 0.85×10^4 ergs per cycle. If no magnetic field was applied to R the specific heat was so much smaller that a temperature of 0.3° K was reached within six minutes.

IV. Nuclear demagnetization and nuclear orientation[1].

82. Possibilities of nuclear demagnetization. The lowest temperatures attainable by adiabatic demagnetization of paramagnetic salts are probably of the order of 10^{-3} °K, the limit being determined by the interactions between the ions, see Sect. 4. Dilution may reduce the interactions, but it decreases at the same time the specific heat per unit volume. The suggestion was made independently by GORTER[2] and by KURTI and SIMON[3] that essentially lower temperatures might be reached by adiabatic demagnetization of substances containing atoms with a nuclear magnetic moment.

Suppose we have a nucleus with spin $I\hbar$ and magnetic moment μ_n. Each of the electronic levels of the atom is split by a magnetic field into $2I+1$ sublevels, a distance $\mu_n H/I$ apart. In order to obtain an appreciable difference in population of the sublevels (hence a noticeable decrease of entropy) $\mu_n H/I$ must be of an order of magnitude comparable with kT. Since μ_n is always at least a factor 10^3 smaller than the electronic magnetic moment, values of H/T of roughly 10^7 oersteds per degree are needed. At present, fields of 10^7 oersteds cannot be made for any reasonable length of time. For this reason successful nuclear demagnetization experiments can only be expected from a cascade demagnetization technique (see Sect. 80) in which the first stage is a paramagnetic salt and the second one is the nuclear substance.

In the cascade demagnetizations performed thus far (described in Sect. 80) the starting temperature of the second stage was always of the order of 0.1° K. The highest fields available at the present time (see Sect. 22) are of the order of 10^5 oersteds so that the starting temperature of a nuclear demagnetization must be at least as low as 0.01° K. At this temperature the problem of the heat transfer between the stages (thermal valve "open") has not yet been solved in an satisfactory way. The heat conduction of the valve itself is not the main problem, but the thermal contacts between the valve and the stages (see Sect. 80). The experimental values found for the coefficient of Eq. (68.1) suggest that the area of contact in the case of metal vanes in a paramagnetic salt pill (or in compressed dielectric powder containing nuclear spins) must be at least 10^5 cm.² at 0.01° K. If liquid helium is used as an intermediate the situation is not much improved, due to the KAPITZA layer (see Sect. 71). The conditions at the second stage are more favourable if a metal can be used for the nuclear substance, but then care must be taken to avoid the influence of eddy current heating due to the variations of the field.

Another difficulty may arise from the relaxation time between the nuclear spin system and the lattice. POUND showed[4] that a relaxation time of more than an hour occurs in lithium nitrate. This is exceptionally long, but in general the times increase with falling temperature[5]. Metals have shorter relaxation times than dielectric crystals, but it is difficult, at the present time, to make any estimations of values at 0.01° K.

A discussion of the possibilities and outlooks of nuclear demagnetization was given by SIMON in 1939[6], assuming that the problem of the thermal switch

---

[1] Details, seen from the nuclear point of view are given by R. J. BLIN-STOYLE and M. A. GRACE in Vol. 41 of this Encyclopedia.

[2] C. J. GORTER: Phys. Z. **35**, 923 (1934).

[3] N. KURTI and F. SIMON: Proc. Roy. Soc. Lond., Ser. A **149**, 152 (1935).

[4] R. V. POUND: Phys. Rev. **79**, 685 (1950).

[5] B. V. ROLLIN and J. HATTON: Phys. Rev. **74**, 346 (1948).

[6] F. SIMON: C. R. Conf. Magnétisme, Strasbourg, Vol. 3, 1, 1939.

at 0.01° K is solved, and that no prohibitively long relaxation times are encountered. Due to the smallness of nuclear magnetic moments the interactions between the nuclei are much weaker than those between magnetic ions the same distance apart, giving rise to essentially lower values of the characteristic temperature Θ introduced in Sect. 1 (see also Sect. 4). It proves to be possible to have even a high concentration of magnetic nuclei and still a low value of Θ, providing (incidentally) a usefully large specific heat. It is difficult to make an estimate of the actual value of Θ for a given substance, but it may be expected that, in the case of a metal, it does not exceed 10^{-4} or 10^{-5} °K. This is also the order of magnitude of the final temperature reached by a nuclear demagnetization if most of the entropy is removed by the magnetic field.

SIMON calculated that, in the case of indium metal ($I = \frac{9}{2}$), a magnetic field of 30000 oersteds at 0.01° K decreases the magnetic entropy by only a few percents, ΔS being still practically proportional to $(H/T)^2$. The final temperature is then well above Θ and the specific heat does not reach its maximum value, so that the temperature rise for a given heat leak is rather large. In the case of copper ($I = \frac{3}{2}$) the final temperature for 30000 oersteds is about 5Θ, for 100000 oersteds it is $\frac{3}{2}\Theta$ and here the specific heat is half that of the maximum.

It follows from the above that nuclear demagnetization involves some difficult technical problems. It is not surprising that, though the possibility was suggested more than twenty years ago, no successful experiments have been reported. The aims of nuclear demagnetization are the same as those of ordinary demagnetization: the attainment of a new region of low temperatures; the determination of the temperature scale in this region; the cooling of other substances with the nuclear substance (the latter will involve really serious problems); and investigations on nuclear magnetism itself. In this connection it is interesting that the possibilities of both nuclear ferromagnetism[1] and nuclear antiferromagnetism[2] have been predicted from theory.

At the end of this section we want to mention some experiments by ROLLIN and HATTON[3], and by POUND and PURCELL[4,5]. They made use of dielectric crystals with a long relaxation time between the nuclear spin system and the lattice. By demagnetizing the substance they succeeded in obtaining a low temperature of the spin system alone. It is obvious that these experiments, though they are of considerable theoretical interest, have no importance for the development of the demagnetization technique as described in the present article. PURCELL and POUND succeeded even, by suddenly reversing the magnetization, to obtain a state in which the occupations of the higher nuclear levels were larger than those of the lower ones; a state that can be described with the help of a negative nuclear spin temperature.

A spin system with such a negative temperature has some remarkable properties. The gradual restoration of the thermal equilibrium with the lattice does not take place via $T_n = 0$, but via $T_n = \infty$. During the whole process there is a heat flow from the spin system to the lattice, so that one should consider the negative temperatures as "higher than infinity", rather than as "lower than absolute zero". The most interesting conclusion is probably that, even for these negative temperatures, the law of unattainability of absolute zero remains valid.

[1] H. FRÖHLICH and F. R. N. NABARRO: Proc. Roy. Soc. Lond., Ser. A **175**, 382 (1940).
[2] C. G. B. GARRETT: J. Chem. Phys. **19**, 1154 (1951).
[3] B. V. ROLLIN and J. HATTON: Phys. Rev. **74**, 346 (1948).
[4] R. V. POUND: Phys. Rev. **81**, 156 (1951).
[5] E. M. PURCELL and R. V. POUND: Phys. Rev. **81**, 279 (1951).

83. The production of nuclear orientation. In connection with the difficulties encountered in experiments on nuclear demagnetization the general interest shifted, in recent years, to investigations on nuclear orientation. This subject is related to nuclear demagnetization as ordinary paramagnetism to adiabatic demagnetization of paramagnetic salts.

The results obtained in these experiments are more in the field of nuclear physics than in that of low temperatures. For this reason there is a separate contribution devoted to the subject in vol. XLI of this Encyclopedia, by R. J. BLIN-STOYLE and M. A. GRACE. In the present article we restrict ourselves to a short survey, leaving out all the details of nuclear technique.

In the following we distinguish carefully between two types of orientation, viz. nuclear "polarization" and nuclear "alinement". In the first case the orientation is parallel and the substance shows a net magnetic moment. In the second type there is an antiparallel orientation, involving a decrease of entropy but not a resulting magnetic moment.

The most obvious method to obtain polarized nuclei is the application of a large magnetic field. It was shown in Sect. 82 that, in order to obtain an appreciable effect, fields of the order of 10^5 oersteds are needed at $0.01°$ K. This method is now generally known as the "brute force method". It has been applied in only one experiment by DABBS, ROBERTS and BERNSTEIN[1] on In^{115} nuclei. A nuclear polarization of 2.1% was obtained with a field of 11150 oersteds at $0.04°$ K.

Nuclear orientation can be obtained in a much easier way by making use of magnetic ions with a nuclear moment. The coupling between the nuclear spin and the electronic moment may give rise to a hyperfine splitting of the order of $0.01°$ or $0.1°$ K. Hence, if the electronic moments are oriented at $0.01°$ K, the nuclear orientation follows automatically. As was pointed out in Sect. 31 there are two types of hyperfine splitting; magnetic interaction between the nuclear and electronic magnetic moments, and electric interaction between the electric quadrupole moment of the nucleus and the gradient of the electric field produced by the electrons. The possibility of nuclear orientation resulting from the first kind of hyperfine structure was suggested independently by GORTER and ROSE. The method was discussed later in more detail by BLEANEY. POUND showed that also the second kind of hyperfine splitting may lead to a nuclear orientation.

The original idea of GORTER[2] and ROSE[3] was to orient the electronic moments at about $0.01°$ K with the help of a magnetic field. A considerable effect can be obtained at this temperature with a few hundred oersteds. Since the equivalent field of the electrons at the nucleus is 10^5 or 10^6 oersteds this entails an appreciable nuclear polarization.

The simplest technique is the demagnetization of a paramagnetic salt with a nuclear spin from a large field to a field of a few hundred oersteds. It is obvious, however, that no high degree of nuclear polarization can be reached in this way, because the hyperfine splitting itself sets a limit to the lowest temperature (see Sects. 4 and 31). Most of the electron spin entropy is removed at the initial temperature, but only very little of the nuclear entropy, and it is the latter that counts.

This difficulty can be obviated by cooling the nuclear paramagnetic salt with the help of a larger quantity of a salt without a nuclear spin. A better method is to use a mixed crystal with a large number of "cooling" ions and a

---

[1] J. W. T. DABBS, L. D. ROBERTS and S. BERNSTEIN: Phys. Rev. **98**, 1512 (1955).

[2] C. J. GORTER: Leiden Commun. Suppl. No. 97d; Physica,'s-Grav. **14**, 504 (1948).

[3] M. E. ROSE: Phys. Rev. **75**, 213 (1949).

small amount of ions with a nuclear moment. Eventually the magnetic interactions of the cooling ions can be reduced further by dilution with non-magnetic
ions.

It should be emphasized that the Gorter-Rose method as described thus
far is based on the assumption that the final field of a few hundred oersteds is
the dominant factor in orienting the electron spins. As Bleaney[1,2] remarked
this is often not true. Suppose we have a paramagnetic salt diluted to such an
extent that the cooperative effects (see Sect. 53) begin well below 0.01° K. The
Stark splitting is of the order of a few tenths of a degree, so that, at about
0.01° K, all the ions are in the lowest Kramers doublets. This gives rise to
strong orienting forces in the direction of the symmetry axis of the Stark effect
(see Sect. 30). The forces are much stronger than those arising from a magnetic
field of a few hundred oersteds. Under these circumstances the Gorter-Rose
method in its original form cannot be applied, but it is obvious that an orientation
along the symmetry axis occurs in zero magnetic field, giving rise to a nuclear
alinement. Most of the investigations on nuclear orientation performed thus
far have been based on this effect. The experiments must be carried out with
a single crystal or with a number of equally oriented single crystals, and the
results are most surveyable if all the ions in the lattice have the same symmetry
axis.

Pound's method of nuclear orientation[3], quoted above, requires large field
gradients in order to obtain appreciable effects. They may be produced by
asymmetric electron clouds as they occur in homopolar bonds. The method gives
rise to alinement and the experiments must be carried out with single crystals.
Paramagnetic ions are needed for cooling only. No successful experiments have
been reported thus far with this method.

All the methods described here can provide important results from the point
of view of nuclear physics, but it should be realized that only the brute force
method may lead to successful nuclear demagnetizations.

84. Detection of nuclear orientation. Two methods are available for the
detection of nuclear orientation. The first one can only be applied in the case of
polarization. It is based on the fact that the cross section of the interaction between a neutron and a nucleus is different as to whether their spins are parallel
or antiparallel[4,5]. If a beam of polarized neutrons passes through a nuclear
sample the absorption depends on the degree of polarization of the nuclei. (De
Vries[6] showed that also in the case of non-polarized neutrons there is a second
order difference in absorption by polarized and by unpolarized nuclei.)

The method has been applied successfully in Oak Ridge, both in the experiments with the brute force method quoted in Sect. 83, and in investigations
with the Gorter-Rose method on Mn^{55} and Sm^{149}. The neutron absorption in
Mn^{55} was small[7], so that it was difficult to detect a change in transmission of the
beam, but the compound nucleus Mn^{56} is gamma radioactive with a half-life of
2.6 hours. After the irradiation with neutrons the sample was taken from the
cryostat and the intensity of the gamma radiation was counted. The Sm^{149}

---

[1] B. Bleaney: Proc. Phys. Soc. Lond., A **64**, 315 (1951).

[2] B. Bleaney: Phil. Mag. **42**, 441 (1951).

[3] R. V. Pound: Phys. Rev. **76**, 1410 (1949).

[4] M. E. Rose: Nucleonics **3**, No. 6, 23 (1948).

[5] M. E. Rose: Phys. Rev. **75**, 213 (1949).

[6] O. J. Poppema: Thesis, Groningen 1954, Chap. II.

[7] S. Bernstein, L. D. Roberts, C. P. Stanford, J. W. T. Dabbs and T. E. Stephenson: Phys. Rev. **94**, 1243 (1954).

sample[1] was cooled with iron ammonium alum. The sample was used as a neutron polarizer, and the degree of polarization of the transmitted beam was analyzed with a single crystal of magnetized magnetite.

The second detection method makes use of radioactive nuclei. SPIERS[2] showed that the directional distribution of gamma radiation emitted by oriented nuclei is anisotropic. STEENBERG[3,4,5] and COX and TOLHOEK[6,7] derived expressions for many types of transitions and for the state of polarization of the emitted radiation. Since the directional distributions are even functions of $\cos \vartheta$ (ϑ is the angle with the axis of the orientation) the method can be applied in the case of alinement.

A complication arises from the requirement that the half-life of the nuclei must be sufficiently long to carry out the investigations, preferably longer than a month. This means that almost no direct γ-emitters are available, but only nuclei with a preceding β-emission or K-capture. One might think that the preceding transition could disturb the orientation. Usually, however, this effect is small and under certain circumstances corrections may be applied.

The combination of requirements of paramagnetism, hyperfine structure and radioactivity with a half-life of reasonable length imposes an appreciable restriction to the choice of nuclei for these experiments. Nevertheless this method has yielded the largest number of successful experiments.

The most detailed investigations were made with Co^{60}. In Oxford[8-11] experiments were made with a mixed crystal of the composition (1% Co, 12% Cu, 87% Zn) $Rb_2(SO_4)_2 6H_2O$. The copper ions acted as the cooling agent. The gamma-ray intensity was measured in the K_1 and K_2 directions (see Sect. 40) and anisotropies up to 33% were found. The linear polarization of the gamma radiation was also investigated[12].

The first Leiden investigations[13,14] were performed with a crystal of (3.5% Co, 96.5% Zn) $(NH_4)_2(SO_4)_2 6H_2O$. The crystal was embedded in chromium potassium alum for thermal insulation. An anisotropy of 15% was found. In later experiments[15,16] the chromium alum was removed. One percent of cobalt and 10 or 20% of copper were added to the zinc. Complete angular diagrams could be measured in the $K_1 - K_2$ plane, giving values of J_a/J_r between 0.80 and 1.15

[1] L. D. ROBERTS, S. BERNSTEIN, J. W. T. DABBS and C. P. STANFORD: Phys. Rev. 95, 105 (1954).

[2] J. A. SPIERS: Nature, Lond. 161, 807 (1948).

[3] N. R. STEENBERG: Phys. Rev. 84, 1051 (1951).

[4] N. R. STEENBERG: Proc. Phys. Soc. Lond. A 65, 791 (1952).

[5] N. R. STEENBERG: Proc. Phys. Soc. Lond. A 66, 391 (1953).

[6] H. A. TOLHOEK and J. A. M. COX: Physica, 's-Grav. 18, 357, 359, 1257 and 1262 (1952).

[7] H. A. TOLHOEK and J. A. M. COX: Physica, 's-Grav. 19, 101, 673 (1953).

[8] J. M. DANIELS, M. A. GRACE and F. N. H. ROBINSON: Nature, Lond. 168, 780 (1951).

[9] B. BLEANEY, J. M. DANIELS, M. A. GRACE, H. HALBAN, N. KURTI and F. N. H. ROBINSON: Phys. Rev. 85, 688 (1952).

[10] M. A. GRACE and H. HALBAN: Physica, 's-Grav. 18, 1227 (1952).

[11] B. BLEANEY, J. M. DANIELS, M. A. GRACE, H. HALBAN, N. KURTI, F. N. H. ROBINSON and F. E. SIMON: Proc. Roy. Soc. Lond., Ser. A 221, 170 (1954).

[12] G. R. BISHOP, J. M. DANIELS, G. GOLDSCHMIDT, H. HALBAN, N. KURTI and F. N. H. ROBINSON: Phys. Rev. 88, 1432 (1952).

[13] C. J. GORTER, O. J. POPPEMA, M. J. STEENLAND and J. A. BEUN: Leiden Commun. No. 287b; Physica, 's-Grav. 17, 1050 (1951).

[14] C. J. GORTER, H. A. TOLHOEK, O. J. POPPEMA, M. J. STEENLAND and J. A. BEUN: Leiden Commun. Suppl. No. 104a; Physica, 's-Grav. 18, 135 (1952).

[15] O. J. POPPEMA, J. A. BEUN, M. J. STEENLAND and C. J. GORTER: Leiden Commun. No. 291a; Physica, 's-Grav. 18, 1235 (1952).

[16] O. J. POPPEMA, M. J. STEENLAND, J. A. BEUN and C. J. GORTER: Leiden Commun. No. 298b; Physica, 's-Grav. 21, 233 (1955).

at the lowest temperatures (here J_r is the intensity in the absence of alinement and J_a is the intensity when alinement is present). Recently investigations were also published on the circular polarization of the gamma radiation[1].

Experiments on nuclear polarization of Co^{60}, using the GORTER-ROSE method, were made in Oxford[2] with the help of crystals of the composition (0.5% Co, 99.5% Mg)$_3$ Ce$_2$(NO$_3$)$_{12}$ 24H$_2$O. Very low temperatures could be reached by demagnetizing the cerium (see Sect. 48) and the external polarizing field was applied in the direction of small g-value of the cerium, thus influencing the temperature very little. Anisotropies up to 50% were found in an external field of about 400 oersteds.

Alinement experiments on Co^{58} were made in Oxford[3,4] with a rubidium tutton salt of the same composition as that used for the investigations with Co^{60}. Anisotropies up to 20% were measured at the lowest temperatures. The polarization of the emitted gamma radiation was also investigated[5].

Experiments on Co^{56} were made in Leiden[6,7]. The composition of the crystals was (1% Co, 20% Cu, 79% Zn) (NH$_4$)$_2$ (SO$_4$)$_2$6H$_2$O. Six different gamma rays were investigated and values of J_a/J_r (see above) in the K_2 direction were found up to 1.12.

Measurements on Mn^{54} were made in Oxford[8]. A crystal of Ce$_2$ Mg$_3$(NO$_3$)$_{12}$·24H$_2$O was used in which small part of the Mg ions was replaced by the Mn. It was found that the anisotropy exhibits a maximum of 28% at $T^* = 0.01°$. Below this temperature it decreases, the value at $T^* = 0.003°$ being 21%. This effect was ascribed to the influence of the magnetic field produced by the cerium ions at the position of the Mn. For this reason an external magnetic field of 1000 oersteds was applied in the direction of small g-value of the cerium ions (see Sect. 48). In this way an anisotropy of 90% was reached at the lowest temperatures. Also the linear polarization of the gamma rays was investigated[9].

Ce^{141} and Nd^{147} were investigated at the National Bureau of Standards[10]. Cerium magnesium nitrate crystals were used containing either some Ce^{141} or a small amount of Nd^{147}. Anisotropies up to 12% were found for the cerium, and up to 39% for the neodymium.

For further discussion of the results given here and for the computation of nuclear quantities from the data, we refer to the contribution of R. J. BLIN-STOYLE and M. A. GRACE in Vol. XLI of this Encyclopedia.

The following nuclear quantities can be derived from nuclear orientation experiments. From the shape of the directional distribution of the gamma radiation one can determine the multipole order of the transition and often also the spin

[1] J. C. WHEATLEY, W. J. HUISKAMP, A. N. DIDDENS, M. J. STEENLAND and H. A. TOL-HOEK: Leiden Commun. No. 301b; Physica, 's-Grav. **21**, 841 (1955).

[2] E. AMBLER, M. A. GRACE, H. HALBAN, N. KURTI, H. DURAND, C. E. JOHNSON and H. R. LEMMER: Phil. Mag. **44**, 216 (1953).

[3] J. M. DANIELS, M. A. GRACE, H. HALBAN, N. KURTI and F. N. H. ROBINSON: Phil. Mag. **43**, 1297 (1952).

[4] M. A. GRACE and H. HALBAN: Physica, 's-Grav. **18**, 1227 (1952).

[5] G. R. BISHOP, J. M. DANIELS, G. GOLDSCHMIDT, H. HALBAN, N. KURTI and F. N. H. ROBINSON: Phys. Rev. **88**, 1432 (1952).

[6] L. J. GALLAHER, CH. WHITTLE, J. A. BEUN, A. N. DIDDENS, C. J. GORTER and M. J. STEENLAND: Leiden Commun. No. 298c; Physica, 's-Grav. **21**, 117 (1955).

[7] O. J. POPPEMA, J. G. SIEKMAN, R. VAN WAGENINGEN and H. A. TOLHOEK: Leiden Commun. Suppl. No. 109b; Physica, 's-Grav. **21**, 223 (1955).

[8] M. A. GRACE, C. E. JOHNSON, N. KURTI, H. R. LEMMER and F. N. H. ROBINSON: Phil. Mag. **45**, 1192 (1954).

[9] G. R. BISHOP, J. M. DANIELS, H. DURAND, C. E. JOHNSON and J. PEREZ: Phil. Mag. **45**, 1197 (1954).

[10] E. AMBLER, R. P. HUDSON and G. M. TEMMER: Phys. Rev. **97**, 1212 (1955).

change accompanying the transition. This often enables one to establish the spins of the levels in the decay scheme. If the spin of the initial nucleus is known, and if also H and T are known, then the magnetic moment of the initial nucleus can be calculated from the magnitude of the gamma anisotropy. Alternately, if H and μ_n are known, the directional distribution may be used as a thermometer. By measuring the direction of the linear polarization of the emitted gamma radiation one can distinguish between the electric or the magnetic character of the transition. In the case of polarized nuclei circular polarized gamma radiation is emitted, and from the sense of the circular polarization the sign of μ_n can be determined.

Suggestions have been made[1,2] to determine the nuclear gyromagnetic ratio of oriented nuclei by applying a radio frequency magnetic field. An experimental problem seems to be, however, to avoid spurious heating by non-resonant electron spin-spin absorption during the search for the nuclear resonance.

General references.

ALLEN, J. F., N. KURTI, K. MENDELSSOHN and F. SIMON: Low temperature physics; four lectures. London: Pergamon Press 1952.
AMBLER, E., and R. P. HUDSON: Rep. Progr. Physics 18, 251 (1955).
CASIMIR, H. B. G.: Magnetism and very low temperatures. Cambridge: University Press 1940.
GARRETT, C. G. B.: Magnetic cooling. Cambridge (Mass.) 1954.
KLERK, D. DE: Thesis, Leiden 1948.
—, and M. J. STEENLAND: Progress in low temperature physics (ed. GORTER), p. 273. Amsterdam 1955.
STEENLAND, M. J.: Thesis, Leiden 1952.
VLECK, J. H. VAN: Ann. Inst. Poincaré 10, 57 (1947).

The paper of AMBLER and HUDSON came too late for discussion in this article. The same is true for the results communicated at the Paris low temperature conference, 1955.

In general, publications which came out until early 1955 have been incorporated. Some later papers, of which the author knew before publication (mainly from the Kamerlingh Onnes Laboratory) have also been reviewed.

---

[1] H. A. TOLHOEK and S. R. DE GROOT: Physica, 's-Grav. 17, 82 (1951).
[2] N. BLOEMBERGEN and G. M. TEMMER: Phys. Rev. 89, 883 (1953).

Superconductivity. Experimental Part.

By

B. SERIN.

With 43 Figures.

I. Introductory survey.

The growth of our knowledge of superconductivity has reached the point where it is possible to arrange almost all the experimental facts into a fairly simple logical pattern. Our purpose in this treatise is to describe in detail each element of this pattern.

A consecutive detailed examination of somewhat artificially bounded areas, such as we contemplate, has an inherent weakness. There is the danger that the non-specialist may develop misconceptions about important matters just because of the narrow view that is available to him. Such misconceptions in principle enjoy only a temporary existence and are eventually corrected as the full picture emerges. However, the task of revising our conceptions is never pleasant and at times is most difficult. The best policy is to avoid generating misconceptions from the beginning. To accomplish this last objective, we propose to present in this chapter a few of the fundamental facts about superconductivity in bold outline. We hope thereby to help the reader build a basic vocabulary in this subject and to provide him with a frame of reference from which to view the detailed and specialized matters presented in the following chapters of this article.

1. Disappearance of electrical resistance. Superconductivity was discovered by KAMERLINGH ONNES[1] in 1911. ONNES was engaged at that time in measuring the electrical resistance of various metals at temperatures near the absolute zero. The resistance was measured in the usual way by determining, with a sensitive potentiometer, the drop in potential across a sample when a given current was passing through the metal. The samples were immersed in a bath of liquid helium and could thus be maintained at temperatures between about 2° and 4° K.

The resistance of mercury underwent extraordinary changes which could not have been foreseen. For example, the resistance of one specimen was 0.08 ohms at a temperature somewhat above 4° K. Upon decreasing the temperature below 4° K [2], the resistance dropped precipitously so that it was less than 3×10^{-6} ohms at 3° K. Increasing the temperature to above 4° K restored the resistance to its earlier value of 0.08 ohms. Later measurements[3] demonstrated that, for sufficiently small measuring currents, the resistance fell in a temperature interval of the order of 0.01° K. The temperature at which the sudden drop in resistance occurs is termed the *transition temperature*, T_c. It soon became clear that the resistance below the transition temperature was immeasurably small. The most refined measurements (see Sect. 4) have proved that this resistance is at most 10^{-11} of

[1] H. KAMERLINGH ONNES: Leiden Comm. **1911**, 122b, 124c.
[2] We give only approximate temperatures, since the temperature scale used at that time was incorrect.
[3] H. KAMERLINGH ONNES: Leiden Comm. **1913**, 133a.

its value above T_c. Thus we conclude with ONNES[1] that "Mercury has passed into a new state which on account of its extraordinary electrical properties may be called the *superconductive state*".

The recognition of the superconductive state as a distinct phase of matter has since been completely justified by experiment. So far, twenty-two metallic elements have been found to pass into this remarkable phase. The transition temperatures range from as low as 0.4° K to as high as 11° K. A large number of alloys and compounds also become superconductive. Probably white tin most closely approximates ideal superconductive behavior. Some of the results of the careful measurements by DE HAAS and VOOGD[2] of the resistive transition of a single crystal of pure tin are shown in Fig. 1. In the limit of zero measuring current the width of the transition is about 0.001° K.

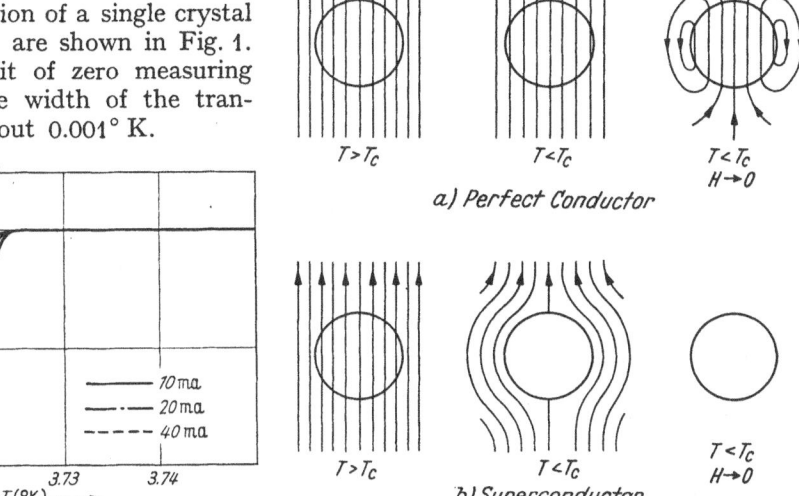

Fig. 1. Resistive transition of pure tin (after DE HAAS and VOOGD[2]).

Fig. 2a and b. Comparison of the behavior of a perfect conductor and a superconductor in a magnetic field.

2. MEISSNER effect. The magnetic properties of a metal in the superconductive state are just as unusual as the electrical properties. This fact was not realized, however, until 1933. Until then, there seems to have been tacit agreement that the magnetic properties could be deduced directly from the property of infinite conductivity. MEISSNER and OCHSENFELD[3] put this inference to an experimental test and found it to be false.

In this crucial experiment a small test coil was used to measure the magnetic field about a single crystal of tin in the form of a long cylinder. The cylinder axis was transverse to the direction of an applied uniform magnetic field. The field was kept constant as the temperature was lowered through the transition point. Fig. 2a shows the result to be expected on the basis of infinite conductivity. Above the transition temperature, the field inside the cylinder is the same as outside, since the magnetic susceptibility of the normal metal is negligibly small. When the temperature is lowered below T_c, the identical situation should prevail because there has been no change in the total magnetic flux threading the sample. It is only by changing the external field (e.g. by reducing it to zero) that a super-conducting current should be induced in the cylinder. The field distribution

[1] H. KAMERLINGH ONNES: Leiden Comm. **1913**, Suppl No. 34.
[2] W. J. DE HAAS and J. VOOGD: Leiden Comm. **1931**, 214c.
[3] W. MEISSNER and R. OCHSENFELD: Naturwiss. **21**, 787 (1933).

actually observed is shown in Fig. 2b. When the temperature is lowered below the transition point, the magnetic field lines near the sample are warped in the manner necessary to make the magnetic induction vanish inside the cylinder. Furthermore, upon reducing the external field to zero, there is no superconducting current induced in the cylinder.

The vanishing of the magnetic induction inside a substance in the super-conductive phase[1] is now recognized as a second fundamental property of the phase. This characteristic is usually termed the MEISSNER effect, and stated concisely by the expression, $B = 0$. Like many physical laws, this one describes ideal behavior, from which all actual substances depart in various degrees. We discuss the experimental situation fully in Chapter II.

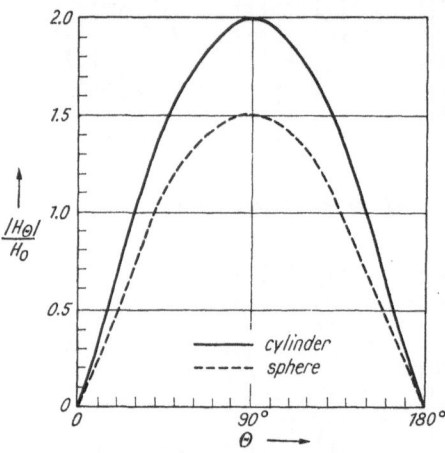

Fig. 3. Magnetic field on the surfaces of a super-conductive cylinder and sphere. The field is transverse to the cylinder axis.

When a sample is in a static external magnetic field, the condition $B = 0$ inside is maintained by superconducting currents flowing on the surface. Thus only a super-conductive substance could be expected to exhibit the MEISSNER effect. The current distribution is uniquely determined by the external field distribution and the shape of the sample. Superconducting currents cannot exist in a simply-connected superconductor in the absence of an external magnetic field. Changes in flux are of no consequence, since no flux ever threads the sample. It is only for multiply-connected shapes, such as a ring, that currents can persist without an external field being present.

The above discussion may be summarized by stating that infinite conductivity is a necessary but not a sufficient condition for the existence of the MEISSNER effect. The existence of the latter shows that there is an essential qualitative difference between the superconductive state and a state in which the electrical conductivity approaches an infinite limit.

The relation

$$B = H + 4\pi M = H + 4\pi\chi H$$

between the magnetic induction B, the magnetic field H, the magnetization M, and the susceptibility χ, permits a formal description of the magnetic properties of a superconductor in which the surface currents are ignored. We see that we can have $B = 0$, if $M = -H/4\pi$, or $\chi = -1/4\pi$. Thus it is often stated that a superconductor has the properties of a perfect diamagnetic material. The major advantage of this viewpoint is that it is possible to take over *in toto* the well known magnetostatic description of normal substances. For a substance having the very large susceptibility of a superconductor, we expect the magnetic properties to be extremely dependent on specimen shape. In Fig. 3, we show the fields at the surfaces of a superconductive cylinder and of a sphere which have been placed in an initially uniform magnetic field. The only simple case occurs when the specimen has the form of a long cylinder with its axis parallel to the field. Under these circumstances, the field everywhere on the surface (except in the immediate neighborhood of the ends) equals the applied field.

---

[1] Actually, the induction is finite in a very thin layer about 10^{-5} cm. thick at the surface of the metal (see Chapter IV). The effects associated with this penetration are negligibly small in samples of macroscopic dimensions.

3. Threshold field. Three years after he discovered superconductivity, KAMERLINGH ONNES[1] observed that the resistance of a superconductor could be restored to its value in the normal state by the application of a large magnetic field. For simplicity, we consider initially only those observations made on cylindrical specimens longitudinal to the applied field direction. TUYN and ONNES[2] showed that under these conditions the resistance increases rapidly in a very small field interval. The field value at which the jump in resistance occurs is termed the *threshold field*, H_c. This field is zero at the transition temperature and increases as the temperature is decreased below this point, according to the approximate relation

$$H_c = H_0 \left[1 - (T/T_c)^2 \right] . \tag{3.1}$$

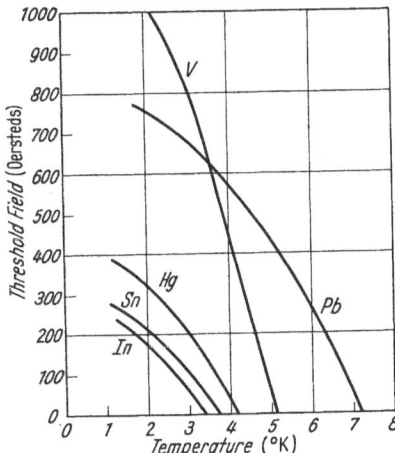

Fig. 4. Threshold field curves of several superconductive elements.

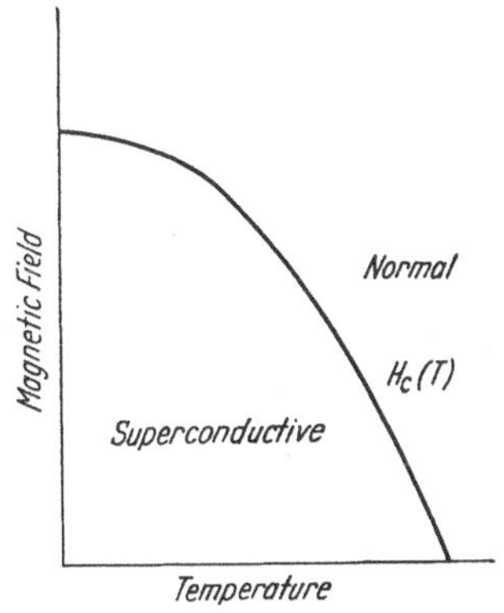

Fig. 5. Phase diagram in the H-T plane.

Here, T is the temperature and H_0, the threshold field at the absolute zero. The two numbers, H_0 and T_c, are characteristic of each superconductive substance. Fig. 4 shows the threshold field curves of several of the elements. We emphasize that Eq. (3.1) is only approximately true. The observed deviations from this expression have acquired more significance in recent years.

After the discovery of the MEISSNER effect, it was found that the magnetization also abruptly increases from its value $(-H/4\pi)$ to zero, when the magnetic field, H, reaches the threshold value. Experience has shown that determining the threshold field from measurements of the change in magnetization is preferable to observing the jump in resistance (see Sect. 10). For example, the magnetization may be determined by having the specimen inside a long coil connected in series with a ballistic galvanometer. Both sample and coil are in an applied uniform magnetic field. The deflection of the galvanometer, produced by suddenly removing the sample from the coil, is proportional to the total magnetic moment of the specimen. Methods of this type are used almost exclusively today.

GORTER and CASIMIR[3] proved that the threshold field curve provides us with the phase diagram shown in Fig. 5. The curve of H_c as a function of temperature, T, divides the $H - T$ plane into two regions. Any point below the curve specifies values of magnetic field and temperature for which any small volume is in the *superconductive phase*; the points above define the states of the *normal phase*. The curve itself defines the unique values of H and T for which the two

---

[1] H. KAMERLINGH ONNES: Leiden Comm. **1914**, 139f.

[2] W. TUYN and H. KAMERLINGH ONNES: Leiden Comm. **1925**, 174a.

[3] C. J. GORTER and H. G. B. CASIMIR: Physica, Haag **1**, 306 (1934).

phases are in equilibrium. It is possible to describe the energetics of the phase transition in terms of the shape of the threshold field curve. Thus, as we shall show in Chapter III, threshold field measurements provide a convenient basis for determining several of the thermodynamic properties of the two phases.

For other sample and field geometries, the magnetic behavior becomes quite complicated when superconductivity is destroyed by a field. Consider, for example, the case of a long cylinder with axis transverse to the field direction. We see from Fig. 3 that the field at the equator is twice the value of the applied field. Thus the field at this point reaches the threshold value when the applied field is $H_c/2$. In circumstances such as these, the specimen passes into a mixed phase of normal and superconductive regions, termed the *intermediate state*. The relative amount of normal phase increases as the external field is further increased, and when the threshold value is reached, the last traces of the superconductive regions disappear. Thus the superconductive property is destroyed over an appreciable range of values of the applied magnetic field.

The dependence on measuring current of the breadth of the resistive transition can be qualitatively described in terms of the intermediate state produced by the magnetic field of the current. Moreover, at temperatures below the transition point, the current in a specimen cannot exceed a critical value without destroying the superconducting property. This phenomenon is called the Silsbee effect[1]. It, too, is a consequence of the magnetic field of the current, and there is a direct connection between the critical current value and the threshold field[2].

II. Electrical and magnetic properties of macroscopic superconductors.

The experiments designed to set an upper limit for the value of the electrical resistivity of the superconductive phase are described first in this chapter. This is followed by a detailed consideration of the properties of a ring, and a brief examination of the physics of superconducting circuits. We then discuss the magnetic behavior of superconductors arising from the Meissner effect. A major fraction of all the experiments done since 1933 have had roots in the Meissner effect, and it occupies a correspondingly prominent place in this disquisition. The chapter closes with a discussion of the resistive transition, a table of the known superconductive elements, and some brief examinations of isolated topics.

4. Persistent currents. For certain shapes of superconducting specimens, the magnetic properties arising from the very large conductivity may overshadow those properties arising from perfect diamagnetism. Such a situation occurs in coils and rings. We consider a shorted coil of wire in the normal state placed in an external magnetic field. When the coil is cooled below the transition point, the magnetic flux through the hole remains essentially the same as before. Thus, if the magnetic field is subsequently changed, a current must be induced in the coil in keeping with Faraday's law. This current flows on the surfaces of the superconducting wires and is superimposed on the shielding surface currents which maintain the induction zero inside the wires. For convenience, we term the induced current a total current and the shielding currents, Meissner. These matters are discussed in detail in following sections, but the qualitative description just presented is adequate for our immediate needs.

For a closed circuit of resistance R, the current I, in the absence of an applied voltage, decays in time according to the relation

$$I = I_0\, e^{-Rt/L}$$

---

[1] F. B. Silsbee: Bull. Bur. Stand. **14**, 301 (1917).
[2] W. Tuyn and H. Kamerlingh Onnes: Leiden Comm. **1926**, 174a.

where I_0 is the initial current value, L the self-inductance of the circuit, and t the time. The current falls to half its initial value in a relaxation time,

$$\tau = \frac{L}{R}.$$

Thus, a lower limit for the resistance of a circuit may be deduced from a determination of the upper limit for the decay time of an induced current. This was the basis of a series of experiments that ONNES performed with coils and rings. In his early investigations[1], the circuit consisted of a coil with a very large number of turns of fine lead wire. A closed superconductive circuit was formed by fusing together the wires at the ends of the coil. The coil was cooled below the transition point in a large magnetic field and the field was then removed. The total current induced thereby produced a magnetic field about the coil, which was detected by observing the deflection of a compass needle placed outside the cryostat[2]. A change in the magnitude of the induced current results in a corresponding change in the deflection of the needle. The induced currents remain undiminished in magnitude for such long times that ONNES named them *persistent* currents.

In a typical run, no change in the current could be detected after intervals of several hours. Since the precision of the current measurements was about 1%, the rate of decay of the current was judged to be at most 1% per hour. This first crude result yielded a relaxation time greater than 100 hours! By contrast, when the coil was lifted out of the liquid helium bath to raise the temperature of the lead above the transition point, the current was "destroyed instantaneously". In other words, the relaxation time was certainly less than one second. The relaxation time of 100 hours corresponds to an upper limit for the resistivity of superconductive lead of about 10^{-16} ohm cm.—a number which should be compared with the value 10^{-9} ohm cm. for the residual resistivity of pure copper or silver at liquid helium temperatures. Two later ingenious experiments[3] provided conclusive proof that current (in the familiar sense of the word) was indeed flowing through the coil, by demonstrating that a persistent current can be initiated by a battery and can be made to disappear by breaking the circuit[4]. Finally, we must note that in the foregoing experiments, whenever the persistent current was interrupted at temperatures less than T_c, the coil was left with a magnetic moment which was perhaps 5% of the moment with current present. These observations puzzled the early investigators; we discuss such anomalies in Sect. 7β.

The upper limit for the resistivity of superconductive lead was reduced by another order of magnitude by ONNES and TUYN[5], several years later. Two rings of lead were immersed in liquid helium. The larger, outer ring was fixed, whereas the smaller, inner ring was suspended by a torsion fiber. The inner

---

[1] H. KAMERLINGH ONNES: Leiden Comm. **1914**, 140b, 140c.

[2] The field of the lead coil actually was compensated by the field of a second coil outside the cryostat. The current in the latter was adjusted until there was no net deflection of the compass needle.

[3] H. KAMERLINGH ONNES: Leiden Comm. **1914**, 141b.

[4] ONNES used a mechanical switch made of lead to make and break a circuit. The switch had no contact resistance. R. HOLM and W. MEISSNER [Z. Physik **74**, 715 (1932)] showed that there is no resistance at a clean contact between two superconductors, even when they are different metals.

[5] H. KAMERLINGH ONNES and W. TUYN: Leiden Comm. **1924**, Suppl. No. 50a. In this same communication, ONNES describes a similar experiment involving a hollow lead sphere. His interpretation of this experiment was incorrect. We discuss it in Sect. 9. A more complete description of these experiments was given by W. TUYN; Leiden Comm. **1929**, 198.

ring was thus free to rotate in the outer one. With the planes of the rings coincident, persistent currents where induced in both rings. The inner ring was then rotated through an angle of about 30°. In this position, the forces between the currents produce a torque on the inner ring, and a resulting twist in the fiber. A change in the magnitudes of the currents consequently results in a change in the torsion. On the basis of this experiment, Onnes estimated that the relaxation time was longer than 2000 hours. Grassmann[1] performed a simplified version of the preceding experiment, and achieved extraordinary precision. He was able to conclude that the resistivity of superconductive lead was at most 10^{-20} ohm cm. With this investigation, interest in this field ceased.

The experiments with presistent currents furnish the basis for the inference that the electrical resistivity of superconductors is identically zero in static fields[2]. This conclusion is clearly an extrapolation of our experience. But it has proved so fruitful that it is now accepted as a basic property of these substances. We assume infinite conductivity for superconductors in all subsequent considerations.

5. Superconducting ring. Because of its infinite conductivity there can be no electric field associated with a current in a superconductor[3].

With this fact in mind, let us consider a superconductive body with a hole in the presence of a magnetic field as shown schematically in Fig. 6. Under these circumstances, Faraday's law of induction becomes:

$$\text{The total magnetic flux through the body} = \text{constant} = \Phi_i. \qquad (5.1)$$

We define Φ to be the calculated flux through the body due to the distorted applied magnetic field surrounding the superconductive material. Φ_i is the value of Φ immediately after the body becomes superconducting. The field is distorted, of course, by the shielding surface currents.

Upon changing the applied field from its initial value, so that Φ differs from Φ_i, an additional source of flux must appear if (5.1) is to be satisfied. The total persistent current supplies this flux. For example, if the field is reduced, such a current will flow on the surface of the hole in the direction shown in Fig. 6. It follows from (5.1) that the magnitude of this current, I, is determined by the relation

$$L\,I + \Phi = \Phi_i, \qquad (5.2)$$

where L is the self-inductance of the body calculated on the basis that all currents are superficial. This relation is valid, so long as the sum of the distorted applied field and the magnetic field of the current does not exceed the threshold value anywhere on the superconductive surface. Eq. (5.2) was verified directly in the somewhat crude experiments of Grayson Smith and Wilhelm[4], who determined with test coils the magnetic field about a superconducting loop, with and without a persistent current flowing.

For the case of a superconducting ring, we see that its total magnetic moment is the sum of the moment due to the current, I, and the diamagnetic moment of the Meissner currents. When the diameter of the ring is much larger than the wire diameter, the diamagnetic moment is very much smaller than the moment of the persistent current. As a result, the magnetic properties are dominated by the persistent current, so that the behavior of the ring depends critically on its initial state.

---

[1] P. Grassmann: Phys. Z. **37**, 569 (1936).
[2] In very high frequency fields, superconductors exhibit resistivity; see Sect. 21.
[3] This statement is true only for quasi-static fields.
[4] H. Grayson Smith and J. O. Wilhelm: Proc. Roy. Soc. Lond., Ser. A. **157**, 132 (1936).

For example, if the ring is cooled below the critical point in zero magnetic field, and a field is subsequently applied,

$$I = - \Phi/L , \qquad (5.3)$$

with the result that the ring behaves like a diamagnetic body. On the other hand, if the ring is cooled in a field and the field is then decreased to zero,

$$I = + \Phi_i/L .$$

Thus the ring is left with a persistent current, which gives it the equivalent of a paramagnetic moment. The current and its associated moment persist as long

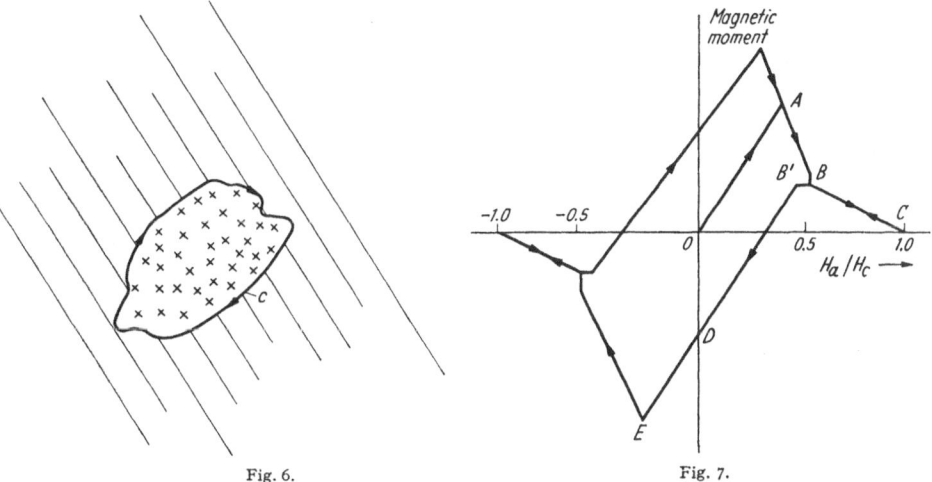

Fig. 6.   Fig. 7.

Fig. 6. Schematic diagram of a superconductor with a hole. The field in the hole is into the paper, and the arrows denote the direction of superconducting current flow when the applied field is reduced.

Fig. 7. Magnetic moment due to the persistent current in a superconductive ring (after SHOENBERG[1]).

as the temperature is not increased, so that we often say that the ring has a frozen-in magnetic moment in such circumstances.

SHOENBERG[1] measured the total moment as a function of field, by determining with a SUCKSMITH balance the force exerted on a ring by a slightly inhomogeneous magnetic field. He has discussed these measurements in great detail in his book *Superconductivity* [1]. We merely state the results, and point out some of their important features.

In Fig. 7 we show how that part of the magnetic moment due to the total current varies with the applied field for the case of a ring cooled in zero field. The diameter of the ring was about four times the diameter of the wire. Along OA, the current in the ring increases in accordance with (5.3). SHOENBERG was able to determine the slope of OA so accurately as to verify that L must have the value for a superficial current and not the value for a current flowing through the whole cross section of the wire. This result confirms our view that the total current flows only on the surface of a superconductor.

We recall that the magnetic field lines about a wire carrying a current are concentric circles about the wire axis. As a result, when the field of the current is added to the distorted applied field, the total field along the outer rim of

---

[1] D. SHOENBERG: Proc. Roy. Soc. Lond., Ser. A **155**, 712 (1936).

the ring exceeds the field on the inner rim (see Fig. 8a). Thus, the magnetic field reaches the threshold value first on the ring's outer rim. This occurs at point A in Fig. 7. With further increases in the applied field, the persistent current decreases in such a manner as to maintain the total field on the outer rim just equal to the threshold value until the point B is reached. In the region AB the ring is in the superconductive state. At B, the field along the inner rim reaches the threshold value, H_c, and the ring then passes into an intermediate state which persists until the applied field reaches H_c, when the ring becomes normal.

Upon decreasing the applied field from above the threshold value, the portion BC is retraced, but at B' the ring becomes superconducting again. Thus, a further decrease in field induces a current in the ring which flows in the opposite direction to that of the current originally induced by the field. Along $B'E$, the field on the inner rim equals H_c. At point E, the field at the outer rim reaches H_c—and so on through the remainder of the loop.

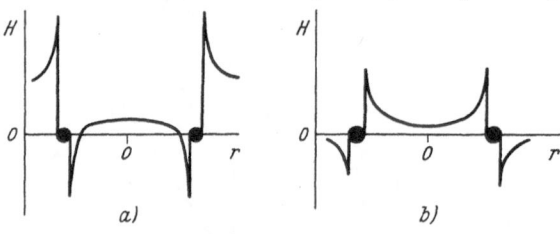

Fig. 8a and b. Magnetic field distribution in the central plane of a super-conductive ring. a) Field applied with ring superconductive (after Dolecek and de Launay[2]). b) Field frozen in and the applied field then reduced to zero.

We see that the ring behaves irreversibly since we cannot return to the origin except by warming above the transition point. The persistent current at D (external field zero) is

$$I = H_c\, a/2,$$

where a is the radius of the wire. These currents can be surprisingly large. For example, for $a \sim 1$ cm. and $H_c \sim 100$ Oe; $I = 50$ abamp. = 500 amp.!

The finer details of the curve in Fig. 7, such as the discontinuity at B and the non-coincidence of B and B', can be understood only in terms of the complete solution of the magnetic field distribution about the ring. This solution is complicated. The general case has been worked out by de Launay[1]; and Dolecek and de Launay[2] have pointed out that in cases where the wire diameter is comparable to the ring diameter, the hysteresis loops differ markedly from the one shown in Fig. 7.

Two magnetic field distributions about a superconducting ring are shown in Fig. 8. Fig. 8a illustrates the case in which the ring has been cooled below T_c in zero field, and a field, H_a, applied; whereas, Fig. 8b illustrates the case in which the ring has been cooled in a field and the field then reduced to zero. We observe that in the former case, even though the total flux through the ring is zero, there is a non-zero field at the center of the ring. In the latter case, even though the frozen-in flux is large, the field at the center of the ring is quite small.

After considering the complex behavior of a superconducting ring, we can appreciate the even greater complications involved in understanding the properties of the coils Onnes studied in his early experiments. The difficulties originate in the intricate magnetic interactions between the many superconductive wires of the coil. It seems impossible to unravel this case. However, the coil experiments do furnish, without ambiguity, an upper limit for the resistance of superconductors.

[1] J. de Launay: Naval Research Laboratory Report No. P3441, April 1949.
[2] R. L. Dolecek and J. de Launay: Phys. Rev. **76**, 445 (1949).

6. Superconducting circuits. Superconducting circuits may be treated by the same basic principle used in the previous section. This is the requirement of conservation of the total magnetic flux through a multiply-connected super-conductor. Since in most cases an external magnetic field is not present, problems can usually be analyzed exclusively in terms of the self and mutual inductances of the circuit elements[1].

These circuits have been discussed in detail by von Laue[2]. We consider a particular example to illustrate the method of analysis. The circuit in our example is shown in Fig. 9. It consists of two coils, connected in parallel, which have self-inductances L_1 and L_2 and a mutual inductance L_{12}. This parallel combination is connected in series with a large resistance, r, a battery, E, and a switch, S.

In the normal state, the coil resistances differ, so that the coils respectively carry currents I_1^0 and I_2^0. The temperature is now lowered until the coils are superconducting. We assume that the resistance, r, is so large that the total current flowing from the battery is unchanged by the superconductive transition. As a result, in the superconductive state the current in each coil remains equal to its initial value, I_1^0 or I_2^0.

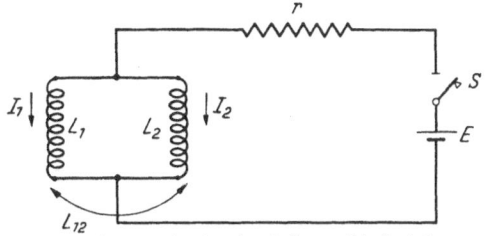

Fig. 9. Superconductive circuit discussed in Sect. 6.

We now open the switch S. A net current can no longer flow through the parallel arrangement and the currents in the coils must change. In a normal circuit the currents go to zero. However, in the superconductive case, a persistent current can flow around the two coils which together form a closed circuit.

The magnitude of the persistent current may be calculated from the principle of flux conservation. If I_1 and I_2 are the currents flowing in the coils after the switch is open, we have

$$L_1 I_1 + L_{12} I_2 - L_2 I_2 - L_{12} I_1 = L_1 I_1^0 + L_{12} I_2^0 - L_2 I_2^0 - L_{12} I_1^0. \qquad (6.1)$$

Since no current flows out of the parallel combination, the further restriction

$$I_1 + I_2 = 0 \qquad (6.2)$$

must be satisfied. Eqs. (6.1) and (6.2) may be solved to yield

$$I_1 - I_1^0 = -\frac{(L_2 - L_{12})}{L_1 + L_2 - 2L_{12}} I;$$

$$I_2 - I_2^0 = -\frac{(L_1 - L_{12})}{L_1 + L_2 - 2L_{12}} I;$$

where $I = I_1^0 + I_2^0$. I is the total current carried by the coils before the switch was opened.

It is interesting to consider the case $L_1 = L_2$. Under these circumstances,

$$I_1 = -I_2 = \frac{I_1^0 - I_2^0}{2}.$$

We observe that if $I_1^0 = I_2^0$, there is no persistent current. This situation corresponds to the case in which the battery is connected only after the coils are superconducting. Since the circuit is reactively symmetrical, it is clear that the coils must

---

[1] The values of the inductances must be those for surface currents.
[2] M. von Laue [2], pp. 8-12.

carry identical currents. When the battery is subsequently disconnected, the same symmetry insures that no persistent current is induced.

The foregoing discussion should illuminate some of the experiments of Onnes[1]. Furthermore, Justi and Zickner[2] investigated the electrical behavior of just such a parallel combination for various values of L_1, L_2 and L_{12}, and found agreement betwen theory and experiment.

An interesting application of superconducting circuits occurs in the superconducting galvanometer. As instruments, the early designs[3] suffered from a lack of sensitivity, and were useful almost solely for investigating the properties of superconducting circuits. More recently, Pippard and Pullan[4] constructed a galvanometer of greater utility. This instrument is noteworthy for its small self-inductance (about 1 μh), and will detect 10^{-5} amp, corresponding to an e.m.f. of 10^{-12} v. It is particularly suited for the measurement of thermo-electric voltages generated at very low temperatures.

7. Superconducting ellipsoid. α) *Theoretical description.* For a superconductor without holes, the requirement that the magnetic induction, B, vanish inside insures that it will enclose no net flux when placed in a magnetic field. Thus, no total current can be induced in the specimen[5], and its magnetic behavior is quite simple compared with the cases discussed in the preceding two sections.

As we indicated previously, there are two equivalent ways of describing the present situation. In the first, we use as variables the magnetic induction, B, the magnetic field, H, and the surface density of Meissner current, j_s. The following conditions prevail about the specimen:

Inside: $B = H = M = 0$,

Surface: $j_s \neq 0$,

Outside: $B = H = H_a +$ (field of j_s).

M is the magnetization (i.e. the magnetic moment per unit volume) and H_a is the applied field, which we take to be uniform in the absence of the specimen. In the second mode of description, we have:

Inside: $B = 0$, $H \neq 0$, $M \neq 0$,

Surface: $j_s = 0$,

Outside: $B = H = H_a +$ (field of M).

The former corresponds to the physical state of a specimen. The latter description gives the same field outside the specimen, but the presence of H and M inside are convenient fictions. We prefer the second description, nevertheless, because it is more familiar.

The only general specimen shape for which the magnetic description is simple is the ellipsoid. Thus, we consider an ellipsoid with one of its axes parallel to the direction of the applied field. Under these circumstances, inside the specimen, B, H, and M are all uniform and parallel to the applied field direction[6].

[1] H. Kamerlingh Onnes: Leiden Comm. **1914**, 141b.

[2] E. Justi and G. Zickner: Phys. Z. **42**, 257 (1941).

[3] H. Grayson Smith and F. G. A. Tarr: Trans. Roy. Soc. Canada (3) **29** (III), 23 (1935). H. Grayson Smith, K. C. Mann and J. O. Wilhelm: Trans. Roy. Soc. Canada (3) **30** (III), 13 (1936).

[4] A. B. Pippard and G. T. Pullan: Proc. Cambridge Phil. Soc. **48**, 188 (1952).

[5] We describe here the ideal case.

[6] J. A. Stratton: Electromagnetic Theory, §§ 3.27 and 4.18. New York: McGraw-Hill-Book Co. 1941.

Furthermore, inside [1]

$$H = H_a - 4\pi D M,\qquad (7.1)$$

where D is a number which depends on the shape of the body and is called the demagnetizing coeffficient [2]. D varies between zero and unity. Upon adding the fundamental requirement,

$$B = 0 = H + 4\pi M,\qquad (7.2)$$

inside the specimen, (7.1) and (7.2) yield

$$M = -\frac{H_a}{4\pi(1-D)},\qquad (7.3)$$

$$H = \frac{H_a}{(1-D)}.\qquad (7.4)$$

The total dipole moment of the specimen is antiparallel to the applied field direction and equals MV, where V is the volume of the body. To find the field outside, we may replace the body by a dipole of this total moment located at its center. The field is then the sum of the applied and the dipole fields.

We recall that the normal component of B and the tangential component of H are continuous at all points on the specimen surface. Since B and H inside the specimen have been determined,

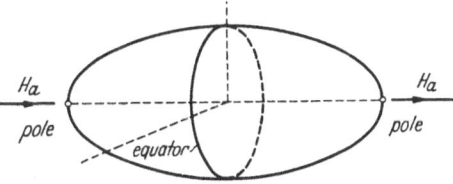

Fig. 10. Definitions of pole and equator for an ellipsoid.

the field on the surface at the two poles and the points of the equator may be deduced. (Consult Fig. 10 for a definition these terms.) At the poles

$$H_{\text{outside}} = B_{\text{inside}} = 0,$$

whereas, for points on the equator

$$H_{\text{outside}} = H_{\text{inside}} = \frac{H_a}{(1-D)}.\qquad (7.5)$$

Furthermore, the field direction at the specimen is clearly everywhere tangential to its surface; and the field varies in magnitude from zero at the poles to a maximum on the equator. Since in general $D>0$, the field on the equator usually exceeds the applied field. These considerations apply only so long as the specimen is superconducting. Most investigations have been restricted to studies of spheres and of long cylinders with axes either longitudinal or transverse to the direction of the applied field. The demagnetizing coefficients for the three cases are:

Longitudinal cylinder [3]: $D = 0,$

Transverse cylinder [3]: $D = \frac{1}{2},$

Sphere: $D = \frac{1}{3}.$

---

[1] Henceforth, we use only scalar magnitudes when all the vector quantities have parallel directions.

[2] E. C. STONER [Phil. Mag. **36**, 803 (1935)] gives historical references and an outline of the derivation of (7.1).

If an axis of the ellipsoid does not coincide with the field direction, an equation similar to (7.1) holds for each rectangular component of the field, with a different value of D (D_x, D_y, D_z) for each direction. The three values satisfy the relation, $D_x + D_y + D_z = 1$.

STONER also gives extensive tables of values of D for ellipsoids of revolution.

Tables for the general ellipsoid may be found in the paper of J. A. OSBORN: Phys. Rev. **67**, 51 (1945).

[3] These values are for cylinders of infinite length.

Fig. 11 shows the dependence of the magnetization on applied field as defined by (7.3). The diamagnetic moment of the sample increases linearly with slope $[4\pi(1-D)]^{-1}$. This can continue only until the applied field reaches the value $(1-D)\,H_c$, where H_c is the threshold field; for at this point the field on the equator attains the threshold value [cf. Eq. (7.5)].

A longitudinal cylinder presents no problems. The magnetic field is the same at all points on its surface and is equal to the applied field. Thus, when the applied field is H_c, the cylinder as a whole passes into the normal phase and the magnetic moment drops abruptly to zero. For $D>0$, on the other hand, the field on the equator is H_c, when the applied field is only $(1-D)\,H_c$. In such cases the specimen passes into the intermediate state.

The macroscopic magnetostatic description of the intermediate state was derived by PEIERLS[1] and by LONDON[2]. They assumed that the magnetization increases linearly from its minimum value, $-H_c/4\pi$, to zero when the applied field increases from $(1-D)\,H_c$ to H_c. This behavior is illustrated by the dashed lines in Fig. 11. Thus,

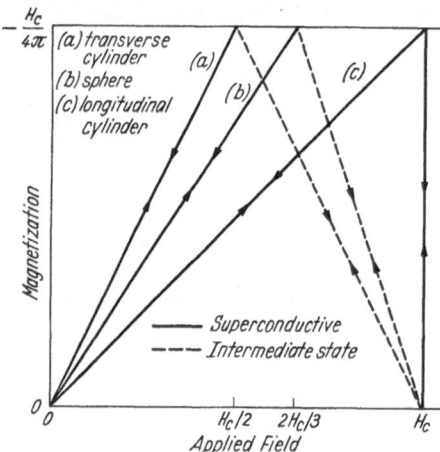

Fig. 11. Variation of the magnetization of various superconductive ellipsoids with the applied field.

$$M = -\frac{1}{4\pi D}\,(H_c - H_a).\qquad(7.6)$$

From (7.1) and (7.2) it follows that inside the specimen

$$H = H_c,\qquad(7.7)$$

$$B = H_c - \frac{1}{D}\,(H_c - H_a).\qquad(7.8)$$

We note that B increases linearly from zero to H_c when H_a changes from $(1-D)\,H_c$ to H_c. In addition, it is essential to observe that the area under all possible magnetization curves of the type shown in Fig. 11 is $-H_c^2/8\pi$.

The boundary conditions on $\boldsymbol{B}$ and $\boldsymbol{H}$ on the specimen surface enable us to deduce the behavior of the field at the poles and on the equator as functions of H_a. As in the superconductive case, the field at the poles equals the magnetic induction inside. This is given by (7.8) provided $(1-D)\,H_c \leqq H_a \leqq H_c$. In the same interval of the applied field, the field on the equator is constant and equal to H_c. When H_a exceeds H_c, the field everywhere equals H_a. The dependence of the fields at these points on H_a is shown in Fig. 12.

To account for this macroscopic behavior, it is presumed that microscopically a specimen in the intermediate state is a mixture of normal and superconductive regions. The microscopic magnetic induction is zero in the superconductive regions and equal to H_c in the normal regions. The normal regions occupy a fraction

$$x = B/H_c\qquad(7.9)$$

of the total volume, where B is the average macroscopic induction. Clearly, the magnetization of the specimen is provided by the superconductive regions, whose total contribution is

$$M = -(1-x)\,H_c/4\pi.\qquad(7.10)$$

[1] R. PEIERLS: Proc. Roy. Soc. Lond., Ser. A **155**, 613 (1936).
[2] F. LONDON: Physica, Haag **3**, 450 (1936).

Substituting into (7.10) from (7.9) and (7.8), we obtain an expression for M identical to (7.6). These remarks apply only to specimens of smallest dimension larger than about one millimeter. For smaller specimens, size effects associated with the detailed structure of the intermediate state become important.

Everything that has been said to this point could have also been derived from a postulate of infinite conductivity, provided the specimen was initially superconductive in zero applied field. The break with such a view comes with the postulate that, on the basis of the MEISSNER effect, we expect ellipsoidal specimens to behave reversibly when the applied field is changed. In concrete

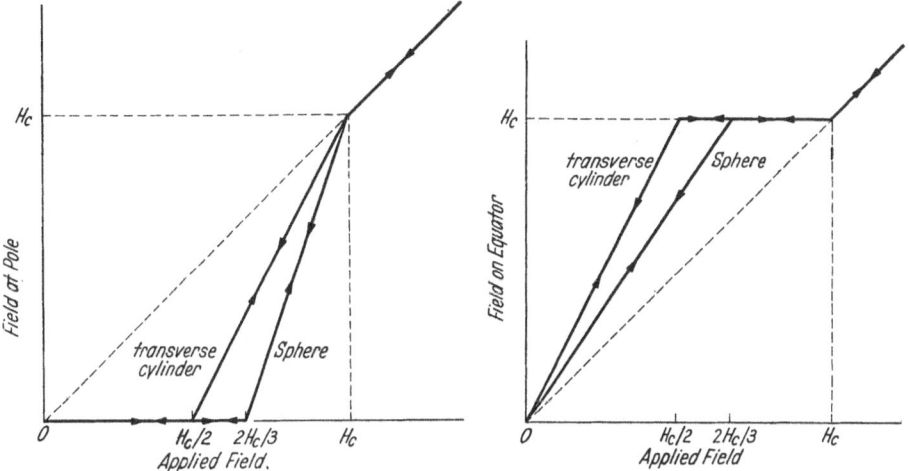

Fig. 12. Magnetic fields at the poles and on the equators of a sphere and a transverse cylinder.

terms, we expect the curves in Figs. 11 and 12 to be retraced when H_a is altered in any way. A perfect conductor cannot behave in this fashion.

The case of the longitudinal cylinder is clear-cut, since it can exist in only two states—superconductive or normal. A perfectly conducting cylinder would be left with a large frozen-in paramagnetic moment after the field is decreased from above H_c to zero, instead of with the zero moment to be expected from the MEISSNER effect. The postulate of reversibility extended to other geometries implies a restriction on the form of the intermediate state. The superconductive regions in this state are then limited to the general shape of laminae parallel to the field, since they cannot approach the form of closed rings which trap magnetic flux.

β) *Experimental verification.* The discovery of the magnetic shielding property of superconductors by MEISSNER and OCHSENFELD was followed by three years of widespread and intense activity and interest in this field. One gains the impression from the literature that the formal description of the magnetic properties discussed in the preceding section was developed independently of the experimental investigations which were being conducted at the same time. The subject developed so rapidly that it is impossible unambiguously to assign priority to any one investigator as being responsible for clarifying a given topic. For discussion, we have selected a few particular investigations from the mass of work. The examples were chosen solely because they seem most suitable for illustrating the more important information to be gained from this fruitful period. A more complete set of references is to be found in SHOENBERG[1].

[1] D. SHOENBERG [1], pp. 51—55.

The very first measurements of the dependence of the magnetic moments of superconductive specimens on field clearly indicated that the observed behavior could not be described exclusively in terms of infinite conductivity. The results were somewhat obscured, however, by the presence of frozen-in magnetic moments. That is to say, when the applied magnetic field was reduced to zero after having exceeded the threshold value, the specimen retained a paramagnetic moment in the absence of the field. The value of this moment was perhaps 10 to 15% of the maximum diamagnetic moment in the superconductive state. While a frozen-in moment of this magnitude is considerably smaller than that to be expected from perfect conductivity, it is still large enough to be disturbing. For a short period, the current view was that the magnetic induction in super-conductors was not precisely zero. But it soon became clear that frozen-in moments in specimens of ellipsoidal shape can be attributed to secondary causes such as polycrystallinity, strains, and impurities.

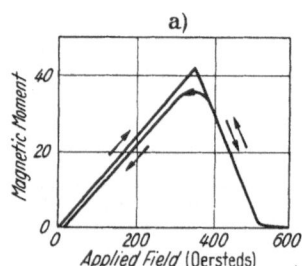

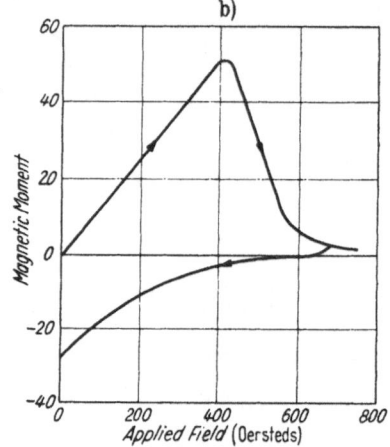

Fig. 13a and b. Magnetization curves of: a) a pure lead sphere, and b) an impure lead sphere at 4.2° K (after SHOENBERG[1]).

These points are effectively illustrated by the measurements of the magnetic moments of small lead spheres made by SHOENBERG[1]. The magnetic moment was determined by measuring the body force on a specimen produced by a slightly inhomogeneous field. The experimental results are shown in Fig. 13a. The observed behavior compares very favorably with the calculated behavior shown in Fig. 11. There is the initial increase in diamagnetic moment with applied field, followed by the discontinuous change in slope when the field reaches $\frac{2}{3}$ of the threshold value and the vanishing of the moment[2] when H_c is reached. The original curve is followed closely when the applied field is reduced, except for the effects of a small frozen-in moment of less than 3%. To illustrate that the slight irreversibility in these observations was attributable to secondary causes, SHOEN-BERG repeated the measurements with a sphere made of lead containing 1.5% bismuth as impurity. These observations are shown in Fig. 13b. There are several features to be noted. The change in slope of the magnetization curve at the maximum occurs gradually rather than discontinuously, and the super-conductive property does not disappear at a definite field value. Instead, the magnetization decreases very slowly as the applied field is increased above 600 oersteds. If the turn-over point is taken to be $\frac{2}{3}$ of the threshold value, the

---

[1] D. SHOENBERG: Proc. Roy. Soc. Lond., Ser. A **155**, 712 (1936).

[2] We neglect the magnetization in the normal phase. Because of the extremely small susceptibility of normal metals, the resulting magnetic moments are not detectable by the methods usually employed for measuring the relatively huge moments of superconductors.

threshold field is 615 oersteds, which agrees well with the value obtained by extrapolating the sharply descending portion of the magnetization curve. The threshold field of the pure sphere at the same temperature was 520 oersteds. Thus the effect of impurity in this case is to increase appreciably the average value of the threshold field. Upon reducing the field, a hysteresis loop of large area results. The frozen-in moment is almost 50% [1]! This extreme irreversible behavior is more characteristic of superconductive rings than of solid ellipsoidal specimens. Since, in this case, a small amount of impurity produces drastic changes in the magnetic properties, we infer that the slight irreversible behavior exhibited by nominally pure specimens can be attributed to minute amounts of impurity of both physical and chemical origin.

MENDELSSOHN'S [2] measurements of the magnetization of mercury cylinders provide another example of how closely it is possible to approach ideal behavior in practice. The cylinder axes were parallel to the applied field. A coil of wire was arranged so that it could be moved rapidly from a position in which it enclosed the sample to one far removed from the specimen. The magnetic moment of a specimen was determined by measuring the voltage impulse induced in the coil when it was moved between the two positions. A typical result is shown in Fig. 14; it is clear that it is an excellent approximation to the ideal behavior illustrated in Fig. 11.

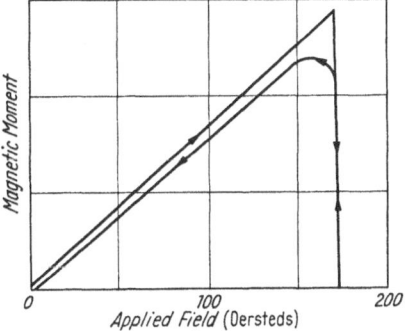

Fig. 14. Magnetization curve of a mercury cylinder at 3.1° K (after MENDELSSOHN [2]).

As a final example, we mention that DE HAAS and GUINAU [3] measured the magnetic field on the equator and at the poles of a sphere. These observations are in agreement with the behavior described in the preceding section and illustrated in Fig. 12. Bismuth wires were used for the field measurements. Since at low temperatures the resistivity of bismuth increases very rapidly with increasing magnetic field, small wires of this metal may conveniently be used for localized field measurement. This technique is obviously superior to the test coils employed in earlier investigations.

The few experiments we have described, together with the many others done at that time and since, constitute the basis for the generally accepted view that ideal superconductive behavior (as described in the preceding section) may be realized in pure, undistorted single crystals of ellipsoidal shape.

The non-ideal magnetic properties of SHOENBERG'S impure lead sphere have come to be regarded as typical of the behavior of alloys. These properties are the persistence of some superconductive regions in very high magnetic fields, and the generation of large frozen-in moments with their associated irreversible magnetic transitions. Furthermore, many substances exhibit an incomplete MEISSNER effect. If they are cooled below the transition point in the absence of a field, it is found that an appreciable amount of flux passes through the sample when a field is applied. As a result, the magnetic induction is not zero inside the specimen even under these circumstances. While we have stated that such

[1] This very large moment may be the consequence of insufficient annealing of the specimen, cf. Chapter VII.
[2] K. MENDELSSOHN: Proc. Roy. Soc. Lond., Ser. A **155**, 558 (1936).
[3] W. J. DE HAAS and O. A. GUINAU: Physica, Haag **3**, 182, 534 (1936). — Leiden Comm. 241 a, b.

behavior is typical of alloys, we wish to emphasize that it is possible to prepare particular alloys which approach ideal behavior. On the other hand, certain of the superconductive elements are notorious for their non-ideal properties. These substances occur in columns IVa and Va in the periodic table (Ti, V, Nb, Ta, etc.). The elements in periods II, III and IVb (In, Sn, Hg, Pb, etc.) generally exhibit excellent behavior, with tin coming closest to possessing ideal magnetic properties. The properties of alloys are discussed more completely in Chapter VII.

In the foregoing discussion we have neglected effects associated with possible supercooling and superheating in the phase transition induced by the magnetic field. These phenomena are known to occur in a large variety of phase transitions, and they also have been observed in the superconductive transition. Supercooling and superheating are connected with the problem of the nucleation of one phase in a matrix of the second phase. We discuss this matter in detail in Sect. 25.

8. Non-ellipsoidal Specimens. The magnetostatic problem of finding the field distribution about a specimen of non-ellipsoidal shape cannot be solved in general. It is clear, nevertheless, that the magnetization of the specimen is not uniform. If we consider a non-uniform magnetization to be the primary characteristic of such shapes, we can include in this same category two additional cases which lend themselves readily to mathematical analysis. These are the hollow cylinder and the hollow sphere. For the case of either the hollow sphere or the hollow infinite cylinder placed in a uniform applied field, one can prove that the magnetostatic boundary conditions cannot be satisfied by assuming the material of the specimen to be uniformly magnetized.

The non-uniform magnetization has no consequences so long as the specimen is superconductive, since the magnetic induction is made to vanish in the walls by an appropriate surface current density. The external magnetic field about a superconductive hollow sphere or cylinder is identical with that about the corresponding solid specimen of equal exterior dimensions. As a result the total magnetic moment of a hollow sample is the same as that of a solid specimen. Therefore, we expect the magnetic behavior of a virgin hollow specimen to be identical with that of the corresponding solid one until the advent of the intermediate state.

Non-uniform magnetization does produce changes in the intermediate state of such hollow specimens. We see that it is no longer possible in principle for these shapes to have the usual type of intermediate state described in Sect. 7α, since such a state is characterized by a uniform magnetization. Peierls[1] has shown that a hollow or non-ellipsoidal specimen may be expected to consist of relatively large superconductive regions and regions having the usual mixed structure. However, it has not been possible to derive the detailed structure of particular cases.

The magnetic behavior of short, solid tin cylinders observed by Shoenberg[2] provides indirect evidence for the existence of a non-uniform intermediate state in non-ellipsoidal specimens. The most direct evidence was obtained by Gittleman[3], who studied the magnetic field distribution on the inner and outer surfaces of a long hollow tin cylinder in a transverse applied field. Bismuth wire probes were used for the field measurements. The intermediate state structure deduced by Gittleman is shown in Fig. 15. We note that there are large superconductive

[1] R. Peierls: Proc. Roy. Soc. Lond., Ser. A **155**, 613 (1936).
[2] D. Shoenberg: Proc. Cambridge Phil. Soc. **33**, 260 (1937); cf [1] pp. 34ff.
[3] J. Gittleman: Phys. Rev. **92**, 561 (1953).

regions in the neighborhood of the equatorial points and regions of the usual mixed state near the poles. Several years earlier, KOCH[1] had, on the basis of crude theoretical arguments, suggested that in a hollow sphere a similar structure exists in which the superconductive region is a closed band about the equator.

Non-ellipsoidal specimens have been observed to deviate markedly from ideal behavior in several other characteristic ways. When the specimen is in the intermediate state, it may take as long as one half hour for the field distribution about the sample to reach a steady value after the applied magnetic field is changed[2, 3, 4], whereas in a solid ellipsoidal specimen, the corresponding time is at most a few seconds. When the applied field is reduced to zero from above the threshold value, a non-ellipsoidal specimen may exhibit a large frozen-in magnetic moment[4], and the magnetization curve is, of course, irreversible. This frozen-in moment is not attributable to the effect of impurities, but is associated with the presence of closed superconducting rings which trap magnetic flux in the specimen. The situation is clearest in a hollow sphere, where a superconductive band around the equator could behave just like a ring. Presumably, similar structures are established in the hollow cylinder and in solid non-ellipsoidal specimens. Finally, we mention that SERIN, GITTLEMAN and LYNTON[5] showed that in hollow cylinders with walls smaller than a critical size, specimens may make the transition into the intermediate state before the applied field reaches one half the threshold value. Below the critical size, the thinner the wall, the smaller the field at which the transition occurs.

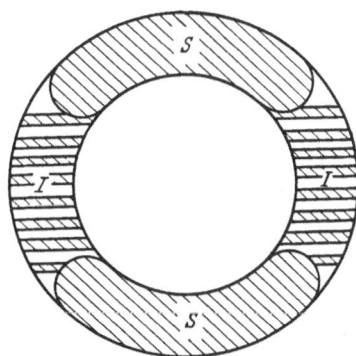

Fig. 15. Intermediate state of a hollow tin cylinder in a transverse field (after GITTLEMAN).

9. Nature of superconducting currents. ONNES and TUYN[6] performed an experiment with a hollow sphere similar to the experiment with two rings described in Sect. 4. A hollow lead sphere, suspended from a torsion fiber, was free to turn inside a fixed lead ring. The specimens were cooled below the transition point in a large magnetic field and the field was then removed. The torque on the sphere was measured as a function of its angular position, and it was found that the observations could be interpreted only by assuming that the magnetic moment maintained a fixed direction in the sphere and rotated with it. On the basis of this measurement, ONNES and TUYN concluded that persistent currents flowing on a superconductor are rigidly bound to the material of the metal. In view of the discussion in the preceding section, this conclusion must be regarded as incorrect. What was observed was the frozen-in moment associated with a ring of superconducting material in the hollow sphere. This point is clearly illustrated in the more recent measurements of FRITZ, GONZALEZ and JOHNSTON[7]. CONDON and MAXWELL[8] measured the torque on a *solid* superconductive tin sphere oscillating in a magnetic field. The observed torque was less than one

[1] K. M. KOCH: Z. Physik **121**, 488 (1943).
[2] See footnote 3, p. 226.
[3] K. MENDELSSOHN and R. B. PONTIUS: Nature, Lond. **138**, 29 (1936).
[4] J. BABISKIN: Phys. Rev. **85**, 104 (1952).
[5] B. SERIN, J. GITTLEMAN and E. A. LYNTON: Phys. Rev. **92**, 566 (1953).
[6] See footnote 5, p. 215.
[7] J. J. FRITZ, O. D. GONZALEZ and H. L. JOHNSTON: Phys. Rev. **80**, 894 (1950).
[8] E. U. CONDON and E. MAXWELL: Phys. Rev. **76**, 578 (1949).

percent of the value to be expected if the current were rigidly fixed to the sphere, and these results were confirmed by Houston and Muench[1] in a similar experiment.

On the basis of these later experiments with solid specimens, we must conclude that the magnetic moment of a superconductive specimen, and therefore the associated surface current density, is directly coupled to the external magnetic field and not to the body of the specimen. This result is to be expected from the Meissner effect[2], since, as can be seen from the expression

$$B = H + 4\pi M,$$

B can be zero only if H and M have opposite directions. The foregoing experiments provide, therefore, additional confirmation of the Meissner effect. Substantiation is provided also by the experiments of Houston and Squire[3] and Wexler and Corak[4] in which it was shown that there is no detectable e.m.f. generated between the equator and the pole of a superconductive sphere rotating very rapidly in a magnetic field parallel to the axis of rotation.

The gyromagnetic ratio (ratio of magnetic moment to angular momentum) of superconducting currents was measured by Kikoin and Goobar[5], and the experiment was repeated recently by Pry, Lathrop and Houston[6]. In the latter version, a superconducting tin sphere suspended from a torsion fiber was set into torsional oscillations in a magnetic field which was reversed at the end of every half period of vibration. When the steady state of oscillation is reached, the angular momentum provided by the reversal in magnetization of the sphere just makes up for the angular momentum lost by damping in each half period. The gyromagnetic ratio found is just $(-e/2\,mc)$, which proves that the diamagnetism is associated with the motion of electrons and has no possible direct connection with the electron spin. On the basis of quantum theory, Broer[7] has shown that this type of measurement can be explained solely in terms of the conservation of the total angular momentum of the metal—electrons and lattice atoms. As a result, experiments of this type cannot, in principle, provide information about the detailed microscopic nature of the diamagnetic currents.

10. Resistive transition. The resistance of a superconductive wire, in a magnetic field parallel to the wire axis, abruptly changes from zero to its normal value when the field reaches a critical value which depends on the temperature. De Haas and Engelkes[8] showed that in tin (at a given temperature) the field value which completely restores the resistance of a wire is the same as the field needed to restore the permeability of a tin sphere to unity. Thus, in tin the critical field determined from either the restoration of resistance or the vanishing of magnetization is identical with the threshold value.

De Haas, Voogd and Casimir-Jonker[9] investigated the resistive transition in more complicated cases. Their results for the resistive transition of tin wires of circular cross section with axes transverse to the applied magnetic field are shown in Fig. 16. Under these conditions, we expect the first trace of resistance to appear at one half the threshold value, and then to increase approximately

[1] W. V. Houston and N. Muench: Phys. Rev. **79**, 967 (1950).
[2] See footnote 8, p. 227.
[3] W. V. Houston and C. F. Squire: Phys. Rev. **76**, 685 (1949).
[4] A. Wexler and W. S. Corak: Phys. Rev. **78**, 260 (1950).
[5] I. K. Kikoin and S. V. Goobar: J. Phys. USSR. **3**, 333 (1940).
[6] R. H. Pry, A. L. Lathrop and W. V. Houston: Phys. Rev. **86**, 905 (1952).
[7] L. J. F. Broer: Physica, Haag **13**, 473 (1937).
[8] W. J. de Haas and A. D. Engelkes: Leiden Comm. 247d. — Physica, Haag **4**, 325 (1937).
[9] W. J. de Haas, J. Voogd and J. M. Casimir-Jonker: Leiden Comm. 229c. — Physica, Haag **1**, 281 (1933).

linearly with field until the normal value is reached at the threshold field. It is clear from Fig. 16 that this does not happen. The shape of the curve of resistance as a function of field is very sensitive to the magnitude of the measuring current; the curves approach linear behavior only as the measuring current approaches larger values. Furthermore, the first trace of resistance appears when the applied field is 0.6 H_c rather than one half the threshold value. The diameter of the wires used in the experiment were only about 0.25 mm. As we shall see in Sect. 24β, the anomalies exhibited by these small specimens are associated with the detailed structure of the intermediate state of the wires. DE HAAS et al. also investigated the transition of wires of ellipsoidal rather than circular cross section. For ellipsoidal specimens, resistance first appears at an applied field about equal to the value necessary to have the field at the equator equal H_c, but there are systematic deviations. In view of the 0.6-anomaly in wires of circular cross section, this last result is not surprising.

The tin lattice is quite anisotropic. It was natural, therefore, for DE HAAS, VOOGD and CASIMIR-JONKER to determine whether the direction between the applied field and the crystal axes had any effect on the magnetically induced resistive transition. No effects were observed within the precision of measurement. Several years previously, DE HAAS and VOOGD[1] had determined that the relative orientation of the measuring current and the crystal axes had no effect on the transition temperature of tin as determined by the abrupt disappearance of resistance at T_c. Recently, CROFT, OLSEN-BÄR and PO-WELL[2] repeated this type of experiment with gallium. Gallium crystals have the

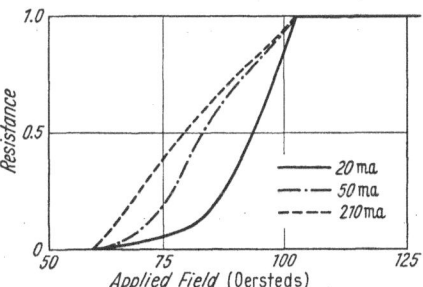

Fig. 16. Resistive transition of a 0.25 mm wire at 2.92° K in a transverse field (after DE HAAS et al.).

largest anisotropy in resistance of all the metals. The transition temperatures of two specimens, each with current flowing in the direction of one of the extremes of resistance, were measured. The upper limit for the difference in the transition temperatures between the two was 0.002° K. On the basis of these experiments, we conclude that anisotropy has no appreciable effect on the macroscopic behavior of superconductors.

The magnetically induced resistive transition of a wire longitudinal to the field can be used, in principle, to determine the threshold value. Although the practical situation is excellent in tin and many other superconductors, in some elements and a good many alloys resistance measurements may give false results. The deception comes about because a few thin threads of superconductive metal running the length of an otherwise normal wire will result in the wire's still exhibiting zero resistance. Such a configuration can result in an apparently greater transition temperature and larger threshold field which are characteristic of the threads rather than the bulk of the specimen. For this reason, magnetic measurements which are governed by the average volume properties of a specimen are generally preferred for the determination of threshold fields and transition temperatures.

11. The superconductive elements. Over the years, an increasing number of elements, alloys and compounds have been found to be superconductive. A review

[1] W. J. DE HAAS and J. VOOGD: Leiden Comm. **1931**, 214c.
[2] A. J. CROFT, M. OLSEN-BÄR and R. W. POWELL: Phil. Mag. **45**, 123 (1954).

of the sources of data on transition temperature and threshold field values for the elements has been prepared recently by EISENSTEIN[1]. In Table 1 we list the most likely values for the transition temperatures of the elements. These values have been taken, with the exception of the three cases noted, from EISENSTEIN.

Table 1. *Transition temperatures of the elements.*

Element	T_c (°K)	Element	T_c (°K)	Element	T_c (°K)	Element	T_c (°K)
Al	1.197	Nb	8.70	La	5.4	Hg [2]	4.173
Ti	0.387	Tc	11.2	Hf *	0.37 (?)	Tl	2.392
V	4.89	Ru	0.47	Ta	4.38	Pb	7.2
Zn	0.905	Cd	0.560	Re	1.699	Th	1.37
Ga	1.103	In	3.396	Os	0.71	U [3]	0.8
Zr	0.546	Sn [2]	3.729				

* There is some doubt about the superconductivity of Hf; see: R. A. HEIN, Phys. Rev. **102**, 1511 (1956).

The transition temperatures listed in the table are for specimens having the distribution of isotopic masses which occur in nature. This stipulation is necessary because the transition temperature is a sensitive function of the average isotopic mass. The effect of isotopic constitution on superconductive properties is discussed in the next chapter. Superconductive alloys and compounds are discussed in Chapter VII.

III. Thermodynamic properties of the normal and superconductive phases.

In this chapter, we discuss first the experimental determinations of the specific heats of various superconductors. We shall see that the form of the temperature dependence of the electronic specific heat changes radically when a metal changes from the normal phase to the superconductive. This is a most important finding, because it clearly indicates that the energy distribution of the electrons undergoes a fundamental change in character when the metal passes from one phase to the other.

We then treat the thermodynamic description of the phase transition. This theory reveals the essential connection between the thermal and mechanical properties of the two phases on the one hand, and the magnetic threshold field curve of the superconductive phase on the other. The latter discussion is interlaced with references to those experimental results which bear on the theory.

12. Specific heat of superconductors. The relation

$$\gamma\,T + A\left(\frac{T}{\Theta}\right)^3,\qquad (12.1)$$

describes the dependence of the specific heat of normal metals on the absolute temperature, T. The first term is contributed by the conduction or "free" electrons, and the second term comes from the crystal lattice[4]. γ is a constant of a metal and has values in the neighborhood of 10^{-4} cal./mole (°K)^2. Θ is the DEBYE temperature of the lattice and $A = 464$ cal./mole °K.

---

[1] J. EISENSTEIN: Rev. Mod. Phys. **26**, 277 (1954).
[2] Author's values.
[3] J. E. KILPATRICK, E. F. HAMMER and D. MAPOTHER: Phys. Rev. **97**, 1634 (1955).
[4] Cf. the detailed treatment of normal metals in the article contributed by P. H. KEESOM and N. PEARLMAN in vol. XIV of this Encyclopedia.

Since the lattice contribution to the specific heat of the superconductive phase is the same as in the normal (cf. Chapter VIII), it is a great convenience to imagine that the lattice contribution has been subtracted from the total specific heat of each phase, thereby leaving the electronic residue. Henceforth, we shall assume that this has been done, so that the term specific heat (unless otherwise qualified) will mean the electronic contribution alone. On this basis, the specific heat of the normal phase is

$$C_n = \gamma\, T = \gamma\, T_c\, t \qquad (12.2)$$

where $t = T/T_c$.

As a practical matter, it is not always possible to separate the electronic component from the total specific heat with great precision. To obtain a good value for the electronic specific heat, it is desir- able to have the lattice contribution as small as possible. This will occur in those metals having large values of the DEBYE tempera- ture, Θ. Unfortunately, many of the super- conductive metals in periods II, III and IVb of the table of the elements (i.e. those having almost ideal superconducting properties) are mechanically soft and so have relatively small values of Θ. Of this group, the two metals having the largest Θ are tin ($\Theta \sim 185°$ K) and aluminum ($\Theta \sim 420°$ K). The elements in peri- ods IVa and V, being hard, have relatively large DEBYE temperatures, but it is difficult to obtain specimens which approach ideal super- conductive behavior.

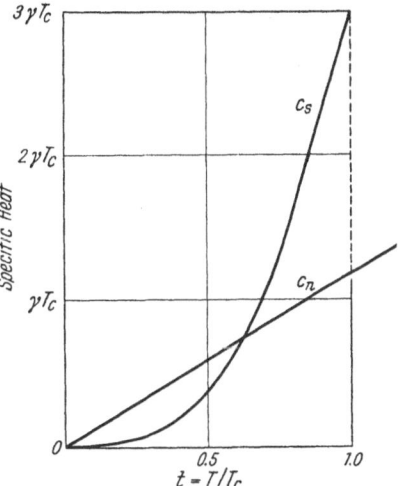

Fig. 17. The approximate specific heats of the normal and superconductive phase are shown as a function of the reduced temperature.

One of the most significant facts about super- conductivity is that in passing from the nor- mal phase into the superconductive phase, the specific heat jumps discontinuously from a value, $\gamma\, T_c$, in the former phase to a value of about $3\gamma\, T_c$ in the latter. The discontinuous increase in the specific heat in passing through the transition point was first seen in tin by KEESOM and VAN DEN ENDE[1]. A short time later, KEESOM and KOK[2] measured the magnitude of this jump and found it to be 0.0024 cal./mole °K. They observed that there is no latent heat associated with the transition, and also found the same jump and the absence of a latent heat in thallium[3]. Because of the discontinuity in specific heat and lack of latent heat of transition, superconductors provide an ideal practical example of a second order phase transition.

After the initial jump, the specific heat of superconductors decreases more rapidly with decreasing temperature than the specific heat of a normal metal. The most careful work prior to World War II was done by KEESOM and VAN LAER[4], who measured tin. In this substance, the superconductive specific heat varies approximately as T^3; in fact,

$$C_s \sim 3\gamma\, T_c\, t^3. \qquad (12.3)$$

The behavior of C_s and C_n is illustrated in Fig. 17. We note that while initially $C_s > C_n$, for temperatures less than $t \sim 1/\sqrt{3}$, C_s rapidly gets smaller than C_n.

[1] W. H. KEESOM and J. N. VAN DEN ENDE: Leiden Comm. 1932, 219b.
[2] W. H. KEESOM and J. A. KOK: Leiden Comm. 1932, 221e.
[3] W. H. KEESOM and J. A. KOK: Leiden Comm. 1934, 230c.
[4] W. H. KEESOM and P. H. VAN LAER: Leiden Comm. 252b. — Physica, Haag 5, 193 (1938).

The variable, $t = T/T_c$, is quite useful in comparing the behavior of different superconductors with various transition temperatures, and we will use it in all subsequent considerations.

We cannot state too emphatically that the T^3-dependence of specific heat and the expression (12.3) are meant to be only very crude statements which roughly fit the data of all superconductors. Systematic deviations from this dependence are already evident in the measurements of KEESOM and VAN LAER on tin. The deviations are shown in Fig. 18 as a function of temperature. Despite the bad scatter of the data, they exhibit a clear systematic departure from t^3-dependence. Since the war, principally due to improvements in the technique

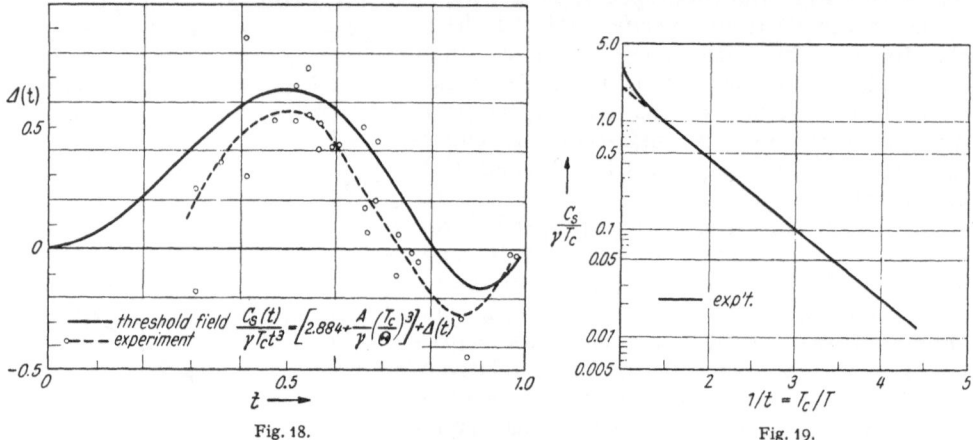

Fig. 18.   Fig. 19.

Fig. 18. Deviations from the t^3-law of the specific heat. O-tin (after KEESOM and VAN LAER). Solid line calculated from the threshold field data.

Fig. 19. Specific heat of superconductive vanadium (after CORAK et al.[2]).

of resistance thermometry, many very precise specific heat data have been obtained, and these also show departures from the t^3-law in several substances. Such deviations are quite evident in the measurements of niobium reported by BROWN, ZEMANSKY and BOORSE[1]. Most significant is the recent work of CORAK, GOODMAN, SATTERTHWAITE and WEXLER[2], whose measurements of the specific heat of vanadium are shown in Fig. 19. The data fit the expression

$$a\, e^{-b/t}$$

with $a = 9.17$, $b = 1.50$, much better than (12.3).

Such an exponential dependence suggests that a gap of the order of kT_c exists in the electronic energy spectrum. As the temperature is raised, an increasing number of electrons are excited across the gap. As several attempts at a theoretical treatment of superconductivity have suggested the existence of such a gap, it seems likely that the variation with temperature of the specific heat of superconductors is of the exponential form[3], provided only that t is small enough. The fit of the tin data of KEESOM and VAN LAER to an exponential temperature dependence is certainly no worse than the fit to T^3, except in the

---

[1] A. BROWN, M. W. ZEMANSKY and H. A. BOORSE: Phys. Rev. **92**, 52 (1953).

[2] W. S. CORAK, B. B. GOODMAN, C. B. SATTERTHWAITE and A. WEXLER: Phys. Rev. **96**, 1442 (1954).

[3] The form is probably $(a t^n e^{-b/t})$, but it is practically impossible to determine the value of n because of the familiar domination of the exponential.

immediate neighborhood of the transition point, where there are pronounced deviations from the exponential. Recent more precise measurements of the specific heat of tin over a more extended temperature range by CORAK and SATTERTHWAITE [1] have revealed that the exponential form is correct except near T_c.

13. Thermodynamics of the phase transition. α) *Free energy difference between the phases.* The specific heat is closely related to the magnetic threshold curve of superconductors. Before the discovery of the MEISSNER effect, it was occassionally suggested that the magnetically induced transition from the superconductive to the normal phase could be treated by thermodynamics. RUTGERS [2], starting from the standard treatment of second order phase transitions, and reasoning by analogy, arrived at essentially the same expressions we will presently derive unambiguously from thermodynamics. Moreover, these expressions were experimentally verified. At that time, however, a superconductor was regarded as a perfect conductor. This viewpoint completely excluded the possibility of a thermodynamic treatment, since it was obvious that the magnetic transition of a perfect conductor is, in principle, irreversible. It is clear that the state of the subject was reaching a crisis in 1933, and the time was ripe for the experiment of MEISSNER and OCHSENFELD. With the discovery of the perfect shielding property, it became evident that superconducting currents differ in an essential way from usual conduction currents, and that the phase transition in a magnetic field could be (and is) reversible. The first clear thermodynamic treatment of the phase transition was derived by GORTER and CASIMIR [3].

In applying thermodynamics, we assume that a given specimen, in the absence of a magnetic field, has a free energy [4] $g_s(T, p)$ when in the superconductive phase and $g_n(T, p)$ when in the normal, where p is the pressure and T, the absolute temperature. Neglecting small pressure effects, at any temperature below the transition point, g_s must be less than g_n, since the superconductive phase is stable. When a superconductive specimen is placed in a magnetic field, H_a, its total free energy is the sum of g_s and the magnetic work done on the specimen; that is to say,

$$G_s(T, p, H_a) = g_s(T, p) - \int_0^{H_a} \mu(H_a)\, dH_a, \qquad (13.1)$$

where μ is the total magnetic moment of the specimen [5]. Since μ is negative, the effect of the field is to increase the free energy of the superconductive phase. We can neglect the magnetic energy of a normal metal, so that

$$G_n(T, p, H_a) = g_n(T, p). \qquad (13.2)$$

The two phases can exist in equilibrium below the transition point when $G_s = G_n$, and the normal phase is stable when $G_s > G_n$.

To avoid the extraneous effects due to geometry, it is simplest to consider initially the case of a long cylinder with its axis longitudinal to the field direction. We know that under these circumstances the specimen remains superconducting until the applied field attains the threshold value, H_c, at which point it passes into the normal phase. Thus, $G_s(T, p, H_c) = G_n(T, p, H_c)$ and we see from (13.1) and (13.2) that

$$g_n(T, p) - g_s(T, p) = \frac{V H_c^2(p, T)}{8\pi}, \qquad (13.3)$$

[1] W. S. CORAK and C. B. SATTERTHWAITE: Phys. Rev. **99**, 1660 (1954).
[2] See P. EHRENFEST: Leiden Comm. **1933**, Suppl. 75b.
[3] C. J. GORTER and H. G. B. CASIMIR: Physica, Haag **1**, 306 (1934).
[4] We use the GIBBS' free energy $U - TS + pV$.
[5] See E. C. STONER: Phil. Mag. **23**, 833 (1937) for a discussion of this expression for magnetic work.

where V is the volume of the specimen. Although (13.3) has been derived for a longitudinal cylinder, it applies quite generally. If we consider any other ellipsoidal shape, then, despite the complications of the intermediate state, we recognize that magnetic work is done only on those portions of the specimen which remain superconductive in any given magnetic field. Since we showed in Sect. 7α that the area under the magnetization curve of any ellipsoid is $VH_c^2/8\pi$, this amount of magnetic work is needed to convert a specimen completely from the super-conductive to the normal phase. Eq. (13.3) describes these cases, too, so that the result is clearly independent of shape, and $H_c^2/8\pi$ is the difference in free energy *per unit volume* of the two phases. This result having been established, Gorter and Casimir showed that it has the following most interesting consequence: Any infinitesimal superconductive volume becomes unstable with respect to a transition to the normal phase when the magnetic field on its *surface* reaches the threshold value. In all this discussion we have neglected the effects of any surface energy which may be present at an interface between the two phases; these are discussed in Chapter V.

It is in the sense of the foregoing discussion that we say that the threshold field curve provides the phase diagram illustrated in Fig. 5. The curve divides the H-T plane into two regions: one specifying the states of the superconductive phase and the other, the normal phase, with the curve itself defining the unique values of H and T for which the two phases are in equilibrium. The diagram refers to any small superconductive volume, and H is the total magnetic field on its surface. For describing the states of the whole of a particular body, the appropriate variable is the applied field, H_a. The phase diagram is now more complicated, there being regions corresponding to intermediate states as well as to superconductive and normal ones.

β) Specific heat and latent heat differences. Differentiation of (13.3) with respect to the temperature yields

$$S_n(T) - S_s(T) = - \frac{V H_c}{4\pi} \frac{d H_c}{d T}, \tag{13.4}$$

where S is the entropy[1]; and a second differentiation gives

$$C_n(T) - C_s(T) = - \frac{V T}{4\pi} \left[H_c \frac{d^2 H_c}{d T^2} + \left(\frac{d H_c}{d T} \right)^2 \right], \tag{13.5}$$

where C is the specific heat[1].

We see from these expressions that many of the thermodynamic properties of the two phases can be derived from the magnetic threshold curve. Some of the deductions are independent of its detailed shape. For example, at the transition temperature, the threshold field is zero and the slope of the curve is finite. Thus we see from (13.4) that the entropy difference is zero at this temperature, and there is no latent heat of transition. It also follows from (13.5) that there is a discontinuous increase in specific heat on passing from the normal phase to the superconductive phase at the transition temperature. Both these phenomena are observed, as we pointed out in the previous section.

At all temperatures above the absolute zero, the slope of the curve is negative, so that the entropy of the normal phase is always greater than the superconductive entropy. This means that the superconductive phase is a state of greater order than the normal phase. Furthermore, heat is absorbed (i.e. there is a latent heat of transition) when the phase transition from the superconductive to the normal state is induced by a magnetic field. On the other hand, if the magnetic transition is carried out adiabatically, the temperature of the specimen is decreased.

---

[1] The pressure is assumed constant.

As the absolute zero is approached, the slope of the threshold field curve approaches zero, and the entropies of both phases tend toward zero, in agreement with NERNST's theorem. The entropy difference is thus zero at T_c and at $0°$ K; in between, it rises to a maximum at $t \sim 1/\sqrt{3}$.

The detailed comparison between the measured specific heats, latent heats and threshold field values and the predictions of (13.4) and (13.5) constitute the best test of the reversibility of the magnetic transition. It has been found to be reversible in numerous investigations; in addition to the references to be found in SHOENBERG[1], we list a recent contribution by DOLACEK[2].

The thermodynamic treatment of the intermediate state requires no additional assumptions; it is necessary only to add together the contribution which each phase makes to the properties of the mixture. We refer the reader to SHOENBERG[3] for a detailed discussion.

14. The threshold field curve. α) *General remarks.* The magnetic threshold curve of any superconductor is the parabola,

$$h = \frac{H_c}{H_0} = (1 - t^2), \tag{14.1}$$

to a first approximation, where H_0 is the threshold field at absolute zero. From (13.3) it follows that $H_0^2 V/8\pi$ is the difference in internal energy of the two phases at $0°$ K.

KOK[4] first showed that by taking

$$C_n - C_s = \gamma T_c t - 3\gamma T_c t^3,$$

the integration of (13.5) yields a parabolic threshold field curve, with

$$T_c^2 = \frac{H_0^2 V}{2\pi\gamma}. \tag{14.2}$$

Thus, threshold field measurements of the superconductive phase provide a convenient tool for determining the specific heat of the *normal* phase. Eq. (14.2) is correct only if the threshold field curve is parabolic over the entire temperature range. When the curve is non-parabolic, γ may be still obtained from the data for the lowest temperatures. Since at sufficiently low temperatures, $S_s \ll S_n$,

$$S_n - S_s \approx S_n = \gamma T = -\frac{V H_c}{4\pi} \frac{dH_c}{dT}. \tag{14.3}$$

DAUNT and MENDELSSOHN[5] first used (14.3) to derive γ from threshold measurements. The equation can also be immediately integrated to yield

$$h = \frac{H_c}{H_0} = (1 - 4\pi\gamma\, T_c^2\, t^2/H_0^2\, V)^{\frac{1}{2}} \approx (1 - 2\pi\gamma\, T_c^2\, t^2/H_0^2\, V). \tag{14.4}$$

From (14.2) we see that $H_0^2 V/2\pi\gamma\, T_c^2$ is about unity, so that terms of the order of t^4 have been neglected in the expansion of the square root. We conclude from (14.4) that the magnetic threshold curve is, in general, parabolic at sufficiently low temperatures, so that the slope of the curve of h *vs* t^2 in this region may be used to calculate γ. A review of the values of γ obtained from measurements on superconductors has been made recently by DAUNT[6].

[1] D. SHOENBERG [1], pp. 60—65.
[2] R. L. DOLACEK: Phys. Rev. **96**, 25 (1954).
[3] D. SHOENBERG [1], pp. 65—71.
[4] J. A. KOK: Physica, Haag **1**, 1103 (1934).
[5] J. G. DAUNT and K. MENDELSSOHN: Proc. Roy. Soc. Lond., Ser. A **160**, 127 (1937).
[6] J. G. DAUNT [3], Chapter XI.

The magnetic threshold curve may also be viewed from the standpoint of the two-fluid models. In these models the free energy of the superconductive phase is written in terms of an order parameter[1] which varies with temperature in such a way as to make the free energy a minimum. The simplest model is that of GORTER and CASIMIR[2], in which

$$g_s = \frac{(1 - \omega) H_0^2 V}{8\pi} - \frac{1}{2} (1 - \omega)^\alpha \gamma T^2. \tag{14.5}$$

Since,

$$g_n = \frac{H_0^2 V}{8\pi} - \frac{1}{2} \gamma T^2,$$

the two expressions may be combined with (13.3) to determine the magnetic threshold field at any temperature. In (14.5), ω is the order parameter. α is an adjustable parameter which can have values between zero and unity. For $\alpha = \frac{1}{2}$, the equilibrium values of ω are

$$(1 - \omega) = t^4, \tag{14.6}$$

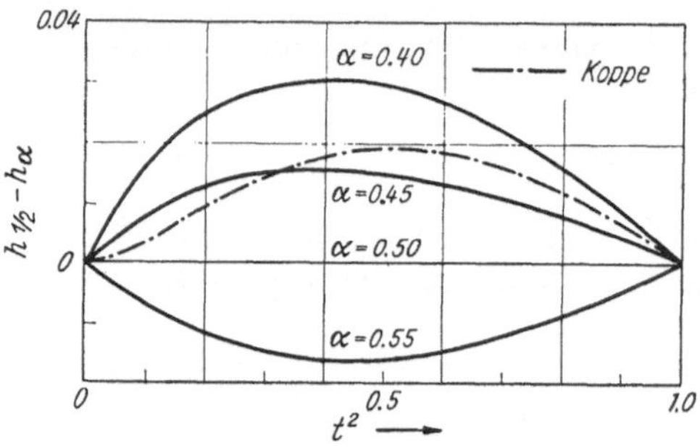

Fig. 20. Deviation from a parabolic threshold field curve for different values of α in the two-fluid model (after MARCUS and MAXWELL[4]).

and the resulting threshold field curve is parabolic. When α has any other value, the magnetic threshold curves are non-parabolic; the deviations from the parabolic form are illustrated in Fig. 20. The free energy function in the two fluid model of KOPPE[3] cannot be written in any simple form, but it predicts a unique non-parabolic threshold field curve which is also illustrated in Fig. 20.

The magnetic threshold data of tin, thallium, mercury and indium have been compared very carefully with the predictions of the foregoing models by MARCUS and MAXWELL[4] and MAXWELL and LUTES[5]. They find that the curve for mercury is parabolic (see also Ref. 9 on p. 237), whereas the curves of the other three superconductors are not. The mercury data can thus be fitted to the GORTER model with $\alpha = \frac{1}{2}$, and contradict the KOPPE model which predicts a non-parabolic curve. The data of the remaining three elements cannot be fitted exactly either to the first by appropriately choosing α nor to the second, although the data approach closely to the case of $\alpha = 0.38$.

We must remark that the deviations of the data from even the parabolic form are quite small, rarely exceeding 10%; and the deviations from the more complicated forms are even smaller, perhaps a few percent. The latter are revealed only after the most precise experimentation. Because of these circumstances, it would appear that at least at high temperatures, the simple GORTER model with $\alpha = \frac{1}{2}$ provides a sufficiently accurate basis for thinking qualitatively about superconductivity from the two-fluid viewpoint. This standpoint is strengthened by the realization that the physical picture of the microscopic nature of the two fluids (or, what is the same thing, of the order parameter) is at present not entirely clear.

---

[1] See BARDEEN, Sect. 4. Unless otherwise qualified, references to this author advert to his article in this volume.

[2] See C. J. GORTER [3], Chapter I.

[3] H. KOPPE: Ann. Phys. 6, 405 (1947).

[4] P. M. MARCUS and E. MAXWELL: Phys. Rev. 91, 1035 (1953).

[5] E. MAXWELL and O. S. LUTES: Phys. Rev. 95, 333 (1954).

We have also compared the data with the predictions of the GINSBURG energy gap model[1], treating σ (which is proportional to the gap energy) as an adjustable parameter. Although $\sigma = 2.75$ results in an approximately parabolic magnetic threshold curve, no value of σ yields a curve coming as close to the data for tin, indium and thallium as the curves for either of the two-fluid models.

β) Isotope effect. In 1950, MAXWELL[2] and REYNOLDS, SERIN et al.[3] discovered that the transition temperature of a superconductive element depends on the isotopic mass. This dependence was searched for in lead on two previous occasions[4] with negative results. At the earlier times, the only isotopes available were those occurring in nature as a result of radioactive decay. Since World War II, however, with the growth of the atomic energy establishment, stable isotopes of a large number of elements have become relatively abundant. The first observations were made on mercury specimens, and a surprisingly large effect of lattice mass on the superconductive properties was found. This effect is strikingly illustrated in Fig. 21, where the threshold fields of various isotopes of mercury are shown as a function of the temperature, near the transition temperatures. SERIN, REYNOLDS and NESBITT[5] established that in mercury the relationship between transition temperature, T_c, and isotopic mass, M is

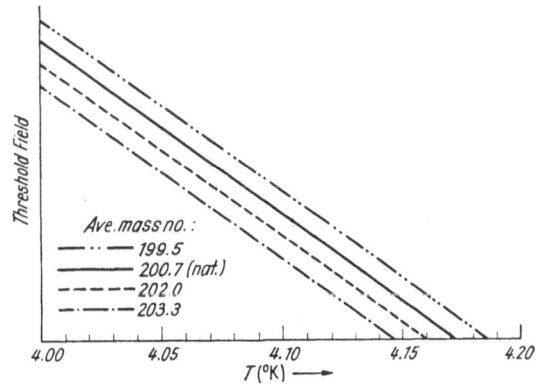

Fig. 21. Isotope effect in mercury (after REYNOLDS et al.[3]).

$$M^{\frac{1}{2}} T_c = \text{const}. \qquad (14.7)$$

The dependence on $M^{\frac{1}{2}}$ is approximately true in tin[6], and thallium[7], but lead[8] seems to have the exceptionally high exponent of 0.73.

The extension of the observations to low temperatures by REYNOLDS, SERIN and NESBITT[9] revealed that the magnetic threshold curve of mercury is parabolic. The only effect of isotopic mass is to change the threshold field at 0° K, H_0, in the same proportion as the change in T_c, so that the ratio H_0/T_c remains constant for the various isotopes. It follows from (14.2) that the electronic specific heat constant of the normal metal, γ, is thus independent of lattice mass. The fact that H_0 is proportional to $M^{-\frac{1}{2}}$ means that the internal energy difference between the normal and superconductive phases at absolute zero is inversely proportional to the mass.

LOCK, PIPPARD and SHOENBERG[6] determined the threshold field curves of isotopes of tin. The range of masses available in this element is quite large,

[1] See BARDEEN, Sect. 5.
[2] E. MAXWELL: Phys. Rev. **78**, 477 (1950).
[3] C. A. REYNOLDS, B. SERIN, W. H. WRIGHT and L. B. NESBITT: Phys. Rev. **78**, 487 (1950).
[4] H. KAMERLINGH ONNES and W. TUYN: Leiden Comm. **1922**, 160b. — E. JUSTI: Phys. Z. **42**, 325 (1941).
[5] B. SERIN, C. A. REYNOLDS and L. B. NESBITT: Phys. Rev. **80**, 761 (1950).
[6] J. M. LOCK, A. B. PIPPARD and D. SHOENBERG: Proc. Cambridge Phil. Soc. **47**, 811 (1951).
[7] E. MAXWELL: Nat. Bur. Stand. Circular **1952**, 519.
[8] M. OLSEN-BÄR: Nature, Lond. **168**, 245 (1951).
[9] C. A. REYNOLDS, B. SERIN and L. B. NESBITT: Phys. Rev. **84**, 691 (1951).

resulting in relatively very large isotopic effects, so that it is possible to arrive at extremely accurate conclusions concerning the thermodynamic parameters. As we mentioned in the previous section, the threshold field curve of tin is *not* parabolic. Despite this fact, the ratio H_0/T_c is a constant for the isotopes. Furthermore, the reduced field $h = H_c/H_0$ is the same function of the reduced temperature, t, for all the masses. These two facts constitute a quite general proof that γ is independent of isotopic mass, as can readily be seen from an examination of (14.4). From their data, Lock *et al.* estimated that the difference in γ between $^{116.2}$Sn and $^{123.8}$Sn is such that $\Delta\gamma/\gamma < 1.4 \times 10^{-3}$. They also found that the relation between transition temperature and mass is $M^{0.46} T_c = $ const. in this element. These results were confirmed by Serin, Reynolds and Loh-

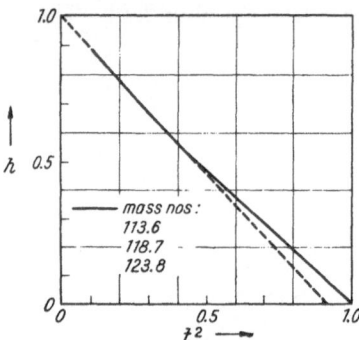

mass nos.:
113.6
118.7
123.8

Fig. 22. Isotope effect in tin (after Serin et al.[1]).

man[1] and by Maxwell[2]. The threshold field data of Serin *et al.* for the tin isotopes are shown in Fig. 22. The universality of the relationship between reduced field and reduced temperature is evident, as is the deviation from parabolic behavior.

The isotope effect is of great significance because it is one of the few experimental findings that gives a direct clue as to the microscopic mechanism which results in the extraordinary superconductive phase in metals. On the basis of this effect, we must conclude that superconductivity comes about primarily as the result of some strong interaction between the electrons and the lattice waves of a metal[3].

γ) *Relationship to the specific heat.* Since the magnetic threshold curves of several superconductors are not parabolic, we expect that their specific heats deviate from a strict T^3-dependence. For example, the deviation of the specific heat calculated from the magnetic threshold data of tin is shown in Fig. 18. The field measurements can give no reliable information about the specific heat at the lowest temperatures, but the deviations shown for $t > 0.3$ cannot be ignored. In fact, when the precise specific heat data for tin and vanadium which have been just recently obtained (see Sect. 12) are used to calculate the threshold fields, the calculated fields are in complete agreement with the magnetic measurements[4, 5]. Thus, we see that careful magnetic measurements can reliably provide insight into some of the finest details of the properties of superconductors.

The material of this section has recently been reviewed by Serin[6] in somewhat greater detail.

15. Mechanical effects. Sizoo and Onnes[7] first observed that the transition temperature of tin is lowered by pressure and increased by tension. The effect of pressure is very small—about $(-5) \times 10^{-5}$ °K/atm. in tin. Probably because of the inherent difficulty of these measurements, this field languished for many

[1] B. Serin, C. A. Reynolds and C. Lohman: Phys. Rev. **86**, 162 (1952).
[2] E. Maxwell: Phys. Rev. **86**, 235 (1952).
[3] See Bardeen, Chapter V.
[4] C. B. Satterthwaite, W. S. Corak, B. B. Goodman and A. Wexler: Phys. Rev. **99**, 1660 (1954).
[5] W. S. Corak and C. B. Satterthwaite: Phys. Rev. **99**, 1660 (1954).
[6] B. Serin [*3*], Chapter VII.
[7] G. J. Sizoo and H. Kamerlingh Onnes: Leiden Comm. **1925**, 180b.

years until KAN, SUDOVSTOV and LASEREV[1] performed an ingenious experiment. Their specimens were placed in a closed container of water, which was then frozen. Pressures of about 2000 atm. were produced as the result of the large expansion which occurs when water solidifies. The measurements showed that an effect of pressure on T_c of about the same magnitude and sign as in tin occurs in indium and mercury. Thallium seems to be an exception, since the effect has about the same magnitude but opposite sign at low pressure. Recent work by CHESTER and JONES[2] at pressures near 13 000 atm. showed that at these higher pressures, the coefficient of thallium, while small, does seeem to have the customary neg-ative sign. They also measured tin at pressures of 11 000 to 18 000 atm., and obtained results which agree with the earlier measurements. In addition CHESTER and JONES made the remarkable discovery that bismuth is a superconductor at pressures above 20 000 atm. with a transition temperature of about $7° K$. This field has been reviewed recently by SQUIRE[3].

The change in transition temperature with pressure is accompanied by a corresponding change of threshold field with pressure at temperatures below T_c. In tin, near the transition temperature, $\partial H_c/\partial p \sim (-7.5) \times 10^{-3}$ Oe/atm. The effect of relatively small one-dimensional compression and tension on the thresh-old fields of tin has been carefully investigated by GRENIER, SPONDLIN and SQUIRE[4]. They find that the effect of stress is to change the threshold field at $0° K$, H_0, in the same proportion as the change in T_c. As a result, the magnetic threshold curves of a specimen under various stresses are all parallel to the zero stress curve. Recently, MUENCH and RORSCHACH[5] measured $(\partial H_c/\partial p)$ for tin as a function of pressure (~ 1900 atm.) and found H_0/T_c to be independent of stress in this case also.

Once it is established that the magnetic threshold field depends on pressure, eq. (13.6) connecting the free energy difference to the threshold field may be manipulated according to familiar principles of thermodynamics to derive rela-tions between several of the properties of the two phases[6]. In particular, it can be shown that the volume of the superconductive phase should be larger than the normal phase at all temperatures below the transition point. The difference in volume rises from zero at T_c (since the transition is second order) to a fractional change of a few parts in 10^{-7} at $0° K$. We also expect the compressibility to be smaller in the superconductive phase than in the normal, the relative change being about 10^{-5} at all temperatures. Furthermore, the difference between the thermal expansion coefficient of the normal and superconductive phases should be about 10^{-7} per $°K$. This difference about equals the coefficient in the normal phase, but a coefficient of such magnitude is difficult to measure.

Several of the small effects associated with the change in threshold field with pressure have been observed in recent years. LASAREV and SUDOVSTOV[7] measured the difference in volume of the two phases by observing the change in curvature of a bimetallic strip made of tin and non-superconducting brass when a magnetic field exceeding the threshold value was applied. The observed volume changes and the measured values of H_c and $(\partial H_c/\partial p)$ are in good agreement with the relationship between these quantities given by thermodynamics.

[1] L. S. KAN, A. I. SUDOVSTOV and B. G. LASEREV: J. Exp. Theor. Phys. USSR. **18**, 825 (1948).
[2] P. F. CHESTER and G. O. JONES: Phil. Mag. **44**, 1281 (1953).
[3] C. F. SQUIRE [3], Chapter VIII.
[4] C. GRENIER, R. SPONDLIN and C. F. SQUIRE: Physica, Haag **19**, 833 (1953).
[5] N. L. MUENCH and H. E. RORSCHACH jr.: Phys. Rev. **99**, 668 (1955), Abstract Y 6.
[6] See BARDEEN, Sect. 3.
[7] See SHOENBERG [1], pp. 76—77.

Landauer[1] measured the elastic moduli of tin by determining the resonant frequency ($\sim$50 kc/sec.) of a composite oscillator made up of a tin specimen cemented to a quartz crystal oscillator. A difference in moduli between the superconductive and normal states was observed which indicated that the velocity of sound is smaller in the superconductive than in the normal phase. At $3.7°$ K the relative change in Young's modulus is 4×10^{-6} and in the shear modulus, 6×10^{-6}. The changes increase with decreasing temperature; and, in particular, the shear modulus change was found to increase five-fold between 3.7 and 3° K. Thermodynamic theory does not treat these moduli, so that the measurements cannot be compared directly with the theory. Welber[2], using the same technique, confirmed this change in tin, and reported a relative change in Young's modulus of lead of about 1×10^{-4}. The shear modulus of tin has been measured statically by Grassmann and Olsen[3]. In this ingenious experiment a fine tin wire was used as a torsion fiber, but while the torsion constant was found to increase in going from the superconductive to the normal state, the magnitude of the effect was found to be much smaller than that observed by Landauer. Grassmann and Olsen report that the change in shear modulus vanishes at T_c and increases to only about 3.5×10^{-6} at 0° K. This disagreement is at present unresolved.

Recently, Bömmel[4] measured the attenuation of ultrasonic pulses in normal and superconductive lead. The attenuation for frequencies of 9 to 27 Mc/sec. was found to decrease sharply when the specimen passed from the normal to the superconductive phase. The change in attenuation increased rapidly with frequency. A similar effect was found in tin by Mackinnan[5]. This effect is not related to the changes in elastic properties which we have already discussed, but is probably connected with the rapid decrease of the number of "normal" electrons in the superconductive phase as the temperature is reduced below T_c (see Sect. 18).

The only attempt to measure the difference in the coefficient of thermal expansion between the superconductive and normal phases was made by McLennan, Allen and Wilhelm[6]. They reported that they could not detect a discontinuity in the coefficients of lead and of Rose's metal near their transition points. A reexamination of the data by Westerfield[7] indicates that these do not definitely preclude the existence of discontinuities.

For completeness, we mention that plastic deformation has been found to cause greatly increased values of T_c, H_c and dH_c/dT in bulk superconductive specimens. These effects are probably related to the anamalous properties exhibited by very thin metal films evaporated on glass at 4° K. Such films also have higher critical temperatures and larger threshold field values than the bulk superconductor. Moreover, some bismuth films have been found to be superconducting. Annealing generally is found to reduce the anomalous behavior of films. We refer the reader to Squire[8] for detailed references.

[1] J. K. Landauer: Phys. Rev. 96, 296 (1954).

[2] B. Welber: Phys. Rev. 98, 1196 (1955), Abstract W7.

[3] P. Grassmann and J. L. Olsen: Helv. phys. Acta 28, 24 (1955).

[4] H. E. Bömmel: Phys. Rev. 96, 220 (1954).

[5] L. Mackinnan: Phys. Rev. 98, 1181 (1955), Abstract QA9.

[6] J. C. McLennan, J. F. Allen, J. O. Wilhelm: Trans. Roy. Soc. Canada (3) 25 (III), (1931).

[7] E. C. Westerfield: Phys. Rev. 55, 319 (1939).

[8] C. F, Squire, [3], Chapter VIII,

IV. Penetration of a magnetic field into a superconductor.

In the treatment given in Chapter II of the properties of macroscopic super-conductors, it was necessary to distinguish scrupulously between total currents and what we called MEISSNER currents. The former are induced in multiply-connected bodies to maintain the total flux through the body constant, and the latter are the shielding surface currents which maintain the induction zero inside the superconductive material. This distinction is artificial, since it is clear that both currents have the same intrinsic nature. We adopted it so as to be able to treat problems using two limiting forms of MAXWELL's equations—the first being the limit of infinite conductivity and the second, of perfect diamagnetism. We repeatedly emphasized that these two formulations are distinct and cannot be made to overlap within MAXWELL's electrodynamics.

Shortly after the discovery of the MEISSNER effect, F. and H. LONDON[1] developed an electrodynamics of superconductivity which includes, within a single framework, the properties of both singly and multiply connected bodies. Speaking qualitatively, this formulation embodies both infinite conductivity and perfect diamagnetism. Actually, LONDON's formulation is unnecessarily complicated for the discussion of most properties of macroscopic superconductors, and the less elegant formulation of Chapter II is preferable because of its simplicity, but this in no way detracts from LONDON's accomplishment.

In addition to providing a unified description of the electromagnetic behavior of superconductors, LONDON's equations in their final form predict several observable properties not contained in the cruder formulations. The most outstanding of these is the penetration of the magnetic field into the surface of a superconductor to a distance of the order of 10^{-6} cm. This result agrees with our intuitive feeling that the induction cannot decrease discontinuously to zero on passing through such a surface. In addition, the theory predicts that super-conductors exhibit resistivity in high frequency alternating electric fields, and that thin films should have much larger threshold fields than the bulk specimens of the same metal. We discuss the first two of these predictions in this chapter, as well as the experiment which demonstrated that a static electric field does not penetrate into a superconductor. We defer the discussion of the properties of films until the next chapter. We shall see that these predictions of LONDON's theory have all been qualitatively verified, but it has become fairly clear in recent years that the theory is inadequate to give a quantitative description of superconductors.

The experiments performed in this field up to 1952 have been exhaustively described and analyzed by SHOENBERG[2]. As a result, we shall treat this topic with as much brevity as is possible without impairing comprehensibility, and discuss the more recent developments at greater length.

16. Measurements of penetration depth. α) *General remarks*. Measurements of the penetration of the magnetic field into a superconductor are generally expressed in terms of the length, λ, which occurs in LONDON's theory. This is the distance from the surface of a massive superconductor in which the field falls to $1/e$ of its value on the surface[3]. λ essentially determines the scale of the decay curve of the field and occurs in all solutions of LONDON's equations (including those for very small specimens) in some dimensionless combination with the space

---

[1] See F. LONDON [4] for a comprehensive account; see also BARDEEN, Chapter III.
[2] D. SHOENBERG [1], Chapter V and pp. 179—214.
[3] Cf. BARDEEN, Sect. 10.

variables. Shoenberg[1] has pointed out that treating λ as such a scale factor enables one to treat problems independently of the detailed solutions to the electromagnetic equations. Nevertheless, we shall use the solutions given by London's theory, but with the recognition that these solutions may be only a qualitative guide to the actual field penetration.

β) *Colloids.* The first experiments which clearly demonstrated magnetic field penetration effects were performed by Shoenberg[2] on colloids of mercury. The specimens were in the form of a large number of very small spheres suspended in chalk. The total magnetic moment, μ, of a specimen was measured as a function of the magnetic field at various temperatures. The results were expressed in terms μ/μ_∞, where μ_∞ is the moment of a spherical specimen containing the same total mass of mercury as the colloid.

We confine our discussion to the specimen having particles of radius about 2×10^{-6} cm., since its magnetization curve showed practically no supercooling, no hysteresis and no evidence of an intermediate state. The magnetic moment of a *single* sphere of radius a in an applied field H_a, is [Bardeen, eq. (11.14)]

$$- H_a a^5/30 \lambda^2.$$

Since the specimen is made up of a large number of spheres of various radii,

$$\mu = - H_a \sum_i a_i^5/30 \lambda^2.$$

The magnetic moment of the corresponding massive sphere is

$$\mu_\infty = - \frac{H_a}{4 \pi (1 - \frac{1}{3})} \sum_i \frac{4}{3} \pi a_i^3.$$

Thus,

$$\frac{\mu}{\mu_\infty} = \frac{1}{15 \lambda^2} \frac{\sum_i a_i^5}{\sum_i a_i^3} = \frac{1}{15} \frac{\bar{a}^2}{\lambda^2}, \tag{16.1}$$

where the definition of $\bar{a}^2$ is obvious from the equation.

The measured values of μ/μ_∞ as a function of temperature are shown in Fig. 23. It is clear that appreciable penetration has occurred since the maximum value of μ/μ_∞ is only 0.01. However, it is not possible to deduce λ, because $\bar{a}^2$ in (16.1) is not known, but it is possible to deduce $\lambda(T)/\lambda(T_0)$ from these experiments, where $\lambda(T_0)$ is the value of λ at some standard temperature. We discuss the form of the temperature variation of λ in Sect. 18. Also shown in Fig. 23 are the measured values of the threshold fields of the colloid specimen. The fields are clearly very much greater than for a bulk specimen; the interpretation of these observations is discussed in the next chapter.

It is important to note that the transition temperature of the mercury colloid deduced from Fig. 23 is 4.15° K, which is very close to the value, 4.17° K, of large specimens.

γ) *Thin wires.* Désirant and Shoenberg[3] measured the magnetic moments of composite specimens containing about 100 thin mercury wires in glass capillary tubes. The diameter of the wires was about 10^{-3} cm., and the wire size in any specimen was the same within about 10%. The wire radii are large compared

[1] D. Shoenberg [1], pp. 139—142.
[2] D. Shoenberg: Proc. Roy. Soc. Lond., Ser. A **175**, 49 (1940).
[3] M. Désirant and D. Shoenberg: Proc. Phys. Soc. Lond. **60**, 413 (1948).

to λ so that the effect of penetration is too small to be measured directly with any accuracy; rather, what was measured was the change in λ with temperature. For a wire of radius $a \gg \lambda$ the magnetic field penetration is equivalent in effect to removing a sheath of thickness λ from the surface of the specimen. As a result,

$$\frac{\mu}{\mu_\infty} = 1 - \frac{2\lambda}{a};$$

and

$$\frac{\mu(T) - \mu(T_0)}{\mu_\infty} = \frac{2[\lambda(T) - \lambda(T_0)]}{a}. \qquad (16.2)$$

We see from (16.2) that measurements μ/μ_∞ as a function of T determine $[\lambda(T) - \lambda(T_0)]$ in this experiment. Combining these measurements with the colloid measurements, which determine $\lambda(T)/\lambda(T_0)$, it is possible to deduce the absolute value of $\lambda(T_0)$. For mercury, this method yielded $\lambda_0 = \lambda(0°\,\mathrm{K}) = 7.6 \times 10^{-6}$ cm. This value is now regarded as being nearly twice the true one, probably because the wires in the composite specimen had a slight spread of transition temperatures.

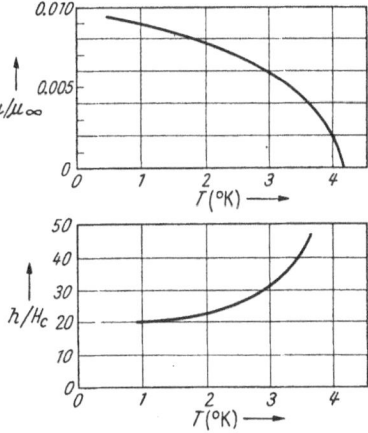

δ) *Thin films.* Lock[1] measured the magnetization in a longitudinal field of thin films of tin, indium and lead. The films were deposited on thin sheets of mica by evaporation in vacuum. Their thickness was about 10^{-5} cm., and composite specimens consisting of a few hundred mica sheets were used. The magnetic moment per unit area of film is [Bardeen, eq. (10.7)]

$$\mu = \frac{-2aH_a}{4\pi}[1 - (\lambda/a)\,\mathrm{Tan}\,a/\lambda],$$

where $2a$ is the film thickness, so that

$$\frac{\mu}{\mu_\infty} = 1 - (\lambda/a)\,\mathrm{Tan}\,a/\lambda. \qquad (16.3)$$

Taking,

$$\lambda = \lambda_0(1 - t^4)^{-\frac{1}{2}}, \qquad (16.4)$$

Fig. 23. μ/μ_∞ and h/H_c as functions of temperature for a mercury colloid (after Shoenberg).

Lock found that the data for a number of film thicknesses taken at several temperatures could be fitted very well to the expression (16.3) by appropriately choosing λ_0. The values of λ_0 obtained in this way are given in Sect. 18. The range of film sizes was too small, however, uniquely to establish (16.3) as the correct law of penetration; other penetration laws can be found which fit the data equally well.

ε) *Large cylinders.* 1. Low frequency methods. Due to the change in penetration depth with temperature, the inductance of a coil with a superconductive core should change very slightly with temperature. Casimir[2] first tried to observe this change, but because of a technical flaw in experimental design, his measurements gave a negative result. The experiment was repeated recently by Laurmann and Shoenberg[3] at a frequency of 70 c/sec. with cylindrical specimens of tin and mercury of about 1 cm. diameter. They managed to overcome the many technical difficulties inherent in this experiment and obtained

[1] J. M. Lock: Proc. Roy. Soc. Lond., Ser. A **208**, 391 (1951).
[2] H. G. B. Casimir: Physica, Haag **7**, 887 (1940). — Leiden Comm. 261c.
[3] E. Laurmann and D. Shoenberg: Proc. Roy. Soc. Lond., Ser. A **198**, 560 (1949).

results for the temperature variation of $[\lambda(T) - \lambda(T_0)]$ and for λ_0 in good agreement with those obtained by the other methods.

Shalnikov and Sharvin[1] performed an interesting variant of this experiment, in which the temperature of a tin specimen in a constant magnetic field was varied at 4 c/sec. Due to the change in field penetration with temperature, an alternating emf was induced in a coil surrounding the specimen. The values of λ near T_c deduced from this investigation are several times greater than those found by Laurmann and Shoenberg. The excessively large values are probably attributable to the poor surface condition of the specimen.

2. High frequency reactance. As part of this extensive series of measurements of the surface impedance of metals at microwave frequencies, Pippard[2] determined the change of λ with temperature. The simplest observations to interpret are those dealing with the change in resonant frequency of a system on passing from the superconductive to the normal phase. At sufficiently low temperatures, the frequency change is proportional to

$$\delta - \lambda,$$

where δ is the electromagnetic skin depth of the normal phase. In tin, δ is independent of temperature, so that once the constant of proportionality is determined, the observations may be used to evaluate $[\lambda(T) - \lambda(T_0)]$. The results obtained in this way agree well with those given by the other methods.

17. Non-penetration of the static electric field. The theory, as originally formulated, left to experiment the task of answering the question of whether the electric field penetrates to a depth λ into a superconductor or terminates on surface charges. The answer was obtained by H. London[3], who looked for a small change in the capacity of a condenser when its plates became superconducting. The plates were made of mercury separated by a thin mica sheet. Although there were several technical difficulties, the expected effect if penetration occurred was four times the error of measurement. No change in capacity was detected, so that the present form of the theory was adopted in which no static electric fields can exist inside a superconductor.

18. Temperature dependence of λ. In Sect. 16δ we remarked that the data of Lock can be fitted very well to theory, if it is assumed that the temperature variation of the penetration depth is given by

$$\lambda = \frac{\lambda_0}{(1 - t^4)^{\frac{1}{2}}}. \qquad (18.1)$$

The results of all the other investigations discussed in Sect. 16 are also in accord with (18.1), within the limits of experimental error, so that it is now generally accepted that to a first approximation, λ varies with temperature in this way. The values of λ_0 for several metals are given in Table 2.

Table 2. *Values of the penetration depth at 0° K.*

Metal	λ_0 (cm)	Reference (Footnote of p. 243)
In	6.4×10^{-6}	1
Sn	5.0	1
Hg	3.8 (∥); 4.5 (⊥)	3
Pb	3.9	1

Comparing (18.1) with (14.6) which gives the temperature dependence of the order parameter, ω, in Gorter's two-fluid model, we see that

$$\left(\frac{\lambda_0}{\lambda}\right)^2 = \omega. \qquad (18.2)$$

[1] A. I. Shalnikov and Yu. V. Sharvin: J. Theor. Exp. Phys. USSR. **18**, 102 (1948).
[2] A. B. Pippard: Proc. Roy. Soc. Lond., Ser. A **191**, 385 (1947).
[3] H. London: Proc. Roy. Soc. Lond., Ser. A **155**, 102 (1936).

Furthermore, LONDON's theory gives [cf. BARDEEN, eqs. (7.13) and (10.3)]

$$\lambda^2 = \left(\frac{m \, c^2}{4 \, \pi \, n_s \, e^2} \right), \qquad\qquad (18.3)$$

where m is the mass, and n_s is the density of superconducting electrons. Adopting the convention of the two-fluid model in which all the electrons are "superconducting" at $0°$ K, we see that (18.2) and (18.3) are in accord with the view that ω is the fraction of "superconducting" electrons at any temperature, t. Conversely, $(1 - \omega)$ is the fraction of "normal" electrons at any t.

The quantity, m, in (18.3) must be taken to be the "effective" mass of the electrons. From the measurements of λ_0, it is thus possible to use (18.3) to evaluate m/n_{s0}, where $n_{s0} = n_s(0°$ K$)$. The ratio, $\left[\dfrac{m}{n_{s0}} \Big/ \dfrac{m_e}{n_e} \right]$ (where m_e is the normal electron mass and n_e, the electron density determined from the valence) is about 0.3 for indium, tin and mercury and 0.7 for lead. From the foregoing very naive standpoint, these results may be taken to mean that only a small fraction of the electrons became "superconducting" or that the effective mass is large.

In the previous chapter we noted that the data for the temperature variation of both the threshold field and the specific heat of superconductors are not in complete agreement with the GORTER model. On the other hand, the determinations of the temperature variation of the penetration depth are in apparent agreement with the model. This discrepancy can probably be explained away on the basis that penetration depth measurements are extremely difficult and the precision of measurement required to reveal small deviations from t^4-behavior was not attained.

19. Magnetic field dependence of λ. PIPPARD[1] measured the variation of the penetration depth in tin with magnetic field. The experimental results are shown in Fig. 24; $\lambda(H_c)$ is the value of λ at the threshold field and $\lambda(0)$ is the value in no field. The remarkable fact to emerge from this investigation is the very small change in λ (at most a few percent) produced by fields near the threshold value.

PIPPARD points out that if the thermodynamic effect of the field were confined to the penetration depth at the surface, the change in entropy density produced by the field in this small layer is enormous. (For example, near the criti-

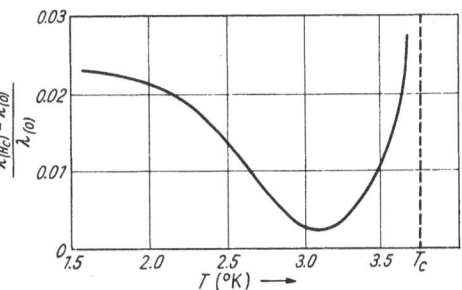

Fig. 24. Magnetic field dependence of the penetration depth (after PIPPARD[1]).

cal temperature, the change in entropy density in the layer at the threshold field would be about one quarter of the difference between the entropy densities of the normal and superconductive phases.) Such a situation seems quite unrealistic. PIPPARD therefore concludes from the almost negligible change in λ that the entropy change must be distributed in a layer of thickness a which is considerably larger than the penetration depth, thereby resulting in a much smaller change in entropy density.

According to the two-fluid model, the increased entropy density is associated with an increase in the fraction of normal electrons in this layer. PIPPARD therefore considers a simple model in which the order parameter in the surface layer, a,

[1] A. B. PIPPARD: Proc. Roy. Soc. Lond., Ser. A **203**, 210 (1950).

is constant, but depends on the field at the surface as well as the temperature. As a result, the value of ω in the layer differs from its value in the remainder of the specimen. It is assumed that in the bulk of the specimen the order parameter has its usual equilibrium value.

Minimization of the free energy calculated from the foregoing model determines ω in the layer, and this result combined with (18.2) serves to determine the dependence of λ on field. It is found that a must be about $20\lambda_0$ (or about 10^{-4} cm.) to obtain qualitative agreement between the predictions of the model and the measurements. Bardeen (Sect. 30) gives a detailed discussion of this experiment.

Implicit in the foregoing approach is the assumption that there is a long range order in the superconductor which prohibits the parameter, ω, from changing rapidly in distances small compared with 10^{-4} cm. Speaking qualitatively, this means that the density of superconducting electrons (and therefore the associated wave-functions of the superconductive state) also must be slowly varying within the same distance (cf. Bardeen, Sect. 6). It should be remarked that the foregoing range of order must be a maximum occurring in bulk materials rather than a minimum necessary for establishing the ordered superconductive phase. This observation is the result of the realization that thin films and colloids having dimensions at least two orders of magnitude smaller than 10^{-4} cm. do exhibit superconducting properties and have the same transition temperatures as large specimens.

20. Dependence of λ on purity. The variation of penetration depth in tin with the mean free path for an electron in the normal phase as determined by Pippard[1], is shown in Fig. 25. The free path was decreased by alloying indium with the tin. A small amount of impurity rapidly decreased the normal electrical conductivity of tin and the maximum indium concentration used in these measurements was 3%. As can be seen from Fig. 25, this small amount of indium increases the penetration depth in tin by more than a factor two. The thermodynamic parameters of the impure specimens, however, differ only slightly from those of pure tin. For example, the transition temperature of the 3% indium specimen was 3.63° K, whereas pure specimens have $T_c = 3.73$° K. There are corresponding small changes in the threshold field values.

Pippard concludes that these observations of large changes in the penetration depth unaccompanied by correspondingly large changes in the thermodynamic parameters constitute a contradiction to London's theory. According to the theory [as can be seen from (18.3)] λ depends only on m/n_s, i.e. the effective mass and density of the superconducting electrons. Since the latter are very little affected by impurity, as is evidenced by the small change in the thermodynamic parameters, the observed large changes in λ with impurity are unexplained. Pippard therefore developed his non-local description of the relationship between the magnetic field and current in superconductors which is discussed in great detail by Bardeen (Sects. 16 to 26). According to this theory, the relationship between current density and field is not a point relation, but an integral relation involving the field in a region of linear dimensions about equal to the range of order surrounding the point. The effect of impurity scattering is to decrease the range of order. The data are in good agreement with the theory.

The foregoing experiments also served to clear up somewhat the mystery of the observed anisotropy of λ in tin[2]. The dependence of the penetration

---

[1] A. B. Pippard: Proc. Roy. Soc. Lond., Ser. A **216**, 547 (1953).
[2] A. B. Pippard: Proc. Roy. Soc. Lond., Ser. A **203**, 98 (1950).

depth on the angle between the current and the tetragonal axis is shown in curve (b) of Fig.26. SHOENBERG[1] has shown that if LONDON's equations are generalized to describe anisotropic bodies, the predicted anisotropy in λ is contradicted by the observed anisotropy in tin. However, on the basis of the non-local theory, the anisotropy in λ in pure specimens is a reflection of the anisotropy in the FERMI surface for the electrons of the metal[2], rather than a direct reflection of crystalline anisotropy. By contrast, as can be seen from curve (a) in Fig.26,
in impure specimens the observed anisotropy is very small. This result is also a qualitative consequence of the decrease in the range of order by impurity.

This set of experiments seems to indicate clearly that LONDON's equations must be modified. However, they will undoubtedly continue to serve as a useful qualitative guide to penetration effects.

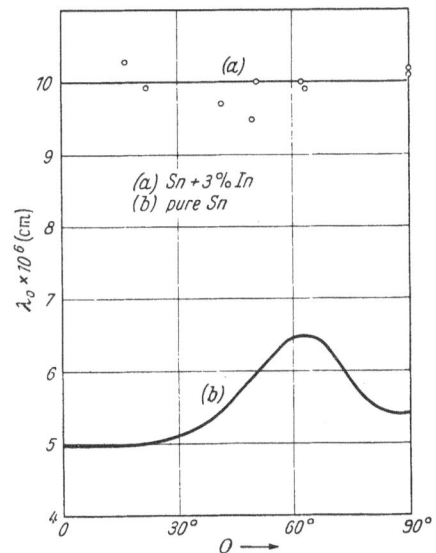

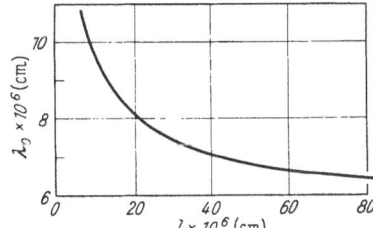

Fig. 25. Variation of λ_0 with mean free path, l (after PIPPARD).

Fig. 26. Variation of λ_0 with the angle, ϑ, between the current and the tetragonal axis of pure tin, and tin + 3% indium (after PIPPARD [2]).

21. High frequency resistance of superconductors.

According to LONDON's equations, a changing electric field can exist in a superconductor, and the metal then exhibits resistance. It can readily be shown[3] that, in the case of alternating fields, the normal current density, j_n, is related to the superconducting current density, j_s, by

$$\frac{|j_n|}{|j_s|} = \left(\frac{\lambda}{\delta}\right)^2. \tag{21.1}$$

Here, δ is the skin depth as it is usually defined for normal metals—

$$\delta = c/(4\pi^2 f \sigma)^{\frac{1}{2}},$$

where f is the frequency and σ is the conductivity. We shall see that the conductivity is a rather complicated concept in this theory, but for qualitative considerations, the normal value may be used.

Eq. (21.1) forms a basis for understanding readily the two limiting cases of very low and very high frequencies. At low frequencies, $\delta \gg \lambda$, so that the total current is the superconducting current, and the familiar description is obtained in which the electric field is zero and the magnetic field penetrates to a depth, λ. At very high frequencies, $\delta \ll \lambda$, so that the normal current completely dominates

[1] D. SHOENBERG [1], pp. 189—191.
[2] A. B. PIPPARD: Proc. Roy. Soc. Lond., Ser. A 224, 273 (1954).
[3] F. LONDON [4], pp. 30—32.

the situation, and we expect no difference in properties between the superconductive and normal phases. This result agrees with the observed identical optical properties of normal and superconductive metals (see Chapter VIII). We also see from (21.1), that when $\delta \sim \lambda$ the two currents have about equal magnitudes, and we expect some intermediate type of behavior. For tin this last condition obtains at frequencies of about 10^3 Mc/sec.

The first measurements were made by H. LONDON[1] who observed the rate at which heat was developed in superconductive tin by fields of frequency 1500 Mc/sec. It was demonstrated clearly that considerable resistance was present below T_c. More recently, PIPPARD[2, 3] has measured the surface impedance of tin and mercury at 1200 Mc/sec. and of tin at 9400 Mc/sec. FAWCETT[4] also

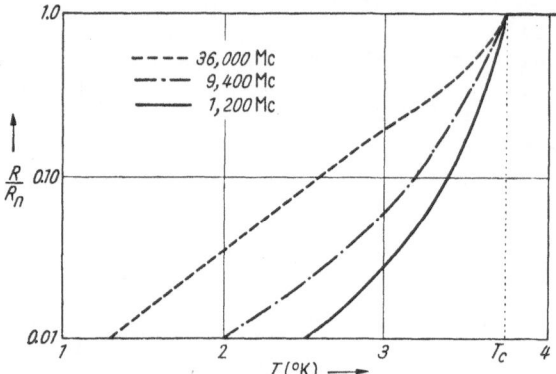

Fig. 27. Dependence of the surface resistance of tin on temperature, at various frequencies (after PIPPARD[2, 3] and FAWCETT[4]).

measured the surface resistance of tin at 36000 Mc/sec. The results of these observations on tin are shown schematically in Fig. 27. It is to be noted that the transition temperature is independent of frequency. Furthermore, the resistance decreases very rapidly with temperature, and it appears that even at the highest frequency there would be no resistance at $0°$ K. We also mention that, at the higher frequencies, the surface resistance exhibits the same type of anisotropy as the penetration depth.

The detailed interpretation[5] of the resistivity measurements is very difficult for two reasons. The first is that only the "normal" electrons contribute to the conductivity in the superconductive state, so that one must use a two-fluid model to calculate δ. The second source of difficulty is that the theory of the conductivity of even the normal phase is most complicated because the mean free path of the electrons in the normal metal is very large compared to the skin depth. As a result, the rather complicated theory of the anomalous skin effect[6] must be used to describe the normal conductivity. Thus, the problem becomes one of grafting a two-fluid model on to an already complicated theory of conductivity. At best, what has been achieved is qualitative agreement between theory and experiment. In particular, we mention that the observed dependence of the surface resistance on frequency does not agree with the theory (cf. BARDEEN, Sect. 34).

We remarked that at the highest microwave frequency so far used for measurement, 4×10^4 Mc/sec., it appears that the surface resistance of tin is vanishingly small at $0°$ K. On the other hand, at optical frequencies greater than 10^7 Mc/sec. (wavelengths of about $20\,\mu$), the resistance of superconductors is a constant, independent of the temperature and equal to the resistance of the normal metal. It is not clear at present how the transition from the first type of behavior to

[1] H. LONDON: Proc. Roy. Soc. Lond., Ser. A **176**, 522 (1940).
[2] A. B. PIPPARD: Proc. Roy. Soc. Lond., Ser. A **191**, 370 (1951).
[3] A. B. PIPPARD: Proc. Roy. Soc. Lond., Ser. A **203**, 98 (1950).
[4] E. FAWCETT: Proc. Phys. Soc. Lond. A **66**, 1071 (1953).
[5] In addition to the original papers, see D. SHOENBERG [1], pp. 197—206.
[6] G. E. H. REUTER and E. H. SONDHEIMER: Proc. Roy. Soc. Lond., Ser. A **195**, 336 (1948).

the second will take place as the frequency is increased above 10^4 Mc/sec. It is possible that as frequency is increased the surface resistance at all temperatures will continue to increase until it becomes obvious that the resistance is finite at $0°$ K (cf. Fig. 27). Or, it is possible that at a well defined threshold frequency, a constant surface resistance will appear. Further experiments are clearly necessary.

22. HALL effect. ONNES and HOF[1] first attempted to observe a HALL e.m.f. in superconductive tin and lead. The experiment was inconclusive because they could just barely detect the effect in the normal metal in very high magnetic fields. More recently, LEWIS[2] showed that the HALL voltage in superconductive vanadium was certainly less than $\frac{1}{8}$ of the voltage generated in the normal metal. As discussed by BARDEEN (Sect. 9) there is an excellent theoretical basis for believing that no HALL voltage can be generated in superconductors.

V. Phenomena associated with the surface energy between the superconductive and normal phases.

In the discussion of the intermediate state presented in Chapter II, we postulated that this state consisted of a uniform mixture of superconductive and normal regions. While the relative concentrations of the two phases were specified, nothing was said about the absolute sizes of the regions. Shortly after the nature of the intermediate state became clear, LANDAU[3] developed a detailed theory of this state which predicted the sizes of the individual superconductive and normal regions. Essential to this theory is the existence of an additional free energy at every interface separating the two phases, which we call a positive surface energy. LONDON[4] has shown that a positive surface energy is also necessary to insure that macroscopic specimens exhibit the MEISSNER effect. It may be shown that in the absence of a surface energy (or for a negative surface energy) the state of lowest magnetic free energy of a superconductive specimen in any field, however small, is one in which the specimen is subdivided into an infinitely fine mixture of superconductive and normal layers. There clearly would be no MEISSNER effect under these circumstances. Since perfect diamagnetism is an essential property of a superconductor, we must assume that a positive interphase surface energy exists. This assumption precludes the proliferation of normal and superconductive layers because a very large total surface free energy becomes associated with this process. As a result, the state of the specimen exhibiting the MEISSNER effect is energetically favored over the state in which it is subdivided into layers.

In this chapter, in addition to discussing the experiments which reveal the structure of the intermediate state, we present the findings concerning the phenomena of supercooling and superheating. We also discuss the problems associated with the propagation of phase boundaries in superconductive metals, and, finally, consider the properties of thin films. All of these phenomena are connected in some way with surface energy.

23. The characteristic length, $\varDelta$. To determine the detailed form of the intermediate state, one must find that structure of normal and superconductive layers for which the total free energy is an absolute minimum. The calculation

[1] H. KAMERLINGH ONNES and K. HOF: Leiden Comm. **1914**, 142b.
[2] H. W. LEWIS: Phys. Rev. **92**, 1149 (1953).
[3] L. D. LANDAU: Phys. Z. Sowjet. **11**, 129 (1937).
[4] F. LONDON [4], pp. 125—129. See also BARDEEN, Sect. 27.

reduces to finding the best compromise between the volume free energy decrease which can be brought about by increasing the number of layers and the accompanying energy increase resulting from the increased number of interphase boundaries. The free energy difference between the normal and superconductive phases is $H_c^2/8\pi$ per unit volume. If the surface energy per unit area is α, we see that the theory must involve a characteristic length, Δ, given by

$$\Delta = \frac{8\pi\alpha}{H_c^2}. \tag{23.1}$$

The same length is important also for the description of the other phenomena we consider in this chapter, and most measurements reduce to a determination of Δ.

In recent years, several phenomenological theories of the interphase boundary energy have been developed. They are all based on the two-fluid model of superconductors. The surface energy at a superconductive-normal interface is associated with the gradual change of the order parameter, ω, from zero in the normal phase to the appropriate temperature dependent equilibrium value in the superconductive phase. We refer the reader to BARDEEN (Sects. 28 and 29) for a detailed discussion of the theories.

24. Intermediate state. α) *Observation of structure.* LANDAU calculated the intermediate state structure for an infinite plate of thickness L, with its surface normal to the applied magnetic field. The domain structure was supposed to consist of alternating laminae of superconductive and normal phase running parallel to the field. According to the theory, the total thickness, a, of two neighboring laminae is a slowly varying function of the field. For calculating orders of magnitude, we take

$$a \sim 10\sqrt{L\Delta}. \tag{24.1}$$

The exact expression for a is given by BARDEEN (Sect. 32). The period of the structure, a, is the sum of the thickness, a_n, of a normal layer and the thickness, a_s, of a superconductive layer. As the field is increased, a_n increases at the expense of the superconductive regions until, at the threshold field, all the latter have vanished. LANDAU[1] later formulated a variant of this theory in which it was proposed that as the normal layers approached the surface of the specimen they began to break up into ever smaller branches, so that the surface consisted of an infinitely fine mixture of superconductive and normal metal. Experiment has shown that the earlier, unbranched model is correct, so that we shall have no more to say about the branching model.

The first evidence that the laminar structure exists was obtained by SHAL-NIKOV[2], who employed a bismuth wire to measure the magnetic field in a narrow gap between two hemispheres. Use was made of the fact that the resistance of bismuth is approximately proportional to the square of the field. We remind the reader that in a specimen in the intermediate state, the average field is B. But, according to the laminar model, the field is the threshold value, H_c, in the normal layers and zero in the superconductive layers, with the normal regions occupying a fraction, B/H_c, of the specimen. Because of the non-linear dependence of resistance on field, it is possible to tell whether the field in the gap is everywhere B or oscillates between zero and H_c. In the former case the resistance of the bismuth wire would be proportional to B^2, whereas in the latter it would be proportional to BH_c. The two cases are distinguishable, because in the

[1] L. D. LANDAU: J. Phys. USSR. **7**, 99 (1943).
[2] A. I. SHALNIKOV: J. Phys. USSR. **9**, 202 (1945).

intermediate state $BH_c > B^2$. SHALNIKOV found that the rms field seen by the wire was indeed greater than B, and thereby substantiated the discontinuous structure of the intermediate state.

The first direct observations of the laminar structure were made by MESKOVSKY and SHALNIKOV[1]. For these beautiful experiments, bismuth magnetic field probes were made which were small enough to resolve the discontinuous changes in field which occurred when the probe was moved across the laminae. A typical probe was $5 \times 10\,\mu$ in cross section and 0.15 mm. long! The field distribution in a gap between two tin hemispheres and on the surface of a tin sphere were investigated. Discontinuous field changes in both the gap and on the surface were observed, conclusively establishing the existence of a non-uniform structure. The form of normal and superconductive regions was shown to be very complicated, and clearly does not approach the laminar model envisioned by LANDAU.

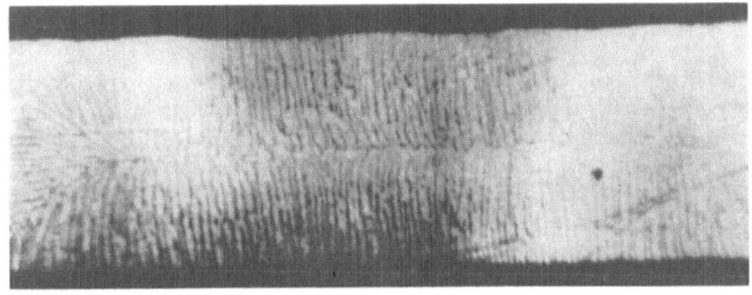

Fig. 28. Powder pattern on a tin plate, 1.3 cm.×4.5 cm.×0.32 cm. thick, in the intermediate state. $T = 1.93°$ K; $H_a/H_c = 0.82$ (Photograph, courtesy of Dr. A. L. SCHAWLOW.)

Nevertheless, for a sphere of 4 cm. diameter in a field $H_a/H_c \sim 0.7$, the period of the laminar structure is about 0.2 mm. If we take $a = 0.2$ mm. and $L = 4$ cm. in (24.1) to obtain a crude value for $\varDelta$, we find $\varDelta \sim 10^{-4}$ cm. The measurements also revealed that the structures in increasing and decreasing fields were different.

An even more elegant method for observing the structure of the intermediate state has been developed recently by SCHAWLOW et al.[2]. In this method, niobium powder is spread over the surface of a flat specimen of a superconductor having a lower transition temperature. Because of the high transition temperature of niobium, the powder remains superconductive when the specimen is in the intermediate state. Consequently, the powder is pushed away from the normal regions and congregates on the superconductive regions on the specimen surface. The resultant powder pattern is readily photographed.

Fig. 28 shows a pattern obtained on a tin plate 1.3 cm.×4.5 cm.×0.32 cm. thick. The domains clearly approximate closely the laminar shape, and moreover, their spacing varies with field in the manner predicted by LANDAU. The patterns can therefore be used to calculate $\varDelta$ as a function of temperature. SCHAWLOW[3] finds that $\varDelta$ decreases from 10×10^{-5} cm. at $3°$ K to 2.5×10^{-5} cm. at $1.25°$ K.

β) Resistance and magnetization measurements of small specimens. Fig. 16 of Chapter II shows how the resistance of a cylindrical tin wire (of 0.25 mm.

[1] A. G. MESHKOVSKY and A. I. SHALNIKOV: J. Phys. USSR. **11**, 1 (1947). — J. Exp. Theor. Phys. USSR. **17**, 851 (1947). See also D. SHOENBERG [I], pp. 104—110.
[2] A. L. SCHAWLOW, B. T. MATTHIAS, H. W. LEWIS and G. E. DEVLIN: Phys. Rev. **95**, 1344 (1954).
[3] A. L. SCHAWLOW: Int. Conf. Low Temp. Phys., Paris 1955.

diameter) varies with the strength of a transverse magnetic field. We remarked that resistance first appeared when the applied field was $0.58\,H_c$ rather than exactly $\frac{1}{2}\,H_c$. The intermediate state first appears at $\frac{1}{2}\,H_c$ in cylinders of large diameter in transverse fields, and we expect a trace of resistance to appear at the same field value.

This apparent anomaly has its origin in the small size of the wire and the laminar structure of the intermediate state. For large specimens, we have assumed that the intermediate state appears when the magnetic field reaches the threshold value anywhere on the surface. This assumption is valid only so long as the free energy associated with the internal interphase surfaces is negligible compared with the volume free energy of the specimen. From (24.1) it may be shown readily that the surface energy contribution will be negligible when $d > 100\,\Delta$, where d is the smaller dimension of the specimen. When the surface energy is not negligible, its effect is to delay the appearance of the intermediate state until the field reaches a larger value than is predicted when surface energy effects are ignored. Only in a larger field can the free energy of the superconductive state exceed the free energy of the intermediate state, with the result that the latter state is stable. Since $100\,\Delta \sim 10^{-2}$ cm., it is not surprising that the wire of Fig. 16 does not enter the intermediate state until $0.58\,H_c$. The particular value 0.58 is associated with the particular diameter of the wire, and, in fact, we expect the ratio $\varrho = H_a/H_c$, (where H_a is the applied field at which resistance first appears) to depend on the wire diameter.

Andrew[1] first investigated the dependence of ϱ, for small tin and mercury wires, on wire size and temperature. He found that ϱ was independent of temperature but that for tin wires in transverse fields, the ratio increased monotonically from $\varrho = 0.54$ for 0.105 cm. wire diameter to $\varrho = 0.67$ for 0.0027 cm. diameter, indicating that ϱ approaches 0.5 in very large specimens. For the larger specimens the effect of current was similar to that shown in Fig. 16. In increasing field, the curves were more concave for smaller measuring currents, as if in the limit of zero current, the transition in a transverse field would become identical with the longitudinal transition. For the smaller specimens, a limiting current curve was found, smaller currents resulting in no further concavity. For the $30\,\mu$ diameter wires, the curve was practically linear and there was scarcely any dependence on current. These results suggest that in large specimens the laminae are parallel to the cylinder axis for small measuring currents, and then turn normal to the axis as the current increases. For small diameter specimens, the laminae apparently are always normal to the axis, even for small currents.

The dependence of ϱ on wire diameter found by Andrew is in qualitative agreement with Landau's model[2].

Recently, Lutes and Maxwell[3] investigated the resistance transition of extremely small tin wires in transverse fields. They found that at sufficiently low temperatures, wires of 1.2×10^{-4} cm. diameter made a direct transition from the superconductive to the normal phase without passing through the intermediate state. The transition occurred at $1.69°$ K when ϱ was 0.67.

This observation probably can be explained on the basis that the free energy of the intermediate state of such small specimens still exceeds the free energy of the superconductive phase in fields approaching values for which the normal phase becomes stable. For a cylinder in a transverse field, the normal phase is stable relative to the superconductive phase when $\varrho > 1/\sqrt{2}$. The discrepancy

[1] E. R. Andrew: Proc. Roy. Soc. Lond., Ser. A **194**, 80 (1948).
[2] E. R. Andrew: Proc. Roy. Soc. Lond., Ser. A **194**, 98 (1948).
[3] O. S. Lutes and E. Maxwell: Phys. Rev. **97**, 1718 (1955).

between the observed value of 0.67 and this larger value is, at present, not understood.

Because of the delay in appearance of the intermediate state, the magnetization curves of small specimens deviate from the ideal macroscopic curves illustrated in Fig. 11. Fig. 29a shows a typical magnetization curve calculated for a small cylinder in a transverse field on the basis of a detailed model of the intermediate state[1]. DÉSIRANT and SHOENBERG[2] made a careful study of the magnetization curves of transverse cylinders of various radii, and qualitatively verified the existence of the unusual features shown in Fig. 29a. A typical measured curve is shown in Fig. 29b. The measurements reveal that in increasing fields, the magnetization drops abruptly with the appearance of the intermediate state at fields for which ϱ is appreciably greater than one half. Moreover, the magnetization vanishes at fields noticeably smaller than the threshold value, as determined from measurements in a longitudinal field. The difference between H_c and the field at which the magnetization disappears increases with decreasing specimen diameter. When the field is reduced from above the threshold value, the curve is retraced for the most part, except for a slight hysteresis which has two features. We note first that the "horn" of the magnetization curve $(a-b)$ is not retraced. Secondly, the return transition to the intermediate state $(c-d)$ occurs discontinuously at a field considerably less than the

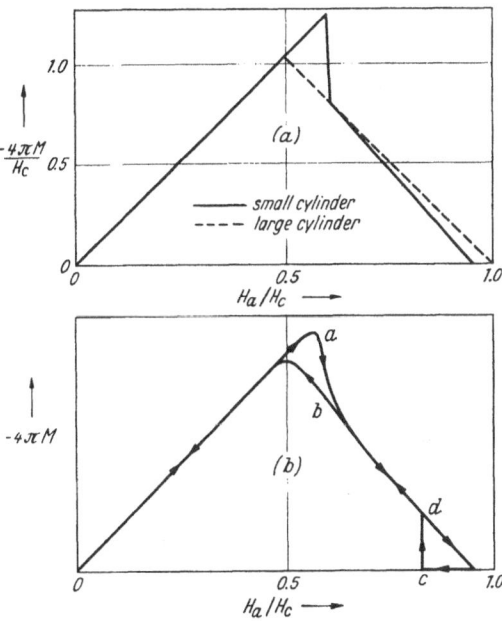

Fig. 29a and b. a) Magnetization curve of a small transverse cylinder as calculated by ANDREW[1], contrasted with the behavior of a large cylinder. b) Magnetization measurements of DÉSIRANT and SHOENBERG[2] on a 1.3×10^{-2} cm. diameter tin cylinder at 3.0° K.

value at which the magnetization vanishes. The latter feature is a consequence of the supercooling of the specimen; we discuss supercooling in the next section.

The foregoing resistance and magnetization data were analyzed in considerable detail in an attempt to obtain values of $\varDelta$. At best, only qualitative agreement was obtained between the data and the theoretical calculations which were based, for the most part, on the branching model. Since this model seems now to be out of date, and recent work (see Sect. 26) has provided more reliable information about $\varDelta$, we do not discuss these calculations here[3].

γ) SILSBEE *effect and paramagnetic effect.* A current,

$$i_c = \frac{H_c\, a}{2},\tag{24.2}$$

flowing in a long cylindrical wire of radius a, produces a magnetic field equal to the threshold value on the surface of the wire. Soon after the discovery of

---

[1] See footnote 2, p. 252.

[2] M. DÉSIRANT and D. SHOENBERG: Proc. Roy. Soc. Lond., Ser. A **194**, 63 (1948).

[3] See ANDREW [Proc. Roy. Soc. Lond., Ser. A **194**, 98 (1948)] and D. SHOENBERG [*1*], pp. 117—120.

the threshold field, SILSBEE suggested that the magnetic field of the current itself is responsible for restoring the resistance of superconductive wires carrying currents in excess of i_c (cf. Sect. 3). This case differs from the corresponding phenomenon which occurs in superconductive rings. In a ring, when the field at the surface tends to exceed the threshold value, the persistent current adjusts itself so that the field just equals H_c, and the specimen remains superconductive (cf. Sect. 5). However, when the current in a wire is maintained constant by an external source, resistance is restored for all currents greater than i_c. For currents exceeding this threshold value, the wire is in an intermediate state.

LONDON[1] derived a model for the intermediate state in the presence of a current which is illustrated in Fig. 30. According to the model, the resistance $\frac{R}{R_n}$ is zero for $i < i_c$ and jumps discontinuously to $R_n/2$ when $i = i_c$, where R_n is

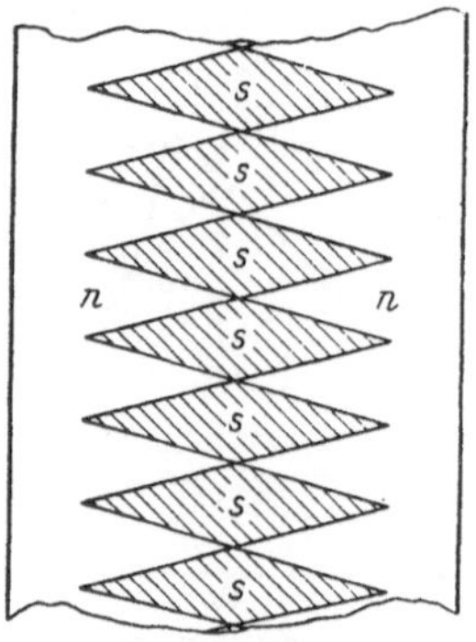

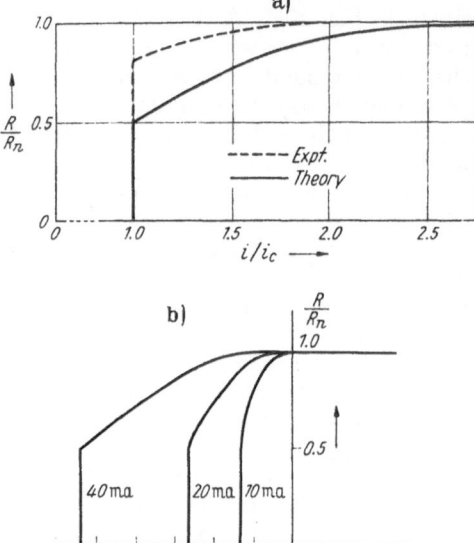

Fig. 30.

Fig. 31.

Fig. 30. Intermediate state of a wire carrying a current (after LONDON).

Fig. 31 a and b. a) Resistance of a tin wire as a function of current. The experiment by SCHUBNIKOW and ALEKSEYEVSKY [2] was on a 0.01 cm. diameter wire at 1.95° K; $H_o = 218$ Oe. b) Theoretical resistive transitions of a 0.01 cm. tin wire carrying various currents.

the resistance in the normal phase. As i is increased further, the resistance approaches R_n asymptotically. The data for a thin tin wire[2], as illustrated in Fig. 31a, are only in qualitative agreement with these predictions. The resistance at i_c jumps to about $0.8 R_n$ rather than to $R_n/2$, and the approach to R_n is more rapid than predicted by the model—all resistance being restored when $i \sim 2i_c$. SCOTT[3] found that the fraction of resistance discontinuously restored at i_c varied with wire diameter in indium wires. The smaller the wire, the larger was the initial jump. However, for increasing currents SILSBEE's condition (24.2) still holds, independently of wire size[4]. KUPER[5] attributes the larger size of the jump at i_c to the scattering of the conduction electrons at the normal-super-conducting interfaces in the intermediate state. The scattering contributes an additional resistance. The theory is in fair agreement with the data.

[1] F. LONDON [4], pp. 120—124; also BARDEEN, Sect. 33.
[2] L. W. SCHUBNIKOW and N. E. ALEKSEYEVSKY: Nature, Lond. 138, 804 (1936).
[3] R. B. SCOTT: J. Res. Nat. Bur. Stand. 41, 581 (1948).
[4] See A. B. PIPPARD [Phil. Mag. 41, 243 (1950)] for a theoretical justification of this finding.
[5] C. G. KUPER: Phil. Mag. 43, 1264 (1952).

LONDON'S model also may be used to determine how the resistance of a wire carrying a fixed current approaches zero as the temperature is reduced below the transition point. Theoretical curves are shown in Fig. 31 b; i_c increases as the temperature is lowered, so that the resistance decreases. When i_c exceeds the current in the wire, the resistance drops discontinuously to zero. The theory agrees only qualitatively with experiment (cf. Fig. 1), but it is clear that, because of the magnetic field of a measuring current, the resistance of a superconductor can vanish abruptly at T_c only in the limit of zero current.

STEINER and SCHOENECK[1] first observed that a superconductive rod in a weak longitudinal magnetic field can exhibit unusual magnetic properties when carrying large currents. When the current exceeds a certain minimum value, the longitudinal magnetic flux in the rod instead of being smaller than the flux in the normal phase, actually is greater. This phenomenon is termed the "paramagnetic effect", because the rod apparently behaves like a paramagnetic substance. The effect is shown, perhaps most strikingly, in the data obtained by MEISSNER, SCHMEISSNER and MEISSNER[2] for tin and mercury specimens. Some of the tin data are shown in Fig. 32. The specimen was surrounded by a coil which was connected to a ballistic galvanometer, and the galvanometer deflection produced by reversing the longitudinal field was observed as the temperature was

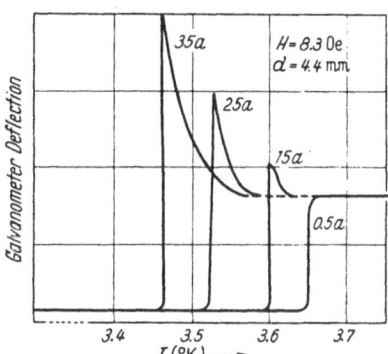

Fig. 32. Paramagnetic effect in tin (after MEISSNER et al.[2]).

lowered through the transition point. As can be seen in Fig. 32, deflections as much as twice the deflection in the normal phase are observed because of the paramagnetic effect.

For a given field and rod, the maximum additional flux in the paramagnetic state is a linear function of the current, with an intercept on the current axis which defines a minimum current, I_0, for the effect to be observed. I_0 also depends linearly on the magnetic field and specimen diameter, according to the relation

$$I_0 = I_g + \gamma\, dH,$$

where I_g and γ are characteristic constants of the metal, d is the diameter and H the longitudinal field. For tin, $I_g = 1.2$ amp and $\gamma = 0.17$ amp/mm. Oe. At the point of maximum flux the magnitude of the total magnetic field on the specimen surface is just the threshold value.

The foregoing observations have been verified by THOMPSON and SQUIRE[3] and SHIBUYA and TANUMA[4].

The paramagnetism is undoubtedly associated with the intermediate state produced by the unusual magnetic field distribution about the specimen. MEISSNER[5] has proposed a theory of the effect which is in qualitative agreement with experiment, but the theory does not account for the characteristic current, I_g.

25. Supercooling and superheating. The existence of a positive surface energy (frequently called a surface tension) between the two phases leads us to expect

[1] K. STEINER and H. SCHOENECK: Phys. Z. **44**, 346 (1943).
[2] W. MEISSNER, F. SCHMEISSNER and H. MEISSNER: Z. Physik **130**, 529 (1951).
[3] J. C. THOMPSON and C. F. SQUIRE: Phys. Rev. **96**, 287 (1954).
[4] Y. SHIBUYA and S. TANUMA: Phys. Rev. **98**, 938 (1955).
[5] H. MEISSNER: Phys. Rev. **97**, 1627 (1954).

superconductors to exhibit supercooling and superheating[1]. These effects are observed often in the more familiar phase transitions. For example, in the liquid-vapor transition the liquid can usually be superheated above the equilibrium boiling temperature before boiling occurs; and the vapor, supercooled below this same temperature before liquid condenses. In superconductivity, the terms are used less rigidly, because usually the temperature is maintained constant, and the magnetic field is varied about the threshold value. When the normal phase persists in fields less than H_c, we say the metal is supercooled; and when the superconductive phase persists in fields greater than H_c, we say it is superheated.

Ever since 1925, supercooling has been observed in occasional investigations of the superconductive transition. To illustrate the form that this phenomenon takes, we show in Fig. 33 a magnetization curve of a tin cylinder in a longitudinal magnetic field. This curve is typical of many the author has obtained incidentally to other investigations. We note that when the applied field was reduced below the threshold value, the specimen persisted in the normal phase until $H_a/H_c = S = 0.7$. At this point, there was an abrupt transition to the superconductive state. Abrupt transitions are characteristically associated with supercooling and superheating, and the zero breadth and unusual speed of the transitions provide positive criteria for recognizing these phenomena. A value of S of about 0.9 is generally observed in tin. Large supercoolings are common in aluminum; Shoenberg[2] observed values of S as small as 0.37 in spherical specimens of this metal.

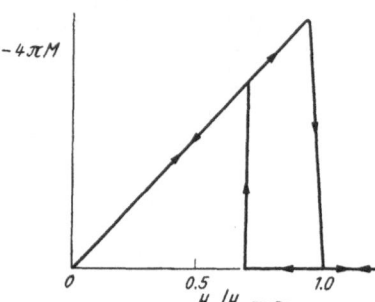

Fig. 33. Magnetization curve illustrating supercooling.

Superheating is not so commonly observed. Faber[3] did find it up to $S \sim 1.02$ in an occasional tin specimen. However, Garfunkel and Serin[4] demonstrated that considerable superheating can be observed under suitable conditions. In their experiments, a short coil was wound on the central portion of a superconductive tin cylinder which was placed in a longitudinal magnetic field. By sending a current through the coil, the field over the central section was made locally larger than the field over the remainder of the specimen. The magnetic moment of the specimen was measured, and it was found that the field at the center consistently had to exceed the threshold value before that section became normal. At temperatures near the transition point, the central section superheated until S was 1.17. The ends of the specimen could not be superheated, presumably because of the large local fields present there resulting from demagnetization effects. The inability of the ends to superheat explains why specimens as a whole rarely do so.

In order for the phase transition to proceed, a small *stable* region of the new phase must first form in the matrix of the existing phase. Once such a nucleus forms, it can grow until the new phase fills the whole specimen. Supercooling and superheating reflect the difficulty of the nucleation process, which is intimately

[1] This observation was made first by H. London: Proc. Roy. Soc. Lond., Ser. A, **152**, 650 (1935).

[2] D. Shoenberg: Proc. Cambridge Phil. Soc. **36**, 84 (1940).

[3] T. E. Faber: Proc. Roy. Soc. Lond., Ser. A **214**, 392 (1952).

[4] M. P. Garfunkel and B. Serin: Phys. Rev. **85**, 834 (1952).

connected with the interphase surface energy. In fact, it can be shown that if the surface energy is everywhere positive, a stable nucleus can never form[1].

We remind the reader than an analogous situation exists in supercooled vapors, and that in vapors, dust and ions serve as nuclei. Similarly in superconductors, flaws in the metal act as nuclei. The connection between flaws and supercooling was conclusively demonstrated by FABER[2]. In these beautiful experiments, a tin rod with several short coils around it at various points along its length was placed in a longitudinal magnetic field and slightly supercooled. By sending a current through a given coil, the field in a particular region of the specimen was further decreased below the threshold value until the superconductive phase rapidly grew out of this region, filling the whole specimen. The degree of supercooling varied greatly from point to point along the rod, suggesting that nucleation in a given region occurs at a particular flaw. The supercooling of the rod as a whole is governed by the weakest flaw in the rod, which explains why a small supercooling is generally observed. The minimum S for a flaw in tin was 0.45.

The flaws do not appear to be associated with surface conditions, impurity or crystal boundaries. FABER established that the flaws in tin usually lie at the surface and are between 10^{-4} and 10^{-3} cm. in size. But, if the surface layer is removed by electropolishing, new flaws are exposed, indicating that they are uniformly distributed throughout a specimen. The flaws are usually unaffected by heating the specimen to room temperature and cooling again, but handling the specimen changes them. FABER and PIPPARD[3] suggest that flaws are regions where the crystal lattice is distorted by a network of dislocations.

According to FABER, the flaws provide local regions in which the interphase surface tension is negative. These regions are superconductive when the specimen is supercooled and serve as stable nuclei. However, even though nuclei are present, they are still prevented from growing by the positive surface tension of the bulk of the metal until the field is reduced well below the threshold value. By making a simple model of a flaw, FABER showed that the amount a given nucleus supercools is determined by its size and shape and the surface energy parameter, Δ. For any shape, however, $(1 - S^2)$ is proportional to Δ. The observations are in good agreement with this model. Even though the amount of supercooling varies in magnitude from flaw to flaw, it depends on temperature in the same way for all. The different magnitudes are not very interesting because they presumably are associated with the various sizes and shapes of nuclei. However, the universal temperature dependence is due to temperature the dependence of Δ, and these data are in good agreement with other determinations of this dependence. Despite this agreement, the measurements do not provide very reliable estimates of Δ.

Recently, FABER and MAPOTHER et al.[4] have observed similar effects in aluminum rods in which the supercooling was exceptionally large; particular regions supercooled to $S = 0.02$. This large supercooling is probably characteristic of pure, undistorted crystals.

The foregoing observations make it desirable to re-examine briefly the concept of an ideal superconductor. Clearly, a specimen which was ideal in the

---

[1] D. SHOENBERG [1], pp. 123—124.

[2] See footnote 3, p. 256.

[3] T. E. FABER and A. B. PIPPARD [3], Chapter IX. The material in this and the following section is reviewed in this reference.

[4] T. E. FABER and D. E. MAPOTHER, J. F. COCHRAN and R. E. MOULD in separate contributions to Int. Conf. Low Temp. Phys. (Houston 1953). See also reference 3.

sense of being perfect in structure, would not exhibit ideal magnetic properties. Because of supercooling and superheating, the magnetization curve of such a specimen would exhibit a large hysteresis, and the phase transitions would be highly irreversible. In order for the transitions to be ideal in the sense of being reversible, a specimen must have at least one very weak flaw. Such a flaw would provide the nucleus by way of which the specimen could reach the states of thermodynamic equilibrium.

26. Kinetics of the phase transition. Delays in the attainment of equilibrium in the superconductive transition have often been observed. They can be particularly long in the intermediate state, where the induction in a specimen can sometimes change slowly over a period of one half hour after the external magnetic field is altered (cf. Sect. 8). Such observations are difficult to analyze, because of the complex configuration of the two phases in the intermediate state. Recently, Faber[1, 2] has measured the rate of phase propagation in a long cylindrical rod in a longitudinal magnetic field. There is no intermediate state under these circumstances, and it has proved possible to deal with the transient behavior in some detail.

We consider first the case which arises when the specimen is initially superconductive and the applied field, H_a, is suddenly increased above the threshold value, H_c, inducing a transition to the normal phase. Under these circumstances, the superconductive phase contracts radially until it disappears. The motion of the phase boundary involves the establishment of the applied magnetic field in the normal phase, and this process is retarded by eddy currents. Several early investigators[3, 4] suggested that phase propagation is associated with eddy current effects, but the problem has been treated rigorously only recently by Pippard[5].

As a result of the eddy currents, the magnetic field inside the specimen is less than the applied field. Pippard assumes that the phase boundary moves at such a rate that the field strength on it is maintained precisely at the threshold value through the action of the induced currents. At any instant, the field varies from H_c on the boundary to the applied value, $H_a > H_c$, on the surface. When the currents have died away, the transition is complete, and the field in the specimen is everywhere H_a. As a result of solving·the electromagnetic equations, subject to these boundary conditions, Pippard finds that the superconductive phase vanishes in a time,

$$\tau = \pi r_0^2 \sigma \frac{H_a}{H_a - H_c}, \tag{26.1}$$

in a cylinder of radius r_0, and normal conductivity, σ. The surface of the specimen is assumed to be maintained at constant temperature by contact with liquid helium. Under these circumstances, the transition can take place isothermally because the thermal conductivity of the normal phase is large enough to conduct in the latent heat absorbed in the transition. The total Joule heat developed is $(H_a^2 - H_c^2)$ per unit volume, and this is small compared to the latent heat in these experiments. We note that the Joule heat vanishes as $H_a \to H_c$.

Faber[1] measured the rate of collapse of the superconductive phase in tin specimens as a function of r_0, σ, H_a and H_c. The specimen was surrounded by

[1] T. E. Faber: Proc. Roy. Soc. Lond., Ser. A **219**, 75 (1953).
[2] T. E. Faber: Proc. Roy. Soc. Lond., Ser. A **223**, 174 (1954).
[3] K. Mendelssohn and R. B. Pontius: Physica, Haag **3**, 327 (1936).
[4] H. Grayson Smith and K. C. Mann: Phys. Rev. **54**, 766 (1938).
[5] A. B. Pippard: Phil. Mag. **41**, 243 (1950).

a search coil which was connected to a galvanometer of very short response time, and the voltage pulse generated in the coil as the flux penetrated the specimen was recorded. The conductivity was varied by alloying the tin with very small amounts of indium. On the whole, the measurements are in agreement with (26.1). The discrepancy between theory and experiment is small at low temperature, but near T_c the observed transition time is about 20% too long. The detailed shape of the observed pulse is qualitatively similar to the shape predicted by theory. As a result of these observations, there is little doubt that the speed of the transition to the normal phase is principally controlled by eddy currents. The small discrepancies between theory and experiment are not understood.

We note that for the purest tin specimens, $\sigma \sim 1$ emu, so that $\pi r_0^2 \sigma = 0.031$ for a specimen of 1 mm. radius. When the applied field is raised to 1.1 H_c, according to (26.1) the transition takes place in 0.34 sec. The time will, of course, be longer in thicker specimens and shorter in impure specimens of the same radius. The experiments also provide evidence that the relaxation time for the destruction of superconductivity is less than 10^{-7} sec.

The reverse transition into the superconductive phase is considerably more complicated. As a result of his measurements of the speed of the phase propagation FABER[1] has been able to arrive at a fairly accurate picture of how the transition proceeds. The measurements were made on long tin rods which had several search coils distributed along them. The coils were connected to a string galvanometer. The rods were slightly supercooled in a longitudinal magnetic field, and the phase transition was then induced at one end of the rod by further reducing the field there. The passage of the superconductive phase down the rod was timed by noting the intervals between the pulses induced in the successive search coils by the flux expelled from the rod.

From the shape of the induced pulses, FABER deduced that the phase transition proceeded in three fairly distinct steps. Initially, a filament of superconductivity grows longitudinally along the surface of the rod with a uniform velocity of about 10 cm./sec. The filament then expands sideways until a superconductive sheath about 5×10^{-3} cm. thick forms around the specimen in approximately 0.1 sec. Finally, the sheath expands inward until the last traces of flux have been expelled from the specimen, probably through small gaps in the sheath. This last process is slow, and takes at least several seconds.

Surface energy effects play a large role in the propagation of the initial superconductive filament. If it were not for surface energy, the filament could become very thin compared to the penetration depth, and advance without displacing any magnetic flux. Under such circumstances, no eddy currents would impede its movement, and the filament would move with extreme speed. However, the surface tension results in a filament about 10^{-3} cm. thick. When a superconductive filament of this thickness moves down a rod, it pushes the magnetic flux ahead of it out of the metal, and eddy currents are induced about the leading edge. The currents impede its motion. Because the currents flow in a region which may be small compared with the electronic mean free path, the appropriate "anomalous" conductivity must be used.

Analysis of the propagation according to this model yields an optimum thickness for the filament, which allows it to travel with a maximum velocity. Under completely anomalous conditions the speed of propagation in tin is

$$v = 1.6 \times 10^{-3} (\varDelta - \lambda)^2 \left(\frac{H_c - H_a}{H_c}\right)^3 . \tag{26.2}$$

---

[1] T. E. FABER: Proc. Roy. Soc. Lond., Ser. A **223**, 174 (1954).

The observed dependence of velocity on applied field is in good agreement with (26.2). Thus, the velocity measurements may be used to evaluate $(\varDelta - \lambda)$. $\varDelta$ may then be calculated from the known variation of λ with temperature (cf. Sect. 18). The values of $\varDelta$ obtained in this way are shown in Fig. 34; they are in good agreement with the values found by Schawlow from his intermediate state measurements (cf. Sect. 24a). We note that $\varDelta$ varies with temperature in very much the same way as the penetration depth; in fact, $\varDelta \sim 5\lambda$.

The mechanism by which the last traces of flux escape from the specimen, thereby completing the Meissner effect, is briefly discussed in Chapter VII.

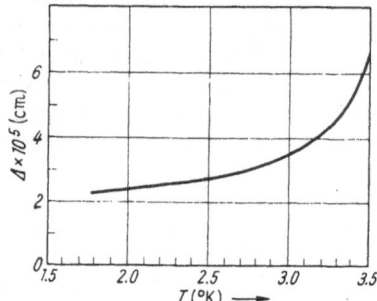

Fig. 34. Dependence of $\varDelta$ for tin on temperature (after Faber).

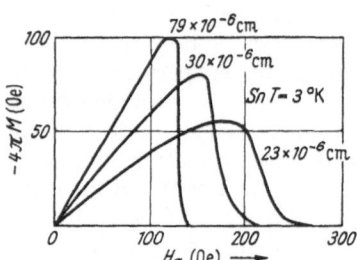

Fig. 35. Magnetization curves of thin tin films (after Lock[3]).

27. Films and colloids. The magnitude of the magnetization of a film in a longitudinal magnetic field is appreciably smaller than $H_a/4\pi$ because of the penetration of the field into the specimen. As a result, when the applied field is increased to the threshold value, H_c, the magnetic work done per unit volume is less than $H_c^2/8\pi$, and it is necessary to increase the field above this value before the phase transition takes place in the film. We must also consider the possibility that the surface free energy between the superconductive phase and vacuum, α_s, may be different from the surface energy, α_n, at an interface between the normal phase and vacuum. When this surface energy term is included, it can be shown that the threshold field, h, of a film of thickness, $2a > \lambda$, is given by the relation

$$\frac{h}{H_c} = 1 + \frac{\beta + \lambda}{2a}, \tag{27.1}$$

where $\beta = 8\pi(\alpha_n - \alpha_s)/H_c^2$; *provided* it is assumed that the magnetization curve is linear right up to the field, h, where it drops sharply to zero.

The increase in threshold value with decrease in size of small specimens has been observed in numerous investigations[1, 2]. The behavior of mercury colloids has already been illustrated in Fig. 23. For thick films the measured threshold values of h do depend on a in the way indicated by (27.1), and when this expression is used to evaluate β, it is found[2] that β is about equal to λ.

However, Lock[3] has shown that the magnetization curves of thin films are non-linear. This behavior is illustrated in Fig. 35, where it is clear that deviations from linearity become very pronounced in the thinnest films. When this happens,

[1] R. B. Pontius: Phil. Mag. 24, 787 (1937). — E. T. S. Appelyard, J. R. Bristow, H. London and A. D. Misener: Proc. Roy. Soc. Lond., Ser. A 172, 540 (1939). — D. Shoenberg: Proc. Roy. Soc. Lond., Ser. A 175, 49 (1940). — N. Alexeevski: J. Phys. USSR. 9, 305 (1945).
[2] E. R. Andrew: Proc. Phys. Soc. Lond. A 62, 88 (1949).
[3] J. M. Lock: Proc. Roy. Soc. Lond., Ser. A 208, 391 (1951).

the arguments leading to (27.1) break down, and no deductions can be made about the value of β from threshold measurements alone. PIPPARD[1] has shown, however, that the area, A, under the magnetization curve of a film, is given by

$$\frac{A}{H_c^2/8\pi} \gtrless 1 + \frac{\beta}{a}. \qquad (27.2)$$

This result is independent of the shape of the curve. The equality holds when the magnetization is reversible and the inequality when it is irreversible. Thus, the magnetization curve may be used to find an upper limit for β. On this basis, LOCK[2] determined that β in tin is considerably smaller than λ. PIPPARD[1] finds that the values of β calculated from the magnetization curves of mercury colloids are about ten times smaller than LOCK's, and he suggests that β may be zero or even slightly negative.

The reasons for the observed threshold values being larger than those predicted by (27.1) with $\beta = 0$ are not very clear. The large values arise from the decrease of the slope of the magnetization curve from its initial value as the field is raised. The nature of this falling off has been discussed only qualitatively by PIPPARD, who suggests that it is caused by the appearance of normal islands in the film below the threshold field.

The hysteresis exhibited by the magnetization curves of mercury colloids has been explained fairly well by PIPPARD[3] on the basis of a simple two-fluid model. Hysteresis is only exhibited by the larger particles of about 10^{-4} cm. radius and not by particles of 5×10^{-6} cm. radius.

VI. Thermal effects.

28. Thermal conductivity. Heat conduction in metals is a very complicated phenomenon. Several mechanisms contribute to the thermal resistance, and their relative importance varies with the particular substance, its purity and temperature[4]. The situation becomes more intricate with the transition into the super-conductive phase, because the several mechanisms are differently affected by the transition. Fortunately, it is possible to find cases in which each of the mechanisms predominates so that their effects can be separately estimated. We shall concern ourselves, for the most part, in this section with illustrating these limiting cases.

This field has been reviewed recently by OLSEN and ROSENBERG[5] and by MENDELSSOHN[6]. We refer the reader to them for detailed references, since we shall confine our attention mostly to a few recent papers which seem particularly illuminating.

The thermal conductivity is usually determined by the standard method, in which the temperature is measured at two points along a rod when it is heated at one end and in contact with a heat sink at the other end. Above $1°$ K, the helium bath serves as the heat sink; and helium gas thermometers are generally

---

[1] A. B. PIPPARD: Proc. Cambridge Phil. Soc. **47**, 617 (1951).

[2] See footnote 3, p. 260.

[3] A. B. PIPPARD: Phil. Mag. **43**, 273 (1952).

[4] Heat conduction of metals at normal temperatures has been treated by G. LEIBFRIED in vol. VII, part 1, and at low temperatures by P. G. KLEMENS in vol. XIV of this Encyclopedia.

[5] J. L. OLSEN and H. M. ROSENBERG: Adv. Physics, Phil. Mag. Suppl. **2**, 28 (1953).

[6] K. MENDELSSOHN [3], Chapter X.

used. Below $1°$ K, the specimen is cooled by a paramagnetic salt, which subsequently serves as the heat sink, and the temperatures are determined by carbon film resistance thermometers.

Before discussing particular cases, it is helpful to note that the total thermal conductivity of a metal may be written as,

$$\varkappa = \varkappa_e + \varkappa_g,$$

where $\varkappa_e$ is the conductivity of the electrons and $\varkappa_g$, of the lattice. In subsequent sections, we place the additional subscript n or s after each symbol to denote, respectively, the conductivity of the normal and superconductive phases.

α) *Normal phase.* We will briefly survey heat conduction in normal metals before discussing the properties of superconductors. The observations are usually interpreted in terms of the theory of Makinson[1]. At temperatures low compared to the Debye temperature, Θ, the electronic thermal conductivity, $\varkappa_{en}$, is limited by two scattering processes, and the theory gives

$$\frac{1}{\varkappa_{en}} = \alpha T^2 + \frac{\varrho_0}{L_0 T}, \qquad T < \frac{\Theta}{10}. \tag{28.1}$$

The first term is the thermal resistance resulting from the scattering of electrons by the vibrations of the lattice. α is proportional to Θ^{-2}. The second term arises from impurity scattering. ϱ_0 is the residual electrical resistivity of the metal, and L_0, the Lorentz number, equals 2.44×10^{-8} watt-ohms/deg.[2]. $\varkappa_{en}$ has a maximum at

$$T_{max} = \left(\frac{\varrho_0}{2 \alpha L_0} \right)^{\frac{1}{3}}.$$

Since the addition of impurities to a metal increases ϱ_0, it has the effect of both decreasing $\varkappa_{en}$ and shifting the maximum to higher temperatures. The position of the maximum also depends on the Debye temperature of the metal, and in general, T_{max} increases with increasing Θ. Lattice scattering predominates above T_{max}, and below it impurity scattering is relatively more important.

The theory gives for the lattice thermal conductivity

$$\frac{1}{\varkappa_{gn}} = \frac{A}{T^2} + \frac{B}{T^3} \frac{1}{\zeta}, \qquad T < \frac{\Theta}{10}. \tag{28.2}$$

The first term comes from scattering by the electrons, and it is the dominant term above $1°$ K. A is proportional to Θ^2. The second term is due to scattering by the crystal boundaries, where ζ is the mean free path of the waves for boundary scattering and B is a constant. This term becomes increasingly important as the temperature is reduced below $1°$ K. Impurity scattering of the lattice waves is negligible because at these temperatures their average wavelength is very large compared to the size of an impurity scattering center.

In pure metals $\varkappa_{en} \gg \varkappa_{gn}$, so that the total conductivity, $\varkappa_n$, is given by (28.1). However, with decreasing purity, the contribution of $\varkappa_{gn}$ to $\varkappa_n$ becomes of increasing relative importance.

Hulm[2] has measured the thermal conductivity of tin, indium, tantalum and mercury specimens. The data for a few specimens of tin and mercury are shown in Figs. 36 and 37 to illustrate particular aspects of the foregoing discussion.

[1] R. E. B. Makinson: Proc. Cambridge Phil. Soc. **34**, 474 (1938).
[2] J. K. Hulm: Proc. Roy. Soc. Lond., Ser. A **204**, 98 (1950).

We note that the maximum thermal conductivity occurs just below 4° K in the pure tin specimen; the conductivity then approaches the linear variation resulting from impurity scattering alone. As impurity is added, the maximum shifts to a higher temperature, the conductivity is reduced and the shape of the curve changes as lattice scattering becomes relatively more important. Finally, in the specimen, $Sn + 4\%$ Hg, the lattice scattering mechanism remains large down to 2° K. In pure mercury on the other hand the maximum is not reached at 1° K, indicating that lattice scattering predominates. The addition of 1.2% In to the mercury results in a barely perceptible maximum at 3° K. The thermal

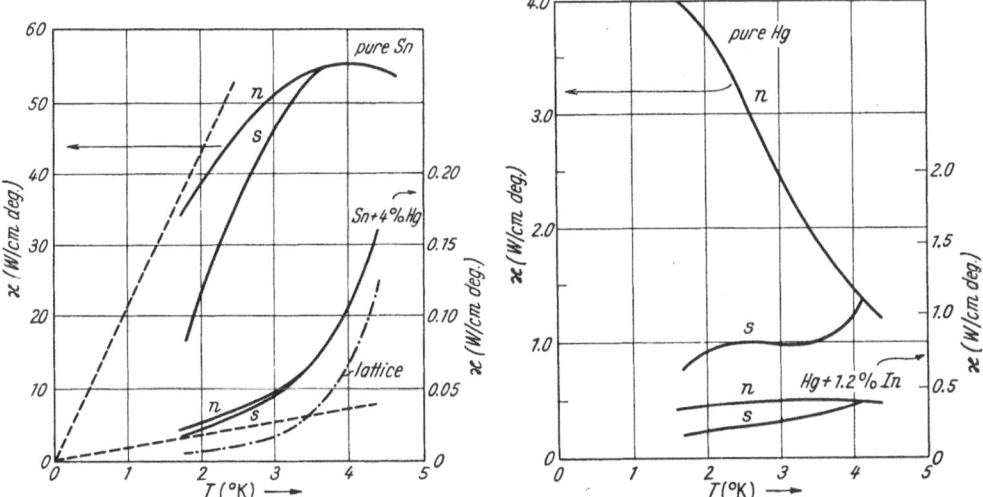

Fig. 36. $\varkappa$ for two tin specimens (after HULM).   Fig. 37. $\varkappa$ for two mercury specimens (after HULM).

conductivity of indium varies with temperature in the same way as does tin, and lead[1] behaves similarly to mercury.

β) *Superconductive phase.* No detailed theory of heat conduction in superconductors has yet been developed. However, the two-fluid model is quite useful for analyzing the experimental findings. According to this model, the superconductive fluid takes no part in the conduction process. It can neither carry a thermal current nor scatter lattice waves. Only the normal electrons are available for these two functions, and their number rapidly diminishes as the temperature is reduced below T_c. As a result, we expect $\varkappa_{es}$ to be smaller than $\varkappa_{en}$, and $\varkappa_{gs}$ to be larger than $\varkappa_{gn}$. However, since $\varkappa_{en}$ is very much greater than $\varkappa_{gn}$ in pure metals, we can assume that unless the change in $\varkappa_g$ is extremely large, the total conductivity of the superconductive phase $\varkappa_s$ will be smaller than $\varkappa_n$. This assumption is borne out by the data for pure metals, and is clearly illustrated in the particular cases shown in Figs. 36 and 37. In certain alloys, however, $\varkappa_s$ is larger than $\varkappa_n$; to illustrate this point the data obtained by OLSEN[2] for $Pb + 10\%$ Bi are shown in Fig. 38. Recent work has established that there is no discontinuity in the thermal conductivity at T_c, even though the slopes of the curves for the normal and superconductive states can be very different at this temperature.

---

[1] W. J. DE HAAS and A. RADEMAKERS: Physica, Haag **7**, 922 (1940). — Leiden Comm. 261e.

[2] J. L. OLSEN: Proc. Phys. Soc. Lond. A **65**, 518 (1952).

The changes in thermal conductivity which occur upon the transition into the superconductive phase can be better appreciated by plotting the ratio, $\varkappa_s/\varkappa_n$, as a function of temperature. Examination of the ratio curves shown in Fig. 39, reveals that tin and mercury behave very differently.

Those tin specimens for which $\alpha T^2 < \varrho_0/L_0 T$ fall on the same curve, and the data are in fair agreement with the approximate theoretical curve proposed by Heisenberg[1]. The theory is based on Koppe's two-fluid model, and gives

$$\frac{\varkappa_{es}}{\varkappa_{en}} = \frac{2t^2}{1+t^4}, \qquad t > 0.4.$$

The systematic deviation from this curve of the data for the more impure specimens is due to the appreciable lattice conductivity component in both the normal and superconductive phases. By subtracting the electronic component, $\varkappa_{gs}$ can be obtained. In Sn + 4 % Hg, $\varkappa_{gs}/\varkappa_{gn}$ is greater than unity and increases

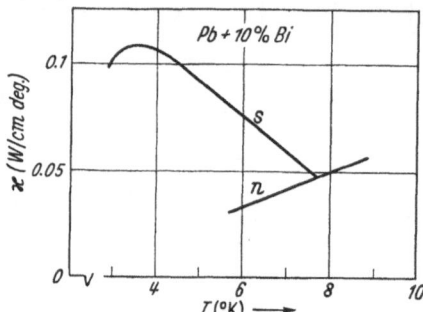

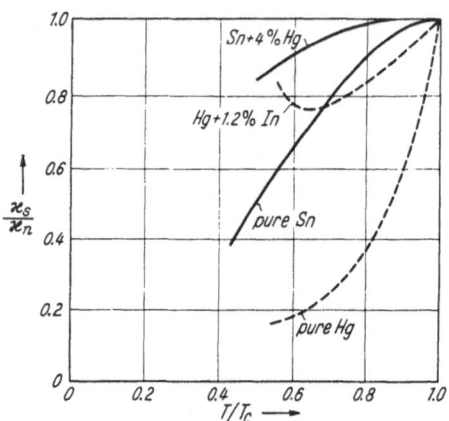

Fig. 38. Thermal conductivity of Pb + 10% Bi (after Olsen).

Fig. 39. $\varkappa_s/\varkappa_n$ for several specimens of tin and mercury (after Hulm[2]).

monotonically with decreasing temperature over the whole range of measurement. Hulm[2] suggested that an even greater relative increase in $\varkappa_{gs}$ could explain the behavior of those alloys in which $\varkappa_s > \varkappa_n$. Olsen's measurements of heat transport in those lead-bismuth alloys in which $\varkappa_s$ is indeed greater than $\varkappa_n$ are in fair agreement with this suggestion. It has been given considerable substance by the recent extensive determinations made by Laredo[3] of the thermal conductivity of tin in the temperature range from 0.5 to 4° K. He found that for Sn + 3 % In, $\varkappa_s > \varkappa_n$ from 0.5 to 2° K, with a maximum in $\varkappa_s$ occuring at about 1° K. The increase in $\varkappa_s$ can be very well explained as due to the relatively large contribution of $\varkappa_{gs}$ in this temperature region[4].

The conductivity of pure mercury, though still electronic, is mainly determined by the scattering of electrons by lattice waves. The ratio, $\varkappa_s/\varkappa_n$, for the purer specimens shown in Fig. 39, decreases roughly as t^5, as the temperature becomes less than T_c. The effect of adding impurity to mercury is to make the ratio curve tend to approach the curve for tin. In the very impure specimens,

[1] W. Heisenberg: *Two Lectures*, Cambridge University Press (1949).

[2] J. K. Hulm: Proc. Roy. Soc. Lond., Ser. A **204**, 98 (1950).

[3] S. J. Laredo: Proc. Roy. Soc. Lond., Ser. A **229**, 473 (1955).

[4] K. Mendelssohn [see e.g., K. Mendelssohn and J. L. Olsen: Proc. Phys. Soc. Lond. A **63**, 2 (1950)] suggested an ingenious circulation mechanism, analogous to the superfluid flow which occurs in liquid He II, to account for the large values of $\varkappa_s/\varkappa_n$ which occur in some alloys. Although the matter is not completely clear, recent evidence does not seem to favor this mechanism.

however, the electronic conductivity is so much reduced, that the lattice conductivity becomes important and causes the ratio to increase at low temperatures. No explanation has been given for the t^5-variation of $\varkappa_s/\varkappa_n$ in superconductors (such as mercury) in which lattice scattering predominates.

Experiments by DAUNT and HEER[1] and GOODMAN[2] gave the first qualitative indication that $\varkappa_s/\varkappa_n$ becomes extremely small in pure metals below 1° K. For example, in pure tin, $\varkappa_s/\varkappa_n \sim 10^{-3}$ at 0.2° K. OLSEN and RENTON[3] overcame several of the technical difficulties inherent in the foregoing work, and clearly established that $\varkappa_s$ in pure lead varies as T^3 between 0.4 and 0.9° K. LAREDO[4] also observed the cubic dependence in tin, and, moreover, he found that specimens of widely differing impurity content had about the same thermal conductivity below 0.5° K. The same T^3-law has been observed at sufficiently low temperatures in specimens of lead, thallium, tin, indium, niobium and tantalum by MENDELSSOHN and RENTON[5]. The cubic temperature dependence is a consequence of the fact that all the electrons are superconductive at these very low temperatures. As a result, $\varkappa_{es} \to 0$, and the lattice conduction is impeded only by boundary scattering. This scattering mechanism is unaffected by either the superconductive transition or impurities, with the result that

$$\varkappa_s = \varkappa_{gs} = \varkappa_{gn} = \frac{\zeta T^3}{B},$$

if T is small enough. The values found for the mean free path, ζ, for boundary scattering are about equal to the specimen diameters of tin rods.

LAREDO[4] finds that in the temperature range $t = 0.15$ to 1.0, the electronic contribution to the thermal conductivity, $\varkappa_{es}$, is in only fair agreement with the predictions of the HEISENBERG-KOPPE two-fluid model. He also finds that $\varkappa_{es}$ is anisotropic in tin, the conductivity along the tetragonal axis being greater than perpendicular to it.

In concluding this section, we mention that lead wires have been used with success as thermal switches below 1° K. Owing to the small value of $\varkappa_s$, the switch is "open" when the wire is superconductive; it is "closed" when made normal by a magnetic field exceeding H_c.

γ) *Intermediate state.* DE HAAS and RADEMAKERS[6] showed that when the superconductivity of a lead rod at 5° K was destroyed by a transverse magnetic field, the thermal conductivity increased linearly from its superconducting to its normal value as the field was increased from $\frac{1}{2} H_c$ to H_c. This linear change is in accord with a model of the intermediate state in which the layers are perpendicular to the cylinder axis, with the n-layers having conductivity, $\varkappa_n$, and the s-layers, $\varkappa_s$. HULM[7] showed that the magnetic transitions of tin rods in a longitudinal field are sharp ,and occur at H_c.

An entirely new and unexpected transverse transition was first observed by MENDELSSOHN and OLSEN[8] in a slightly impure lead specimen at about 3° K. Instead of increasing monotonically from the superconductive to the normal value with increasing field, the thermal conductivity first decreased sharply at $\frac{1}{2} H_c$, and then, after passing through a minimum, increased to the normal

[1] J. G. DAUNT and C. V. HEER: Phys. Rev. 76, 854 (1949).
[2] B. B. GOODMAN: Proc. Phys. Soc. Lond., Ser. A 66, 217 (1953).
[3] J. L. OLSEN and C. A. RENTON: Phil. Mag. 43, 946 (1952).
[4] S. J. LAREDO: Proc. Roy. Soc. Lond., Ser. A 229, 473 (1955).
[5] K. MENDELSSOHN and C. A. RENTON: Proc. Roy. Soc. Lond., Ser. A 230, 157 (1955).
[6] W. J. DE HAAS and A. RADEMAKERH: Physica, Haag 7, 922 (1940). — Leiden Comm. 261 e.
[7] J. K. HULM: Proc. Roy. Soc. Lond., Ser. A 204, 98 (1950).
[8] K. MENDELSSOHN and J. L. OLSEN: Phys. Rev. 80, 859 (1950).

value at H_c. At 5° K, the specimen exhibited the usual linear behavior. Minima have since been observed in pure lead[1] and in tin and indium[2] at about 2° K. The depth of the minimum seems to increase with decreasing temperature and a very large effect has been found recently in tin at 0.5° K by Laredo and Pippard[3]. The same deepening of the minimum is observed in lead down to 1° K, but Olsen and Renton[4] found that the relative depth began to decrease below this temperature. In general the transition is irreversible, the minimum being less deep when the field is reduced from above H_c than in the initial transition from zero field, and in many investigations no minimum at all is found in the transition from high fields.

The data of Mendelssohn and Olsen are shown in Fig. 40. Sladek[5] observed minima in the longitudinal field transition of rods of alloys of indium containing

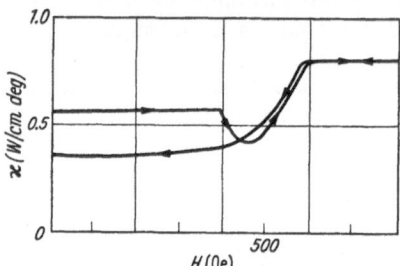

Fig. 40. Thermal conductivity of a slightly impure lead specimen in a transverse magnetic field (after Mendelssohn and Olsen).

more than 15% thallium. He showed that this effect was associated with the persistence in the specimens of thin superconductive filaments in fields exceeding the threshold value.

Recently, it has become clear that the minimum in the transverse transition can be explained without invoking new mechanisms of thermal conduction. Cornish and Olsen[6] considered a crude model in which it was assumed that the electrons and lattice waves had virtually independent temperature distributions, and on this basis, obtained fair agreement with experiment. The effect, at least at low temperatures, has since been treated with considerable rigor by Laredo and Pippard[3]. They use a laminar model of the intermediate state of the rod with the layers transverse to the axis. As in any such model, the major impedance to the flow of heat results from the poor conductivity of the superconductive layers. Heat is conducted in them, for the most part, by the lattice. By an ingenious analysis, Laredo and Pippard demonstrate that the mean-free path, ζ, for lattice scattering is much smaller in a layer than in the bulk superconductor. In the latter, ζ about equals the specimen diameter, whereas in the former, it equals the lamina thickness. Thus, the effective lattice conductivity of the superconductive layers is reduced by a factor of about five below the value in a bulk specimen. This increased impedance of the s-layers results in the rapid decrease of thermal conductivity when a specimen enters the intermediate state. As the field is further increased, the s-layers become progressively thinner, so that their impedance is gradually reduced and the thermal conductivity of the specimen slowly increases. The lattice conductivity in the layers is so much reduced that the electronic component of conductivity cannot be neglected. Taking this into account, the final calculated result is in very good agreement with the data obtained in increasing magnetic field. The reduction in depth of the minimum in decreasing field is not explained, although it is suggested that the difference arises because the structure of the intermediate state depends on whether the magnetic field is increasing or decreasing.

[1] R. T. Webber and D. A. Spohr: Phys. Rev. **84**, 384 (1951).
[2] D. P. Detwiler and H. A. Fairbank: Phys. Rev. **88**, 1049 (1952).
[3] S. J. Laredo and A. B. Pippard: Proc. Cambridge Phil. Soc. **51**, 368 (1955).
[4] J. L. Olsen and C. A. Renton: Phil. Mag. **43**, 946 (1952).
[5] R. J. Sladek: Phys. Rev. **97**, 902 (1955).
[6] F. H. J. Cornish and J. L. Olsen: Helv. phys. Acta **26**, 369 (1953).

29. Thermoelectric effects. Many experiments have shown that no thermo-electric emf, E, is developed in a circuit containing two superconducting metals with junctions at different temperatures (see e.g., STEINER and GRASSMANN[1]). This means that the absolute thermoelectric power, $e = dE/dT$, of all super-conductors is zero. As a result, the absolute thermoelectric power of a normal metal can be determined by measuring the emf developed by thermocouples formed between the metal and a superconductor.

In their measurements on the couple, indium-lead (the lead being super-conducting), KEESOM and MATTHIJS[2] found evidence to suggest that e did not fall abruptly to zero at the transition temperature of indium, but decreased gradually over a temperature interval of about $1°$ K, reaching zero at T_c. CASIMIR and RADEMAKERS[3] repeated the experiment more carefully with a tin-lead couple, and reported that the thermoelectric power of tin showed an unusual decrease at about $0.15°$ K above the transition temperature, which, they suggested, "fore-shadowed" the onset of superconductivity.

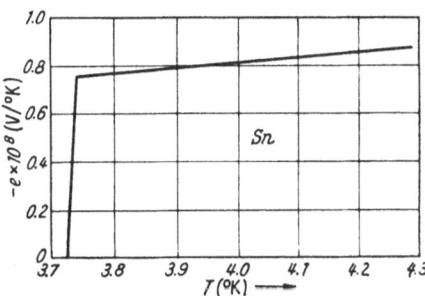

Fig. 41. Thermoelectric power of tin as a function of temperature (after PULLAN[4]).

This matter has been settled recently by PULLAN[4], who unambiguously demon-strated that the thermoelectric power of tin drops abruptly to zero at the transition temperature in contradiction to the earlier observations. A tin-lead couple was used, but the temperature difference between the junctions in these experiments was only about $0.01°$ K. Thus, the thermo-electric power could be found by dividing the measured emf by this small temperature difference, rather than by the usual less desirable procedure of differentiating the experimental curve of E as a func-tion of T. The extremely small emf's developed were measured with a supercon-ducting galvanometer. PULLAN's results are shown in Fig. 41; it is clear that the thermoelectric power falls to zero in a temperature interval of about $0.01°$ K, which is just the spread in the mean temperature of the specimen.

PULLAN also found that the thermoelectric power of tin is unaffected by small magnetic fields, in agreement with the more extensive measurements of STEELE[5]. The latter experiments have been analyzed in considerable detail by SHOENBERG[6].

The difference between the THOMSON heats of two metals is given by

$$\sigma_1 - \sigma_2 = T \frac{de_{12}}{dT},$$

where e_{12} is the thermoelectric power between them. Since $de_{12}/dT = 0$ between superconductors, this relation shows that the THOMSON coefficients of all super-conductive metals are equal. Furthermore, NERNST's theorem requires that σ be zero at $0°$ K, so that it is reasonable to expect the THOMSON heat to vanish over the whole superconducting temperature range. This expectation was

[1] K. STEINER and P. GRASSMANN: Phys. Z. **36**, 527 (1935).
[2] W. H. KEESOM and C. J. MATTHIJS: Physica, Haag **5**, 1 (1938). — Leiden Comm. 250d.
[3] H. B. G. CASIMIR and A. RADEMAKERS: Physica, Haag **13**, 33 (1947). — Leiden Comm. 270d.
[4] G. T. PULLAN: Proc. Roy. Soc. Lond., Ser. A **217**, 280 (1953).
[5] M. C. STEELE: Phys. Rev. **81**, 262 (1951).
[6] D. SHOENBERG [1], pp. 86—94.

confirmed experimentally by Daunt and Mendelssohn[1], who showed that the induction of a large persistent current produced no detectable change in the temperature distribution in an unequally heated lead ring. They found that σ for lead was less than 4×10^{-9} V/°K.

VII. Superconductive alloys and compounds.

30. Many alloys and compounds become superconductive at low temperatures. Alloys of the superconducting elements, either with each other or with non-superconducting metals, are known to be superconducting in a large range of concentrations. Shoenberg[2] lists about 40 alloy systems which fall into this category. Before World War II, many chemical compounds were also found to be superconductive[3]. Since the war, the properties of compounds have been intensively investigated and numerous new superconductors have been discovered. In Table 3 we present some examples of compounds as they have been classified and summarized in recent papers. No attempt has been made to achieve completeness or a logical scheme of classification; the main purpose is to convey the large number and diverse types of compounds which become superconductive above 1° K.

Table 3. *Superconductive compounds above 1° K.*

Class	Example	Approximate No. in class	References
Bi-compounds	LiBi	10	4
(Metals) + (non-metallic elements)	NbN	30	5
Ni As-structure	PdSb	4	6
Mo and W alloys	Mo_3Os	10	7

Recently, Matthias[7] found the empirical correlation shown in Fig. 42 between the transition temperature of a superconductor and its number of valence electrons per atom. This suggests that optimum conditions for the occurrence of superconductivity seem to exist for 5 and 7 valence electrons per atom.

A good many compounds exhibit reasonably sharp magnetic transitions in a longitudinal field.

Alloys and the "hard" superconductors, on the other hand, tend to exhibit diffuse transitions (cf. Sect. 7β). Small magnetic fields begin to penetrate gradually into the specimen and some superconductive threads persist in very large fields. This persistence of threads makes resistance measurements unreliable, because the specimen exhibits zero resistance in very high fields when the bulk of the specimen is in the normal phase. Since the threads are thin, the resistive transition is very sensitive to the strength of the measuring current. Furthermore, when the field is lowered from a high value, the specimen is left with a large frozen-in magnetic moment.

The foregoing paragraph accurately describes the properties of the majority of cases. However, recent work has revealed important exceptions in particular

[1] J. G. Daunt and K. Mendelssohn: Proc. Roy. Soc. Lond., Ser. A **185**, 225 (1946).
[2] D. Shoenberg [*1*], pp. 230—231.
[3] D. Shoenberg [*1*], pp. 228—229.
[4] B. T. Matthias and J. K. Hulm: Phys. Rev. **87**, 799 (1952).
[5] G. F. Hardy and J. K. Hulm: Phys. Rev. **93**, 1004 (1954).
[6] B. T. Matthias: Phys. Rev. **92**, 874 (1953).
[7] B. T. Matthias: Phys. Rev. **97**, 74 (1955).

systems and in dilute alloys, and has also provided clues as to the cause of the non-ideal behavior. We briefly review this work below.

WEXLER and CORAK[1] determined the B-H curves of several specimens of vanadium which differed from each other in mechanical hardness. In the softest specimen at any given temperature, the initial penetration of the magnetic field occurred at a sharply defined value, indicating that the superconductivity of a substantial fraction of the material was destroyed at a well defined field. As the field was further increased, the induction gradually rose to $B = H$, showing that superconductive regions tended to persist in the specimen. WEXLER and CORAK suggest that the sharp penetration fields correspond closely to the equilibrium fields which would be manifested by pure, unstrained specimens. With increasing specimen hardness, the point of initial penetration became less and less well defined and the transition temperature was lowered. Increased specimen hardness could be correlated with an increased content of absorbed nitrogen and

oxygen. As a result, it is proposed that the magnetic properties exhibited by the hard superconductors are due to mechanical strain arising from either mechanical work or interstitially located impurities such as carbon, nitrogen and oxygen. SHOENBERG[2] had earlier observed ideal magnetic behavior in a very pure thorium wire, even though thorium is in the "hard" group. It is suggested that the structure of thorium and its large atomic volume reduce the strains introduced by interstitial impurities.

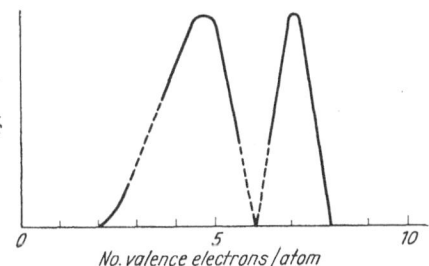

Fig. 42. Empirical correlation between the transition temperatures of compounds and their number of valence electrons per atom (after MATTHIAS).

The superconducting properties of indium-thallium solid solutions in the composition range from pure indium to 50% thallium were studied by STOUT and GUTTMAN[3]. The magnetic induction and electrical resistance of long cylindrical specimens in longitudinal fields were measured at various temperatures. Up to 10% thallium, the magnetic induction in the specimens at a given temperature jumped from zero to H at a sharply defined field, and resistance was restored at substantially the same field. Thus, these specimens exhibited ideal superconductive properties. As the concentration of thallium was increased above 10%, the flux penetration occurred over a wider range of field, and resistance appeared only after practically all the flux had penetrated, indicating that alloy effects became evident only at fairly high concentrations. The transition temperatures of the alloys were all smaller than T_c for pure indium. Along these same lines, LOHMAN and SERIN[4] investigated the transition temperatures of dilute solid solutions of antimony, bismuth, cadmium, indium, lead, mercury and zinc in tin. All specimens exhibited sharp transitions, and in all cases the effect of the impurity was initially to lower the transition temperature.

Recently, LYNTON, SERIN and ZUCKER[5] have extended the measurements on the tin solutions to low temperatures. They find in addition to the initial decrease of transition temperature, that the electronic specific heat constant, γ,

[1] A. WEXLER and W. S. CORAK: Phys. Rev. 85, 85 (1952).
[2] D. SHOENBERG: Proc. Cambridge Phil. Soc. 36, 84 (1940).
[3] J. W. STOUT and L. GUTTMAN: Phys. Rev. 88, 703 (1952).
[4] See B. SERIN [3], Chapter VII.
[5] E. A. LYNTON and B. SERIN: Int. Conf. on Low Temp. Phys., Paris 1955.

of the normal phase shows a small, gradual increase for all impurities, of valence
both higher and lower than tin.

In the discussion of Sect. 26 of the transition from the normal to the superconductive phase of tin, we stated that a superconductive sheath initially formed
on the surface of a rod. The sheath only occupies about 1/5 of the total volume
of a specimen, and we indicated that the flux trapped inside the sheath slowly
escapes through small gaps in it. PIPPARD[1] has suggested that the superconductive
metal on either side of a gap is prevented from coalescing by the necessarily
large value of the range of coherence, $\xi \gg \lambda$, in pure tin. In impure tin, when
$\xi \sim \lambda$, the gaps are assumed to be able to close, thereby permanently trapping
flux (λ is the penetration depth).

To check the effect of impurity on flux trapping, PIPPARD performed an
extensive series of experiments on the amounts trapped in rods of pure tin and
of tin alloyed with indium up to 3%. The
procedure was to place a rod in a transverse magnetic field which was increased
above the threshold value and then reduced to zero. The specimen was then
rapidly rotated through 180°, and the emf
induced in a search coil was observed in
order to obtain a measure of the amount
of flux trapped.

Significant results were obtained only after the specimens had been annealed for at
least about 20 days, apparently to homogenize the alloys. The proportion of trapped
flux in pure tin was only about 0.1%, and it
increased steadily as the indium concentration was increased. In addition to this effect,

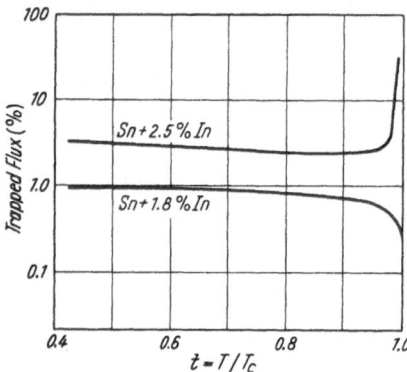

Fig. 43. Trapped flux as a function of the temperature of tin-indium alloys (after PIPPARD[1]).

it seemed that a change of behavior occurred between 2.3 and 2.5% of indium,
which is illustrated in Fig. 43. For indium concentrations less than 2.3%, the
proportion of trapped flux tends to zero as the temperature approaches T_c.
For greater indium concentrations, the trapping rises to a value of about 50%
at the transition temperature. PIPPARD suggests, therefore, that the apparent
change of behavior at 2.3% indium marks the beginning of spontaneous coalescence.
This interpretation of the experiments has recently been questioned by BUD-
NICK, LYNTON and SERIN[2]. They find that as a result of continued annealing
(up to about 100 days) the rise in trapped flux in impure tin specimens near T_c
becomes progressively less pronounced. It appears that even the most impure
specimens would show no rise after sufficient annealing. They therefore conclude that there is no direct evidence for the existence of spontaneous coalescence,
but that in well annealed specimens, any flux trapped can, at all temperatures,
for the most part escape when the external field is reduced to zero.

VIII. Diverse properties unchanged in the superconductive transition.

31. X-ray and neutron diffraction patterns. KEESOM and ONNES[3] observed
the X-ray diffraction patterns exhibited by lead above and below the transition
temperature and found no change in the patterns. Thus any changes in the

[1] A. B. PIPPARD: Phil. Trans. Roy. Soc. Lond. A **248**, 97 (1955).
[2] J. I. BUDNICK, E. A. LYNTON and B. SERIN: Phys. Rev. **103**, 286 (1956).
[3] W. H. KEESOM and H. KAMERLINGH ONNES: Leiden Comm. **1924**, 174b.

lattice spacing upon passing between the normal and superconductive phases are extremely small.

Recently, WILKINSON, SHULL, ROBERTS and BERNSTEIN[1] measured the coherent and incoherent scattering of neutrons by the electrons in vanadium, lead and niobium above and below their transition points, and found that in no case was there a change of the coherent scattering or the diffuse background. This result clearly indicates that there is no detectable change in the electronic distribution with advent of superconductivity. Examination of the nuclear scattering in lead and niobium showed that there were no pronounced changes in the atomic lattice vibrations at the transition temperature[2]. These same authors report that the total thermal neutron cross section for tin in the normal and superconducting states is the same within one percent.

32. Interaction with electrons. McLENNAN, McLEOD and WILHELM[3] measured the absorption by thin lead films of electrons having energies of a few Mev. The advent of superconductivity produces no change in the absorption coefficient.

MEISSNER and STEINER[4] determined that the transmission coefficient of tin foils for electrons of about 10 ev energy is the same in the superconductive and normal phases.

Recently, STUMP and TALLEY[5] measured the lifetime of positrons in superconductive lead and tin. The lifetime in lead appreciably increased when the lead passed from the normal to the superconductive state, but no change in lifetime was found in tin. At the present writing, the interpretation of these measurements is obscure.

33. Interaction with radiation, and field emission. The photoelectric current coming from lead illuminated with ultraviolet light was measured by McLENNAN, HUNTER and McLEOD[6], and it was found that the superconductive transition caused no detectable change in current.

Within the error of measurement of 0.2%, GOMER and HULM[7] observed no difference in the field emission current from tantalum at temperatures above and below T_c.

DAUNT, KEELEY and MENDELSSOHN[8] measured the reflectivity of lead and tin for infra-red radiation of about 10μ wavelength. There is no measurable difference in the reflectivities of the superconductive and normal phases. This result has been verified for tin by RAMANTHAN[9].

WEXLER[10] found that the amount of visible light transmitted by thin lead films is unaffected by the advent of superconductivity.

I should like to thank warmly my colleagues, Drs. E. A. LYNTON and P. LINDENFELD and my wife, BERNICE SERIN for their many helpful comments on this article.

[1] M. K. WILKINSON, C. G. SHULL, L. D. ROBERTS and S. BERNSTEIN: Phys. Rev. **97**, 889 (1955).

[2] This is equivalent to finding no change in the DEBYE temperature, Θ.

[3] J. C. McLENNAN, J. H. McLEOD and J. O. WILHELM: Trans. Roy. Soc. Canada (3) **23** (III), 264 (1929).

[4] W. MEISSNER and K. STEINER: Z. Phys. **76**, 201 (1932).

[5] R. STUMP and H. E. TALLEY: Phys. Rev. **96**, 904 (1954).

[6] J. C. McLENNAN, R. G. HUNTER and J. H. McLEOD: Trans. Roy. Soc. Canada (3) **24** (III), 3 (1930).

[7] R. GOMER and J. K. HULM: J. Chem. Phys. **20**, 1500 (1952).

[8] J. G. DAUNT, T. C. KEELEY and K. MENDELSSOHN: Phil. Mag. **23**, 264 (1937).

[9] K. G. RAMANTHAN: Proc. Phys. Soc. Lond. A **65**, 532 (1952).

[10] A. WEXLER: Phys. Rev. **70**, 219 (1946).

Bibliography.

[1] Shoenberg, D.: Superconductivity, 2nd ed. Cambridge: University Press 1952.
[2] Laue, M. von: Theory of Superconductivity. New York: Academic Press 1952.
[3] Gorter, C. J.: Progress in Low Temperature Physics. Amsterdam: North-Holland Publishing Company 1955.
[4] London, F.: Superfluids, vol. 1. New York: John Wiley and Sons 1950.

References Appended in Proof, March 1956.

Chap. II.

Sect. 24γ.

(1) Bedard, F., and H. Meissner: Measurements of contact resistance between normal and superconducting metals. Phys. Rev. **101**, 31 (1956).

Chap. III.

Sect. 14γ.

(2) Worley, R. D., M. W. Zemansky and H. A. Boorse: Heat capacities of V and Ta in the normal and superconducting phases. Phys. Rev. **99**, 447 (1955).

Sect. 15.

(3) Grenier, C.: The anisotropy of the effect of the elastic deformation on the super-conductivity of tin. C. R. Acad. Sci., Paris **238**, 2300 (1954); **240**, 2302 (1955).
(4) Garber, M., and D. E. Mapother: Effect of hydrostatic pressure on the superconducting transition of tin. Phys. Rev. **94**, 1065 (1954).
(5) Muench, N. H.: Effects of stress on superconducting Sn, In, Tl and Al. Phys. Rev. **99**, 1814 (1955).
(6) Hatton, J.: Effect of pressure on the superconducting transition of thallium. Phys. Rev. **100**, 1784 (1955).
(7) MacKinnon, L.: Relative absorption of 10 Mc/sec. longitudinal sound waves in a superconducting polycrystalline tin rod. Phys. Rev. **100**, 655 (1955).
(8) Bömmel, H. E.: Ultrasonic attenuation in superconducting and normal-conducting tin. Phys. Rev. **100**, 758 (1955).
(9) Pippard, A. B.: Ultrasonic attenuation in metals, Phil. Mag. **46**, 1104 (1955).
(10) Pippard, A. B.: Thermodynamics of a sheared superconductor. Phil. Mag. **46**, 1115 (1955).

Chap. IV.

Sect. 16 ε (2).

(11) Faber, T. E., and A. B. Pippard: The penetration depth and hf resistance of super-conducting Al. Proc. Roy. Soc. Lond., Ser. A **231**, 336 (1955).

Sect. 21.

(12) Grebenkamper, C. J., and J. P. Hagen: High frequency resistance of metals in the normal and superconducting state. Phys. Rev. **86**, 673 (1952).
(13) Grebenkamper, C. J.: Superconductivity of V at 24,000 Mc/sec. Phys. Rev. **96**, 316 (1954).
(14) Grebenkamper, C. J.: H-f resistance of Sn and In in the normal and superconducting state. Phys. Rev. **96**, 1197 (1954).
(15) Fawcett, E.: The surface resistance of normal and superconducting tin at 36 kMc/sec. Proc. Roy. Soc. Lond., Ser. A. **232**, 519 (1955).
(16) Blevins, G. S., W. Gordy and W. H. Fairbank: Superconductivity at millimeter wave frequencies. Phys. Rev. **100**, 1215 (1955).
(17) Biondi, M. A., M. P. Garfunkel and A. O. McCoubrey: Millimeter wave absorption in superconducting aluminum. Phys. Rev. **101**, 1427 (1956).
N.B. The foregoing two references [(16), (17)] report a new observation. For frequencies, such that $h\nu \lesssim kT_c$ (i.e. frequencies > 77 kMc/sec. in the case of Sn, and > 22 kMc/sec. in the case of Al) the metal seems to have the same residual resistance as in the normal state, at temperatures at which there is already complete dc superconductivity. For a given frequency, the temperature has to be reduced below the usual transition point before the surface resistance begins to decrease. The higher the frequency, the lower the temperature to start the decrease in surface resistance.

Sect. 22.

(18) LEWIS, H. W.: Search for a HALL effect in a superconductor II. Phys. Rev. **100**, 641 (1955).

Chap. V.

Sect. 24α.

(19) SCHAWLOW, A. L.: Structure of the intermediate state of superconductors. Phys .Rev. **101**, 573 (1956).

Sect. 24γ.

(20) GRASSMANN, P., and L. RINDERER: Critical values of the current in superconducting Pb-Bi alloy in external magnetic field. Helv. phys. Acta **27**, 309 (1954).
(21) RINDERER, L.: Destruction of superconductivity by the current carried and an applied transverse magnetic field. Z. Naturforsch. **10**a, 174 (1955).
(22) MEISSNER, H.: Paramagnetic effect in superconductors II. Phys. Rev. **101**, 31 (1956).

Sect. 26.

(23) FABER, T. E.: The phase transition in superconductors IV, Al. Proc. Roy. Soc. Lond., Ser. A **231**, 353 (1955).
(24) GALKIN, A. A., and P. A. BEZUGLYI: The kinetics of the destruction of superconductivity by a magnetic field. Zh. eksp. teor. Fiz. **28**, 463 (1955).

Chap. VI.

Sect. 28β.

(25) PHILLIPS, N. E.: Thermal conductivity of In-Tl alloys. Phys. Rev. **100**, 1719 (1955).

Sect. 28γ.

(26) RENTON, C. A.: Effect of a magnetic field on the heat conductivity of a superconductor. Phil. Mag. **46**, 47 (1955).

Chap. VII.

Sect. 30.

(27) MATTHIAS, B. T., and E. CORENZWIT: Superconductivity of Zr Alloys. Phys. Rev. **100**, 626 (1955).
(28) ZHURAVLEV, N. N., and G. S. ZHDANOV: Superconducting Bi-Rh compounds. Zh. eksp. teor. Fiz. **28**, 228 (1955); also ALEKSEEVSKI, N. E., G. S. ZHDANOV and N. N. ZHURAVLEV: Zh. eksp. teor. Fiz. **28**, 237 (1955).
(29) GLOVER III, R.: An empirical rule for the position of superconductors in the periodic table. Z. Physik **140**, 494 (1955).
(30) TEASDALE, T. S.: Permanent magnetic moments of a superconductive sphere. Phys. Rev. **99**, 1248 (1955).
(31) DOIDGE, R. P.: The transition to superconductivity. Phil. Trans. Roy. Soc. Lond. **248**, 553 (1956).

Chap. VIII.

Sect. 32.

(32) ALBERS-SCHÖNBERG, H., and E. HEER: Directional correlation measurements in super-conducting metals. Helv. phys. Acta **28**, 389 (1955).

Sect. 33.

(33) McCRUM, N. G., and C. A. SHIFFMAN: The optical constants of tin below the super-conducting transition temperature. Proc. Phys. Soc. Lond. A **67**, 368 (1954).

Theory of Superconductivity.

By

J. Bardeen.

With 20 Figures.

I. Introduction.

1. Although superconductivity falls into the domain where one would expect ordinary non-relativistic quantum mechanics to be valid, it has proved to be extremely difficult to obtain an adequate theoretical explanation of this remarkable phenomenon. In spite of the large amount of excellent experimental and theoretical work devoted to the problem, there remain major unsettled questions. However, the area in which the answers are to be found has been narrowed considerably. There are very strong indications, if not quite a proof, that superconductivity is essentially an extreme case of diamagnetism rather than a limit of infinite conductivity. The isotope effect indicates that the superconducting phase arises from interactions between electrons and lattice vibrations.

That the magnetic properties come from orbital motion of electrons and not from electron spins is shown by a measurement of the g-value from the gyromagnetic effect. Kikoin and Goobar[1], and more recently Pry, Lathrop and Houston[2], from observations of the angular momentum picked up by a sphere when a magnetic field is switched on, have found that the g-value is close to unity, as expected for orbital motion.

Let us first consider the nature of the electromagnetic properties: Is superconductivity, as the name implies, a limit of infinite conductivity in which the electrons are not scattered or is it a limit of perfect diamagnetism ($B=0$) as is indicated by the Meissner effect? These two aspects are very closely related. If the conductivity is infinite, the magnetic field in the interior of a massive specimen cannot change when the external field is changed, but the field need not be zero if the specimen is cooled in the presence of an external field. If the diamagnetic aspects are assumed basic, one must show why a current flowing in a ring is metastable and does not decay. Since the discovery of the Meissner effect in 1933 and the London [16] phenomenological description which followed shortly afterwards, it has generally been assumed those aspects usually associated with infinite conductivity are a consequence of the magnetic properties. The supercurrents are then always associated with and determined by the magnetic field. In other words, in the presence of a magnetic field the stable condition is that with currents, flowing near the surface, which prevent the penetration of the field.

When a current, I, flows in a ring, there is a one-parameter family of solutions determined by the flux through the ring or by the current flow. The complete current distribution is determined by this parameter. The lowest state

[1] I. K. Kikoin and S. V. Goobar: C. R. Acad. Sci. URSS. **19**, 249 (1938). — J. Phys. USSR. **3**, 333 (1940).
[2] R. H. Pry, A. L Lathrop and W V. Houston: Phys. Rev. **86**, 905 (1952).

corresponds to $I = 0$, but states with $I \neq 0$ are metastable and persist indefinitely. A possible explanation is that in the multiply-dimensional phase space of all of the electrons, there is only one unique path which leads to states of lower energy. Fluctuations are unlikely to lead to this path.

There exists, however, no rigorous proof of the metastability of such current distributions, and some recent theories, such as those of BORN and CHENG and of HEISENBERG and KOPPE have taken the viewpoint that persistent currents exist independently of the magnetic field, and that it is the stability or metastability of such currents which is the basic property. In a discussion of a one-dimensional model, FRÖHLICH[1] has also suggested that currents in the absence of magnetic fields may be metastable. Opposing the view that currents are really stable is a proof of BLOCH[2], extended by BOHM[3] to many-electron wave-functions, that in the state of lowest energy the current density must vanish. It is possible that entropy considerations may favor distributions of microscopic current loops, so that states with currents flowing are thermodynamically stable, as suggested by HEISENBERG [9].

An interesting experiment which indicates that the diamagnetic property is basic is a measure of the damping and period of a superconducting sphere oscillating in a magnetic field[4]. One expects no eddy current damping from either model. However, there should be a change in period from torques introduced by undamped eddy currents if the conductivity is infinite. On the diamagnetic model, the currents are always associated with the magnetic field and stay fixed in space as the sphere rotates. There is then no additional torque and no change in period, and this is what was found experimentally.

We adopt here the viewpoint that the diamagnetic aspects are basic, and show how they might follow as a consequence of quantum theory, along the lines suggested originally by F. LONDON [14], [15]. He showed that the LONDON equation $c \operatorname{curl} \Lambda \boldsymbol{j} = -\boldsymbol{H}$ follows if the wave functions are so rigid that they are not modified at all by a magnetic field. While many of the qualitative consequences of this equation of the LONDON's have been confirmed, the theory has not received a really good quantitative check. PIPPARD [20] has suggested on empirical grounds a modified form of the theory in which the current density at a point depends on the integral of the vector potential over a region surrounding the point. We shall show that when first order changes of the wave functions produced by the magnetic field are taken into account, one is led to a "nonlocal" version of the theory similar to that suggested by PIPPARD.

Another major question concerns the nature of the interactions responsible for the thermal transition and thermal properties. The isotope effect (SERIN[5] p. 237) shows rather conclusively that superconductivity arises from interactions between electrons and lattice vibrations, and theories based on this idea have been proposed independently by FRÖHLICH [4] and the author [1]. FRÖHLICH's theory, developed without knowledge of the isotope effect, gave a relation between critical temperature, T_c, and isotopic mass,

$$\sqrt{M}\, T_c \sim \text{const}$$

---

[1] H. FRÖHLICH: Proc. Roy. Soc. Lond., Ser. A **223**, 296 (1954).

[2] Quoted by L. BRILLOUIN: Proc. Roy. Soc. Lond., Ser. A **152**, 19 (1935).

[3] D. BOHM: Phys. Rev. **75**, 502 (1949).

[4] W. V. HOUSTON and N. MUENCH: Phys. Rev. **79**, 967 (1950). Closely related experiments are those of E. U. CONDON and E. MAXWELL: Phys. Rev. **76**, 578 (1949); **79**, 967 (1950) and FRITZ, GONZALEZ and JOHNSON: Phys. Rev. **79**, 967 (1950); **80**, 894 (1950).

[5] Unless otherwise qualified, references to this author advert to his preceding article in this volume.

which is close to that found empirically. Because of mathematical difficulties involved, neither of these theories is very satisfactory.

Anything approaching a rigorous deduction of superconductivity from the basic equations of quantum theory is a truly formidable task. The energy difference between normal and superconducting phases at the absolute zero is only of the order of 10^{-8} ev per atom. This is far smaller than errors involved in the most exacting calculations of the energy of either phase. One must neglect terms or make approximations which introduce errors which are many orders of magnitude larger than the small energy difference one is looking for. One can only hope to isolate the physically significant factors which distinguish the two phases. For this, considerable reliance must be placed on experimental findings and the inductive approach.

A great deal can be learned from the thermal and electrical properties: Specific heat, thermal conductivity, electrical conductivity observed in the penetration depth at microwave frequencies, and other properties which give information about the excited states of superconductors. A powerful method of attack in the analysis of properties of materials at low temperatures is to consider the nature of the elementary excitations; e.g. LANDAU's rotons in liquid helium II or BLOCH's spin waves in ferromagnetics[1]. The individual particle model gives a satisfactory account of excited states of electrons in normal metals. At $T = 0°$ K, all the levels are filled up to the FERMI level, E_F, and none of the higher states are occupied. Elementary excitations correspond to raising an electron from an occupied state to a higher unoccupied state. The fraction of electrons which are thermally excited at a temperature T is of the order of kT/E_F, and the average excitation energy is of the order of kT. This gives a thermal energy proportional to T^2 and a specific heat proportional to T, as observed in normal metals. The thermal properties of superconductors indicate that there are excited electrons similar to those of the normal phase, but that the number is greatly reduced when $T < T_c$. Since the transition is of second order, the number must be substantially the same just below T_c. Thus the elementary excitations in the superconductor, as in the normal metal, are probably those of individual particles. This is essentially the basis for the various two-fluid models which have been suggested to account for the thermal properties. The "normal" component corresponds to the excited electrons. The number of low-lying excited states and thus of excited electrons is greatly reduced in a superconductor.

It has been suggested that the electrons form some sort of condensed state in the superconducting phase such that a finite energy $\varepsilon \sim kT_c$ is required to excite an electron near $T = 0°$ K. This has been called the "energy gap" model. This would lead to a specific heat and thermal conductivity varying as $\exp(-\varepsilon/2kT)$ at very low temperatures. While there is some evidence[2] that this is the case, the question is still open. According to the GORTER-CASIMIR two-fluid model, which fits the data for a number of superconductors at moderate temperatures, the specific heat varies as T^3. If this latter model is correct at $T \to 0$ there is reduced density of low-lying states, but no true energy gap.

It can be shown that the energy gap model, originally introduced to account for the thermal properties, if taken literally gives the MEISSNER effect, and in fact leads to a theory similar to PIPPARD's modification of the LONDON equation for the current density in a magnetic field. Thus the essential task of the microscopic theory is to show why there are few low-lying excited states in the superconducting phase.

---

[1] GINSBURG [5] has advocated this approach to superconductivity.
[2] See SERIN, Sect. 12 and 27.

Since the BLOCH individual particle approximation accounts satisfactorily for the properties of normal metals, it has been thought that superconductivity arises from one of the terms neglected in this theory. One of these is the correlation between the positions of the electrons due to COULOMB forces, used in the HEISENBERG theory [9]. It was suggested that electrons with energies near the FERMI surface form a lattice and so tend to keep apart and reduce to the long range COULOMB energy between electrons. Another is magnetic interactions between electrons, as suggested by WELKER[1]. A third is electron-phonon interactions, originally introduced to account for scattering of electrons and thus resistance. They also contribute to the energy of both normal and superconducting phases, and are presumably responsible for the transition.

It has been shown recently by PINES and the author [2] why, as is indicated experimentally by the isotope effect, electron-phonon interactions are more important than COULOMB interactions. The reason is essentially that the COULOMB interaction between electrons is a screened interaction of relatively short range. The long range part of the interaction gives rise to plasma oscillations which are of such high frequency that they are not normally excited and also to coupled electron-ion oscillations which are just the lattice vibrations of long wavelength. The remaining screened interaction is sufficiently weak so that it can be treated by perturbation methods, and thus does not have a marked effect on the wave functions. On the other hand the criterion for superconductivity, as given for example in FRÖHLICH's theory, is essentially that the electron-phonon interaction be so large that it cannot be treated by perturbation theory. This means that the wave functions may be modified greatly by the interaction. Mathematical methods for treating such large interactions are lacking, so that we still do not have a satisfactory picture of the superconducting phase.

Partly because of the difficulties involved in developing a fundamental microscopic theory, considerable effort has been devoted to the development of phenomenological theories. These include two-fluid models to describe the thermal properties, equations such as those of the LONDON's to account for the electrodynamic properties, and theories of boundary effects to derive properties of the intermediate state and related phenomena. Most of these theories are still on a rather insecure foundation and have not been subjected to convincing quantitative checks. The only relations one can be really sure of are those based on thermodynamics.

In Chap. II we give a brief outline of the thermodynamic relations, and then discuss some of the two-fluid models which have been proposed; Chap. III is devoted to the LONDON phenomenological theory and generalizations of these equations which have been proposed by PIPPARD, Chap. IV to boundary effects and the intermediate state, including the LANDAU theory [13], and Chap. V to the microscopic theories. The latter is concerned mainly with the formulation of the problem of calculating electron-phonon interactions. An outline is given of FRÖHLICH's theory and other attempts, none very successful, for calculating the interaction energy in normal and superconducting states.

II. Thermodynamic properties and two-fluid models.

a) Thermodynamic relations.

2. Temperature dependence of critical field. The critical field is determined by thermodynamic considerations. Exclusion of flux gives an increase in magnetic energy, and when this increase more than compensates for the lower free energy

---

[1] H. WELKER: Z. Physik **114**, 525 (1939).

of the superconducting phase in zero field, a transition to the normal state occurs. The MEISSNER effect shows that there is a unique state of a superconductor under given conditions of temperature, pressure and external applied magnetic field, so that one should be able to derive thermodynamic relations concerning the transition parameters. As a matter of fact, thermodynamics was applied with good results prior to the discovery of the MEISSNER effect, first by KEESOM[1], and later by RUTGERS[2] and by GORTER[3]. The reason for this success became apparent only after the discovery that the transition really is reversible.

Our treatment is patterned closely after that in the excellent book of SHOEN-BERG [23], who has also contributed to the theory. It is most convenient to take the pressure, P, and the external field, H_a, as independent variables. We shall restrict the discussion at present to massive specimens for which we may assume that the field is zero in the interior. Boundary effects and thin films are discussed in Chap. IV. One then deals with the GIBBS free energy,

$$G = U - TS + PV - \int_0^{H_a} M(H_a)\, dH_a,$$
(2.1)

where U, the internal energy, and S, the entropy, are assumed independent of H_a, and $M(H_a)$ is component of the magnetic moment in the direction of H_a. If the free energy depends on a parameter x, the equilibrium value is such that

$$\left(\frac{\partial G}{\partial x}\right)_{T, P, H_a} = 0.$$
(2.2)

The entropy, volume and magnetic moment are given by:

$$S = -\left(\frac{\partial G}{\partial T}\right)_{P, H_a},$$
(2.3)

$$V = \left(\frac{\partial G}{\partial P}\right)_{T, H_a},$$
(2.4)

$$M = -\left(\frac{\partial G}{\partial H_a}\right)_{T, P}.$$
(2.5)

It will be most convenient to consider first a long rod parallel to the field for which the demagnetizing coefficient is zero. The magnetic moment is then

$$M = -\frac{VH_a}{4\pi}.$$
(2.6)

If $G_s(0)$ is the free energy in absence of an external field,

$$G_s(H_a) = G_s(0) + \frac{1}{8\pi} V H_a^2.$$
(2.7)

The critical field, H_c, is that for which $G_s(H_c)$ becomes equal to the free energy G_n of the normal metal, assumed independent of H_a;

$$G_n = G_s(0) + \frac{1}{8\pi} V H_c^2.$$
(2.8)

This result must be independent of the shape of the body, and it is true that the area under the magnetization curve is independent of shape:

$$-\int_0^{H_1} M(H_a)\, dH_a = \frac{1}{8\pi} V H_c^2.$$
(2.9)

[1] W. H. KEESOM: Rapp. et Disc. 4e. Congr. Phys. Solvay, p. 288.
[2] A. J. RUTGERS: Physica, Haag 1, 1055 (1934); 3, 999 (1936).
[3] C. J. GORTER: Arch Mus. Teyler 7, 378 (1933).

Here H_1 is sufficiently large to bring the specimen to the normal state. Surface effects and any small dia- or paramagnetism of the normal state are neglected. It is also assumed that the transition takes place reversibly.

Relations between various thermodynamic quantities at the transition can be obtained by taking appropriate derivatives of (2.7) along the transition curve $H_c(T)$. The entropy difference between normal and superconducting states is:

$$S_n - S_s = - \frac{V H_c}{4\pi} \frac{dH_c}{dT}. \qquad (2.10)$$

The heat, Q, absorbed in going from the superconducting to normal states is

$$Q = T(S_n - S_s) = - T \frac{V H_c}{4\pi} \frac{dH_c}{dT}. \qquad (2.11)$$

The difference in specific heats obtained from the relation $C = T(\partial S/\partial T)$ is given by

$$C_s - C_n = \frac{T V_m}{4\pi} \left(H_c \frac{d^2 H_c}{dT^2} + \left(\frac{dH_c}{dT}\right)^2 \right), \qquad (2.12)$$

where V_m is the specific volume.

In zero applied field, the transition is of second order. There is no latent heat $(Q=0)$, but there is a discontinuity in specific heat, ΔC, at the transition point:

$$\Delta C = \frac{V_m T_c}{4\pi} \left(\frac{dH_c}{dT}\right)^2. \qquad (2.13)$$

A comparison of observed values of the left and right sides of (2.12), known as RUTGERS' relation, is given in Table 1, taken from SHOENBERG.

3. Pressure and volume variations. There are small but observable changes of H_c and T_c with pressure. These can be related, by means of thermodynamic relations, with the small volume difference, neglected so far, between normal and superconducting phases and also with differences in thermal expansion and compressibility between the two phases. We suppose that H_c in (2.8) is a function of P and T, and apply (2.4) to find

Table 1. *Test of* RUTGERS' *Relation* [Eq. (2.12)]. (From SHOENBERG [23], p. 62).

Metal	T_c °K	$(dH_c/dT)_{T=T_c}$	$\Delta C \times 10^3$ calc. cal/°K	$\Delta C \times 10^3$ obs. cal/°K
Pb	7.22	200	10	12.6
Ta	4.40	320	9.4	8.2, 9
La	4.37	1000	190.00	13.9
Sn	3.73	151	2.61	2.4, 2.9
In	3.37	146	2.08	2.3
Tl	2.38	139	1.47	1.48
Al	1.20	177	0.71	0.46

$$V_n - V_s(0) = \frac{V H_c}{4\pi} \left(\frac{\partial H_c}{\partial P}\right)_T + \frac{H_c^2}{8\pi} \left(\frac{\partial V}{\partial P}\right)_T. \qquad (3.1)$$

Since

$$G_s(H_a) = G_s(0) + \frac{V}{8\pi} H_a^2, \qquad (3.2)$$

we have

$$V_s(H_c) - V_s(0) = \left(\frac{\partial V}{\partial P}\right)_T \frac{H_c^2}{8\pi}, \qquad (3.3)$$

and, from (3.1), the change in volume ΔV, at the transition is

$$\Delta V = V_n - V_s(H_c) = \frac{V H_c}{4\pi} \left(\frac{\partial H_c}{\partial P}\right)_T. \qquad (3.4)$$

On combining (3.4) with the thermodynamic relation

$$\left(\frac{\partial H_c}{\partial P}\right)_T = - \left(\frac{\partial H_c}{\partial T}\right)_P \left(\frac{\partial T}{\partial P}\right)_{H_c}, \qquad (3.5)$$

and making use of (2.11), we find the Clausius-Clapeyron relation:

$$\left(\frac{\partial P}{\partial T}\right)_{H_e} = -\frac{Q}{\Delta V}. \tag{3.6}$$

The left hand side is the change in pressure required to keep the critical field at the same value, H_c, as the temperature is changed. Further derivatives of (3.4) with respect to T and to P give the change of the thermal expansion coefficient, $\Delta\alpha$, and the change in compressibility, ΔK, at the transition, where

$$\alpha = \frac{1}{V}\frac{\partial V}{\partial T}, \qquad K = -V\frac{\partial P}{\partial V}. \tag{3.7}$$

For the transition in zero applied field, $H_c = 0$, the expressions reduce to:

$$\Delta\alpha = \frac{1}{4\pi}\left(\frac{\partial H_c}{\partial T}\right)_P \left(\frac{\partial H_c}{\partial P}\right)_T, \tag{3.8}$$

$$\Delta K = \frac{K^2}{4\pi}\left(\frac{\partial H_c}{\partial P}\right)^2. \tag{3.9}$$

There are also relations due to Ehrenfest which apply to all second-order transitions:

$$\frac{dT_c}{dP} = \frac{V_m T_c \Delta\alpha}{C_n - C_s} = \frac{\Delta K}{K^2 \Delta\alpha}. \tag{3.10}$$

A much more complete discussion of these relations is given in Shoenberg's book [23].

b) Two-fluid models.

4. Gorter-Casimir and related theories. Various two-fluid models have been suggested to account for the thermal properties of superconductors. They are based on two general assumptions: (1) There is a condensed state, the energy of which is characterized by some sort of order parameter; (2) all of the entropy comes from excitations of individual particles similar to those of the normal metal[1]. The number of excited electrons depends on the temperature as well as on the order parameter. An order parameter of some sort is required to give a second-order transition such that the condensation energy varies from a maximum at $T = 0°$ K to zero at the transition temperature. The excited electrons account not only for the entropy and part of the specific heat, but also for the thermal conductivity, a.c. resistance and viscosity of electrons in the superconducting phase.

The first and best-known two-fluid model is that of Gorter and Casimir [8] which in the usual form leads to a specific heat varying as T^3. Koppe [11] derived a particular form of the two-fluid model on the basis of the Heisenberg theory. However, Koppe's theory does not depend on the interaction assumed to be responsible for the condensation, and may well have more general validity. Ginsburg's theory[2] is based on the energy gap model in which it is assumed that a minimum energy $\varepsilon \sim k T_c$ is required to excite an electron from the condensed phase. Rather general formulations which include the others as special cases have been discussed by Koppe [11], Bender and Gorter[3] and Marcus and Maxwell[4].

---

[1] The naive interpretation as two independent fluids is not justified.
[2] W. L. Ginsburg: J. exp. theor. Phys. USSR. **14**, 134 (1946) and [5].
[3] P. L. Bender and C. J. Gorter: Physica, Haag **18**, 597 (1952).
[4] P. M. Marcus and E. Maxwell: Phys. Rev. **91**, 1035 (1953).

The choice of the order parameter is somewhat arbitrary. We shall follow MARCUS and MAXWELL and others and take a parameter ω which varies from unity at $T=0°$ K to zero at $T=T_c$ and which is such that the condensation energy relative to the normal metal is $-\beta\omega$, where β is a parameter, characteristic of the metal, given by

$$\beta = \frac{V_m H_0^2}{8\pi}.$$ (4.1)

Here H_0 is the critical field at $T=0°$ K. The HELMHOLTZ free energy of the normal phase may be expressed in the form:

$$F_n = U(0) - \tfrac{1}{2}\gamma T^2 + F_L(T)$$ (4.2)

where γT is the electronic specific heat and F_L is the contribution of lattice vibrations. In going to the superconducting phase, it is assumed that any change in $U(0)+F_L$ is accounted for by the term $-\beta\omega$ and that $(\tfrac{1}{2})\gamma T^2$ is reduced by a factor $K(\omega)$, so that

$$F_s = U(0) - \beta\omega - \tfrac{1}{2}\gamma T^2 K(\omega) + F_L(T).$$ (4.3)

The reduction factor K may depend on T as well as on ω.

If superconductivity arises from interactions between electrons and lattice vibrations, the condensation energy may appear as a reduction in zero-point energy of the oscillations. If the predominant wave lengths involved are so short that the oscillations are not excited at low temperatures, as appears to be the case, the temperature dependent terms in $F_L(T)$ will not be affected by the transition.

The various theories differ in what is taken for $K(\omega)$. To agree with experiment, $K(\omega)$ must approach zero as $\omega \to 1$ (corresponding to very low temperatures) and, in order to have a second order transition, must approach unity as $\omega \to 0$ (corresponding to $T=T_c$).

GORTER and CASIMIR made the *ad hoc* assumption that

$$K(\omega) = (1-\omega)^\alpha,$$ (4.4)

which is a simple function satisfying both limiting values. They found best agreement with experiment by taking $\alpha = \tfrac{1}{2}$, the value which leads to an electronic specific heat varying as T^3 and a parabolic critical field curve. As will be discussed in the following, MARCUS and MAXWELL find that a smaller value of α gives a better fit to the critical field curves of several elements, so that it is probably best to leave α as a parameter to be determined empirically.

The equilibrium value of ω is that which makes (4.3) a minimum:

$$\omega_e = 1 - \left(\frac{T}{T_c}\right)^{\frac{2}{1-\alpha}} = 1 - t^{\frac{2}{1-\alpha}}.$$ (4.5)

where $t = T/T_c$ is the reduced temperature and T_c is the critical temperature:

$$T_c = \sqrt{\frac{2\beta}{\alpha\gamma}} = \sqrt{\frac{V_m H_0^2}{4\pi\alpha\gamma}}.$$ (4.6)

The entropy is

$$S_s = \gamma T (1-\omega)^\alpha = \gamma T\, t^{\frac{2\alpha}{1-\alpha}},$$ (4.7)

and the electronic specific heat is

$$C_s = \frac{\gamma T (1+\alpha)}{1-\alpha}\, t^{\frac{2\alpha}{1-\alpha}}.$$ (4.8)

The critical field is obtained from the difference between F_n and F_s:

$$h^2 = \left(\frac{H_c}{H_0}\right)^2 = 1 - \frac{1}{\alpha}\, t + \frac{1-\alpha}{\alpha}\, t^{\frac{2}{1-\alpha}}. \qquad (4.9)$$

For the special value $\alpha = \tfrac{1}{2}$ these reduce to:

$$\left.\begin{array}{ll} \omega_e = 1 - t^4, & s = \gamma\, T_c\, t^3, \\[4pt] c_s = 3\gamma\, T_c\, t^3, & h = 1 - t^2. \end{array}\right\} \qquad (4.10)$$

MAXWELL[1] finds that of the three parameters, only β varies with isotopic mass; the other two quantities, α and γ, remain constant.

One of the great successes of the GORTER-CASIMIR theory is the prediction of the way the penetration depth, λ, varies with temperature. According to the LONDON theory, λ^2 is inversely proportional to the concentration of electrons responsible for superconductivity, n_s. In the two-fluid model, it is assumed that n_s is proportional to ω, so that for $\alpha = \tfrac{1}{2}$:

$$\lambda^{-2} \sim \omega_e \sim 1 - t^4,$$

or

$$\lambda = \frac{\lambda_0}{\sqrt{1 - t^4}}. \qquad (4.11)$$

where λ_0 is the penetration depth at $t = 0°$ K. This predicted variation is in agreement with observation.

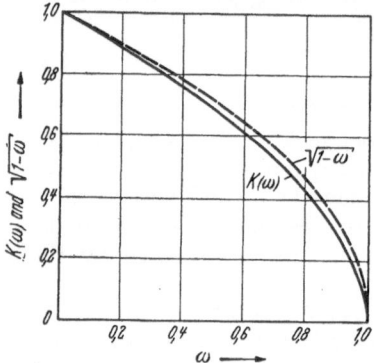

Fig. 1. Comparison of $K(\omega)$ from KOPPE's theory and $\sqrt{1-\omega}$ (after KOPPE [11]).

KOPPE's expression [11] for $K(\omega)$ is rather complicated, although it is based on a rather simple idea. It is assumed that the condensation takes place in momentum space, and that it occurs over a fraction ω of the FERMI surface. States outside of the condensed area of the FERMI surface are used to form the wave function of the condensed state, and so are no longer available to form excited states of individual particles. Thus it is assumed that the density of available states above the FERMI surface is reduced by a factor $1 - \omega$, while the density of states below the FERMI surface is unchanged. This latter assumption does not appear to be a reasonable one, because one would expect that states both above and below the FERMI surface would be used to form the condensed state, so that the density of both would be reduced.

A plot of $K(\omega)$, as calculated by KOPPE, is compared in Fig. 1 with one of the GORTER-CASIMIR function, (4.4). They are close together except when ω is close to unity (very low temperatures). The KOPPE theory differs from the GORTER-CASIMIR theory in that it gives an exponential temperature dependence as $t \to 0$, as would be expected for a model with an energy gap to the excited states. GOODMAN[2] has given an interpretation of the theory in terms of a model of this sort, in which the energy gap varies from $\sqrt{\pi^2/6}\, k\, T_c$ at $T = 0$ to zero at $T = T_c$. As GOODMAN points out, this interpretation may be more reasonable than KOPPE's. KOPPE's theory gives a variation of thermal conductivity with temperature in reasonably good agreement with observed values for metals in which the mean free path is limited by impurity scattering.

[1] E. MAXWELL: Phys. Rev. 87, 1126 (1952).
[2] B. B. GOODMAN: Proc. Phys. Soc. Lond. A 56, 217 (1953).

5. Energy gap models. A two fluid model based directly on an energy gap model is that if GINSBURG [5]. He assumes that the free energy of the super-conducting phase may be written

$$F_s = U(0) - \beta - \frac{1}{2}\gamma_s T^2 e^{-\frac{\varepsilon}{2kT}} + F_L(T),\qquad(5.1)$$

where γ_s and ε are chosen so as to give a second order transition at $T = T_c$. One may express γ_n and γ_s in terms of β, ε and T_c.

The conditions

$$F_n(T_c) = F_s(T_c) \quad \text{and} \quad \left(\frac{\partial F_n}{\partial T}\right)_{T_c} = \left(\frac{\partial F_s}{\partial T}\right)_{T_c}\qquad(5.2)$$

give

$$\gamma_s = \frac{4\beta}{\sigma T_c^2}\,e^{\sigma},\qquad(5.3)$$

$$\gamma_n = \frac{2\beta}{T_c^2}\left(1 + \frac{2}{\sigma}\right),\qquad(5.4)$$

where

$$\sigma = \frac{\varepsilon}{2kT_c}.\qquad(5.5)$$

The critical field curve is given by:

$$\left(\frac{H_c}{H_0}\right)^2 = 1 - \left(\frac{T}{T_c}\right)^2\left\{\frac{2+\sigma}{\sigma} - \frac{2}{\sigma}e^{\sigma\left(1-\frac{T_c}{T}\right)}\right\}.\qquad(5.6)$$

GINSBURG finds a reasonably good fit with the nearly parabolic critical field curve of Hg if $\sigma = 2.75$.

It should be noted that β in (5.1) is not multiplied by a factor, such as ω in (4.3), which goes to zero as $T \to T_c$. Further, it is necessary to take $\gamma_s > \gamma_n$, which does not appear reasonable if the term is to be interpreted as coming from excited electrons. The particular choice of parameters to make the transition one of second order seems to be an artificial one. However, the theory can be interpreted in a more reasonable way. One may assume that the energy ε required to create an excited electron and "hole" decreases linearly with the temperature, so that the temperature dependent free energy of excitation in the superconducting phase is

$$-\frac{1}{2}\gamma_n T^2 e^{-\frac{\varepsilon_0}{2kT}\left(1-\frac{T}{T_c}\right)},\qquad(5.7)$$

where ε_0 is the value of ε at $T = 0°$ K. In order to get a second order transition, one may assume that the condensation energy [given by $-\beta\omega$ in (4.3)] decreases with increase in number of excited electrons and is equal to

$$-\beta\left(1 - \left(\frac{T}{T_c}\right)^2 e^{-\frac{\varepsilon_0}{2kT}\left(1-\frac{T}{T_c}\right)}\right),\qquad(5.8)$$

which goes to zero as $T \to T_c$. The sum of (5.7) and (5.8) is of the form assumed by GINSBURG,

$$-\beta - \frac{1}{2}\gamma_s T^2 e^{-\frac{\varepsilon_0}{2kT}},\qquad(5.9)$$

where

$$\gamma_s = \left(\gamma_n - \frac{2\beta}{T_c^2}\right)e^{\frac{\varepsilon_0}{2kT_c}} = \frac{4\beta}{\sigma T_c^2}e^{\sigma},\qquad(5.10)$$

as given by (5.3) and (5.4).

This interpretation also has its difficulties. One would not expect a linear decrease in ε with T. Further, (5.7) is not quite the correct form for the free energy of excited electrons in an energy gap model. One should have

$$
F = -2kT \int_{E_F + \frac{\varepsilon}{2}}^{\infty} \log\left(1 + e^{\frac{E_F - E}{kT}}\right) N(E)\, dE , \tag{5.11}
$$

where $N(E)$ is the density of states and E_F the FERMI energy[1].

There are various ways one might develop a more satisfactory theory for the energy gap model. It would be desirable to introduce an order parameter. This could, perhaps, be taken to be the number of excited electrons, n_{exc}, and one could assume that the condensation energy decreases with increasing n_{exc} and would go to zero at a particular value of $n_{exc}(T_c)$ which would be equal to the number in the normal phase at the critical temperature. Another possibility would be to take the relative value of the energy gap as an order parameter. It would be desirable to develop such a theory if experiment or theory indicates that there is a true energy gap.

The GORTER-CASIMIR theory, for example in the prediction of the variation of penetration depth with temperature, is most successful at the higher temperatures, near T_c. Perhaps the correct theory will yield something like the GORTER-CASIMIR model at high temperatures $(T > 0.5\, T_c)$ an energy gap model at low temperatures $(T < 0.5\, T_c)$.

III. LONDON theory and generalization.

a) LONDON theory.

6. Introduction. The phenomenological theory of H. LONDON and F. LONDON [16], developed soon after the discovery of the MEISSNER effect, is based on the diamagnetic approach in that it gives a unique relation between current and magnetic field. On the other hand, it is closely related to the infinite conductivity approach, since the allowed current distributions represent a particular class of solutions for electron motion in the absence of scattering.

The LONDON theory gives a complete and consistent electrodynamics of superconductors which has been applied to a wide variety of problems and which has been very successful in correlating and predicting results of experiments. One outstanding success was the prediction, prior to observation, of a penetration depth, together with a correct estimate ($\sim 10^{-6}$ cm.) of its order of magnitude. Nevertheless, the theory has not received a good quantitative check, and in fact in at least one case (anisotropy of the penetration depth in tin [20]) appears to be in direct contradiction with experiment[2]. The LONDON equations probably represent only an idealized limiting case of more complicated equations required for actual superconductors. As such, they will continue to be very useful, even though the solutions may not be in complete quantitative agreement with experiment.

---

[1] See, for example, A. H. WILSON: The Theory of Metals, (2nd Ed.), p. 329, Cambridge 1954.

[2] PIPPARD (private communication) states that the interpretation of the anisotropy experiments is in doubt because the correction from the real part of the surface impedance is not negligible at the frequency employed. The extrapolation procedure used to get the low temperature limit does not eliminate the correction terms.

PIPPARD [20], on empirical grounds, has proposed a generalization of the LONDON equations in such a way as to give a non-local relation between current density and magnetic field. The current density at a point is determined by the field in a region of $\sim 10^{-4}$ cm. surrounding the point. While the details of his theory may not be correct, there is good experimental and theoretical justification for a generalization of this sort (Sect. 26). PIPPARD's theory has not yet been developed to give a complete electrodynamics of superconductors.

The basis for PIPPARD's theory is his concept of coherence. By this is meant that the range of order or the wave functions of the condensed superconducting phase extend over rather large regions of space, of the order of 10^{-4} cm. in pure material. Experimental evidence for such a coherence range comes from (1) the abruptness of the transition in zero field, which indicates the absence of local fluctuations and shows cooperation of large numbers of electrons, (2) the fact that the observed change in penetration depth with field is small, which shows that the order parameter can not change over regions less than $\sim 10^{-4}$ cm., and (3) the large boundary energy between normal and superconducting phases, which shows that the transition region is spread out over $\sim 10^{-4}$ cm. The experiments on which these remarks are based have been carried out for the most part on tin.

An estimate of the coherence length can be obtained in a general way from the uncertainty principle[1], if it is assumed that the interactions responsible for superconductivity are between states with energies within $\Delta E \sim k\, T_c$ of the FERMI surface, E_F. The states involved then lie in a thin shell in k-space with a thickness of the order of

$$\Delta k \sim \frac{\Delta E}{E_F}\, k_F \sim 10^{-4} \times 10^8 \sim 10^4 \text{ cm.}^{-1}. \tag{6.1}$$

The smallness of the range Δk indicates that the wave functions extend over large distances in real space. The uncertainty relation,

$$\Delta k\, \Delta x \sim 1, \tag{6.2}$$

gives a range $\Delta x \sim 10^{-4}$ cm.

That wave functions extending over large regions of space are favorable for a large diamagnetism has been known, of course, for a long time, and has been discussed by SLATER[2], KLEIN and LINDHARD[3] and others. The condition is essentially that the extent of the wave function be greater than the penetration depth, so that electrons cannot be localized within the penetration depth[4]. Diamagnetism is a quantum phenomenon; in classical theory electrons move under the influence of a magnetic field in such a way that there is no net diamagnetism. Quantum effects, as shown by LANDAU, give a small net diamagnetism for a degenerate electron gas. Electrons in bound states give a susceptibility proportional to the square of the radius of the orbit. Superconductivity is an extreme case in which the field changes markedly over the size of the wave packets which represent the system.

F. LONDON [14] showed that the LONDON equation relating current density and field follow from a quantum theoretic approach if it is assumed that the wave functions are not modified at all by a magnetic field. Since the penetration depth does not vary much with field, a linear theory should be satisfactory, but in calculating the current density, one *should* include first order changes of the

---

[1] J. BARDEEN reference [27], p. 5, and A. B. PIPPARD, reference [20].
[2] J. C. SLATER: Phys. Rev. **51**, 195 (1937); **52**, 214 (1937).
[3] O. KLEIN and J. LINDHARD: Rev. Mod. Phys. **17**, 305 (1945).
[4] J. BARDEEN: Phys. Rev. **81**, 829 (1951).

wave functions produced by the magnetic field. As pointed out in the introduction, the energy gap model, introduced to account for thermal properties, leads to the MEISSNER effect and to an expression for current density similar to that proposed by PIPPARD. In Sect. 26 we give arguments pro and con for the PIPPARD and LONDON versions of the theory.

Since the original version of the LONDON theory has been presented very adequately and very completely in at least two books, one by F. LONDON [15] and the other by M. VON LAUE [25], we shall give here only a brief account of the essential features and important solutions. We then give LONDON's argument which shows how such a theory might follow from quantum mechanics. This will be followed by a discussion of PIPPARD's version of the theory and a derivation of the latter from the energy-gap model.

7. Basic equations of LONDON theory[1]. The LONDON equations relate the current density at a point with the electric and magnetic fields. In the formulation of the theory, a distinction is made between the supercurrent, j_s, related to the diamagnetic properties of the condensed phase, and the normal current, j_n, which presumably comes from motion of excited individual particles. The total current density is the sum

$$j = j_s + j_n.\tag{7.1}$$

It is assumed that j_n obeys OHM's law:

$$j_n = \sigma E.\tag{7.2}$$

The equations peculiar to the theory are those which relate j_s with the magnetic and electric fields:

$$\text{(I)} \qquad c \operatorname{curl} \Lambda j_s + H = 0,\tag{7.3}$$

$$\text{(II)} \qquad \frac{\partial}{\partial t}(\Lambda j_s) - E = 0.\tag{7.4}$$

Here H is the total field (which might better be called B) resulting from the currents in the body as well as the external field. The parameter Λ is a constant characteristic of the material.

We shall assume for simplicity that the permeability and dielectric constant are both equal to unity so that we need not distinguish between B and H and E and D. MAXWELL's equations in GAUSSIAN units are then:

$$\left.\begin{aligned}
c \operatorname{curl} H &= 4\pi j + \frac{\partial E}{\partial t},\\
c \operatorname{curl} E &= -\frac{\partial H}{\partial t},\\
\operatorname{div} H &= 0,\\
\operatorname{div} E &= 4\pi \varrho.
\end{aligned}\right\}\tag{7.5}$$

Eqs. (7.1) through (7.5) are the basic equations of the LONDON theory.

Eqs. (7.1) and (7.2) may be combined with (7.3) and (7.4) to give:

$$c \operatorname{curl} \Lambda j = -H - \sigma \Lambda \frac{\partial H}{\partial t},\tag{7.6}$$

$$\frac{\partial}{\partial t}(\Lambda j) = E + \sigma \Lambda \frac{\partial E}{\partial t}.\tag{7.7}$$

---

[1] For more complete discussions, see references [15] and [25].

Eqs. (7.6) and (7.7) have the advantage that only the total current density appears rather than j_s and j_n. The separation of the total current into the two components may be somewhat artificial.

In anisotropic media, Λ is to be interpreted as a tensor. The general theory for such media has been developed most completely by von Laue[1]. He writes the London equations in the form

$$c \operatorname{curl} \boldsymbol{G} = -\boldsymbol{H}, \tag{7.8}$$

$$\frac{d\boldsymbol{G}}{dt} = \boldsymbol{E}, \tag{7.9}$$

where $\boldsymbol{G}$ is interpreted as the momentum density of the supercurrents. For anisotropic media,

$$G_i = \sum_j \Lambda_{ij}\, i_{sj} \tag{7.10}$$

where Λ_{ij} is a tensor.

Von Laue[2] has also suggested how the theory might be generalized if a non-linear theory is required. In this case, (7.10) might be replaced by a general functional relation between $\boldsymbol{G}$ and j_s. This version was stimulated by Heisenberg's theory, which indicated that non-linear effects might occur at high fields. Present indications are that non-linear effects are small.

A perfect conductor consisting of an electron gas subject to no scattering leads to (II) but not to (I). The Londons added (I) to the earlier "acceleration" theory of Becker, Sauter and Heller[3] to account for the Meissner effect. The derivation is as follows. Let $\boldsymbol{v}(x, y, z, t)$ be the average drift velocity of the electron gas. The particle acceleration is then given by the Lorentz force:

$$\frac{d\boldsymbol{v}}{dt} = \frac{\partial \boldsymbol{v}}{\partial t} + \boldsymbol{v} \cdot \operatorname{grad} \boldsymbol{v} = -\frac{e}{m}\left(\boldsymbol{E} + \frac{1}{c}\boldsymbol{v} \times \boldsymbol{H}\right). \tag{7.11}$$

This equation may be written in the form:

$$\frac{\partial \boldsymbol{v}}{\partial t} + \operatorname{grad} \frac{v^2}{2} + \frac{e}{m}\boldsymbol{E} = \boldsymbol{v} \times \left(\operatorname{curl} \boldsymbol{v} - \frac{e}{cm}\boldsymbol{H}\right). \tag{7.12}$$

The curl of this equation is

$$\frac{\partial}{\partial t}\left(\operatorname{curl} \boldsymbol{v} - \frac{e}{mc}\boldsymbol{H}\right) = \operatorname{curl}\left[\boldsymbol{v} \times \left(\operatorname{curl} \boldsymbol{v} - \frac{e}{mc}\boldsymbol{H}\right)\right]. \tag{7.13}$$

Thus if

$$\operatorname{curl} \boldsymbol{v} - \frac{e}{mc}\boldsymbol{H} = 0 \tag{7.14}$$

at time $t = 0$, it is equal to zero at all times. Eq. (I) corresponds to restricting the solutions to this particular group[4]. Since the current density is

$$\boldsymbol{j}_s = -n_s e \boldsymbol{v}, \tag{7.15}$$

---

[1] M. v. Laue: Ann. Phys., Lpz. (6) **3**, 31 (1948); reference [25].
[2] M. v. Laue: Ann. Phys., Lpz. (6) **5**, 197 (1949), reference [24], Chap. 20, also J. Geiss: Ann. Phys., Lpz. (8) **9**, 40 (1951). — Z. Physik **129**, 449 (1951).
[3] R. Becker, F. Sauter and G. Heller: Z. Physik **85**, 772 (1933).
[4] It has been pointed out by J. Lindhard, Phil. Mag. **44**, 916 (1953), that the London equations do not represent a particular solution of the acceleration equations when the Fermi distribution of velocities of the electron gas is taken into account. If a magnetic field is applied to a superconductor of finite size, the field will eventually penetrate into the interior; the time for penetration increases with size. These considerations emphasize that the diamagnetic approach is most likely the correct one.

where n_s is the density of electrons, (7.3) is identical with (7.14) if one takes

$$\Lambda = \frac{m}{n_s e^2}. \tag{7.16}$$

Except for the nonlinear term, grad $\frac{1}{2} v^2$, which can be shown to be negligibly small, (7.12) subject to condition (7.14) is identical with (II).

By taking the curl of (II), one sees immediately that the time derivative of (I) is equal to zero. The additional restriction is that (I) itself rather than its time derivative be satisfied.

We now believe that (I) is actually the more basic equation, and is the one which would be derived first from a microscopic theory. How nearly, then, can one obtain (II) from (I)? The time derivative of (I) is the curl of (II). That the divergence of (II) is equal to zero follows from the equation of continuity. Since space charge in a metal is negligible,

$$\operatorname{div} \boldsymbol{j}_s = 0, \quad \operatorname{div} \boldsymbol{E} = 0. \tag{7.17}$$

In an infinite medium, (II) itself would then have to be satisfied. In a finite medium, all one can say is that

$$\frac{\partial}{\partial t} (\Lambda \boldsymbol{j}_s) - \boldsymbol{E} = \operatorname{grad} \boldsymbol{\Phi}, \tag{7.18}$$

where

$$\nabla^2 \boldsymbol{\Phi} = 0. \tag{7.19}$$

Thus one cannot quite derive (II) from (I). Both equations are required for a complete description.

Eq. (I) does imply that under steady state conditions, the electric field must vanish within a superconductor. In a simply connected body, this solution is unique; there are no currents except in the presence of an external magnetic field. In a multiply connected body, such as a ring, there are different solutions corresponding to different stationary currents flowing around the ring, even in the absence of an external field. The magnetic field which gives the supercurrent is produced by the supercurrent itself. This is also true of a current flowing in a superconducting wire between contacts with normal metals; the current flow gives a magnetic field which in turn makes the supercurrent flow in the wire. There are no normal currents flowing in the wire, and the electric field is zero. Solutions for simply connected bodies are discussed in Sect. 11, for multiply connected in Sect. 13.

An electric field can exist within the penetration depth at high frequencies (generally in the microwave range). This produces a normal current flow and dissipation which can be observed.

8. Energy-momentum theorems. The Londons have generalized the energy-momentum theorems of electrodynamics so as to apply to superconductors. To interpret the results, they make use of a two-fluid model in which there are super and normal current densities $\boldsymbol{j}_s$ and $\boldsymbol{j}_n$ and charge densities ϱ_s and ϱ_n. As discussed in Sect. 4, a naive interpretation of these as referring to two independent fluids is probably not justified, and this must be kept in mind when evaluating the significance of the resulting equations.

Maxwell's equations lead to the well-known energy theorem:

$$\operatorname{div} \frac{c}{4\pi} (\boldsymbol{E} \times \boldsymbol{H}) + \frac{\partial}{\partial t} \left[\frac{1}{8\pi} (E^2 + H^2) \right] = -\boldsymbol{j} \cdot \boldsymbol{E}. \tag{8.1}$$

From the relations between $\boldsymbol{j}_n$ and $\boldsymbol{j}_s$ and $\boldsymbol{E}$, one finds

$$\boldsymbol{j} \cdot \boldsymbol{E} = \frac{\partial}{\partial t}\left(\frac{1}{2}\Lambda j_s^2\right) + \frac{1}{\sigma}j_n^2. \tag{8.2}$$

The first term is the rate of increase of the "kinetic energy" of the superfluid:

$$\text{K.E. density} = \tfrac{1}{2}\Lambda j_s^2, \tag{8.3}$$

the second the dissipation of energy by resistance to normal current flow. The K.E. term represents the increase in energy of the system with increase in current density, and need not all be real kinetic energy. Because the supercurrent flows in a thin film, $\boldsymbol{j}_s$ is large and the K.E. term is of the same order as the magnetic energy. Eq. (8.1) states that electromagnetic energy flowing into a volume goes into increasing the energy of the electromagnetic field, into the K.E. of the super-currents or into JOULE heat. The K.E. represents a reversible work which is recovered when the magnetic field is decreased.

The LONDONS also derived a stress tensor for superconductors analogous to the MAXWELL tensor in electrodynamics. The MAXWELL tensor is defined by

$$T_{ik} = \frac{1}{8\pi}\left[(E^2 + H^2)\,\delta_{ik} - 2\,(E_i\,E_k + H_i\,H_k)\right], \tag{8.4}$$

where i and $k = 1, 2, 3$ and δ_{ik} is the unit tensor. The physical interpretation is in terms of momentum flow; T_{ik} is the i-component of the current density of the k-component of momentum. A consequence of MAXWELL's equation is that

$$\varrho\,\boldsymbol{E} + \frac{1}{c}\,(\boldsymbol{j}\times\boldsymbol{H}) + \frac{\partial}{\partial t}\left(\frac{\boldsymbol{E}\times\boldsymbol{H}}{4\pi c}\right) = -\,\text{Div}\,\boldsymbol{T} \tag{8.5}$$

where $\text{Div}\,\boldsymbol{T}$ is a vector with components

$$(\text{Div}\,\boldsymbol{T})_k = \sum_i \frac{\partial T_{ik}}{\partial x_i}. \tag{8.6}$$

The first two terms on the right represent the force density on matter of charge density ϱ and current density $\boldsymbol{j}$, as follows from the LORENTZ force. The third term may be interpreted as the time rate of change of momentum density in the electromagnetic field. The tensor $\boldsymbol{T}$ thus represents a stress whose divergence is equal to the time rate of change of momentum (matter plus field) per unit volume.

Further progress can be made if one uses the two-fluid model to divide super-current and normal current, so that the LORENTZ force is

$$\varrho\,\boldsymbol{E} + \frac{1}{c}\,(\boldsymbol{j}\times\boldsymbol{H}) = \varrho_s\,\boldsymbol{E} + \frac{1}{c}\,(\boldsymbol{j}_s\times\boldsymbol{H}) + \varrho_n\,\boldsymbol{E} + \frac{1}{c}\,(\boldsymbol{j}_n\times\boldsymbol{H}). \tag{8.7}$$

It is assumed that the conservation law applies to the supercurrent:

$$\text{div}\,\boldsymbol{j}_s = -\frac{\partial \varrho_s}{\partial t}. \tag{8.8}$$

With help of a tensor $\boldsymbol{S}$ defined by

$$S_{ik} = \Lambda\,(j_{si}\,j_{sk} - \tfrac{1}{2}j_s^2\,\delta_{ik}), \tag{8.9}$$

it is then possible to write (8.5) in the form

$$\varrho_n\,\boldsymbol{E} + \frac{1}{c}\,(\boldsymbol{j}_n\times\boldsymbol{H}) + \frac{\partial}{\partial t}\left(\frac{\boldsymbol{E}\times\boldsymbol{H}}{4\pi c}\right) + \frac{\partial}{\partial t}\,(\Lambda\,\varrho_s\,\boldsymbol{j}_s) = -\,\text{Div}\,(\boldsymbol{T}+\boldsymbol{S}). \tag{8.10}$$

The interpretation is made that $\Lambda \varrho_s \boldsymbol{j}_s$ represents the momentum density of the supercurrent. The stress tensor is now $\boldsymbol{T} + \boldsymbol{S}$; the London tensor $\boldsymbol{S}$ is added to the Maxwell tensor $\boldsymbol{T}$.

The left hand side of (8.10) vanishes under steady state conditions, since $\boldsymbol{E}$ and $\boldsymbol{j}_n$ both vanish. This means that the Div $(\boldsymbol{T} + \boldsymbol{S})$ vanishes, so that there are no body forces acting on the superfluid. There are, however, surface forces. Since $\boldsymbol{T}$ is continuous across the surface, while $\boldsymbol{S} = 0$ outside the body, and is thus discontinuous at the surface, it is only the latter which contributes to the surface force. If ν is a direction normal to the surface, the force per unit area, is just $S_{\nu i}$.

9. Hall effect in a superconductor. If one assumes that the diamagnetic approach to superconductivity is the correct one, there can be no Hall field in a simply connected specimen. Since there is a unique distribution for the diamagnetic currents in a static external field, there is no way they could die down so as to give up energy to an external circuit. This argument does not apply to a metastable current in a superconducting ring in zero external field. To the extent, however, that the currents in the ring are also of diamagnetic origin, there would be no Hall effect in this case either.

One might expect a Hall voltage if the magnetic field exerts a transverse force on current elements so as to give a Hall electric field:

$$\boldsymbol{E} = -R(\boldsymbol{j} \times \boldsymbol{H}), \tag{9.1}$$

where R is the Hall constant. We have seen in the previous section that when the London stresses are included, there are no net body forces acting on the superfluid, and thus one would expect no Hall effect.

From a quantum point of view, superconductivity occurs when the wave functions of the electrons extend over distances large compared with penetration depth. This means that one cannot localize electrons within the penetration depth, and one cannot localize the force due to a magnetic field as one would for a classical stream of particles.

No Hall effect has been observed in a superconductor. The most recent search for one to date (1955) was by H. W. Lewis[1].

b) Solutions of the London equations.

10. Penetration depth. The equations which apply to a superconductor in a static field are:

$$\boldsymbol{E} = 0,$$
$$c \operatorname{curl} \Lambda \boldsymbol{j}_s = -\boldsymbol{H},$$
$$c \operatorname{curl} \boldsymbol{H} = 4\pi \boldsymbol{j}_s.$$

The latter two may be combined to give,

$$\nabla^2 \boldsymbol{H} = \boldsymbol{H}/\lambda^2, \tag{10.1}$$

and

$$\nabla^2 \boldsymbol{j}_s = \boldsymbol{j}_s/\lambda^2. \tag{10.2}$$

where

$$\lambda = \sqrt{\Lambda c^2/4\pi}, \tag{10.3}$$

is the penetration depth, of the order of 5×10^{-6} cm. in most superconductors.

---

[1] H. W. Lewis: Phys. Rev. **92**, 1149 (1953); **100**, 641 (1955).

It has been shown quite generally by von Laue [15], [25] that the solutions of (10.1) are such that the field drops exponentially to zero in the interior of a massive specimen.

The case of a plane boundary can be treated quite simply. Let the superconductor occupy the half space $x > 0$, and let $H_y = H_0$, $H_x = H_z = 0$ at the boundary $x = 0$. Then $H_x = H_z = 0$ everywhere, and H_y depends only on x. The equation for H_y is:

$$\frac{d^2 H_y}{dx^2} = \frac{H_y}{\lambda^2} .$$

(10.4)

The solution which vanishes as $x \to \infty$ is:

$$H_y = H_0 e^{-\frac{x}{\lambda}} .$$

(10.5)

The currents are in the z-direction, with

$$j_z = \frac{c}{4\pi} \frac{\partial H_y}{\partial x} = -\frac{c H_0}{4\pi \lambda} e^{-\frac{x}{\lambda}} .$$

(10.6)

Thus the currents are confined to a thin layer near the surface.

The solution for a massive specimen in an external field is similar. The field is nearly parallel to the surface and drops off exponentially toward the interior.

With different boundary conditions (10.4) may be used to obtain a solution for a slab with a field parallel to the surfaces. If the faces are at $y = \pm a$, so that the thickness is $2a$, the solution of (10.4) which gives a field H_0 at each surface is

$$H = H_0 \frac{\mathrm{Cos}\,(y/\lambda)}{\mathrm{Cos}\,(a/\lambda)} .$$

(10.7)

The magnetic moment per unit volume is:

$$I = \frac{M}{V} = \frac{1}{8\pi a} \int_{-a}^{+a} (H - H_0)\, dy .$$

(10.8)

With use of (10.7), this becomes

$$I = -\frac{H_0}{4\pi} \left(1 - \frac{\lambda}{a} \mathrm{Tan}\, \frac{a}{\lambda} \right) .$$

(10.9)

Measurements of Lock[1] on thin films of tin and indium are in agreement with (10.9). However, the smallest values of a/λ in his experiments were not much less than unity, so that a really critical test of the London penetration law was not obtained.

11. Solutions for sphere and circular cylinder. Simple solutions of London's equations which frequently are used are those for a circular cylinder with a field parallel to the axis and a sphere in a uniform external field. The equation for the field, H, which is parallel with the axis of the cylinder, is:

$$\frac{d^2 H}{dr^2} + \frac{1}{r} \frac{dH}{dr} - \frac{H}{\lambda^2} = 0 ,$$

(11.1)

where r is the radial distance from the axis. The solution which gives a field H_0 at the surface of a cylinder of radius a is:

$$H = H_0 \frac{I_0(r/\lambda)}{I_0(a/\lambda)} ,$$

(11.2)

---

[1] J. M. Lock: Proc. Roy. Soc. Lond., Ser. A **208**, 391 (1951).

where I_0 is the BESSEL function of imaginary argument. The magnetic moment per unit volume is:

$$I = \frac{1}{2\pi a^2} \int_0^a (H - H_0)\, r\, dr = -\frac{H_0}{4\pi}\left(1 - \frac{2\lambda}{a}\,\frac{I_1(a/\lambda)}{I_0(a/\lambda)}\right). \tag{11.3}$$

To treat a sphere of radius a in a uniform external field, H_0, we introduce spherical coordinates, r, ϑ, φ. Outside the sphere, the field is that of a dipole of magnetic moment M located at the center of the sphere plus the external field:

$$H_r = \left(H_0 + \frac{2M}{r^3}\right)\cos\vartheta, \tag{11.4}$$

$$H_\vartheta = \left(-H_0 + \frac{2M}{r^3}\right)\sin\vartheta. \tag{11.5}$$

To get the field inside the sphere, we introduce the current density which is of the form:

$$j_\varphi = f(r)\sin\vartheta, \qquad j_r = j_\vartheta = 0. \tag{11.6}$$

From the equation

$$\lambda^2\, \text{curl curl }\boldsymbol{j} = \boldsymbol{j} \tag{11.7}$$

we find the following differential equation for $f(r)$:

$$\frac{d^2f}{dr^2} + \frac{2}{r}\frac{df}{dr} - \left(\frac{2}{r^2} + \frac{1}{\lambda^2}\right)f = 0. \tag{11.8}$$

The solution which is well-behaved at the origin is of the form

$$f(r) = \frac{cA}{4\pi r^2}\left(\text{Sin}\,\frac{r}{\lambda} - \frac{r}{\lambda}\,\text{Cos}\,\frac{r}{\lambda}\right), \tag{11.9}$$

where A is a constant to be determined.

From the first LONDON equation, $4\pi\lambda^2\,\text{curl}\,\boldsymbol{j} = -c\,\boldsymbol{H}$, we obtain the following expressions for H_r and H_ϑ

$$H_r = \frac{2\lambda^2 A}{r^3}\left[\text{Sin}\,\frac{r}{\lambda} - \frac{r}{\lambda}\,\text{Cos}\,\frac{r}{\lambda}\right]\cos\vartheta, \tag{11.10}$$

$$H_\vartheta = \frac{\lambda^2 A}{r^3}\left[\left(1 + \frac{r^2}{\lambda^2}\right)\text{Sin}\,\frac{r}{\lambda} - \frac{r}{\lambda}\,\text{Cos}\,\frac{r}{\lambda}\right]\sin\vartheta, \tag{11.11}$$

The values of M and A are determined by the requirement that H_r and H_ϑ be continuous at $r = a$:

$$M = -\frac{H_0 a^3}{2}\left(1 - \frac{3\lambda}{a}\,\text{Cot}\,\frac{a}{\lambda} + \frac{3\lambda^2}{a^2}\right), \tag{11.12}$$

$$A = -\frac{3H_0 a}{2\,\text{Sin}\,(a/\lambda)}. \tag{11.13}$$

When the radius of the sphere is very small, the magnetic moment is

$$M = -\frac{H_0 a^5}{30\lambda^2}. \tag{11.14}$$

SHOENBERG[1] has applied these results to analysis of measurements of the magnetic properties of mercury colloids with particle size of the order of 10^{-6} to 10^{-5} cm. Because of the range of particle sizes, it was not possible to test (11.12)

---

[1] D. SHOENBERG: Nature, Lond. **143**, 434 (1939). — Proc. Roy. Soc. Lond., Ser. A **175**, 49 (1940); reference [23], p. 143.

in a quantitative way, although information about the penetration depth could be obtained from the temperature variation in the range where (11.14) is applicable.

An interesting solution is that for a body of axial symmetry rotating about its axis, which was first obtained by BECKER[1] and coworkers on the basis of the acceleration theory. If the system starts from rest with no current flow, this solution is essentially that obtained from the LONDON theory ([15], p. 78). We have noted that the LONDON theory picks out one unique solution from a variety of solutions allowed by the acceleration theory. The solution is such that nearly all of the electrons follow the motion of the positive ions, so that there is no current in the interior. Electrons within a penetration depth of the surface lag behind, to give a net current flowing near the surface. This current is such as to produce a uniform magnetic field in the interior of just such a magnitude as to give a LARMOR frequency equal to the frequency of rotation:

$$H = \frac{2mc}{e}\,\omega. \tag{11.15}$$

This field, of the order of 10^{-4} gauss for an angular frequency of 10^3 sec.$^{-1}$ is probably sufficiently large to be detected in a careful experiment. In a coordinate system which rotates with the body the CORIOLIS force just balances, to the first order in ω, the force due to the magnetic field. This, of course, is the basis of the LARMOR theorem.

The surface currents for the case of a sphere of radius R are given by ([15], p. 81):

$$\left.\begin{aligned}
j_e &= j_r = 0, \\
j_\varphi &= -\frac{3n_s e \omega R}{\operatorname{Sin}\beta R}\cdot\frac{1}{\beta r}\left(\operatorname{Cos}\beta r - \frac{1}{\beta r}\operatorname{Sin}\beta r\right)\sin\vartheta \approx -\frac{3n_s e \omega}{\beta}\,e^{-\beta(R-r)}\sin\vartheta,
\end{aligned}\right\} \tag{11.16}$$

where $\beta = 1/\lambda$. The magnetic moment, M, is:

$$M = \frac{mc}{e}R^3\omega\left(1 - \frac{3}{\beta R}\operatorname{Cot}\beta R + \frac{3}{\beta^2 R^2}\right) \approx \frac{mc}{e}R^3\omega. \tag{11.17}$$

12. Wire carrying a current. Another simple solution is that for the current and field distribution in long straight wire of circular cross-section carrying a current parallel with the axis. We introduce cylindrical coordinates, z, r, ϑ. The only nonvanishing component of current density is $j_z(r)$ and of magnetic field is $H_\vartheta(r)$, both of which depend only on the radial distance r. Outside of the cylinder,

$$H_\vartheta = \frac{J}{2\pi c r}, \tag{12.1}$$

where J is the total current.

The equation for $j_z(r)$ is

$$\frac{d^2 j_z}{dr^2} + \frac{1}{r}\frac{dj_z}{dr} - \frac{j_z}{\lambda^2} = 0. \tag{12.2}$$

The solution is of the form

$$j_z = \frac{J}{2\pi a \lambda}\,\frac{I_0(r/\lambda)}{I_1(a/\lambda)}, \tag{12.3}$$

where a is the radius of the wire, and I_0 and I_1 are BESSEL's functions of imaginary argument. Since $I_0(x)$ approaches infinity as $e^{-x}/\sqrt{x}$, the current is confined to

---

[1] R. BECKER, F. SAUTER and G. HELLER: Z. Physik **85**, 772 (1933).

a thin layer of the order of the penetration depth next to the surface. The field inside the wire can be obtained from j_z by using the LONDON equation:

$$H_\theta = -\frac{4\pi\lambda^2}{c}\left(-\frac{\partial j_z}{\partial r}\right) = \frac{J}{2\pi c a}\frac{I_1(r/\lambda)}{I_1(a/\lambda)}.\tag{12.4}$$

It should be noted that (12.1) and (12.4) are equal at the boundary $r = a$.

It is of interest to determine how the current flows from a normal conductor into a superconductor. As far as the normal conductor is concerned, the super-conducting boundary at the interface is an equipotential surface. The current density is normal to the interface. Since current flow across the boundary

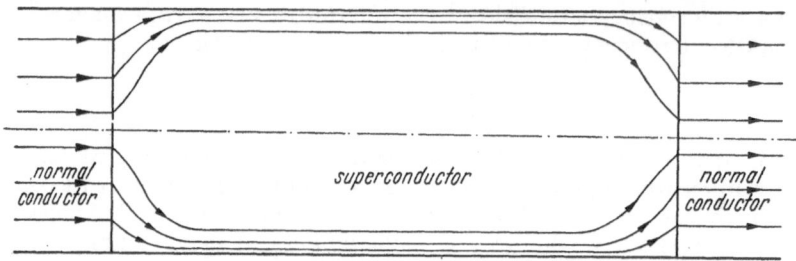

Fig. 2. Current flow from normal metal to superconductor in a rectangular bar (after LONDON [15], p. 37).

must be continuous, this gives a boundary condition for flow in the supercon-ductor. The mathematical problem of determining flow within the supercon-ductor is then to find an appropriate solution of (10.2) in which the normal component of current is specified at the boundaries of normal regions and is equal to zero at a free surface.

Such solutions have been given by LONDON ([15], p. 37) for flow from a normal to a superconducting region in a rectangular bar and by VON LAUE ([25], Chap. 8) for the corresponding flow in a bar of circular cross-section. Fig. 2, illustrates the flow pattern for LONDON's case.

13. Flow in multiply connected bodies[1]. A multiply connected body is one in which there exist loops which cannot be continually deformed to a point with-out passing out of the body. The simplest example is a ring. While the first LONDON equa-tion (I) gives a unique solution for a simply connected body in a static external field, it does not give a unique solution for multi-ply connected bodies. This allows for the possibility of persistent currents. The second Eq. (II) implies that such currents are stable in time. From the dia-magnetic approach, one might expect to derive an equation analogous to (I). The problem of showing that persistent currents are metastable and do not decay in time would then remain. This problem is discussed in Sect. 14. In the present section, we shall discuss the consequences of LONDON equations (I) and (II).

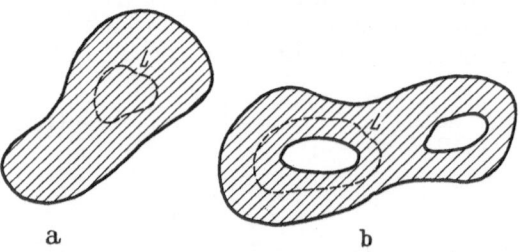

a   b

Fig. 3a and b. Loops for integration in (a) simply, and (b) multiply connected bodies.

---

[1] This section is based to a large extent on reference [15], p. 73—78.

First consider a simply connected body (Fig. 3 a) and apply STOKES' theorem to (I) for a loop entirely within the body. This gives

$$\iint_A \boldsymbol{H} \cdot d\boldsymbol{A} = -c \oint_L \Lambda \boldsymbol{j}_s \cdot d\boldsymbol{s}, \tag{13.1}$$

where $d\boldsymbol{A}$ is an element of area of the cap and $d\boldsymbol{s}$ a line element of the loop L. It follows that

$$\Phi_L = \iint_A \boldsymbol{H} \cdot d\boldsymbol{A} + c \oint_L \Lambda \boldsymbol{j}_s \cdot d\boldsymbol{s} = 0. \tag{13.2}$$

The quantity Φ_L is called the "fluxoid", (to distinguish it from the flux as given by the first term). Eq. (13.2) implies that the fluxoid through any loop L in a simply connected body is zero.

If the loop surrounds a hole in a multiply connected body (Fig. 3 b), we can not use (I), which applies only within the superconductor, but we can get an expression for the fluxoid by applying STOKES' theorem to MAXWELL's equation, $c \operatorname{curl} \boldsymbol{E} = -\partial \boldsymbol{H}/\partial t$, and then using II. STOKES' theorem gives

$$\frac{d}{dt} \iint_A \boldsymbol{H} \cdot d\boldsymbol{A} = -c \oint_L \boldsymbol{E} \cdot d\boldsymbol{s}. \tag{13.3}$$

Replacing E by the use of (II), we find

$$\frac{d}{dt} \iint_A \boldsymbol{H} \cdot d\boldsymbol{A} = -c \oint_L \frac{\partial}{\partial t} (\Lambda \boldsymbol{j}_s) \cdot d\boldsymbol{s}, \tag{13.4}$$

or

$$\frac{d}{dt} \left\{ \iint_A \boldsymbol{H} \cdot d\boldsymbol{A} + c \oint_L \Lambda \boldsymbol{j}_s \cdot d\boldsymbol{s} \right\} = \frac{d\Phi_L}{dt} = 0. \tag{13.5}$$

Thus

$$\Phi_L = \text{const.} \tag{13.6}$$

Since the currents are confined to a thin layer near the surface, one may take the loop L in the interior of a massive specimen where $\boldsymbol{j}_s = 0$. The fluxoid is then simply the total flux which passes through the hole and through the penetration region adjacent to the hole, and is independent of the particular loop chosen as long as it passes once and only once around the hole. The fluxoid defined in this way is then a quantity associated with the hole.

It has been shown by LONDON that (I) leads to a unique solution under static conditions if the fluxoids $\Phi_1, \Phi_2 \ldots \Phi_n$ through each hole of a multiply connected body are specified. These will in general be determined by the past history of the specimen.

In the derivation of (13.6), it was not necessary to assume static conditions, and in fact it applies to time varying external fields. Thus, when the external field is changed, the current distribution will change in such a way that the fluxoids remain constant. This is what is found experimentally.

c) The LONDON approach to superconductivity.

14. Introduction of vector potential. F. LONDON has stressed the explanation of superconductivity from a diamagnetic point of view and has indicated how an equation of the form (I) might follow from quantum theory[1]. This is the

---

[1] F. LONDON: Proc. Roy. Soc. Lond., Ser. A **152**, 24 (1935). — Phys. Rev. **74**, 562 (1948) and reference [*15*], p. 142.

approach we have emphasized in this article and believe to be correct, although it is likely that (I) requires modification along the lines suggested by PIPPARD.

According to the diamagnetic description, there is just one stable current distribution in an isolated simply connected body in an external field. Any fluctuations will be about this stable distribution. Except at very high frequencies, the currents change adiabatically with changes in external field, so that there is no dissipation.

Electric fields exist only when the external field is changing, and then only within the penetration depth of the magnetic field. At very high frequencies, these fluctuating electric fields may give rise to a dissipative flow described in terms of an ordinary conductivity for the superconducting phase, such as is given by the two-fluid model. It is also possible that there may be dissipation associated with relaxation effects of the supercurrents. We shall not be concerned with such high frequency behavior in this section.

According to the diamagnetic description, the supercurrents are always associated with and determined by the magnetic field. A persistent current flowing in a ring is a metastable phenomenon. In this case the magnetic field which gives the current flow is that of the currents themselves. The entire current distribution is determined uniquely by the fluxoid through the ring. Thus the metastable currents represent a one-parameter family in the phase-space of all of the electrons. Almost all random fluctuations are likely to increase rather than decrease the free energy. It is unlikely that the point representing the system in phase space will find the one path which leads down hill. While this is the most likely explanation for the metastability of persistent currents, it has not yet been put on a firm quantitative basis.

LONDON noted that Eq. (I) may be expressed in the form:

$$c \Lambda \boldsymbol{j}_s = - \boldsymbol{A}, \tag{14.1}$$

where $\boldsymbol{A}$ is the vector potential, such that

$$\boldsymbol{H} = \operatorname{curl} \boldsymbol{A} \tag{14.2}$$

The curl of (14.1) gives (I). Eq. (14.1) holds only for a particular choice of gauge. First, it is required that

$$\operatorname{div} \boldsymbol{A} = 0. \tag{14.3}$$

This insures that $\operatorname{div} \boldsymbol{j}_s = 0$. One can always add to $\boldsymbol{A}$ the gradient of a scalar, φ, without changing $\boldsymbol{H}$. If φ satisfies LAPLACE's equation,

$$\nabla^2 \varphi = 0, \tag{14.4}$$

(14.3) is also satisfied. If (14.4) holds in an infinite medium, φ must be a constant so that $\boldsymbol{A}$ is determined uniquely. A further condition is required to specify $\boldsymbol{A}$ in a body of finite dimensions.

First consider an isolated simply connected body. It is required that the normal component of $\boldsymbol{j}_s$ vanish at the surface, and this requires that the gauge be chosen so that $A_\perp = 0$. Suppose that in a particular gauge, denoted by a prime, $A'_\perp \neq 0$ at the surface. Then one may change the gauge by adding grad φ so that

$$A_\perp = A'_\perp + \operatorname{grad}_\perp \varphi = 0 \tag{14.5}$$

at the surface. Eq. (14.5) specifies the normal derivative of φ at the surface, and this determines, up to a constant, a unique solution of LAPLACE's equation,

(14.4). Thus the conditions div $\boldsymbol{A} = 0$ and $A_\perp = 0$ determine a unique choice of gauge. With this choice, $\boldsymbol{A}$ goes to zero in the interior of a massive specimen. If any other choice is made, (14.1) would have to be written in a different manner.

In a multiply connected body we still require that

$$\nabla^2 \varphi = 0 \quad \text{and} \quad \frac{\partial \varphi}{\partial n} = 0 \tag{14.6}$$

at the surface, but φ is not yet completely determined. We may apply STOKES' theorem to (14.2) for a loop around a hole, as in Fig. 3. This gives

$$\Phi = \iint_A \boldsymbol{H} \cdot d\boldsymbol{A} = \oint_L \boldsymbol{A} \cdot d\boldsymbol{s}. \tag{14.7}$$

We may suppose that the loop is taken in the interior where $\boldsymbol{H} = 0$. Since curl $\boldsymbol{A} = 0$, we may take in this region

$$\boldsymbol{A} = \operatorname{grad} \varphi \tag{14.8}$$

where φ may now be a multiple valued function whose gradient is single valued. The line integral of grad φ around the loop is then not zero, and φ may be chosen so that

$$\oint_L \operatorname{grad} \varphi \cdot d\boldsymbol{s} = \Phi. \tag{14.9}$$

If the fluxoids for each hole are specified, grad φ is determined uniquely.

If current is introduced at the boundary, the normal component of current no longer vanishes, but its value determines $A_\perp$. If the values of $A_\perp$ are given over the surface of a simply connected body, the gauge is determined uniquely.

Thus for a particular and uniquely determined choice of gauge, the LONDON equation (I) may be expressed in the form (14.1). It is this form which is related to the derivation from quantum theory.

15. LONDON derivation from quantum theory (cf. [15], Chap. E). The general expression for the current density for a system of N particles described by the wave function $\Psi(\boldsymbol{r}_1, \boldsymbol{r}_2, \ldots \boldsymbol{r}_N)$ in a magnetic field with vector potential $\boldsymbol{A}(\boldsymbol{r}_\alpha)$ is:

$$\boldsymbol{j}(\boldsymbol{r}) = \sum_{\alpha=1}^{N} \int \left\{ \frac{e\hbar}{2im} [\Psi^* \operatorname{grad}_\alpha \Psi - \Psi \operatorname{grad}_\alpha \Psi^*] - \frac{e^2}{mc} \boldsymbol{A}(\boldsymbol{r}_\alpha) \Psi^* \Psi \right\} \times \\ \times \delta(\boldsymbol{r} - \boldsymbol{r}_\alpha) d\boldsymbol{r}_1 \ldots d\boldsymbol{r}_N. \tag{15.1}$$

We may assume that the current density vanishes in the absence of a magnetic field, when $\Psi = \Psi_0$ and $\boldsymbol{A} = 0$,

$$\boldsymbol{j}_0(\boldsymbol{r}) = \sum_{\alpha=1}^{N} \int \frac{e\hbar}{2im} [\Psi_0^* \operatorname{grad}_\alpha \Psi_0 - \Psi_0 \operatorname{grad}_\alpha \Psi_0^*] \delta(\boldsymbol{r} - \boldsymbol{r}_\alpha) d\boldsymbol{r}_1 \ldots d\boldsymbol{r}_N = 0. \tag{15.2}$$

LONDON points out that an equation of the form (14.1) follows if it is assumed that the wave functions are so rigid that they do not change at all when a magnetic field is applied. One may then take $\Psi = \Psi_0$ and find

$$\boldsymbol{j}(\boldsymbol{r}) = -\frac{\varrho(\boldsymbol{r}) e^2}{mc} \boldsymbol{A}(\boldsymbol{r}), \tag{15.3}$$

where $\varrho(\boldsymbol{r})$ is the particle density:

$$\varrho(\boldsymbol{r}) = \sum_{\alpha=1}^{N} \int \Psi^* \Psi \, \delta(\boldsymbol{r} - \boldsymbol{r}_\alpha) \, d\boldsymbol{r}_1 d\boldsymbol{r}_2 \ldots d\boldsymbol{r}_N. \tag{15.4}$$

This rigidity must be a quantum effect. It can be shown quite generally (VAN LEEUWEN's theorem) that a classical system exhibits no diamagnetism. A well-known consequence of quantum theory is that an atom or molecule in a uniform field has a diamagnetic moment proportional to the square of the radius of the orbit. The large diamagnetism of superconductors is undoubtedly associated with a large extent of wave functions or range of order. The diamagnetism is, of course, so great that the field drops to zero in the interior.

It is undoubtedly going too far to assume such a rigidity that the wave functions are not changed at all by a magnetic field. One should take into account at least the first order changes in the wave functions produced by the magnetic field, and we shall consider this problem in Sect. 16. If this were done, the expression for current density would be gauge invariant.

The LONDON expression (14.1) is not gauge invariant but one can give a reasonable argument for the particular choice of gauge required by the theory, $\operatorname{div} \boldsymbol{A} = 0$ and $A_\perp = 0$ on the surface[1]. Let $\boldsymbol{A}$ be the vector potential for an arbitrary choice of gauge. Terms in the HAMILTONian which involve the magnetic field are

$$H_m = \frac{1}{2m} \sum_{\alpha=1}^{N} \left(\boldsymbol{p}_\alpha + \frac{e}{c} \boldsymbol{A}(\boldsymbol{r}_\alpha) \right)^2, \tag{15.5}$$

in which $-e$ is the charge of an electron. Consider the class of wave functions of the form

$$\Psi = e^{\frac{ie}{c\hbar} \sum_\alpha \varphi(\boldsymbol{r}_\alpha)} \Psi_0(\boldsymbol{r}_1, \dots \boldsymbol{r}_N). \tag{15.6}$$

The exponential factor is introduced when a gauge transformation $\boldsymbol{A} \to \boldsymbol{A} + \operatorname{grad} \varphi$ is made, and is required if the gauge is to be chosen arbitrarily. For a particular choice of gauge, a particular value of $\varphi(\boldsymbol{r}_\alpha)$ will be required to represent the zero-order wave function. The function may be chosen to make W_m, the first-order change in energy resulting from the magnetic field, a minimum.

$$
\left.
\begin{aligned}
W_m &= \int \left\{ \Psi^* H_m \Psi - \sum_{\alpha=1}^{N} \Psi_0^* \frac{p_\alpha^2}{2m} \Psi_0 \right\} d\tau \\
&= \frac{1}{2m} \sum_{\alpha=1}^{N} \int \Psi_0^* \left[-\frac{ie\hbar}{c} \operatorname{div}(\boldsymbol{A}_\alpha + \operatorname{grad} \varphi_\alpha) - \right. \\
&\quad \left. - \frac{2ie\hbar}{c} (\boldsymbol{A}_\alpha + \operatorname{grad} \varphi_\alpha) \operatorname{grad}_\alpha + \frac{e^2}{c^2} (\boldsymbol{A}_\alpha + \operatorname{grad} \varphi_\alpha)^2 \right] \Psi_0 \, d\boldsymbol{r}_1 \dots d\boldsymbol{r}_N.
\end{aligned}
\right\} \tag{15.7}
$$

After an integration by parts, (15.7) may be written

$$W_m = \frac{1}{c} \int (\boldsymbol{A} + \operatorname{grad} \varphi) \boldsymbol{j}_0(\boldsymbol{r}) \, d\tau - \frac{e}{2mc^2} \int (\boldsymbol{A} + \operatorname{grad} \varphi)^2 \varrho_0(\boldsymbol{r}) \, d\tau, \tag{15.8}$$

where $\boldsymbol{j}_0(\boldsymbol{r})$ and $\varrho_0(\boldsymbol{r})$ are the current and particle densities for the wave function Ψ_0.

We shall take $\boldsymbol{j}_0 = 0$ and assume for simplicity that $\varrho_0(\boldsymbol{r}) = \mathrm{const}$. The condition that W_m be invariant for small changes $\varDelta \varphi$ of φ is then

$$\delta W_m = -\frac{e \varrho_0}{m c^2} \int (\boldsymbol{A} + \operatorname{grad} \varphi) \cdot \operatorname{grad} \varDelta \varphi \, d\tau = 0. \tag{15.9}$$

An integration by parts gives (for a simply connected body)

$$\int \varDelta \psi \operatorname{div}(\boldsymbol{A} + \operatorname{grad} \varphi) \, d\tau - \int \varDelta \varphi (\boldsymbol{A} + \operatorname{grad} \varphi) \cdot d\boldsymbol{s} = 0 \tag{15.10}$$

---

[1] J. BARDEEN: Phys. Rev. **81**, 469 (1951).

in which the second integral is over the surface. If this equation is to be satisfied for all $\Delta \varphi$, we must have

$$\operatorname{div}(\boldsymbol{A} + \operatorname{grad} \varphi) = 0, \tag{15.11}$$

$$(\boldsymbol{A} + \operatorname{grad} \varphi)_\perp = 0 \tag{15.12}$$

on the surface. Choosing φ in this way is equivalent to taking the London choice of gauge and $\Psi = \Psi_0$. This means that the London choice is the proper one to take if first order changes in wave functions produced by the magnetic field are neglected.

d) Pippard's non-local modification of the London equation.

16. Reasons for modification. In the introduction to this chapter, we mentioned Pippard's concept of coherence [20] and the nature of the evidence that the wave functions or range or order describing the superconducting phase extend over large regions of space ($\sim 10^{-4}$ cm.). If this is the case, one might expect that the relation between current density and field might not be a point relation, involving only differentials, but an integral relation involving the field in a region about the point in question extending over distances of the order of the coherence distance[1]. Pippard's proposed modification of the London equation (I) is of this sort.

Experimental evidence indicating that such a relation is required is not extensive but is nevertheless quite suggestive. First is the change in penetration depth with indium concentration in alloys containing zero to 3% indium (Serin, Chap. VII). A decrease in penetration depth by a factor of about two is observed, although there is hardly any change at all in critical temperature. Pippard believes that the change in penetration depth is due to a reduction in mean free path of the electrons with added indium and a consequent decrease in coherence distance. Second is the variation in penetration depth of single crystals of tin with orientation[2]. The penetration depth is a maximum when the angle, ϑ, between the crystal axis and the tetrad axis is about 60°, and decreases for angles on either side (Serin, Sect. 19). Such a change cannot be accounted for by assuming that the parameter Λ in the London equation is a tensor, for this would give a monotone change with angle. Pippard had observed a corresponding change in the high frequency resistance of normal tin, which again cannot be accounted for by simply making the conductivity a tensor, but has been explained by the theory of the anomolous skin effect. In the latter, the mean free path is larger than the skin depth so that the electric field acting on an electron varies over the free path. The current density then depends on an integral of the electric field over a region of the order of the free path. A third is the dependence of penetration depth on the parameters of the metal, which appears to work out better with Pippard's version of the theory, as will be discussed later (Sect. 26).

---

[1] Strictly, the London Eq. (I) is not a point relation, since the current density at a point depends on the magnetic field in a region surrounding the point. In an appropriate gauge, the current density *is* proportional to the vector potential, but the latter is determined by an integral of the field extending over a large area. As we shall discuss in Sect. 26, Schafroth and Blatt have argued that (I) is valid only when the order extends without limit. Pippard's coherence distance is a length significant from an energetic point of view. Appreciable energy would be required to confine the wave packets describing the superconducting state to a region extending over less than a coherence distance. For example, the width of the boundary between normal and superconducting phases in the intermediate state is of the order of the coherence distance. The actual extent of the ordered ground state in the superconducting phase may be and probably is much greater than the coherence distance.

[2] See, however, footnote 2, p. 284.

17. Chamber's derivation of current density in anomalous skin effect. Pippard [20] obtained his phenomenological relation between current and field by analogy with the theory of the anomalous skin effect. He used the theory in a form derived by Chambers[1]. Since Chambers' derivation is quite simple, we shall reproduce it here. It is assumed that the electric field, $E(r)$, is a function of position, r. We are interested in calculating the steady state current density at a point, which we shall take for simplicity to be the origin. Electrons at the origin will have come from various distances after making the last collision. The probability that an electron at a distance r will travel to the origin without making a collision is $e^{-r/l}$. The net current can be determined from the shape of the displaced Fermi surface for electrons at the origin. It is assumed that the Fermi surface is on the average undisplaced after a collision, so that we want to determine the shape from the changes in momentum which have taken place after the last collision. We follow the motion of electrons with velocities on the Fermi surface. Consider electrons travelling from r to $r - dr$ toward the origin. The net change in velocity normal to the Fermi surface, and thus in the direction $-r$ is

$$d\boldsymbol{v}(r) = -\frac{e\,\boldsymbol{r}\cdot\boldsymbol{E}\,dr}{m\,v_0\,r} = -\frac{e\,\boldsymbol{r}\,(\boldsymbol{r}\cdot\boldsymbol{E})}{m\,v_0\,r^2}\,dr, \qquad (17.1)$$

where v_0 is the velocity at the Fermi surface. To get the average change at the Fermi surface we must multiply by the probability $e^{-r/l}$ that the electron will reach the origin without collision. If $\Delta\boldsymbol{v}_F$ is the net displacement of the Fermi surface, the net current density is

$$\boldsymbol{j}(0) = -\frac{3n\,e}{4\pi}\int \Delta\boldsymbol{v}_F\,d\Omega, \qquad (17.2)$$

where n is the concentration of electrons and $d\Omega$ is the element of solid angle. With

$$\Delta\boldsymbol{v}_F = e^{-r/l}\,d\boldsymbol{v} \qquad (17.3)$$

and the element of volume

$$d\tau = r^2\,dr\,d\Omega, \qquad (17.4)$$

we find

$$\boldsymbol{j}(0) = \frac{3n\,e^2}{4\pi\,m\,v_0}\int \frac{\boldsymbol{r}\,(\boldsymbol{r}\cdot\boldsymbol{E})\,e^{-r/l}}{r^4}\,d\tau. \qquad (17.5)$$

Since for this model the conductivity is

$$\sigma = \frac{n\,e^2\,l}{m\,v_0}, \qquad (17.6)$$

the expression for $\boldsymbol{j}$ may be written:

$$\boldsymbol{j} = \frac{3\sigma}{4\pi\,l}\int \frac{\boldsymbol{r}\,(\boldsymbol{r}\cdot\boldsymbol{E})\,e^{-r/l}}{r^4}\,d\tau. \qquad (17.7)$$

In a non-static situation, E would be evaluated at the retarded time $\left(t - \frac{r}{v_0}\right)$. The correction is important only at extremely high frequencies. An equivalent expression for the current density for the anomalous skin effect was obtained earlier by Reuter and Sondheimer[2] by a different method.

A question arises as to how to treat the boundary. If the scattering at the surface is completely random, electrons leaving the surface will have no net

---

[1] The derivation is given in reference [21], p. 18.
[2] G. E. H. Reuter and E. H. Sondheimer: Proc. Roy. Soc. Lond., Ser. A **195**, 336 (1948).

momentum parallel with the surface. An equivalent distribution would be obtained in an infinite medium if E is set equal to zero everywhere outside of the surface. This amounts to integrating (17.7) over the actual volume. If scattering is specular, the situation is more complicated. A plane boundary can be treated by the method of images. If the medium occupies the half-space $x > 0$, one may assume that $E(-x, y, z) = E(x, y, z)$ and then integrate over the entire space. In the model used by REUTER and SONDHEIMER it was assumed that a fraction p are specularly reflected and a fraction $(1-p)$ diffusely scattered. Experimental evidence favors $p = 0$.

In application of (17.7) with $p = 0$ to an actual problem, it would be required that div $j = 0$ in the interior and that the normal component of j vanish at a free surface. It is a consequence of (17.7) that div $j = 0$ in an infinite medium, provided that div $E = 0$. In order to satisfy this condition in a finite region and with $E = 0$ outside the region, it is necessary to require that $E_\perp = 0$ at a free surface so that there is no discontinuity in the lines of flux.

It is not at all evident, however, that when the integration is carried out in this manner $j_\perp$ would be zero for a body of arbitrary shape. In assuming that the correct way to introduce a boundary is to set $E = 0$ outside the surface, the implicit assumption was made that $j_\perp = 0$ but this may not be true. If it is not, another term would have to be added to correct for the current flow into the surface. There is no difficulty for a plane boundary, and this is the only case for which explicit solutions have been obtained. We shall see that similar problems arise in fitting the boundary conditions in PIPPARD's expression for the diamagnetic current in a superconductor.

18. PIPPARD theory. In direct analogy with CHAMBERS' expression for the current in the anomalous skin effect, PIPPARD[1] assumes that the LONDON equation (I) may be replaced by

$$j(r) = -\frac{3}{4\pi \xi_0 c \Lambda} \int \frac{R(R \cdot A(r')) e^{-R/\xi}}{R^4} d\tau'. \tag{18.1}$$

Here ξ_0 is a parameter of the metal of the dimensions of a length and is presumably related to the coherence distance in the pure metal and $R = |r - r'|$. The parameter ξ is closely related to the mean free path, l. PIPPARD finds that he can fit the dependence of penetration depth on mean free path in tin-indium alloys if he assumes that

$$\frac{1}{\xi} = \frac{1}{\xi_0} + \frac{1}{\alpha l}, \tag{18.2}$$

where α, adjusted empirically, is 0.80. The gauge is chosen so that div $A = 0$ and $A_\perp = 0$.

Penetration phenomena have been computed from (18.1) only for the case of a plane boundary in a semi-infinite medium, which is the one of most practical interest. When combined with

$$\operatorname{curl} H = \operatorname{curl} \operatorname{curl} A = -\nabla^2 A = \frac{4\pi j}{c}. \tag{18.3}$$

(18.1) becomes

$$\nabla^2 A = \frac{3}{\xi_0 \Lambda c^2} \int \frac{R(R \cdot A') e^{-R/\xi}}{R^4} d\tau'. \tag{18.4}$$

To treat the problem of a plane boundary, we shall take the x-axis normal to the surface, j and A in the z direction and H in the y direction. Eq. (18.4)

---

[1] A. B. PIPPARD: Physica, Haag **19**, 765 (1953); references [20] and [21].

then reduces to the following integral equation for $A_z(x)$:

$$\frac{d^2 A_z(x)}{dx^2} = \frac{3\pi}{\xi_0 \Lambda c^2} \int_0^\infty k\left(\frac{x-t}{\xi}\right) A_z(t)\, dt,$$ (18.5)

where

$$k(v) = \int_1^\infty \left(\frac{1}{s} - \frac{1}{s^3}\right) e^{-s|v|}\, ds.$$ (18.6)

The solution of (18.5) is obtained in a manner similar to the one used by REUTER and SONDHEIMER for the corresponding equation for the anomalous skin effect.

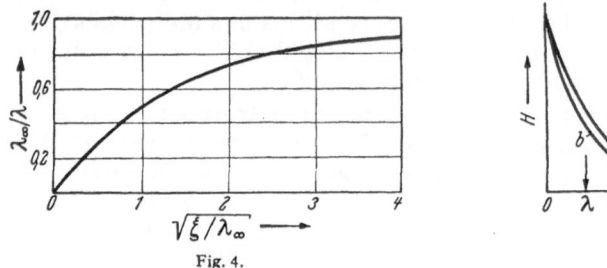

Fig. 4.

Fig. 5.

Fig. 4. Plot of λ_∞/λ versus $\sqrt{\xi/\lambda_\infty}$ (after PIPPARD [20]).

Fig. 5. Penetration of magnetic field at a plane boundary according to (a) LONDON theory and (b) PIPPARD theory with specular reflection at surface (after PIPPARD [20]).

The penetration depth for the assumption of random scattering at the surface is:

$$\lambda = \frac{\int_0^\infty H_y\, dx}{H_y(0)} = \frac{\pi\xi}{\int_0^\infty \log\left(1 + \frac{\beta \varkappa(t)}{t^2}\right) dt},$$ (18.7)

where

$$\beta = \frac{3\pi \xi^3}{\xi_0 \Lambda c^2}$$ (18.8)

and

$$\varkappa(t) = \frac{2}{t^3}\{(1 + t^2)\arctan t - t\}.$$ (18.9)

As we shall discuss later (Sect. 24), (18.9) is closely related to the FOURIER transform of (18.1). Limiting expressions for β small and β large are:

$$\lambda = \sqrt{\frac{\xi_0 \Lambda c^2}{4\pi \xi}}, \qquad\qquad \beta \text{ small,}$$ (18.10)

$$\lambda = \frac{3^{\frac{1}{6}}(\xi_0 \Lambda c^2)^{\frac{1}{3}}}{2\pi^{\frac{4}{3}}} = \lambda_\infty, \qquad \beta \text{ large.}$$ (18.11)

A plot of λ_∞/λ as a function of $\sqrt{\xi/\lambda_\infty}$ is given in Fig. 4. Experimental values derived from data on the tin-indium alloys are shown on the curve. Values of ξ_0, α and λ_∞ as adjusted empirically to give the fit to the theoretical curve are $\xi_0 = 1.2 \times 10^{-4}$ cm., $\alpha = 0.80$ and $\lambda_\infty = 5.28 \times 10^{-4}$ cm. PIPPARD assumes that ξ cannot exceed ξ_0, [as is implicit in (18.2)], so that the limit λ_∞ is not attained. The minimum value of λ as $l \to \infty$ is 5.72×10^{-6} cm.

The field distribution has not been calculated for random scattering, but PIPPARD [20] has given the solution for the simpler case of specular reflection. The vector potential is then given by an integral of the form

$$A(x) = C \int_0^\infty \frac{\cos x\,t\,dt}{t^2 + \beta\varkappa(t)}, \qquad (18.12)$$

where C is a constant. The integral was evaluated by expressing the integrand, $1/(t^2 + \beta\varkappa(t))$, as a sum of functions whose FOURIER transforms are known. The resulting curve for $H = dA/dx$ is compared in Fig. 5 with the exponential decrease given by the LONDON theory. Parameters are adjusted to make the penetration depth the same for both. The value of β was chosen to be 2000, so that $\lambda = 0.0432\xi_0$. It is interesting to note that there is a small reversal of the sign of the field. Since the maximum negative value is only about 3% of the value at the surface, it would be difficult to observe.

e) Derivation of diamagnetic properties from energy gap model.

19. Perturbation theory method. In the energy gap model, as described in Chap. II, it is assumed that the superconducting phase differs from the normal phase in that an extra energy ε is required to excite an electron. Otherwise, excited electrons in the superconducting phase are assumed similar to those of the normal phase. We have mentioned that this model accounts in a reasonable way for the temperature variation of specific heat, thermal conductivity, electrical conductivity as observed in the skin depth at microwave frequencies, and viscosity of the electron gas as observed from attenuation of ultrasonic sound waves. We shall show in this section that the model also accounts for the diamagnetic properties and leads to a phenomenological theory very similar to that suggested by PIPPARD and described in Sect. 18.

The calculation is made by a perturbation theoretic method in which first-order changes in wave functions produced by the magnetic field are used to calculate the current density. In order to simplify the calculation, it is assumed that the medium is infinite. Sources of the field may be introduced by inserting current sheets in the interior. This method has been applied by KLEIN [12] and by SCHAFROTH[1] to the calculation of the diamagnetic properties of an electron gas. The problems involved in applying results calculated for an infinite medium to one with a finite boundary are similar to those encountered in CHAMBERS' derivation of the anomalous skin effect as discussed in the preceding section.

In order to carry out the perturbation theory calculation, three things are needed, the energies of the excited states, the matrix elements of the magnetic interaction which connect these with the ground state, and the current densities associated with the excited states. These are obtained from the corresponding quantities for the normal state as follows. We assume that the ground state of the superconductor (the one which exists at $T = 0°$ K) is some sort of condensed phase which cannot be described simply in terms of one-electron wave functions. Excited states are assumed similar to individual particle excitations of the normal state, in which an electron is excited from a state k_0 below the FERMI surface to a state k above the FERMI surface. For simplicity, we take a degenerate free-electron gas in which the excitation energies in the normal state are:

$$W_n(k_0 \to k) = \frac{\hbar^2}{2m}(k^2 - k_0^2) \qquad \text{(normal)}. \qquad (19.1)$$

---

[1] M. R. SCHAFROTH: Helv. phys. Acta **24**, 645 (1951).

We assume that excitation energies in the superconducting phase differ from these by an additive energy

$$W_s(\boldsymbol{k_0} \to \boldsymbol{k}) = W_n(\boldsymbol{k_0} \to \boldsymbol{k}) + \varepsilon \qquad \text{(superconducting)}. \tag{19.2}$$

We assume further that the expressions for the matrix elements and current densities of the excited states in the superconducting phase are exactly the same as those of the normal phase for the corresponding transition. While these assumptions cannot be completely correct, they have good empirical justification from the success of the two-fluid model and very likely lead to results of the right general sort.

The calculation is carried out in the following manner. The magnetic terms in the Hamiltonian of a free-electron gas are

$$\frac{1}{2m} \sum_i \left(\boldsymbol{p}_i + \frac{e}{c}\boldsymbol{A}(\boldsymbol{r}_i)\right)^2 = \frac{1}{2m} \sum_i \left\{ \boldsymbol{p}_i^2 + \frac{e}{c}(\boldsymbol{p}_i \cdot \boldsymbol{A} + \boldsymbol{A} \cdot \boldsymbol{p}_i) + \frac{e^2}{c^2}A^2 \right\}. \tag{19.3}$$

Since we are keeping only linear terms in $\boldsymbol{A}$, we may omit the terms in A^2. This term would have to be kept if we were calculating energy rather than current, or if we were taking a general gauge. We shall choose the gauge in $\boldsymbol{A}$ so that div $\boldsymbol{A} = 0$ and also such that $\boldsymbol{A} = 0$ when the magnetic field is zero. The magnetic interaction then becomes:

$$\sum_i \left\{ -\frac{ie\hbar}{mc}\boldsymbol{A} \cdot \text{grad}_i \right\}. \tag{19.4}$$

We shall be interested in vector potentials which are everywhere finite and can be resolved into Fourier components. This excludes, for example, a magnetic field which is uniform everywhere and for which electrons would make circular orbits no matter how weak is the magnetic field. One cannot, of course, expect to derive properties of the circular orbits from a perturbation-theoretic approach. The diamagnetic properties of a free-electron gas can be derived from the circular orbits, but they are not essential. If there is a mean free path which prevents the electrons from making complete orbits, one would expect a perturbation-theoretic approach to be valid, and it does indeed lead to the usual Landau formula regardless of the mean free path (Sect. 22)[1].

Let the one electron functions normalized to unit volume of the normal state be denoted by

$$\psi = e^{i\boldsymbol{k} \cdot \boldsymbol{r}_l}. \tag{19.5}$$

The perturbed wave function, $\Psi(\boldsymbol{r}_1, \boldsymbol{r}_2 \dots \boldsymbol{r}_n)$ may be expressed in the form:

$$\Psi = \Psi_0 + \sum_{\boldsymbol{k}, \boldsymbol{k_0}} a_{\boldsymbol{k}, \boldsymbol{k_0}} \Psi_{\boldsymbol{k}, \boldsymbol{k_0}}, \tag{19.6}$$

where Ψ_0 is the ground state wave function and $\Psi_{\boldsymbol{k}, \boldsymbol{k_0}}$ the excited state many-electron wave function in which one electron is excited from $\boldsymbol{k_0}$ to $\boldsymbol{k}$. The coefficients of the expansion for the normal state are

$$a_{\boldsymbol{k}, \boldsymbol{k_0}} = -\frac{\int \psi_{\boldsymbol{k}}^* \left\{ -\frac{ie\hbar}{mc}\boldsymbol{A} \cdot \text{grad} \right\} \psi_{\boldsymbol{k_0}} d\tau}{W_n(\boldsymbol{k_0} \to \boldsymbol{k})}. \tag{19.7}$$

The current density is:

$$\boldsymbol{j}(\boldsymbol{r}) = \sum_{\boldsymbol{k_0}} \left\{ \sum_{\boldsymbol{k}} \left(a_{\boldsymbol{k}, \boldsymbol{k_0}}^* \boldsymbol{j}_{\boldsymbol{k}, \boldsymbol{k_0}} + a_{\boldsymbol{k}, \boldsymbol{k_0}} \boldsymbol{j}_{\boldsymbol{k_0}, \boldsymbol{k}} \right) - \frac{e^2}{mc}\boldsymbol{A}(\boldsymbol{r}) \psi_{\boldsymbol{k_0}}^* \psi_{\boldsymbol{k_0}} \right\}, \tag{19.8}$$

---

[1] The de Haas-van Alphen terms, of course, do not appear in this approximation.

where
$$\boldsymbol{j}_{\boldsymbol{k},\,\boldsymbol{k}_0} = \frac{i\,e\,\hbar}{m\,c}\,\{\psi_{\boldsymbol{k}}^*\,\mathrm{grad}\,\psi_{\boldsymbol{k}_0} - \psi_{\boldsymbol{k}_0}\,\mathrm{grad}\,\psi_{\boldsymbol{k}}^*\}. \tag{19.9}$$

For our free electron model,
$$\boldsymbol{j}_{\boldsymbol{k},\,\boldsymbol{k}_0} = -\frac{e\,\hbar}{m\,c}\,(\boldsymbol{k} + \boldsymbol{k}_0)\,e^{i\,(\boldsymbol{k}_0 - \boldsymbol{k})\cdot\boldsymbol{r}}. \tag{19.10}$$

We assume that the corresponding expressions for the superconducting phase are similar except that $W_s(\boldsymbol{k}_0 \to \boldsymbol{k})$ replaces $W_n(\boldsymbol{k}_0 \to \boldsymbol{k})$.

The expression for the current density may be expressed for general one-electron functions in terms of appropriately defined density matrices. If we substitute (19.7) and (19.9) into (19.8) we find

$$\boldsymbol{j}(\boldsymbol{r}) = \sum_{\boldsymbol{k}_0}\left\{\sum_{\boldsymbol{k}} \frac{e^2\,\hbar^2}{2\,m^2\,c}\left[\frac{\int \psi_{\boldsymbol{k}}^*(\boldsymbol{r}')\,\boldsymbol{A}(\boldsymbol{r}')\cdot\nabla'\,\psi_{\boldsymbol{k}_0}(\boldsymbol{r}')\,d\tau'}{W_s(\boldsymbol{k}_0 \to \boldsymbol{k})}\,(\psi_{\boldsymbol{k}}^*(\boldsymbol{r})\,\nabla\,\psi_{\boldsymbol{k}}(\boldsymbol{r}) - \right.\right.$$
$$\left.\left. - \psi_{\boldsymbol{k}}(\boldsymbol{r})\,\nabla\,\psi_{\boldsymbol{k}_0}^*(\boldsymbol{r})) + \text{comp. conj.}\right] - \frac{e^2}{m\,c}\,\boldsymbol{A}(\boldsymbol{r})\,\psi_{\boldsymbol{k}_0}^*(\boldsymbol{r})\,\psi_{\boldsymbol{k}_0}(\boldsymbol{r})\right\}. \tag{19.11}$$

Expressions of the form $\nabla\varrho(\boldsymbol{r}',\boldsymbol{r})$ occur in (19.11), where $\varrho(\boldsymbol{r}',\boldsymbol{r})$ is a density matrix obtained by summing over a shell of constant energy, or, in the free electron model, of $|\boldsymbol{k}| = \text{const}$;

$$\varrho(\boldsymbol{r}',\boldsymbol{r}) = \sum_{|\boldsymbol{k}|=\text{const}} \psi_{\boldsymbol{k}}^*(\boldsymbol{r}')\,\psi_{\boldsymbol{k}}(\boldsymbol{r}). \tag{19.12}$$

We shall make use of (19.12) in estimating the effect of impurity scattering on the current density and also in a derivation for a different model.

20. Calculation of Fourier components. It is convenient to use Fourier transforms and to derive a relation between the Fourier component $\boldsymbol{j}(\boldsymbol{q})$ of the current density for the wave vector $\boldsymbol{q}$ and the corresponding component $\boldsymbol{A}(\boldsymbol{q})$ of the vector potential. When the gauge in $\boldsymbol{A}$ is chosen so that $\mathrm{div}\,\boldsymbol{A} = 0$, $\boldsymbol{A}(\boldsymbol{q})$ and $\boldsymbol{q}$ are perpendicular and $\boldsymbol{j}(\boldsymbol{q})$ is parallel with $\boldsymbol{A}(\boldsymbol{q})$. The relation may be written in the form
$$4\pi\boldsymbol{j}(\boldsymbol{q}) = -c\,K(q)\,\boldsymbol{A}(\boldsymbol{q}) \tag{20.1}$$

where $K(q)$ is a function of $|\boldsymbol{q}|$ to be determined. The Fourier transforms are defined by a series or integral as follows:
$$\boldsymbol{A}(\boldsymbol{r}) = (2\pi)^{\frac{3}{2}}\sum\boldsymbol{A}(\boldsymbol{q})\,e^{i\,\boldsymbol{q}\cdot\boldsymbol{r}} = (2\pi)^{-\frac{3}{2}}\int\boldsymbol{A}(\boldsymbol{q})\,e^{i\,(\boldsymbol{q}\cdot\boldsymbol{r})}\,d\boldsymbol{q}, \tag{20.2}$$

with corresponding expressions for $\boldsymbol{j}(\boldsymbol{r})$. Since $\mathrm{div}\,\boldsymbol{A} = 0$,
$$\boldsymbol{q}\cdot\boldsymbol{A} = 0. \tag{20.3}$$

The matrix element for the Fourier component $\boldsymbol{q}$ is
$$a_{\boldsymbol{k}+\boldsymbol{q},\,\boldsymbol{k}} = -(2\pi)^{\frac{3}{2}}\frac{i\,e\,\hbar}{m\,c}\,\frac{\boldsymbol{A}(\boldsymbol{q})\cdot\boldsymbol{k}}{W_s(\boldsymbol{k}\to\boldsymbol{k}+\boldsymbol{q})} = a_{\boldsymbol{k},\,\boldsymbol{k}+\boldsymbol{q}}^* \tag{20.4}$$

and
$$\boldsymbol{j}_{\boldsymbol{k},\,\boldsymbol{k}+\boldsymbol{q}} = -\frac{e\,\hbar}{2\,m}\,(2\boldsymbol{k}+\boldsymbol{q})\,e^{i\,\boldsymbol{q}\cdot\boldsymbol{r}} = \boldsymbol{j}_{\boldsymbol{k}+\boldsymbol{q},\,\boldsymbol{k}}^*. \tag{20.5}$$

When these expressions are inserted in (19.8), we find
$$\boldsymbol{j}(\boldsymbol{q}) = 2\sum_{\boldsymbol{k}}\left\{-\frac{e^2}{m\,c}\,\boldsymbol{A}(\boldsymbol{q}) + \frac{e^2\,\hbar^2}{2\,m^2\,c}\sum_{\boldsymbol{q}}\frac{(2\boldsymbol{k}+\boldsymbol{q})\,(\boldsymbol{A}^*(\boldsymbol{q})\cdot\boldsymbol{k}+\boldsymbol{A}(\boldsymbol{q})\cdot\boldsymbol{k})}{W_s(\boldsymbol{k}\to\boldsymbol{k}+\boldsymbol{q})}\right\}. \tag{20.6}$$

The sum over $\boldsymbol{k}$ is over occupied states, that over $\boldsymbol{q}$ over states such that $\boldsymbol{k}+\boldsymbol{q}$ is unoccupied. Actually, the latter restriction can be removed, since the expression

$\sum\limits_{q} \cdots$ changes sign if k and $k+q$ are interchanged as initial and final states, and the sum of the two terms would vanish. In the corresponding calculation of Klein, he allowed $k+q$ in the second sum to be unrestricted, but it will be more convenient for us to keep it as a sum over unoccupied states. We are required to evaluate sums of the form

$$S = 2 \sum_k \frac{A(q) \cdot k \, f(k) \, (1 - f(k+q)) \, (2k+q)}{(\hbar^2/2m) \, ((k+q)^2 - k^2) + \varepsilon}, \qquad (20.7)$$

where $f(k)$ is the probability that the state k is occupied. At $T = 0°$ K, we assume $f(k) = 1$ for $|k| < k_F$ and $f(k) = 0$ for $|k| > k_F$. To carry out the summa-

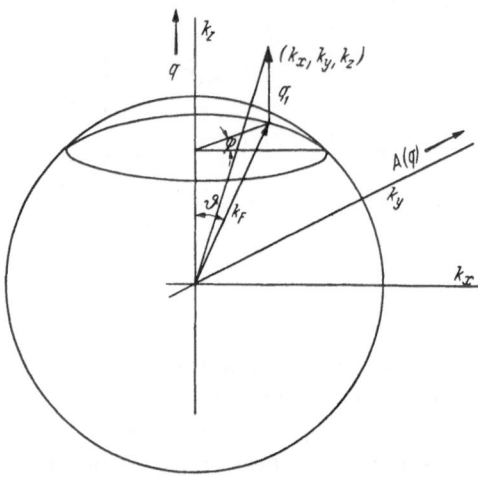

tion, we introduce polar coordinates in k-space with the polar axis in the direction of q and $A(q)$ directed along the y-axis. As indicated in Fig. 6, we take as variables the polar angles ϑ and the distance q_1, measured parallel to the polar axis from the Fermi surface to the point:

$$\left.\begin{aligned} k_z &= k_F \cos\vartheta + q_1, \\ k_y &= k_F \sin\vartheta \sin\varphi, \\ k_x &= k_F \sin\vartheta \cos\varphi. \end{aligned}\right\} \quad (20.8)$$

The Jacobian of the transformation from k_x, k_y, k_z to ϑ, φ, q, is $k_F^2 \sin\vartheta \times \cos\vartheta$. The only nonvanishing component of S is that in the k_y direction, i.e. parallel to $A(q)$.

Fig. 6. Coordinate system for integration over k-space.

The sum S may be written as the following integral.

$$S = \frac{3n\,A(q)}{4\pi k_F^3} \int\limits_0^{\pi/2} d\vartheta \int\limits_0^{2\pi} d\varphi \int\limits_{\substack{-q \\ -2k_F\cos\vartheta}}^{0} \frac{2 k_F^4 \sin^3\vartheta \cos\vartheta \sin^2\varphi \, dq_1}{\{(\hbar^2/2m) \, [(k_F\cos\vartheta + q_1 + q)^2 - (k_F\cos\vartheta + q_1)^2] + \varepsilon\}}, \quad (20.9)$$

where n is the total number of electrons in unit volume. The lower limit is $-q$ or $-2k_F\cos\vartheta$, whichever is most positive. This insures that the initial point is within the Fermi sphere and that when q is added, the final state is outside the sphere. The integration over φ is immediate and that over q_1 can be obtained in terms of elementary integrals. If we introduce $u = \cos\vartheta$, we are left with the integral

$$S = -\frac{3m\,n\,k_F\,A(q)}{2\hbar^2 q} \int\limits_0^1 u(1-u^2) \log\left\{\frac{k_F u + \frac{q}{2} + \frac{m\varepsilon}{\hbar^2 q}}{\left|k_F u - \frac{q}{2}\right| + \frac{m\varepsilon}{\hbar^2 q}}\right\} du. \qquad (20.10)$$

In the expression for $j(q)$, there are two terms equivalent to S. The value of $K(q)$ defined by (20.1) is

$$K(q) = \frac{4\pi n e^2}{m c^2} \left[1 - \frac{3k_F}{2q} \int\limits_0^1 u(1-u^2) \log\left\{\frac{k_F u + \frac{q}{2} + \frac{m\varepsilon}{\hbar^2 q}}{\left|k_F u - \frac{q}{2}\right| + \frac{m\varepsilon}{\hbar^2 q}}\right\} du. \right] \qquad (20.11)$$

When $\varepsilon = 0$, this leads to the Landau expression for the diamagnetic susceptibility of a free electron gas. We shall see that when $\varepsilon \sim k T_c$ we get a Meissner effect with an expression for $j(q)$ similar to that proposed by Pippard.

The integral in (20.11) may be evaluated explicitly. We want

$$I = \int_0^1 u(1 - u^2) \log \frac{u + \alpha + \beta}{|u - \alpha| + \beta} \, du, \tag{20.12}$$

where we have introduced the notation

$$\alpha = \frac{q}{2 k_F}, \qquad \beta = \frac{m \varepsilon}{\hbar^2 q k_F}. \tag{20.13}$$

We are interested only in small values of q such that $\alpha \ll 1$. We may therefore expand I in a power series in α, and find after some analysis the following result:

$$\left. \begin{aligned} I = 2\alpha &\left\{ -(\beta - \beta^3) \log \frac{1 + \beta}{\beta} + \frac{2}{3} + \frac{\beta}{2} - \beta^2 \right\} + \\ &+ 2\alpha^3 \left\{ \frac{1}{3(1 + \beta)} + \beta \log \frac{1 + \beta}{\beta} - (1 + 2\beta) \right\} + O(\alpha^5). \end{aligned} \right\} \tag{20.14}$$

This gives to the order α^2

$$\left. \begin{aligned} K(q) = \frac{3}{2\lambda_0^2} &\left\{ \beta(1 - \beta^2) \log \frac{1 + \beta}{\beta} - \frac{\beta}{2} + \beta^2 - \right. \\ &\left. - \alpha^2 \times \left[\beta \log \frac{1 + \beta}{\beta} - (1 + 2\beta) + \frac{1}{3(1 + \beta)} \right] \right\}, \end{aligned} \right\} \tag{20.15}$$

where

$$\lambda_L^2 = \frac{m c^2}{4 \pi n e^2} \tag{20.16}$$

is the square of the LONDON expression for the penetration depth.

The limit $\beta \to 0$ corresponds to LANDAU diamagnetism:

$$K(q) = \frac{\pi n e^2 q^2}{m c^2 k_F^2}. \tag{20.17}$$

The diamagnetic susceptibility is

$$4 \pi \chi_d = -\frac{K(q)}{q^2} = -\frac{2 \pi n}{E_F} \left(\frac{e \hbar}{2 m c} \right)^2, \tag{20.18}$$

where $E_F = \frac{\hbar^2 k_F^2}{2m}$ is the energy at the FERMI surface.

In the superconducting case, for which $\beta \neq 0$, the term in α^2 is negligible. Since, with $\varepsilon \sim k T_c$, β is small for the q important for ordinary penetration phenomena, $\lambda_L q \sim 1$, a good approximation for $K(q)$ in this range is[1]

$$K(q) = \frac{3 m \varepsilon}{2 \lambda_L^2 \hbar^2 q k_F} \log \left(1 + \frac{\hbar^2 q k_F}{m \varepsilon} \right). \tag{20.19}$$

In the limit of β large (q very small), $K(q)$ approaches the value corresponding to the usual LONDON theory:

$$K(q) = \frac{1}{\lambda_L^2} \left(1 - \frac{9}{8 \beta} + \cdots \right) \quad \text{as} \quad \beta \to \infty. \tag{20.20}$$

21. Alternative derivation. Before discussing the application of (20.19) to penetration phenomena, we shall give an alternative derivation based on somewhat different assumptions which leads to a theory almost identical with that of PIPPARD. Instead of assuming that the energy of excited states is increased

---

[1] J. BARDEEN: Phys. Rev. **97**, 1724 (1955).

by ε in going from the normal to the superconducting phase, we simply omit from the expansion transitions in which the energy difference between initial and final states is less than ε. This again means that the energy of the lowest excited state considered is ε above the ground state, but there are differences in the matrix elements and density of states for excitations of equivalent energy for the two models. The present calculation is also of interest because it shows which excitations are most important in cancelling the term proportional to $\boldsymbol{A}$ in the current density.

We shall make use of the formulation (19.11) for current density, which uses density matrices, and express the current as an integral of the vector potential in ordinary space so that it can be compared directly with Pippard's equation. The average over an energy surface of $\psi_k^*(\boldsymbol{r}') \psi_k(\boldsymbol{r})$ is

$$\langle \psi_k^*(\boldsymbol{r})' \psi_k(\boldsymbol{r}) \rangle = \frac{1}{2} \int_{-1}^{1} e^{ikRu}\, du = \frac{\sin kR}{kR}, \tag{21.1}$$

where

$$R = |\boldsymbol{r} - \boldsymbol{r}'|. \tag{21.2}$$

The gradient with respect to $\boldsymbol{r}$ is

$$\nabla\left(\frac{\sin kR}{kR}\right) = \frac{kR \cos kR - \sin kR}{kR^2}\, \nabla R. \tag{21.3}$$

After some reduction, the current density may be expressed in the form

$$\boldsymbol{j}(\boldsymbol{r}) = \frac{2e^2}{\pi^4 mc} \int \frac{(G(R)\,\nabla R)\, \boldsymbol{A}' \cdot \nabla' R\, d\boldsymbol{r}'}{R^4} - \frac{ne^2\, \boldsymbol{A}(\boldsymbol{r})}{mc}, \tag{21.4}$$

where

$$G(R) = \int_0^{k_F} dk_0 \left\{ \int_0^{k_0 - \Delta k} + \int_{k_0 + \Delta k}^{\infty} \right\} \frac{k f(k)\, k_0 f(k_0)\, dk}{k_0^2 - k^2} \tag{21.5}$$

and

$$f(k) = kR \cos kR - \sin kR. \tag{21.6}$$

The integral over k_0 is over initial occupied states, that over k over final states. We have summed over all final states with the exception of those for which $|\boldsymbol{k} - \boldsymbol{k}_0| < \Delta k$. The value of Δk is chosen so that

$$\varepsilon = \hbar^2 k_F \Delta k / m. \tag{21.7}$$

As we have mentioned earlier, summing over all final states is equivalent to summing over unoccupied final states, since the integrand is antisymmetric in k_0 and k. The range of integration is shown in Fig. 7. This is not changed by the omission of terms with $|\boldsymbol{k} - \boldsymbol{k}_0| < \Delta k$, since the restriction is symmetric in k and k_0.

To evaluate $G(R)$, first consider the integral of the following over the contour, C, of Fig. 8

$$\int_C \frac{2k\, e^{ikR}}{k_0^2 - k^2}\, dk. \tag{21.8}$$

The contour C, runs along the real axis, with semicircles of radius ϱ above the real axis around the poles at $-k_0$ and $+k_0$, and returns along a semicircle at infinity. Since the latter integral vanishes, and there are no poles inside the contour, the sum of the integrals along the real axis and the small semicircles must vanish. If we take $\varrho = \Delta k$, the integral along the real axis is just the integral over k with omission of regions such that $|\boldsymbol{k} - \boldsymbol{k}_0| < \Delta k$, and we can determine its value from integrations around the small semicircles. If we assume Δk is small and keep only terms to the first order, we find in this way that

$$\left\{ \int_0^{k_0-\Delta k} + \int_{k_0+\Delta k}^\infty \right\} \frac{k \sin k R}{k_0^2 - k^2}\, dk = \left(-\frac{\pi}{2} + R\,\Delta k \right) \cos k_0 R + \frac{\Delta k}{2 k_0} \sin k_0 R. \quad (21.9)$$

A derivative with respect to R gives the value of the integral with $k^2 \cos kR$ in the numerator. Thus we find

$$G(R) = \int_0^{k_F} k_0 f(k_0)\, dk_0 \left\{ -\frac{\pi}{2}\left(-k_0 R \sin k_0 R - \cos k_0 R \right) + \Delta k \left(\frac{df(k_0)}{dk_0} + \frac{f(k_0)}{2 k_0} \right) \right\}. \quad (21.10)$$

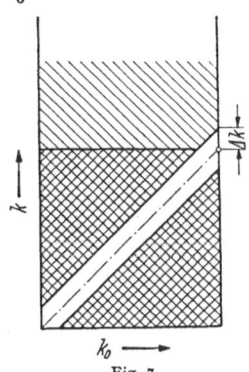

The term independent of Δk cancels the term proportional to $\boldsymbol{A}$ in $\boldsymbol{j}(\boldsymbol{r})$ and leaves the small LANDAU term. The integrand oscillates rapidly about zero and gives a contribution only for $R \approx 0$.

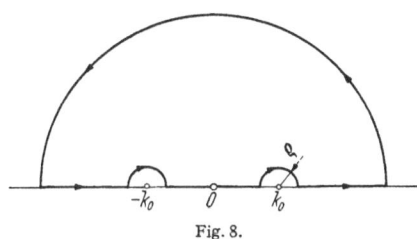

Fig. 7.   Fig. 8.

Fig. 7. Range of integration for integral in Eq. (21.5). Since the integrand is an odd function of k and k_0, integrations over the two cross-hatched regions add to zero.

Fig. 8. Contour in complex plane for integral (21.8).

It is the term proportional to Δk which gives the result we are looking for. This term is

$$\Delta k \int_0^{k_F} k_0 f(k_0) \left\{ \frac{df(k_0)}{dk_0} + \frac{f(k_0)}{2 k_0} \right\} dk_0 = \frac{1}{2}\, k_F\, [f(k_F)]^2\, \Delta k. \quad (21.11)$$

The explicit expression for $[f(k_F)]^2$ is

$$[f(k_F)]^2 = k_F^2 R^2 \cos^2 k_F R + \sin^2 k_F R - 2 k_F R \sin k_F R \cos k_F R. \quad (21.12)$$

We are interested in large values of $k_F R$ where the trigonometric terms are oscillating rapidly. To a good approximation we may replace $\cos^2$ and $\sin^2$ by the average value $\frac{1}{2}$ and omit the third term which averages to zero. Thus we find for $k_F R$ large,

$$\langle [f(k_F)]^2 \rangle = \tfrac{1}{2}\left(1 + k_F^2 R^2 \right) \approx \tfrac{1}{2} k_F^2 R^2. \quad (21.13)$$

The current density is:

$$\boldsymbol{j}(\boldsymbol{r}) = -\frac{3 e^2 n\, \Delta k}{2 \pi^2 m c} \int \frac{\boldsymbol{R}\,(\boldsymbol{R} \cdot \boldsymbol{A}(\boldsymbol{r}'))\, d\tau'}{R^4}. \quad (21.14)$$

This is of the same form PIPPARD assumed for the limit $\xi \to \infty$. Comparing the coefficients, we find for the relation between his ξ_0 and our Δk:

$$\xi_0 = \frac{\pi}{2 \Delta k}. \quad (21.15)$$

This is of the order of magnitude one would estimate from the uncertainty relation between an extension of the wave functions in space, ξ_0, and a significant wave vector difference, Δk:

$$\xi_0 \Delta k \sim 1. \quad (21.16)$$

Pippard's version of (21.14) for a pure metal has an extra factor of $\exp(-R/\xi_0)$ in the integrand. This insures the expression for the current density approaches the usual London expression when A varies very slowly. The criterion is that the Fourier components of A have wave vectors q such that $q\xi_0 \ll 1$. This is also true of our version of the theory as given by (20.20) and probably would also be true of (21.14) in a higher approximation. Thus the integrand in (21.14) requires a correction of the sort introduced by Pippard, but the variation with R may be different than a simple exponential.

Regardless of the details of the theory, it is a likely possibility that matrix elements and energies of excited states differ between normal and superconducting phases only for excitation energies of the order of kT_c. That they are similar for high energies is shown by the fact that there is no difference in reflective power in the infrared at a wavelength of 10μ[1]. Since it is the integral over the entire range of energies which cancels the term proportional to $A(r)$ in the current density, the cancellation must be nearly complete for Fourier components $q \sim 10^5$ or 10^6 cm.$^{-1}$ if only low energy excitations ($\Delta k \sim 10^4$ cm.$^{-1}$) are changed. This would mean that the correct theory in the penetration range is non-local, and that the London theory requires modification along the lines suggested by Pippard. It may be that excited states of higher energy than kT_c are affected by the transition. A natural energy which enters into the lattice-vibration theory is the phonon energy, $k\Theta$ which is of the order of 10 to 100 times larger than kT_c. The relevant Δk might then be of the order of 10^5 or 10^6 cm.$^{-1}$ rather than $\sim 10^4$ cm.$^{-1}$, and these would correspond to distances in real space of the order of the penetration depth. The term proportional to the vector potential might then not be cancelled as completely, and one might get a theory more like the London theory. This point is discussed at greater length in Sect. 26.

It is quite possible, of course, that the particular integral relation suggested by Pippard is not correct, but it is quite interesting to see that it follows from a simple model. We shall show in the next section how the kernel is affected by impurity scattering.

22. Effect of mean free path. The relation (18.1) for the diamagnetic current, suggested by Pippard, involves a factor $e^{-R/\xi}$ in the integral, where ξ is a parameter approximately equal to the mean free path, l, for impurity scattering. We shall show that a similar factor is to be expected from the perturbation theory approach of Sect. 19. The expression for the current density, (19.11) involves the density matrices (19.12) summed over a constant energy surface. In Sect. 19, 20 and 21 we have use free-electron wave functions which give (21.1) for the average of $\psi_k^*(r')\,\psi_k(r)$ over the surface $|k| = \mathrm{const}$. We shall consider here how this result is modified by impurity scattering and how the integral for the current density is in turn affected.

One would like ideally to determine the exact wave functions for electrons moving in a metal with a random distribution of impurity centers, and then determine an average of $\psi^*(r')\,\psi(r)$ over an energy surface. This is a hopeless task. One would expect coherence of the excited state, although not necessarily

---

[1] J. Daunt, T. C. Keeley and K. Mendelssohn: Phil. Mag **23**, 264 (1937). In measurements of surface resistance of superconducting Sn at millimeter wave frequencies, G. S. Blevins, W. Gordy and W. M. Fairbank, Phys. Rev. **100**, 1215 (1955), find that there is a marked increase in absorption when $h\nu > kT_c$, but indications are that the surface resistance at very low temperatures may remain substantially below that of the normal state even when $h\nu$ is considerably greater than kT_c. Similar results have been reported by M. P. Garfunkel, M. A. Biondi and A. D. McCoubrey [33] from measurements on Al.

the ground state, wave functions would be destroyed over a distance of a mean free path, so that a factor similar to that suggested by PIPPARD is not unreasonable.

That such a factor really does come in is indicated by the following calculation. Suppose that the scattering centers are randomly distributed in a slab of width w normal to the x-direction and that there are no centers outside of the slab, as illustrated in Fig. 9. The solutions outside of the slab are then plane waves. If it is assumed that scattering is incoherent, we can calculate $\langle \psi^*(\boldsymbol{r}')\,\psi(\boldsymbol{r})\rangle_k$ exactly by use of general scattering theory, provided that $\boldsymbol{r}$ and $\boldsymbol{r}'$ are outside of the slab.

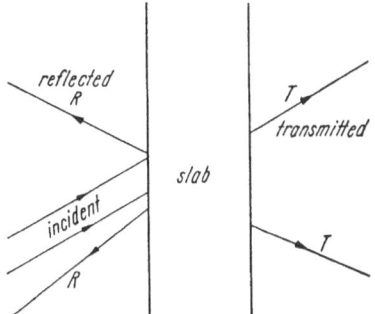

Fig. 9. Incident, reflected and transmitted waves for slab with random distribution of scattering centers.

A complete orthogonal set of wave functions is obtained from (a) incident waves coming from the left in region 1 and (b) incident waves coming from the right in region 2. Associated with each incident wave will be waves scattered to the left and right and a transmitted wave which represents the undeflected wave. Wave functions of the types (a) and (b) may be written as follows:

(a)   in (1)
$$\psi = e^{i\boldsymbol{k}\cdot\boldsymbol{r}} + \sum_{\boldsymbol{k}'} R_{\boldsymbol{k}\boldsymbol{k}'}\, e^{-i\boldsymbol{k}'\cdot\boldsymbol{r}} \sqrt{\frac{\cos\vartheta_k}{\cos\vartheta_{k'}}}\,,$$

in (2)
$$\psi = \sum_{\boldsymbol{k}'} T_{\boldsymbol{k}\boldsymbol{k}'}\, e^{i\boldsymbol{k}'\cdot\boldsymbol{r}} \sqrt{\frac{\cos\vartheta_k}{\cos\vartheta_{k'}}}\,,$$
(22.1)

(b)   in (1)
$$\psi = \sum_{\boldsymbol{k}'} T_{\boldsymbol{k}\boldsymbol{k}'}\, e^{-i\boldsymbol{k}'\cdot\boldsymbol{r}} \sqrt{\frac{\cos\vartheta_k}{\cos\vartheta_{k'}}}\,,$$

in (2)
$$\psi = e^{-i\boldsymbol{k}\cdot\boldsymbol{r}} + \sum_{\boldsymbol{k}'} R_{\boldsymbol{k}\boldsymbol{k}'}\, e^{i\boldsymbol{k}'\cdot\boldsymbol{r}} \sqrt{\frac{\cos\vartheta_k}{\cos\vartheta_{k'}}}\,.$$
(22.2)

The cosine factors normalize the scattered waves to the same flux as the incident wave. In all cases, the x-components of $\boldsymbol{k}$ and $\boldsymbol{k}'$ are restricted to positive values, so that a positive sign in the exponent corresponds to a wave moving from left to right, a negative sign from right to left. Density matrices can be constructed from this complete set.

In computing the density matrices, we shall assume that the scattered waves add in random phase and average to zero. If both $\boldsymbol{r}$ and $\boldsymbol{r}'$ are in the left region,

$$\sum_{\substack{|\boldsymbol{k}|=\text{const}}} \psi_{\boldsymbol{k}}^*(\boldsymbol{r}')\,\psi_{\boldsymbol{k}}(\boldsymbol{r}) = \sum_{\substack{|\boldsymbol{k}|=\text{const} \\ k_x > 0}} \{e^{i\boldsymbol{k}\cdot(\boldsymbol{r}'-\boldsymbol{r})} + e^{-i\boldsymbol{k}\cdot(\boldsymbol{r}'-\boldsymbol{r})}\}. \qquad (22.3)$$

The sum on the right is over positive values of k_x. It is equivalent to summing one of the two terms over all k_x, positive and negative, and is thus exactly the same as the free electron value. A different result is obtained if $\boldsymbol{r}$ is on the left and $\boldsymbol{r}'$ on the right of the slab

$$\sum \psi_{\boldsymbol{k}}^*(\boldsymbol{r}')\,\psi_{\boldsymbol{k}}(\boldsymbol{r}) = \sum T_{\boldsymbol{k}\boldsymbol{k}} \{e^{i\boldsymbol{k}\cdot(\boldsymbol{r}'-\boldsymbol{r})} + e^{-i\boldsymbol{k}\cdot(\boldsymbol{r}'-\boldsymbol{r})}\}. \qquad (22.4)$$

There is now an extra factor, $T_{\boldsymbol{k}\boldsymbol{k}}$, whose square is the probability that the electron pass through the slab without being scattered. Since the probability that an electron goes a distance R without being scattered is $e^{-R/l}$, it is reasonable to suppose in the general case of a random distribution of scattering centers, the density matrix is reduced by a factor $e^{-R/2l}$, where $R=|\boldsymbol{r}-\boldsymbol{r}'|$.

According, we shall assume that when scattering centers give a mean free path, l, (21.1) is changed to

$$\langle \psi_k^*(r') \, \psi_k(r) \rangle_s = \frac{\sin kR}{kR} \, e^{-\frac{R}{2l}}. \tag{22.5}$$

The subscript s implies the value when scattering is taken into account. The calculation of diamagnetic properties proceeds in exactly the same way as that of Sect. 21. The expression for $f(k)$, (21.6), is changed to

$$f_s(k) = kR \cos kR - \left(1 + \frac{R}{2l}\right) \sin kR. \tag{22.6}$$

The integral for $G(R)$, (21.10), is changed to

$$\left. \begin{aligned} G_s(R) &= e^{-\frac{R}{l}} \int_0^{k_F} k_0 f_s(k_0) \, dk_0 \times \\ &\times \left\{ -\frac{\pi}{2} \left(k_0 R \sin k_0 R - \left(1 + \frac{R}{2l}\right) \cos k_0 R\right) + \Delta k \left(\frac{df_s(k_0)}{dR_0} + \frac{f_s(k_0)}{2k_0}\right) \right\}. \end{aligned} \right\} \tag{22.7}$$

The terms independent of Δk give, with neglect of terms of order $(k_F l)^{-2}$, just the small LANDAU diamagnetism. Thus a finite mean free path has a negligible effect on ordinary diamagnetism. The terms proportional to Δk give, to the same order, in place of (21.14)

$$j(r) = - \frac{3e^2 n \Delta k}{2\pi^2 m c} \int e^{-\frac{R}{l}} \frac{R(R \cdot A(r')) \, d\tau'}{R^4}. \tag{22.8}$$

This is exactly of the form proposed on phenomenological grounds by PIPPARD. As indicated in Sect. 21, a correction to the integrand is required of the sort which would replace l by ξ as suggested by PIPPARD [Eq. (18.2)].

f) Non-local theories.

23. Electrons with small effective mass. It has been suggested by the author[1] that a gas of electrons of small effective mass will obey the LONDON equations. The LANDAU expression[2] for the diamagnetic susceptibility of a degenerate electron gas may be written in the form [cf. Eq. (20.18)]:

$$\chi_s = - \frac{n_s}{2E_s} \left(\frac{e\hbar}{2m_s c}\right)^2, \tag{23.1}$$

where n_s is the concentration of electrons, E_s is the FERMI energy and m_s an effective mass. A large diamagnetism is obtained if m_s is sufficiently small.

It was first proposed that a small effective mass in superconductors might arise from small periodic displacements of the ions such as to produce BRILLOUIN zone boundaries with small energy gaps at the FERMI surface. The associated reduction of energies of electrons near the FERMI surface would stabilize the structure. A very large number of boundaries would be required for a FERMI surface of arbitrary shape. The period in k-space would be very small and in ordinary space very large compared with the interatomic distance. The energy reductions would be small, so that the transition would occur only at low temperatures. The zones would not be completely filled, so that the electrical properties would be described by electrons or holes of very small effective mass, depending on whether or not the zone is nearly empty or nearly filled. For an energy reduction of $\Delta E \sim kT_c$ the effective mass is of the order of $(\Delta E/E_F)m \sim$

[1] J. BARDEEN: Phys. Rev. **59**, 928A (1941); **81**, 829 (1951).
[2] L. D. LANDAU: Z. Physik **64**, 629 (1930).

$10^{-4}\,m$, where E_F and m are the FERMI energy and mass of electrons in the normal metal. The maximum energy E_s of the electrons or holes would be of order ΔE. Each zone would accommodate only a small number of electrons, but with a very large number of zones one might expect that n_s is of the order of the number of electrons in the normal metal with energies within ΔE of the FERMI surface. Thus we expect

$$\frac{n_s}{E_s} \sim \alpha\, \frac{n}{E_F}\,, \tag{23.2}$$

where α is perhaps of the order of 0.1. Thus the susceptibility is increased by a ratio $\sim (m/m_s)^2$ which is of the order of 10^8. Since ordinary susceptibilities are of the order of 10^{-7} to 10^{-6}, one might expect to get a value of $|\chi_d|$ larger than $(4\pi)^{-1}$ and thus a very large diamagnetism[1].

With the discovery of the isotope effect, the theory was modified and it was proposed that the small mass comes from reduction of energy of electrons near the FERMI surface brought about by interactions of electrons with lattice vibrations. There would be no true zone structure, but again most electrons with energies within ΔE of the FERMI surface might be involved and n_s would be of a similar order of magnitude.

While it is now believed that an adequate description of superconductivity cannot be obtained from an individual particle model, it is of interest to investigate the properties of a degenerate electron gas in which n_s and E_s are treated as independent variables, and are *not* related as they would be if the electrons were in a single BRILLOUIN zone. A large diamagnetism is not obtained from a single zone regardless of how small m_s is made unless E_s is an impossibly large value.

A calculation has been given by KLEIN [12] in which it is assumed that $|\mathbf{k}|$ is the same for all electrons. While one could integrate $|\mathbf{k}|$ over a FERMI distribution, it is more convenient to follow KLEIN and take $|\mathbf{k}| = k_s$ as a representative value for the electron distribution. The expression derived by KLEIN for $K(q)$ for such an electron gas, and which follows from the theory of Sect. 20, is

$$K(q) = \frac{1}{2\lambda_s^2} \left\{ 1 - \frac{1}{2}\left(\frac{2k_s}{q} - \frac{q}{2k_s}\right) \log \frac{1 + \frac{q}{2k_s}}{\left|1 - \frac{q}{2k_s}\right|} \right\}, \tag{23.3}$$

where λ_s is the penetration depth given by

$$\lambda_s^2 = \frac{c^2\, m_s}{4\pi\, e^2\, m_s}\,. \tag{23.4}$$

Since we are assuming that the ratio n_s/m_s is of the order of n/m, λ_s will be of the order of λ_0.

Limiting values of (23.3) for q/k_s very large and very small are:

$$K(q) = \frac{1}{\lambda_s^2} \qquad \text{for} \qquad \frac{q}{k_s} \gg 1\,, \tag{23.5}$$

$$K(q) = \frac{2}{3\lambda_s^2}\left(\frac{q}{2k_s}\right)^2 \qquad \text{for} \qquad \frac{q}{k_s} \ll 1\,. \tag{23.6}$$

---

[1] E. N. ADAMS II [Phys. Rev. **89**, 633 (1953)] has given a critical discussion of the LANDAU-PEIERLS susceptibility when a large diamagnetism results from a small effective mass in the BLOCH theory. He concludes that neglected terms from off-diagonal interband transitions may actually be quite large, and the result is therefore uncertain.

The first limit corresponds to the ordinary LONDON theory with a penetration depth λ_s. Important values of q for ordinary penetration phenomena are of the order of λ^{-1} or of the order of 2×10^5 cm.$^{-1}$. The value of k_s is of order

$$k_s \sim \frac{E_s}{E_F} k_F \sim 10^{-4} k_F \sim 10^4 \text{ cm.}^{-1}. \tag{23.7}$$

Thus the first limit applies for the usual penetration phenomena and one may expect the ordinary LONDON theory to be valid in this range. There is, however, a small penetration of field in a massive specimen. The second limit corresponds to a susceptibility equal to:

$$\chi = \frac{\chi_s}{1 - 4\pi \chi_s}, \tag{23.8}$$

where χ_s is the analogue of (23.1) for this model;

$$\chi_s = -\frac{n_s}{6 E_s} \left(\frac{e \hbar}{2 m_s c} \right)^2. \tag{23.9}$$

The difference of a factor of 3 between (23.1) and (23.9) comes from the difference between a distribution of k_s over a FERMI distribution and a uniform k_s. According to (23.8) $4\pi |\chi|$ is less than unity no matter how large is $|\chi_s|$, so that there is some penetration of field into the interior[1].

In order to have a uniform density of electrons all the way up to the surface, it is necessary to assume that the boundary conditions are such that the phases of the various electrons at the surface are distributed at random. This is implicitly assumed when calculations made for an infinite medium are applied to a bounded medium.

It should be noted that a large diamagnetism occurs when the wavelength of the electron waves is large compared with the penetration depth. The wave functions of the electrons then extend over distances large compared with the penetration depth. An extreme case is an ideal EINSTEIN-BOSE gas of charged particles. Below the condensation temperature, a significant fraction of the electrons are in the very lowest state, and the wave function for this state extends throughout the volume. This would correspond in the above example to a limit $k_s \to 0$ and for this one gets just the ordinary LONDON theory with no penetration of field in a massive specimen, even if one assumes an ordinary mass for the electrons [12]. The diamagnetic and thermal properties of a charged EINSTEIN-BOSE gas have been worked out most completely by SCHAFROTH[2].

24. Comparison between different versions of non-local theories. Two different formulations of non-local theories, based on somewhat different assumptions, have been given in Sects. 20 and 21. A third, which leads to the usual LONDON theory for infinite correlation length, has been suggested by SCHAFROTH and BLATT[3].

The versions of Sects. 20 and 21 cannot be compared directly because that of Sect. 20 used FOURIER transforms while that of Sect. 21 did not. Because it has not been possible to get a simple expression in coordinate space for the former it is easiest to compare the FOURIER transforms. The expression for $K_p(q)$ as defined by (20.1) for PIPPARD's relation (18.1) is

$$K_p(q) = \frac{3\pi \xi}{\xi_0 \Lambda c^2} \cdot \frac{2}{\xi^3 q^3} \{ (1 + \xi^2 q^2) \text{ arc tan } \xi q - \xi q \}. \tag{24.1}$$

[1] H. FRÖHLICH: Nature, Lond. **168**, 280 (1951).
[2] M. R. SCHAFROTH: Phys. Rev. **100**, 463 (1955).
[3] M. R. SCHAFROTH and J. BLATT: Phys. Rev. **100**, 1221 (1955).

We have given essentially this result in connection with PIPPARD'S expression (18.7) for the penetration depth, which applies to the case of random scattering at the surface:

$$\lambda = \frac{\pi}{\int\limits_0^\infty \log\left(1 + \frac{K_p(q)}{q^2}\right) dq}.$$

(24.2)

Since impurity scattering was not included in the discussion in Sect. 20, we should compare the expression derived there, (20.19), with the limiting value of (24.1) for $\xi \to \infty$. For infinite free path, (24.1) becomes:

$$K_p(q) = \frac{3\pi^2}{\xi_0 \Lambda c^2 q} = \frac{3\Delta k}{2\lambda_L^2 q};$$

(24.3)

thus $K_p(q)$ is inversely proportional to q. Except for the slowly varying logarithmic term, this is also true of (20.19). Thus the theories are very similar. If we make use of the definition of Δk in (21.7), we may write (20.19) in the form

$$K(q) = \frac{3\Delta k}{2\lambda_L^2 q} \log\left(1 + \frac{q}{\Delta k}\right).$$

(24.4)

Thus the only difference is in the logarithmic term. Since $\Delta k \ll q$ for ordinary penetration phenomena, the argument of the logarithm is a large number. The term varies slowly with q and may be treated as a constant to a good approximation. Thus the theory of Sect. 20, which is probably more realistic than that of 21, also leads to a theory very close to that proposed by PIPPARD, but with a somewhat larger coefficient in front of the integral. If one takes $q \sim \lambda^{-1}$ in the logarithm, the value of PIPPARD'S ξ_0 turns out to be

$$\xi_0 = \frac{\pi}{2\Delta k \log\left(1 + \frac{1}{\lambda \Delta k}\right)}.$$

(24.5)

Theoretical values of ξ_0 will be compared with those deduced from observation in Sect. 25.

SCHAFROTH and BLATT[1] have assumed a form equivalent to the following:

$$K_{SB}(q) = \frac{q}{\lambda_0^2(q + \mu)},$$

(24.6)

where μ is a parameter such that μ^{-1} is of the order of correlation length. This reduces to the LONDON expression as $\mu \to 0$, corresponding to infinite correlation distance. The form (24.6) was suggested because they believe that $K(q)q$ must approach zero as $q \to 0$ when the correlation length is finite[2]. They are able to fit PIPPARD'S data on the variation in penetration depth in tin-indium alloys by taking $\lambda_0 = 5.9 \times 10^{-6}$ cm. and $\mu = 0.31\,l^{-1}$, where l is the mean free path. It should be noted that (24.6) gives an infinite penetration of flux for a half-space with a plane boundary. In other words, $\int\limits_0^\infty H(x)\, dx$ does not converge. They state that this must occur in a system with a finite correlation distance[3]. Details of the calculation are not available as this is written.

25. Boundary conditions, gauge invariance. The perturbation theory derivation of the current density was carried out for an infinite medium. A problem

---

[1] M. R. SCHAFROTH and J. BLATT: Phys. Rev. **100**, 1221 (1955).

[2] See the discussion in Sect. 26.

[3] The correlation distance is defined roughly as the maximum distance over which the momenta of two particles of the fluid are correlated.

remains as to how to apply the results to a body of finite dimensions. We shall assume for simplicity that the body is simply connected; the extension to multiply connected bodies is similar to that of the ordinary London theory as discussed in Sect. 13 [1]. The problem of introducing a boundary is closely related to that of gauge invariance. The condition $\text{div}\,\boldsymbol{A}=0$, on which all of our expressions have been based, determines the gauge uniquely in an infinite medium. If one adds $\text{grad}\,\varphi$ to $\boldsymbol{A}$, the condition on φ is

$$\nabla^2\varphi=0, \tag{25.1}$$

and the Fourier components of φ must vanish. However, it is not true that φ must vanish (up to a constant) for a finite body. One is at liberty to specify either φ or $\partial\varphi/\partial n$ on the boundary. The theory should be formulated in such a way that $\text{div}\,\boldsymbol{j}=0$ everywhere and $j_\perp=0$ at a free surface. More generally one would like to formulate the theory so that $j_\perp$ could be specified on the surface. In an infinite medium, the condition $\text{div}\,\boldsymbol{A}=0$ insures that $\text{div}\,\boldsymbol{j}=0$ for any relation of the form (20.1).

Pippard has suggested that for random scattering at the surface, one should carry out the integration in (18.1) over the volume occupied by the body; this corresponds to setting $\boldsymbol{A}=0$ outside of the body. Although he does not state so explicitly, presumably one should take the London choice of gauge with $A_\perp=0$ at the free surface. The lines of the vector field of $\boldsymbol{A}$ will then be parallel to the surface, and there will be no violation of $\text{div}\,\boldsymbol{A}=0$ at the surface. This insures that $\text{div}\,\boldsymbol{j}=0$ inside. This choice determines $\boldsymbol{A}$ uniquely. It might happen, however, that $j_\perp$ is *not* equal to zero at the surface for this choice, and thus the proper boundary conditions would not be satisfied. Fortunately, this difficulty does not arise for such simple but important cases as penetration at a plane surface, a cylinder in an axial field, and a sphere in a uniform external field.

The physical argument for taking $\boldsymbol{A}=0$ outside the surface for random scattering at the surface is similar to that for the anomalous skin effect, as discussed in Sect. 17. Electrons then enter the surface in random directions as if they come from a field-free space. The perturbation theory derivation leads to a similar result, as indicated by the argument of Sect. 22. If there is random scattering, the density matrix for two points inside the surface is the same as for an infinite medium, but the matrix would, of course, vanish for one point inside and the other outside the surface. Thus one should carry the integration over the actual volume. Since a derivative of the density matrix is involved, and it is discontinuous at the surface, the possibility that a surface integral should be added is not excluded. Such a surface integral would be required in any case to satisfy the boundary condition when $j_\perp$ is specified at the surface. No surface integral is required if the integral over the body satisfies the proper condition, e.g. $j_\perp=0$ at the surface. If the integral does give a net current flowing into the surface, the flow away from the surface cannot be completely random, and one cannot expect to satisfy the conditions by taking $\boldsymbol{A}=0$ outside of the surface. In this case, a surface integral must be added.

The only solution that has been obtained so far is that for penetration at a plane boundary. In this case one may take $\boldsymbol{A}$ and $\boldsymbol{j}$ both parallel with the boundary. The penetration depth, as derived by Pippard by a method similar to that used for the anomalous skin effect, is given by (24.2)

The solution for specular reflection at a plane boundary can be obtained by introducing an infinite plane current sheet as a source in an infinite medium.

---

[1] The fluxoid theorem is no longer valid, but the solution is uniquely determined by the total currents flowing around the loops.

If the surface is the plane $x = 0$, the current sheet is such that $H_y = H_0$ for $x = +0$, and $H_y = -H_0$ for $x = -0$. Electrons crossing the boundary from the side $x < 0$ will then correspond to specular reflection of electrons coming from $x > 0$. One may analyze the source field into its FOURIER components, $A_0(q)$, and determine the components of the total vector potential $A(q)$ from the MAXWELL equation

$$(q^2 + K(q)) A(q) = q^2 A_0(q). \qquad (25.2)$$

In this way, one derives the following for the penetration depth (see Sect. 18):

$$\lambda = \frac{2}{\pi} \int_0^\infty \frac{dq}{q^2 + K(q)}. \qquad (25.3)$$

A similar expression was derived by REUTER and SONDHEIMER from the corresponding theory for the anomalous skin effect.

With PIPPARD'S $K_p(q)$ for the limit $\xi = \infty$, as defined by (24.3), the integral (25.3) gives:

$$\lambda_\infty = \frac{8}{9} \frac{3^{\frac{1}{6}}}{(2\pi)^{\frac{1}{3}}} (\xi_0 \lambda_L^2)^{\frac{1}{3}} \qquad \text{(specular reflection)}. \qquad (25.4)$$

The corresponding expression for random scattering, as given by (18.11) is

$$\lambda_\infty = \frac{3^{\frac{1}{6}}}{(2\pi)^{\frac{1}{3}}} (\xi_0 \lambda_L^2)^{\frac{1}{3}} \qquad \text{(random scattering)}. \qquad (25.5)$$

The expressions for the penetration depth are almost identical; the numerical factors differ by a ratio $\frac{8}{9}$.

The integral (25.3) cannot be evaluated explicitly for $K(q)$ given by (24.4), but a good approximation can be obtained if q in the logarithm is replaced by an average value, λ^{-1}. The only difference is then an extra factor of the logarithm in the radical. With the numerical factor for random scattering, (24.4) gives

$$\lambda_\infty = \frac{3^{\frac{1}{6}}}{(2\pi)^{\frac{1}{3}}} \left[\xi_0 \lambda_L^2 / \log\left(1 + \frac{1}{\lambda_\infty \Delta k}\right) \right]^{\frac{1}{3}}. \qquad (25.6)$$

FABER and PIPPARD[1] have made estimates of λ for tin and aluminum from observed values of the skin resistance at high frequencies and have compared these estimates with observed penetration depths. Theory indicates that

$$\lambda_L^2 = \frac{1}{4\sqrt{3}\,\pi^2 v_0\, \omega^2 \Sigma_\infty^3}; \qquad (25.7)$$

where v_0 is the velocity at the FERMI surface, ω is the angular frequency and Σ_∞ the skin conductance of the normal metal at frequency ω. At high frequencies $\omega^2 \Sigma_\infty^3$ approaches a constant characteristic of the metal. To estimate ξ_0 or Δk it is assumed that [see Eqs. (21.7) and (21.15)]:

$$\xi_0 = \frac{\pi}{2\Delta k} = \frac{a\,\hbar\, v_0}{k\, T_c}, \qquad (25.8)$$

where a is parameter of order unity. When (25.7) and (25.8) are inserted in (25.5), v_0 drops out and we find

$$\lambda_\infty = \frac{1}{2\pi\, \omega^{\frac{2}{3}} \Sigma_\infty} \left(\frac{a\,\hbar}{k\, T_c}\right)^{\frac{1}{3}}, \qquad (25.9)$$

which is the expression used by FABER and PIPPARD. They also used the specific heat to estimate the density of states at the FERMI surface, and thus to find v_0.

---

[1] T. E. FABER and A. B. PIPPARD: Proc. Roy. Soc. Lond., Ser. A **231**, 53 (1955).

An estimate then could be obtained for ξ_0. They assume that in the pure metal, $\xi = \xi_0$, and correct λ_∞ to get the actual penetration depth for $T = 0°$ K. Results of their calculation are given in Table 2. FABER and PIPPARD used (25.5) and got the best fit with $a = 0.15$.

Table 2. *Comparison of calculated and observed values of penetration depth at $T = 0°$ K for tin and aluminum, after* FABER *and* PIPPARD.

Metal	$10^4 \frac{\hbar v_0}{kT_c}$ cm.	$10^4 \xi_0$ cm. $a = 0.15$	$10^6 \lambda_L$ cm. Eq. (25.7)	$10^6 \lambda_\infty$ cm. Eq. (25.5)	$10^6 \lambda$ cm. Fig. 4	$10^6 \lambda$ cm. (obs.)
Tin · · · · ·	1.4	0.21	3.5	4.15	5.25	5.1
Aluminum	8.2	1.23	1.6	4.45	4.75	4.9

If (25.6) with the extra logarithmic factor is used, the effect is to multiply a by the value of the logarithm. This would give $a = 0.35$ for tin and $a = 0.70$ for aluminum. The theoretical value of a corresponding to $\varepsilon = kT_c$ is $\pi/2$, as given by (21.15). The rough order of magnitude agreement is certainly satisfactory.

PIPPARD's theory predicts that the penetration depths of tin and aluminum should be about the same, even though the values given by the LONDON theory in its simplest form differ by a factor of two. The large value of ξ_0 for aluminum compensates for the small value of λ_L.

26. Discussion of phenomenological theories. PIPPARD [20] has given the following experimental verifications of his version of the phenomenological equations of superconductivity. The theory accounts for (1) the variation of penetration depth, λ, of tin-indium alloys with mean free path, (2) the anisotropy of λ for tin, particularly the maximum at an intermediate angle, (3) the fact that λ is considerably larger than that given by the LONDON expression, and (4) the relative values of λ for tin and aluminum (Sect. 25). There are, of course, many things not accounted for as yet. Perhaps the most important is the variation of λ with temperature, which is explained so well by the usual LONDON theory when combined with the GORTER-CASIMIR two-fluid model (Sect. 4). It is not as yet certain that penetration phenomena in thin films and other bodies of small dimensions can be explained as well with the PIPPARD version as they are with the LONDON version.

There are two fundamental lengths which enter the theory: ξ_0, which is associated by means of the uncertainty relation with the energy kT_c,

$$\hbar v_0/\xi_0 \sim kT_c, \qquad (26.1)$$

and λ, the penetration depth, which is associated similarly with an energy

$$\varepsilon_\lambda \sim \hbar v_0/\lambda \sim \xi_0 kT_c/\lambda. \qquad (26.2)$$

It is indicated from the derivations from the energy-gap models of Sects. 19 and 21 that a theory similar to PIPPARD's will follow if $\xi_0 \gg \lambda$ and if excitations with an energy ε_λ are essentially the same as those of a normal metal. If only excited states with energy $\sim kT_c$ are involved in the normal-superconducting transition, so that only these states have their energies and matrix elements changed appreciably, one is led to a theory of the PIPPARD type.

To get a theory more similar to LONDON's, it would be necessary either to have an energy gap of the order of ε_λ or to have matrix elements corresponding to excited states of this energy markedly reduced by the transition. While the former does not occur, the latter is not out of the question. The natural energy which comes into the lattice-vibration theory of superconductivity is the phonon energy, $\hbar\omega$, which is 10 to 100 times kT_c. Estimates of T_c from current theories have been far too high, and it is not known why T_c is as small as it is. One possibility is that electron-phonon interactions involving energies $> kT_c$ come in equally in both the normal and superconducting phases, and that it is only states of lower energy which are involved in the transition. Another is that states of much higher energy are involved and have their wave functions changed markedly, but that the energy difference is nevertheless very small, so that the transition occurs at a low temperature. If the latter were true, one might be led to a theory more like that of LONDON. Studies in the millimeter range of wave lengths should give a good indication of which picture is correct.

SCHAFROTH[1] has argued that one cannot have a true MEISSNER effect, in the sense of a finite penetration depth in a massive specimen, if the correlation length is finite. Since the LONDON and PIPPARD theories do give a MEISSNER effect, they argue that neither theory can apply to real systems. The proof is based on the "rotating bucket theorem"[2]. This theorem concerns the thermal equilibrium motion of a fluid when the container is rotated. A correlation length is defined from the thermal average, over the available states of the system at temperature T, of the product of the momenta of two particles separated by a distance R. The product is defined in a quantum mechanical way so that there is no difficulty from the uncertainty principle, but the classical interpretation is valid only when it is not violated. The correlation length is the distance beyond which the momenta are essentially uncorrelated. It is shown that for sufficiently slow rotational velocities and sufficiently large volumes, the fluid will rotate with the container (i.e. have its classical moment of inertia) if the correlation length is smaller than the dimensions of the container.

The connection with superconductivity is made essentially as follows. Suppose that the superconductor is in the form of a long circular cylinder, and imagine fictitious sources of magnetic field within the body which have no other interaction with the electrons than the magnetic one. The sources are taken such as to produce a uniform magnetic field within the body. This could be done, for example, by having an imaginary uniformly charged cylinder in coincidence with, but not interacting with the superconductor. Now let the charged cylinder rotate. According to the LONDON theory, as indicated in Sect. 11, the electrons of the superconductor will rotate so as to neutralize the current almost everywhere, and leave a uniform magnetic field. In the rotating superconductor, the positive ions provide the current and the electrons rotate around with them. In the present case, we suppose that the ions of the superconductor remain fixed; it is the fictitious charges and electrons which rotate. The uniform magnetic field causes rotation of the electrons at just the LARMOR frequency, ω_0.

Now consider a superconducting cylinder rotating with angular frequency ω_0 in no magnetic field. In a coordinate system rotating with the superconductor, there is a CORIOLIS force which to first order in ω_0 acts like a uniform magnetic field. This, of course, is just the basis of the LARMOR theorem. Now apply the "rotating bucket" theorem to this system. If there is a finite correlation length, the electrons will move with the container, and thus will not rotate relative to the rotating coordinate system.

Translating to the equivalent case of a non-rotating cylinder in a uniform magnetic field, the electrons will not rotate in the field, but will remain substantially at rest. Thus we have a contradiction. The only solution allowed by the

[1] M. R. SCHAFROTH: Phys. Rev. **100**, 502 (1955).
[2] J. M. BLATT, S. T. BUTLER and M. R. SCHAFROTH: Phys. Rev. **100**, 481 (1955).

London equations in a uniform magnetic field is the one for which the electrons rotate with the Larmor frequency, and such a rotation in a system of finite correlation length is not allowed by the "rotating bucket" theorem. The objection applies not only to the usual London equations, but to any system which gives a true Meissner effect, such as that of Pippard.

It is perhaps dangerous to base the argument on a fictitious system, but if we accept the result, we must either abandon a perfect Meissner effect or admit an infinite correlation length. Perhaps one might picture a limiting case of a superconductor of finite correlation length as a metal broken up into non-interacting regions separated by this insulating boundaries. Even though there is a good Meissner effect in each region, there would be some penetration of flux through the specimen as a whole. The smaller the regions, the greater would be the penetration of flux. Thus, to get a true Meissner effect in a massive specimen, the ordered ground state must extend throughout the volume.

In real superconductors, the correlation length, L, may not be infinite, but it may be very large, so as to give a nearly perfect Meissner effect. If L is large compared with the other basic lengths which enter into the theory, Pippard's coherence distance, ξ_0, and the penetration depth, λ, one would expect the Pippard or London type of equation to hold as a close approximation. In a pure metal, one would expect L to be of the order of the mean free path or larger, and thus may be as much as 10^{-2} cm, which is indeed large compared with ξ_0. In well-prepared alloys which exhibit a Meissner effect, L is probably also quite large.

One should distinguish between L, which refers to the range of the order in the ground state, and the mean free path, l, which refers to the elementary excitations (excited electrons). The former may be much larger than the latter. The range of order will persist as the temperature is raised above $T=0°$ K until the critical temperature is approached.

At present it is uncertain whether the correct "ideal" theory is of the Pippard or London types, with some arguments in favor of each. Schafroth and Blatt, as mentioned in Sect. 24 have been able to account for the mean free path effects in tin-indium alloys with a theory which reduces to the London theory when the correlation length is infinite. If this view is adopted, there remain the problems of anisotropy of the penetration depth of tin and of the magnitude of the penetration depth. The London theory probably represents a limiting case which is never actually reached; it may be that the Pippard limit is not attained either, and that actual metals will require an intermediate theory.

The Pippard theory has not been developed to give a complete electrodynamics of superconductivity. For relatively slow changes, corresponding to angular frequencies ω much less than a critical frequency, ω_c, given by

$$\omega_c = k\,T_c/\hbar \sim v_0/\xi_0, \tag{26.3}$$

we may expect that Pippard's equation (18.1) will be valid if A is considered to be a function of time. A time varying magnetic field gives rise to an electric field, $E = \dfrac{1}{c}\dfrac{\partial A}{\partial t}$. The time derivative of (18.1) then gives an equation analogous to the London equation (II):

$$\frac{\partial j_s}{\partial t} = -\frac{3}{4\pi\Lambda\xi_0}\int \frac{R(R\cdot E)\,e^{-\frac{R}{\xi}}\,d\tau}{R^4} \qquad (\omega\ll\omega_c). \tag{26.4}$$

For frequencies of the order of ω_c, one might expect quantum effects to come in, since the quantum energy will then be of the order of the energy gap. When

$\omega \gg \omega_c$, the excitations are assumed similar to those in the normal state, so that we may expect that CHAMBERS' expression (17.5) for the anomalous skin effect will be valid. This expression may be written in a form analogous to (26.4):

$$ \boldsymbol{j} = \frac{3}{4\pi \Lambda v_0} \int \frac{\boldsymbol{R}(\boldsymbol{R} \cdot \boldsymbol{E})\, e^{-\frac{R}{l}}\, d\tau}{R^4} \qquad (\omega \gg \omega_c). \qquad (26.5) $$

When $\boldsymbol{E}$ varies with time, the retarded time, $t - \dfrac{R}{v_0}$, is to be taken in the integrand. One might expect that (26.4) and (26.5) will give similar results for $\omega \sim \omega_c$, and this is true. Consider the current generated by a pulse $\boldsymbol{E}(\boldsymbol{r}, t)$ lasting for a time

$$ \Delta t = 1/\omega_c = \xi_0/v_0. \qquad (26.6) $$

We may estimate from (26.5) the current, $\boldsymbol{j}(\Delta t)$, just after the pulse and from (26.4) the average rate of increase of current during the pulse. We expect to find

$$ \frac{\partial \boldsymbol{j}_s}{\partial t} \sim \frac{\boldsymbol{j}(\Delta t)}{\Delta t} \sim \frac{v_0}{\xi_0}\, \boldsymbol{j}(\Delta t), \qquad (26.7) $$

and this is indeed just the relation indicated by (26.4) and (26.5). It should be noted that the critical time is that taken for an electron at the FERMI surface to go a distance ξ_0.

It would be desirable to write (18.1) in an obviously gauge invariant way. For an infinite medium, this can be done and one has, corresponding to (I):

$$ \operatorname{curl} \boldsymbol{j} = -\frac{3}{4\pi \xi_0 c \Lambda} \int \frac{\boldsymbol{R}(\boldsymbol{R} \cdot \boldsymbol{H})\, e^{-\frac{R}{\xi}}\, d\tau}{R^4}. \qquad (26.8) $$

This may be combined with div $\boldsymbol{j} = 0$ to get the current density for the infinite medium. Eqs. (26.4) and (26.7) were suggested by LONDON in a discussion following a paper of PIPPARD[1]. The difficulty comes in applying the equation to a finite body with the condition of random scattering at the surface. One might think one could integrate (26.9) over the body, which would correspond to setting $H = 0$ outside the body in the infinite medium. However, this is not the same thing as setting $\boldsymbol{A} = 0$ outside, with the additional condition that $A_\perp = 0$. For the latter, $\boldsymbol{H}$ is also zero outside, but there is an added surface integral which comes from the infinite magnetic field associated with the discontinuity of $\boldsymbol{A}$ at the surface. This infinite field presumably has the effect in the infinite medium of introducing an effective random scattering at the surface. The surface integral is simply expressed in terms of $\boldsymbol{A}$, but not in terms of $\boldsymbol{H}$, so that the result is not obviously gauge invariant. The theory is, of course, actually gauge invariant because there is a prescription for specifying the gauge to be used in a unique way.

IV. Boundary effects; the intermediate state.

a) Theory of boundary energies.

27. Introduction. Before much was known about the structure of the intermediate state, LANDAU [13] proposed a theory based on alternating lamina of normal and superconducting regions, as illustrated in Fig. 10a for a plate oriented normal to the field. The field H is equal to the critical field H_c in the normal regions, and the relative widths of the regions are such as to give the proper

---

[1] A. B. PIPPARD: Physica, Haag **19**, 765 (1953).

average magnetization as determined by the demagnetizing factor of the specimen. In a subsequent theory, LANDAU[1] suggested that the lamina may branch out, as in Fig. 10, in such a way as to make the field equal to H_c everywhere at the surface. Experiments and theory both indicate that the earlier unbranched model is the correct one for most cases of practical interest[2]. The theory is discussed in Sect. 32.

An important parameter in LANDAU's theory is the energy of the boundary between normal and superconducting phases. The energy is defined relative to an ideal boundary in which there is an abrupt transition between the two regions and in which the field $H = H_c$ everywhere in the normal region and $H = 0$ in the superconducting region. The position of the ideal boundary is taken such that the net flux is the same as for the actual boundary. Present evidence indicates that there is actually a gradual change from one region to the other, so that the transition region is spread out. The energy per unit area of boundary may be written in the form:

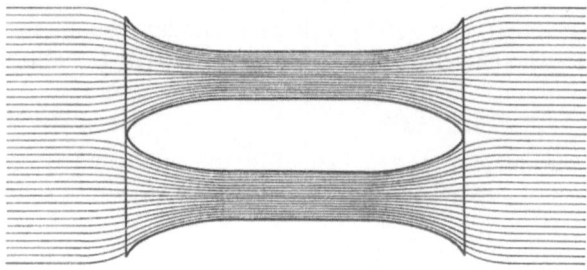

a.) *Unbranched model*

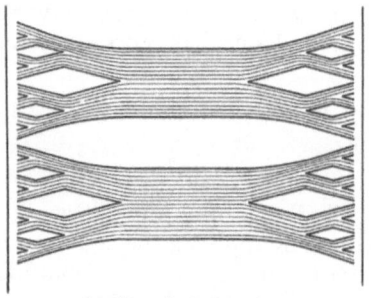

b) *Branched model*

Fig. 10a and b. Intermediate state for slab with magnetic field normal to face, according to LANDAU. a) Unbranched model. b) Continually branched model.

$$\alpha_{ns} = \frac{\Delta H_c^2}{8\pi}, \quad (27.1)$$

where Δ has the dimensions of a length and is of the order of the boundary width. Some authors have defined the energy relative to an ideal boundary in which the field penetrates into the superconducting region with a penetration-depth λ as it would at a free surface. This energy is larger than (27.1) by $\lambda H_c^2/8\pi$, or is equal to

$$\alpha_{ns}^{(1)} = \frac{\Delta_1 H_c^2}{8\pi}, \quad (27.2)$$

where

$$\Delta_1 = \Delta + \lambda. \quad (27.3)$$

LONDON ([15], p. 128) has pointed out that one must have $\Delta > 0$ in order to observe a MEISSNER effect. Otherwise, the net free energy could be lowered by introduction of thin normal laminae parallel to the field. Consider, for example, a slab in a field parallel to the plane of the slab. In an external field, H, there is an increase in magnetic energy of the superconducting phase of $H^2/8\pi$ per unit volume. Now suppose that the material breaks up into a series of normal and

---

[1] L. D. LANDAU: J. Phys. USSR. **1**, 99 (1943).

[2] D. SHOENBERG [24] has given an excellent summary of the extensive work done in this field in the USSR.

superconducting laminae, such that the thickness of the superconducting laminae is smaller than the penetration depth and the thickness of the normal laminae is small compared with the superconducting. There would then be a large penetration of field, with a decrease in magnetic energy of $\sim H^2/8\pi$ without much change in the energy in zero field. If d is the thickness of the slab, the number of boundaries introduced is of the order of $2d/\lambda$. The layered structure would be favored for a field $H \sim H_c$ unless

$$\frac{2d}{\lambda}\,\alpha_{ns}^{(1)} > \frac{dH_c^2}{8\pi} \tag{27.4}$$

or

$$\Delta_1 > \lambda/2.$$

The criterion for stability against introduction of a single thin normal lamina in the interior is the more stringent condition

$$\Delta > 0. \tag{27.5}$$

Observed values of Δ are generally much larger than λ.

The boundary energy is also of importance in supercooling phenomena. A superconducting nucleus must be above a critical size before it is stable.

Also to be considered is the energy per unit area of the free surface of normal and superconducting phases, designated by α_n and α_s. A significant difference between these would have an influence on the transition of thin films. PIPPARD[1] has shown that the difference is probably small compared with $\lambda H_c^2/8\pi$. For a body of arbitrary shape, the difference in free energy between the normal phase, F_n, and the superconducting phase, F_s, in a magnetic field is (see Sect. 2):

$$F_n - F_s = \int \mathbf{H} \cdot d\mathbf{M} = -\int \mathbf{M} \cdot d\mathbf{H}, \tag{27.6}$$

where the integration extends from zero field to the field at which the specimen becomes normal and the magnetic moment, M, vanishes. The second integral follows from the first by an integration by parts, since M vanishes at both limits. Thus $F_n - F_s$ is just the area of the plot of $-M$ versus H. If we include the surface energies,

$$F_n = V f_n + A\,\alpha_n, \quad F_s = V f_s + A\,\alpha_s, \tag{27.7}$$

where V is the volume, A the surface area, and f_n and f_s refer to unit volume. Since

$$f_n - f_s = \frac{H_c^2}{8\pi}, \tag{27.8}$$

$$-\int \mathbf{M} \cdot d\mathbf{H} = \frac{V H_c^2}{8\pi} + A(\alpha_n - \alpha_s). \tag{27.9}$$

PIPPARD has used this expression to estimate $\alpha_n - \alpha_s$ from SHOENBERG's data on mercury colloids and from data of LOCK and others on thin films[2]. As stated above, he found that the difference is very small and is perhaps zero within the accuracy of the method.

Theories of the interphase boundary energy have been based on the two fluid model and the concept of an order parameter which gives the effective concentration of superconducting electrons, n_s. The order parameter is assumed to change gradually from a temperature-dependent equilibrium value on the superconducting side of the boundary to zero on the normal side. The width of

---

[1] A. B. PIPPARD: Proc. Cambridge Phil. Soc. **47**, 617 (1951).
[2] A summary is given in [23], p. 171.

the transition region is of the order of Δ. GINSBURG and LANDAU [6] have proposed a phenomenological extension of the LONDON equations to take into account a space variation of the order parameter. It is assumed that n_s is given by the square of an effective wave function, Ψ, and that there is an extra term in the energy proportional to $|\text{grad } \Psi|^2$. The coefficient of this term is evaluated in terms of the critical field, H_c, and the penetration depth, λ, so that there are no undetermined parameters. Since the free energy is expanded in a power series in n_s, or Ψ^2, the theory is presumably valid only near the transition temperature, T_c, where n_s is small. The author[1] has extended the theory by use of the GORTER-CASIMIR two-fluid model so as to apply to all temperatures. This theory is discussed in Sect. 28 and applied to the calculation of boundary energies in Sect. 29.

Independently of the GINSBURG-LANDAU theory, PIPPARD[2] proposed a qualitative theory of boundary energies which is also based on a space-variation of an order parameter, but which differs in some important aspects. PIPPARD suggests that the width of the boundary region, and thus Δ, is determined by the coherence distance in the superconducting phase. In pure metals, Δ is assumed to be of the order of ξ_0, as estimated from the uncertainty relation (21.16). In alloys, Δ is assumed to be of the order of the mean free path, l. Up to the spring of 1955, there were no experiments to show a dependence of Δ on l. Actually, because of the way λ depends on l, the PIPPARD and GINSBURG-LANDAU theories do not lead to very different predictions.

In Sect. 30 we discuss the change in penetration depth with field as resulting from changes in the order parameter, or n_s, and from true non-linear terms. In Sect. 31 we discuss transitions in thin films and other small specimens as affected by boundary effects and by changes in the order parameter.

28. GINSBURG-LANDAU theory and extensions. Following a general theory for phase transitions of the second kind proposed by LANDAU and LIFSHITZ[3], GINSBURG and LANDAU assume that near the transition temperature, T_c, the free energy difference between superconducting and normal phases may be expanded in a power series in an order parameter ω, defined so that $\omega = 0$ in the normal phase and $\omega = 1$ in the superconducting phase at $T = 0^\circ$ K (see Sect. 4)

$$f(T) = F_c(T) - F_n(T) = a(T)\,\omega + \tfrac{1}{2} b(T)\,\omega^2 + \cdots \qquad (28.1)$$

The equilibrium value of ω is that which makes $f(T)$ a minimum. Keeping only terms to the second order, we find:

$$\omega_e = -\frac{a}{b}. \qquad (28.2)$$

For unit volume of material, the equilibrium value of $f(T)$ is

$$f_e(T) = -\frac{H_c^2}{8\pi} = -\frac{a^2}{2b}. \qquad (28.3)$$

GINSBURG and LANDAU identify ω with the square of an effective wave function, Ψ, defined so that $|\Psi|^2$ is equal to the concentration of superconducting electrons, n_s. We shall use a different normalization and assume, as indicated above, that $\omega = |\Psi|^2 = 1$ at $T = 0^\circ$ K, so that

$$n_s = n_0\,|\Psi|^2, \qquad (28.4)$$

---

[1] J. BARDEEN: Phys. Rev. **94**, 554 (1954).
[2] A. B. PIPPARD: Proc. Roy. Soc. Lond., Ser. A **203**, 210 (1950).
[3] L. D. LANDAU and E. M. LIFSHITZ: Statistical Physics, p. 204. Oxford 1940.

where n_0 is the value of n_s at $T = 0°$ K. Since λ^2 is inversely proportional to n_s,

$$\omega_e(T) = |\Psi_e|^2 = \frac{\lambda_0^2}{\lambda(T)^2}. \tag{28.5}$$

From (28.2), (28.3) and (28.5), we find

$$a = -\frac{H_c^2}{4\pi}\frac{\lambda^2}{\lambda_0^2}. \tag{28.6}$$

$$b = \frac{H_c^2}{4\pi}\frac{\lambda^4}{\lambda_0^4}. \tag{28.7}$$

Thus the expression for $f(T)$ is completely determined to the second order in ω in terms of measurable quantities.

Prior to his learning about the GINSBURG-LANDAU theory, the author[1] made an independent estimate of the boundary energy on the basis of the model of electrons of small effective mass outlined in Sect. 23. It was assumed that a wave function for the boundary could be obtained by multiplying each of the slowly-varying one-particle functions which describe the superconducting electrons by a function $\Psi(r)$ which varies from 0 to 1 across the boundary. The density of superconducting electrons is then proportional to Ψ^2. The concentration of normal electrons was presumed to vary in such a way as to keep the total electron density constant. The free energy difference between normal and superconducting phases at $T = 0°$ K was taken to be:

$$f(\Psi) = -\frac{H_c^2}{8\pi}\Psi(r)^2. \tag{28.8}$$

An additional energy density which comes from the variation in $\Psi(r)$ as determined from the effective mass concept is:

$$\frac{n_s}{2m_s}\hbar^2 |\operatorname{grad} \Psi|^2. \tag{28.9}$$

An expression for $\Psi(r)$ was found by minimizing the boundary energy, as obtained from (28.8), (28.9) and the terms representing magnetic energy. Except for the expression representing the free energy difference, the theory is similar to the earlier theory of GINSBURG and LANDAU and leads to nearly equivalent results.

More recently, the author has used the GORTER-CASIMIR two-fluid model to obtain for the free energy difference an expression which should be valid throughout the temperature range. Near $T = T_c$, the theory approaches the GINSBURG-LANDAU theory and near $T = 0°$ K the author's earlier theory described in the preceding section. If the parameter $\alpha = \frac{1}{2}$, Eqs. (4.2), (4.3) and (4.4) give for the free energy difference:

$$f(t, \omega) = \frac{H_0^2}{4\pi}\left\{t^2\left(1 - \sqrt{1 - \omega}\right) - \frac{1}{2}\omega\right\}, \tag{28.10}$$

where H_0 is the critical field at $0°$ K and $t = T/T_c$ is the reduced temperature. When $t \ll 1$ (28.10) approaches (28.8) and when $t \sim 1$ and $\omega = |\Psi|^2$ is small, (28.10) approaches (28.1). Since most of the theory can be carried through for a general $f(t, \omega)$, we will not restrict the function to the form (28.10) until detailed calculations of boundary energies are made in Sect. 29.

GINSBURG and LANDAU treat $\Psi(r)$ as an effective wave function[2], so that the energy density in a magnetic field defined by the vector potential $A(r)$ is taken to be

$$\frac{n_0}{2m_s}\left|-i\hbar\operatorname{grad}\Psi + \frac{eA}{c}\Psi\right|^2, \tag{28.11}$$

---

[1] J. BARDEEN: Phys. Rev. **81**, 1070 (1951).
[2] They suggest that the density matrix $\varrho(r, r') \sim \Psi^*(r)\Psi(r')$.

where m_s is an effective mass of the superconducting electrons with charge $-e$. This is just the kinetic energy term for a gas with n_0 electrons in a state defined by the wave function $\Psi(r)$, which might be interpreted as the ground state of an equivalent Einstein-Bose gas. As mentioned in Sect. 23, such a gas does obey the London equations. We may assume that, in a simply connected body, Ψ is real when the standard London choice of gauge is made: div $A = 0$, $A_\perp = 0$. Taking into account the energy density of the field, $H^2/8\pi$, we may write for the Helmholtz free energy difference:

$$F = \int \left\{ \frac{n_0}{2m_s} \left| -i\hbar \operatorname{grad} \Psi + \frac{e}{c} A \Psi \right|^2 + \frac{H^2}{8\pi} + f(\Psi) \right\} d\tau - \int_0^{H_a} M \cdot dH_a. \quad (28.12)$$

We are treating A (and thus H) as an arbitrary function to be adjusted to make F a minimum.

We shall treat only the case of a plane boundary for which we may assume, as we have in other cases, that the magnetic field is in the z-direction and that the boundary is normal to the x-direction. The only component of A is $A_y(x) = A(x)$ and both A and Ψ depend only on x. The magnetization per unit volume is $(H - H_a)/4\pi$. Thus the free energy difference per unit area reduces to:

$$F = \int_{-\infty}^{+\infty} \left\{ \frac{n_0 \hbar^2}{2m_s} \left[\left(\frac{d\Psi}{dx} \right)^2 + \left(\frac{eA\Psi}{\hbar c} \right)^2 \right] + \frac{1}{8\pi} \left(\frac{dA}{dx} \right)^2 + f(\Psi) + \frac{H_a^2}{8\pi} - \frac{H_a}{4\pi} \frac{dA}{dx} \right\} dx. \quad (28.13)$$

The problem is to find the functions $\Psi(x)$ and $A(x)$ which make F a minimum subject to appropriate boundary conditions. The differential equations for Ψ and A as derived by the usual variational procedures are:

$$\frac{d^2\Psi}{dx^2} = \frac{m_s}{n_0 \hbar^2} \frac{df}{d\Psi} + \frac{e^2 A^2}{\hbar^2 c^2} \Psi, \quad (28.14)$$

$$\frac{d^2 A}{dx^2} = \frac{4\pi e^2 n_0 \Psi^2}{m_s c^2} A. \quad (28.15)$$

These are a pair of coupled nonlinear equations in A and Ψ which are to be solved subject to appropriate boundary conditions. In Sect. 29 they are applied to the phase boundary and in Sect. 30 to the boundary at a free surface.

The Ginsburg-Landau theory leads to the usual London theory when the effective wave function is a constant. If a non-local theory, such as Pippard's, is correct it is necessary to modify the equations. The following appears to be a natural way to generalize the theory. For simplicity, we consider only the one-dimensional case, which leads to equations analogous to (28.14) and (28.15). We suppose the current density is given by

$$j(x) = -\frac{n_0 e^2}{m_s c} \int_{-\infty}^{+\infty} K(x - x') A(x') \Psi(x) \Psi(x') dx'. \quad (28.16)$$

The kernel $K(x - x')$ is normalized so that

$$\int_{-\infty}^{+\infty} K(x - x') dx' = 1. \quad (28.17)$$

If $K(x-x')$ is a δ-function, the LONDON expression is obtained. We may replace (28.13) by the following equation for the free energy, F:

$$F = \int\limits_{-\infty}^{+\infty} \left\{ \frac{n_0 \hbar^2}{2m_s} \left(\frac{d\Psi}{dx}\right)^2 + \frac{1}{8\pi} \left(\frac{dA}{dx}\right)^2 + f(\Psi) + \frac{H_a^2}{8\pi} - \frac{H_a}{4\pi} \frac{dA}{dx} \right\} dx + \\ + \frac{1}{2c} \int\limits_{-\infty}^{+\infty} \int\limits_{-\infty}^{+\infty} K(x-x')\, A(x)\, A(x')\, \Psi(x)\, \Psi(x')\, dx\, dx'. \qquad (28.18)$$

This leads to the following integro-differential equations for Ψ and A:

$$\frac{d^2\Psi}{dx^2} = \frac{m_s}{n_0 \hbar^2} \frac{df}{d\Psi} + \frac{e^2}{\hbar^2 c^2} A(x) \int\limits_{-\infty}^{+\infty} K(x-x')\, \Psi(x')\, A(x')\, dx', \qquad (28.19)$$

$$\frac{d^2A}{dx^2} = \frac{4\pi n_0 e^2}{m_s c^2} \Psi(x) \int\limits_{-\infty}^{+\infty} K(x-x')\, \Psi(x')\, A(x')\, dx'. \qquad (28.20)$$

No calculations have been made as yet with use of these generalized equations.

29. Energy of the boundary between normal and superconducting phases. We shall assume that the boundary is near the plane $x=0$, and that the superconducting side corresponds to $x>0$ and the normal phase to $x<0$. The field in the normal region is the critical field $H=H_c$. Our problem is to solve (28.14) and (28.15) subject to the boundary conditions. As illustrated in Fig. 11, the effective wave function Ψ increases gradually from zero to the equilibrium value Ψ_e with increasing x across the boundary, while A is everywhere negative, with a uniform positive slope for $x\ll0$, corresponding to a constant field $H=H_c$ and A approaches zero for $x\gg0$. The boundary conditions are then:

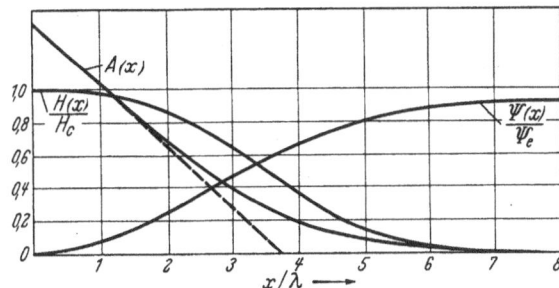

Fig. 11. Variation of A, H and Ψ across a boundary between normal and superconducting regions, according to a modification of the GINSBURG-LANDAU theory. (Calculated for $s=0.2$, $T=0°$ K.)

Normal region, $x\ll0$,

$$\left.\begin{array}{c} \Psi = \dfrac{d\Psi}{dx} = f(\Psi) = 0; \\[2mm] \dfrac{dA}{dx} = H = H_c. \end{array}\right\} \qquad (29.1)$$

Superconducting region, $x\gg0$,

$$\left.\begin{array}{c} \Psi = \Psi_e; \\[2mm] A = \dfrac{dA}{dx} = \dfrac{d\Psi}{dx} = 0; \\[2mm] f(\Psi) = -\dfrac{H_e^2}{8\pi}. \end{array}\right\} \qquad (29.2)$$

Note that the integrand of the expression (28.13) for the free energy difference is zero at both limits, so that $F=0$ for an ideal boundary as defined in Sect. 27.

There is an integral of (28.14) and (28.15) which can be used to aid in the solution of the equations and also to simplify the expression for F:

$$\frac{n_0 \hbar^2}{2 m_s} \left(\frac{d\Psi}{dx}\right)^2 + \frac{1}{8\pi} \left(\frac{dA}{dx}\right)^2 = f(\Psi) + \frac{e^2 n_0 \Psi^2 A^2}{2 m_s c^2} + \frac{H_c^2}{8\pi}. \tag{29.3}$$

The constant of integration has been chosen to fit the boundary conditions. With use of (29.3), (28.13) and taking $H_a = H_c$, we may write:

$$\alpha_{ns} = 2 \int_{-\infty}^{+\infty} \left\{ f(\Psi) + \frac{e^2 n_0 \Psi^2 A^2}{2 m_s c^2} + \frac{H_c}{8\pi}\left(H_c - \frac{dA}{dx}\right)\right\} dx. \tag{29.4}$$

Since the equations to be solved for Ψ and A are nonlinear, a general solution has not been obtained, but solutions have been found for some limiting cases and a numerical integration has been obtained which applies in the low temperature limit. The equations to be solved, (28.14) and (28.15), may be simplified by use of the following reduced variables:

$$U = \frac{\Psi}{\Psi_e}, \quad \xi = \frac{x}{\lambda}, \quad V = \frac{e A \lambda}{\hbar c}, \quad s = \frac{e H_c \lambda^2}{\hbar c}, \tag{29.5}$$

in which λ is the usual penetration depth defined by

$$\lambda^2 = \frac{m_s c^2}{4\pi e^2 n_0 \Psi_e^2}. \tag{29.6}$$

With this notation (28.14) and (28.15) become

$$\frac{d^2 U}{d\xi^2} = \frac{4\pi s^2}{H_c^2} \frac{df}{dU} + V^2 U, \tag{29.7}$$

$$\frac{d^2 V}{d\xi^2} = U^2 V. \tag{29.8}$$

We shall use the Gorter-Casimir expression (28.10) for f for which

$$\Psi_e^2 = 1 - t^4. \tag{29.9}$$

If we replace ω in (28.10) by $U^2 \Psi_e^2$ we obtain

$$\frac{4\pi}{H_c^2} f(U) = \frac{1}{1-t^2}\left\{ t^2\left(1 - \sqrt{1 - U^2\Psi_e^2}\right) - \frac{1}{2} U^2 \Psi_e^2\right\}, \tag{29.10}$$

since for this model, $H_c = (1-t^2) H_0$. It follows from (29.10) that:

$$\frac{4\pi}{H_c^2} \frac{df}{dU} = \frac{1+t^2}{1-t^2}\left\{ 1 - \frac{t^2}{\sqrt{1 - U^2\Psi_e^2}}\right\} U, \tag{29.11}$$

which result may be substituted in (29.7). Limiting forms of the equation for low and high temperatures are

$$\frac{d^2 U}{d\xi^2} = (V^2 - s^2)\, U \qquad (t \to 0), \tag{29.12}$$

$$\frac{d^2 U}{d\xi^2} = [V^2 - 2 s^2 (1 - U^2)]\, U \qquad (t \to 1). \tag{29.13}$$

The second of these is equivalent to the one used by Ginsburg and Landau.

One useful limiting case is that for which the dimensionless parameter $s \ll 1$. The width of the transition region is then large compared with the penetration depth. To a good approximation one may neglect the magnetic terms on the superconducting side of the boundary, since V is very small where U is appreciable. Before giving the general solution for small s which applies at all temperatures, we shall first make the further restriction that the temperature is small so that (29.12) can be applied. If V^2 is neglected in comparison with s^2 for $\xi \to 0$, the appropriate solution of (29.12) is

$$U = \sin s\,\xi$$

which joins smoothly with the solution $U = 1$ in the body of the superconducting region at $\xi = \pi/(2s)$ or at $x = \pi\,\lambda/(2s)$. We neglect the magnetic field terms in (29.4) and find

$$\alpha_{ns} = \frac{H_c^2}{4\pi} \int_0^{\frac{\pi\lambda}{2s}} \left(1 - \sin^2 \frac{s\,x}{\lambda}\right) dx = \frac{\pi}{2}\frac{\lambda H_c^2}{8\pi s} = \frac{\hbar c H_c}{16\,\lambda e}\begin{pmatrix}s \to 0\\ t \to 0\end{pmatrix}. \qquad (29.14)$$

The corresponding expression derived by GINSBURG and LANDAU from (29.13) is

$$\alpha_{ns} = \frac{4}{3}\frac{\lambda H_c^2}{8\pi s} = \frac{\hbar c H_c}{6\pi\,\lambda e} \qquad \begin{pmatrix}s \to 0\\ t \to 1\end{pmatrix}. \qquad (29.15)$$

It is interesting to note that (29.14) and (29.15) differ by only about 20%, so that the expression for α_{ns} in terms of λ and H_c is relatively insensitive to temperature.

A general expression for $s \ll 1$ which applies for all t can be obtained by neglecting magnetic terms in the superconducting region and terms involving U in the normal region. The expression for the boundary energy is then

$$\alpha_{ns} = 2 \int_0^\infty \left(f(\Psi) + \frac{H_c^2}{8\pi}\right) dx. \qquad (29.16)$$

The corresponding limiting expression for the integral (29.3) is

$$\frac{n_0 \hbar^2}{2 m_s}\left(\frac{d\Psi}{dx}\right)^2 = f(\Psi) + \frac{H_c^2}{8\pi}, \qquad (29.17)$$

so that (29.16) becomes

$$\alpha_{ns} = \frac{n_0 \hbar^2}{m_s}\int_0^\infty \left(\frac{d\Psi}{dx}\right)^2 dx = \frac{n_0 \hbar^2}{m_s}\int_0^{\Psi_e} \frac{d\Psi}{dx}\, d\Psi. \qquad (29.18)$$

Eq. (29.17) may now be used to find an expression for $d\Psi/dx$ in terms of Ψ. With use of (29.10), we obtain,

$$\frac{d\Psi}{dx} = \sqrt{\frac{m_s H_0^2}{4\pi n_0 \hbar^2}}\left(\sqrt{1 - \Psi^2} - t^2\right). \qquad (29.19)$$

When (29.19) is inserted into (29.18) and the integration is carried out, we get an expression for α_{ns} which may be written in the form:

$$\alpha_{ns} = \frac{\lambda H_c^2}{8\pi}\frac{(\Psi_e^{-1}\arcsin \Psi_e - t^2)}{s(1 - t^2)} \qquad (s \ll 1). \qquad (29.20)$$

The coefficient of $\lambda H_c^2/(8\pi)$ approaches $\pi/2$ as $t \to 0$ and $\tfrac{4}{3}$ as $t \to 1$, and so agrees in these two limits with (29.14) and (29.15). Since the coefficient varies slowly with t, either limiting form will probably give reasonably good results for all t.

A numerical solution valid for all s has been obtained for the low temperature limit, $t \ll 1$. It is convenient to express α_{ns} in the form

$$\alpha_{ns} = g(s) \frac{\lambda H_c^2}{8\pi s},\qquad(29.21)$$

so that Δ, defined by (27.1), is given by

$$\frac{\Delta}{\lambda} = \frac{g(s)}{s}.\qquad(29.22)$$

A plot of the numerical factor $g(s)$ as a function of s is given in Fig. 12. Between $s = 0.1$ and $s = 1.0$, $g(s)$ varies almost linearly with s,

$$g(s) \approx (1.1 - 1.6s)\qquad(29.23)$$

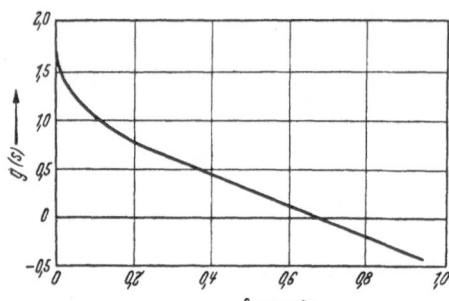

Fig. 12. Plot of $g(s)$ versus s (after Bardeen, reference 1 p. 324).

and becomes negative for s larger than 0.68. Ginsburg and Landau find that Δ becomes negative for s larger than 0.5 in the high temperature limit. A variational solution which gives approximate values for all s and t has been given by Lewis[1].

The plot of Fig. 11 which gives the variation of H and V across the boundary for $s = 0.2$ was obtained from the numerical solution valid for $t \ll 1$.

In Table 3 are listed values of s_0 and Δ_0 for metals for which H_0 and λ_0 are known. All values apply to the low temperature limit. It is to be noted that s_0 is quite small for all metals listed, so that the width of the boundary region is large compared with λ, and values of Δ_0/λ_0 are correspondingly large. The temperature dependence of Δ/λ is not very great, since s varies rather slowly with temperature:

$$\frac{s}{s_0} = \frac{\lambda^2 H_c}{\lambda_0^2 H_0} = \frac{1-t^2}{1-t^4} = \frac{1}{1+t^2}.\qquad(29.24)$$

Perhaps the most reliable experimental estimates of Δ are based on analysis of the structure of the intermediate state of a slab in a field normal to the plane of the slab. According to the theory of Landau, to be discussed in Sect. 32, the width of the domains depends on the boundary energy and on the dimensions of the specimen. An estimate for tin made in this way by Schawlow and Lewis[2] is in good agreement with the theoretical value given in Table 3, and the temperature variation is about as predicted. However, a vanadium specimen appeared to have an enormously larger boundary energy.

Table 3. *Values of $s_0 = e H_0 \lambda_0^2/c$ for various metals*

Metal	λ_0 cm. $\times 10^{-6}$	H_0 gauss	s_0
Al	4.9	106	0.039
Hg	4.5	415	0.125
In	6.4	270	0.17
Pb	3.9	800	0.19
Sn	5.0	305	0.115

b) Applications to specific problems.

30. Change in penetration depth with field. Pippard[3] has observed a small but significant change of penetration depth of tin with field, as indicated by the dotted line of Fig. 13. It was found that $\Delta\lambda/\lambda$ is proportional to the square

[1] H. W. Lewis: Phys. Rev. **99**, 669 (1955).
[2] A. L. Schawlow: Phys. Rev. **101**, 573 (1956). — H. W. Lewis (to be published).
[3] A. B. Pippard: Proc. Roy. Soc. Lond., Ser. A **203**, 210 (1950).

of the field, H. The values plotted apply when the field is critical, $H = H_c$. Observations were made by a microwave method; the reactive part of the surface impedance was measured as a function of a static applied field.

The observed change is a minimum at about $3°$ K and increases on each side to values between 2 and 3%. This suggests that two effects are operative, one important near $T = T_c$ and the other at low temperatures. PIPPARD himself suggested that the change near $T = T_c$ arises from changes with field of the order parameter, or n_s, near the surface in such a way as to allow a greater penetration of field and a consequent lower free energy. In order that the predicted change be as small as observed, PIPPARD found it necessary to assume that the change in order extends to a depth of $\sim 10^{-4}$ cm. This is one piece of evidence for a coherence distance of this order. As we shall see later in this section, the GINSBURG-LANDAU theory predicts an even smaller change than that observed.

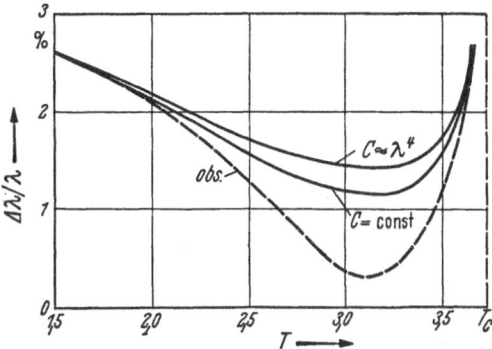

Fig. 13. Relative change in penetration depth between $H = 0$ and $H = H_c$ as a function of temperature. Comparison of semi-empirical theory, adjusted to fit near $T = 0°$ K and $T = T_c$, and PIPPARD's observations.

The author[1] has suggested that the increase $\Delta \lambda / \lambda$ at low temperatures comes from true non-linear terms which would appear in a more exact version of the LONDON theory. These would presumably come from second order changes in the wave functions which would give terms in the expression for the current density which are quadratic in the field. If the free energy, F_s, of a superconducting slab of thickness W is expanded a power series in the applied field, H, parallel to a face, terms to the fourth order are:

$$F_s = F_{s0} + (W - 2\lambda) \frac{H_a^2}{8\pi} - 2C H_a^4, \tag{30.1}$$

in which the coefficient C may depend on the temperature. Penetration terms for both faces of the slab have been included. The magnetic moment per unit area of surface is

$$M = -\frac{\partial F}{\partial H_a} = \frac{H_a}{4\pi} \{-W + 2\lambda + 32\pi C H_a^2\}. \tag{30.2}$$

The effective penetration depth is

$$\lambda_{\text{eff}} = \lambda + 16\pi C H_a^2, \quad \text{or} \quad \left(\frac{\Delta\lambda}{\lambda}\right)_1 = \frac{16\pi C H_a^2}{\lambda}. \tag{30.3}$$

This change, quadratic in the field, is in addition to that resulting from changes in λ with changes in order parameter. If it is assumed that C is independent of temperature, $\Delta\lambda / \lambda$ for $H_a = H_c$ varies as

$$\left(\frac{\Delta\lambda}{\lambda}\right)_1 \sim \frac{(1 - \tau^2)^2}{\sqrt{1 - \tau^4}}. \tag{30.4}$$

If the vector potential rather than the field is a better measure of the magnitude of the non-linear term, one might expect C to vary as λ^4, since according to the LONDON theory, A contains an extra factor of λ:

$$A = -\lambda H = -\lambda H_a e^{-\frac{x}{\lambda}}.$$

[1] J. BARDEEN: Phys. Rev. 87, 192 (1952); 94, 554 (1954).

If this variation is assumed, there is an extra factor of $(1-t^4)^{-2}$, giving

$$\left(\frac{\Delta\lambda}{\lambda}\right)_1 \sim \frac{(1+t^2)^{-2}}{\sqrt{1-t^4}}.$$ (30.5)

Both (30.4) and (30.5) indicate a rapid drop in $\Delta\lambda/\lambda$ with increase in temperature so that it is reasonable to suppose that a true non-linear effect is responsible for the change in penetration depth at low temperatures.

PIPPARD made use of the GORTER-CASIMIR two-fluid model to estimate $\Delta\lambda$ resulting from changes in the order parameter. He assumed that the change takes place uniformly in a slab of depth a adjacent to the surface, and found that for $H_a = H_c$:

$$\left(\frac{\Delta\lambda}{\lambda}\right)_2^2 = \frac{\lambda_0}{2a}\frac{t^4}{(1-t^2)^2},$$ (30.6)

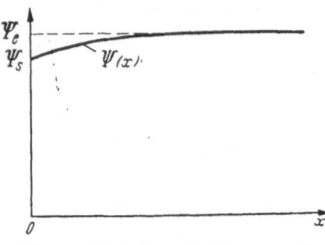

Fig. 14. Variation of $\Psi(x)$ near free surface in an applied magnetic field.

which gives a very rapid rise as $t\to 1$. To fit the high temperature part of the observed change, PIPPARD found that a value of $a\sim 10^{-4}$ cm. was required. This is in accordance with other estimates of the range of coherence.

Attempts to fit the entire observed curves by combining (30.4) or (30.5) with (30.6) are shown in Fig. 13. Since parameters were adjusted to get the best fit at both high and low temperatures, not very much significance can be attached to the agreement, but it is gratifiyng that the minimum occurs at about the right place. The observed decrease in $\Delta\lambda/\lambda$ in the low temperature part of the curve appears to be more rapid than given by either (30.4) or (30.5).

The change in penetration depth with field has also been estimated from the GINSBURG-LANDAU theory as modified by use of the two-fluid model. As illustrated in Fig. 14, when a field is applied there is a decrease in the effective wave function, Ψ, from an equilibrium value, Ψ_e, in the interior to a value Ψ_s at the surface, $x=0$. This will allow a greater penetration of field and a consequent lower magnetic energy at the expense of an increased energy from the change in Ψ. The problem is to determine $\Psi(x)$ so as to make the overall free energy, F, as defined by (28.12) a minimum. While this can be done in a straight forward way by starting from (28.12), the calculation can be carried out much more simply in the limit of s small, or a coherence distance large compared with λ. One may then assume that where the field is appreciable, Ψ does not depart significantly from Ψ_s, so that

$$\frac{\Delta\lambda}{\lambda} = \frac{\Psi_e - \Psi_s}{\Psi_e} = -\frac{\Delta\Psi_s}{\Psi_e},$$ (30.7)

where $\Delta\Psi_s$ is a negative quantity.

Further, in calculating the change in free energy from the change in Ψ, we may assume that

$$\Delta\Psi(x) = \Psi(x) - \Psi_e$$

is small. Thus for an applied field, H_α

$$\left.\begin{aligned}
\delta F &= \delta \int_0^\infty \left\{\frac{n_0 \hbar^2}{2m_s}\left(\frac{d\Psi}{dx}\right)^2 + f(\Psi)\right\} dx - \Delta\lambda \frac{H_a^2}{8\pi} \\
&= \delta\Psi_s \left\{-\frac{n_0 \hbar^2}{m_s}\left(\frac{d\Psi}{dx}\right)_s + \frac{\lambda}{\Psi_e}\frac{H_a^2}{8\pi}\right\} + \int_0^\infty \delta\Psi\left\{-\frac{n_0 \hbar^2}{m_s}\frac{d^2\Psi}{dx^2} + \frac{df}{d\Psi}\right\} dx.
\end{aligned}\right\}$$ (30.8)

This gives the two equations:

$$\left(\frac{d\Psi}{dx}\right)_s = \frac{m_s}{n_0\,\hbar^2}\,\frac{\lambda}{\Psi_e}\,\frac{H_a^2}{8\,\pi}\,, \tag{30.9}$$

$$\frac{d^2\Psi}{dx^2} = \frac{m_s}{n_0\,\hbar^2}\,\frac{df}{d\Psi}\,. \tag{30.10}$$

A first integral of (30.10) which satisfies the boundary condition for large x is [cf. (29.17)]:

$$\left(\frac{d\Psi}{dx}\right)^2 = \frac{2\,m_s}{n_0\,\hbar^2}\,(f(\Psi) - f(\Psi_e)) \approx \frac{m_s}{n_0\,\hbar^2}\,(\varDelta\Psi)^2\,\frac{d^2f}{d\Psi^2}\,. \tag{30.11}$$

From (29.10) and (29.11), we find for $\Psi = \Psi_e = \sqrt{1 - t^4}$

$$\frac{d^2f}{d\Psi^2} = \frac{H_0^2}{4\,\pi}\,\frac{1 - t^4}{t^4}\,. \tag{30.12}$$

An equation for $\varDelta\Psi_s$ is obtained by comparing (30.11) evaluated at the surface and (30.9). The final result is that for $H = H_c$,

$$-\frac{\varDelta\Psi_s}{\Psi} = \frac{\varDelta\lambda}{\lambda} = \frac{s_0\,t^2}{2\,(1 + t^2)^2}\,, \tag{30.13}$$

where s_0 is the parameter defined by (29.5) and listed in Table 3. This result agrees with a corresponding calculation of GINSBURG and LANDAU [6] in the limit $s \to 0$, $t \to 1$. It is in only rough agreement with PIPPARD's observations on tin. With $s_0 = 0.115$, the maximum value $(t = 1)$ is 0.015, only about one half or one third of the observed value, and, perhaps more serious, the predicted rise in $\varDelta\lambda/\lambda$ near $t = 1$ is not as rapid as observed. PIPPARD's expression (30.6) based on a definite range of order gives a better fit.

31. Transitions in thin films and other small specimens. A study of transitions in thin films or other small specimens should provide a good test of an order parameter theory such as that of GORTER and CASIMIR. If the dimensions are sufficiently small, the order parameter, ω, will not vary over the specimen, but in a strong applied field may depart from the equilibrium value, ω_e, for zero field. The total free energy in an applied field H_a is sum of the free energy in zero field plus the magnetic energy:

$$F(H_a, \omega) = F(0, \omega) - \int_0^{H_a} M(H, \omega)\,dH \tag{31.1}$$

where $M(H, \omega)$ is the magnetic moment for an order parameter ω. We suppose that ω is defined as in Sect. 4, and that the penetration depth, λ, increases with decreasing ω. This means that $-M(H, \omega)$ is zero for $\omega = 0$ and increases with ω. The value of ω chosen to make $F(H_a, \omega)$ a minimum will depend on the field, and will change so as to allow a greater penetration in strong fields. This will produce a rounding of the magnetization curve and a higher critical field.

Other effects are also predicted [6]. When the dimensions are small, the transition in a field may be of second rather than first order. The order parameter may decrease gradually to zero as the field is increased, so that there is no latent heat when the transition point is reached. When the dimensions are larger, above a critical value, the transition is of first order, but hysteresis is predicted. The critical field observed on increasing H from below in the superconducting phase is larger than that predicted for decreasing H from the normal phase.

Unfortunately, the experimental results are ambiguous. While some effects similar to those predicted are observed, it is not at all certain that they are really due to a change in order parameter. Perhaps the best experiments are those of Lock[1] on the magnetization curves of thin films of tin, indium and lead deposited on mica. A typical set of data for thin films of tin of various thicknesses, all at 3° K, is illustrated in Fig. 15. There is considerable rounding of the curves near the critical field. Lock attempted to account for his data in terms of a change with field of the order parameter, ω, of the Gorter-Casimir theory. While a qualitative agreement was found, the observed rounding was much larger than predicted, particularly for the thicker films. A change of penetration depth of 3% with field, the maximum observed by Pippard for a massive specimen, would give an almost negligible rounding of the magnetization curve of a thick specimen. Pippard has suggested that because of the structure of the films, the coherence distance, and thus the boundary energy, may be much less than for a massive specimen. The boundary energy may actually become negative, so that normal nuclei can be formed before the critical field is reached. An argument against this point of view is that, according to Pippard's theory and measurements on alloys, a decrease in coherence distance gives an increase in penetration depth. Analysis of Lock's measurements, however, give $\lambda_0 = 5 \times 10^{-6}$ cm., about the same as observed for massive specimens. Thus the cause of the rounding is in doubt.

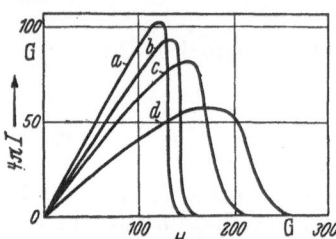

Fig. 15. Magnetization of tin films at 3° K. Thickness of films (units of 10^{-6} cm.): (a) 79, (b) 55, (c) 30, (d) 23. (After Lock, reference 1).

Abrikosov[2] has extended these concepts in order to account for experimental results of Zavaricky[3] on unannealed films of tin and thallium. He suggests that if the boundary energy is negative, superconducting regions can persist at fields above H_c. For a thick film and large s, superconductivity is not completely destroyed until the field is greater than $2sH_c$, which is greater than H_c when $s > \frac{1}{2}$. According to the Ginsberg-Landau theory, the boundary energy becomes negative when $s > \frac{1}{2}$. He also suggests that the "hard" superconductors, which do not show a Meissner effect may be accounted for on the same basis.

Pippard[4] has discussed the consequences of an equation of the form (31.1) in a qualitative way. He has worked out the theory for small spheres, also making use of the Gorter-Casimir model, and applied it to Shoenberg's observations on colloidal mercury. Again, while there was some qualitative agreement, a really clean cut test of the order-parameter theory was not obtained, in part because of the large range in size of the colloidal particles.

Ginsburg[5] has given a rather complete theory for thin films, also based on (31.1), but with use of the free energy expression (28.1), valid near $T = T_c$, for $F(0, \omega)$.

Since the detailed expressions are rather complicated, they will not be given. The qualitative behavior of the plots of $F(H, \omega)$ for thin films or small particles is illustrated in Fig. 16 and 17, taken from Pippard's paper. As shown in Fig. 16, the total free energy is the sum of $F(0, \omega)$, shown as curve (a), and the magnetic contribution, proportional to H^2. This sum is shown as curve (b). The magnetic term is zero at $\omega = 0$ and increases to a maximum at $\omega = 1$. The sum of the two

[1] J. M. Lock: Proc. Roy. Soc. Lond., Ser. A **208**, 391 (1951).
[2] A. A. Abrikosov: Dokl. Akad. Nauk SSSR. **86**, 489 (1952).
[3] For references see the review by Shoenberg [24].
[4] A. B. Pippard: Phil. Mag. **43**, 273 (1952).
[5] W. L. Ginsburg: Dokl. Akad. Nauk SSSR. **83**, 385 (1952).

terms may, as illustrated in the figure, go through a maximum and a minimum with increasing ω. This is true when the size of the specimen is above a critical value. A series of curves for different relative values of H for such a specimen is illustrated in the right hand diagram of Fig. 17. On the other hand, if the size is very small, the magnetic term is small and there is no maximum, as illustrated on the left.

The equilibrium value of ω is the one which makes $F(H, \omega)$ a minimum. For the very small specimen, ω decreases from ω_e, appropriate for $H = 0$, to zero as H is increased. In this case the transition will be of second order; there will be no latent heat. For the larger specimens, ω decreases as H increases, but will change abruptly from a finite value to zero when a critical field is reached.

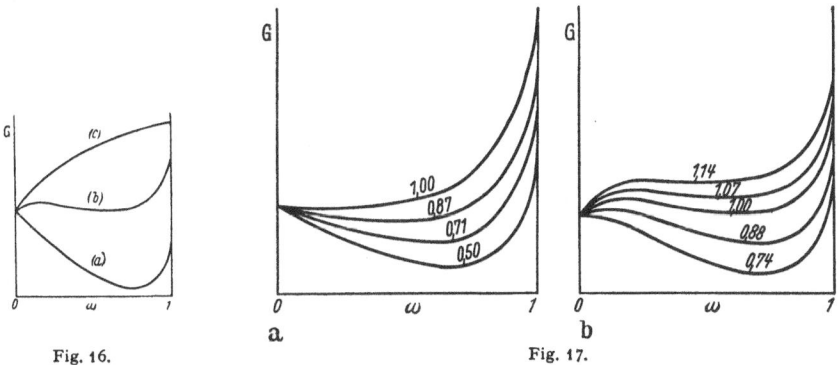

Fig. 16.   Fig. 17.

Fig. 16. GIBBS' function for a small superconducting particle in a magnetic field as a function of the order parameter, schematic). (a) In absence of field, (c) field contribution, (b) total. (After PIPPARD, reference 4, p. 334).

Fig. 17a and b. GIBBS' function versus order parameter for different relative values of H. a) Very small particle. b) Larger particle. (After PIPPARD, reference 4 p. 334).

Hysteresis may be expected if the transition takes place by a change in H. Because of the maximum, the transition to $\omega = 0$ would not take place on increasing field until the upper curve is reached and on decreasing field until the lower curve is reached. Hysteresis which might be accounted for in this way has been observed both with thin films and small spheres.

PIPPARD finds that the critical radius, a, of a sphere for hysteresis effects to occur is given in terms of λ_0, the penetration depth at $0°$ K and the reduced temperature, $t = T/T_c$, as follows

$$a = \sqrt{\frac{21}{8}} \frac{t \lambda_0}{\sqrt{1 - t^2}}. \tag{31.2}$$

According to GINSBURG, the critical film thickness, $2a$, for hysteresis effects is, near $t = 1$, given by

$$a = \frac{\sqrt{5}}{2} \lambda_0 = 1.12 \lambda_0.$$

As pointed out in Sect. 2, Eq. (2.9), the area under the magnetization curve depends only on the volume of the specimen and is independent of shape. This assumes that the transition is reversible and that the difference in surface energies, $\alpha_n - \alpha_s$, is negligible. PIPPARD has shown, as noted in Sect. 27, that the latter is small. If there were no rounding of the magnetization curve, the transition field would be determined by the magnetic moment, as given by expressions in Sect. 11 for specimens of simple shape. For a film of thickness $2a$, the transition field, H_T, is given in terms of the critical field, H_c for a massive specimen by

$$\left(\frac{H_T}{H_c}\right)^2 = \frac{1}{1 - \frac{\lambda}{a}\,\mathrm{Tan}\,\frac{a}{\lambda}}. \tag{31.3}$$

For a cylindrical wire of radius a in a longitudinal field,

$$\left(\frac{H_T}{H_c}\right)^2 = \frac{1}{1 - \left(\frac{2\lambda}{a}\right)\frac{I_1(a/\lambda)}{I_0(a/\lambda)}}. \tag{31.4}$$

For a sphere of radius a,

$$\frac{3}{2}\left(\frac{H_T}{H_c}\right)^2 = \frac{1}{1 - \frac{3\lambda}{a}\,\mathrm{Cot}\,\frac{a}{\lambda} + \frac{3\lambda^2}{a^2}}. \tag{31.5}$$

Relations of this sort were first given by Ginsburg[1]. Rounding of the magnetization curves will increase the transition field above these values, as will also a positive value for $\alpha_n - \alpha_s$.

H. London[2] was the first to point out that small specimens should have higher critical fields than massive ones, and later a more complete theory was developed by von Laue[3]. However, the criterion for the transition used by these authors differs from the one given in the preceding section. They assumed that destruction of superconductivity takes place by gradual motion of a normal boundary from the surface to the interior. The width of the boundary is assumed negligible, and the boundary energy neglected[4]. The criterion for stability of the interface may then be expressed in terms of a *critical current density* which cannot be exceeded:

$$j = \frac{c\,H_c}{4\pi\lambda}. \tag{31.6}$$

For a thin film of thickness $2a$, this would give a transition field

$$\frac{H_T}{H_c} = \mathrm{Cot}\,\frac{a}{\lambda}. \tag{31.7}$$

As the boundary moves in, effectively decreasing a, a larger and larger field would be required. No matter how large the field, a superconducting core would remain, contrary to observation. The difficulty is the neglect of boundary energies. It can be shown ([15], p. 136) that if $\Delta > 0$, formation of such a boundary is energetically unfavorable, and the transition will take place abruptly[5].

c) The intermediate state.

32. Landau theory. As early as 1937, when little was known about the intermediate state, Landau [13] suggested a laminar structure of alternating normal and superconducting domains. Later experiments have verified this predicted structure. Detailed calculations were made for a plate oriented normal to the

---

[1] W. L. Ginsburg: J. Phys. USSR. **9**, 305 (1945).
[2] H. London: Proc. Roy. Soc. Lond., Ser. A **152**, 650 (1935).
[3] M. v. Laue: Ann. Phys., Lpz. **32**, 71, 253 (1938).
[4] P. M. Marcus [Phys. Rev. **88**, 373 (1952)] has given a rather complete discussion of phase transition in cylinders from this point of view.
[5] For a discussion of the questions involved, see M. von Laue [Ann. Phys., Lpz. **10**, 296 (1952)] and articles of F. London, F. Beck and others immediately following.

field. The suggested domain structure for an unbranched model is illustrated in Fig. 10a. The field in the normal regions of width a_n is equal to the critical field, H_c. The field in the interior of the superconducting regions of width a_s drops to zero. The relative thickness of the domains is such as to carry the flux through the plate. For an external field H,

$$H a = H (a_n + a_s) = H_c (a_n + 2\lambda).\qquad(32.1)$$

LANDAU determined the shape of the regions by requiring that the field at the boundary of the superconducting region be everywhere equal to H_c.

The total free energy per unit area of plate surface is the sum of two terms, a boundary energy, which for a plate of width L is

$$\frac{2L\,\Delta}{a}\,\frac{H_c^2}{8\pi}\qquad(32.2)$$

and a magnetic energy, independent of L and proportional to a, which, including both front and back surfaces, may be written:

$$2\,a\,\psi(\eta)\,\frac{H_c^2}{8\pi},\qquad(32.3)$$

Table 4. *Values of $\psi(\eta)$.*

η	$\psi(\eta)$	η	$\psi(\eta)$
0.1	0.0055	0.6	0.0182
0.2	0.0136	0.7	0.0128
0.3	0.0195	0.8	0.0065
0.4	0.0224	0.9	0.0020
0.5	0.0221		

where $\eta = H/H_c$. The function $\psi(\eta)$ was given in the form of a complicated integral. Limiting expressions for large and small η are

$$\psi(\eta) = \frac{\eta^2}{\pi}\log\frac{1}{2\eta}\qquad\qquad \eta \ll 1,\qquad(32.4)$$

$$\psi(\eta) = \frac{\log 2}{\pi}(1-\eta)^2\qquad 1-\eta \ll 1.\qquad(32.5)$$

A numerical evaluation by LIFSHITZ and SHARVIN[1] is given in Table 4.

The value of a is determined so that the total free energy, F, the sum of (32.2) and (33.3) is a minimum:

$$F = \frac{H_c^2}{4\pi}\left(\frac{L\,\Delta}{a} + a\,\psi(\eta)\right).\qquad(32.6)$$

This gives

$$a = \sqrt{\frac{L\,\Delta}{\psi}}\qquad(32.7)$$

and the corresponding value of F is

$$F = \frac{H_c^2}{2\pi}\sqrt{\psi\,L\,\Delta}.\qquad(32.8)$$

As noted by LIFSHITZ and SHARVIN, the thickness of the layers can be quite large. For $\eta = \frac{1}{2}$ and $L = 2$ cm., $a_s \approx a_n \approx 0.06$ cm. for tin with $\Delta = 1.5 \times 10^{-4}$ cm.

The field in the normal regions near the surface can be a good deal smaller than critical with this model. In the example cited above, the field at the surface in the center of the normal region is only $0.73\ H_c$. Since it was difficult to see what might keep such regions from going superconducting, LANDAU later proposed a branched laminar model (Fig. 10b), with continual branching such that the average field is everywhere equal to H at the surface. LANDAU found the following expression for the free energy for a model with such repeatedly branched layers:

$$F' = 0{,}277\,H_c^2\,(L\,\Delta^2)^{\frac{1}{3}}\,\eta^{\frac{1}{2}}(1-\eta)^{\frac{2}{3}}.\qquad(32.9)$$

---

[1] E. M. LIFSHITZ and Y. V. SHARVIN: Dokl. USSR. Akad. Nauk **79**, 783 (1951).

LIFSHITZ and SHARVIN noted that for reasonable values of L and Δ ($L \sim 1$ cm., $\Delta \sim 10^{-4}$ cm.), F' is considerably larger than F, so that the unbranched model is favored for specimens of normal dimensions. Since F goes as $L^{\frac{1}{3}}$ while F' goes as $L^{\frac{1}{2}}$, the repeatedly branched model is favored for very large values of L/Δ, but they are so large that they are not likely to occur. Some branching, is possible, however, and an intermediate model in which one or two branches occur might give a still lower free energy.

The unbranched model has been used by SCHAWLOW and LEWIS[1] to estimate boundary energies.

33. Destruction of superconductivity by currents. Interesting intermediate state phenomena occur when superconductivity is gradually destroyed by current flow. The simplest and best understood case is that of a long straight cylindrical wire of radius a carrying a current J. It is observed (Fig. 18) that about two-thirds of the normal resistance is restored suddenly when a critical current

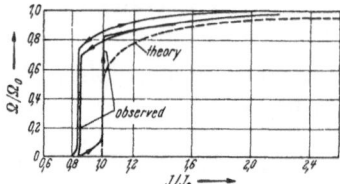

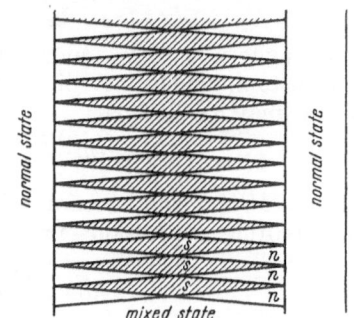

Fig. 18. Resistance of indium as a function of current according to LONDON's theory and according to measurements of SCOTT (after LONDON [15], p.123).

Fig. 19. Intermediate state in a superconducting wire in which a current flows (after LONDON [15], p. 120).

is reached, and that the remaining resistance is restored gradually as the current is increased indefinitely. According to SILSBEE's hypothesis, which is true only approximately, the first restoration of resistance occurs when the field at the circumference of the wire due to the current flowing in the wire reaches the critical value, H_c. A theory which accounts for most of the observed facts, at least in a qualitative way, has been given by LONDON ([15], p. 120). We shall give an outline of LONDON's theory and then discuss briefly the "paramagnetic" effect which occurs when there is a magnetic field applied parallel with the axis in addition to the current.

LONDON suggested that when the current is above the critical value, an outer shell $R < r < a$ becomes normal, while the interior, $r < R$ goes into an intermediate state, as illustrated in Fig. 19. The intermediate state is assumed to consist of a stack of thin discus-shaped superconducting regions imbedded in a normal matrix. Because of symmetry, the magnetic field has only one component, $H_\varphi(r)$. In the normal region, $H = H_c = \text{const.}$ In the intermediate region, $r > R$, $H_\varphi > H_c$. The electric field has only one component, E_z. Under static condition, curl $\boldsymbol{E} = 0$, so that

$$\text{curl}_\varphi \, \boldsymbol{E} = \frac{\partial E_z}{\partial r} = 0. \tag{33.1}$$

Thus E_z must be independent of r. Current flow in the intermediate state is nearly parallel with the axis and normal to the boundaries of the superconducting regions.

---

[1] Reference 2, p. 330.

The problem is to determine how the thickness, $w(r)$, of a discus-shaped region varies with the radial distance r. It must be such that the current density, j, is consistent with the MAXWELL equation

$$\text{curl } \boldsymbol{H} = \frac{4\pi j}{c},$$

or

$$j_z(r) = \frac{c}{4\pi} \frac{1}{r} \frac{\partial}{\partial r}(r H_\varphi) = \frac{c H_c}{4\pi r}. \tag{33.2}$$

If σ is the normal conductivity, the effective conductivity for axial flow is:

$$\sigma_{\text{eff}}(r) = \frac{w(0)}{w(0) - w(r)} \sigma. \tag{33.3}$$

The current density is therefore

$$j_z(r) = \sigma_{\text{eff}} E_z = \frac{w(0)\,\sigma E_z}{w(0) - w(r)} = \frac{c\,H_c}{4\pi r}, \tag{33.4}$$

and

$$\frac{w(r)}{w(0)} = 1 - \frac{4\pi\sigma E_z r}{c H_c}. \tag{33.5}$$

Thus $w(r)$ goes to zero at a distance

$$R = \frac{c H_c}{4\pi\sigma E_z}. \tag{33.6}$$

The current density between R and a is σE_z so that the total current in the wire is

$$\left.\begin{aligned}
J &= \pi\sigma E_z (a^2 - R^2) + \frac{c H_c R}{2} \\
&= \pi\sigma E_z a^2 + \frac{c^2 H_c^2}{16\pi\sigma E_z}.
\end{aligned}\right\} \tag{33.7}$$

The equation may be solved to express E_z as a function of J. The final result may be written in the form

$$\frac{\Omega}{\Omega_0} = \frac{1}{2}\left\{ 1 + \sqrt{1 - \left(\frac{J_c}{J}\right)^2} \right\}, \tag{33.8}$$

where Ω/Ω_0 is the resistance of the wire relative to that of the normal state and

$$J_c = a c H_c \tag{33.9}$$

is the critical current according to SILSBEE's hypothesis.

Eq. (33.8) indicates that one-half of the resistance should be restored when the critical current is reached. As shown in Fig. 18, the critical current is close to (33.9), but the jump in resistance is rather more than one-half. There is some hysteresis; on decreasing, the current drops to about 0.85 critical before the resistance disappears. It is perhaps incorrect to assume that discus-shaped regions have negligible thickness. In a more exact theory, it would be necessary to take boundary effects into account[1].

A rather unusual intermediate state phenomenon occurs when a longitudinal magnetic field is present along with a large current flowing in the cylinder. It has first been observed by STEINER[2] and confirmed later by others that the average flux density in the superconducting wire may be much larger than that in the applied field. This "paramagnetic" effect is not as yet completely understood, although it is almost certain that it is observed in a rather complex intermediate state phenomenon which does not involve anything basically new.

---

[1] H. KOPPE, Ann. Phys. Lpz. 6, 375 (1949) and C. G. KUPER, Phil. Mag. 43, 1264 (1952).

[2] K. STEINER and H. SCHOENECK: Phys. Z. 38, 887 (1937). — K. STEINER: Z. Naturforsch. 4a, 271 (1949).

The paramagnetic effect is observed when the current is above a critical value, J_0, which depends linearly on the applied magnetic field:

$$J_0 = J_g + 2\pi a \gamma H_a,\qquad(33.10)$$

where J_g and γ are constants for a material at a fixed temperature. The maximum flux through the cylinder occurs for a current slightly larger than J_0. At this point of maximum flux, the total field at the outside of the cylinder is equal to H_c:

$$H_a^2 + H_\varphi^2 = H_c^2\qquad(33.11)$$

where H_φ is the field from the current, J,

$$H_\varphi = \frac{2v}{a c}.\qquad(33.12)$$

The resistance of a cylinder in the paramagnetic state is qualitatively similar to that of the intermediate state of a long cylinder carrying a current in the absence of an external field, as discussed earlier in this section.

It has been definitely established that the added flux through the cylinder comes from circular currents flowing around the cylinder[1]. The effect is destroyed if the cylinder is slotted to prevent such currents from flowing. The combination of circular plus axial flow gives flow lines following helical paths around the axis.

A theory of the paramagnetic effect somwhat along the lines of LONDON's theory has been given by H. MEISSNER[2]. The combination of H_φ and H_a at the surface will give flux lines which spiral around the surface. MEISSNER suggests that the superconducting domains of the intermediate state will follow more or less and be elongated along the lines of flux. The conductivity would then be highly anisotropic, with lower conductivity in a direction parallel with the field. The current would then follow helical paths and give the paramagnetic flux. While the theory is in qualitative and even semi-quantitative agreement with experiment, it does not yield a critical current (J_g). Further developments will probably require a discussion of boundary energies.

34. Kinetics of phase transitions and high frequency effects. As is the case for many phase transitions, the transition between the normal and superconducting phases occurs by nucleation and growth[3]. Because of the large boundary energies involved, a relatively large nucleus must be formed before it is stable and will grow. Various aspects of the problem of nucleation and growth have been studied at a number of laboratories, and some theoretical work has been devoted to the problem. There is an excellent review of the subject by FABER and PIPPARD ([7], Chap. IX, p. 159), in which extensive references to the literature may be found.

Both supercooling and superheating are observed. Actually, it is more convenient to vary the magnetic field than the temperature, so that "supercooling" refers to a metal remaining in the normal state when the magnetic field is reduced to a value lower than H_c and "superheating" to a metal remaining in the superconducting state as the field is increased above H_c. Usually supercooling is more

---

[1] MEISSNER, SCHMEISSNER and MEISSNER: Z. Physik **130**, 521 (1951); **130**, 529 (1951); **132**, 529 (1952). — Phys. Rev. **90**, 709 (1953). Other experiments on the effect are those of T. S. TEASDALE and H. E. RORSCHACH jr.: Phys. Rev. **90**, 709 (1953) and J. C. THOMPSON and C. F. SQUIRE: Phys. Rev. **96**, 287 (1954).

[2] H. MEISSNER: Phys. Rev. **97**, 1627 (1954); **101**, 31 (1956).

[3] T. E. FABER: Proc. Roy. Soc. Lond., Ser. A **214**, 392 (1952).

marked than superheating. This is because there usually exist local regions where the field is abnormally high at which normal nuclei may start growing. This was demonstrated by GARFUNKEL and SERIN [1] in experiments on a rod in a longitudinal field. An additional coil was placed near the center of the rod so that the field could be increased locally from below to above H_c. With this geometry, which avoids a large local field near the ends of the rod, considerable superheating was observed.

The velocity of propogation of the normal-superconducting boundary has been studied most extensively by FABER [2]. The studies were made for the most part by placing a number of detecting coils along a rod, so the propagation of a phase boundary along the rod could be studied. The rod was supercooled in a field a little below H_c. A superconducting nucleus could be started by suddenly decreasing the field locally by means of an auxiliary coil, and this spreads out along the rod in the order of a few seconds. The velocity of propagation is determined mainly by eddy current damping. The theory, worked out independently by PIPPARD [3] and by LIFSHITZ [4] accounts in a satisfactory way for the data. The basis of the theory is to consider the energy balance between the free energy released, when a fresh volume of normal metal is released, and the energy absorbed by the eddy currents. The latter is proportional to the velocity of propagation and the former to $H_c^2 - H^2$.

FABER [5] suggests that a superconducting nucleus spreads out in the form of a thin sheath of thickness d adjacent to the surface of the rod. According to theory, the maximum velocity of propagation of the sheath along the rod is obtained when the thickness has an optimum value given by

$$d_{\mathrm{opt}} = \frac{3}{4} \frac{(\varDelta - \lambda) H_c}{H_c - H} . \qquad (34.1)$$

The velocity for this thickness is:

$$v = C \frac{(H_c - H)^3}{H_c^3 (\varDelta - \lambda)^2} , \qquad (34.2)$$

where C is a constant which can be roughly estimated from the eddy current damping theory. FABER has used this result to estimate relative values of $\varDelta - \lambda$ from his experimental data.

A large amount of experimental and theoretical work has been devoted to the study of superconductors under high frequency fields. Some has involved small amplitude fields; the surface impedance is measured at microwave frequencies. A review article of PIPPARD [21] gives a summary of this work together with references to the literature. Another aspect has been the study of the kinetics of the phase transition, for which large amplitude fields in all regions of the spectrum are of interest. The theory of the destruction of superconductivity by alternating fields of large amplitude has been discussed by LIFSHITS [6]. We shall give here only a very brief summary of the theoretical aspects of work on surface impedance, because the subject is treated at greater length elsewhere in this series.

[1] M. P. GARFUNKEL and B. SERIN: Phys. Rev. **85**, 834 (1952).

[2] T. E. FABER: Proc. Roy. Soc. Lond., Ser. A **214**, 392 (1952); **219**, 75 (1953); **223**, 174 (1954).

[3] A. B. PIPPARD: Phil. Mag. **41**, 243 (1950).

[4] I. M. LIFSHITZ: Ž. eksper. teor. Fiz. **20**, 834 (1950).

[5] T. E. FABER: Proc. Roy. Soc. Lond., Ser. A **223**, 174 (1954); reference [7], p. 176.

[6] I. M. LIFSHITS: Dokl. Akad. Nauk SSSR. **90**, 363 (1953). — I. M. LIFSHITS and M. I. KAGANOV: Dokl. Akad. Nauk SSSR **90**, 529 (1953).

The surface impedance, $\boldsymbol{Z}$, is defined as the ratio of the complex quantity, $E_0(\omega)$, representing the alternating electric field of frequency ω at the surface to the integrated complex current density $J(x)$:

$$\boldsymbol{Z} = R + iX = \frac{E_0(\omega)}{\int\limits_0^\infty J(x)\,dx}. \tag{34.3}$$

Interpretation of data on superconductors has generally made use of the two-fluid model. The electric field which comes from the time variation of the magnetic field in the penetration region, acts on the normal component and gives a loss. The problem was first considered by H. LONDON[1]; later PIPPARD[2] pointed out that in most experiments, the mean free path is larger than the penetration depth, and gave a semi-quantitative theory to take this into account. The mathematical theory of the "anomalous skin effect" was developed more completely by REUTER and SONDHEIMER[3] and by MAXWELL, MARCUS and SLATER[4].

While the two-fluid model accounts in a qualitative way for the resistive part of the impedance, R, and its variation with temperature difficulties arise when a quantitative fit of the observed data is attempted. PIPPARD[5] has worked out, in part by means of dimensional analysis empirical laws which fit the observed data in different temperature ranges. At relatively low temperatures, where R in the superconducting phase is less than 5% of that in the normal phase, the data can be fitted by

$$R = A(\omega)\,\frac{t^4(1-t^2)}{(1-r^4)^2}, \tag{34.4}$$

where t is the reduced temperature. The frequency dependence is contained in the factor $A(\omega)$, which is found empirically to vary as $\omega^{\frac{3}{2}}$. The usual sort of two-fluid model, such as that of the original version of the LONDON theory, predicts a variation proportional to ω^2. It appears that some other relaxation effect may be coming in to alter the frequency dependence. After consideration of various possible mechanisms, FABER and PIPPARD[6] consider the most likely one may be a relaxation process in the superconducting state with a time constant of the order of the time it takes a phonon to travel across a coherence distance, ξ_0. This time is of the order of 10^{-9} sec., so that the relaxation would appear in the right frequency range. Another possibility is that the $\omega^{\frac{3}{2}}$ variation is a transition range between an ω^2 variation at lower frequencies and a slower variation at higher frequencies.

LANDAU has suggested that "normal" electrons may be bound at low temperatures in very large orbits, and that this would give a very high dielectric constant, of the order of 10^9. If this were the case, there would be appreciable displacement currents within the penetration depth at microwave frequencies. A report of a theory of ABRIKOSOV, who has extended the REUTER-SONDHEIMER theory to include displacement currents, and applications to microwave data of HAJKIN on thin films is included in SHOENBERG's review [24]. PIPPARD [21] believes that the experiments can be interpreted in other ways, and there is as yet no convincing evidence in favor of a large dielectric constant. Since there is good

---

[1] H. LONDON: Proc. Roy. Soc. Lond., Ser. A **176**, 552 (1940).
[2] A. B. PIPPARD: Proc. Roy. Soc. Lond., Ser. A **191**, 385 (1947).
[3] G. E. H. REUTER and E. H. SONDHEIMER: Proc. Roy. Soc. Lond., Ser. A **195**, 336 (1948).
[4] E. MAXWELL, P. M. MARCUS and J. C. SLATER: Phys. Rev. **76**, 1332 (1949).
[5] A. B. PIPPARD: Proc. Roy. Soc. Lond., Ser. A **203**, 195 (1950).
[6] T. E. FABER and A. B. PIPPARD: Proc. Roy. Soc. Lond., Ser. A **231**, 53 (1955).

evidence that the superconducting wave functions extend over large volumes, a large dielectric constant is a possibility and should be kept in mind in analysis of microwave data.

V. Electron-phonon interactions.

a) Introduction.

35. Microscopic theories. The BLOCH theory, in which it is assumed that each electron moves independently in a periodic potential determined by the ions and an averaged charge density of the valence electrons, gives a good qualitative and in some cases quantitative explanation of the electrical properties of normal metals, but fails to account for superconductivity. Most attempts to give a microscopic theory of superconductivity have made use of interactions omitted from the BLOCH theory. These include correlations between the positions of the electrons brought about by COULOMB interactions, magnetic interactions between electrons and interactions between electrons and phonons. While all of these interactions are undoubtedly important for a complete theory, the isotope effect shows that the main one responsible for the transition is the electron-phonon interaction.

Prior to the discovery of the MEISSNER effect, it was thought that superconductivity was simply infinite conductivity, and that it would be necessary to show why the electrons in the superconducting state are not scattered in such a way as to give resistance. Some of the more recent theories such, as those of HEISENBERG and of BORN and CHENG, also have attempted to account for superconductivity in terms of the stability of currents.

A major stumbling block to all such theories is a theorem of BLOCH that the lowest state is one of zero current (Sect. 1). BLOCH's theorem does not apply to diamagnetic currents. There can be a net current density in the lowest state in the presence of a magnetic field. LONDON's approach, which we believe to be correct, is based on the idea that all supercurrents are diamagnetic in origin. In the case of a persistent current flow in a ring, the current itself gives a magnetic field which in turn produces the supercurrents. While LONDON has given some qualitative arguments to show why such currents should be metastable, no real proof has been given, and probably cannot be without discussion of a specific model.

We shall give a brief description of HEISENBERG's theory [9] because it may contain some elements of truth, although the basic assumption that COULOMB interactions between electrons are responsible for superconductivity is not correct. HEISENBERG[1] attempted to show that electrons with energies near the FERMI surface may at low temperatures condense into electron lattices of low density moving in different directions. These electrons can be described roughly by wave packets formed from states with wave vectors within Δk of the FERMI surface, $|\boldsymbol{k}| = k_F$. The spread of the wave packet is of order $\Delta x = 1/\Delta k$. The kinetic energy required to localize the electron is of order $\hbar^2 k_F \Delta k/m$, where m is an effective mass. The gain in COULOMB energy obtained on formation of a lattice of such wave packets was estimated to be very roughly or order

$$- e^2 \Delta k \log \frac{k}{\Delta k}.$$

This will more than compensate for the increase in kinetic energy if Δk is sufficiently small. A more accurate estimate of the COULOMB energy was made later

---

[1] W. HEISENBERG: Z. Naturforsch. **2**a, 185 (1947); **3**a, 65 (1948), also [9] and [11].

by KOPPE[1]. A difficulty in these calculations is that the screening of the fields of the individual electrons by other electrons is not taken into account. If a screened COULOMB field of short range were used, the gain in COULOMB energy by formation of such an electron lattice would be negligible.

Since electrons near the FERMI surface are travelling in all directions, a lattice must be formed from a group of electrons in the same region of k-space, all moving in the same direction. A moving electron lattice would give a net current, which HEISENBERG argued, would be thermodynamically stable. Ordinarily, the supercurrents in different domains would be in random directions and so give no macroscopic current. The MEISSNER effect was explained by the effect of a magnetic field on the distribution of supercurrents. General theoretical objections against a theory of this sort have been given by LONDON ([15], p. 142). Some of the detailed predictions of the theory are not in accord with observation. Perhaps the most important is a maximum current density which approaches zero as $T \to 0° K$. This would imply that at low temperatures there should be a marked increase in penetration depth with field, which has not been found experimentally. On the other hand, we have seen (Sect. 5) that predictions of KOPPE's two-fluid model, based rather loosely on the theory, are in at least rough agreement with observation.

Another theory based on COULOMB interactions and persistent currents is that of BORN and CHENG[2]. It was suggested that superconductivity occurs in metals with overlapping energy bands in which the lower band is nearly full. An attempt was made to show that, below a critical temperature, the lowest free energy occurrs with an asymmetric distribution of electrons in k-space, with more electrons in some corners of the BRILLOUIN zones than others. This appears to be a violation of BLOCH's theorem that the state of lowest energy has zero current.

A mathematical formulation for treating many particle wave functions in the theory of metals has been suggested by TISZA[3], with a view toward application to the problem of superconductivity. His functions are "super" BLOCH functions which represent the coordinated motion of a group of electrons with a net momentum. The theory was not developed very far, but presumably would be useful for a theory in which persistent currents play a dominant role. We believe that the objections raised by LONDON to all such theories are valid.

Another interaction which has been suggested as being responsible for superconductivity is the magnetic interaction between electrons. Such interactions can be taken into account in the HARTREE approximation by including the magnetic fields of the electron currents in a self-consistent manner. This is of course essential when the diamagnetism is large, and has been done in the discussion of Chap. III. Electron currents are determined by the magnetic field and these currents also contribute to the field. There is no evidence, however, that it is necessary to take specific magnetic interactions between individual electrons into account. WELKER[4] once attempted to base a theory of superconductivity on magnetic exchange interactions.

Another possibility, which we now believe to be correct, is that motion of the ions is involved in the transition to the superconducting state. The author[5]

[1] H. KOPPE: Ann. Phys., Lpz. 1, 405 (1947). — Z. Naturforsch. 3a, 1 (1948); 4a, 74 (1949); 6a, 284 (1951); also [11].
[2] M. BORN and K. C. CHENG: Nature, Lond. 161, 968, 1017 (1948). — J. Phys. Radium 9, 249 (1948). — K. C. CHENG: Nature, Lond. 163, 247 (1949).
[3] L. TISZA: Phys. Rev. 80, 717 (1950). Also see J. M. LUTTINGER: Phys. Rev. 80, 727 (1950).
[4] H. WELKER: Z. Physik 114, 525 (1939).
[5] J. BARDEEN: Phys. Rev. 59, 928 (A) (1941).

once suggested that there are small periodic displacements of the lattice in such a way as to produce a very large unit cell in real space and a fine-grained BRILLOUIN zone structure in k-space. The displacements were assumed to be such as to produce small energy gaps near the FERMI surface so that the energies of the occupied states are lowered. It is known that some alloys (for example, the γ-phase alloys) take up a complicated structure which gives planes of discontinuity near the FERMI surface. It was supposed that the same sort of thing could occur in many metals at low temperatures, not matter how complicated the FERMI surface, if the zone structure is very fine-grained. First rough estimates indicated that the energy decrease of electrons near the FERMI surface might be sufficient to compensate for the energy required to displace the ions, but more careful estimates made later showed it too small by an order of magnitude or more. Most favorable metals are those with a large interaction between electrons and lattice and thus a large resistivity in the normal state. The diamagnetic properties were accounted for by the very small effective mass of electrons and holes with energies near the FERMI surface (see Sect. 24). Since the best estimates seemed to indicate that transitions of this sort are not to be expected, the details of the theory were never published. Some features were retained in a later theory[1] based on a dynamic interaction between electrons and lattice vibrations, which was suggested by the isotope effect.

Without having prior knowledge of the isotope effect, FRÖHLICH [4] proposed a theory of superconductivity based on electron-phonon interactions. While such interactions had long been used to account for thermal scattering of electrons and thus the resistivity of normal metals, it had not been recognized that they would also give a contribution to the energy. FRÖHLICH calculated the interaction energy by use of second-order perturbation theory. He showed that if the interaction is sufficiently large, the energy at the absolute zero would be lowered if a thin shell of electrons adjacent to the FERMI surface of the normal metal were displaced outward a small distance in k-space. He presumed that this shell distribution represents the superconducting state. There is considerable doubt about the details of the theory, because the criterion for superconductivity, the condition that the shell distribution have a lower energy then the normal one, is essentially the same as the condition that the interactions be so large that perturbation theory cannot be applied. It is believed that the basis of the theory is correct, and that the criterion gives a good indication for the occurrence of superconductivity, but that better mathematical methods are required to give a reliable picture of the nature of the superconducting state[2]. We shall discuss FRÖHLICH's theory in more detail in Sect. 42.

Since an adequate mathematical theory of superconductivity based on electron-phonon interactions has not been given, we shall devote most attention in this Chapter to the formulation of the problem. Both FRÖHLICH and the author started from the BLOCH theory in which it is assumed that each electron moves independently in a periodic potential field. Vibrational coordinates and interactions between electrons and vibrations were introduced exactly as is done in the theory of conductivity. The strength of the interactions was estimated empirically from the high temperature resistivity.

---

[1] J. BARDEEN: Phys. Rev. **79**, 167 (1950); **80**, 567 (1950); **81**, 829, 1070 (1951); **82**, 978 (1951); also [1].

[2] FRÖHLICH himself has emphasized the need for new mathematical methods; H. FRÖHLICH: Physica, Haag **19**, 755 (1953); reference [30], p. 909. There are illuminating discussions by BOHR. HEISENBERG and others following the first of these, and there are also interesting discussions following the second.

There are two objections to this formulation: First, the COULOMB interactions between electrons should be introduced at the start; second, displacements of the electrons brought about by electron-phonon interactions have an important effect on the vibrational frequency and also on the effective matrix element for the interaction. An important part of the problem is to show how these should be determined from first principles. Starting from a formulation which includes COULOMB interactions between electrons, we shall show that the usual BLOCH theory should be a reasonably good starting point to develop a theory of super-conductivity. We also show why electron-phonon interactions have a larger influence on the wave functions than COULOMB interactions, even though the interaction energies are much smaller. Our treatment, given in Sects. 37 to 41, follows closely one of PINES and the author [2].

36. Importance of screening in metals. An essential point to be remembered is that COULOMB interactions in a metal are screened interactions. This applies to the interactions between electrons and ions as well as to the interactions between electrons. This was recognized in early calculations of the electron-phonon interaction by HOUSTON[1] and by NORDHEIM[2]. The potential energy of an electron at a distance r from an ion was taken to be:

$$v(r) = -\frac{Z e^2}{r} e^{-\alpha r},$$

where α is a screening constant, estimated from a FERMI-THOMAS model. To cal-culate the change in potential resulting from a lattice vibration, it was assumed that the ions move rigidly under the displacements. In a later calculation, the author[3] determined the screening by a HARTREE self-consistent field method and applied the results to a calculation of the resistivity of monovalent metals More recently, NAKAJIMA[4] has derived nearly equivalent results by use of field theoretic methods.

It is also important to take the screening into account in a calculation of the vibrational frequency. Entirely erroneous results would be obtained if the response of the electrons to the motion of the ions were not included. The HARTREE self-consistent field method has been extended by TOYA[5] to derive an expression for the vibrational frequency. Equivalent results follow from NAKAJIMA's derivation.

Prior to NAKAJIMA's work, FRÖHLICH[6] and KITANO and NAKANO[7] inde-pendently used similar field theoretic methods to determine the effect of electron motion on the vibrational frequency by starting from the BLOCH HAMILTONian in which COULOMB interactions between electrons are not explicitly introduced.

From a field-theoretic point of view, there is an interaction between electrons brought about by virtual emission and absorption of phonons, and it is this interaction which is believed to be responsible for superconductivity [4]. There is also a phonon self-energy, which can be quite large when the interaction is strong enough to give superconductivity. This means physically that one must take the electron motion into account in a derivation of the phonon frequency ([1], p. 264).

---

[1] W. V. HOUSTON: Phys. Rev. **34**, 279 (1929); **88**, 1321 (1952).
[2] L. W. NORDHEIM: Ann. Phys., Lpz. **9**, 607 (1931).
[3] J. BARDEEN: Phys. Rev. **52**, 688 (1937).
[4] S. NAKAJIMA: Buss. Kenkyu **65**, p. 116 (1953); reference [*30*], p. 916.
[5] T. TOYA: Buss. Kenkyu **59**, 179 (1952).
[6] H. FRÖHLICH: Proc. Roy. Soc. Lond., Ser. A **215**, 291 (1952).
[7] Y. KITANO and H. NAKANO: Progr. Theor. Phys. **9**, 370 (1953).

The reason that the Bloch Hamiltonian is reasonably satisfactory for most problems in the theory of metals, including superconductivity, is that the Coulomb interactions are screened out within a distance of the order of the interparticle spacing. To give an example, Abrahams[1] has estimated the collision cross-section and mean free path for screened electrons in the alkali metals. He finds that the scatterings possible are so greatly restricted by the exclusion principle that the mean free path for electron-electron collisions is greater than that for electron-phonon interactions at practically all temperatures.

Bohm and Pines[2] have shown that the long range part of the Coulomb interaction leads to a coherent motion of the electrons which can be described in terms of plasma oscillations. These are of such high frequency that they are not normally excited. There remains a short range Coulomb interaction between the individual electrons. Pines and the author [2] have extended this theory to take the motion of the ions into account. In the combined collective motion of electrons and ions, there are high frequency modes corresponding to the plasma oscillations of an electron gas and low frequency modes corresponding to longitudinal lattice vibrations. Expressions for the electron-phonon interaction and vibrational frequency derived in this treatment are nearly equivalent to those found by the Hartree self-consistent field method. The collective treatment is not applicable to phonons of short wave length; for these, the Nakajima formulation is probably the most satisfactory.

As pointed out be Fröhlich, a canonical transformation can be used to eliminate the electron-phonon interaction, from the Hamiltonian, and one is left with an interaction between electrons which corresponds to one he had derived earlier by perturbation theory methods. When the electron-phonon interaction is large, this procedure breaks down for a small number of terms with small energy denominators. These terms are not important for calculating the matrix element for the interaction and vibrational frequencies, but they are just the terms important for superconductivity. Since they cannot be treated by perturbation theory methods, they can have a large effect on the wave functions.

The general plan of this Chapter is as follows. In Sect. 37, the Hamiltonian is derived in a form suitable for a field-theoretic treatment. The canonical transformation which eliminates the linear terms of the electron-phonon interaction and the Nakajima method for deriving the shielded interaction and phonon frequency is given in Sect. 38, and is followed by the collective treatment in which plasma coordinates are introduced in Sect. 39. Convergence of the expansion and the criterion for superconductivity are discussed in Sect. 40. The remaining sections are concerned with attempts which have been made to calculate the electron-phonon interaction energy when the interaction is strong enough to give superconductivity.

b) Formulation of the electron-phonon interaction problem.

37. Derivation of the Hamiltonian. In order to formulate properly the problem of calculating the interactions between electrons and phonons in a metal, we derive in this section an expression for the Hamiltonian in a form sufficiently general for our purpose. Coulomb interactions between electrons and motions of the ions are included from the start, but several approximations are made in order to simplify the equations. These amount essentially to neglect of anisotropic effects not believed to be important for the superconductivity problem. It is

[1] E. Abrahams: Phys. Rev. **95**, 839 (1954).

[2] D. Bohm and D. Pines: Phys. Rev. **82**, 625 (1951); **85**, 338 (1952); **92**, 609 (1953). — D. Pines: Phys. Rev. **92**, 626 (1953).

assumed that lattice waves are either longitudinal or transverse, and that electrons interact only with the longitudinal component. This is a valid approximation for waves of long wavelength, but is not correct for short waves except for certain directions of travel. We also assume, as is often done in the BLOCH theory, that the matrix elements for the electron-phonon and COULOMB interactions depend only on the wave vector difference between initial and final states. In the calculation of COULOMB interactions, approximations are made which amount to treating the valence electrons as a free electron gas.

We assume a monatomic crystal of N atoms of valence Z, so that there are $n = ZN$ valence electrons. Positions of the valence electrons are denoted by $r_i (i = 1, 2, \ldots n)$ of the ions by $R_j (j = 1, 2, \ldots N)$.

The total HAMILTONian of the crystal is the sum of four terms, the kinetic energy of the electrons, the electron-ion interaction energy, the COULOMB interaction between electrons and the HAMILTONian for the ions, including kinetic energy, COULOMB and exchange interactions:

$$H = \sum_i \frac{p_i^2}{2m} + \sum_{i,j} v(r_i - R_j) + \sum_{i \neq j} \frac{e^2}{|r_i - r_j|} + H_{\text{ion}}. \qquad (37.1)$$

As written, the COULOMB interaction energies of the separate terms are very large, but these large terms tend to cancel in the sum. Included is a large negative contribution from the second term, which represents the interaction between each electron and the sum of the COULOMB fields of all of the ions, and large positive contributions from the COULOMB interactions between the electrons and between the ions. In order to avoid dealing with these large energies, we suppose that there is subtracted from the electron-ion interaction, the interaction of each electron with a uniform positive sea, from the electron-electron interaction, the self-energy of a uniform negative sea, and from the ion-ion interaction, the self-energy of a uniform positive sea. Since the sum of these three terms adds to zero, the total energy is unchanged. The ion-ion interaction energy less the energy of a uniform positive sea is equivalent to the energy of the ions in a uniform negative sea, including the self-energy of the negative charge.

We want to modify this HAMILTONian by introducing phonon coordinates to represent the ion motion and by introducing occupation numbers of a set of BLOCH functions to represent the electron wave function. The transformed HAMILTONian will then contain creation and destruction operators for the electrons.

The BLOCH functions, $\psi_k(r)$, are a set of one-particle functions for the electrons which apply to a crystal with the ions fixed in equilibrium positions. They may be defined by a HARTREE approximation or by a HARTREE-FOCK approximation in which effects of electron exchange are included. We shall use an even simpler approximation here and assume that the density of valence electrons is uniform, so that the effective potential, $V(r)$, in which the electrons move is that of the ions in equilibrium positions compensated by a uniform negative charge. If $v(r - R_j^0)$ is the potential of the ion at the equilibrium position R_j^0,

$$V(r) = \sum_j v(r - R_j^0) + \text{compensating charge}. \qquad (37.2)$$

The BLOCH equation for the one-particle functions is

$$\left(\frac{p^2}{2m} + V(r)\right)\psi_k(r) = E_k \psi_k. \qquad (37.3)$$

The electrons are described in an extended zone scheme, so that the wave vector k is not necessarily in the first BRILLOUIN zone. The designation of the wave vector for a particular state presumably would be chosen so that the approximations concerning matrix elements mentioned in the preceding paragraph would be most nearly valid. This implies that an electron in state k is treated much like a free electron with the same wave vector. As usual, periodic boundary conditions are introduced to get a discrete set of k-values. We shall omit spin-orbit interactions, and where necessary indicate the spin by an index s which can take on the values $\pm\frac{1}{2}$.

We describe the electron wave function in second quantization by giving the occupation numbers for this set of BLOCH functions. Creation and destruction operators, c_{ks}^{*}, c_{ks} are defined in the usual way and obey the commutation relations for FERMI particles

$$[c_{ks}^{*}, c_{k's'}]^{+} = c_{ks}^{*} c_{k's'} + c_{k's'} c_{ks}^{*} = \delta_{kk'} \delta_{ss'}. \tag{37.4}$$

Except where required for clarity, we shall omit the spin index. The number of electrons in the state k, s is

$$n(k, s) = c_{ks}^{*} c_{ks}. \tag{37.5}$$

The normal modes of vibration of a crystal lattice consist of waves which may be designated by a wave vector, $\varkappa$, taking on values in the first BRILLOUIN zone. If there is one atom per unit cell, there are N distinct values of $\varkappa$ and for each $\varkappa$ three independent waves corresponding to different directions of polarization, designated by $\sigma = 1$, 2 or 3. The departure, δR_j, of an ion from its equilibrium position can be expanded in terms of the coordinates, $q_{\varkappa\sigma}$, of the normal modes:

$$\delta R_j = R_j - R_j^0 = \frac{1}{\sqrt{NM}} \sum_{\varkappa\sigma} q_{\varkappa\sigma} \, \boldsymbol{\varepsilon}_{\varkappa\sigma} \, e^{i\varkappa \cdot R_j^0}. \tag{37.6}$$

The direction of polarization, $\boldsymbol{\varepsilon}_{\varkappa\sigma}$, is taken in the same sense for $\varkappa$ and $-\varkappa$, so that the reality requirement is $q_{-\varkappa} = q_{\varkappa}^{*}$. The mass of an atom is denoted by M.

In a more correct formulation, part of the problem would be to determine the directions of polarization going with a given $\varkappa$, and for this it is necessary to take into account the displacement of the electrons associated with the wave. As stated earlier, we simplify the problem by assuming that the waves are either longitudinal or transverse, and that electron motion affects only the longitudinal wave. This implies that the frequencies of the transverse waves are determined by motion of the ions in a fixed negative sea. It is known from the work of FUCHS that the elastic constants for shear of the monovalent metals can be determined accurately in this way[1]. It is probably a good approximation for long waves, but less good for short waves which really have both longitudinal and transverse components. We shall be concerned here only with longitudinal waves, and shall use $q_{\varkappa}$ without explicit designation of σ to indicate this component.

When phonon coordinates are introduced, the HAMILTONian for the ions compensated by a uniform sea of negative charge may be written,

$$H_{\text{ion}} = H_{\text{ph}} + H_{\text{tr}} + H_{\text{ion-ion}}, \tag{37.7}$$

where

$$H_{\text{ph}} = \tfrac{1}{2} \sum_{\text{zone}} (p_{\varkappa}^{*} p_{\varkappa} + \Omega_{\varkappa}^{2} q_{\varkappa}^{*} q_{\varkappa}) \tag{37.8}$$

represents the longitudinal phonons, H_{tr} the transverse phonons and $H_{\text{ion-ion}}$ the interaction energy of the ions in equilibrium positions. Since $\Omega_{\varkappa}$ as defined

---

[1] Cf. the article of G. LEIBFRIED in vol. VII, part 1 of this Encyclopedia.

includes only ion-ion interaction terms, it is not the true frequency. To get the correct frequency, $\omega_{\varkappa}$, one must include the shielding of the ions from electron motion, and part of our task is to show this should be done.

The electron-ion interaction terms may be expanded in the $q_{\varkappa}$. To terms of the first order, we have

$$\sum_{i,j} v(r_i - R_j) = \sum_{i,j} v(r_i - R_j^0) - \frac{1}{\sqrt{NM}} \sum_{\varkappa} \varepsilon_{\varkappa} \cdot \nabla_r v(r - R_j^0) q_{\varkappa} e^{i\varkappa \cdot R_j^0}. \qquad (37.9)$$

The first term may be combined with the kinetic energy of the electrons to give:

$$H_{el} = \sum_i \left\{ \frac{p_i^2}{2m} + V(r_i) \right\} = \sum_{k,s} c_{ks}^* c_{ks} E_k. \qquad (37.10)$$

To express the second term of (37.9) in terms of the creation and destruction operators, we need the matrix elements:

$$v_{\varkappa}^i = -\frac{1}{\sqrt{NM}} \int \psi_{k+\varkappa}^* \left\{ \sum_j \varepsilon_{\varkappa} \cdot \nabla_r v(r - R_j^0) e^{i\varkappa \cdot R_j^0} \right\} \psi_k \, d\tau. \qquad (37.11)$$

We make the simplifying assumption that $v_{\varkappa}^i$ depends only on $\varkappa$ and is independent of k. The general selection rule for a matrix element connecting the electron states k' and k is:

$$k' = k \pm \varkappa. \qquad (37.12)$$

It may happen that $k' - k$ lies outside of the first Brillouin zone. In this case there may be matrix elements for $q_{\varkappa}$ of the reduced wave vector of $\varkappa$. These correspond to the Umklapp processes of Peierls. We shall take this possibility into consideration by allowing the $\varkappa$ in $v_{\varkappa}^i$ to run out of the first zone, and remember that in the corresponding $q_{\varkappa}$, $\varkappa$ represents the reduced vector in the first zone. The interaction term may then be written:

$$H_{int} = \sum_{\varkappa k s} c_{k+\varkappa, s}^* c_{ks} q_{\varkappa} v_{\varkappa}^i = \sum_{\varkappa} q_{\varkappa} v_{\varkappa}^i \varrho_{-\varkappa}, \qquad (37.13)$$

where

$$\varrho_{\varkappa} = \sum_{ks} c_{k-\varkappa, s}^* c_{ks}; \qquad \varrho_{-\varkappa} = \sum_{ks} c_{k, s}^* c_{k-\varkappa, s}. \qquad (37.14)$$

In (37.13), the sum is not restricted to the first Brillouin zone, but runs over all $\varkappa$. Similarly, in (37.14), the sum is over all k, for we have used the extended zone scheme to represent the electron wave functions. Note that from the definitions we have used, $(v_{\varkappa}^i)^* = v_{-\varkappa}^i$.

Finally, the Coulomb interactions between the electrons may be expressed in the form

$$H_{coul} = \tfrac{1}{2} \sum_{\varkappa} M_{\varkappa}^2 \varrho_{-\varkappa} \varrho_{\varkappa}, \qquad (37.15)$$

where we have again assumed that the matrix elements depend only on the wave vector difference between initial and final states. For free electrons,

$$M_{\varkappa}^2 = \frac{4\pi e^2}{\varkappa^2}. \qquad (37.16)$$

Our final Hamiltonian is now in the form:

$$H = H_1 + H_{tr} + H_{ion\text{-}ion}, \qquad (37.17)$$

where

$$H_1 = H_{el} + H_{ph} + H_{int} + H_{coul}. \tag{37.18}$$

Our problem now is to transform (37.18) so as to determine the phonon frequencies and the effective matrix element of the electron-phonon interaction.

38. Simplified derivation of vibrational frequencies and interaction potential.
As discussed in Sect. 36, it is necessary in a calculation of the interaction potential and vibrational frequencies to take into account the motion of the electrons which tend to shield the ions. In the following section, we shall show how these may be determined by appropriate canonical transformations of the HAMILTONian. Since much of the physics of the problem is buried in the formalism when this method is used, we shall give first a simplified approximate treatment of the problem.

This simple derivation follows closely one given in [2] and parallels in some respects one given earlier by BOHM and STAVER[1]. The interaction potential, $v_\varkappa$, can be written as the sum of two terms, one, $v_\varkappa^i$ due to the motion of the ions and a second, $v_\varkappa^\varrho$, due to the motion of the electrons:

$$v_\varkappa = v_\varkappa^i + v_\varkappa^\varrho. \tag{38.1}$$

In the author's 1937 paper (reference 3, p. 346), $v_\varkappa^\varrho$ was determined by a HARTREE self-consistent field method in which it was assumed that the wave functions of the individual electrons change adiabatically with the ion motion. In the simplified derivation, we use the FERMI-THOMAS approximation.

It follows from the HAMILTONian (37.18) that the equation of motion for a longitudinal vibration with wave vector $\varkappa$ is:

$$\frac{d^2 q_\varkappa}{d t^2} + \Omega_\varkappa^2 q_\varkappa = -v_{-\varkappa}^i \varrho_\varkappa \tag{38.2}$$

where $\varrho_\varkappa$ is the FOURIER component of electronic density fluctuation. A part, $\varrho_\varkappa^0$, consists of random fluctuations which would exist in the absence of ion motion, and which average to zero; and another, $\delta \varrho_\varkappa$, gives the coherent motion in response to the vibration. The electron potential, $v_\varkappa^\varrho$ is related to $\delta \varrho_\varkappa$ by POISSON's equation:

$$\varkappa^2 v_\varkappa^\varrho q_\varkappa = 4\pi e^2 \delta \varrho_\varkappa = \varkappa^2 M_\varkappa^2 \delta \varrho_\varkappa, \tag{38.3}$$

where $M_\varkappa^2$ is defined by (37.16). The equation of motion may be written

$$\frac{d^2 q_\varkappa}{d t^2} + \omega_\varkappa^2 q_\varkappa = -v_{-\varkappa}^i \varrho_\varkappa^0, \tag{38.4}$$

where $\omega_\varkappa$ is the actual frequency, given by

$$\omega_\varkappa^2 = \Omega_\varkappa^2 + M_\varkappa^{-2} v_{-\varkappa}^i v_\varkappa^\varrho. \tag{38.5}$$

In the FERMI-THOMAS approximation, the electron density is proportional to $(E_F - \delta V(r))^{\frac{3}{2}}$ where E_F is the FERMI energy and $\delta V(r)$ the fluctuating potential from combined ion and electron motion.

Thus

$$\delta \varrho(r) = -\frac{3}{2} \frac{n}{E_F} \delta V(r). \tag{38.6}$$

---

[1] D. BOHM and T. STAVER: Phys. Rev. **84**, 836 (1951).

The FOURIER components are

$$\delta \varrho_{\varkappa} = - \frac{3}{2} \frac{n}{E_F} v_{\varkappa}. \tag{38.7}$$

By use of (38.1) and (38.3), we find

$$v_{\varkappa}^{\varrho} = \frac{- v_{\varkappa}^i}{1 + \dfrac{\varkappa^2 E_F}{6\pi e^2 n}}. \tag{38.8}$$

For small $\varkappa$, both $v_{\varkappa}^i$ and $v_{\varkappa}^{\varrho}$ are inversely proportional to $|\varkappa|$, but the shielding is such that the sum, $v_{\varkappa}$, is proportional to $|\varkappa|$:

$$v_{\varkappa} = \frac{\varkappa^2 E_F v_{\varkappa}^i}{6\pi e^2 n + \varkappa^2 E_F}. \tag{38.9}$$

Eqs. (38.5) and (38.8) for $\omega_{\varkappa}^2$ and $v_{\varkappa}^{\varrho}$ are similar to those derived by the self-consistent field method, and also to those derived by the canonical transformation to be discussed in the following sections.

39. Elimination of linear terms by a canonical transformation. In Sect. 36 we noted that several authors have taken into account the effect of electron motion on the vibrational frequencies by a canonical transformation which eliminates from the HAMILTONian terms linear in the phonon coordinates. We shall follow here, with some modifications given in [2], the treatment of NAKA-JIMA in which COULOMB interactions between electrons are included from the start. While similar to the self-consistent field method, it goes beyond the simple adiabatic approximation for treating ion motion. NAKAJIMA writes the HAMIL-TONian in a form equivalent to the following:

$$\left. \begin{aligned} H_1 = \sum_{ks} E_k c_{ks}^* c_{ks} + \tfrac{1}{2} \sum_{\text{zone}} (p_{\varkappa}^* p_{\varkappa} + \omega_{\varkappa}^2 q_{\varkappa}^* q_{\varkappa}) + \sum_{\varkappa} v_{\varkappa} q_{\varkappa} \varrho_{-\varkappa} + \\ + \tfrac{1}{2} \sum_{\varkappa} M_{\varkappa}^2 \varrho_{\varkappa} \varrho_{-\varkappa} + \sum_{\varkappa} (v_{\varkappa}^i - v_{\varkappa}) q_{\varkappa} \varrho_{-\varkappa} + \tfrac{1}{2} \sum_{\text{zone}} (\Omega_{\varkappa}^2 - \omega_{\varkappa}^2) q_{\varkappa}^* q_{\varkappa}. \end{aligned} \right\} \tag{39.1}$$

A canonical transformation is made to eliminate the linear term in $q_{\varkappa}$ in the first line. The effective interaction, $v_{\varkappa}$, is chosen in such a way as to eliminate the linear terms in $q_{\varkappa}$ in the second line and $\omega_{\varkappa}^2$ is chosen so that there are no diagonal terms in $q_{\varkappa}^* q_{\varkappa}$ in the second line. It should be noted that the term which cancels $v_{\varkappa}^i - v_{\varkappa} = -v_{\varkappa}^{\varrho}$ represents electron response to ion motion. The transformed HAMILTONian is expanded in a series in the generating function, S, which is linear in the $v_{\varkappa}$:

$$H_1' = e^{-iS/\hbar} H_1 e^{iS/\hbar} = H_1 + \frac{i}{\hbar} [H_1, S] - \frac{1}{2\hbar^2} [[H_1, S], S] + \cdots. \tag{39.2}$$

The expressions in square brackets represent commutators; for example

$$[H_1, S] = H_1 S - S H_1. \tag{39.3}$$

For S we take

$$S = i \sum_{\varkappa k} c_k^* c_{k-\varkappa} \{ f(k, \varkappa) q_{\varkappa} - i g(k, \varkappa) p_{-\varkappa} \}. \tag{39.4}$$

The required commutators are:

$$\left[\sum E_k c_k^* c_k, S \right] = - \sum (E_k - E_{k-\varkappa}) c_k^* c_{k-\varkappa} \{ f(k, \varkappa) q_{\varkappa} - i g(k, \varkappa) p_{-\varkappa} \}; \tag{39.5}$$

$$\left[\tfrac{1}{2} \sum (p_{\varkappa}^* p_{\varkappa} + \omega_{\varkappa}^2 q_{\varkappa}^* q_{\varkappa}), S \right] = i \sum c_k^* c_{k-\varkappa} \{ -i\hbar p_{-\varkappa} f(k, \varkappa) + \hbar \omega_{\varkappa}^2 q_{\varkappa} g(k, \varkappa) \}; \tag{39.6}$$

$$[\varrho_{-\varkappa'} \varrho_{\varkappa'}, S] = \varrho_{-\varkappa'} [\varrho_{\varkappa'}, S] + [\varrho_{-\varkappa'}, S] \varrho_{\varkappa'}, \tag{39.7}$$

$$[\varrho_{\varkappa}, S] = i \sum_{k \varkappa' k'} \{ \delta_{k k'} c_{k-\varkappa}^* c_{k'-\varkappa'} - \delta_{k'-\varkappa', k-\varkappa} c_{k'}^* c_k \} \{ f(k', \varkappa') q_{\varkappa'} - i g(k', \varkappa') p_{-\varkappa'} \}. \tag{39.8}$$

In order to simplify the notation, we have omitted the spin index, s. The terms linear in $q_\varkappa$ in (39.5) and (39.6) are of the form $c_k^* c_{k-\varkappa} q_\varkappa$. The coefficients f and g are chosen so that the first order terms (39.5) and (39.6), cancel the zero order term in $q_\varkappa$ from the third term in the first line of (39.1) and so that the coefficient of $p_{-\varkappa}$ vanishes. This procedure gives:

$$g(k, \varkappa) = \frac{-\hbar f(k, \varkappa)}{E_k - E_{k-\varkappa}}, \tag{39.9}$$

$$f(k, \varkappa) = \frac{\hbar(E_k - E_{k-\varkappa}) v_\varkappa}{(E_k - E_{k-\varkappa})^2 - \hbar^2 \omega_\varkappa^2}. \tag{39.10}$$

The interaction $v_\varkappa$ is now chosen so that the diagonal component of the coefficient of $c_k^* c_{k-\varkappa} q_\varkappa$ in (39.7) cancels the corresponding coefficient of the second term of the second line of (39.1). NAKAJIMA considers only those terms in (39.8) for which $\varkappa = \varkappa'$ so that

$$[\varrho_\varkappa, S] = i \sum_k \{n(k - \varkappa) - n(k)\} \{f(k, \varkappa) q_\varkappa - i g(k, \varkappa) p_{-\varkappa}\} \tag{39.11}$$

may be treated as a c-number in (39.7). This is equivalent to a HARTREE approximation, since diagonal exchange terms in (39.7) are then neglected.

A complete expansion of the commutator (39.7) is:

$$[\varrho_{-\varkappa'} \varrho_{\varkappa'}, S] = i \sum \{c_{k'}^* c_{k'-\varkappa'} (c_{k-\varkappa'}^* c_{k-\varkappa} - c_k^* c_{k-\varkappa+\varkappa'}) + \\
+ (c_{k-\varkappa'}^* c_{k-\varkappa} - c_k^* c_{k-\varkappa+\varkappa'}) c_{k'}^* c_{k'-\varkappa'}\} \{f(k, \varkappa) q_\varkappa - i g(k, \varkappa) p_{-\varkappa}\}. \tag{39.12}$$

The spin index of each pair of c's is the same. Our problem is to find the diagonal components of the coefficients of terms of the form $c_k^* c_{k-\varkappa} q_\varkappa$. In addition to those for which $\varkappa = \varkappa'$, k' arbitrary, which give (39.11), there are the following combinations:

(a) $k' = k$,   (b) $k' = k - \varkappa$,   (c) $k = k' - \varkappa'$,   (d) $k' = k - \varkappa + \varkappa'$.

The sum of these gives

$$\tfrac{1}{2} \sum M_{\varkappa'}^2 [\varrho_{-\varkappa'} \varrho_{\varkappa'}, S] = i \sum_{k, \varkappa} \left\{ \sum_{k', s} M_\varkappa^2 [n(k' - \varkappa) - n(k')] f(k', \varkappa) + \right.$$
$$+ \sum_{\varkappa'} M_{\varkappa'}^2 [n(k - \varkappa - \varkappa') - n(k - \varkappa')] f(k, \varkappa) + $$
$$\left. + \sum_{k'} M_{k'-\varkappa}^2 [n(k') - n(k' - \varkappa)] f(k', \varkappa) \right\} c_k^* c_{k-\varkappa} q_\varkappa + $$
$$+ \text{corresponding terms in } g(k, \varkappa). \tag{39.13}$$

The first sum in the curly brackets is over both spin states, the other two sums, which represent exchange terms, are only over the spin which is parallel with that in $c_k^* c_{k-\varkappa}$.

When exchange terms are included, it is no longer advantageous to introduce $v_\varkappa$, because one cannot eliminate the linear terms in $q_\varkappa$ and $p_{-\varkappa}$ in the second line with $f(k, \varkappa)$ and $g(k, \varkappa)$ defined by (39.9) and (39.10). Instead, there is an infinite system of equations to be solved:

$$\sum_{k'} (2M_\varkappa^2 - M_{k-k'}^2)(n(k' - \varkappa) - n(k')) f(k', \varkappa) - (W_{k-\varkappa} - W_k) f(k, \varkappa) + $$
$$+ (E_k - E_{k-\varkappa}) f(k, \varkappa) + \hbar^2 \omega_\varkappa^2 g(k, \varkappa) + v_\varkappa^i = 0,$$
$$\sum_{k'} (2M_\varkappa^2 - M_{k-k'}^2)(n(k' - \varkappa) - n(k')) g(k', \varkappa) - (W_{k-\varkappa} - W_k) g(k, \varkappa) + $$
$$+ (E_k - E_{k-\varkappa}) g(k, \varkappa) + \hbar f(k, \varkappa) = 0, \tag{39.14}$$

where W_k is the exchange energy of an electron in the state k:

$$W_k = - \sum_{\varkappa'} M_{\varkappa'}^2 \, n \, (k - \varkappa'). \qquad (39.15)$$

The factor of 2 multiplying $M_\varkappa^2$ in (39.14) takes account of the sum over both spin states.

When the distribution of electrons in k-space is symmetric, $n(k) = n(-k)$, the solutions satisfy the equations:

$$f(\varkappa - k, \varkappa) = - f(k, \varkappa); \qquad g(\varkappa - k, \varkappa) = g(k, \varkappa). \qquad (39.16)$$

The direct term in the equation for $g(k, \varkappa)$ then vanishes

$$\sum_{k'} M_\varkappa^2 \left[n(k - \varkappa') - n(k') \right] g(k', \varkappa) = 0. \qquad (39.17)$$

Even with this simplification, the equations cannot be solved readily when the exchange terms are included in the sums over k' in (39.14). It can be shown that these exchange terms are important only for short wavelengths.

The other exchange terms simply add an energy W_k to the individual particle energy E_k. This energy, if included, would make an important difference at long wavelengths. In the usual theory of an electron gas, it is known that the exchange energy, W_k, leads to a very small density of states at the FERMI surface and to a low temperature specific heat which is much smaller than observed. BOHM and PINES have shown that when the screening of the fields of the electrons are taken into account in the collective description, the exchange terms are greatly reduced in magnitude, and no longer have an important effect on the density of states and specific heat. We shall show in Sect. 40 that when the electron-lattice interaction is calculated using the collective description, no exchange terms at all appear for long wavelengths and for short wavelengths the exchange terms are reduced in magnitude because of the shielded interaction. We shall omit the exchange terms in the subsequent discussion, but remember that they might make an appreciable contribution for short wavelengths.

When exchange terms are omitted, the linear terms in the second line of (39.1) are eliminated if

$$2 \sum_k M_\varkappa^2 \left(n(k - \varkappa) - n(k) \right) f(k, \varkappa) = \hbar (v_\varkappa^i - v_\varkappa), \qquad (39.18)$$

or, substituting for $f(k, \varkappa)$:

$$\left. \begin{aligned}
v_\varkappa^i - v_\varkappa &= - 2 M_\varkappa^2 \, v_\varkappa \sum_k \frac{(E_k - E_{k-\varkappa}) \, (n(k) - n(k-\varkappa))}{(E_k - E_{k-\varkappa})^2 - \hbar^2 \omega_\varkappa^2} \\
&= 2 M_\varkappa^2 \, v_\varkappa \sum_k \frac{n(k) - n(k - \varkappa)}{E_{k-\varkappa} - E_k + \hbar \omega_\varkappa},
\end{aligned} \right\} \qquad (39.19)$$

which can be solved to give $v_\varkappa$. The factor of two in (39.18) and (39.19) is for electron spin. As a result of (39.17), terms in $p_{-\varkappa}$ vanish for a symmetric distribution in k-space. Eq. (39.19) is equivalent to the author's 1937 result if $\hbar \omega_\varkappa$ in the denominator is neglected. This term has a negligible effect on the interaction.

Elimination of diagonal terms in $q_\varkappa^* \, q_\varkappa$ gives the following for $\omega_\varkappa^2$:

$$\Omega_\varkappa^2 - \omega_\varkappa^2 = - 2 \, S \, v_{-\varkappa}^i \, v_\varkappa \sum_k \frac{n(k - \varkappa) - n(k)}{E_{k-\varkappa} - E_k + \hbar \omega_\varkappa}. \qquad (39.20)$$

The $\varkappa$ on the left corresponds to the reduced vector in the first BRILLOUIN zone; the sum S is over all $\varkappa$ which correspond to this same reduced wave vector. Eq. (39.20) is equivalent to:

$$\Omega_\varkappa^2 - \omega_\varkappa^2 = \mathsf{S}\, M_\varkappa^{-2}\, v_{-\varkappa}^i\, (v_\varkappa^i - v_\varkappa) \tag{39.21}$$

and thus to (38.5). Principal parts are to be taken in sums over vanishing energy denominators.

Eq. (39.21) has been tested for the alkali metals [2]. The matrix element $v_\varkappa$ was calculated in the limit of $\varkappa$ small from the 1937 expression. Calculated values of $\omega_\varkappa$ were found to be in reasonable agreement with values deduced from observed elastic constants. This agreement gives added confidence in the general method of procedure.

40. Collective description of electron-ion interaction. BOHM and PINES (see Sect. 36) have taken the long range part of the COULOMB interaction into account by introducing extra coordinates which describe the collective motion of the electron gas as plasma oscillations. Coordinates of the individual electrons are retained, so that there are more coordinates than are required for the system. It is therefore required that the system wave function satisfy certain supplementary conditions. This treatment has been extended by PINES and the author [2] to take the motion of the ions into account. In addition to the plasma oscillations, there are coupled electron-ion motions which correspond to longitudinal sound waves. We shall give a brief outline of this theory, which is patterned closely after one of PINES[1] for treating interactions between electrons in the absence of ion motions.

The collective description is used only for oscillations with wave vectors $|\varkappa| < \varkappa_c$, where $\varkappa_c$ is a critical value derived by PINES, usually somewhat less than that corresponding to the FERMI surface. Interactions of electrons with phonons with $|\varkappa| > \varkappa_c$ is probably treated best by self-consistent field methods.

An additional set of variables, $P_\varkappa$, $Q_\varkappa$ ($|\varkappa| < \varkappa_c$) is introduced to describe the plasma oscillations. These are first introduced by adding to the HAMILTONian, H_1, as defined by (37.18), a kinetic energy term:

$$H = H_1 + \frac{1}{2} \sum_{|\varkappa| < \varkappa_c} P_\varkappa^* P_\varkappa - i \sum_{|\varkappa| < \varkappa_c} \left(\frac{4\pi e^2}{\varkappa^2}\right)^{\frac{1}{2}} P_\varkappa \varrho_\varkappa . \tag{40.1}$$

It is required that the system wave function satisfy the subsidiary conditions

$$P_\varkappa \Psi = 0. \tag{40.2}$$

Thus for wave functions satisfying (40.2), the energies of the HAMILTONian, H, with the additional variables, is the same as the energies of H_1. A series of canonical transformations is made from the $P_\varkappa$, $Q_\varkappa$ defined above to variables which represent the plasma oscillations.

First consider the canonical transformation generated by

$$S = \sum_{|\varkappa| < \varkappa_c} (-i\, M_\varkappa \varrho_{-\varkappa} + u_\varkappa\, q_{-\varkappa})\, Q_\varkappa , \tag{40.3}$$

where $M_\varkappa$ is defined by (37.15) and $u_\varkappa$ is to be determined. This transformation gives new plasma and phonon coordinates which represent combined electron

---

[1] D. PINES: Phys. Rev. **92**, 626 (1953); also article in Solid State Physics, Vol. I, editors F. SEITZ and D. TURNBULL, p. 367 New York 1955.

and ion motions. The new phonons describe a motion in which the field of the ions is shielded by motion of the electrons. The subsidiary condition (40.2) is transformed to:

$$e^{-iS/\hbar} P_{\varkappa} e^{iS/\hbar} \Psi = [P_{\varkappa} - i M_{\varkappa} \varrho_{-\varkappa} + u_{\varkappa} q_{-\varkappa}] \Psi = 0. \tag{40.4}$$

With approximations identical with those made by Pines, the Hamiltonian (40.1) is transformed to:

$$\left. \begin{aligned} H = &\sum_k E_k c_k^* c_k + \frac{1}{2} \sum_{|\varkappa| < \varkappa_c} \{P_{\varkappa}^* P_{\varkappa} + (\Omega_{\varkappa}^2 - u_{\varkappa}^2) q_{\varkappa}^* q_{\varkappa}\} + \\ &+ \frac{1}{2} \sum_{|\varkappa| < \varkappa_c} \{P_P^* P_{\varkappa} + (\omega_P^2 + u_{\varkappa}^2) Q_{\varkappa}^* Q_{\varkappa}\} + \sum_{|\varkappa| < \varkappa_c} \{v_{\varkappa}^i - i M_{\varkappa} u_{\varkappa}\} q_{\varkappa} \varrho_{-\varkappa} + \\ &+ \sum_{|\varkappa| < \varkappa_c} u_{\varkappa} \varrho_{\varkappa}^* Q_{\varkappa} - \sum_{\substack{|\varkappa| < \varkappa_c \\ k}} M_{\varkappa} \frac{\hbar \varkappa}{m} \cdot \left(k - \frac{1}{2} \varkappa\right) c_k^* c_{k-\varkappa} Q_{\varkappa} + \\ &+ \sum_{|\varkappa| > \varkappa_c} \{P_{\varkappa}^* P_{\varkappa} + \Omega_{\varkappa}^2 q_{\varkappa}^* q_{\varkappa} + 2 v_{\varkappa}^i q_{\varkappa} \varrho_{-\varkappa} + M_{\varkappa}^2 \varrho_{-\varkappa} \varrho_{\varkappa}\}, \end{aligned} \right\} \tag{40.5}$$

where $\omega_P = \sqrt{4\pi n e^2/m}$ is the plasma frequency. It has been assumed that $E_k = \hbar^2 k^2/(2m)$. The mass to be used is somewhat uncertain. A physical argument of Mott (unpublished) suggests that when the plasma frequency is large compared with the energy at the surface of the first Brillouin zone, the ordinary electron mass, not the effective mass in the first zone, should be used. Perhaps the most important approximation made in deriving (40.5) is the "random phase" approximation of Bohm and Pines.

The significance of the various terms in (40.5) is as follows. Those in the first line represent the energies of the individual electrons, the phonon field and the plasma field. The first term in the second line represents the electron-phonon interaction, the second the plasma-phonon interaction and the third the plasma-electron interaction. The last line represents terms for $|\varkappa| > \varkappa_c$ for which collective coordinates are not introduced. The last term in this sum represents a shielded Coulomb interaction between the individual electrons.

On comparing the electron-phonon interaction with that introduced in the self-consistent field method, we see that

$$v_{\varkappa} = v_{\varkappa}^i - i M_{\varkappa} \varrho_{-\varkappa}, \tag{40.6}$$

so that

$$v_{\varkappa}^{\varrho} = - i M_{\varkappa} \varrho_{-\varkappa}. \tag{40.7}$$

The $u_{\varkappa}$ are to be determined in such a way that the new plasma and phonon variables will be uncoupled and will represent independent oscillations. It is required that there be no coupling via the subsidiary conditions as well as in the Hamiltonian. The value of $u_{\varkappa}$ and the phonon frequency $\omega_{\varkappa}$ are determined by carrying out a canonical transformation which eliminates to a given order the electron-phonon interaction terms in (40.5). It is required that to the same order there be no coupling between phonons and electrons in the transformed subsidiary conditions, and this will be the case if the phonon variables no longer appear in the conditions in this order.

The canonical transformation is just that defined by (39.4) with $f(k, \varkappa)$ and $g(k, \varkappa)$ defined by (39.9) and (39.10). For $|\varkappa| < \varkappa_c$, $v_{\varkappa}$ is defined by (40.6), for $|\varkappa| > \varkappa_c$ by the method of Sect. 39. The transformed HAMILTONian is:

$$
\begin{aligned}
H = &\sum_{k,s} E_k c_k^* c_k + \sum_{|\varkappa| < \varkappa_c} \left\{ \frac{1}{2} P_\varkappa^* P_\varkappa + \omega_\varkappa^2 q_\varkappa^* q_\varkappa) + \frac{1}{2} \left(P_\varkappa^* P_\varkappa + (\omega_P^2 + u_\varkappa^2) Q_\varkappa^* Q_\varkappa \right) \right\} - \\
&- \sum_{\substack{|\varkappa| < \varkappa_c \\ k,s}} M_\varkappa \frac{\hbar \varkappa}{m} \cdot \left(k - \frac{1}{2} \varkappa \right) c_k^* c_{k-\varkappa} Q_\varkappa - \frac{1}{2} \sum_{|\varkappa| < \varkappa_c} v_{-\varkappa} \varrho_\varkappa c_k^* c_{k-\varkappa} g(k, \varkappa) + \\
&+ \sum_{|\varkappa| < \varkappa_c} \frac{1}{2} (P_\varkappa^* P_\varkappa + \omega_\varkappa^2 q_\varkappa^* q_\varkappa) + \sum_{|\varkappa| > \varkappa_c} \frac{1}{2} M_\varkappa^2 \varrho_\varkappa \varrho_{-\varkappa} - \\
&- \frac{1}{2} \sum_{\substack{|\varkappa| > \varkappa_c \\ k,s}} v_{-\varkappa}^i \varrho_\varkappa c_k^* c_{k-\varkappa} g(k, \varkappa) + \text{higher order terms.}
\end{aligned}
\qquad (40.8)
$$

The requirement that $q_\varkappa$ no longer appear in the subsidiary condition gives the following for $u_\varkappa$:

$$
u_\varkappa = -2 i M_\varkappa v_\varkappa \sum_k \frac{n(k - \varkappa) - n(k)}{E_{k-\varkappa} - E_k + \hbar \omega_\varkappa}. \qquad (40.9)
$$

The expression for $\omega_\varkappa (|\varkappa| < \varkappa_c)$ is:

$$
\omega_\varkappa^2 = \Omega_\varkappa^2 - u_\varkappa^2 + 2 v_\varkappa^2 \sum_k \frac{n(k - \varkappa) - n(k)}{E_{k-\varkappa} - E_k + \hbar \omega_\varkappa}. \qquad (40.10)
$$

The factors of two come from summing over spins. These expressions give values for $\omega_\varkappa$ and $v_\varkappa$ identical with those of NAKAJIMA. No exchange effects appear in the plasma treatment.

The result of the canonical transformation is to replace the electron-lattice interaction by an interaction between electrons described by

$$
H_2 = - \sum_{\substack{|\varkappa| < \varkappa_c \\ k}} \{ v_{-\varkappa} \varrho_\varkappa c_k^* c_{k-\varkappa} g(k, \varkappa) - \sum_{\substack{|\varkappa| > \varkappa_c \\ k}} v_{-\varkappa}^i \varrho_\varkappa c_k^* c_{k-\varkappa} g(k, \varkappa). \qquad (40.11)
$$

The sum over spins has been carried out. There is an important difference between the terms for $|\varkappa| < \varkappa_c$ for which plasma coordinates were introduced and those for $|\varkappa| > \varkappa_c$ treated by the NAKAJIMA transformation. The shielded interaction constant, $v_{-\varkappa}$, appears for the former while the unscreened interaction constant, $v_{-\varkappa}^i$ appears for the latter. The difference is particularly important for $\varkappa$ small, since $v_{-\varkappa}^i$ becomes infinite while $v_\varkappa$ approaches zero as $\varkappa \to 0$. If it were correct to use $v_{-\varkappa}^i$ for all $\varkappa$ it would mean that small values of $\varkappa$ give the most important contribution to the interaction energy. If, as the collective description indicates, one should use $v_{-\varkappa}$ for $\varkappa$ small, this is not true. We shall discuss this question further in the following section.

The interaction H_2 is similar to one derived by FRÖHLICH[1] without explicit introduction of COULOMB interactions. As he noted, the diagonal part of H_2, called E_2, represents an interaction between electrons in k-space nearly equivalent to a part of the ordinary second order perturbation theory energy he had earlier called "E_2" and had used as the basis for his theory of superconductivity [4]. For $|\varkappa| < \varkappa_c$,

$$
E_2 = \hbar^2 \sum_{k,\varkappa} \frac{|v_\varkappa|^2 \, n(k) \, (1 - n(k - \varkappa))}{(E_k - E_{k-\varkappa})^2 - \hbar^2 \omega_\varkappa^2}. \qquad (40.12)
$$

The second order energy, W, is the sum of E_2 and the change in zero point energy of the oscillators.

[1] H. FRÖHLICH: Proc. Roy. Soc. Lond., Ser. A **215**, 291 (1952).

In the usual perturbation theory derivation, virtual transitions are considered in which a phonon of wave vector $\varkappa$ is emitted and an electron is scattered from k to $k - \varkappa$. Summing over both spins, we find

$$W = 2 \sum \frac{|V_\varkappa|^2 \, n\,(k)\,(1 - n\,(k - \varkappa))}{E_k - E_{k-\varkappa} - \hbar\,\omega_\varkappa}. \tag{40.13}$$

The squared matrix element,

$$|V_\varkappa|^2 = |v_\varkappa|^2 \, q_\varkappa \, q_{-\varkappa} \tag{40.14}$$

is to be averaged over zero-point oscillations:

$$\langle q_\varkappa \, q_{-\varkappa} \rangle = \frac{\hbar}{2\,\omega_\varkappa}. \tag{40.15}$$

It may be verified that

$$W = \tfrac{1}{2}\,(\Omega_\varkappa^2 - \omega_\varkappa^2)\,\langle q_\varkappa \, q_{-\varkappa} \rangle + E_2. \tag{40.16}$$

The first term is essentially FRÖHLICH's energy E_1. The conclusion is that the usual second order perturbation theory, in which one uses the actual shielded matrix element appropriate to resistivity calculations, $V_\varkappa = v_\varkappa \, q_\varkappa$ and the actual vibrational frequency, $\omega_\varkappa$, is satisfactory[1]. The COULOMB terms introduce some difference for $|\varkappa| > \varkappa_c$ because of the difference between $v_\varkappa^i$ and $v_\varkappa$, but for these $\varkappa$ the shielding is not very important.

In (40.8) there is also a weak interaction between electrons and plasma, which may be eliminated in a manner similar to that used for electron-phonon interaction. This also leads to an interaction between electrons, actually considerably larger in magnitude than E_2, and one which can not be treated by perturbation theory methods, implying that the wave function derived from the collective model are not accurate. Because of the large energy of the plasma quanta, this interaction probably plays no role in superconductivity.

There remains in the final HAMILTONian a shielded COULOMB interaction between electrons, represented by $\sum_{|\varkappa| > \varkappa_c} \tfrac{1}{2} M_\varkappa^2 \, \varrho_\varkappa \, \varrho_{-\varkappa}$. This also may be treated by perturbation theory.

We shall discuss the convergence of the canonical transformation which eliminates first order terms in the electron-phonon interaction and leads to the interaction H_2 in the following section.

We shall not discuss here the complication introduced by the subsidiary condition on the wave function, which condition transforms to

$$(P_\varkappa - i\,M_\varkappa \, \varrho_{-\varkappa})\,\Psi = 0. \tag{40.17}$$

The problems introduced by this condition are similar to those in the absence of ion motion, and they have been discussed by BOHM and PINES[2]. The subsidiary condition is usually just ignored in getting an approximate wave function for the HAMILTONian, but it is doubtful whether this is really justified.

41. Convergence of the canonical transformation. The canonical transformation, S (39.4), may be thought of as introducing a new set of BLOCH functions which depend on the vibrational coordinates, and a new set of vibrational coordinates which depend on the electron coordinates. The expansion (39.2) of

[1] This analysis answers objections raised by G. WENTZEL: Phys. Rev. **83**, 168 (1951) and W. KOHN and VACHASPATI: Phys. Rev. **83**, 462 (1951). For earlier discussions of the problem, see K. HUANG: Proc. Phys. Soc. Lond. A **64**, 867 (1951) and also [1].

[2] D. BOHM and D. PINES: Phys. Rev. **92**, 609 (1953). — D. PINES: Phys. Rev. **92**, 626 (1953). See also questions raised by N. I. ADAMS jr.: Phys. Rev. **98**, 1130 (1955) and subsequent comment by PINES (submitted to Physical Review).

the new Hamiltonian in a power series in S will converge rapidly if a small number of terms are omitted from S. These are the terms for which the energy denominators are small. We shall show (1.2) that the omitted terms do not contribute appreciably to the matrix element and vibrational frequency, $v_\varkappa$ and $\omega_\varkappa$. It is, however, just these terms which are important for superconductivity. In his analysis of the problem, Fröhlich[1] suggested omitting these terms from the canonical transformation and treating them separately, and we shall follow this procedure here.

The expansion coefficients for a given electron state, $\boldsymbol{k}$, are $\hbar^{-1} f(\boldsymbol{k}, \varkappa) q_\varkappa$. For the present purpose, we may omit the small term $\hbar^2 \omega_\varkappa^2$ in the denominator of $f(\boldsymbol{k}, \varkappa)$ [Eq. (39.10)], and take

$$\frac{1}{\hbar} f(\boldsymbol{k}, \varkappa) q_\varkappa = \frac{v_\varkappa q_\varkappa}{E_{\boldsymbol{k}-\varkappa} - E_{\boldsymbol{k}}}. \tag{41.1}$$

The expansion will converge rapidly if

$$\sum_\varkappa \frac{|v_\varkappa|^2 \langle q_\varkappa q_{-\varkappa}\rangle}{(E_{\boldsymbol{k}-\varkappa} - E_{\boldsymbol{k}})^2} = \sum_\varkappa \frac{|V_\varkappa|^2}{(E_{\boldsymbol{k}-\varkappa} - E_{\boldsymbol{k}})^2} \ll 1. \tag{41.2}$$

We shall omit from the expansion those $\varkappa$ for which $|E_{\boldsymbol{k}-\varkappa} - E_{\boldsymbol{k}}| < \Delta E$. We shall show that ΔE may be taken sufficiently large to satisfy (41.2) and at the same time sufficiently small so that the omitted terms have a negligible effect on the calculation of $v_\varkappa$ and $\omega_\varkappa$.

An order of magnitude estimate may be obtained by assuming that $|V_\varkappa|^2$ is independent of $\varkappa$. The sum (41.2) then becomes:

$$|V_\varkappa|^2 \left\{ \int_0^{E_{\boldsymbol{k}}-\Delta E} \frac{N(E)\,dE}{(E-E_{\boldsymbol{k}})^2} + \int_{E_{\boldsymbol{k}}+\Delta E}^{E_{\max}} \frac{N(E)\,dE}{(E-E_{\boldsymbol{k}})^2} \right\} \ll 1, \tag{41.3}$$

which is nearly equivalent to:

$$N(E_{\boldsymbol{k}}) |V_\varkappa|^2 \ll \Delta E. \tag{41.4}$$

This condition sets a lower limit on ΔE. When summed by principal parts, the omitted terms do not contribute abnormally to $v_\varkappa$ and $\omega_\varkappa$; the error introduced is of the order of $\Delta E/E_F$. For most metals, the left hand side of (41.4) is of the order of the phonon energy, $\hbar\omega$. Since $\hbar\omega \sim 10^{-4} E_F$, we may choose ΔE large compared with $\hbar\omega$ and at the same time small compared with E_F.

It is the terms for which $\Delta E \ll \hbar\omega_\varkappa$ which are important for superconductivity. The criterion for superconductivity, which will be discussed later (Sect. 42) is essentially that the interaction be so large that the canonical expansion does not converge if these terms are included. These terms then require separate treatment, and they should provide the basis for an adequate theory.

c) Calculation of interaction energy.

42. Fröhlich's shell distribution. In his original article on a theory of superconductivity based on interaction between electrons and lattice vibrations, Fröhlich [4] assumed that electrons in the superconducting state take up a modified distribution in $\boldsymbol{k}$-space. He showed that if the interaction is sufficiently large, and if the interaction energy is calculated by use of second-order perturbation theory, the shell distribution illustrated in Fig. 20 will have a lower energy than the normal Fermi distribution.

---

[1] H. Fröhlich: Proc. Roy. Soc. Lond., Ser. A **215**, 191 (1952).

FRÖHLICH's theory is based on the energy E_2 of (40.12). It should be noted that this energy is independent of the excitation of the vibrational modes. FRÖHLICH's original definition of E_2 which we shall call E_{2F} included only the part of E_2 which represents an interaction between electrons in k-space:

$$E_{2F} = -\sum_{k,\varkappa} \frac{|v_\varkappa|^2 \, n(k) \, n(k-\varkappa)}{(E_k - E_{k-\varkappa})^2 - \hbar^2 \omega_\varkappa^2}. \tag{42.1}$$

This interaction is a repulsion if $|E_k - E_{k-\varkappa}| < \hbar\omega_\varkappa$; it is an attraction if $|E_k - E_{k-\varkappa}| > \hbar\omega_\varkappa$. The shell distribution will have lower energy if the decrease in interaction energy more than makes up for the increase in FERMI energy. The displacement in energy is of the order of $\hbar\omega_\varkappa$. FRÖHLICH assumed that $v_\varkappa$ is proportional to $\varkappa$ and neglected Umklapp processes. He expressed his results in terms of a parameter, F, which in our notation is equal to:

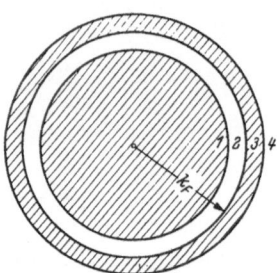

$$F = \frac{|V_\varkappa|^2 N(E)}{\hbar\omega_\varkappa}, \tag{42.2}$$

where $N(E)$ is the density of states of one spin at the FERMI surface. Since both $|V_\varkappa|^2$ and $\omega_\varkappa$ are assumed proportional to $\varkappa$, F is independent of $\varkappa$. For a metal of valence Z, the condition that the shell distribution have a lower energy than the normal one is[1]

$$F > \frac{2}{(4Z)^{\frac{1}{3}}} \qquad (4Z > 1). \tag{42.3}$$

This gives a criterion for superconductivity.

The value of F may be estimated empirically from the high-temperature resistivity. The exact relation depends in part on the particular assumptions made. The author[2] expressed an almost equivalent criterion in the forms:

$$\frac{\hbar}{\tau} > 2\pi\, kT; \tag{42.4}$$

$$10^{-6}\, \varrho\, n > 1 \qquad \left(\begin{array}{l} \varrho = \text{resistivity in esu at } 0°\text{ C} \\ n = \text{no. electrons cm.}^3. \end{array} \right) \tag{42.5}$$

In (42.4), τ is the relaxation time for high temperature conductivity (τ^{-1} is proportional to T). This simple criterion does, in fact, distinguish pretty well between superconductors and non-superconductors.

As we have noted earlier, FRÖHLICH's theory is in doubt because the criterion is essentially the same as the condition that perturbation theory break down. The difference in energy between the shell distribution and the normal distribution comes mainly from interactions between the shell and the gap (regions 2 and 3 in Fig. 20). These are states for which

$$\Delta E = |E_k - E_{k-\varkappa}| < \hbar\omega_\varkappa. \tag{42.6}$$

This condition emphasizes the dynamic nature of the interactions responsible for superconductivity, and is believed to be valid even though other features of the theory may not be. The condition for non-convergence of the perturbation

[1] A factor of two was omitted from Eq. (3.19) of FRÖHLICH's paper.
[2] J. BARDEEN: Phys. Rev. **80**, 567 (1950).

theory expansion for these terms is approximately

$$\sum_{\Delta E < \hbar \omega_{\varkappa}} \frac{|V_{\varkappa}|^2}{(\hbar \omega_{\varkappa})^2} > 1 \tag{42.7}$$

or

$$|V_{\varkappa}|^2 N(E) > \hbar \omega_{\varkappa}, \tag{42.8}$$

which is almost identical with (42.3).

The difference in energy per unit volume between the normal and the shell distribution is of the order of

$$\beta N(E) (\hbar \omega_m)^2 \tag{42.9}$$

where ω_m is the maximum frequency of the DEBYE spectrum and β is a numerical factor somewhat less than unity. Since ω_m^2 is inversely proportional to the mass M of the atom, the energy difference, and thus

$$\frac{H_c^2}{8 \pi} \sim \frac{1}{M}. \tag{42.10}$$

Since H_c is proportional to T_c, the isotope effect, $\sqrt{M} T_c \sim$ const, is predicted by the theory.

A major difficulty with the theory is that the predicted energy differences are too large. The natural energy which enters the theory is the phonon energy, $\hbar \omega$, which is much larger than $k T_c$. The numerical factor, β, is not sufficiently small to get agreement with observed energy differences, which are of the order of $N(E) (k T_c)^2$.

FRÖHLICH pointed out that in the true lowest state, more than one shell might be displaced from the FERMI surface. A very complete analysis of this problem, including the thermal properties, has been made by KITANO ([30], p. 919). He finds that over a certain range of the interaction parameter, F, an odd number of shells is most stable, over the remainder, an even number is the stable configuration. A phase transition, if one occurs at all, is of the first kind with a latent heat, and the transition temperature does not vary in the correct way with isotopic mass. Further, the specific heat in the superconducting state is nearly proportional to T, contrary to observation. Since the perturbation theory interaction on which these results are based is of doubtful validity, the significance of these results, other than to emphasize this fact, is questionable. Features of FRÖHLICH's theory which are likely to remain valid are (1) the criterion for superconductivity, (42.3) or (42.8) and (2) the dynamic nature of the interaction, indicated by (42.6).

It has not been shown from the perturbation theory approach how one might account for the electromagnetic properties of the superconducting state. SCHAFROTH[1], in fact, has calculated the diamagnetic properties from the model by a method similar to that of KLEIN discussed in Chap. III. The free energy of an infinite system in a magnetic field was expanded in a series in increasing powers of the interaction parameter, and the behavior of $K(q)$ in the limit $q \to 0$ was investigated. The definition of $K(q)$ is given by (20.1); it is the factor which relates the FOURIER components of current and vector potential. A perfect MEISSNER effect is obtained if $K(q)$ approaches a finite limit >0 as $q \to 0$. SCHAFROTH was able to show that $K(q) \to 0$ as $q \to 0$ in every order of the perturbation theory expansion. However, the possibility of a large, although not perfect

---

[1] M. R. SCHAFROTH: Helv. phys. Acta **24**, 645 (1951). — Nuovo Cim. **9**, 291 (1952).

diamagnetism is not ruled out. This may be all that is required to account for the experimental data.

Buckingham[1] has used the interaction E_2 in an attempt to account for anomalies in the low temperature specific heats of the normal phases of Na and Be. The interaction leads to changes in density of states near the Fermi surface and thus to departures of the electronic specific heat from the usual linear law. It is not all certain, however, that the predicted terms actually occur in normal metals. The author has shown [1] that the states for normal metals may be defined in such a way that the density of states in energy is hardly affected at all by the electron-phonon interaction. A better picture of the normal phase wave functions for large interactions is required before definite predictions regarding the electronic specific heat can be made.

43. Variational method. The author (see Sect. 35) used an approach somewhat different from that of Fröhlich, but the end results and the basic difficulties are much the same. The approach was more of an empirical one. It was assumed that the states of the individual electron in the normal phase include effects of weak coupling to virtual states of high energy, and that the main difference between superconducting and normal phases comes from virtual transitions in which energy difference between initial and excited states is small ($\sim \Delta E$). In the one-particle approximation, the result is a decrease in energy of electrons near the Fermi surface by $\sim \Delta E$. Since the number of electrons affected is $\sim N(E)\Delta E$, the total interaction energy per unit volume is $\sim N(E)\Delta E^2$. If it is assumed that these interactions do not occur in the normal phase, but only in the superconducting phase, this would represent the energy difference between the two phases at $T = 0°$ K. About the right order of magnitude is obtained if $\Delta E \sim kT_c$, but, as in Fröhlich's theory, estimated values of ΔE are more like an average phonon energy, $\hbar \omega$, so that the estimated energy difference is too large. It was suggested that a probable reason for this is that a large part of this interaction energy occurs in the normal phase; only a small part represents the true energy difference.

The value of ΔE was estimated by including only states which are coupled closely together by the interaction, the condition being that the interaction energy per electron of the states included be of the order of ΔE. The wave functions were taken to be of the form

$$\Psi_k = N \left(\psi_k + \sum_{k'} b_{kk'} q_{kk'} \psi_{k'} \right), \tag{43.1}$$

where N is a normalizing factor and $q_{kk'}$ is the amplitude of a vibrational mode which connects the Bloch states k and k'. The linear combinations are between states with energies within $\pm \Delta E$ of the Fermi surface, as illustrated by regions 2 and 3 of Fröhlich's shell structure (Fig. 20). Half of the states (unprimed k) are initial states; the other half (primed k) virtual states. Since, when the interaction is large, all states occur with a probability of about 50%, with $N = 1/\sqrt{2}$, it does not make much difference which states are taken as the initial states. They might be taken in region 2 (corresponding to Fröhlich's "normal" distribution), with k in 2 and k' in 3 or (corresponding to the shell distribution) with k in 3 and k' in 2. The coefficients $b_{kk'}$ are chosen by a variational method to give a minimum energy. When the linear combinations are formed, there are two states for each k, one having a lower energy, and the other a higher energy than the initial state. Since there are twice as many states included as there

---

[1] M. J. Buckingham: Nature, Lond. **168**, 281 (1951).

are electrons, only the low energy states need be occupied. The final result for the interaction energy is similar to that of FRÖHLICH; the isotope effect is predicted, but the magnitude of the energy difference is too large[1].

To account for magnetic properties, it was suggested[2] that electrons near the FERMI surface behave as if they have a very small effective mass,

$$m_{\text{eff}} \sim \frac{\Delta E}{E_F}\, m \sim 10^{-4}\, m,\qquad\qquad (43.2)$$

and that this gives a large diamagnetism and thus the MEISSNER effect (see Sect. 24).

As we have already mentioned, one major difficulty with the theory is that the interaction energy which occurs in the normal state for electrons with energies within ΔE of the FERMI surface is not taken into account properly. Probably associated with this difficulty is that the dynamic nature of the interaction (42.6) does not come explicitly into the theory. Undoubtedly it is necessary to go beyond the one-particle approximation to get a satisfactory description. Further, we need a better picture of what really constitutes the difference between superconducting and normal wave functions.

44. Other methods for calculation of electron-phonon interaction energies.
Attempts have been made to calculate interaction energies by use of a theory of TOMONAGA for a one-dimensional FERMI-DIRAC gas with interactions and also by use of a BLOCH-NORDSIECK transformation. These attempts are discussed in [1]. Neither one has been very successful. The TOMONAGA theory is not only restricted to one-dimension, but the basic approximation of the theory is not valid when the interactions are sufficiently strong to satisfy the criterion for superconductivity. The recoil of an electron on emission or absorption of a phonon is neglected in the BLOCH-NORDSIECK transformation. This is a good approximation only for phonons of very long wavelength (small ϰ).

The basis of the TOMONAGA theory[3] is to approximate the operator for the kinetic energy of the particles in a FERMI DIRAC distribution by an expression involving creation and destruction operators for electron-hole pairs. It is valid when the excitations are not too large and when the interactions between particles are of long range. The theory is related to the plasma oscillation theory of BOHM and PINES which we have used in Sect. 40.

The TOMONAGA theory was applied to the superconductivity problem independently by WENTZEL[4] and by DRESDEN[5]. WENTZEL attempted to show that the lattice would become unstable when the criterion for superconductivity is satisfied. He suggested that it might be necessary to introduce COULOMB interactions explicitly into the theory in order to bring about stability. As discussed in [1], we do not agree with this conclusion; rather we believe, as stated above, that the approximation of the kinetic energy is invalid for strong interactions.

Since we believe that some essential features and difficulties of the superconductivity problem are involved, a brief review will be given here. The treatment follows closely that given in [1]. Let dE/dn represent the spacing between the levels of the individual electrons near the FERMI surface of the one-dimensional model. Let a_n^* and a_n represent creation and destruction operators, as defined by

---

[1] For discussion of the problems involved, see reference [1], reference [27], p. 5 and J. BARDEEN: Phys. Rev. **82**, 978 (1951).

[2] J. BARDEEN: Phys. Rev. **81**, 5 (1951).

[3] S. TOMONAGA: Progr. Theor. Phys. **5**, 544 (1950).

[4] G. WENTZEL: Phys. Rev. **83**, 168 (1951).

[5] M. DRESDEN: Reference [27], p. 21.

TOMONAGA, for electron-hole pairs separated by n levels and corresponding to an excitation energy $n(dE/dn)$. The HAMILTONIAN for the electrons alone is then approximated by

$$H_e = \sum_n \hbar \Omega_n a_n^* a_n, \tag{44.1}$$

where $\hbar \Omega_n$ represents the energy

$$\hbar \Omega_n = n \frac{dE}{dn}. \tag{44.2}$$

The HAMILTONIAN H_e is that for a set of harmonic oscillators of frequency Ω_n, and can by a change of variables be written in the form:

$$H_e = \sum_n \left\{ \tfrac{1}{2} \left(P_n^2 + \Omega_n^2 Q_n^2 \right) - \tfrac{1}{2} \hbar \Omega_n \right\}. \tag{44.3}$$

We suppose there is a lattice vibration of frequency ω_n which connects electron states separated by n levels. The HAMILTONIAN for the vibrational modes alone is:

$$H_L = \sum_n \tfrac{1}{2} \left(p_n^2 + \omega_n^2 q_n^2 \right). \tag{44.4}$$

Finally the interaction terms are given by

$$H_I = \sum_n c_n q_n Q_n. \tag{44.5}$$

The complete HAMILTONIAN, the sum of (44.3), (44.4) and (44.5), is that of a set of coupled harmonic oscillators. It can be reduced to diagonal form by an appropriate transformation. The frequencies are given by the roots of

$$(\lambda_n^2 - \Omega_n^2)(\lambda_n^2 - \omega_n^2) = C_n^2, \tag{44.6}$$

the solution of which is

$$\lambda_n = \frac{1}{\sqrt{2}} \sqrt{\Omega_n^2 + \omega_n^2 \pm \sqrt{(\Omega_n^2 - \omega_n^2)^2 + 4 C_n^2}}. \tag{44.7}$$

One root becomes imaginary when

$$C_n > \Omega_n \omega_n. \tag{44.8}$$

This condition is essentially the same as the criterion for superconductivity for this model. If V_n is the matrix element for the interaction for zero point oscillations,

$$C_n^2 = \frac{4 n}{\hbar^2} V_n \Omega_n \omega_n, \tag{44.9}$$

so that (44.8) becomes:

$$\frac{n V_n^2}{\hbar \Omega_n} > \frac{\hbar \omega_n}{4}, \tag{44.10}$$

or

$$N(E) V_n^2 > \hbar \omega_n / 4. \tag{44.11}$$

Except for a numerical factor, this is of the same form as FRÖHLICH's criterion for superconductivity (42.3).

If $\omega_n < \Omega_n$, it is the vibrational mode which is pushed down by the interaction and becomes unstable. On the other hand, if $\Omega_n < \omega_n$, it is the electrons which become unstable. This latter condition corresponds to $\Delta E < \hbar \omega$, i.e., that the difference in energies of the electron states be less than the phonon energy. As we have stated in Eq. (42.6), it is virtual transitions of this sort which are believed to give superconductivity. Actually, of course, the electrons do not become unstable; the apparent instability simply means that the description of the electrons in terms of an equivalent set of oscillators is inadequate.

The excitation of the electrons can not be considered small, and Tomonaga's theory is not applicable. A marked change in electron wave functions is indicated, but the theory cannot tell what will actually occur.

Another method, discussed in [1], is to eliminate the interaction terms by means of a Bloch-Nordsieck transformation. The basis of the approximation is neglect of recoil of the electron during emission or absorption of a phonon. This can be expected to be valid only for interaction with phonons of long wavelength, so that the velocity of the electron is not changed very much. The interaction need not be small.

The kinetic energy operator, $p^2/2m$ is replaced by a term linear in the momentum, $V \cdot p$, in which V represents a constant average velocity of the electron. The net result of the transformation is to replace the electron-phonon interaction terms by an interaction between electrons. The most important term is:

$$U(r_1, r_2, \ldots, r_n) = - \sum_{k,i,j} \frac{|v_\varkappa|^2 \cos[\varkappa \cdot (r_i - r_j)]}{\omega_\varkappa^2 - (V \cdot \varkappa)^2}, \qquad (44.12)$$

where $v_\varkappa$ and $\omega_\varkappa$ are the interaction potential and vibrational frequency as defined in Sect. 38. The diagonal terms of (44.12) give just the usual second-order perturbation theory expression for the interaction energy. This result gives some justification for use of the perturbation theory expression for large interactions, provided only that the phonon wavelengths are long. However, it is undoubtedly a poor approximation to keep only diagonal terms when the interaction is large. One should get a better solution of the Schrödinger equation in which the interaction between electrons is given by (44.12).

The nature of the interaction (44.12) has been discussed by Singwi[1]. Electrons near the Fermi surface move much more rapidly than the velocity of sound, S. One may consider the emission of phonons as a Cerenkov radiation, or as a "bow wave" of a projectile in air moving faster than the velocity of sound. The disturbance is confined to a wake, the angle of which is approximately $S/V \sim 10^{-3}$ radians. Carrying out the summation in (44.12) by taking principal parts, Singwi finds that the interaction energy of two electrons is indeed zero except when one is in the wake of the order. The interaction is positive (repulsive) and is a maximum at the boundary of the wake, where it becomes singular. Bohm and Staver[2] had suggested earlier that the wake nature of the interaction might be important. They proposed that chains of electrons might be formed in the superconducting state, with one electron following in the wake of another. Singwi also discussed this possibility. One difficulty with this picture arises from the uncertainty principle. As we have discussed earlier, there is good evidence that the wave functions of electrons in the superconducting state spread out over large regions, and it is difficult to picture them as representing localized and relatively non-interacting "chains".

The nature of the emission of sound waves from an electron in a metal has been discussed by Klein[3] from a somewhat different point of view. He suggests that what might be important for superconductivity is the relative value of the energy uncertainty, $\Delta\varepsilon$, as calculated from the relaxation time, and the excitation energy, ε, of the electron from the Fermi surface. The proposed criterion for superconductivity is $\Delta\varepsilon/\varepsilon \sim 1$ for ε small, of the order of kT_c, and $\Delta\varepsilon/\varepsilon$ should then decrease toward zero as ε increases. However, the usual conductivity

[1] K. S. Singwi: Phys. Rev. 87, 1044 (1952). — K. S. Singwi and B. M. Udgaonkar: Phys. Rev. 94, 38 (1954). Some errors in the first of these are corrected in the second.
[2] D. Bohm and T. Staver: Phys. Rev. 84, 836 (1951).
[3] O. Klein: Ark. Fysik 5, 459 (1952).

theory gives an increase in $\Delta\varepsilon/\varepsilon$ as ε increases. Klein suggests that perhaps the compensation of the ionic motion is not as complete as predicted by the theory, as a result of a small part of the electron density remaining stationary as the ions move. This would give a small contribution to $v_\varkappa$ varying as $1/\varkappa$ rather than as $\varkappa$, and would thus give a large scattering probability for $\varkappa$ very small. This is given as a purely *ad hoc* assumption, with no physical basis.

45. Fröhlich's one-dimensional model. Fröhlich[1] has investigated a one-dimensional Fermi-Dirac gas of free electrons interacting with lattice displacements. He finds that the vibration corresponding to the wave vector which connects states at the top of the Fermi distribution may become unstable so as to create a finite sinusoidal displacement of the lattice and an energy gap in the electron states at the Fermi surface. There are just enough electrons to fill the states below the gap. These features are reminiscent of the author's early theory based on a three-dimensional model with energy gaps at the Fermi surface, as discussed briefly in Sect. 35. There are, however, some differences. In the one-dimensional case, an energy gap can be formed even though the interaction is fairly weak. Another difference is that Fröhlich considers the slow motion of the lattice wave together with the electrons through the crystal[2]. This corresponds to a displacement of the entire system of electrons plus lattice wave in **k**-space. There is a current flow associated with a displacement of this sort. This differs from a displacement of the electrons in **k**-space, with the lattice distortion kept fixed. Since the band is completely filled, this latter would not lead to a new state and there would be no current flow. In contrast, a moving sinusoidal lattice wave carries the electrons along with a resultant current flow.

The calculation of the energy of the system was made by an adiabatic self-consistent field method. Fröhlich finds the following for the energy gap, W

$$W = 8E_F\,e^{-\frac{3}{2ZF}},\qquad(45.1)$$

where E_F is the Fermi energy, Z the number of free electrons per atom and F the interaction parameter defined in an analogous way to that for the three-dimensional case (Sect. 42). In order that the approximations of the theory be valid, it is required that

$$E_F > W > \hbar\omega_{max},\qquad(45.2)$$

where $\hbar\omega_{max}$ is the maximum phonon energy. If V is the velocity with which the system moves, the energy is

$$E = n\left(\frac{1}{3}E_F - \frac{1}{8}\frac{W^2}{E_F} + \frac{1}{2}(m+m_1)V^2\right),\qquad(45.3)$$

where m_1 is an effective mass defined by

$$m_1 S^2 = \frac{3W^2}{4E_F Z F}.\qquad(45.4)$$

Here S is the velocity of sound. Condition (45.2) implies that m_1 will generally be larger than m.

Since the whole configuration moves with only a single degree of freedom, Fröhlich suggests that scattering to decrease the current would be unlikely, so that the current should be stable.

[1] H. Fröhlich: Proc. Roy. Soc. Lond., Ser. A **223**, 296 (1954).
[2] The general theory of coupled electron-lattice motion of this sort has been discussed by L. Brillouin: Proc. Nat. Acad. Sci U.S.A. **41**, 401 (1955).

KUPER[1] has calculated the thermal properties of the one-dimensional model. He finds that the energy gap decreases with increasing temperature, and goes to zero at a critical temperature, T_c. However, the approximations made in the theory are not valied unless T_c is larger than the DEBYE temperature, Θ_D, for the model. In actual superconductors, of course, T_c is much smaller than Θ_D. He looked into the question of stability of currents corresponding to the disdlacements discussed above, but was unable to come to a definite conclusion.

The model used is an unrealistic one, which makes it difficult to tell how much these results mean when applied to actual superconductors. The instability of the lattice for relatively weak interactions is true only for the one-dimensional case. As mentioned in Sect. 35, it appears that in three dimensions, the energy gained by the electrons when a large number of energy gaps is formed at the FERMI surface is considerably less than the energy required to deform the lattice. However, it is very likely true that there also exists for the general case a current flow in which the whole configuration of the superconducting state is displaced in k-space. It should be remembered, however, that the wave packets for the superconducting state extend over large volumes of real space, so that any such currents will also extend over large volumes. It is this feature which makes it difficult for the electrons to respond to a magnetic field as do classical electrons, and which leads to a large diamagnetism. While currents of this sort probably do not play much of a role in the MEISSNER effect, they may be important for persistent currents in wires of small cross-section, where the current density does not vary much over the area.

46 Concluding remarks. While there is some qualitative understanding of the nature of the superconducting state, we still do not have a good mathematical theory, or even a good physical picture of the difference between the normal and superconducting states. A superconductor is an ordered phase in which quantum effects extend over large distances in space, distances of the order of 10^{-4} cm. in pure metals. It is this large extent of the wave packets which undoubtedly accounts for the remarkable magnetic properties. As is the case for other second-order phase transitions, a superconductor is probably characterized by some sort of an order parameter which goes to zero at the transition point. However, experimental evidence for an order parameter is inconclusive, and we do not have any understanding at all of what the order parameter represents in physical terms.

The isotope effect shows that superconductivity arises from interactions between electrons and lattice vibrations, and theory indicates that, when the electron-lattice interaction is large, one can expect a marked change in the electron wave functions. Better mathematical methods for treating large interactions are required. The TOMONAGA intermediate coupling theory has been applied with success to the polaron problem[2] of an electron moving in an ionic crystal, and there is hope that some such method might be applied to electrons in a metal.

One of the major difficulties is to isolate the interactions responsible for superconductivity. The energy difference between the normal and superconducting phases is only a very small part of the total electron-phonon interaction energy. Theory indicates that the interactions responsible are those for which the difference between the energies of the electron states, ΔE, is less than the phonon energy, $\hbar\omega$. However, even if we consider only these, the energy involved in

---

[1] C. G. KUPER: Proc. Roy. Soc. Lond., Ser. A **227**, 214 (1955).

[2] T. D. LEE, F. E. Low and D. PINES: Phys. Rev. **90**, 297 (1953). — F. E. Low and D. PINES: Phys. Rev. **91**, 193 (1953). — T. D. LEE and D. PINES: Phys. Rev. **92**, 883 (1953). — H. FRÖHLICH: Adv. in Phys. **3**, 325 (1954).

the phase transition is still only a small fraction of the total. It is quite possible that the significant values of ΔE are of the order of $k T_c$ but, if so, we do not know why $k T_c$ is so much smaller than $\hbar\omega$. A possible explanation is that relatively long wavelength phonons are involved. This would be expected if the inter-action energy is calculated by a self-consistent field method, such as that of NAKAJIMA, rather than by the collective description. It will be recalled that $v_{\varkappa}^i$, which becomes large for small $\varkappa$, appears in H_2 (40.11) for those terms calculated by the NAKAJIMA method, while the screened interaction, $v_{\varkappa}$, appears for those terms treated by the collective description. Although the plasma treatment we have discussed may not be completely valid, one would expect on physical grounds that only screened interactions actually occur. Evidence that short wavelength phonons are important is that the transition temperature of thin films does not vary much with film thickness, even when the thickness is only a few atom layers [24].

It is probable that the vibrational modes are not affected very much by the phase transition. The energy difference is only a small fraction of the total zero-point energy of the modes. While it is possible that a small number of the modes might be affected to a large extent, this is not likely because one would expect that a large fraction of the modes participate in the transition. If this conclusion is correct, one should be able to treat the vibrational coordinates by perturbation theory methods, if not the electrons[1]. In this case, one could, by an appropriate canonical transformation, replace the electron-phonon inter-action by an interaction between electrons. Thus one would take an interaction such as that given by H_2 (40.11) seriously, and attempt to get a good description of the electron wave functions for a HAMILTONian with this interaction term. It would not be a satisfactory approximation to keep only the diagonal terms as is done in perturbation theory. In this way one would replace the electron-phonon interaction problem by the still difficult problem of treating a FERMI-DIRAC gas with interactions so large that they cannot be treated by perturbation theory methods.

We would expect to find as a consequence of an adequate theory a justification for the energy-gap model. An essential difference between the normal and superconducting state appears to be that in the latter a finite energy, ε, is required to excite an electron. The magnetic properties can be determined by perturbation theory methods, as discussed in Chap. III. A non-local theory, perhaps similar to that suggested by PIPPARD, would probably result; the LONDON theory would represent only a limiting case not actually attained. Relaxation processes at high frequencies would depend on the details of the model.

A framework for an adequate theory of superconductivity exists, but the problem is an extremely difficult one. Some radically new ideas are required, particularly to get a really good physical picture of the superconducting state and the nature of the order parameter, if one exists.

The author is indebted to A. B. PIPPARD and J. M. BLATT for stimulating correspondence about some of the controversial questions treated here and to J. R. SCHRIEFFER for aid with the manuscript.

General references.

[1] BARDEEN, J.: Rev. Mod. Phys. 23, 261 (1951). A review of attempts which have been made to calculate electron-phonon interaction energies for application to superconductivity.
[2] BARDEEN, J., and D. PINES: Phys. Rev. 99, 1140 (1955). Formulation of the electron-phonon interaction problem, including effects of COULOMB interactions.

[1] See the discussion by J. BARDEEN, reference [30], p. 913.

[3] BURTON, E. F., and others: The Phenomenon of Superconductivity. Toronto 1934.

[4] FRÖHLICH, H.: Phys. Rev. **79**, 845 (1950). The basic paper of FRÖHLICH's theory.

[5] GINSBURG, W. L.: Fortschr. Phys. **1**, 101 (1953). A review of theories of superconductivity, with consideration of the spectrum of elementary excitations. Good bibliography, particularly of work done in USSR.

[6] GINSBURG, W. L., and L. D. LANDAU: Ž. eksper. teor. Fiz. **20**, 1064 (1950). An extension of the LONDON phenomenological theory to take into account a space variation of the order parameter. — GINSBURG, W. L.: Nuovo Cim., Ser. II **2**, 1234 (1955).

[7] GORTER, C. J. (editor): Progress in Low Temperature Physics, vol. I. New York 1955.

[8] GORTER, C. J., and H. B. G. CASIMIR: Phys. Z. **35**, 963 (1934). — Z. techn. Phys. **15**, 539 (1934). — Physica, Haag **1**, 306 (1934). Thermodynamic relations and the two-fluid model.

[9] HEISENBERG, W.: Two Lectures, Cambridge, 1948. One of the lectures is a review of the HEISENBERG-KOPPE theory.

[10] KOPPE, H.: Fortschr. Phys. **1**, 420 (1954). A review of the phenomenological theory.

[11] KOPPE, H.: Ergebn. exakt. Naturw. **23**, 283 (1950). A review, based in large part on the HEISENBERG-KOPPE theory.

[12] KLEIN, O.: Ark. Mat., Astronom. Fys. Ser. A **31**, No. 12 (1944). — KLEIN O., and J. LINDHARD: Rev. Mod. Phys. **17**, 305 (1945). Calculation of diamagnetic properties of an electron gas with applications to superconductivity.

[13] LANDAU, L. D.: Phys. Z. Sowjet **11**, 129 (1937). Unbranched model of intermediate state.

[14] LONDON, F.: Une conception nouvelle de la supraconductibilité. Paris 1937. Review of phenomenological theory.

[15] LONDON, F.: Superfluids, vol. I. New York 1950. Macroscopic theory of superconductivity. A basic source for the present article.

[16] LONDON, H., and F. LONDON: Proc. Roy. Soc. Lond., Ser. A **149**, 71 (1935). — Physica, Haag **2**, 341 (1935). Basic papers of the LONDON phenomenological theory.

[17] MEISSNER, W.: Ergebn. exakt. Naturw. **11**, 219 (1932).

[18] MEISSNER, W.: Handbuch der Experimentalphysik, vol. 11 (pt. 2), 204 (1935).

[19] MENDELSSOHN, K.: Rep. Progr. Phys. **12**, 270 (1949).

[20] PIPPARD, A. B.: Proc. Roy. Soc. Lond., Ser. A **216**, 547 (1953). Basis for PIPPARD's non-local phenomenological theory.

[21] PIPPARD, A. B.: Adv. Electronics a. Electron Physics **6**, 1—45 (1954).

[22] SERIN, B.: Superconductivity, Experimental Part, in this volume.

[23] SHOENBERG, D.: Superconductivity, 2nd Ed. Cambridge 1952. An excellent introduction to the subject, with a very complete bibliography.

[24] SHOENBERG, D.: Nuovo Cim. **10**, 459 (1953). A review of work on superconductivity in the USSR.

[25] LAUE, M. VON: Theorie der Supraleitung, 2. Aufl. Berlin-Göttingen-Heidelberg 1949. English translation by L. MEYER and W. BAND, New York, 1952. A very complete account of the phenomenological theory.

Conference Proceedings (since 1949).

[26] 1949. International Conference on Low Temperatures, M.I.T., Cambridge, Mass.

[27] 1951. Low Temperature Symposium, National Bureau of Standards, Circular 519, Washington, 1952.

[28] 1951. Oxford Conference on Low Temperatures.

[29] 1953. LORENTZ-KAMERLINGH ONNES Conference, Physica, Haag **19**, No. 9 (Sept. 1953). Short papers followed by discussions.

[30] 1953. International Conference on Theoretical Physics, Kyoto and Tokyo, 1954.

[31] 1953. International Conference on Low Temperature Physics, Houston, Texas.

[32] 1955. Ninth Congress of the International Institute of Refrigeration, Paris.

[33] 1955. Conference on Low Temperature Physics and Chemistry, Baton Rouge, La. USA.

Liquid Helium.

By

K. MENDELSSOHN.

With 101 Figures.

Introduction.

The phenomenon of superfluidity, like that of superconductivity occupies a unique position in the pattern of our known physical world. While at first there was a tendency to regard the unusual effects which were discovered as a limiting aspect of the properties of aggregate matter at very low temperatures, it is now quite clear that they have a more profound significance. The fact that these highly ordered states should make their appearance at temperatures which are two or three orders of magnitude smaller than the condensation of gases into the liquid and solid states may be nothing more than an accident due to the particular physical conditions obtaining on the surface of the earth. It is not at all impossible that in the universe as a whole the aggregation of matter may proceed more generally according to a pattern in which ordering of velocities takes precedence over ordering of positions. Thus, when discussing the behaviour of liquid helium, it is well to remember that the phenomena discovered so far may represent only a very limited aspect of a new pattern of assemblies of inter-acting particles. The peculiar analogy between superconductivity, an aggreg-ation of charged light particles, obeying FERMI-DIRAC statistics, and liquid helium, an assembly of uncharged atoms, following BOSE-EINSTEIN statistics, seems to emphasize the fundamental nature of the new state. The fact that these two rather dissimilar assemblies should follow the same pattern indicates that the pattern itself must be remarkably general. It probably has the same kind of generality as a crystal which may be brought together by a variety of forces such as ionic, VAN DER WAALS or exchange interaction and may be composed of quite different atoms but nevertheless has always the same basic properties.

The lack of completeness in our knowledge of superfluidity and our inability to understand the significance of its pattern make a systematic survey difficult. So far we do not know whether all essential phenomena have been observed and it is impossible to assess the relative importance of those which have been observed. In its short history the accepted ideas about liquid helium have changed profoundly on more than one occasion because new evidence had been obtained or old evidence had been regarded in a new light. In several instances the following up of some chance observation, sometimes disregarded or forgotten for a decade, has altered the picture completely. In these circumstances it would be unduly presumptuous to base an account of liquid helium on what appears essential to the author at the time of writing. The information has therefore been presented first in the form of a fairly detailed historical survey and a number of subsequent chapters in which the knowledge of the various phenomena has been brought up to date. It is hoped that in this way no observational fact which might gain particular importance in future work has been omitted. Even so, the reader is warned that for serious work on the subject he should have recourse to the considerable number of detailed summary articles and to original papers.

The literature on liquid helium is very large and cannot be easily subdivided into important and less important contributions. For the reasons mentioned above it is always possible that some early and obscure paper may contain information of considerable value which has been disregarded by later workers. In order to make the present article manageable, references have been restricted to work mentioned in the text. For the remainder, reference should be made to summaries with exhaustive literature index which have been listed at the end of this article.

A. Historical survey.

The lines of the helium spectrum were first seen by a number of observers of the sun's atmosphere in 1868, and in the following years they were ascribed to a new element which had not yet been found on earth. The first terrestrial occurrence of the element was discovered by RAMSAY who in 1895 separated a small quantity of the gas from uranium bearing minerals. Five years later, he and TRAVERS showed that helium failed to be condensed in liquid hydrogen and therefore had a lower boiling point than the latter. From a number of experiments in which samples of helium were compressed and expanded at low temperatures and from measurements of the gas isotherms the boiling point was estimated by various authors to lie below 6 °K. In only one observation, made by KAMERLINGH ONNES, was there reliable evidence of a mist of liquid drops.

1. First liquefaction and solidification. One obstacle to large scale liquefaction was the scarcity of the new element. The abundance of helium in the earth's atmosphere is about 0.0005 volume percent and its separation from the air requires considerable quantities of liquid hydrogen. Monazite, from which the gas for the first liquefaction was obtained contains about 1 to 2 cm^3 of gas per gm. It was only after the large scale extraction of helium from certain well gases that it became generally available.

In the first successful liquefaction of helium in 1908, KAMERLINGH ONNES used the conventional type of LINDE-HAMPSON cycle based on the JOULE-KELVIN effect. For experiments on a smaller scale this method was supplemented by helium liquefaction due to desorption cooling and by single adiabatic expansion, both these methods being due to SIMON. In 1934 KAPITZA liquefied helium by cooling the gas in a reciprocating expansion engine and this method has more recently been adapted by COLLINS to a form of helium liquefier which is commercially available[1].

On the same day on which KAMERLINGH ONNES first liquefied helium [1], he tried, by reducing the vapour pressure above the liquid, to reach the triple point. This and subsequent attempts of the same kind failed, and it became clear that helium under its own saturation pressure will remain liquid at all temperatures below the critical point. The main object of this work was to reach as low a temperature as possible and to determine the vapour pressure curve over the region investigated. The actual temperatures reached by pumping off the vapour are subject to a small degree of uncertainty, considering the temperature scale used by the authors. The early results were re-calculated by KEESOM on the basis of the "1932 scale" and yield the following picture. In his first liquefaction, on July 10th, 1908, KAMERLINGH ONNES reached a temperature of 1.72 °K. In this and the following three attempts in 1909, 1910 and 1919 mechanical

---

[1] For more details cf. Vol. XIV, articles by DAUNT and COLLINS, this Encyclopedia.

pumps were used and the temperatures attained were 1.38, 1.04 and 1.00 °K respectively. Using diffusion pumps he reached in 1922 a vapour pressure of 0.013 mm Hg, corresponding to a temperature of 0.83 °K, and ten years later, Keesom succeeded in pumping helium down to 0.71 °K.

The success of the magnetic cooling method in the following year, 1933, diminished interest in the attainment of very low temperatures with helium, but the pumping off experiments had clearly demonstrated that down to less than a seventh of its critical temperature, helium retained its liquid state of aggregation. This did not, of course, preclude the possibility of a triple point below that temperature and for a satisfactory solution of the problem, the melting curve had to be investigated. In 1926 Keesom [2] used a cryostat whose temperature could be changed and into which a strong-walled capillary containing helium under pressure was immersed. He found that at the boiling temperature the capillary was blocked at a pressure of 128 kg/cm² but free at 126 kg/cm² and concluded that the melting pressure of helium must lie between these values. Since the melting pressures in the temperature range where the liquid is stable are not excessively high, the same author constructed a cryostat embodying a pressure container made from glass which permitted visual observation of the melting process [3]. In it an iron stirrer could be moved by an external magnet and it was observed that, as the melting curve was passed, the stirrer lost its mobility. This was the only visual indication that solidification had taken place since solid helium turned out to be perfectly transparent.

2. The diagram of state. The melting curve above 2.5 °K was found to rise rapidly with temperature and extrapolation of this section to lower temperatures would not preclude the possible existence of a triple point. However, below this temperature, the measured values of the melting curve showed a surprising deviation to higher pressures from any such extrapolation (Fig. 1). Indeed, Tammann [4] showed that the relation between the melting pressure (in atmospheres) and the absolute temperature can be expressed as:

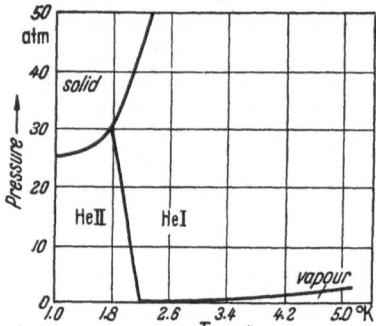

Fig. 1. Phase diagram of liquid helium.

$$T - 1 = \log (p - 24.0) \qquad (2.1)$$

thereby indicating that even at absolute zero a pressure of 24 atmospheres would be required to force helium into the solid state.

The realisation that helium has a diagram of state which differs essentially from that of any other substance by the fact that solid and vapour phase cannot co-exist provided the first indication of the unique position of this substance. Formerly it had been supposed that owing to the symmetry of the constituent particles, helium would serve as an ideal model substance for investigations of the solid, and particularly of the liquid, state. The unusual shape of the melting curve and the obvious absence of a triple point showed clearly that helium was not a very representative liquid and that some new factor, which was not operative in "normal" substances, had to be taken into account.

An indication of this new factor was actually apparent in Kamerlingh Onnes' first experiment of July 10th, 1908. He had then made a rough determination of the liquid density and found it to be about 0.15, an astonishingly low value. The great number of new facts discovered then and shortly afterwards seem to have, however, overshadowed the smallness of the absolute value and an explanation of the unusual equilibrium between kinetic energy and inter-action forces had to wait until 1923. At that time BENNEWITZ and SIMON discussed the deviation from TROUTON's rule in hydrogen with reference to the zero point energy. SIMON [5] now applied these considerations to the case of helium in which he postulated the zero point energy to be so high that it would prevent solidification. Expressing the extension of TROUTON's rule as

$$(L + E_0)/T = \text{const} \tag{2.2}$$

where L is the latent heat of evaporation and E_0 the zero point energy, he deduced for the latter in the case of helium a value of 64 cal/mol. His postulate that the high zero point energy of the substance prevents its solidification under satura-tion pressure emphasized the unique position of liquid helium. It also explained that this unique position is directly due to the influence of the quantum prin-ciple. Six years later, WOHL [6] correlated the measured density of the liquid with the very much higher estimate of its density from the gas kinetic diameter of the helium atom and the application of the law of corresponding states. The introduction into this law of a quantum parameter such as was introduced into the extension of TROUTON's rule proved of great success in DE BOER's prediction of the vapour pressure curve of the light helium isotope.

It is as well to separate this general evidence for the fact that liquid helium is a "quantum fluid" from the anomalous behaviour which has attracted so much attention. Such evidence as exists at present suggests that failure to crystal-lize under its own vapour pressure owing to the influence of the zero point energy may be a phenomenon which is shared by both isotopes. The anomalous trans-port properties, on the other hand, may, possibly for quantum statistical reasons, be confined to the heavier isotope.

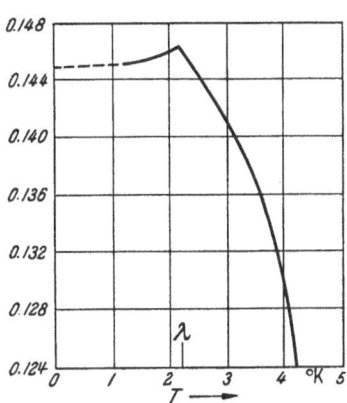

Fig. 2. Density of liquid helium.

3. The lambda-phenomenon. The discovery of the lambda-phenomenon, as the anomalous be-haviour of He⁴ in its liquid state has been called, has been a gradual process. This is not surprising if one realizes that the observed phenomena were to-tally unexpected and have no counterpart in the behaviour of any other liquid. The way in which the liquid density varies with temperature was first investigated by KAMERLINGH ONNES in 1911. Measurements were carried out between 1.5 and 4.3 °K which yielded the somewhat sur-prising result of a density maximum near 2.2 °K. The experiments were repeated with greater ac-curacy in 1924 in which this maximum was well defined [7] (Fig. 2). While the care takento eliminate errors due to a possible anomaly in the expansion of the glass vessel indicate the importance which KAMERLINGH ONNES attached to the result, its significance was not yet understood at the time. The analogy with the density maximum in water was clearly tempting enough to exclude other explanations.

The first realisation that the density maximum might indicate a more pro-found change in the liquid arose from a determination of the latent heat of va-porisation by DANA and KAMERLINGH ONNES in 1926 [8]. They found that this quantity which between 1.5 and 3.5 °K is in first approximation independent of temperature showed a slight minimum and that this minimum co-incided with the anomaly in the density curve (Fig. 3). The authors suggested that the two anomalies were possibly a sign of some discontinuous change taking place in the liquid at this temperature. Thus, in the last year of his life, Kamerlingh Onnes reported the suspected existence of two states of liquid helium.

At the same time Dana and Kamerlingh Onnes had carried out determinations of the specific heat of the liquid but their publication only lists values above 2.6 °K. It appears that work was also carried out by them at a somewhat lower temperature. Since, however, they had reason to believe that some of their results were falsified by secondary causes, they did not include in their report very high values of the specific heat obtained in this region.

A full investigation of the problem and a clear demonstration of the two states of the liquid was left to Keesom and his co-workers. First the dielectric constant of the liquid was measured and a change similar to that in the density

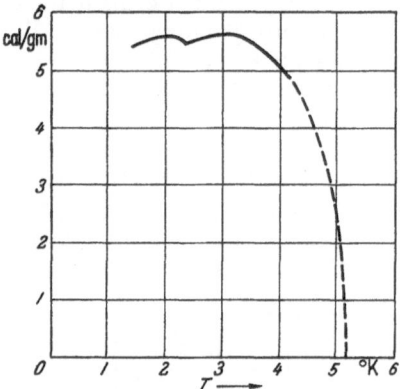

Fig. 3. Latent heat of evaporation.

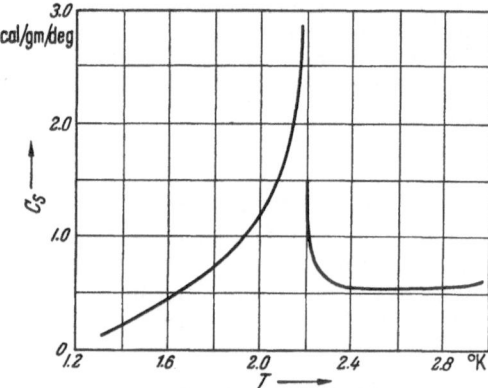

Fig. 4. Specific heat of liquid helium.

was found [9]. These observations were followed by the most important investigation on the static properties, the determination of the specific heat between 1.3 and 4.1 °K by Keesom and Clusius [10]. It revealed a large anomaly, somewhat resembling in shape the inverted Greek letter lambda from which the phenomenon has derived its name. Particular attention was devoted to the peak of the anomaly and it was found that, while a discontinuity occurs in the specific heat at about 2.19 °K, there is no latent heat connected with this transition (Fig. 4). The sharpness of the transition was estimated by later experiments of Keesom and Miss Keesom [11] to be within a few thousandths of a degree.

4. Helium I and Helium II. Using the same apparatus, Keesom and Clusius also investigated the position of the transition, called the lambda-point, as the pressure is changed. By taking cooling curves of the calorimeter filled with helium under more than saturation pressure, they observed that the lambda-point moved with rising pressure to progressively lower temperatures. The melting curve is reached at a temperature of 1.75 °K where the melting pressure is about 30 atmospheres. It is significant that this is the region in which the melting curve looses its downward slope and becomes temperature independent. The liquid region in the diagram of state of helium is therefore divided by the lambda-line into two completely separated regions. Following an early suggestion by Keesom and Wolfke, the two forms of the liquid above and below the lambda-line have been named liquid helium I and liquid helium II (Fig. 1).

Further work, especially by KEESOM and Miss KEESOM, to which we shall refer later has shown that all along the lambda-line the same condition holds as at saturation pressure, namely that the two liquids are not separated by a latent heat. This means that under equilibrium conditions liquid helium I and II can never be co-existent. Occasional reports by later observers who claimed to have seen a liquid-liquid boundary are probably erroneous.

The discovery of the lambdy-transition in liquid helium led EHRENFEST [12] to consider this type of transformation in more general terms. He proposed a distinction between different types of transformation according to the discontinuities in the derivatives of the thermal potential. He defined the order of a transformation by whether discontinuities will occur in the first, second or higher derivatives of the potential. Thus a change involving a latent heat, such as melting, has to be considered as of first order whereas the lambda-transformation is of the second order since no discontinuity exists in the thermal energy but only in the specific heat. For the change of the lambda-point with pressure this then leads to

$$\left(\frac{dp}{dT}\right)_\lambda = \frac{\Delta c_p}{T v \Delta a_p} \tag{4.1}$$

and

$$\left(\frac{dp}{dT}\right)_\lambda = \frac{\Delta (\partial v/\partial T)_p}{\Delta (\partial v/\partial p)_T} \tag{4.2}$$

where Δc_p and Δa_p are the discontinuities in the specific heat and the coefficient of expansion at the transition between helium I and II.

Since then, a number of careful measurements leading to the entropy diagram and the diagram of state of liquid helium have been carried out which will be discussed in detail later. This work did not lead to the discovery of any salient new facts, but it emphasized the curious position of the phase equilibrium between liquid and solid helium at low temperatures. According to the third law of thermodynamics the entropy of the liquid as well as of the solid phase must become zero at absolute zero. The lambda-anomaly in the specific heat of the liquid now indicates a rapid loss of entropy within a few tenths of a degree below the lambda-point. Quite apart from the interesting question about the way in which order is established in the liquid in this region, the entropy loss must make itself felt in the shape of the melting curve. The variation of the melting pressure with temperature, which according to the CLAUSIUS-CLAPEYRON equation is the ratio of entropy change to volume change, will become zero as the entropy difference between solid and liquid phase disappears. Therefore, as SIMON [13] has pointed out, the change in slope of the melting curve is closely bound up with the lambda-phenomenon since at these temperatures the entropy of the liquid drops to a value which is not far from that of the solid entropy.

At absolute zero, and effectively in the last 1.5° above it, the melting process of helium bears no similarity to that observed normally. Solidification will not take place on cooling but solely on the application of external pressure. The melting heat disappears and the phase change becomes a purely mechanical process at which no thermal change takes place.

5. Super heat conduction. What has been said about the discovery of the lambda-phenomenon is *a fortiori* true for the discovery of the anomalous transport effects. The appearance of a transformation in the liquid phase was unexpected and surprising, but the thermal effects showed at least some resemblance to transitions observed in the solid state. The transport effects, however, have no counterpart at all in physical observations, if we except the equally enigmatic phenomenon of superconductivity. It may appear astonishing that it took more than 25 years after the first liquefaction of helium for these striking phenomena to be recognized. However, as we shall see, curious facts were observed and recorded and other features of the experiments, which appear striking in retrospect, were passed over without comment. It is simply that the obvious conclusions to be drawn from these observations would make no sense, just as the now established effects still do not fit into the known pattern of the behaviour of aggregate matter.

A very conspicuous change takes place in the liquid when it is cooled through the lambda-point by pumping off the vapour. It must have been seen by a great number of observers many times and was, in fact, recorded in 1932 by McLennan, Smith and Wilhelm [14] who write: "... the liquid was watched closely as the triple point was approached. When a pressure of 38 mm was reached, the appearance of the liquid underwent a marked change, and the rapid ebullition ceased instantly. The liquid became very quiet and the curvature at the edge of the meniscus appeared to be almost negligible."

This sudden cessation of boiling is indeed quite a dramatic effect and has since been used generally to demonstrate the lambda-point to a large audience. Neither McLennan and his co-workers nor anyone else tried to interpret the effect, and it was not until the enormous increase in the heat conduction was found directly that the obvious connection between the two phenomena was realised. This failure to perceive the true reason for the change in the aspect of the liquid is clearly due to the fact that no mechanism was known by which the heat conductivity in a dielectric liquid can suddenly increase up to a million times.

The idea of a large heat conduction only occurred when the conclusion had become quite inescapable. In their calorimetric experiments Keesom and Miss Keesom [11] set out to determine the sharpness of the lambda-point from a plot of the actual observations in which the temperature rise of the calorimeter on heating was recorded. It was then found that at the lambda-point not only the rate at which heat is taken up changed but also the way in which it was taken up by the liquid (Fig. 5). Below the lambda-point the calorimeter temperature became instantly steady as soon as the heating current was switched off. Above the lambda-point over-heating was very evident. Since the shape of such records is a standard test in calorimetry for the equalisation of temperature, it now became clear that there existed a rapid change in heat conduction at the lambda-point. The magnitude of the change was, however, not yet appreciated as is evident from the fact that the first apparatus designed for a determination of the heat conduction proved completely unsuitable. In this first attempt Keesom and Miss Keesom [15] tried to measure the temperature difference at the faces of a flat disc of liquid. This method worked with helium I but when the apparatus was filled with helium II, the temperature difference proved unmeasurably small. Only when a narrow capillary had been substituted for the disc shaped chamber, could measurements be taken. Determinations at 1.4 and 1.75 °K yielded a value of approximately 190 cal/degree · cm which by far exceeds the heat conductivity of any other substance. It is about 200 times larger than that of copper at room temperature and three million times that of helium I.

Although the extreme magnitude of the heat conduction as discovered in these experiments in 1935 and 1936 only constitutes part of the heat flow phenomenon, it provided the stimulus for research into the transport phenomena. An important additional fact which had escaped notice in these first experiments was observed in the following year by Allen, Peierls and Uddin in Cambridge [16]. They, too, measured the heat conduction of helium II enclosed in a capillary but found that, besides being very large, it also depended on the temperature gradient. A little later the authors themselves called this result in doubt and possibly caused by a complicating effect. However, later work established that the value of the heat flow is not only influenced by the temperature gradient but also by the dimensions of the apparatus in which it is measured. The concept of "heat conductivity" in the accepted sense as a constant ratio of heat current density to temperature gradient has thus lost its usefulness when dealing with liquid helium II. Limiting the discussion to one capillary size and

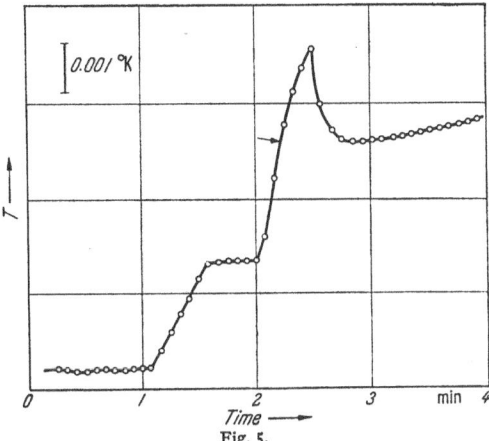

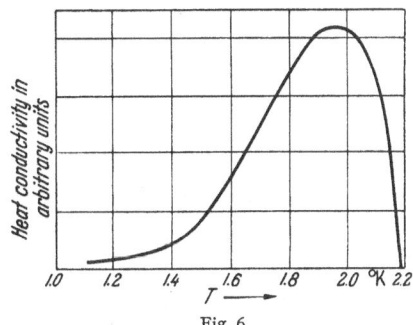

Fig. 5.

Fig. 6.

Fig. 5. The change in heat conduction on passing the lambda-point as shown by thermometer readings (in arbitrary units) in a measurement of the specific heat. The arrow marks the position of the lambda-point.

Fig. 6. Temperature dependence of the heat flow in liquid helium II under constant temperature gradient.

constant temperature gradient, it was found that on cooling below the lambda-point the heat conduction rises rapidly to a maximum at about 2 °K and then falls off again to lower temperatures (Fig. 6).

6. The thermo-mechanical effects. The complicating feature which made the interpretation of these results doubtful and which was traced shortly afterwards by ALLEN and JONES [17] turned out to be another unexpected effect which again is characteristic for helium II only. The heat conduction apparatus used in Cambridge consisted of a thermally insulated glass reservoir embodying a heater which communicated with the helium bath through a glass capillary (Fig. 7). The heat flow through the capillary was measured by applying a heating current and comparing the height of the meniscus in the reservoir with the level of the bath. Since the heated end of the liquid had a higher vapour pressure than the bath, the level in the reservoir was depressed and the degree of separation of the menisci therefore acted as a sensitive differential thermometer.

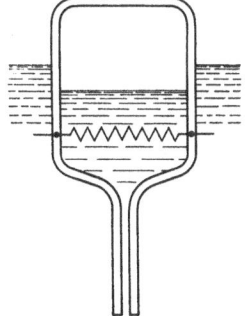

Fig. 7. Arrangement for measuring the heat conduction of liquid helium II (diagrammatical) used by ALLEN, PEIERLS and UDDIN.

At small heat currents the separation of the levels was very small and sometimes almost non-existent. This could be understood in terms of very large heat flow under small temperature gradients. However, under certain conditions of temperature gradient and absolute temperature, there also appeared to be cases in which the level in the reservoir showed a definite *rise* above the bath level. In terms of vapour pressure differences this would have meant that on supplying heat to the helium in the reservoir its temperature fell, and this was clearly non-sensical. The experiment was therefore varied by opening the top of the reservoir so that no difference in vapour pressure was possible. Repeating the heat flow experiment under these conditions (Fig. 8), yielded the astonishing result that on supplying heat, the meniscus in the reservoir rose above the bath level. The authors were able to enhance the effect very much by shining a light on a tube closely packed with emery powder which carried a fine nozzle projecting above the level of the helium bath. A free jet of liquid helium was then seen to rise as high as 30 cm into the vapour space above the bath level. This striking demon--

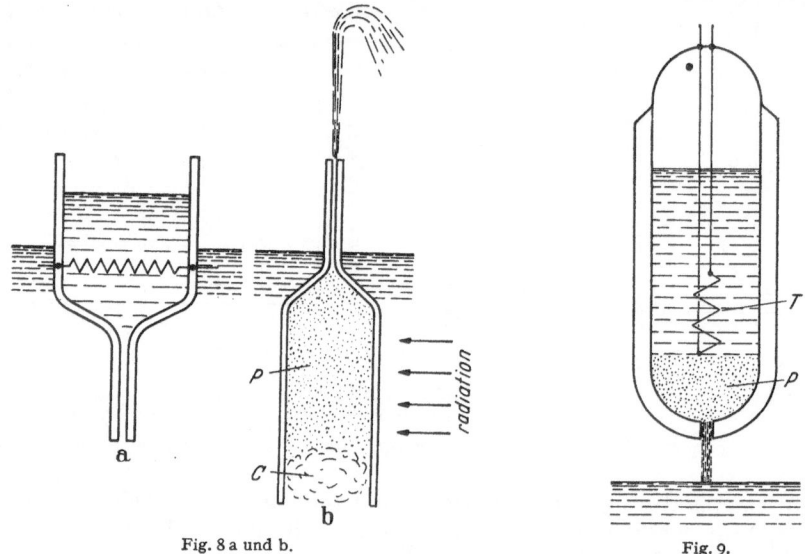

Fig. 8 a und b.　　　　　　　　　　　　　　　　　　　Fig. 9.

Fig. 8 a and b. The thermo-mechanical ("fountain") effect. (a) First observation, (b) fountain produced in a tube filled with fine powder (P) and closed by a cotton wool plug (C).

Fig. 9. The mechano-caloric effect.

stration earned the phenomenon the name "fountain-effect" which is still fre-quently used. However, the conditions in this demonstration experiments are somewhat complex and tend to obscure the true nature of the phenomenon which is presented in a much clearer form by the first experiment. We therefore prefer the descriptive term "thermo-mechanical effect" which now has come into general use.

The meaning of these observations is that as heat is supplied to the reservoir, a flow of liquid towards this supply of heat will take place. As the column of liquid at the heated end of the capillary rises, its weight will push helium back through the capillary into the bath and a dynamic equilibrium is established in which the thermo-mechanical flow is balanced by the return flow. The reason why a so much greater height of liquid column could be established in the fountain experiment, was evidently due to the return flow being largely inhibited by the flow resistance of the powder filled tube. Further experiments with a powder filled tube in which the height of the column of liquid was carefully measured showed that for constant temperature difference the reaction force showed a dependence on absolute temperature which was not unlike that of the heat conduction.

The discovery of the thermo-mechanical effect immediately suggested the possible existence of another phenomenon which would be its thermo-dynamical counterpart. The thermo-mechanical effect shows that in liquid helium II the establishment of a temperature difference will cause the appearance of a difference in fluid pressure. The question therefore arose whether the establishment of a pressure difference will set up a corresponding difference in temperature. A search for such a "mechano-caloric effect" was made in the following year by DAUNT and MENDELSSOHN in Oxford [18] (Fig. 9) who observed that flow of helium II from a higher to a lower level is indeed accompanied by a temperature gradient. Their experiment was carried out with a small Dewar vessel which was completely closed except for a small orifice at the bottom. The lower part of the vessel was filled with closely packed jeweller's rouge which formed a plug, P, with many fine channels and above this plug a resistance thermometer, T, was fitted. When the vessel was partly immersed in the bath of helium II the meniscus inside it adjusted itself eventually to the same height as the bath level and the temperature inside the vessels was the same as that of the helium bath. On withdrawing the Dewar vessel from the bath, liquid helium was seen to run out through the orifice at the bottom and it was observed that at the same time the temperature inside the vessel rose slightly. Conversely, if, starting from the equilibrium position, the vessel was lowered further into the bath so that liquid flowed in through the plug a fall in temperature was noted. This showed that the heat content of the fluid which had

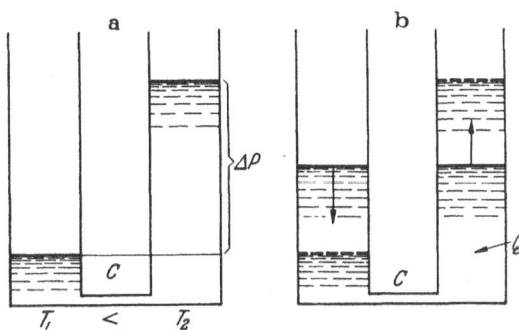

Fig. 10 a and b. The (a) mechano-caloric and (b) thermo-mechanical effects in two volumes of liquid helium II connected by a capillary link C.

passed through the plug was lower than under starting conditions and that the heat content of the fluid which was left behind rose accordingly. The powder plug was thus found to act as an "entropy filter" and rough estimate from the first experiment indicated that the entropy of the liquid which had passed through the filter was very small and possibly zero.

7. Superfluidity. While the high heat flow in helium II was the first indication of the existence of anomalous transport phenomena, the discovery of the thermo-mechanical effects was preceded by that of another unexpected phenomenon. On various occasions it had been noted that small leaks became noticeable in evacuated containers when these were immersed in helium II and it had, in fact, been suggested by KEESOM and KEESOM in 1932 that the viscosity of the liquid might decrease below the lambda-point. The first measurements of the viscosity were carried out three years later by WILHELM, MISENER and CLARK [19] in Toronto who made observations of the damping of an oscillating cylinder in liquid helium at four temperatures above and one just below the lambda-point. In their experiments the damping increased very appreciably from 4.2 to 2.4 °K and was then found to be lower again at 2.2(?) °K. BURTON [20] deduced from these data viscosity values of 110 micropoise at 4.2 °K, rising to 270 micropoise just above and then falling to 33 micropoise just below the lambda-point. This interpretation of the data has been called in doubt by a number of authors who drew attention to the likely occurrence of turbulence, but it seemed clear that the change at the lambda-point wa real. In 1938 KEESOM and MacWOOD [21] repeated the Toronto experiments, using an oscillating disc method which permitted somewhat easier interpretation. They found a gradual drop with falling temperature in helium I to be followed by a discontinuity at the lambda-point with a higher value for the viscosity in helium II and a subsequent rapid decrease as the temperature was lowered further. More recent measurements which will be discussed below have cleared up the discrepancies between these results in helium I and at the lambda-point. The observations of Keesom and MacWood confirm, however, the drop in viscosity below the lambda-point found by the Canadian workers.

Early in 1938 two short papers appeared in the same issue of "Nature", by Kapitza [22] and by Allen and Misener [23] respectively, in which the flow of helium II through narrow channels was described. In both cases the liquid was flowing from a raised glass reservoir under gravity back into the helium bath. The link between reservoir and bath used by Kapitza was the gap between two optically flat plates whereas Allen and Misener studied the flow through capillaries. The width of the flow channel was varied in the former experiment by inserting spacers in the gap and in the latter two different capillaries were employed. It was in these observations that a new striking phenomenon of liquid helium II was revealed which has become known as "superfluidity", a term suggested by Kapitza in this first paper. He found that when in his arrangement the glass plates were opposed without an intervening spacer so that the gap, as determined from the optical fringes, was of the order of 5×10^{-5} cm, the flow of helium I could just be detected over several minutes. In the helium II region, however, the whole reservoir was emptied within a few seconds. The drop in viscosity on passing the lambda-point was estimated to be at least 1500 times. A similar result was obtained by Allen and Misener who observed moreover in these and subsequent experiments that the flow velocity was greatly independent of the pressure difference and of the diameter of the capillary. Indeed, for the finer capillaries it was found that this velocity increased with decreasing capillary diameter.

8. Film transfer. In the same volume of "Nature" in which the discovery of superfluidity was reported, and again in the same issue two letters dealing with still another strange phenomenon in liquid helium II were published. The authors, Daunt and Mendelssohn [24] in Oxford and Kikoin and Lasarew [25] in Kharkov respectively described observations on the helium film. The first indication of a peculiar transport effect in liquid helium was observed even before the lambda-point was discovered. In 1922, when Kamerlingh Onnes [26] made an attempt to reduce further the temperature to which liquid helium could be cooled by pumping off the vapour, he employed an arrangement in which two concentric Dewar vessels were used (Fig. 11). The object of the experiment was to shield the liquid in the inner vessel thermally from the influx of radiation by the liquid in the outer vessel. He therefore expected the liquid surrounding the inner vessel to evaporate more rapidly and the meniscus in the inner vessel to remain higher. He found, however, that the liquid levels in both vessels fell at the same rate as vapour was drawn off. Moreover, when by shining a lamp on the cryostat the surrounding liquid was evaporated more rapidly and the outer meniscus fell below the inner one, the two levels re-adjusted themselves again to the same height once the lamp was removed. A similar adjustment in the opposite direction was observed after the outer level had been raised by scooping liquid from the inner into the outer vessel. Kamerlingh Onnes believed this effect to be a distillation phenomenon and the experiment was not repeated for another 16 years.

In 1932 Closs and Mendelssohn [27] reported in some detail on a disturbing effect in calorimetric measurements at helium temperatures which they traced to the evaporation of a layer of helium from the surface of the calorimeter. They noted that the effect only occurred when the calorimeter had been cooled to below ~2 °K. Another disturbing effect was observed by ROLLIN and SIMON [28], also in attempts at pumping down to a low temperature a cryostat filled with liquid helium II. They found that the rate of evaporation was much higher than was to be expected when allowance was made for the known sources of heat influx into the cryostat. From a series of experiments in which this effect was investigated they came to the conclusion that the inner wall of the tube connecting the cryostat with the pumping line was covered with a film of liquid helium. This work was carried out in 1936 in conjunction with experiments revealing the anomalously high heat conduction of liquid helium II. It was therefore only natural that these authors at first ascribed the high rate of evaporation from their cryostat to a high heat conduction of the helium film on the walls of the

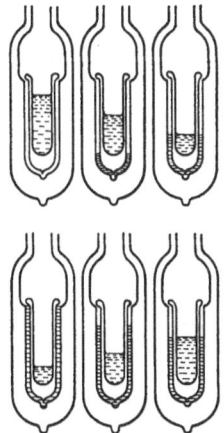

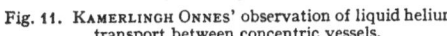

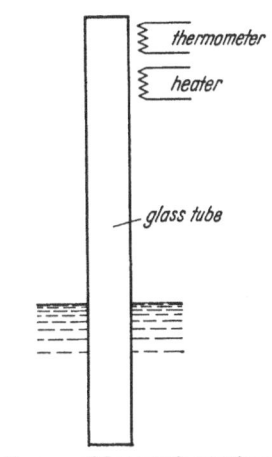

thermometer

heater

glass tube

Fig. 11. KAMERLINGH ONNES' observation of liquid helium transport between concentric vessels.

Fig. 12. KIKOIN and LASAREW's experiment on the helium film.

connecting tube. However, this process of heat influx into cryostats containing helium II is quite different as was demonstrated by the experiments of DAUNT and MENDELSSOHN as well as by the subsequent work of ROLLIN and SIMON [29].

The observation of KIKOIN and LASAREW followed a pattern somewhat similar to those of ROLLIN and SIMON. They used a glass tube whose lower end dipped into a bath of liquid helium II while at its upper end a heating coil and a thermometer were attached (Fig. 12). In the helium II region the upper end of the tube had always the same temperatures as the lower one when no heating took place. The same was true for small heating currents but at a critical value of the current, which increased with falling temperature, the temperature at the upper end of the tube rose rapidly. This observation corroborated the idea of a surface film of helium II and the critical currents were regarded as the heat input necessary to evaporate the film completely. However, these authors, too, explained their observations as due to a very high heat conduction of the film and thus failed to recognize the true nature of the film phenomenon. In fact, they regarded their results as a refutation of KAPITZA's idea of convection currents in the liquid, pointing out that the film was far too thin to allow for such a convection process.

The experiments of DAUNT and MENDELSSOHN, on the other hand, were designed in the first place to repeat KAMERLINGH ONNES' observation of a re-adjustment of levels in liquid helium which had never been reported again in the intervening sixteen years. Their first apparatus [30] consisted of two small glass vessels on top of each other which were joined by a co-axial glass tube. Some liquid was introduced into each vessel and the variation with time of the menisci

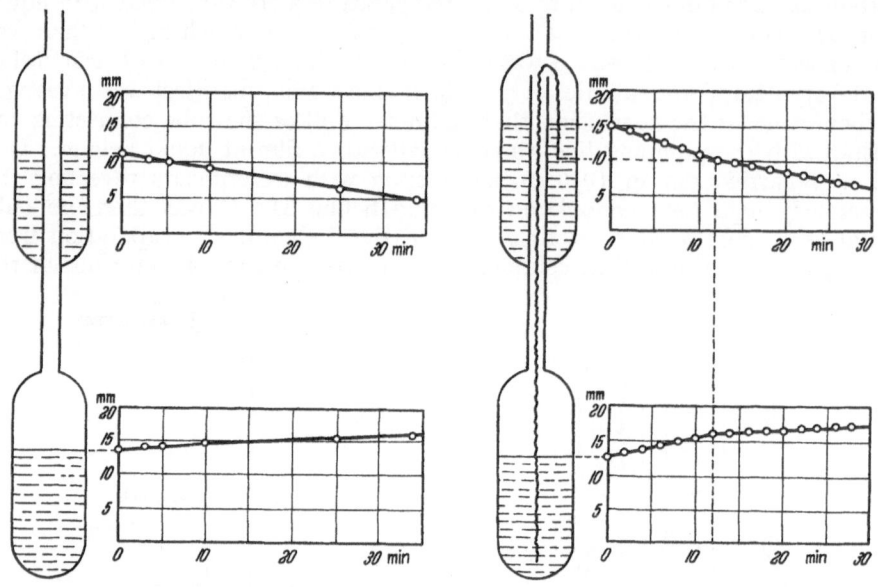

Fig. 13. DAUNT and MENDELSSOHN's first observation of transfer of liquid helium II through the film.

in the two vessels was watched. The result was disappointing since no rapid re-adjustment, such as had been seen by KAMERLINGH ONNES, did take place. The effect was found again, but it was so small that it almost escaped detection.

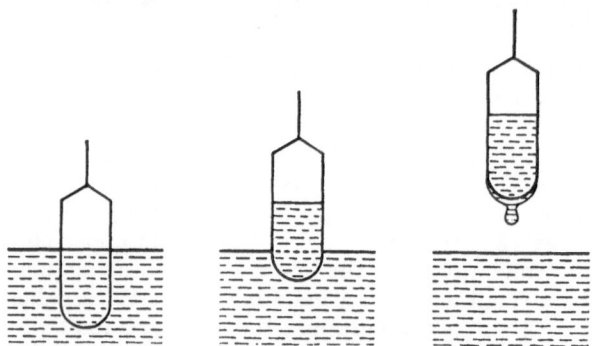

Fig. 14. Transfer into and out of beakers.

In the course of 30 min the upper level dropped by 7 mm while the lower one rose, but only by 3 mm (Fig. 13). The gain of liquid in the lower vessel could either be accounted for by distillation as suggested by KAMERLINGH ONNES or by transport of liquid along the surface of the connecting glass tube. In order to decide this, the authors increased the solid surface connecting the two volumes of liquid helium by inserting a number of wires into the apparatus. An increased loss from the upper vessel coupled with a proportionate gain in the lower vessel which lasted as long as the wires dipped into the liquid demonstrated that the flow took place in a film along the solid surface.

For a better study of the phenomenon, named "transfer effect" by the authors, DAUNT and MENDELSSOHN used the simple device of a small cylindrical glass beaker which was suspended on a fine glass fibre and could be lowered and raised with respect to the helium bath (Fig. 14). This has proved to be a very convenient type of experiment and has been used since in a great number of investigations on transport along the film. On lowering the empty beaker partly into the liquid, it was observed to fill up gradually until the meniscus of helium inside was at the same height as the level of the bath. After raising the beaker slightly the process was reversed, helium now passing from the beaker into the bath until again equalisation of the levels had taken place. Finally the beaker was lifted completely out of the bath, and it was then seen that drops of helium formed at the

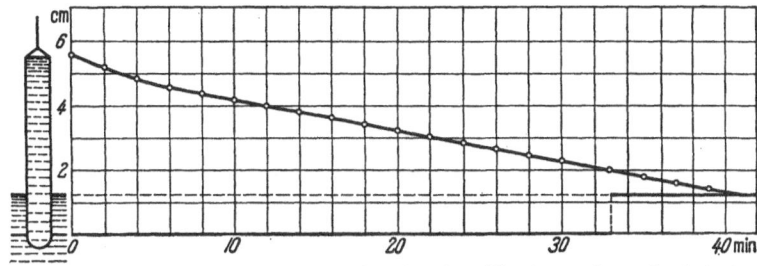

Fig. 15. Film transfer of helium II from a beaker plotted against time. The change of outer level at minute 33 did not affect the rate of transfer.

bottom of the outside surface of the beaker which grew and fell back into the bath at regular intervals.

On timing the rate at which liquid helium was transferred back into the bath from a filled and raised beaker, DAUNT and MENDELSSOHN found that it did not change appreciably throughout the process of emptying. In their first as in all subsequent experiments it was noted that the transfer is slightly higher within a few millimetres of the rim but then settles down to a steady rate which is uninfluenced even by sudden changes in the relative heights of the menisci. The beaker experiment thus showed that the transfer of helium along the film is independent of the potential difference, of the length of the path and of the height of the intervening barrier since these all change in the course of the experiment (Fig. 15).

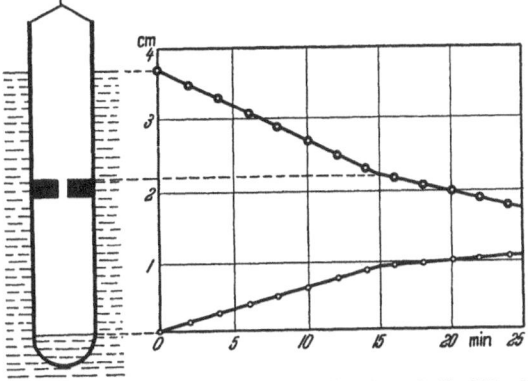

Fig. 16. The limiting effect of a constriction above the liquid level on the film transfer.

The experimental conditions were then slightly varied by introducing a constriction of the walls inside the beaker. As the empty beaker was lowered partly into the bath, liquid began to collect inside at the same rate as in the previous experiment. However, in this case the vessel holding the bath itself was also fairly narrow so that its level fell noticeably as the meniscus inside the beaker rose (Fig. 16). In agreement with the previous experiment the transfer rate remained constant until the bath level had fallen to the height of the constriction inside the beaker. From then onward the transfer proceeded at a reduced rate, the ratio between this and the original rate being the same as the diameter of the constriction to the inside diameter of the unconstricted beaker.

The last observation seemed to indicate that the transfer is limited by the minimum width of the connecting surface above the upper level. Another experiment was, however, carried out to demonstrate that the result obtaines had not been simulated by thermal effects. A beaker in the form of an unsilvered Dewar vessel was constructed, the inner section of which was made up of a wider cylindrical vessel on top and a narrower one at the bottom. Heat exchange in this arrangement could only take place by evaporation through the surface and, if the transfer was limited by this, it should have been proportional to the squares of the upper and lower diameters of the inner vessel. The result of the experiment showed, however, that the rates of transfer were strictly proportional to the diameters and not to their squares (Fig. 17).

While these and other observations showed that at any given temperature the rate of transfer per centimetre width of the connecting surface was completely

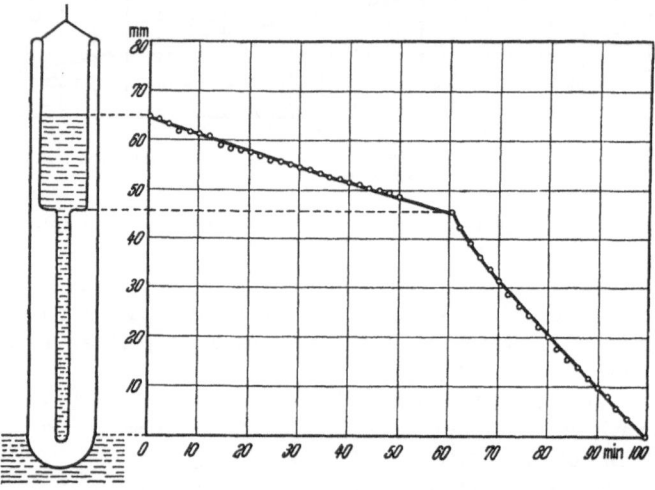

Fig. 17. Experimental proof that the film transfer is proportional to the width of the connecting solid perimeter.

uninfluenced by the conditions of the experiment, a strong dependence of this rate on temperature was found. From the value zero at the lambda-point this rate rises to a practically temperature independent value of 7.5×10^{-5} cm³/sec per cm width of the connecting surface. The nature of this surface appeared to have no effect on the transfer since beakers of copper and aluminium yielded the same value of the rate as glass.

Using different methods, KIKOIN and LASAREW [31] as well as DAUNT and MENDELSSOHN [30] made rough determinations of the thickness of the helium film, both experiments yielding a value of $\sim 3 \times 10^{-6}$ cm. The latter authors also found that a thermo-mechanical effect exists if instead of being joined by a capillary, the two volumes of liquid helium of different temperature are connected through the film.

As, early in 1938, the work on superfluidity in Cambridge and the film experiments in Oxford proceeded, it became increasingly clear that there existed a certain resemblance between the film transfer and the flow phenomena in the finest capillaries. The work on the flow of bulk liquid through capillaries and slits gave results of considerable complexity which, however, showed signs of becoming somewhat simpler as the width of the flow channel was reduced. The flow phenomena then gradually approached the unusual but intrinsically simple pattern of the film transfer. The impression was gained that by using increasingly narrow capillaries, a particular type of flow was "filtered" out from the complex transport phenomena of which the bulk liquid was capable. The film, serving as an exceedingly fine capillary exhibited the pattern of this "superflow" in the simplest and most clear cut form. These observations and the conclusions drawn from them led to a phenomenological model of two interpenetrating fluids of the same substance but with different hydrodynamical properties which has proved of great value as a working hypothesis in designing experiments with liquid helium II.

9. Structure models for He II. The discovery of the lambda-point and in particular the large anomaly in the specific heat necessarily led to speculations concerning the structure of the liquid below this temperature. The rapid fall in entropy below the lambda-point, signifying a large increase in the state of order in helium II, drew again attention to the failure of helium to exhibit a triple point. The structural significance of the lambda-point was first discussed by KEESOM in 1932 who compared the specific heat anomaly in helium with those of similar shape found in ammonium salts and solid methane [32]. Mentioning three possible reasons for the energetic change, a transition inside the atoms, a change in their state of motion or a spatial re-arrangement, he rejected the first two in favour of the third. He pointed out that the fraction (0.42) of the critical temperature

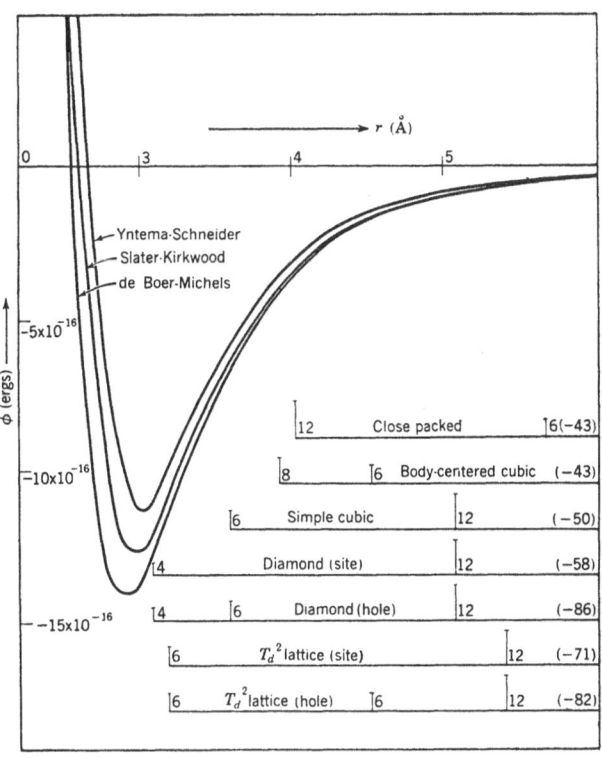

Fig. 18. Interaction energy of two helium atoms as a function of their distance r.

at which the lambda-transition takes place is the same as that at which the triple point of hydrogen occurs which in turn is close to the reduced triple points of neon, nitrogen, argon, and oxygen. This evidence made it attractive to search for a structural model of a quasi-crystalline nature which might be brought into accord with the fact that the aspect of the substance remained that of a liquid. KEESOM visualized some form of liquid crystal in which small crystalline regions of variable size and shape would account for a high degree of space order, allowing at the same time the substance as a whole to retain its liquid aspect. Four years later the question was taken up in some detail by F. LONDON [33] who considered in more general terms the state of helium at absolute zero. Taking into account the large zero point energy of the substance, he compared cubic face-centred, simple cubic and diamond structures and their potential energies as a function of the atomic volume. He showed that, while in the absence of zero point motion the cubic face-centred lattice would be the most stable, under the actual conditions of helium the diamond lattice is the most favourable of the three. These considerations have retained much of their usefulness in spite of the fact that they were based on crystal models, but it is perhaps characteristic of the trend of thought at the time that LONDON avoided the term "liquid" in the title of his paper, referring to "condensed" helium. In the following year, 1937, FRÖHLICH [34] pointed out that the diamond lattice might be considered as a body-centred cubic lattice in which only half the sites are occupied, the vacancies also forming a diamond lattice. This seemed to offer a model in analogy to a binary alloy of atoms and holes in which the lambda-anomaly played the part of an order-disorder transformation.

The idea of some sort of ordered structure of the liquid naturally suggested X-ray analysis, but experiments of this nature on helium present particular difficulties as the scatter produced by any container will overshadow the weak scattering of the liquid. The problem was solved in 1938 by KEESOM and TA-CONIS [35] who obtained diffraction patterns from irradiation of free jets of liquid helium I and II. The result showed unequivocally that the X-ray pattern did not undergo any change at the lambda-point, yielding a single ring at a scatter-ing angle of 28°. The diamond structure proposed by LONDON should have given rings at angles of 23 and 38°, and while KEESOM and TACONIS could show that the observed pattern would be compatible with a hypothetical lattice of the space group $T_d 2$, the X-ray investigation cast severe doubt on a crystalline structure of helium II.

10. BOSE-EINSTEIN condensation. F. LONDON in particular pointed out that a non-localized structure in condensed helium would, because of the high zero point energy, be energetically more favourable than a crystal. Indeed, liquid helium II, instead of being close to a solid crystal, is, owing to its low density, much closer to a gas than an ordinary liquid. It was this gas-like nature combined with the high degree of order of helium II which led LONDON to his important theory. In 1938, in the same volume of NATURE in which superfluidity and the film flow were announced, he published a note [36] drawing attention to a possible connection of the lambda-point with a curious condensation phenomenon postu-lated by EINSTEIN in 1925. In his treatment of an ideal gas obeying BOSE statistics the latter had shown that for any given molar volume there exists a critical temperature below which a finite value of momentum cannot be given to all molecules. This means that at temperatures below this critical value a certain fraction of the molecules has passed into the lowest energy state with momentum zero in which they will have ceased to contribute to the pressure. The condensation phenomenon of the BOSE-EINSTEIN gas is thus a particular aspect of the general phenomenon of gas degeneracy which had already been postulated by NERNST so as to make his heat theorem applicable to non-condensed systems.

Two years after EINSTEIN's publication certain doubts as to the correctness of his derivation had arisen and since there appeared to be no gas in which the degeneracy would not be overshadowed by ordinary condensation, no further attention was paid to this hypothetical momentum condensation.

In his first publication LONDON pointed out that EINSTEIN's condensation process in the ideal gas would be accompanied by a peak in the specific heat at the temperature where, on cooling, particles begin to pass into the state of zero momentum. This is a third-order transformation in which neither the energy nor the specific heat exhibit a discontinuity (Fig. 19). The fact that the specific heat anomaly in helium is of a somewhat different nature, being of the second order, need not be surprising in view of the difference between a liquid with strongly interacting atoms and the ideal gas of EINSTEIN's model. On the other hand, the condensation temperature for an ideal gas with the atomic mass and density of liquid helium would be at 3.14 °K which is remarkably close to the actual lambda-temperature. LONDON himself was fully aware of the very rough approximation of his model but felt that the influence of the statistics was of such significance as to relegate the deviations from the ideal gas to a place of secondary importance. He and many others have attempted to account for the influence of interaction, and in one of his early papers F. LONDON [37] showed qualitatively that one may expect a decrease in the density of the lowest energy states. Such a loosening up of the lowest states is necessary to account for the greater steepness of the specific heat function, as observed in liquid helium, in comparison with that of the ideal gas.

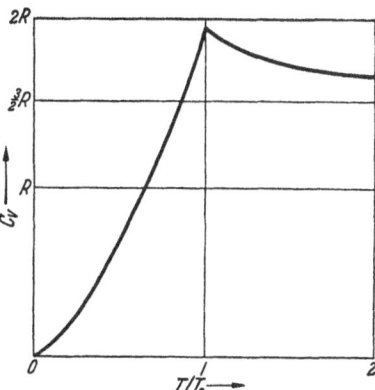

Fig. 19. Specific heat (at constant volume) of an ideal BOSE-EINSTEIN gas.

Whether or not the EINSTEIN condensation process will eventually prove to be the true explanation of the lambda-phenomenon, LONDON's idea gave to the theoretical approach of the helium problem an entirely new turn. The search for a model which was highly ordered in co-ordinate space had become sterile. Even if the particular model chosen by LONDON should fail to be satisfactory, its emphasis on order in the space of velocities has had a profound influence on all subsequent theories.

11. The two-fluid model. The immediate result of LONDON's work was somewhat unexpected, not least to the author himself, in that it led to a phenomenological treatment which, though of doubtful physical significance, proved eminently successful as a working hypothesis. TISZA who was in close contact with LONDON's original work formulated a macroscopic description of helium as a condensing BOSE-EINSTEIN gas which has since become known as the "two-fluid-model" [38]. As a container with liquid helium I is cooled below the lambda-point, condensation of atoms into a state of zero momentum is supposed to begin at this temperature. There will be no separation of the new "phase" in co-ordinate space since the condensation process shall only affect the velocity and not the position of the helium atoms in the lowest state. Helium II is accordingly considered to be a mixture of two completely interpenetrating fluids which have different heat content but consist of the same type of particle, namely helium atoms.

Avoiding the difficult problem of a rigorous treatment of an interacting BOSE-EINSTEIN fluid, TISZA showed that by making certain assumptions his model would not only provide a suitable framework for the tangled phenomena observed in liquid helium but that, in addition, it was capable of predicting new effects [39]. The assumptions concern the behaviour of the condensed and the thermally excited fluids respectively. These fluids are in TISZA's model distinguished by different hydrodynamical behaviour in addition to the difference in heat content. While the uncondensed "normal" fluid is supposed to retain the properties of an ordinary liquid or vapour, the condensed "superfluid" fraction of helium II is meant to be incapable of taking part in dissipation processes.

Hence, an oscillating disc in helium II will experience friction by the normal fluid while a fine capillary will allow superfluid to pass without experiencing friction. The widely differing values for the viscosity of helium II which had been obtained with these two methods can thus be, at least qualitatively, explained by the model.

Similarly, an interpretation could be found for the thermo-mechanical effect. Since in the model the temperature of a volume of liquid helium II simply means a certain relative concentration of the two fluids, a change in this concentration will be registered as either a cooling or a heating. The anomaly in the specific heat, being due to the "evaporation" of the BOSE-EINSTEIN condensate, is therefore according to TISZA the heat required to excite helium atoms from the superfluid into the normal state. As heat is supplied to one of two volumes of liquid which are connected by a capillary, the temperature of this volume is raised or, in other words, the relative concentration of normal fluid is increased. This causes superfluid from the other vessel to be drawn through the capillary towards the supply of heat in order to balance the difference in concentration. Flow of superfluid through the capillary is non-dissipative and therefore unimpeded, whereas a counterflow of normal fluid would be subject to friction and will be negligibly small if the capillary is sufficiently fine. There will thus exist a net flow of helium from the cold to the heated container as had actually been observed. The process has been likened

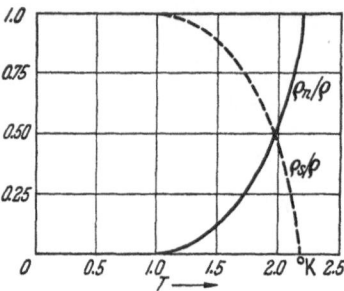

Fig. 20. The ratio of normal and superfluid densities in the two-fluid model of helium II.

to osmotic diffusion, the capillary or powder plug taking the place of a semipermeable membrane. The obvious conclusion from this explanation of the thermo-mechanical effect which was drawn by TISZA was the prediction of the inverse effect, namely that helium forced through a fine capillary should be richer in superfluid and thereby exhibit a drop in temperature. It is to be noted that the publication of TISZA's prediction actually preceded the discovery of the mechano-caloric effect mentioned above.

The anomalously high heat transport in helium II also fell in well with the assumptions of the two-fluid model. The case to be considered is very similar to that of the thermo-mechanical effect, except that the link between the two vessels is not a fine capillary but a tube wide enough to permit the passage of normal fluid without undue friction. The supply of heat will again cause an increase in the normal concentration, demanding flow processes in the liquid in order to restore the concentration balance. However, in this case the flow of superfluid in the direction towards the heater will be compensated for by a counterflow of normal fluid in the opposite direction. The energy to be supplied by the heater to each unit mass of superfluid excited into normal fluid amounts to the total thermal energy at this temperature since the energy of the BOSE-EINSTEIN condensate is zero. The counter-current in liquid helium II therefore appears as a peculiar internal convection mechanism carrying a very large thermal energy. It seems, moreover, plausible that this complex process of heat transport may well be the cause of the observed dependence of the thermal conductivity of helium II on the temperature gradient.

The most far reaching prediction arising out of TISZA's model was his anticipation of thermal waves in the liquid, a phenomenon which has since become known as "second sound". The formalism of two interpenetrating fluids of different entropy led to a wave equation for inhomogeneities of temperature instead of the dissipative equation of heat conduction. He suggested therefore that a disturbance of concentration equilibrium between the two fluids might be equalized by a wave propagation of this disturbance rather than by its diffusion. The wave motion to be expected would thus have a certain resemblance to that of acoustic sound but with the significant difference that no appreciable variations in the liquid density should occur. Their place would be taken by variations in the relative density of the two fluids, i.e. by variations in the temperature. In helium II the relevant parameter for the dissipation of a heat impulse is according to this view not the heat conductivity of the substance but the velocity of the thermal wave in it. On the basis of his model, TISZA postulated that this velocity should rise from zero at the lambda-point to a maximum at about 1.5 °K, falling again to lower temperatures.

The undoubted success of the two-fluid model as proposed by TISZA has often resulted in a tendency to ascribe to it greater physical reality than could be claimed for it. Quite apart from the fact that on the atomistic scale a distinction of "atoms I" and "atoms II" is quantum-mechanically hardly permissible, other difficulties must arise. The idea that at the absolute zero helium should consist entirely of atoms with zero momentum leaves one of the outstanding features of the substance, its high zero point energy, unaccounted for. For the same reason the model for the thermo-mechanical effect is somewhat misleading. Here equalisation of the concentration difference is visualized as osmotic diffusion through a semi-permeable capillary. However, this clearly cannot take place if in addition to the normal fluid being immobilized through friction, the superfluid is accorded zero momentum. These difficulties can be avoided if momentum is ascribed to the superfluid, but then the already vague connection between the property of superfluidity and the BOSE-EINSTEIN condensate is weakened still further.

On the other hand, it is interesting to note that since its first formulation the two-fluid model has in one form or another been part of all subsequent theories. This may simply be due to the fact that it is a phenomenological description which fits the observational data well and for this reason will be in accord with any successful theory. This is certainly the least which can be said for it and while the two-fluid model may have the danger of lacking physical reality, its formalism has undoubtedly provided the most useful basis of experimental research.

12. The thermo-mechanical cycle. The ideas of F. LONDON and TISZA were utilized immediately by H. LONDON [40] in a generalized form which proved to be of great value for experimental work, avoiding at the same time the inconsistencies arising out of any specific model. H. LONDON's approach was purely thermodynamical and as such independent of any model theory. He treated the observed phenomenon of the thermo-mechanical effect as a reversible cycle similar to that presented by a thermo-electric circuit. Using again the scheme for the thermo-mechanical effect, the system can be regarded as a reversible heat engine where in the heated reservoir the rise of the liquid column produces a pressure difference ΔP between the two volumes of helium II which differ in temperature by ΔT (see Fig. 10b). Taking ΔS to be the difference in entropy between the liquid passing through the capillary and that doing work in the return path from the higher to the lower level, H. LONDON arrives at the general relation

$$\frac{\Delta P}{\Delta T} = \varrho \, \Delta S \qquad (12.1)$$

where ϱ is the density of liquid helium at this temperature. The heat of transport Q which is supplied to one reservoir and liberated in the other is then given by

$$Q = T \Delta S. \tag{12.2}$$

So far the treatment merely takes account of the observed thermo-mechanical effect and is independent of any theory. Assuming the flow through the capillary to be carried out by a fluid of zero entropy which corresponds to zero THOMSON heat in the thermo-electric analogy, the two equations become

$$\frac{\Delta P}{\Delta T} = \varrho S \tag{12.3}$$

and

$$Q = TS \tag{12.4}$$

respectively, where S is the total entropy of the liquid. H. LONDON raised the question whether or not S appearing in the heat of transport would include the phonon entropy, a problem which was to receive a certain amount of attention later.

Such data as existed on the thermo-mechanical effect as well as the observation of the mechano-caloric effect showed the ΔS was quite large but were no sufficiently accurate to decide whether it had the value S. H. LONDON's paper was published at the outbreak of war in 1939 when cryogenic research had been suspended in Holland and England where most of the work on helium had been carried out. Quantitative confirmation of his formula was however produced two years later by KAPITZA.

13. Thermal counterflow. In 1941 KAFITZA published two papers containing a large amount of experimental observations on liquid helium II. The first paper [41] dealt mainly with the mechanism of heat transport in capillaries and its relation to mass flow. KAPITZA showed that while the heat flow in his capillary was very large, as had been observed in the Leiden experiments, it could be drastically reduced if the liquid in the capillary was agitated. Both forced flow of liquid along the capillary as well as the introduction of rotary motion by means of a co-axial stirrer in the capillary had this effect. In a number of very ingeneous experiments he then demonstrated the existence of a counter current in the capillary. A thermally insulated and closed bulb which contained a heater and thermometer was attached to a capillary whose other end was opposed by a vane (Fig. 21). On supplying heat, a rise in temperature in the bulb was recorded and this was always accompanied by a force acting on the vane. By displacing the vane slightly sideways, KAPITZA could show that heat flow in the capillary was coupled with mass flow which emerged from the open end of the capillary as a jet. He also carried out experiments in which the total reaction

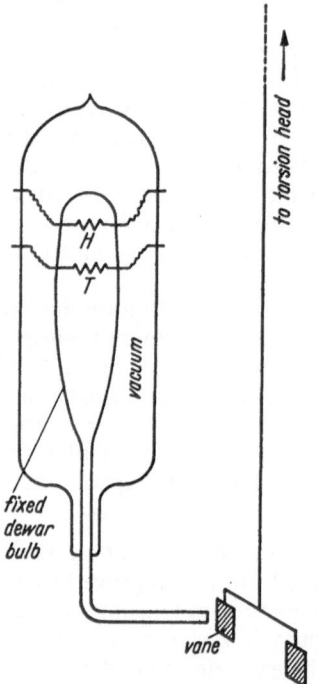

Fig. 21. KAPITZA's apparatus for demonstrating the existence of counter flow in helium II.

of the jet was measured, and it became clear that a large amount of the supplied heat had been turned into kinetic energy.

These observations agreed well with the ideas put forward by TISZA and H. LONDON but the work of these authors is not mentioned and was, owing to the war, possibly unknown to KAPITZA. It is significant that the mechanism of heat transport which he visualized made no use of a two-fluid model. Instead he suggested the possibility of two spatially separated mass currents, inflow into the bulb of a surface layer on the inner perimeter of the tube and outflow through the centre of the tube. The difference in heat content between the two currents is accounted for by the VAN DER WAALS forces of the capillary wall on the surface layer of liquid.

KAPITZA's second paper [42] which was written seven months later, dealt with the flow of helium II through a narrow slit under the influence of a temperature difference. It was, in fact, a quantitative study of the mechano-caloric effect under closely adiabatic conditions. The quantities measured were the heat of transport Q and the thermo-mechanical pressure difference ΔP corresponding to a difference in temperature ΔT. The work was thus a verification of H. LONDON's equations and showed that with considerable accuracy ΔS is equal to the total entropy of liquid helium II. KAPITZA concluded from his experiments

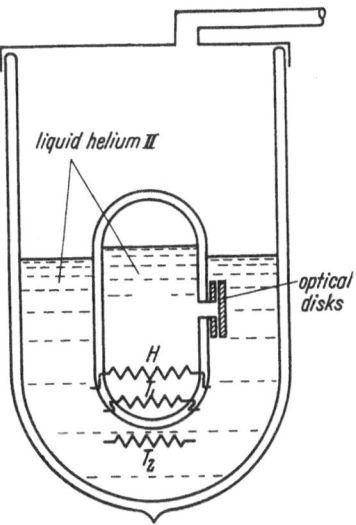

Fig. 22. Apparatus for the study of thermal effects in helium II.

that the entropy of liquid helium flowing through the narrow slit was zero. He mentioned that this had been suggested by TISZA and H. LONDON but ascribed the true explanation to a new theory of liquid helium by LANDAU [43] which was published simultaneously with his own paper. At the same time he corrected his earlier model of surface flow and substituted for it LANDAU's new two-fluid model.

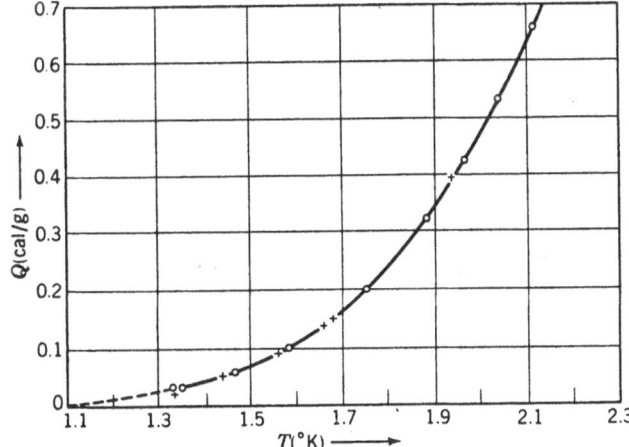

Fig. 23. The heat of transport derived from the thermo-mechanical effect (circles) and the mechano-caloric effect (crosses) plotted together with TS (full line) derived from specific heat measurements.

14. Phonons and rotons. The dates of KAPITZA's two papers suggest that LANDAU's theory was formulated early in 1941. In his opening paragraph LANDAU criticizes TISZA's two-fluid model, first regarding the basic idea of an ideal BOSE gas which he consideres not applicable to a liquid and secondly pointing out that the model would not yield superfluidity. In LANDAU's theory the problem of

accounting for the interaction forces between the helium atoms is avoided by
treating the liquid as a quasi-continuum. In a way this treatment can be compared with the quantum theories of the specific heat of a solid. There are accordingly modes of vibration corresponding to sound waves which can be excited,
but these do not constitute all forms of motion of which a liquid is capable.
Provision has also to be made for the modes of vortex motion. Thus LANDAU
constructs the energy spectrum of a liquid from two types of excitations; to the
phonons of the solid body he added a spectrum of "rotons" by which term
he defined the elementary excitations of the vortex spectrum.

He argued that in a quantum liquid no continuous transition between
the states of potential motion (curl $v = 0$) and of vortex motion (curl $v \neq 0$)

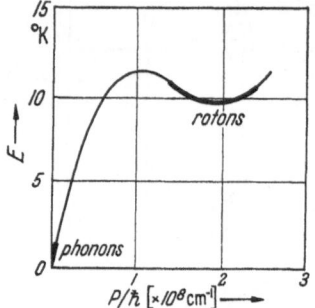

Fig. 24. Thermal excitations in the
LANDAU theory.

can exist and that there is an energy gap between the
lowest levels of the phonon and roton spectra. From
purely dimensional arguments it follows that the gap
should be of the order of

$$\Delta E \sim m^5/\varrho^2 \hbar^2 \qquad (14.1)$$

where m is the mass of the helium atom and ϱ the
density of the liquid. It is perhaps significant that
LANDAU did not postulate from first principles a
lower ground level for the phonon than for the roton
spectrum. He merely pointed out that if superfluidity is to result from the model, this has to be
the case.

As LANDAU pointed out, the phenomena to be expected from his theoretical
model are identical with a two-fluid description, and similar to TISZA he devides
the total density of the liquid ϱ into two temperature dependent parts ϱ_n and ϱ_s
which correspond to the normal and superfluid states so that always $\varrho_n + \varrho_s = \varrho$.
He was, however, careful to stress that the two-fluid aspect cannot be regarded
as more than a convenient way of saying that the liquid as a whole is capable
of two types of motion simultaneously. The convenient terms of "superfluid"
and "normal" were thus not associated by LANDAU with interpenetrating atomic
fluids but with the effective masses of the respective types of motion. Of these
one has the normal behaviour of any usual liquid while the other is superfluid.
The existence of the energy gap and the lower position of the phonon states in
relation to the rotons leads to superfluidity. There is no interaction between
the two types of flow and thus no friction.

The specific heat of the liquid is made up of two parts corresponding to the
energy spectrum of phonons and rotons. The former, which alone are excited
at low enough temperatures lead to a T^3 term which changes over to a steeper
rise as the rotons begin to appear with rising temperature. The only measurements of the specific heat at temperatures below 1 °K which were available at
the time were some preliminary observations of SIMON and PICKARD. These
turned out to give values far in excess of the true ones determined later by
different authors, and LANDAU therefore wondered whether the onset of the roton
excitation might occur at exceedingly low temperatures. In fact, the phonon
entropy quoted by him (calculated in 1940 by A. MIGDAL) was later found to be
in fair agreement with the measured values.

In many respects the consequences of LANDAU's theory turned out to be
identical with TISZA's two-fluid model, and H. LONDON's equation to which

due reference is made in LANDAU's paper is also derived by the latter. Significant discrepancies, however, arise in the behaviour of liquid helium predicted by the two theories for very low temperatures. According to TISZA the "superfluid" retains a DEBYE entropy whereas LANDAU's superfluid is completely without excitations, phonons as well as rotons contributing to the normal liquid. For instance, the heat of transport Q in H. LONDON's second equation should become negligibly small well below 1 °K in TISZA's model but retain the value corresponding to the phonon entropy in the LANDAU theory. A similar difference in behaviour was to be expected in the velocity of second sound. This phenomenon was also predicted by LANDAU and although he had been anticipated by TISZA, whose second paper, which postulated the existence of thermal waves, was written more than two years earlier, it seems likely that owing to war conditions LANDAU was ignorant of this part of the work. The formalism of LANDAU's theory led to two different equations for the propagation of sound which is evidently the reason for the Russian workers giving the name of "second sound" to thermal waves. Indeed, the first and unsuccessful attempt to generate and detect second sound waves was made with acoustic apparatus by SHALNIKOV and SOKOLOV (probably in 1941). TISZA's prediction led to a drop of the second sound velocity to zero at absolute zero, while LANDAU expected the velocities of first and second sound u_1 and u_2 to be represented by the ratio

$$u_1/u_2 = \sqrt{3} .$$
(14.2)

15. Second sound. The failure to observe second sound acoustically was explained in 1944 by LIFSHITZ [44] who pointed out that a quartz oscillator will radiate second sound a million time less intensely than first sound, and that a suitable generator for second sound would be a body whose temperatur changes periodically. Using such a thermal generator, PESHKOV [45] was able to demonstrate in the same year the existence of standing thermal waves for the first time. In 1946, having further perfected his elegant technique, PESHKOV [46] reported accurate measurements of the velocity of second sound between the lambda-point and 1.1 °K. The velocity rises sharply as the temperature is lowered below T_λ and reaches a maximum of 20.3 m/sec at about 1.65 °K (Fig. 25). On further cooling a gradual decrease was observed with the velocity becoming apparently independent of temperature at the end of the available range. These experiments did not as yet permit a decision between the treatments of TISZA and LANDAU, but two years later PESHKOV [47], by extending the range of his experiments to 1.03 °K, was able to show a slight increase in the velocity at the lowest temperature. The question was, however, definitely settled in 1949 by PELLAM and SCOTT [48] who studied second sound pulses in magnetically cooled helium. While, owing to the difficulty of the technique, these workers were unable to determine the actual temperatures at which the readings were taken, they found that by cooling they could raise the velocity of second sound to 34 m/sec. This spectacular increase left little doubt that, as regards the propagation of second sound, the prediction of LANDAU appeared to be the more probable one.

Besides second sound, LANDAU suggested another experiment. This was carried out in 1946 by ANDRONIKASHVILI [49] and it may be considered as a further demonstration of the two-fluid model of helium II. The apparatus used consisted of a closely spaced stack of circular aluminium discs which were suspended by means of a torsion fibre in a bath of liquid helium (Fig. 26). The period of oscillation of the stack was measured as the temperature of the bath was varied and was found to decrease with falling temperature. The explanation for this phenomenon is given by the different hydrodynamical behaviour of the superfluid and normal parts of the liquid. Whereas the former does not follow the motion of the stack, the normal fluid is dragged along with it in the narrow gaps between the discs. As the temperature is lowered below the lambda-point and the concentration of superfluid increases, the total moment of inertia of the stack decreases because

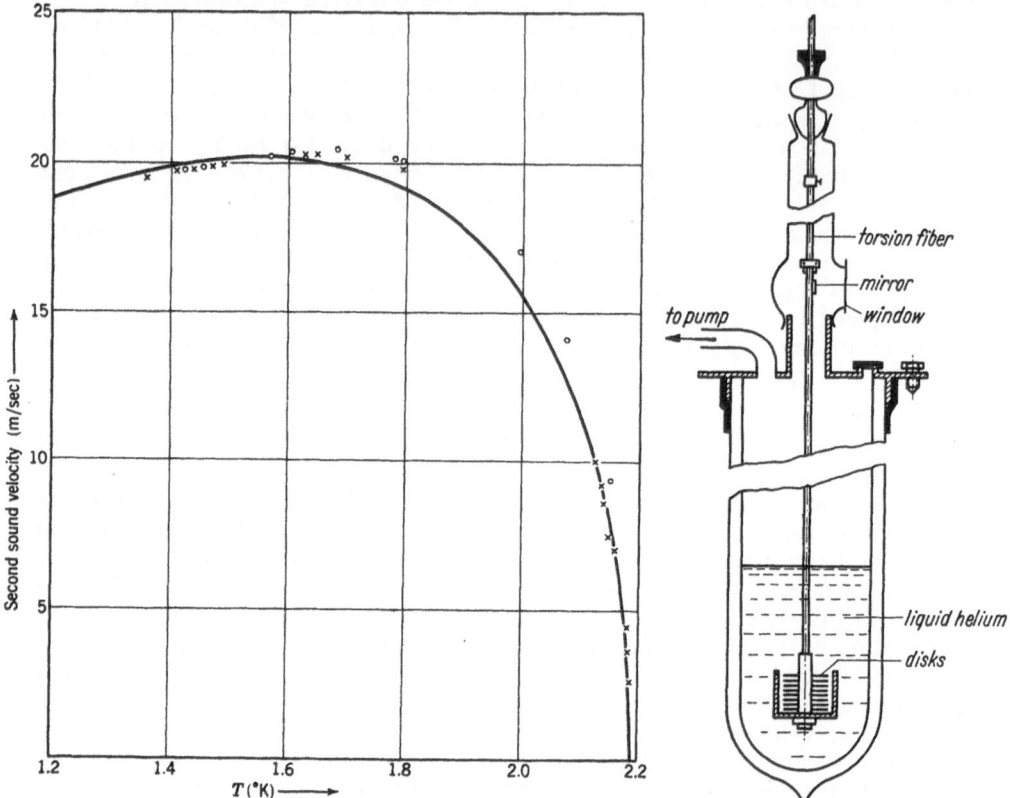

Fig. 25. Early measurements of the velocity of second sound by Peshkov (crosses) and Lane (circles) in comparison with Tisza's model (full line).

Fig. 26. Andronikashvili's apparatus for measuring the density of the normal fluid in helium II.

progressively less helium will take part in the oscillations. The observations thus lead to a direct determination of the normal concentration ϱ_n as a function of temperature (Fig. 27).

16. Analogy with superconductivity. At about the same time when Androni-kashvili made his measurement of the normal constituent of helium II, Daunt and Mendelssohn carried out an experiment designed to investigate the behaviour of the superfluid. In 1942 these authors had drawn attention to a far reaching similarity between the phenomena of superfluidity and electrical super-conductivity [50]. In particular they emphasized the existence in both cases of a temperature dependent critical velocity of transport below which dissipation is completely absent. In superconductors the lack of dissipation is exemplified by the existence of persistent currents and by the fact that in a current-potential measurement the voltage across a superconductor drops to zero when the current becomes sub-critical. The observations on helium films, where the superfluid properties are most clearly exhibited, had all been made at the critical flow rate, and DAUNT and MENDELSSOHN [51] therefore devised an experiment in which the rate of film transfer was limited to less than the critical value. The arrangement was an analog to the current potential measurement on a superconductor and simply consisted of two concentric beakers which could be lowered or raised in a

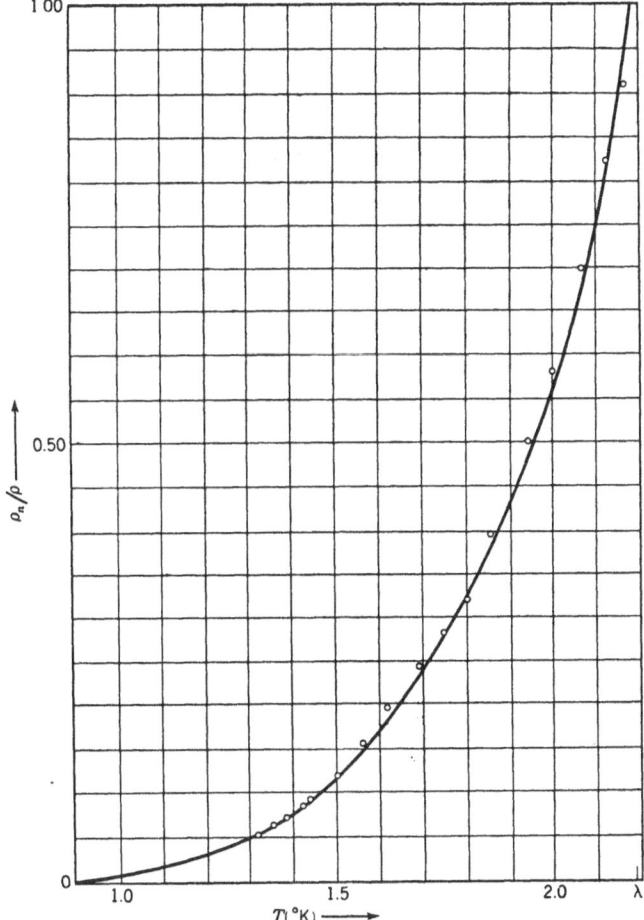

Fig. 27. The normal concentration ϱ_n/ϱ as function of temperature.

bath of liquid helium (Fig. 28). In equilibrium the levels in the inner and outer beakers will adjust themselves by film transfer to the same height as the bath level. If the beaker was raised, helium had to pass by film flow from the inner into the outer beaker and from there into the bath. It was found that in this process the levels in the two beakers stayed at the same height and the identical result was observed when the flow was reversed. In the analog the difference between the levels of inner beaker and bath represents the battery, the total flow of helium the current, the outer beaker a resistor limiting the flow over the rim of the inner beaker to below the critical value, and the inner beaker represents the superconductor. The complete absence of a pressure difference between inner and outer beaker thus corresponds to the absence of a potential difference across a superconductor. The result showed that below the critical velocity in helium the transport of mass is carried by pure potential flow since kinetic energy is preserved in the passage through the liquid from the descending to the ascending film.

In the ten years which have elapsed since these experiments were carried out there has been a large volume of experimental work as well as of theoretical speculation on the helium problem. Many points at issue have been clarified and much detail has been added to our knowledge of the phenomena. In particular the question of the critical velocity and the appearance of friction has been further investigated, as well as the phenomena of second sound and the viscosity.

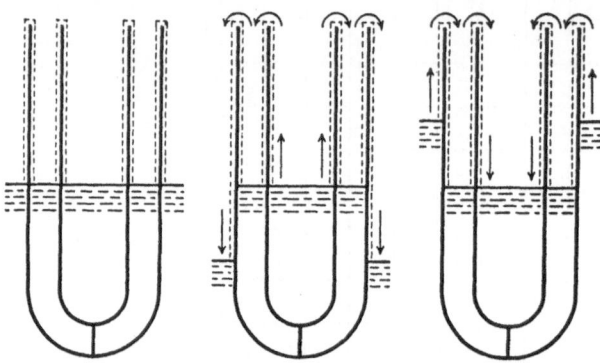

Fig. 28. Film transfer under zero pressure difference.

However, the impression is gained that no outstanding new discoveries have been made which will rank with those described here. Perhaps it should be mentioned here that the specific heat below 1 °K has been measured accurately by KRAMERS, WASSCHER and GORTER [52] in 1952 who could show that below ~0.6 °K it obeys a T^3 law. A theory which combines certain features of the work of F. LONDON and LANDAU but goes further in its interpretation of the basic fact has lately been developed by FEYNMAN [53]. A discussion of these developments will be given in the following sections.

17. Liquid He³. There is, however, another field of research which has had the most profound effect on the helium problem and which has yielded results of equal importance with any of those mentioned above. This is the work on the light isotope of helium with the atomic weight 3. In contradistinction to He⁴ which obeys BOSE-EINSTEIN statistics, He³ has an odd number of nucleons and is therefore subject to FERMI-DIRAC statistics. In view of F. LONDON's suggestion that the lambda-phenomenon may be due to the condensation of momenta in a BOSE-EINSTEIN fluid, the difference in statistics gives special significance to experiments on liquid He³.

He³ was first observed as the product of nuclear bombardment of Li⁶ with protons carried by OLIPHANT, KINSEY and RUTHERFORD in 1933. Its natural occurrence was not discoverod until 1939 when ALVAREZ and CORNOG investigated atmospheric helium mass-spectrographically with a cyclotron. They found for the isotopic ratio He³/He⁴ a value of ~10⁻⁷ which was decreased by a further power of ten in the case of helium gas obtained from wells. Since the war a large number of experiments of various kinds have been carried out on atmospheric helium in which the He³ content had been enriched by thermal diffusion. However, the release of nuclear energy in reactors has also led to the production of small amounts of pure He³. The bombardment of lithium with neutrons produces tritium which under β-decay with a mean life of about ten years turns into He³. Since 1948 quantities sufficient for experiments at low temperatures have become available but the nature of the source has, until fairly recently, confined these experiments to the Los Alamos and Argonne laboratories.

In the ten years preceding the first liquefaction of He³ a number of speculations concerning its state at absolute zero were made which all favoured the idea that this substance would not liquefy by cooling alone. It was predicted that at 0 °K the vapour pressure of He³ would be different from zero and that the liquid state would only be realized under external pressure. In 1948 DE BOER [54] extended the law of corresponding states to cases in which the quantum deviations become appreciable. He and LUNBECK [55] plotted the reduced critical temperatures of the rare gases as well as H₂ and N₂ against the parameter which takes care of the quantum effects in DE BOER's theory. Extrapolation to the case of He³ then led to positive values for the critical constants of this substance, suggesting the existence of a critical point between 3.1 and 3.5 °K. A few months later the

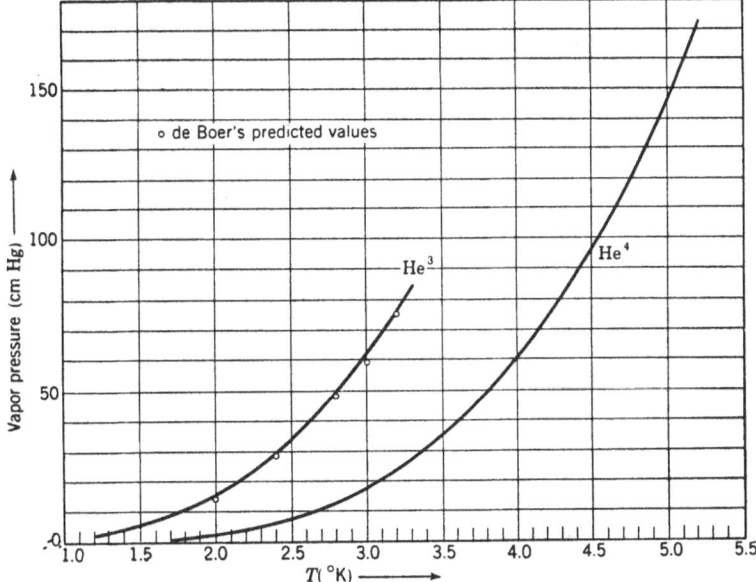

Fig. 29. Vapour pressure curves of He³ and He⁴.

first liquefaction of He³ was carried out by SYDORIAK, GRILLY and HAMMEL [56] at Los Alamos who found the critical point at 3.34 °K. They also measured the vapour pressure between 1.2 °K and the critical point, and their work was repeated and extended to 1.02 °K by ABRAHAM, WEINSTOCK and OSBORN [57] at the Argonne Laboratory. The experimental values obtained in these investigations were found to be in excellent agreement with the predictions of DE BOER and LUNBECK (Fig. 29).

In the first experiment in Los Alamos the He³ gas was liquefied in a steel capillary which was subsequently replaced by a glass tube. In the latter He³ could be visually observed as a colourless liquid of small surface tension. A sudden reduction of the vapour pressure was seen to cause violent boiling of the liquid similar to that observed in He I which indicated that, at least at this temperature, He³ does not exhibit the high heat conduction of superfluid He II. A direct test of superfluidity was made in 1949 at the Argonne Laboratory [58]. Liquid He³ was forced under its vapour pressure through a narrow channel into an evacuated vessel. The result of this experiment (Fig. 30) shows a monotonically decreasing rate of flow through the "superleak" with falling temperature. Using He⁴ in the same apparatus resulted in a strong rapid increase of the flow rate at the lambda-point due to the onset of superfluidity. These observations thus showed that no superfluidity took place in liquid He³ above 1.05 °K. The range of the validity of this conclusion was extended in the following year by Daunt and Heer [59] to much lower temperatures. These authors measured the heat influx into a magnetically cooled vessel containing He³ to He⁴ mixtures of different concentration. If the mixture is superfluid, film transport along the connecting tube and re-condensation of the evaporated film will cause a strong influx of heat into the apparatus. Daunt and Heer could thus determine the lambda-temperature for any given concentration of He³ by the onset of film transport (Fig. 31). The highest concentration of 89% He³ yielded superfluidity at 0.38 °K and the authors concluded that according to their extropolation pure He³ would not be superfluid above 0.25 °K and probably remain normal even at absolute zero.

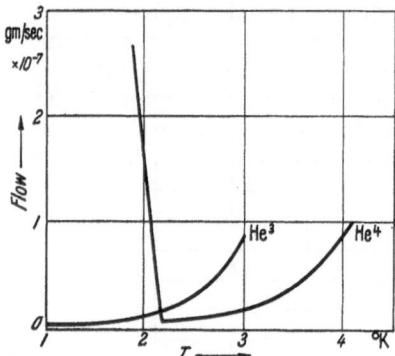

Fig. 30. Outflow of liquid He³ and He⁴ through a narrow capillary as function of temperature. The onset of superfluidity in He⁴ at 2.19 °K is marked.

The failure of liquid He³ to show superfluidity thus gave added significance to F. London's idea of interpreting the lambda-phenomenon as the Bose-Einstein condensation process of an interacting fluid. This is further emphasized by the similarity in general behaviour of the two isotopes in the liquid state. De Boer's treatment of He³ and He⁴ vapour pressures using the same type of equation with success for both isotopes is an example of this similarity. Another

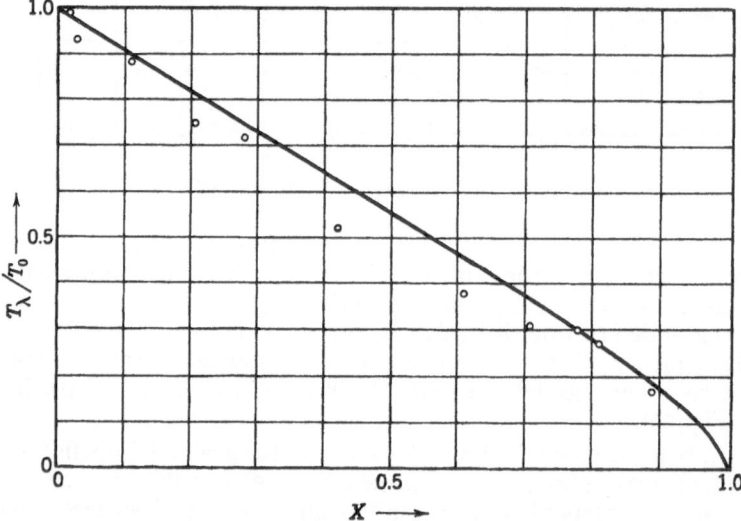

Fig. 31. The onset of superfluidity as function of He³ concentration and temperature.

is provided by a comparison of the temperature functions of the densities of liquid and vapour for He³ and He⁴. The values for the former were derived by Grilly, Hammel and Sydoriak [60] from differential measurements of the liquid and vapour densities and have since been confirmed and extended by Kerr [61] in direct determinations. The curve for the liquid density of He⁴ shows the kink which originally led to the discovery of the lambda-point (Fig. 32). However, it is also apparent that the variation in density due to it does not alter the density function radically and is rather in the nature of a second order effect. This means that in spite of the difference of statistics the two liquids are comparable in their physical behaviour. The fact that in such similar liquids the one which obeys BOSE-EINSTEIN statistics shows superfluidity while this property is lacking in the FERMI-DIRAC liquid, clearly lends weight to F. LONDON's theory. At the present state of development of LANDAU's general theory of quantum fluids, there is no reason to suppose that liquid He³ should behave differently from He⁴ in respect to superfluidity. LANDAU has ascribed superfluidity to the ad hoc assumptions that the lowest levels in the energy spectrum are phonon and not roton excitations and there is no indication in his theory that this relative position may be inverted under the influence of statistics.

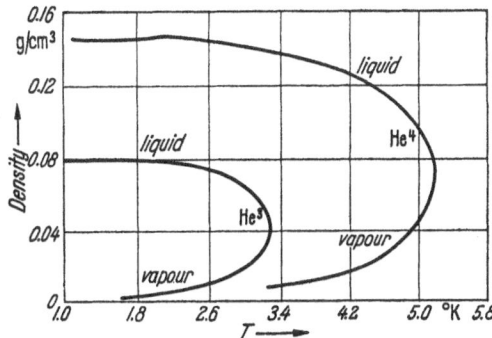

Fig. 32. Densities of He³ and He⁴ as function of temperature.

18. The melting curve of He³. The equilibrium curve between liquid and solid He³ has been investigated at the Argonne Laboratory [62] down to a temperature of 0.16 °K. Between the highest measured value at 1.5 °K and 0.5 °K the melting curve follows a square law from which the equilibrium pressure P can be evaluated as

$$P = 26.8 + {} \atop + 13.1\ T^2\ \text{atm.} \Bigg\} (18.1)$$

This would indicate that, like the heavier isotope, He³ under its saturation pressure must remain liquid at absolute zero and that for its solidification an external pressure of very much the same order as in He⁴ is required. The experimental points of the melting curve below ∼0.5 °K, however, depart from the square law and become temperature independent at a value of just under 30 atm (Fig. 33). This is somewhat similar to the change of direction exhibited by the

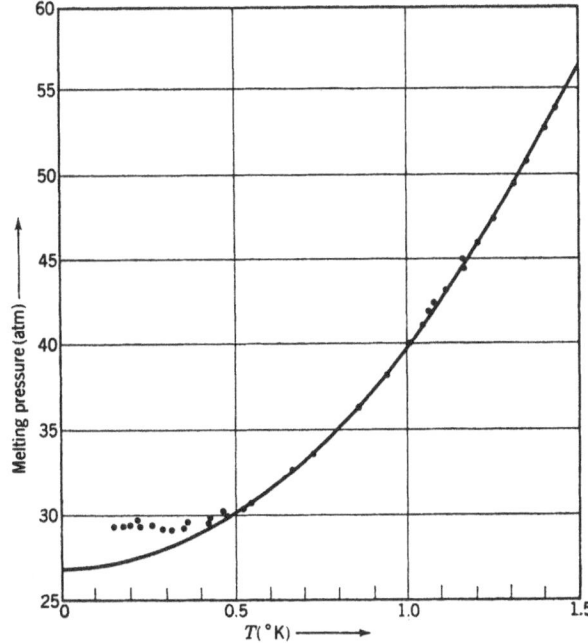

Fig. 33. Melting curve of liquid He³.

melting curve of He⁴ in that region of the diagram of state where the lambda-curve intersects the melting curve. Since the behaviour of the He⁴ melting pressure is in that case due to the rapid drop in entropy of the liquid phase, i.e. to the lambda-phenomenon, the existence of a lambda-point in He³ in the region between 0.5 and 1 °K might be suspected. However, the experiments on the failure of He³ to become superfluid make any such explanation in close analogy to He⁴ unlikely. For some time, therefore, the diviation of the measured points from the square law were ascribed to a rather trivial technical reason, namely that below 0.5 °K the thermal contact between the capillary containing the He³ and the paramagnetic salt which acted as coolant as well as thermometer had become insufficient. Thus, the temperature of the substance could have remained at 0.5 °K while the recorded temperature of the salt had decreased. In 1954 SYDORIAK [63] drew attention to another technical difficulty of the method which might give a very different significance to the deviation observed at the lowest temperatures. The method used for determining the melting pressures consisted of recording, with changing pressure, the readings of two manometers at different ends of a capillary filled with He³. The capillary passed through the cryostat and as soon as the melting pressure was attained at the coldest spot along the capillary, the latter became blocked with solid He³ and the readings of the two manometers became independent. The identification of the pressure at which the blockage occurred with the melting pressure at the coldest spot of the capillary is, however, only justified as long as the temperature coefficient of the melting pressure does not change sign. Should the melting curve pass through a minimum, then the capillary will block at a place corresponding to more than the minimum temperature and the substance can be in the liquid state at the coldest spot.

19. Spin alignment in liquid He³. The possibility that the melting curve of He³ might exhibit such a minimum arose from considerations of spin degeneracy in the solid and liquid phases of the substance. The He³ nucleus has a spin of $\frac{1}{2}$ and the state of the substance at absolute zero must therefore be one in which the spins are aligned. In 1950 POMERANCHUK [64] suggested that in the solid state

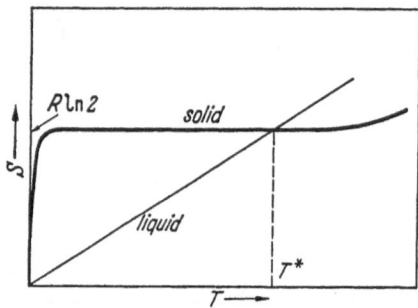

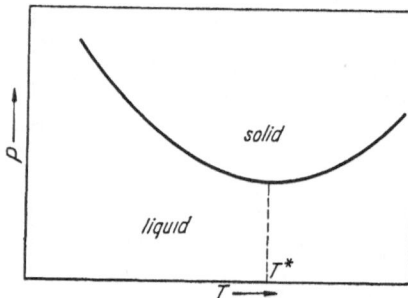

Fig. 34. Suggested entropy diagram of liquid and solid He³.   Fig. 35. Suggested melting curve of He³.

the exchange interaction leading to spin alignment would be very small so that ordering of the spins may only occur at temperatures corresponding to the order of the coupling energy between neighbouring nuclear dipoles, which is about 10^{-7} °K. In the liquid phase, on the other hand, the exchange energy leading to spin coupling can be expected to be very much larger than in the solid so that alignment may take place at rather higher temperatures. Even allowing for an appreciable entropy due to phonon and roton excitations in the liquid, the total entropy of the liquid phase may therefore decrease below that of the solid at not too low a temperature (Fig. 34). When this happens, the sign of the temperature function of the equilibrium curve must change sign (Fig. 35). The possiblity of a minimum in the melting curve of liquid He³ thus cannot be excluded and the observed deviation from the square law may, in fact, indicate the existence of such a minimum.

Spin alignment in liquid He³ should, of course, be noticeable in the magnetic susceptibility, and this effect has therefore been investigated. In analogy with the approximation to an ideal gas, which led to the correct order of magnitude of the BOSE-EINSTEIN condensation temperature in the heavier isotope, the change in the susceptibility might have been expected at easily accessible temperatures. For a FERMI-DIRAC gas of the atomic mass and density of liquid He³ the degeneracy temperature works out to about 5 °K. The first experiments carried out in the temperature region above 1 °K did not, however, indicate any spin alignment, and it was clear that the ideal gas approximation does not lead to the correct degeneracy temperature. Very recently W. M. FAIRBANK and co-workers [65], working in close co-operation with the late F. LONDON at Duke University have extended these measurements to 0.2 °K and were able to show that

the expected effect does occur (Fig. 36). They found a striking departure of the susceptibility of liquid He³ from which a degeneracy temperature of ∼0.45 °K, that is about ten times lower than the ideal gas approximation, was deduced. An incidental result of some interest was that the spin-lattice relaxation time found in this work was rather long, varying between 30 and 200 seconds in different experimental arrangements.

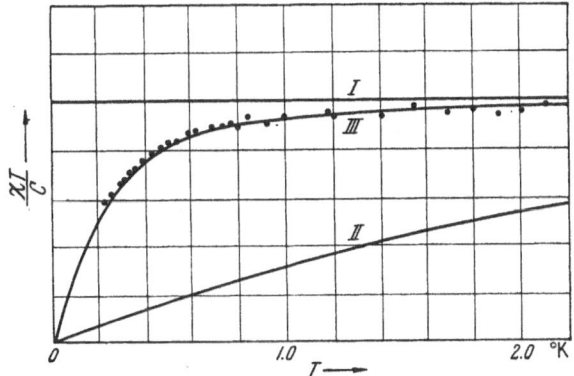

Fig. 36. Nuclear alignment in liquid He³. χ nuclear susceptibility C normalizing CURIE constant. Curve I: CURIE's law; curve II: degeneracy temperature $T_v = 5$ °K; curve III: $T_v = 0.45$ °K.

The entropy change due to spin alignment should also be noticeable in the specific heat of liquid He³. The first determination by DAUNT and co-workers indicated a linear relation with temperature. In 1954 ROBERTS and SYDORIAK [63] measured the specific heat down to 0.5 °K and in 1955 the experiments were extended to 0.23 °K by the team at the Argonne Laboratory [66]. The results, while not permitting unambiguous analysis, show definite evidence of a spin contribution. Like the specific heat of liquid He⁴ above the lambda-point, that of He³ is roughly proportional to the absolute temperature in the region above 1.4 °K. This similarity seems to be one of the properties of the liquid helium phase which does not depend on statistics and, although no theoretical basis exists for this specific heat function, extrapolation to the absolute zero may not be too unreasonable. Below 1.4 °K the measured specific heat of liquid He³ is larger than this extrapolation, the excess amounting to more than 300% at 0.25 °K (Fig. 37). The method of separation between the thermal excitations and the heat due to the destruction of spin alignment is far too crude to draw conclusions as to the exact temperature function of the spin entropy. However, as far as the order of temperatures is concerned, the anomaly in the specific heat is in agreement with that in the susceptibility, and it seems justified to assume that it is connected with the deviation of the melting curve from the square law.

Practically nothing is as yet known about the physical properties of solid He³. However, if POMERANCHUK's prediction [64] that the spin entropy of the solid will only vanish at about 10^{-7} °K is correct, a minimum in the melting curve at ∼0.5 °K has to be expected. The temperature coefficient of the melting pressure should then change again in the neighbourhood of 10^{-7} °K.

For the sake of clarity we have omitted in this historical account the earliest work on dilute solutions of He³ in He⁴ which preceded the first liquefaction of pure He³ by about one year. The first experiment in this field was carried out by DAUNT, PROBST and JOHNSTON [67] who demonstrated that He³ is not carried along in superflow. It was shown that, when He II passes along a solid surface by film flow or through a narrow slit, He³ impurity will not take part in this flow and is thus filtered out. Shortly afterwards it was found that this non-participation in superflow is also true for the bulk liquid and that He³ is therefore carried with the flow of the normal component in the two-fluid description. If, for instance, heat is supplied to the liquid, He³ will move with the heat current and its distribution over the available volume of bulk liquid becomes inhomo geneous. This phenomenon has led to considerable errors in early determinations of the partial pressure above a solution of given concentration. It has also been the basis of a method for the separation of the helium isotopes [68].

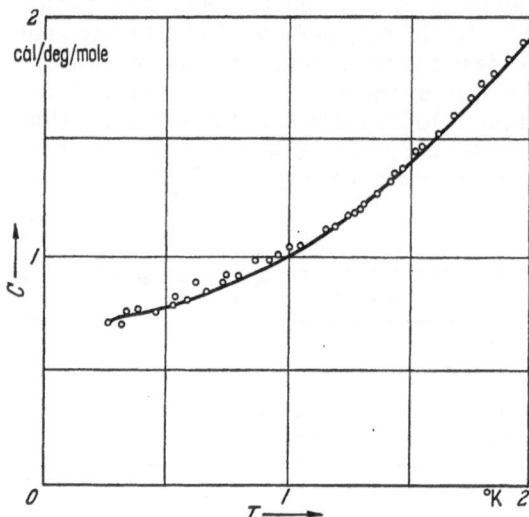

Fig. 37. Specific heat of liquid He³ under its saturated vapour pressure.

B. The diagram of state.

The diagram of state of liquid helium is unique in that the vapour pressure and melting curves do not intersect. The absence of a triple point is, however, not the only characteristic feature. The fact that two liquid modifications exist which are separated by the lambda-curve but which pass into each other without discontinuity in the energy creates two distinct points in the diagram. These are the intersections of the lambda-curve with the vapour pressure and melting curves. Since there is a very rapid drop in entropy with falling temperature as the lambda-curve is traversed, the effect on the solid-liquid equilibrium is pronounced. The three features of the phase diagram which have received particular attention are therefore:

1. the solid-liquid equilibrium,

2. the variation of density with pressure and temperature of the liquid phase itself, and

3. the liquid-vapour equilibrium.

20. The melting curve. Accurate determination of the equilibrium pressures between the solid and the liquid phase between 1.0 and 1.8 °K were carried out by SWENSON [69] who used the blocked capillary method. He found that between 1.0 and 1.5 °K the data could be represented by the relation

$$P = 0.053\, T^8 + 25.00 \text{ atm} \qquad (20.1)$$

which leads to

$$\frac{dP}{dT} = 0.425\, T^7 \text{ atm/deg} \qquad (20.2)$$

for the interval between 1.0 and 1.4 °K. Combining his data with the earlier work of KEESOM, SWENSON obtains for the melting pressure between 2.0 and 4.0 °K the dependence

$$P = 15.45 \, T^{1.57} - 8 \text{ atm}. \tag{20.3}$$

By extrapolation to absolute zero the equilibrium pressure there is derived as

$$P_0 = 25.00 \, (\pm 0.05) \text{ atm}. \tag{20.4}$$

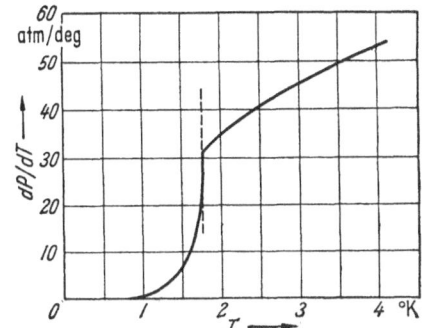

Fig. 38. Temperature coefficient of the melting curve.

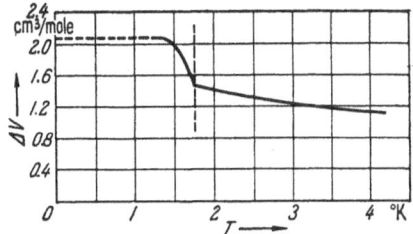

Fig. 39. Volume difference between liquid and solid helium as function of temperature.

Thus this value only exceeds by 1 atm the early extrapolation made by TAMMANN and quoted in Eq. (2.1).

The temperature at which the lambda-curve intersects the equilibrium curve between solid and liquid was determined from these data as 1.77 ($\pm$0.005) °K and the pressure as 30.0 ($\pm$0.1) atm.

The slope of the melting curve between this temperature and 1.0 °K decreases by more than a factor of 100 as shown in Fig. 38 indicating that in this temperature region the entropy difference between the liquid and the solid phases disappears rapidly. As was pointed out by SIMON [70], the disappearance of the entropy difference requires a fortiori the disappearance of the melting heat ($T\Delta S$) and since the latter is given by $\Delta U + P\Delta V$, this can occur in two different ways Firstly, both the energy difference between the phases ΔU as well as the volume difference ΔV can become zero which would mean that the two states of helium would be identical. For this there is no indication from the experimental evidence. The other way of reaching zero melting heat is for ΔU becoming equal to $P\Delta V$ as absolute zero is approached. This relation was predicted by SIMON in 1934 and its

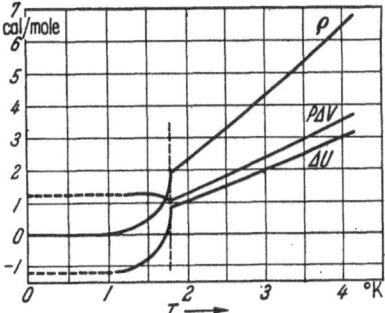

Fig. 40. The melting heat ϱ, $P\Delta V$, and the energy difference ΔU between liquid and solid helium.

correctness appears clearly from determinations of the volume difference between 1.2 and 1.9 °K carried out by SWENSON. Combining his results with those obtained at higher temperatures in Leyden, the curve shown in Fig 39 was obtained which extrapolates to the value of 2.07 ($\pm$0.06) cm³/mole for ΔV at absolute zero.

The melting heat was calculated from the data of Figs. 38 and 39, by the CLAUSIUS-CLAPEYRON equation and was found to be of the form 0.021 T^8 cal/mole for the temperature interval between 1.0 to 1.4 °K. The temperature variation

26*

of the melting heat together with that of $P\Delta V$ and ΔU is shown in Fig. 40. ΔU passes through zero at 1.72 °K which is the temperature at which a tangent

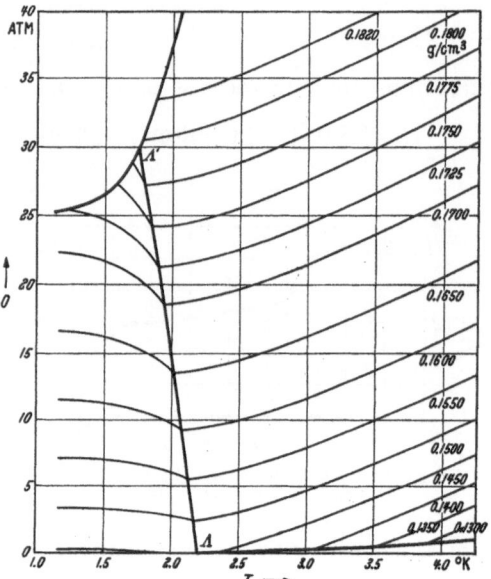

Fig. 41. Isopycnals of liquid helium.

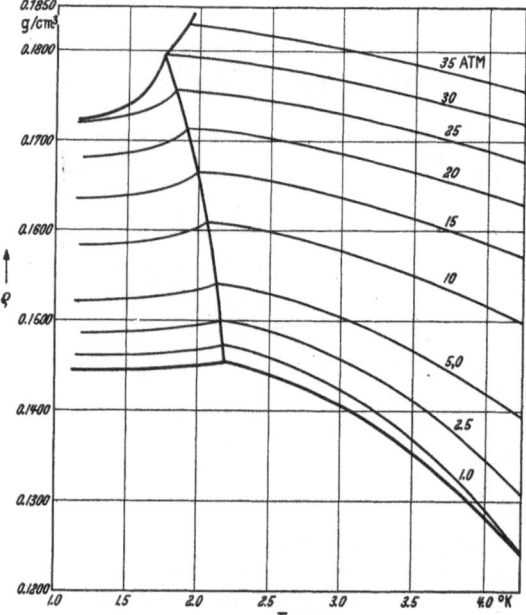

Fig. 42. Isobars of liquid helium.

drawn through the origin of the diagram of state will touch the melting curve. At absolute zero ΔU has the value of -1.2 cal/mole. The invariance with temperature of ΔU below 1 °K leads to the purely mechanical nature of the melting process in helium at very low temperatures to which reference has already been made in Sect. A.

21. Density variation. The equation of state of the liquid phase of helium was investigated by KEESOM and Miss KEESOM [71] in 1933 who determined the isopycnals between 1.15°K and the boiling point and under pressures up to 35 atm. A diagram of smoothed curves for different densities against pressure and temperature is given in Fig. 41. The isopycnals which cross the lambda-curve are made up of two branches which form an angle at this curve and therefore lead to discontinuities in $(\partial p/\partial T)_v$ and in $(\partial v/\partial T)_p$. An interesting feature of the observations was that an isopycnal at high pressure which reaches the melting curve from the helium I region showed pronounced supercooling of the liquid state below the melting temperature. No such supercooling was ever observed in the isopycnals which reach the solid-liquid equilibrium curve from the helium II region. From their results the authors also derived the isobars which are shown in Fig. 42. These curves, too, show two branches, characteristic for helium I and II which meet at an angle at the lambda-curve.

22. Vapour pressure. The first determination of the pressure of the saturated helium vapour against a gas thermometer was carried out by KAMERLINGH ONNES [72] in 1910. Since then a great number of very careful and detailed measurements have been made, particularly by KEESOM and his school. The accurate knowledge of the vapour pressure curve of helium is mainly important as an aid to temperature measurement and for this reason has received much attention. A description of the various attempts to link up the vapour pressure with primary thermometers or to derive it from thermodynamical formulae on the basis of measured values of other thermodynamic functions lies outside the scope of this article.

The question of a discontinuity at the intersection of the lambda-curve with the vapour pressure curve has been discussed by KEESOM [73]. He has shown that the meeting of two branches of the heat of vaporisation at the lambda-point such as is suggested by the measurements of KAMERLINGH ONNES and DANA requires two branches of the vapour pressure curve to meet at the lambda-point with the same tangent. KEESOM has pointed out that deviations from a smooth function of the vapour pressure due to this anomaly will be too small to falsify the determination of this quantity as they are smaller than the accuracy of measurement.

In 1955 equations for the vapour pressure curve were proposed by VAN DIJK and DURIEUX [74] in Leiden and by CLEMENT, LOGAN and GAFFNEY [75] in Washington. Both formulae agree with the latest experimental determinations to within about $\pm 0.002°$. The equation of the last named authors reads

$$\left.\begin{aligned} \ln P = I - \frac{A}{T} + B \ln T + \frac{1}{2} C T^2 - \\ - D \left[\frac{\alpha \beta}{\beta^2 + 1} - \frac{1}{T}\right] \arctan (\alpha T - \beta) - \ln \frac{T^2}{1 + (\alpha T - \beta)^2} \end{aligned}\right\} \quad (22.1)$$

in which, if the pressure is measured in mm Hg at 20° C, the constants have the following values: $I = 4.6202_5$, $A = 6.399$, $B = 2.541$, $C = 0.00612$, $D = 0.5197$, $\alpha = 7.00$ and $\beta = 14.14$.

C. Entropy.

Even without considering the lambda-phenomenon, knowledge of the entropy diagram of liquid helium is of particular interest. Since it became clear that helium under its saturation pressure would remain in the liquid state even at absolute zero, the question of how the state of perfect order, which any substance in internal equilibrium must exhibit at zero temperature, would be established. The obvious way of determining the entropy is to measure the specific heat down to sufficiently low temperatures so that no appreciable entropy changes might be expected and extrapolation to absolute zero become feasible. Using the measured specific heats above 1 °K and a number of other determinations such as the entropy changes on melting and evaporation, an entropy diagram has been established which is shown in Fig. 43. Above the lambda-point the entropy of the liquid under saturation conditions varies roughly linear with the absolute temperature and extrapolates with fair approximation to zero at absolute zero. With increasing pressure the entropy decreases in this region as would be expected.

Just above the lambda-point there is a decrease in the entropy which corresponds to the rise in the specific heat with falling temperature in the same temperature region. This is followed by a very rapid fall in entropy in the helium II region. The position regarding the pressure is now reversed, the entropy of the liquid rises as it is compressed. The possible reasons for this rapid increase in the state of order of the liquid as it is cooled below the lambda-point has been the subject of much speculation, a summary of which has been given in part A of this article.

Points of particular interest beyond the data given in Fig. 43 are the exact nature of the entropy variation below 1 °K and the determination of the entropies of the two constituents in the two-fluid model. We will therefore discuss separately determinations of the specific heat at low temperatures, as well as measurements of the entropy difference between superfluid and normal constituents as determined by measurements of the heat of transport and of the thermo-mechanical effect.

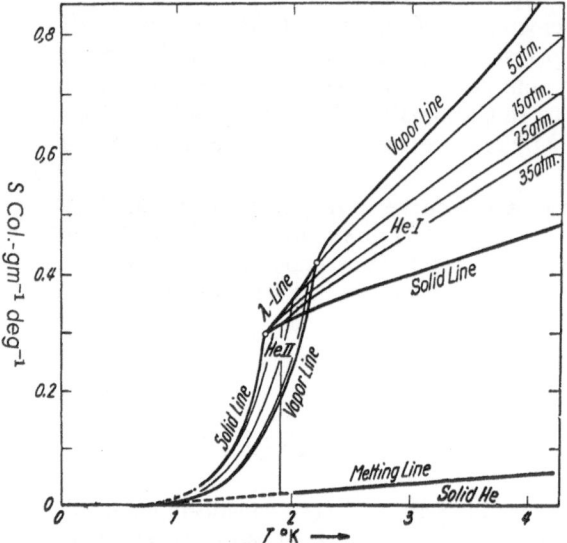

Fig. 43. Entropy diagram of helium.

23. Specific heats. Even before any theoretical model for liquid helium II had been proposed and simply by regarding the specific heat anomaly as the excitation of unspecified degrees of freedom it was assumed that this excitation would commence at a finite temperature only. Consequently, at temperatures below this onset the specific heat it was expected to rever to a simple cubic temperature dependence, characteristic of phonon excitation.

Preliminary experiments, mentioned in part A, were carried out by Simon and Pickard [76] in Oxford but, as appeared later, the results were in error. It should be pointed out here that accurate determinations of the specific heat, and in particular that of liquid helium, in the temperature region below 1 °K are extremely difficult. The substance has to be cooled by the adiabatic de-magnetization of a paramagnetic salt with which it has to be kept in good thermal contact. Since make and break thermal contacts in this region are difficult, the salt will remain in contact with the substance during the measurement of specific heat. This means that the heat capacity of the salt has also to be known very accurately and that its own heat capacity, particularly at the lowest temperatures is likely to outweigh by far that of the substance under investigation. Since usually the magnetic susceptibility of the salt is used as thermometer, excellent thermal contact between salt and sample has to be maintained throughout the course of the measurement. In the case of helium II there is the added difficulty that film flow and re-condensation along the tube by which the liquid has been filled into the calorimeter will cause a large influx of heat which cannot be tolerated in a temperature region where heat capacities are, on the whole, small and where the maintenance of a good vacuum is exceptionally difficult. It is thus not surprising that reliable data of the entropy of helium II at very low temperatures have not been obtained until fairly recently.

Simon and Pickard overcame the difficulty presented by the heat leak due to film transport by employing a sealed capsule containing the paramagnetic salt and helium gas under high pressure at room temperature. On cooling, the helium liquefied and formed a completely enclosed sample for calorimetric measurements.

In 1941 Keesom and Westmijze [77] published a short note on measurements between 0.4 and 1.5 °K. They gave no details as to the method used and merely state that between 0.6 and 1.5 °K the specific heat could be represented by

0.023 T^6 cal/g/degree. They suggested that between 0.4 and 0.6 °K the power dependence might be somewhat smaller, but that their values were considerably lower than those obtained by SIMON and PICKARD. The temperature interval between 0.6 and 1.6 °K was also covered by the work of HULL, WILKINSON and WILKS [78] who used the capsule method and were the first to give a detailed account of their experiments. They state that below 1.4 °K the specific heat can be represented by 0.024 $T^{6.2}$ cal/g/degree.

A very careful study ranging from 1.9 down to 0.2 °K has been made by KRAMERS, WASSCHER and GORTER [52]. They used a calorimeter which was

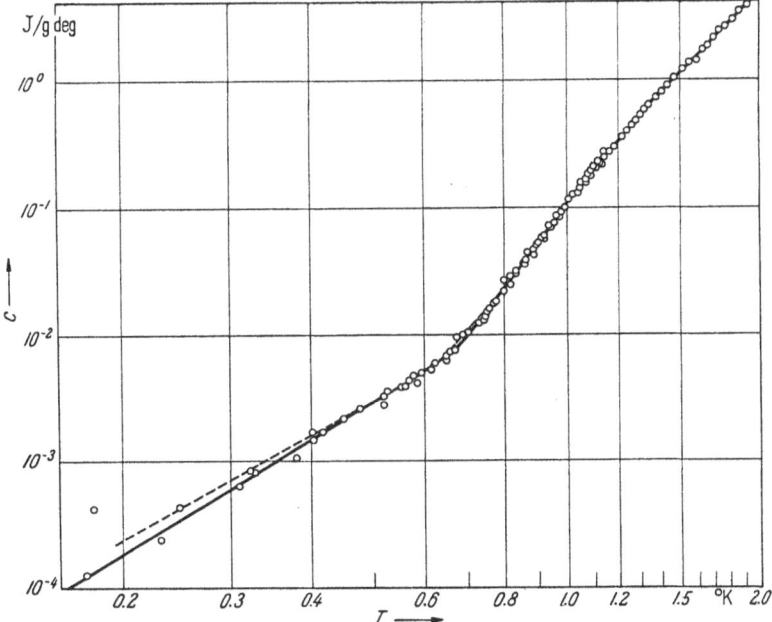

Fig. 44. Specific heat of liquid helium.

connected to the helium supply by a fine capillary through which the liquid was admitted. The results of this work are shown in the logarithmic plot of Fig. 44. The most striking feature is the fairly sharp change in steepness in the neighbourhood of 0.7 °K. Below this temperature the results can be represented by 0.0235 ($\pm$ 0.0015) T^3 joule/g/degree. Above this temperature there is somewhat better agreement with HULL, WILKINSON and WILKS than with KEESOM and WESTMIJZE whose values appear to be slightly too low. The slight curvature of the upper branch of the values which is particularly noticeable at the highest temperatures makes it clear that in this region the specific heat cannot be accurately represented by a single power function.

These results appear to reveal clearly the phonon entropy which gives rise to the cubic temperature function below 0.6 °K and which had been postulated by a number of authors. The absolute value is in good agreement with the theoretical prediction which postulates for the phonon specific heat

$$C = \frac{16}{15} \pi^5 \frac{k^4 T^3}{h^3 \varrho u_1^3} \tag{23.1}$$

and then gives 0.021 T^3 joule/g/degree with the measured value of the sound velocity u_1. The onset of extra excitations above 0.7 °K corresponds in LANDAU's theory to the appearance of rotons and in TISZA's two-fluid model to the evaporation in velocity space of the BOSE-EINSTEIN condensate. The form of the expected specific heat dependence on temperature is thus similar in the two theories but, as has been mentioned in part A, the significance of the two contributions in the specific heat function for the problem of superfluidity is fundamentally different. In LANDAU's theory the superfluid is not only free of the roton but also of the phonon entropy whereas according to TISZA the superfluid should retain its phonon entropy. A decision as to whether or not phonon entropy is carried by the superfluid cannot therefore be decided on the basis of specific heat measurements alone but requires a separate determination of the entropy of the normal fluid. Such data are provided by the measurements at low enough temperatures of the thermo-mechanical effect and of the heat of transport.

Before, however, discussing these determinations a short survey must be given of the measurements of the specific heat of helium above 1 °K which have been carried out recently. HERCUS and WILKS [79] in Oxford have measured the specific heat under different pressures between those of the saturated vapour and the melting pressure in the temperature interval between 1 and 2 °K. They found satisfactory agreement with the only other direct determination of the specific heat under pressure which were a few data obtained by KEESOM and CLUSIUS [10] under 19 atm in 1932. A more detailed comparison could be made between their results and the isopycnal data of KEESOM and Miss KEESOM [71]. Bearing in mind that this comparison is based on the second differential of a smoothed curve, they concluded that here, too, the agreement was satisfactory. On the other hand, their values at saturated pressure did not agree too well with those of KRAMERS, WASSCHER and GORTER, being on the whole 10% too high. Quite recently the specific heat in this region has been re-measured by KAPADNIS [80] who found satisfactory agreement with the values of KRAMERS, WASSCHER and GORTER and there thus exists a possibility that the higher values of HERCUS and WILKS may be in error. It is at present uncertain to which degree their specific heats and derived entropy values under pressure are affected.

24. The thermo-mechanical effect. The thermo-dynamical derivation of the relation between the thermo-mechanical pressure difference and the temperature rise corresponding to it by H. LONDON has already been mentioned in Sect. 12. Assuming strict reversibility, $\Delta p/\Delta T = \varrho \, \Delta S$ in which ΔS is the entropy difference between the bulk liquid and that part of the liquid which has flown in the direction of the higher temperature through the capillary link. It is to be expected according to the two-fluid model that the superfluid should be free of some or all excitations and the entropy difference might therefore be equal to the total or almost the total entropy of the liquid. According to TISZA's theory ΔS should correspond to that part of the heat content of the liquid which is taken up in anomalous excitations above 0.7 °K. LANDAU's theory, on the other hand postulates that ΔS is strictly equal to the total measured entropy of the liquid.

The problem to be solved by measurement can therefore be divided up into two parts. First, whether the general concept of the two-fluid model is correct, i.e. whether ΔS in LONDON's equation is of the order of S, and secondly whether ΔS is strictly equal S. The early measurements on the thermo-mechanical effect by ALLEN and REEKIE [81] were not sufficiently adiabatic to solve the first question. Using the arrangement shown in Fig. 22, KAPITZA [42] measured the thermo-mechanical pressures and corresponding temperature differences between 1.3 and 2.1 °K and found that they agreed with the total entropy according to the existing specific heat measurements. Similar conclusions were reached by MEYER and MELLINK [82] who made measurements of the same kind. While the second part of the problem could not be solved by these experiments because neither the specific heat at low temperatures was well enough known nor did the measurements of the thermo-mechanical data extend low enough, they established $\Delta S \approx S$ well over the rest of the temperature range. The work of MEYER and MELLINK, in particular, has shown that this relation holds to temperatures close to the lambda-point, provided reversible conditions could be maintained in the capillary link.

Observations have now been extended to 0.8 °K by PESHKOV [83] and, by measuring the integrated thermo-mechanical pressure between 1 °K and temperatures ranging down as far as 0.1 °K by BOTS and GORTER [84]. These latter authors in their first experiments found thermo-mechanical pressures yielding values of ΔS much below the entropies determined in specific heat determinations. Recent repetition of their measurements [85] indicates, however, that the entropy of the superfluid does not contain the phonon contribution. The result is thus in agreement with the observations of the heat of transport described below.

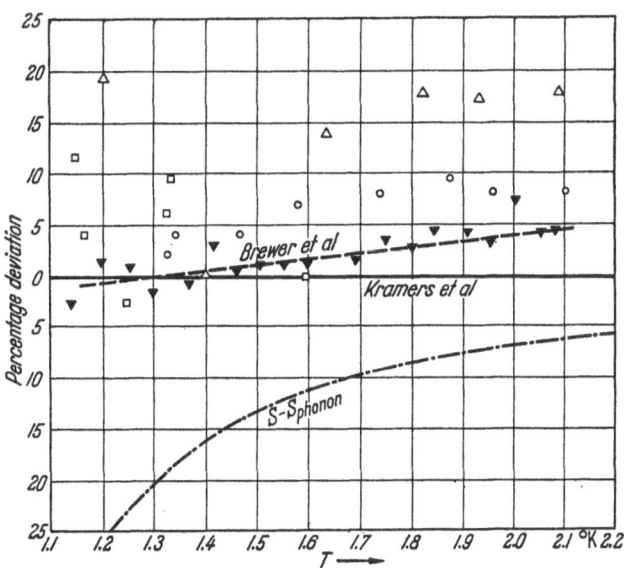

Fig. 45. Percentage deviation of measured heats of transport from the entropy calculated from specific heat measurements. ○ KAPITZA, △ CHANDRASEKHAR and MENDELSSOHN, □ PESHKOV, ▼ BREWER, EDWARDS and MENDELSSOHN.

25. The heat of transport. The first measurement of the heat which has to be supplied to unit mass of superfluid in order transform it into bulk liquid was made by KAPITZA [42], and he found that $Q_T \approx T S$, where S is taken from integration of the measured specific heat. CHANDRASEKHAR and MENDELSSOHN [86] used the arrangement shown in Fig. 93 to measure the heat of transport. They insured thus reversibility by having the film as connecting link between the two volumes of helium. In their work as in all other experiments of this kind, the heat supplied to an adiabatically separated vessel was determined together with the mass of helium transferred into the vessel by this quantity of heat. These experiments led to values of ΔS which were in better agreement with the specific heat measurements of HERCUS and WILKS [79] than with those of KRAMERS, WASSCHER and GORTER [52]. Since the former appear to be erroneously high, the question must arise whether heat of transport measurements using the film as connecting link may contain a term, so far unaccounted for, due to an additional energy required to turn the film into bulk liquid.

Further measurements of the heat of transport, but using bulk liquid on both sides of the slit, have recently been made by BREWER, EDWARDS and MENDELSSOHN [87]. Their values are in good agreement with the accepted specific heat data indicating with a fair degree of accuracy that ΔS is strictly equal S. A composite graph of the relevant results is given in Fig. 45 in which the percentage deviation of the heat of transport from the values of KRAMERS, WASSCHER and GORTER has been plotted against the absolute temperature. The curve for the value of Q_T to be expected if the superfluid were to contain phonons is also included. It is clear that in the neighbourhood of 1.4 °K the deviation of the heat of transport thus calculated from that corresponding to the total entropy becomes appreciable. The results of PESHKOV and of CHANDRASEKHAR and MENDELSSOHN show too much scatter to be useful, with the additional difficulty that the latter ones are consistently too high. At the relevant low temperatures, the values given by KAPITZA are in good agreement with the specific heats, but they are too high over the rest of the temperature range to be fully convincing. The latest results obtained in Oxford show a systematic deviation from the values of KRAMERS, WASSCHER and GORTER, but this deviation is not large enough to cast serious doubt on the agreement of the measured heat of transport with that derived from the specific heat. At 1.2 °K the difference between the heats of transport with and without phonons is 30% whereas the percentage deviation of the values of BREWER, EDWARDS and MENDELSSOHN is never greater than $\pm 3\%$ between this temperature and 1.7 °K.

Considering these results in conjunction with those on the velocity of second sound at low temperatures, there can be little doubt that the superfluid lacks not only the entropy due to the anomalous excitations but also that due to phonons. While this cannot be taken a proof for the correctness of the LANDAU theory, it certainly is in disagreement with the model proposed by TISZA.

D. Superfluidity.

Short mention of the discovery of superfluidity has already been made in Sect. 7. The subsequent publication of the full investigation by ALLEN and MISENER [88] contained all the basic facts of the flow of helium II through capillaries varying in diameter from 1 mm to 10^{-5} cm. The narrowest tubes had been made in a very ingeneous way by drawing down an alloy tube containing a great number of fine stainless steel wires. In this manner a great number of fine channels of approximately the same and uniform diameter were formed. They summarized their results as follows:

1. The dependence of the velocity of flow on pressure became less

(a) as the radius of the capillary was reduced, or

(b) as the capillary was lengthened, or

(c) as the temperature was lowered.

2. In the largest capillaries at a temperature close to the lambda-point, an approximation to laminar conditions of flow was observed.

3. In all capillaries at low pressures, the velocity increased with decreasing temperature, but the reverse held at higher pressures in large capillaries.

4. At all temperatures there was a minimum in the relation between the radius of the channel and the velocity at constant pressure. For channels smaller than 5×10^{-4} cm in width, the velocity increased rapidly with decreasing channel size.

5. At pressure above 50 dynes/cm² in the narrowest channels, the velocity was completely independent of pressure at all temperatures. The curve between the pressure independent velocity and the temperature bears a strong resemblance both in magnitude and shape to that for mobile surface films of helium II above the hydrostatic surface of the liquid.

6. In channels less than 10^{-3} cm in width, at temperatures close to the lambda-point, and at low pressures, the flow became laminar with evidence of an exceedingly small viscosity.

On the basis of these observations the authors concluded that two different flow mechanisms appeared to be operating simultaneously, frictionless superflow and ordinary viscous flow. The critical velocity of superflow was tentatively ascribed as ocurring along the walls which seemed then reasonable since the flow rate was found to be directly proportional to the radius. Fig. 46 gives the dependence of the mean velocity on pressure head, and it can be seen that there is a gradual change from potential flow in the narrowest channels to a more complex pattern, due to the appearance of dissipative flow. In capillaries of the order of 10^{-3} cm or more diameter, viscous flow is becoming so important that the characteristics of superflow are completely swamped. It has therefore become customary to discuss the observations on fine and wide channels separately, and we will follow this system since it permits clarification of a somewhat involved pattern of results. The discussion is further complicated by the fact that flow in helium II can be due to either a hydrostatic or a thermomechanical pressure difference. Since here the nature of the flow channel seems to be significant, we will discuss the two types of pressure side by side for the same arrangement.

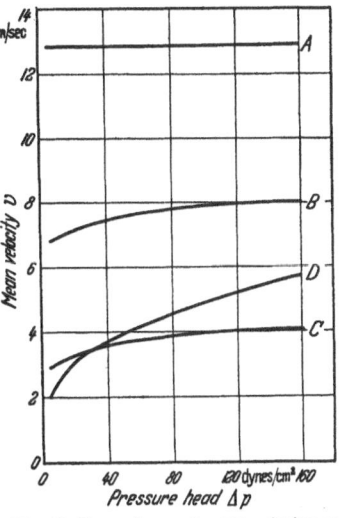

Fig. 46. Dependence of mean velocity on pressure head at 1.2 °K. Channel width: $A = 1.2 \times 10^{-4}$ cm; $B = 7.9 \times 10^{-5}$ cm; $C = 3.9 \times 10^{-4}$ cm; $D = 5.0 \times 10^{-3}$ cm.

Finally, it should be mentioned that ALLEN and MISENER also measured the flow of helium II through closely packed powder forming very narrow channels. The results differed, however, fundamentally from those obtained with capillaries and this type of flow will therefore be treated separately.

26. Narrow channels. ALLEN and MISENER'S curves for the temperature dependence of the velocity in channels of 1.2 and 7.9×10^{-5} cm diameter is shown in Fig. 47 from which it can be seen that there is indeed a close similarity between them and the transfer rate of the saturated film. The readings were taken under a pressure head of 160 dynes/cm². In spite of the fact that, as is evident from Fig. 46, the flow in the larger of the two channels is already slightly pressure dependent, the temperature function is still unaffected. For comparison the curves for capillaries of 5×10^{-3} cm and 4.38×10^{-2} cm diameter have been included which show clearly in which way the narrow channels differ from the wider ones. Later work has, on the whole, confirmed the type of temperature dependence observed by ALLEN and MISENER in narrow channels.

The fact that in the finest channels the flow is pressure independent immediately leads to the concept of a critical velocity by which the flow is limited and which, as is clear from Fig. 47, will become temperature independent at the lowest temperatures. In view of the very similar phenomena in the film, the idea of such a critical velocity has been frequently discussed and suggestions have been made by various authors as to relation between the channel width d and the critical velocity v_c. On mainly dimensional arguments it has been postulated that

$$m\, v_c\, d \sim \hbar \tag{26.1}$$

where m is the mass of the helium atom. This relation has been found approximately correct numerically in the case of the saturated film and ALLEN and MISENER'S narrowest channels which yield for the product $v_c d = 0.8 \times 10^{-4}$ and 1.6×10^{-4} cm²/sec respectively. Since the ratio of d-values in the two cases is only 5, this can hardly be taken as proof for the relation. Indeed further work on channels between 10^{-5} and 10^{-4} cm diameter has led to such widely varying results for the product $v_c d$ that it must at present appear to be incorrect, since the divergence of values can probably not be explained by errors in experimentation. A certain improvement has been achieved by making, as was suggested by MOTT, v_c vary with $d^{-\frac{1}{2}}$ instead of with d^{-1}, but even so, large discrepancies remain.

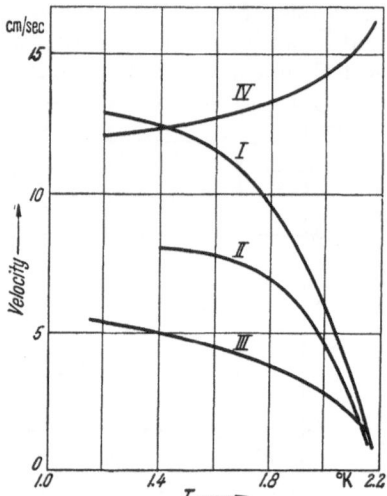

Fig. 47. Variation of flow velocity with temperature. Pressure head = 160 dynes/cm². Capillary size: $I = 1.2 \times 10^{-5}$ cm; $II = 7.9 \times 10^{-5}$ cm; $III = 5.0 \times 10^{-3}$ cm; $IV = 4.38 \times 10^{-3}$ cm.

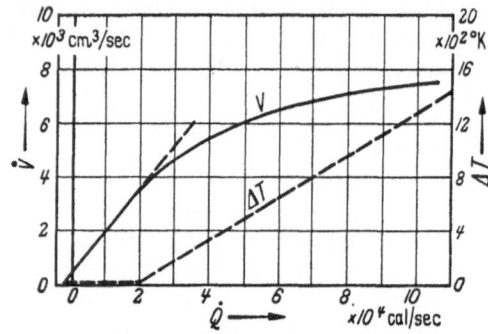

Fig. 48. Rate of flow $\dot{V}$ and temperature difference ΔT in thermo-mechanical transport through a narrow channel as function of the heat input $\dot{Q}$.

The chief difficulty in establishing and testing a relation of the above kind is the small range of sizes available for observation. In the neighbourhood of 10^{-5} cm the accurate determination of channel width becomes very difficult and thus does not permit reliable evaluation of results. On the other hand, in channels much larger than 10^{-4} cm the flow becomes very pressure dependent, as can be seen from Fig. 46 and consequently the determination of a critical velocity, or even its existence, becomes problematic. A decision as to the correct form of the dependence of velocity on channel size will thus have to wait until a method has been found which allows to define and measure critical velocities in wide channels. The recent work on the determination of the thermal resistance in wide capillaries mentioned in part F (Sect. 32) may possibly offer such an opportunity.

Corroberative evidence for the existence of a critical velocity in channels of 10^{-5} to 10^{-4} cm width was furnished by KAPITZA [42] who measured the flow rate under a thermo-mechanical pressure gradient. He used the arrangement shown in Fig. 22. On supplying heat to the inside of the bulb, liquid helium was drawn into through the slit between the optically flat discs. The width of this could be changed by the insertion of spacers and was measured with optical fringes. The rate of volume flow was measured for different rates of heat input, and a typical result at a constant ambient temperature is shown in Fig. 48. For the lowest heat inputs there is a roughly linear rise of volume flow which, however, at a certain value of Q begins to lag behind the impressed heat input. This behaviour suggests strongly the existence of a critical velocity up to which the volume flow will without friction follow the heat flow. KAPITZA also measured the temperature difference ΔT between the inside of the bulb and the helium bath. He noted no increase in temperature inside the bulb up to the critical heat input (corresponding to the critical velocity just defined) when a gradual rise of the temperature in the bulb over that of the bath was observed. Considering H. LONDON's equation according to which any change of liquid level in the bulb must be accompanied by a change in temperature, the last mentioned result of KAPITZA's must appear strange. However, the temperature changes due to

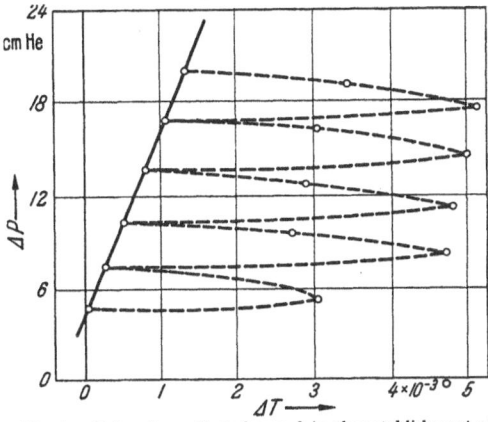

LONDON's equation were, under the conditions of his experiment probably too small to be detected. The rise in ΔT at the critical heat input should therefore be considered as a surge in temperature far above the value of ΔT corresponding to the thermo-mechanical pressure difference existing between bulb and bath.

The existence of these surges was subsequently demonstrated by MEYER and MELLINK [82] in Leiden who used a similar arrangement buth with a more sensitive thermometer. They found, as is shown in Fig. 49, relaxation effects which were particularly noticeable in narrow slits and close

Fig. 49. Relaxation effect observed in the establishment of a thermo-mechanical pressure difference.

to the lambda-point. The full line represents the true variation of pressure with temperature difference under equilibrium conditions which is in agreement with the LONDON equation. However, the way in which these equilibrium conditions could be realised depended on the rate and manner of heat supply. Only for

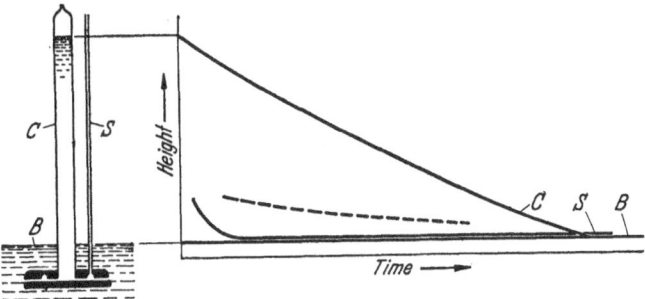

Fig. 50. Outflow through a slit (1 micron) under a hydrostatic pressure difference with intermediate pressure measured in a static tube. Level in reservoir C, in static tube S and in bath B.

small rates of heat input was the full line followed without deviation. On larger (supercritical) rates of heat input a rapid rise in temperature, indicated by the broken lines, was noted. On switching off the heat, the temperature in the bulb decreased again until a point on the equilibrium line was reached.

The existence of critical velocities was clearly shown in experiments in Oxford in which in the same flow channel the flow under hydrostatic and thermo-mechanical pressure differences was investigated. Of a number of devices used that employed by BOWERS and MENDELSSOHN [89] gave the most satisfactory results. The object of their experiments was not only to measure the value and pressure dependence of the flow rate but also the intermediate pressure at an arbitrary place along the flow channel. Their apparatus which is shown in Fig. 50 is similar to that used by Kapitza in the discovery of superfluidity. It was a cylindrical reservoir to whose lower end a flat glass disc was attached. This disc was opposed by another optically flat glass plate so that helium could flow in and out of the reservoir through an annular channel of about 1 micron width. An annular groove was cut into the upper plate, about half way along the flow channel, to which a narrow "static tube" for pressure measurement at this point in the flow channel was attached.

As the reservoir was filled and withdrawn from the bath, its level C was dropping at an almost constant rate, indicating superflow with a critical velocity. The level S in the static tube dropped immediately to the height of the bath level B and stayed there throughout the experiment, the small level difference with the bath being due to surface tension in the narrow tube. This showed that there was no pressure gradient inside the flow channel and, since the static tube was placed at an arbitrary position, one can only conclude that the whole pressure drop in a channel carrying superflow must occur at the end.

Flow under a thermo-mechanical pressure difference was produced by closing the top of the reservoir and supplying energy to a heating coil inside the reservoir. As heat was supplied, the level inside the reservoir rose, showing that liquid was drawn in under a thermo-mechanical pressure difference. The level in the static tube, however, again remained at the height of the bath level. This demonstrated that the flow channel failed to maintain a thermo-mechanical as well as a hydrostatic pressure difference. Since in the last experiment the pressure difference between bath and bulb is accompanied by a temperature difference, the behaviour of the static tube suggests that in thermo-mechanical superflow, too, there is no temperature gradient in the flow channel and that the temperature difference is concentrated at one end of the channel.

In these experiments, too, critical flow rates were found. Three different criteria for the critical rates were observed which agreed with each other within the experimental error. The first criterion was similar to Kapitza's shown in Fig. 48. There was a linear rise of the flow rate with $\dot{Q}$ and a fairly sharp departure of the curve from linearity at the critical heat input. Secondly, a phenomenon akin to the relaxation effect by Meyer and Mellink was observed. When the rise of level in the reservoir with heat input was observed under increasing values of $\dot{Q}$, a change of behaviour occurred at the critical rate. Below this rate the rise stopped immediately when the heating was switched off, whereas above it the level continued to rise for a while after heating was discontinued. This indicated that beyond a critical rate superfluid could not pass through the flow channel sufficiently fast to keep pace with the heat supply and overheating took place. The third criterion was observed in the static tube. Its level, which remained with the meniscus of the bath for small flow rates, fell below the bath level when a certain flow rate was exceeded. The explanation for this behaviour is possibly that above the critical rate the helium column in the reservoir lags behind the thermo-mechanical equilibrium pressure and that the helium in the flow channel now flows to an effectively lower pressure.

Measured values for these three criteria are given in Fig. 51 together with the flow rates under three hydrostatic pressure differences. Since the pressure dependence of the flow is weak, the three curves differ little and the critical values obtained under thermo-mechanical pressure are all well within this spread.

Using essentially the same apparatus Swim and Rorschach [90] at the Rice Institute have accurately measured the pressure dependence for gaps between the plates of 2.4 and 4.3 microns. They employed a tall reservoir which allowed them to go up to pressure heads of 2×10^3 dyne/cm^2 and a set of curves for the narrower gap is shown in Fig. 52. Expressing the volume flow rate in terms of $(\Delta p)^n$, they found that n varied between 0.33 and 0.36 from 1.4 to 2.1 °K in the case of the wider slit and between 0.27 and 0.28 between 1.4 and 1.77 °K for the narrower one. They also confirmed the behaviour of the level in the static tube observed by BOWERS and MENDELSSOHN. WINKEL [91] and others in Leiden also have determined a few critical velocities in slits between 0.43 and 3.1 microns. They used various criteria, but readings were only taken at three different temperatures. These suggest a falling off of the velocities with decreasing temperature between 1.9 and 1.5 °K.

BOWERS and WHITE [92] in Oxford investigated the flow of helium II through etched copper membranes with an average channel size of 1 micron. Nothing was known concerning the actual shape of these channels. The work was limited to flow under hydrostatic pressures since the good heat conduction of the membranes did not

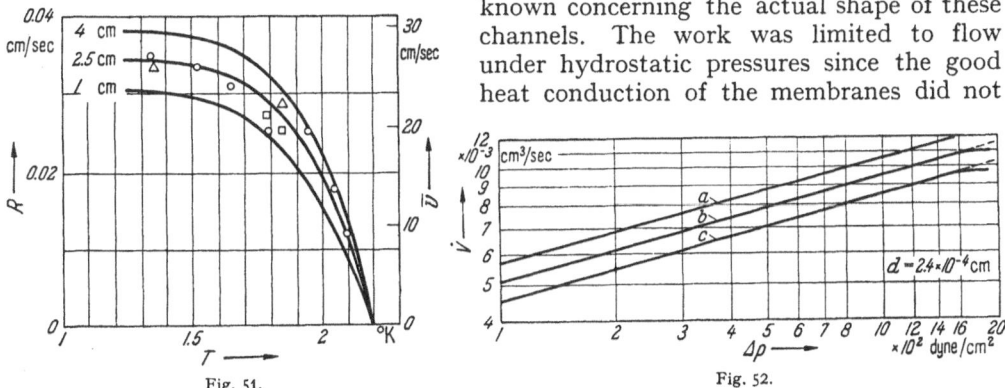

Fig. 51.

Fig. 52.

Fig. 51. Critical rates determined by three different criteria in thermo-mechanical flow and flow rates under hydrostatic pressure heads as function of temperature for a 1 micron slit. ($\bar{v}$ average flow velocity.)

Fig. 52. Flow rates under hydrostatic pressure through 2.4 micron slit (a) 1.40 °K, (b) 1.64 °K and (c) 1.77 °K

permit the setting up of a temperature difference. The pressure dependence of the flow was found to be similar to that of slits between discs, n having a value of 0.2 between 1.2 and 1.9 °K, rising to almost 0.3 at 2.1 °K. The work with membranes permitted an extension of the measurement of the intermediate pressure. In the experiments with slits the ratio of the normal flow resistances of the two parts of the annular gap amounted to 4:1 with the total pressure drop in superflow being concentrated in the part with the higher normal resistance. Geometry did not allow in this arrangement to make the normal resistances more nearly equal. However, in the work with membranes, two of these could be selected which had a ratio of 4:3 in the normal flow resistances. Here the full pressure drop under superflow was again taken up by the membrane with the slightly higher flow resistance, irrespective of whether it was placed between reservoir and static tube or between the bath and static tube.

27. Flow through packed powder. In their fundamental experiments on superflow, ALLEN and MISENER [88] also used a tube which was closely packed with jeweller's rouge, i.e. finely powdered Fe_2O_3, the grain size of which was estimated as being of the order of 10^{-5} cm. The size of the individual flow channels between the grains was thus about 10^{-6} cm which is smaller than the finest wire tubes used by the same authors and approximating the thickness of the film. On the other hand, it is clear that the geometry of the channels in such a powder must be complex.

Flow observations were made between the lambda-point and 1.1 °K, and it was found that for constant pressure head the flow rate changed with temperature in much the same way as in wire tubes or in the film. The pressure dependence of the flow, however, was marked and but for a temperature very close to the lambda-point the flow rate was always proportional to the square root of the pressure. This is characteristic of turbulent flow which evidently, as the lambda-temperature is approached, changes into laminar flow. For the latter region the coefficient of viscosity was estimated as being of the order of 10^{-8} poise.

In view of the striking difference in behaviour between the tubes filled with wires and those filled with powder, the latter were further investigated by CHANDRA-SEKHAR and MENDELSSOHN [93] who used a static tube and also extended the work

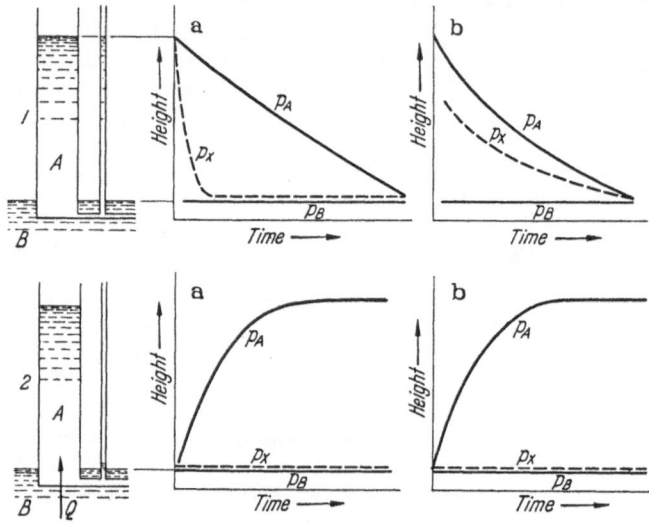

Fig. 53a and b. Superflow under (1) hydrostatic, and (2) thermo-mechanical pressure in (a) slits, and (b) powder packed tubes. (P_x level in static tube.)

to flow under thermo-mechanical pressure differences. Their observations under hydrostatic pressure completely confirmed the results of ALLEN and MISENER. On the other hand, it was noted that when liquid helium was drawn into the reservoir by the application of a heat current, the level in the static tube remained at the height of the bath meniscus. Thus under a thermo-mechanical pressure difference the behaviour of the powder tube was identical with that of a fine slit, suggesting no dissipation inside the flow channels and the existence of a thermo-mechanical pressure difference at the entrance of the powder plug. Care was taken that in flow under hydrostatic as well as under thermo-mechanical pressures the same velocity range was investigated. The results on slits and powder tubes, taken together, thus reveal an inconsistency in the flow mechanism which has as yet not been resolved. A schematic diagram of this inconsistency is shown in Fig. 53 in which (a) and (b) refer to slits and powder tube respectively while (1) and (2) denote hydrostatic and thermo-mechanical pressures.

In thermo-mechanical flow through the powder tube the same criteria for critical velocities as in slits were observed. The sudden onset of friction when a critical flow rate is exceeded is even more pronounced than in the observation on slits. Fig. 54 shows a plot of the rate of volume flow against heat input in which the sharp departure from linearity is clearly marked. When comparing the critical rates obtained in experiments with thermo-mechanical flow with the flow rates under varying hydrostatic pressure heads it was found that they corresponded to a pressure difference of ~ 1200 dyne/cm^2 at all temperatures (Fig. 55). Since it was noted that in each case the pressure drop occurred at that end of the powder tube which was connected to the reservoir, experiments have recently been carried out in which heat could be supplied either to the reservoir or to the bath. It was found that the pressure drop now always took place at the warm end of the powder tube.

In the work of CHANDRASEKHAR and MENDELSSOHN, too, a comparison between the results and the expected viscosity in helium II has been made. The power law by which the hydrostatic pressure is related to the flow velocity

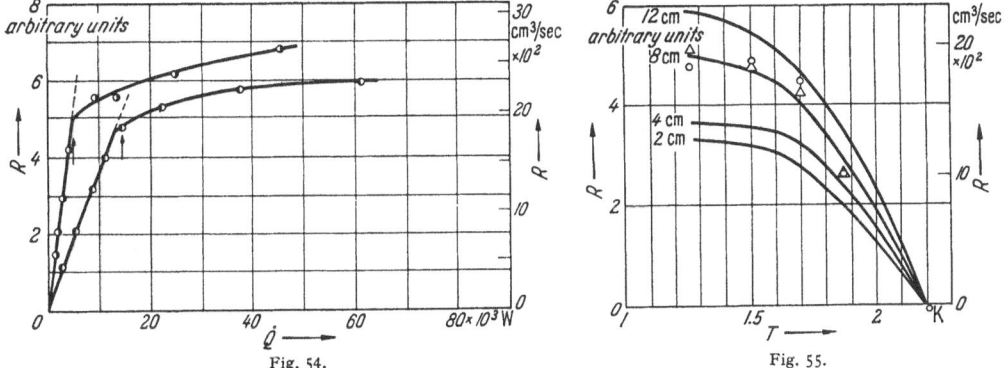

Fig. 54.   Fig. 55.

Fig. 54. Thermo-mechanical flow rates through a powder packed tube against heat input at ◑ 1.27 °K and ◑ 1.52 °K.

Fig. 55. Critical rates determined by two different criteria in thermo-mechanical flow, and flow rates under hydrostatic pressure heads as function of temperature for a tube with packed powder.

suggested in these experiments also the existence of turbulent flow. Comparing the flow of helium I through the same tubes with that below the lambda-point a coefficient of viscosity was obtained which was 100 times smaller than the measured viscosity of the normal fluid in the helium II region. The authors therefore concluded that the flow observed by them in the powder tubes was superflow in which the onset of a small degree of dissipation led to the appearance of turbulence in the superfluid in hydrostatic flow.

It must thus be concluded that while some, admittely inadequate, analysis has been possible for narrow and wide capillaries, the behaviour of powder tubes cannot at present be explained at all.

28. Wide capillaries. The temperature variation of the flow rate in wide capillaries at constant hydrostatic pressure as observed by ALLEN and MISENER has already been included in the data given in Fig. 47. Their work was supplemented by some further measurements on wide capillaries made by JOHNS, GRAYSON SMITH and WILHELM [94] in Toronto. These authors worked in the border region between helium I and II and, by interpreting their results as due to a sum of a viscous and a superfluid term, deduced the coefficient of viscosity between the boiling point and 1.8 °K. The lack of data between 2.25 and 3.4 °K as well as the fact that the numerical value of their viscosity points in the helium I region is much too high, however, make the interpretation of these result doubtful.

The only other investigation on wide capillaries is a careful study made by ATKINS [95] in 1951 on four diameters, ranging between 2.6×10^{-3} and 4.4×10^{-2} cm and using different lengths. Corrections were applied for the thermo-mechanical effect, film flow, end effects and acceleration due to changing velocity. A typical graph of results is given in Fig. 56, where the mean velocity is plottet against the pressure gradient at a temperature of 1.22 °K. The pressure dependence of the velocity is complex, but it is clear that the data for vanishing pressure gradients for the three larger capillaries do not allow a decision whether there is a finite intercept on the velocity axis for zero pressure gradient. The author therefore concludes that for capillaries of 8×10^{-3} cm or more the critical velocity, if it exists, cannot exceed 1 cm/sec. The results admit the possibility of a critical velocity of the order of 3 cm/sec in the finest capillary of 2.6×10^{-3} cm diameter.

ATKINS has tried to interpret his results on the basis of the GORTER-MEL-LINK theory of mutual friction (see also part F, Sect. 32) assuming that the superfluid can at best pass through the capillary with the critical velocity. The normal fluid is retarded by friction against the wall and against the superfluid. This leads to the following equation for the average mean flow velocity

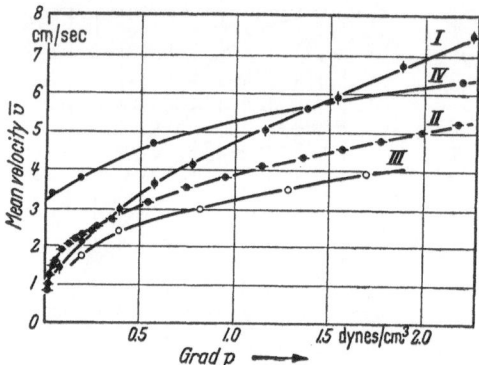

Fig. 56. Mean flow velocity under hydrostatic pressure heads at 1.22 °K for capillaries of diameter: $I = 4.4 \times 10^{-3}$ cm, $II = 2.03 \times 10^{-3}$ cm, $III = 8.15 \times 10^{-3}$ cm and $IV = 2.62 \times 10^{-3}$ cm; and of length: $I = 48.6$ cm, $II = 46.6$ cm, $III = 8.03$ cm and $IV = 7.76$ cm.

$$\bar{v} = \frac{\varrho_s}{\varrho} \left(- \operatorname{grad} \frac{p}{C \varrho \varrho_n} \right)^{\frac{1}{3}} + \left. + r^2 \operatorname{grad} \frac{p}{8 \eta_n} \right\} \quad (28.1)$$

where ϱ, ϱ_s and ϱ_n are the total, the superfluid and the normal densities, r the radius of the capillary, η_n the coefficient of viscosity of the normal fluid, and C the constant of mutual friction. The author concludes from a comparison of his results with the equation that, while in rough outline the observations can be explained by an equation of this kind, the detailed agreement is not too good. Considering the comparison of the thermal conductivity data to which the reader is referred (Sect. 32) with the simple form of the mutual friction theory, the existing disagreement cannot be considered surprising.

Summarizing the observations on flow through channels of varying size, it must appear evident that, while it is possible to account in an extremely qualitative manner for the phenomena, far too little information is available to test the existing theories rigorously. As in the case of thermal conduction the interplay of the flow mechanisms of the two-fluid model only permit a reasonably clear assessment of the relevant factors when one of the mechanisms is predominant. It also seems certain that until methods are found to separate the flow mechanisms in capillaries of medium diameter by experimental or theoretical means, little progress can be expected.

A note should be added on very recent and as yet unpublished experiments in Oxford and Philadelphia by which the range of observation of superflow has been increased. The work is still in progress but the results obtained so far appear to be sufficiently significant to deserve mention. In these experiments the flow of helium II through porous glass has been investigated. The samples were intermediate products in the making of boro-silicate glass supplied by the Corning Glass Works. The average pore size is estimated to be of the order of 10^{-7} cm. This is well below the thickness of the saturated film. In these samples the onset of superfluidity occurred at temperatures between 1.4 and 1.9 °K, varying from specimen to specimen. The onset itself is very sharp and the nature of the flow appears to be complex. This is the first instant in which the onset of superfluidity of the bulk liquid had been found depressed below the normal lambda-point at 2.19 °K. It is worthy of note that the onset of superfluidity in these samples of porous glass occurs at roughly the same temperature at which superfluidity ests in in subsaturated films of about 10^{-7} cm thickness. On purely dimensional reasons MENDELSSOHN [96] has suggested that

$$T_\lambda \sim \text{const} \left(\frac{1}{l^2} - \frac{1}{d^2} \right) \tag{28.2}$$

where l is the mean free path in helium I and d the distance in helium II over which the momentum of a helium atom will remain unchanged. The latter has therefore the meaning of a length characteristic of the lambda-phenomenon. If the size of the liquid domain is decreased below the value of d, the lambda-temperature must decrease. Work on unsaturated films has suggested that the value of d may be of the order of 10^{-6} cm. The depression of the onset of superfluidity in the passage of helium II through porous glass is thus in good agreement with these considerations, whose basis is simply the assumption that the energy of the lambda-anomaly is that which is required to shorten the mean free path in helium II to the mean free path in helium I.

E. Viscosity.

The discoveries described in Sect. 7 will make it clear that, while in helium I a coefficient of viscosity may be expected to have a well defined meaning, the phenomena in helium II can hardly be described in such terms. The discrepancy of results in the early work with helium II, which by the use of different methods extended to a factor 10^6 in the measured values of the viscosity, requires a quite different approach to the problem. The phenomena of superfluidity were therefore being treated separately in part D and the present part, apart from the data on helium I, will be limited to the determination of the viscosity of the normal component of helium II and of its concentration in dependence on temperature.

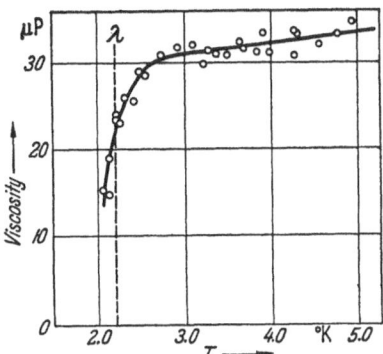

Fig. 57. Viscosity of liquid helium I.

29. Helium I. In addition to the work listed in part A, some observations on the viscosity have been made by JOHNS and others [97] in Toronto, using capillary flow. They have been briefly discussed together with the experiments on superfluidity in Sect. 28. Flow through capillaries was also used by BOWERS and MENDELSSOHN [98] whose results are shown in Fig. 57. They differ from the earlier work in Leiden with oscillating discs in that they yield almost temperature independent values of the viscosity coefficient between 5 and 2.7 °K which are followed by a fairly rapid decrease from about 30 micropoise at the latter temperature to about 22 micropoise at the lambda-point. Moreover, there was no sharp change at the lambda-point, the transition in the viscosity being quite continuous. These discrepancies are, however, not due to the different methods of measurement used, since more recent experiments by DE TROYER, VAN ITTER-

Beek and van den Berg [99], also using oscillating discs, have completely confirmed the data obtained with capillaries. The magnitude and temperature dependence of the viscosity of helium I is therefore well established.

It is worthy of note that in contra-distinction to ordinary liquids, the viscosity of helium I does not decrease with falling temperature. The absolute value, too, is very small, only about three times that of the gas. These features emphasize the gas-like properties of the liquid which are due to its low density, caused in turn by the high zero point energy. A change to a behaviour more similar to liquids could indeed be observed in measurements carried out in Leiden [100] in which the viscosity was determined at elevated pressures. Another interesting feature is the drop between 2.7 °K and the lambda-point which co-incides with the rise in the same temperature region of the specific heat in preparation for the lambda-phenomenon. No theory exists for the liquid in this region, but if the model of Bose-Einstein condensation is considered, then this temperature range must correspond to the anomalous increase of the specific heat of the ideal gas over the value $\frac{3}{2}R$ due to the filling up of the states of low energy. The possibility has also been considered that in this region fluctuations might cause a mixture of helium I with small localized clusters of helium II. Experiments have therefore been carried out with very fine capillaries [101] designed to decide whether or not in this temperature range the behaviour of the flow departs in any way from the classical viscous pattern. The results have, however, been negative, showing no departure from ordinary viscous flow.

30. Helium II. The early experiments with oscillating discs showed that in the helium II region the viscosity falls with decreasing temperature monotonically. It was, however, pointed out by Landau and also by Tisza that under the assumption of the two-fluid model the apparent viscosity measured in this way will have little meaning. Since it is only the normal component which interacts with the disc while the superfluid is not dragged along with it, the square of the damping is a measure of $\eta_n \varrho_n$. The relevant quantity is thus the coefficient of viscosity of the normal constituent η_n and in order to derive this from the observed damping, knowledge of the relative concentration of the normal component in dependence of temperature is necessary.

The experiment of Andronikashvili [49] which permits a direct measure of the normal concentration by the dragging along of fluid in the motion of a stack of closely packed discs has been described in part A (Sect. 15) and the results obtained by him are shown in Fig. 27. In the region accessible to him, between 1.3 °K and the lambda-point the ratio ϱ_n/ϱ was found to be equal to $(T/T_\lambda)^{5,6}$. These data have since been supplemented by derived values of the normal concentration from measurements of the velocity of second sound. A composite logarithmic plot of ϱ_n/ϱ against the absolute temperature for the whole region between 0.1 °K and the lambda-point is given in Fig. 58. As can be seen, there is a more rapid drop with temperature between 0.6 and 1.0 °K which in the Landau theory corresponds to the excitation of rotons. This is followed at lower temperatures by a function T^4 which corresponds to phonon excitations. The significance of the appearance of phonon excitations in the normal rather than in the superfluid concentration and its bearing on the theories of Landau and Tisza is discussed in other sections. For the purpose of viscosity determinations in helium II the data of Andronikashvili are so far sufficient since the work has as yet not been extended to temperatures below 1 °K.

In addition to the measurements of de Troyer [99] and others whose work has been mentioned in the previous section, determinations of the coefficient

of viscosity of the normal constituent have been carried out by ANDRONIKASH-
VILI and by HOLLIS-HALLET [102]. All these authors have used oscillating discs and,
as Fig. 59 shows, their results are in good agreement. With decreasing temperature η_n falls from its value at the lambda-point which is identical with the viscosity of helium I at this temperature to a flat minimum in the neighbourhood
of 1.7 °K. Below this temperature a steep rise occurs, leading to a value of over

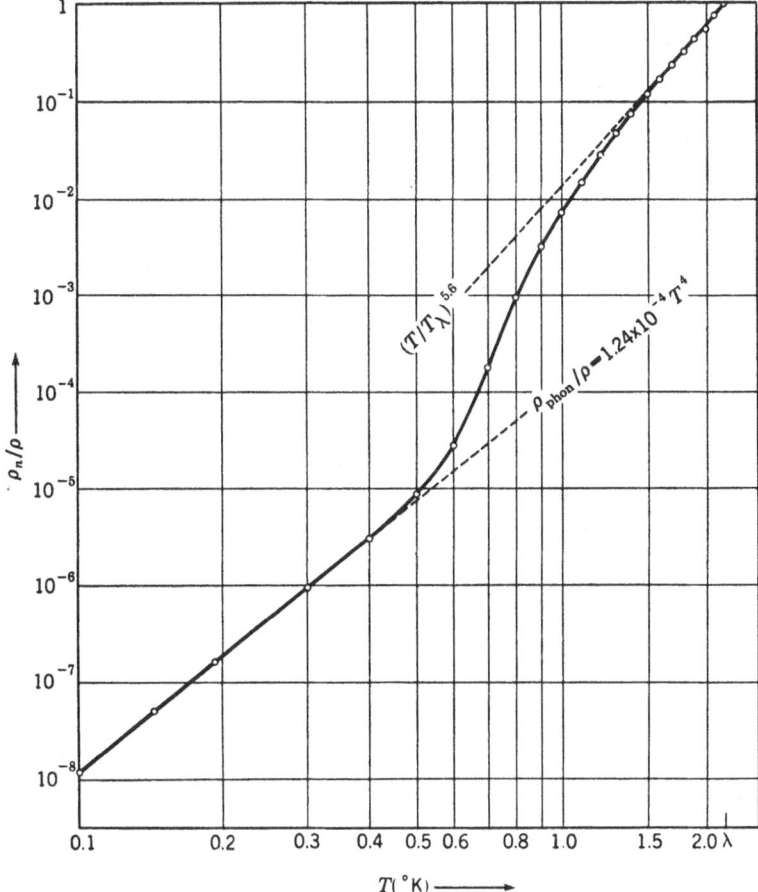

Fig. 58. The concentration of normal fluid as a function of temperature.

35 micropoise at the lowest measured temperature just below 1.3 °K. This value
is already slightly higher than the viscosity of helium I at the boiling point and
the trend of the curve suggests that it is still rising.

HOLLIS-HALLET [103] has also made measurements with a viscometer in which
the drag of a rotating on a stationary cylinder was observed. The values obtained
in this manner show fair agreement with the oscillating disc between the lambda-
point and 1.5 °K. They do not, however, follow the very steep rise below this
temperature, reaching only about 17 micropoise at 1.1 °K. It is not clear whether
this discrepancy is true or whether it is due to to the incorrect evaluation of the
correction terms. It has to be remembered, however, that in this temperature
region the normal constituent is already very diluted, amounting to no more
than 2 or 3% of the liquid volume. More information may possibly be obtained

from careful counterflow experiments such as used to measure the thermal conduction. With sufficient constancy of temperature and accurate measurement
of small temperature differences it should be possible to carry these into the
subcritical region where an unambiguous determination of the normal viscosity
will be feasible.

The theory of the normal viscosity in helium II has been worked out by
LANDAU and, particularly, by KHALATNIKOV [*104*]. Their approach is based
on the energy spectrum
proposed by LANDAU, and
for the viscosity problem
the collisions of these excitations are treated similar to those between par-
ticles in a gas. Since, as
is seen from Fig. 27, the
concentration of normal fluid
is falling rapidly below the
lambda-point, the density
of excitations below 1.9 °K
is small enough to treat them

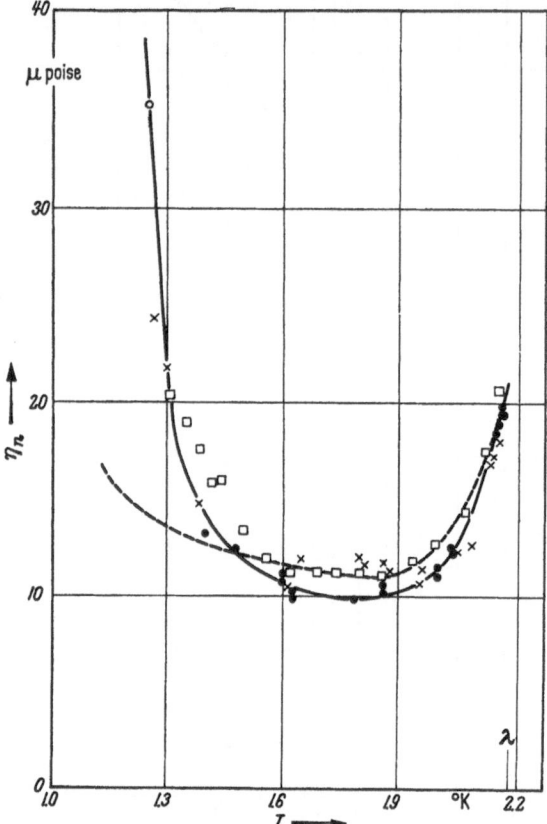

Fig. 59. Viscosity of the normal component of helium II. ● HOLLIS-
HALLETT, □ ANDRONIKASHVILI and × DE TROYER *et al.* The broken
curve is measured with a rotating cylinder viscometer.

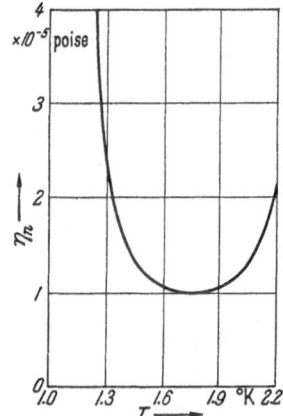

Fig. 60. Viscosity of the normal component
of helium II according to the theory of
LANDAU and KHALATNIKOV.

as an ideal gas. Since for the non-ideal gas the interaction between the excitations is becoming important, the theory cannot be applied to the region
between 1.9 °K and the lambda-point. The gas of excitations is composed of
phonons and rotons, both of which can take part in the processes leading
to dissipation, and the coefficient of normal viscosity η_n is made up of contributions due to both types of excitation. Assumptions have to be made concerning the collision properties of rotons, and it is postulated that they will behave
as heavy particles whose momentum exceeds that of a phonon by about 50 times
at 1 °K. The mean free path of a roton will thus be mainly determined by roton-
roton collisions and, while admittedly nothing is known about roton-roton interaction, it can be made plausible that the roton viscosity will be independent of
temperature. At higher temperatures the mean free path of the phonons will

also be determined by collision with rotons, but as the roton concentration decreases the phonon mean free path becomes greater. In this way evaluation of the phonon contribution to the normal viscosity leads to a rapid rise below 1.6 °K which is in good agreement with the observations on oscillating discs. It has been suggested that subtraction of the calculated phonon contribution from the observed viscosity in the region between 1.3 and 1.9 °K leads to a constant term of about 10 micropoise which might be considered as representing the postulated temperature independent roton viscosity.

F. Heat conduction.

31. Helium I. The first work by KEESOM and Miss KEESOM [15] yielded only a single value at 3.3 °K of 6×10^{-5} cal/degree $\cdot$ cm $\cdot$ sec. Two further investigations between the lambda-point and the boiling point were carried out 15 years later by GRENIER [105] at the Rice Institute and by BOWERS and MENDELS-
SOHN [106] which gave results in good agreement with each other as well as with KEESOM's value. As shown in Fig. 61 the heat conduction rises linearly with the absolute temperature and there is no noticable de-
viation from this behaviour in the vicinity of the lambda-point. This

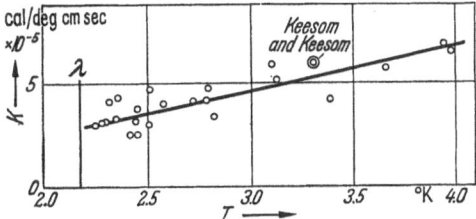

Fig. 61. Heat conductivity of helium I.

latter fact is remarkable in view of the appreciable changes in the viscosity, the attenuation of sound and, above all, in the specific heat in this region.

An estimate of the value of the heat conduction of a gas with the viscosity and the specific heat of liquid helium I leads to the correct order of magnitude. This fact, as well as the linear dependence on temperature emphasizes the si-
milarity of liquid helium with a gas, which is due to the high zero point energy and has been discussed earlier (Sect. 10). It has to be remembered that in this simple gas kinetic model the heat conduction is proportional to both the specific heat and the viscosity. In the region below 2.6 °K where these properties change in preparation for the lambda-point, the specific heat rises with falling temperature while the viscosity drops. It is thus just possible that the invariance of the heat conduction with temperature in this region may be due to the cancel-
ling out of anomalous effects of opposite sign.

32. Helium II. Before discussing in detail the heat conduction data below the lambda-point and their significance, it is necessary to define clearly what is meant under the heading of heat *conduction*. As has been discussed above, the high heat transport in helium II can conveniently be explained as due to the independent motion of the superfluid and normal constituents of the liquid. These allow for a number of transport effects in which heat is carried, including the thermo-dynamic cycle discussed by H. LONDON. However, in the latter the flow of heat is accompanied by a net flow of mass. Experiments of this kind, as for instance determinations of the heat of transport have to be considered separately, and we will concern ourselves here *only with heat transport processes in which the net flow of mass is zero*.

It had been postulated by TISZA and LANDAU and conclusively demonstrated by KAPITZA [41] that in a tube filled with helium II whose closed end is heated, a countercurrent of liquid will be set up in which the superfluid component moves in the direction of the heat supply, while the normal component moves away from it. The amount of heat transported in this way is the product of the total entropy of the liquid at this temperature and its absolute temperature. Since the motion of the superfluid will be free of friction, the resistance to this counter current should be determined entirely by the friction encountered by the normal constituent.

The simplest assumption to make for this process is that the normal constituent will move through the tube with laminar flow and, once the viscosity of the normal component is known, it should be possible to account for the thermal resistance in a quantitative manner. However, the first measurements showed that important additional assumptions would be necessary to bring the two-fluid model into agreement with the experimental results. ALLEN, PEIERLS and UDDIN [16] in their experiments which led eventually to the discovery of the thermo-mechanical effect noted that the heat current density, besides having anomalously high values, also was dependent on the temperature gradient. Since it was known that the results were falsified by the thermo-mechanical pressure head, they were not analysed in detail, but it is clear that they would not conform with the model sketched above. The whole question was investigated in great detail in the following years by KEESOM, Miss KEESOM and SARIS [107] and by KEESOM, SARIS and MEYER [108] who used an arrangement without a free surface of liquid helium at the warm end. Instead

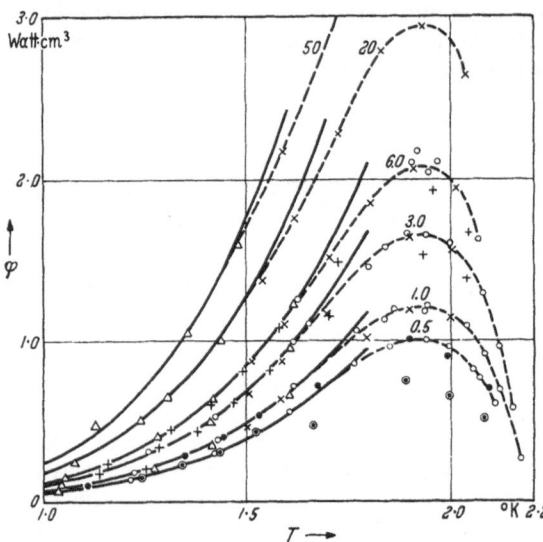

Fig. 62. Heat current density as function of temperature for different temperature gradients. The numbers indicate the temperature gradient × 10³ in deg/cm.

of reading the temperature difference by the differential vapour pressure, phosphorbronze thermometers were employed so that the apparatus conformed with the conditions mentioned above for a true heat conduction measurement.

A set of data obtained by the last named authors is given in Fig. 62. The diameter of the tubes in which the heat conduction was determined varied between 0.032 and 0.16 cm and the length between 1 and 35 cm. The values plotted against the temperature are the heat current densities $\dot{Q}/A$, where A is the cross section of the tube. Comparing the results with the heat conductivity equation

$$\dot{Q} = K A \Delta T/l \qquad (32.1)$$

where l is the length of the tube, it appears that the heat conduction K is independent of the length of the tube and that $\dot{Q}$ is proportional to A. On the other hand, it is clear that the relation between heat current density and temperature gradient is not that required by Eq. (32.1). The results can instead be represented by a proportionality of $\dot{Q}/A$ to the cube root of the temperature gradient. The heat transport in these capillaries can thus be neither explained by a classical heat conduction process nor by the simple application of the two-fluid model, assuming dissipation by laminar flow of the normal component.

Regarding the general shape of the temperature dependence of the curves, they agree with those obtained by ALLEN and GANZ [109] who used a capillary tube of similar diameter and determined with it the dependence of the heat conduction on pressure as is shown in Fig. 63. In both these investigations it was found that for any given temperature gradient the heat current density was a 5-th power function of the absolute temperature between the lower limit of measurement ($\sim$1 °K) and about 1.63 °K. Beyond this temperature the curve flattens out and, after passing through a maximum (at about 1.9 °K under saturation pressure), tends to zero at the lambda-point. For the lower temperature region KEESOM, SARIS and MEYER could represent their data by the equation

$$\dot{Q}/A = 0.623\, T^5\, (d\,T/dl)^{\frac{1}{3}} \quad (32.2)$$

where $\dot{Q}$ is expressed in watts.

It should be noted that in the work at different pressures not only the lambda-point is shifted to lower temperatures, but the absolute value of the heat conduction is appreciably reduced. This is evidently due to increased dissipation in the normal fluid under rising density.

The cube root dependence of the heat current density on the temperature gradient is in curious disagreement with the observations of ALLEN and REEKIE [81] who measured the heat flow as well as the thermo-mechanical pressure in a tube filled with fine powder.

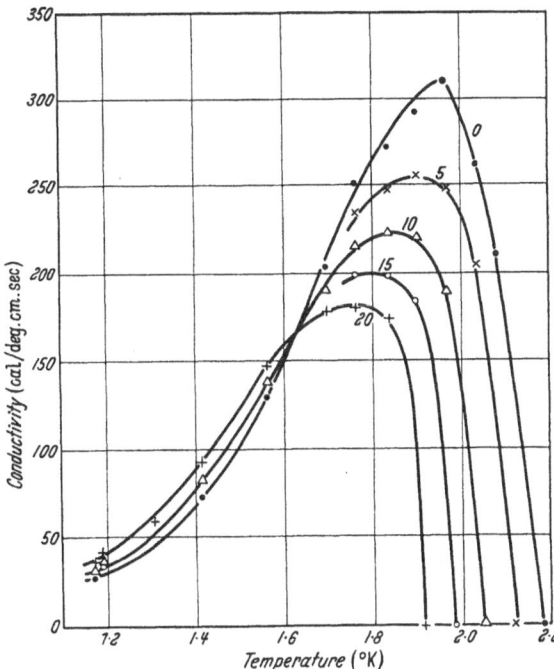

Fig. 63. Heat conductivity as a function of temperature and pressure (in atm.).

It was estimated that the channels between the individual grains would form a network of capillaries of the order of 10^{-3} cm. The shape of the temperature dependence was not markedly different from that shown in Fig. 62, but it was found that, at least for the lower temperatures, the heat flow was strictly proportional to the temperature gradient. The same result was obtained by KEESOM and DUYCKAERTS [110] who measured the heat flow through a fine slit between two spherical glass surfaces. The width of this slit could be varied between 0.7 and 15 microns and its length was about 0.25 cm. In addition to finding proportionality between $\dot{Q}$ and $\varDelta T$ at the lower temperatures and for the finest slits, they also observed that as the slit width was increased, this pattern changed over into the cube root dependency.

The entirely different nature of the heat conduction in capillaries and powder, and particulary the change over with channel size observed by KEESOM and DUYCKAERTS, indicates that the processes determining the heat resistance must be different in both cases. The case of narrow channels can be reasonably represented by the simple two-fluid model, but that of the wider capillaries cannot be fitted in.

In order to account for these discrepancies, Gorter [111] suggested in 1948 a modification of the two-fluid model which provided for dissipation to occur by mutual friction between the normal and superfluid constituents of the liquid. The velocity dependence of the friction force F was chosen in such a way that the observations on wider capillaries would be satisfied by

$$F = C \varrho_s \varrho_n (v_s - v_n)^3 \qquad (32.3)$$

where v are velocities and ϱ densities, the suffixes referring to the superfluid and normal concentrations.

Since then a considerable amount of work, practically all of it in Leiden, has been carried out in order to test this model for the heat conduction process in helium II and to establish the value of the constant C. The apparatus usually employed for these determinations as well as for those of the thermo-mechanical effect is a thermally insulated vessel whose only communication with the helium bath is by means of a narrow slit. In the diagram shown in Fig. 22 the slit is produced by pressing an optically flat plate over the orifice of the vessel which has a flat polished rim. For wider gaps spacers between rim and plate are inserted. The width of the slit is determined by gas flow or by an optical interference method. Energy can be supplied to the inside of the vessel by means of a resistance heater, and the difference in temperature between vessel and bath is determined by differential thermometers. Since it is often desirable in this work to make observations under small temperature differences, particular care has to be taken to ensure great constancy in the temperature of the helium bath during the course of an experiment. This experimental difficulty is probably the cause for a good deal of uncertainty in the results, especially in those involving small heat inputs and temperature differences. Mellink [112], Mellink and Meyer [113] and Hung, Hunt and Winkel [114] have made experiments not only with different slit width but also with various lengths of slit and from the latter they have concluded that the friction process must occur throughout the length of the flow channel and not just at its ends.

Assuming that in such an arrangement as shown in Fig. 22 the thermo-mechanical pressure difference between bath and vessel remains constant, the net flow of mass must be zero. In terms of the two-fluid model this means that the same amount of normal fluid will leave the vessel as superfluid is entering it, so that

$$\varrho_n v_n + \varrho_s v_s = 0. \qquad (32.4)$$

Since it can be taken from Kapitza's work [42] that the total entropy of the liquid is taken up by the superfluid (the question of the phonon entropy can be ignored in the temperature region of these experiments), the heat transported is

$$\dot{Q} = \varrho \dot{V} S T \qquad (32.5)$$

where $\dot{V}$ is the volume of superfluid passing in unit time. As according to Eq. (32.4) the same volume returns in normal form with viscous dissipation

$$\dot{V} = \frac{G}{\eta} \Delta p \qquad (32.6)$$

η being the viscosity of the normal constituent, G the Poiseuille constant and Δp the thermo-mechanical pressure. According to H. London's equation, the latter is equal to $\varrho S \Delta T$, ΔT being the temperature difference between vessel and bath. We thus obtain for the thermal resistance

$$\frac{\Delta T}{\dot{Q}} = \frac{1}{G}\frac{\eta}{\varrho^2 S^2 T}. \tag{32.7}$$

Comparing the observed values of the heat conduction in narrow channels, for which (32.7) should be applicable, it is found that in gaps larger than 1 micron, the order of magnitude is that given by the formula. For narrower channels, however, the heat conduction is much higher than can be explained. LONDON and ZILSEL [115], who in 1948 first discussed this question on the basis of KAPITZA's results and those of MEYER and MELLINK, pointed out that for a 0.3 micron gap the observed heat conduction exceeds the calculated one by a factor of 2.5×10^2. This discrepancy is disturbing but recent experiments by DELSING [116] on the viscosity of the normal component in very narrow slits suggest that it may be due to some extent to experimental difficulties. He has pointed out that in the region fo the discrepancies the heat conduction along the walls of the vessel becomes of the same order of magnitude or larger than that through the slit itself.

As the width of the channel through which the heat flow takes place is increased, the viscous drag on the returning normal fluid will diminish, and such dissipation as may be due to mutual friction should eventually become dominant in the determination of the heat conductivity. H. LONDON's formula for the thermo-mechanical effect presupposes complete reversibility and the appearance of friction must therefore decrease the pressure difference corresponding to a given ΔT in LONDON's formula. For a mutual friction which is proportional to the third power of the relative velocity the modification of Eq. (32.7) reads

$$\frac{\Delta T}{\dot{Q}} = \frac{1}{G}\frac{\eta}{\varrho^2 S^2 T} + \frac{C \varrho_n}{S^4 T^3 \varrho_s^3}\dot{Q}^2. \tag{32.8}$$

At present there exists a good deal of evidence for a friction term which is proportional to the cube of the velocity, besides that on which the original postulate was based. On the other hand, the existing data are not all equally convincing. It has to be remembered that in a process of the kind in which a counterflow is supposed to occur, a number of dissipation phenomena besides those listed above may make their appearance. Nothing has so far been said about the possible occurence of either turbulence in the channel or of dissipation in end effects. The latter would for a classical fluid add a further term to Eq.(32.8) which sould be proportional to $\dot{Q}$.

Another, rather unsatisfactory, aspect of GORTER's original model is that it assumes the existence of friction over the whole velocity range. The phenomenon of superfluidity, when first discovered, suggested immediately the disappearance of friction and while the phenomena could for some time be explained by a small value of the constant C in Eq. (32.3), this became untenable as more measurements were made. Moreover, the behaviour of film transfer showed clearly the existence in helium II of frictionless transport of mass and, while admittedly the present theories do not go far enough to make predictions, it would nevertheless appear strange if this phenomenon were to be confined to films. The flow experiments carried out in Oxford on narrow slits, porous membranes and packed powder which have been described in part D have indeed given clear evidence that in channels filled with bulk liquid frictionless transport will take place as long as a certain critical velocity is not exceeded. Similar indications were already given by the early experiments of ALLEN and MISENER and of KAPITZA. The possible size dependence of this critical velocity, the correct function of which is still

unknown but would appear to produce a decrease with increasing width of channel, may be the reason for the apparently satisfactory agreement of the early
heat conduction data with Eq. (32.3). To allow for a critical velocity in GORTER's
mutual friction model, Eq. (32.3) has been modified to read

$$F = C \varrho_s \varrho_n [(v_s - v_n) - v_c]^3 \qquad\qquad (32.9)$$

v_c being the critical velocity. If v_c is small in wider channels the deviation from
Fq. (32.3) may have been to small to be observed. It is therefore encouraging
that in the very latest work carried out in Leiden by WINKEL, DELSING and
POLL [117] on narrow slits between 0.43 and 3.1 microns definite indication of
the existence of critical velocities
has been obtained.

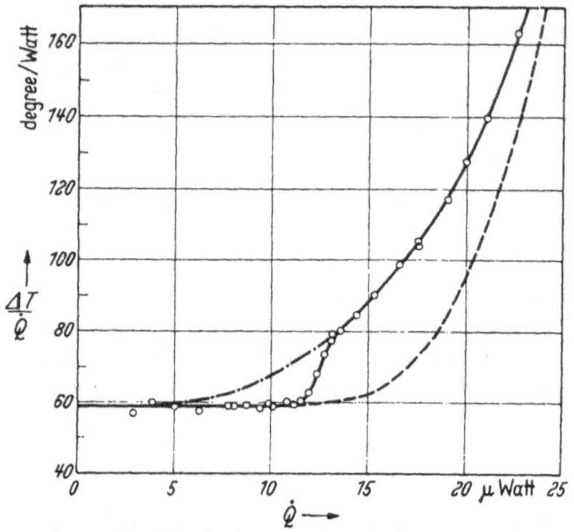

Fig. 64. Heat resistance as a function of heat input (in arbitrary units) in a capillary of 5×10^{-3} cm diameter. Full line:
experimental results; ———— Eq. (32.9); — · — · — Eq. (32.3).

Thus the whole subject of heat
conductivity in helium II is one in
which sufficient information to form
a clear and unambiguous picture is
still lacking. On the other hand, it
appears that from latest work a fairly
consistent pattern is emerging. Under
these circumstances it may be permissible to quote some very recent results obtained by BREWER, EDWARDS
and MENDELSSOHN [118] since they
seem to contain a new factor which
may dissolve the discrepancies mentioned earlier. The se authors have
measured the thermal resistance of
a capillary of 50 micron diameter.
They have used an arrangement
which allowed the ambient temper-
ature to be monitored to less than 10^{-5} degrees and were thus able to investigate the heat conduction under extremely small temperature gradients.
They then found that even in such a very wide capillary the thermal resistance
was at first quite independent of the temperature gradient. In this region Eq.(32.7)
was well satisfied numerically, too, since η was found in good agreement with
viscosity measurements. The surprising feature was, however, the onset of friction which is shown in Fig. 64 where the heat resistance $(\varDelta T/\dot{Q})$ is plotted against
the heatin put $\dot{Q}$. At a certain critical heat input which corresponds in the two-
fluid model to a critical velccity, the thermal resistance rises quite suddenly
and almost discontinuously. From then onwards there is a gradual increase in
the thermal resistance which, when analysed would correspond to friction nearer
to a fourth rather than to third power of the velocity. It would certainly be
premature to discuss this latter feature of the result because, as mentioned
above, friction, once it appears, may be of a very complex nature. The significant phenomenon is that the gradually rising thermal resistance does not
merely indicate onset of friction above a critical value of $\dot{Q}$ but that it extrapolates
back to the heat conduction at zero temperature gradient. In other words,
leaving aside for the moment the exact power law of the friction force, an equation of the form (32.9) would have led to the broken curve in Fig. 64. Instead,
friction follows a law of the form indicated in Eq. (32.3), again with reservations

regarding the exponent but with the important proviso that the law is not followed to zero velocity.

Whereas formerly, even when a critical velocity was assumed, it was always thought that friction was added to the existing flow process beyond a certain velocity value, the mechanism suggested by Fig. 64 is totally different. It appears that the system is capable of exhibiting two quite distinct patterns of energy transport. In the first, which corresponds to the region of constant thermal resistance, the superfluid constituent suffers no dissipation whereas the second possible mechanism implies dissipation in the superfluid for all velocities. The results suggest that these distinct flow patterns cannot exist simultaneously but that they change over into each other at a critical value of $\dot{Q}$. The aspect of energy transport in helium II thus becomes closely analogous to the phenomenology of superconductors where there also exist two conduction mechanisms which at the threshold value of current or field change into each other discontinuously. It is perhaps significant that in some of the experiments a hysteresis was observed in that the critical value of $\dot{Q}$ was found to be higher for ascending than for decreasing heat input. While these experiments may thus bring the question of heat conduction in helium II nearer to solution, it should be emphasized that much further work and corroboration is clearly required before the pattern suggested here can be accepted as definitely correct.

33. Experiments below 1 °K. The results mentioned in the previous section all refer to the temperature range above 1 °K which is accessible with the conventional type of helium cryostat. However, a separate discussion of the heat conductivity in the magnetic temperature range is justified not only in view of the different techniques employed, but mainly because the phenomena observed differ fundamentally from those in the rest of the helium II region.

The first determination was made by KÜRTI and SIMON [119] in 1938. These authors had observed that in a capsule filled with paramagnetic salt and partly filled with liquid helium, the heat exchange was poor at low temperatures. They concluded that the thermal conductivity of the liquid was not anomalously high as above 1 °K and made some experiments to measure its value between 0.2 and 0.5 °K. The results of this work showed qualitatively that in this temperature range the heat conductivity is very much smaller than would be expected from an extrapolation of the work above 1 °K. The actual values, ranging between 0.2 and 2 cal/degree · cm · sec are probably somewhat too small. Subsequently, estimates of the heat conductivity were made by a number of authors from data obtained during demagnetization experiments, but a full study has been made only fairly recently by H. A. FAIRBANK and WILKS [120].

The apparatus used is shown in Fig. 65. It consisted of a strong walled metal capsule C which extended into a tube G carrying the capillary in which the heat conductivity was measured. The capsule contained the pill of paramagnetic salt P and was filled at room temperature with helium gas under high pressure and then sealed off. At low temperature the helium condensed, filling G and part of C. Energy was supplied by the heater H and the temperature drop along the capillary was determined with two carbon thermometers T_1 and T_2. The whole measuring section was enclosed in a sealed-off vacuum jacket J. The result, obtained on two capillaries, A of 0.8 and B of 0.29 mm diameter, are shown in Fig. 66.

Above 0.6 °K the heat conductivity rises suddenly more sharply, and in this region it was also found to become dependent on the temperature gradient.

Altogether, the phenomena are here of the same kind as discussed in the preceding section. This change over corresponds to the rise of the specific heat observed at the same temperature and is clearly due to the onset of excitation other than phonons. Below 0.6 °K the heat conduction is independent of the temperature gradient and follows the temperature variation of the specific heat. The difference in heat conduction between the two capillary sizes is due to the mean free path being of the order of the diameter and thus the effect is due to boundary scattering, as has been observed in dielectric solids at low temperatures. The results seem to be in agreement with the theory of Landau and Khalatnikov in that the mean free path significant in heat conduction and viscosity becomes very long at low temperatures. This is important for the assessment of second sound phenomena at the lowest temperatures which are discussed in another section. In this way the data

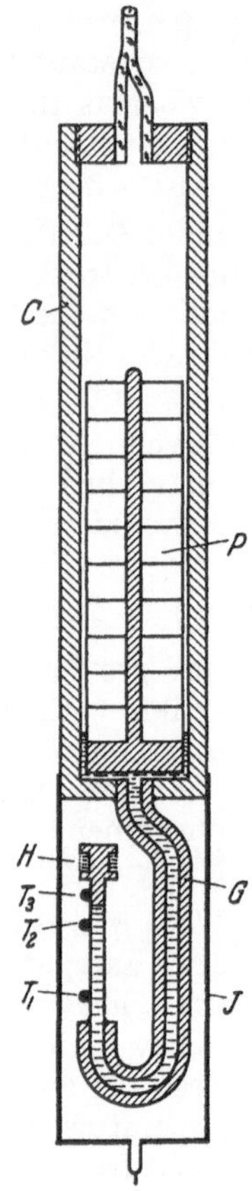

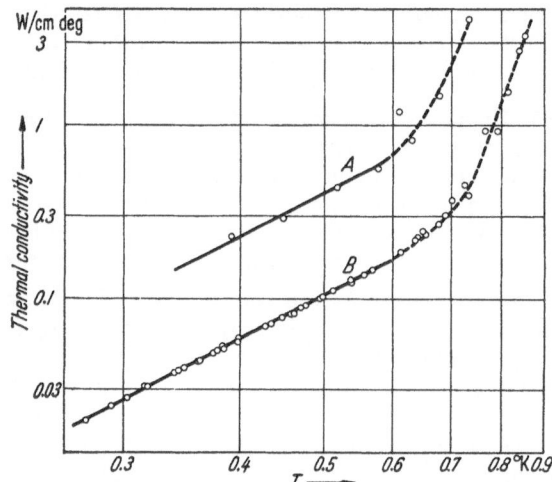

Fig. 65. Apparatus for measuring the heat conduction of liquid helium below 1 °K.

Fig. 66. Heat conductivity of liquid helium. Capillary diameter: $A = 0.80$ mm; $B = 0.29$ mm.

obtained here add weight to the assumption that observations on second sound at the lowest temperatures have to be treated with caution since they may be largely falsified by mean free path effects.

34. Boundary effects. Kapitza [42] noted that in a metal block freely suspended in liquid helium II which was fitted with internal heater and thermometer, the temperature rose well above that of the helium bath. From this he deduced that the boundary between the metal and the liquid must offer an additional thermal resistance. His results showed that this boundary resistance does not depend on the nature of the solid surface and that it occurs within 10^{-3} cm of the solid surface. He also found that the heat transport through the boundary increased roughly proportional to the third power of the temperature for small temperature differences between solid and helium bath.

Since then the boundary resistance has been observed by a number of authors and theoretical interpretations have been given by Gorter, Taconis and Been-

AKKER [121] and by KHALATNIKOV [122]. The former authors have considered the phenomena to be expected in the liquid itself close to the solid surface. They formed an estimate of the temperature difference in the liquid in the direction perpendicular to the surface which has to be maintained in order to allow a transformation of superfluid into normal fluid at a rate sufficient for the total heat flow. KHALATNIKOV's explanation is that of a contact resistance which will occur at any boundary and which is only particularly noticeable in helium II because of the high heat conduction of the latter. He assumes that the heat transfer between metal and liquid takes place by the radiation of sound waves and postulates that above as well as below 0.6 °K the heat transfer coefficient should be proportional to T^3.

H. A. FAIRBANK and WILKS [120] have used the arrangement shown in Fig. 65 for an investigation of the boundary resistance. They added a third thermometer T_3 which was kept thermally in good contact with the heater by means of a block of very pure copper housing both. On supplying a large amount of heat, they noticed that the temperature difference between T_3 and T_2 was several times bigger than that between T_2 and T_1 which indicated that most of the heat resistance occurred at the boundary surface. Measuring the heat flow through the boundary between the copper surface and the helium for a temperature difference ΔT, they obtained the relation shown in Fig. 67. The temperature gradient was varied by an order of magnitude in these observations, and the heat flow was found to be proportional to the gradient. The variation of the heat

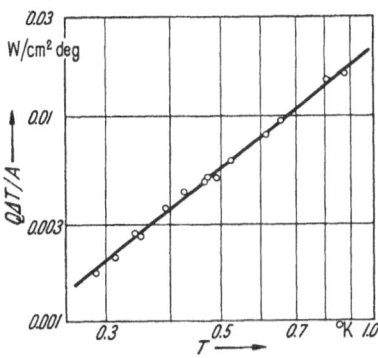

Fig. 67. Heat flow Q through a boundary area A between liquid helium and copper for a temperature difference ΔT as a function of temperature.

transport through the boundary was found to be proportional to T^2 and there was no indication of a change at 0.6 °K. From this the authors conclude that the type of process considered by KHALATNIKOV is more likely to be responsible for the boundary resistance than a transformation from superfluid to normal.

G. Wave propagation.

35. First sound. Since the discovery of thermal waves in liquid helium II it has become customary to refer to the phenomenon of ordinary sound waves, i.e. to the propagation of density variations as "first", in contradistinction to "second" sound. The velocity of first sound was first measured in 1938 by FINDLAY, PITT, GRAYSON SMITH and WILHELM [123] at a frequency of 1.338 Mc/sec. Readings were taken between the normal boiling point at 4.2 °K and a lower limit at 1.76 °K which showed at first with falling temperature a rise in the velocity, leading to a maximum at about 2.5 °K. This was followed by a decrease and a sharp minimum at the lambda-point. In the helium II region the velocity was then found to increase again with falling temperature. The same authors repeated their experiments under pressures up to 5.55 atm and obtained results in which the minimum at the lambda-point was even more pronounced. This early work has since been supplemented by observations of PELLAM and SQUIRE [124], using 15 Mc/sec and by ATKINS and CHASE [125], using 14 Mc/sec. In both these investigations pulse techniques were employed, whereas the early experiments had been carried out with standing waves. The last mentioned authors extended the measurements to 1.2 °K, and it was found that over most of the range covered by the three investigations the results are in good agreement. A graph of the values is given in Fig. 68.

A number of authors have discussed the variation of the first sound velocity at the lambda-point according to the EHRENFEST relations for transitions of the second order. KEESOM already has pointed out that on this basis a discontinuity

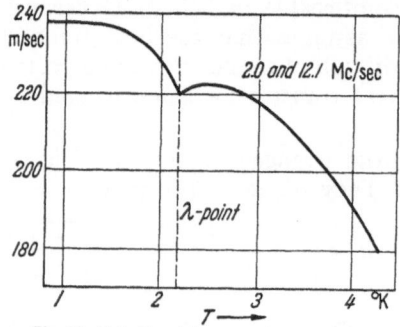

Fig. 68. Velocity of first sound as a function of temperature.

in the velocity of sound at the lambda temperature has to be expected. ATKINS and CHASE have made a careful investigation in this temperature region and found indeed a very sharp minimum at the lambda-point as is shown in Fig. 69. However, as mentioned by ATKINS the velocity of first sound obeys the relation

$$u_1^2 = \gamma V/\beta \qquad (35.1)$$

β being the isothermal compressibility and γ the ratio of the specific heats, which both vary rapidly at this temperature. He estimated from the known data on the specific heat and the coefficient of expansion that the discontinuity in u_1 should amount to 2.5% but stated that the observed rapid variation does not permit a decision of this question on the basis of the available results. TISZA

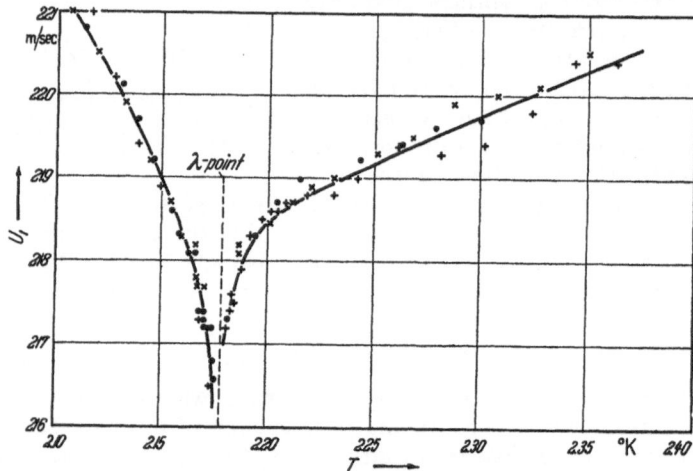

Fig. 69. Velocity of first sound at the lambda-point.

has drawn attention to the possibility that at the lambda transition the derivatives of the GIBBS free energy may not only become discontinuous but in fact reach infinite values. This would allow the velocity of first sound to become zero in a very small interval of temperature.

ATKINS and CHASE have extrapolated their results of the velocity of first sound to absolute zero an arive at a value of (237 ± 2) m/sec. This is of some interest in view of LANDAU's formula (14.2) relating the values of the velocities of first and second sound at absolute zero.

While the velocity of first sound only offers an indirect test of theoretical models of helium II, its attenuation has had wider applications in this field. PELLAM and SQUIRE used the pulse technique which has served to determine the temperature dependence of u_1 also to measure the attenuation coefficient over the same range of temperatures. They compared their results with the classical equation for the attenuation coefficient.

$$\alpha = \frac{8\pi^2 \eta v^2}{3\varrho\, u_1^3} + \frac{2\pi^2 (\gamma - 1)\, K v^2}{\varrho\, u_1^3 C_p} \tag{35.2}$$

in which the first term accounts for the attenuation due to viscosity and the second for that due to thermal conduction. C_p is the specific heat at constant pressure, v the frequency, η the coefficient of viscosity, and K the thermal conductivity. Substituting the known values, the authors found excellent agreement between the boiling point and 3 °K, but below this temperature α departed from the theoretical curve, rising to a very sharp maximum at the lambda-point. This anomaly they attributed to transformations of small localized regions of liquid from the I-state to the II-state in the neighbourhood of the lambda-temperature. This suggestion is similar to the explanation advanced

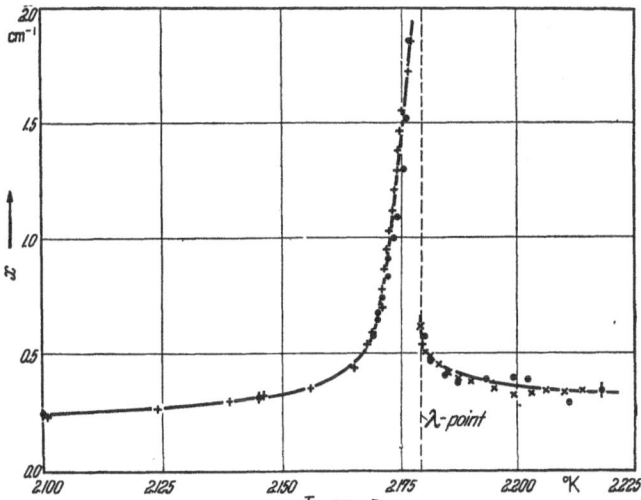

Fig. 70. Attenuation of first sound at the lambda-point.

by KEESOM in trying to account for the anomalously high specific heat just above the lambda-point. The problem has been discussed again in some detail by PIPPARD who, on the basis of the attenuation results, has suggested that the size of the local regions would be of the order of 10^3 atoms.

Another anomaly observed by PELLAM and SQUIRE is the rise of the attenuation coefficient at the lower end of their temperature range, below 1.7 °K. As they pointed out, this rise cannot be brought into agreement with Eq. (35.2) which should lead in this region to a falling value of α with decreasing temperature. In 1950 KHALATNIKOV [126] extended his treatment of the viscosity of the normal component of the LANDAU two-fluid model to the problem of attenuation of first sound and postulated a rapid increase in α as the temperature is lowered below 2 °K. He considered inelastic collisions in the gas of phonon and roton excitations and found that the two most important collision processes were those between phonons and between phonons and rotons. He calculated the relaxation times characteristic for the establishment of equilibrium by these collision mechanisms and applied the results of this work to prediction of the attenuation at lower temperatures, using the absolute values obtained by PELLAM and SQUIRE.

The theory was tested experimentally by ATKINS and CHASE who carried out measurements down to 1.2 °K and later by CHASE and HERLIN [127] who, using a magnetic cryostat, extended the range to 0.1 °K. The former authors also investigated in great detail the attenuation in the close proximity of the lambda-temperature. They confirmed and extended the work of the earlier authors finding a very sharp peak, their results are given in Fig. 70. CHASE and HERLIN [22] observed two peaks in a broad maximum at about 0.9 °K. It was then thought that these two peaks were connected with the two relaxation phenomena of the KHALATNIKOV theory. Very recent work with similar equipment also carried out at the Massachusetts Institute of Technology seems, however, to suggest that the detailed structure of the maximum may not be so well defined as was originally thought and that the two peaks may be due to spurious effects.

While it thus appears that such detailed features may not be justified by either experiment or theory, there can be no doubt that in general outline the attenuation of first sound can be reasonably well understood assuming the model proposed by KHALATNIKOV. A composite curve for the temperature variation of the coefficient of attenuation based on the results of a number of different workers is given in Fig. 71. Measurements under pressure have been made recently by NEWELL and WILKS [128] and these authors also mention work below

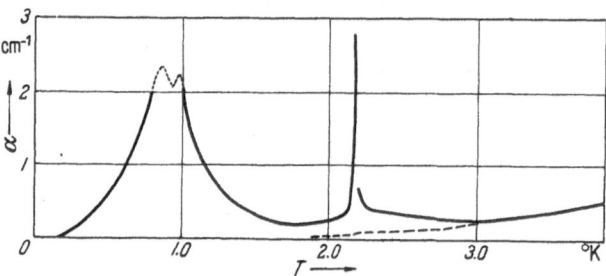

Fig. 71. Attenuation of first sound as a function of temperature.
--- Eq. (35.2).

1 °K under saturated vapour pressure. They also found a maximum in the attenuation at about 0.9 °K but again not the double peak mentioned by CHASE and HERLIN.

36. Second sound. A good deal of the work on second sound and its theoretical implication has already been mentioned in Sect. 15. In particular the significance of the observation by PELLAM and SCOTT [48] that below 1 °K the second sound velocity rises again has been discussed with reference to the theories of TISZA and LANDAU. The value of 34 m/sec at an unspecified low temperature reported by these authors left no doubt about an appreciable increase over the value of about 20 m/sec which had been observed by a number of authors in the region between 1 and 1.8 °K. The subsequent work has been mainly concerned with measurements below 1 °K which were designed to investigate more accurately and in greater detail the effect observed by PELLAM and SCOTT. ATKINS and OSBORN [129] made velocity determinations down to a recorded temperature of 0.1 °K. Experimentally these observations are very difficult and their work was the first attempt at measuring the rise in u_2 in a more than qualitative manner. They found that, as the temperature is lowered below 1 °K, u_2 rises at first gradually to about 50 m/sec until at about 0.4 °K a steep increase to 120 m/sec was observed which was followed by a further gradual rise to 150 m/sec at 0.1 °K. Three years later DE KLERK, HUDSON and PELLAM [130] investigated the same temperature region with an improved cryostat. While they observed the same general features in the variation of u_2 with temperature as ATKINS and OSBORN, the rapid rise was found to be rather less steep than observed in the earlier work and was displaced from 0.4 to about 0.6 °K (Fig. 72). Moreover, at 0.1 °K u_2 had a measured value 40 m/sec higher than that reported by ATKINS and OSBORN. There thus appears to be little doubt that in the first of these measurements the temperature of the liquid helium carrying the second sound pulses was consistently higher than that of the paramagnetic salt which acted as coolant and whose temperature was recorded.

Since the rise in u_2 found by PELLAM and SCOTT appeared to confirm LANDAU'S theory, it was disappointing that already the observations of ATKINS and OSBORN yielded at their lowest temperature a value of u_2 which was slightly in excess of LANDAU'S prediction of $u_1/\sqrt{3}$. This discrepancy was made worse by the measurements of DE KLERK, HUDSON and PELLAM which showed that LANDAU'S prediction was exceeded by at least 40%. It has been suggested by a number of authors

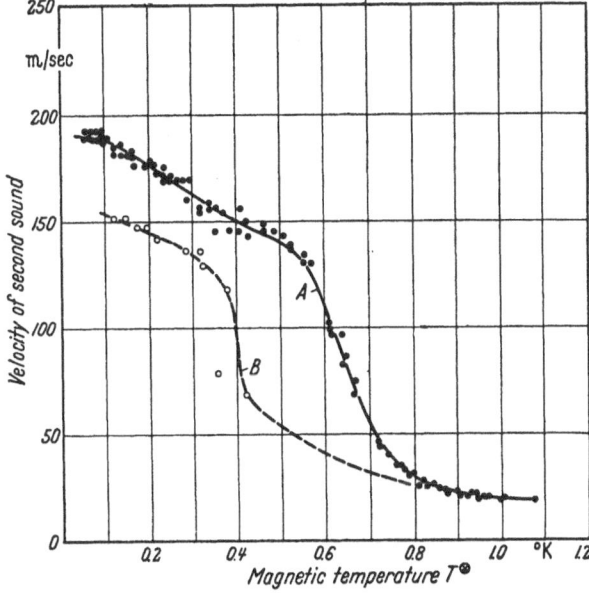

that the reason for this discrepancy is less likely to be due to the inadequacy of the LANDAU theory but to the fact that the observations made below 1 °K may not all correspond to true second sound. OSBORN observed shock wave propagation in second sound at higher temperatures, and this phenomenon has been suggested as the reason for the high observed values of u_2 at the lowest temperatures. Another possible cause for the discrepancy might be found in the long mean free path of the phonons.

Fig. 72. Velocity of second sound below 1 °K. Curve A: DE KLERK, HUDSON and PELLAM. Curve B: ATKINS and OSBORN.

The whole question has been examined recently by KRAMERS et al. in Leiden [131]. They have measured the propagation of heat pulses in tubes of different sizes and also found a high value of u_2, of the order of 200 m/sec, at the lowest temperatures. The general shape of the temperature variation is similar to that observed by DE KLERK, HUDSON and PELLAM, and they find it convenient to discuss the observed phenomena in three temperature regions, that below 0.5 °K, that between 0.5 and 0.7 °K and that above 0.7 °K. These intervals correspond to the two regions of gradual change with temperature at either end of the range and the region of rapid

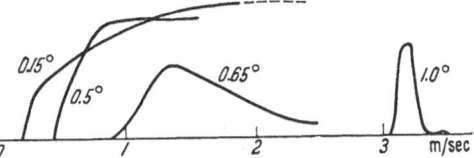

Fig. 73. Pulse shapes in second sound measurements at different temperatures.

change between them. KRAMERS found that the shape of the received pulse changes profoundly as the temperature is lowered and the three regions correspond roughly to different types of propagation. Pulse shapes at different temperatures are given in Fig. 73.

Above 0.7 °K the pulse is still reasonably sharp but broadens as the temperature is lowered. This corresponds to the damping predicted for this region by KHALATNIKOV. Here the propagation can still be regarded as that of second sound waves. Below 0.5 °K the original shape of the pulse is completely lost. Only the front can be observed which, at the lowest temperatures proceeds with a velocity close to that of first sound. In this region the phonon mean free path has become so long that in the tubes used in the experiments true second sound cannot be expected. The increase in phonon mean free path observed has been compared with that postulated by KHALATNIKOV and found to be less rapid with falling temperature than predicted. It has been suggested that this shorter mean free path may be due to the presence of minute quantities of the light isotope He³. The intermediate region between 0.5 and 0.7 °K in which the rapid increase of the observed u_2 takes place denotes the transition from

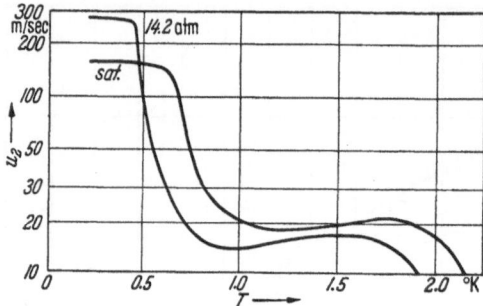

Fig. 74. Velocity of second sound at the vapour pressure and at 14.2 atm.

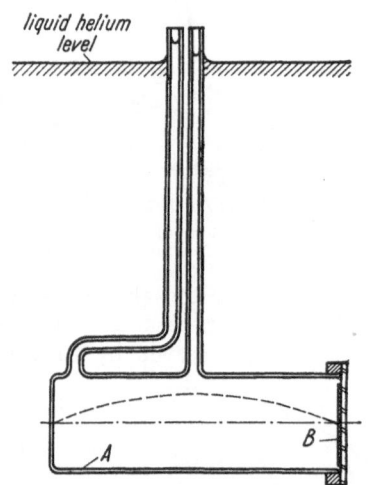

Fig. 75. Thermal pilot tube for second sound investigation. *A* cavity, *B* heater.

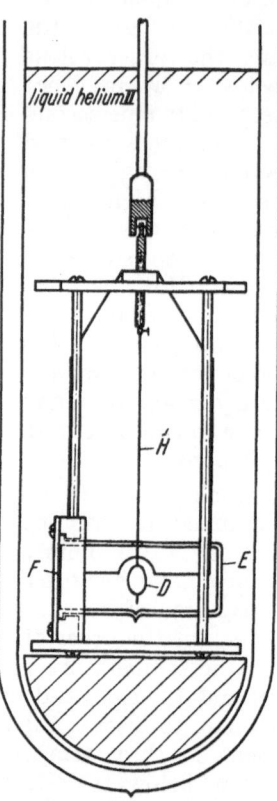

Fig. 76. RAYLEIGH disc for second sound experiments. *D* disc in the form of a galvanometer mirror, *E* cavity, *F* heater, *H* suspension.

second sound to the type of propagation characteristic of the region of long phonon mean free path. The pulse, while its shape can still be recognized is broadened so much that a definition of the true speed of propagation of the original signal becomes impossible.

Further information on the type of two-fluid model to which helium II corresponds can be obtained from measurements of second sound under pressure. According to LANDAU's theory the superfluid is to be free of all excitations, phonon excitations being as well as rotons part of the normal fluid. It has already been mentioned that a sharp rise in the second sound velocity must occur according to this model in the temperature region where the phonon entropy becomes predominant. Since under pressure this will happen at a lower temperature the sharp rise in u_2 should accordingly be also displaced. Moreover, according to LANDAU'S Eq. (14.2) the second sound velocity must be proportional to that of first sound at absolute zero and since the latter increases with pressure, the temperature functions of u_2 at different pressures should cross over at very low temperatures.

Determinations of the velocity of second sound under different pressures have been made by HERLIN and his co-workers [132], and some of the most significant results have been summarized in Fig. 74. For clarity's sake only the curves under vapour pressure and 14.2 atm are shown. They exhibit quite definitely the two effects which were expected from the theory.

The experiments mentioned so far were all carried out either with standing waves or pulses, both generated by means of heaters. KÜRTI and McINTOSH [133] have recently described a method in which second sound is produced through the alternating magnetization and de-magnetization of a paramagnetic salt. They have demonstrated the feasibility of the method at ordinary helium temperatures and pointed out that, since no irreversible heat is generated, it should be particularly useful in the region below 1 °K.

PELLAM and his co-workers [134] have applied two other devices to the detection of second sound; the pitot tube and the RAYLEIGH disc. In both cases second sound was generated by a heater plate at one end of a cylindrical resonant cavity. In the case of the pitot tube which is shown in Fig. 75, manometric tubes were attached at the centre and at the end of the cavity opposed to the heater. When second sound at a wavelength of twice the length of the cavity was generated an increase in pressure at the end, i.e. at the node was observed. The arrangement for the RAYLEIGH disc is shown in Fig. 76 and in the case of this device a number of quantitative measurements have been made. These have been discussed on the basis of the two-fluid model and found to be consistent with it. The results could be well represented by the sum of the torques t_n and t_s due to the normal and superfluid components in the liquid when

and
$$\left.\begin{array}{l} t_n = \tfrac{4}{3} r^3 \varrho_n v_n^2 \\[4pt] t_s = \tfrac{4}{3} r^3 \varrho_s v_s^2 \end{array}\right\} \qquad (36.1)$$

where ϱ and v are the densities and velocities of the two components.

H. The saturated film.

37. Film thickness. The first observation on the film thickness by KIKOIN and LASAREW [31] and by DAUNT and MENDELSSOHN [135] respectively were made by measuring the amount of liquid helium required to cover a known surface with film. The former authors used a cylinder of large area (see Fig. 77) which ended in two fine tubes. One of these served to suspend the cylinder above the helium bath while the second one dipped into the liquid. A heater was wound on to the lower tube and a thermometer on the

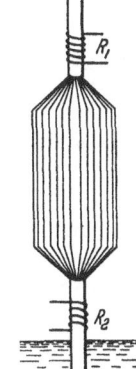

Fig. 77. KIKOIN and LASAREW's apparatus for measuring the film thickness.

upper one. When no heat was supplied, the cylinder was covered with the film and the thermometer registered the same temperatures as the bath. On supplying heat, the film was evaporated from the cylinder and, when the heater was switched off again, liquid from the bath had to supply the film coverage. Consequently a slight drop of the bath level occurred at this stage in the experiment and from this the thickness of the film was estimated as 2 to 3×10^{-6} cm.

DAUNT and MENDELSSOHN [30] used an apparatus consisting of a tube whose lower end dipped into liquid helium II while the upper end was at room temperature. The temperature gradient along this tube was controlled by successive baths of liquid He I and liquid hydrogen. The bottom of the tube was drawn out into a fine capillary and into it a few mm³ of liquid had been condensed. Into the liquid dipped a fine copper wire which in turn was attached to a scroll of polished copper sheet of large surface. As is shown in Fig. 78 the scroll could be lifted up to room temperature by

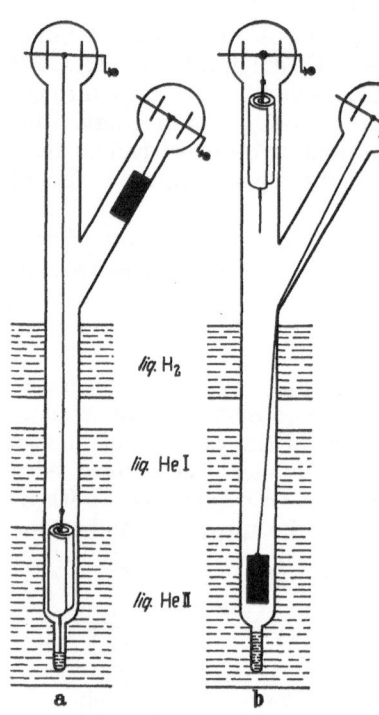

means of a winch. When this was done the film evaporated and, since the temperature along the tube was controlled, had to condense in the capillary. In order to allow for the vapour displaced by the scroll a lump of copper of equal mass was lowered into its position. The results which were obtained between 1.59 and 2.14 °K showed some scatter with an average of 3.5×10^{-6} cm but no defined dependence on temperature. The authors also found that their method yielded no film thicker than 10^{-7} cm just above the lambda-point.

JACKSON and his co-workers [136] at Bristol improved this work by a great number of experiments, using an elegant optical method. They used stainless steel mirrors covered partly with one and partly with three layers of barium stearate. The degree of polarization of monochromatic light reflected from the mirror was investigated, matching the two areas with and without the helium film being the actual measurement. The determined quantity was therefore the rotation of a nicol prism and from this the thickness of the film had to be deduced. While it may be assumed that in first approximation the change in the

Fig. 78 a and b. Method of DAUNT and MENDELSSOHN for measuring the film thickness.

angle of rotation is directly proportional to the thickness of the film, absolute calibration is a rather more difficult problem. The method is thus particularly suited to investigations of the variation of the film thickness with temperature, with height above the liquid meniscus and with the state of motion. All these questions have been explored and the results of measurement of height and temperature variation are given in Fig. 79. There is good agreement with the work of DAUNT and MENDELSSOHN as far as the temperature dependence is concerned since between 1.1 °K and the lambda-point the film thickness is fairly constant. In all the earlier measurements of JACKSON it was also found that there was a rapid decrease in the film thickness at the lambda-temperature above which films of only about ten atomic layers were found. Quite recently indications of a thick film even above the lambda-point were noted, but this result requires clearly further corroboration.

In their first experiments the Bristol workers used simple steel plates which were later supplemented by steel mirrors forming the outside of a flow beaker so that the thickness of the stationary as well as that of the moving film could

be measured. Care was taken to limit heat influx along the suspension by inter-
posing a beaker filled with liquid helium (Fig. 80) and the level inside the steel
beaker was read on a communicating glass tube. It was then found that the
thickness of the film and its dependence on height above the liquid meniscus
was the same whether the film was stationary or in motion. The arrangement
was further varied by using a steel beaker whose diameter was different in the
upper and in the lower parts. In this way it was possible to observe the thickness
of the film flowing at the critical or a sub-critical rate of volume flow, and it
was again found that the thickness was the same in both cases.

Trying to express the variation of film thickness d with height h above the liquid meniscus as

$$d = \frac{d_0}{h^{1/z}} \qquad (37.1)$$

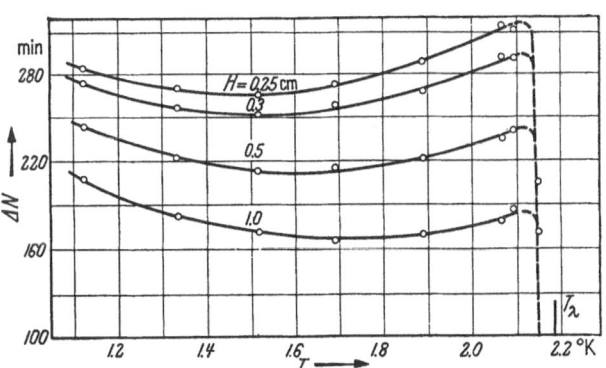

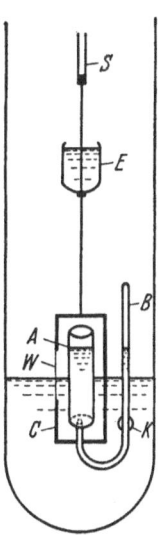

Fig. 79. Film thickness for different heights above the liquid level as a function of temperature. $\varDelta N$ = Nicol rotation.

Fig. 80. Apparatus of JACKSON and HENSHAW for measuring simultaneously film thickness and transfer rate.

where d_0 is the thickness at $h = 1$ cm, JACKSON found z to change both with
temperature and with height. At present it can only be said that the value of
z seems to vary between 2 and 9. Taking into account the calibration difficulties,
d_0 as determined with the optical method appears to be of the order of 2×10^{-6} cm.

A more indirect method for measuring the film thickness was employed by
ATKINS [137] who made use of the level oscillations following the emptying of a helium reservoir by film flow. The apparatus and a typical result are shown in Fig. 81.
A reservoir which is in good thermal contact with the helium bath can be filled
or emptied by film transfer and the level in the capillary attached to the reservoir
is observed. As the meniscus in the capillary drops to the level of the bath,
oscillations due to the kinetic energy of the film are observed which, thanks to
the lack of dissipation, are greatly undamped. From the period the moving
mass of helium can be calculated and, making certain assumptions, the thickness
of the film can be deduced. Since the assumptions extend into a model of the
flow mechanism in the film about which nothing is known, the value of the
results was mainly in corroboration of observations obtained with the more
direct methods. In this respect it is satisfactory to note that the value of d_0
obtained by ATKINS as 2×10^{-6} cm is in good agreement with JACKSON'S work
and that the value of z which was found to be ~ 7 is also compatible with it,
although the large uncertainty here does not permit to regard the values as in
agreement.

A third method which to some extent is similar to that used by DAUNT and MENDELSSOHN was employed by BOWERS [138]. He also measured the amount of helium deposited on a metal foil of large surface to which a fine wire which dipped into the helium bath was attached. In his arrangement the mass of helium in tbe film was, however, determined by weighing with a very sensitive micro-balance. The result obtained in this way, assuming the film to have uniform thickness over the whole surface, is in satisfactory agreement with that of DAUNT and MENDELSSOHN but yields a thickness three times larger than that found by JACKSON and ATKINS when reduced to the value d_0. The method is not very convenient for a determination of z since integration over the height of the foil is required. The value found for z was 2.

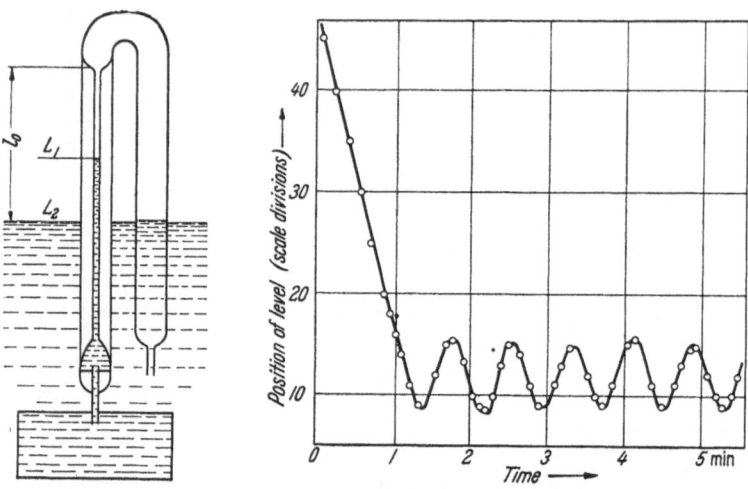

Fig. 81. ATKINS' method for determining the film thickness through oscillations.

Theories of the helium film fall into two groups, those explaining it as an ordinary adsorption due to the VAN DER WAALS forces and the others which consider the film as an integral part of the lambda-phenomenon. The former type, as proposed by SCHIFF [139] and by FRENKEL [140], might appear to be in disagreement with the widely observed fact that the thick helium film vanishes at the lambda-point, since there is no reason why the same amount should not be adsorbed on the solid wall in the helium I region as below the lambda-point. However, the possibility cannot be excluded that the thick film in the helium II region owes its existence only indirectly to the lambda-phenomenon. It has been suggested that quite generally such thick films will be in equilibrium with the saturated vapour but that they may be stripped off by evaporation if the temperature of the adsorbing surface is raised very slightly over that of the liquid phase. Below the lambda-point this loss will be made good immediately by superflow from the helium bath whereas in the helium I region such a re-stocking cannot take place because the absence of superflow will permit even a small heat leak to raise the temperature of the adsorbing surface. It is for this reason that further research on the possible persistence of the film into the helium I region is of considerable importance. The inverse cube potential due to the VAN DER WAALS forces, when superimposed on the graviational potential yields a value for z of 3 which, while not in disagreement with experiment, cannot be taken as a confirmation in view of the wide variation of the observational results.

The same is true for the theories of Bijl, de Boer and Michels [*141*] and of Temperley [*142*] which regard the film as directly due to the lambda-phenomenon. In the former, the zero point energy of mean free path in an ideal Bose-Einstein condensate is minimized together with the gravitational energy which then leads to an equilibrium film thickness of the form

$$d = \left(\frac{h^2}{8\,m^2 g}\right)^{\frac{1}{2}} h^{-\frac{1}{2}} \qquad (37.1)$$

where h is the quantum constant, m the mass of the helium atom, and g the gravitational acceleration. While the factor $(h^2/8\,m^2 g)^{\frac{1}{2}}$ is of the observed magnitude of the film thickness, the experiments are clearly not good enough to make a distinction between values of $z = 2$ or $z = 3$. In addition, there are grave objections to the theory in this simple form which when accounted for would lead to a different value for z.

38. Transfer rates. The experiments of Daunt and Mendelssohn had yielded a fairly simple pattern for the film transfer which agreed well with that theoretically postulated for potential flow. Their findings have been summarized in Sect. 8 and comparatively little has to be added in order to describe our present knowledge of the transfer phenomena. However, for a curious reason the possible influence of a number of factors on the film flow has been investigated in some detail.

Repetition of the original experiments by Atkins [*143*] in Cambridge and by de Haas and van den Berg [*144*] in Leiden revealed a behaviour of film transfer which was completely at variance with the work of Daunt and Mendelssohn [*30*] and was of a much more complex character. In particular these authors observed much higher transfer rates and a strong dependence on the level difference, the height of the intervening barrier and on the shape of the transfer vessel. While Atkins ascribed his results mainly as due to his departure from the ordinary test tube form of the transfer vessel employed, the Leiden authors thought that complete absence of radiation was responsible for their findings. Both these influences were subsequently investigated in Oxford [*145*], [*146*] without success, the observations merely confirming the original experiments. Vessels similar to those employed by Atkins gave the same pressure independent transfer rates as test tube type beakers. Using an arrangement in which the transfer vessel could be completely shielded from incident radiation during part of the experiment, it was found that the flow rates were the same with and without radiation and identical with the values found by Daunt and Mendelssohn. As is evident from Fig. 82 the filling up of the transfer beaker by film transport during the interval when radiation was excluded had proceeded at exactly the same rate as when visual observations had been made.

The discrepancies were ultimately solved by Bowers and Mendelssohn [*146*] who traced the high and pressure dependent transfer rates to the presence of impurities in the liquid helium used in the experiments. An analysis of techniques employed in Cambridge and Leiden suggested that contamination of the cryostats with solidified gases (air or hydrogen) might have occurred and transfer from a test tube type beaker of helium with varying degrees of air impurity was measured. It could be shown that, whereas with pure helium the outflow was slow and pressure independent, subsequent addition of increasing amounts of air contamination produced effects identical with those observed in Cambridge and Leiden. The result of this experiment is shown in Fig. 83 where (1) denotes uncontaminated transfer, (a) transfer over a layer of air adsorbed on the glass at about 90 °K and (2),(3) and (4) transfer with increasing air contamination. Still higher contamination did not increase the transfer above that shown in curve (4). The deposit of solid air on the walls of the beaker was, on reaching curve (4), still too small to be seen, but further addition of air resulted in a visible deposit whose aspect suggested a granular nature with a grain size of the order of the wavelength of visible light.

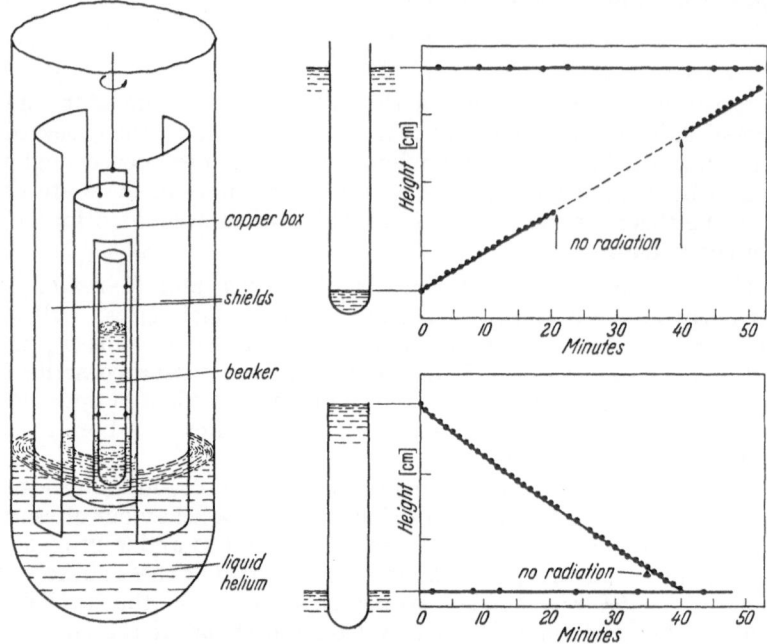

Fig. 82. Test on the influence of radiation on film transfer.

Further experiments with contamination by hydrogen and neon· yielded similar results. Subsequent experiments by ATKINS [148] and by DE HAAS and VAN DEN BERG in which special care had been taken to work with very pure helium failed

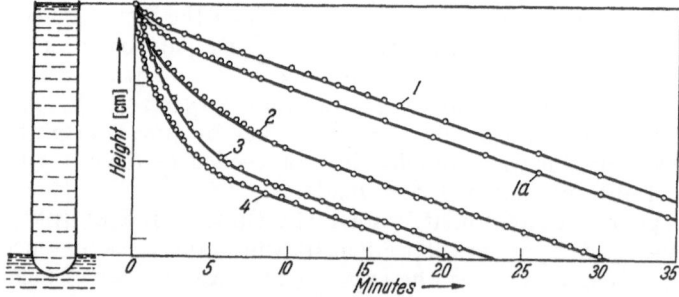

Fig. 83. Influence of contamination with air on the transfer rate. 1 clean beaker, 1a adsorbed air, 2, 3 and 4 successive deposits of solid air.

to exhibit the anomalously high and pressure dependent film flow rates found by them earlier, and it can thus be assumed that the original pattern of film transfer observed by DAUNT and MENDELSSOHN is the fundamental and significant one.

The influence of impurities on the rate of transfer raises the question of the importance of the substrate on which the helium II film is deposited. The problem was first investigated in DAUNT and MENDELSSOHN's experiments as it was noted that transfer along drawn copper wires was, per unit connecting perimeter, larger than along the glass wall of the beaker. They therefore measured the flow rate from a carefully polished copper beaker and found that it was the same as on glass. The higher rate along the wire was ascribed to flow of helium in the surface cracks caused by drawing and scraping. However, subsequent experiments in Oxford [149] on nickel and platinum surfaces and similar work by BOORSE and DASH [150] at Columbia on copper, lead, iron, stainless steel and lucite revealed transfer rates which were often much higher than those on glass although in most cases the general form of the dependence on temperature was the same. Thus, the question arose whether there existed any direct influence of the chemical nature of the substrate on the film transfer or whether the observed effects were simply due to an increase in the geometrical solid perimeter produced by surface cerrations and irregularities. BOORSE and DASH had already shown that transfer over a machined copper surface which was twice that over glass was further increased three times by etching. While this observation did not exclude the possibility of a higher intrinsic transfer rate on copper, it clearly showed the far reaching effect of surface roughness. A decrease of the transfer rate over perspex was observed by CHANDRASEKHAR [151] when the surface of the plastic was carefully polished. Finally, he and MENDELSSOHN [152] found that the transfer over the surface of a beaker of stainless steel which had been given the finest optical polish was identical with that over glass. These authors could also show that once the same beaker had been brought to red heat which destroyed the extreme surface smoothness, the transfer rate had increased. All this work leaves little doubt that there is no appreciable effect of the chemical nature of the substrate on the film transport and that such variations as have been observed are caused merely by surface roughness. This is not really surprising, since it has to be assumed that the helium atoms in direct contact with the solid surface will be in a state which differs greatly from that of the liquid. The VAN DER WAALS forces of the substrate will cause the first layers of helium to be of much higher density than that of the liquid phase, resembling a compressed solid layer of helium. The boundary of the liquid film in which the transfer occurs is therefore not provided by the substrate but by this layer of solid helium and it may be assumed that no first order effects due to the substrate can occur in the film.

Quite apart from any significance for the understanding of the lambda-phenomenon, the film transfer, and in particular the manner in which it is influenced by the conditions of the substrate, is of considerable practical importance. It may be recalled that ROLLIN and SIMON [28] postulated the existence of the film from the large heat influx into helium cryostats, and the subsequent realisation that it was the transport of mass and not the heat conduction which caused this effect made it an important consideration in the construction of helium containers. A study of the old experiments of KAMERLINGH ONNES [26] and of KEESOM [153] which were designed to produce the lowest temperatures which can be reached by pumping off the vapour above a helium II bath indicates that in both cases the film transfer was far in excess of the rate on clean glass. KAMERLINGH ONNES noted a striking speed of re-adjustment of helium levels in his concentric Dewar beakers. This and the speed at which helium evaporated in KEESOM's experiment suggests that in both cases the glass surfaces were heavily contaminated. BLAISSE, COOKE and HULL [154] and later AMBLER and KÜRTI [155] investigated means of reducing the heat influx due to film transfer by suitable construction of the tubes through which helium II cryostats are pumped. At the time of the first mentioned work the influence of contamination was unknown and while in the research of the latter authors precautions were taken to avoid impurities in the helium, it is always difficult to be certain that the impurity content in any such experiment was negligible. In particular, when the cryostat connections under investigation are metal tubes, it is usually quite impossible to say whether high film flow is due to irregularities of the solid surface or to contamination.

In spite of the amount of work done on transfer over smooth and irregular surfaces, no clear assessment of the relevant factors in the latter case has been possible. The simplest explanation is clearly that, owing to cerrations, the actual surface perimeter is larger than the measured one and that it is covered with a film of the same thickness as, say, on glass. Some, but not all, observations can be explained in this way. The next step is to assume that bulk liquid is formed in the surface cracks and that in addition to ordinary film transfer capillary flow can occur in channels filled with liquid. Reasonable assumptions concerning the size of the cracks and knowledge of the surface tension would normally allow an estimate of the height to which liquid can rise in these cracks. However, in liquid helium II account has also to be taken of the thermo-mechanical effect. If, as certainly will be the case in tubes leading into helium cryostats, the upper end of the solid surface is warmer than that in contact with the bulk liquid, an additional thermo-mechanical level height must be added to that pro-

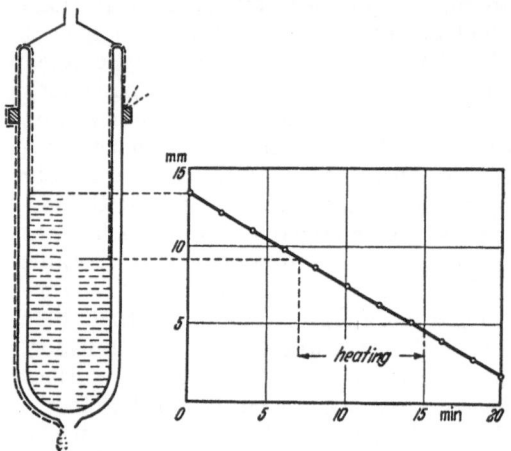

Fig. 84. Experimental proof that evaporation of the film does not effect the transfer rate.

duced by surface tension. This is possibly the explanation for the very high transfer rates encountered not only in the experiments of KAMERLINGH ONNES and KEESOM but also in a great number of helium cryostats used by various authors since. It has been suggested by a number of authors that film transfer under a temperature gradient may be higher than in gravitational outflow from a beaker, but so far there exists little reliable evidence for such an effect. High transfer rates which have occasionally been observed when a temperature gradient was present may thus be due to flow of bulk liquid in small cracks and under the influence of a thermo-mechanical pressure head.

That transfer towards a higher temperature is not necessarily greater than under gravity was demonstrated in the early experiments of DAUNT and MENDELSSOHN [30]. Since it had at first been suggested that the high evaporation rates of helium II cryostats were due to a high thermal conductivity of the film, the latter was investigated in the following apparatus. A small Dewar vessel (Fig. 84) was fitted with an external metal ring in which heat could be produced by means of eddy currents. The vessel was lifted out of the helium bath, and the transfer from it was observed by the drop of the meniscus as well as by counting the liquid drops falling from the vessel. When the ring was heated by gradually increasing the induced eddy current, it was noted that the number of drops decreased and finally disappeared completely, showing that the film had been completely evaporated. Since the rate at which the meniscus dropped was the same with or without heating, it is clear that no appreciable amount of heat was conducted along the film into the Dewar vessel where it would have caused additional evaporation of liquid. It should be remarked, however, that in this experiment heat was supplied to the outer surface of the vessel and not to its rim which means that, while it proves the low heat conduction of the film, it does not exclude the possibility of higher transfer rates under a thermo-mechanical pressure.

Surface irregularities which cause ordinary surface tension rise in cracks are probably responsible for a slightly increased transfer rate which is almost invariably observed when the liquid level is within a few millimetres from the rim of a beaker. This feature was noted [30] in the very first beaker experiment (Fig. 15). As already mentioned, the transfer rate was found to be practically constant for the residual length of the beaker. DAUNT and MENDELSSOHN investigated a possible change of rate with height in a separate experiment and noted that there appeared to exist a slight increase in transfer with increasing level difference, but that the effect was of the same order of magnitude as the accuracy of their observation. A similar effect, also very slight, was found by AT-KINS [154] who suggests a relation between variation of rate with the change of film thickness with height.

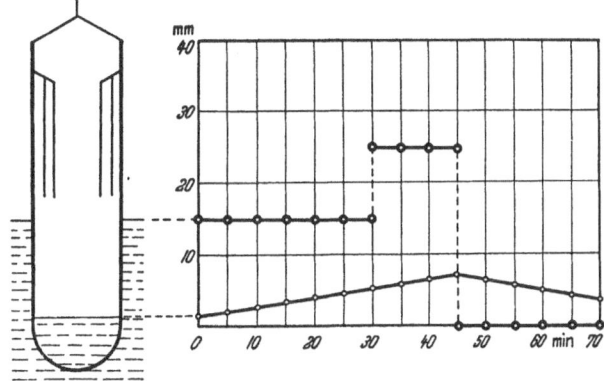

Fig. 85. Formation of bulk liquid from the film. Drops formed between minutes 30 and 45.

However, it has to be kept in mind that his value of z in Eq. (37.1) is anomalously high and that the usual thickness determinations would demand a very much increased variation of transfer rate over the length of the beaker. Re-examining the question of height dependence of rate, MENDELSSOHN and WHITE [149] could not detect a definite variation in observations on carefully shielded beakers except for a drop of about 10% when the levels approached to within 3 mm. This effect was subsequently explained by ESELSON and LASAREW [156] who showed that it is due to insufficient heat contact between the inside of the beaker and the bath which results in thermo-mechanical pressure difference. Under more isothermal conditions the effect was not observed.

The appearance of liquid drops at the bottom of a test tube filled with helium II shows that at some stage in the transfer process bulk liquid will form out of the film. DAUNT and MENDELSSOHN [30], using a beaker with a funnel shaped inset as shown in Fig. 85, observed that drops of liquid would fall off the tip of the funnel if the latter was below the height of the bath level. From this they concluded that bulk liquid can be formed at constrictions below the higher level. The question was further investigated by CHANDRASEKHAR and MENDELSSOHN [86] who employed a beaker with staggered diameters below which were arranged glass skirts of different width. The film could thus originate at three different perimeters AA, BB, CC and the maximum diameters were arranged in the sequence $DD > AA > EE > BB > CC$ (see Fig. 86). The three stages of the experiment show that liquid drops will always form when the film has to flow over a perimeter which is narrower than the original one. It is also evident that the film will re-form from the bulk liquid even below the higher meniscus since no drops were observed on DD when BB was the perimeter at which the film originated. The nature of the bulk liquid so formed has been demonstrated in a striking observation by HAM and JACKSON [157]. They employed a conical beaker of stainless steel in which the inner diameter at A was the same as the outer diameter at B and the inner diameter at B equal to the outer diameter at C as shown in Fig. 87. Using the optical method for the determination of film thickness, they observed that when the filled

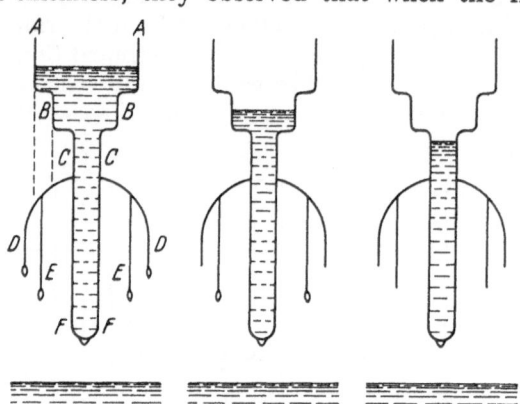

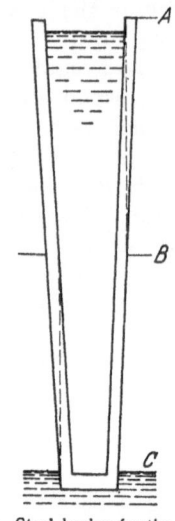

Fig. 86. Experiments on the formation of bulk liquid from the film.

Fig. 87. Steel beaker for the observation of liquid helium drops forming out of the film.

beaker was raised out of the helium bath drops, which appeared as bright spots, could be seen running down the outer surface. The number of drops increased towards the lower end of the beaker because of the diminishing outer perimeter.

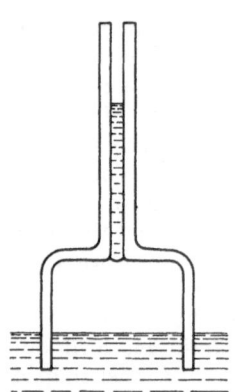

Fig. 88. Formation of helium II under surface tension in a capillary.

No drops were observed between A and B at which stage the outer circumference of the beaker was equal to the originating perimeter. As the experiment proceeded and the inner level in the beaker fell, fewer drops were observed until finally they ceased completely when the inner level had reached B. The authors, comparing their observations with FRENKEL's theory for drops on a wetted surface found in agreement with it that the smallest drops did not run but appeared stationary.

Bulk liquid can form from the film above the meniscus of the helium bath in cavities of such a size and at such height that due to normal surface tension the liquid will be stable. Although KELVIN already pointed out that a fine capillary suspended in the vapour above the level of a liquid should fill up with liquid under equilibrium conditions, the process is not observed as distillation under isothermal conditions is far too slow.

In helium II the conditions are different as the film can transfer liquid along the walls of the vessel and over the suspension into the capillary. LANE and DYBA [158] showed that a capillary to whose lower end a wide tube was attached which in turn dipped into a bath of liquid helium II would fill up with liquid from the bath (Fig. 88). The height of the liquid corresponded to half the height to which a freely suspended capillary of equal diameter could be filled up with helium I, allowing for the slight change of surface tension with temperature. The reason is that in the case of helium I the liquid column in the capillary is held up by a "skin" at the top and at the bottom whereas in helium II the lower skin is "pierced" by the film. Another observation by the same authors could be explained equally well. They noticed that the gap between two plates, separated by spacers, only one of which dipped into the bath, would also fill up. Here the film had evidently to form bulk liquid

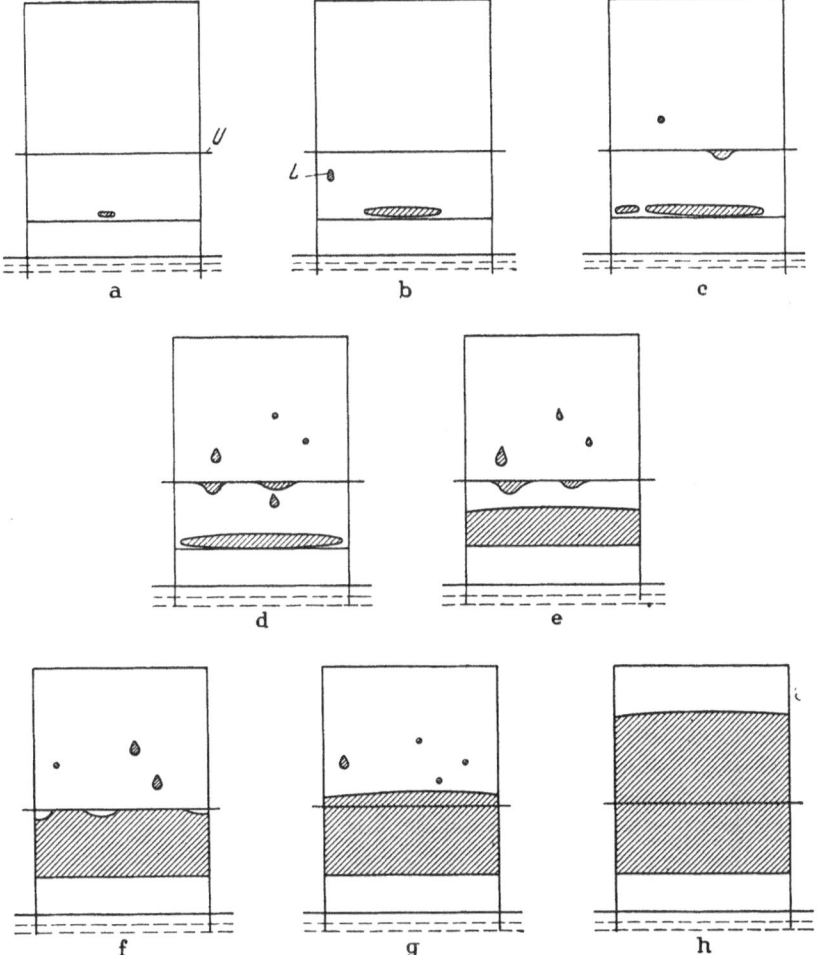

Fig. 89 a—h. Successive stages in the formation of bulk liquid in a gap of 12 microns width between parallel glass plates. U spacer wire, L a particle of dust.

at the spacers. On the other hand, it had been noticed by others that even if there were no spacers, liquid would form between closely opposing plates. These phenomena were further studied visually in some detail by McCrum [159] who used an interference method in order to observe the formation of bulk liquid in a gap of only 30 microns or less between two parallel plates which were opposed without spacers. It was then found that if the gap between the plates was completely free of dust or any kind of bridging material, no liquid would form even well below the surface tension height. The presence of bridging material which sometimes was in such small specks that it was not visible could, on the other hand, be noticed quite clearly owing to the drops of helium formed at these centres. A typical example of one of these observations is given in Fig. 89 where the gradual formation of bulk liquid from the film in time intervals of a few seconds is illustrated. Shortly after dipping the longer plate into the bath, a drop of liquid is seen to form at the lower end of the gap. Subsequently liquid appears on a wire acting as a spacer and on particles of dust until finally the whole gap fills up to the surface tension height.

One of the first experiments carried out by various authors on the film was a measure of the dependence of transfer rate on temperature. In view of what has been said about the difficulty of obtaining a cleam and smooth surface it is not surprising that the absolute values differ somewhat from one investigation to another. They agree, however, in the general shape of the dependence which shows a gradual rise from the lambda-point, becoming almost independent of temperature at about 1.4 °K (Fig. 90). The most

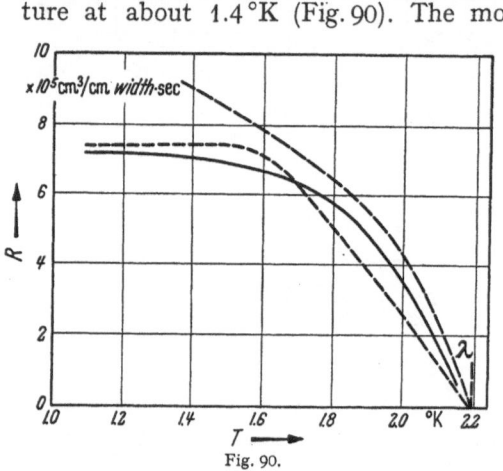
Fig. 90.

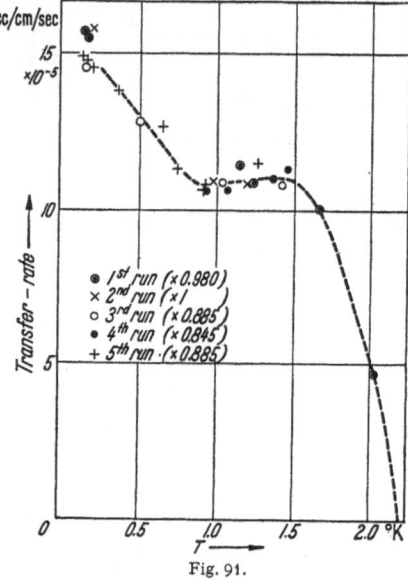
Fig. 91.

Fig. 90. Film transfer rates over glass surfaces (in cm³ of liquid per second for each cm of width of the connecting perimeter Full curve: Mendelssohn and White, ---- Daunt and Mendelssohn, — — — Webbers et al.

Fig. 91. Film transfer rate below 1 °K. The results have been normalized to the numerical value of the second run.

recent measurements by Mendelssohn and White between 1.1 °K and the lambda-point which have been carried out under fairly stringent conditions yielded for the transfer rate

$$R = A \left[1 - \left(\frac{T}{T_\lambda} \right)^n \right]. \qquad (38.1)$$

R is measured in cm³/sec per cm of perimeter and n has the value of ~ 7. The value for A is about 7.5×10^{-5} which is identical with that found originally by Daunt and Mendelssohn. These figures all refer to smooth glass surfaces, but it should be noted that the rate is somewhat higher for backed out glass. This is probably due to the evaporation of water from small cracks and a consequent increase in effective perimeter.

The transfer rate below 1 °K has been measured by Ambler and Kürti [160] who observed a marked increase by about 30% in the transfer rate between about 0.7 °K and 0.15 °K. No detailed measurements were made between the lambda-point and 1 °K, but the shape of the curve in this region seems to agree qualitatively with the above mentioned measurements. The apparatus used by these authors was a glass beaker with a constriction from which the outflow was observed. Pills of paramagnetic salt used to cool the helium were arranged inside and outside this beaker. The result is extremely interesting since the temperature at which the rise in the transfer rate occurs roughly coincides with that at which the anomalous excitations begin to play a role. The observed effect thus should be considered together with the change in the specific heat function and the rise in the second sound velocity which occur in the same temperature region. Unfortunately, the observed transfer rates in the region between 1 °K and 1.5 °K where they can be compared with the conventional experiments differed in individual runs by about 15% and were altogether higher than the accepted ones (Fig. 91). The possibility that the results may be influenced by surface contamination cannot therefore be excluded. The only other investigation to temperatures in the magnetic range has recently been carried out by

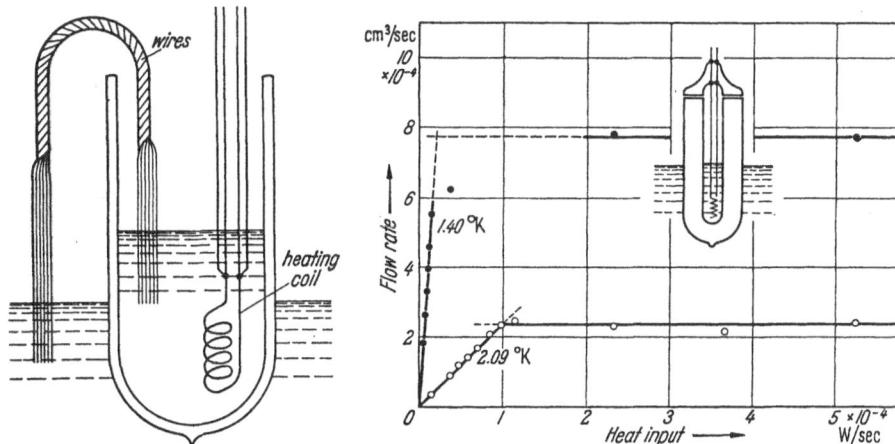

Fig. 92. Thermo-mechanical effect with the film as connecting link.

Fig. 93. Thermo-mechanical film transfer under adiabatic conditions.

WARING [161] at Yale whose values are always higher than those found by AMBLER and KÜRTI but who finds hardly any dependence on temperature below 0.9 °K. WARING determined the transfer rate by measuring the evaporation from a vessel held at the low temperature which was connected to the region of higher temperatures by a glass capillary. The author suggests that the discrepancy between his and the earlier results may be due to the fact that he observed transfer under a thermal rather than under a gravitational potential. It is clear that further investigation is required to form a clear picture of these effects.

Finally, the occurence of the thermo-mechanical effect with the film as a connecting link has to be mentioned. It was first observed by DAUNT and MENDELSSOHN [18] who found that in a small Dewar partly immersed into helium II the inner level rose when heat was applied to the inside of the vessel. The effect could be much enhanced [162] by adding a large connecting perimeter in the form of a bundle of wires as shown in Fig. 92. From a quantitative assessment of the circulation process in the vapour and through the film it was concluded that viscous return flow in the film is negligible. The same effect was studied by CHANDRASEKHAR and MENDELSSOHN [86] who used a Dewar vessel with a closed top into which the film could pass but which effectively inhibited outflow of vapour. Using this highly adiabatic arrangement it was found that up to a certain limit of transfer rate the inflow was exactly proportional to the heat current (Fig. 93). Beyond this critical rate the transfer did not increase further as the heat input was raised. The experiments show that film transfer under a thermomechanical pressure difference exhibits the same features as found in gravitational flow. There is a sharply defined critical velocity (given by the transfer rate) and below this the flow is free of friction.

J. The unsaturated film.

39. Film formation. The existence of a thick film of helium on all surfaces in contact with the liquid has naturally raised the question as to its origin. As was already mentioned in the previous chapter, the two types of existent theories favour either a pure adsorption phenomenon or some mechanism directly connected with the lambda-phenomenon. KISTEMAKER [163] in Leiden was the first to study the adsorption of helium to concentrations approaching saturation. The results were not very accurate and his work has been followed by a number of other investigations in different laboratories, using various substrates. In spite of much effort it is evident that our present knowledge of the film formation is most unsatisfactory.

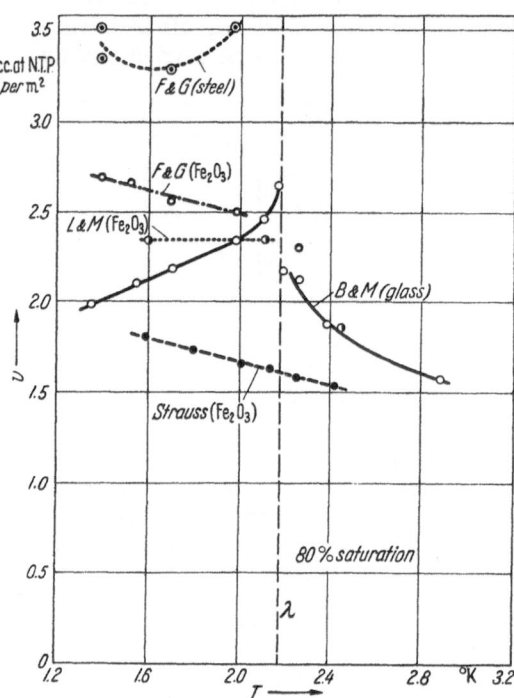

Fe$_2$O$_3$ as substrate has been used by LONG and MEYER [164] and by STRAUSS [165] in Chicago as well as by FREDERIKSE and GORTER [166] in Leiden. The latter authors also used a steel surface. Glass was used by BREWER and MENDELSSOHN [167] and by TJERKSTRA, HOOFTMAN and VAN DEN MEIDENBERG [168] and aluminium by BOWERS [169]. Since the layer closest to the substrate will be highly compressed, it is convenient to express the results in cm^3 of helium adsorbed per m^2.

Fig. 94. Results on the adsorption of helium at 80% saturation as a function of temperature. F. & G. FREDERIKSE and GORTER, L. & M. LONG and MEYER, B. & M. BREWER and MENDELSSOHN.

The great divergence of the results both in magnitude and temperature dependence is shown in Fig. 94 in which the adsorbed amount at 80% saturation is plotted for temperatures below and above the lambda-point. While the absolute value of the coverage varies up to a factor of 2, the coverage is seen to exhibit a decrease or an increase or no change with rising temperature below the lambda-point. Above the lambda-point the adsorbed amount always decreases with rising temperature, as would normally be expected, but here again there are differences. In some observations the adsorption curve shows a break at the lambda-point while in others it remains continuous. There is little reason to make the substrate alone responsible for the observed discrepancies. In jeweller's rouge falling as well as temperature independent coverages were found in the helium II region, with and without break at the lambda-point. In glass BREWER and MENDELSSOHN observed a rise of coverage with temperature and the same is true for KISTEMAKER's work. However, TJERKSTRA, HOOFTMAN and VAN DEN MEIDENBERG found no dependence on temperature.

In these circumstances it is hardly worth while to reproduce and discuss the adsorption isotherms obtained by the different authors, and the reader is referred to the original publications. Since it would be difficult to ascribe the wide divergence of results entirely to faults in experimentation, one may suspect that in the formation of the helium film some factor is operative which as yet has not been taken into account. If this should be so, it must seem unlikely that the process of film formation is one of ordinary adsorption. It has been suggested by BREWER and MENDELSSOHN that it may be energetically more favourable in the helium II region to form clusters of helium rather than to add further layers beyond a given coverage. In this case the nature of the experimental arrangement may have an influence on the result.

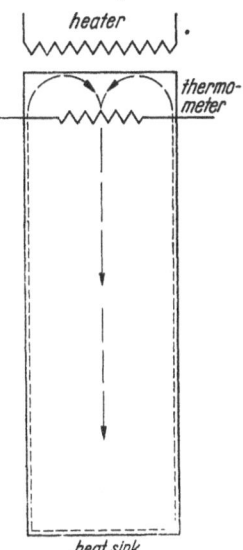

40. Transfer rates. In spite of the discrepancies in the adsorption measurements, recent, work on the onset of superfluidity in sub-saturated helium films has yielded consistent results. Two entirely different methods have been used. In the first, the flow through a superleak is measured. The superleak forms the link between two containers with helium gas below saturation but of different pressure. Gas flow through the superleak is very small and practically all of the mass transport is made up of film transfer over the perimeter of the leak. BROWN and MENDELSSOHN [170] employing superleaks made of wires drawn down in a metal tube could show that transfer takes place at sub-saturation pressures. LONG and MEYER [171] made careful and quantitative experiments with this method using two different procedures. In the first one, the chamber receiving the helium flowing through the superleak was at a much lower pressure than that in which the gas was stored under sub-saturation pressure. In the second procedure, the pressure in both chambers was almost the same. The results obtained

Fig. 95. Thermal method for determining the transfer of the unsaturated helium film.

in these experiments were, however, completely different. In the first procedure the onset of superfluidity was depressed to temperatures below the lambda-point as the pressure was reduced below the saturation value. The second procedure yielded superfluidity at the lambda-point for all concentrations down to a film thickness of 1.5 statistical layers. Since the authors felt that the second procedure represented more closely equilibrium conditions, they assumed the results obtained with the latter as the valid ones.

The other method [172] of investigating the flow properties of sub-saturated films is based on a measure of the heat transported in the circulation process involving the film and the gas phase. The apparatus used is shown in Fig. 95 and consists in a closed thermally insulated tube, the lower end of which is attached to a heat sink of controlled temperature while the upper one bears a thermometer and a heater. The tube is filled with gas of the desired pressure below saturation, and its inner walls are then covered with the unsaturated film corresponding to this concentration. As heat is supplied, some of the film is evaporated at the top and the helium will return to the heat sink through the gas phase, setting up a convection current. When this current reaches its critical value, i.e. when the film is completely evaporated, the temperature at the top rises suddenly and the critical rate of mass transport by means of the subsaturated film is given as

$$r_c = \frac{\dot{Q}_c}{L + T \varDelta S} \tag{40.1}$$

where $\dot{Q}_c$ is the heat input at the moment when the temperature rises, L the latent heat of liquid helium and ΔS the entropy difference between the helium

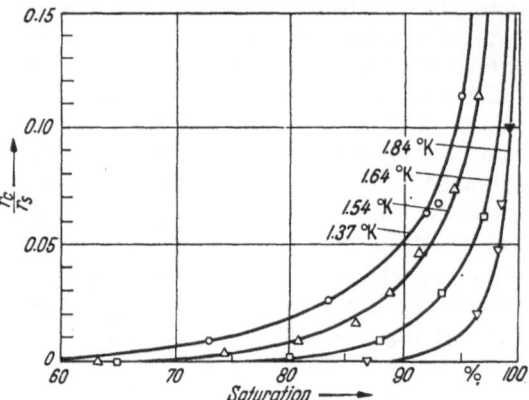

Fig. 96. Critical transfer rate of the unsaturated film at different temperatures as a function of percentage saturation. r_c/r_s is the ratio of observed rate to the rate of the saturated film.

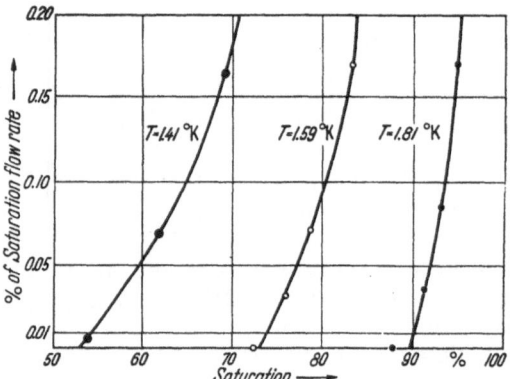

Fig. 97. The onset of superfluidity at different temperatures.

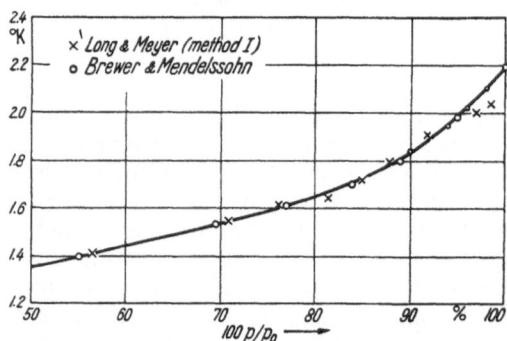

Fig. 98. Onset temperatures of superfluidity as a function of percentage saturation. The large circles refer to glass as substrate and the small ones to German silver.

flowing in the film and that of the bulk liquid. This might be taken as identical with the total entropy. Fig. 96 shows how for four different temperatures the critical rate varies with concentration. Since in this region the transfer rate of the saturated film also varies with temperature, the plotted rates are reduced values (r_c/r_s) where r_s refers to the saturated film. As is evident, film transfer is very much reduced as the pressure is lowered below saturation. Even at 90% saturation r_c is only 5% of r_s at 1.37 °K and at 1.84 °K the same concentration is necessary to produce the first onset of superfluidity. The sharpness of the onset can be demonstrated well by this method because, as quantities of heat in this temperature region go, L is very large which makes the measurement very sensitive. Fig. 97 which shows the transport rates of the sub-saturated film at values below 0.2% r_s leaves no doubt that the onset of superfluidity is discontinuous and, for a given concentration, will occur at a well defined temperature.

The latter observation is thus in accord with LONG and MEYER'S procedure 1 and not with procedure 2. The reason for this is not quite clear but may possibly be connected with the large discrepancies in the adsorption measurements. If there should exist some anomalous feature in the formation of the helium II film, it is conceivable that in procedure 2 the superleak was, in fact, filled with bulk liquid. It is therefore encouraging to note that LONG and MEYER'S values for the onset of superfluidity as obtained with procedure 1 are in good agreement with the values found by BREWER and MENDELSSOHN [172] using the thermal method described above. The results of both investigations are shown in Fig. 98, the crosses referring to flow through a superleak and the circles to the thermal method.

An interesting feature of the results is that observations with different substrates, glass and german silver, which are represented by circles of different size gave the same values for the onset temperatures of superfluidity. On the other hand, the transfer rates of the saturated film over these substrates differed

by a factor of two. This is an additional indication that high transfer rates on rough surfaces are not necessarily caused by the increase in the geometrical perimeter alone. In fact, it was found that at 1.53 °K the value of r_c was the same for german silver and glass up to a concentration of 93% when the rate over german silver began to rise much more steeply (Fig. 99). It can thus be concluded that in the present case a transport process which was additional to that provided by the film was coming into play at high concentrations and it seems likely that this was superflow of bulk liquid which had formed in cracks.

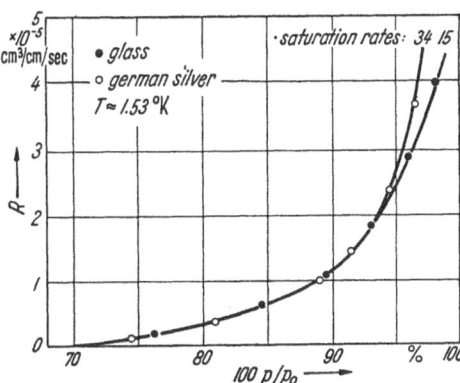

Fig. 99. Critical transfer rates of the unsaturated film on glass and German silver as a function of percentage saturation.

In view of a number of theoretical suggestions, it would be interesting to establish a relation between film thickness and rate of transfer. The unsatisfactory state of the adsorption measurements makes it unfortunately quite impos-

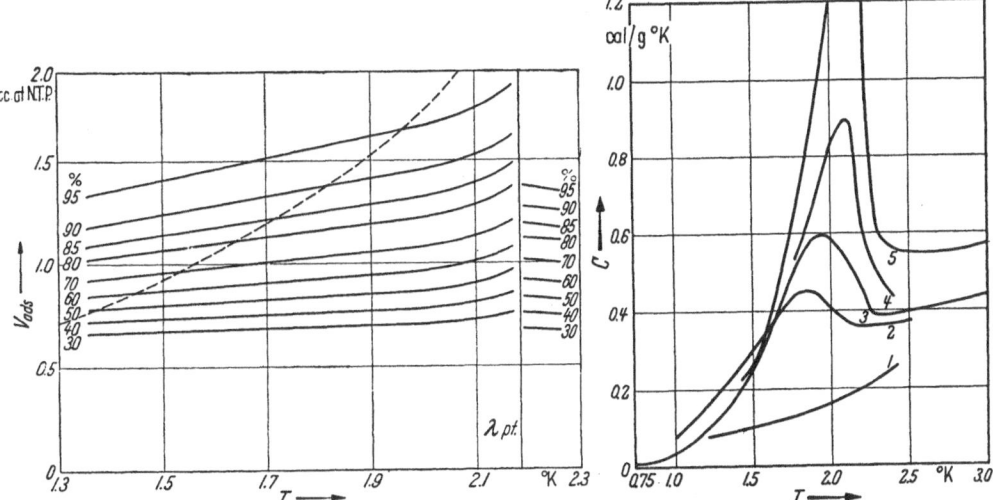

Fig. 100. Adsorption of helium on glass for different percentage saturations as a function of temperature. The dotted line indicates the onset of superfluidity observed with the thermal method.

Fig. 101. Specific heat of adsorbed helium as a function of temperature. The estimated number of layers are (1) 3—4, (2) 5—6, (3) 7—9, (4) 9—12, (5) bulk liquid.

sible to link up the values of r_c with reliable estimates of the number of layers in the film. BREWER and MENDELSSOHN [168] have, on the basis of their adsorption isothermals and their flow rate measurements constructed a diagram in which the onset of superfluidity of sub-saturated films is plotted together with the coverage on glass against the temperature. As will be seen from Fig. 100, the isothermals all break at the lambda-point whereas the onset of superfluidity is strongly temperature dependent. Careful measurement of one of the isothermals in the region where it is intersected by the superfluidity curve yielded no anomaly in the adsorption at this value. Unless one can assume some such process as the formation of clusters instead of homogeneous coverage which might explain the break at the lambda-point, the results of Fig. 100 must appear inconsistent.

However, even allowing for the uncertainty in the film thickness d, the experiments on subsaturated films seem to indicate that the product vd, where v is the critical velocity, is not, as has sometimes been suggested, constant but increases with d.

41. Specific heat. The specific heat of sub-saturated layers of helium has been measured by FREDERIKSE [173] in Leiden. The substrate was jeweller's rouge and the specific heat curves shown in Fig. 101 were obtained for different coverages. It has been suggested that owing to the strong compression at the actual surface of the substrate, the film here is much denser and that the actual coverage given by the author in layers ought to be re-assessed on the basis that the first four statistical layers will actually form the compressed single layer. It is interesting that the specific heat due to these first four statistical layers does not, in fact, show an anomaly, whereas thicker coverages all show anomalies which with increasing concentration become larger and move to higher temperatures.

K. Theoretical Appendix.

An account of the two important theoretical approaches to the helium problem, that of F. LONDON's BOSE-EINSTEIN condensation and that of LANDAU's energy spectrum of phonons and rotons, has been given in Sect. A and the significance of experimental results has been discussed under the various sub-headings. The essence of the two-fluid model has also been explained. Summaries of the main features of the theories and their elaborations are given in this appendix.

42. The ideal BOSE-EINSTEIN gas. Condensation in momentum space will take place if the number of atoms is larger than

$$N_c = 2.612 \, V \left(\frac{2\pi \, m \, k \, T}{h^2} \right)^{\frac{3}{2}} \tag{42.1}$$

or if the temperature is lower than

$$T_c = \frac{h^2}{2\pi \, m \, k} \left(\frac{N}{2.612 \, V} \right)^2 \tag{42.2}$$

or if the molar volume (V/N) is smaller than

$$v_c = \frac{1}{2.612} \left(\frac{h^2}{2\pi \, m \, k T} \right)^{\frac{3}{2}}. \tag{42.3}$$

The fraction n_0 of atoms in the ground state below the condensation temperature is

$$n_0 = N \left[1 - \left(\frac{T}{T_c} \right)^{\frac{3}{2}} \right]. \tag{42.4}$$

For the thermo-dynamic functions F. LONDON has derived the two branches, those below the critical temperature T_c denoted by $(-)$ and those above T_c denoted by $(+)$.

For the total energy the relations are obtained as

$$E_- = 0.770 \, R \, T \left(\frac{T}{T_c} \right)^{\frac{3}{2}} \tag{42.5}$$

and

$$E_+ = \frac{3}{2} R T \left[1 - 0.4618 \left(\frac{T_c}{T}\right)^{\frac{3}{2}} - 0.0226 \left(\frac{T_c}{T}\right)^3 - 0.0020 \left(\frac{T_c}{T}\right)^{\frac{9}{2}} - \cdots \right]. \quad (42.6)$$

From this the specific heats are calculated as

$$C_{v_-} = 1.926 \, R \left(\frac{T}{T_c}\right)^{\frac{3}{2}} \quad (42.7)$$

and

$$C_{v_+} = \frac{3}{2} R \left[1 + 0.231 \left(\frac{T_c}{T}\right)^{\frac{3}{2}} + 0.045 \left(\frac{T_c}{T}\right)^3 + 0.007 \left(\frac{T_c}{T}\right)^{\frac{9}{2}} + \cdots \right]. \quad (42.8)$$

The free energies are

$$A_- = - 0.513 \, R T \left(\frac{T}{T_c}\right)^{\frac{3}{2}} \quad (42.9)$$

and

$$A_+ = - \frac{3}{2} R T \left[\ln \left(\frac{T}{T_c}\right)^{\frac{3}{2}} + 0.0265 + 0.3075 \left(\frac{T_c}{T}\right)^{\frac{3}{2}} + \right. \\ \left. + 0.0075 \left(\frac{T_c}{T}\right)^3 + 0.0003 \left(\frac{T_c}{T}\right)^{\frac{9}{2}} + \cdots \right] \quad (42.10)$$

and the entropies

$$S_- = 1.284 \, R \left(\frac{T}{T_c}\right)^{\frac{3}{2}} \quad (42.11)$$

and

$$S_+ = \frac{3}{2} R \left[\ln \left(\frac{T}{T_c}\right) + 1.0265 - 0.1537 \left(\frac{T_c}{T}\right)^{\frac{3}{2}} - \right. \\ \left. - 0.0150 \left(\frac{T_c}{T}\right)^3 - 0.0015 \left(\frac{T_c}{T}\right)^{\frac{9}{2}} - \cdots \right]. \quad (42.12)$$

For the pressures one obtains

$$P_- = 1.342 \, k T \left(\frac{2\pi m k T}{h^2}\right)^{\frac{3}{2}} \quad (42.13)$$

and

$$P_+ = \frac{R T}{V} \left[1 - 0.4618 \left(\frac{V_c}{V}\right) - 0.0226 \left(\frac{V_c}{V}\right)^2 - 0.0020 \left(\frac{V_c}{V}\right)^3 - \cdots \right] \quad (42.14)$$

where the critical volume is $N v_c$, v_c having been defined by Eq. (42.3).

LONDON points out that these power series equations are unsuitable to decide the value of the functions close to the critical temperature and that it is impossible to say whether or not they will be discontinuous at this point. However, he shows independently that C_{v_+}, when calculated differently, converges towards the value of C_{v_-} as given by Eq. (42.7). Extending the calculation to the temperature dependence of the specific heat at the condensation point, he obtains for

$$\lim_{T \to T_c} \left(\frac{\partial C_{v+}}{\partial T}\right)_v = - 0.78 \frac{R}{T_c} \quad (42.15)$$

and for

$$\lim_{T \to T_c} \left(\frac{\partial C_{v-}}{\partial T}\right)_v = + 2.88 \frac{R}{T_c} \quad (42.16)$$

and thus for the discontinuity of the temperature coefficient of the specific heat

$$\Delta \left(\frac{\partial C_v}{\partial T}\right) = 3.66 \left(\frac{R}{T_c}\right). \quad (42.17)$$

Although no condensation in co-ordinate space takes place, it is interesting to see from Eq. (42.13) that the pressure below T_c only depends on the temperature and not on the volume. It is therefore analogous in its behaviour to that of a saturated vapour.

Whereas under condition of constant volume the condensation of the ideal Bose-Einstein gas has no discontinuities in either energy or specific heat [as shown by (42.17) only the derivative of the latter with respect to the temperature becomes discontinuous], the process can be a transition of the first order when carried out under constant pressure. From (42.13) it follows that the (P, T)-diagram contains a transition line with the critical value of the pressure as given by this equation. For pressures larger than this value the volume changes discontinuously to zero from $N v_c$ [see Eq. (42.3)]. This change is accompanied by a latent heat of condensation

$$Q = T_c \Delta S = \tfrac{5}{2} 0.513 R T_c.$$ (42.18)

43. The non-ideal Bose-Einstein gas. While the possibility offered by the Bose-Einstein condensation phenomenon to account for a rapid loss of entropy without invoking ordering processes in co-ordinate space, such as crystallization, was attractive, its difficulties were immediately recognized. F. London himself in his first papers has stressed the differences between an ideal gas and a liquid but has pointed out that for an ideal gas with the mass of the helium atom the values of T_c and S_c from (42.2) and (42.11), (42.12), being 3.14 °K and $1.28 R$ respectively, are remarkably close to the lambda-point T_λ and the entropy S_λ of liquid helium which are 2.19 °K and $0.8 R$. He has therefore made attempts to account for the interaction forces in order to see whether closer agreement would result by making plausible assumptions. His early but rather intuitive efforts were reasonably successful. They have been followed by numerous attempts of a great number of authors to produce more rigorous derivations, none of which is, however, sufficiently convincing to provide a firm basis for a theory of an interacting Bose-Einstein liquid. The fundamental difficulties involved in any approach of this kind have been stressed by Feynman.

London's first consideration of the non-ideal state was based on a qualitative notion how interaction, in particular the exchange forces, would influence the ordering process. He suggested that the density of states at the lowest energy would be reduced, and later assumed the possible appearance of an energy gap Δ above the ground state.

An energy spectrum of this type was actually used by A. Bijl, de Boer and Michels who wrote the energy momentum relation as

$$E = \Delta + \frac{p^2}{2m'},$$ (43.1)

where m' is an "effective" mass. Suitable choice of Δ and m' as functions of the density can be made by reference to the experimental data. Generalized energy relations for this model have been derived by Dingle and by F. London. The latter has pointed out that adaptation to the real case observed in liquid helium would require $\Delta \sim 4 k T_\lambda$ which permits evaluations of the thermo-dynamical functions as follows.

The normal density

$$\varrho_n = \varrho \left(\frac{T}{T_\lambda} \right)^{\frac{3}{2}} e^{\frac{\Delta}{k T_\lambda} - \frac{\Delta}{k T}}$$ (43.2)

and, calculated per gram, for the energies below and above the lambda-point

$$e_- = \frac{kT}{m'}\left(\frac{T}{T_\lambda}\right)^{\frac{3}{2}}\left(\frac{3}{2} + \frac{\Delta}{kT}\right)e^{\frac{\Delta}{kT_\lambda} - \frac{\Delta}{kT}} \tag{43.3}$$

and

$$e_+ = \frac{kT}{m'}\left(\frac{3}{2} + \frac{\Delta}{kT}\right). \tag{43.4}$$

For the specific heat at constant volume

$$c_{v-} = \frac{k}{m'}\left(\frac{T}{T_\lambda}\right)^{\frac{3}{2}}\left(\frac{15}{4} + \frac{3\Delta}{kT} + \frac{\Delta^2}{k^2 T^2}\right)e^{\frac{\Delta}{kT_\lambda} - \frac{\Delta}{kT}} \tag{43.5}$$

and

$$c_{v+} = \frac{3k}{2m'} \tag{43.6}$$

which leads at the lambda-point to the discontinuity

$$\frac{(c_{v-} + c_{v+})}{c_{v+}} = \frac{2}{3}\left(\frac{3}{2} + \frac{\Delta}{kT_\lambda}\right)^2 \tag{43.7}$$

and for the entropies

$$s_- = \frac{k}{m'}\left(\frac{T}{T_\lambda}\right)^{\frac{3}{2}}\left(\frac{5}{2} + \frac{\Delta}{kT}\right)e^{\frac{\Delta}{kT_\lambda} - \frac{\Delta}{kT}} \tag{43.8}$$

and

$$s_+ = \frac{k}{m'}\left\{\ln\left[\frac{m'}{\varrho}\left(\frac{2\pi m' kT}{h^2}\right)^{\frac{3}{2}}\right] + \frac{5}{2}\right\}. \tag{43.9}$$

Chosing for Δ/k the value 8.8 °K and for m' the mass of the helium atom multiplied by 9.1, LONDON has evaluated Eqs. (43.2) and (43.8), and has shown that the agreement with the observed temperature functions of the normal density and of the entropy is quite satisfactory. The specific heat, when evaluated in the same way from (43.5) and (43.6), yields a discontinuity at the lambda-point, but the temperature dependence both above and below this temperature is not in good agreement and the size of the discontinuity evaluated from (43.7) is much smaller than the experimental value.

44. Phonons. When it was realized that helium even at absolute zero would remain in the liquid state, the question of thermal excitations near absolute zero in this liquid was discussed by various authors. It is generally assumed that, although longitudinal as well as shear waves might occur, it would only be the former which are excited at the lowest temperature. The experimental attempts at determining the phonon contribution to the thermal energy of liquid helium by various types of measurement have been mentioned in the earlier sections. Knowledge of the velocity of first sound or of the compressibility of helium will allow the phonon contribution to be evaluated from the DEBYE theory. This yields for the energy

$$E_p = \frac{4\pi^5 k^4 T^4}{15\varrho h^3 u_1^3} \tag{44.1}$$

for the entropy

$$S_p = \frac{16\pi^5 k^4 T^3}{45\varrho h^3 u_1^3} \tag{44.2}$$

and for the specific heat

$$C_p = \frac{16\pi^5 k^4 T^3}{15\varrho h^3 u_1^3}. \tag{44.3}$$

45. LANDAU's theory. In its early form the theory of LANDAU considered a spectrum of phonon excitations which is devided from "roton" excitations, i.e. elementary excitations of vortex motion by an energy gap Δ which is of the order of kT_λ. While criticizing the arguments used by BIJL, he postulated a similar energy momentum relation for the rotons as proposed by Bijl, de Boer and Michels for all excitations, which is given in Eq. (43.1). Thus, assuming that the rotons obey Bose-Einstein statistics, the thermo-dynamic relations will be similar to those of Sect. 43.

When attempting to fit the results of the velocity of second sound obtained by Peshkov to the predictions of the theory, Landau noted that the originally proposed energy momentum relation did not give the correct result. He therefore suggested a modified energy spectrum as shown in Fig. 24 in which the roton momenta are clustered around a certain value p_0 so that the energy momentum relation in this neighbourhood can be written as

$$E = \Delta + \frac{(p - p_0)^2}{2m'}. \qquad (45.1)$$

He points out that under these conditions it is no longer permissible to make a qualitative distinction between phonon and roton excitations, and he proposes to distinguish rather between excitations of long wave (small p) and of short wave ($p \approx p_0$). He stresses that in the modified model the postulate of superfluidity as a necessary consequence of the energy spectrum will be retained as well as the macroscopic hydrodynamics developed on its basis. The energy relations, on the other hand, are different, and are given as follows:

For the free energy of the "rotons"

$$F_r = - \frac{2 m'^{\frac{1}{2}} (k\,T)^{\frac{3}{2}} p_0^2}{(2\pi)^{\frac{3}{2}} \varrho\, \hbar^3} e^{-\Delta/k\,T} \qquad (45.2)$$

for their entropy

$$S_r = \frac{2 (k\,m')^{\frac{1}{2}} p_0^2 \Delta}{(2\pi)^{\frac{3}{2}} \varrho\, T^{\frac{1}{2}} \hbar^3} \left(1 + \frac{3\,k\,T}{2\Delta} \right) e^{-\Delta/k\,T} \qquad (45.3)$$

and for their specific heat

$$C_r = \frac{2 m'^{\frac{1}{2}} p_0^2 \Delta^2}{(2\pi)^{\frac{3}{2}} \varrho\, k^{\frac{1}{2}} T^{\frac{3}{2}} \hbar^3} \left[1 + \frac{k\,T}{\Delta} + \frac{3}{4} \left(\frac{k\,T}{\Delta} \right)^2 \right] e^{-\Delta/k\,T}. \qquad (45.4)$$

For the normal density due to the rotons the relation

$$\frac{(\varrho_n)_r}{\varrho} = \frac{2 m'^{\frac{1}{2}} p_0^4}{3 (2\pi)^{\frac{3}{2}} \varrho\, (k\,T)^{\frac{1}{2}} \hbar^3} e^{-\Delta/kT} \qquad (45.5)$$

is derived. The three constants of the theory are given as $\Delta/k = 9.6°$, $p_0/\hbar = 1.95 \times 10^8$ cm^{-1} and $m' = 0.77$ times the mass of the helium atom.

Literature references.

[1] Kamerlingh Onnes, H.: Commun. Phys. Lab. Univ. Leiden No. 108 (1908).
[2] Keesom, W. H.: Commun. Phys. Lab. Univ. Leiden No. 184b (1926).
[3] Keesom, W. H.: Proc. Roy. Acad. Amst. 29, 1136 (1926).
[4] Tammann, G.: Ann. Phys. 82, 240 (1927).
[5] Simon, F.: Nature, Lond. 133, 529 (1934). — Z. Physik 16, 183 (1923).
[6] Wohl, K.: Z. phys. Chem. Abt. B 2, 77 (1929).
[7] Kamerlingh Onnes, H., and J. D. A. Boks: Commun. Phys. Lab. Univ. Leiden No. 170b (1924).
[8] Dana, L. I., and H. Kamerlingh Onnes: Commun. Phys. Lab. Univ. Leiden No. 179c (1926).
[9] Wolfke, M., and W. H. Keesom: Commun. Phys. Lab. Univ. Leiden, No. 190a (1928).
[10] Keesom, W. H., and K. Clusius: Commun. Phys. Lab. Univ. Leiden, No. 219e (1932).
[11] Keesom, W. H., and Miss A. P. Keesom: Physica, Haag 2, 557 (1935).
[12] Ehrenfest, P.: Commun. Phys. Lab. Univ. Leiden Suppl. No. 75b (1933).

[13] SIMON, F.: Nature, Lond. **133**, 529 (1934).
[14] McLENNAN, J. C., H. D. SMITH and J. O. WILHELM: Phil. Mag. **14**, 161 (1932).
[15] KEESOM, W. H., and Miss A. P. KEESOM: Physica, Haag **3**, 359 (1936).
[16] ALLEN, J. F., R. PEIERLS and M. Z. UDDIN: Nature, Lond. **140**, 62 (1937).
[17] ALLEN, J. F., and H. JONES: Nature, Lond. **141**, 243 (1938).
[18] DAUNT, J. G., and K. MENDELSSOHN: Nature, Lond. **143**, 719 (1939).
[19] WILHELM, J. O., A. D. MISENER and A. R. CLARK: Proc. Roy. Soc. Lond., Ser. A **151**, 342 (1935).
[20] BURTON, E. F.: Nature, Lond. **135**, 265 (1935).
[21] KEESOM, W. H., and G. E. MacWOOD: Physica, Haag **5**, 737 (1938).
[22] KAPITZA, P.: Nature, Lond. **141**, 74 (1938).
[23] ALLEN, J. F., and A. D. MISENER: Nature, Lond. **141**, 75 (1938).
[24] DAUNT, J. G., and K. MENDELSSOHN: Nature, Lond. **141**, 911 (1938).
[25] KIKOIN, A. K., and B. G. LASAREW: Nature, Lond. **141**, 912 (1938).
[26] KAMERLINGH ONNES, H.: Commun. Phys. Lab. Univ. Leiden No. 159 (1922).
[27] CLOSS, J. O., and K. MENDELSSOHN: Z. phys. Chem. Abt. B **19**, 291 (1932).
[28] ROLLIN, B. V.: Actes 7th Int. Congr. Froid **1**, 187 (1936).
[29] ROLLIN, B. V., and F. SIMON: Physica, Haag **6**, 219 (1939).
[30] DAUNT, J. G., and K. MENDELSSOHN: Proc. Roy. Soc. Lond., Ser. A **170**, 423 (1939).
[31] KIKOIN, A. K., and B. G. LASAREW: Nature, Lond. **142**, 289 (1939).
[32] KEESOM, W. H.: Commun. Phys. Lab. Univ. Leiden Suppl. No. 71e (1932).
[33] LONDON, F.: Proc. Roy. Soc. Lond., Ser. A **153**, 576 (1936).
[34] FRÖHLICH, H.: Physica, Haag **4**, 639 (1937).
[35] KEESOM, W. H., and K. W. TACONIS: Physica, Haag **5**, 270 (1938).
[36] LONDON, F.: Nature, Lond. **141**, 643 (1938).
[37] LONDON, F.: J. Phys. Chem. **43**, 49 (1939).
[38] TISZA, L.: Nature, Lond. **141**, 913 (1938).
[39] TISZA, L.: C. R. Akad. Sci., Paris **207**, 1035, 1186 (1938).
[40] LONDON, H.: Proc. Roy. Soc. Lond., Ser. A **171**, 484 (1939).
[41] KAPITZA, P. L.: J. Phys. USSR. **4**, 181 (1941).
[42] KAPITZA, P. L.: J. Phys. USSR. **5**, 59 (1941).
[43] LANDAU, L.: J. Phys. USSR. **5**, 71 (1941); **11**, 91 (1947).
[44] LIFSHITZ, E.: J. Phys. USSR. **8**, 110 (1944).
[45] PESHKOV, V.: J. Phys. USSR. **8**, 131 (1944).
[46] PESHKOV, V.: J. Phys. USSR. **10**, 389 (1946).
[47] PESHKOV, V.: J. Exp. Theor. Phys. **18**, 951 (1948).
[48] PELLAM, J. R., and R. B. SCOTT: Phys. Rev. **76**, 869 (1949).
[49] ANDRONIKASHVILI, E. L.: J. Phys. USSR. **10**, 201 (1946).
[50] DAUNT, J. G., and K. MENDELSSOHN: Nature, Lond. **150**, 604 (1942). — MENDELSSOHN, K.: Proc. Phys. Soc. Lond. A **57**, 371 (1945).
[51] DAUNT, J. G., and K. MENDELSSOHN: Nature, Lond. **157**, 389 (1946).
[52] KRAMERS, H. A., J. WASSCHER and C. J. GORTER: Physica, Haag **18**, 329 (1952).
[53] FEYNMAN, R. P.: Phys. Rev. **91**, 1291, 1301 (1953); **94**, 262 (1954).
[54] BOER, J. DE: Physica, Haag **14**, 139 (1948).
[55] BOER, J. DE, and R. J. LUNBECK: Physica, Haag **14**, 510 (1948).
[56] SYDORIAK, S. G., E. R. GRILLY and E. F. HAMMEL: Phys. Rev. **75**, 303, 1103 (1949).
[57] ABRAHAM, B. M., D. W. OSBORNE and B. WEINSTOCK: Phys. Rev. **80**, 366 (1950).
[58] OSBORNE, D. W., B. WEINSTOCK and B. M. ABRAHAM: Phys. Rev. **75**, 988 (1949).
[59] DAUNT, J. G., and C. V. HEER: Phys. Rev. **79**, 46 (1950).
[60] GRILLY, E. R., E. F. HAMMEL and S. G. SYDORIAK: Phys. Rev. **75**, 1103 (1949).
[61] KERR, E. C.: Phys. Rev. **96**, 551 (1954).
[62] OSBORNE, D. W., B. ABRAHAM and B. WEINSTOCK: Phys. Rev. **82**, 263 (1951); **85**, 158 (1952).
[63] ROBERTS, T. R., and S. G. SYDORIAK: Phys. Rev. **93**, 1418 (1954).
[64] POMERANCHUK, I.: J. Exp. Theor. Phys. **20**, 1919 (1950).
[65] FAIRBANK, M., W. D. ARD and G. K. WALTERS: Phys. Rev. **95**, 566 (1954).
[66] OSBORNE, D. W., B. M. ABRAHRM and B. WEINSTOCK: Bull. Inst. Int. Froid, Suppl. Annexe, **3**, 11 (1955).
[67] DAUNT, J. G., R. E. PROBST and H. L. JOHNSTON: J. Chem. Phys. **15**, 759 (1947).
[68] SOLLER, T., W. M. FAIRBANK and A. D. CROWELL: Phys. Rev. **91**, 1058 (1953).
[69] SWENSON, C. A.: Phys. Rev. **79**, 626 (1950).
[70] SIMON, F., and C. A. SWENSON: Nature, Lond. **165**, 829 (1950).
[71] KEESOM, W. H., and Miss A. P. KEESOM: Commun. Phys. Lab. Univ. Leiden **1933**, Nos. 224d, e.
[72] KAMERLINGH ONNES, H.: Commun. Phys. Lab. Univ. Leiden **1911**, Nos. 119, 124b.

[73] Keesom, W. H.: Commun. Phys. Lab. Univ. Leiden Suppl. **1933**, No. 75a.
[74] Dijk, H. van, and M. Durieux: Bull. Inst. Int. Froid, Suppl. Annexe **1955**, 3, 595.
[75] Clement, J. R., J. K. Logan and J. Gaffney: Bull. Inst. Int. Froid, Suppl. Annexe **1955**, 3, 601.
[76] Bleaney, B., and F. Simon: Trans. Faraday Soc. **35**, 1205 (1939).
[77] Keesom, W. H., and W. K. Westmijze: Physica, Haag **8**, 1044 (1941).
[78] Bull, R. A., K. R. Wilkinson and J. Wilks: Proc. Phys. Soc. Lond. A **64**, 379 (1951).
[79] Hercus, G. R., and J. Wilks: Phil. Mag. **45**, 1163 (1954).
[80] Kapadnis, D. G.: Thesis, Leiden 1956.
[81] Allen, J. F., and J. Reekie: Proc. Cambridge Phil. Soc. **35**, 114 (1939).
[82] Meyer, L., and J. H. Mellink: Physica, Haag **13**, 197 (1947).
[83] Peshkov, V.: J. Exp. Theor. Phys. **27**, 351 (1954).
[84] Bots, G. J. C., and C. J. Gorter: Phys. Rev. **90**, 1117 (1953).
[85] Bots, G. J. C.: Bull. Inst. Int. Froid, Suppl. Annexe **1955**, 3, 44.
[86] Chandrasekhar, B. S., and K. Mendelssohn: Proc. Phys. Soc. Lond. A **68**, 857 (1955).
[87] Brewer, D. F., D. O. Edwards and K. Mendelssohn: Proc. Phys. Soc. Lond. A **68**, 93 (1956).
[88] Allen, J. F., and A. D. Misener: Proc. Roy. Soc. Lond. A **172**, 467 (1939).
[89] Bowers, R., and K. Mendelssohn: Proc. Roy. Soc. Lond. A **213**, 158 (1952).
[90] Swim, R. T., and H. E. Rorschach: Phys. Rev. **97**, 25 (1955).
[91] Winkel, P., A. M. G. Delsing and J. D. Poll: Physica, Haag **21**, 331 (1955).
[92] Bowers, R., and G. K. White: Proc. Phys. Soc. Lond. A **64**, 558 (1951).
[93] Chandrasekhar, B. S., and K. Mendelssohn: Proc. Roy. Soc. Lond., Ser. A **218**, 18 (1953).
[94] Johns, H. E., J. O. Wilhelm and H. Grayson Smith: Canad. J. Res. A **17**, 149 (1939).
[95] Atkins, K. R.: Proc. Phys. Soc. Lond. A **64**, 833 (1951).
[96] Mendelssohn, K.: Phys. Soc. Cambridge Conf. Rep. **1947**, 35.
[97] Burton, E. F.: Nature, Lond. **142**, 72 (1938).
[98] Bowers, R., and K. Mendelssohn: Proc. Roy. Soc. Lond., Ser. A **204**, 366 (1950).
[99] Troyer, A. de, A. van Itterbeek and G. J. van den Berg: Physica, Haag **17**, 50 (1951).
[100] Tjerkstra, H. H.: Physica, Haag **19**, 217 (1953).
[101] Champeney, D. C.: Proc. Phys. Soc. Lond. A (in print).
[102] Hollis Hallet, A. C.: Proc. Roy. Soc. Lond., Ser. A **210**, 404 (1952).
[103] Lee, K. S., and A. C. Hollis Hallett: Bull. Inst. Int. Froid., Suppl. Annexe **1955**, 3, 71.
[104] Landau, L., and I. M. Khalatnikov: J. Exp. Theor. Phys. **19**, 637, 709 (1949).
[105] Grenier, C.: Phys. Rev. **83**, 598 (1951).
[106] Bowers, R.: Proc. Phys. Soc. Lond. A **65**, 511 (1952).
[107] Keesom, W. H., A. P. Keesom and B. F. Saris: Physica, Haag **5**, 281 (1938).
[108] Keesom, W. H., B. F. Saris and L. Meyer: Physica, Haag **7**, 817 (1940).
[109] Allen, J. F., and E. Ganz: Proc. Roy. Soc. Lond., Ser. A **171**, 242 (1939).
[110] Keesom, W. H., and G. Duyckaerts: Physica, Haag **13**, 153 (1947).
[111] Gorter, C. J., and J. H. Mellink: Physica, Haag **15**, 285 (1949).
[112] Mellink, J. H.: Physica, Haag **13**, 180 (1947).
[113] Meyer, L., and J. H. Mellink: Physica, Haag **13**, 197 (1947).
[114] Hung, C. S., B. Hunt and P. Winkel: Physica, Haag **18**, 629 (1952).
[115] London, F., and P. R. Zilsel: Phys. Rev. **74**, 1148 (1948).
[116] Delsing, A. M. G.: Bull. Inst. Int. Froid., Suppl. Annexe **1955**, 3, 28; Discussion.
[117] Winkel, P., A. M. G. Delsing and J. D. Poll: Physica, Haag **21**, 331 (1955).
[118] Brewer, D. F., D. O. Edwards and K. Mendelssohn: Phil. Mag., (in print).
[119] Kürti, N., and F. Simon: Nature, Lond. **142**, 207 (1938).
[120] Fairbank, H. A., and J. Wilks: Proc. Roy. Soc. Lond., Ser. A **231**, 545 (1955).
[121] Gorter, C. J., K. W. Taconis and J. J. M. Beenakker: Physica, Haag **17**, 841 (1951).
[122] Khalatnikov, I. M.: J. Exp. Theor. Phys. **22**, 687 (1952).
[123] Findlay, J. C., A. Pitt, H. G. Smith and J. O. Wilhelm: Phys. Rev. **54**, 506 (1938); **56**, 122 (1939).
[124] Pellam, J. R., and C. F. Squire: Phys. Rev. **72**, 1245 (1947).
[125] Atkins, K. R., and C. E. Chase: Proc. Phys. Soc. Lond. A **64**, 826 (1951).
[126] Khalatnikov, I. M.: J. Exp. Theor. Phys. **20**, 243 (1950).
[127] Chase, C. E., and M. A. Herlin: Phys. Rev. **97**, 1447 (1955).
[128] Newell, J. A.: Bull. Inst. Int. Froid, Suppl. Annexe **1955**, 3, 80.
[129] Atkins, K. R., and D. V. Osborn: Phil. Mag. **41**, 1078 (1950).
[130] Klerk, D. de, R. P. Hudson and J. R. Pellam: Phys. Rev. **89**, 326 (1953).

[*131*] KRAMERS, H. C., T. VAN PESKI-TINBERGEN, J. WIEBES, F. A. W. VAN DEN BURG and C. J. GORTER: Physica, Haag **20**, 743 (1954).
[*132*] MAURER, R. D., and M. A. HERLIN: Phys. Rev. **82**, 329 (1951).
[*133*] KÜRTI, N., and J. MCINTOSH: Bull. Inst. Int. Froid, Suppl. Annexe **1955**, 3, 84.
[*134*] PELLAM, J. R., and W. B. HANSON: Phys. Rev. **85**, 216 (1952).
[*135*] DAUNT, J. G., and K. MENDELSSOHN: Nature, Lond. **142**, 475 (1938).
[*136*] BURGE, E. J., and L. C. JACKSON: Proc. Roy. Soc. Lond., Ser. A **205**, 270 (1951).
[*137*] ATKINS, K. R.: Proc. Roy. Soc. Lond., Ser. A **203**, 119 (1950).
[*138*] BOWERS, R.: Phys. Rev. **91**, 1016 (1953).
[*139*] SCHIFF, L. I.: Phys. Rev. **59**, 839 (1941).
[*140*] FRENKEL, J.: J. Phys. USSR. **2**, 365 (1940).
[*141*] BIJL, A., J. DE BOER and A. MICHELS: Physica, Haag **9**, 655 (1941).
[*142*] TEMPERLEY, H. N. V.: Proc. Roy. Soc. Lond., Ser. A **198**, 438 (1949).
[*143*] ATKINS, K. R.: Nature, Lond. **161**, 925 (1948).
[*144*] HAAS, W. J. DE, and G. J. VAN DEN BERG: Rev. Mod. Phys. **21**, 524 (1949).
[*145*] BROWN, J. B., and K. MENDELSSOHN: Proc. Phys. Soc. Lond. A **63**, 1312 (1950).
[*146*] BOWERS, R., and K. MENDELSSOHN: Proc. Phys. Soc. Lond. A **63**, 1318 (1950).
[*147*] ATKINS, K. R.: Proc. Roy. Soc. Lond., Ser. A **203**, 240 (1950).
[*148*] BERG, G. J. VAN DEN, and W. J. DE HAAS: Physica, Haag **17**, 797 (1951).
[*149*] MENDELSSOHN, K., and G. K. WHITE: Proc. Phys. Soc. Lond. A **63**, 1328 (1950).
[*150*] BOORSE, H. A., and J. G. DASH: Phys. Rev. **82**, 851 (1951).
[*151*] CHANDRASEKHAR, B. S.: Phys. Rev. **86**, 414 (1952).
[*152*] CHANDRASEKHAR, B. S., and K. MENDELSSOHN: Proc. Phys. Soc. Lond. A **65**, 226 (1952).
[*153*] KEESOM, W. H.: Commun. Phys. Lab. Univ. Leiden **1932**, No. 219a.
[*154*] BLAISSE, B. S., A. H. COOKE and R. A. HULL: Physica, Haag **6**, 231 (1939).
[*155*] AMBLER, E., and N. KÜRTI: Phil. Mag. **43**, 1307 (1952).
[*156*] ESELSON, B. N., and B. G. LASAREW: J. Exp. Theor. Phys. **23**, 552 (1952).
[*157*] HAM, A. C., and L. C. JACKSON: Phil. Mag. **44**, 214 (1953).
[*158*] LANE, C. T., and R. V. DYBA: Phys. Rev. **92**, 829 (1953).
[*159*] MCCRUM, N. G.: Phil. Mag. **45**, 1302 (1954).
[*160*] AMBLER, E., and N. KÜRTI: Phil. Mag. **43**, 260 (1952).
[*161*] WARING, R. K.: Phys. Rev. **99**, 1704 (1955).
[*162*] DAUNT, J. G., and K. MENDELSSOHN: Proc. Phys. Soc. Lond. A **63**, 1305 (1950).
[*163*] KISTEMAKER, J.: Physica, Haag **13**, 81 (1947).
[*164*] LONG, E. A., and L. MEYER: Phys. Rev. **76**, 440 (1949).
[*165*] STRAUSS, unpublished, but cf. E. A. LONG and L. MEYER: Adv. in Phys. **2**, 1 (1953).
[*166*] FREDERIKSE, .H. P. R., and C. J. GORTER: Physica, Haag **16**, 402 (1950).
[*167*] TJERKSTRA, H H., F. J. HOOFTMAN and C. J. N. VAN DEN MEIJDENBERG: Physica, Haag **19**, 935 (1953).
[*168*] BREWER, D. F., and K. MENDELSSOHN: Phil. Mag. **44**, 340 (1953).
[*169*] BOWERS, R.: Phil. Mag. **44**, 485 (1953).
[*170*] BROWN, J. B., and K. MENDELSSOHN: Nature, Lond. **160**, 670 (1947).
[*171*] LONG, E. A., and L. MEYER: Phys. Rev. **79**, 1031 (1950).
[*172*] BOWERS, R., D. F. BREWER and K. MENDELSSOHN: Phil. Mag. **42**, 1445 (1951).
[*173*] FREDERIKSE, H. P. R.: Physica, Haag **15**, 860 (1949).

Books and reviews on liquid helium.

ALLEN, J. F.: Liquid Helium, a lecture in: Low temperature physics. London: Pergamon Press 1952.
ATKINS, K. R.: Wave Propagation and Flow in liquid Helium II. Phil. Mag. Suppl. **1**, 169 (1952).
DAUNT, J. G.: Properties of Helium Three at Low Temperatures. Phil. Mag. Suppl. **1**, 209 (1952).
DAUNT, J. G., and R. S. SMITH: The Problem of Liquid Helium—some recent aspects. Rev. Mod. Phys. **26**, 172 (1954).
DINGLE, R. B.: Theories of Helium II. Phil. Mag. Suppl. **1**, 111 (1952).
KEESOM, W. H.: Helium. Amsterdam: Elsevier 1942.
LONDON, F.: Superfluids, vol. II. New York: Wiley 1954.
FEYNMAN, R. P.: Application of quantum mechanics to liquid helium; PELLAM, J. R.: RAYLEIGH disks in liquid helium II; HALLET, A. C. H.: Oscillating disks and rotating cylinders in liquid helium II; HAMMEL, E. F.: The low temperature properties of helium three; and J. J. M. BEENAKKER and K. W. TACONIS: Liquid mixtures of helium three and four. Articles in Progress in Low Temperature Physics edited by C. J. GORTER. Amsterdam: North Holland Publishing Co. 1955.

Sachverzeichnis.

(Deutsch-Englisch.)

Bei gleicher Schreibweise in beiden Sprachen sind die Stichwörter nur einmal aufgeführt.

30

Wärmewellen 2. Art in flüssigem Helium II, *second sound in liquid Helium II* 388, 393, 394, 430, 434, 435, 437, 458.

Wärmezufuhr, *heat supply* 57.

Wärme, zugeführte, *heat of transport* 390, 391, 409, 410.

Wechselwirkungsenergien, Variations-methode zu ihrer Berechnung, *variational method of calculating interaction energies* 362.

WEISSsche Konstante, WEISS *constant* 82, 84.

WEISSsches Molekülfeld, WEISS *molecular field* 128, 129 f.

Widerstandsthermometer aus Kohle, *resistance thermometers of carbon* 185.

— aus Phosphorbronze, *resistance thermometers of phosphorbronze* 185, 189, 190.

Widerstandsübergang von Supraleitern, *resistive transition of superconductors* 228.

Wirbelbewegung, *vortex motion* 392.

Wirbelströme, *eddy currents* 258.

— in Magneten, *eddy currents in magnets* 73, 75, 76, 203.

Wismut, supraleitendes, *superconducting bismuth* 239.

Wismutdraht-Technik, *bismuth wire technique* 225, 226, 250.

Zerstörung der Supraleitung durch Ströme, *destruction of superconductivity by currents* 338.

Zustandsdiagramm, *diagram of state* 372.

Zustandssumme, *partition function* 41, 42, 80, 84.

Zwei-Flüssigkeiten-Modell, *two-fluid model* 236, 244, 250, 263, 276, 280, 387, 388, 389, 392, 426.

zweiter Hauptsatz der Wärmelehre, *second law of thermodynamics* 54, 58.

Zwischenzustand bei Supraleitern, *intermediate state of superconductors* 197, 214, 222, 250, 254, 265, 321, 322, 336.

— bei Supraleitern, hohle Proben, *intermediate state of superconductors, hollow specimens* 226.

Subject Index.

(English-German.)

Where English and German spelling of a word is identical the German version is omitted.

ENCYCLOPEDIA OF PHYSICS

EDITED BY

S. FLÜGGE

VOLUME XIV

LOW TEMPERATURE PHYSICS I

WITH 215 FIGURES

Springer-Verlag Berlin Heidelberg GmbH
1956

HANDBUCH DER PHYSIK

HERAUSGEGEBEN VON

S. FLÜGGE

BAND XIV

KÄLTEPHYSIK I

MIT 215 FIGUREN

Springer-Verlag Berlin Heidelberg GmbH
1956

ISBN 978-3-662-38851-8   ISBN 978-3-662-39773-2 (eBook)
DOI 10.1007/978-3-662-39773-2

Softcover reprint of the hardcover 1st edition 1956

Contents.

The Production of Low Temperatures Down to Hydrogen Temperature[1].

By

J. G. Daunt.

With 101 Figures.

Preface.

The purpose of this article is to give the fundamental physical principles involved in the many techniques for the production of low temperatures down to temperatures attainable with liquid hydrogen. In carrying through this aim, emphasis is laid on the evolution and establishment of new ideas and methods. However, it is not my purpose to detail the technological and mechanical developments attendant on each process of refrigeration, which can be found more suitably in engineering publications. In other words, each process of refrigeration is treated at the stage in which it was or is a problem in physics laboratories; but those aspects of the techniques which are concerned with their engineering or commercial development are omitted.

The article is subdivided into nine chapters which are concerned respectively with: A. Gas refrigerating machines; B. Compressed vapor refrigerating machines; C. D. and E. Cooling by JOULE-THOMSON expansion and the liquefaction of air and hydrogen by the LINDE method; F. and G. Cooling by isentropic expansion and the liquefaction of air, etc., by the CLAUDE method; H. Single adiabatic expansion for hydrogen liquefaction; I. Heat exchangers and regenerators. Although the refrigerating processes used in gas refrigerating machines and in compressed vapor machines are capable of producing deep refrigeration in, for example, the liquefaction of air, the detailed discussion of the physical principles involved in them which is given in chapters A and B is included to provide a thermodynamic background for the rest of the work as well as to recognize their own intrinsic interest.

No attempt has been made to recognize all the many significant centers of low temperature research and technology, be they academic or commercial. Instead reference is given only to those which by their detailed publications in readily accessible learned Journals have notably added to the general knowledge of low temperature production.

I am indebted to many previous authors from whose works on low temperature production I have gained and relayed valuable information. Since the reader may wish to consult these works also, I append my list herewith together with brief comments concerning each.

[1] J. A. EWING: The Mechanical Production of Cold. Cambridge Univ. Press 1908. A classic giving detailed references to 19th century work.
[2] M. and B. RUHEMANN: Low Temperature Physics. Cambridge Univ. Press 1937. Chap. I, Part I, gives an historical picture of the development of low temperature physics.

---

[1] A condensed bibliography of the most essential books on the subject is given at the end of the Preface.

[3] H. LENZ: Handbuch der Experimentalphysik, vol. IX/I, p. 47. 1929. Liquefaction of gases and its thermodynamical foundation. Good account of early work on JOULE-THOMSON expansion.

[4] W. MEISSNER: Handbuch der Physik, vol. XI, p. 272. 1926. Production of low temperatures and the liquefaction of gases. Excellent survey of the field to 1926.

[5] M. RUHEMANN: The separation of gases. Oxford Press 1940. Chap. V gives good outline of refrigeration to low temperatures.

[6] J. A. VAN LAMMEREN: Technique of Low Temperatures. Springer 1941. Besides detailing methods of production of low temperatures, this gives a useful chapter on cryostats and a full bibliography.

[7] M. DAVIES: The physical principles of gas liquefaction and low temperature rectification. Longmans Green & Co. 1949. A short but authoritative book from a modern standpoint.

[8] H. HAUSEN: Wärmeübertragung in Gegenstrom, Gleichstrom and Kreuzstrom. A detailed account of interchangers, regenerators, etc.

[9] R. PLANK: Handbuch der Kältetechnik, vol. I. Springer 1954. Excellent historical article.

[10] H. J. MACINTIRE and F. W. HUTCHINSON: Refrigeration Engineering. Wiley & Sons Inc. 1937. Good text from engineering point of view of refrigeration to − 50° C.

Of course the above list makes no pretense at completeness regarding publications on low temperature production, no more than does the total of references given in the text of the article. It represents only a personal view of the major contributions which present aspects of the field of fundamental physical significance.

A. Refrigeration by gas (air) machines.

Refrigeration by gas engines, in which the gas has generally been air, has been carried out by two distinct classes of machine (a) the "open-cycle" machine in which atmospheric air circulates in the low pressure circuit of the machine at atmospheric pressure and (b) the "closed-cycle" machine in which the same mass of gas is circulated in a completely closed circuit, in general everywhere at pressures above atmospheric.

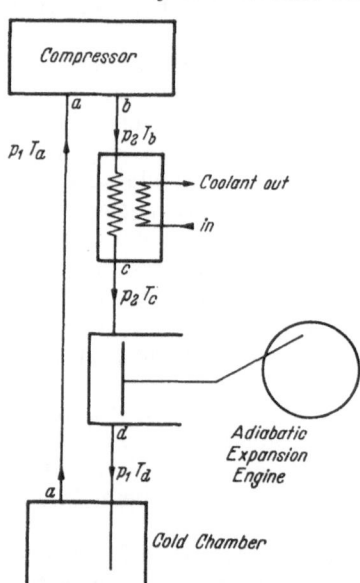

Fig. 1. Flow diagram for open-cycle gas (air) refrigerating machine.

1. The open-cycle gas refrigerating machines. The pioneering work of practical development of open-cycle machines was due largely to GIFFORD (1873) and to J. COLEMAN and J. and H. BELL (1877)[1]. A diagrammatic sketch of such a machine is given in Fig. 1. The gas (air) is adiabatically compressed in the compressor from pressure p_1 to p_2. It is subsequently cooled to temperature T_c, ideally at the compressor pressure, p_2, in the cooler which may have water as the coolant fluid. Thence the gas passes to the expansion engine in which it is expanded adiabatically, doing external work. The mechanical work obtained from the expansion engine is coupled back to the compressor, often by having compressor and expander on the same shaft. The cool gas at low pressure, p_1, and temperature, T_d, passes from the expansion engine to the cold chamber which it maintains refrigerated. Gas from the cold chamber returns to the low pressure side of the compressor at a temperature T_a, which is approximately that at which the cold chamber is maintained.

---

[1] See R. PLANK: Handbuch der Kältetechnik, vol. I., Berlin: Springer 1954.

Diagrammatically the operation of the machine may be followed in the indicator diagram of Fig. 2. Here the adiabatic path $a \to b$ represents the adiabatic compression from p_1 to p_2 (the letters a, b, etc. are also marked at the corresponding points of Fig. 1). The path $b \to c$ is the isobaric cooling from temperature T_b to T_c, during which heat Q_2 is given off. The path $c \to d$ is the adiabatic expansion down to pressure p_1 and path $d \to a$ represents the passage through the cold chamber in which the gas extracts heat, Q_1, from the chamber. For each mole of gas circulated:

$$Q_2 = C_{p_2} (T_b - T_c) \atop Q_1 = C_{p_1} (T_a - T_d).} \qquad (1.1)$$

Since in both adiabatic processes the ratio of the pressures, p_2/p_1, is the same, then:

$$\frac{T_a}{T_b} = \frac{T_d}{T_c}. \qquad (1.2)$$

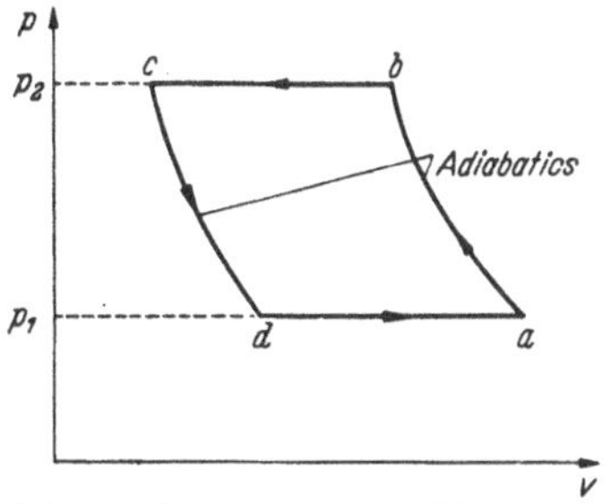

Fig. 2. Indicator diagram of idealized operation of an open-cycle gas refrigerating machine.

Combining this with eq. (1.1) and assuming, as for a perfect gas, that $C_{p_1} = C_{p_2}$ we get:

$$\frac{Q_1}{Q_2} = \frac{T_a}{T_b} = \frac{T_d}{T_c} \qquad (1.3)$$

and the coefficient of performance, ξ, is given by:

$$\xi = \frac{Q_1}{Q_2 - Q_1} = \frac{T_a}{T_b - T_a} = \frac{T_d}{T_c - T_d} = \frac{(T_a - T_d)}{(T_b - T_a) - (T_c - T_d)}. \qquad (1.4)$$

This coefficient of performance is less than the thermodynamical coefficient, ξ^{ideal}, which would be given by a reversible CARNOT engine transferring its heat from the cold chamber at temperature T_a to the cooling fluid in the cooler at T_c. ξ^{ideal} is:

$$\xi^{ideal} = \frac{T_a}{T_c - T_a} \qquad (1.5)$$

and the relative thermodynamic efficiency η_{rel} for the gas refrigerating machine is therefore:

$$\eta_{rel} = \frac{\xi}{\xi^{ideal}} = \frac{T_c - T_a}{T_b - T_a}. \qquad (1.6)$$

In an actual air open-cycle machine the temperature of the cold chamber, T_a might be, say, 0° C and the temperature, T_c, of the cooling fluid in the cooler, say, 24° C. This would give $\xi^{ideal} = 11.4$. For actual running conditions with $p_1 = 1$ atmosphere and $p_2 = 4$ atmosphere T_d, the temperature of the air after expansion, would be about $-65°$ C, which from eq. (1.4) yields $\xi = 2.3$. The relative efficiency, η_{rel}, is therefore ≈ 0.2. In fact[1] the coefficients of performance, ξ, are about 0.5 to 0.7. The discrepancies here are due to mechanical inefficiencies in the machine, etc. More important, however, is the fact that the absorption of heat, Q_1, in the cold chamber takes place over a wide range of temperature, the cold gas entering the chamber from the expansion engine at temperatures much less than that of the chamber. The isobaric expansion thereby

_____

[1] See J. A. EWING: The Mechanical Production of Cold. Cambridge Univ. Press 1908.

taking place in the cold chamber cannot do useful external work, which thereby introduces considerable irreversibility.

This irreversibility introduced by the isobaric expansion in the cold chamber can be reduced by reducing the value of T_d, the temperature of the gas issuing from the expansion engine. To effect this, smaller compression ratios, $r = p_2/p_1$, should be employed. The effect of lowering r on the value of ξ can be seen in the following way:

In an adiabatic compression from pressure p_1 to p_2 the work done per mole of perfect gas is:

$$W_{ad} = C_p T_1 \left[\left(\frac{p_2}{p_1} \right)^{\frac{\gamma-1}{\gamma}} - 1 \right]. \tag{1.7}$$

Where T_1 is the temperature at initial pressure p_1 and where γ is the ratio C_p/C_v. For the engine of Figs. 1 and 2 the *net* work, W, is the difference in the work of adiabatic compression and expansion and is given by:

$$W = C_p (T_a - T_d) \left[\left(\frac{p_2}{p_1} \right)^{\frac{\gamma-1}{\gamma}} - 1 \right]. \tag{1.8}$$

An alternative expression for the coefficient of performance, ξ, therefore is obtained from eqs. (1.1) and (1.8):

$$\xi = \left[\left(\frac{p_2}{p_1} \right)^{\frac{\gamma-1}{\gamma}} - 1 \right]^{-1}. \tag{1.9}$$

From this latter expression for ξ it will be seen that by *decreasing* the compression ratio, $r = p_2/p_1$, the value of ξ can be increased. Lower values of r must be counterbalanced by increased compressor and expansion engine speeds in order to obtain the same net refrigeration. This is mechanically disadvantageous since greater frictional losses are thereby introduced.

The low coefficients of performance associated with air refrigerating machines as described above have resulted in their being largely replaced by compressed vapor refrigerating machines, which, as is shown in chapter B, have much higher efficiencies. The air refrigerating machine in general is only retained where the convenience of having air as the refrigerating fluid is paramount, as is the case in some shipboard applications and, as has been recently introduced, in "comfort cooling" in aircraft. In the latter the same rotary compressor may be used for the cooling systems as is used at higher altitudes for a heating cycle.

In order to improve the coefficient of performance of the gas refrigerating machines it is clearly necessary to avoid the loss of useful work in isobaric expansion in the cold chamber and to make the process of compression more economical of energy by carrying it out quasi-isothermally instead of adiabatically. A close approach to such desired isothermal processes of heat evolution and heat absorption has been made recently by KÖHLER and JONKERS[1] in a closed-cycle gas refrigerating machine to be described in Sect. 5.

2. The vortex tube. The vortex tube was originally devised by RANQUE[2] and the first publication of a systematic experimental study of it was due to HILSCH[3].

---

[1] J. W. L. KÖHLER and C. O. JONKERS: Philips Techn. Rev. **16**, 69, 105 (1954).

[2] G. J. RANQUE: Bull. bi-mensuel Soc. Française de Phys. June 2, 1933, p. 112. Publication bound with J. Phys. Radium (7) **4** (1933). See also, for example, U.S. Patent No. 1952281. Dec. 6, 1932.

[3] R. HILSCH: Z. Naturforsch. **1**, 203 (1946). English translation in Rev. Sci. Inst. **18**, 108 (1947).

Further experimental studies have been made by JOHNSON[1], by ELSER and HOCH[2], by MACGEE[3] and by others[4]. A typical arrangement of a vortex tube is given in Fig. 3. Essentially the vortex tube consists of a cylindrical tube into which a high velocity gas jet is introduced from a tangential nozzle placed approximately in the center of the cylinder. A circular iris, I, is placed in the tube just to one side of the plane of the jet nozzle (in Fig. 3 this is to the right hand side and close to the nozzle, N) so that gas passing from the region of high pressure through the iris comes from the central region of the tube. A turbulent screwlike flow of gas then takes place in a direction away from the iris (to the left in Fig. 3) and its exit from the tube is partially restricted by the valve, V. As a result of this restriction

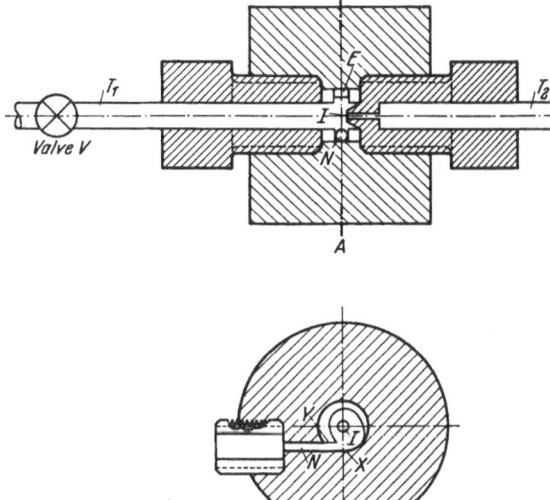

Section A–A

Fig. 3. Construction of Vortex tube. T_1 and T_2 are the side tubes mounted in the body B. N is the nozzle, tangential to the tube and I the iris in tube T_2.

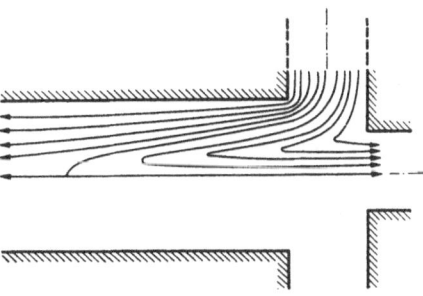

Fig. 4. Flow diagram (neglecting rotation) of the air in the warm tube (Tube T_1 of Fig. 3) of a Vortex tube. [After J. J. VAN DEEMTER: Appl. Sci. Res. A 3, 174 (1952).]

in the valve, a fraction, μ, of the gas is forced to flow back through the iris and this is found to be cooled. The fraction, $(1-\mu)$, passing through the valve is heated. A diagram showing the gas flow paths, but omitting the rotational motions, is shown in Fig. 4.

In its practical construction, as is shown in Fig. 3, the gas entering the tube at N is guided into a spiral flow by the eccentric ring, E, which is shown at x and y in the section of Fig. 3. The tube itself is made by attaching side tubes, T_1 and T_2, to the central block, B, with threaded fittings. These side tubes are made from thin walled poor thermally conducting material to minimize heat transfer between the hot and cold ends of the whole system[5].

In all HILSCH's experiments the gas used was air introduced to the nozzle at a temperature, T_N, which was approximately ambient.

He studied three tubes of different sizes. Noting the nozzle diameter by d_N, the sizes and other data are given in Table 1. The choice of d_I, the iris diameter, was made to secure approximately the maximum cooling effect. In all tubes the lengths of the side tubes were about 50 times the tube diameter, d_T.

[1] A. F. JOHNSON: Canad. J. Res. F **25**, 299 (1947).
[2] K. ELSER and M. HOCH: Z. Naturforsch. 6a, 25 (1951).
[3] R. MACGEE jr.: Refrig. Engng. **58**, 975 (1950).
[4] For a full bibliography see W. CURLEY and R. MACGEE jr.: Refrig. Engng. **59**, 166 (1951).
[5] See M. P. BLAHER: J. Sci. Instrum. **27**, 168 (1950) for a neat construction using plastics.

Table 1. *Dimensions of* Hilsch's[1] *Vortex Tubes.*

Tube No.	d_N mm.	d_T mm.	d_I mm.	Gas flow for input pressure = 11 atm. m.³/h.
1	1.1	4.6	2.6	7.0
2	2.3	9.6	4.2	30.5
3	4.1	17.6	6.5	97.0

Fig. 5 shows his results for tube No. 2 for the temperature of the emergent cold gas as a function of the value of μ, which is the fraction equal to the mass of cold gas divided by the total mass flow to the nozzle. The results shown in Fig. 5 are for four different input pressures, p_N, from 1.5 to 10 atm.

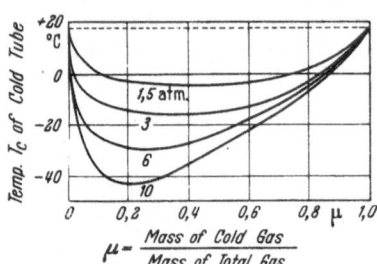

$$\mu = \frac{\text{Mass of Cold Gas}}{\text{Mass of Total Gas}}$$

Fig. 5. Plot of Hilsch's measurements of exit temperature of cold tube of Vortex tube (Hilsch's Tube No. 2) as function of the fractional mass flow, μ, through the cold tube. The four curves are for four different pressures, as marked, of the gas entering the nozzle. [After R. Hilsch: Z. Naturforsch. 1, 203 (1946).]

It will be seen that at the higher pressures considerable coolings up to 60° C could be obtained for suitable choice of μ. The value of μ experimentally is varied by varying the setting of the value V on the side tube T_1.

If T_c and T_H are the temperatures of the emergent cold and hot streams then the energy balance must give:

$$\mu = \frac{T_H - T_N}{T_H - T_c} \tag{2.1}$$

where T_N is the initial temperature of the gas entering the nozzle.

The original paper should be consulted for further experimental detail. Johnson[2] quoted that his results were in general agreement with those of Hilsch. Elser and Hoch[3] made measurements with CO_2, CH_4, A and He gas well as air and found qualitatively similar results. Their results showed that the total temperature difference, $T_H - T_C$, produced for $\mu = 0.5$ decreased with increasing atomicity of the molecule, the value for $T_H - T_c$ for CO_2 and CH_4 being essentially the same.

Theoretical studies of the cooling produced in the vortex tube are numerous, having been made for example by ter Haar and Wergeland[4], Burkhardt[5], Prins[6], Webster[7], Fulton[8], Sheper[9], van Deemter[10,11], A completely rigorous theoretical treatment of the complex turbulent motion of the gas in the vortex tube is clearly a formidable problem, particularly since the velocity profile within the tube is as yet experimentally undetermined. Qualitatively however, the cooling can be understood as follows. The rotating air stream within the tube produces a radial pressure gradient increasing as one moves from

[1] R. Hilsch: Z. Naturforsch. 1, 203 (1946).
[2] A. F. Johnson: Canad. J. Res. F 25, 299 (1947).
[3] K. Elser and M. Hoch: Z. Naturforsch. 6a, 25 (1951).
[4] D. ter Haar and H. Wergeland: Forh. Kong. Norske. Vid. Selskat. 20, 55 (1947).
[5] G. Burkhardt: Z. Naturforsch. 3a, 46 (1948).
[6] J. A. Prins: Nederl. Tijdschr. Natuurk. 14, 241 (1948).
[7] D. S. Webster: Refrig. Engng. 58, 163 (1950).
[8] C. D. Fulton: Refrig. Engng. 58, 473 (1950).
[9] G. W. Sheper: Refrig. Engng. 59, 985 (1951).
[10] J. J. van Deemter: Appl. Sci. Res. A 3, 174 (1952).
[11] See W. Curley and R. MacGee jr.: Refrig. Engng. 59, 166 (1951) for a fuller bibliography.

the axis towards the tube wall. The effect of turbulence in this pressure field is to create adiabatic mixing and tends to produce an adiabatic temperature distribution with the colder gas therefore being in the axial region of the tube A completely adiabatic distribution would however not be realized due to the effect of the thermal conductivity of the gas tending to reduce the radial temperature gradient and due to the effect of non-uniform angular velocity. This latter effect has been described by VAN DEEMTER as follows: "When the angular velocity is not uniform a second mechanism is interfering, namely a flow of mechanical energy radially outward. By turbulent friction (eddy viscosity) the inner layers of fluid try to force the outer layers to move with uniform angular velocity and therefore work is done by the fluid in the center on the fluid in the region of the wall. Due to this transfer of energy the outer region is heated at the cost of the inner region. A non uniform angular velocity therefore involves a deviation from the adiabatic temperature distribution.

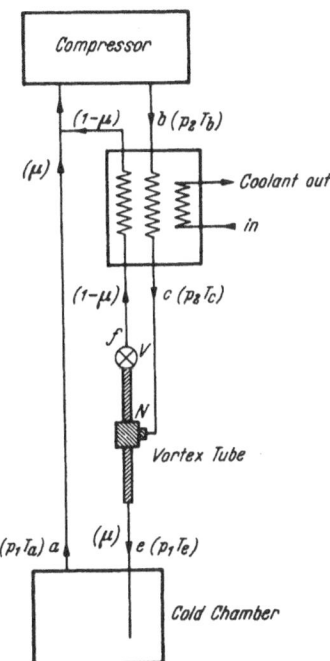

Fig. 6. Flow diagram for possible open-cycle refrigerating machine using Vortex Tube expansion.

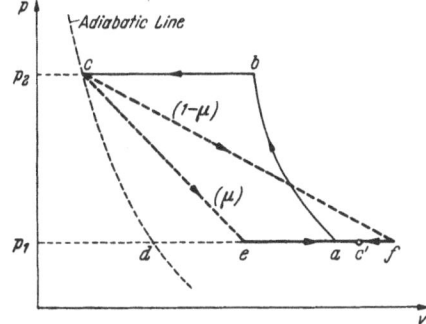

Fig. 7. Indicator diagram of idealized operation of open-cycle refrigerating machine using Vortex Tube expansion.

3. The coefficient of performance of the vortex tube. In order to discuss the coefficient of performance of the vortex tube used as an open-cycle gas refrigerating machine, a possible mode of operation is given in Fig. 6, showing the circuit and in Fig. 7, giving the cycle on an indicator diagram. First the gas is adiabatically compressed from pressure p_1 and temperature T_a to pressure p_2 and temperature T_b, following the adiabatic path $a \rightarrow b$ of Fig. 7. The work of compression W_c for one mole of a perfect gas is:

$$W_c = C_p T_a \left[\left(\frac{p_2}{p_1} \right)^{\frac{\gamma-1}{\gamma}} - 1 \right] = C_p (T_b - T_a). \tag{3.1}$$

The gas then passes through the cooler, ideally at constant pressure p_2 until it is cooled to the temperature T_c of the coolant. This gives the path $b \rightarrow c$ on Fig. 7. On leaving the cooler it enters the nozzle, N, of the vortex tube. In the tube the gas divides, the cold fraction in amount μ passing at a temperature T_e and pressure p_1 into the cold chamber. This is represented by the path $c \rightarrow e$ of Fig. 7. Since the cooling process is not exactly adiabatic the point e on the indicator diagram is at a higher temperature than the point d, which is on the adiabatic through c at pressure p_1. The fraction $(1-\mu)$ of heated gas leaves the valve, V, of the vortex tube at temperature T_f and pressure p_1 and this is represented by the path $c \to f$. Note that f is at larger volume than a, since $T_f > T_c \lessgtr T_a$. This fraction $(1-\mu)$, cooled to the temperature T_c by passing through the cooler, returns to the compressor input. This is represented by the path $f \to c'$. c' is not coincident with a unless $T_a = T_c$. This non-coincidence of c' and a results in a somewhat larger work of compression than that given in eq. (3.1).

The cold fraction μ of the gas entering the cold chamber absorbs heat, Q_1, isobarically at pressure p_1, and leaves at the temperature T_a of the material in the cold chamber. This is given by the path $e \to a$ of Fig. 7. Also for one mole of total gas passing through the compressor,

$$Q_1 = \mu\, C_p (T_a - T_e).\tag{3.2}$$

If one neglects the excess work of compression needed to change the fraction $(1-\mu)$ of the gas from c' to a, then the coefficient of performance is:

$$\xi_{\text{vortex}} = \frac{Q_1}{W} = \frac{\mu\,(T_a - T_e)}{(T_b - T_a)}.\tag{3.3}$$

The correct value of ξ, allowing for the fact that c' and a are not coincident is smaller than the value of eq. (3.3) by

$$100\,(1-\mu)\,\frac{T_c - T_a}{T_a} \text{ percent}\tag{3.4}$$

which for the application considered later is only a correction of less than 5 %.

It is of interest to compare the value of (ξ_{vortex}) for the vortex tube given by eq. (3.3) with the value for the coefficient of performance of an open-cycle gas refrigerating machine using an isentropic expansion engine. The latter is given in eq. (1.4). It is to be noticed that if in computing ξ for the isentropic expander machine one neglected to use the work given out by the expander, then the value of ξ would be

$$\xi' = \frac{(T_a - T_d)}{(T_b - T_a)}.\tag{3.5}$$

This is larger than ξ_{vortex} for the vortex tube cycle, (a) by a factor $(1/\mu)$, because the vortex tube only uses a fraction μ of the gas passing through the cold chamber and (b) by an extra amount due to the fact that $T_e > T_d$, because the vortex expansion is not isentropic.

Table 2. *Coefficient of performance etc., for actual vortex-tube refrigerating cycle using air.* Take T_c (Temperature of coolant) = (temperature of nozzle) = 20° C; $p_1 = 1$ atm.; $p_2 = 11$ atm.

1. Min. observed temperature T_e, of cold gas from vortex tube (Hilsch's Tube No. 3)	$-48°$ C
2. μ for min. T_e	0.40
3. Temperature drop, $(T_c - T_e)_{\text{obs}}$, for vortex tube	68° C
4. Temperature drop, $(T_c - T_d)$, for ideal isentropic expansion	144° C
5. Coefficient of perf. ξ_{vortex} for vortex tube cycle. Assume $T_a = 0°$ C. Eq. (3.3)	0.07
6. Coefficient of perf., ξ', for gas refrigerating (adiabatic expansion) machine, *not* using work of expansion. Assume $T_a = 0°$ C. Eq. (3.5)	0.45
7. Full. coefficient of perf. ξ, for gas refrigerating (adiabatic expansion) machine. Assume $T_a = 0°$ C. Eq. (1.4)	0.96
8. Efficiency η_{rel} of vortex tube cycle relative to gas refrigerating (adiabatic expansion) machine	7.3 %

ξ_{vortex} can be computed from HILSCH's experimental observations on his tube No. 3. The results of this computation are given in Table 2. It is assumed that the vortex tube cycle is as in Fig. 6 with $T_c = 20°$ C, $T_a = 0°$ C and $p_1 = 1$ atm., $p_2 = 11$ atm. Then the observed temperature, T_e, of the cold gas emerging from the vortex tube would have a minimum value of $-48°$ C for $\mu = 0.40$. For adiabatic compression, eq. (3.3) makes the coefficient of performance $\xi_{\text{vortex}} = 0.07$. This is to be compared with the coefficients of performance ξ' and ξ for a gas refrigerating (isentropic expansion) machine operating between the same temperatures T_c and T_a. ξ' is for the latter machine *not* using the work of expansion. One finds $\xi' = 0.45$ and $\xi = 0.97$, showing that the vortex tube cycle has much poorer coefficient of performance than either of these. Its relative efficiency, $\eta_{\text{rel}} = \xi_{\text{vortex}}/\xi$, relative to the gas refrigerating machine is then 7.3%. Since the gas refrigerating machines described earlier have poor coefficients of performance compared with, for example, compressed vapor refrigerating machines, it is unlikely that the vortex tube will have serious practical application in refrigeration, except where extreme simplicity is a vital requirement.

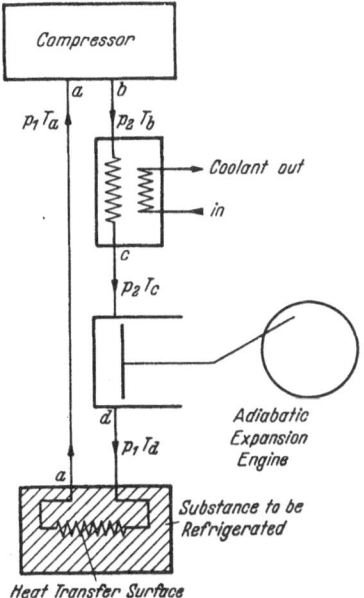

Fig. 8. Flow diagram for closed-cycle gas refrigerating machine.

4. The closed-cycle gas refrigerating machines.
The early development of closed-cycle machines using air as the working gas was due to GORRIE[1] (1845) and A. KIRK[2] (1861) and later to ALLEN and to WINDHAUSEN[3]. The system, which is essentially a STIRLING air engine reversed, is exactly analogous to that of an open-cycle gas refrigerating machine, as described above. The difference between the two types lies in the fact that in the closed-cycle system the same mass of gas is continually recycled, generally everywhere at pressures above atmospheric. One of the advantages of the closed-cycle is that *dry* air can be used, so avoiding the frosting difficulties encountered in the open-cycle machines. Also smaller compressors and expansion engines may be used, thus reducing frictional losses. A schematic diagram of the closed-cycle system is given in Fig. 8 which is identical with that of Fig. 1 for the open-cycle system except that the cold chamber is replaced by a heat transfer surface (generally a long pipe or pipes carrying the refrigerating gas) in contact with the substance to be refrigerated. In the system developed for example by ALLEN, known as the "ALLEN dense-air system", air was the refrigerating gas and it worked between pressures $p_1 = 4.5$ atm. and $p_2 = 16.5$ atm.

Although relatively low compression ratios could therefore be used in the closed-cycle system with at the same time reasonably small compressor and expansion engine, the coefficients of performance of such machines, prior to the

[1] J. GORRIE: US Patent 8080. May 1851. See also W. SIEMENS: Min. Proc. Instn. Civ. Engrs. **68**, 179 (1882).

[2] A. KIRK: Min. Proc. Instn. Civ. Engrs. **37**, 244 (1873/74).

[3] See J. A. EWING: Mechanical Production of Cold. Cambridge Univ. Press 1908 and R. PLANK: Handbuch der Kältetechnik, vol. I. Berlin: Springer 1954.

new Philips machines[1], were very small. Their theoretical values are given in eqs. (1.4), (1.6) and (1.9) being formally the same as for the open-cycle machines. The obsolescence of the traditional closed-cycle system therefore is due also to its poor performance.

The Philips closed-cycle gas refrigerating machine described by Köhler and Jonkers[1] has been able to approach more closely the ideal thermodynamical efficiency to be expected largely by carrying out the absorption of heat at the lower temperature approximately isothermally. This obviates the irreversibilities introduced by the isobaric expansion in the passage of the gas through the heat transfer surface. To enable such an isothermal expansion to be carried out the expansion engine must operate, both as to its input as well as its output, at the lower temperature. The possibility of low temperature operation of such engines was developed, initially by Claude[2], at a date much later than that of the traditional gas refrigerating machines described above. The early gas machines therefore owe their major inefficiency to the high temperature operation of the expansion engine.

5. The Philips gas refrigerating machine. A new development of the closed-cycle has been described in detail by Köhler and Jonkers[1] of the Philips Industries. Since this enables in a single stage expansion engine temperatures to be maintained which are low enough to liquefy air under atmospheric pressure with fair efficiency, a detailed description of it may be of value.

In principle the significant difference between this and the earlier closed-cycle refrigerating machines described above is due to the introduction of a heat exchanger between the compressor and the expansion engine. Fig. 9 is a basic diagram of this system and is to be compared with Fig. 8 of the earlier systems. By the use of the interchanger it is possible to maintain the expansion engine at the temperature T_1 of the refrigeration and so enables almost all the cold produced by the expansion to be available essentially isothermally at T_1.

It is possible to consider ideally the system illustrated in Fig. 9 as consisting of:

a) *Isothermal* compression to pressure p_2 of the gas at the temperature T_2, the isothermalism being effectively maintained by the cooler. (For this reason the compressor and cooler are grouped as one unit as indicated by the broken rectangular box marked T_2 enclosing them.)

b) Cooling of the high pressure gas from T_2 to T_1 by passage through the interchanger with consequent reduction in volume.

c) *Isothermal* expansion from pressure p_2 to p_1 in the expansion engine with consequent extraction of heat from the substance to be refrigerated. (Here also the isothermalism of the combined processes of expansion and of extraction of heat from the substance to be refrigerated via the heat transfer surface is diagrammatically indicated by the enclosing of the expansion engine and of the heat transfer surface of Fig. 9 inside one broken rectangular box marked T_1.)

d) Passage of the low pressure return gas back through the interchanger from temperature T_1 to T_2.

If one may assume the interchanger to be perfect such that no temperature differences exists between the ingoing and outgoing gas and such that no pressure differences are needed maintaining the gas flow through it, then this idealized cycle may be represented by the indicator diagram of Fig. 10. In this the path $a \to b$ represents the compression from p_1 to p_2 along the isotherm T_2. Path

[1] J. W. L. Köhler and C. O. Jonkers: Philips Techn. Rev. **16**, 69, 105 (1954).
[2] G. Claude: Liquid Air, Oxygen and Nitrogen. Paris 1913. See also section G.

$b \rightarrow c$ represents the cooling form T_2 to T_1 in passage through the interchanger at constant pressure p_2, path $c \rightarrow d$ the isothermal expansion in the expansion engine at temperature T_1, and path $d \rightarrow a$ the return path through the interchanger at constant pressure p_1.

The heat extracted, Q_1, at T_1 from the substance to be refrigerated is then given by:

$$Q_1 = \int_c^d p \, dv \qquad (5.1)$$

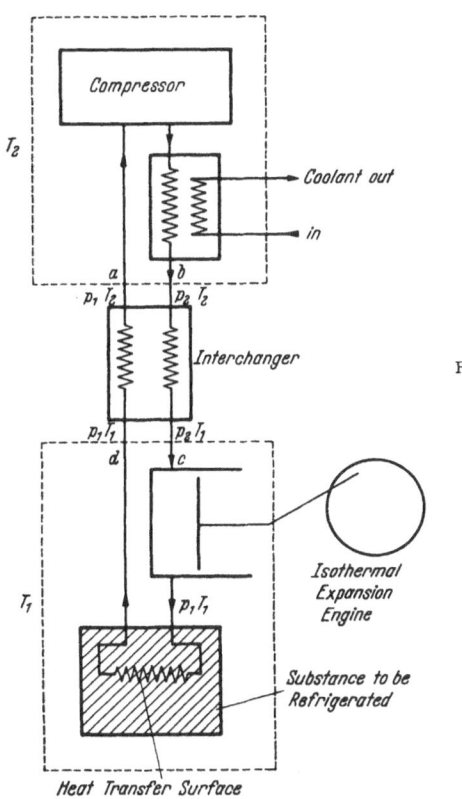

Fig. 9. Flow diagram for closed-cycle gas refrigerating machine using low temperature expansion engine and interchanger. Diagram represents basic features of the Philips gas refrigerating machine.

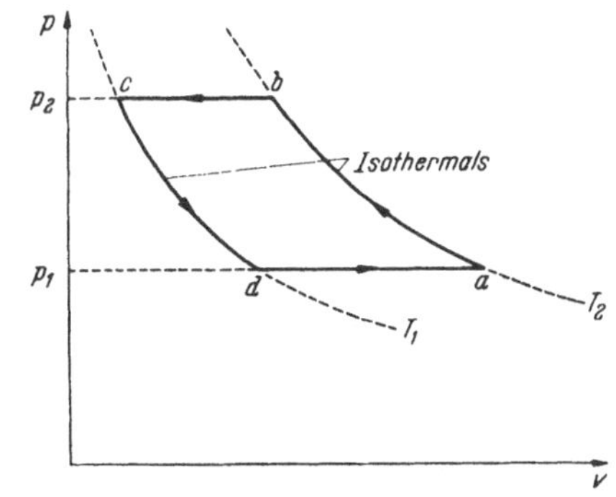

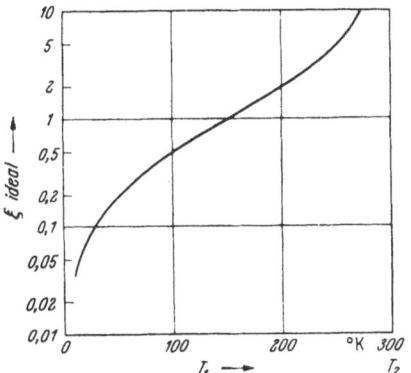

Fig. 10. Indicator diagram for idealized operation of closed-cycle refrigerating machine of Fig. 9.

Fig. 11. Coefficient of performance of ideal (CARNOT) refrigerating machine, having upper temperature, T_2, of 300° K, as function of refrigeration temperature T_1.

which for a perfect gas is:

$$Q_1 = R\,T_1 \ln(p_2/p_1) \qquad (5.2)$$

and the work done, W, during the complete cycle, again assuming a perfect gas as the working substance, is:

$$W = \int_a^b p \, dv - \int_d^c p \, dv = R(T_2 - T_1) \ln(p_2/p_1). \qquad (5.3)$$

Eqs. (5.2) and (5.3) therefore give for the coefficient of performance, ξ^{ideal}:

$$\xi^{\text{ideal}} = \frac{Q_1}{W} = \frac{T_1}{T_2 - T_1} \qquad (5.4)$$

which is the value for a CARNOT system. Fig. 11 shows the variation of ξ^{ideal} with variation of the refrigerating temperature T_1, when the upper temperature

T_2 is maintained at $300°$ K, i.e. approximately ambient temperature. This clearly shows the rapid fall off of the ideal coefficient of performance with diminishing T_1.

In the Philips machine the compressor and the expander are combined together in one unit of the so-called "displacer" type, and is illustrated in Fig. 12. Here *1* is the main piston moving in the cylinder *2*. The displacer piston *3* moves coaxially with the main piston with a phase difference, φ, which is determined by the crankshaft connections. (See Fig. 14 for constructional detail of this.)

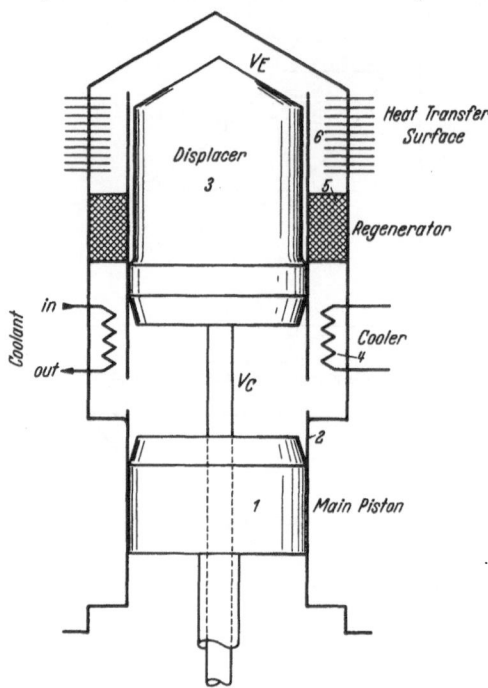

The combined movement of the main piston and the displacer cyclically compresses the gas in the "compressor" space V_c, transfers it through the cooler *4*, the regenerator *5*, and the heat transfer surface *6*, expands it in the "expander" space V_E and then transfers it back, via *6*, *5* and *4*, to the space V_c. In comparing this with the idealized diagram of Fig. 9 the significant difference lies in the use of a regenerator rather than an interchanger. (See chapter I for more detail concerning the operation of a regenerator.) This is occasioned by the fact that the gas transfer between spaces V_c und V_E alternates in direction cyclically rather than being composed of two continuous counterflows.

The variations of volume of the spaces V_c and V_E can be seen in Fig. 13 in which the positions of the pistons are shown as a function of the shaft angle, α. The constant volumes taken up by the two pistons are shown by the shaded areas. In the lower part of Fig. 13 the full curve gives the variation of the compressor volume V_c which shows its sinusoidal variation with α and points up the phase angle φ with respect to the expansion volume V_E.

Fig. 12. Schematic diagram of Philips gas refrigerating machine using displacer type engine. [After J.W. L. KÖHLER and C. O. JONKERS: Philips techn. Rev. 16, 105 (1954).]

For harmonic variations of the volumes V_c and V_E as function of shaft angle, α, write:

$$V_E = \tfrac{1}{2} V_0 (1 + \cos \alpha) \tag{5.5}$$

$$V_c = \tfrac{1}{2} w V_0 [1 + \cos (\alpha - \varphi)] \tag{5.6}$$

where w is the ratio of the maximum value of V_c to the maximum value of V_E and is approximately unity.

For a perfect gas these volume variations result in pressure variations in the system as a whole given by[1]:

$$p = p_{max} \frac{1 - \delta}{1 + \delta \cos (\alpha - \vartheta)} \tag{5.7}$$

[1] See RINIA, DU PRÉ, DE BREY and VAN WEENEN: Philips Techn. Rev. 8, 2, 129 (1946), 9, 97, 125 (1947).

where, neglecting dead volumes in the system,

$$\delta = \frac{\sqrt{\tau^2 + w^2 + 2\tau w \cos\varphi}}{\tau + w} \tag{5.8}$$

τ being the ratio, T_2/T_1, of the temperature of the isothermal compression to the temperature T_1 of the isothermal expansion.

The angle ϑ is the phase angle between the pressure and volume of the gas in the expansion and is given by

$$\tan\vartheta = \frac{w\sin\varphi}{\tau + w\cos\varphi} . \tag{5.9}$$

Using these relations between p and V_c and V_E it can be shown[1] that the coefficient of performance, ξ, is the same as that for the ideal CARNOT cycle given in eq. (5.4). Thus the performance of the ideal machine is, as would be expected, unaffected by whether the cycle is made along isotherms and isobars, as in Fig. 10, or around a smooth curve determined by the harmonic variations of V_c and V_E.

The heat, Q_1, absorbed by the expander at the low temperature, T_1, per cycle is:

$$\left. \begin{array}{l} Q_1 = \pi\bar{p}\,V_0 \times \\[2mm] \times \dfrac{\delta}{1 + \sqrt{1-\delta^2}} \sin\vartheta \, . \end{array} \right\} \tag{5.10}$$

It is to be noted that since δ and ϑ are determined uniquely by the choice of $\tau(=T_2/T_1)$, w and φ, Q_1 can be increased only by increase of the average pressure $\bar{p}$ and the volume V_0. The *rate* of heat absorption,

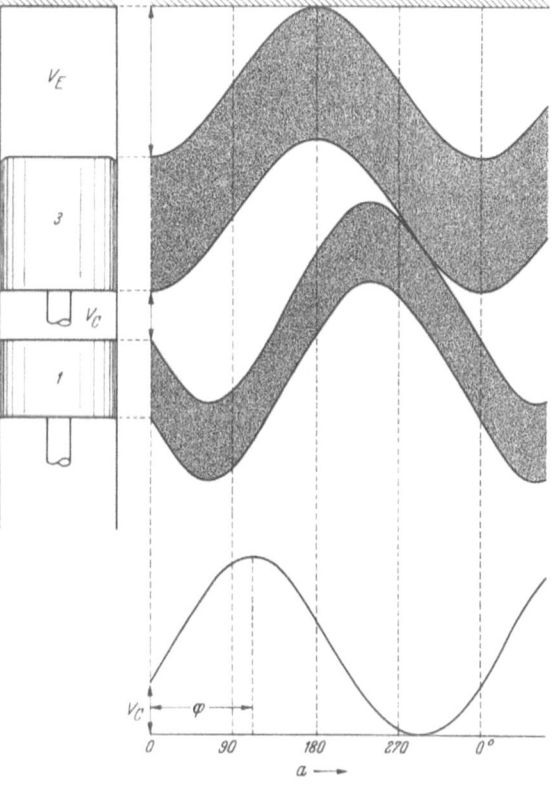

Fig. 13. Diagram showing the volumes V_E and V_C of the expander space and the compressor space respectively of the displacer type engine of the Philips refrigerating machine as a function of the crank angle, α. The constant volumes taken up by the two pistons are shown by the shaded areas. The lower curve gives the variation of the compressor space. [After J. W. L. KÖHLER and C. O. JONKERS: Philips techn. Rev. 16, 105 (1954).]

of course being proportional to the number of cycles for second, is governed also by the speed of rotation of the shaft. The Philips machine ran at 1440 revs/min.

The question of the determination of the temperature ratio, $\tau = T_2/T_1$, is of interest. In a simple adiabatic expansion, as for example in the earlier gas refrigerating machines of Figs. 1 and 8.

$$\tau = \frac{T_2}{T_1} = r^{\frac{\gamma-1}{\gamma}} \tag{5.11}$$

where r is the compression ratio, $r = p_2/p_1$. The value of τ therefore cannot be greater than $r^{0.4}$, the value for monatomic gases. In the Philips machine using

---

[1] J. W. H. KÖHLER and C. O. JONKERS: Philips Techn. Rev. **16**, 69 (1954).

the displacer-type engine the connection between τ and r is more complex. Eq. (5.7) gives for r

$$r = \frac{p_{\max}}{p_{\min}} = \frac{1 + \delta}{1 - \delta} \tag{5.12}$$

which determines δ for a given r value. Now eq. (5.8) determines τ as a function of δ, w and φ. The final temperature T_1 of the refrigeration for prior choice of

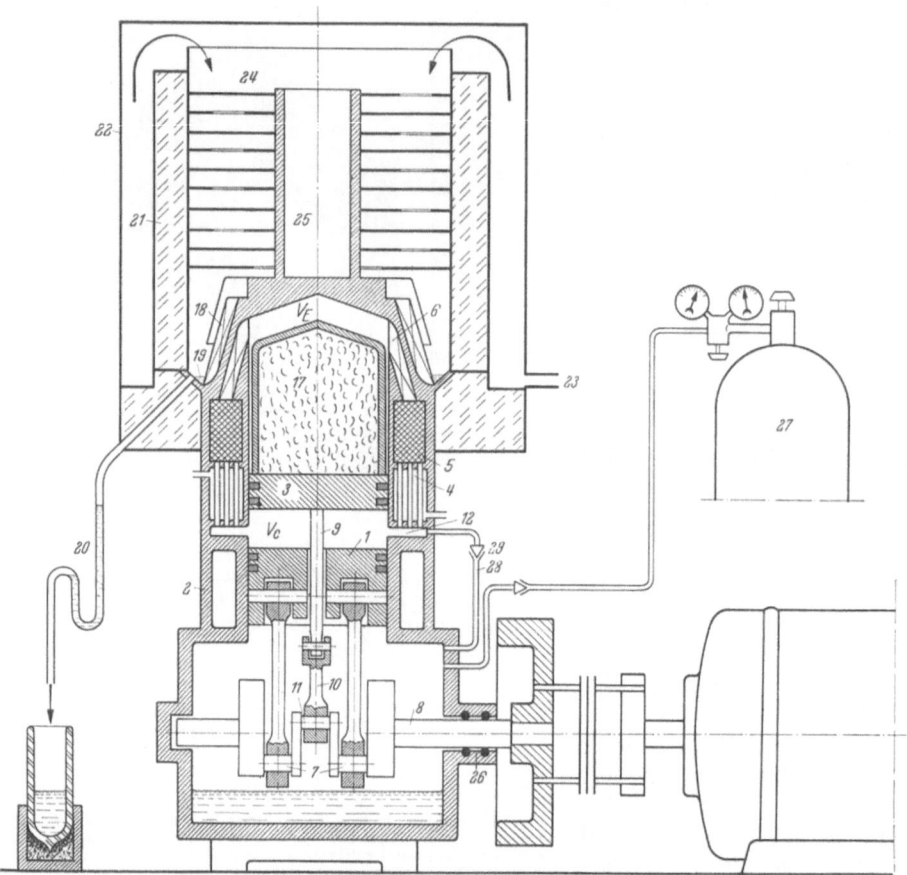

Fig. 14. Detailed drawing of construction of Philips gas refrigerating machihe. [After J. W. L. Köhler and C. O. Jonkers: Philips techn. Rev. 16, 105 (1954).]

the constructional parameters w and φ is given as a function of r by combining eq. (5.8) and (5.12); in the reported machine, $\tau = T_2/T_1$ was approximately 3.8 whereas $r \approx 2.2$.

In practice the ideal coefficient of performance is not realized and there is a lower limit to the temperature T_1 of refrigeration owing to the inevitable losses and irreversibilities introduced in actual operation. The main factors influencing the deviations from the ideal performance are:

1. Regenerator inefficiency. Although the regenerator efficiency can be made high ($\sim 99\%$), its deviation from 100% efficiency accounts for both the lowest limit to the value of T_1 and the upper limit to the value of $\bar{p}$ and hence to the value of Q_1 [see eq. (5.10)].

2. The so-called "adiabatic loss" due to the nonisothermalism of compression and expansion.

3. Frictional losses both mechanical and in the gas flow.

4. Insulation losses.

For a detailed discussion of these inefficiency factors the original papers should be consulted.

A detailed drawing of the Philips gas refrigerating machine is given in Fig. 14, which is largely self-explanatory. The refrigerant used is hydrogen or helium. One novel feature concerns the displacer piston, 3. Its lower end, 3, which is in contact with the cylinder, is at the high (ambient) temperature and here therefore the rings are mounted. The

Fig. 15. Photograph of (left) the cooler, (center) the regenerator and (right) the cylinder head with heat transfer surface (fins) of the Philips gas refrigerating machine. In right foreground is one unit of the regenerator, consisting of a ring of fine compressed metal wire. Matchbox gives scale. [Courtesy of J. W. L. KÖHLER and C. O. JONKERS: Philips techn. Rev. 16, 105 (1954).]

upper surface of the displacer piston, in thermal contact with cold expansion space, V_E, is thermally isolated from the lower end, 3, by means of the tall dome, 17, which is made of heat insulating material.

The cooler 4, which is water cooled, the regenerator 5 and the heat transfer surface 6 are shown also in Fig. 15. The cooler consists of a large number of parallel pipes carrying the gas, over which water flows. The regenerator consists of a pile of rings made from a mass of fine metal wire, the rings being mounted in the annular space between two nylon cylinders. The heat transfer surface consists of fins or slots cut both inside and outside of the copper cylinder head.

For use as an air liquefier, the cylinder head is surrounded by an insulated can, 24 (see Fig. 14) in which atmospheric air condenses on the copper finned surface, 18, of

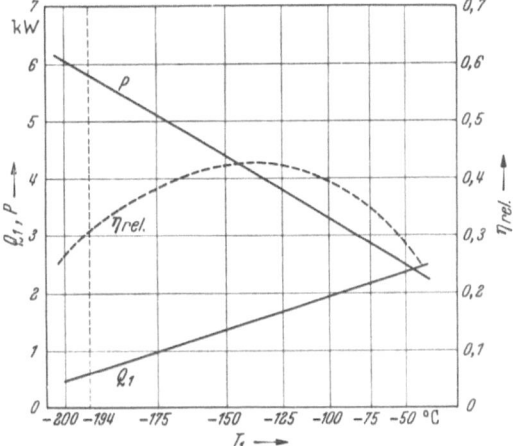

Fig. 16. Performance data on Philips gas refrigerating machine as function of refrigerating temperature, T_1. Q_1, is the heat extracted by machine at T_1, P is shaft power and η_{rel} the relative efficiency compared with an ideal CARNOT refrigerator. [After J. W. L. KÖHLER and C. O. JONKERS: Philips techn. Rev. 16, 105 (1954).]

the cylinder head. The liquid air that collects is led away via the drain pipe, 20.

When operated in this way as an air liquefier with a shaft power of 5.8 kW and when fed with dry air it produced 6.6 liters liquid per hour. This corresponds to an efficiency of 0.88 kW-hr/liter liquid. As will be seen from Tables 12, 13 and 15, this compares very favorably with other methods of air liquefaction, particularly for such a small scale unit. The ratio of the observed coefficient of performance, ξ, under these conditions to the ideal CARNOT coefficient of performance, ξ^{ideal}, is approximately 0.3.

The future possibilities opened up by this machine are manifold. First its high efficiency, η_{rel}, relative to an ideal CARNOT system in the temperature range $-50°$ C to $-200°$ C together with its mechanical simplicity makes it most valuable in this range of operation. (The observed values of $\eta_{rel} \equiv \xi/\xi^{ideal}$ are shown in Fig. 16.) This is just the temperature range beyond the satisfactory operation of single stage compressed vapor machines and in consequence the Philips machine may find many uses where at present multistage or cascade compressed vapor machines are employed. (See chapter B.)

B. Compressed vapor refrigerating machines.

6. The simplified cycle of operations. A schematic diagram of a typical single stage compressed vapor refrigerator is given in Fig. 17. The purpose of such a refrigerator is to withdraw heat from a low temperature, T_1, by means of evaporation of a suitable liquid in an "evaporator". In PERKINS'[1] original machine sulphuric ether $[(C_2H_5)_2O]$ was the liquid to be evaporated. This was also the working substance actually used in the improved machine of J. HARRISON (1857)[2]. The first introduction of ammonia as the working substance, a substance still in wide use today, was due to LINDE[3] in 1876, and that of sulphur dioxide to PICTET[4] also at about the same date.

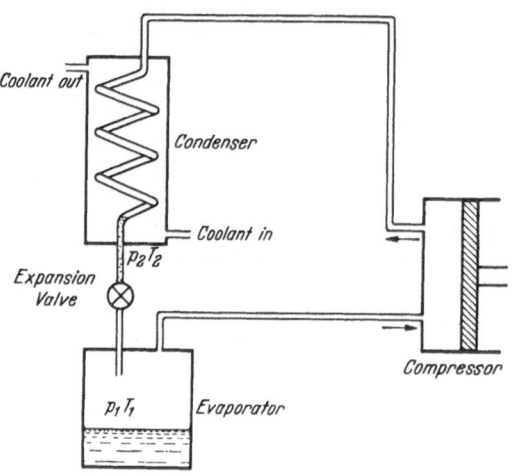

Fig. 17. Flow diagram for idealized compressed vapor refrigerating machine.

As shown in Fig. 17 the complete refrigerator consists of: 1. an evaporator which absorbs the heat from its low temperature surroundings. In an household refrigerator for example the evaporator consists of the tubing coiled around the freezing compartment. 2. A compressor, used to liquefy the vapor in the condenser by compression alone, 3. a condenser, which absorbs the heat of compression of the vapor and which maintains a temperature sufficiently low to allow condensation of the vapor under the pressure produced by the compressor. And 4. an expansion valve which allows the compressed liquid to expand approximately isenthalpically into the evaporator, an isenthalpic expansion being one in which the enthalpy $(H = U + pV)$ remains constant.

In order to show the simplest cycle of operations performed by such a refrigerator, first consider a "wet-compression" machine, in which the vapor entering the compressor contains a certain fraction of liquid and in which on compression

---

[1] J. PERKINS: English Patent No. 6662, 1834. See also "Mechanical production of cold" by J. A. EWING. Cambridge Univ. Press 1908.

[2] See, for example, R. PLANK: Handbuch der Kältetechnik, vol I. Berlin: Springer 1954.

[3] See, for example, K. v. LINDE: English Patent No. 1458 (1876) and see also R. PLANK: Handbuch der Kältetechnik, vol. I. Berlin: Springer 1954.

[4] R. PICTET: C. R. Acad. Sci. Paris **85**, 1214, 1220 (1877) and see also R. PLANK: Handbuch der Kältetechnik, vol. I. Berlin: Springer 1954.

the vapor is just exactly saturated. The cycle of operations is shown thermodynamically on the entropy versus temperature diagram given in Fig. 18, which
is a schematic entropy diagram for ammonia. Such diagrams were introduced
by MOLLIER[1]. In Fig. 18 the ordinates give the absolute temperature, T (degrees KELVIN), starting from the absolute zero, $T = 0$, and the abscissae the
entropy, S. Since a thermodynamic evaluation of the performance of the refrigerator depends only on differences in entropy involved, the zero for the entropy
is not of importance. In Fig. 18, the full lines are representative of typical isobars for various pressures, p, such that $p_3 > p_2 > p_1$. The heavy broken curve
represents the boundary between the liquid and vapor states, i.e. it gives the
locus of points representing saturated vapor states or 100% liquid states.

In the wet-compression machine suppose the wet vapor at p_1 and T_1 entering
the compressor from the evaporator is represented by the point a, and let an
ideally adiabatic compression raise the vapor to p_2 and T_2 to the point c, where
it is exactly saturated. Condensation in the condenser at temperature T_2, now
takes the state of the working substance to the point d. In passing through the
expansion valve, the substance undergoes isenthalpic expansion, as in JOULE-
THOMSON expansion, and follows the curve
$d \to e$ on the entropy diagram, back to T_1 and
p_1. Since such an expansion is irreversible,
an increase in the entropy must take place
as indicated schematically in Fig. 18. The
fraction ε of the substance still remaining
liquid after the expansion is given by:

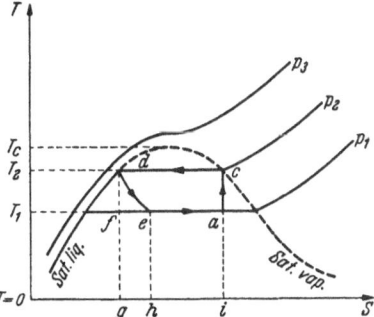

$$\varepsilon = \frac{ae}{af}. \qquad (6.1)$$

Finally by evaporation of the liquid in the
evaporator, the percentage of vapor increases to its initial value as is indicated
by the state of the substance moving from
point e to point a. This completes the cycle
of operations.

Fig. 18. $T-S$ diagram showing thermodynamical
cycle of operations for idealized compressed vapor
refrigerating machine using "wet" compression
(T_c is critical temperature of working fluid).

7. The efficiency and actual cycles of operations. Let the amount of heat
absorbed from the surroundings at the low temperature T_1 be denoted by Q_1
and the heat given out at the high temperature T_2 by Q_2. Then from the first
and second laws of thermodynamics, a reversible CARNOT machine working between T_1 and T_2 would be the most efficient machine and would be such that:

$$Q_1^{\max} = Q_2 \frac{T_1}{T_2} \qquad (7.1)$$

and the ideal maximum coefficient of performance, $\xi^{\max}$, may be defined by the
ratio of the heat absorbed at the low temperature to the net work done, neglecting
all losses, would be given by:

$$\xi^{\max} = \frac{Q_1^{\max}}{W} = \frac{Q_1^{\max}}{Q_2 - Q_1^{\max}} = \frac{T_1}{T_2 - T_1}. \qquad (7.2)$$

In the cycle of operations for the idealized wet-compression compressed vapor
refrigerator described above, the operation, which is not reversible and hence
which must reduce its coefficient of performance below that of a reversible engine

---

[1] R. MOLLIER: Z. ges. Kälteind. **3** (1896), which paper gives the $T-S$ diagram for CO_2.

working between the same two temperatures, is the isenthalpic expansion at the expansion valve, as represented by the line $d \rightarrow e$ in Fig. 18. In this cycle of operations, the heat absorbed, Q_1 in the evaporator at the constant temperature T_1 is:

$$Q_1 = T_1(S_a - S_e) = \text{area } (aehia) \qquad (7.3)$$

where the subscripts refer to the corresponding points on the diagram of Fig. 18. Also the heat Q_2, given up to the condenser, at the constant temperature T_2 is:

$$Q_2 = T_2(S_c - S_d) = \text{area } (cdgic). \qquad (7.4)$$

For an ideal reversible machine the expansion of the working substance from the condenser into the evaporator would be made reversibly and adiabatically and would be represented by the line $d \rightarrow f$. In this case, the maximum heat absorbed at T_1 would be:

$$Q_1^{\text{max}} = T_1(S_a - S_f) = \text{area } (afgia), \qquad (7.5)$$

and from the figure it is seen that:

$$\frac{Q_1^{\text{max}}}{Q_2} = \frac{\text{area } (afgia)}{\text{area } (cdgic)} = \frac{T_1}{T_2}. \qquad (7.6)$$

in agreement with eq. (7.1). The lack of reversibility of the actual cycle of operations therefore involves a loss in refrigeration given by:

$$Q_1^{\text{max}} - Q_1 = \text{area } (efghe) \qquad (7.7)$$

which is dependent on the relative values of T_1 and T_2, as is discussed in more detail later. Moreover the relative efficiency η_{rel} of the actual cycle relative to the ideal reversible thermodynamic engine can be expressed by:

$$\eta_{\text{rel}} = \frac{\xi}{\xi^{\text{max}}} < \frac{Q_1}{Q_1^{\text{max}}} < \frac{ae}{af} \qquad (7.8)$$

the inequality sign being due to the fact that for the irreversible machine an amount of work ($=$ area $efghe$) must be done on it in addition to the amount ($=$ area $acdfa$) required by the reversible machine.

It will readily be seen from Fig. 18 that, as the value of the temperature T_1 of refrigeration is reduced the ratio ae/af must diminish. Consequently one may conclude that a) from eq. (7.1) as T_1 is reduced, Q_1^{max} is reduced and b) from eq. (7.8) as T_1 is reduced the value of η_{rel} also is reduced. To refrigerate at lower temperature therefore becomes more wasteful of energy.

In considering the efficiency of compressed vapour refrigerators it is necessary also to consider the "dry-compression" type of machines[1]. This "dry-compression" is said to exist if the compression starts from the saturation line and ends in the superheated vapor region. In the entropy versus temperature diagram of Fig. 19, this dry compression is represented by the vertical line $a \rightarrow c$; going from pressure p_1 of the saturated vapour at the temperature T_1 of the evaporator

---

[1] The pioneer thermodynamic analysis was carried out by K. v. LINDE (Z. ges. Kälteind. 28. Jan. 1895 and Z. VDI 2. Febr. 1895) and may be studied in detail in many textbooks on Refrigeration Engineering as for example: J. A. EWING: ,,The Mechanical Production of Cold", Cambridge Univ. Press 1908; M. DAVIES: "The Physical Principles of Gas Liquefaction and Low Temperature Rectification", Longmans, Green & Co. 1949; R. G. OWENS and F. OPHULS: ASRE Refrigerating Data Book, 7th Ed. 1951, Part I, p. 3 and p. 11.

to some pressure p_2. In the ideal case, the compression is considered adiabatic (i.e. isentropic) and hence the compression line $a \to c$ drawn vertical.

The subsequent cycle of operations for the machine is similar to that for the wet-compression machine and can be followed in Fig. 19. The cooling of the vapor in the condenser follows the path of constant pressure $c \to c' \to c'' \to d$. At c' the first drops of liquid form in the condenser. The path $c'' \to d$ represents cooling of the liquid below the temperature of first condensation. The path $d \to e$, as before, represents the isenthalpic expansion at the valve into the evaporator, while the final path $e \to a$ represents evaporation of the liquid in the evaporator with the associated production there of cold. This completes the cycle of operations.

From the diagram, it will be seen that the heat absorbed, Q_1, in the evaporator at the constant temperature is:

$$Q_1 = T_1(S_a - S_e) = \text{area } (aehia) \quad (7.9)$$

whereas for an ideal reversible machine, in which the change from temperature T_2 to T_1 would be isentropic, following the line $d \to f$, the heat absorbed in the evaporator would be:

$$Q_1^{\text{max}} = T_1(S_a - S_f) = \text{area } (afgia). \quad (7.10)$$

As before therefore, the relative efficiency would be:

$$\eta_{\text{rel}} = \frac{\xi}{\xi^{\text{max}}} < \frac{Q_1}{Q_1^{\text{max}}} < \frac{ae}{af} \quad (7.11)$$

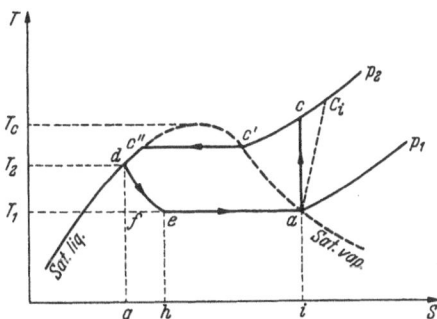

Fig. 19. $T-S$ diagram showing thermodynamical cycle of operations for idealized compressed vapor refrigerating machine using "dry" compression. (T_c is critical temperature of working fluid.)

and the excess work that would have to be done on the machine over that for the corresponding reversible one would be represented by the area of $efghe$.

Again this result for η_{rel} emphasizes the need for small temperature differences, $(T_2 - T_1)$, in order to maintain the term ae/af as large as possible; i.e. the smaller the temperature differences the more nearly reversible does the refrigerator become.

The exact values therefore for the theoretical coefficient of performance for any machine can be calculated, as indicated above, from graphical integration of appropriate areas in the entropy and temperature diagram for the working substance used once the operating conditions are known.

It should be remembered that were the cycle of operations truly a reversible CARNOT cycle, then the coefficient of performance would be independent of the working substance. It is the isenthalpic expansion at the throttling valve, which constitutes the main deviation of the cycle from a CARNOT cycle and hence which introduces the importance of a study of the properties of the working substance in calculation of the coefficient of performance.

For tables and graphs of the entropy as a function of temperature, as well as other thermodynamic data, for ammonia, a widely used refrigerator working substance, the references given in Table 3 should be consulted. This table also gives references for similar data for other working substances. Other thermal data of importance in the design of compressed vapour refrigerators, e.g. critical temperature, critical pressure, etc. for most of the common working substances are given also in Table 3.

Table 3. *Data on fluids used in compressed vapor refrigerators.*

T_B = boiling point; T_c = critical temperature; p_c = critical pressure; T_F = freezing point.

		T_B °K	T_c °K	p_c atm.	T_F °K	Reference to thermodynamic data
Trichlorotrifluoroethane (Freon 113)	$C_2Cl_3F_3$	320.7	487.2	33.6	238.2	1, 2
Methylene Chloride or Dichloromethane (Carrene No. 1)	CH_2Cl_2	313.7	489.4	45.6	176.5	10
Trichloromonofluoromethane (Freon 11 or Carrene No. 2)	CCl_3F	296.8	470.7	36.4	162.2	1, 2
Ethyl Chloride	C_2H_5Cl	285.7	460.4	51.9	134.5	
Dichloromonofluoromethane (Freon 21)	$CHCl_2F$	281.8	451.4	51.0	138.3	1, 2
Dichlorotetrafluoroethane (Freon 114)	$C_2Cl_2F_4$	276.7	419.2	32.2	179.1	1, 2
Sulphur Dioxide	SO_2	263.1	430.3	77.7	200.3	1, 5
Methyl Chloride	CH_3Cl	249.4	416.2	65.9	169.6	1, 3
Dichlorodifluoromethane (Freon 12)	CCl_2F_2	243.1	384.2	39.6	118.3	1, 2, 4
Ammonia	NH_3	239.8	405.5	111.5	195.5	1, 6, 11
Monochlorodifluoromethane (Freon 22)	$CHClF_2$	232.5	369.1	48.7	113.3	1, 2
Propane	C_3H_8	230.8	368.7	44.0	83.0	1, 8
Propylene	C_3H_6	226.1	364.3	43.9	87.9	1
Carbon Dioxide	CO_2		304.2	73.0	216.6	1
Ethane	C_2H_6	184.8	305.5	47.2	101.2	1, 13
Nitrous Oxide	N_2O	183.6	309.6	71.7	170.8	1
Ethylene	C_2H_4	169.3	282.8	58.9	103.8	1, 7, 9, 12
Methane	CH_4	111.7	190.6	45.8	90.5	1, 7, 12, 14

8. Use of the $H-S$ Mollier Diagram. It is possible, to obtain an evaluation of the coefficient of performance more readily by use of one of the so-called "Mollier diagrams", which gives a plot of the enthalpy H against the entropy, S [15]. Fig. 20 represents diagrammatically such a Mollier diagram, the full lines marked p_1, p_2, and p_3 being typical isobars ($p_3 > p_2 > p_1$) and the light broken lines being typical constant temperature lines ($T_3 > T_2 > T_1$).

The heavy broken curve is the boundary curve enclosing under it the two-phase heterogeneous region. It is to be noticed that in the two phase region the isobars and the isotherms are linear and coincident. Moreover, since quite generally, $(\partial H/\partial S)_p = T$, the slope of the isotherms (or isobars) in the twophase region gives the absolute temperature directly. Table 3 gives references for such diagrams (and for $p-H$ diagrams) for ammonia and other working substances.

[1] ASRE Refrigerating Data Book, 7th Ed., Part II, p. 105 and ff. 1951.

[2] Publications of Kinetic Chemicals Inc. Wilmington Delaware. (Colored Charts may be obtained from this company.)

[3] R. and H. Chemicals Dept. E. I. du Pont de Nemours Co. Wilmington. Delaware. (Charts may be obtained from this company.)

[4] ASRE Circular No. 12. Publ. Amer. Soc. Refrig. Engng. 40 W. 40 St. New York, N. Y.

[5] D. F. Rynning and C. O. Hurd: Trans. Amer. Inst. Chem. Engr. 41, 465 (1945).

[6] Nat. Bur. Stand., Circular 1923, No. 142.

[7] W. H. Keesom and D. J. Houthoff: Leiden Comm. Suppl. 65a, b (1928).

[8] Dana, Jenkins, Burdick and Timm: Refrig. Engng. 12, 403 (1926).

[9] R. York and E. F. White: Trans. Amer. Inst. Chem. Engr. 40, 227 (1944).

[10] R. W. Waterfill: Industr. Engng. Chem. 24, 616 (1932).

[11] H. J. MacIntire and F. W. Hutchinson: Refrigerating Engineering. New York: J. Wiley & Sons 1950.

[12] W. H. Keesom, A. Bijl and L. A. J. Monte: Leiden Comm. Suppl. 108b (1954) and Appl. Sci. Res. 4, 25 (1954).

[13] Barkelew, Valentine and Hurd: Trans. Amer. Inst. Chem. Engr. 43, 25 (1947).

[14] C. S. Matthews and C. O. Hurd: Trans. Amer. Inst. Chem. Engr. 42, 55 (1946).

[15] Such diagrams were first introduced by R. Mollier: Z. VDI 48, 271 (1904). Diagrams of $p-H$, which are also of value in determining the characteristics of refrigerators, etc., and which were also introduced by Mollier, are referred to also as Mollier diagrams.

The cycle of operations of both a dry-compression or wet-compression vapor refrigerator can be followed in detail by means of the $H-S$ diagram of Fig. 21, on which only two isobars are represented, p_1 corresponding to the pressure at which the working substance is evaporated in the evaporator (i.e. the compressor input pressure) and p_2 corresponding to compressor output pressure. In a wet-compression cycle (corresponding to the cycle of Fig. 18) the state of the working substance at the compressor input is given by the point a on the isobar, p_1. Adiabatic compression takes the working substance to c, which is vertically above a on the p_2 isobar, and which represents saturated vapor. Condensation in the condenser takes the substance to d, which being on the saturation curve represents saturated liquid. The isenthalpic expansion (constant H) through

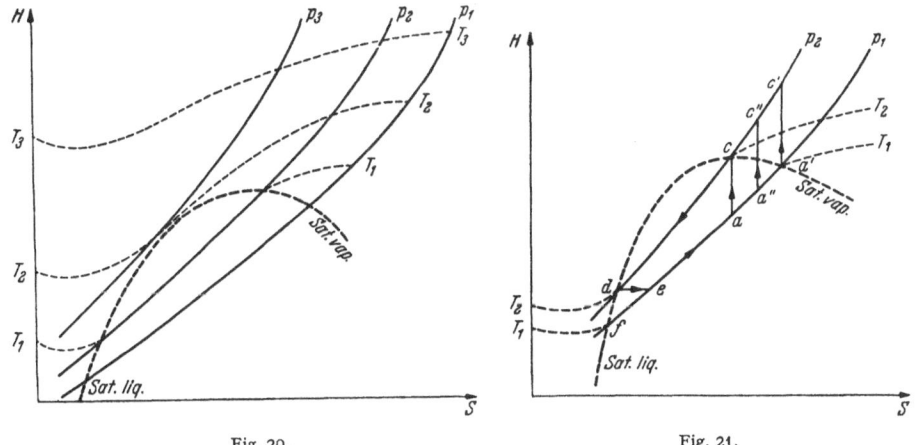

Fig. 20.   Fig. 21.

Fig. 20. Schematic diagram of $H-S$ MOLLIER diagram, showing typical isobars and isotherms. The heavy broken curve marks the boundary of the two-phase heterogeneous region.

Fig. 21. $H-S$ MOLLIER diagram showing thermodynamic cycle of operations for idealized compressed vapor refrigerating machine. (The unprimed letters are for "wet" compression; the single primed letters are for "dry" compression and the double primed letters are for mixed compression.)

the throttling valve is represented by the horizontal path $d \rightarrow e$, and the subsequent evaporation at constant temperature in the condenser is represented by the line $e \rightarrow a$.

The coefficient of performance, ξ, can be obtained, if it assumed that there are no losses, in the following manner.

The heat, Q_1, absorbed in the evaporator at constant pressure and temperature must by definition be:

$$Q_1 = H_a - H_e \qquad (8.1)$$

and the heat, Q_2, given up to the condenser is:

$$Q_2 = H_c - H_d \qquad (8.2)$$

where the subscripts refer to the value of the enthalpy, H, at the appropriately lettered point.

Remembering that $H_d = H_e$ and that the net work done on the machine, W, (neglecting dissipational losses), is given by $W = Q_2 - Q_1$, we obtain for the coefficient of performance:

$$\xi = \frac{Q_1}{W} = \frac{H_a - H_e}{H_c - H_a}. \qquad (8.3)$$

These values of H can be immediately determined by consulting the MOLLIER diagram for the working substance of interest, once the working pressures are known.

In the case of a dry-compression refrigerating machine, which is the one usually adopted in practice, the cycle as plotted on the MOLLIER diagram is essentially the same as that described above, except that now the starting point, a', must be on the saturated vapor line, and that the end of the compression must be represented by c' on the p_2 isobar vertically above a'. The value of the coefficient of per-

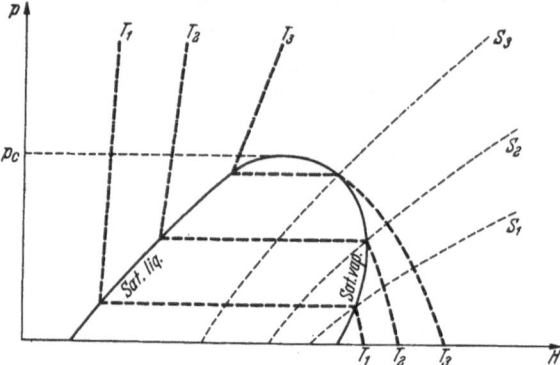

Fig. 22. Schematic diagram of p—H MOLLIER diagram, showing typical isotherms and lines of constant entropy (S). p_c is the critical pressure of the fluid.

formance, ξ, is given by eq. (8.3) with the appropriate primed values substituted. Again note that a knowledge of p_1 and p_2 is still sufficient to determine ξ. If the machine is not an ideal wet-compression or dry compression type, but is a mixture of the two types, then the cycle of operations would be represented on Fig. 21 by $a'' \to c'' \to d \to e \to a''$, where the compression $a'' \to c''$ is again assumed isentropic. In this case to calculate ξ, either the temperature corresponding to the point c'' must be known, or alternatively the "wetness" of the vapor entering the compressor must be known in order to fix the point a''. In practice, as is discussed in more detail in the below, the compression is never ideally isentropic and hence both the above facts must be known as well as p_1 and p_2 in order to evaluate ξ.

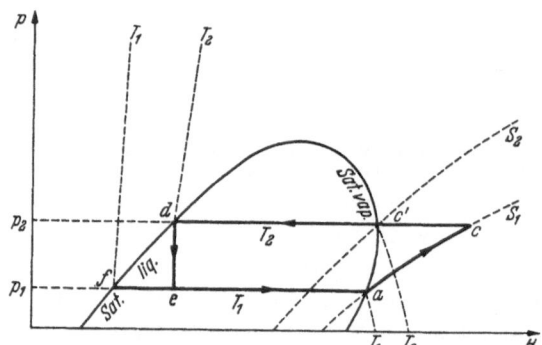

Fig. 23. p—H MOLLIER diagram showing thermodynamical cycle of operations for idealized compressed vapor refrigerating machine using "dry" compression.

9. Use of the $p-H$ MOLLIER Diagram.
The $p-H$ thermodynamic diagram, introduced also be MOLLIER[1], is of great practical value in assessing the theoretical coefficients of performance of compressed vapor refrigerators. The diagrams, which plot the pressure p, against the enthalpy, H, include lines of constant temperature, constant entropy and constant volume. A simplified picture of such a diagram is given in Fig. 22. The full drawn dome-like line encloses the two phase (liquid and vapor) heterogeneous region, the maximum of which gives the critical pressure, p_c. The heavy broken lines are those for constant temperature; $T_1 < T_2 < T_3$. Inside the heterogeneous region the isotherms are horizontal. The light broken lines are those for constant entropy; $S_1 > S_2 > S_3$.

A dry-compression cycle of a compressed-vapor refrigerator can be followed on such a MOLLIER diagram as is shown in Fig. 23. Starting with saturated

---

[1] R. MOLLIER: Z. VDI **48**, 271 (1904). For references to such diagrams for common refrigerating substances see Table 3.

vapor at the evaporator temperature, T_1, and pressure, p_1, at the point a, adiabatic compression carries the substance along the isentrop from a to c, where c corresponds to the intersection with the isobar, p_2, which is the condenser pressure. The influx at constant pressure, p_2, to the condenser from the compressor is given by the horizontal path $c \rightarrow c'$ and the subsequent condensation at constant pressure in the condenser at temperature, T_2, by the horizontal path $c' \rightarrow d$. The isenthalpic expansion through the valve is given by the vertical line $d \rightarrow e$ and the subsequent evaporation at constant pressure p_2 in the evaporator by the line $e \rightarrow a$. The fraction ε liquefied in the evaporator is given by

$$\varepsilon = \frac{a\,e}{a\,f}. \tag{9.1}$$

To compute the theoretical coefficient of performance it is necessary to know only T_1 and T_2 whence the values of H_a, H_c, and H_e can be immediately read from the diagram. These values inserted into eq. (8.3) give ξ_{theor}. For further detail on the use of these $p - H$ diagrams in practice, OPHULS' article in the ASRE Refrigerating Data Book[1] may be consulted.

10. Practical single stages systems, their working substances and efficiencies. In practice single stage compressed vapor refrigerators, as have been discussed above, may be put to a variety of uses for the extraction of heat at low temperature. For example the domestic refrigerator and air conditioning systems perform the extraction of heat at about $0°$ C; the so-called domestic "deep-freezer" at about $-15°$ C. Extraction of heat at still lower temperatures, however, by such machines or in similar multistage and cascade types is of more interest here, as for example is performed in the liquefaction of other substances and in the precooling stages of air liquefiers, whether of the LINDE or Cascade types.

Since in single stage compressed vapor refrigerators it is convenient to have the temperature, T_2, at which heat is absorbed from the working substance by the condenser only a little higher than ambient temperature ($20°$ C), the following criteria can be imposed in making a choice of working substance suitable for the machine:

a) Ambient temperature (T_2) must be less than the critical temperature, T_c, in order to liquefy the vapor by compression, and furthermore the pressure, p_2, required for this liquefaction must not be abnormally high.

b) It is convenient to have the temperature of the evaporator, T_1, around the boiling point, T_B, of the working substance, so that the pressure, p_1, of evaporation may be approximately atmospheric. This is, however, not a strict requirement. The only essential requirement here is that T_1 shall be larger than the temperature of the triple point, T_{Tr}, in order to avoid solidification of the working substance.

c) The substance must be chosen to give a small value for the compression ratio p_2/p_1. This is necessary because the work of compression increases rapidly with the compression ratio. This question is intimately connected with the choice imposed on the value of the evaporator temperature T_1.

d) Within the limitations set out in items a), b), and c) above, the choice of working substance is finally decided by a compromise between the thermodynamic properties (i.e. high coefficient of performance, etc.) and practical considerations governing the use of the machine (for example some substances with good thermodynamic properties are very corrosive or alternatively highly toxic when released from machine).

---

[1] F. OPHULS: ASRE Refrigerating Data Book, 7th Ed., Part. I, p. 11. 1951.

Table 4. *Characteristics of refrigerants used in compressed vapor machines operating between the following temperatures: Evaporation — 15° C; Condenser 30° C* [1].

	Ammonia NH₃	Carbon Dioxide CO₂	Sulphur Dioxide SO₂	Methyl Chloride CH₃Cl	Freon 12 CCl₂F₂
1. Condenser pressure (atm.) at 30.0° C (p_2 of Fig. 21)	11.52	70.8	4.52	6.34	7.36
2. Evaporator pressure (atm.) at — 15° C (p_1 of Fig. 21)	2.33	22.7	0.805	1.42	1.80
3. Compression ratio, r	4.95	3.13	5.63	4.46	4.09
4. Enthalpy [2] (cal/g.) of saturated vapor (leaving evaporator) at — 15° C (H_a of Fig. 21)	341.0	77.1	102.0	108.5	43.8
5. Enthalpy [2] (cal/g.) of liquid entering evaporator at — 15° C (H_e of Fig. 21)	77.3	46.3	23.4	25.9	15.4
6. Net refrigerative effect, Q_1 (cal/g.) · [($H_a - H_e$) of Fig. 21]	263.7	30.8	78.6	82.6	28.4
7. Coefficient of performance ξ (dry compression, neglecting losses) [3]	4.7	2.88	4.24	4.82	4.65
8. Refrigeration in cals per liter of piston displacement	306	1880	517	292	195
9. Practical hazard	toxic	high pressure	very toxic corrosive	toxic flamable	none

The above considerations led to the early adoption for machines operating with evaporator temperatures above about — 50° C of the following long used working substances: ammonia, carbon dioxide and sulphur dioxide [4]. A comparison between these substances can be made from Table 4 which sets out their various characteristics when used in compressed vapor refrigerators operating between 30° C and —15° C. From the table it will be seen that the coefficients of performance and the compression ratios for NH_3 and SO_2, as well as for the more recent substances used, CH_3Cl and CCl_2F_2, are about the same, whereas for CO_2 their values are somewhat smaller.

From the point of view of the smallness of the necessary piston displacement per unit of refrigeration, CO_2 is outstanding. CO_2 also has the advantage of being

Table 5. *Theoretical coefficients of performance (i.e. losses neglected) in an ammonia vapor compression refrigerator (dry compression) with condenser temperature at 30° C for various values of the evaporator temperature, T_1. Also given are the evaporator pressures, the compression ratio, and the efficiency, η_{rel} relative to a* CARNOT *cycle operating between the same temperatures.*

Temperatute of evaporator T_1 °C	Pressure in evaporator p_1 (atm.)	Compression ratio, r p_2/p_1	Theoretical coefficient of performance ξ	Theoretical efficiency relative to CARNOT cycle η_{rel}
0	4.23	2.7	8.1	0.89
— 10	2.87	4.0	5.6	0.85
— 20	1.88	6.1	4.1	0.815
— 30	1.18	9.75	3.15	0.78
— 40	0.71	16.2	2.45	0.735
— 50	0.39	28.5	1.9	0.685

---

[1] Some of this data is taken from L. S. Morse. 7th Int. Cong. Refrig. **3**, 718 (1937).

[2] Enthalpy measured from — 40° C.

[3] For ideal CARNOT cycle operating between these temperatures, $\xi^{max} = 5.75$. A value of $\xi = 1$ corresponds to an efficiency of 0.212 "tons" per h.p.

[4] See E. Griffiths and J. H. Asbery: Proc. Brit. Assoc. Refrig. Mar. 1925 for references to early literature.

non-toxic and non-combustible. It is used therefore where these practical assets are of importance. However the high pressures needed in the CO_2 cycle and its poorer coefficient of performance compared with the other substances listed in Table 4 make it to be not of general application. For more detailed comparisons of various working substances and of their relative efficiencies see KEESOM[1] and PLANK[2].

The very high value of the calculated (neglecting losses) efficiency of the ammonia machine relative to the ideal CARNOT efficiency ($\eta_{rel} = 0.82$) is to be noted[3]. It is this factor that has made the compressed vapor machine preferable to old-types of gas (air) refrigerating machines.

In reaching to temperatures lower than the example of $-15°$ C given above with single stage machines the coefficient of performance, as noted above, diminishes and the compression ratio increases. Table 5 and Fig. 24 show the way the theoretical (neglecting losses) value of the coefficient of performance, ξ, of a compressed vapor refrigerator using ammonia in dry compression varies with the lowering of the evaporator temperature, T_1, from $0°$ C to $-50°$ C. For these calculations, as for Table 4, the condenser temperature, T_2, was arbitrarily fixed at $30°$ C, a figure commonly used in practice. It is clear from the Table 5 that ξ goes to quite low values as T_1 is reduced, and moreover the theoretical efficiency, η_{rel}, relative to a CARNOT cycle operating between the same temperatures also decreases as T_1 decreases. For wet compression machines the values of ξ are slightly larger than those given here for the dry compression.

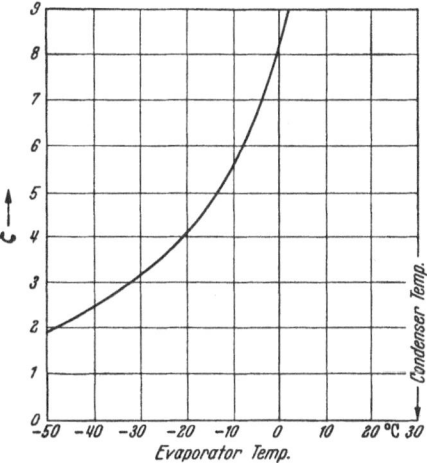

Fig. 24. The theoretical (i.e. loss-free) value of the coefficient of performance, ξ, as a function of the evaporator temperature, T_1, for a dry cycle ammonia compressed vapor refrigerating machine. The condenser temperature, T_2, is taken to be $30°$ C.

It should be noted also from Table 5 that the compression ratio, r, for the lower temperatures of evaporator operation reaches values (≈ 30 for $T_1 = -50°$ C) which are much too high for single stage compressors. For operation therefore to $-50°$ C and below, multi-stage systems are necessary, as are described later. For single-stage systems the compression ratio must be less than about 8 or 9 for reasonable practical efficiency and consequently for evaporator operation below about $-20°$ C, substances more favorable than ammonia must be looked for.

A further reason for choosing a more favorable substance lies in the choice of the evaporator pressure, p_1. If this is below atmospheric pressure, as it is for ammonia below $-33°$ C, there is the practical disadvantage of the possibility of air leaking into the apparatus and degrading its performance and of the necessity for large volume compressors, with consequent increased frictional losses. In order to overcome this second problem, at least partially, suitable substances must be sought which have lower boiling points than ammonia. This is also

---

[1] W. H. KEESOM: Leiden Comm. Suppl. **76a** (1933).

[2] R. PLANK: Z. ges. Kälteind. **47**, 81 (1940).

[3] In refrigerating engineering the common unit of refrigerating capacity is the "ton", derived from the average rate of heat absorption required to freeze 2000 lbs. (1 ton) of ice from water at the melting point every twenty four hours. The "ton" therefore is equivalent to 840 cals/sec. or 200 b.t.u./min. It is common also to express the coefficient of performance, ξ, in "tons" per horse-power. A value of $\xi = 1$ corresponds to 0.212 "tons"/h.p.

the criterion for choice of substances giving lower compression ratios, as has been pointed out by DAVIES[1], as follows:

The vapor pressure of the many substances of interest can be approximately expressed by

$$\ln p = A - \frac{L}{RT} \qquad (10.1)$$

where L is the latent heat of evaporation per mole. This equation is valid only over a relatively small range of temperature, over which L may be assumed constant. Near the critical temperature where $L \to 0$, the equation cannot be satisfactorily applied. The compression ratio, r, therefore is given by:

$$\ln r = \ln\left(\frac{p_2}{p_1}\right) = -\frac{L}{R}\left(\frac{T_1 - T_2}{T_1 T_2}\right). \qquad (10.2)$$

For a given T_1 and T_2, therefore, r will be diminished by choice of substances with decreasing values of their latent heats of evaporation. Since, according to TROUTON's rule the ratio of L (cal/mole) to the boiling point, T_B, in degrees K, is approximately constant and equal to the value 21, low L values correspond to low boiling points. Therefore, from eq. (10.2), the compression ratios, r, should diminish with diminishing T_B.

For the boiling points of various substances suitable for compressed vapor refrigerators, Table 3 should be consulted in which the substances are arranged consecutively in order of their boiling points, the highest boiling point being at the top of the list. The six substances listed, all having boiling points above that of SO_2, are most suitable for relatively high temperature refrigeration, such as is encountered in air conditioning, refrigerated transportation, etc. For the others, Table 6 lists both the evaporator pressure, p_1, and the compression ratio, r, for operation in dry compression between 30° C and $-50°$ C. It is clear from this table that the lower boiling point substances have compression ratios which would allow single-stage machines to be succesfully operated. However, in practice, as is outlined below, two-stage machines are more economical for operation to $-50°$ C and below.

Table 6. *Evaporator pressure, p_1; compression ratios, $r = p_2/p_1$, for the lower boiling point substances used in compressed vapor machines operating in dry compression between 30° C and $-50°$ C.*

Substance		p_1 (atm.)	r
Methyl Chloride	CH_3Cl	0.27	23.7
Freon 12	CCl_2F_2	0.375	19.5
Ammonia	NH_3	0.39	29.5
Freon 22	$CHClF_2$	0.64	18.6
Propane	C_3H_8	0.69	15.2
Propylene	C_3H_6	0.90	13.9
Carbon Dioxide	CO_2	6.73	10.5
Ethane	C_2H_6	5.45	8.4
Nitrous Oxide	N_2O	6.45	9.8

It is noted that the last two substances listed in Table 3, namely ethylene and methane have critical points well below 30° C and hence cannot be included in the comparison of Table 6. These substances are of great value in cascade vapor refrigeration where they can be operated with condenser temperatures well below their critical points.

Table 3 includes many working substances for relatively high temperature evaporation which have been introduced within the past two and one half decades. For example dichlorodifluoromethane (CCl_2F_2), known as Freon 12,

---

[1] M. DAVIES: The physical principles of gas liquefaction and low temperature rectification. London: Longmans, Green & Co. 1949.

was introduced[1] in 1930; and many organic chlorofluoride refrigerants have been subsequently developed which have been given the generic name of Freon[2]. For evaporator temperatures down to about $-15°$ C. Methyl Chloride[3] and Freon 12 compare very favorably with ammonia, as is shown in Table 4. The virtue of Freon 12, besides its general lack of practical hazards, is that its thermodynamic properties are very similar to those of NH_3 and hence it can be used to replace NH_3 in existing machines with standard positive compressors without any radical changes being necessary. It is to be noted that Freon 22 ($CHClF_2$) has a still lower boiling point than Freon 12 (see Table 3) and Freon 13 ($CClF_3$) (not listed in Table 3) has a boiling point of 192° K and both therefore are suitable in cascade vapor compression refrigeration for operation to about $-70°$ C.

This discussion of single stage compressed vapor refrigerators, has been up to now a theoretical one in which losses have been neglected. Before closing, therefore, some comment should be made on the principal sources of loss, which reduce the coefficients of performance below the theoretical values, and which are as follows:

a) Losses in the condenser. In order for the heat of condensation to flow to the condenser coolant a temperature difference must exist between the working substance of the refrigerator and the coolant. Hence the condenser temperature, T_2, of the working substance must be higher than for an ideal reversible system.

b) Losses in the evaporator. For similar reasons to those stated in a) above, the temperature, T_1 of the working substance in the evaporator must be less than that of the material being cooled. This again means η_{rel} is diminished.

c) Loss in the piping etc. Frictional losses in the piping between the components of the machine and in the compressor passages must occur and these represent an extra load on the compressor.

d) Unwanted heat transfer. In the piping, unwanted heat transfer may occur due to imperfect insulation which tends to reduce the refrigerative effort.

e) Mechanical losses in compressor and motor.

f) Non-ideality of compression. In compression heat transfer occurs between the gas and the compressor walls and this means that in the initial period of compression, when the gas is below ambient temperature, the gas receives unwanted heat. This heating of the gas, in excess of that due to the compressional work done on it, in practice more than counterbalances the outflow of heat through transference to the compressor walls at the final period of compression. As a result the compression is not isentropic, but results in a net entropy increase of the gas. This is illustrated in Fig 19 in which the path $a \rightarrow c$ represents the ideal adiabatic compression and in which the path $a \rightarrow c_i$ represents the irreversible actual non-isentropic compression. Since $H_{c_i} > H_c$, the coefficient of performance, as given by eq. (8.3) is reduced below the theoretical value.

The losses due to non-ideality of compression can be reduced by multistage compression, which also allows a closer approach to thermodynamic perfection by the use of multiple expansion valves (see below). The other losses are inherent and in general reduce the actual coefficient of performance to between 0.6 and 0.8 times the theoretical value of ξ.

---

[1] T. MIDGLEY and A. L. HENNE: Industr. Engng. Chem. 22, 542 (1930). See also R. J. THOMPSON: Industr. Engng. Chem. 24, 620 (1932) for further early description.

[2] See for example T. MIDGLEY: J. Industr. Engng. Chem., Feb. 1937, and publications of Kinetic Chemicals (E. I. duPont de Nemours and Company).

[3] See also publications of R. and H. Chemical Dept. E.I. duPont de Nemours and Company.

11. Multistage systems for operation to −50° C and below. One difficulty generally associated with single stage vapor compression refrigerators, certainly those using ammonia or similar working substances, was that for operation to − 50°C the compression ratio becomes uncomfortably high. It is convenient there-fore to carry out the com-pression in several stages, by which one also gains from the thermodynamical point of view[1]. In a multi-stage compressor system, however, it is also possible to gain greater thermody-namic efficiency by intro-ducing also multiple expan-sion. It is at the isenthalpic expansion where irrever-sibilities are inevitably in-troduced, and it is thermo-dynamically axiomatic that if these irreversibilities are introduced by a sum of a succession of small temper-ature drops rather than by one large one, the total irreversibility is reduced. In such a multi-stage com-pression, multiple expan-sion system, the vapor fraction appearing after each expansion is returned at its equilibrium pressure to a junction between stages of compression.

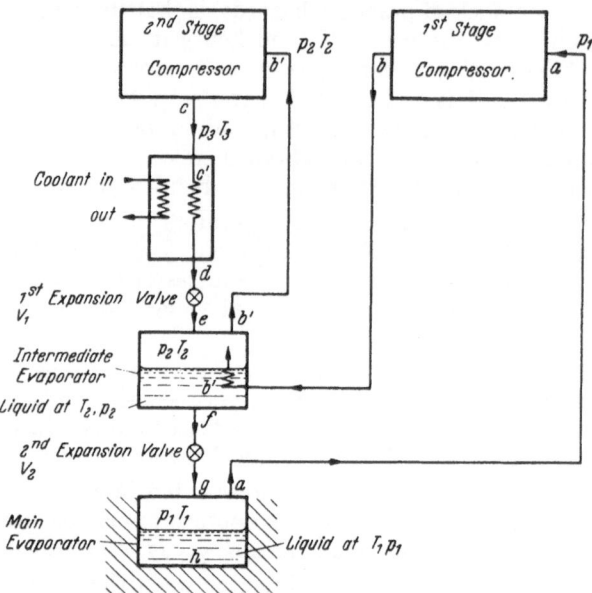

Fig. 25. Flow diagram for idealized two-stage compressed vapor refrigerat-ing machine using also two-stage expansion.

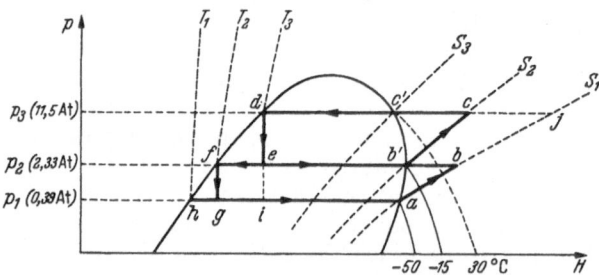

Fig. 26. p—H MOLLIER diagram showing thermodynamic cycle of operations for idealized two-stage compressed vapor refrigerating machine of Fig. 25.

The multi-stage com-pression, multiple expan-sion, vapor compression machine is illustrated in principle by the two stage system of Fig. 25. The whole output of the second stage at pressure p_3 at c is led into the condenser at c' and liq-uefied at d at the condenser temperature, T_3. On passing the first expansion valve V_1, it is partly liquefied in the intermediate evaporator at p_2 and T_2. The remaining vapor fraction passes via b' back to the input of the second stage of compression at pressure, p_2. The liquid is further expanded through the second valve, V_2, to p_1 and T_1 and the liquid fraction collected in the main evaporator where it can absorb heat from the material to be refrigerated. The evaporating liquid, leaving at a at pressure p_1 is fed to the input side of the first stage of the com-pressor. The output from this first stage at pressure p_2 is cooled to the saturation temperature T_2 by passing it through the intermediate evaporator at b'.

[1] See for example N.R. SPARKS: Theory of Mechanical Refrigeration. New York: McGraw Hill (1938) for a discussion of the relative efficiencies of multi-stage compression, single ex-pansion, systems.

The cycle of operations can be followed on the $p-H$ MOLLIER diagram in Fig. 26 (here the lower case letters correspond to the equivalent points on Fig. 25). The cycle on this diagram is: Second stage adiabatic compression (*all* the gas) from $b' \to c$; condensation of *all* the gas in condenser from $c \to c' \to d$; expansion through V_1 from $d \to e$; return vapor at intermediate pressure, p_2, from $e \to b'$; liquid in the intermediate evaporator corresponds to the point f. Next for the expansion at valve V_2 the path is $f \to g$; the return vapor is from $g \to a$; the adiabatic compression of the fraction x of the gas in the first stage of the compressor is from $a \to b$ and finally the gas comes back to b' by cooling in the intermediate evaporator.

The numerical values of pressure and temperature noted on Fig. 26 are for a typical NH_3 system.

It should be noted that, whereas all the gas passes through the second stage of compression, only a fraction x passes through the first. Evaluation of x is made as follows:

At the expansion at V_1 from $d \to e$, a fraction, y, of liquid is formed given by:

$$y = b' e/b' f. \tag{11.1}$$

This however is *not* the fraction, x, of substance subsequently passing on through V_2 from f to g. This latter fraction, is given by the heat balance equation for the adiabatically isolated intermediate evaporator, which is obtained as follows: The heat developed in the intermediate evaporator, Q_2, by cooling of the fraction x from $b \to b'$ is:

$$Q_2 = x(H_b - H_{b'}) \tag{11.2}$$

and this must equal to the heat extracted by evaporation in the intermediate evaporator in amount:

$$Q_2 = (y - x)(H_{b'} - H_f). \tag{11.3}$$

By combining eqs. (11.1), (11.2) and (11.3) we get

$$x = \frac{H_{b'} - H_e}{H_b - H_f} = \frac{H_{b'} - H_d}{H_b - H_f}. \tag{11.4}$$

In addition the fraction, z, of the substance passing through V_2 which is thereby liquefied is

$$z = \frac{H_a - H_g}{H_a - H_h} = \frac{H_a - H_f}{H_a - H_h} \tag{11.5}$$

so that the final fraction, ε, of *all* the gas flow out of the compressor (second stage) which is liquefied in the main evaporator is:

$$\varepsilon = x z. \tag{11.6}$$

To compute the overall coefficient of performance, ξ, of this two-stage machine, we note that the heat absorbed, Q_1, at the low temperature, T_1, is:

$$Q_1 = x(H_a - H_g) \tag{11.7}$$

and the heat, Q_3, finally given out in the condenser at T_3 is:

$$Q_3 = H_c - H_d. \tag{11.8}$$

The work done, W, therefore is:

$$W = Q_3 - Q_1 = (H_c - H_d) - x(H_a - H_g) \tag{11.9}$$

and the coefficient of performance, ξ is given by:

$$\frac{1}{\xi} = \frac{W}{Q_1} = \frac{H_c - H_d}{H_a - H_g} \cdot \frac{1}{x} - 1. \tag{11.10}$$

For the conditions indicated in Fig. 26 for the two-stage multiple expansion NH_3 machine (i.e. $T_1 = -50°$ C, $T_2 = -15°$ C and $T_3 = 30°$ C) the following values are found for no loss conditions:

$$x = 0.735; \quad z = 0.89; \quad \varepsilon = 0.655; \quad \xi = 2.26; \quad \eta_{rel} = 0.81. \tag{11.11}$$

It is of interest to compare this for a theoretical single-stage system, also using NH_3, operating between the same end temperatures, namely $-50°$ C and $30°$ C. Here it is assumed that the compression ratio of 29.5 can be maintained and that the system is also loss free. Such a system would be represented on Fig. 26 by the cycle $a \rightarrow j \rightarrow d \rightarrow i \rightarrow a$; and the fraction liquefied, ε, at T_1 would be ai/ah. For this cycle:

$$\varepsilon = 0.74; \quad \xi = 1.9; \quad \eta_{rel} = 0.685. \tag{11.12}$$

It is to be noted therefore that the fraction liquefied is greater here than in the two-stage system, indicating that per cal. of refrigeration at T_1 more substance must be circulated per minute in the two-stage system than for the single-stage machine. However, the two-stage machine is about 20% more efficient theoretically and approaches closely to the ideal CARNOT cycle.

Machines of this kind employing as working substance NH_3 and CO_2 and Freon are used for refrigeration to $-50°$ C and below[1], the first being due to LINDE in 1898. Three-stage compressor multiple expansion machines have been built for still lower temperature operation down to $-75°$ C.

12. Cascade compressed vapor systems and the liquefaction of air. In reaching for lower temperatures by vapor compression refrigeration historically the quest was to obtain final evaporator temperatures sufficiently low to liquefy air, nitrogen or oxygen by compression only. The critical temperatures of these so-called "permanent gases" are (cf. Table 8) respectively 132.5° K, 126° K and 154.3° K. Evaporator temperatures below $-147°$ C were therefore desirable. As has been discussed above, to approach lower evaporator temperature working substances with lower boiling points than NH_3, SO_2, etc. are required, such as are provided by ethylene and methane (cf. Table 3). Since these substances have themselves critical temperatures well below ambient, being 282.8° K for ethylene and 190.6° K for methane, it is necessary to provide condenser temperatures for their vapor compression cycles by the evaporators of other vapor-compression machines operating to higher temperatures. This is the so-called "cascade" system.

To make the most general illustration of the cascade system of refrigeration and gas liquefaction, KEESOM's model[2] for N_2 liquefaction will be described. In this system four substances are operated in cascade, the fourth (substance D of Fig. 27) being N_2. As shown in Fig. 27 the first cycle, using substance A, produces liquid A in evaporator A by the vapor-compression refrigeration process. The cold gas A leaving evaporator A exchanges its heat with the warm gas B leaving compressor B. Gas B is subsequently condensed by thermal contact with liquid A in evaporator A. Evaporator A therefore is also the condenser B

---

[1] See for example H. E. REX: Refrig. Engng. **58**, 566 (1950).
[2] W. H. KEESOM: Leiden Comm. Suppl. **76a** (1933).

for substance B. It is possible therefore to use as substance B a gas with a critical temperature only just greater than the temperature in evaporator A. After passing the throttling valve V_B, substance B collects as liquid in evaporator B, which in turn serves as condenser C for substance C. Finally the temperature maintained in evaporator C is sufficiently low for substance D to be condensed therein by compression alone. After condensation in the condensing or precooling spiral D in evaporator C, liquid D collects at lower pressure in reservoir D,

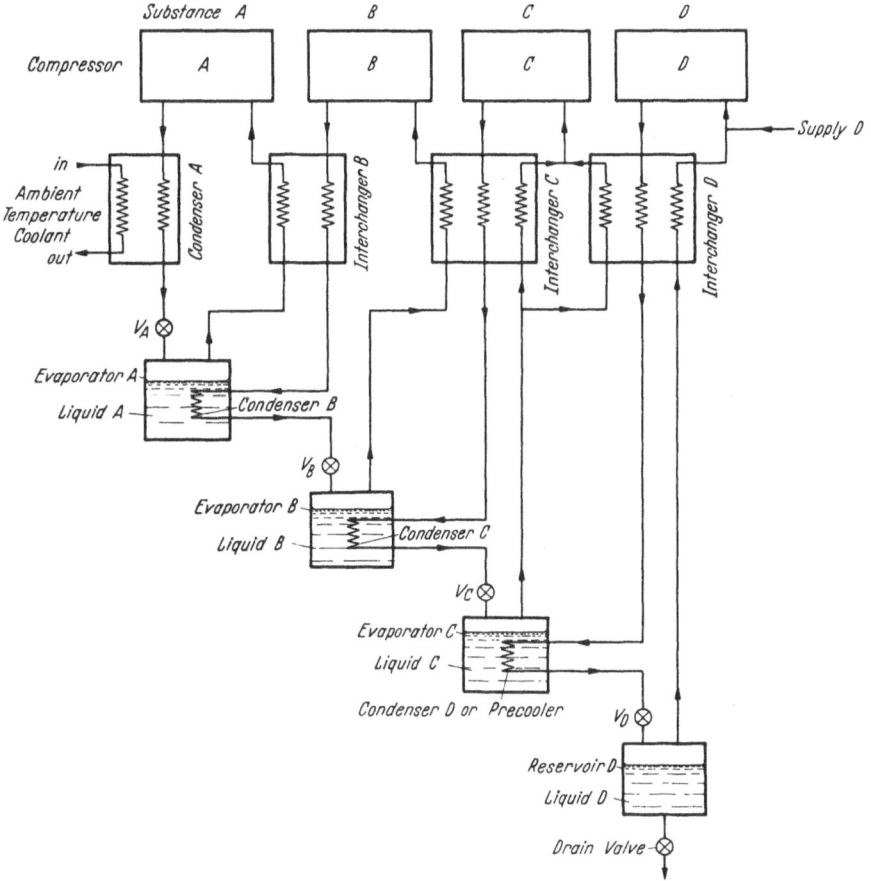

Fig. 27. Schematic flow diagram for idealized cascade compressed vapor refrigerating (and/or liquefying) machines using four different working substances in cascade.

wherefrom it may be withdrawn through the drain valve. It is to be noted that the efficiency will be enhanced greatly by use of the interchangers, B, C, and D whereby the incoming warm gases B, C, and D are cooled before reaching their respective condensers. Their introduction in the cascade process was due to KAMERLINGH-ONNES[1].

It will be seen therefore that each cycle of the system is a vapor-compression refrigerator, the efficiencies of which, as we have seen above, approach closely those of ideal CARNOT machines.

[1] H. KAMERLINGH-ONNES: Leiden Comm. **14** (1894); **87** (1903). — Leiden Comm. Suppl. **35** (1913). See also C. A. CROMMELIN: Leiden Comm. Suppl. **45** (1922).

In the system investigated by Keesom[1], which was previously successfully operated in practice[2], the substances *A*, *B*, *C*, and *D* were respectively ammonia (NH_3), ethylene (C_2H_4), methane (CH_4) and nitrogen (N_2). One of the advantages of this choice was that no substance had to evaporate at less than atmospheric pressure, thus allowing standard type compressors to be employed. Details of the temperatures and pressures of condensation and evaporation of the various substances in this system are given in Table 7.

Table 7 also shows the main features, giving working substances, temperatures und pressures (where known) of the historical landmarks in the development of the cascade process of air liquefaction[3]. Pictet's first attempt in 1877[4] is noted here. Using SO_2 and CO_2 in a double cascade system he claimed to have reached 133° K in the CO_2 evaporator. High pressure air at about 500 atm. passed in thermal contact with this evaporator and was expanded at a valve. No liquid was formed, only a jet of mist. It must be concluded, since the critical temperature and pressure for air are 132.5° K and 37.2° atm. (cf. Table 8), that the temperature of the CO_2 bath was much higher than 133° K. The Polish school under

Table 7. *Data on early cascade compressed vapor refrigerators for air, O_2 or N_2 liquefaction.*

Designer	Pictet[5]	Wroblewski and Olszewski[6]	Dewar[7]	K-Onnes[8]	Keesom[9]
Substance A	SO_2	+	+	CH_3Cl	NH_3
Condensed A at Temperature: °K	248			294	298
Condensed A at Pressure: (atm.)	2.75			15	10.2
Evaporated A at Temperature: °K	208	~175	195	186	240
Evaporated A at Pressure: (atm.)	≪1	~0.07	1	0.26	1
Substance B	CO_2	C_2H_4	C_2H_4	C_2H_4	C_2H_4
Condensed B at Temperature: °K	~208	~175	~195	~186	242
Condensed B at Pressure: (atm.)	~2.5	1.4	~3.7	3	19
Evaporated B at Temperature: °K	145 (?)	~123	~145	128	169
Evaporated B at Pressure: (atm.)	≪1	~0.13	<1	0.35	1
Substance C	not employed	not employed	not employed	O_2	CH_4
Condensed C at Temperature: °K				128	172
Condensed C at Pressure: (atm.)				17	24.7
Evaporated C at Temperature: °K				80	112
Evaporated C at Pressure: (atm.)				0.26	1
Substance D	air	O_2	O_2	air	N_2
Condensed D at Temperature: °K	not condensed*	~125	~145	80	114.6
Condensed D at Pressure: (atm.)	~350	up to 100	up to 75	1	18.6

[1] W. H. Keesom: Leiden Comm. Suppl. **76a** (1933).
[2] A. Huguenin: Festschrift zum 70. Geburtstag von Prof. A. Stodola, p. 272, Zürich 1929.
[3] See also R. Plank: Handbuch der Kältetechnik, Vol. I. 1954.
[4] R. Pictet: C. R. Acad. Sci. Paris **85**, 1214, 1220 (1877).
[5] R. Pictet: C. R. Acad. Sci. Paris **85**, 1214, 1220 (1877).
[6] K. Olszewski: Ann. Phys. u. Chem. **31**, 58 (1887). See also Phil. Mag. **39**, 188 (1895).
[7] J. Dewar: Proc. Roy. Inst., June **1886**. — Phil. Mag. **39**, 298 (1895).
[8] H. Kamerlingh-Onnes: Leiden Comm. **14** (1894); **87** (1903). — Leiden Comm. Suppl. **35** (1913). See also C. A. Crommelin: Leiden Comm. Suppl. **45** (1922).
[9] W. H. Keeson: Leiden Comm. Suppl. **76a** (1933).
+ Bath of CO_2 "snow" and ether was employed.
* An attempt to liquefy the O_2 by expansion at a valve after cooling to about 143° K was made. Only a "mist" emerged.

WROBLEWSKI and OLSZEWSKI[1] pioneered in the use of ethylene in the cascade, and employed a triple cascade system with a final ethylene evaporator at about 125° K by which O_2 was condensed under pressure. They were successful in obtaining sufficient quantities of liquid air, O_2 and N_2 to enable them to establish a considerable body of data concerning these liquids. DEWAR[2] gave brief details of a similar O_2 liquefier also using ethylene as the lowest temperature evaporator.

The first large scale air liquefier using the cascade system was due to KAMER-LINGH-ONNES[3] in 1894, which machine produced about 14 liters liquid per hour and remained in routine service for many years. Technically, however, it suffered from the drawback of using pressures much less than one atmosphere in all three evaporators, necessitating large-displacement pumps.

For the quadruple cascade system of KEESOM and HUGUENIN, using NH_3, C_2H_4, CH_4 and N_2 all evaporating at atmospheric pressure, the practical effi-ciency was about 0.54 kW-hr per kg. of N_2 liquefied. This corresponds to about 2.5 times the energy required to liquefy one liter of N_2 by an ideal reversible thermodynamic cycle. This value of 0.54 kW-hr/kg.-N_2 is much lower than the values obtained for air or N_2 liquefaction using LINDE or CLAUDE systems (cf. Tables 12 and 15) and it reflects the high efficiency of the vapor-compression type of refrigeration.

C. Cooling of gases by the JOULE-THOMSON effect.

13. Discussion of the cooling produced by the JOULE-THOMSON effect in gases. In 1852 JOULE and THOMSON[4] reported on their first observations made on the

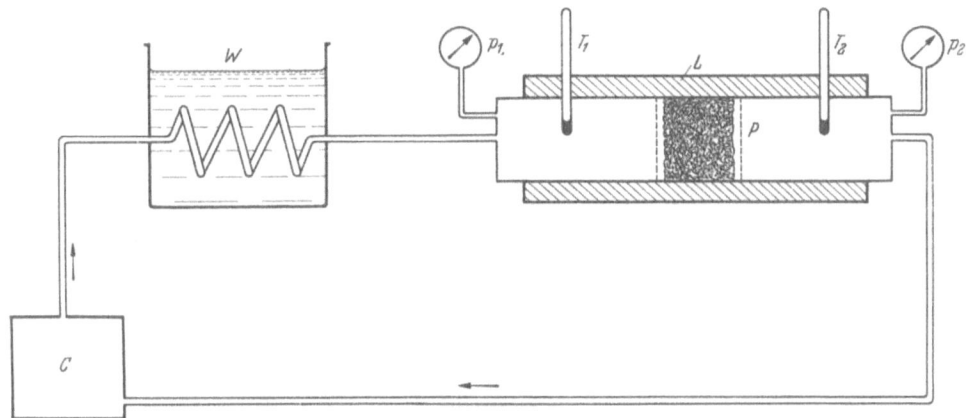

Fig. 28. Schematic diagram of apparatus used by JOULE and THOMSON for measurement of temperature changes at isenthalpic expansion (i.e. the porous-plug experiment).

senthalpic expansion of a gas streaming through a nozzle at approximately room temperature. Subsequent experimental arrangements used by them[5] for ex-periments on isenthalpic expansion employed a porous plug instead of a nozzle, as is diagrammatically shown in Fig. 28. The gas under investigation was circulated

---

[1] K. OLSZEWSKI: Ann. Phys. u. Chem. **31**, 58 (1887). See also Phil. Mag. **39**, 188 (1895) for a review of their work.

[2] J. DEWAR: Proc. Roy. Inst., June **1886**. — Phil. Mag. **39**, 298 (1895).

[3] H. KAMERLINGH-ONNES: Leiden Comm. 14 (1894); 87 (1903). — Leiden Comm. Suppl. **35** (1913). See also C. A. CROMMELIN: Leiden Comm. Suppl. **45** (1922).

[4] J. P. JOULE and W. THOMSON: Phil. Mag. **4**, 481 (1852). (William THOMSON assumed the name of Lord KELVIN in 1892.)

[5] J. P. JOULE: Sci. Pap. **2**, 216.

by the compressor, C, first through the water-cooling tank, W, to remove the heat of compression, then through the porous plug, P (originally of cotton), which was well insulated thermally from the surroundings by the lagging, L. After a little time a steady state was reached with the pressures p_1 and p_2 and the temperatures T_1 and T_2 attaining time-independent values.

Consider a unit mass of gas passing in this process from the high pressure p_1 to the low pressure p_2. At p_1 and T_1 let one mole of gas have volume v_1 and at p_2 and T_2 let the molar volume be v_2. Then the work ΔW done by the gas in expanding through the plug is:

$$\Delta W = p_2 v_2 - p_1 v_1 \qquad (13.1)$$

and since ideally no heat is considered to flow in or out due to the thermal isolation of the plug (adiabatic conditions) it must follow from the first law of thermodynamics that

$$\Delta W = U_1 - U_2 \qquad (13.2)$$

where U_1 and U_2 are the internal energies per mole at p_1, T_1 and p_2, T_2. Hence:

$$H_1 = U_1 + p_1 v_1 = U_2 + p_2 v_2 = H_2, \qquad (13.3)$$

i.e. the process is one at constant enthalpy.

It can be shown[1] thermodynamically moreover that the differential change in temperature across the plug per unit pressure difference, i.e. the so-called JOULE-THOMSON coefficient, α_H, is given by:

$$\alpha_H = \left(\frac{\partial T}{\partial p}\right)_H = \frac{1}{C_p}\left[T\left(\frac{\partial v}{\partial T}\right)_p - v\right] \qquad (13.4)$$

where C_p is the molar specific heat of the gas at constant pressure. Or, if $\Delta T = T_1 - T_2$ and $\Delta p = p_1 - p_2$ are relatively small,

$$\Delta T = \frac{1}{C_p}\left[T\left(\frac{\partial v}{\partial T}\right)_p - v\right]\Delta p. \qquad (13.5)$$

For a mole of a perfect gas $pv = RT$, whence $T\left(\frac{\partial v}{\partial T}\right)_p = v$ and hence the change in temperature, ΔT, in the JOULE-THOMSON experiment would be zero. JOULE and THOMSON found that for many gasses, i.e. air, CO_2, O_2, N_2, ..., the temperature, T_2, at the outlet side of the plug was less than that, T_1, at the inlet side. For air, for example, in the temperature range, $T = 273°$ K to $373°$ K, and the pressure range 1 to 6 atm. they found that:

$$\Delta T = 0.276 \left(\frac{273}{T}\right)^2 \Delta p. \qquad (13.6)$$

On the other hand for hydrogen they found that a heating effect took place (i.e. $T_2 > T_1$) rather than the cooling observed for other gases. Subsequent work[2] has shown that *all* gases show both heating and cooling effects depending

---

[1] See for example J. K. ROBERTS and A. R. MILLER: Heat and Thermodynamics, p. 105. London: Blackie & Son 1951.

[2] Some representative experimental work has been done by the following authors. K. OLSZEWSKI: Ann. Phys. 7, 818 (1902) on H_2. — F. E. KESTER: Phys. Rev. 21, 260 (1905) on CO_2. — K. OLSZEWSKI: Phil. Mag. 13, 722 (1907) on air and N_2. — W. P. BRADLEY and C. F. HALE: Phys. Rev. 29, 258 (1909) on air. — J. DALTON: Leiden Comm. 109c (1909) on air. — F. NOELL: Forsch. Ing.-Wes. 184 (1916) on air. — L. C. HOXTON: Phys. Rev. 13, 438 (1919) on air. — E. S. BURNETT: Phys. Rev. 22, 590 (1923) on CO_2. — J. R. ROEBUCK: Proc. Amer. Acad. Arts Sci 60, 537 (1925) on air. — N. EUMORFOPOULOS and J. RAI: Phil. Mag. 2, 961 (1926) on air. — H. HAUSEN: Forsch. Ing.-Wes. 274 (1926) on air. — J. R. ROEBUCK: Proc. Amer. Acad. Arts Sci. 64, 287 (1930) on air. — J. R. ROEBUCK and H. OSTERBERG: Phys. Rev. 43, 60 (1933) on He. — J. R. ROEBUCK and H. OSTERBERG: Phys. Rev. 46, 785 (1934) on A. — J. R. ROEBUCK and H. OSTERBERG: Phys. Rev. 48, 450 (1935) on N_2. — J. L. ZELMANOV: J. Phys. USSR. 3, 42 (1940) on He. — ROEBUCK, MURRELL

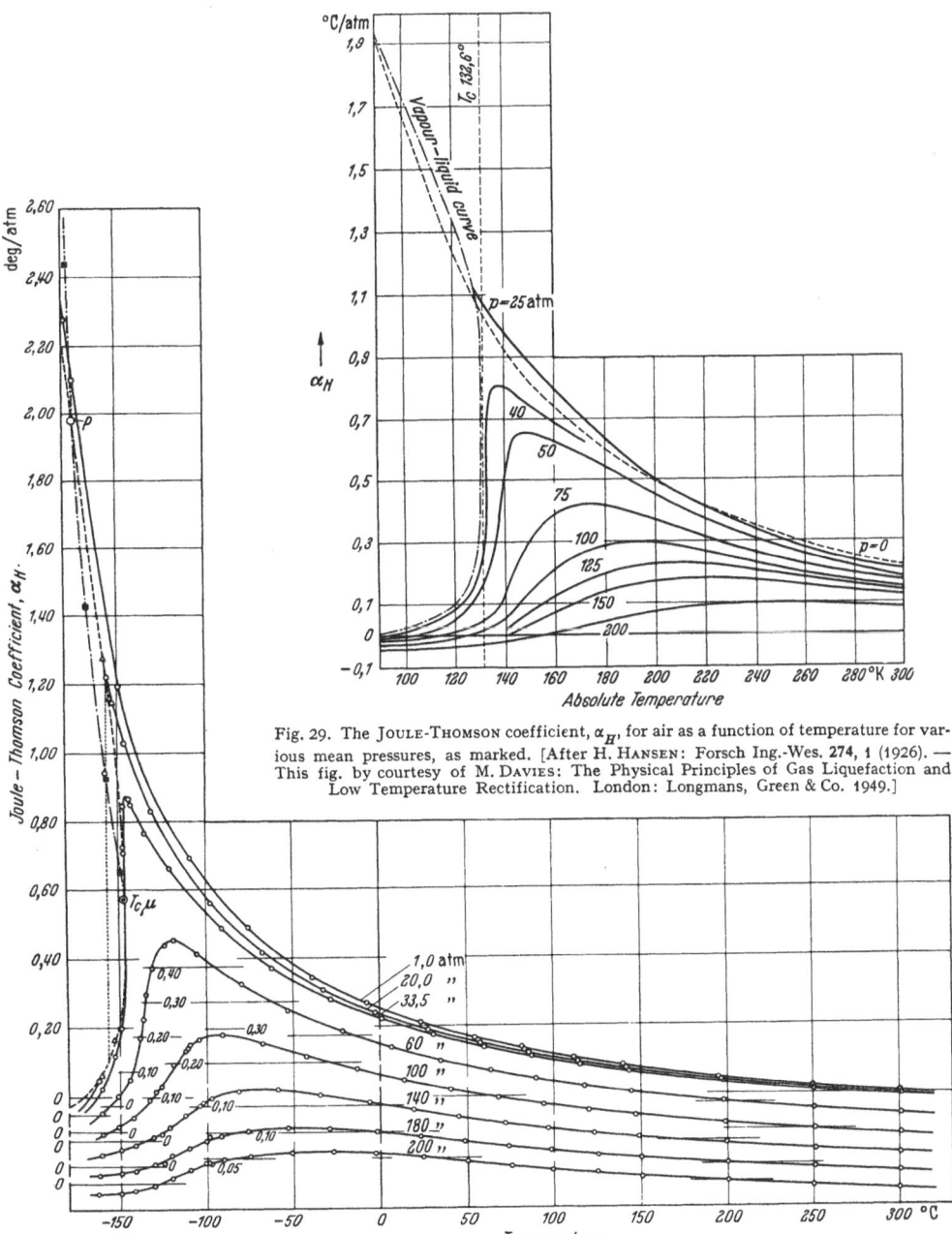

Fig. 29. The JOULE-THOMSON coefficient, α_H, for air as a function of temperature for various mean pressures, as marked. [After H. HANSEN: Forsch Ing.-Wes. 274, 1 (1926). — This fig. by courtesy of M. DAVIES: The Physical Principles of Gas Liquefaction and Low Temperature Rectification. London: Longmans, Green & Co. 1949.]

Fig. 30. The JOULE-THOMSON coefficient, α_H for N_2 as a function of temperature for various pressures, as marked. For the lower curves the zero has been shifted, as shown, to clarify the reading. [After J. R. ROEBUCK and H. OSTERBERG: Phys. Rev. 48, 450 (1935).]

and MILLER: J. Amer. Chem. Soc. 64, 400 (1942) on CO_2. — JOHNSTON, BEZMAN and HOOD: J. Amer. Chem. Soc. 68, 2367 (1946) on H_2. — JOHNSTON, SWANSEN and WIRTH: J. Amer. Chem. Soc. 68, 2373 (1946) on D_2. — CHARNLEY, ISLES and TOWNLEY: Proc. Roy. Soc. Lond., Ser. A 218, 133 (1953) on N_2, C_2H_4, CO_2, N_2O. — For a detailed review of the experimental work up to 1929 see H. LENZ, Handbuch der Experimentalphysik, vol. 9/1, p. 47, 1929 and A. EUCKEN, Handbuch der Experimentalphysik, vol. 8/1, p. 511, 1929.

on the absolute values of the input pressure and temperature. Curves showing the Joule-Thomson coefficient α_H, as a function of temperature for various pressures, for air, due to Hausen[1] and for N_2 due to Roebuck and Osterberg[2] are shown in Figs. 29 and 30. It will be seen that quite generally the cooling effect, i.e. the numerical value of α_H, decreases with increasing pressure. The behavior with variation of T is more complicated showing a maximum at low pressures near the critical temperature T_c and indicating that the position of this maximum shifts to higher temperatures as the pressure is increased. It is to be noticed (cf. Fig. 29) that for example in air at $p = 200$ atm., α_H is negative (i.e. there is a heating effect) if the initial temperature is below about $153°$ K. This behavior for air is typical for all gases. It is also observed that at sufficiently high temperatures, α_H is negative for all values of the pressures. (This is illustrated by Fig. 31 which gives the variation of α_H with temperature for air as measured by Roebuck[3].) The limiting temperature where for $p = 0$, α_H changes from positive to negative values is called the "Inversion Temperature", T_i. Above T_i, α_H is always negative for all values of the pressure p. The inversion temperatures, T_i, of many commonly used gases are given in Table 8 and from eq. (13.5)

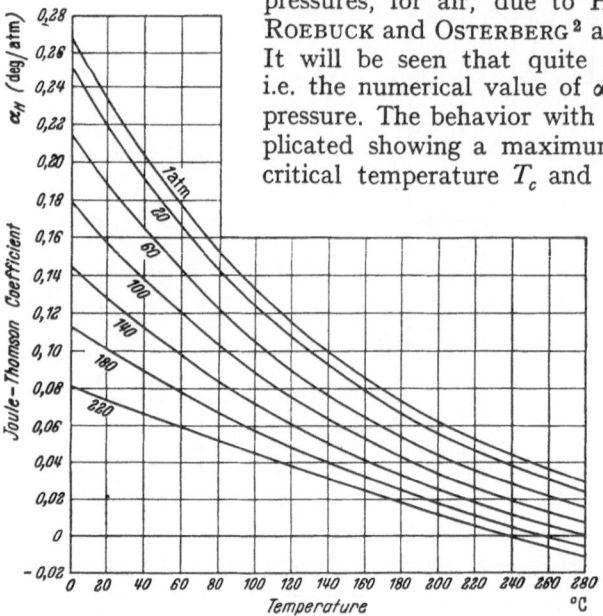

Fig. 31. The Joule-Thomson coefficient, α_H, for air as a function of temperature for various pressures, as marked. This shows the high temperature behavior where α_H tends to negative values. [After J. R. Roebuck: Proc. Amer. Acad. Arts Sci. 64, 287 (1930).]

Table 8.

Substance		Boiling point °K	Critical temperature °K	Critical pressure atm.	Inversion temperature °K	$r = \dfrac{RT_c}{p_c v_c}$	Triple point °K	Triple pressure Cm.Hg
Helium	He³	3.2	3.34	1.15				
Helium	He⁴	4.2	5.19	2.26	∼50.5(?)	3.27		
Hydrogen	H₂	20.4	33.2	12.98	204.6	3.13	13.96	5.4
Deuterium	D₂	23.7	38.8	17.4			18.73	12.86
Tritium	T₂	25.0	43.7*	20.8*			20.6	16.2
Neon	Ne	27.2	44.4	25.9		3.25	24.57	32.4
Nitrogen	N₂	77.3	126.0	33.5	621	3.42	63.1	9.29
Air (21% O₂)		80.1	132.5	37.2	603	3.42		
Carbon Monoxide CO		81.1	134.1	34.6				
Argon	A	87.4	151.1	48.0	723	3.43	83.9	51.2
Oxygen	O₂	90.1	154.3	49.7	893	3.42	54.4	0.12
Krypton	Kr	121.3	210.1	54.0		3.44	104	
Xenon	Xe	164.0	289.7	58.2		3.58	133	

[1] H. Hausen: Forsch. Ing.-Wes. **274**, 1 (1926).
[2] J. R. Roebuck and H. Osterberg: Phys. Rev. **48**, 450 (1935).
[3] J. R. Roebuck: Proc. Amer. Acad. Arts Sci. **64**, 287 (1930).
* Calculated. See E. F. Hammel: J. Chem. Phys. **18**, 228 (1950).

it will be seen that at T_i

$$T_i \left(\frac{\partial v}{\partial T} \right)_{p=0} = v. \tag{13.7}$$

From Table 8, the values of T_i for the gases originally used by JOULE and THOMSON (air, O_2, N_2) except H_2 are seen to be well above room temperature. For H_2 and He for example, however, $T_i = 204.6°$ K and $T_i \approx 50.5°$ K[1] respectively, and hence heating effects would be observed in JOULE-THOMSON experiments at room temperature.

14. The inversion curve. A calculation of the inversion temperature and an analysis of the way in which α_H changes from positive to negative values for variations of the parameters p and T can be made from eq. (13.7), provided the equation of state for the gas under consideration is known. As a first approximation, it is instructive to make such calculations using the VAN DER WAALS equation of state as representative of the gas, namely:

$$\left(p + \frac{a}{v^2} \right) (v - b) = RT \tag{14.1}$$

where a and b are the terms due to the attractive and repulsive components respectively of the mutual interactions between the molecules.

Now the condition that the JOULE-THOMSON effect should vanish, i.e. that $\alpha_H = 0$, is from eq. (13.7) given by

$$\left(\frac{\partial T}{\partial v} \right)_p = \frac{T}{v} \tag{14.2}$$

where the value of $(\partial T/\partial v)_p$ for the VAN DER WAALS gas can be obtained directly by differentiation of eq. (14.1). This leads, after a little work, to the result:

$$p = \frac{2}{3b} RT - \frac{1}{3a} RT^2 \tag{14.3}$$

as a first approximation valid for T large.

If "reduced" pressures, π, and temperatures, τ, are used such that $\pi = p/p_c$; $\tau = T/T_c$ and $r = RT_c/p_c v_c$ where p_c, v_c and T_c are the critical pressure, volume and temperature respectively, eq. (14.3) can be rewritten as:

$$\pi = 2r\tau - \tfrac{1}{9} r^2 \tau^2 \tag{14.4}$$

where again its validity is restricted to high values of τ only. When τ tends to the value 1, the relationship is considerably complicated. However, eq. (14.3) and (14.4) serve to show that the states of the gas, as represented by points on a p, T-diagram, for which the JOULE-THOMSON effect is zero, lie on an approximately parabolic curve. A graphical representation of this curve is given in Fig. 32 in which the broken curve is the locus of points for which $\alpha_H = 0$ for a gas obeying a VAN DER WAALS equation. Every point beneath the curve represents a state of the gas for which a cooling takes place in the JOULE-THOMSON effect ($\alpha_H > 0$), whereas for every point above the curve a heating takes place. The intercept with the τ axis at $\pi = 0$ at the high temperature side gives the inversion temperature, as previously defined. In reduced temperatures it will be seen that $\tau_i = 18/r$ for a VAN DER WAALS gas, as also is obtained from eq. (14.4). This serves to explain the curve of Fig. 31 for which, at temperatures higher than the inversion

---

[1] Owing to the paucity of adequate data this value is not too accurate. See W. H. KEESOM: Helium. Amsterdam: Elsevier 1942.

temperature, α_H is negative for all values of p. Fig. 32 also shows that there is a lower inversion temperature given when $\tau \approx 2.2/r$ for a VAN DER WAALS gas, a result which does not appear in eq. (14.4) due to its very approximate nature at low τ. For a VAN DER WAALS

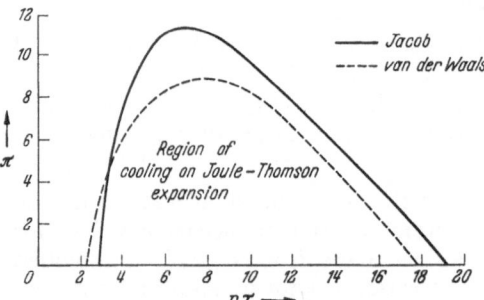

Fig. 32. The "reduced" inversion curve, giving the reduced pressure, π, as a function of the reduced temperature τ. [After M. JACOB: Phys. Z. 22, 65 (1921)]. The broken curve gives the reduced inversion curve for a VAN DER WAALS Gas. (This fig. by courtesy of M. DAVIES: The Physical Principles of Gas. Liquefaction and Low Temperature Rectification. London: Longmans, Green & Co. 1949.)

Table 9. *Values of the critical pressure, p_c, critical temperature, T_c, and r $(= RT_c/p_c v_c)$ used by* JACOB [Phys. Z. **22**, 65 (1926)] *for his reduced inversion curve.*

Substance	p_c atm,	T_c °K	r
CO_2 . .	72.4	304	3.53
C_2H_4 .	50.0	283	3.42
O_2 . .	49.7	154	3.42
air . .	37.2	132.5	3.42
N_2 . .	33.4	126	3.42
H_2* . .	12.8	33.2	3.27
He** .	2.26	5.19	3.27

gas therefore held at any temperature between $\tau = 2.2/r$ and $\tau = 18/r$ there is a cooling in the JOULE-THOMSON effect only for pressures less than some value

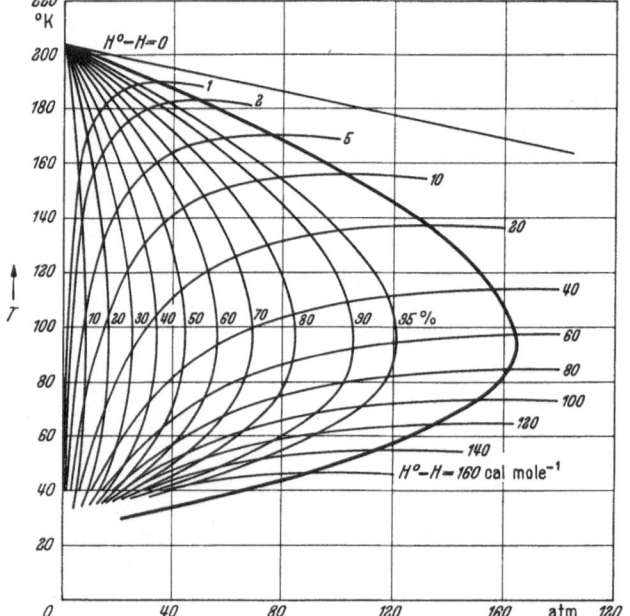

Fig. 33. Inversion curve (heavy full curve) for normal H_2. [After WOOLLEY, SCOTT and BRICKWEDDE: J. Res. Nat. Bur. Stand. 41, 379 (1948).] The light curves which cross the inversion curve horizontally are curves of constant refrigeration per mole of gas. The light curves roughly parallel to the inversion curve are lines at which the refrigeration is the marked percentage of the 100% on the inversion curve.

determined by the "inversion curve" of Fig. 32.

In practice, real gases cannot be represented by a reduced VAN DER WAALS equation of state. The experimental values of the inversion of the JOULE-THOMSON effect for a number of gases, namely CO_2, C_2H_4, O_2, air, N_2, H_2, have been collated by JACOB[1] and his results show that all these gases quoted except H_2 are describable by one reduced inversion curve as shown by the full curve in Fig. 32. The actual values of p and T can be obtained from this curve by using the values of p_c, T_c and r for each gas as tabulated in Table 9. JACOB describes the curve by the equation

$$\left. \begin{aligned} \pi = 23.37 - 1.174\, r\, \tau \\ -178.6/r^2\, \tau^2. \end{aligned} \right\} (14.5)$$

---

* MEISSNER [Z. Physik **18**, 12 (1923)] used the following values: $p_c = 12.80$ atm., $T_c = 33\ 18°$ K, $r = 3.276$, obtained by KAMERLINGH-ONNES, CROMMELIN and CATH [Leiden Comm. **151**c (1917)]. WOOLLEY, SCOTT and BRICKWEDDE [J. Res. Nat. Bur. Stand. **41**, 379 (1948)] give the following values: $p_c = 12.98$ atm.; $T_c = 33.19°$ K, and $r = 3.13$.

** These data come from ZELMANOV's work [J. Phys. USSR. **3**, 43 (1940)].

[1] M. JACOB: Phys. Z. **22**, 65 (1921).

The results of the measurements on argon by ROEBUCK and OSTERBERG[1] also fit JACOB's general reduced inversion curve.

The gases of very low atomic weight deviate most from the classical VAN DER WAALS law of corresponding states, due to quantum effects (see DE BOER[2]). Hence, whereas CO_2, C_2H_4, O_2, N_2, A and air all give an inversion curve approximately the same as found by JACOB, hydrogen and helium are markedly different and their inversion curves are shown separately in Figs. 33 and 34, as obtained from the work of WOOLLEY et al[3] and ZELMANOV[4].

15. Some physical considerations. It is of interest to consider the physical

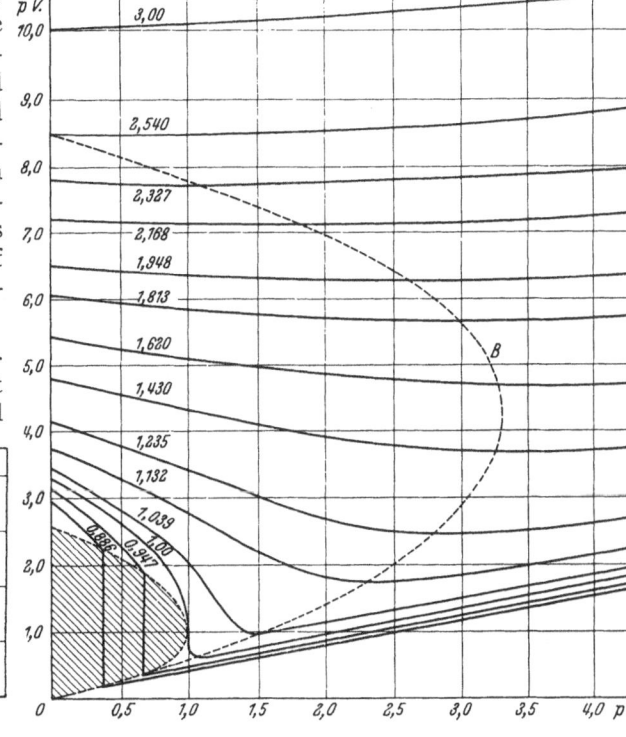

Fig. 34.

Fig. 35.

Fig. 34. The inversion curve for He, in reduced units (cf. Table 9). [After ZELMANOV: J. Phys. USSR. **3**, 43 (1940).] The reduced inversion curve for all gases other than H_2 and He due to JACOB [Phys. Z. **22**, 65 (1926)] is also shown. [The argon points are measurements by ROEBUCK and OSTERBERG: Phys. Rev. **46**, 785 (1934).]

Fig. 35. Typical isotherms on a pv versus p plot for a fluid in arbitrary units. (This diagram by courtesy of J. K. ROBERTS and A. R. MILLER: Heat and Thermodynamics. London: Blackie & Son Ltd. 1951.) The broken curve gives the locus of the minima, the upper intersection of which with the pV axis gives the BOYLE point. The shaded area gives the two phase heterogeneous region.

processes which lead to the inversion of the JOULE-THOMSON effect. In order to do this, the thermodynamic evaluation of α_H, as given in eq. (13.4), needs further development, as follows:

The enthalpy, $H = U + pv$ by definition, is given by:

$$\left. \begin{aligned} \left(\frac{\partial H}{\partial p}\right)_T &= \left(\frac{\partial U}{\partial p}\right)_T + p\left(\frac{\partial v}{\partial p}\right)_T + v \\ &= T\left(\frac{\partial S}{\partial p}\right)_T + v \quad (\text{Since } dU + p\,dv = T\,dS) \\ &= -T\left(\frac{\partial v}{\partial T}\right)_p + v \quad (\text{MAXWELL equation}) \\ &= -C_p \cdot \alpha_H . \end{aligned} \right\} \quad (15.1)$$

[1] J. R. ROEBUCK and H. OSTERBERG: Phys. Rev. **46**, 785 (1934).

[2] J. DE BOER et al.: Physica, Haag **14**, 139, 149, 520 (1948).

[3] WOOLLEY, SCOTT and BRICKWEDDE: J. Res. Nat. Bur. Stand. **41**, 379 (1948).

[4] J. L. ZELMANOV: J. Phys. USSR. **3**, 43 (1940). This work superceeds the earlier work of KEESOM and HOUTHOFF: Leiden Comm. Suppl. **65f** (1928).

One may write the JOULE-THOMSON coefficient, α_H, therefore, as:

$$\alpha_H \cdot C_p \equiv \left(\frac{\partial T}{\partial p}\right)_H \cdot C_p = -\left(\frac{\partial U}{\partial p}\right)_T - \left(\frac{\partial (pv)}{\partial p}\right)_T. \qquad (15.2)$$

The first term on the right hand side of eq. (15.2) would be zero for a perfect gas, since by JOULE's law the internal energy, U, of a perfect gas is a function of T only. Finite values of α_H, therefore, due to finite values of this first term, can be described physically as being due to deviations from JOULE's law. Now deviations from JOULE's law are due to the work done against the attractive forces between the molecules on expansion and are always such that an expansion produces a decrease in the internal energy. Hence $-(\partial U/\partial p)_T$ is always positive (or zero) and in consequence the process due to the deviations from JOULE's law always produces a cooling effect in the JOULE-THOMSON expansion.

The second term on the right hand side of eq. (15.2) would be zero for a gas obeying BOYLE's law. The deviation of a real gas from BOYLE's law can be such that $\left(\partial (pv)/\partial p\right)_T$ can be positive or negative dependent on the conditions, as is shown in Fig. 35, which gives a plot of isotherms on a pv versus p diagram typical for all gases[1]. The broken curve on the graph of Fig. 35 gives the locus of points where $\left(\partial (pv)/\partial p\right)_T = 0$, and the temperature corresponding to the isotherm which is horizontal at $p = 0$ [i.e. such that $\left(\partial (pv)/\partial p\right)_T = 0$ for $p = 0$] is called "The BOYLE Temperature", T_{Boyle}, of the substance. It will be seen therefore that for all temperatures above the BOYLE Temperature, T_{Boyle}, the term $-\left(\partial (pv)/\partial p\right)_T$ is always negative, and must contribute to a heating on JOULE-THOMSON expansion. For $T > T_{Boyle}$ therefore, whether there is a cooling or heating in JOULE-THOMSON expansion depends on the relative magnitudes of the two terms, one the cooling due to deviations from JOULE's law and the other the heating due to deviations from BOYLE's law. On the inversion curve where there is neither heating nor cooling, these two opposing effects exactly cancel out[2]. Moreover the negative values of $-\left(\partial (pv)/\partial p\right)_T$ can be regarded as being due to the compressibility being smaller than that of a perfect gas and hence as being conditioned by the repulsive forces between the molecules. For further details of the relative contribution of these two effects with air see HAUSEN[3].

D. Theory of gas liquefaction using JOULE-THOMSON expansion.

16. The ideal cycle of operations. In considering the operation of a machine for producing liquefied gases, attention should be paid to the fact that such a machine must possess twofold functions, first it must supply the gas to be liquefied and secondly it must refrigerate the gas sufficiently to produce the desired change of phase. It so happens that in most machines for gas liquefaction, the working substance used to provide the refrigerative effort of the machine is identical with the gas to be liquefied. Such identity between the refrigerative working substance and the gas liquefied tends to cause confusion of the two functions of a liquefier. In the following discussion of the theory of gas liquefaction, therefore, care will be taken to distinguish clearly between these two functions.

For purposes of providing some standard of reference in calculating the efficiency and in assessing the optimum mode of operation of liquefiers, it is con-

---

[1] See J. K. ROBERTS and A. R. MILLER: Heat and Thermodynamics, p. 105. London: Blackie & Son 1951.

[2] For a VAN DER WAALS gas, the inversion temperature, T_i, defined as that where $\alpha_H = 0$ for $p \to 0$, is such that $T_i = 2T_{Boyle}$.

[3] H. HAUSEN: Z. techn. Phys. **7**, 444 (1926).

venient first to consider an ideal cycle for the refrigerative function of such a machine. Clearly if the gas to be liquefied has a critical temperature either above ambient temperature or at least above a temperature which can be maintained readily, liquefication by compression is directly possible and this can be carried out if necessary with the help of a compressed vapor refrigerator or with a cascade system of compressed vapor refrigerators, as have been described in detail earlier. It will be sufficient therefore to limit our considerations to gases which have critical temperatures at relatively low temperature. In common practice this limitation narrows our considerations to the liquefication of air, hydrogen, and helium. (For a tabulation of critical temperatures see Table 8).

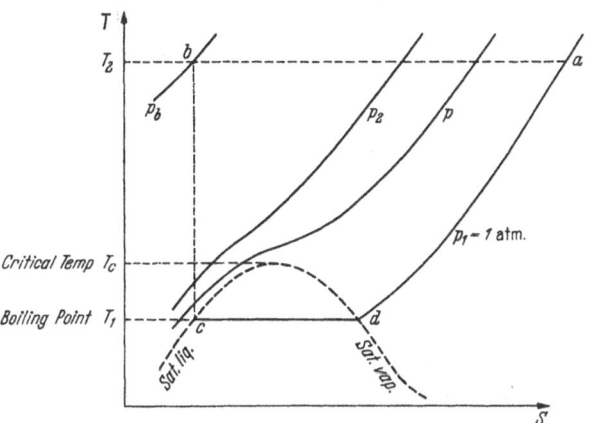

Fig. 36. $T-S$ diagram for a typical fluid showing isobars. The cycle $a \rightarrow b \rightarrow c \rightarrow d$ is for theoretical refrigeration by ideal isentropic expansion.

Consider the entropy, S, versus temperature, T (degrees absolute) diagram for the gas under consideration to be represented diagrammatically by Fig. 36, where the full lines are the isobars $(p_2 > p_1)$ and the broken curve is the limit of the heterogeneous two phase region. (Liquid and vapor). Suppose further that the temperature, T_2, at which compression of the gas is to take place is, as shown in Fig. 36, well above the critical temperature, T_c, and that T_1 is the boiling point of the gas considered at $p_1 = 1$ atm. Now the following reversible refrigerative cycle using the gas as working substance can, in theory, be postulated. First, starting at T_2 and p_1 (the point "a"), the gas is compressed isothermally to b. Then the gas isentropically and reversibly is expanded from b to c, such that the point c represents the condition of liquid saturation at the temperature T_1. Then at constant temperature T_1 and constant pressure, p_1, the liquefied gas is evaporated in an evaporator until a saturated vapor is produced at d. This process of evaporation absorbs heat from the surroundings, at the low temperature T_1, in amount proportional to the latent heat of vaporization of the liquid. Finally the gas is allowed to warm up to the temperature T_2, at constant pressure p_1, the heat for this coming from the surroundings. A representative block diagram for such an ideal machine is given in Fig. 37, which is self-explanatory and for which the heat absorbed at the low temperature, T_1, is designated Q_1; the heat evolved during isothermal compression at T_2 is designated Q_2 and the heat absorbed in isobaric warming from T_1 to T_2 is designated Q_3. By consideration of the diagram of Fig. 36, it is clear that

$$\left.\begin{aligned} Q_1 &= L_{T_1} = H_d - H_c \\ Q_2 &= T_2(S_a - S_b) \\ Q_3 &= H_a - H_d \end{aligned}\right\} \tag{16.1}$$

where L_{T_1} is the latent heat of evaporation at T_1 and where H is the enthalpy corresponding to the subscripted point.

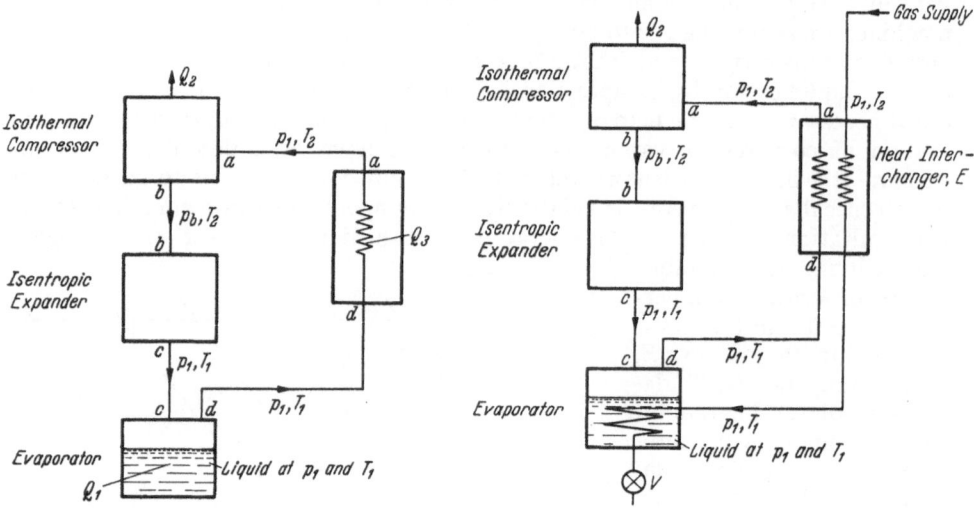

Fig. 37. Schematic flow diagram for theoretical refrigeration by the ideal isentropic expansion cycle of Fig. 36.

Fig. 38. Schematic flow diagram of idealized liquefaction process using the refrigeration by ideal isentropic expansion.

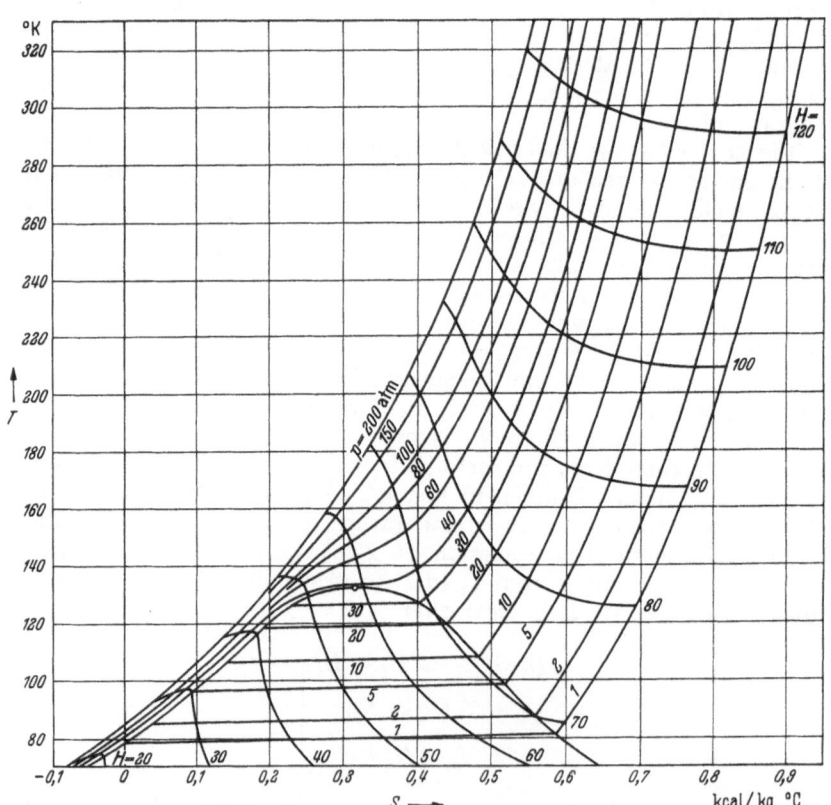

Fig. 39. $T—S$ diagram for air. [After Hausen: Forsch. Ing.-Wes. 274, 1 (1926).]

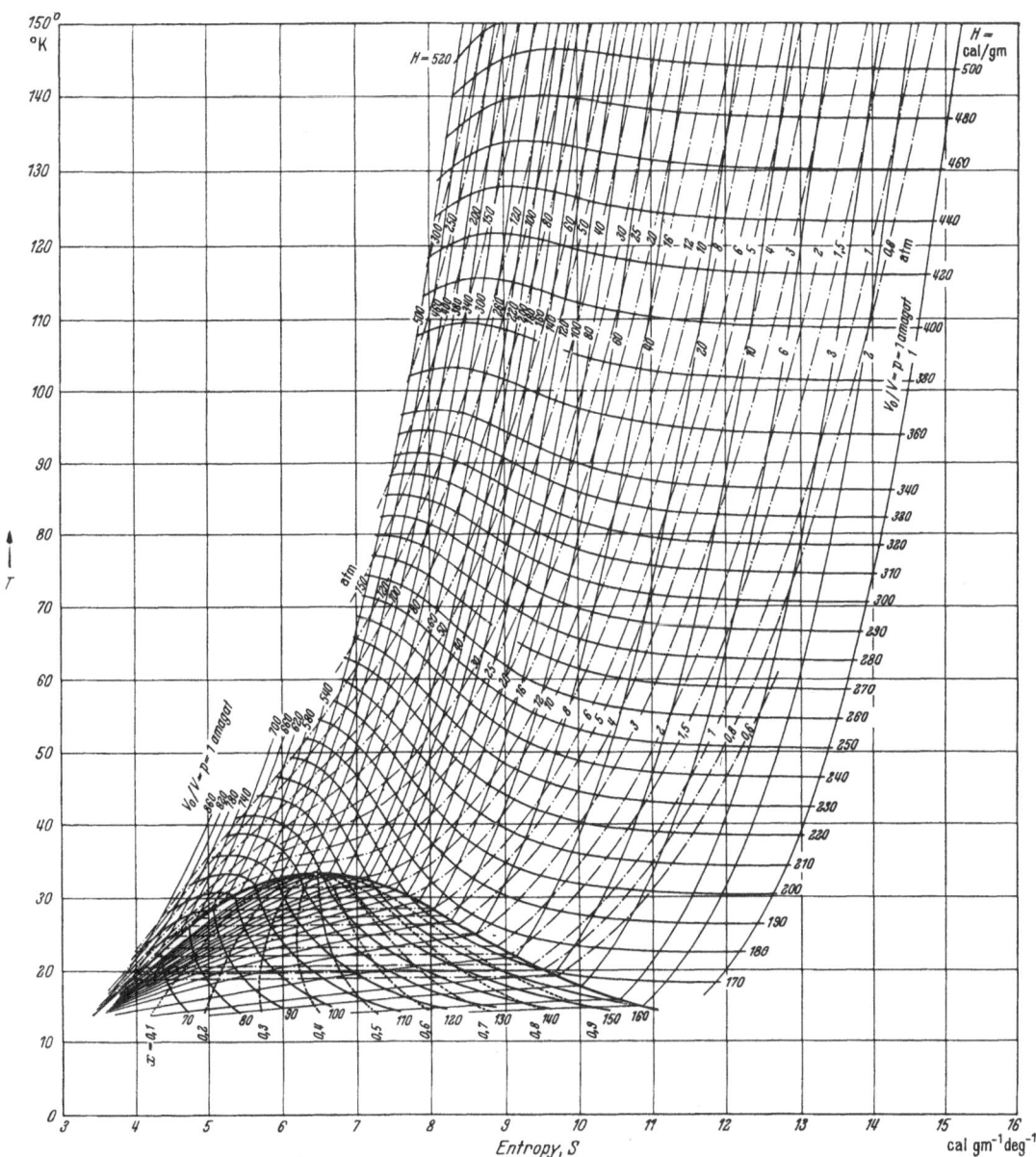

Fig. 40. $T-S$ diagram for H_2. [After Woolley, Scott and Brickwedde: J. Res. Nat. Bur. Stand. 41, 379 (1948).]

The net work done on one mole of the working substance circulated in this reversible refrigerative cycle is

$$
\left.
\begin{aligned}
W &= Q_2 - (Q_1 + Q_3) \\
&= T_2 (S_a - S_b) - (H_a - H_c) \\
&= T_2 \Delta S - \Delta H
\end{aligned}
\right\} \tag{16.2}
$$

where ΔS and ΔH are the differences in entropy and enthalpy respectively between the gas at the high temperature, T_2, and the saturated liquid at the low temperature, T_1, when the pressure p_1 is the same. The numerical values of ΔS and ΔH therefore are determined once p_1 and T_2 are fixed.

The ideal reversible refrigerative cycle considered above can be supposed to perform the refrigeration for a gas liquefier. To complete the machine as a liquefier, it is necessary to introduce the gas to be liquefied. Let it be supposed that

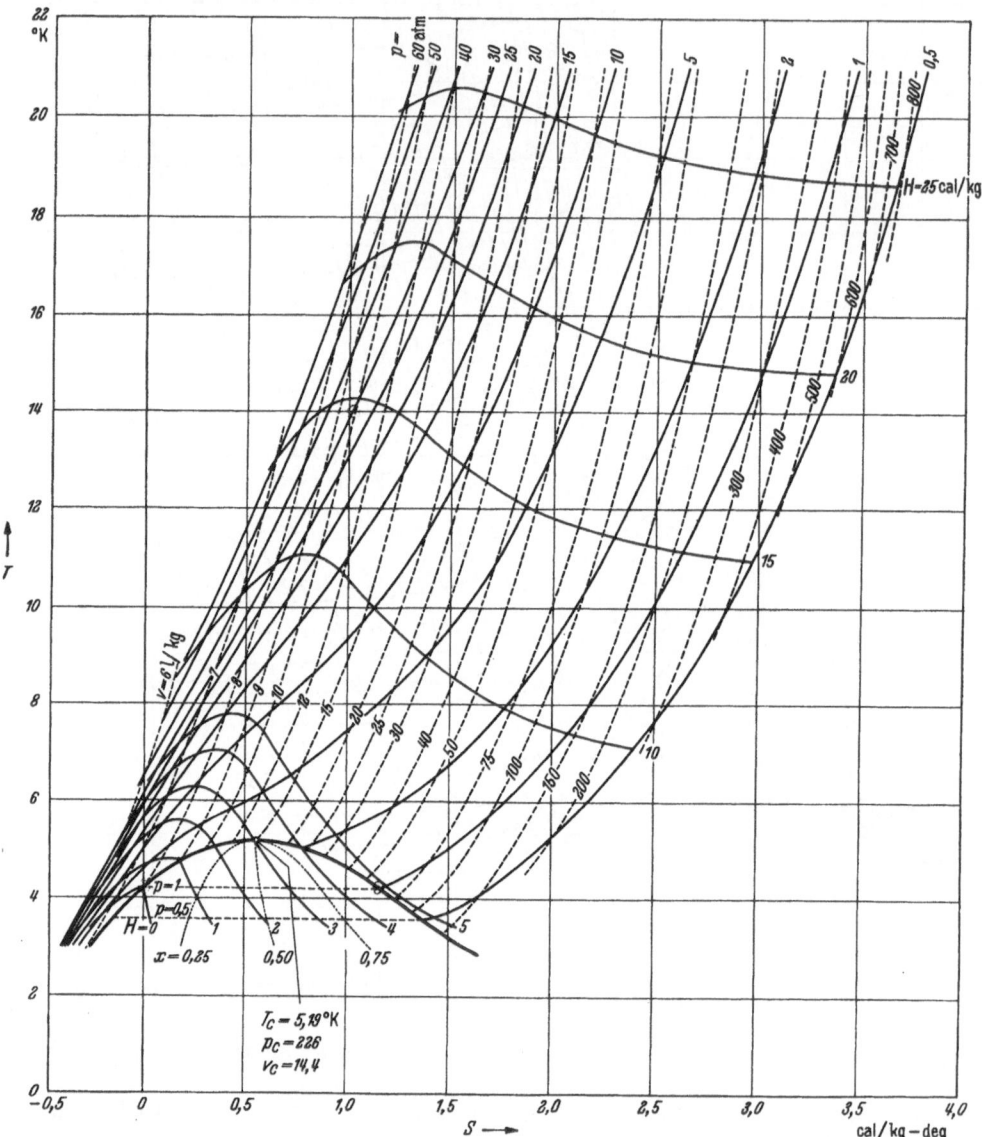

Fig. 41. $T-S$ diagram for He. [After Zelmanov: J. Phys. USSR. **8**, 129 (1944).]

the *same* gas is used both as working substance in the refrigerative cycle and as gas to be liquefied. The gas to be liquefied can, in principle, be introduced as shown in the block diagram of Fig. 38, first entering the ideal heat interchanger E at p_1 and T_2. If the gas flow to be liquefied is the *same* number of moles per unit time as that circulated in the refrigerator section, then it will leave the ideal interchanger at p_1 and T_1, having given up an amount of heat Q_3 per mole to the working substance. On leaving the interchanger at p_1 and T_1 the gas passes into thermal contact with the liquid working substance in the evaporator and for every mole condensed here liberates an amount of heat Q_1 to the evaporator. The liquid finally is drained off through the valve V. With such a system therefore for every mole of working substance circulated in the refrigerator section,

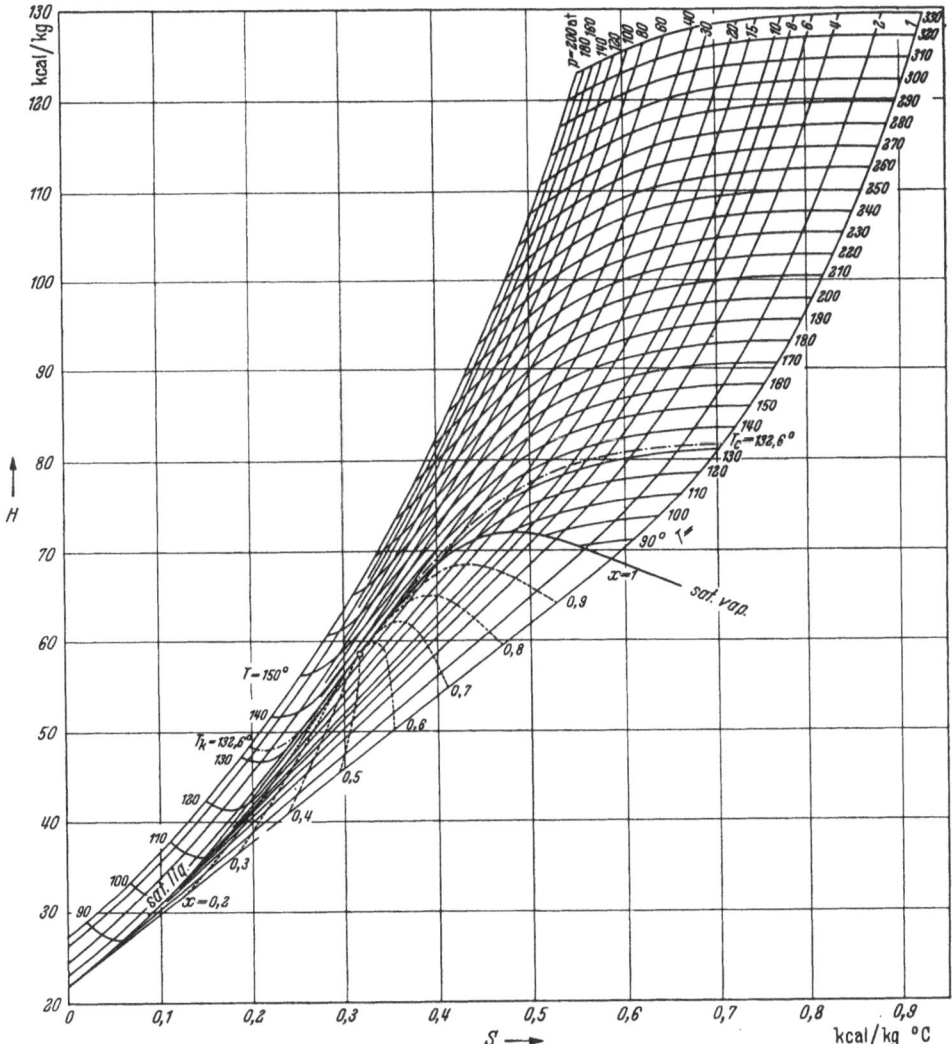

Fig. 42. H—S diagram for air. [After HAUSEN: Forsch. Ing.-Wes. 274, 1 (1926).]

one mole is liquefied in the liquefier section. The minimum amount of work to liquefy one mole of gas using the combined ideal machine of Fig. 38, is therefore given by eq. (16.2); i.e.

$$W_{\min} = T_2 \, \varDelta S - \varDelta H \qquad (16.3)$$

where $\varDelta S$ and $\varDelta H$ have the same significance as before.

Numerical evaluations of W_{min}, can be obtained from the known values for ΔS and ΔH. The measurements are conveniently presented in the form of $T - S$ and $H - S$ diagrams. The $T - S$ diagrams for air, hydrogen and helium are given in Fig. 39, 40 and 41 and the $H - S$ diagram for air in Fig. 42[1]. For fuller bibliography for such thermodynamic data, see Table 10. Table 11 shows the values

Table 10. *Brief Bibliography for Thermodynamic Data.*

Air	$T - S$ and $H - S$ diagrams.
	H. Hausen: Forsch. Ing.-Wes. **274**, 1 (1926).
	V. C. Williams: Trans. Amer. Inst. Chem. Engr. **39**, 93 (1943).
	J. H. Ruston: Refrig. Engng. **53**, 28 (1947).
	L. C. Claiton and D. B. Crawford: Trans. Amer. Soc. Mech. Engrs. **71**, 885 (1949).
N₂	$T - S$ and $H - S$ diagrams.
	W. H. Keesom and D. J. Houthoff: Leiden Comm. Suppl. **65**c (1928).
	R. W. Miller and J. D. Sullivan: U.S. Bur. Mines Techn. Paper **1928**, No. 424.
	L. C. Claiton and D. B. Crawford: Trans. Amer. Soc. Mech. Engrs. **71**, 885 (1949).
	Lunbeck, Michels and Wolkers: Appl. Sci. Res. A3, 197 (1951/53).
	$p - H$ diagram.
	W. H. Keesom, A. Bijl and L. A. J. Monté: Leiden Comm. Suppl. **108**a (1954).
O₂	$T - S$ diagram.
	R. W. Miller and J. D. Sullivan: U.S. Bur. Mines Techn. Paper **1928**, No. 424.
	L. C. Claiton and D. B. Crawford: Trans. Amer. Soc. Mech. Engrs. **71**, 885 (1949).
H₂	$T - S$ diagram and tables.
	Woolley, Scott and Brickwedde: J. Res. Nat. Bur. Stand. **41**, 379 (1948).
	W. H. Keesom and D. J. Houthoff: Leiden Comm. Suppl. **65**d (1928).
He⁴	$T - S$ diagram.
	J. L. Zelmanov: J. Phys. USSR. **8**, 129 (1944).
	W. H. Keesom and D. J. Houthoff: Leiden Comm. Suppl. **65**e (1928).
	W. H. Keesom, A. Bijl and L. A. J. Monté: Appl. Sci. Res. **4**, 25 (1954) and Leiden Comm. Suppl. **108**c (1954).
	Tables.
	S. W. Akin: Trans. Amer. Soc. Mech. Engrs. **72**, 751 (1950).

of W_{min} as obtained from the above data, together with the other pertinent data for the three gases mentioned. Such an ideal process could not be realized in practice, not only because of the irreversibility introduced in expansions, etc., but also because the pressure, p_b, required for example for air in the above ideal cycle would be approximately 10^5 atm.

---

[1] $T - S$ and $H - S$ diagrams for air due to Hausen: Forsch. Ing.-Wes. **274**, 1 (1926). — $T - S$ diagram for H_2 due to Woolley, Scott and Brickwedde: J. Res. Nat. Bur. Stand. **41**, 379 (1948). — $T-S$ diagram for He due to Zelmanov: J. Phys. USSR. **8**, 129 (1944). — For thermodynamic properties of O_2 and N_2 see for example U.S. Bur. Mines Paper 424, 1928.

Table 11. *Minimum work required for liquefaction and other data for air, H_2 and He.*

Substance	p_1 atm.	T_2 °K	ΔS [1] cal/mole-°K	ΔH [2] cal/mole	W_{min} cal/mole	liquid density	W_{min} kW-hr/liter liquid
Air	1	293	26.1	2.88×10^3	4.77×10^3	0.88	0.167
Hydrogen (normal)	1	293	25.2	1.86×10^3	5.51×10^3	0.07	0.224
Helium (He[4]) . . .	1	293	27.0	1.46×10^3	6.45×10^3	0.122	0.235

17. The actual LINDE cycle of operations. Instead of using an isentropic expansion, as is demanded in the ideal liquefier described above, to produce the cooling effect, one very important method due in 1895 to LINDE[3] and HAMPSON[4] independently made use of the cooling obtained in isenthalpic JOULE-THOMSON expansion. The first liquefiers based on this principle were for the liquefaction of air using air also as the "working substance" for the refrigerative

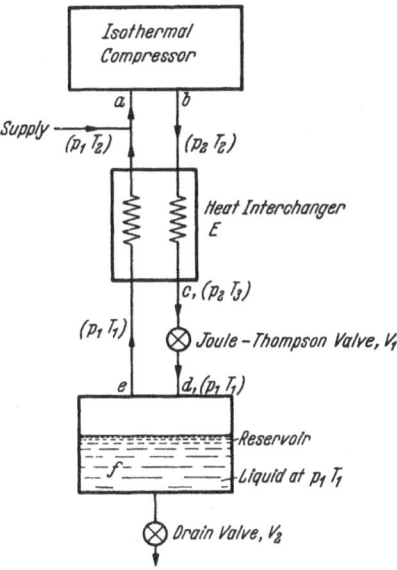

Fig. 43. Schematic flow diagram for simple LINDE type air liquefier.

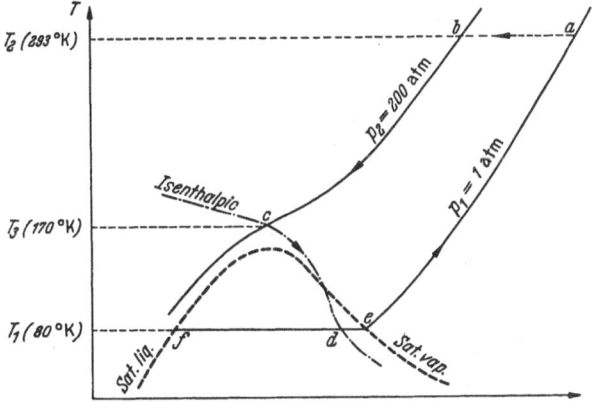

Fig. 44. $T-S$ diagram showing the thermodynamic cycle of operations for the simple LINDE air liquefier of Fig. 43.

process, and the success of their operation was dependant on the use of heat interchangers, such as had been discussed much earlier by SIEMENS[5]. A schematic diagram of the simplest form of "LINDE" *air* liquefier is given in Fig. 43, and the $S-T$ diagram showing the cycle of operations is given in Fig. 44, the lettered points on both diagrams indicating the state of the air at the various places of the cycle. Such a system for air liquefication can be considered without serious loss of theoretical generality. The process is carried through as follows: First the air enters the compressor at p_1 and T_2 and leaves the compressor at p_2 and T_2. In practice the compression itself is not isothermal, but conditions can be made to approach an isothermal one by employing a multi-stage compressor with appropriate intercoolers between stages and with a final after cooler. The

---

[1] See curves of Figs. 39, 40 and 41.

[2] For tabulations of enthalpy see for example: for *air* HAUSEN [Forsch. Ing.-Wes. **274** (1926)]. For H_2 see WOOLLEY et al [J. Res. Nat. Bur. Stand. **41**, 379 (1948)]). For He see S. W. AKIN [Trans. Amer. Soc. Mech. Engrs. **72**, 751 (1950)]. See also Table 10.

[3] K. v. LINDE: German patent 88 824 and Z. ges. Kälteind. **4**, 23 (1897). See also The Engineer. Nov. 13. and 20. 1896.

[4] HAMPSON: May 1895. English Patent, 10 165.

[5] W. SIEMENS: English Patent, No. 2064. 1857. See also Min. Proc. Inst. Civ. Engrs. **68**, 179 (1882).

compression therefore takes the air from state a to state b of Fig. 44. It then passes through the high pressure side of the heat interchanger E to the valve V_1, where it is allowed to expand to pressure p_1 isenthalpically. This JOULE-THOMSON expansion at V_1 produces a cooling, provided the initial temperature, T_2 is below the inversion temperature (cf. Sect. 14). Since the inversion temperature for air is 603° K, an initial temperature of 293° K is satisfactory in an air liquefier. (For hydrogen and helium liquefiers based on the cooling produced in JOULE-THOMSON expansion, lower valves of T_2 must be employed as discussed in chapter E.) The cooled air, after leaving the valve V_1, passes via the reservoir into the return low pressure side of the heat interchanger E, to the compressor and on its way gives up its cold to the high pressure air from the compressor. During the starting conditions therefore the cooling at the valve V_1 is regenerative due to the action of the heat interchanger, and the valve V_1 will continue to cool until a steady state is reached with a certain fraction ε of the air flow being liquefied during the JOULE-THOMSON expansion.

When the steady (liquefying) state is reached the high pressure gas enters the heat interchanger at p_2 and T_2 (point b of Fig. 44) and leaves it at p_2 and T_3 (point c). The air expands isenthalpically through the valve V_1 from p_2, T_3 to p_1, T_1 passing along the isenthalpic line from c to d. At d therefore the fraction of air in the liquid state is given by

$$\varepsilon = \frac{de}{ef} \tag{17.1}$$

where e and f represent the saturated vapor state and the saturated liquid state at the temperature T_1. This fraction ε of liquid collects in the reservoir at T_1 and p_1, ready to be drained off when required through the valve, V_2. The remaining quantity of gas $(1-\varepsilon)$, represented by the point e, on Fig. 44, enters the low pressure side of the heat interchanger at T_1 and p_1 and leaves it at T_2 and p_1, having given up its cold to the downcoming high pressure air.

To calculate the work required in the above process to liquefy one mole of air, one must suppose, as in the ideal cycle of operations, that the liquefied air in amount ε is evaporated in the reservoir, so absorbing heat Q_1 at the low temperature, and that *all* the air in warming up to the temperature T_2 gains heat Q_3 in the process. Under such a supposition the refrigerative function of the machine is completely cyclic, having the same mass of air passing all points per unit time. Then, as in the considerations of the ideal cycle of operations (cf. Fig. 38), one can suppose that air, in amount ε is cooled from T_2 to T_1 by liberating heat to the interchanger and then is liquefied at T_1, evolving to the evaporator an amount of heat of Q_1. Under these circumstances, we have, per ε moles of air liquefied:

$$\left.\begin{aligned}
&\text{Heat given out on isothermal compression: } Q_2 = T_2\,(S_a - S_b)\\
&\text{Heat absorbed in evaporating } \varepsilon \text{ moles at } T_1: Q_1 = \varepsilon L = \varepsilon\,(H_e - H_f)\\
&\text{Heat given out to interchanger (path } b \rightarrow c): Q_4 = H_b - H_c\\
&\text{Heat absorbed in interchanger (path } e \rightarrow a): Q_3 = H_a - H_e
\end{aligned}\right\} \tag{17.2}$$

Whence the net work done, W, is given by:

$$\left.\begin{aligned}
W &= Q_2 - Q_1 + Q_4 - Q_3\\
&= T_2\,(S_a - S_b) - \varepsilon\,(H_e - H_f) + (H_b - H_c) - (H_a - H_e)
\end{aligned}\right\} \tag{17.3}$$

and whence the net work done per mole of air liquefied is

$$W_{\text{molar}} = W/\varepsilon. \tag{17.4}$$

The values of the entropy and enthalpy at the various points can be obtained from the thermodynamic diagrams once T_1, T_2 and T_3 are known. Typical values for the temperatures are: $T_1 = 80°$ K, $T_2 = 293°$ K, $T_3 = 170°$ K and $p_1 = 1$ atm., $p_2 = 200$ atm. For these values

$$W_{\text{molar}} = 32.7 \times 10^3 \text{ cal/mole liquid}$$

giving the energy required to liquefy one mole of liquid air. This is to be compared to the value given in Table 11 of $W_{\text{min}} = 4.77 \times 10^3$ cal/mole as the minimum energy of the ideal cycle required to liquefy one mole of air. The discrepancy is partly due to the relatively low pressure (~ 200 atm.) used in an actual cycle and partly due to the fact that the "LINDE" system adopts an irreversible JOULE-THOMSON expansion rather than a reversible isentropic expansion.

It is of interest to note that the work of compression, which is the directly measured external work required in the liquefying process is given by

$$W_{\text{compr}} = Q_2/\varepsilon = T_2(S_a - S_b)/\varepsilon \tag{17.5}$$

per mole of air liquefied[1]; and this work of compression evaluated under the typical conditions stated above ($T_2 = 293°$ K, $p_1 = 1$ atm, $p_2 = 200$ atm., etc.) is:

$$W_{\text{compr}} = 35.4 \times 10^3 \text{ cal/mole liquid},$$

i.e., the work of compression is only very little larger than the theoretical energy required to liquefy one mole of air, and it can in consequence be regarded as the energy required to liquefy one mole of air. However, in practice, no compression is isothermal, as is assumed in this calculation, so that further allowance must be made for non-ideality of the compression process. In practice the work of compression is at least 1.7 times the calculated isothermal work required, still further reducing the efficiency of the LINDE process as compared with the ideal cycle of operations.

To calculate the theoretical operations of hydrogen and helium liquefiers, similar general methods as that used above are used, which will become apparent after the technical operations for such liquefiers are known (see chapter E).

18. Use of the $H-S$ diagram and conditions for maximum liquid yields. The $H-S$ diagram giving the enthalpy, H, as a function of the entropy, S, for various isobars and isotherms has already been of use in calculating the efficiencies of compressed vapour refrigerators (see chapter B). Such a diagram is also of value in discussing the efficiency of LINDE air liquefiers (and hydrogen and helium liquefiers).

Consider the simple LINDE type air liquefier as shown diagrammatically in Fig. 43 and as described above. The cycle of operations can be followed on the $H-S$ chart as shown in Fig. 45. In this chart, as described previously (cf. Figs. 20 and 21), the heavy broken curve represents the boundary of the heterogeneous two-phase region, the full curves are typical isobars ($p_2 > p > p_1$), and the light broken curves are typical isotherms ($T_2 > T_3 > T_1$). It is to be noticed that inside the heterogeneous region the isobars and isotherms are linear and coincident, having a slope equal to the absolute temperature. Now the cycle

---

[1] If the compressed air is treated as a perfect gas, then for isothermal compression $W_{\text{compr}} = R T_2 \ln (p_2/p_1)/\varepsilon$.

of the LINDE air liquefier is as follows: The point a represents the gas at T_2 and p_1, i.e. at the input to the compressor. Isothermal compression takes the representative point to b at T_2 and p_2. In practice $T_2 \approx 293°$ K and p_1 and p_2 approximately 1 and 200 atm respectively. The path b to c represents the cooling of the input gas as it passes down the interchanger and at c the gas is isenthalpically expanded from p_2 and T_3 to p_1 and T_1. This isenthalpic expansion at the valve is represented by the horizontal line (constant H), $c \rightarrow d$. The point d therefore gives a measure of the amount of gas liquefied, the fraction ε, liquefied at the expansion being (ed/ef). The liquefied fraction is represented by the point f, being saturated liquid at T_1 and p_1. The fraction remaining gaseous is represented by the point e, being saturated vapour at T_1 and p_1. This gaseous fraction returns via the interchanger to the compressor, as represented in the diagram by the path $e \rightarrow a$.

If one supposes, as is implicitly assumed in the above graphical representation, that a) the thermal isolation of the interchanger and low temperature parts of the machine is perfect, so that no heat is lost or gained therein and b) that the interchanger is perfect so that the gas emerging at low pressure regains the temperature T_2, then considering the liquefaction system, the enthalpy, H, is conserved. This is so since by hypothesis no heat is lost or gained and at the JOULE-THOMSON expansion H is invariant. Hence, the total enthalpy H_b of the gas entering the interchanger must be the same as the sum of the enthalpy H_a of the gas emerging from the interchanger and of the enthalpy H_f of the liquid produced. Since the liquid fraction is taken to be ε, this enthalpy balance equation can be written:

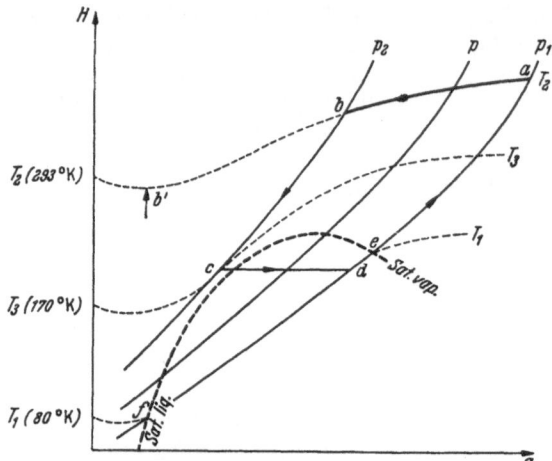

Fig. 45. $H-S$ diagram showing the thermodynamic cycle of operations for the simple LINDE air liquefier of Fig. 43.

or

$$H_b = (1 - \varepsilon) H_a + \varepsilon H_f \\ \varepsilon = \frac{H_a - H_b}{H_a - H_f}. \tag{18.1}$$

This equation enables, ε, the so-called "liquefaction coefficient" to be readily evaluated from a $H-S$ diagram provided the working pressures and temperatures are known.

In practice the temperature of compression T_2 is fixed by considerations of convenience, being approximately a little above ambient temperature, and the input pressure p_1 also fixed, being generally approximately 1 atm., a pressure at which it is convenient to supply make-up gas from a gasholder. In consequence H_a and H_f are predetermined. The liquefaction coefficient, ε therefore, as is seen from equation (18.1), depends only on the value of H_b, becoming larger as H_b is made smaller. This result is of interest in that the liquefaction coefficient is independent of the conditions at the expansion and is determined instead by the conditions at the high temperature input end of the interchanger.

The conditions at the input (high pressure) side of the interchanger (point b of Fig. 45) can be examined theoretically, since thermodynamically quite generally it is known that:

$$\left(\frac{\partial H}{\partial S}\right)_T = T - v\left(\frac{\partial T}{\partial v}\right)_p . \tag{18.2}$$

First this equation shows that for a perfect gas, for a mole of which $pv = RT$, $(\partial H/\partial S)_T = 0$; i.e., isothermals on the $H-S$ diagram are straight lines parallel to the S axis for perfect gases, a condition which is approximated at high values of S and T for real gases. For such a case therefore no liquefaction can be obtained since the enthalpies before and after isothermal compression would be the same [cf. eq. (18.1)]. Such a conclusion is supported by the previous considerations of the JOULE-THOMSON effect given in chapter C.

Secondly this equation enables the condition for maximum liquefaction coefficient, ε, to be obtained. As is indicated by eq. (18.1), ε is a maximum when H_b is a minimum.

The isotherm at T_2, as is shown in Fig. 45, has a minimum at the point marked b^1, and at this point $(\partial H/\partial S)_{T_2} = 0$. By eq. (18.2) therefore this minimum occurs when:

$$T_2 = v\left(\frac{\partial T}{\partial v}\right)_p . \tag{18.3}$$

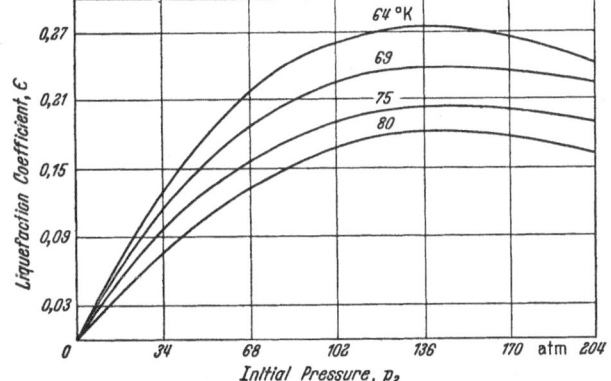

Fig. 46. The computed liquefaction coefficient, ε, for hydrogen liquefaction using a simple LINDE circuit as a function of pressure, p_2. The four curves are for four precooling temperatures, T_2. [Curves due to JOHNSTON, BEZMAN and HOOD: J. Amer. Chem. Soc. 68, 2367 (1946).]

If reference is now made to eq. (13.4) it will be seen that the condition specified by eq. (18.3) immediately above is also the condition for zero JOULE-THOMSON coefficient. In other words for maximum ε, T_2 must be the inversion temperature corresponding to the pressure at the high pressure stage of the compressor.

Another way of stating this conclusion, which was first arrived at by MEISSNER[1], is that the maximum liquefaction coefficient is obtained when p_2 and T_2 correspond to a point actually on the inversion curve. For air this would mean that for $T_2 \approx 293°$ K maximum liquefaction occurs for $p_2 = 440$ atm. In practice values of p_2 only about half of this value are employed. For hydrogen, using the data given by WOOLLEY et al[2] and assuming $T_2 = 80°$ K one obtains $p_2 = 157$ atm. For helium using the data of ZELMANOV[3] and assuming $T_2 = 15°$ K, one obtains $p_2 = 31$ atm. In practice the choice of working pressure is not too critical since the maximum in the curve of ε versus p_2 is a broad one. This is illustrated in Fig. 46, taken from the measurements of JOHNSTON et al[4] on H_2, where plots of the liquefaction coefficient, ε, versus p_2 are made for a number values of the temperature T_2. It will be seen from the curve for $T_2 = 80°$ K, that the maximum value of $\varepsilon = 18.4\%$ at $p_2 = 140$ atm. On the other hand for $p_2 = 100$ atm. ε is still as high as 16.6%.

[1] W. MEISSNER: Z. Physik 18, 12 (1923).
[2] WOOLLEY, SCOTT and BRICKWEDDE: J. Res. Nat. Bur. Stand. 41, 379 (1948).
[3] ZELMANOV: J. Phys. USSR. 3, 43 (1940).
JOHNSTON, BEZMAN and HOOD: J. Amer. Chem. Soc. 68, 2367 (1946).

19. Increasing the yield by precooling. Considerable increases in the lique-faction coefficient, ε, and consequent decreases in the power requirement per liter liquefied, can be obtained by precooling of the high pressure gas before it enters the LINDE circuit.

A diagram of such a system is given in Fig. 47. The high pressure gas leaves the compressor at p_2 and T_3, enters the heat inter-changer, E_1, and cools in the refrigerator evaporator to a temperature T_2. This refrigerator evaporator in the case of air liquefaction might be the evaporator of an ammonia machine (cf. chapter B) or might be the lowest temperature evaporator of a cascade refrigerator. In the case of hydrogen liquefaction the refrigerator evaporator is a bath of liquid $(N_2 + O_2)$ mixture boiling under reduced pressure; whereas for helium liquefaction it would be a bath of liquid hydrogen also boiling under reduced pressure. The gas leaving the refrigerator evaporator at p_2 and T_2 then enters the LINDE circuit which consists of the interchanger E_2 and the JOULE THOMSON expansion valve, V_1. The remainder of the diagram is self-explanatory.

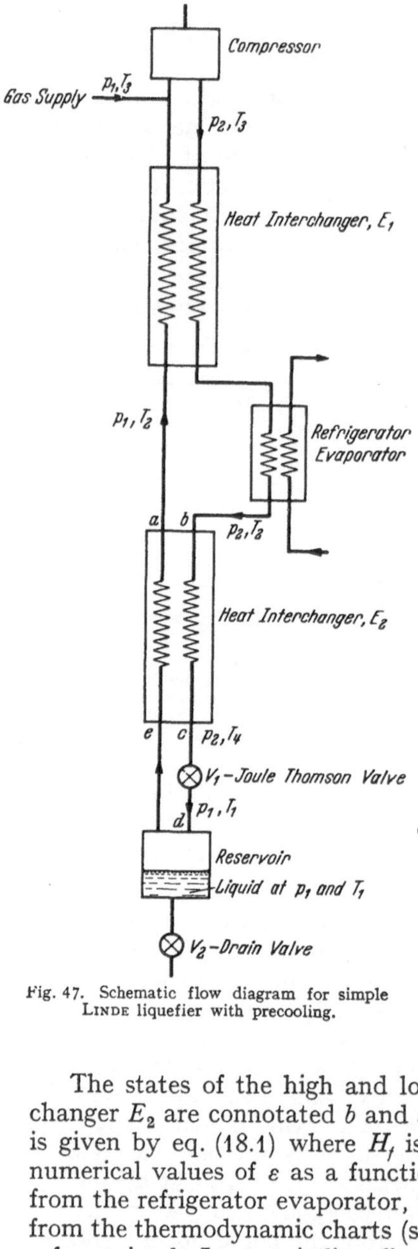

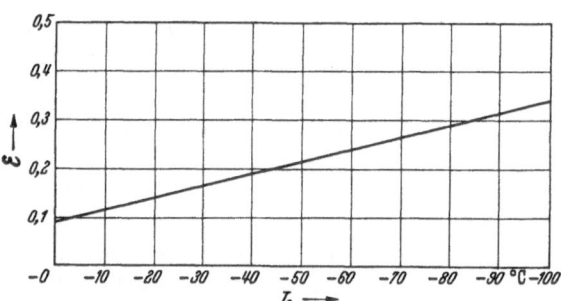

Fig. 47. Schematic flow diagram for simple LINDE liquefier with precooling.

Fig. 48. The theoretical liquefaction coefficient, ε, for air liquefaction, using a simple LINDE system with precooling, as a function of the precooling temperature, T_2, for constant input pressure, p_2, of 200 atm.

The states of the high and low pressure gas at the warm end of the inter-changer E_2 are connotated b and a respectively; then the liquefaction coefficient is given by eq. (18.1) where H_f is the fluid enthalpy at p_1 and T_1. To obtain numerical values of ε as a function of the temperature T_2 of the gas emerging from the refrigerator evaporator, the enthalpy values, H_a and H_b must be taken from the thermodynamic charts (see Table 10 for references). We have computed ε for a simple LINDE air liquefier with precooling for the following typical conditions, namely: $p_1 = 1$ atm., $T_1 = 80°$ K, $p_2 = 200$ atm. The results are shown in Fig. 48 which plots ε as a function of the precooling temperature T_2.

In practice the precooling is frequently performed in an air liquefier by an ammonia refrigerator so precooling the air to about $-45°$ C[1]. With cascaded

[1] R. LINDE: Z. VDI **65**, 1357 (1921).

compressed vapor refrigerators, using for example Freon 22 and Freon 13, lower precooling temperatures can be reached. The high efficiency of these compressed vapor refrigerators allows a great gain to be made in the total power required for the air liquefaction. A table showing the computed (theor.) and some typical observed (pract.) efficiencies of precooled and non-precooled air liquefiers is given in Table 12. The difference between the theoretical and practical efficiencies

Table 12. *Air liquefaction. No. of kW-hrs per liter liquid*[1].
Fixed data: $p_2 = 200$ atm.; $p_1 = 1$ atm; $T_1 = 80.1°$ K.

Temperature of precooling (T_2) °K	Ideal reversible process kW-hr/liter	Simple LINDE			Two-stage LINDE	
		Theor kW-hr/liter	Pract kW-hr/liter	(percent) ε	Theor kW-hr/liter	Pract kW-hr/liter
293	0.167	1.05	2.04	9	0.71	1.05
228		0.61	1.09	20	0.40	0.73
200		0.50	0.83	24.5		

being due to heat losses in imperfect insulation, imperfect heat interchangers and to non-isothermal compression. The latter source of loss can be minimized by multiple compression in many stages, but as mentioned previously it is difficult to get less than about 1.7 times the ideal isothermal values of work of compression. The losses due to imperfect insulation are strongly dependent on the physical size of the machine. For larger machines the surface to volume ratio diminishes and so larger machines suffer less loss from this source. As an example of this Table 13 (taken from "Industrial Gases" by H. C. GREENWOOD,

Table 13. *Outputs and efficiency of simple* LINDE *liquid air plants.* (Taken from "Industrial Gases" H. C. GREENWOOD. Publ. Van Nostrand. 1919.)

Liters of liquid air per hour without precooling	—	—	12.5	35	70
Liters of liquid air per hour with precooling	0.75	5	20	50	100
Power kW	2.6	14.2	39	78	142
kW—hr/liter without precooling	—	—	3.12	2.22	2.04
kW—hr/liter with precooling	3.45	2.85	1.95	1.56	1.43

pub. VAN NOSTRAND 1919) gives the outputs and efficiencies of various sizes of LINDE liquid air plants. It will be seen from the table that for the machines with precooling the 100 liter/hr one is approximately twice as efficient as the 5 liter/hr model.

The question of interchanger efficiency is taken up in chapter I.

In hydrogen liquefiers employing the simple LINDE circuit precooling is always necessary, in order, as explained earlier, to get the gas below its inversion temperature of 204.6° K. In practice, this is done by cooling the high pressure gas with a mixture of liquid O_2 and N_2 boiling under reduced pressure. The minimum temperature possibly available ranges from 54.4° K for pure O_2 to 63.1° K for pure N_2, these being the triple points of O_2 and N_2 respectively.

---

[1] Data taken from R. LINDE: Z. VDI **65**, 1357 (1921). — H. LENZ: Handbuch der Experimentalphysik, vol. 9/1, p. 127. 1929; see also M. DAVIES: Gas Liquefaction and Rectification. London: Longmans 1949. — M. RUHEMANN: Gas Separation. Oxford Press 1940.

Curves showing the liquefaction coefficient ε for hydrogen, as computed from Joule-Thomson effect measurements of Johnston et al[1], plotted as a function of the high pressure, p_2 for various values of the precooling temperature, T_2, have already been given in Fig 46. These curves, for T_2 equal to 64° K, 69° K, 75° K and 80° K show clearly the marked gain in the yield as the precooling temperature is reduced.

Similar curves, shown in Fig. 49, for T_2 between 70° K and 38° K, have been prepared by Keyes et al[2] from earlier thermodynamic data on H_2 by Keesom and Houthoff[3]. These show more markedly the profound gain to be obtained

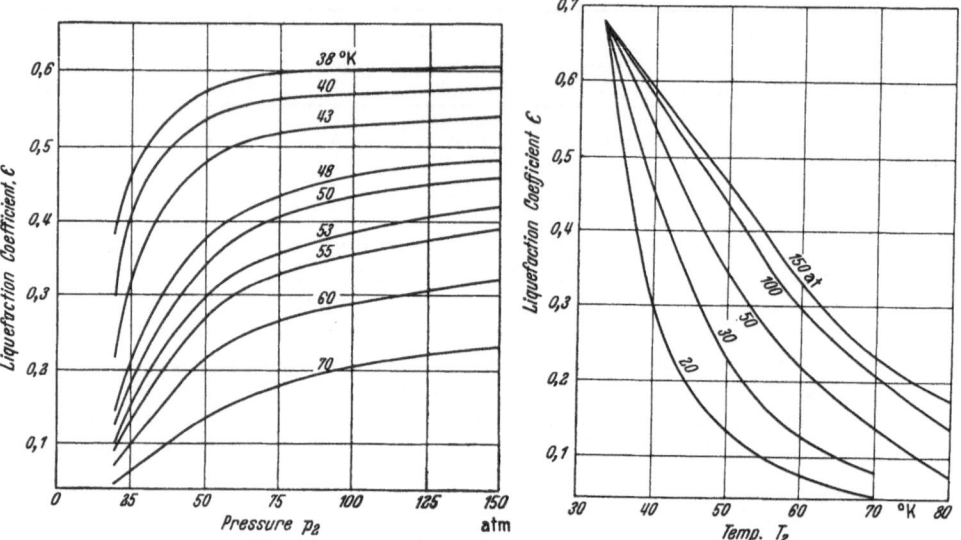

Fig. 49. The theoretical liquefaction coefficient, ε, for hydrogen liquefaction, using a simple Linde system, as a function of the input pressure, p_2, for various precooling temperatures, T_2. [Curves due to Keyes, Gerry and Hicks: J. Amer. Chem. Soc. 59, 1426 (1937).]

Fig. 50. The theoretical liquefaction coefficient, ε, for hydrogen liquefaction, using a simple Linde system, as a function of the precooling temperature, T_2, for various values of the input pressure, p_2. [Curves due to Keyes, Gerry and Hicks: J. Amer. Chem. Soc. 59, 1426 (1937).]

by precooling to lower temperatures. Although the thermodynamic data on which Keyes's curves are based are not so accurate as the more recent data quoted (see Table 10 also), it is of interest to present his calculations of the liquefaction coefficient, ε, for hydrogen as a function of the precooling temperature, T_2 for various values of the input pressure p_2. These are shown by the curves of Fig. 50.

A more detailed estimate of the effect of the precooling temperature on the hydrogen liquefaction coefficient can be gained from the graph of Fig. 33. The heavy line in this graph is the inversion curve. The light curves which cross the inversion curve horizontally are curves of constant $(H_a - H_b)$ where H_a and H_b are respectively the enthalpies of the low pressure and high pressure gas at the precooler temperature, T_2. $(H_a - H_b)$ is approximately equal to the amount of refrigeration per mole of gas available for liquefaction. It will be seen from the curves how markedly this increases as the precooling temperature T_2 is reduced. If one supposes that the working parameters p_2 and T_2 are always adjusted to give a point on the inversion curve (maximization of yield) then for $T_2 \approx 96°$ K,

[1] Johnston, Bezman and Hood: J. Amer. Chem. Soc. 68, 2367 (1946).
[2] Keyes, Gerry and Hicks: J. Amer. Chem. Soc. 59, 1426 (1937).
[3] Keesom and Houthoff: Leiden Comm. 65 (1928).

$(H_a - H_b) = 60$ cal/mole whereas for a temperature, $T_2 = 66°$ K, $(H_a - H_b)$ has double this value.

In this Fig. 33 the light curves which are roughly parallel to the inversion curve and converge with it at the inversion temperature of 204.6° K are lines showing the pressure at which $(H_a - H_b)$ has reached the marked fraction of its maximum value for the given temperature. The inversion curve itself is the 100% curve of this family. It is interesting to note that in the usual range of precooling temperatures, $T_2 = 60°$ K to 90° K, for hydrogen almost 95% of the maximum possible refrigeration is obtained when the pressure is only about 75% of the optimum inversion pressure.

20. Increasing the efficiency by double expansion. The two-stage LINDE circuit. An improvement in air liquefaction efficiency due to LINDE was the introduction of the two-stage circuit, whereby the pressure is reduced by two isenthalpic expansions in succession, rather than all in one expansion as in the simple LINDE circuit. At all isenthalpic expansions irreversibility is introduced and hence inefficiencies. However if the thermodynamic path from the high temperature and pressure to the low temperature and pressure is made in a series of isenthalpic steps the total irreversibility introduced is made smaller the greater the number of steps.

A diagram of the two-stage air liquefier is given in Fig. 51. This is simplified by not including any precooling. For such a modification see Fig. 47. The states of the air at the various points in the machine can be followed by comparison with the $T-S$ diagram of Fig. 52 in which the lettered symbols refer to the corresponding points of Fig. 51. The compression is made up in two main stages

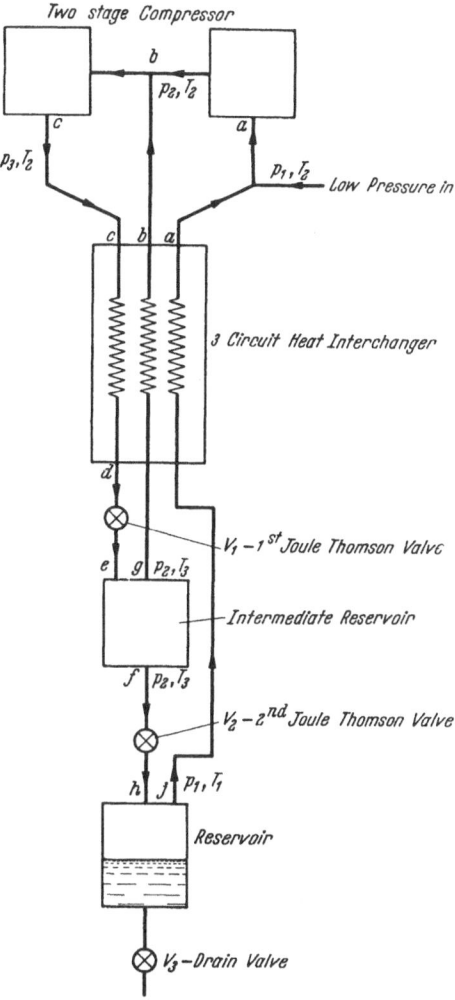

Fig. 51. Schematic flow diagram for two-stage LINDE air liquefier (without precooling) using double JOULE-THOMSON expansion.

so that the intermediate pressure return gas, at p_2, can feed into the compressor between stages. After passing through the heat interchanger the high pressure gas at pressure p_3 expands in the first JOULE-THOMSON valve, V_1, to an intermediate pressure p_2 which in practice is about 40 to 50 atm. ($p_3 \approx 200$ atm,). This first expansion is given on Fig. 52 by the isenthalpic path $d \rightarrow e$. A fraction x of this expanded air is further expanded through the valve, V_2 into the reservoir at atmospheric pressure (p_1). The remaining fraction $(1 - x)$ returns via the heat interchanger to the compressor at the intermediate pressure p_2. That fluid which expands at V_2 (path $f \rightarrow h$ of Fig. 52) partially liquefies with a liquefaction coefficient, say, y. The non-liquefied fraction $(1-y)$ returns through the interchanger which is constructed to allow heat exchange between all three gas streams simultaneously.

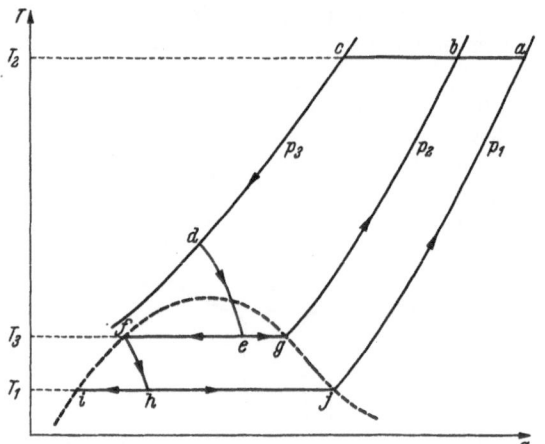

Fig. 52. $T-S$ diagram showing the thermodynamical cycle of operations for the idealized two stage LINDE air liquefier of Fig. 51.

The overall liquefaction coefficient, ε, therefore, defined as the liquefied fraction of the *total* gas stream is

$$\varepsilon = x\,y. \qquad (20.1)$$

As before we can assume that for an ideal machine no enthalpy losses occur, so that the enthalpy of the gas entering the interchanger at c must be equal to the sum of the enthalpies of the gas streams emerging at a and b together with the enthalpy of the liquid produced. This gives:

$$\left.\begin{aligned} H_c = x\,(1-y)\,H_a + \\ + (1-x)\,H_b + \\ + x\,y\,H_i. \end{aligned}\right\} \quad (20.2)$$

For given values of H_a and H_i, which are fixed, since they are the enthalpy of the low pressure gas at the temperature of compression and the enthalpy of the liquid respectively, eq. (20.2) shows that the liquefaction coefficient ε depends not only on the high pressure p_3 but also on the intermediate pressure p_2 and on the fraction x of the fluid passing through valve V_2. In order to obtain values of ε, or more importantly, the efficiency as expressed in kW-hr work needed per liter liquefied, eq. (20.2) must be solved by numerical appeal to the $H-S$

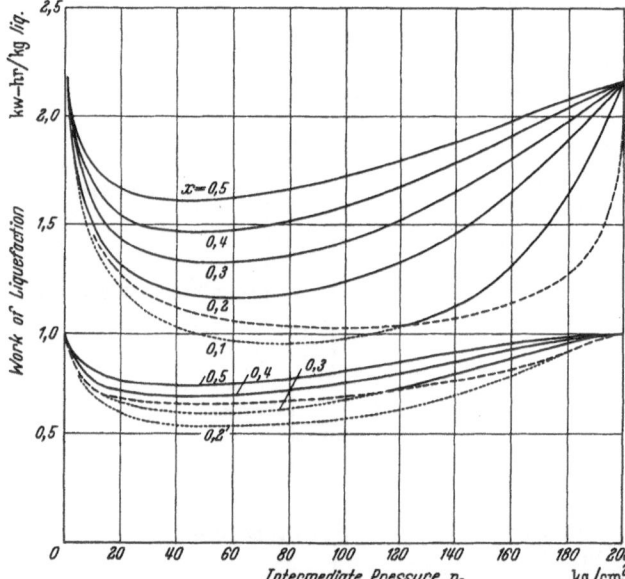

Fig. 53. The practical efficiency of two-stage LINDE air liquefiers, expressed, as the work of liquefaction in kW-hrs per kg. liquid produced, as a function of the intermediate pressure, p_2 for various values of the fraction x of fluid passing to the second JOULE-THOMSON valve. The initial high pressure p_3 is taken to be 200 atm. The upper family of curves is for no precooling, the lower family for precooling to a temperature of $-50°$ C. The dashed lines give the theoretical lower limits for each family of curves. [Curves due to H. LENZ: Handbuch der Experimentalphysik, Bd. 9/1, p. 127, 1929).]

diagram for air. In Fig. 53 the practical efficiency, i.e. the work of liquefaction in kW-hr/kg. liquid, is shown for a two-stage LINDE air liquefier operating at a high pressure (p_3) of 200 atm., the efficiency being plotted as a function of the intermediate pressure p_2 for various values of the parameter x. Two families of curves are shown, one for a system without precooling and one for a system with an ammonia precooling system, precooling to $-50°$ C. The broken curves indicate the theoretical limits for each family of curves, this limit marking the conditions at which the liquefier no longer is self-cooling.

It will be seen from Fig. 53 that for the non-precooled system all curves tend to the same limit, namely 2.3 kW-hr/kg. liquefied, for p_2 tending to zero. This is the limit corresponding to a simple LINDE circuit (cf. Table 12). In practice, the two-stage LINDE circuit will operate with $p_2 \approx 50$ atm. and $x \approx 0.2$, yielding a practical efficiency of about 1.15 kW-hr/kg. liquefied which is approximately twice as efficient as a simple LINDE circuit operating at the same high pressure, $p_3 = 200$ atm. It is to be noted, also as is clear from Table 12, that for systems with precooling the efficiencies are much higher than those without precooling.

One way of viewing the appreciable gain in efficiency for the two-stage system is to consider the work of compression. In the two-stage system only a fraction $x(1-y)$ of the gas has to be compressed all the way from one atmosphere pressure to the high pressure, p_3. The most significant fraction $(1-x)$ of the gas has only to be compressed from the intermediate pressure p_2 to the final value p_3. The heat of compression from pressure p_i to pressure p_f per mole of a *perfect* gas is:

$$W_{\mathrm{comp}} = RT \ln (p_f/p_i) \qquad (20.3)$$

and for air at room temperatures this is a satisfactory approximation. Eq. (20.3) clearly shows the marked influence of the compression ratio.

As far as the writer is aware the two-stage LINDE circuit has not been used for hydrogen or helium liquefaction.

E. Practical aspects of gas liquefaction using only JOULE-THOMSON expansion.

21. Early air liquefiers. Fig. 54 gives a scale drawing of the first HAMPSON air liquefier. Dry air at 200 atm. pressure enters the machine

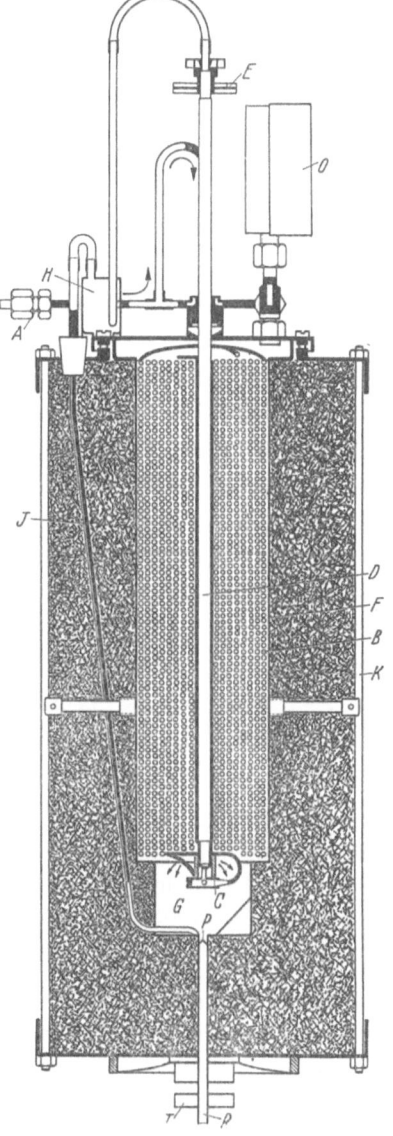

Fig. 54. Section through HAMPSON air liquefier. (Fig. from G. CLAUDE: Liquid Air, Oxygen and Nitrogen. Philadelphia: Blakiston 1913.)

at A and passes through the copper tubes of the interchanger, B[1]. This is a closely packed spiral of copper tubing wound in the annular space between the tubes D and F. The JOULE-THOMSON expansion takes place at the valve C which is controlled by the knob E. The low pressure gas (at about 1 atm). after entering

[1] For very complete details of the exact sizes of tubing, dimensions of the apparatus see, for example, K. OLSZEWSKI: Ann. Phys. **10**, 768 (1903).

the volume G, passes up through the interstices between the copper tubing in the interchanger and exits therefrom to the atmosphere. In the steady state the liquid air, collected in G, can be drained off through the tube R. The whole unit, which is only about 20 inches tall, is enclosed in an insulating jacket, K.

Due to its simplicity of construction numbers of these machines were used in laboratories for occasional production of liquid air. It took but 10 minutes to reach the liquefying state and with a 5 h.p. compressor produced about 1 liter liquid per hour. This indicates an efficiency of 3.8 kW-hr per liter liquid which, by comparison with other systems given in Table 12 is somewhat poor even for a small scale liquefier.

Fig. 55 shows a diagram of a LINDE air liquefier which was exhibited at the Paris Exhibition of 1900. This is a precooled two stage liquefier. Air from the

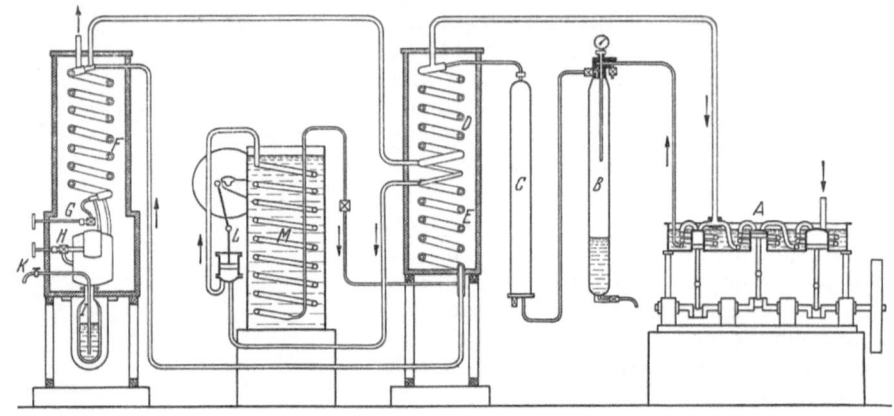

Fig. 55. Diagram of early two-stage LINDE air liquefier with ammonia refrigerated preccooling for 8 liters/hr production.
(Fig. from G. CLAUDE: Liquid Air, Oxygen and Nitrogen. Philadelphia: Blakiston 1913.)

compressor A at 200 atm. pressure passes through the dryer and purifier tubes B and C into the precooling interchanger $D-E$. This is cooled to about $-50°$ C by interchanging heat with the auxiliary ammonia compressed vapor refrigerator $L-M$. The cooled compressed air then passes into the three channel interchanger F, composed of 3 concentric tubes wound into a helix, and expands to 40 atm. pressure at the valve G and so cools to about 130° K. The second JOULE-THOMSON expansion to 1 atm. pressure takes place at the valve H and the liquid air froms here at 80° K. The fraction of the gas returning at 40 atm. passes back to the interstage junction of the compressors via all the interchangers. The return gas at 1 atm. pressure only passes back through the interchanger F, at the warm end of which it issues to the atmosphere. This machine yielded an output of 8 liters liquid per hour.

22. More recent air liquefiers. A detailed description of the more recent types of air liquefiers using the LINDE system of liquefaction is beyond the scope of this article. Those which have been marketed commercially have been developments of the basic two-stage LINDE system illustrated in Fig 55. In general, however, present practice is to produce either pure liquid oxygen or pure liquid nitrogen by low temperature rectification of air. Small air rectifiers suitable for laboratory operation have been constructed in general using the method of isentropic expansion for cooling (cf. chapters F and G), as for example the CLAUDE-HEYLANDT system (Sect. 32) and the low pressure system (Sect. 36 and 37).

Very large scale air rectification is now an important industry and plants capable of production of up to 120 tons of liquid O_2 per day have been reported[1]. These so-called "tonnage oxygen plants[2]" do not use JOULE-THOMSON expansion for cooling however, but instead use turbo-expansion engines (cf. Sect. 36) together with regenerators rather than interchangers[3]. The introduction of the use of regenerators and reversing interchangers, as detailed in chapter I, has allowed great simplification of the air purification problem with the consequent realization of large scale operation.

23. Large hydrogen liquefiers. The first liquefaction of hydrogen was made by J. DEWAR at the Royal Institution, London, in 1898[4]. He precooled the high pressure hydrogen gas below its inversion temperature by passing it through a coiled tube immersed in liquid air boiling under reduced pressure. From there, the gas passed through a simple LINDE circuit, the liquefaction taking place through JOULE-THOMSON expansion at a valve No detailed description of this apparatus was given, although similar machines were built by the British-Oxygen Co., one being purchased in 1904 by the Bureau of Standards[5].

The first detailed description of a hydrogen liquefier, embodying essentially the principles used in DEWAR's liquefier was given by M. W. TRAVERS in 1901[6]. A drawing of the liquefier is given in Fig. 56. TRAVERS' own description of this apparatus was partly as follows: "The hydrogen from the compressor under a pressure of 200 atm. before entering the liquefier passes through a coil A, which is cooled to $-80°$ C in a mixture of solid carbonic acid and alcohol. It then enters the coil contained in the chamber B, which is continually replenished with liquid air during an experiment. The lower portion of this coil passes into the chamber C, which is closed and communicates through the pipe ff with an exhaust pump; liquid flows continuously from B into C through a pin valve, controlled by a lever b, and boiling under a pressure of 100 mm. of mercury lowers the temperature to $-200°$ C. The gas now passes into the refrigerator-coil D which is enclosed in the vacuum vessel H, and, expanding at the valve E, passes upwards through the interstices of the coil and the annular space F, surrounding B and C, to the outlet G, where it can return through W and R and the cock i to the main supply pipe N. The liquid which separates from the gas is ultimately collected in the vacuum vessel K, which can easily be removed . . ."

"In constructing the apparatus the coil D was wound on the thin steel tube c which contains the valve rod . . . The coil itself consists of 80 feet of drawn copper tube 5/64 inch internal and 9/64 external diameter; in winding it the spirals ran alternately away from and towards the central tube and great care was taken to preserve a uniform external diameter of $2\frac{3}{4}$ inches. The coils were carefully spaced and fixed in position with solder as each layer was wound. The length of the refrigerator-coil D was 7 inches . . ."

[1] See for example I. ROBERTS: Refrig. Engng. **60**, 950 (1952).

[2] See for example ref. 1 above; A. M. CLARK: Bull. Inst. Internat. Froid. Annexe **1954**, p. 2, 39. — R. SCHLATTERER: Bull. Inst. Internat. Froid. Annexe **1954**, p. 2, 21. — B. H. VAN DYKE: Steel **123**, 103 (1948). — Chem. Eng. News. **54**, 126 (1947).

[3] For example, the so-called LINDE-FRANKL system as reported by HOCHGESAND, Mitt. Forsch. Anst. GHH Konzern. **4**, Part 1 (1935), and by J. WUCHERER, Iron Coal Tr. Rev. **159**, 723 (1949).

[4] J. DEWAR: J. Chem. Soc. **73**, 529 (1898). — C. R. Acad. Sci. Paris **126**, 1408 (1898). — Proc. Roy. Soc. Lond. **63**, 256 (1898).

[5] F. G. BRICKWEDDE: Ohio State Univ. Eng. Exp. Station, News **18**, No. 3, 30 (1946).

[6] M W. TRAVERS: Phil. Mag. **1**, 411 (1901), see also „Experimental Study of Gases", p. 198, New York: MacMillan & Co. 1901, and Encyclopedia Britannica **14**, 184 also by TRAVERS.

The hydrogen gas, which was prepared by the action of dilute sulphuric acid on commercial granulated zinc, was stored in a gas holder. After compression

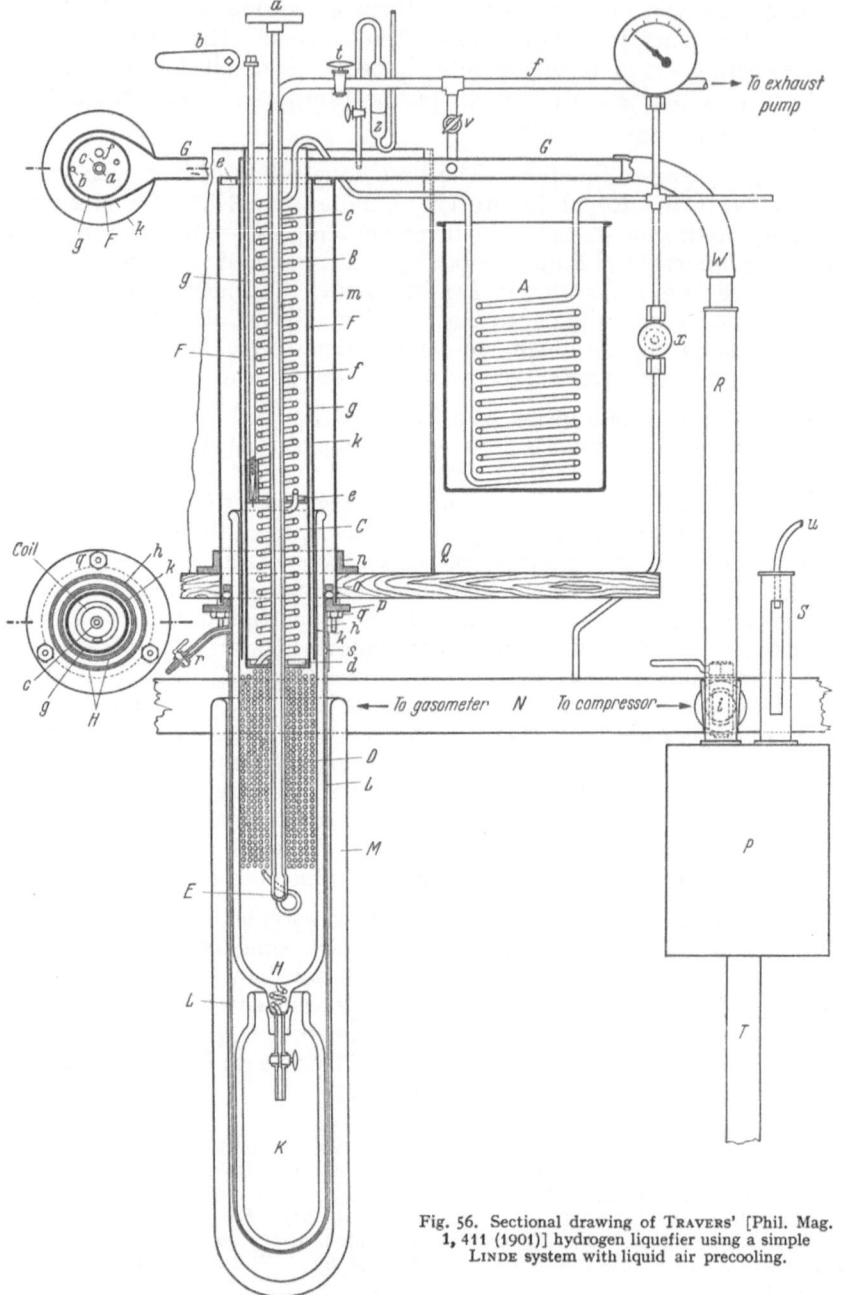

Fig. 56. Sectional drawing of Travers' [Phil. Mag. 1, 411 (1901)] hydrogen liquefier using a simple Linde system with liquid air precooling.

it passed through a cylinder containing lumps of solid caustic potash to remove traces of water and other impurities.

No statement was made by Travers concerning the rate of liquefaction with this equipment.

Since the successful operation of these early hydrogen liquefiers[1], many others of large and small capacity have been built in many institutions and commercial organizations, too numerous to list. However, there have not been many detailed publications on the constructional features of the larger machines. Some of the most notable detailed publications have been as follows:

1. H. KAMERLINGH-ONNES' first hydrogen liquefier in Leiden[2] built in 1906. This reference gives a detailed drawing of the machine which followed closely the design of TRAVERS and used again a HAMPSON-type interchanger for the main hydrogen circuit. Operated at a pressure of between 180 and 200 atm. the machine liquefied hydrogen at a rate of 4 liters/hour. Considerable detail is also given concerning the purification system for the H_2 gas.

2. H. KAMERLINGH-ONNES' second hydrogen liquefier in Leiden[3] which was completed in 1912. This is a similar but improved version of the first liquefier. The compressor flow was 40 m[3] gas per hour with an input pressure of between 150 and 200 atm. The liquefaction rate was 13 liters liquid per hour. A similar machine was built in Toronto about 1921[4].

3. W. MEISSNER[5] described a hydrogen liquefier installed at the Physikalisch-Technische Reichsanstalt in Berlin in 1928. This machine using a flow of 37 m.[3] per hour of gas with an input pressure of 175 atm produced 10 liters liquid per hour.

4. JONES, LARSEN and SIMON[6] described the hydrogen liquefaction installation at the Clarendon Laboratory, Oxford, in 1948. This reference gives also a full and detailed account of the many auxilliary apparatus needed in a complete liquid hydrogen production plant as well as detail of the liquefier itself. Unfortunately no detail is given concerning the construction of the heat interchangers. This plant is noteworthy in having provision for ortho→para conversion during liquefaction by passing the gas over charcoal at 75° K. This point is discussed in detail in Sect. 25. The gas flow through the liquefier is 50 m.[3]/hour with an input pressure of from 165 to 175 atm. and the production rate is 13 liters liquid per hour.

5. K. CLUSIUS[7] has described his hydrogen liquefier installed in Zürich in 1953. Details of the interchangers and a scale drawing are given. With a flow of 23 m.[3]/hr and input pressure 160 atm. the liquefaction rate is 7 liters/hour.

6. R. SPOENDLIN has described[8] a LINDE type machine set up at the CNRS, Bellevue, France, which not only produces liquid hydrogen but also in cascade, in the same unit, liquid helium. He has also given a detailed description[9] of the HAMPSON type heat interchangers employed and a detailed analysis of the thermodynamic losses incurred at each stage of liquefier circuit. The gas flow for hydrogen production only is 21 m.[3]/hour at an input pressure of 150 atm., yielding a liquefaction rate of 7 liters per hour.

---

[1] For description of others see for example K. OLSZEWSKI: Ann. Phys. 10, 773 (1903). — Bull. int. Acad. Cracovie 1908, 389; 1912. 1. See also LILIENFELD: Z. kompr. flüss. Gase 13, 186 (1911).

[2] H. KAMERLINGH-ONNES: Leiden Comm. 94 (1906).

[3] H. KAMERLINGH-ONNES (Posthumous publication): Leiden Comm. 158 (1926). — C. A. CROMMELIN: Leiden Comm. Suppl. 45 (1922).

[4] J. C. McLENNAN: Roy. Soc. Can. Trans. 15, 31 (1931). — J. C. McLENNAN and G. M. SHRUM: Roy. Soc. Can. Trans. 16, 181 (1922).

[5] W. MEISSNER: Phys. Z. 29, 610 (1928).

[6] JONES, LARSEN and SIMON: Research 1, 420 (1948).

[7] K. CLUSIUS: Z. Naturforsch. 8, 479 (1953).

[8] R. SPOENDLIN: J. Res. CNRS. 28, 1 (1954).

[9] R. SPOENDLIN: J. Res. CNRS. 3, 309 (1951).

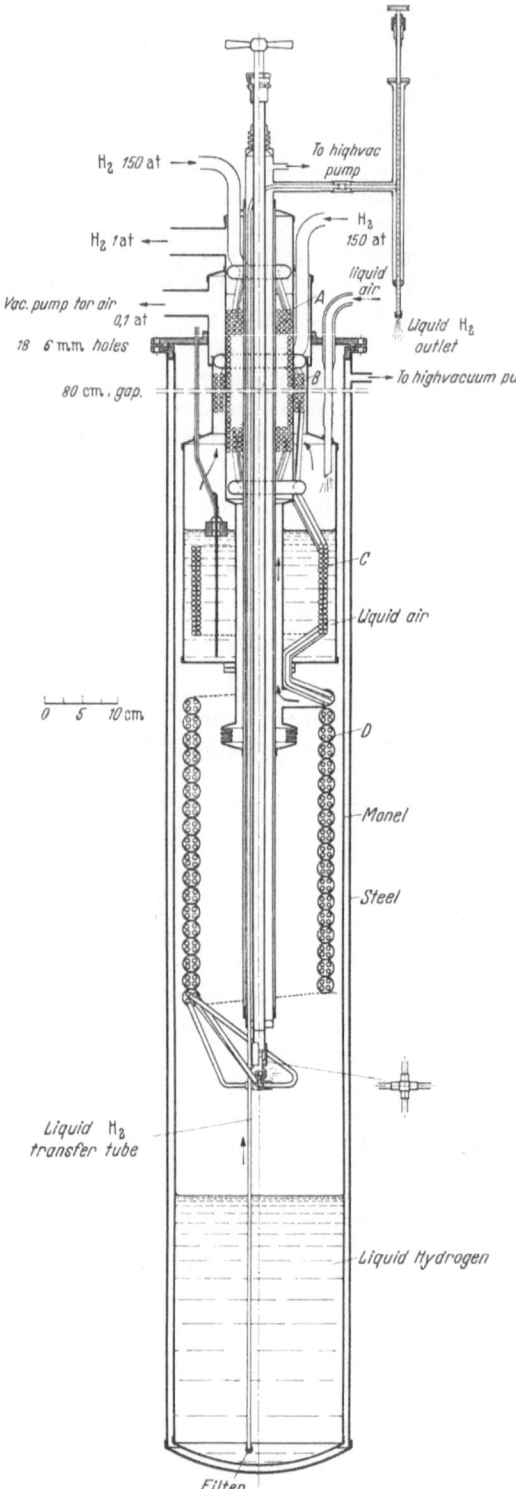

H₂ 150 at

To highvac
pump

H₂
150 at

H₂ 1 at

liquid
air

Vac. pump for air
0,1 at

A

18 6 mm holes

Liquid H₂
outlet

80 cm, gap.

B

To highvacuum pump

C

Liquid air

0   5   10 cm

D

Monel

Steel

Liquid H₂
transfer tube

Liquid Hydrogen

Filter

7. The third hydrogen liquefier at the Leiden Laboratory under the direction of C. J. Gorter has been described[1] in some detail in 1953. A scale drawing of the machine is given in Fig. 57 which is self-explanatory. The pertinent data on the heat interchangers are as follows: Interchanger A, 12 parallel 6 mm. o.d., 6 mm. i.d., 7 m. long. 1st layer 5, 2nd layer 4, 3rd layer 3. Interchanger B, 4 parallel 6 mm. o.d., 4 mm. i.d., 10 m. long. 1st layer 3 parallel crossing halfway to 2nd layer, 2nd layer 3 parallel crossing halfway to 1st layer. Interchanger C, 4 parallel 6 mm. o.d., 4mm. i.d., 4 m. long. Interchanger D, 4 parallel 6 mm. o.d., 4 mm. i.d., 10 m. long inside a 22 mm. o.d., 20 mm. i.d. tube. The machine is designed for a flow of 120 m.³/hr with an input pressure of about 200 atm. and with a liquefaction rate of 40 liters liquid per hour. The estimated efficiency is given as 5.3 kW-hrs/liter liquid, of which figure 40.6% is used for the energy of compression, 48.1% represents the free energy of the liquid air used (1.5 liters liquid air per liter of liquid hydrogen) and the remaining 11.3% being accounted for by the energy to run the vacuum pump for precooling purposes. It has also been run with a flow of 220 m.³/hr, producing 75 liters liquid H₂ per hour *inside* the machine[2].

The pertinent data concerning the liquefiers referred to above are summarized, where known, in Table 14. Here the input pressure (p_2 of chapter D), the input

[1] De Modernisering van het Kamerlingh Onnes Laboratorium te Leiden. 1953.
[2] Private communication from Prof. K. W. Taconis.

Fig. 57. Sectional drawing of the 1953 Linde-type hydrogen liquefier of the University of Leiden. (Drawing by courtesy of Prof. C. J. Gorter from "De Modernisering van het Kamerlingh-Onnes Laboratorium te Leiden".)

temperature to the LINDE circuit (T_2 of chapter D), the flow rate, the liquefaction rate and the observed coefficient of liquefaction, ε, are tabulated.

Table 14. *Some published data concerning H_2 liquefiers in practice.*

Location	Reference	Date	Input pressure (p_2) atm	Input temperature to LINDE circuit (T_2) °K	Gas flow rate m.³/hr	lique-faction rate liter/hr	ε † %
University College London	TRAVERS[1]	1901	200	73	—	—	—
Leiden University	K-ONNES[2]	1906	180—200	—	20**	4	16.8
Cracow	OLSZEWSKI[3]	1912	200	—	—	—	—
Leiden University	K-ONNES[4]	1912	150—200	*	40**	13	27.2
PTR Berlin	MEISSNER[5]	1928	175	—	37	10	22.7
Johns Hopkins University	BLANCHARD and BITTNER[6]	1942	150	63	34	8.4	20.8
Oxford University	JONES, LARSEN and SIMON[7]	1948	165—175	66	50	13	21.8
Zürich	CLUSIUS[8]	1953	160	65	23	7	24.4
CNRS. Paris	SPOENDLIN[9]	1954	150	63	21	7	27.8
Leiden University	GORTER[10]	1953	200	~63	120	40	28

24. Problems of gas purification and the KAPITZA H_2 circuit. The main impurities present in hydrogen gas are oil vapor (from the compressor, if used), water vapor, oxygen and nitrogen. The relative abundance of each impurity will depend on the source and treatment of the gas. Hydrogen obtained by electrolysis, a method frequently employed in liquefaction plants, tends to be of higher purity than commercial grade technical hydrogen. The oil vapor and water vapor can readily be removed by a cold trap before the gas enters the liquefier. In addition some plants pass the gas over drying plants, such as $CaCl_2$, etc., at room temperature. A very satisfactory cold trap has been described by BLANCHARD and BITTNER[11].

Fig. 58 shows the construction of this high pressure dryer which consists of a long rather open winding HAMPSON-type interchanger. The top of the "jacket" tube is cooled in a tank containing dry ice and alcohol. The input gas enters the jacket tube at the bottom and the water and other condensible impurities such as oil vapor in the gas condense out on the outer walls of the return spiral of the interchanger, since in going up the tube the return spiral is progressively

---

† The liquefaction coefficient, ε, given in this table may in some instances refer to the rate of provision of liquid H_2 *outside* the liquefier. For such situations the transfer loss must be known, before comparison of ε with theory can be made.

* Vapor pressure of 2 mm for precooling bath stated only.

** Compressor displacement only quoted.

[1] TRAVERS: Phil. Mag. **1**, 411 (1901).
[2] K-ONNES: Leiden Comm. **94 f** (1906).
[3] OLZEWSKI: Krakauer Anz. **1912**.
[4] K-ONNES: Leiden Comm. **158** (1926); Suppl. **45** (1922).
[5] MEISSNER: Phys. Z. **29**, 610 (1928).
[6] BLANCHARD and BITTNER: Rev. Sci. Instrum. **13**, 394 (1942).
[7] JONES, LARSEN and SIMON: Research **1**, 420 (1948).
[8] CLUSIUS: Z. Naturforsch. **8**, 479 (1953).
[9] SPOENDLIN: J. Res. CNRS. **28**, 1 (1954).
[10] GORTER: De Modernisering van het KAMERLINGH-ONNES Laboratorium te Leiden. 1953.
[11] E. R. BLANCHARD and H. W. BITTNER: Rev. Sci. Instrum. **13**, 394 (1942).

cooler. At the top the dry cool gas enters the return spiral via a strainer. The condensed water drains to the bottom of the jacket tube where it may be drained off.

For a flow of H_2 of 34 m.³/hr at 165 atm. Blanchard and Bittner used for the jacket monel tube $1\frac{3}{4}''$ o.d., 0.22'' wall, 82 cm. long, and for the return spiral two parallel soft copper tubes $\frac{3}{16}''$ o.d.; 0.030'' wall, each 7 m. long.

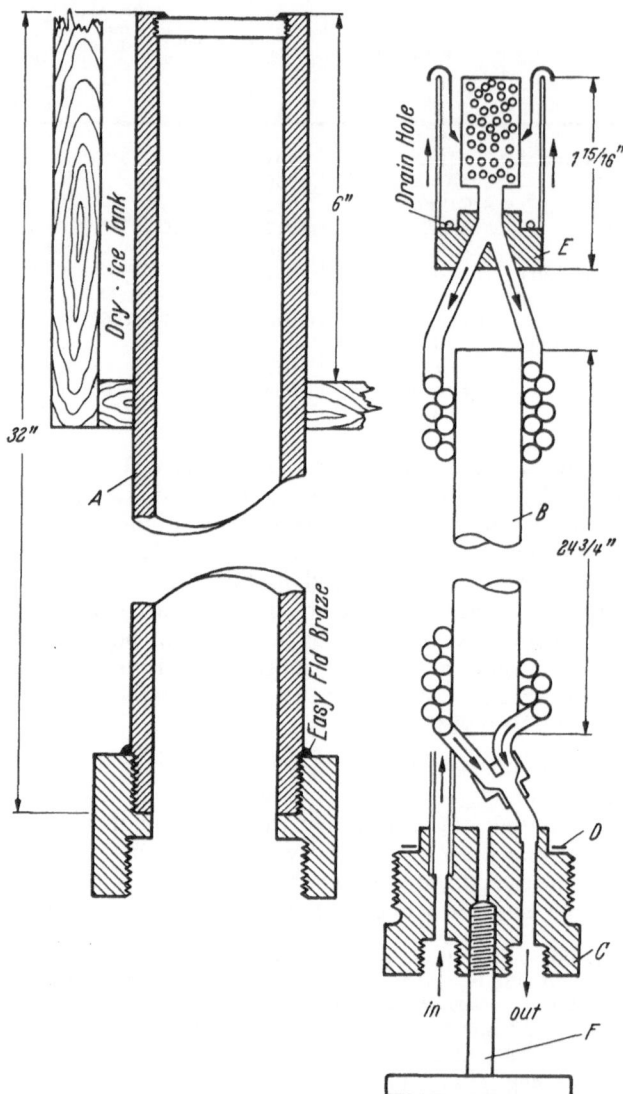

Fig. 58. Section views of purifying system for high pressure hydrogen due to Blanchard and Bittner. [Rev. Sci. Instrum. 13, 394 (1942).]

Oxygen can also readily be removed from the gas stream by passing it over hot copper shavings or a suitable catalytic purifier.

The removal of the nitrogen from the gas is more difficult. It is not unusual for these impurities to amount to about 0.5% by volume. This would mean therefore for each liter of hydrogen liquefied, with a liquefaction coefficient, ε, of say 25%, 3.14 m.³ of H_2 gas must pass through the machine which would deposit, with the 0.5% impurity content, approx. 20 gms of solid N_2, etc. For liquefying, say, 10 liters H_2 therefore the machine would accumulate about 100 to 200 cm.³ of solid impurity which could very well block or partially block the high pressure tubes and valves and, still more importantly, this impurity would "plate" out on the inside surfaces of the interchangers so causing inefficient thermal transfer. This problem can be solved in various ways. MEISSNER[1] included a special trap in his hydrogen liquefier which collected the impurities and which could be warmed at intervals with an electric heater without warming up the liquefier itself and on warming the trap was flushed out. Alternatively, an absorption purifier consisting of activated charcoal at liquid nitrogen temperatures can be included in the high pressure H_2 line just before the liquefier, as has been reported for example by JONES, LARSEN and SIMON[2] and by HOOD and GRILLY[3]. KAMERLINGH-ONNES instituted a third method[4], namely prepurification of the H_2 gas by means of a trap cooled with liquid air, boiling under reduced pressure, *before* the gas is passed into the low pressure gas holder. KAMERLINGH-ONNES estimated that by this means the impurity content is reduced to less than $\frac{1}{20}$%.

A quite different arreangement to deal with the purity problem was introduced by KAPITZA and COCKROFT in 1932[5]. Their hydrogen liquefier consisted of two separate circuits, one a closed circuit of very pure hydrogen which served as the refrigerant, and a second of commercial technical hydrogen (purity 99.5%) which was condensed in a low pressure circuit. A schematic diagram of the apparatus is given in Fig. 59. The closed high purity circuit, having a total content of

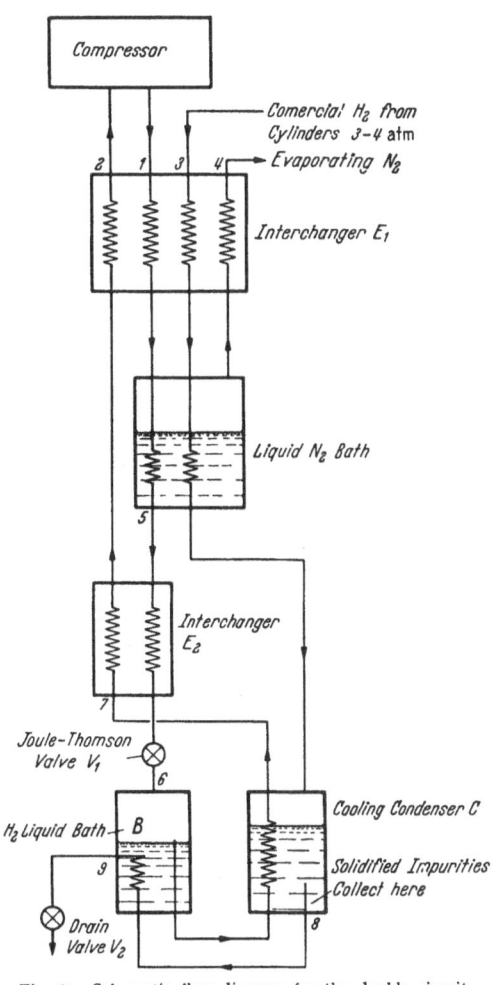

Fig. 59. Schematic flow diagram for the double circuit hydrogen liquefier of KAPITZA and COCKROFT. [Nature, Lond. 129, 224 (1932).]

about 0.7 m.³ of H_2 at NTP, included the compressor which forced the gas at 160 to 170 atm. pressure at *1* into the first heat interchanger E_1. This gas reached the temperature of liquid N_2 on leaving the liquid N_2 bath at *5*. (For simplicity only one liquid N_2 bath is shown in Fig. 59. In practice, two liquid N_2 baths were used, the first one with the liquid N_2 boiling at atmospheric pressure at

[1] W.-MEISSNER: Phys. Z. 29, 610 (1928).
[2] JONES, LARSEN and SIMON: Research 1, 420 (1948).
[3] C. B. HOOD and E. R. GRILLY: Rev. Sci. Instrum. 23, 357 (1952).
[4] H. KAMERLINGH-ONNES: Leiden Comm. 94f. (1906).
[5] P. KAPITZA and J. D. COCKROFT: Nature, Lond. 129, 224 (1932).

about $77°$ K and the second boiling under reduced pressure at about $65°$ K.
A heat interchanger was placed between the two baths. In this way the load
on the pump reducing the pressure over the liquid N_2 was reduced well below
that which would occur using only one liquid N_2 bath.) At 5 the pure H_2 entered
the heat exchanger E_2 of the simple LINDE circuit and at 6 the isenthalpic expansion at the JOULE-THOMSON valve, V_1, took place. The liquid formed collected in the H_2 liquid bath, B, until it was half full; from here the remaining
gas flowed through the spiral inside the cooling condenser C and back at 7 into
the low pressure return circuit.

Technical H_2 from cylinders with purity about 99.5% after passing through
a reducing valve entered the second H_2 circuit at 3 at a pressure of from 3 to
4 atm. After being cooled by passing through interchanger E_1 and the liquid
N_2 baths, this H_2 passed into the cooling condenser, C, where it liquefied at 3
to 4 atm. This cooling condenser was filled with wire gauze to help liquefaction.
The impurities solidified in C and fell into the bottom part of C which was of
ample volume. To draw off the liquid H_2 the liquid flowed through the draw-off
tube at 8, through a further cooling spiral 9 in the liquid H_2 bath B and out
through the drain valve, V_2.

With this arrangement, by using relatively large diameter tubes in the condensing circuit, the impurities caused no blocking troubles and it was reported
that this machine at Cambridge University produced about 4 liters liquid H_2
per hour. No detailed data on construction however were given.

BLANCHARD and BITTNER[1] however have given a very full constructional
description, including data on the heat interchangers, of a similar two-circuit H_2
liquefier which they constructed at Johns Hopkins University. This machine
had a flow in the high purity H_2 refrigeration circuit of 34 m.³/hr at an input
pressure of 150 atm., and yielded 8.4 liters liquid per hour when the precooling
temperature was $63°$ K (cf. Table 14). The construction of a similar machine
has been briefly reported by HUFFMAN[2].

25. Liquid neon circuits. It was suggested by CLUSIUS[3] that many of the
disadvantages of high pressure LINDE type liquefiers for liquid hydrogen could
be avoided by using instead a neon liquefier and then liquefying the hydrogen
in a separate low pressure condensing system. It is to be noted that the boiling
point, under atmospheric pressure, of Ne is $27.2°$ K whereas the critical point
of H_2 is $33.2°$ K (cf. Table 8). A somewhat similar system has been constructed
and reported on by HOOD and GRILLY[4]. It consisted of a high pressure LINDE-
type neon liquefier, similar in essential details to the hydrogen liquefiers described
above and a separate neon-hydrogen converter. The liquefier had an input
flow rate of approximately 100 m.³ Ne per hour at from 140 to 170 atm. input
pressure and produced about 17 liters liquid per hour with a precooling temperature of $71°$ K ($\varepsilon = 20\%$). This liquid Ne was then transferred into the "converter" which was a separate unit and at the same time H_2 gas at 6.6 atm. was
condensed in the converter, thus producing 27.5 liters/hour of liquid hydrogen.
In this way the plant produced about 20% more liquid H_2 per hour than it would
do producing liquid hydrogen directly in the same LINDE-type liquefier. As
the authors stated, it would have been preferable to make the two gas system
all into one unit and this may account for their subsequent abandonment of
the neon circuit.

[1] E. R. BLANCHARD and H. W. BITTNER: Rev. Sci. Instrum. **13**, 394 (1942).
[2] H. M. HUFFMAN: Chem. Rev. **40**, 1 (1947).
[3] K. CLUSIUS: Z. ges. Kälteind. **39**, 94 (1932).
[4] C. B. HOOD and E. R. GRILLY: Rev. Sci. Instrum. **23**, 357 (1952).

26. Ortho$\rightarrow$para-H_2 conversion. If normal-H_2, i.e. 75% ortho-H_2 and 25% para-H_2, is liquefied and stored in storage Dewars, the ortho$\rightarrow$para conversion takes place relatively rapidly[1, 2] and the heat of conversion would result in a rapid loss of the stored liquid hydrogen. Measurements of this effect have been made for example by LARSEN, SIMON and SWENSON[3] and GRILLY[4]. The latter author for example found that a 25 liter capacity Dewar, which had a basic evaporation rate for fully converted liquid H_2 of 22 cm.[3] liquid/hour, freshly filled with liquid H_2 of 70% ortho-H_2 content lost 10 liters liquid in the first 4 days; whereas the same vessel filled with liquid H_2 of 32% ortho-H_2 content only lost 4 liters in the equivalent period. Very distinct gain in storage economy is to be made if ortho-para conversion is made to take place during or before the liquefaction process. In the Oxford University H_2 liquefier described by JONES, LARSEN and SIMON[5] provision for such conversion is made by having one of the precooling stages of the liquefier include a conversion catalyst. In this liquefier the high pressure gas passes over a charcoal catalyst contained in a vessel about 1.5 liters volume and maintained at 75° K (the charcoal was activated carbon grade C.S. supplied by the British Carbo-Norit Union Co.). In this way the exit hydrogen was converted to about 50% ortho-H_2 content. In a system described by GRILLY[6] the catalytic system consisted of a circular bundle of seven brass cylindrical tubes, (each of 3,42 liter volume, 89 cm. long, 7.00 cm. o.d., 0.318 cm. wall) connected in series and mounted in a Dewar at 78° K. The first two tubes contained charcoal and the others a chrome alumina catalyst (20% Cr_2O_3 on an Al_2O_3 carrier supplied by the Harshaw Chemical Co.). Approximately 18 m.[3]/hr (NTP) normal H_2 at 4 atm. pressure could be passed through this and the effluent H_2 was of 51.8% ortho-H_2 content. In GRILLY'S arrangement this effluent gas was fed into the low pressure feed line of the H_2 liquefier system.

27. Small hydrogen liquefiers. Many small hydrogen liquefiers have been made, used and reported on. They have proved of extreme value where liquid hydrogen is required only in small quantity, particularly since they can be operated from commercial high-pressure cylinders of H_2 without the need for expensive compressors. It would not be practical to list all those which have been described but the following gives a representative selection of them. First NERNST[7] used one producing approximately $\frac{1}{2}$ liter liquid per hour. Subsequent descriptions, for example, have been given by LATIMER[8]; LATIMER, BUFFINGTON and HOENSHEL[9]; RUHEMANN[10]; KEYES, GERRY and HICKS[11]; AHLBERG, ESTERMANN and LUNDBERG[12]; FAIRBANKS[13]; DE SORBO, MILTON and ANDREWS[14].

---

[1] E. CREMER and M. POLANYI: Z. phys. Chem., Abt. B **21**, 459 (1933).

[2] SCOTT, BRICKWEDDE, UREY and WAHL: J. Chem. Phys. **2**, 454 (1934).

[3] LARSEN, SIMON and SWENSON: Rev. Sci. Instrum. **19**, 266 (1948).

[4] E. R. GRILLY: Rev. Sci. Instrum. **24**, 1 (1953).

[5] JONES, LARSEN and SIMON: Research **1**, 420 (1948).

[6] E. R. GRILLY: Rev. Sci. Instrum. **24**, 1 (1953).

[7] W. NERNST: Z. Elektrochem. **17**, 735 (1911). See J. E. LILIENFELD: Z. kompr. flüss. Gase **13**, 165 (1911).

[8] W. M. LATIMER: J. Amer. Chem. Soc. **44**, 90 (1922).

[9] LATIMER, BUFFINGTON and HOENSHEL: J. Amer. Chem. Soc. **47**, 1571 (1925).

[10] M. RUHEMANN: Z. Physik **65**, 67 (1930).

[11] KEYES, GERRY and HICKS: J. Amer. Chem. Soc. **59**, 1426 (1937).

[12] AHLBERG, ESTERMANN and LUNDBERG: Rev. Sci. Instrum. **8**, 422 (1937).

[13] H. A. FAIRBANKS: Rev. Sci. Instrum. **17**, 473 (1946).

[14] DE SORBO, MILTON and ANDREWS: Chem. Rev. **39**, 403 (1946).

It is considered sufficient to detail the description of only one of these machines, since they all are very similar. A diagram of the general arrangement of Ahlberg, Estermann and Lundberg's H_2 liquefier is shown in Fig. 60. High pressure hydrogen from a number of cylinders initially at about 150 atm. pressure is first passed through a $CaCl_2$ drying tube and then into the purifying precooling unit A before passing into the liquefier B. The first heat interchanger I in A consists of 2 feet of $\frac{1}{8}''$ o.d., 0.030'' wall, copper tubing pushed inside approximately 2 feet of $\frac{3}{16}''$ i.d., $\frac{1}{16}''$ wall, lead tubing and wound into a helix of a few turns. The copper tube carries the high pressure gas; the annular region inside the lead tubing the low pressure. The four tubes containing the charcoal

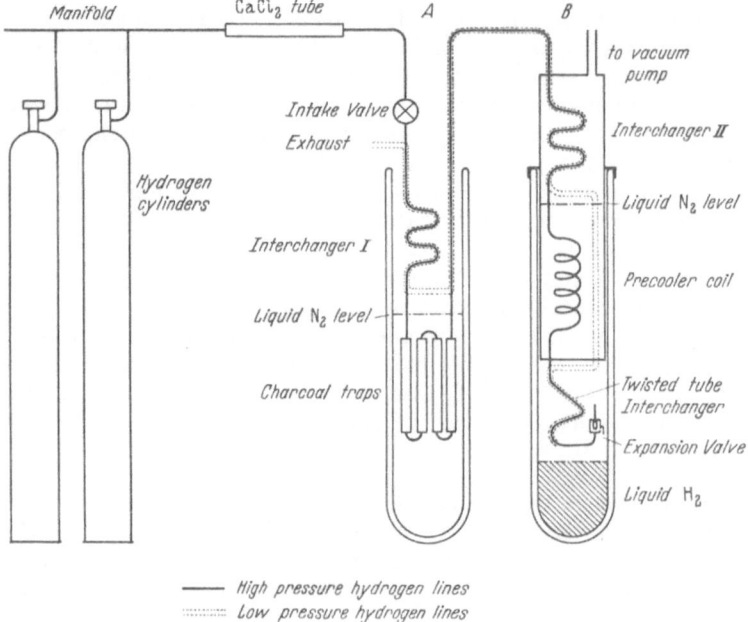

Fig. 60. General arrangement of the small scale Linde-type hydrogen liquefier of Ahlberg, Estermann and Lundberg. [Rev. Sci. Instrum. 8, 422 (1937).]

are of $\frac{1}{2}''$ o.d. heavy brass, 10'' long, connected in series. The whole unit A fits in a Dewar $2\frac{5}{8}''$ i.d., 24'' long containing liquid N_2 boiling under atmospheric pressure.

The liquefier unit is shown in detail in Fig. 61. Interchanger II consists of 5 feet of double tubing exactly like interchanger I. The high pressure gas then passes through a pre-cooling coil, PC, made of 20 feet of $\frac{1}{8}''$ o.d., 0.030'' wall, copper tubing. This coil is immersed in liquid N_2 contained in a closed space made with a German-silver tube $2\frac{1}{4}''$ o.d., $\frac{1}{32}''$ wall and 16'' long; and the liquid N_2 can be made to boil under reduced pressure by pumping through the exit tube, VP. After passing through the precooling spiral, PC, the H_2 gas enters the final Linde circuit at the interchanger, TT.

This is a "Twisted tube" interchanger following the design of G. F. Nelson, as described by Bichowsky[1], and was made as follows: a 20'' length of $\frac{1}{8}''$ o.d., 0.015'' wall, annealed German-silver tube was rolled flat (with about $\frac{1}{10}$ mm.

---

[1] F. R. Bichowsky: J. Ind. Chem. Soc. 14, 62 (1922).

inside clearance) and twisted around its axis to form a spiral of about 1 cm. pitch. This was then inserted in a thin lead tube, $\frac{3}{16}''$ i.d., 0.015'' wall, and this sheath formed the low pressure return. After passing the TT interchanger the gas expands at a needle valve and returns via the low pressure circuit to the atmosphere. The liquefier unit is contained in a Dewar approximately $2\frac{1}{2}''$ i.d. which serves to collect the liquid H_2 formed on expansion at the valve and a vacuum jacketed syphon, S, is provided for withdrawal of the liquid H_2.

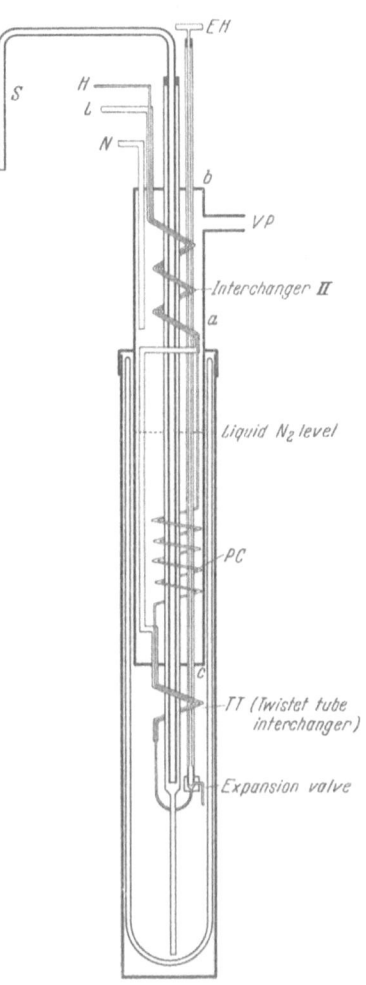

Fig. 61. Section through liquefier column of the small scale hydrogen liquefier of Fig. 60. [Due to AHLBERG, ESTERMANN and LUNDBERG: Rev. Sci. Instrum. **8**, 422 (1937).]

In operation, if the liquefier Dewar was first precooled with liquid N_2, the machine began liquefying about 5 min after N_2 temperatures had everywhere been reached. For liquefaction the flow was maintained at about 5 m.^{3} (NTP) per hour and yielded about 1 liter per hour when the precooling temperature was about 69° K. This corresponds to a liquefaction coefficient, $\varepsilon = 16\%$. If a stronger pump were to have been used on the liquid nitrogen precooling bath, so that the precooling temperature were about 63° K (triple pt. of N_2), then ε would have increased to about 24%.

28. Combined H_2 and He liquefiers. Small LINDE-type hydrogen liquefiers have been built in combination with small helium liquefiers from time to time in order to provide an inexpensive cryostat for low temperatures. Reports describing such equipment have been made for example by ROLLIN[1], SEILER[2] and SCHALLAMACH[3]. Since the hydrogen liquefying circuits in these machines were essentially similar to that described in Sect. 27 above, further comment is unecessary.

Large combination LINDE-type hydrogen and helium liquefiers have been described by ASHMEAD[4] and by SPOENDLIN[5]. In ASHMEAD's machine in Cambridge University no provision was made for withdrawal of free liquid H_2, the machine being essentially a helium liquefier. SPOENDLIN's machine at the Bellevue Laboratories has already been discussed briefly in Sect. 23 above.

F. Theory of gas liquefaction using isentropic expansion.

29. Comparison of isentropic with isenthalpic expansions. The method of gas liquefaction based on the cooling by JOULE-THOMSON or isenthalpic expansion, described in earlier sections, cannot in principle be as efficient as an equivalent

[1] B. V. ROLLIN: Proc. Phys. Soc. Lond. **48**, 18 (1936).
[2] K. SEILER: Ann. Phys. 39, 129 (1941).
[3] A. SCHALLAMACH: J. Sci. Instrum. 20, 195 (1943).
[4] J. ASHMEAD: Proc. Phys. Soc. Lond. **63**, 504 (1950).
[5] R. SPOENDLIN: J. Res. CNRS. **28**, 1 (1954).

system using isentropic expansion for cooling, due to the inevitable thermo-dynamic irreversibilities accompanying the former process. Any irreversibility introduced into the cooling cycle must decrease the efficiency and in an isenthalpic expansion the rate of change of entropy with pressure is given by:

$$\left(\frac{\partial S}{\partial p}\right)_H = -\frac{v}{T}. \tag{29.1}$$

The total irreversible loss of heat is given by $\int T\,dS$, the limits of integration being determined by the initial and final temperatures and pressures. On the other hand in an isentropic expansion, in which the gas is allowed to do external work, there is by definition no increase in the entropy. Methods therefore, such as those introduced by Claude[1] for the liquefaction of air, in which cooling is obtained by expansion accompanied by the doing of external work are in principle superior.

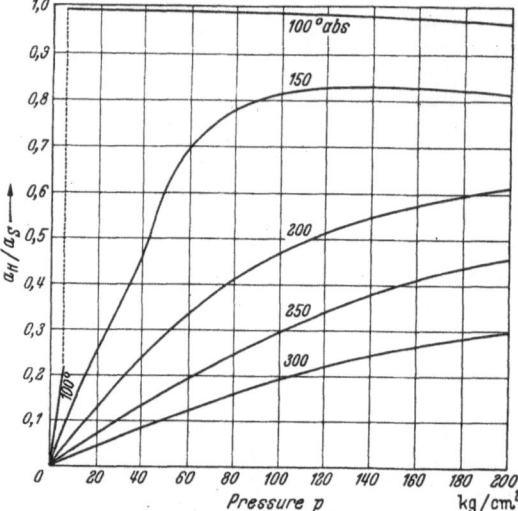

Fig. 62. Plot of the ratio α_H/α_S for air versus pressure for various isotherms, where α_H and α_S are the cooling coefficients in isenthalpic and isentropic expansion respectively. [The curves are due to H. Hausen: Z. techn. Phys. 7, 450 (1926).]

It is of interest to compare the coolings produced by the two expansion methods in more detail, particularly in order to gain an insight into the reasons underlying the remarkably high efficiencies obtained in isenthalpic processes. Writing the First and Second Laws of thermodynamics as:

$$T\,dS = C_p\,dT - T\left(\frac{\partial v}{\partial T}\right)_p dp \tag{29.2}$$

we obtain for the case of isentropic expansion $(dS = 0)$, an immediate evaluation of the coefficient of cooling,

$$\alpha_S = \left(\frac{\partial T}{\partial p}\right)_S = \frac{T}{C_p}\left(\frac{\partial v}{\partial T}\right)_p. \tag{29.3}$$

This is compared with the Joule-Thomson coefficient of cooling α_H, given in eq. (13.4) by:

$$\alpha_H = \left(\frac{\partial T}{\partial p}\right)_H = \frac{T}{C_p}\left(\frac{\partial v}{\partial T}\right)_p - \frac{v}{C_p} \tag{29.4}$$

whence we get:

$$\alpha_H = \alpha_S - \frac{v}{C_p} \tag{29.5}$$

indicating that as p is increased or T decreased, so diminishing v, the Joule-Thomson cooling coefficient, α_H, approaches the reversible isentropic cooling coefficient, α_S. Moreover, as the critical temperature is approached, C_p tends to an infinite value; hence in regions near critical isenthalpic expansions also must be highly efficient—a point already observed in discussion of compressed vapor refrigeration.

[1] G. Claude: Air liquide, Oxygène, Azote, Gaz rares. Paris 1926.

The relation between α_H and α_S can also be written as

$$\frac{\alpha_H}{\alpha_S} = 1 - \frac{v}{T}\left(\frac{\partial T}{\partial v}\right)_p \tag{29.6}$$

and a plot of this relationship is given in Fig. 62, which shows (α_H/α_S) as a function of pressure for air for various values of T. It will be seen first, that (α_H/α_S) tends to relatively high values as the temperature is reduced, and secondly that for readily approachable temperatures pressures between 50 and 150 atm. give the high values of (α_H/α_S). This latter fact accounts for the high efficiencies of the two-stage LINDE circuit, in which the pressure after the first expansion is not permitted to drop below about 40 atm.

From these results, therefore, one can conclude that for expansions in regions where the fluid is near the critical point (or *a fortiori* when the expanded fluid is liquid) or for expansions between relatively high pressures, the isenthalpic process is highly efficient and due to its mechanical simplicity it offers here as good a system as the isentropic expansion engine. However, where the pressure drops on expansion are large and/or take place from relatively high temperatures, the isentropic expansion is far preferable despite its technical complexity.

30. The actual CLAUDE cycle of operations. A diagrammatic sketch of the typical system introduced by CLAUDE[1] for air liquefaction in 1902

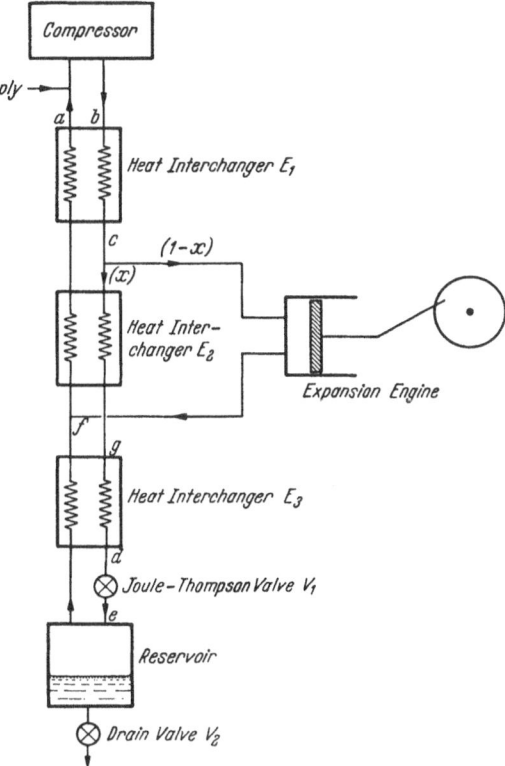

Fig. 63. Schematic flow diagram for CLAUDE's air liquefier using an adiabatic expansion engine.

is given in Fig. 63 and the steps in the operation can be followed in the $T-S$ diagram of Fig. 64. The air is first compressed at about room temperature, T_2, from pressure p_1 to p_2. In CLAUDE's earlier machines $p_1 = 1$ atm. and $p_2 \approx 40$ atm. This compression is represented by the isothermal path $a \rightarrow b$ in Fig. 64. The compressed air passes through the first heat interchanger, E_1, and at the exit of E_1 it splits into two streams, a fraction x passing down through the next heat interchanger E_2 and the remainder, $(1-x)$, going to an expansion engine. This point of division is marked at c on the $T-S$ diagram, and the temperature T_c was about 190° K in the CLAUDE machine. The expansion engine may be a lubricated reciprocating engine, as originally employed by CLAUDE[1, 2], an unlubricated reciprocating engine as used for example also by CLAUDE[2] and later by COLLINS[3] or a turbine

[1] G. CLAUDE: C. R. Acad. Sci. Paris **134**, 1568 (1902).

[2] G. CLAUDE: Liquid air, Nitrogen and Oxygen. Paris 1926.

[3] S. C. COLLINS: Rev. Sci. Instrum. **18**, 157 (1947). See also COLLINS, NASON and CANADY: Refrig. Engng. **59**, No. 12 (1951).

as reported by HOCHGESAND[1] and KAPITZA[2]. In general the work output of this expansion engine is fed back to the main compressor, although, as we shall see later, the power developed here is only a small fraction of the power used by the compressor. The detailed consideration of these various expanders is given in chapter G below.

In principle, it would be possible to produce liquid air in the expander directly. In practice, however, this produced technical difficulties and hazards and is not adopted. In the CLAUDE system the temperature reached on expansion is a little above the boiling point of air and the process of expansion is illustrated on the $T-S$ diagram by the path $c \rightarrow f$. If the expansion were truly isentropic, the path $c \rightarrow f$ would be a vertical line at constant entropy. Some irreversibilities are introduced, however, and in our diagram we have accounted for these by the slight slope of the line $c \rightarrow f$. In the CLAUDE system the actual liquefaction is completed by isenthalpic expansion at a JOULE-THOMSON valve, V_1, the final heat interchanger E_3 together with the valve V_1 forming a simple LINDE liquefaction circuit.

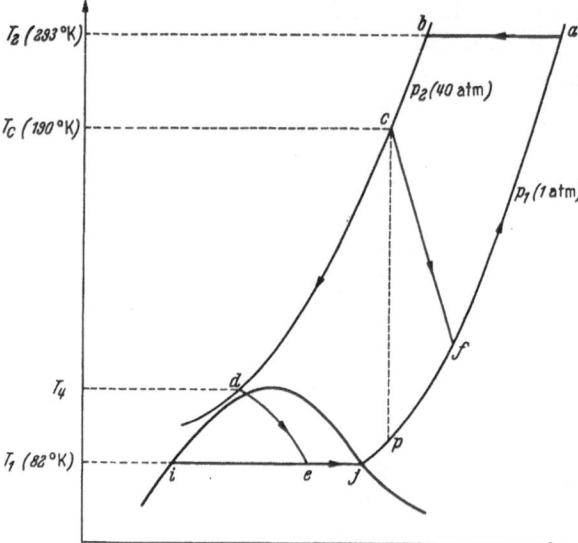

Fig. 64. $T-S$ diagram showing the thermodynamic cycle of operations for the CLAUDE air liquefier of Fig. 63.

The low pressure cold expanded gas from the expansion engine joins the main low pressure circuit at f, returns up through the heat interchanger E_2 thereby regeneratively cooling the fraction x of high pressure gas which passes into the final LINDE circuit at g. If one supposes that of the air expanding at the JOULE-THOMSON valve V_1 a fraction y is liquefied, then by appeal to Fig. 64 one sees that

$$y = \frac{je}{ji} \tag{30.1}$$

and moreover the overall liquefaction coefficient ε must be given by:

$$\varepsilon = x y. \tag{30.2}$$

The remaining part of the diagram of Fig. 63 is self explanatory.

One can regard the cooling occurring by means of the expansion engine in the CLAUDE system as a way of replacing the precooling (ammonia cycle) system in a LINDE liquefier. Moreover by choice of the value of the high pressure p_2 and the fraction x, this "precooling" temperature can be varied between wide limits.

31. The liquefaction coefficient and the efficiency. Ideally, the enthalpy of the air at p_2 and T_2 entering the machine at b (see Figs. 63 and 64) must be equal to the sum of the enthalpy of the fraction of air at p_1 and T_2 returning at a, the enthalpy of the liquid formed at p_1 and T_1 and the work produced by the

[1] B. C. P. HOCHGESAND: Mitt. Forsch. Anst. GHH Konzern **4**, Part I (1935).
[2] P. L. KAPITZA: J. Phys. USSR. **1**, 7 (1939).

expanding air in the expansion engine. This enthalpy balance can be written symbolically as:

$$H_b = (1 - \varepsilon)\, H_a + \varepsilon H_i + W \qquad (31.1)$$

where the subscripts indicate the states of the gas, where ε is the overal liquefaction coefficient and where W is the change in enthalpy of the air expanded in the expansion engine. By appeal to Fig. 64 remembering that a fraction $(1 - x)$ passes through the expansion engine, we have:

$$W = (1 - x)\,(H_c - H_f). \qquad (31.2)$$

Eqs. (31.1) and (31.2) are not sufficient to allow a unique computation of ε, since in principle x is an adjustable parameter. In order to complete the computation, it is necessary to consider the heat balance in any one of the heat interchangers. For simplicity consider the first interchanger, E_1, which on the high pressure side carries the whole gas stream. For this, assuming a perfect heat interchanger, (see chapter I for methods of allowing for interchanger imperfections), the heat extracted from the high pressure gas must be equal to that taken up by the low pressure gas, or symbolically:

$$H_b - H_c = (1 - \varepsilon)\, C_{p_1}\,(T_2 - T_c). \qquad (31.3)$$

Combining eqs. (31.1) and (31.2) we may write:

$$\varepsilon = \frac{(H_a - H_b) + (1 - x)\,(H_c - H_f)}{H_a - H_i}. \qquad (31.4)$$

A simultaneous use of this and eq. (31.3) enables T_c and hence ε to be evaluated for various values of x.

It is of interest to compare this liquefaction coefficient with that for a simple LINDE circuit operating between the same initial and final temperatures and pressures connotated by the subscripts a and b. For the simple LINDE circuit, [see eq. (18.1)]

$$\varepsilon = \frac{H_a - H_b}{H_a - H_i} \qquad (31.5)$$

which is smaller than that for the CLAUDE system [eq. (31.4)] by an amount $(1 - x)\,(H_c - H_f)/(H_a - H_i)$. For the CLAUDE system with $p_2 = 40$ atm., $T_c = 190°$ K and $x = 0.2$, eq. (31.4) leads to a value of the ideal efficiency of 0.33 kW-hr/liter liquid.

Some typical values of H_a, H_b, T_c, ε, etc. due to DAVIES[1] for practical operation of a CLAUDE air liquefier may be of interest. Some of these numerical values, where appropriate, have been included in parentheses in Fig. 64.

Typical operational conditions.

High pressure, $p_2 = 40$ atm. Initial temperature $T_2 = 293°$ K. Low pressure, $p_1 = 1$ atm. Liquefaction temperature $T_1 = 82°$ K. Temperature T_c at entrance to expansion engine = 200° K. Temperature T_f at exit to expansion engine = 110° K. Temperature T_a at exit of interchanger $E_1 = 288°$ K. Temperature of low pressure gas entering interchanger $E_1 = 170°$ K.

Resultant data.

$H_a =$	119 cal/kg.	$x =$	0.30
$H_b =$	118 cal/kg.	$y =$	0.43
$H_i =$	23 cal/kg.	$\varepsilon =$	0.14
$H_c =$	93.2 cal/kg.	$W =$	11.7 cal/kg. of entrant air.
$H_f =$	76.5 cal/kg.		

---

[1] M. DAVIES: The Physical Principles of Gas Liquefaction and Low Temperature Rectification. London: Longmans, Green & Co. 1949.

By assuming that the actual work of compression W_{comp} is 1.7 times the computed *isothermal* work of compression, the above data yield: $W_{comp} = 0.99$ kW-hr/liter liquid. This is to be compared with the work done by the expansion engine per liter *liquefied*, $W/\varepsilon = 0.09$ kW-hr/liter liquid. It is interesting to note that the work obtained from the expansion engine is only a small fraction (of the order of 10%) of the total work of compression. The markedly increased efficiencies of CLAUDE air liquefiers over those of simple LINDE circuits is not due therefore to the power fed back from the expansion engine.

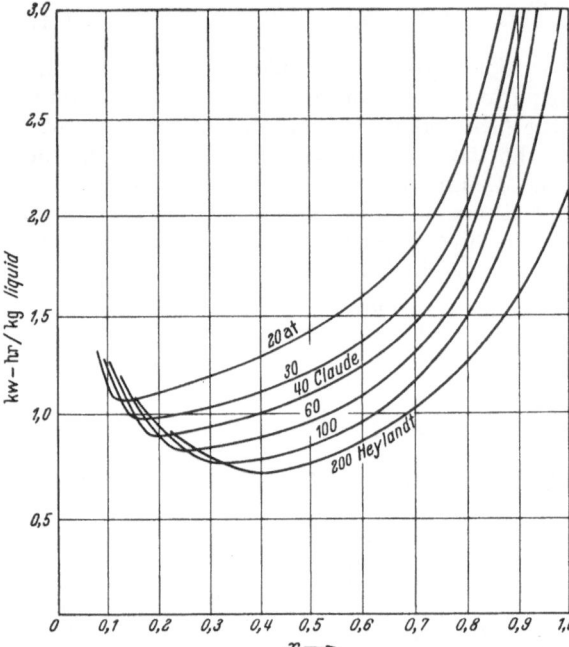

Computations of the practical efficiencies to be expected from CLAUDE air liquefying systems have been made by LENZ[1], using as basis the above equations. In his computations moreover he has taken into account typical interchanger inefficiencies, heat losses and irreversibilities in expansion, so that they represent an approach to "practical" conditions. The results are summarized in Fig. 65 in which the power consumption in kW-hr per kilogram of air liquefied is plotted as a function of $(1 - x)$, the fraction of gas passing through the expansion engine, for various values of the high pressure, p_2. It will be seen that each curve passes through a well defined minimum and that the value of $(1 - x)$ at which the minimum occurs gets smaller as

Fig. 65. The practical efficiencies of air liquefiers using the CLAUDE system of Fig. 64 giving the number of kW-hr per kg liquid air produced as a function of the fraction x of gas passing directly to the JOULE-THOMSON expansion valve for various values of the input pressure, p_2. [Curves due to H. LENZ: Handbuch der Experimentalphysik, Bd. 9/1, S. 135. 1929.)

the initial pressure, p_2, is increased. This is further illustrated by Table 15, given by DAVIES[2], which gives the value of $(1 - x)$, the temperature T_c at which the gas enters the expansion engine and the energy consumption in kW-hr/liter liquid for some practical CLAUDE systems, operating at their maximum efficiency, i.e. operating at about the minima of the curves of Fig. 65. This table indicates the additional fact that not only is the value of $(1 - x)$ decreased as p_2 is increased, but also that T_c increases with increasing p_2 for maximal efficient operation.

By comparison with Table 12 it will be seen that the practical efficiencies of the CLAUDE and CLAUDE-HEYLANDT liquefaction methods are about the same order as those for the two-stage LINDE circuit with precooling. By adopting the more thermodynamically efficient method of isentropic expansion, therefore, the CLAUDE system rivals and can even improve on the precooled two-stage LINDE circuit without having to adopt the extra process of precooling with an ammonia machine.

[1] H. LENZ: Handbuch der Experimentalphysik, Bd. 9/1, p. 135. 1929.

[2] M. DAVIES: Physical Principles of gas liquefaction and low temperature rectification. London: Longmans, Green & Co. 1949.

Table 15. *Conditions for operation of* Claude-*type air liquefier at maximum efficiency*
(due to Davies*)* [1].

	Turboexpansion machine (Kapitza[2])			Original Claude machine			Claude-Heylandt machine
High Pressure, p_2, (atm.)	6.5	20	30	40	60	100	200
Fraction $(1-x)$ of gas passing expansion engine		0.88	0.84	0.80	0.75	0.68	0.60
Temperature T_c of gas entering expansion engine. °K	115	155	173	191	215	241	293
Power required (theor) kW-hr/liter liquid.	—	—	—	0.33	—	—	0.32
Power required (pract) kW-hr/liter liquid.	1.5	1.08	1.00	0.90	0.83	0.78	0.70

32. The Claude-Heylandt air liquefier. Table 15 sets out the values of the fraction $(1-x)$ of gas passing throug the expansion engine and of the temperature T_c of the gas as it enters the expansion engine as a function of the high pressure p_2 for maximal efficient operation. It will be seen that there are two limiting cases for the Claude system, the first being when T_c becomes equal to room temperature and the second being when $(1-x)$ tends to 100%. The first limiting case provides the basis of the Heylandt air liquefier. The second limiting case is for low pressure operation with the expansion engine working at very low temperature. Such low pressure machines were adopted in the early nineteen thirties in air rectifiers using the Linde-Fränkl circuit, for which turbine expansion engines were used[3].

In the Claude-Heylandt air liquefier a fraction of about 60% of the gas passes straight into the expansion engine at room temperature. Here it expands to about 150° K before returning to the low pressure side of heat interchanger E_2. This arrangement has two advantages, first the first interchanger E_1 can be eliminated altogether and second the relatively high temperature operation of the expansion engine minimizes lubrication and insulation difficulties. Finally, as will be seen from Table 15, the efficiency is the highest of all Claude type systems.

33. The low pressure air liquefier. The second extreme case of Claude system operations is where the fraction $(1-x)$ of the gas passing through the expansion engine is made very large ($\sim 100\%$), which requires for maximal efficient operation relatively low pressures for the high pressure, p_2, compressor output and low temperature, T_c, at the entrance to the expansion engine. Although, as mentioned in Sect. 32 above, such low pressure machines were used by the Linde Company[3,4] early in the nineteen thirties, the first detailed report on an air liquefier operating on these principles was due to Kapitza[5]. This latter machine operated as follows: The high pressure air entered the machine at 6.5 atm. and after passing through the heat interchanger system split into two streams,

---

[1] M. Davies: The Physical Principles of Gas Liquefaction and Low Temperature Rectification. London: Longmans Green & Co. 1949.
[2] P. L. Kapitza: J. Phys. USSR. **1**, 7 (1939). This was a small machine. For similar machines on a larger scale the power required can be made to approach 1.1 kW-hr/liter liquid.
[3] R. Linde: Z. ges. Kälteind. **41**, 183 (1934).
[4] B. C. P. Hochgesand: Mitt. Forsch. Anst. GHH. Konzern **4**, Part I (1935). See also: J. Wucherer: Iron and Coal Trades Rev. **159**, 723 (1949).
[5] P. L. Kapitza: J. Phys. USSR. **1**, 7 (1939).

one fraction, $(1 - x)$, containing by far the larger mass, passed through a turbine which acted as the expansion engine. The expanded air at about 1.6 atm cooled the fraction x of air which did not pass through the expansion engine and this then was liquefied by compression. The liquid was then let down to lower pressure through a valve, V. (Follow path $e \rightarrow g$ on the $T - S$ diagram of Fig. 67.) Diagrams of the machine, together with a $T - S$ diagram for the operation, are given in Figs. 66 and 67 respectively, on which the approximate operating temperatures and pressures are shown.

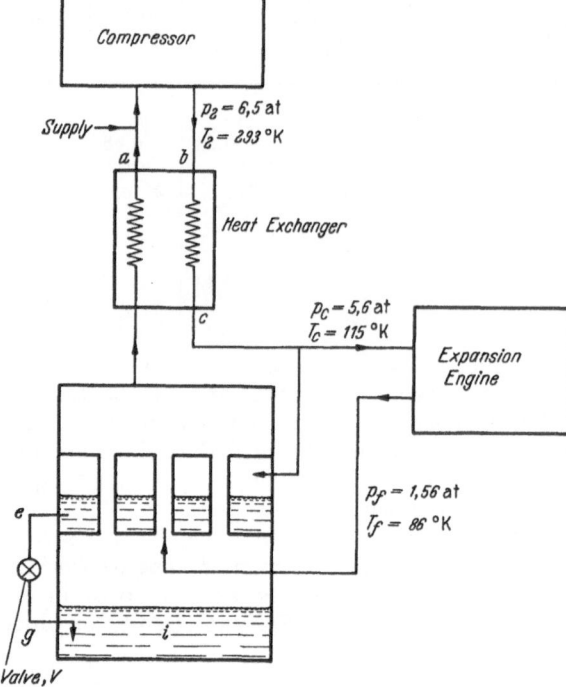

Fig. 66. Schematic flow diagram of KAPITZA's low pressure air liquefier using turbine expansion engine.

KAPITZA's original machine was a small one producing about 35 liters of liquid air per hour and it showed an overal efficiency of about 1.5 kW-hr/liter liquid. It was considered that by larger scale operation and by use of centrifugal turbo-type compression, rather than the reciprocating type compressor actually employed, this efficiency figure could be reduced to about 0.97 kW-hr/liter liquid, a figure comparable with the other CLAUDE type systems.

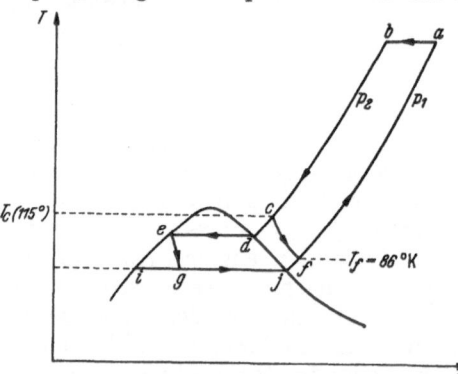

Fig. 67. $T - S$ diagram showing thermodynamical cycle of operations in the low pressure air liquefier of Fig. 66.

One great advantage of low pressure operation of this type air liquefier is reflected in the much lower initial costs for low pressure compressors, etc. and this makes this system attractive. A further and perhaps more important advantage lies in the fact that low pressure operation permits the use of heat regenerators or reversing heat interchangers (see chapter I) instead of the more conventional heat interchanger, and this enables one to dispense with the usual extensive and technically unpleasant equipment for the removal of water vapor and CO_2 from the air supply.

G. Practical aspects of gas liquefaction using expansion engines.

34. Air liquefaction.—Early work with reciprocating expansion engines. The earliest attempts at air liquefaction using reciprocating expansion engines coupled with heat interchangers were reported by COLEMAN[1] and SOLVAY[2]. The first

[1] J. J. COLEMAN: Min. Proc. Inst. Civ. Engrs. **68** (1882).
[2] E. SOLVAY: C. R. Acad. Sci. Paris **121**, 1141 (1895).

successful work, however, was by CLAUDE[1], using a system which has been considered in chapter F above. The CLAUDE system was further improved by HEYLANDT who introduced the scheme of operating the input to the expansion engine at room temperature.

The first type of CLAUDE air liquefier, diagrammatically sketched in Fig. 63, used a single reciprocating expansion engine lubricated with petrol ether of

density of from 0.675 to 0.700. The temperatures at the input and at the output of the expansion engine were about 200° K and 110° K. From 110° K down, the liquefaction was completed by a simple LINDE circuit. Although, as CLAUDE found, it is possible to arrange the expansion such that air liquefies in the engine, and so to eliminate the final LINDE circuit, this was not adopted because (1) liquefaction in the engine caused severe knocking and even mechanical breakdown and (2) the power output from the expansion engine fell markedly and with it the liquefaction rate. This was because the vapor expelled on the exhaust stroke would precool the incoming gas almost to liquefaction and then the possible volume changes on subsequent expansion would be very small.

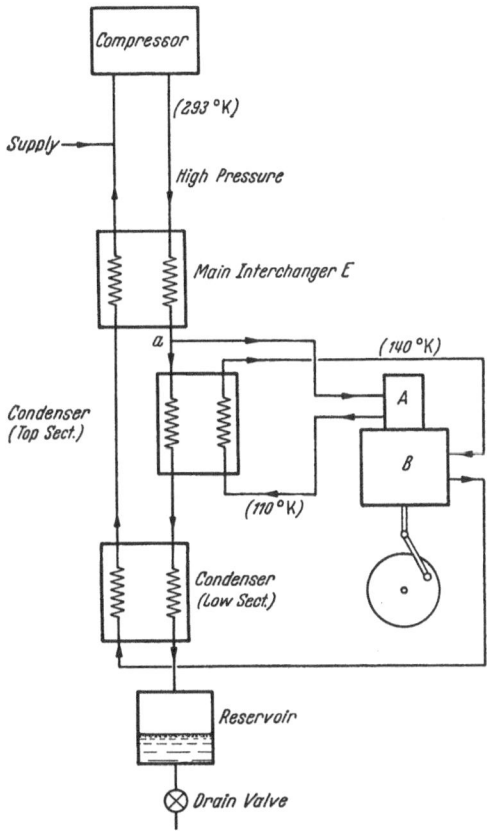

Fig. 68.   Schematic flow diagram of CLAUDE's two-stage expansion type air liquefier.

Two notable improvements were later introduced by CLAUDE. First, he found (1912) that by making the piston rings in the expansion engine of dry de-fatted leather, he could operate the engines without the petrol ether lubrication and greatly reduce cylinder wear. Secondly, he introduced a two-stage expansion engine with liquefaction under pressure. This is diagrammatically illustrated in Fig. 68. The high pressure air after passing through the main interchanger, E, divides at a, part going to the high pressure expansion cylinder A and part into the top section of the condenser. The gas leaving the high pressure expander at about 110° K, passes up through the medium pressure channels of the top section of the condenser emerging at about 140° K. Thence it goes to the second low pressure expansion cylinder, B. After expansion in B the gas passes up through the low pressure channels of the lower section of the condenser. The high pressure gas in the condenser condenses in this lower section and fills the reservoir. In practice CLAUDE made cylinders A and B coaxial, so that a single piston worked for both.

A more detailed drawing of this machine is given in Fig. 69. Here C is the compressor, A the water-cooled aftercooler, B a potassium hydroxide tower for

[1] G. CLAUDE: C. R. Acad. Sci. Paris **134**, 1568 (1902). See also G. CLAUDE: Liquid Air, Oxygen and Nitrogen. Paris 1926.

drying the air and removing the CO_2, E is the main interchanger, L and L' the top and lower sections of the condenser, and F the two-stage expansion engine and D the reservoir.

35. The adiabatic efficiency of the expansion engine. In all expansion engines irreversibilities occur so that the expansion is not truely isentropic. This has been illustrated in the $T-S$ diagram of Fig. 64 for an air liquefier. Here the actual expansion is indicated by the path $c \rightarrow f$, whereas an ideal isentropic

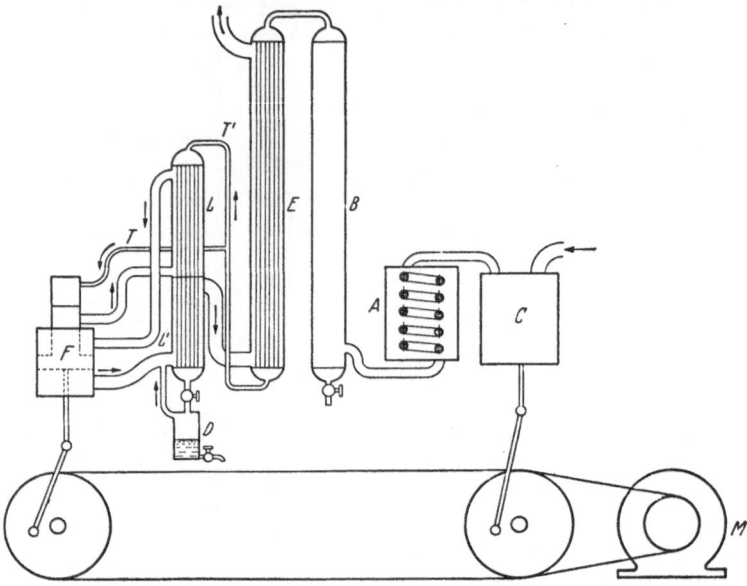

Fig. 69. Diagram of two-stage expansion type air liquefier due to Claude. (Courtesy of Claude: Liquid Air, Oxygen and Nitrogen. Philadelphia: P. Blakiston & Son & Co. 1913.)

expansion would have been the vertical path (constant S) $c \rightarrow p$. It is customary to define the adiabatic efficiency[1], η_{ad}, of the expansion by:

$$\eta_{ad} = \frac{H_c - H_f}{H_c - H_p}. \tag{35.1}$$

Since the work delivered by an expansion engine per unit mass of gas is equivalent to the amount of heat removed from the gas, times the mechanical efficiency of the engine, the adiabatic efficiency, η_{ad}, of eq. (35.1) is the ratio of the heat removed in practice to the maximum possible removable by isentropic expansion in an engine of the same mechanical efficiency.

In the engines employed by Claude, adiabatic efficiencies of from 60 to 70% were obtained.

36. Turbine expansion engines for low pressure air liquefiers. The use of a turbine type expansion engine for air liquefaction was suggested as long ago as 1898 by Rayleigh[2]. Attempts at its construction were made by Thrupp[3]. The first working turbine machines were probably made by the Linde Eismaschinen Company in low pressure air liquefiers and rectifiers in the early nineteen

---

[1] This is called by Kapitza [J. Phys. USSR. **1**, 7 (1939)] the "technical efficiency".
[2] Rayleigh: Nature, Lond. **58**, 199 (1898).
[3] Thrupp: English Patent 26767, 1898.

thirties although little detail of these were available[1]. Some detail of these impulse type turbines, however, were published by HOCHGESAND[2] in 1935 and later in 1947 by SWEARINGEN[3] in the USA.

The first detailed description of a turbine expansion engine in an air liquefier was given by KAPITZA[4] who used a low pressure system, as has been briefly described in Sect. 33. A diagram of this equipment is given in Fig. 70. Here the air entering at *1* is compressed in the two stage compressor, *2*, giving 9.5 to 10 m.^{3} air per minute at about 9 atm. pressure. The compressed air passes through the water cooled aftercooler, *3*, and the oil purifier, *4*, to the valve system

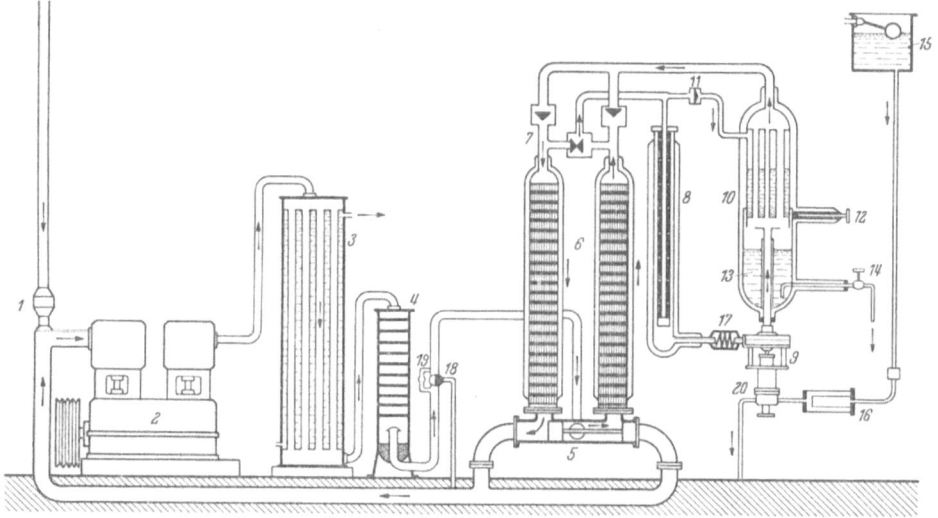

Fig. 70. Diagram of turbine expansion type air liquefier due to KAPITZA. [P. L. KAPITZA: J. Phys. USSR. 1, 24 (1939).]

5 of the regenerators *6*. The regenerators (see chapter I for more detail concerning regenerators) consisted of a pair of vacuum isolated columns packed with flat metallic ribbons 50 mm. wide, 0.1 mm. thick which were covered with "bumps". The valve system 5 at the input and 7 at the output of the regenerators caused the high pressure flow to pass alternately up through first the left hand and then the right hand regenerator column, alternating every 25 to 27 sec. The low pressure return alternates also through the columns in the reverse direction. This arrangement replaces the usual counterflow heat interchanger and permits air containing moisture and CO_2 to be used without predecontamination of these impurities, since they are condensed during the passage of the high pressure air up one column and swept out during the subsequent low pressure flow down the same column. As will be discussed in more detail later on, the operation of a simple reversing regenerator system such as this does not completely remove the contaminants. Indeed, as reported by KAPITZA, the operation of the machine had to be interrupted at rather frequent intervals to remove solid CO_2 accumulated in the turbine expansion engine.

[1] R. LINDE: Z. ges. Kälteind. 41, 183 (1934). See also M. RUHEMANN: Separation of Gases. Oxford Univ. Press 1940.

[2] B. C. P. HOCHGESAND: Mitt. Forsch. Anst. GHH. Konzern 4, Part I (1935).

[3] J. S. SWEARINGEN: Trans. Amer. Inst. Chem. Engr. 43, 85 (1947).

[4] P. KAPITZA: J. Phys. USSR. 1, 7, 29 (1939). See also M. M. LEVITIN and O. A. STETZ-KAYOV: Avtogennoe. Delo 12, Nr. 5, 25 (1941).

It is clear that the hydraulic resistance in each column of the regenerator system must be the same for optimum performance; and, since the low pressure return also passes through the same resistances, it is clear that a reversing regenerator system is limited in practice to low pressure circuits.

The compressed gas passes from the regenerators, 6, to a filter and a stabilizer 8, which consists of a tower containing a few kilograms of activated charcoal. This serves not only to filter impurities but also to stabilize the temperature fluctuations which take place at the cold ends of the regenerator columns during the alternation of the direction of gas flow. From 8 the compressed gas at about 5.6 atm. pressure enters the turbine expansion engine 9 at about 115° K. Here the gas expands to a pressure of 1.56 atm. doing work against the water brake, 20, and reaches a temperature of 86° K. The expanded gas then passes through

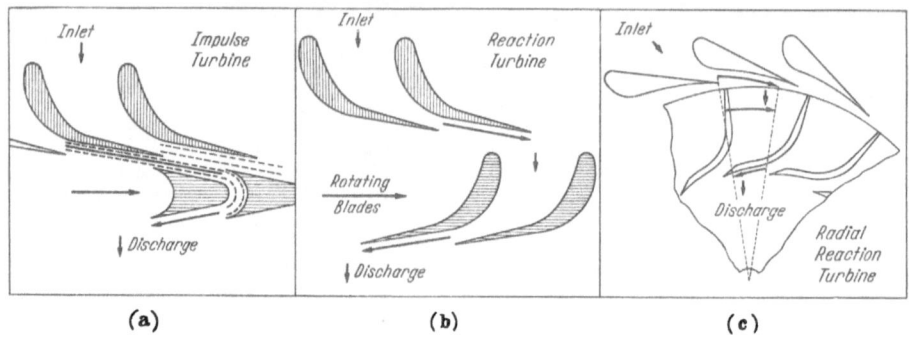

Fig. 71a—c. Schematic diagrams of (a) an axial flow impulse turbine, (b) an axial flow reaction turbine and (c) a radial flow reaction turbine. [Figure due to J. S. Swearingen: Trans. Amer. Inst. Chem. Engr. 43, 85 (1947).]

the condenser 10, back through the regenerator and to the compressor. A fraction of the high pressure gas flow passes through the check valve 11, rather than going to the expansion engine, and is liquefied in the condenser, 10. The liquid air is drawn off through the valve 12 from the condenser into the reservoir 13, whence it can be tapped through the drain valve, 14.

The machine yielded about 34 liters liquid air per hour with an efficiency of 1.5 kW-hr/liter. The power developed in the water brake was about 4 kW.

The turbine was a radial flow impeller type turbine consisting of a straight bladed monel wheel, 8 cm. diameter, 250 g weight, running at 40000 rev/min. and passing about 600 kg. air/hr, with a claimed adiabatic efficiency of 83 %[1]. Although Kapitza in his two papers was the first to give considerable detail regarding the design and construction of the turbine for the air liquefier, particularly concerning the superiority of the radial flow type, which he adopted over more conventional axial flow turbines, attention here will be given to the more recent design data given by Swearingen[2].

Swearingen reported in 1947 on a turbine expansion engine for air liquefiers (and/or oxygen producers) built under the auspices of the US National Defense Research Committee with the following properties: Flow capacity about 3100 kg. air/hr (i.e. about five times larger than Kapitza's turbine), inlet pressure 7 atm., outlet pressure 1.4 atm., inlet temperature 121° K and adiabatic efficiency about 83 %.

---

[1] H. Hausen [Z. ges. Kälteind. 48, 24 (1941)] has reinterpreted Kapitza's data to give an adiabatic efficiency of from 76.5 to 78.5%.

[2] J. S. Swearingen: Trans. Amer. Inst. Chem. Engr. 43, 85 (1947).

In considering the choice of turbine, SWEARINGEN followed KAPITZA in choosing a radial flow type. Schematic diagrams of (1) an axial flow impulse turbine, (2) an axial flow reaction turbine and (3) a radial flow reaction turbine are given in Fig. 71 a, b and c, respectively. The gas in the axial flow impulse turbine (type a) has to make a U turn at high velocity in the rotor blades and this reduces its efficiency considerably. This loss is avoided in the axial flow reaction turbine (type b). In this type only about half the energy is spent in the primary nozzles, the remaining half being taken up by the rotor nozzles which accept the primary air stream without loss since they move with the same speed as the primary jets. In both types, a and b, however, the gas suffers considerable radial force and this may result in lossy operation due to the tendency to expell air outwards radially. In the radial flow reaction turbine (type c) the flow through the rotor nozzles is inwards radially against the centrifugal forces and the work necessary for this flow is directly converted into the rotary mechanical energy of the wheel. The efficiency therefore is high.

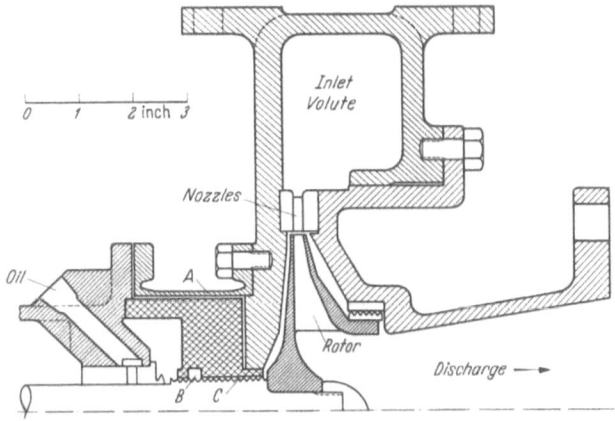

Fig. 72. Expansion turbine for air liquefier due to SWEARINGEN. [Trans. Amer. Inst. Chem. Engr. **43**, 85 (1947).]

A drawing of SWEARINGEN's turbine is given in Fig. 72 in which the wheel is $6\frac{7}{8}''$ diameter and runs at 22000 rev/min.

The construction of other turbine expansion engines, some for use with large liquefiers (or rectifiers), has been briefly reported, for example, by RUSHTON and STEVENSEN[1], by KOTTAS[2], by BLEYLE, HINCKLEY and JEWETT[3], and by WUCHERER[4].

BOSE[5] has proposed an air liquefier using a multiple turbine expansion engine in which the air expands to a pressure much less than one atmosphere and which on return to room temperature must be pumped back to atmospheric pressure before re-entering the compressor. No technical details of the proposed machine are, however, included.

37. Low pressure air liquefiers with reciprocating expansion engines. The use of the "ringless-piston" type of reciprocating expansion engine for gas liquefiers was first introduced by KAPITZA[6] in 1934. In this engine, used for helium liquefaction, the stainless steel piston was a relatively loose fit in the phosphor-bronze cylinder (clearance about 1.5 thousandths inches on a $1\frac{1}{4}$ inch diameter), the gas escaping past the piston acting as the lubricant. In order to secure correct alignment and equalization of pressure, the piston was grooved with a series of

[1] J. H. RUSHTON and E. P. STEVENSON: Trans. Amer. Inst. Chem. Engr. **43**, 61 (1947).
[2] H. KOTTAS: Refrig. Engng. **59**, 762 (1951).
[3] BLEYLE, HINCKLEY and JEWETT: See A. D. Little Inc. reprint.
[4] J. WUCHERER: Bull. Inst. Internat. Froid. Annexe **1954**, p. 2, 69.
[5] A. BOSE: Indian J. Phys. **23**, 433 (1949).
[6] P. KAPITZA: Nature Lond. **133**, 208 (1934). — Proc. Roy. Soc. Lond., Ser. A **147**, 189 (1934).

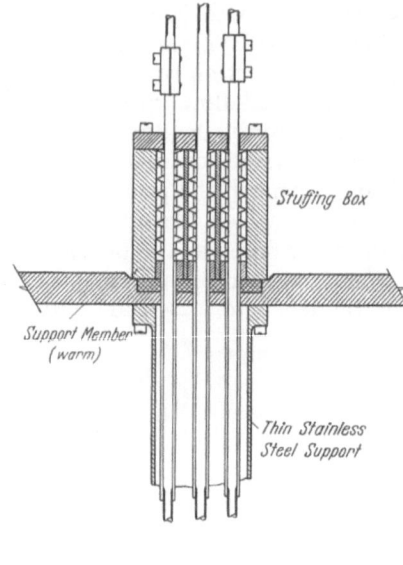

Stuffing Box

Support Member (warm)

Thin Stainless Steel Support

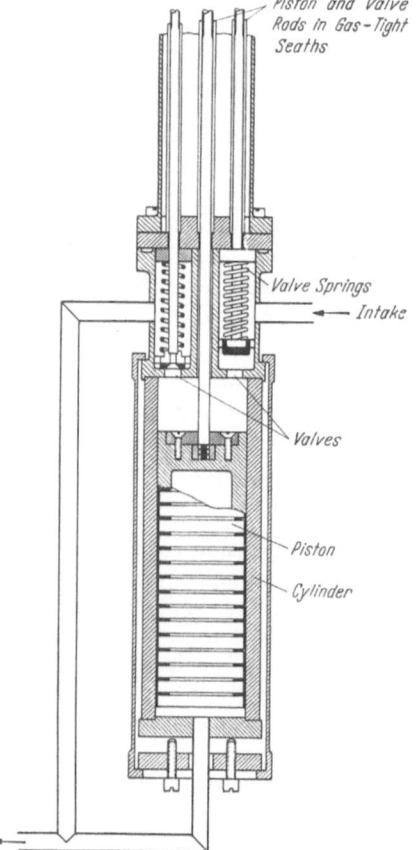

Piston and Valve Rods in Gas-Tight Seaths

Valve Springs

Intake

Valves

Piston

Cylinder

Exhaust

Fig. 73. Reciprocating expansion engine for helium liquefier due to COLLINS. (Fig. From A. LATHAM jr. and H. O MCMAHON: Refrig. Engng. June 1949.)

equally spaced periferal grooves and was loosely joined to the piston rod, so that it was self-aligning. To minimize gas escape past the piston, the power stroke was made much more quickly than the return stroke.

COLLINS[1] improved KAPITZA's expansion engine by making the piston and cylinder of hardened nitrided nitralloy, so enabling the clearance to be reduced to between 0.3 and 0.4 thousandths of an inch on a one inch diameter for his helium liquefier. This reduced gas loss and enabled adiabatic efficiencies of up to 80% to be obtained. COLLINS and co-workers[2] used similar, but larger, ringless-piston expansion engines in low pressure air liquefiers and oxygen generators. They have reported in this work expansion engines with nitrided ringless pistons of 4 inches diameter (clearance between 0.5 and 1.0 thousandths inches). A subsequent type of ringless-piston expansion engine constructed by COLLINS for air liquefaction and/or oxygen generation used a bronze cylinder coated with hard chromium plating on its inner surface and a mild steel piston fitted with a plastic sleeve of micarta. On a single cylinder expansion engine of this latter type, having bore and stroke of 6.5 and 7 inches respectively, the radial clearance was 0.001 inches at room temperature. At the operating temperature (about $140°$ K) this clearance becomes almost closed. Such expansion engines have given as high as 85% adiabatic efficiencies when operating between 12.5 and 1.5 atm. pressures at inlet temperatures of about $140°$ K.

The complete expansion engines, as developed by COLLINS, were similar in general design whether they were for air or helium liquefaction and their layout is illustrated in Fig. 73, which shows his helium liquefier engine. The engine

[1] S. C. COLLINS: Rev. Sci. Instrum. **18**, 157 (1947).
[2] S. C. COLLINS: Rev. Sci. Instrum. **18**, 157 (1947). — COLLINS, NASON and CANNADAY: Refrig. Engng. **59**, No. 12 (1951). Also private communication from Professor S. C. COLLINS.

is single acting with the piston rod always in tension. The valves are seated by spring pressure and opened by rods also in tension. The piston and valve rods are sealed by stuffing boxes maintained at room temperature, some distance from the engine. Detail of the stuffing boxes is given in Fig. 74.

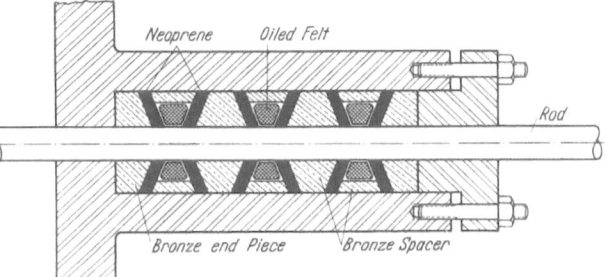

Fig. 74. Enlarged drawing of stuffing box used in expansion engine of Fig. 73. (Fig. from A. LATHAM jr. and H. O. McMAHON: Refrig. Engng. June 1949.)

Another type of nonlubricated expansion engine has been constructed by COLLINS[1] for an air liquefier or oxygen-generator which used a flexible diaphram.

It is of interest to study the flow diagram of an air liquefier used by COLLINS using the micarta covered ringless piston type expansion engine described above. The actual machine constructed by COLLINS[2] incorporated a rectification column so that it delivered liquid nitrogen rather than liquid air at a rate of 30 to 40 liters per hour. We illustrate here in Fig. 75 the flow diagram associated with the air liquefaction part of the process only. Compressed gas at 13 atm. pressure from the compressor, after passing through a water cooled aftercooler, but not otherwise purified of water vapor and CO_2 impurities, enters the reversing interchanger, I, via the reversing valves V_1. This reversing interchanger (see chapter I for fuller details) and the sections A and B of the reversing interchanger II consist of two channels of equal surface area and flow resistance. The

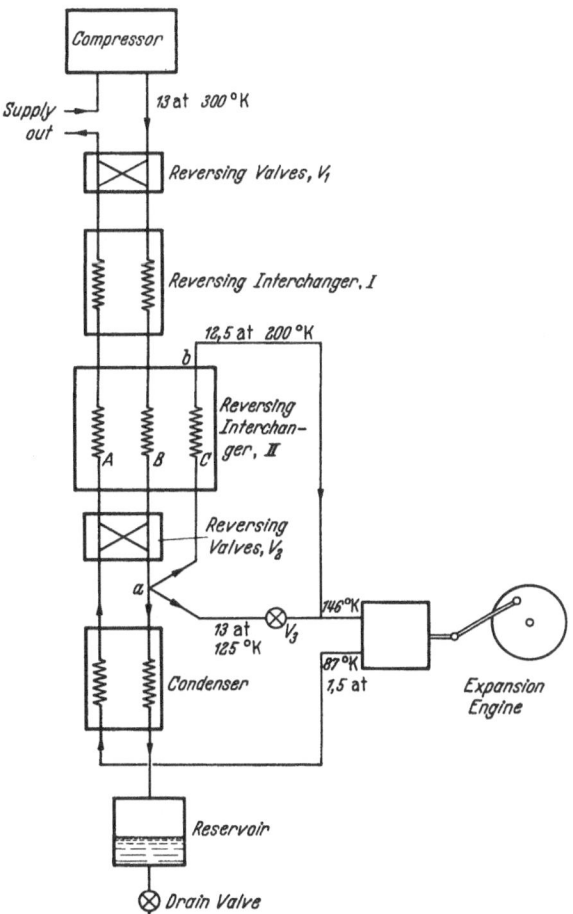

Fig. 75. Schematic flow diagram for low pressure air liquefaction using reversing interchangers due to COLLINS.

compressed gas is forced to flow alternately through one channel and through the other with period of about 3 min., by action of the reversing valves V_1 and V_2.

[1] S. C. COLLINS: Rev. Sci. Instrum. **18**, 157 (1947).

[2] COLLINS, NASON and CANNADAY: Refrig. Engn. **59**, No. 12. Also Private Communication from Professor S. C. COLLINS.

(V_1 is a piston type mechanically operated valve; V_2 is an assembly of four simple check valves which respond to the pressure changes produced by the action of valve V_1.) During one half cycle the compressed gas desposits the water vapor near the warm end and CO_2 near the cold end of the channel through which it passes. During the other half cycle, the return low pressure flow through this channel flushes out these impurities with it. By use of the reversing interchanger, instead of the more usual type of interchanger, therefore it is not necessary to purify the air coming from the compressor. This great virtue possessed also, in part, by the regenerator (as briefly outlined in Sect. 36 above), of automatically purifying the air makes low pressure air liquefaction a very advantageous system; whereas in high pressure (~ 200 atm.) systems such reversing interchangers are largely inapplicable due to the difficulty of making channels which at the sametime do not hold too much air[1] and do not restrict the low pressure flow unduly.

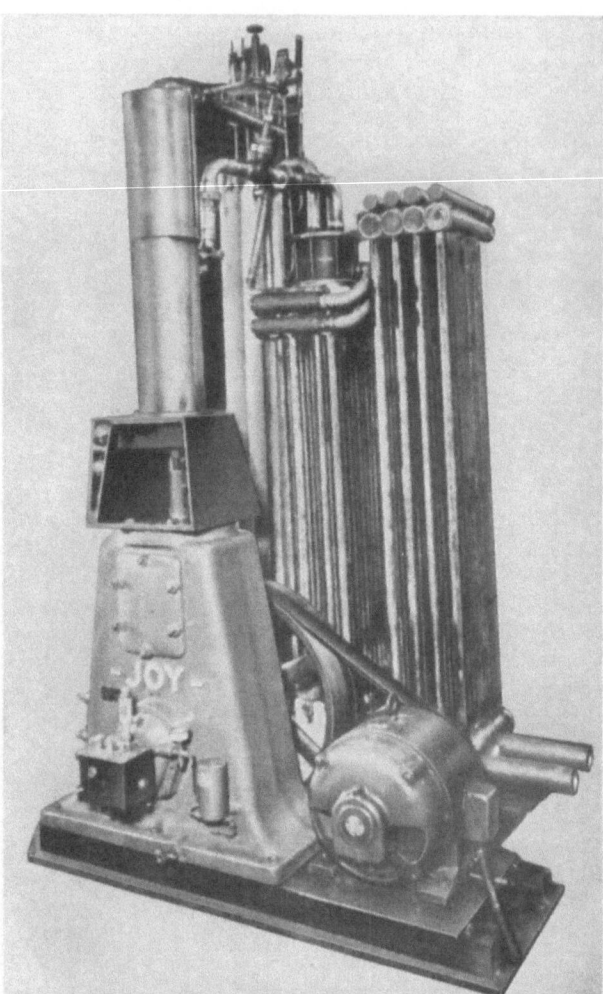

Fig. 76. Photograph of low pressure air rectifier using reversing interchangers and reciprocating expansion engine due to COLLINS. (Courtesy of Professor S. C. COLLINS.)

In reversing interchangers and regenerators the warm end is always placed at the lower elevation, so that free water condensed moves to the warm end, rather than forming a partial block of ice at the colder end.

Referring again to Fig. 75 the compressed gas after emerging at 13 atm. pressure and at $125°$ K from the reversing valves V_2 at the cold end of the reversing interchanger II subdivides at a into three streams, as follows: one stream containing the major fraction of the gas passes to the expansion engine via a restricting valve, V_3, which produces a pressure drop from 13 atm. to 12.5 atm.; one stream passes into the condenser and the third stream goes into the third

[1] On reversal, the air instaneously in any one channel must be reversed in flow direction. If the quantity of air thus reversed in flow is large compared with the total flow, serious inefficiency is introduced.

channel, C, of the interchanger II. This third stream after emerging at b from this interchanger at 200° K and at 12.5 atm. returns to the expansion engine. It is to be noted that this stream through channel C always flows in one direction, regardless of the directions of flow in channels A and B.

The gas which enters the expansion engine at 12.5 atm. pressure and about 146° K (this temperature being the resultant of the two streams, one at 125° K and the other at 200° K), expands to 1.5 atm. and falls in temperature to 87° K. This cooled gas passes up through the condenser and then to the return paths through the reversing interchangers. The liquid condensed under pressure in the condenser, falls into the reservoir whence it may be withdrawn through the drain valve.

The condenser consists of a brass tube 7.5 ft long, 6 inch o.d. and 5.75 inch i.d. filled with 130 copper tubes 7 feet long and 0.25 inch o.d. The tubes are soldered into tube sheets at each end, the tube sheets being soldered to the inner walls of the brass tube. The ends of the brass tubes are then closed. The high pressure air to be condensed flows into the space outside the bundle of copper tubes. The low pressure cooling gas from the expansion engine flows up through the copper tubes. A photo of the complete machine (including rectifying column) is given in Fig. 76.

As is discussed in detail in chapter I, the purpose of splitting the reversing interchanger into two sections I and II was to introduce the extra channel C to the cooler unit. This extra channel is to provide extra cooling for the incoming gas, since under normal conditions the cooling provided by the return low pressure gas alone would be sufficient to reduce the high pressure gas flow to only 145° K. This temperature is insufficient to freeze out all the CO_2 impurity from the air. By use of the extra cooling in channel C the high pressure gas emerges at a at 125° K, a temperature sufficiently low to permit continuous operation of the machine without its being clogged with CO_2 impurity. This ingenious scheme makes the operation of the machine much easier than, for example, the low pressure air liquefier of KAPITZA which, as reported earlier, had to be partially warmed up at intervals to flush out CO_2 impurity.

A similar low pressure oxygen machine using COLLINS' reversing interchangers and a reciprocating expansion engine has been described by LOBO[1].

38. Liquefaction of other gases using expansion engines. Hydrogen liquefiers have been constructed in which the cooling is performed by a separate helium gas cycle, the helium passing through ringless-piston expansion engines[2]. A photograph of such a machine, capable of giving 18 liters liquid H_2 per hour, is given in Fig. 77.

This machine embodies a closed helium system using two expansion engines at two temperature levels. The hydrogen of commercial purity (or Ne, N_2, CO, A, O_2, etc.) to be liquefied enters the unit at low pressure (approximately 0.2 atm. overpressure), passes through a heat interchanger into a condenser where all the flow is condensed. No external purification is required; however, final purification is accomplished by freezing out the impurities in a duplex interchanger within the liquefier. When one side of the duplex interchanger becomes clogged, the flow is switched to the other side and the first side is then derimed. Such a method of liquefaction of H_2 by low pressure condensation has the advantages of safety and in purification requirements over the more usual high pressure JOULE-THOMSON type hydrogen liquefier.

[1] W. E. LOBO: Chem. Ind. **59**, 53 (1946).
[2] A. D. LITTLE Inc. Hydrogen liquefier specifications. 1953.

The first use of expansion engines in helium liquefaction was performed by Kapitza[1] who used a single repricating expansion engine with a non-lubricated ringless piston, after precooling the helium gas with liquid nitrogen. Detail of this and other helium liquefiers is given in the following article on Helium Liquefaction by S. C. Collins. Subsequent construction and development of such machines using two expansion engines has been done notably by Collins[2]. Isolated machines have been made by Meissner[3], Lane[4], Daunt et al[5]. The design

Fig. 77. Photograph of hydrogen liquefier using closed-cycle helium refrigeration (with reciprocating expansion engines). (Courtesy of A. D. Little Inc.)

due to Collins has been marketed by A. D. Little Inc. More recently Collins[6] has made a helium liquefier using three ringless piston expansion engines, capable of giving 30 liters liquid per hour.

A new type of expansion engine in which an elastic metal bellows replaces the conventional cylinder and piston has been introduced and used by Long and Simon[7] for helium liquefaction.

[1] P. L. Kapitza: Nature, Lond. 133, 208 (1934). — Proc. Roy. Soc. Lond., Ser. A 147, 189 (1934).
[2] S. C. Collins: Rev. Sci. Instrum. 18, 157 (1947). — Science, Lancaster, Pa. 116, 289 (1952).
[3] W. Meissner: Phys. Z. 43, 261 (1952).
[4] C. T. Lane: Rev. Sci. Instrum. 12, 326 (1941).
[5] Nicol, Smith, Heer and Daunt: Rev. Sci. Instrum. 24, 16 (1953).
[6] S. C. Collins: Science, Lancaster. Pa. 116, 298 (1952).
[7] H. M. Long and F. E. Simon: Nature, Lond. 172, 581 (1953). Appl. Sci. Res. A 4, 237 (1954). — Z. Kältetechn. 6, 150 (1954).

H. Liquefaction by single isentropic expansion.

39. The methods of gas liquefaction using isentropic expansion described in chapter G above were concerned with methods of continuous gas liquefaction and in consequence the expansion engines were such as to handle a continuous flow of gas. A technically simpler method of gas liquefaction is by use of a single adiabatic expansion to liquefy the gas. This was first tried, unsuccessfully, for oxygen by CAILLETET[1] in 1877 and again unsuccessfully by OLSZEWSKI[2] for H_2 in 1887. By a single rapid expansion of the gas from the lowest attainable temperatures they were able only to produce a visible mist of liquid, but were unable to collect free liquid by this meanss.

The successful application of this method is due to SIMON and his co-workers[3] using in the first instance helium. Since its introduction in 1932 by SIMON this so-called "Expansion method" has been widely and extensively used for the production of liquid helium in small quantities[4]. In 1935 SIMON, COOKE, and PEARSON succeeded in liquefying hydrogen by the same method[5].

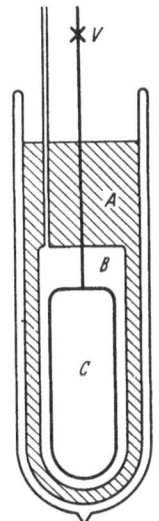

The principle of the method is illustrated by the diagram of Fig. 78. The gas to be liquefied is compressed isothermally to high pressure (between 100 and 200 atm.) into the chamber C which is kept at low temperature by the bath A. For hydrogen liquefaction bath A would contain oxygen or an oxygen-nitrogen mixture evaporating under reduced pressure. For helium liquefaction bath A would contain hydrogen evaporating under reduced pressure. Thermal contact between C and A is maintained by means of exchange gas in the vacuum jacket B which surrounds C. Before allowing the gas in C to expand by opening the valve V (which is at room temperature), the exchange gas is pumped out of B, so that C is adiabatically isolated. On adiabatic expansion of the gas in C, it is cooled and partially liquefied, the liquid remaining in the vessel C.

Fig. 78. Schematic diagram of H_2 or He liquefier using the single expansion method of SIMON. [Z. ges. Kälteind. **39**, 89 (1932).]

The success of this single expansion method of liquefaction is dependent on obtaining a sufficiently small heat capacity of the strong walled container, C, relative to the heat capacity of the gas which it contains. This condition only holds at very low temperature where the specific heats of solids are small and hence the method is applicable in practice to liquefaction of helium and hydrogen only. This explains the lack of success of CAILLETET's experiment with oxygen.

In Table 16, due to PICKARD and SIMON[6], are given the values at two different temperatures of the heat capacities of a steel vessel of 150 cm.³ volume capable of withstanding 100 atm. pressure and the heat capacities of the helium which it would hold at that pressure and at those temperatures. It is clear from the table that at the lower temperature (10° K), the heat capacity of the vessel is negligibly small and hence that essentially all the refrigeration of the expansion will be useful in cooling the gas. At the higher temperature the reverse is true.

[1] L. CAILLETET: C. R. Acad. Sci. Paris **85**, 1213 (1877).

[2] K. OLSZEWSKI: Ann. Phys. u. Chem. **31**, 58 (1887). — Wiener Ber. 95, 1 (1887). See also Phil. Mag. **39**, 188 (1895).

[3] F. SIMON: Z. ges. Kälteind. **39**, 89 (1932). Also Phys. Z. **34**, 232 (1932). — F. SIMON and J. E. AHLBERG: Z. Physik **81**, 816 (1933).

[4] See article on "Helium Liquefaction" below by S. C. COLLINS for fuller details.

[5] SIMON, COOKE and PEARSON: Proc. Phys. Soc. **47**, 678 (1935).

[6] G. L. PICKARD and F. E. SIMON: Proc. Phys. Soc. **60**, 405 (1948).

Table 16. *Heat capacity of 150 cm.³ steel vessel and of helium gas in it at 100 atm.*

	Room temperature cal/°C	10° K cal/°C
Steel vessel, C_S	50	0.1
Helium gas at 100 atm. C_H	2	20
Ratio, C_S/C_H	25	0.005

Table 17. *Heat capacity of 144 cm.³ steel vessel and of hydrogen gas at 150 atm. pressure.*

	Room temperature cal/°C	50° K cal/°C
Steel vessel, C_S	23	3.3
Hydrogen gas at 150 atm. C_H	4.3	12
Ratio, C_S/C_H	5.3	0.28

In Table 17 a similar table is computed, largely from data given by Simon, Cooke and Pearson[1], for hydrogen gas and its container. (This container was of specially high tensile strength steel and allowed a safety factor of 2.) Here it will be seen that at 50° K, a possible starting temperature for the expansion, the heat capacity of the vessel is about one quarter of that of the gas it contains. Although this is not a negligible fraction, as it was for the similar ratio for helium at 10° K, it still allows appreciable liquefaction by this method.

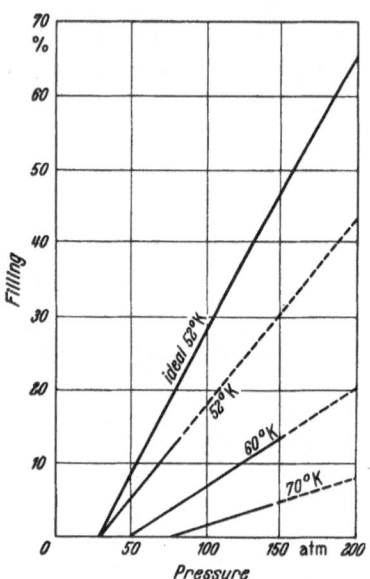

Fig. 79. Percentage filling of vessel in H_2 single expansion type liquefier as function of initial pressure due to Simon, Cooke and Pearson. [Proc. Phys. Soc. **47**, 678 (1935).] The heavy full curve is theoretical (neglecting effects of heat capacity of container) for starting temperature of 52° K. The other curves give observations up to 15% filling and extrapolations to higher pressures for starting temperatures of 52° K, 60° K and 70° K.

To compute the yield of liquid, suppose that the heat capacity of the vessel can indeed be neglected. Per mole of gas adiabatically expanded from initial pressure of p_i and initial temperature T_i to final conditions p_f and T_f, let a fraction x remain as liquid. Then for the isentropic change:

$$S_i = x S_l + (1 - x) S_g \qquad (39.1)$$

where S_i, S_l and S_g are the entropies per mole of gas at p_i and T_i, per mole of liquid at p_f and T_f, and per mole of gas at p_f and T_f. If V_i is the molar volume of the gas at p_i and T_i, and V_l the molar volume of the liquid at p_f and T_f, then the ideal yield, y_{ideal}, defined as the ratio of the volume of liquid obtained to the total volume of the vessel, C, is:

$$y_{ideal} = \frac{x V_l}{V_i} = \frac{V_l}{V_i} \frac{(S_g - S_i)}{(S_g - S_l)} . \qquad (39.2)$$

Taking for hydrogen values which could be realized in practice, namely $p_i = 200$ atm.; $T_i = 52°$ K; $p_f = 1$ atm.; $T_f = 20.4°$ K (the boiling point) eq. (39.2) gives $y_{ideal} = 0.7$.

In the case of helium the results are more striking. For $p_i = 150$ atm., $T_i = 15°$ K, $p_f = 1$ atm., $T_f = 4.2°$ K, one gets from eq. (39.2) $y_{iddal} = 1.0$. Ideal yields of more than 100% can theoretically be computed, since under conditions of high enough initial pressure and low enough T_i the molar volume, V_i, of the gas may be smaller than the molar volume, V_l, of the liquid.

Experimental measurements have been made of the yield actually obtained in this expansion method of hydrogen liquefaction by Simon, Cooke and Pearson[1]. Some of their results are shown in Fig. 79. For comparison the value of

---

[1] Simon, Cooke and Pearson: Proc. Phys. Soc. **47**, 678 (1935).

y_{ideal} is given for $T_i = 52°$ K. The actual yields obtained in these preliminary experiments did not exceed about 15 %. However, if a linear extrapolation of their data is made (as seems reasonable from the corresponding data on helium[1]), then for $T_i = 52°$ K and $p_i = 200$ atm., $y_{actual} \approx 0.45$, which would represent a very practical and useful operation.

SIMON and AHLBERG[2] have suggested adding to the yield by use of JOULE-THOMSON cooling. It was suggested not to expand the gas as is usually done through a valve at room temperature outside the croystat, but through an adjustable valve situated on the chamber C. Then the JOULE-THOMSON cooling in the gas flowing past the valve could be made to contribute to the refrigeration if the expanded gas were made to flow in a tube soldered to the outside of the chamber C. Some experiments on this have been reported[3], but further detail is awaited.

I. Heat interchangers and regenerators.

40. The significance of interchangers and regenerators. In nearly all apparatus for the production of low temperatures, as has been described above, heat interchangers or regenerators play a significant rôle. In the development of the gas refrigerating machine to its modern importance, as realized in the Philips machine[4], the inclusion of a heat regenerator was fundamental to its success. A regenerator rather than a counter-current heat exchanger was employed because the gas flow from the compressor volume to the expansion volume takes place in alternating directions. In the progress towards lower temperatures in the use of compressed vapor machines, the introduction of the use of counter-current heat interchangers by KAMERLINGH-ONNES[5] enabled the cascade compressed vapor cycle to be successfully operated and thereby marked the beginning of large scale air liquefaction by such a process.

The success of LINDE[6] and HAMPSON[7] in first liquefying air by the JOULE-THOMSON expansion method, was due to their employement of counter-current heat exchangers, which, as previously envisaged by SIEMENS[8], acted as a regenerating agent in the initial cooling down process. The CLAUDE[9] system of gas liquefaction using isentropic expansion depends on the use of counter-current heat interchangers. In the more recent *low* pressure air liquefaction and rectification plants, also using cooling by isentropic expansion, as introduced by FRÄNKL[10], regenerators rather than counter-current heat interchangers are employed. By allowing the input and output streams of air to alternate successively through two metal-foil packed regenerators, it was found that not only could the cold of the output stream be adequately transferred to the warmer input stream, as in a counter-current interchanger, but also that prepurification of the input stream was largely unnecessary. The H_2O and CO_2 impurities were deposited on the large metal surface of any one regenerator column when the input stream

[1] G. L. PICKARD and F. E. SIMON: Proc. Phys. Soc. **60**, 405 (1948).
[2] F. SIMON and J. E. AHLBERG: Z. Physik **81**, 816 (1933).
[3] SIMON, COOKE and PEARSON: Proc. Phys. Soc. **47**, 678 (1935).
[4] J. W. L. KÖHLER and C. O. JONKERS: Philips techn. Rev. **16**, 69, 105 (1954).
[5] H. KAMERLINGH-ONNES: Leiden Comm. **14** (1894); **87** (1903); Leiden Comm. Suppl. **35** (1913).
[6] K. v. LINDE: Z. ges. Kälteind. **4**, 23 (1897).
[7] HAMPSON: English Patent 10165. 1895.
[8] W. SIEMENS: English Patent 2064. 1857.
[9] G. CLAUDE: Liquid air, oxygen and nitrogen. Paris 1913.
[10] M. FRÄNKL: German Patent. 490878. 1928.

passed through it and then were subsequently flushed out into the atmosphere by the passage of the output stream through the same channel.

A still more recent development, also made in the progress of low pressure air liquefaction and rectification machines by COLLINS[1], has been the introduction of the reversing heat interchanger. This is a cross between a regenerator system, which by definition must be a reversing system (i.e. the gas flow in any one tower must alternate in direction), and a counter-current heat exchanger. In it the gas streams alternate between the two sections of the interchanger, with the purpose of avoiding prepurification of the air. The impurities, as with the regenerator, are deposited in one section when the input stream passes through it and are flushed out when the flow is reversed and the output stream passes through the same section.

Some detail of the properties of heat interchangers and regenerators as applied to low temperature production is given in this chapter. Their significance, as briefly outlined above is fundamental and can be summed up as being of twofold importance. First their regenerating action is basic, such that in starting up a low temperature producer the first relatively small coolings are regeneratively used in cooling the input gas streams and secondly they serve the vital role of absorbing from the emergent gas streams all or nearly all of the cold carried by them, thereby adding materially to the efficiency of the production of cold.

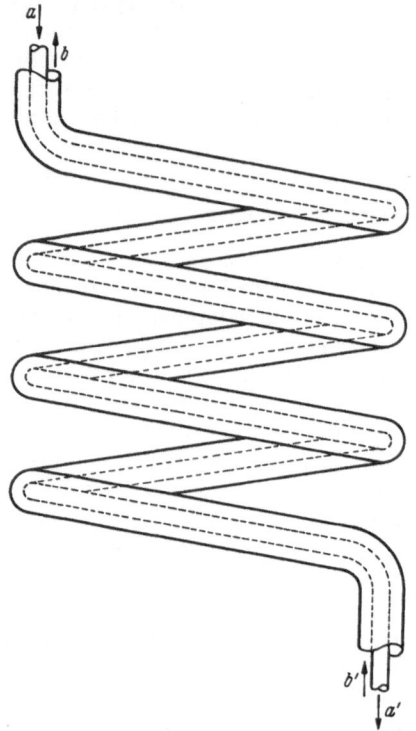

Fig. 80. Schematic diagram of LINDE type counter-current heat interchanger.

41. The LINDE and HAMPSON type interchangers. These two simple types of heat interchangers will first be discribed as representative of many other more complex systems. The LINDE interchanger[2], as used in low temperature production, is a counter-current system, being constructed by mounting one tube inside another and then winding the ensemble into a helix. See Fig. 80. Very often the inner tube, $a \to a'$ of Fig. 80, is made to carry the high pressure input (warmer) gas and the annular region between the two tubes, $b' \to b$ of Fig. 80, the return low pressure (cooler) gas. Such a system has been illustrated in Sect. 21. In general, if the two pressures are widely different, it is preferable to canalize the high pressure flow through the annular region. This is so, because in either arrangement the heat transfer suface is the same (being the wall of the *inner* tube) but in the second arrangement only *one* wall, namely the inner wall of the inner tube, contributes in establishing the pressure drop in the low pressure return. Clearly pressure drops in the low pressure flow are more serious than in the high pressure flow and consequently this arrangement is advantageous

[1] S. C. COLLINS: Chem. Eng. **53**, 106 (1946).

[2] K. VON LINDE: Z. ges. Kälteind. **4**, 23 (1897).

in minimizing it. If the annular region is used for the high pressure flow, it is also quite common practice[1] to wind and solder down a spiral of wire of very coarse pitch along the length of the inner tube before it is inserted in the outer tube. This serves to increase the heat transfer surface and to encourage turbulent high pressure flow.

The LINDE type interchanger although simple to construct suffers from the disadvantage of all counter-current interchangers composed of bundles of parallel tubes, namely that the length of the low pressure return is the same as that of the high pressure line, so introducing appreciable pressure drops in the low pressure flow. Moreover the LINDE type also introduces, in the wall of the outer tube, a weight of metal which contributes in general a negligible factor to the heat transfer surface.

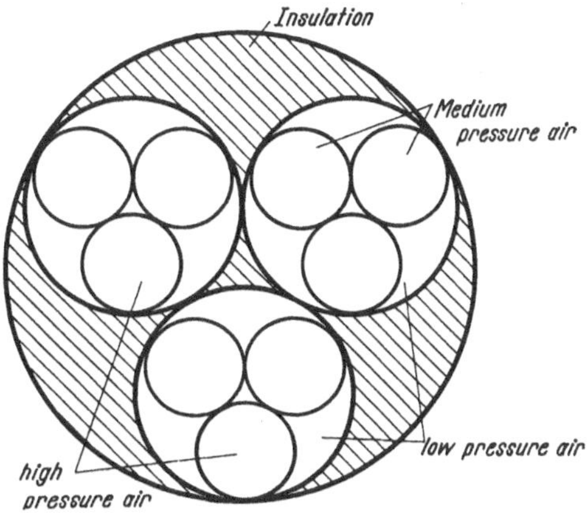

Fig. 81. Section for LINDE type counter-current heat interchanger for three gas flows (high, intermediate, and low pressure).

The exchangers for the LINDE two-stage air liquefiers (see Sect. 21) required three gas streams, one for the high pressure input flow and two for the return flows at intermediate pressure and at low pressure. Again a bundle of tubes was used, as shown in Fig. 81.

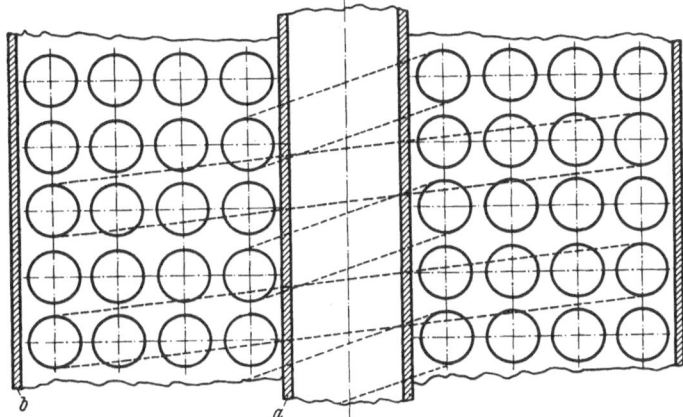

Fig. 82. Schematic drawing for HAMPSON type heat interchanger.

The disadvantages of the LINDE type interchangers are largely absent in the HAMPSON type interchanger[2], illustrated diagrammatically in Fig. 82. In this, the high pressure input gas flows through a tube wound in a succession of spiral pancakes one on top of the other. For a description of this winding see TRAVERS' statements in Sect. 23 and the article by COOK[3]. A more complex high pressure system uses many high pressure tubes in parallel also wound in the same way together. A detailed description of such a method of construction

[1] See for example. S. C. COLLINS and F. G. KEYES: J. Phys. Chem. **43**, 5 (1939).
[2] HAMPSON: English Patent 10165. 1895.
[3] J. W. COOK: Bur. Stand. Sci. Papers. **17**, No. 419 (1921).

of this multiple tube type has been given by SPOENDLIN[1]. The low pressure return flows up through the annular region between the tubes a and b (see Fig. 82) which fit closely the high pressure complex. This flow must take place through the interstices occuring between the high pressure tubes and is essentially at right angles to the high pressure flow. For this reason one may regard the HAMPSON interchanger as of a cross-flow type. In general, in order to regularize the interstices between the high pressure tubes, the latter before winding have a coarse-pitch helix of wire or, for example, nylon[1] wound over them. This serves to keep the tubes apart. Other more elaborate methods have been adopted where, for example[2], the high pressure tubes are wound in layers with spacers between layers. Further details of the possible modifications of construction of such interchangers are given in Sect. 48.

In the HAMPSON type interchanger the low pressure return path is much shorter than the high pressure path and hence the pressure drop on the low pressure side can be made small. Moreover nearly all the metal in the interchanger is useful as heat transfer surface between the warm and cold streams. One disadvantage lies in its requirement of high precision of construction. If the spacings are not uniform, the temperature distribution becomes non-uniform with consequential diminished efficiency of the interchanger.

42. The heat transfer coefficient. Consider the transference of heat from one fluid to another through a wall, generally metal, bounding the two fluids. Such a wall would be formed by the tubing carrying the high pressure in an interchanger, as described above. In general, the fluids will be moving past the walls in steady streaming. The temperature variations near the wall will be as shown in Fig. 83. Let the temperatures within the bulk of each fluid be T and T' ($T > T'$). Then heat will flow from fluid F to fluid F', at a rate say $\dot{q}$ per cm.[2] of wall surface. There will be temperature drops, ΔT, and $\Delta T'$, at the wall-fluid boundaries associated with this flow of heat. Also there will be a temperature drop, ΔT_w across the wall given by:

$$\dot{q} = \frac{\lambda_w \Delta T_w}{t} \tag{42.1}$$

where λ_w and t are respectively the heat conductivity and thickness of the wall.

The transfer of heat from fluid to wall, or vice versa, is described in terms of a heat transfer coefficient, α, such that for small temperature differences:

$$\dot{q} = \alpha \Delta T = \alpha' \Delta T' \tag{42.2}$$

where α and α' are the heat transfer coefficients which give the rate of heat transference per unit temperature difference per cm.[2] of transfer surface[3]. The term α, the evaluation of which is discussed in Sect. 46, is a function of a number of parameters of the fluid, such as the thermal conductivity, kinematic viscosity, density, specific heat at constant pressure. It is also a function of the shape and

---

[1] R. SPOENDLIN: J. Res. CNRS. **15**, 1 (1951).

[2] See, for example, the design of W. F. GIAUQUE used by J. G. DAUNT and H. L. JOHNSTON. [Rev. Sci. Instrum. **20**, 122 (1949).]

[3] In this section and in Sect. 46 some basic information on heat transfer between fluids and solids is given. For more detail the reader is referred to the many texts on this subject, including for example W. H. McADAMS: Heat Transfer. New York: John Wiley & Sons. 1942. — R. C. L. BOSWORTH: Heat transfer Phenomena. New York: John Wiley & Sons 1952. — M. JACOB and G. A. HAWKINES: Elements of Heat Transfer and Insulation. New York: John Wiley & Sons 1950. — Also reference is recommended to the very complete monograph by H. HAUSEN: Wärmeübertragung in Gegenstrom, Gleichstrom und Kreuzstrom. Berlin: Springer 1950.

size of the boundary wall and of the fluid velocity past the wall. In the latter connection it should be emphasized that for all conditions of heat transfer treated here it will be assumed that the fluid flows are turbulent. The fluid mixing accompanying turbulent flow assures that all parts of the fluid contact the wall, so improving the thermal transfer.

In general, differences of the material (if all metal) and of surface state of the wall have a negligible influence on α. However, if the wall should get "plated" with impurities condensed out of the gas flow, such a oil vapor, CO_2, etc., the thermal transfer is reduced. This is a serious problem in the practical operation of heat interchangers.

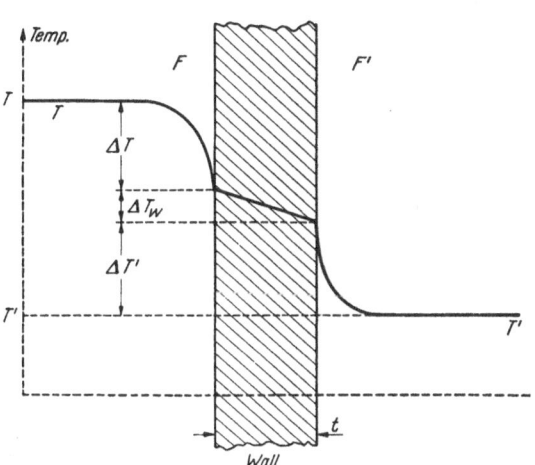

Fig. 83. Schematic plot of temperature between two gases flowing on either side of a metal wall.

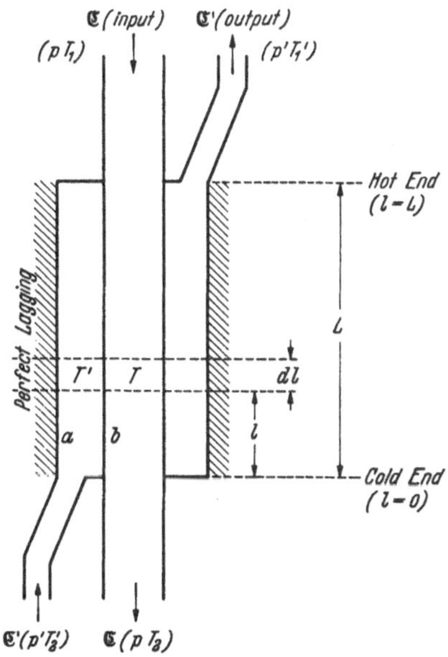

Fig. 84. Schematic diagram for simple LINDE type heat interchanger.

The overall temperature difference between the two bulk fluids is given by:

$$T - T' = \Delta T + \Delta T_w + \Delta T' \tag{42.3}$$

which combined with eqs (42.1) and (42.2) serves to define an overall heat transfer coefficient, $\varkappa$, such that

$$\dot{q} = \varkappa (T - T') \tag{42.4}$$

and

$$\frac{1}{\varkappa} = \frac{1}{\alpha} + \frac{t}{\lambda_w} + \frac{1}{\alpha'}. \tag{42.5}$$

In general the value of the metallic heat conductivity, λ_w, is so large that the second term on the right hand side can be neglected, yielding:

$$\frac{1}{\varkappa} \approx \frac{1}{\alpha} + \frac{1}{\alpha'}. \tag{42.6}$$

43. The temperature distribution in simple interchangers. In order to illustrate the principles of computation of the thermal properties of interchangers, a simple type is first considered. Let us have a simple LINDE type counter-current heat interchanger as illustrated in Fig. 84. Let it consist of two concentric tubes, a and b; b carrying the input gas downwards from the higher temperature region and the annular space between a and b carrying the return gas flow upwards from the lower temperature region. In the steady state, the temperatures T_1 and T_2 $(T_1 > T_2)$ at the entrance and exit of the input stream and the temperatures T_1' and T_2' $(T_1' > T_2')$ at the exit and entrance respectively of the return stream are constant.

For simplicity suppose:

a) The outer wall of tube a is perfectly lagged so that no heat may flow through it.

b) That the conduction of heat down the material of the walls of tubes a and b is negligible.

c) That the overall transfer coefficient, $\varkappa$, across tube b between the two gas flows is the same along the whole length of tube b and that the tube b is of uniform perimeter along its length.

After the preliminary calculations have been carried out, consideration will be given to the effect of relaxing conditions a), b) and c) above.

At a distance l from the colder end of tube a, let the bulk gas temperatures of the input, and output flows in the steady state be T and T' respectively.

Let $\dot{m}$ and $\dot{m}'$ be the mass flow rates of the input and output gases and C_p and $C_{p'}$, the specific heats at the constant pressures p and p' respectively. Then define quantities $\mathfrak{C}$ and $\mathfrak{C}'$ such that:

$$\mathfrak{C} = \dot{m}\,C_p; \quad \mathfrak{C}' = \dot{m}'\,C_{p'}. \tag{43.1}$$

The heat flow, $d\dot{q}$, between the two gas flows over the length l to $l+dl$ is:

$$d\dot{q} = \varkappa\,(T - T')\,P\,dl \tag{43.2}$$

where $\varkappa$, as before, is the overall heat transfer coefficient [eq. (42.6)] and where P is the effective perimeter[1] for heat transfer of tube b.

Also considering the gain or loss of heat in the gas along the length dl, we have:

$$\left. \begin{aligned} d\dot{q} &= \mathfrak{C}\,\frac{dT}{dl} \cdot dl = \mathfrak{C}\,dT, \\ d\dot{q} &= \mathfrak{C}'\,\frac{dT'}{dl}\,dl = \mathfrak{C}'\,dT'. \end{aligned} \right\} \tag{43.3}$$

Integrating eq. (43.2) between $l=0$ and $l=l$, we have:

$$\varkappa\,P\,l = \int_0^l \frac{d\dot{q}}{T - T'}. \tag{43.4}$$

Also under the restrictions a) and b) above concerning external heat fluxes, one gets from the heat balance between the two gas streams between points $l=0$ and $l=l$:

$$\mathfrak{C}\,(T - T_2) = \mathfrak{C}'\,(T' - T_2') = \dot{q} \tag{43.5}$$

and from the heat balance over the whole length, L, of the interchanger

$$\mathfrak{C}\,(T_1 - T_2) = \mathfrak{C}'\,(T_1' - T_2') = \dot{Q}. \tag{43.6}$$

Eqs. (43.5) and (43.6) implicitly assume $\mathfrak{C}$ and $\mathfrak{C}'$, i.e. C_p and $C_{p'}$, to be independent of temperature. Such an assumption is in general only a crude first approximation. It is made here for simplicity, the more detailed treatment being left to the reader.

---

[1] For further discussion of the term "effective" see Sect. 45.

One may re-express eq. (43.5) in the following way, which will prove convenient

$$(T - T') = (T_2 - T_2') - \left(\frac{\mathfrak{C}' - \mathfrak{C}}{\mathfrak{C}'}\right)(T_2 - T) \qquad (43.7)$$

which by differentiation yields:

$$d(T - T') = \left(\frac{\mathfrak{C}' - \mathfrak{C}}{\mathfrak{C}'}\right)dT. \qquad (43.8)$$

Combining eq. (43.8) with eq. (43.3), one gets:

$$d\dot{q} = \frac{\mathfrak{C}'\mathfrak{C}}{\mathfrak{C}' - \mathfrak{C}}\, d(T - T') \qquad (43.9)$$

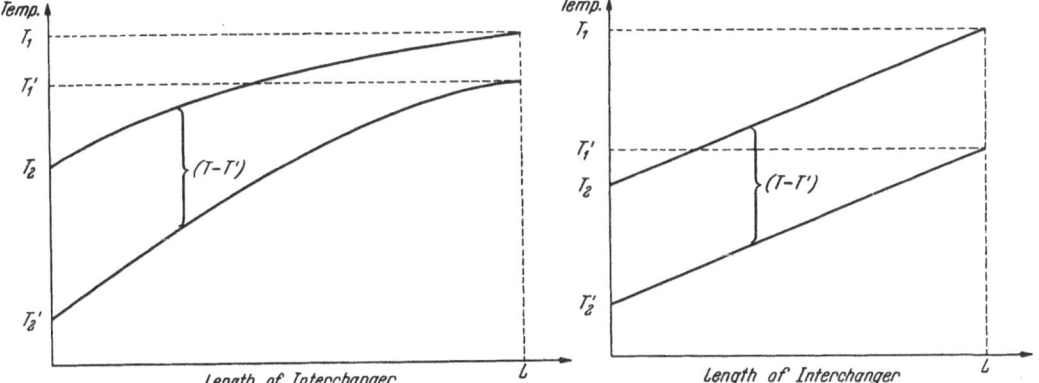

Fig. 85. Temperature along length of interchanger of Fig. 84 as measured from cold end. Case of $\mathfrak{C} > \mathfrak{C}'$

Fig. 86. Temperature along length of interchanger of Fig. 84 as measured from cold end. Case of $\mathfrak{C} = \mathfrak{C}'$.

which can be substituted in eq. (43.4) to permit integration of the latter, with the result that:

$$(T - T') = (T_2 - T_2') \exp\left[-\left(\frac{1}{\mathfrak{C}'} - \frac{1}{\mathfrak{C}}\right)\varkappa\, Pl\right] \qquad (43.10)$$

which gives the variation of $(T - T')$ along the length of the interchanger. At the extreme end, $l = L$, eq. (43.10) gives:

$$(T_1 - T_1') = (T_2 - T_2') \exp\left[-\left(\frac{1}{\mathfrak{C}'} - \frac{1}{\mathfrak{C}}\right)\varkappa\, PL\right]. \qquad (43.11)$$

A diagram showing the exponential form of eq. (43.10) and (43.11), under the assumption that $\mathfrak{C} > \mathfrak{C}'$, is given in Fig. 85. As shown by the figure, $(T - T')$ diminishes exponentially as L increases. To reduce the difference in temperatures, $(T_1 - T_1')$, at the warmer to zero, i.e. in order that the emergent gas stream should take exactly the input gas input temperature, clearly an infinitely long exchanger would be necessary.

If the total heat capacities per sec. of gas flow, $\mathfrak{C}$ and $\mathfrak{C}'$ are *equal*, then the exponent of e in eqs. (43.10) and (43.11) is zero, whence

$$(T - T') = (T_1 - T_1') = (T_2 - T_2'), \qquad (43.12)$$

a situation illustrated in Fig. 86, in which the temperature difference $(T - T')$ between the two streams remains constant.

44. Some practical considerations of gas flows in interchangers used in low temperature producers. In general, the output flows which we have to consider in low temperature technology are smaller than the input flows, measured in g/sec. In liquefiers for example this is certainly true, for some fraction of the input flow is liquefied and this fraction does not therefore return back through the interchangers. There may be situations, however, where the two mass flows, $\dot m$ and $\dot m'$, are equal, such as may occur in a refrigerator. To determine therefore, in the extreme case of $\dot m = \dot m'$, whether $\mathfrak{C}$ is greater or less than $\mathfrak{C}'$ [c.f. the definitory eq. (43.1)], a study must be made of the variation of C_p with pressure.

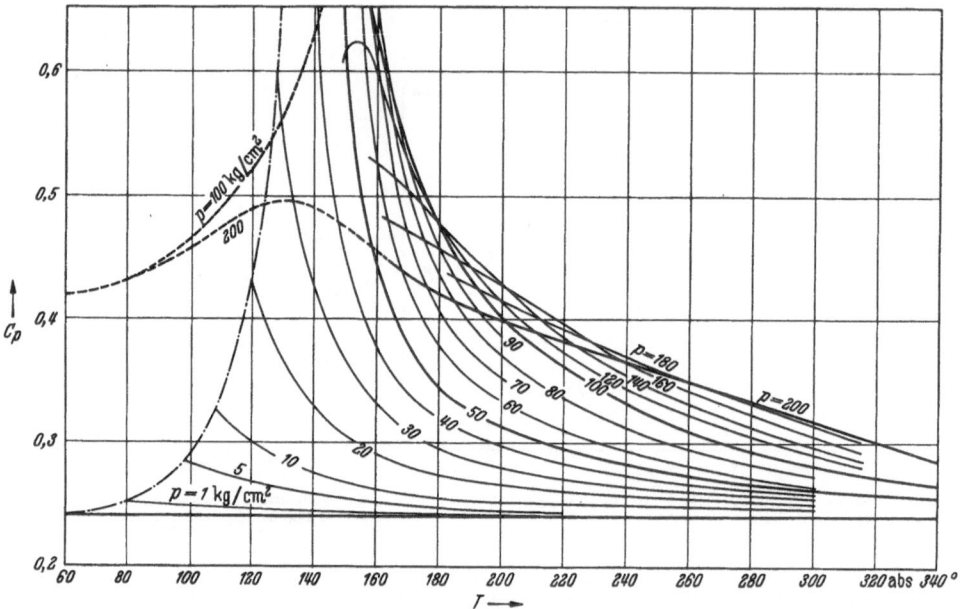

Fig. 87. Specific heat, C_p, at constant pressure for air as function of absolute temperature for various isobars. [Curves due to H. Hausen: Wärmeübertragung in Gegenstrom, Gleichstrom and Kreuzstrom. Berlin: Springer 1950.)

For a perfect gas C_p is independent of p. For a real gas, well above the critical temperature where its equation of state can be adequately described in terms of the first two virial coefficients only, it can be shown that for small pressures:

$$\left(\frac{\partial C_p}{\partial p}\right)_T = - T\left(\frac{\partial^2 B}{\partial T^2}\right). \tag{44.1}$$

This term is generally positive, i.e., C_p increasing with increasing p. This can be seen in the graphs of C_p versus T for air at various pressures given in Fig. 87. From this figure it will be seen also that the effect of pressure on C_p becomes much more complex near the critical temperature (T_c for air $= 132.5°$ K). Since interchangers are generally used at temperatures well above T_c, one may assume that in general $C_p > C_{p'}$, where C_p applies to the input (high pressure) gas and $C_{p'}$ to the output (low pressure) gas. At temperatures above about 160° C, helium is an exception to this general rule[1]: for example, at 275° C ($C_{200\,At}/C_{1\,At}$) = 0.99. Below 150° C, $(\partial C_p/\partial p)_T$ for helium is positive but small. For example, for $-50°$ C $< T < 150°$ ($C_{40\,At}/C_{1\,At}$) < 1.001.

---
[1] W. H. Keesom: Helium, p. 86. Amsterdam: Elsevier 1942.

Assuming $C_p > C_{p'}$ however, for most practical applications, one may further assume therefore that for $T \gg T_c$

$$\mathfrak{C} \geq \mathfrak{C}' \qquad\qquad (44.2)$$

since $\dot{m} \geq \dot{m}'$.

The condition (44.2) means that the exponent of e in eqs. (43.10) and (43.11) is negative (or in the extreme case zero). This in turn means that, as is illustrated in Fig. 85, interchangers are convergent, i.e., are such that $(T - T')$ diminishes as one passes from the cooler to the warmer end.

Eq. (43.11) shows that to maintain a given value of the temperature difference, $(T_1 - T_1')$, at the warm end of the interchanger for given T_2 and T_2' at the cold end, the quantity

$$\left(\frac{1}{\mathfrak{C}'} - \frac{1}{\mathfrak{C}}\right) PL \qquad\qquad (44.3)$$

must be held constant. As $\mathfrak{C}'$ approaches $\mathfrak{C}$, therefore, the total length of the interchanger must be increased. This fact illustrates the difficulty of constructing highly efficient interchangers when the total heat capacities per second of gas flow, $\mathfrak{C}$ and $\mathfrak{C}'$, are nearly equal. This fact, pointed out by Jacobs and Collins[1], explains why the interchanger requirements on a Linde type gas liquefier, which have relatively large liquefaction coefficients, are less critical than those, for example, on a Collins-type helium liquefier which has a much smaller value of coefficient of liquefaction. It also illustrates the fact that the constructional requirements on interchangers for low pressure air liquefiers are more stringent than for high pressure systems, since the ratio $C_p/C_{p'}$ for the low pressure system is smaller.

45. Interchanger efficiency. The efficiency of an interchanger can be defined as the ratio of the total heat transfered per second from one stream to the other to the maximum possible heat transferable per second in an infinitely long interchanger.

Confining ourselves to the practical case where

$$\mathfrak{C} \geq \mathfrak{C}'$$

this means that the efficiency, η, is given by:

$$\eta = \frac{\dot{Q}}{\dot{Q}_{max}} = \frac{\mathfrak{C}'(T_1' - T_2')}{\mathfrak{C}'(T_1 - T_2')} = 1 - \frac{(T_1 - T_1')}{(T_1 - T_2')}. \qquad\qquad (45.1)$$

Eq. (45.1) means that η is given by:

$$\eta = 1 - \frac{\text{temp. diff. between flows at warm end}}{\text{max. temp. diff. between ends of interchanger}}. \qquad\qquad (45.2)$$

By use of eqs. (43.5), (43.7) and (43.11), one may obtain from eq. (45.1) an alternative implicit expression for η in terms of parameters which are more readily pre-calculable, as follows:

$$\ln\left[\frac{\mathfrak{C} - \eta\mathfrak{C}'}{(1 - \eta)\mathfrak{C}}\right] = \left(\frac{\mathfrak{C} - \mathfrak{C}'}{\mathfrak{C}\mathfrak{C}'}\right) \varkappa PL \qquad\qquad (45.3)$$

or, putting $\mathfrak{C}' = x\mathfrak{C}$,

$$\ln\left(\frac{1 - \eta x}{1 - \eta}\right) = (1 - x)\frac{\varkappa PL}{\mathfrak{C}'}. \qquad\qquad (45.4)$$

---

[1] R. B. Jacobs and S. C. Collins: J. Appl. Phys. **11**, 491 (1940).

Graphs showing the variation of η with $(\varkappa PL/\mathfrak{C}')$ for various values of $x\ (=\mathfrak{C}'/\mathfrak{C})$ are shown in Fig. 88.

This figure shows how, when x is made smaller, the same efficiency can be obtained with smaller values of L.

In the limit when $\mathfrak{C}=\mathfrak{C}'$, eqs. (45.3) and (45.4) simplify, since $(T-T')$ is constant along the length of the interchanger and one gets:

$$\eta = \frac{\varkappa\, PL}{\varkappa PL + \mathfrak{C}}\,. \tag{45.5}$$

The eqs. (45.3) and (45.5) enable the efficiency of an "ideal" interchanger to be computed from a knowledge of the rates of gas flow, the gas flow pressures (these determine C_p and $C_{p'}$) and of the term $(\varkappa PL)$. An "ideal" interchanger for this purpose is one in which the restrictions a) and b) of Sect. 43 above apply, namely: no heat flow from outside to the interchanger and negligible heat flows down metal parts of the interchanger.

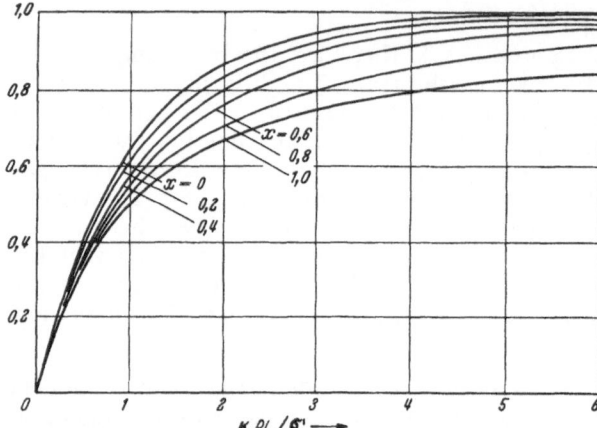

Fig. 88. Efficiency of ideal interchanger of Fig. 84 as function of $(\varkappa PL/\mathfrak{C}')$ for various values of $x\ [x=\mathfrak{C}'/\mathfrak{C}]$. (Curves due to H. HAUSEN: Wärmeübertragung in Gegenstrom, Gleichstrom und Kreuzstrom. Berlin: Springer 1950.)

The term $(\varkappa PL)$ is the product of the average total heat transfer coefficient at the transfer surfaces of the interchanger and of the total effective transfer surface area, A_{eff}, for the whole interchanger.

Of course $\varkappa$ may vary along the length of the interchanger, but such variations can in principle be allowed for by computing

$$(\varkappa\, PL)_{\text{eff}} = \int_0^L \varkappa\, P\, dl. \tag{45.6}$$

It is to be observed that by defining $(\varkappa\, PL)$ in this way one automatically disposes of restriction c) of Sect. 43 above, since by integrating $\varkappa\, P dl$ along the whole length of the interchanger, possible changes in the effective perimeter P of the heat transfer surface with distance l can be allowed for, as well as possible changes in the local value of $\varkappa$.

In principle the value of $\varkappa$ may be calculated using eq. (42.6), by obtaining the values of the individual heat transfer coefficients, α and α', in the manner described in Sect. 46.

For the more complex analysis of interchanger performance when restrictions a) and b) of Sect. 43 above are relaxed, i.e. when allowance has to be made for heat leakage into the interchanger from outside as well as for heat conduction down the material of the interchanger tubing, reference should be made to the detailed monograph of HAUSEN[1].

---

[1] H. HAUSEN: Wärmeübertragung im Gegenstrom, Gleichstrom und Kreuzstrom. Berlin: Springer 1950.

46. Computation of the heat transfer coefficient. In computation of the heat transfer coefficient[1], the dimensional analysis of NUSSELT[2] is employed. The heat transfer coefficient is related to the dimensionless NUSSELT number, Nu, by:

$$\alpha = Nu \cdot \frac{\lambda}{D} \qquad (46.1)$$

where λ is the thermal conductivity of the gas and where D is the hydraulic diameter of the tube through which the gas is flowing and to which it is transferring heat. The hydraulic diameter is defined by:

$$D = \frac{4V}{S} \qquad (46.2)$$

where V is the volume occupied by the fluid per unit length of the channel and where S is the total wetted surface per unit length.

The NUSSELT number, Nu, is a function of other dimensionless quantities—the REYNOLDS number, Re, and the PRANDTL number, Pr, and of the tube geometry, such that:

$$Nu = \text{const } (Re)^x \cdot (Pr)^y \cdot \left(\frac{D}{L}\right)^z \qquad (46.3)$$

where L is the tube length. Even for $L \gg D$ the third term on the right hand side is unity, since from experiment $z \approx 0.05$.

The REYNOLDS number is given by:

$$Re = \frac{4\dot{m}}{\pi \mu D} \qquad (46.4)$$

where $\dot{m}$, as before, is the mass flow through the tube of diameter D per second, and where μ is the viscosity of the gas.

The PRANDTL number is given by:

$$Pr = \frac{\mu C_p}{\lambda} \qquad (46.5)$$

where the terms have their previously mentioned significance.

The PRANDTL number has almost the same numerical value for all gases. (See for example, MCADAMS[3], table 26a, for Pr for a variety of gases at $p = 1$ atm, and $T = 100°$ C.) Moreover it does not change very much in magnitude with change in temperature. For air, for example, $Pr = 0.72$ at $0°$ C and 1 atm., and $Pr = 0.90$ at $-180°$ C and 1 atm.[4].

The heat transfer coefficient, α, therefore can be computed from eqs. (46.1) to (46.5) from known properties (λ, μ and C_p) of the gas flowing, from its flow rate, $\dot{m}$ and from the tube geometry, (D, L), provided the constant and indices x, y and z of eq. (46.3) are known. The latter are determined only by experiment. However, since it is generally assumed that the dimensional analysis of NUSSELT is of general application, it is presumed that the values of x, y and z determined for one gas or for one range of conditions may be used for other similar gases in other ranges of similar conditions. The general texts referred to in Sect. 42

---

[1] See general references given in Sect. 42 for further detail.
[2] W. NUSSELT: Z. VDI **53**, 1750, 1809 (1909). — Phys. Z. **12**, 285 (1911).
[3] W. H. MCADAMS: Heat Transmission, p. 145. New York: McGraw Hill 1951.
[4] See data given by J. A. VAN LAMMEREN (Technik der tiefen Temperaturen, p. 55. Berlin: Springer 1941) for air, H_2 and He at various temperatures.

give the experimental evaluations of x, y and z. For turbulent flow of gases it seems generally agreed[1] that $x = 0.8$, $z \approx 0.05$ and $y \approx 0.4$ for heating of the gas and $y \approx 0.3$ for cooling of the gas.

The small values of y together with the relatively constant value of the PRANDTL number itself allows one to approximate $(Pr)^y$ by a constant over a wide range of conditions and for a great variety of gases. Moreover, as mentioned above, the small value of z means that even where $D \ll L$, the term $(D/L)^z$ can also be approximated by a constant. Hence one may rewrite eq. (46.3) in the following approximate form for gases:

$$Nu = \text{constant } (Re)^{0.8} = \beta \cdot (Re)^{0.8} \tag{46.6}$$

where $\beta = 0.019$ according to HALL and TSOA[1].

An interesting but different formulation has been proposed for hydrogen at low temperature by STARR[2], who put the indices $x = 0.75$ and $y = 1$. This means that from eqs. (46.1) and (46.5) the heat transfer coefficient, α, becomes independent of the heat conductivity, λ, of the gas. The final formula for α then becomes:

$$\alpha = \frac{0.15}{\pi} C_p \mu^{0.25} \frac{\dot{m}^{0.75}}{D^{1.75}} \text{ joule } /^{\circ}\text{K - cm. - sec.} \tag{46.7}$$

where the units are cgs with C_p in joules/g$^{\circ}$- K.

47. Pressure drops in simple interchangers. Since the fundamental bases of heat conductivity and viscosity in gases are both transport processes, of energy and of momentum, it must be concluded that systems showing large total heat transference to the walls should also be systems yielding large resistance to flow. Very high heat transference must in general be associated with significant pressure drops in the gas flow. In the design of an interchanger for low temperature operations it is desirable to minimize, especially in the low pressure flows, these undesirable pressure drops.

In computing the pressure drop, Δp, in turbulent flow of gas through a length dl of tube, again the method of dimensionless analysis may be employed and for simple shapes the FANNING equation is applicable, namely:

$$-dp = f \varrho v^2 \frac{dl}{D} \tag{47.1}$$

where f is the friction factor (a dimensionless number which is a function of the REYNOLDS number) and where v is the velocity of flow. Expressed in terms of mass flow we get:

$$-dp = \frac{16}{\pi^2} f \frac{\dot{m}^2}{\varrho} \frac{dl}{D^5} \tag{47.2}$$

where the friction factor f and its dependence on the dimensionless REYNOLDS number remains to be determined by experiment. BLASIUS[3] gives for turbulent flow:

$$f = \text{const} \cdot (Re)^{-0.25} \tag{47.3}$$

[1] See T. A. HALL and P. H. TSOA [Proc. Roy. Soc. Lond., Ser. A **191**, 6 (1947)] for low temperature measurements. See also B. H. SCHULTZ [Appl. Sci. Res. A **1**, 287, 400 (1947/49)] for theoretical discussion of these factors in their application to heat interchangers.
[2] C. STARR: Rev. Sci. Instrum. **12**, 193 (1941).
[3] H. BLASIUS: Phys. Z. **12**, 1175 (1911).

which, assuming $p = \varrho \frac{RT}{M}$, as for a perfect gas, gives the equation quoted by STARR[1]:

$$- dp = 0 \cdot 021 \, \frac{T \mu^{0.25} \, \dot{m}^{1.75}}{p \, M \, D^{4.75}} \, dl \ \text{g/cm.}^2 \tag{47.4}$$

where cgs units are used, where M is the molecular weight of the gas and where p is in atmospheres. Formula (47.4) is a satisfactory approximation only for temperatures well above the critical temperature.

The following considerations concerning the number of parallel paths in an interchanger are of interest. Suppose that an interchanger be made of n simple LINDE interchangers, as detailed immediately above, in parallel, each carrying $1/n$ of the total required mass flow. Then eq. (47.4) shows that for variation of the number n, the pressure drop, Δp, will remain unchanged provided:

$$n^7 D^{19} = \text{const} \tag{47.5}$$

where D is the hydraulic diameter of the tubing (tube b of Fig. 84) forming the heat interchanging surface.

On the other hand, the total effective heat exchange parameter, increasing which increases the efficiency of the interchanger, is given by $n(\varkappa P L)$, (cf. eq. (45.6)]. Now $\varkappa$ is directly related to the heat transfer coefficient, α, by eq. (42.6) and α as a function of D can in certain circumstances be assumed to be given approximately by eq. (46. 7). By combining these equations, we conclude that *for constant Δp*, the total effective heat exchange parameter, $n(\varkappa P L)$, is proportional to $n^{0.53}$. For more complex interchangers and for other possible choices of the parameters, x, y, and z in the NUSSELT equation, the index 0.53 may alter somewhat. However, this result allows the conclusion to be made that for given Δp, the efficiency of the interchanger is increased by increasing the number of parallel paths. This conclusion is quite general for any interchanger and it emphasizes the need for multiplicity of flow paths in the low pressure side of interchangers, where minimizing Δp is of importance.

48. Practical examples of actual interchangers. The simpler forms of LINDE and HAMPSON interchangers, described in Sect. 41 above, have been elaborated, and some account of these subsequent developments follows herewith.

The two or three tube counter-current interchangers, first used by LINDE, led to the use of bundles of tubes in parallel. Descriptions of such tube-bundle counter-current interchangers have been given by LINDE[2], and experimental tests on the relative merits of such interschangers have been made by JACOBS and COLLINS[3]. The three main types tested by the latter authors consisted of a) three concentric tubes separated by tightly fitting helical wire spacers, which wires were firmly soldered to the tubes touching them. In this type the two gas flows passed through the two annular regions, the inner central tube carrying no flow. b) Seven parallel tubes, all of the same size, all soldered together along their length. This is the type used by KAPITZA[4] in his helium liquefier; the central tube carries the high pressure input gas and the six surrounding tubes carry in parallel the return low pressure gas. c) Seven parallel tubes, all of the same size, pushed into one envelope tube and tightly fitting it. In this all seven tubes

---

[1] C. STARR [Rev. Sci. Instrum. **12**, 193 (1941)] gives this formula in consideration of hydrogen liquefier interchangers. His numerical constant, however, is incorrect by a factor of ten.

[2] R. v. LINDE: Z. ges. Kälteind. **41**, 161 (1934).

[3] R. B. JACOBS and S. C. COLLINS: J. Appl. Phys. **11**, 491 (1940).

[4] P. KAPITZA: Proc. Roy. Soc. Lond., Ser. A **147**, 189 (1934).

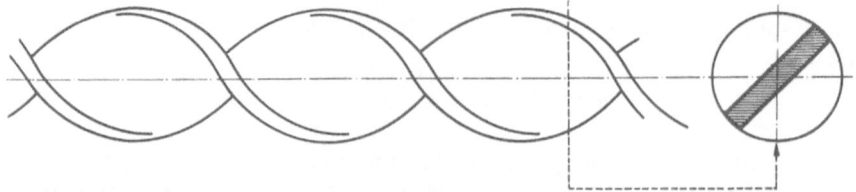

Fig. 89. Diagram of construction of "twisted tube" interchanger due to NELSON [F. R. BICHOWSKI: J. Industr. Engng. Chem. **14**, 62 (1922).]

Fig. 90. Photograph of COLLINS-type interchanger. (Courtesy of A. D. Little Inc.)

carried the low pressure return flow, the high pressure input flow passing down the space between them and the outer envelope. They found for 15 foot lengths of each type interchanger the following efficiencies: type a) 95.7%; type b) 94% and type c) 94.5%.

JACOBS and COLLINS also tested a two tube counter-current type which was devised some time ago by NELSON, called the "twisted tube" interchanger. This was first described by BICHOWSKI[1] and an example of it, used in a small scale hydrogen liquefier, has been described above in Sect. 27. The inner tube, which carries the high pressure flow is flattened and then twisted before being inserted in the outer tube. This construction is illustrated in Fig. 89. JACOBS and COLLINS concluded that its characteristics are similar to the others described and that it can generally be matched by a simpler design. A novel counter-current interchanger for large scale operations built up from perforated plates has been described in detail lay McMAHON, BOWEN and BLEYLE[2].

Some commonly used HAMPSON type interchangers have been described in Sect. 41. A type used by COLLINS[3] in his first helium liquefier has proved very successful, and is diagrammatically illustrated in the subsequent article by S. C. COLLINS. Two coaxial cones of poor conducting metal form the low pressure flow circuit, the low pressure flowing in the

---

[1] F. R. BICHOWSKI: J. Industr. Engng. Chem. **14**, 62 (1922).
[2] McMAHON, BOWEN and BLEYLE, jr.: Trans. Amer. Soc. Mech. Engrs. **72**, 623 (1950).
[3] S. C. COLLINS: Rev. Sci. Instrum. **18**, 157 (1947).

annular region between the cones. The high pressure tube, of brass $\frac{1}{4}''$ o.d., 0.015'' wall, has soldered outside of it a spiral finning made of edgewound copper ribbon 0.010'' × 0.040'' with 30 turns per inch. This finned tube is then wound on the inner cone before the outer cone is placed over it. The low pressure return gas therefore passes at right angles to the high pressure tube, being forced to travel through the spaces between the fins. The arrangement shown in Fig. 90 indicates how cotton cord suitably located ensures that the low pressure gas is guided into the finning.

A variant of the finned interchanger mentioned above has been described by Nicol, Smith, Heer and Daunt [1]. Instead of soldered finning, they formed fins by threading the tube. The found that by passing $\frac{1}{4}''$ o.d., 0.30'' wall, copper tubing through a die mounted in a lathe, threading 0.015'' deep, 28 to the inch, could easily be made. Immediately after the threaded tube feeds out of the die, it was annealed in a torch flame. This threaded tubing was wound in layers on a cylindrical central tube, the outermost layer being finally covered with a cover tube. The low pressure gas flows past the threads, at right angles the high pressure tubes, in annular space between the central tube and the cover tube. As in the previous type interchanger, cotton cords were used to guide the low pressure flow closely around the threading. High efficiencies for this type of interchanger were obtained. A photograph of a more recent six layer model, constructed by Erickson and Daunt, is shown in Fig. 91, which was taken with the cover tube removed.

Fig. 91. Photograph of threaded tube type interchanger due to Nicol, Smith, Heer and Daunt. (Courtesy of Dr. R. A. Erickson.)

A novel type of finned tubing, used in a two tube counter-current interchanger in a helium liquefier has been described by Ashmead [2].

49. The regenerator. Although regenerators for exchange of heat between high temperature gas flows have been in use for a long time, as for example in

[1] Nicol, Smith, Heer and Daunt: Rev. Sci. Instrum. **24**, 16 (1953).
[2] J. Ashmead: Proc. Phys. Soc. Lond. **63**, 504 (1950).

the foundries, their use in low temperature technology is relatively recent. The first suggestion for their low temperature application was due to FRÄNKL[1] in 1928 and soon thereafter they were used by the Linde Company in air rectification units[2].

A schematic diagram of a low temperature regenerator is given in Fig. 92. It consists of two columns, 1 and 2, packed with metal ribbon. The cut in column 2 indicates the metal ribbon "pancakes" piled one on top of the other.

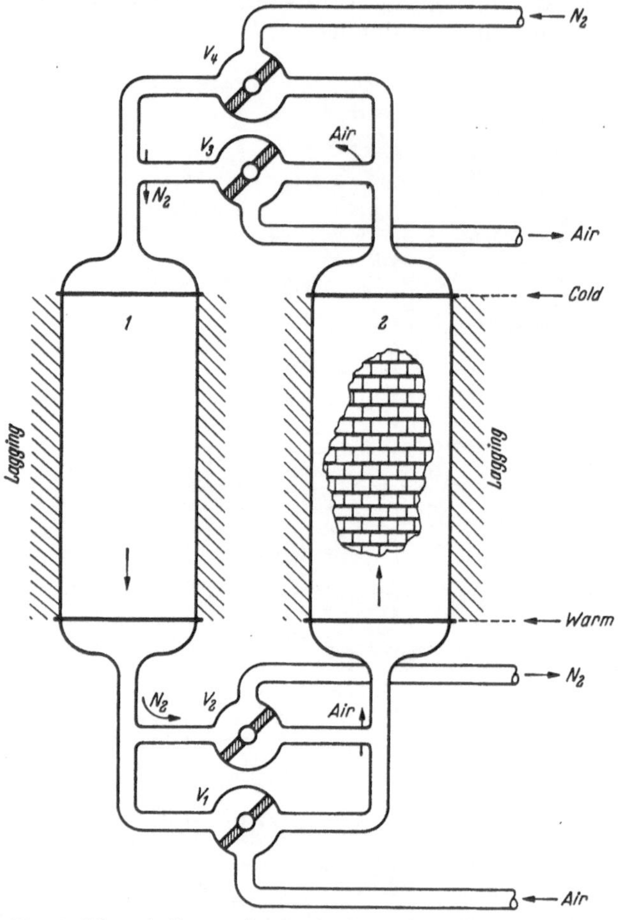

To form a pancake, the metal ribbon is first serrated or crinkled and then wound to diameter to fit the columns, Aluminium, iron or copper may be used for the ribbon, generally about 2 to 3 cm. wide and about 0.5 mm. thick. One method of construction is to wind a pair of ribbons together, one plain and the other saw-tooth form. This results in a pancake of the type shown in Fig. 93; which gives a multiplicity of paths for gas to travel through it parallel to the width of the ribbon. Another construction due to FRÄNKL is to make "rifling" type crinkles and to wind two ribbons together which have opposed rifling. This is shown in Fig. 94. Each pancake is separated by a gauze as each is placed in the regenerator column.

The ratio of surface area of the metal packing to its volume should be greater than 1000 m.$^{-1}$ in order that the dead volume of the column should not be too great, and that the heat transfer from gas to metal be adequate.

The regenerator column therefore consists of a mass of metal of large surface area, past which the gas can flow and interchange heat between metal and gas.

More recently BORCHARD[3] has reported the use of small stones, called "split", of size "between that of a hazel nut and a walnut", in low temperature

Fig. 92. Schematic diagram of regenerator system for exchanging heat between warm air stream and cold N₂ stream in low pressure LINDE-FRÄNKL air rectifier.

[1] M. FRÄNKL: German Patents 490878 and 492431. 1928. US. Patents 1890646. 1932. — 1970299. 1934.
[2] See for example R. LINDE: Z. ges. Kälteind. 41, 183 (1934). — B. C. P. HOCHGESAND: Mitt. Forsch. GHH. Konzern 4, 14 (1935). — J. WUCHERER: Iron Coal Tr. Rev. 159, 723 (1949).
[3] P. BORCHARD: Proc. VIII. Int. Congr. Refrig. London. p. 118 (1951).

regenerator columns instead of metal ribbon. It is reported that this split is better than aluminium filling.

By an arrangement of valves (see V_1, V_2, V_3, V_4, of Fig. 92) gas is made to flow up one column and simultaneously down the other. After a time, τ, the direction of gas flow in each column is reversed by action of the valves, $V_1 - V_4$. In practice, if V_1 and V_2 are the valves at ambient temperature, it is possible to replace the low temperature valves V_3 and V_4 by a system of automatic check valves which respond to the changes in pressure produced by action of the valves V_1 and V_2. The alternation of flow direction is repeated cyclically. In general, in low temperature application the time for each flow period is about 2 to 3 min.

In the regenerator of Fig. 92 the gas streams are not of the same gas, since it is possible to use the regenerator in an air rectifier with

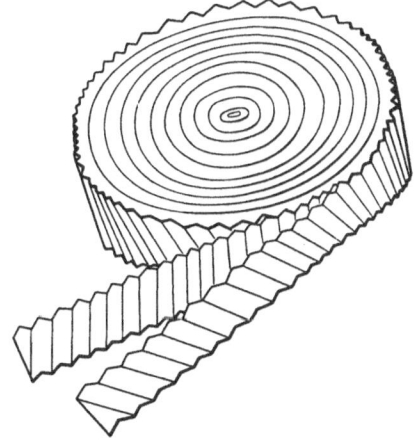

Fig. 93. Schematic section through "pancake" for regenerator column.

Fig. 94. Schematic drawing of "pancake" for regenerator using rifle ribbing due to FRÄNKL. [See also R. LINDE: Z. ges. Kälteind. **41**, 183 (1934).]

air as the input gas and N_2 (or alternately O_2) as the output gas. Of course, the same gas for each flow may be used.

The cold return gas passing through any one column gives up a fraction of its cold to the metal and during the subsequent flow period the warm input gas, flowing in the reverse direction is cooled by transference of part of its heat to the cool metal. In this way the regenerator performs the same function as a countercurrent interchanger. Since both gas streams flow through any one column of the regenerator in turn, the same hydraulic resistance is presented to both streams. This means that for successful operation the pressures of the two streams should not be widely different. This limits the efficient use of regenerators to low pressure liquefaction and rectification systems[1].

A typical regenerator has been described by HAUSEN[2] for an air rectifier with air (or N_2) flow of 10^3 m./hr. The regenerator was 3 m. long and contained 1250 kg. of galvanized iron foil with surface area 860 m.[2]. The input temperature for the air was 20° C and the input temperature of the N_2 at the other end was — 180° C. The warm and cold periods were each of duration 2 min. Some detail of the performance of this regenerator is given below in Sect. 50.

In an extensive paper, LUND and DODGE[3] have given constructional detail for a number of regenerators and have presented the results of their experimental

---

[1] See footnote to p. 84 for the effect of dead volume on the range of operational pressures.
[2] H. HAUSEN: Z. ges. Kälteind. **39**, 1 (1932).
[3] G. LUND and B. F. DODGE: Industr. Engng. Chem. **40**, 1019 (1948).

measurements on the overall heat transfer coefficients and on the pressure drops. Glaser has also reported similar measurements for Fränkl regenerator packings[1].

50. The temperature variations in a regenerator. Let Fig. 95a and b represent a regenerator with input gas entering at steady temperature T_1 and pressure p, and with return gas entering at steady temperature. T_2', at pressure, p'. If $\dot{m}$ and $\dot{m}'$ are the mass rates of flow of the input and output flows respectively, define quantities $\mathfrak{C}$ and $\mathfrak{C}'$, as previously in consideration of counter-current interchangers, by:

$$\mathfrak{C} = \dot{m}\,C_p; \quad \mathfrak{C}' = \dot{m}'\,C_{p'}, \tag{50.1}$$

where C_p and $C_{p'}$, are specific heats at constant pressure of the two streams. For simlicity assume C_p and $C_{p'}$, are temperature independent.

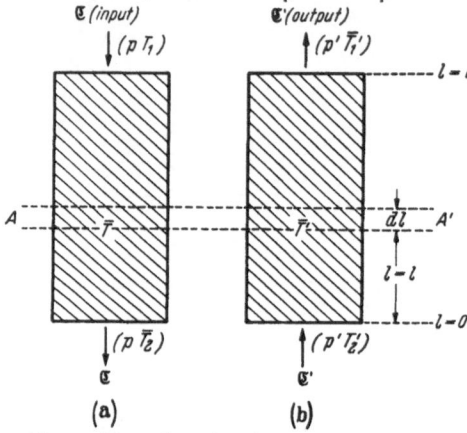

Fig. 95. Schematic section of regenerator columns.

Let the warm period, τ (i.e. the period when warm input gas is flowing through the column of interest), and the cold period, τ' (i.e. the period when cold return gas is flowing) be equal. Then the ratio of the total heat capacity of the input gas which flows in the entire warm period to the total heat capacity of the return gas which flows during the entire cold period is:

$$\frac{\mathfrak{C}\tau}{\mathfrak{C}'\tau'} = \frac{\mathfrak{C}}{\mathfrak{C}'} \tag{50.2}$$

Let the time averaged values[2] of the temperature of the gases emerging from the regenerator columns be $\overline{T_2}$ and $\overline{T_1'}$, and at any equivalent section AA' let the time averaged gas temperatures be $\overline{T}$ and $\overline{T'}$.

At any section AA', T and T', the instantaneous gas temperatures, will be functions of time, as illustrated in Fig. 96.

As the time t of the warm period progresses, T increases. Similarly as the time of a cold period increases, T' decreases.

The curves of Fig. 96 show also the time variation of the surface temperatures, ϑ_s and the mean temperatures, ϑ_m, of the metal ribbon at the same section. (ϑ_m is averaged over the half-thickness of the ribbon.)

In the curves of Fig. 96 and 97 the differences between the surface temperatures, ϑ_s, and the mean temperatures, ϑ_m at any one time have been greatly exaggerated. For metal ribbon, the heat conductivity in it is so great that the differences in temperature across its thickness are very small compared with the temperature differences between metal and gas.

The linear variation of T, T' and ϑ_m with time, shown in Fig. 96, is characteristic of the central regions of the regenerator. Near the ends marked deviations from time linearity occur, as is illustrated in Fig. 97, which shows the temperature variations at the warm end of the column.

These end effects are due to the time independence of T_1 and T_2', the input and return temperatures.

---

[1] H. Glaser: Z. VDI **53**, 925 (1939).
[2] *Time* averages are denoted by a superscipt bar. Averages over configurational space are denoted by a subscript m.

This time variation of temperature at any one location in a regenerator makes the analysis of its performance much more complex than that of countercurrent interchangers, in which all temperatures are steady.

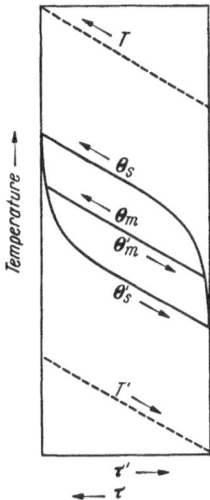

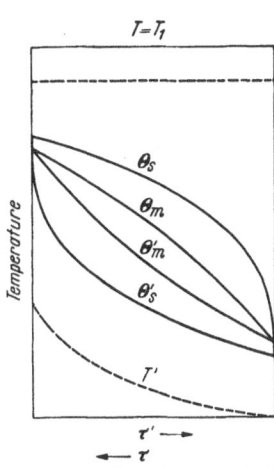

Fig. 96. Temperature as a function of time in hot and cold periods for center sections of regenerator. T and T' are the gas temperatures. ϑ_s and ϑ'_s the metal surface temperatures and ϑ_m the mean metal temperature. (Curves due to H. HAUSEN: Wärmeübertragung in Gegenstrom, Gleichstrom und Kreuzzstrom. Berlin: Springer 1950.)

Fig. 97. Temperature as function of time in hot and cold periods for hot end of regenerator column. T and T' are the gas temperatures. ϑ_s and ϑ'_s the metal surface temperatures and ϑ_m and ϑ'_m mean metal temperatures. (Curves due to H. HAUSEN: Wärmeübertragung in Gegenstrom, Gleichstrom und Kreuzstrom. Berlin: Springer 1950.)

The variation of temperature along the length of a regenerator column is illustrated in Fig 99 which gives the temperature, ϑ_m, of the metal ribbon at various times of the regenerator cycle. The data in these curves were obtained

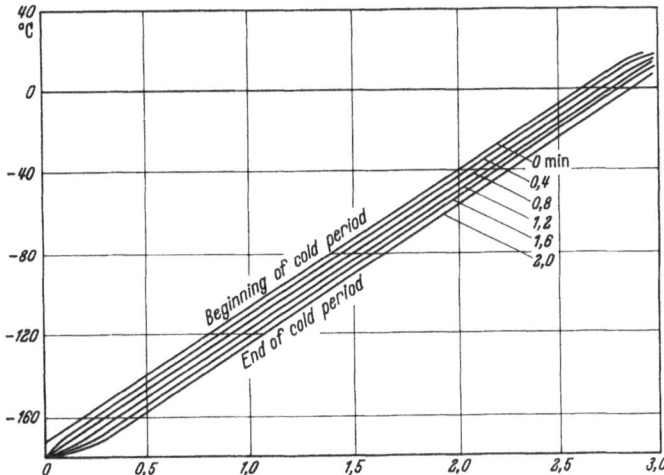

Fig. 98. The mean metal temperature, ϑ_m in a regenerator column as a function of length along the column for various times in the cold period and for the case $\mathfrak{C}=\mathfrak{C}'$. [Curves due to H. HAUSEN: Z. ges. Kälteind. **39**, 1 (1932).]

by HAUSEN[1] for the 3 m. long regenerator detailed in Sect. 49 above and for flows such that $\mathfrak{C}=\mathfrak{C}'$ and $\tau=\tau'$. The gas temperatures, T and T' also follow a linear variation with column length, l, in the central region, the temperature

---

[1] H. HAUSEN: Z. ges. Kälteind. **39**, 1 (1932).

differences $(T - \vartheta_m)$ and $(\vartheta_m - T')$, [cf. Fig. 96] being for this particular regenerator about 1.7° K.

The linear variation of the temperatures with length, l, in the central regenerator region is characteristic of the flow condition $\mathfrak{C} = \mathfrak{C}'$ and $\tau = \tau'$. Note that for countercurrent interchangers the same flow condition also resulted in a linear T (or T') versus l relationship. If $\mathfrak{C}\tau > \mathfrak{C}'\tau'$, the temperature in the central region becomes an exponential function of the column length l, as was the case for the counter-current interchanger. Fig. 99 due to HAUSEN[1] gives the metal temperature, ϑ_m, for the same 3 m. long regenerator of Fig. 98 as a function of length, l, for the flow condition $\mathfrak{C}\tau = 1.05\,\mathfrak{C}'\tau'$. For 0.3 m. $< l < 2.7$ m. HAUSEN gives:

$$\vartheta_m = 31.96 - 192.57 \exp(-0.976\,l + 0.125\,t) \qquad (50.3)$$

where l is in m and time t in min.

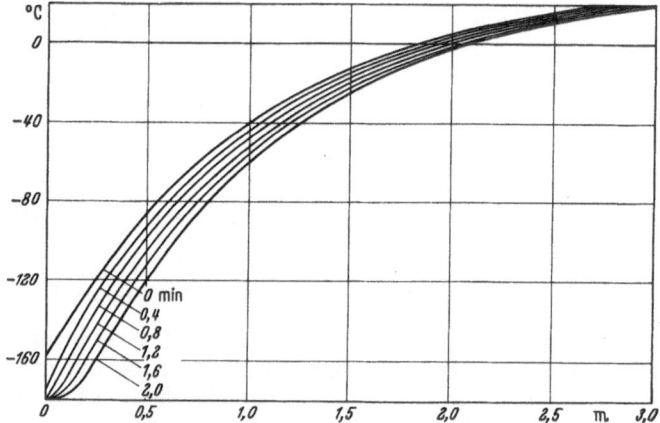

Fig. 99. The mean metal temperature, ϑ_m, in a regenerator column as function of length along the column for various times in the cold period and for the case $\mathfrak{C} = 1.05\,\mathfrak{C}'$. [Curves due to H. HAUSEN: Z. ges. Kälteind. **39**, 1 (1932).]

The gas temperatures T and T' in the same region of length are also exponential functions, such that their average difference $(\overline{T} - \overline{T}')$ also diminishes exponentially as l increases. The variation of average temperature along the length of the regenerator shows qualitatively, therefore, the same behavior as in a counter-current interchanger, for both the flow conditions $\mathfrak{C} = \mathfrak{C}'$ and $\mathfrak{C} > \mathfrak{C}'$. (Assuming $\tau = \tau'$.)

51. Basic equations for regenerator performance. In an article of this length it is only possible to outline briefly the basic equations governing the performance of regenerators. For detailed theoretical treatment the reader is referred to a monograph by HAUSEN[2].

Consider the transfer of heat in the regenerator of Fig. 92 and 95, between the gas and the metal packing. The turbulent flow of gas is in a direction parallel to the metal surface, the metal being essentially planar of thickness, say, δ. The gas temperatures, T and T' in the warm period and cold period respectively, are linear functions of time in the central region of the regenerator column. In consequence, the average rate of transfer of heat from gas to metal, or vice versa, is not only a function of the heat transfer coefficients, α and α' (see Sect. 42

[1] H. HAUSEN: Z. ges. Kälteind. **39**, 1 (1932).
[2] HAUSEN: Wärmeübertragung in Gegenstrom, Gleichstrom und Kreuzstrom, pp. 262 to 452. Berlin: Springer 1950.

for definition), but also of the rate at which heat can diffuse into or out of the metal. The latter is determined by the diffusivity, χ, of the metal, where χ is given by:

$$\chi = \frac{\lambda_m}{\varrho C} \qquad (51.1)$$

where λ_m is the thermal conductivity of the metal and where ϱ and C are the metal's density and specific heat per unit mass. If ϑ is the temperature of the metal along a plane, parallel to the surface and at a distance x from the surface (see Fig. 100), then the rate of change of ϑ with time is given by:

$$\frac{\partial \vartheta}{\partial t} = \chi \frac{\partial^2 \vartheta}{\partial x^2}. \qquad (51.2)$$

Consider the heat transfer in the column between sections at l and $l + dl$ (cf. Fig. 95). At l let the gas temperature during the warm period be T and at $l + dl$ let the temperature of the gas at the same instant be $T + \left(\frac{\partial T}{\partial l}\right) dl$. Then the heat gained by the metal between l and $l + dl$ in a time dt is:

$$dq = \mathfrak{C} \, dt \left(\frac{\partial T}{\partial l}\right)_t dl. \qquad (51.3)$$

On the other hand, we have from eq. (42.2)

$$dq = P \, dl \, \alpha (T - \vartheta_s) \, dt, \qquad (51.4)$$

where P is the area of the heat transfer surface per unit length of column and ϑ_s the surface temperature of the metal. From (51.3) and (51.4)

$$\left(\frac{\partial T}{\partial l}\right)_t = \frac{\alpha P}{\mathfrak{C}} (T - \vartheta_s). \qquad (51.5)$$

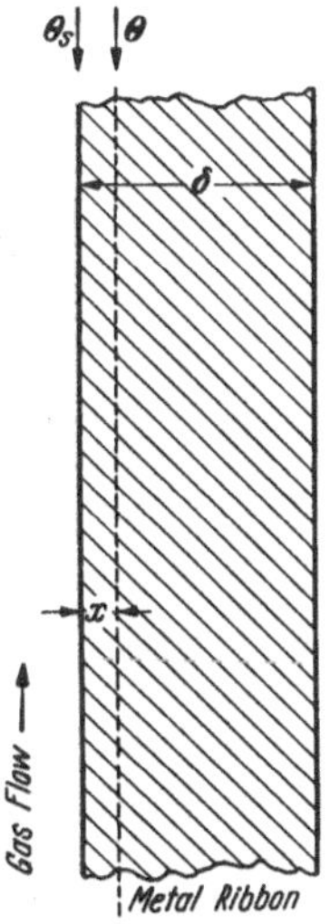

Fig. 100.
Section through metal ribbon of regenerator.

This quantity of heat dq moreover raises the mean temperature ϑ_m of the metal at a rate given by:

$$dq = \tfrac{1}{2} (P \, \varrho \, \delta \, C \, dl) \left(\frac{\partial \vartheta_m}{\partial t}\right)_l dt. \qquad (51.6)$$

Where $\tfrac{1}{2}(P \varrho \delta C \, dl)$ is the heat capacity of the metal between the sections at l and $l + dl$.

Eqs. (51.4), (51.6) combine to give:

$$\left(\frac{\partial \vartheta_m}{\partial t}\right)_l = \frac{2}{\varrho \delta C} \cdot \alpha (T - \vartheta_s). \qquad (51.7)$$

The eqs. (51.2), (51.5) and (51.7) together with similar ones for the cold period are the basic ones from which the regenerator performance can be calculated. It is to be noted that in arriving at these equations we have assumed implicitly that the regenerator columns are "ideal", i.e. that no heat is transferred into or out of the columns through its walls (perfect lagging) and that the heat flow down the material of the column itself is negligible.

Two factors which make the regenerator computations very different from those for counter-current interchangers should be emphasized. First, there is the factor, already noted above, that the mean rate of transfer from gas to metal or vice versa, is not completely defined by the heat transfer coefficients but depends also on the diffusivity and geometry of the metal. To obtain optimum overall heat transference, therefore, metals with high conductivity should be chosen.

Secondly, there is an upper limit to the amount of heat Q_{max} that can possibly be interchanged between the gas streams, which limit is determined by the mass of metal packing. If one calls $\Delta\vartheta_m$ the mean value of the difference between the temperature of the metal at the beginning of one period and that at the end of the same period, then

$$Q_{max} = M C \Delta\vartheta_m \tag{51.8}$$

where M is the mass of the metal packing. A constructional requirement therefore is that

$$\mathfrak{C}'\,\tau'\,(\overline{T_1'} - T_2') < Q_{max}. \tag{51.9}$$

A detailed theoretical treatment for the solution of the basic eqs. (51.2), (51.5) and (51.7) in terms of practical regenerator parameters has been made by HAUSEN[1]. An approximate, but simpler approach has recently been presented by SCHULTZ[2].

52. Reversing interchangers. The reversing interchanger was first developed by COLLINS[3] for use with low pressure air rectifiers, and its possible application for air liquefaction has been described briefly in Sect. 37 above. The type developed by COLLINS is shown in Fig. 101. It consisted of four coaxial copper tubes uniformly spaced by finning made of spiral springs of edgewound copper ribbon. The finning was bonded to adjacent tube walls with soft solder. In the model of Fig. 101 the innermost tube is 0.25" o.d. and the outermost 1.50" o.d. The outermost annulus takes one gas flow, and the two inner annuli in parallel take the other gas flow. The combined area and resistance to flow of the two inner annuli are designed to be the same as that for the outermost annulus. Both gas streams therefore flow through hydraulically identical channels. The use of the passage through the innermost tube is described below.

In use, the gas flows through the two channels are alternated in direction about every three minutes by the use of reversing valves similar to the arrangement shown in Fig 92 for the alternation of gas flows through regenerator columns.

Since any one channel therefore must, in turn, carry either gas stream, the application of such reversing interchangers is limited to low pressure systems, in which the ratio of pressure in the two streams is not too great.

TRUMPLER and DODGE[4] have measured the overall heat transfer and pressure drops in this type of reversing interchanger and have presented detailed data of their results. The theoretical treatment outlined in Sect. 42 to 47, for estimating the characteristics of interchangers is, of course, also applicable to the reversing interchangers.

The general aim in the use of reversing interchangers is similar to that of regenerators, namely to avoid the need for prepurification of the input gas. When used with air liquefiers or rectifiers for example, raw atmospheric air can be used. The water vapor and carbon dioxide impurities are condensed inside one channel of the interchanger when the input gas flows through it and those impurities are flushed out by the subsequent low pressure return flow. The advantage of the reversing interchanger over the regenerator is that it

[1] H. HAUSEN: Wärmeübertragung in Gegenstrom, Gleichstrom und Kreuzstrom. Berlin: Springer 1950.
[2] B. H. SCHULTZ: Appl. Sci. Res. **3**, 173 (1952).
[3] S. C. COLLINS: Chem. Engng. **53**, 106 (1946).
[4] P. R. TRUMPLER and B. F. DODGE: Chem. Engng. Progr. **43**, 75 (1947).

is less bulky, heat storage being essentially eliminated in it. It has an additional advantage when a three channel reversing interchanger is used in, for example, an oxygen or nitrogen producer, that the high purity output stream does not flush the impurities out with it[1].

The operation of reversing interchangers in air rectifiers, which so far are the only equipment in which they have been extensively used[2], is in general such that the return cold gas does not cool the input high pressure gas below about 145° K at the exit of the in-terchanger. At this temper-ature the vapor pressure of CO_2 impurity in the input stream is about 3 mm. Hg, which constitutes an amount sufficient to cause blocking at the expander (cf. Sect. 36 above). In order to bring the CO_2 vapor pressure down to a safe value (about 0.1 mm. Hg) for continuous satisfactory op-eration, an output temper-ature of about 125° K is re-quired. This has been suc-cessfully achieved by use of an "unbalanced passage" in the interchanger, as was sug-gested by TRUMPLER[3]. A typ-ical arrangement of an unbal-anced passage is shown in Fig. 75. Here a fraction of the high pressure gas stream, after passing through the low temperature reversing valves, V_2, divides away from the main stream at a and flows back up the unbalanced pas-sage C of the interchanger in the same direction as the return gas flow. This provides

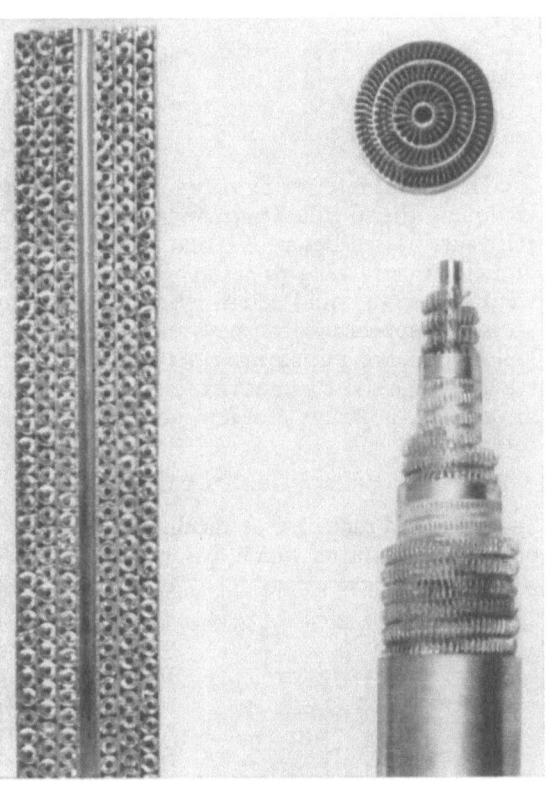

Fig. 101. Photograph showing construction of COLLINS type reversing interchanger. [Chem. Engng. 53, 106 (1946).] (Courtesy of Professor S. C. COLLINS.)

additional cooling for the high pressure input flow. The unbalanced passage terminates at an intermediate temperature, in this case 200° K, part way up the interchanger; and, moreover, the gas flows always in the same direction through it. In the COLLINS-type interchanger of Fig. 101, the unbalanced pas-sage can be provided by causing the unbalanced flow to pass up through the innermost tube.

It is possible also to locate the unbalanced passage in the low pressure gas stream. For a discussion of this, see LOBO and SKAPERDAS[4]. It is in principle possible also to apply the unbalanced passage to the operation of regenerators.

[1] For detailed discussion of this, which is outside the scope of this article, see W. E. LOBO and G. T. SKAPERDAS: Chem. Engng. Progr. 43, 69 (1947).
[2] J. H. RUSHTON and E. P. STEVENSON: Chem. Engng. Progr. 43, 61 (1947).
[3] See reference 4 below.
[4] W. E. LOBO and G. T. SKAPERDAS: Chem. Engng. Progr. 43, 69 (1947).

Helium Liquefiers and Carriers.

By

S. C. COLLINS.

With 24 Figures.

Introduction.

The initial interest in very low temperatures was created chiefly by the desire to liquefy the so-called permanent gases. Hydrogen was first liquefied in 1898 but ten years passed before KAMERLINGH-ONNES and his coworkers at the University of Leiden were able to reduce helium to the liquid state. Fifteen more years went by before liquid helium was produced anywhere else. After 1930, however, with the appearance of new techniques, more and more institutions acquired helium liquefying apparatus. At the present time liquid helium is being produced regularly in almost a hundred scientific laboratories and in quantities which would have seemed fantastic a few years ago.

A. Principles of refrigeration at low temperatures.

1. General remarks. A mechanical refrigerator consists of two distinct assemblages of apparatus which are interconnected to permit the working fluid or refrigerant to flow continuously in a circuit which includes both groups. In the group which occupies the warm zone the refrigerant is compressed and cooled by transfer of heat to a heat sink. Both the enthalpy and the entropy of almost all refrigerants, but not necessarily the temperature, are decreased. In the group which occupies the cold zone, expansion of the refrigerant takes place with a drop in temperature and heat is absorbed at the lower temperature. The enthalpy and the entropy are increased.

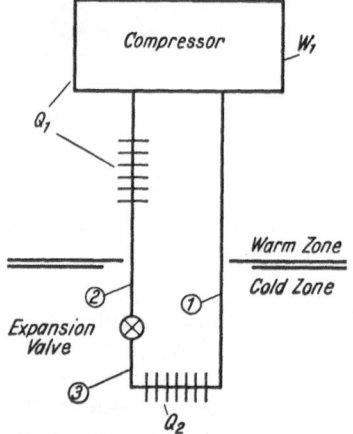

Fig. 1. Simple refrigeration cycle.

A flow diagram of a simple refrigeration cycle is shown at Fig. 1. The heat, Q_1, lost by the refrigerant to the heat sink, is the equivalent of the work, W_1, applied to the compressor shaft, plus the flow work, $-p_1 v_1$, of the low-pressure stream of fluid entering the compressor, plus the flow work, $p_2 v_2$, of the leaving compressed fluid, plus the change in the internal energy, $u_2 - u_1$. This can be written conveniently as

$$Q_1 = h_2 - h_1 + W_1 \tag{1.1}$$

where $h = u + p v$, the specific enthalpy. Similarly a heat balance across the cold zone leads to the relation

$$Q_2 = h_1 - h_2 + W_2. \tag{1.2}$$

It is assumed that steady flow is maintained and that velocities are sufficiently low to make the kinetic energy negligibly small at localities where the thermodynamic properties are to be evaluated.

Ordinarily the expansion occurs at a valve or other throttling device. Consequently no work is done and the heat, Q_2, absorbed by a unit mass of the refrigerant in passing through the cold zone of the refrigerator is exactly equivalent to the decrease in specific enthalpy experienced by the refrigerant during its treatment in the warm zone. Only in a reversible cycle, however, would the gain in entropy in the cold zone equal the decrease of entropy in the warm zone. A convenient and significant efficiency factor, η, for the cold zone can be defined as a ratio of these quantities, that is

$$\eta = \frac{s_1 - s_3}{s_1 - s_2} \tag{1.3}$$

where s is the specific entropy.

2. Choice of refrigerant. All possible refrigerants are equally effective in reversible cycles, but in actual practice the choice of refrigerant profoundly affects the design and operating characteristics of the machinery. This fact is illustrated by the data given in Table 1. Three representative refrigerants, ammonia, air

Table 1.

	T_2	p_1	p_2	W_1	Q_2 or $h_1 - h_2$	η	T_3
	degree	atm.	atm.	cal/gram	cal/gram		°K
Ammonia	300	1	10.5	− 79.1	292.4	0.975	239.8
Air	300	1	10.5	− 47.6	0.65	0.014	297.3
Helium	300	1	10.5	−350.0	−0.75	—	300.6

With counter current heat exchanger.

Ammonia	300	1	10.5	− 79.1	292.4	0.985	239.8
Air	300	1	10.5	− 47.6	0.65	0.049	82
Air	300	1	200	−106.2	8.4	0.268	82
Air (further cooled) .	240	1	200	−106.2	14.4	0.460	82
Helium	300	1	10.5	−350.0	−0.75	—	∞

Cooled with liquid hydrogen. ε

Helium	20	1	35		2.42	0.093	4.2
Helium	16	1	35		3.24	0.155	4.2
Helium	14	1	35		3.53	0.192	4.2
Helium	14	1	30		3.65	0.199	4.2
Helium	14	1	25		3.59	0.195	4.2

and helium have been chosen for use, first in the cycle of Fig. 1 and then in the more complex cycles of Fig. 2 and 3. Reasonable values of temperature and pressure have been assumed for states 1 and 2. Values of W_1 were computed for reversible isothermal compression, while other quantities were read from tables of the properties of the substances considered or were computed.

For ammonia the specific refrigerative capacity, Q_2, and the efficiency are phenomenally high but the minimum temperature, 239.8° K, the boiling point at a lowside pressure of 1 atm. is still a fairly high temperature. For air at the same pressure range the refrigerative capacity is negligibly small, the efficiency impossibly low and the minimum temperature only slightly less than the starting temperature. For helium there is heating instead of cooling.

In the cycle given in Fig. 2 an insulated counterflow heat exchanger has been added to the cold zone just ahead of the expansion valve. Its function is to utilize a part of the refrigerative effect generated at the expansion valve to further cool the compressed fluid below the sink temperature. The numbers in the second part of Table 1 indicate the well-known, but nevertheless remarkable, potentiality of a counterflow heat exchanger in certain processes. It will be noted that in the air system the available refrigeration, $h_1 - h_2$, unchanged in magnitude, is transposed from the 297 to 300° range all the way to the boiling point of liquid air. No additional expenditure of power is involved. The efficiency of the cold zone is higher by virtue of the lower temperature level of refrigeration but is still small. If the pressure level is raised from 10 atm. to 200 atm. (see Table 1), the efficiency moves up to a tolerable value. The decrease of enthalpy with pressure is roughly proportional to the change in pressure while the work expended in compression is a function of pressure ratios. The cycle of Fig. 2 was devised by LINDE and was used successfully by him in the liquefaction of air.

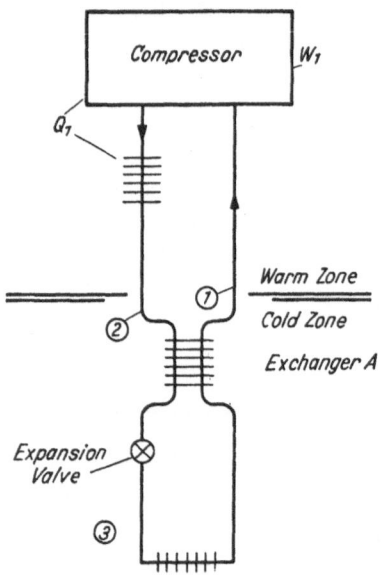

Fig. 2. Refrigerator with counterflow heat exchanger (LINDE cycle).

Helium is still useless as a refrigerant. The addition of a heat exchanger has only served to magnify the heating effect. A way must be found, of course, to bring about a reduction of the enthalpy across the warm zone in order to have a cooling effect in the workless expansion types of cold zone being considered.

3. Temperature-entropy diagram. A temperature-entropy diagram of helium is given in Fig. 4. Constant enthalpy lines and constant pressure lines are superimposed upon the isothermals and isentropics. It will be noted that above 50° K the enthalpy increases as the pressure increases at a given temperature so that upon isenthalpic expansion through a throttling device the temperature rises. In combination with a counterflow heat exchanger the temperature of the helium at the expansion valve may climb so high as to destroy the apparatus. At temperatures below 30°, however, the slope of the isenthalps has definitely changed sign at moderate to low pressures so that the differential, $\left(\frac{\partial h}{\partial p}\right)_T$, becomes negative, (or, as more frequently stated, the JOULE-THOMSON coefficient, $\left(\frac{\partial T}{\partial p}\right)_h$, becomes positive), thereby endowing even a stream of compressed helium with refrigerative value.

4. Liquid hydrogen cooling. The refrigerative effect can be materialized in a helium refrigerator utilizing the cycle of Fig. 3a. It differs significantly from that of Fig. 2. The "warm" zone is extended to include a counterflow heat exchanger, B, and a second stage of cooling of sufficient intensity to assure a temperature lower than the inversion temperature of the isothermal enthalpy change between the chosen pressures. With this condition satisfied a steady state is reached only when the temperature T_3, downstream from the valve, falls to the boiling point of helium under the existing pressure. The refrigerative effect, $h_1 - h_2$, may be usefully applied in condensing a separate stream of helium or in liquefying a fraction of the refrigerant stream itself. In the latter case an amount of gaseous helium equivalent to that liquefied will be injected into the system from an external source at the proper rate to maintain a steady state.

The fraction of the stream, ε, which can be removed as liquid, assuming no leakage of heat from the surroundings, can be expressed in terms of the enthalpy of helium in certain significant states. The enthalpy of the compressed gas entering the cold zone at 2 must be equal to the combined enthalpy of the leaving streams of expanded gas and liquid.

$$h_2 = \varepsilon\, h_3 + (1 - \varepsilon)\, h_1 \qquad (4.1)$$

where h_3 is the specific enthalpy of the liquid phase.

Liquid hydrogen is the only suitable coolant for helium. Its normal boiling point is 20.4° and the triple point temperature is 14.0°. Because of the poor thermal contact between solid hydrogen and its containing walls, heat transfer rates from the helium being cooled to the solid hydrogen are very low. Except in a special case to be described later, hydrogen cooling to temperatures below the triple point is not attempted.

For more efficient utilization of the liquid hydrogen as a coolant a path is provided in exchanger B for the cold hydrogen vapor. It assists in the cooling of the incoming stream of compressed helium.

Referring again to the temperature-entropy chart of Fig. 4, it is apparent that in the low temperature range the enthalpy at a given temperature has a minimum

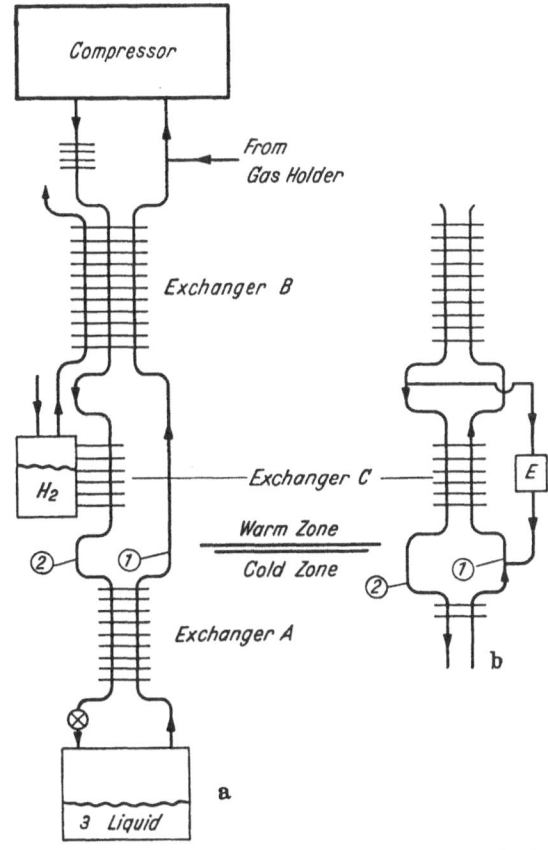

Fig. 3 a and b. Refrigeration cycle with supplementary cooling by liquid hydrogen (a), by external-work machine (b).

value at a definite pressure. The lower the temperature the lower is the pressure at which the minimum occurs. This fact is more obvious in the chart of ZEL-MANOV[1] given in Fig. 5. Once the lowest practicable temperature of precooling is established it is advantageous to select as an operating pressure that value at which the enthalpy is a minimum and, consequently, the potential refrigerative effect a maximum.

5. Cooling by external-work machine. In an alternative method of cooling compressed helium below the inversion temperature an expansion engine is substituted for the hydrogen bath as indicated in Fig. 3b. A fraction of the stream of compressed helium expands adiabatically in the engine and does external work. The temperature decreases chiefly because of the decrease of the internal energy.

---

[1] J. ZELMANOV: J. Phys. USSR. **8**, 135 (1944).

The cold expanded gas discharged by the engine joins other low-pressure helium to absorb heat from the remainder of the stream of compressed helium in exchanger C.

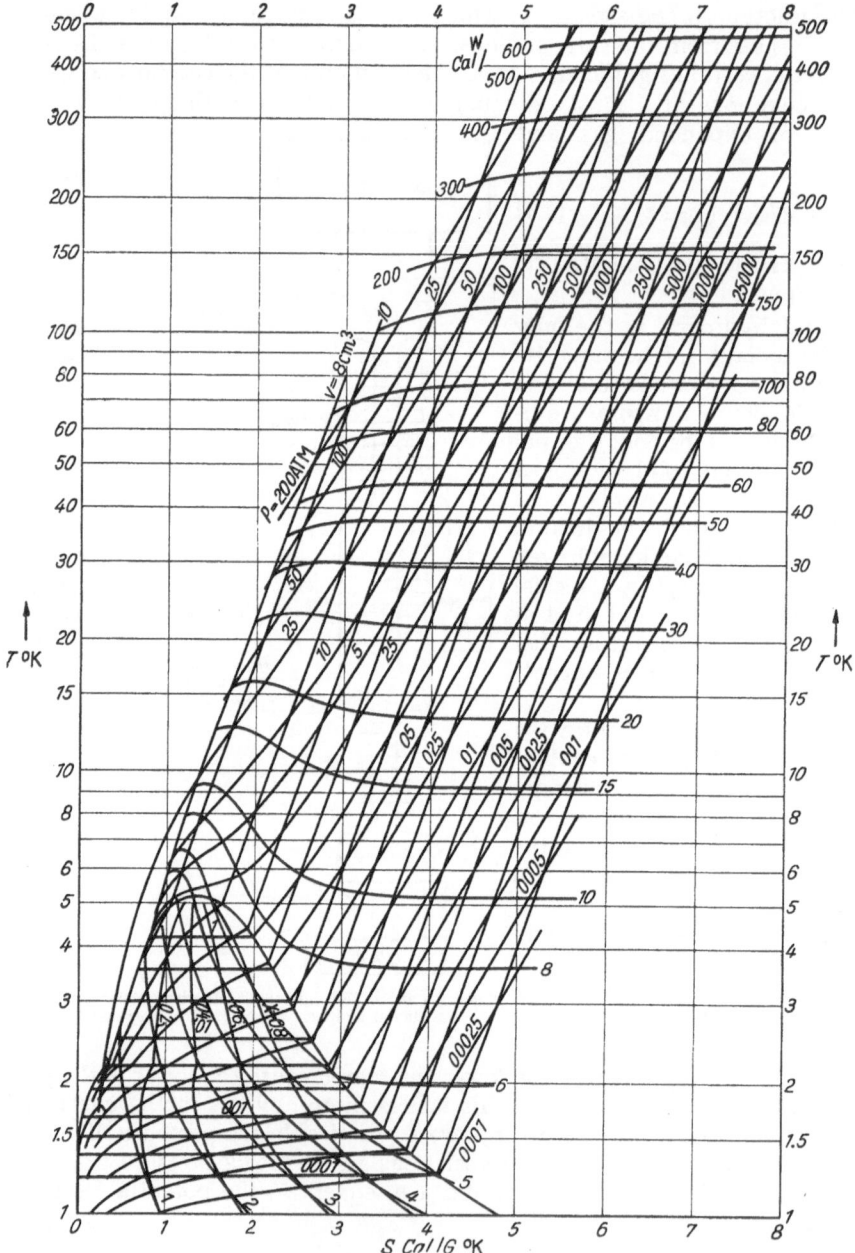

Fig. 4. Temperature-entropy diagram for helium.

According to equation (1.2) the external work, in an adiabatic process, is equal to the decrease in enthalpy which, in turn, if the process is also reversible, equals $\int v\,dp$ by the first law of thermodynamics. For a reversible adiabatic expansion of gas the pressure and volume are related by the equation $pv^{\varkappa} = $ constant, where $\varkappa$ denotes the ratio of the specific heats. The use of this in the integration of the expression above leads to the equation

$$W = -\int v\,dp = \frac{R T_0 \varkappa}{\varkappa - 1}\left[1 - r^{\frac{\varkappa-1}{\varkappa}}\right] \tag{5.1}$$

where r is the ratio of final to initial pressure and T_0 is the temperature of the gas at the beginning of the expansion. With a given pressure ratio the amount of external work done is proportional to the initial temperature. To be useful for liquefying helium, however, the gas must be sufficiently cold at the end of the expansion to cool the cold-zone stream of compressed helium below the inversion temperature.

The refrigeration produced by an expansion engine resides in the cold

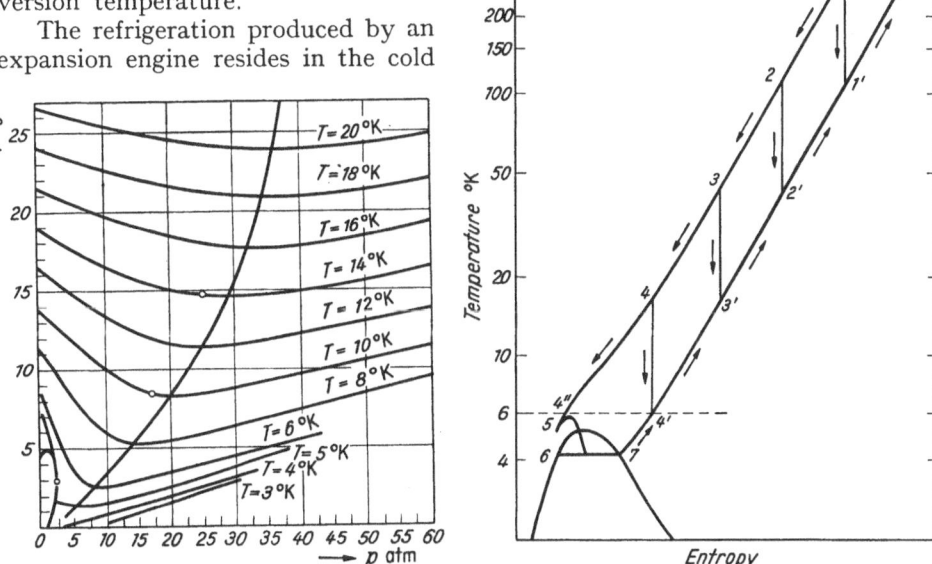

Fig. 5. Variation of enthalpy with pressure at several temperatures.

Fig. 6. T-S diagram of idealized cycle.

gas discharged from the engine. It is usefully applied by allowing the cold gas to absorb heat from the body to be cooled. Whereas $W = -\Delta h$, the available refrigeration equals the product $-c_p \Delta T$ where c_p is the specific heat of the expanded gas and ΔT is the change in temperature during the expansion. The work is exactly equivalent to the available refrigeration only at the inversion temperature for the pressure range involved. Above the inversion temperature the work is the greater. Below the inversion temperature the refrigeration exceeds the work done.

A cold stream of expanded helium is preferable to an evaporating liquid for cooling a stream of compressed helium. Assuming the heat transfer to take place in a counter flow heat exchanger, the loss of entropy by the helium being cooled is more nearly equal to the gain of entropy by the refrigerant and the process is, therefore, more nearly reversible.

6. Idealized liquefier. It is not feasible, however, in a single expansion to have the temperature of the expanding charge fall from 300 to 4.2°. An equivalent result can be achieved by expanding separate charges of compressed helium through a moderate pressure range, beginning at successively lower temperatures. The schematic temperature-entropy diagram of Fig. 6 illustrates a possible arrangement involving expansion at four temperature levels spaced so that the range between 300 and 6° is spanned. The pressure range assumed is 12 atm. to 1 atm. If a higher pressure ratio were used, fewer stages of expansion would be required.

A large stream of helium circulates in the direction of the arrows. A small fraction of the stream is removed as liquid at 6, an equivalent amount of gaseous helium being added to the stream at 0. Isothermal compression (0 to 1) is assumed to occur in the compressor. Cooling of the compressed gas (1 to 5) is accomplished in the counterflow heat exchanger by the transfer of heat to the colder outgoing stream of low-pressure gas (7 to 0). A fraction of the stream of compressed helium at 1 is expanded in an engine to 1′ where it joins the main stream of low pressure gas. The drop in temperature is a result of the external work done. Since helium is an almost perfect gas at higher temperatures, the preferred rate of flow through the first engine (1 to 1′) exactly equals the rate of liquefaction. Under this condition the mass rate of flow in the high pressure channel of the heat exchanger (1 to 2) equals that in the low pressure channel (1′ to 0), and, consequently, the temperature drop from 1 to 2 equals the temperature rise from 1′ to 0. Assuming the heat exchanger to be perfect, no net gain of entropy occurs and thus far the process is reversible. For the next stage of cooling, a second fraction of the stream of compressed helium is split off at 2 and expanded in a second engine to 2′. At lower temperatures the effect of pressure upon the specific heat of helium is not negligible. If the temperature drop from 2 to 3 is to equal the temperature rise from 2′ to 1′, the mass rate of flow through the second engine (2 to 2′) must exceed slightly the rate of liquefaction. By so doing, complete reversibility in this section of the heat exchanger can be closely approached. A third expansion is indicated by the path 3 to 3′ and a fourth by the path 4 to 4′. Finally the remainder of the stream of compressed helium is expanded isenthalpically in a throttling device at 5. Because of the rapidly rising specific heat of the high pressure stream at very low temperatures, the unliquefied portion of the flow through the expansion valve as well as the expanded gas from engines 3 and 4 must exchange heat with a fluid that is appreciably warmer. There is, therefore, a net increase of entropy in this section of the heat exchanger. At the expansion valve, too, a sizable gain in entropy occurs.

Had all of the high pressure stream expanded isentropically as, for instance, by replacing the valve at 5 with an expansion engine, the liquid phase would have been formed within the cylinder. Experience has shown this to be undesirable. The wet cylinder walls cool the incoming compressed gas irreversibly.

If the discharge temperature of engine 4 be assumed to be 6°, that being about as close to saturation as one could safely go, then the "cold zone" or final section of the liquefier would receive compressed helium at 6° and would give up liquid helium at one atm. and a stream of gaseous helium at 1 atm. and 6°. The fraction liquefied, ε, can be computed from equation 4, which becomes in this case

$$h_{4''} = \varepsilon h_6 + (1 - \varepsilon) h_{4'}.$$

The approximate flow distribution for this cycle has been computed to be as follows:

$$\varepsilon = 0.65$$

Flow through expansion valve	= 192 grams per liter helium
Flow through 4th engine	= 154 grams per liter helium
Flow through 3rd engine	= 132 grams per liter helium
Flow through 2nd engine	= 126 grams per liter helium
Flow through 1st engine	= 125 grams per liter helium
Total flow	= 729 grams per liter helium

Work expended isothermal compression of 729 grams of helium at 300°, 1 to 12 atm. −0.314 kwh/liter.

A completely reversible process would require 0.24 kwh/liter.

Whatever the means of getting the temperature below the
inversion point, by liquid hydrogen evaporating at reduced pressure or by cold expanded helium from an engine, in all steady
flow helium liquefiers the cold compressed helium is further cooled
in a counterflow heat exchanger or regenerator spiral and is expanded isenthalpically through a valve.

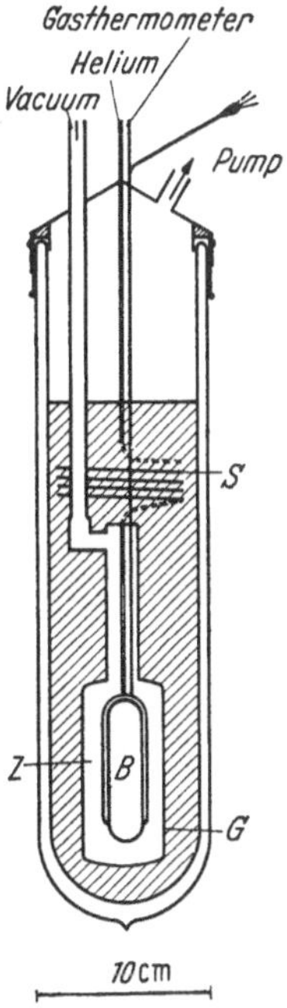

7. SIMON expansion liquefier. There are three distinct types
of helium liquefiers: Namely, steady flow with hydrogen precooling, steady flow with cooling by an external-work machine
and a well-known non-flow process. The first two have been described briefly. The third is SIMON's expansion liquefier[1] which
is shown schematically in Fig. 7. In this liquefier gaseous helium,
cooled in the coil, S, is pumped into a metal container, B, which
is cooled by liquid or solid hydrogen, G. The space Z is filled with
helium at low pressure to make it a conductor. The heat absorbed
by the hydrogen bath must be the sum of the decrease of internal
energy of the helium after it enters the container and the flow
work of compression. The latter term is $\sum mpv$, where m is the
mass of a very small quantity of entering gas and v its specific
volume. If all of the gas enters at the same temperature, T_1, the
total flow work is NRT_1 where N is the number of moles of gas
which enter the container and R is the gas constant. The amount
of hydrogen refrigerant required to absorb the heat produced by
the flow work may actually exceed that needed to bring about
the reduction of the internal energy of the helium. It is a case of comparing
magnitudes of two products, namely, RT_1 and $C_{v\,(\text{mean})}\,(T_1 - T_2)$ where T_2 is
the final temperture.

Fig. 7. SIMON ex-
pansion liquefier.

Because of the low critical pressure of helium it is possible to make
a remarkable reduction of the specific entropy and the specific volume
with reasonable pressures at temperatures attainable with hydrogen. The
density of the compressed gas may
be considerably greater than that of
the liquid phase at one atm. When
the vessel is fully charged at the
lowest practicable temperature, the
container is isolated by evacuating
the space Z and the pressure within
the container is relaxed by allowing
the charge of helium to leak slowly
through a valve. In Fig. 8 are curves
for various temperatures at which
the expansion is started. The coordinates are the percentage of the vessel B filled with liquid after expansion and the pressure of the helium
at the beginning of the expansion.

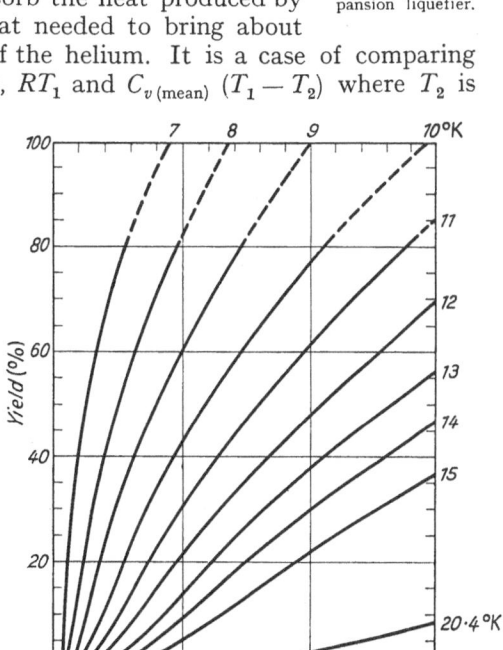

Fig. 8. Liquid yield in single-expansion process.

With starting pressure and temperature of 150 atm. and 11° respectively the container is left about four-fifths full of liquid. Even though heat is transferred

[1] F. SIMON: Phys. Z. **34**, 232 (1933).

with difficulty to solid hydrogen, it is practicable to cool the charge to a temperature as low as 10°.

An all-important factor contributing to the success of the Simon expansion liquefier is the vanishingly small heat capacity of the vessel into which the helium is compressed. To withstand the pressure the walls must be heavy, but at the existing temperature the specific heat is so small that substantially none of the liquid formed is evaporated in cooling the container from the starting temperature to 4.2°. Table 2 gives the heat capacity of a steel container of 150 cm.³ capacity designed to stand 100 atm. together with the heat capacity of the helium it would contain at two temperatures.

Table 2. *Heat capacity (cal/deg) of steel container and helium.*

	Room temperature	10°
Steel container . .	50	0.1
Helium at 100 atm.	2	20
Ratio	25	0.005

At any stage during the expansion the helium left in the vessel will have expanded reversibly. It does work on the gas being pushed out. The over-all expansion is rather efficient and the amount of liquid left is approximately 60% of the quantity possible with a completely reversible expansion. The charging half of the cycle, however, is relatively wasteful of hydrogen because of the negative flowwork.

B. Components of a liquefying system.

8. Compressors. Helium compressors are usually air compressors which have been modified so as to reduce leakage to a minimum and to avoid contamination by air. When a single acting compressor is used the entire crankcase is sometimes made air-tight. With double acting machines having a distance piece between cylinder and crosshead it is only necessary to provide specially-packed piston rod glands. An effort is made to choose a lubricant with very low vapor pressure and great stability toward heat. Silicones have been used to a limited extent. Oil separators are generally necessary to remove entrained oil from the stream of compressed helium. This is especially true of low-pressure liquefiers because of the large specific volume of the compressed helium combined with a relatively large mass rate of flow. Alternate layers of fine steel wool and fiber glass have been found to be especially effective in removing atomized oil.

Oil-free compressors using carbon piston rings and carbon rod packing have long been used successfully in compressing air for various industrial applications. For pumping very dry gases they have not proved entirely satisfactory. The carbon rings wear much more rapidly and the friction is greater. For compressing helium it is generally assumed that the problems of purifying the helium after oil contamination are simpler than those created by oil-less compression.

9. Heat exchangers. In steady flow liquefiers the stream of helium undergoes compression at room temperature, is cooled to a very low temperature and is partially liquefied upon expansion. The unliquefied portion is returned to the compressor for recompression by way of the counterflow heat exchanger. The heat exchanger recovers with remarkable efficiency the refrigeration possessed by the cold expanded gas.

The primary concern in the design of a counterflow heat exchanger is to reduce to a minimum the thermal resistance of the heat path between the two streams of gas. Ordinarily the total thermal resistance is composed of three parts, that due to the relatively stagnant film of gas on the metal surface of each of

the two channels and the resistance of the path through the metal as expressed
by the relation

$$R = \frac{1}{\alpha_1 A_1} + \frac{x}{k_3 A_3} + \frac{1}{\alpha_2 A_2} \qquad (9.1)$$

where R is the total resistance, α_1 and α_2 are the film coefficients of heat transfer
for the channels 1 and 2 respectively, A_1 and A_2 the surface areas exposed to the
gas streams, x the length of path through the metal, k_2 the thermal conductivity
of the metal and A_3 the effective cross sectional area of the metal path. The
effective area, A_3, of the conducting section is very difficult to evaluate in many
heat exchangers but high accuracy is unnecessary because the thermal resistance
of the metal path between the channels is usually negligibly small when compared
with the film resistances. The heat transferred in unit time can be expressed in
terms of R and the temperatures of the two gas streams as

$$q = \frac{t_1 - t_2}{R}. \qquad (9.2)$$

If the efficiency of the heat exchanger is to be high the temperature, t_2, of
the low pressure gas as it leaves the heat exchanger must be very close to t_1, the
temperature of the compressed gas entering the heat exchanger. This means that
R of equation (9.1) must be small and consequently that the film coefficients and
areas of transfer surface must be as large as feasible.

Film coefficients of heat transfer vary widely with the kind of fluid and its
velocity and with the geometry of the path. High values of the heat transfer
coefficient are nearly always attended by large loss of head by the fluid stream
from friction. Many empirical equations have been devised for the calculation
of pressure drop and heat transfer coefficients. For a given fluid in turbulent flow
the equations are generally of the following form[1].

$$\frac{dp}{dL} \sim \frac{V^2 \varrho}{r} \sim \frac{G^2}{r \varrho} \qquad (9.3)$$

and

$$\alpha \sim \frac{C_p G^{0.8} \mu^{0.2}}{r^{0.2}}. \qquad (9.4)$$

Where L refers to length, V the average velocity of the fluid, r the hydraulic
radius which is defined as the cross-sectional area of the conduit divided by the
wetted perimeter, G the mass velocity (lb/hr) (Sq.ft. of cross section), α the film
or surface coefficient from stream to metal wall the density and μ the viscosity.
From these equations it is apparent that practical consideration of the energy
cost of overcoming friction may be necessary in the quest for high film coefficients. Furthermore, it is obvious that less expenditure of energy will be required for a given value of α in the high-pressure conduit than on the low-
pressure side because of the greater density. Because of this fact, more surface
and a larger passageway are usually provided for the low-pressure stream.

The efficiency of a counterflow heat exchanger is defined as the ratio of the
actual rise in temperature of the cold stream during its passage through the heat
exchanger to the maximum increase which could have occurred. Obviously the
temperature of the stream being heated can never exceed that of the incoming
warm stream even though the heat capacity of the latter might be much greater.
If the stream being heated has the larger heat capacity the maximum temperature rise is determined by the quantity of heat which the warmer stream can

---

[1] W. H. McADAMS: Heat Transmission, pp. 128 and 174. New York: McGraw-Hill Company 1942.

provide while being cooled from its initial temperature to the initial temperature of the cold stream. It is often desirable to consider the inefficiency, rather than the efficiency, because it is proportional to the measured ΔT at the warm end of the heat exchanger.

Whenever the product of the mass rate of flow and the specific heat differs for the two streams, the temperature difference between streams cannot remain constant from one end of the heat exchanger to the other, but must be larger at one end, for it is assumed that the heat lost by one stream is equal to the heat gained by the other. In this circumstance it can be shown that equation (9.1) becomes

$$q = \frac{\Delta t - \Delta t'}{R \ln \frac{\Delta t}{\Delta t'}} = \frac{\Delta t_m}{R} \qquad (9.5)$$

where Δt and $\Delta t'$ are the temperature differences at the two ends of the heat exchanger and Δt_m is the so-called logarithmic mean temperatures difference. It is obvious that without changing the thermal resistance of a heat exchanger the temperature approach, Δt, can be made very small at one end of the heat exchanger by reducing the heat capacity of the stream flowing toward that end. Referring first to Fig. 9a in which specific temperatures have been assumed for the case in which both streams possess the same heat capacity and then to Fig. 9b, note the change caused by reduction of the heat capacity of the cold stream to the extent that the quantity of heat removed from the hot stream is only 0.9 its original value and the final temperature is 210° rather than 200°. The efficiency jumps from 0.952, a very low value for liquefiers, to 0.995 which would be considered very high. The temperatures underscored are fixed. Other temperatures follow from the law of conservation of energy, the characteristics of the heat exchanger and the rates of flow. The effect illustrated in Fig. 9 has practical importance in the design of liquefiers. If the fraction of the stream removed as liquid is large, a simple heat exchanger of limited area may have an acceptable efficiency. If the fraction liquefied is small or zero as in a helium refrigerator, the same heat exchanger will be very wasteful of cold. In certain applications, however, an effort is made to enlarge the heat capacity of the outgoing stream by use of nitrogen vapor or hydrogen vapor for instance, with a consequent wastage of cold in order to reduce the ΔT at the cold end of the heat exchanger to a minimum and thereby to cool more effectively the incoming stream.

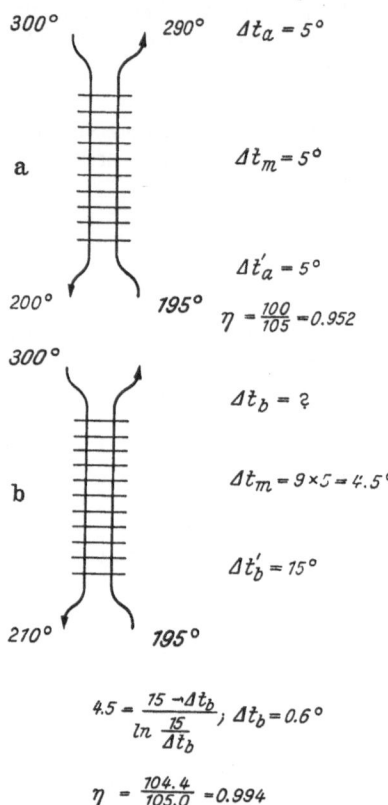

a

$$300° \qquad 290° \quad \Delta t_a = 5°$$

$$\Delta t_m = 5°$$

$$\Delta t'_a = 5°$$

$$200° \qquad 195° \qquad \eta = \frac{100}{105} = 0.952$$

b

$$300°$$

$$\Delta t_b = ?$$

$$\Delta t_m = 9 \times 5 = 4.5°$$

$$\Delta t'_b = 15°$$

$$270° \qquad 195°$$

$$4.5 = \frac{15 - \Delta t_b}{\ln \frac{15}{\Delta t_b}}; \; \Delta t_b = 0.6°$$

$$\eta = \frac{104.4}{105.0} = 0.994$$

Fig. 9. Effect of unbalanced flow in heat exchanger.

10. Types of heat exchangers.
Many types of counter current heat exchangers are in use. In order to have the required area of heat transfer surface and at the same time sufficient longitudinal thermal resistance the heat exchanger is usually made many feet in length. The cross-sectional area is largely determined by the

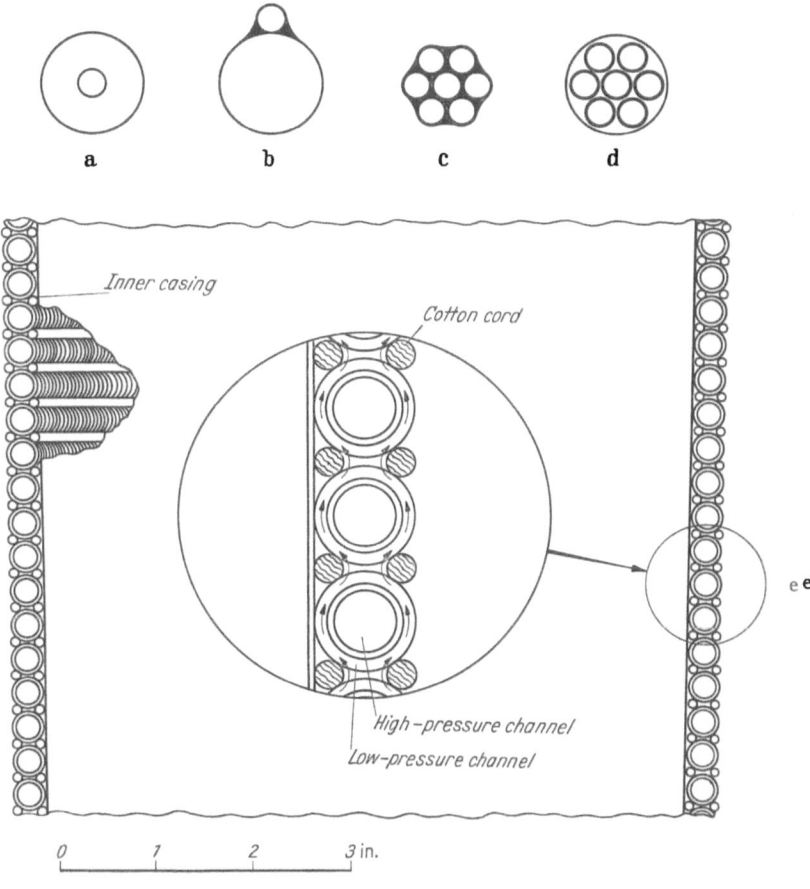

Fig. 10a—f. Several types of heat exchangers.

amount of gas being treated. Resistance to flow is a consideration in the design of a heat exchanger because of the loss of heat it entails on the one hand and the improvement of heat transfer coefficients on the other.

Sectional views of several exchangers are shown in Fig. 10.

Referring to Fig. 10, type A, used by LINDE in liquefying air, is one of the simplest forms of counterflow heat exchanger. One tube is drawn inside a second, so that the bore of the inner tube and the annular space constitute the two channels. It is relatively inferior in performance, since too small a percentage of the total surface takes part in heat transfer. The inner surface of the outer tube contributes much to the loss in heat but generally nothing to the transfer of heat. Type B is an improvement over A in that all wetted surfaces are active in transferring heat. Ordinarily the circumferential temperature gradient is negligibly small, but if the material of which the tube is made is a poor conductor, loss of thermal heat may be serious. Type C is similar to B except that all tubes are of the same diameter. High pressure gas flows through one or two tubes and the low pressure stream through the remaining tubes. The tubes must be bonded together by solder throughout the length. Type D is more easily constructed than C and is more efficient. High pressure gas is caused to flow inside the sheath but outside of the small tubes. The tubes are not soldered together nor are they bonded to the sheath. If extremely high pressure is used the sheath becomes quite heavy, but for pressures of the order of 30 atm. or lower this design is practicable.

In type E the high pressure tubing is so closely finned with very thin copper strip (0.010″) that the external area is tripled. A single helix of the finned tubing completely fills the annulus between two thin-walled stainless steel cylinders. The low pressure gas is compelled to choose a tortuous path through the fins by filling the grooves between adjacent finned tubes with a spiral of cotton cord. Because of the much greater area presented to the low pressure stream, a lower transfer coefficient can be accepted and still have a gain in overall efficiency. A significant characteristic of all of the types mentioned so far is that one or both channels are applied in the form of a helix several inches in diameter. Ordinarily it is possible to use the space within the helix for other elements of the liquefier.

A more compact form of heat exchanger is the HAMPSON type shown in Fig. 10f. The high-pressure stream flows through a number of relatively small tubes connected in parallel. They are quite long but are closely coiled so as to fit within a short cylindrical vessel. The low-pressure stream flows within the cylindrical shell but externally to the tubes. Many ingenious schemes have been devised for forming the coils into an assembly of uniform density and regular outline. If irregularities occur the low pressure stream flows preferentially through the holes where minimum opportunity for heat transfer prevails. Ideally, each tube should wander to and fro across the assembly but this involves construction problems of a serious nature. Essentially the same result is secured by winding the tubes as simple cylindrical helices beginning with a small arbor at the center. By choice of the proper relative diameter of arbor, diameter and length of heat exchanger tube and length of assembly it is possible to build up a multi-layered assembly, each layer containing one more tube than the preceding layer, and yet have all tubes of the same length. It is desirable that the length of all tubes be the same; otherwise the high-pressure stream is not divided equally between the tubes. Presumably the low-pressure stream is distributed uniformly over the cross section of the assembly. If one layer of tubes carries more than its share of the gas, cooling of that share is deficient and the over-all efficiency of the heat exchanger is impaired. The individual tubes are often spaced from each other by wires which are spiraled around the tubes.

The heat exchanger shown in Fig. 10f is noteworthy for its extended surface in both high and low-pressure channels. A ribbon of copper is wound on edge to form a spring for use as annular packing in a nested tube assembly. The spring is soldered to both walls of the annulus. Both channels can be made much shorter for a given efficiency than would be possible with plain tubes. The exchanger tube is still 10 to 20 feet in length, however. This type has been used chiefly for oxygen production.

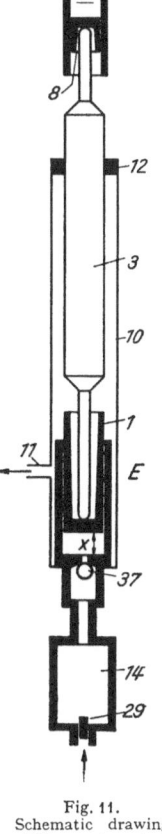

11. Expansion engines. GORRIE[1] was the first to use an expansion engine expressly for creating cold. About 1846 he produced artificial ice by means of a compressed air engine.

SOLVAY[2], 1885 to 1887, attempted to liquefy air by the combination of an air engine with a heat exchanger. The reason for failure is unknown.

KAMERLINGH-ONNES[3], about 1892, attempted to build a hydrogen expansion engine which was to have been used with a heat exchanger for the liquefaction of hydrogen. The project was abandoned because of mechanical difficulties.

CLAUDE[4] in 1902 successfully liquefied air in apparatus cooled by an expansion engine. The important difference between CLAUDE's engine and ordinary steam engines of the day was that the piston was fitted with a leather cup such as is used in a small air pump. The piston could then function without oil.

About 1920 hydrogen engines of the CLAUDE type came into industrial use for the purification of hydrogen. Temperatures lower than can be reached with liquid nitrogen are required. In some of these installations the piston is sealed with nonmetallic piston rings rather than by a leather cup.

KAPITZA[5] in 1934 liquefied helium with apparatus in which a helium engine provided the refrigeration ordinarily supplied by liquid hydrogen. KAPITZA's engine, shown schematically in Fig. 11, is a laboratory model as compared to CLAUDE's industrial engines. The loose-fitting piston, 1, has neither rings nor leather cup. The work is absorbed by the hydraulic mechanism, 2, in a way which permits the piston to complete its power stroke in an uncommonly short time. Leakage of helium by the piston might otherwise be excessive. The piston rod, 3, is made of thin stainless steel tube of a diameter equal to that of the piston. The valve control mechanism is a fairly intricate

Fig. 11.
Schematic drawing of KAPITZA's engine.

assemblage of electromagnets and mechanical linkages. The piston is made of stainless steel and the cylinder of phosphor bronze which has essentially the same coefficient of thermal expansion.

KAPITZA found it necessary to cut small grooves $\frac{1}{4}$ mm. deep and $\frac{1}{4}$ mm. broad at a distance apart of about 5 mm., round the circumference of the piston so that pressure inequalities could not build up along the surface of the piston and cause sideway forces.

[1] J. GORRIE: U.S. Patent 8080. 1851.
[2] E. SOLVAY: C. R. Acad. Sci. Paris **121**, 1141 (1895).
[3] KAMERLINGH-ONNES: Leiden Comm. 23 (1896).
[4] G. CLAUDE: Compt. R. Acad. Sci. Paris **134**, 1568 (1902).
[5] P. KAPITZA: Proc. Roy. Soc. Lond., Ser. A **147**, 189 (1934).

COLLINS in 1941 developed an especially efficient engine for use in oxygen production. In design it is fairly conventional in that it has the usual crankshaft-crosshead arrangement with cam-operated valves. The chief novel features are the very slender piston rod which is in tension during the power stroke and the specially close-fitting piston-cylinder combination. The flexibility of the rod helps to eliminate sideway forces on the piston. The cylinder is made of hardened steel or of a softer metal plated with chromium. For the piston a number of materials have been found suitable, namely, hardened steel, micarta, Bakelite, nylon and leather. The last four substances are used as tightly-fitted sleeves over a steel core. The radial clearance between piston and cylinder is of the order of 0.0001 inch per inch diameter. Leakage of helium is not excessive even at very low engine speed.

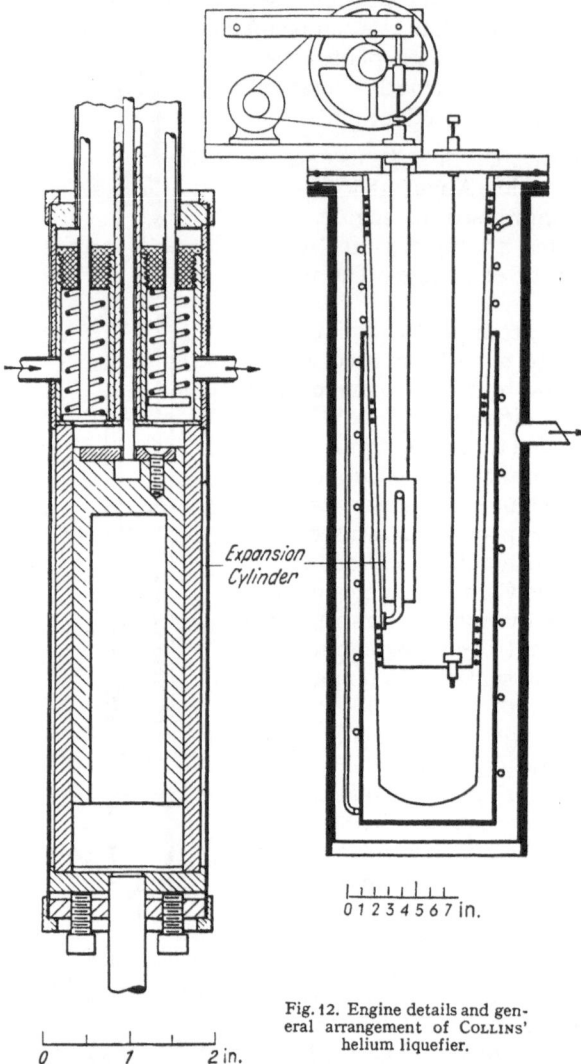

Later this engine [1,2] was adapted for use in helium liquefaction. Multiple cylinders operate at different temperature levels for a closer approach to reversibility. A schematic drawing of one cylinder of the helium engine is given in Fig. 12. Both in principle and in appearance it is quite similar to an 18th century steam engine. Walking beams responding to constant-speed rotating cams control the motion of the piston and the valves.

Fig. 12. Engine details and general arrangement of COLLINS' helium liquefier.

The useful work done by the expanding helium is sometimes absorbed eletrically by belting the engine to an alternating current motor. The motor, turned generator, serves a dual role of work absorber and speed control. In other cases a centrifugal brake has been used to dissipate the work and to control the speed.

LONG and SIMON [3] have liquefied helium by means of a bellows engine shown in Fig. 13. The bellows operates from internal pressure, the central volume of the bellows being displaced by a cup which contains the inlet and outlet valves. The bellows is supported in a vacuum chamber on stainless steel tubing. The motion is transmitted to a crank on the exterior.

[1] S. C. COLLINS: Rev. Sci. Instrum. **18**, 157 (1947).
[2] S. C. COLLINS: Science, Lancaster, Pa. **116**, 289 (1952).
[3] H. M. LONG and F. E. SIMON: Proc. Internat. Inst. Refrig., Sept. 1953.

The advantages of the bellows engine are chiefly the mechanical simplicity and the elimination of the piston leakage problem. The principal disadvantage is the limited life of the bellows, although this may not be as serious as it was originally believed to be.

12. Thermal insulation. Vacuum jackets with highly reflecting surfaces are universally used to minimize leakage of heat into liquid helium and helium apparatus. Many finely divided insulating materials when placed in an evacuated space of about 6 inches thickness compare favorably with the simple high vacuum as far as preventing heat flow is concerned, but their heat capacity is high and the length of time required for equilibrium is great.

For structural elements of the apparatus which must bridge the temperature interval between liquid helium and the surroundings, metal, glass or any one of several plastics may be used. Stainless steel is the favored metal because of its great strength, lack of brittleness and relatively low thermal conductivity. Glass and the plastics have much lower conductivity than stainless steel but, because of their low strength, much thicker sections must be used. Nearly all glasses, in addition to being fragile, have the disadvantage of being permeable to helium so that the vacuum jacket must be pumped out frequently.

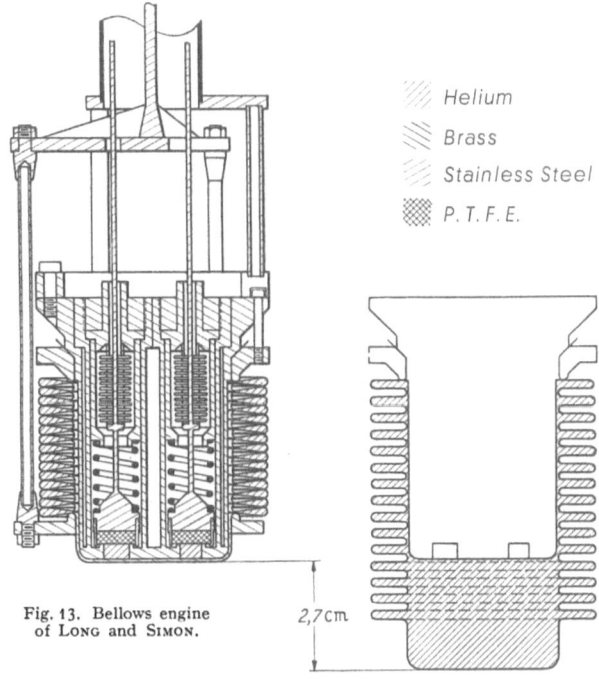

Helium
Brass
Stainless Steel
P. T. F. E.

Fig. 13. Bellows engine of Long and Simon.   2,7 cm

C. Representative helium liquefiers.

13. Steady flow with hydrogen cooling. The newest of a series of helium liquefiers constructed at the University of Leiden[1] is shown in Fig. 14. Helium is compressed to 30 atm. in a threestage compressor and forced to flow through a purification train to the heat exchanger, B, in the liquefier. Here it is cooled by the returning stream of low pressure helium and by the hydrogen vapor being pumped from the main hydrogen bath. Thereafter it is further cooled while flowing through a long coil, C, immersed in the liquid hydrogen. The compressed helium, now at 15°, enters the heat exchanger, A, which ends at the expansion valve, D.

Liquid helium is produced at the rate of 15 liters per hour. Twenty liters of liquid hydrogen are consumed per hour, or 1.33 liters per liter of helium. Although

---

[1] The modernization of the Kamerlingh Onnes Laboratory at Leiden. N. V. De Bataafsche Petroleum Maatschappij.

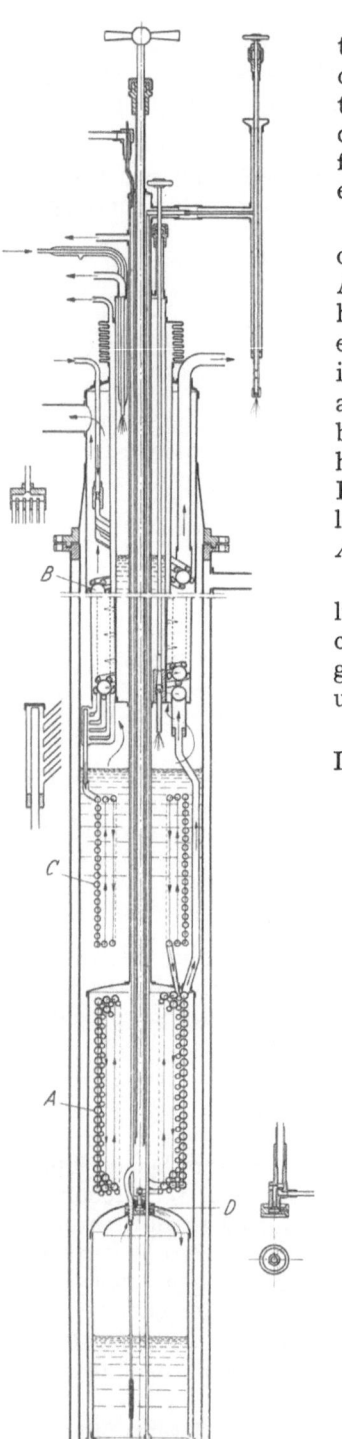

this number is almost twice the liquid hydrogen requirement of perfect apparatus, it differs little from that used by many other liquefiers operating on this cycle. Heat exchanger inefficiency and heat leaks from the surroundings undoubtedly account for the extra hydrogen needed.

In Fig. 15 is shown a sectional view of the liquefier of DAUNT and JOHNSTON[1] at Ohio State University. A metal case 8 inches in diameter and 6 feet high houses the liquefier. Compressed helium at 23 atm. enters Exchanger B of the liquefier. Thisexchanger is div..ded into two parts which are joined in parallel. In one part the high-pressure helium is cooled by returning low-pressure helium, in the other by hydrogen vapor. The heat exchanger, which is of the HAMPSON type, Fig. 10f, is designed for especially low resistance on the low-pressure side. Exchanger A is of the LINDE type, Fig. 10a.

Production of liquid helium is at the rate of 7.5 liters per hour. Twenty liters of liquid hydrogen are consumed in cooling the apparatus from liquid nitrogen temperature to 15°. Thereafter liquid hydrogen is used at the rate of 1.3 liters per liter of liquid helium.

Intermediate in size between the liquefiers of Leiden and Ohio State University is that of the Low

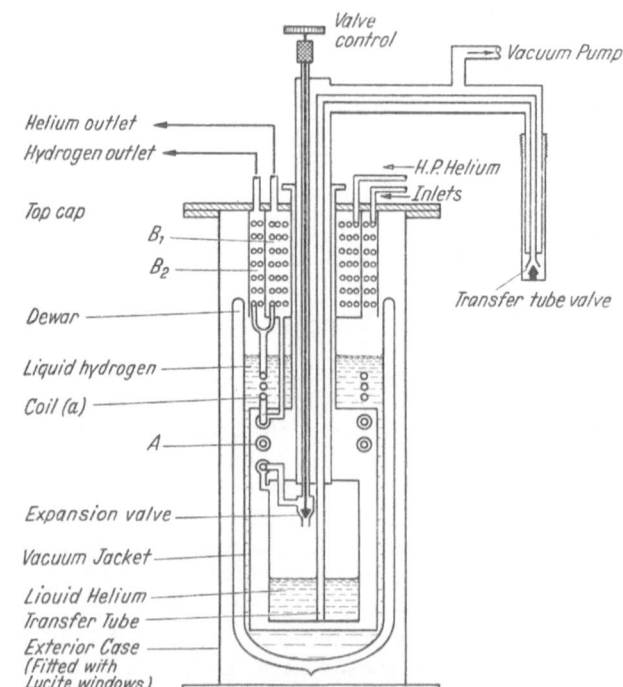

Fig. 14. Sectional view of Leiden liquefier.   Fig. 15. Sectional view of helium liquefier at Ohio State University.

[1] J. G. DAUNT and H. L. JOHNSTON: Rev. Sci. Instrum. 20, 122 (1949).

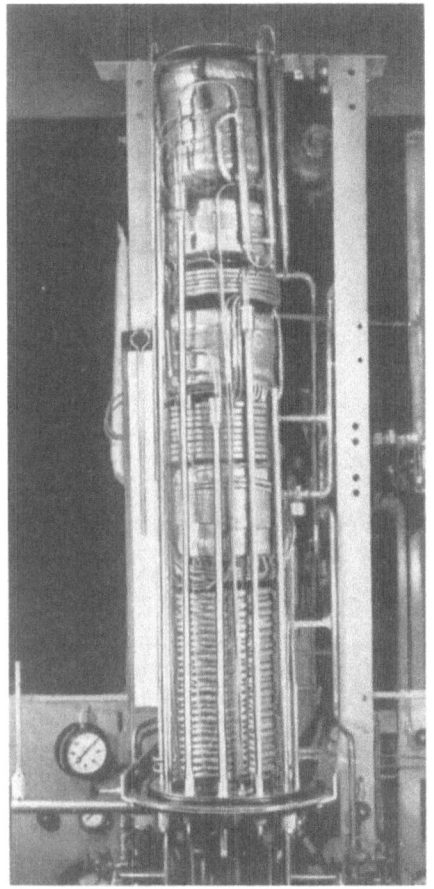

Fig. 16. ASHMEAD liquefier at Cambridge.

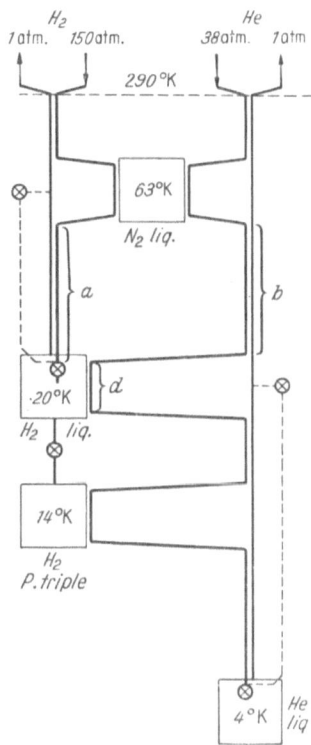

Fig. 17. Flow diagram of Bellevue helium-hydrogen liquefier.

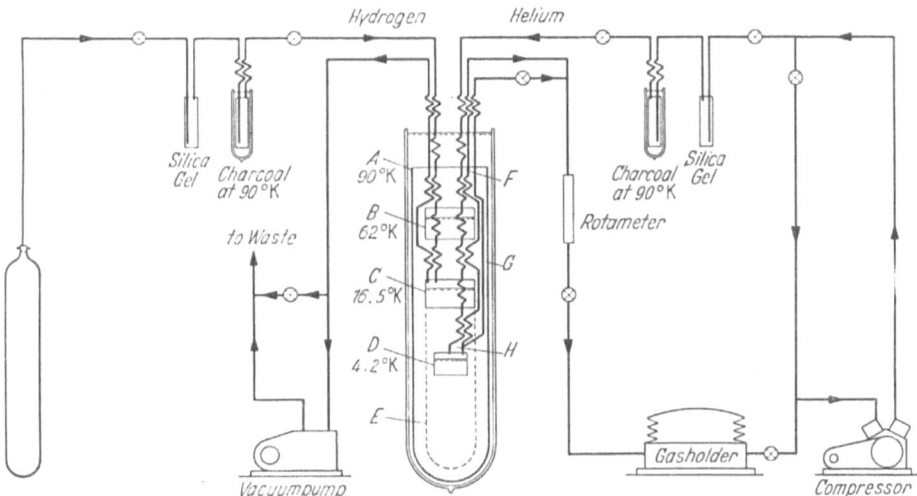

Fig. 18. Flow diagram of helium liquefier at the University of London.

Temperature Laboratory of the General Electric Company[1] at Schenectady. The liquefaction rate ranges from 10 to 14 liters of liquid helium per hour with liquid hydrogen consumption at about 1.3 liters per liter of liquid helium.

Ashmead[2] at Cambridge University employs a closed hydrogen circuit to provide liquid hydrogen continuously (see Fig. 16). Since the hydrogen never leaves the system many of the hazards in the handling of liquid hydrogen are removed. A pumped nitrogen bath at 67° precools the 150 atm. stream of hydrogen. Four-fifths of the liquid hydrogen formed is evaporated at 1.5 atm., the remainder is valved into a second hydrogen bath where the pressure is maintained at 6 cm. Helium at 25 atm. enters the final exchanger at 14°.

The liquefaction rate is 3.8 liters per hour. Twenty-five kilograms of air are used to cool to operating temperature. The starting time is 70 min. The liquid hydrogen rate is 5 liters per hour.

A similar installation but of larger capacity has been constructed by Spoendlin[3] at Bellevue, France. The rate of liquifaction is 7 liters per hour of liquid helium or 7 liters per hour of liquid hydrogen. A flow diagram of the cycle is given in Fig. 17.

Ruhemann[4] produced a miniature liquefier for either hydrogen or helium, precooling compressed hydrogen with liquid air for liquid hydrogen and precooling compressed helium with liquid hydrogen for liquid helium. The experimental chamber was combined with the liquefier and the rate of liquefaction had only to be great enough to compensate for heat leaks. Because of the small scale of operation compressed gases could be taken directly from the cylinder. Chester and Jones[5] at the University of London have extended this technique, combining the hydrogen liquefier, the helium liquefier and a particularly flexible experimental chamber within a single envelope. The starting time is only 30 min. A single cylinder of hydrogen may provide cooling for three days' work. Oxygen is used for cooling at two temperature levels, 90 and 62°. A schematic diagram of the liquefier is given in Fig. 18.

The mammoth hydrogen liquefier of the Cryogenic Engineering Laboratory of the National Bureau of Standards at Boulder, Colorado has been tested as a helium liquefier. Running at two-thirds capacity the liquefaction rate was 120 liters of liquid helium per hour.

14. Steady flow with cooling by external-work machines. A sectional view of Kapitza's[6] helium liquefier is given in Fig. 19. All of the components of the liquefier, heat exchangers, expansion engine, nitrogen container and receiver for the liquid helium are assembled in the highly evacuated space within the case *16*. The engine is lined up along the axis of the cylindrical case, its piston rod *3* passing through the large packed gland *12* to the hydraulic piston *H*. The scheme of heat exchangers is shown in Fig. 20. The compressed helium at 30 atm. is cooled in heat exchanger *A* by outgoing helium and nitrogen vapor, is further cooled to 65° by contact with the ring-shaped nitrogen container in which nitrogen is boiling at reduced pressure and then goes to exchanger *B*. After passing through exchanger *B* the stream, now at 20°, is divided, part going to the engine and part through valve *6*, exchanger *C*, exchanger *D* (corresponding to exchanger *A* in

---

[1] Private communication of M. D. Fiske.
[2] J. Ashmead: Proc. Phys. Soc. Lond. B **63**, 504 (1950).
[3] R. Spoendlin: J. Rech. CNRS. **28**, 1 (1954).
[4] M. Ruhemann: Z. Physik **65**, 67 (1930).
[5] P. F. Chester and G. O. Jones: Proc. Phys. Soc. B **66**, 296 (1953).
[6] P. Kapitza: Proc. Roy. Soc., Lond., Ser. A **147**, 189 (1934).

the conventionalized cycle of Fig. 3) and finally valve *4* to the receiver *5*. The expanded helium leaves the engine at 10° and 2.2 atm. and returns to the compressor by way of exchangers *C*, *B* and *A* in that order.

The purpose of valve *6* is to reduce the pressure from 30 to 17 atm., that being the pressure at which the enthalpy of helium is a minimum at 10°. See Fig. 5.

The liquefier produced 1.7 liters of liquid helium per hour. About 8 liters of liquid nitrogen were used to cool the liquefier and about 3.4 liters

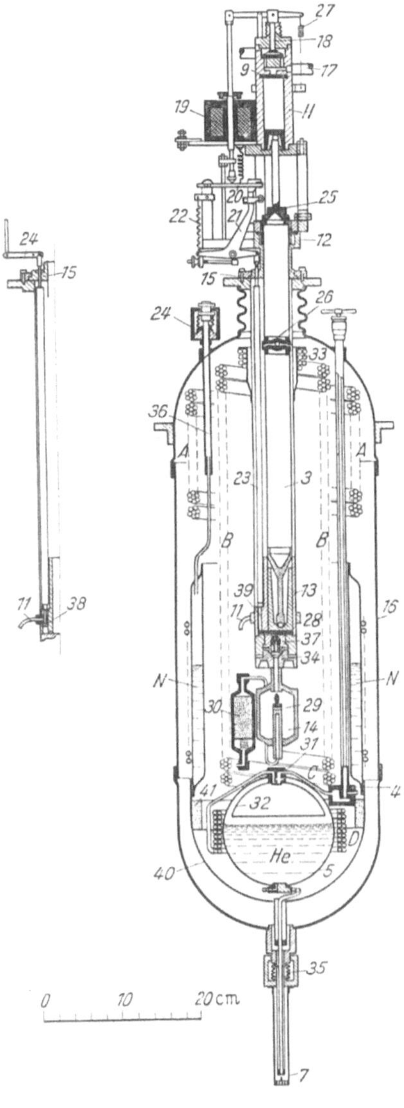

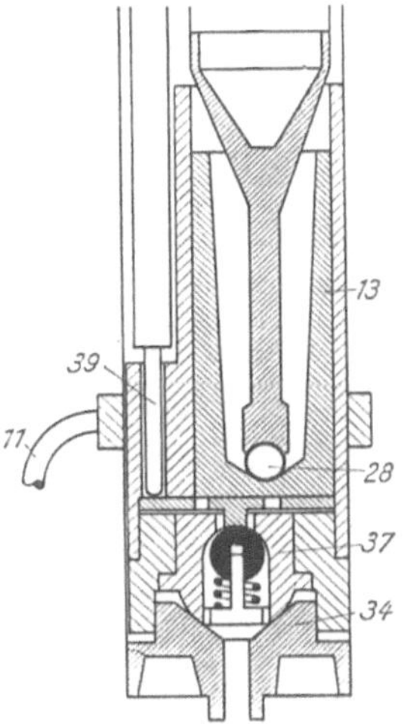

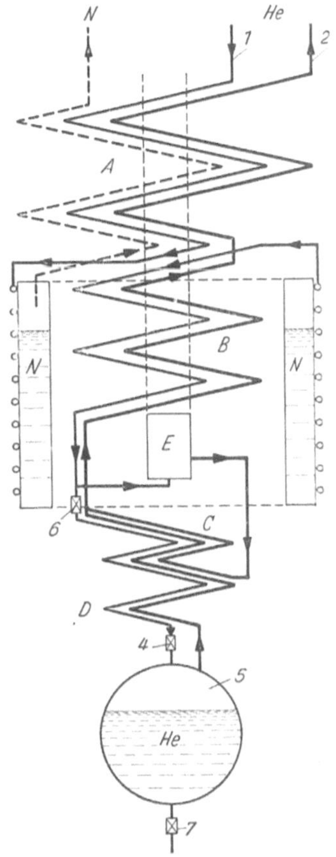

Fig. 19. Sectional view of KAPITZA's liquefier.   Fig. 20. Scheme of heat exchangers.

per hour thereafter, that is about 2 liters of liquid nitrogen per liter of liquid helium.

This liquefier, although an experimental model, performed satisfactorily from 1934 to 1949 with the exception of a few years during the war.

MEISSNER[1] completed a helium liquefier of the KAPITZA type in 1942 at Munich. The general arrangement, which differs somewhat from that of KAPITZA,

---

[1] W. MEISSNER: Phys. Z. **43**, 261 (1942).

is shown in Fig. 21. The exchangers A, B, C and D have identical functions as in KAPITZA's machine. The work is transmitted by the connecting rod to an eccentric disk so that the expansion takes place rapidly. The disk is connected to an electric generator. Valves are controlled positively by cams.

A similar machine was made for the University of Göttingen. KAPITZA liquefiers were also constructed at Kharkov and at Yale University.

Multiple-cylinder engines with appropriate heat exchange systems for liquefaction of helium or hydrogen were introduced by COLLINS[1,2]. Although the

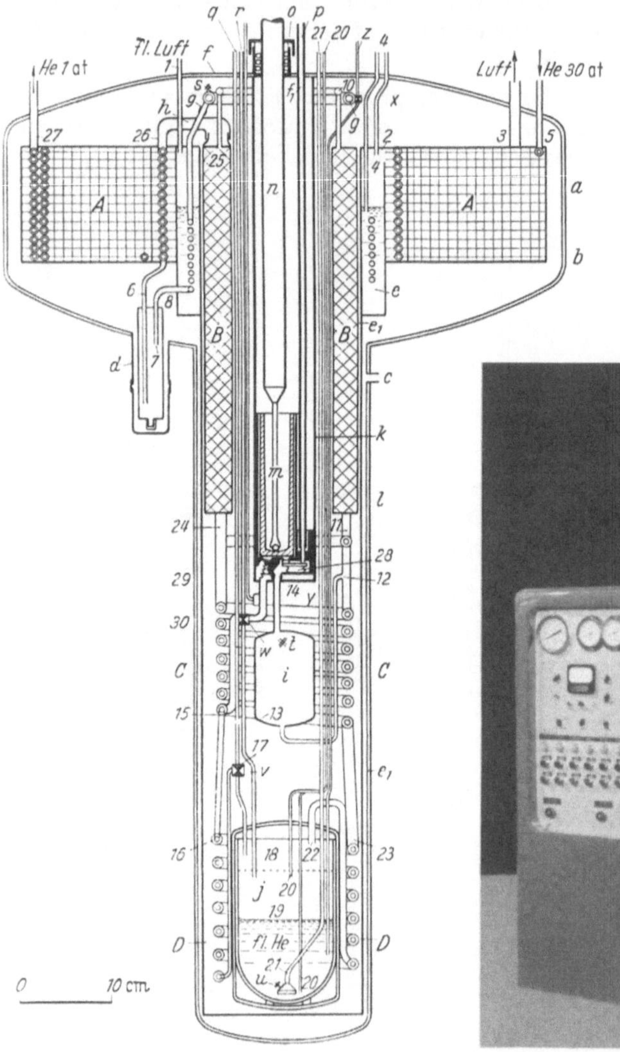

Fig. 21. Sectional view of Munich liquefier.

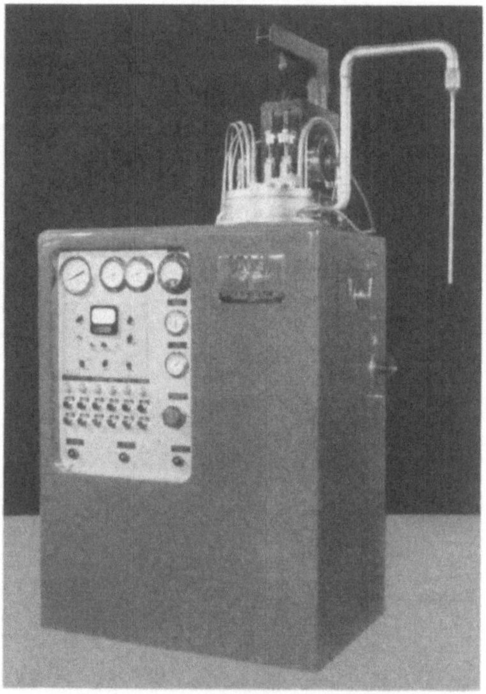

Fig. 22. Helium liquefier (Arthur D. Little, Inc.).

motion of all of the pistons may be controlled from a common crankshaft, the cylinders operate over different temperature ranges. Just as the action of a single cylinder of an engine may obviate the need for liquid hydrogen, a second cylinder which provides refrigeration at a higher temperature enables one to dispense with liquid nitrogen too. Two-cylinder, 3-cylinder and 5-cylinder combinations have been built. All liquefiers so equipped are capable of producing fair yields

[1] S. C. COLLINS: Rev. Sci. Instrum. **18**, 157 (1947).
[2] S. C. COLLINS: Science, Lancaster, Pa. **116**, 289 (1952).

of liquid helium without liquid coolants of any kind, but the yields are much higher when liquid nitrogen is used.

A partial sectional view of a two-cylinder liquefier is given in Fig. 12. Note that the heat exchangers and engine extend downward into a metal Dewar vessel from a heavy plate to which they are attached. The outer wall of the Dewar vessel is a steel cylinder 14 inches in diameter and 5 feet high. It is closed at the bottom and fitted with a flange and o-ring seal at the top. The vacuum jacket contains no apparatus or piping except a single tube which conveys liquid nitrogen over the surface of the radiation shield. The space in which the heat exchangers and engine hang is filled with helium at the pressure existing in the heat exchanger on the downstream side of the expansion valve. Minor leaks from gasketed joints in the engine or mechanical couplings in the piping are of no consequence. Since the temperature is highest at the top there is no convection in the helium atmosphere.

The heat exchanger is of the type shown in Fig. 10 E. Engine connections can be plugged in at appropriate levels. The operating pressure varies from 12 to 14 atm. The rate of liquefaction

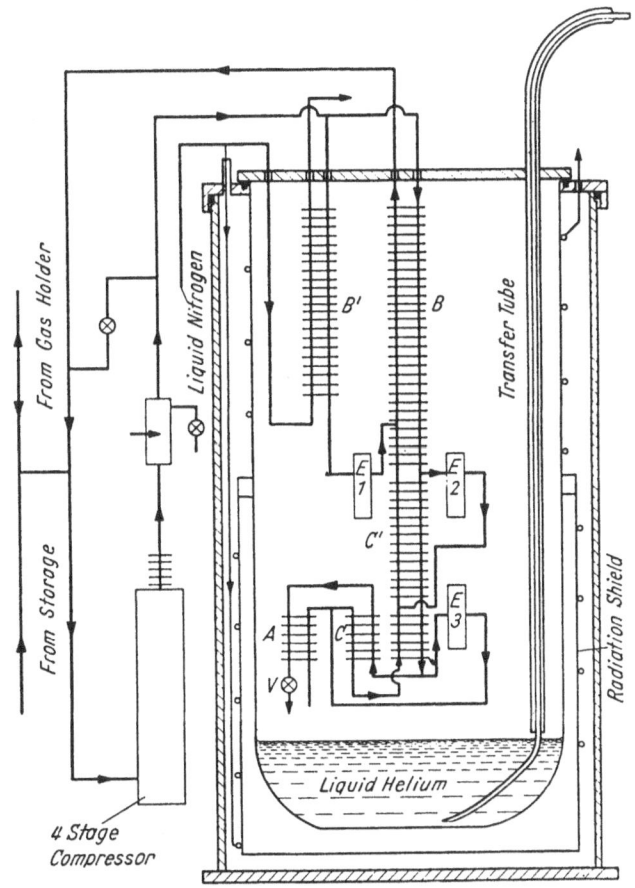

Fig. 23. Schematic diagram of large helium liquefier at the Massachusetts Institute of Technology.

varies from 2 to 8 liters per hour depending upon the amount of compressed helium available and whether liquid nitrogen is used. A commodious experimental chamber with access from the top is provided but is rarely used.

There are now about 80 of these machines in existence. A large majority of them were produced by ARTHUR D. LITTLE, Inc. of Cambridge, Mass. A photograph of the commercial model is shown in Fig. 22.

Fig. 23 is a schematic diagram of a much larger liquefier in which a 3-cylinder engine is used. Again a metal Dewar vessel provides the thermal isolation and the apparatus is immersed in an atmosphere of helium. A HAMPSON type, Fig. 10 F, heat exchanger is employed because of its compactness. It contains about 10000 feet of brass tubing 0.125 inch o.d. and 0.10 inch i.d. There are 162 tubes in parallel coiled into a compact cylinder, 8 inches in diameter and 42 inches long.

The temperature ranges over which the engine works are approximately as follows

1st cylinder	80 to 45°
2nd cylinder	45 to 25°
3rd cylinder	18 to 10°

The rate of liquefaction ordinarily varies from 25 to 28 liters per hour. If no liquid nitrogen is used, the yield is only 10 liters per hour.

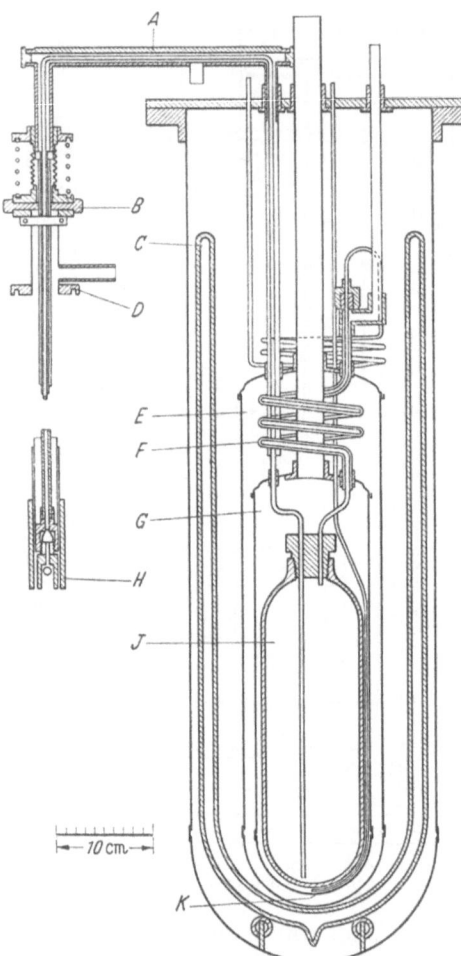

Fig. 24. Sectional view of large expansion liquefier at Oxford.

The starting time is 1.5 to 1.75 hrs. About 40 liters of liquid nitrogen are used during the starting period and approximately one liter of liquid nitrogen per liter of liquid helium thereafter. The helium circulated by the compressor is about 185 cfm. and the pressure varies from 11 to 13 atm. In each cylinder the gas enters at the same high pressure and leaves at the same low pressure of about 1.1 atm.

Brickwedde and Gifford[1] of the National Bureau of Standards and McMahon aof Arthur D. Little, Inc. collaborated in the design and construction of a large liquefier of the Collins type. A higher pressure, 30 atm., is employed. All of the compressed helium passes through the warmer cylinder, the pressure falling from 30 to 15 atm. and the temperature from about 35 to 28°, and 57% expands in the colder cylinder to 1 atm., the temperature falling from about 17° to about 10°. The rate of liquefaction is in the range of 18 to 20 liters per hour.

The liquefier of Long and Simon[2] employs a bellows engine for supplementary cooling. A compressor supplies 15 cubic meters per hour at 20 atm. 70% is thrott led in the regulator valve to 5.5 atm. for use in the engine, while 30% enters the heat exchange system at the higher pressure and eventually flows through the expansion valve.

After reaching a steady state, the liquefier produces 0.4 liter of liquid helium per hour and consumes 1.2 liters of liquid air per hour.

15. Non-flow expansion liquefiers. Although dozens of expansion liquefiers have been built by almost as many different investigators, modifications of Simon's early design, described above in Sect. 7, have been relatively slight.

---

[1] F. G. Brickwedde: Int. Con. Low. Temp. MIT. **1949**.
[2] H. M. Long and F. E. Simon: Proc. Internat. Inst. Refrig., Sept. 1953.

One change, introduced by COOK, ROLLIN and SIMON[1], is the provision for transfer of liquid helium to external vessels. If the valve in the transfer line is used in the later stages of the blow-down, some use can be made of the JOULE-THOMSON effect. The cold gas precools the transport Dewar vessel and may deposit some liquid in the transport vessel even before liquid is formed in the bomb.

SCOTT and COOK[2] of the National Bureau of Standards have reported an efficient liquefier of this type. With an expansion chamber of 410 cm.³ the amount delivered to the external receiver is 310 cm.³. A second liquefaction can be made in 1½ hrs. A later model delivers 1½ liters of liquid helium and uses only 4 liters of liquid hydrogen.

CROFT[3] at the Clarendon Laboratory in Oxford has produced a large expansion liquefier with provision for transferring liquid to other vessels. A longitudinal section of the liquefier is shown in Fig. 24. The transfer siphon, A, terminates in the outlet valve, H. The Dewar vessel, C, contains oxygen to block radiation across the vacuum space, E, and to precool compressed helium. The space G is filled with liquid hydrogen for cooling the bomb, J, and its charge of helium. Almost all of the hydrogen vapor flows out through the heat exchanger, F, and absorbs heat from the incoming helium. During the final stage of cooling the vapor is pumped through the large central tube.

The liquid yields is 1.1 to 1.2 liters in the transport vessel starting at 95 atm. Five liters of liquid hydrogen are used. One hour is required for the first liquefaction and ¾ hr. for subsequent liquefactions.

D. Liquid storage and transport vessels.

16. Glass containers. While glass has been extensively used in the construction of apparatus and for transport vessels, it has the disadvantage of being more or less permeable to helium so that the vacuum in sealed Dewar vessels is impaired after a time. Pyrex is particularly susceptible. The rate of diffusion of helium through glass is very much higher at room temperature than at lower levels. It is important, therefore, that gaseous helium not be allowed to stand in a warm Dewar vessel.

GIAUQUE[4] has described a liquid air-cooled glass flask of 8 liters capacity in which some liquid would be left after 4 days. SYDORIAK and SOMMERS[5] produced a 12-liter Pyrex flask which had an evaporation rate of one-half liter per day. It was necessary to repump the vacuum jacket every few days.

17. Metal containers. Relatively recently storage vessels for liquid helium have come into general use. WEXLER[6,7] described a vessel made of concentric copper spherical shells with a neck of Inconel. It has a capacity of 9.3 liters and the evaporation rate was only 0.09 liter per day. The construction is similar to that of conventional liquid air flasks except that nitrogen shielding is provided and more attention is given to the neck to reduce conduction of heat. Storage containers of this type have been commercially obtainable for a few years.

[1] A. H. COOK, B. V. ROLLIN and F. SIMON: Rev. Sci. Instrum. **10**, 251 (1939).
[2] R. B. SCOTT and J. W. COOK: Rev. Sci. Instrum. **19**, 889 (1948).
[3] A. J. CROFT: J. Sci. Instrum. **29**, 388 (1952).
[4] W. F. GIAUQUE: Rev. Sci. Instrum. **18**, 852 (1947).
[5] S. G. SYDORIAK and H. S. SOMMERS jr.: Rev. Sci. Instrum. **22**, 915 (1951).
[6] A. WEXLER and H. S. JACKET: Rev. Sci. Instrum. **22**, 282 (1951).
[7] A. WEXLER: J. Appl. Phys. **22**, 1463 (1951).

Two such vessels which have been in continuous use in the Cryogenic Laboratory of the Massachusetts Institute of Technology for two years were tested recently for evaporation rate. A standard liquid air flask was tested also after fitting it with an improvised nitrogen-cooled jacket and filling with liquid helium. The results are given in Table 3.

Table 3. *Evaporation rate of liquid helium.*

Vessel	Evaporation per day, liters liquid helium
75-liter helium vessel	0.327
25-liter helium vessel	0.221
25-liter liquid air vessel filled with helium .	0.317

With such extraordinarily low rates of evaporation it becomes practical to store liquid helium and have it continuously available.

The relatively higher rate of escape from the liquid air flask is undoubtedly due mostly to the larger and shorter neck with which it is equipped.

Electrical Conductivity of Metals and Alloys at Low Temperatures.

By

D. K. C. MacDonald.

With 43 figures.

A. Historical summary and general concepts.

1. Introduction. "It is perhaps unfortunate that so much attention has been paid to the resistance of metals, since it is probably one of the least characteristic properties of the substance, and depends on the electronic distribution and the elastic constants in a very complicated way." So said Wilson[1] in the first edition of his classic book "The Theory of Metals". From some points of view this outlook might still be maintained, but we should not forget that one of the most striking and valuable properties of a metal is its ready conduction of electricity. It is true that a complete fundamental understanding of electrical resistance in metals is still lacking today despite unremitting experimental investigation for at least a century; however, the wealth of useful information on metals gained by studies of electrical conductivity appears unequalled by any other comparable measurement. It is of course the rather subtle dependence of the electron scattering on the characteristic parameters of a metal that renders a full theoretical interpretation so difficult; nonetheless, it appears a fortunate fact that so sensitive and informative a parameter can be so readily measured.

2. The work of Drude and Lorentz. Following the discovery by J. J. Thomson in 1897 of the electron as a precise physical entity, it was quickly appreciated by Drude[1] and Lorentz[2] that the electrical properties of a metal must be ascribed to the presence of relatively free electrons. That is to say, an applied electric field would readily transport such electrons through the metal. Drude realized that such a "gas" would also tend to diffuse under a thermal gradient and was then able to explain the observed proportionality of electrical and thermal conductivities in a metal (Wiedemann-Franz law, 1853). Lorentz restated the problem in an elegant fashion based on the type of analysis developed by Boltzmann for the kinetic theory of gases. Introducing a statistical distribution function for the electrons, Lorentz was then able to discuss in a unified manner the electrical conductivity, the thermal conductivity and thermoelectric power of a metal—i.e. the basic electron-transport properties. Thus, despite the title of the present article, we shall not be able to confine ourselves wholly to the question of electrical conductivity nor can we consider alone the conventional low temperature region.

Drude's analysis may be directly compared with the elementary kinetic theory of gases. A uniform, but randomly directed, speed, v, is assigned to the

---

[1] P. Drude: Ann. Phys., Lpz. **1**, 566; **3**, 369 (1900).
[2] H. A. Lorentz: Proc. Amst. Acad. **7**, 438, 585 (1905).

electrons and a corresponding mean free path, l, assumed. The electrical conductivity, σ, is then given by

$$\sigma = \frac{n\,e^2\,l}{m\,v} \qquad (2.1)$$

where n is the number of free (transportable) electrons per unit volume, e is the electron charge, and m the mass of the electron.

[In practical units

$$\sigma = \frac{n\,e^2\,l}{m\,v} \times 10^7 \quad (\Omega\,\text{cm.})^{-1} \qquad (2.2)$$

where n is carrier density/c.c., e is measured in coulombs (1.59×10^{-19} for an electron), m is measured in grams (9.1×10^{-28} for an electron), l is measured in cm., and v is measured in cm./sec.]

The thermal conductivity, K, on the Drude model is given by

$$K = \tfrac{2}{3}\,l \cdot v \cdot C_v \qquad (2.3)$$

where C_v is the specific heat per electron.

The Lorenz number, L, is then defined as:

$$L \equiv \frac{K}{\sigma\,T} = \frac{2\,m\,v^2\,C_v}{3\,e^2\,T} \qquad (2.4)$$

If now we assume that the electrons are a classical ideal "gas", then $C_v = 3\,k/2$ (k Boltzmann's constant), $mv^2 = (3\,kT)$ and thus

$$L = 3\left(\frac{k}{e}\right)^2 \qquad (2.5)$$

[In practical units, if C_v is measured in joules/deg.cm.^{3}:

$$L = \frac{2\,m\,v^2\,C_v}{3\,e^2\,T} \times 10^{-7} \text{ watts.}\,\Omega/\text{deg.}^2,$$

$$= 2.23 \times 10^{-4} \text{ watts.}\,\Omega/\text{deg}^2].$$

This value for L proved to be in rather good overall agreement with experimental results on metals around room temperature. In Lorentz' treatment, the law of scattering by the lattice ions may be generalized (cf. Richardson[1], or e.g. Wilson [I]) and the velocity-distribution enters through the distribution-function. Assuming a hard-sphere atomic model, and classical statistics for the electrons, Lorentz found:

$$\sigma = \sqrt{\frac{8}{3\pi}} \cdot \frac{n\,e^2\,l}{m\,\sqrt{\overline{v^2}}} \qquad (2.6)$$

and

$$K = \sqrt{\frac{8}{3\pi}}\,l \cdot \sqrt{\overline{v^2}} \cdot C_v. \qquad (2.7)$$

Thus, using $C_v = \tfrac{3}{2}n\,k$; $\tfrac{1}{2}m\overline{v^2} = \tfrac{3}{2}\,kT$,

$$L = 2\left(\frac{k}{e}\right)^2. \qquad (2.8)$$

The agreement of L with experiment is now less satisfactory than on Drude's theory, although still within a factor of 2. However, from (2.6) we see that the resistivity $\varrho\,(\equiv 1/\sigma)$ should grow with $\sqrt{T}$ since l would be constant on a hard

---

[1] O. W. Richardson: Electron Theory of Matter. Cambridge University Press 1914.

sphere atomic model: similarly from (2.7) K would also increase with $\sqrt{T}$. Experimentally, the electrical resistance of metals rises more or less linearly with T at normal temperatures; comparison of thermal conductivity is perforce somewhat less direct since the ionic lattice itself will conduct heat, but around room temperature the thermal conductivity of metals is approximately constant or falling slowly with increase of temperature.

At the same time the problem arises of the contribution of the electron "gas" to the specific heat of the metal. Evidently, the overall specific heat should be $3n_a k + \frac{3}{2} n k$ where n_a is the density of atoms. Assuming $n/n_a \approx 1$, we should thus expect a molar heat of about $\frac{9}{2} R \approx 9$ cals/mole, while experimentally the specific heat of metals at room temperatures is about 6 cals/mole, suggesting that the free electrons do not contribute significantly.

As is clear today, there is a common underlying reason for these discrepancies. To a first approximation we may regard the conduction electrons in a metal as energetically insensitive to heat. The specific heat of the electrons is thus very small and their speed $(v$ or $\sqrt{\overline{v^2}})$ may therefore be taken as essentially constant. Consequently, the dependence of electrical resistance on temperature arises almost wholly from the increasing vibrations of the ions, and we may think of the electronic resistance as a most convenient probe responding to the state of order in the metallic structure. If the electron heat were exactly zero, then by (2.7) the thermal conductivity would also vanish. It is the case that in a full analysis today one must carry the approximations to a higher order to derive the thermal conductivity than is necessary for the electrical conductivity.

It is interesting, however, (and not entirely a coincidence) that on the present-day theory L does not alter very much—in fact theory predicts at normal temperatures

$$L = \frac{\pi^2}{3} \left(\frac{k}{e}\right)^2 \tag{2.9}$$

differing very little numerically from DRUDE's value. Consequently, one must evidently beware of regarding the value of the LORENZ number as a very significant test of a theory. Indeed one might say that subsequent theories made it their business—naturally enough!—to produce a LORENZ number in substantial agreement with (2.5) or (2.8).

It is believed today that the unusual attributes of the conduction electrons are due to the requirements of the PAULI exclusion principle operating in the metal. This finds expression through the FERMI-DIRAC statistics of electrons, and to SOMMERFELD[1] goes the credit for first applying these principles to electron transport in metals. SOMMERFELD was closely followed by HOUSTON[2] and BLOCH[3] in applying quantum mechanics to the electron-lattice interaction, and further developments of the modern theory followed rapidly. However, between the DRUDE-LORENTZ period and the work of SOMMERFELD there were not unnaturally a number of different attempts to provide alternative models of electron conduction. Apart from yielding various expressions for electrical and thermal conductivity and the ubiquitous LORENZ number, there was generally also the aim of explaining other related phenomena. We recall that as a result of his systematic researches on the properties of gases H. KAMERLINGH ONNES was able for the first time in 1908 to liquefy helium in Leiden. Helium boils under atmospheric pressure at about 4.2° K and the temperature may quite readily be reduced to

[1] A. SOMMERFELD: Z. Physik **47**, 1 (1928).
[2] W. V. HOUSTON: Z. Physik **48**, 449 (1928). — Phys. Rev. **34**, 279 (1929).
[3] F. BLOCH: Z. Physik **52**, 555 (1928); **53**, 216 (1929); **59**, 208 (1930).

around 1° K by pumping. Energetic research in this new field rapidly produced much valuable work and an early major discovery was that of superconductivity by KAMERLINGH ONNES[1]. This phenomenon was first observed in mercury and has since been found to exist in a number of other metals such as lead, tin, indium, aluminium, and also in compounds. In the latter case, binary superconducting compounds or alloys have been discovered where one component is not a super-conductor (e.g. Cu_3Sn) or even *both* (Au_2Bi).

In addition, less dramatic phenomena such as the *reduction* of resistance which most metals show under pressure seemed unaccountable on the early theories. For if one accepts the simple model of electron-scattering by an assembly of "billiard-ball" atoms one would certainly expect that compression would increase the frequency of collisions so increasing the resistivity. Although theories of these intermediate years between 1905 and 1928 are seldom now considered it seems worth-while to outline them here since for example we have still no established theory for superconductivity and more than a germ of truth may yet lie hidden in these earlier discussions of electron conduction. Indeed, the theory proposed by HEISENBERG[2] just after World War II to account for super-conductivity showed considerable resemblance in its early form to the LINDE-MANN "electron-lattice" model to which we shall refer below.

3. Subsequent theories (THOMSON, LINDEMANN, WIEN). First we recall the dipole theory of J. J. THOMSON[3]. THOMSON assumed that "the atoms of some substances including the metals, contain electrical doublets, i.e. pairs of equal and opposite electrical charges at a small distance apart. ... On this theory the peculiarity of metals is that electrons, not necessarily nor probably those in the doublets, are very easily abstracted by these forces from the atoms when these are crowded together [i.e. in the solid state]". The effect of an applied electric field was to tend to orient these electric doublets and a current could then flow, ... "the electrons passing along the chain of atoms like a company in single file pass-ing over a series of stepping-stones". An experimental physicist today might well be pardoned for looking back with something like nostalgia to a time when fundamental theories were couched in such readily visualized terms as these! THOMSON's paper of 1915 was specifically stimulated by KAMERLINGH ONNES' investigations on superconductivity and indeed THOMSON suggested then that this phenomenon must be regarded as a fatal objection to the free-electron theory. His own theory provided for a superconductive transition if and when the electric doublets lined up spontaneously in what we should presumably denote as a "ferro-electric" transition today. THOMSON worked out his model in some detail showing, for example, that a critical temperature would only exist on his theory if the temperature-coefficient of resistance at normal temperatures exceeded $1/273$. It is interesting to notice that the present much more sophisticated theory of superconductivity of FRÖHLICH[4] and BARDEEN[5] also relates the occurrence or otherwise of the phenomenon in a particular element to the behaviour of the resistivity in the normal state. THOMSON's theory also indicated a proportionality between the resistivity and the thermal energy of the solid which was already known experimentally to fit the facts fairly well in a number of cases.

[1] H. KAMERLINGH ONNES: Commun. Phys. Lab. Leiden Nos. 122b, 124c, 1911.

[2] W. HEISENBERG: Z. Naturforsch. 2a, 185 (1947).

[3] Sir J. J. THOMSON: The Corpuscular Theory of Matter, Chap. 5, (Constable), 1907. — Phil. Mag. 30, 192 (1915).

[4] H. FRÖHLICH: Phys. Rev. 79, 845 1950). — Proc. Phys. Soc. Lond. A 64, 129 (1951). — Nature, Lond. 168, 280 (1951).

[5] J. BARDEEN: Phys. Rev. 80, 567 (1950); 81, 829, 1070 (1951); 82, 978 (1951).

F. A. LINDEMANN[1] (now Lord CHERWELL) in 1915 proposed for the state of the free electrons in a metal a novel conception which still bears a strong air of modernity. After recalling that infra-red data already indicated a free-electron density of the same order as that of the atoms, LINDEMANN detailed the deficiencies of the free-electron "gas" theory. The major problem was that while the observed thermal conductivity appeared to demand a specific heat of normal magnitude, no such specific heat could be observed in calorimetric measurements. The bold suggestion was then made that we should regard the conduction electron assembly as a rigid lattice structure rather than as a "gas". Because of its low mass density, yet having the rigidity of a solid, such a lattice would have a very high sound velocity and a correspondingly high characteristic temperature (about 96000° K for silver). Consequently, the molar heat of such an electron lattice would be negligible at normal temperatures while the high sound velocity ($\sim10^8$ cm./sec.) could still maintain a reasonable thermal conductivity. It seems particularly interesting that the present free-electron "gas" theory based on FERMI-DIRAC statistics leads to very similar parameters—about 64000° K as a characteristic temperature for silver and an electron speed around 10^8 cm./sec.

In 1913 WIEN[2] in a remarkably prophetic paper said „Die Ergebnisse der Strahlungstheorie und die neuere Theorie der spezifischen Wärme haben den Nachweis geliefert, daß die Elektronentheorie der Metalle auf eine wesentlich neue Grundlage gestellt werden muß"[3]. WIEN recognized clearly a number of the vital features which characterize our physical understanding of electron conduction today. He argued that we can only maintain effectively free electrons in an atomic lattice if the electrons have a velocity v which is independent of the temperature and remains unchanged down to the absolute zero of temperature[4]. He deduces that so long as the metal structure is completely regular the conductivity should be infinitely great, referring indeed to KAMERLINGH ONNES' experiments at very low temperatures for confirmation. At higher temperatures the oscillations of the metal atoms will disturb the regular periodicity of the lattice and collisions with the conduction electrons will then take place. Referring to DRUDE's equation, (2.1) above, WIEN points out that we must regard v and n as constants, so long as the metal structure is unaltered, and that only the mean free path, l, will depend on temperature.

WIEN then goes on to consider how the electron scattering will depend on the amplitude of the atomic vibrations. He argues that if n quanta of energy, hv, are distributed amongst a number of atomic oscillators, the scattering will only be independent of the specific mode of distribution if we assume it to depend directly on the square of the atomic vibrations (i.e. on the energy itself). One might perhaps suggest that the "*phonon*" as a useful entity was born with this deduction. It is today established on quantum-mechanical foundations that an electron in a vibrating lattice is scattered by absorption or emission of one quantum of vibrational energy. Since the probability of such a transition is simply proportional to the density of quanta at a given lattice frequency[5] one may visualize

---

[1] F. A. LINDEMANN: Phil. Mag. **29**, 127 (1915).

[2] W. WIEN: Sitzgsber. preuß. Akad. Wiss., Berlin **5**, 184 (1913).

[3] Author's translation: "The results of radiation theory and the recent theory of specific heats have shown that the electron theory of metals must be established on an essentially new foundation."

[4] „Es muß also v eine andere Bedeutung haben als in der bisherigen Theorie; es muß eine Geschwindigkeit sein, die mit der Temperatur nichts zu tun hat, die also auch für $T=0$ unverändert existiert."

[5] More strictly, the probability is proportional to N_q for a lattice transition of the qth mode from N_q to N_q-1 (i.e. absorption by the electron of a quantum) and proportional to N_q+1 for a lattice transition from N_q to N_q+1.

these as individual electron-phonon collisions. The average energy of a lattice oscillator in thermal equilibrium is $h\nu/(e^{h\nu/kT}-1)$, and one could therefore always say of course that the density of quanta or "phonons" of energy $h\nu$ is by definition $1/(e^{h\nu/kT}-1)$. The applicability, however, of the phonon concept in *scattering* processes depends on the foregoing arguments[1].

WIEN himself then assumed that

$$\varrho \propto \int_0^{\nu_{max}} \frac{h\nu\,d\nu}{e^{h\nu/kT}-1} \tag{3.1}$$

i.e.

$$\varrho = C\left(\frac{T}{\Theta}\right)^2 \cdot \int_0^{\Theta/T} \frac{x\,dx}{e^x-1} \tag{3.2}$$

where $k\Theta = h\nu_{max}$, and C is some constant.

Thus at high temperatures, $(\Theta/T \ll 1)$, we may expand the exponential and integrate, yielding:

$$\varrho = C\left(\frac{T}{\Theta}\right). \tag{3.3}$$

The linear temperature dependence is then in good general agreement with experiment at high temperatures. At low temperatures $(\Theta/T \gg 1)$ we may set the integral upper limit as ∞ and hence:

$$\varrho = \frac{\pi^2}{6} C\left(\frac{T}{\Theta}\right)^2. \tag{3.4}$$

Experiment shows, however, that in most metals the decay of resistance at low temperatures is much more rapid.

We can also see broadly from WIEN's theory (as did GRÜNEISEN) that we should now expect the resistance to *diminish* under pressure (in general accord with experiment) since the characteristic lattice frequency, and hence Θ, will increase under compression. There is also of course clear recognition by WIEN that the resistance will in general be a function of a reduced temperature, T/Θ.

4. Foundation of modern electron theory of metals by SOMMERFELD. Despite the physical ingenuity of these "intermediate" theories, it was felt on the whole that the situation as regards the conduction electron assembly remained logically unsatisfactory and SOMMERFELD[2] in 1927 when presenting the foundations of his theory based on FERMI-DIRAC statistics for the electrons said: „Deshalb ist die Idee des ‚Elektronengases' im Metall während der letzten zwanzig Jahre mehr und mehr in Mißkredit gekommen."[3] PAULI's exclusion principle (PAULI-Verbot) postulated that no two electrons in an atom could occupy the same quantum-state (i.e. have the same four quantum numbers) and this was then extended to the assembly of electrons in a molecule. FERMI[4] and DIRAC[5] then made the discerning generalization of the principle to the case of an idealized gas of electrons. The consequences of this "exclusion principle" are that an assembly of otherwise

---

[1] In this article, $e = 2.718\ldots$, whereas e means the elementary electrical charge.
[2] A. SOMMERFELD: Naturwiss. **15**, 824 (1927); **16**, 374 (1928). — Z. Physik **47**, 1 (1928).
[3] Author's translation: "In this way the concept of the electron-gas in metals has become more and more discredited during the past twenty years."
[4] E. FERMI: Z. Physik **36**, 902 (1926).
[5] P. A. M. DIRAC: Proc. Roy. Soc. Lond., Ser. A **112**, 661 (1926).

free particles obeying FERMI-DIRAC statistics should possess a zero-point kinetic energy. The upper bound of linear momentum is simply determined by the linear density of particles i.e. $p_{max} \approx \frac{h}{2d}$ (cf. de BROGLIE's relation $p \approx h/\lambda$), where d is the average distance of separation; thus $v_{max} \approx \frac{h}{2md}$ which because of the small electron-mass signifies a relatively high velocity. If we now set $d \approx 3 \times 10^{-8}$ cm., as a typical interatomic separation we find: $v_{max} \approx 10^8$ cm./sec., while on the classical basis we should have $v \approx \sqrt{\frac{3kT}{m}} \approx 10^7$ cm./sec., for $T = 300°$ K. Corresponding to the momentum $\frac{h}{2d}$, the particles will have an energy $E_0 = \frac{p^2}{2m}$ which on a classical basis would only be attained at a temperature:

$$k\,T_0 \approx E_0 = \frac{p^2}{2m}, \quad \text{i.e.} \quad T_0 \approx \frac{h^2}{8\,m\,k\,d^2},$$

the so-called "degeneracy temperature". For $d \approx 3$ Å this has the value: $T_0 \approx$ 40 000° K. For physical temperatures well below this value the properties of the assembly will almost entirely be determined by the zero-point energy and one sees readily that only that fraction of the particles within a range $\approx kT$ of the maximum energy, E_0, can expect to interchange thermal energy with the environment. Thus the molar energy, U, at a temperature $T(\ll T_0)$ will be in the order of

$$U \approx N E_0 + \left(\frac{kT}{E_0}\right) N \cdot kT$$

$$\approx N E_0 + \left(\frac{T}{T_0}\right) \cdot RT \quad \text{(cf. Fig. 1)}$$

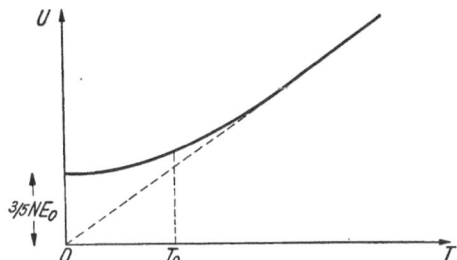

Fig. 1. Energy of FERMI-Dirac "gas" (Sketch only).

while for very elevated temperatures, $T \gtrsim T_0$, we should have simply the classical result $U \approx \frac{3}{2}RT$. Thus at normal temperatures $(T \ll T_0)$ the specific heat will be $C \approx (T/T_0)R$.

In applying the concept of a FERMI-DIRAC gas to the conduction electrons in a metal, SOMMERFELD removed at a stroke the fundamental difficulties which had beset the free-electron gas theory. In terms of DRUDE's formulae: $\sigma = \frac{ne^2l}{mv}$; $K = \frac{2}{3}lvC_v$ we now see that if we are justified in assuming a density, n, of free electrons (and naively we might regard the high FERMI velocity as providing some justification for this) then n and v will essentially be independent of temperature as WIEN was forced to assume. Only l then varies significantly with T. The specific heat is now predicted to be very small because $T/T_0 \ll 1$, as shown by experiment. Finally, the LORENZ number depends on the product: mv^2C_v i.e. on E_0C_v; but $C_v \approx (kT/E_0) \cdot k$, and consequently L retains essentially the same value as in the classical theory again in agreement with experiment. Indeed we now have $L = \pi^2k^2/3e^2$ very close to DRUDE's value of $3\,(k/e)^2$, which was in better numerical accord with experiment than LORENTZ's value of $2\,(k/e)^2$. SOMMERFELD gives (converted to practical units) 2.46×10^{-8} watt Ω/deg.^{2} as the average value for the LORENZ number observed at ambient temperatures for the twelve metals Al, Cu, Ag, Au, Ni, Zn, Cd, Pb, Sn, Pt, Pd, Na, while the FERMI-DIRAC-SOMMERFELD theory gives 2.45×10^{-8} and on DRUDE's theory we have 2.23×10^{-8}. Furthermore if we assume $(1/l) \propto T$ as WIEN did, giving $\varrho \propto T$ in agreement with observation at normal temperatures we then have K constant which is also found in good metals. SOMMERFELD also showed that the magnitude of the thermo-electric effects was of the right order and that they had the appropriate temperature-dependence (again at normal temperatures). The classical theory had failed here very badly for reasons similar to those arising in the specific heat question (we recall that THOMSON himself called the THOMSON heat "the specific heat of electricity").

5. Developments of HOUSTON and BLOCH. The problem now resolves itself into calculating l, the mean free path of electron-scattering by lattice vibrations. HOUSTON[1] first followed WIEN essentially in setting $1/l$ proportional to the mean square amplitude of atomic vibrations, leading, as before, to $\varrho \propto (T/\Theta)$ for $T \gg \Theta$ and $\varrho \propto (T/\Theta)^2$ for $T \ll \Theta$. Shortly after, however, the essential features of the scattering process were recognized[2]. An electron will only be scattered by the lattice vibrating at frequency, ν, if a quantum of energy, $h\nu$, is exchanged between lattice vibrations and conduction electron. The scattering is thus essentially inelastic, although at high temperatures ($kT > h\nu$, i.e. $T > \Theta$) it may be considered elastic, since the energy interchange is then relatively small. It follows immediately that as the absolute zero is approached, the resistance due to thermal vibrations must vanish since both electrons and lattice are rapidly settling into their lowest energy state. That is to say, there will be no resistance arising from the zero-point vibrations of the lattice, a question which had previously been in some doubt.

HOUSTON used theory already developed for the scattering of light waves by a lattice of atoms (cf. also FRENKEL and MIROLUBOW[3]). For the scattering from a single centre, required in the application of the theory to this problem, HOUSTON introduced WENTZEL's theory of the scattering of charged particles by atoms. WENTZEL[4] there assumed that the potential, V, surrounding a single atom, of nuclear charge Z, was given by $V = \dfrac{Z e}{r} \, e^{-r/b}$; b may thus be regarded as a measure of the net extension of the COULOMB field of the nucleus as screened by the outer electrons. In this case the ratio, S, of scattered to incident intensity at angle Ψ is

$$S = \frac{1}{r^2} \cdot \left(\frac{Ze}{4E}\right)^2 \cdot \frac{1}{\left(\mathrm{Sin}\,\dfrac{\Psi}{2} + \dfrac{1}{4 k^2 b^2}\right)^2} \tag{5.1}$$

where E is the energy and k is here the wave-number of the incident particles. On this basis, HOUSTON arrived at the expression:

$$\varrho \propto T \int\limits_0^{\Theta/T} \frac{x^4\,dx}{\left(x^2 + \dfrac{\Theta^2}{T^2} \cdot \dfrac{\pi^2 b^2}{A_0^2}\right)^2 \cdot (e^x - 1)} \tag{5.2}$$

where A_0 is the lattice constant. The second factor on the denominator: $(e^x - 1)$: includes the influence of the PAULI exclusion principle on the scattering[5], and the DEBYE model is assumed for the vibrational spectrum of the lattice.

About the same time BLOCH published a rather full analysis[6], now regarded as classic in this field, treating afresh the whole question of electron scattering

---

[1] W. V. HOUSTON: Z. Physik **48**, 449 (1928). See also SOMMERFELD: Naturwiss. **16**, 374 (1928).

[2] W. V. HOUSTON: Phys. Rev. **34**, 279 (1929). — F. BLOCH: Z. Physik **52**, 555 (1928).

[3] J. FRENKEL and MIROLUBOW: Z. Physik **49**, 885 (1928).

[4] G. WENTZEL: Z. Physik **40**, 590 (1927).

[5] HOUSTON mentions that BLOCH [Z. Physik **52**, 555 (1928)] was first to recognize the importance of the PAULI principle in the electron-ion *scattering* process.

[6] F. BLOCH: Z. Physik **53**, 216, (1929); **59**, 208, (1930).

in a lattice as a quantum-mechanical problem. He adopted a rather simplified model for the electron-ion interaction and obtained the result

$$\varrho \propto \frac{T^5}{\Theta^6} \int\limits_0^{\Theta/T} \frac{x^5\, dx}{(e^x - 1)\,(1 - e^{-x})}.$$ (5.3)

Both BLOCH's and HOUSTON's formulae predict $\varrho \propto T$ for high temperatures and $\varrho \propto T^5$ for sufficiently low temperatures, which is found to be generally the case experimentally. (But see WOODS[1]).

We shall leave until later (Sect. C) a discussion of BLOCH's formula and possible modifications thereof to take better account of the specific character of the ion-electron inter-action.

6. Resistance due to impurities, defects and structural changes. Electrical resistance arises as we have seen whenever the regular periodicity of the ionic lattice is disturbed. We have discussed how this will arise from thermal vibrations and we now wish to consider the influence of static perturbations due to foreign atoms ("impurities") which we might term chemical defects, and due to physical defects in the lattice such as misplaced atoms, grain boundaries and so on. For the most part, these are treated independently in the-oretical discussion, but in general of course both are present in any physical experiment. MATTHIESSEN's rule[2], discovered experimentally, states that the increase of resist-ance due to a small concentration of another metal in solid solution is in general independent of temper-ature. Figs. 2 and 3 illustrate the approximate validity of this rule

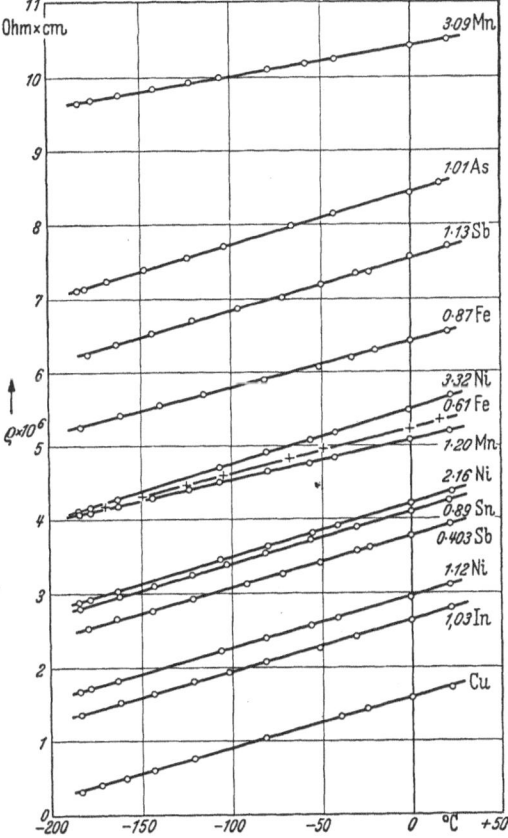

Fig. 2. Electrical resistance of copper and various alloys of copper to illustrate approximate validity of MATTHIESSEN's Rule. (After LINDE.)

both at high and low temperatures and when considering, say, the temperature-variation of resistance in comparison with theory it is standard practice to sub-tract first the so-called "residual resistance" from the observed results. The residual resistance is the limit of the observed resistance when extrapolated to absolute zero (with many metals a measurement at $\sim 4°$ K is quite adequate) which is assumed to be due solely to chemical and physical imperfections. Cor-respondingly the remainder of the resistance (often called the "ideal resistance") is assumed to arise solely from the thermal vibrations of the unperturbed lattice and hence to be characteristic of the ideally pure and physically perfect metal.

[1] S. B. WOODS: Canad. J. Phys. **34**, Feb. (1956).
[2] A. MATTHIESEN: Ann. Phys. u. Chem., Lpz. **7**, 761, 892 (1864).

Let us consider that the electrons in the metal are essentially "free", all having a charge e and (effective) mass m. Then the conductivity, σ, will be given by $\sigma \sim \dfrac{e^2}{m} \sum_{k} \tau(k)$ where $\tau(k)$ is the relaxation-time for electrons with wave-number, k, and the summation is of course to be taken over all occupied levels. Let $\tau_t(k)$ be the relaxation-time for electrons in the state k due to thermal vibrations and τ_{imp} the relaxation-time due to "residual" impurity scattering. We have

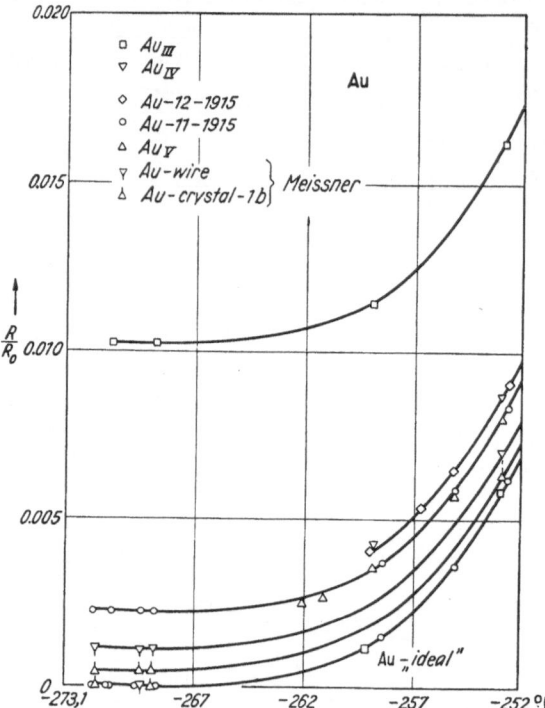

written τ_{imp} as independent of wave-number (although it is easy to generalize) since it has been generally assumed that this is the case for impurity scattering. Recent work on the thermoelectric power of dilute copper alloys[1] casts doubt on the *general* validity of this assumption since it appears necessary to assume a very marked energy-dependence of scattering to account for the results. On the other hand, experiments on cold-worked metal[2] (to increase the number of physical defects without altering the impurity concentration) suggest that the assumption *is* permissible for this type of scattering. Let us then write

$$\frac{1}{\tau(k)} = \frac{1}{\tau_t(k)} + \frac{1}{\tau_{imp}}$$

Fig. 3. Electrical resistance of gold samples of varying purity at low temperatures illustrating Matthiessen's Rule (from van den Berg).

assuming that for an electron in a given state, k, the probabilities of scattering by lattice thermal vibration and by lattice defects are independent and therefore directly additive (Mott and Jones [2], pp. 300 to 301, also discuss the validity of this particular assumption when considering the thermal displacement of an impurity ion). Thus

$$\sigma = \frac{e^2}{m} \sum_{k} \left\{ \frac{1}{\dfrac{1}{\tau_t(k)} + \dfrac{1}{\tau_{imp}}} \right\}. \qquad (6.1)$$

If now $\tau_t(k) = \tau_t$, independent of k, we may write immediately

$$\varrho \equiv \frac{1}{\sigma} = \frac{m}{N e^2} \frac{1}{\tau_t} + \frac{m}{N e^2 \tau_{imp}} = \varrho_{therm} + \varrho_{imp} \qquad (6.2)$$

which is Matthiessen's rule. More generally,

$$\sigma = \frac{e^2 \tau_{imp}}{m} \sum_{k} \left\{ \frac{\tau_t(k)}{\tau_t(k) + \tau_{imp}} \right\}. \qquad (6.3)$$

[1] D. K. C. MacDonald and W. B. Pearson: Proc. Roy. Soc. Lond., Ser. A **219**, 373 (1953). — Acta met. 403, **3**, 392 (1955). — W. B. Pearson: Phil. Mag. **45**, 1087 (1954).
[2] W. B. Pearson: Phys. Rev. **97**, 666 (1955).

If $\tau_t(\mathbf{k}) \ll \tau_{\text{imp}}$ so that thermal scattering is dominant, this reduces to

$$\varrho = \frac{m}{e^2} \left\{ \frac{1}{\sum\limits_{\mathbf{k}} \tau_t(\mathbf{k})} + \frac{1}{\tau_{\text{imp}}} \cdot \frac{\sum\limits_{\mathbf{k}} \tau_t^2(\mathbf{k})}{\left\{ \sum\limits_{\mathbf{k}} \tau_t(\mathbf{k}) \right\}^2} \right\} \tag{6.4}$$

which, if we write $\tau_t(\mathbf{k}) = \bar{\tau} + \delta(\mathbf{k})$ gives

$$\varrho = \varrho_{t,\text{ideal}} + \varrho_{\text{imp}} \left\{ 1 + \frac{\overline{\delta^2(\mathbf{k})}}{\bar{\tau}^2} \right\}. \tag{6.5}$$

Since $\tau_t(\mathbf{k})$ is temperature-dependent, and hence also $\delta(\mathbf{k})$, $\frac{\overline{\delta^2(\mathbf{k})}}{\bar{\tau}^2}$ will in general be a function of temperature. While, if $\tau_{\text{imp}} \ll \tau_t(\mathbf{k})$, so that we are at sufficiently low temperatures for the residual resistance to dominate, we find

$$\varrho = \frac{m}{N e^2 \tau_{\text{imp}}} \left\{ 1 + \frac{\tau_{\text{imp}}}{N} \sum_{\mathbf{k}} \left(\frac{1}{\tau_t(\mathbf{k})} \right) \right\}$$

which if we write again $\tau_t(\mathbf{k}) = \bar{\tau} + \delta(\mathbf{k})$ and assume $\delta(\mathbf{k})/\bar{\tau}$ small gives

$$\varrho = \varrho_{\text{imp}} + \varrho_{\text{therm,id}} \left\{ 1 + \frac{\overline{\delta^2(\mathbf{k})}}{\bar{\tau}^2} \right\}. \tag{6.6}$$

Thus in general MATTHIESSEN's rule will not be exact, although it is a very useful first approximation. One may expect from (6.3) that the total deviation will be greatest when $\tau_t \approx \tau_{\text{imp}}$ (i.e. the thermal component of resistance and the residual component are about equal), but it is also evident that so long as $\tau_t(\mathbf{k})$ is increasing monotonically with decreasing temperature, so also must σ increase and hence ϱ must diminish monotonically. VAN DEN BERG's[1] careful series of measurements of the electrical resistance of a number of metals between $\sim 20°$ K and $\sim 1.5°$ K showed some departures from MATTHIESSEN's rule in copper between $\sim 14°$ K and $\sim 20°$ K, and GRÜNEISEN[2] has also noticed deviations in copper up to $\sim 90°$ K. On the other hand, the *minimum* in the resistance of certain metals at low temperatures, first observed by VAN DEN BERG and DE HAAS on gold, cannot be explained as any obvious departure from MATTHIESSEN's rule. When one comes to make measurements of the electrical resistance produced by lattice defects, strains, etc., it appears very important to the writer that we should avoid relying on MATTHIESSEN's rule wherever possible. That is to say, when we are trying to draw experimental inferences about what are often relatively small changes in the overall resistance due to physical and chemical defects it would appear most desirable that these deductions should be made, if possible, directly from experiments done at sufficiently low temperatures for thermal scattering to be negligible or at least small.

The use of electrical resistance measurements as an indication of the spatial state of order in the lattice is of course a technique of long-standing. Perhaps the most classic example is the observation of the formation of ordered lattice structures in the copper-gold system (see Fig. 4a and b). The difference of resistance even at room temperature in say Cu_3Au between the equilibrium ordered structure and the disordered lattice, preserved by rapid quenching from a high temperature, is evidently very great; it is therefore unnecessary in such cases to

---

[1] G. J. VAN DEN BERG: De elektrische Weerstand van zuivere Metallen... Thesis, Leiden 1938. See also W. J. DE HAAS, J. DE BOER and G. J. VAN DEN BERG: Physica, Haag **1**, 1115 (1933).
[2] E. GRÜNEISEN: Ann. Phys., Lpz. **16**, 530 (1933).

suppress entirely the resistance due to thermal vibrations by making measurements at very low temperatures. For less dramatic structural changes it may be necessary finally to cool the specimen to liquid air, hydrogen, or helium temperatures in order to observe with certainty and accuracy the resistance changes involved; assuming of course a rate of cooling below the *transformation region* sufficient to quench in the particular structural change involved. This, however, is not always possible; in the case of lattice defects generated before melting in the alkali metals[1, 2] it appears that the energy of movement for such defects

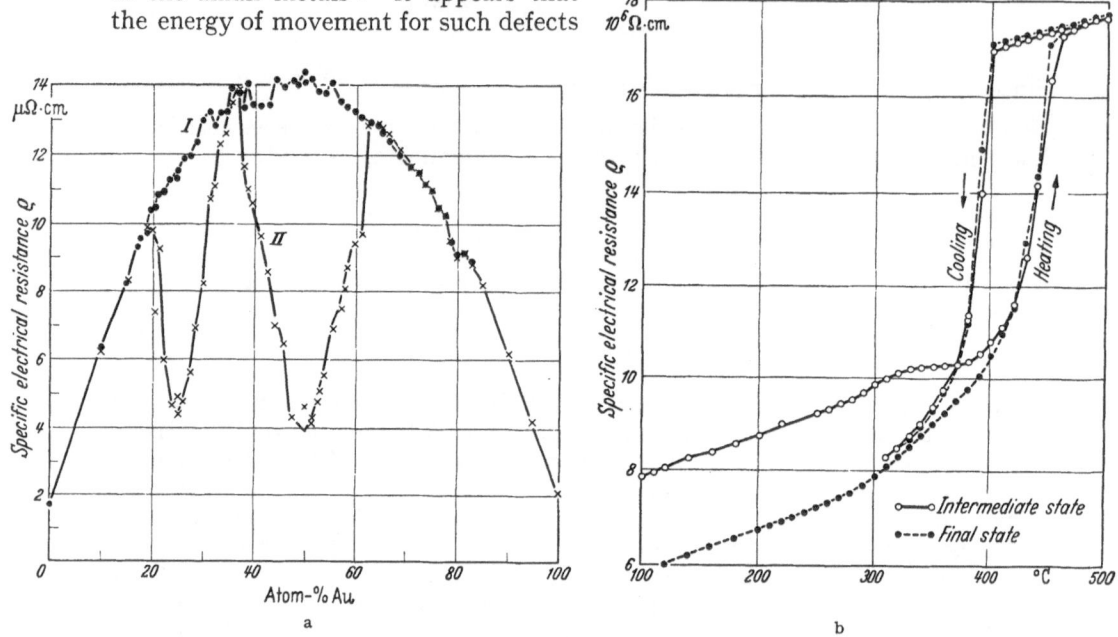

Fig. 4a and b. (a) Specific resistance of gold-copper alloys at 0° C, after Pospisil. I: Quenched disordered alloys. II: Annealed alloys. (b) Variation with temperature of resistance of Au—Cu alloy showing marked influence of ordered lattice structure (after Dehlinger and Graf).

is so small ($\sim$ 1 kcal/mole) that no reasonable rate of quenching can freeze them in. It is therefore necessary to follow the generation of such defects by the characteristic rapid growth of resistance with temperature ($\Delta \varrho \sim e^{-E/kT}$) superimposed on the normal resistance (see Fig. 5a and b). On the other hand, some structural changes appear to affect the resistance only very slightly; in lithium Barrett and Trautz[3] report the occurrence of a partial martensitic transformation of crystal structure from the body-centred cubic structure (common to all the alkali metals at higher temperatures) to a hexagonal close-packed structure. However, there is as yet no evidence of anomalous behaviour arising from this transformation in the specific heat and only secondary evidence of any influence on the variation of resistance with temperature which to the eye appears quite smooth. It appears to be characteristic of such martensitic changes that their onset temperature and degree of completion are strongly influenced by cold work and, presumably, degree of local strain; consequently the overall

[1] D. K. C. MacDonald: Report of Conference on Defects in Crystalline Solids, p. 383, Phys. Soc. Lond. 1955.

[2] L. G. Carpenter: Private communication (I am most indebted to Mr. Carpenter for cordial discussion and correspondence on this problem).

[3] C. S. Barrett and O. R. Trautz: Trans. Amer. Inst. Min. Metallurg. Engrs. 175, 579 (1948). — See also W. B. Pearson: Canad. J. Phys. 32, 708 (1954).

transformation in a given specimen may be "smeared out" over a considerable range of temperature rendering its observation by thermal means very difficult.

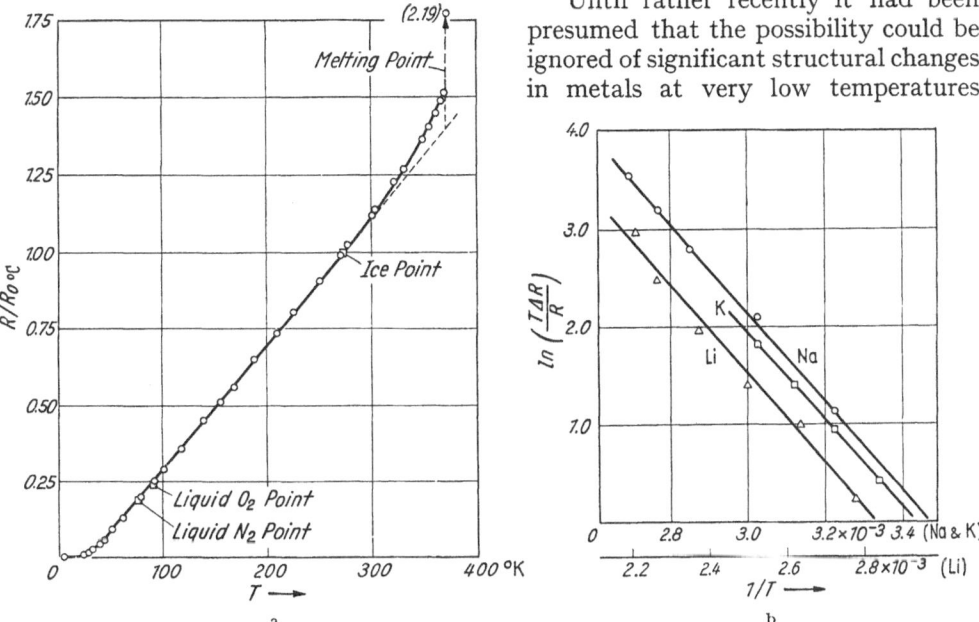

Until rather recently it had been presumed that the possibility could be ignored of significant structural changes in metals at very low temperatures

Fig. 5a and b. (a) Electrical resistance of sodium up to melting point. (b) Excess resistance in alkali metals, presumed due to thermal lattice defects, plotted for deduction of activation energy.

(under normal pressure) on the assumption that the energy change required would be too great for thermal excitation. However, following his work on lithium, BARRETT has found evidence of a similar transformation on cooling sodium, starting about 32° K in strain-free metal, and RAYNE[1] has found a considerable anomalous peak in the specific heat of sodium about 0.85° K which, he suggests, may be due to a martensitic transformation. We recall also the calorimetric measurements of SIMON and PICKARD[2] on sodium which displayed an anomaly[3] in the specific heat in the region of 7° K. BARRETT and his colleagues are at present undertaking a survey of the alkali

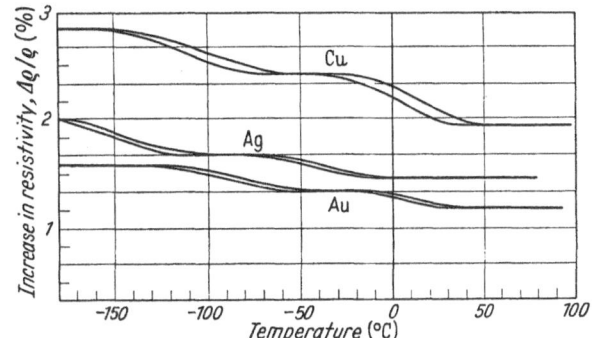

Fig. 6. Variation of resistance with temperature of noble metals after cold work at —183° C (after MANINTVELD).

metals, using a low-temperature X-ray camera, to search systematically for any transformations of this type. The existence of such structural changes at very low temperatures is somewhat disconcerting in making comparison of fundamental

[1] J. RAYNE: Phys. Rev. 95, 1428 (1954).
[2] F. E. SIMON and G. L. PICKARD: Proc. Phys. Soc. Lond. 61, 1 (1948).
[3] But see also: D. H. PARKINSON and J. E. QUARRINGTON: Proc. Phys. Soc. Lond. A 68, 762 (1955).

theory with experiment when one might otherwise have hoped that all perturbing features in the ionic lattice (such as thermal expansion) were quite negligible.

The variation of resistance with rising temperature after cold-work (or irradiation) at a low temperature shows in general a number of "recovery"-processes where the resistance drops markedly; these take place at more or less well-defined temperatures from which activation energies for these processes may be deduced and the specific mechanism thus identified in favourable cases. Broom[1] has given a recent survey written essentially from the experimental point of view, with a broad bibliography, of this very active field and Fig. 6 shows experimental results obtained by Manintveld on the noble metals after extension at −183° C. Two discrete recovery processes are evident and Table 1 shows the activation energies deduced by Manintveld for these processes.

Table 1. *Recovery processes in the noble metals after low temperature deformation.*

Metal	Q_1 (e.v.) "Low"-temperature recovery	Q_2 (e.v.) "High"-temperature recovery
Gold . . .	0.29 ± 0.03	0.69 ± 0.06
Silver . . .	0.18 ± 0.02	0.69 ± 0.07
Copper. . .	0.20 ± 0.03	0.88 ± 0.09

It has been suggested that the higher temperature recovery is due to diffusion of single lattice vacancies and the lower to diffusion of aggregates of vacancies. Table 2, taken from Broom, summarizes data on recovery in a number of metals after cold work and radiation damage.

Table. 2 *Observed activation energies for recovery of resistivity* (after Broom).

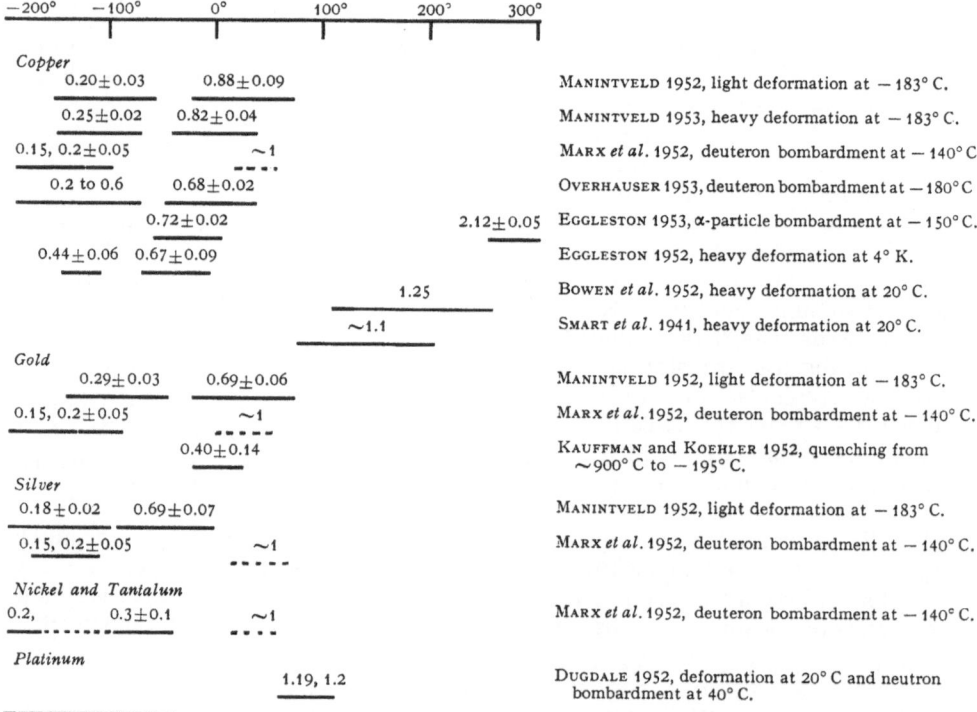

−200°	−100°	0°	100°	200°	300°

Copper
0.20±0.03   0.88±0.09   Manintveld 1952, light deformation at −183° C.
0.25±0.02   0.82±0.04   Manintveld 1953, heavy deformation at −183° C.
0.15, 0.2±0.05   ~1   Marx et al. 1952, deuteron bombardment at −140° C
0.2 to 0.6   0.68±0.02   Overhauser 1953, deuteron bombardment at −180° C
0.72±0.02   2.12±0.05   Eggleston 1953, α-particle bombardment at −150° C.
0.44±0.06   0.67±0.09   Eggleston 1952, heavy deformation at 4° K.
1.25   Bowen et al. 1952, heavy deformation at 20° C.
~1.1   Smart et al. 1941, heavy deformation at 20° C.

Gold
0.29±0.03   0.69±0.06   Manintveld 1952, light deformation at −183° C.
0.15, 0.2±0.05   ~1   Marx et al. 1952, deuteron bombardment at −140° C.
0.40±0.14   Kauffman and Koehler 1952, quenching from ~900° C to −195° C.

Silver
0.18±0.02   0.69±0.07   Manintveld 1952, light deformation at −183° C.
0.15, 0.2±0.05   ~1   Marx et al. 1952, deuteron bombardment at −140° C.

Nickel and Tantalum
0.2,   0.3±0.1   ~1   Marx et al. 1952, deuteron bombardment at −140° C.

Platinum
1.19, 1.2   Dugdale 1952, deformation at 20° C and neutron bombardment at 40° C.

[1] T. Broom: Adv. Physics **3**, 26 (1954).

The scattering of electrons, and the consequent resistance, due to chemical impurities is always due fundamentally of course to the disturbance of the otherwise regular ionic lattice, but may arise specifically in a number of ways. The most obvious source is a heterovalent impurity atom. If for example a divalent atom such as zinc is present in copper we may expect the two valence electrons of the zinc atom to join the other conduction electrons contributed by the copper atoms leaving an isolated divalent ion embedded in the lattice of monovalent copper ions[1]. There will thus be an excess unit positive charge at the site of the zinc ion. If one assumes a simple COULOMB (and therefore long-range) interaction of a conduction electron with such a charge the resistance is theoretically infinite, but in fact there

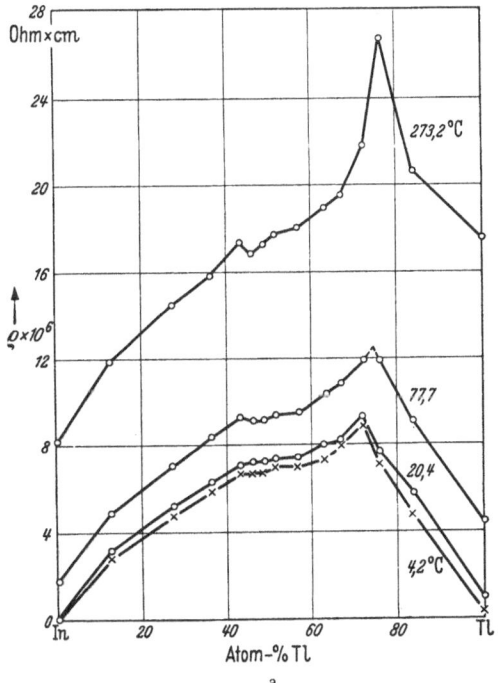

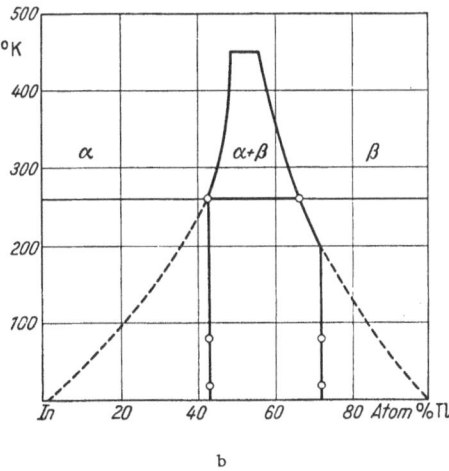

Fig. 7a and b. (a) Specific resistance of indium-thallium alloys (after MEISSNER, FRANZ and WESTERHOFF). (b) Equilibrium diagram data for indium-thallium alloys (after MEISSNER, FRANZ and WESTERHOFF).

is strong "screening" of the foreign ion by the remaining conduction electrons and a short range potential of the type $\frac{e^2}{r} e^{-r/b}$, where $b \sim 10^{-8}$ cm., is a closer approximation[2]. (We shall discuss this "electron-screening" in more detail later in this article.) In dilute alloys the additional resistance will be simply proportional to the atomic concentration, x, of the solute while NORDHEIM[3] has shown that for a general homogeneous (random) solid solution the resistance will be proportional to

Solute in copper	Scattering cross-section
Zn	0.6×10^{-16} cm.^{2}
Sn	5×10^{-16} cm.^{2}
Fe	18×10^{-16} cm.^{2}

$x(1-x)$, and for simple disordered alloys (such as the In-Pb system) this theoretical law is well obeyed. For dilute alloys of copper some electron scattering cross-sections are given in the adjoining table.

If the solute atom is homovalent with the solvent metal (as, for example, with K in Na) no charge difference will result for the alien ion but scattering can

---

[1] We are here simply assuming that this is an adequate description of the copper matrix.

[2] N. F. MOTT: Proc. Cambridge Phil. Soc. **32**, 281 (1936). See also MOTT and JONES [2], p. 292—294.

[3] L. NORDHEIM: Ann. Phys., Lpz. **9**, 641 (1931).

occur in two ways. The internal field of the solute ion will in general differ from that of the parent metal and Mott[1] has also given a theoretical calculation for the scattering cross-section arising in this way. An experimental estimate for the scattering cross-section of K in Na is about 2.5×10^{-17} cm.², and for Ag in Cu about $2.4_5 \times 10^{-17}$ cm.². In addition, dependent on the relative "*size-factor*" of the solute atom, the parent matrix may be locally strained and distorted by the foreign atom; this displacement of the ions will also give rise to electron-scattering. Friedel[2] has recently surveyed in some detail the field of metallic solution, particularly the energetic and thermodynamic questions involved. (Cf. also Pearson[3].)

If and when the introduction of sufficient solute leads to formation of a new

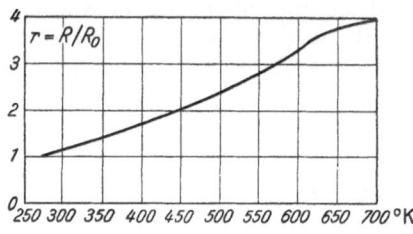
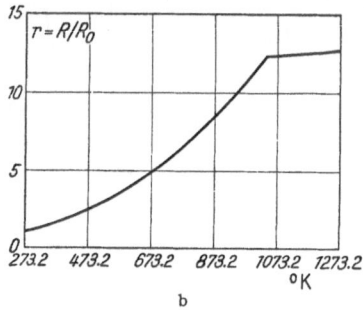

Fig. 8a and b. (a) Electrical resistance of nickel in the neighbourhood of the Curie-point (after Gerlach). (b) Resistance of iron near the Curie-point (after Holborn and Meyer).

phase in the equilibrium diagram we can expect to see a marked departure from the smooth parabolic variation of resistance with solute concentration, x. This is illustrated in Figs. 7a and b where the resistance-measurements of Meissner, Franz and Westerhoff enabled them to construct an equilibrium diagram for the indium-thallium binary system. Changes of magnetic order in a metal can also affect the electrical resistance (cf. Figs. 8a and b), but we shall not here be concerned specifically with magnetic metals or alloys.

7. The anomalous resistance-minimum at low temperatures. Apart from the phenomenon of superconductivity which has defied adequate theoretical interpretation for some forty odd years since its discovery, perhaps the most puzzling feature today in electrical resistance at low temperatures is the occurrence of the *minimum* in certain metals, referred to briefly above. De Haas and van den Berg at Leiden undertook a series of careful and detailed measurements on the electrical resistance of metals below $\sim 20°$ K starting about 1933. Earlier Leiden work and the extensive series of measurements by Meissner and Voigt[4] had in general provided data on many metals at, say, the liquid oxygen ($\sim 90°$ K) and liquid nitrogen ($\sim 78°$ K) points, the boiling and triple points of hydrogen ($\sim 20°$ K and $14°$ K) and in the liquid helium region (say 4 to $1.5°$ K) but the intermediate regions were unexplored. As pointed out in Chapter C of this article, it is just in the region between perhaps 30 and $4°$ K that the most significant comparison may be made with theory, and information derived about the nature of electron scattering. De Haas and van den Berg discovered that the resistance of their samples of gold (and, less markedly, silver) passed through a minimum at temperatures ranging from ~ 7 to $\sim 4°$ K. Figs. 9a and b show some of their

[1] N. F. Mott: Proc. Cambridge Phil. Soc. **32**, 281 (1936). See also Mott and Jones [2], p. 290–291.

[2] J. Friedel: Rapp. 10e Congr. Solvay Phys.

[3] W. B. Pearson: Lattice Spacings and Structures of Metals, Chap. 3. London: Pergamon Press 1956.

[4] W. Meissner and B. Voigt: Ann. Phys., Lpz. **7**, 761, 892 (1930).

results. They noticed that the higher the residual resistance[1] of their specimens the higher the temperature at which the minimum occurred, and were able for their specimens to relate uniquely the resistance at the minimum to its temperature of occurrence and to indicate that for an "ideally pure" specimen the minimum would vanish entirely (see Fig. 10). The phenomenon has since been observed in a number of metals and

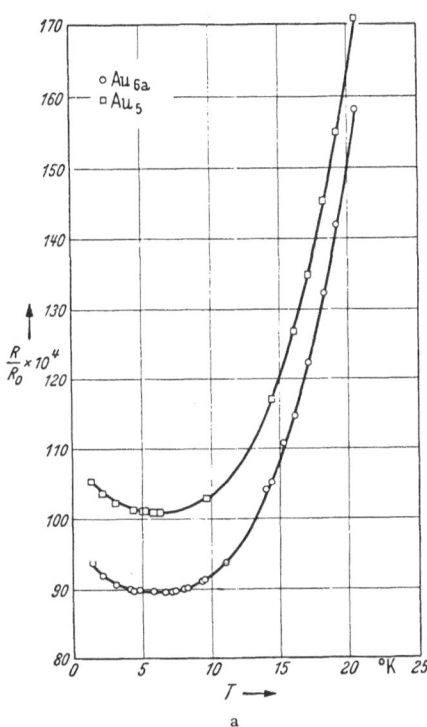

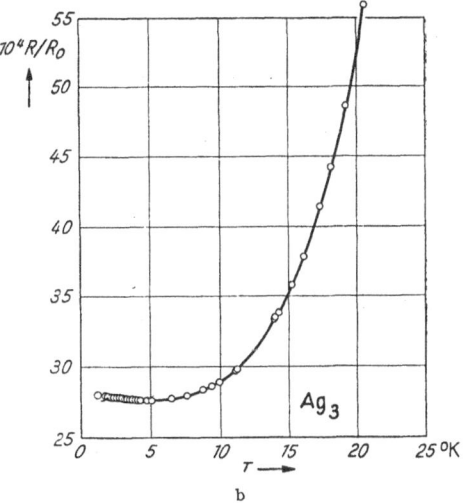

Fig. 9a and b. (a) Electrical resistance of two specimens of gold at very low temperatures (after DE HAAS and VAN DEN BERG). (b) Resistance of a specimen of silver at very low temperatures (after dE HAAS and VAN DEN BERG).

detailed experimental investigations have been undertaken particularly by GERRITSEN and LINDE, MENDOZA and THOMAS, and MACDONALD and PEARSON (The phenomenon obviously calls for co-operative research!). The post-war research has established that different impurity atoms in a given metal will give rise to different variations of the resistance-minimum; the *single* curve of DE HAAS and VAN DEN BERG can probably be ascribed to the presence of a single major residual impurity (perhaps silver) in the relatively pure specimens used by them in their experiments. A rather full investigation of copper as parent metal has shown that a wide variety of solute atoms

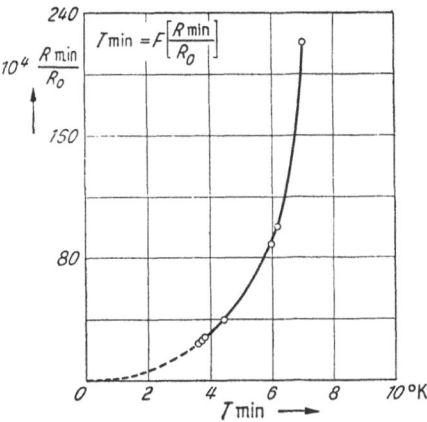

Fig. 10. Minimum resistance plotted against temperature of minimum for various specimens of gold (after DE HAAS and VAN DEN BERG).

can produce a minimum (e.g. Fe; Sn; Ge; Pb; Ga; In) while certain solutes (e.g. Ni; Ag; Au) cause no appreciable effect, merely giving rise to

---

[1] Strictly we can only talk of the resistance at the minimum, but so long as the overall magnitude of the minimum is not too large we can reasonably continue to refer to this as the residual resistance without ambiguity.

temperature-independent scattering in the usual way. It has been shown that striking anomalies in the low temperature thermoelectric power arise concurrently with the resistance minimum and we shall discuss the two phenomena together in more detail in Sect. C.

B. Notes on experimental methods and techniques.

8. Measurement of resistance by bridge methods. An almost endless variety of bridge techniques, stemming essentially from the classical WHEATSTONE Bridge, is available for the accurate measurement of electrical resistance (see, e.g.:

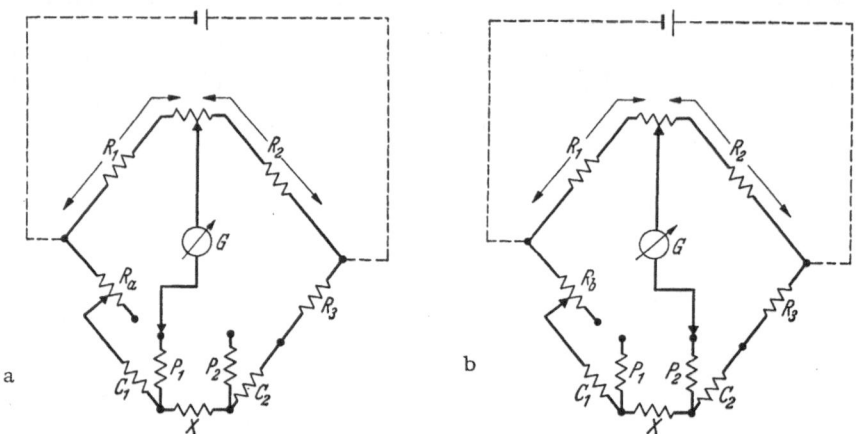

Fig. 11 a and b. (a) First balance of bridge. (b) Second balance of bridge.

v. STEINWEHR[1], MEISSNER[2], CHALMERS and QUARRELL[3]). Two basic advantages are that they employ a null method of observation, the bridge being "balanced" when zero current flows through the indicating galvanometer, and that they are independent of the precise value of the electromotive force used to put them in operation. On the other hand they suffer from the major disadvantage that lead and contact resistances are included in the measurement. Up to a point lead resistance can be compensated for by duplication in another arm of the bridge, or by taking *two* readings of the bridge before and after a suitable lead reversal. A simple example of this is illustrated in Figs. 11a and b. The ratio arms, R_1 and R_2, are assumed adjusted to equality. The resistance to be measured is X shown together with its current leads C_1, C_2 to connect it to the adjacent arms of the bridge and potential leads (P_1 and P_2) attached at appropriate points of the resistance to be measured; the resistances (P_1 and P_2) of these leads will include any contact resistance arising when they are attached to X. Two balances are taken as shown in Fig. 11 a and b, requiring settings of R_a and R_b for balance. Thus

$$R_a + C_1 = X + C_2 + R_3, \tag{a}$$
$$R_b + C_1 + X = C_2 + R_3. \tag{b}$$

Hence

$$X = \frac{R_a - R_b}{2}.$$

---
[1] H. v. STEINWEHR: Handbuch der Physik, Bd. 16, Chap. 17. 1927.
[2] W. MEISSNER: Handbuch der experimentellen Physik, Bd. 11/2, p. 25. 1935.
[3] B. CHALMER sand A. G. QUARRELL: The Physical Examination of Metals, vol. 2, chap. 3. London: Arnold 1941.

We have to assume here that C_1 and C_2 remain unchanged between the two measurements and of course that equality of the ratio arms, R_1 and R_2, is maintained. In many types of measurement this restriction would not be a serious one since we can normally make the resistances C_1 and C_2 small compared with X; consequently any small *change* in C_1 and C_2 which may occur can be made of negligible importance. For many applications also the additional time involved in making *two* blances would cause us no concern.

When, however, we turn to the measurement of the resistance of pure metals at low temperatures special difficulties have to be overcome. The problem to be faced in the low temperature field is clearly brought out when we consider, for example, a measurement of the resistance of a pure metal such as sodium. Sodium has a resistivity at 20° C of about $4 \times 10^{-6}\ \Omega$ cm. Experiments can be made today on samples so pure that the resistivity at $\sim 4°$ K is only about 2×10^{-4} of this value, i.e. about $8 \times 10^{-10}\ \Omega$cm. If we refer to equation (2.2) of chapter A we find, assuming $v \sim 10^8$ cm./sec. and one free electron per atom, that this indicates a mean free path for the electrons in the metal approaching $\frac{1}{5}$ mm. If now we

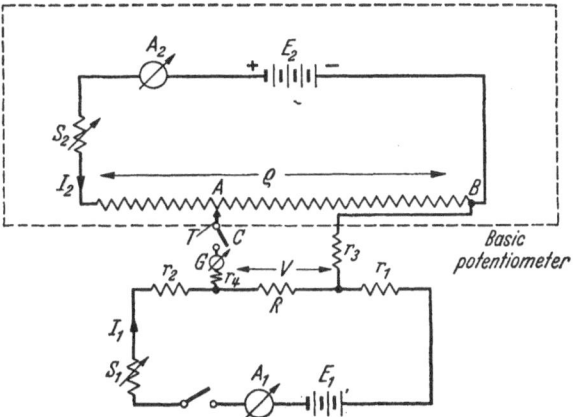

Fig. 12. Basic potentiometer circuit for measuring resistance.

wish our measurement to be essentially independent of surface effects (i.e. "size effects") we must make the diameter of our specimen somewhat larger than this value. Let us choose $\frac{1}{2}$ mm., and assume that there is space in our apparatus for a 10 cm. long specimen. The total resistance, R, we then have to measure with some degree of accuracy is

$$R = \frac{\varrho L}{\pi r^2} \approx 4.1 \times 10^{-6}\ \Omega, \quad \text{i.e.} \quad 4\mu\ \Omega,$$

and it will frequently be desirable that the overall measurement does not occupy more than, say, a few seconds. It is evident immediately that we must adopt methods where problems such as contact resistance (much less lead resistance!) cannot enter. Furthermore, we shall certainly find ourselves restricted in the current that we may employ to measure R because of heat generation in particular. In this case we might use 100 mA (10^{-1} amp.) as our measuring current in which case we shall have to measure 4×10^{-7} volts, with, we might hope, something like 1% accuracy.

9. Measurement of resistance with potentiometers. The basic instrument to meet our demands is, in principle, the potentiometer and an example of its use is shown in Fig. 12[1]. The battery, E_1, supplies to the specimen R a suitable current I_1, which may be adjusted to the required value by the rheostat S_1 and measured on the ammeter, A. The potential difference developed across

---

[1] An authoritative article with references but also, unfortunately, a complete lack of uniformity in diagram notation, is: W. P. WHITE: "Potentiometers for thermoelectric measurements" in "Temperature.—Its Measurement and Control in Science and Industry", Amer. Inst. of Physics. New York: Reinhold 1941.

$R (V = IR)$ will tend to drive a current around the circuit r_4, $\varrho_{(AB)}$, r_3, R when the switch-contact, C is closed, so deflecting the galvanometer, G. Now a fixed current I_2 is being supplied to the resistance-wire ϱ by the battery E_2 and controlled by the rheostat S_2. We can then adjust the resistance-tap, T, on ϱ until no current is observed to flow through G when the contact C is closed. The potential-difference across the part AB of the resistance-wire, ϱ, must then be identical to that being developed across the specimen R. Assuming that the potentiometer has been suitably calibrated, etc., the setting of the slider, T, will indicate the potential-difference in question. Under these idealized conditions of balance, no current is flowing through the resistances r_3 and r_4 which include the contact and lead resistance of these *potential leads* to the specimen. r_1 and r_2 represent the corresponding contact and lead resistances of the *current leads* to the specimen and it is evident that they do not enter into the measurement involved. The only influence of r_1 and r_2 is that, if and when they vary, the current I_1 will alter and S_1 must be adjusted to compensate for this and restore the current through R to its proper value.

In practice, of course, it is not possible either to maintain, or indeed observe, a perfect balance in the galvanometer circuit. What we have to ensure is that we can observe an acceptable approximate balance and that the current drawn by the galvanometer circuit has negligible influence upon the potential developed across R which we wish to measure. Thus let us assume that the slide wire, ϱ, in the potentiometer has a resistance of say $10\,\Omega$ and the overall lead resistances r_3 and r_4 are each $5\,\Omega$. We must then use a galvanometer of appropriate resistance with as high a voltage sensitivity as possible. Thus, for example, the Cambridge instrument with a $20\,\Omega$ coil and a current sensitivity ~ 300 mm./μA (at 1 metre scale distance) requires $100\,\Omega$ critical damping resistance (i.e. an additional $60\,\Omega$ in the circuit) for a response-time ~ 2 secs. If then we assume that we can detect readily when the galvanometer is 0.5 mm. or more off balance this will call for a net driving potential $\sim 2 \times 10^{-7}$ volts which, however, would represent about 50% error in the problem quoted earlier. If the galvanometer were inserted directly in circuit *without* additional damping resistance we should roughly halve this error but the response will now be considerably more sluggish (~ 8 to 10 secs.). A more expensive galvanometer (such as the Leeds and Northrup type H.S.) would improve the state of affairs very considerably giving a sensitivity of about 3×10^{-8} volts/mm. in this circuit with about 5 sec. response-time, with a consequent accuracy of about 5%. It is evident, that in this type of problem the current drawn by the galvanometer ($\sim 10^{-9}$ amps) can have no direct influence on the potential difference developed across the specimen.

The situation can then be further improved by increasing the length of the light beam. It is not at all difficult to provide a 3 metre "throw" for these galvanometers in place of the more conventional 1 metre scale distance. Alternatively one might turn to an ingenious and simple optical deflection multiplier devised recently by Dauphinee[1] (see Figs. 13a and b). Instead of the light beam being simply reflected *once* from the galvanometer mirror to the scale where it is observed, an additional fixed mirror is placed close in front and almost parallel to the galvanometer mirror. The light beam then suffers a number of successive reflections at the galvanometer mirror before finally departing for the scale and the net angular deflecion is consequently increased. Dauphinee has obtained satisfactory results on a suitable galvanometer with an angular amplification of as high as six. The order of the image which can be usefully used depends

---

[1] T. M. Dauphinee: Rev. Sci. Instrum. **26**, 873 (1955).

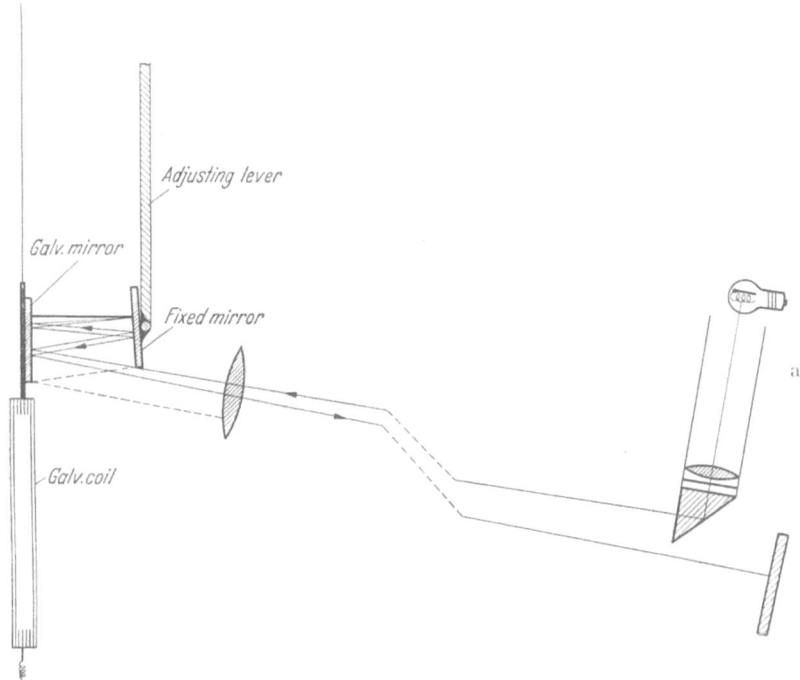

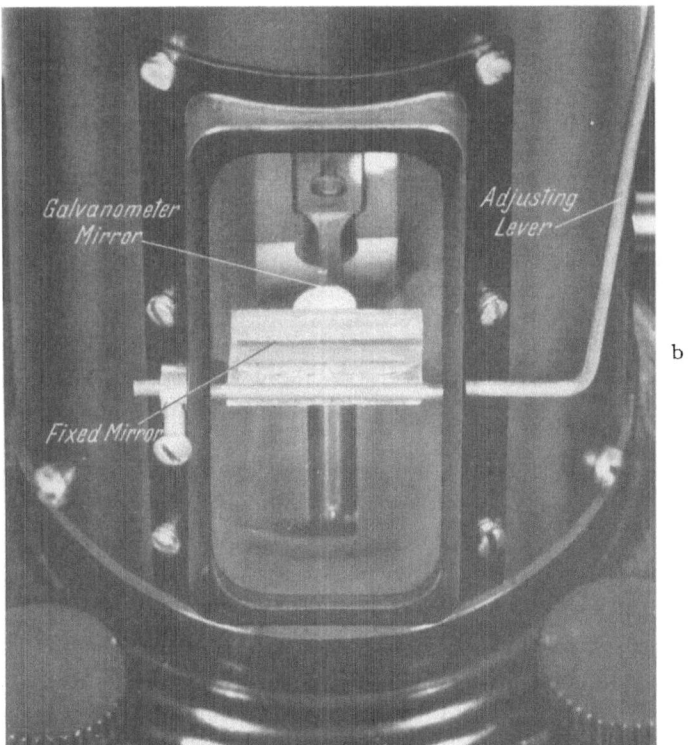

Fig. 13 a and b. Schematic diagram of optical multiplier by DAUPHINEE.

naturally on the size of the mirror in the particular galvanometer under consideration, and with a small mirror one might be limited to an angular gain of 3 or 4.

A defect of the type of potentiometer illustrated in Fig. 12 is the presence of
variable contacts in the galvanometer circuit. When we recall that we are in-
tending to measure an overall potential difference of a few microvolts it is evident
that we must avoid the introduction of any unnecessary sources of thermoelectric
force particularly when they are likely to be variable. White's amusing remarks
are relevant here "... Many of you must have personal recollection of that ma-
jestic battleship among potentiometers, the Fuessner-Wolff of 30 or more
years ago ... Here the resistance of the three switches—six contacts in all—and

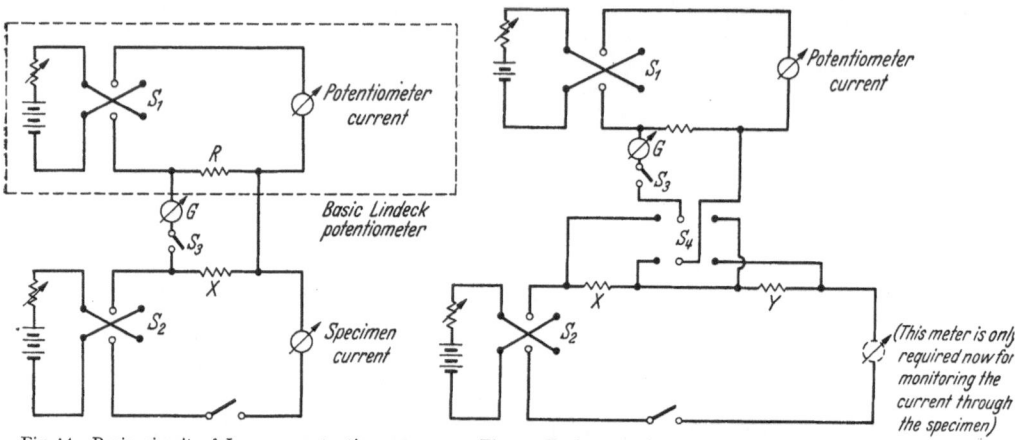

Fig. 14. Basic circuit of Lindeck potentiometer.   Fig. 15. Basic potentiometer circuit for direct comparison
of resistances.

worse than that, their variations, were always in ... the galvanometer circuit ...
Many of you may remember the morning ritual of cleaning and oiling switches
which was needed to get proper results ... the success of potentiometers specially
designed for thermoelectric and other low-voltage work was due not to unpre-
cedented refinement in the construction of switches and other parts, but to
devices or arrangements which put the switches where they could not do harm."
A typical feature of sensitive instruments is the splitting of the potentiometer
bank itself into two parts, a battery circuit and the galvanometer circuit; careful
placing of switches and the variation of resistance elements by suitably chosen
shunt arms rather than by simple series switching enabled Diesselhorst, Wenner
and others to develop instruments capable of reading accurately in the 10^{-7}
to 10^{-8} volt range.

Another approach to this problem is the use of the Lindeck potentiometer
circuit. In this case, rather than maintaining a constant current through the
potentiometer resistance banks and varying their effective resistance, the current
in the potentiometer circuit is *varied* through a fixed resistance thus eliminating
the troublesome switches in the potential measuring circuit.

We must not forget that some unavoidable thermal e.m.fs. will be generated
in the potential leads from the resistance we are measuring and to correct for this
some form of reversing switch may be introduced (we shall mention later a superconducting switch suitable for use in liquid helium). At the same time it will
then be necessary in our present circuit to reverse the current through the potentiometer circuit and then rebalance the system. Fig. 14 shows the essential

circuits using now a LINDECK type of potentiometer. The measurement of resistance may also be reduced to a direct comparison of the potential produced by the specimen resistance with that from a standard resistor in series and Fig. 15 shows a circuit for this purpose, although many variants are possible. The use of high-speed double-pole double-throw "choppers"[1] in suitable cases can enable such potential comparisons to be made rapidly and, in effect, continuously.

If we are to avoid adequately the intrusion of thermal e.m.f. we must also remember, particularly in sensitive measurements, that these will be varying, often quite rapidly. Rapid commutation will therefore be essential, coupled with quick response of the galvanometer. If also we wish to avoid any switch contacts in the galvanometer circuit (cf. S_3 in Figs. 14 and 15), the reversing-switches S_1 and S_2 must be coupled, operate rapidly and almost simultaneously. To achieve this is a matter of some difficulty since only a slight degree of asynchronism will produce a rather large "out-of-balance" surge in the galvanometer circuit. KAPITZA and MILNER[2] used small fast relay-operated switches for this purpose and CHAMBERS has been able to reduce the galvanometer "kick" on reversal to about 1% of that due to either e.m.f. acting alone.

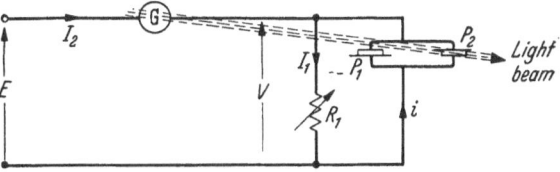

Fig. 16. Basic galvanometer amplifier circuit with strong series negative feed-back.

10. The galvanometer amplifier with negative feed-back. Photo-electric amplification was originally used on the light beam of a galvanometer simply as a means of increasing its sensitivity (e.g., cf. MILNER[3]). Although a relatively simple optical system coupled with selenium cells can provide a large intrinsic amplification only a relatively small fraction of this gain can be used directly, and systematic or random drift is often rather troublesome. The inclusion of strong "negative feed-back", however, in a galvanometer amplifier enables one then to put the "unused" amplification to work in a number of useful ways. By "feeding-back" a large fraction of the output of the amplifying photocell *in opposition* to the input voltage one achieves enhanced stability and linearity, a high input resistance and rapid response of the galvanometer amplifier, while sufficient *net* amplification is retained.

Consider the system of Fig. 16. The voltage E, which we wish to measure, tends to drive a current, say I_2, through the galvanometer G. The galvanometer mirror reflects a strong beam of light on to the twin photo-cells, P_1 and P_2, which are normally illuminated equally. When the galvanometer mirror deflects, one photo-cell (say P_1) receives more light and P_2 receives less, so that a current I_1 is driven through the resistance R_1. Because of the high amplification, $I_1/I_2 \gg 1$ (a ratio of around 2×10^3 can be quite readily achieved with simple selenium cells)[4]. The voltage V then opposes the original voltage E tending to nullify the primary current I_2 through the galvanometer. Since $I_2/I_1 \ll 1$ we may assume to

---

[1] See T. M. DAUPHINEE: Canad. J. Phys. **31**, 577 (1953); cf. also T. M. DAUPHINEE and E. MOOSFR: Rev. Sci. Instrum. **26**, 660 (1955).

[2] P. KAPITZA and C. J. MILNER: J. Sci. Instrum. **14**, 165 (1937).

[3] C. J. MILNER: Proc. Roy. Soc. Lond., Ser. A **160**, 207 (1937). — (P. KAPITZA and C. J. MILNER: J. Sci. Instrum. **14**, 165 (1937) is also of interest in this field.)

[4] Cf. also J. S. PRESTON: J. Sci. Instrum. **23**, 173 (1946), who designed a simple galvanometer amplifier of this type using *parallel* "negative feed-back" to give *low* input resistance, while also envisaging the use of *series* "negative feed-back" in appropriate circumstances.

an excellent first approximation:

$$I_2 \approx 0.$$

Hence

$$E = V = I_1 R_1.$$

It also follows therefore that i, the current in the photo-cell arm, is equal to I_1 and hence finally:

$$E = i R_1. \tag{10.1}$$

We may then insert in the photo-cell arm any suitable secondary current-indicating instrument, usually another galvanometer, to measure i directly. With reasonable attention to design features (contacts, minimization of vibration, etc.)

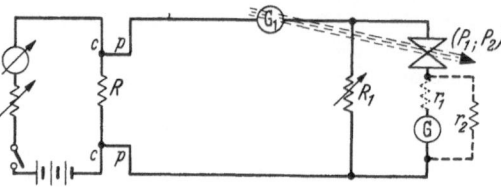

Fig. 17. Circuit of galvanometer amplifier used for resistance measurement.

it is possible, using two inexpensive Cambridge galvanometers of internal resistance $\sim 500\,\Omega$ to work with $R = 20\,\Omega$, or even less in favourable circumstances. This value then produces a net gain in voltage sensitivity over a *single* such galvanometer of about 25; that is to say, the unaided galvanometer has a sensitivity of around 2 mm./μV at 1 metre scale distance and the instrument now shows about 5 cm./μV. We need not concern ourselves appreciably with the linearity or otherwise of the primary galvanometer or of the photo-cell response since the effect of the strong "negative feed-back" is to make the system highly independent of the specific galvanometer and cell parameters. This is clear from the performance equation (10.1). The effective input resistance of the system is very high:

$$R_{\text{in}} \equiv \frac{E}{I_2} = \frac{I_1}{I_2} \cdot \frac{E}{I_1} = M \cdot R \tag{10.2}$$

where $M \sim 10^3$, so that even with R as low as $10\,\Omega$ we should have $R_{\text{in}} \approx 10^4\,\Omega$ and the system may therefore be used with confidence to measure directly the true potential developed across any normal metallic specimen. Indeed we may conveniently regard the instrument as a very rapid self-balancing LINDECK potentiometer, requiring no switch or key in the galvanometer circuit.

The sensitivity of the instrument may be immediately altered over a wide range[1] simply by variation of R.(typically we use a resistance decade box covering 10 to 10000 Ω), and the response-time of the amplifier will only tend to become at all appreciable at the lowest values. With $R \gtrsim 100\,\Omega$ in the typical case, the response-time of the amplifier itself will be very much shorter than that of the unaided primary galvanometer since it is being strongly "driven" by the negative feedback.

In most cases the secondary instrument is again a galvanometer and scale in which case it will require calibration and an accuracy of between $\frac{1}{2}$ and $\frac{1}{10}\%$ should be expected depending on the scale deflection used. It is of course probable that other methods of presenting or recording the final output of the amplifier

---

[1] This is a particularly useful feature in low-temperature investigations. It is quite customary to normalize the resistance to that observed at, say, 0° C; for when we are measuring reasonably pure metals the resistance at this temperature is almost entirely due to thermal scattering and consequently the precise degree of impurity or physical strain present is of little consequence. However, in liquid helium the resistance may drop by as much as 10^4; furthermore we may well be interested in samples of various sizes and thus may sometimes wish to measure resistances as high as about 1 Ω.

would be very useful. It is an advantage of a direct deflection instrument in this type of work that any thermal e.m.fs. in the system are constantly displayed and, even if these are varying quite significantly, the relatively rapid deflection of the instrument enables one to make a reasonably accurate visual correction.

Some further information about design details, etc., of galvanometer amplifiers of this and similar types will be found in the literature[1,2,3,4]. Fig. 17 shows schematically a typical layout for resistance measurement and Figs. 18a and 18b show the apparatus in use and its optical system.

Fig. 18a. Galvanometer amplifier in use with low temperature apparatus.

A practical feature of operation should perhaps be mentioned. If the leads to the photo-cells be incorrectly attached so that their output tends to increase, rather than minimize, the current through the primary galvanometer, the system will be highly unstable,—and very obviously so! This provides in fact a rapid test for the correct sense of connection. However, it is possible occasionally for *oscillatory* instability to arise with the *correct* mode of connection. A crude

[1] J. S. PRESTON: J. Sci. Instrum. **23**, 173 (1946).
[2] D. K. C. MACDONALD: J. Sci. Instrum. **24**, 232 (1947).
[3] B. FRANKENHAEUSER and D. K. C. MACDONALD: J. Sci. Instrum. **26**, 145 (1949).
[4] See also A. V. HILL: J. Sci. Instrum. **25**, 225 (1948). — A. C. DOWNING: Journ. Sci. Instrum. **25**, 230 (1948).

analysis suggests that this may arise if the state of the selenium cells is such that they have a combination of high amplification and some time-delay in response (perhaps of the order of milliseconds). The condition may be cured by shunting a suitable resistance (or even condenser) across the terminals of the primary galvanometer or by changing the photo-cells.

If a very small potential difference or e.m.f. which we wish to measure is generated in a specimen at low temperatures then an obvious source of trouble is the large temperature difference along the connecting leads from the specimen to the measuring apparatus at room temperature. Any inhomogeneity of the leads

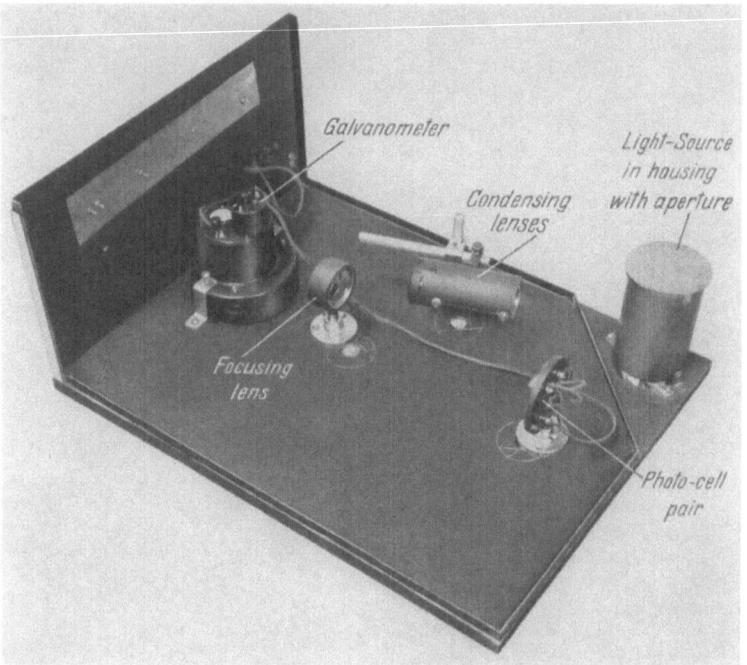

Fig. 18b. Optical system of galvanometer amplifier.

in the temperature gradient will give rise to a "spurious" thermal e.m.f. in these leads; even with carefully matched wires mechanical strain or "kinking" may suffice to produce an undesirably large voltage. There are two ways in which this difficulty can be overcome. We can measure the potential directly by an instrument operating at the low temperature (e.g. in liquid helium) so that no leads have to be brought out to room temperature, or we may use an amplifier or dicriminator operating in liquid helium so that any subsequent "spurious" e.m.f. introduced will not influence the final measurement made at room temperature.

11. The superconducting galvanometer. The first solution was adopted by Pippard and Pullan[1], who designed a superconducting galvanometer for use in liquid helium and used it to measure thermoelectric power in metals below 4° K, particularly in the neighbourhood of the transition point of superconductors where the thermoelectric power is going rapidly to zero. The requirement there

---

[1] A. B. Pippard and G. T. Pullan: Proc. Cambridge Phil. Soc. **48**, 188 (1952). See also H. Grayson-Smith et al.: Trans. Roy. Soc. Canada **29** (III), 23 (1935); **30** (III), 13 (1936).

of a voltage sensitivity $\sim 10^{-12}$ volts was achieved by having a circuit resistance $\sim 10^{-7}\,\Omega$ with a tangent galvanometer of current sensitivity $\sim 10^{-5}$ amp. With such a low circuit resistance (R) it is necessary to keep the effective inductance (L_{eff}) as low as possible if a very long time-constant (L_{eff}/R sec.) is to be avoided. This, in turn, requires that the static magnetic field shall be very small ($\sim 10^{-2}$ gauss) if the desired current sensitivity is to be attained.

The mechanical design calls for considerable care in construction and provision of a vibrationless support, but with the instrument PULLAN[1] obtained very interesting results. He showed in particular that the thermoelectric power of an ideal superconductor vanishes very abruptly at the transition point in contrast to earlier work which had indicated that the thermo-electric power started to fall off some way above the actual superconducting transition temperature.

12. The superconducting reversing switch. The simplest form of discriminator, as the second solution to the problem, is a reversing switch close to the specimen in the "cold bath". The potential to be measured will then be reversed in polarity when the switch is operated while any unwanted e.m.f. generated between the switch and measuring instrument will remain unaffected; the difference of two readings in reversed positions should therefore give the true e.m.f. In our laboratory a number of different designs were tried for a remotely-controlled reversing-switch with mechanical contacts to operate in a liquid helium bath. The major difficulty, however, was that the mechanical contacts always proved unsatisfactory and unreliable in operation. The solution was found in a switch designed by TEMPLETON[2] which used superconductors; switching was provided by stroying the superconductivity at will with a magnetic field generated locally.

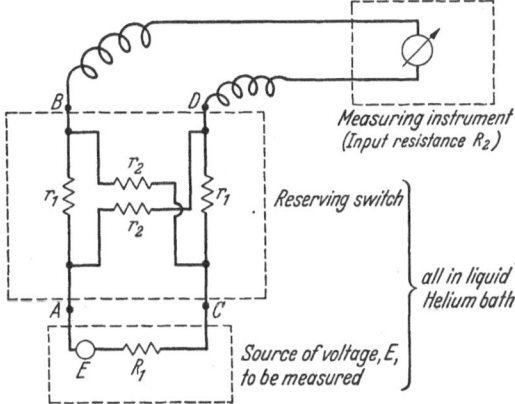

Fig. 19. Circuit of superconducting reversing-switch for low temperature measurements.

Consider the circuit of Fig. 19. The resistance R_1 of the source of e.m.f., E, may be $10^{-3}\,\Omega$ or less; the resistances r_1 and r_2 in the switch are small coils of tantalum wire whose normal (i.e. *non*-superconducting) resistance is about $1\,\Omega$ and a magnetic field of about 60 gauss at $4.2°$ K (generated by about 2.5 mW in a small electromagnet whose core forms the framework of the switch) is required to render them so. The input resistance R_2 of the measuring instrument is presumed to be large compared with $1\,\Omega$. When the magnetic field is applied to the arms r_2, so that the arms r_1 remain superconducting, the polarity of the switch is normal, the source being connected to the measuring instrument effectively via the routes AB; CD. If the field is now applied to the arms r_1, and cut off from r_2, the role of these switch arms is transposed and connection is via AD; CB. With the typical figures given the reversing switch is found to operate very well.

The switch has been used successfully by PEARSON and TEMPLETON[3] for the measurement of small thermoelectric potentials at low temperatures and

---

[1] G. T. PULLAN: Proc. Roy. Soc. Lond., Ser. A **217**, 280 (1953).

[2] I. M. TEMPLETON: J. Sci. Instrum. **32**, 172 (1955).

[3] W. B. PEARSON and I. M. TEMPLETON: Proc. Roy. Soc. Lond., Ser. A **231**, 534 (1955).

particularly in connection with redetermination of the absolute thermoelectric scale. A photograph of the first model made of the switch appears in Fig. 20.

13. The superconducting modulator. Templeton[1] also developed a "superconducting modulator" for use in liquid helium. A section of fine superconducting (tantalum) wire is

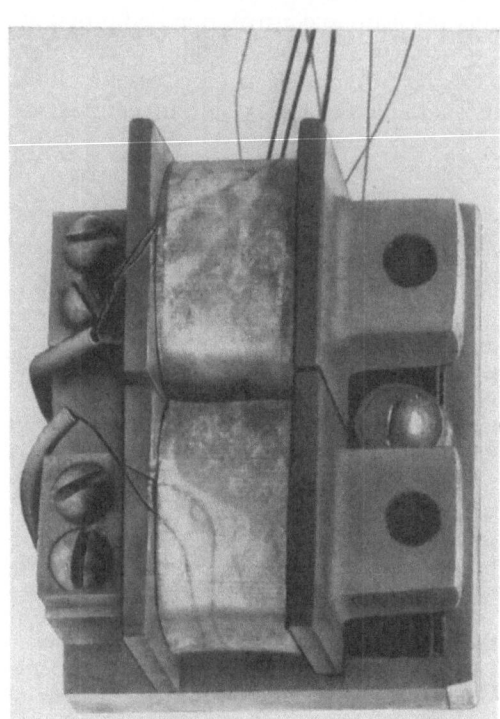

Fig. 20. Superconducting Reversing Switch by Templeton to operate in liquid helium (Linear enlargement about 2×).

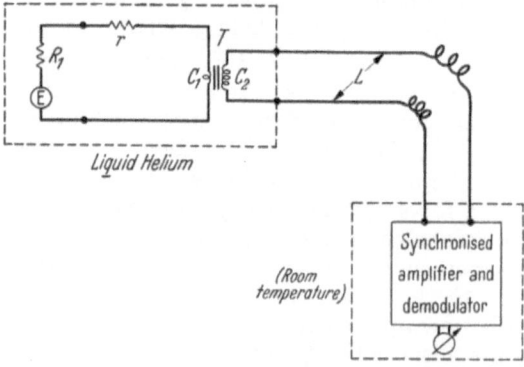

Fig. 21. Schematic diagram of superconducting modulator by Templeton.

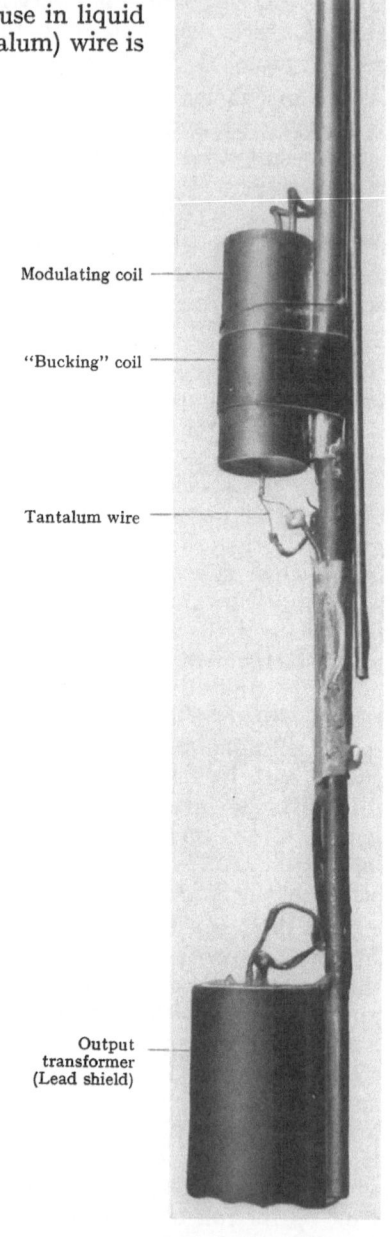

Modulating coil

"Bucking" coil

Tantalum wire

Output transformer (Lead shield)

Fig. 22. Superconducting modulator by Templeton, to operate in liquid helium. (Somewhat less than full size.)

connected in series with the very low resistance source of e.m.f. to be measured and the primary of a miniature audio-frequency transformer with a mu-metal core.

[1] I. M. Templeton: J. Sci. Instrum. **32**, 314 (1955).

The resistance of the tantalum wire in the normal state is about $10^{-1}\,\Omega$, tantalum being chosen because of its small transition field ($\sim$60 gauss) at 4.2° K. The remainder of the primary circuit including the primary of the transformer is made of lead wire.

In the sketch (Fig. 21), the tantalum modulating resistance, r, is normally held very close to the transition point by a steady magnetic field and is then "swung" in and out of superconductivity by a small oscillating magnetic field at 800 c/s. The impedance of the primary coil, C_1, of the transformer, T, is about $4 \times 10^{-3}\,\Omega$ and the step-up ratio is about 100:1; the secondary coil, C_2, of copper has an output impedance of about $10\,\Omega$. As the resistance r is modulated, the voltage E appears across C_1 during each half-cycle when r is superconducting, and the stepped-up output from C_2 at 800 c./s. is then led out to a more or less conventional synchronized amplifier. The essentially d.c. thermal e.m.fs. developed in the output leads, L, are thus quite impotent to affect the measurement, and little trouble is experienced from vibration.

The instrument (see Fig. 22) has been used for measurements of the resistance of single crystals (i.e. "as grown" without any reduction of cross-section in acid or by other means) of copper and dilute alloys whose resistance is extremely low at helium temperatures. A limiting noise figure of about 2×10^{-11} volt has been obtained with this instrument.

14. Preparation of alkali metal specimens. It is frequently assumed in theoretical discussions that the monovalent alkali (group IA) metals (Li, Na, K, Rb, Cs) should conform most closely in practice to the idealized metallic model of almost-free conduction electrons interacting only weakly with the ionic lattice. The group IB (or "noble") metals (Cu, Ag, Au), also assumed to be monovalent in the solid state, are often appealed to as well, but are usually regarded as somewhat less ideal structures for comparison with theory. We shall therefore make a few remarks about the preparation of alkali metal specimens in particular since for the experimenter it is somewhat unfortunate (although presumably no coincidence) that this group of metals is highly reactive.

Lithium is exceptional amongst the alkalis in reacting, when liquid, with ordinary glasses and quartz. This means that unlike the other members of the family it cannot be melted and run into an ordinary glass mould in the usual way, an otherwise excellent method of preparing samples of this metallic group for experiment. One can use with some success for lithium specimens a special borax-coated glass[1]; however, since lithium oxidizes fairly slowly (although the oxide when once formed is rather objectionable) the best procedure is probably to extrude the metal cold through a steel die. Lithium is the "toughest" of the alkali metals but still soft enough that it can be cut with a sharp knife and extruded quite readily. On extrusion it should covered with "Vaseline", "Nujol", medicinal liquid paraffin (e.g. "Stanolax") or some similar inert substance to prevent oxidation. No alkali metal, of course, should be allowed to associate with water, but lithium and sodium can be readily handled in the open air with dry equipment without fear of them catching fire—caesium, at the other extreme, will catch fire practically instantaneously if exposed and further, having a low melting point, ($\sim$27° C) will then melt immediately. Considerable care must therefore be taken in handling caesium and rubidium. For some work the extruded and coated "wire" of lithium (or even a suitable piece cut out directly from the

---

[1] D. K. C. MacDonald and J. E. Stanworth: Proc. Phys. Soc. Lond. B **63**, 455 (1950). [It also appears possible to run the metal into an ordinary glass mould provided that:
   (i) A thin film of medicinal liquid paraffin (e.g. "Stanolax") is present.
   (ii) The lithium is chilled sufficiently rapidly after being run into the mould.]

block of metal) can then be conveniently mounted in a simple ebonite or other plastic frame for support and electrical leads screwed in or attached by small collars. In some cases it may be necessary to suspend the wire more or less freely or perhaps position it lightly by passing it through an over-sized hollow glass tube; if this can be sealed at both ends and kept filled with liquid paraffin the specimen will remain in good condition for lengthy periods.

Alternatively, for electrical or thermal conductivity work, the lithium can be melted in an inert atmosphere (e.g. helium gas) in a stainless steel crucible and then forced into a suitable steel mould for the conductivity of which an appropriate correction will of course then have to be made. It should be noted that any oxide which may form on the surface of the molten metal is quite "tough" and that the surface tension is rather high.

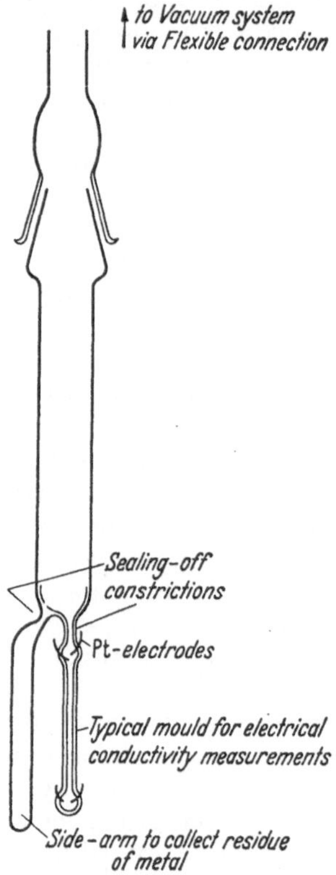

For the remaining metals a convenient procedure for casting into suitable glass moulds (electrodes of platinum, which does not react with the alkalis, are readily fused into soft glass) is to evacuate the mould (e.g. see Fig. 23) and fill with helium gas to 1 atmosphere pressure. The alkali metal to be cast will be in a sealed glass capsule of some sort. This can then be safely broken when *adequately* immersed under benzene which has previously been dried by sodium wire or chips in it. The metal, having a thin protective layer of benzene on it, is then transferred rapidly to the mould and the apparatus immediately evacuated. The mould is then placed in a bath of liquid paraffin which is heated to perhaps 10 to 20° C above the melting temperature of the metal concerned and after some minutes the metal melts; helium admitted up to 1 atmosphere will then force the metal into the mould which is thereafter allowed to cool *slowly*. The sample can later be sealed off with a small gas (or gas and air) flame, or it may be cracked off (under benzene if necessary), sealed with plasticine, "*Q*"-grease or something similar and covered finally with "Durofix" ("Glyptal") or some similar cement. It is unwise to try to melt an alkali metal in the apparatus by applying a flame directly to the glass as the temperature is very difficult to control at all precisely, and all molten alkali metals will react to some extent with glass if excessively heated; furthermore the inert paraffin bath is an excellent safety precaution in case of any breakage.

Fig. 23. Typical glass mould for casting alkali metals.

Instead of running the molten metal directly into a mould as described, it may be *distilled* into the mould after melting and this is generally best done with a controlled electric heater. Also, instead of starting from a supply of the solid metal, it may be produced directly in the apparatus by reaction, e.g., of the chloride with a reagent such as pure magnesium filings. This procedure may be of particular value for rubidium and caesium which are difficult to obtain commercially in a very pure metallic state. The relevant parts of the apparatus should in this case be made of quartz to withstand the heat necessary for the reaction to proceed.

15. The general preparation and handling of dilute alloys. It appears today in the low temperature field that considerable fundamental interest must attach to the scattering of electrons by impurity atoms, particularly in very dilute alloys (say 10^{-3} to 10^{-1} atom%). Some very brief notes on the preparation and handling of such alloys may therefore be of interest; for more detailed information on alloy preparation, etc., a metallurgical text such as that by CHRISTIAN, HUME-ROTHERY and PEARSON[1], or the papers in the literature must be referred to.

The three main factors involved in dilute alloy preparation are:

1. Pure starting metals.
2. Melting and heat treatment without contamination.
3. Achievement of the desired equilibrium (or metastable) state of the alloy for experiment.

α) Pure starting metals. Pure starting metals may be obtained by one or more of the following methods:

(i) chemical preparation,
(ii) electrolytic deposition,
(iii) distillation,
(iv) zone purification.

The last process can yield very high purity metals in those cases where it may be used. Its success, it will be appreciated, depends on the variation with temperature and composition of the solidus and liquidus boundaries for the binary (or more complex) systems involving these impurities in the metal.

β) Melting and heat treatment. This involves generally the use of refractories for crucibles in which alloys can be prepared, preferably by melting, and also an inert atmosphere or vacuum in which the melting is conducted. The crucible material must not react with the molten metals, as otherwise they will pick up oxygen and also the metallic component of the refractory crucible. The presence of inert gases may prevent oxidation and may not be objectionable—or, on the other hand, it may lead to unsound ingots, due to bubbles formed by gas rejected as the metal solidifies. The use of high vacuum conditions *can* be objectionable in removing solvent or solute by vaporization. Preferential removal of either will yield an alloy whose true composition may differ very appreciably from the nominal value as determined by starting percentages.

The choice of reducing, oxidizing or "neutral" conditions for the melting process may be of importance. For example, if the presence of oxygen is not otherwise objectionable a certain amount of oxygen may be of value in keeping an undesirable solute out of solid solution (i.e. as an insoluble oxide), so that it does not contribute to the impurity scattering. On the other hand, alloys made in graphite crucibles or in a hydrogen atmosphere should generally be regarded as prepared under reducing conditions.

γ) Annealing of alloys to required equilibrium or metastable state. The problem of annealing is generally less critical than that of melting as the temperatures involved are lower. Solutes of low solubility may be held in metastable solid solution by annealing at a sufficiently high temperature, with subsequent quenching. The quenching rate must be sufficiently rapid to retain the desired homogeneous metastable state without partial decomposition, and the final temperature must be low enough that the alloy will not "age" or decompose appreciably before use. In some cases, it might be necessary to store the quenched alloy in, say,

---

[1] J. CHRISTIAN, W. HUME-ROTHERY and W. B. PEARSON: Metallurgical Equilibrium Diagrams. London: Institute of Physics 1952.

liquid nitrogen. If contacts have subsequently to be made to a quenched alloy the influence of soldering heat on the stability of the alloy must be considered. This is of particular importance in thermoelectric work where a local inhomogeneity may play a very significant role.

δ) *Physical state of the alloys.* In some fields of experiment it may be desirable to have a specifically monocrystalline specimen. The methods of growth are now well-known:

(i) Crystallization in a predetermined temperature gradient by moving the furnace at a fixed rate relative to the crucible or *vice versa.*

(ii) Withdrawing the crystal (started often from a "seed" crystal) from the melt at a specific rate.

(iii) Strain-annealing.

In shaping single crystals one has of course to be careful not to strain the crystal and in many instances it would appear that chemical or physical electrolytic action to dissolve the crystal to the required size rather than cutting to shape is the only safe approach. Even in this case there is the possibility that the dissolution process may remove one component preferentially from the reacting surface of an alloy. If the metal were subsequently annealed so that a surface excess was then redistributed throughout its volume, this effect might well be important in a dilute alloy.

I am particularly grateful to Drs. T. M. Dauphinee, R. G. Chambers and W. B. Pearson for helpful discussions and comments on this chapter.

C. Comparison of experimental data with theory.

16. The lattice model. If we are to interpret our experiments usefully, we must at some stage decide which theory we intend to adopt as our standard of comparison. The development of the theory of electrical resistance and its comparison with data parallels in many respects the similar problem in specific heats — i.e. the internal energy of solids. The chief differences are that the experimental measurement of electrical resistance is generally considerably easier and more rapid than that of specific heats, while the theory on the other hand is very much more complex[1].

After the foundations of quantum physics were laid at the start of this century Einstein[2], Born and v. Kármán[3], and Debye[4] applied the concepts to the interpretation of specific heat measurements on solids. Einstein[5] himself realized that his bold assumption of a single frequency of atomic vibration in a solid could hardly be expected to be a very precise physical model. However, the broad understanding of the behaviour of the specific heat which immediately followed his work justified entirely the use initially of this relatively crude model. A vital outcome was the recognition of a temperature, Θ, characteristic of each substance. Above this temperature thermal energy "blurs" the *individual* features of any particular lattice structure (and as a consequence we then have the very general classical result $E = 3NkT$) while at lower temperatures the specific heat (together with many other observable parameters) will depend quite critically on the characteristics of the specific solid lattice. Thus the "anomalous" specific

---

[1] The theory of the specific heat of solids is given in some detail in vol. VII of this Encyclopedia by M. Blackman.

[2] A. Einstein: Ann. Phys. Lpz. **22**, 180, 800 (1907).

[3] M. Born and Th. v. Kármán: Phys. Z. **13**, 297 (1912); **14**, 15 (1913).

[4] P. Debye: Ann. Phys., Lpz. **39**, 789 (1912).

[5] A. Einstein: Ann. Phys. **34**, 170 (1911).

heat of diamond which had been found experimentally to be much below the classical value was immediately explained as the result of a high characteristic frequency, v, (as evinced by the extreme hardness of diamond). The resulting Θ ($k\Theta = hv$) lies much above room temperature and consequently the specific heat already lies well below the classical DULONG-PETIT value and we are in the "low-temperature region" for diamond.

DEBYE then adopted an ingenious model which recognized that in a solid there must be a complete gamut of characteristic vibrational modes whose wavelengths would range from the macroscopic dimensions of the crystal down to that of the interatomic distance. The model (variously known as a "jelly" or "quasi-continuum") remained sufficiently general to retain the valuable concept of a single characteristic temperature, Θ_D, for each particular solid. DEBYE's model on the whole accounted very well for the experimental results and in particular for the rate of decay of specific heat at low temperatures where the EINSTEIN model predicted much too rapid a drop[1].

Almost simultaneously with DEBYE, BORN and VON KÁRMÁN considered the precise dynamical analysis of the proper vibrations of an atomic lattice with the interatomic force constants as starting-point. Because of the attractive simplicity and generality of DEBYE's model, it is only of recent years that the BORN-v. KÁRMÁN analysis was pursued and developed, particularly by BLACKMAN[2], with the interest it deserved. There is no doubt today that if we have a sufficient knowledge of the range and character of interatomic forces in any particular solid then the frequency spectrum, specific heat and other allied properties of that solid can be derived with excellent accuracy, although generally with considerable labour, using dynamical lattice theory as developed by BORN and his school. It is furthermore true that certain experimental results which were regarded as "anomalous" when compared with the general DEBYE formulae have been found to be direct consequences of the specific features of the lattice under consideration (although of course an actual "hump" in the variation with temperature of the specific heat cannot be accounted for by lattice theory alone). On the other hand, in metals our knowledge of the precise character of electron-ion and electron-electron interactions is still quite meagre so that a sophisticated analysis is often not warranted. In addition, as with the specific heat problem, it is certainly necessary to understand the major features of the experimental data before carrying refinements or particularization of the model too far.

The analysis of the variation with temperature of electrical resistance closely parallels that of the thermal energy or specific heat, but with added complexity. Not only must we have knowledge of the lattice vibrations but also of the mechanism of electron-ion interaction or "scattering", which consequently includes some specification of the behaviour of the electron assembly itself. The introduction by PLANCK of the concept of zero-point energy of vibration did not affect the understanding of specific heat behaviour in solids but it was not until much later (BLOCH, HOUSTON and SOMMERFELD) that it became clear that zero-point vibrations of the lattice would not in fact contribute to electrical resistance. It appears quite justified to say today that since the period 1927 to 1932 we have understood *broadly* the mechanism of electrical resistance due to lattice vibrations

---

[1] „... der Abfall der Atomwärme bei den untersuchten Elementen Pb, Ag, Zn, Cu, Al und ferner beim KCl bei tiefen Temperaturen *langsamer erfolgt*, als der Formel von EINSTEIN entspricht". — W. NERNST and F. A. LINDEMANN: Sitzgsber. preuß. Akad. Wiss. Berlin **22**, 494 (1911).

[2] M. BLACKMAN: e.g.: Rep. Progr. Phys. **8**, 11 (1941). — Proc. Roy. Soc. Lond., Ser. A **148**, 365, 384 (1935); **149**, 117, 128 (1935). — Proc. Phys. Soc. Lond. A **64**, 681 (1951).

(although the same cannot necessarily be said for certain domains of thermal conductivity and thermoelectricity) while there remain many aspects where numerical agreement and detailed understanding of certain processes leaves much to be desired. Hence we find that, although everyone is agreed (ignoring here the BORN-RAMAN controversy) on the fundamental calculation of the specific heat in simple solids there is no such univeral agreement on the theoretical formula for the variation of resistance in even a "simple" metal.

17. The BLOCH-GRÜNEISEN formula for resistance. The most widely employed expression for the resistivity is the so-called BLOCH-GRÜNEISEN formula which reads

$$\varrho = \frac{KT^5}{M\Theta^6} \cdot \int\limits_0^{\Theta/T} \frac{z^5\,dz}{(e^z - 1)\,(1 - e^{-z})} \qquad (17.1)$$

where M is the atomic weight and K is a constant characteristic of the metal (and its volume).

For T/Θ greater than about 0.5, this gives in close approximation

$$\varrho \approx \frac{K}{4} \cdot \frac{T}{M\Theta^2} \qquad (17.2)$$

as may be checked by expanding the integrand for small values of z.

For $T/\Theta < 0.1$, on the other hand, we have

$$\varrho \approx \frac{KT^5}{M\Theta^6} \int\limits_0^\infty \frac{z^5\,dz}{(e^z - 1)\,(1 - e^{-z})} = \frac{(124.4)\,KT^5}{M\Theta^6}. \qquad (17.3)$$

Thus the ratio of resistance at a low temperature $(T_1 < \sim\Theta/10)$ to that at a "high" temperature $(T_2 > \sim0.5\Theta)$ would be

$$\frac{\varrho_1}{\varrho_2} = \frac{497.6}{\Theta^4} \cdot \frac{T_1^5}{T_2}. \qquad (17.4)$$

Agreement with (17.1) would only be expected at best in a rather idealized monovalent metal, where the conduction electrons may be regarded as free so that their energy is given simply by $E = \frac{1}{2}mv^2$, and the lattice vibrations are sufficiently described by the DEBYE model (i.e. dispersion is ignored). The scattering of the conduction electrons by the vibrating lattice is also very much simplified. The theory is developed on the assumption that once the "static" electron-ion interaction is specified, the net scattering then depends only on the *displacement* of the ion. Consistent with this, the interaction is further supposed localized close to the centre of the ion, while over the remainder of the atomic volume a conduction electron is regarded as truly free. This behaviour corresponds essentially to almost perfect "screening" of the ionic charge by the other conduction electrons in the metal, to which we shall refer again later.

There are certain other approximations involved in the derivation of (17.1) both of a physical and mathematical character, but these are probably less important than the rather gross assumptions already mentioned.

WILSON([1], page 254) has said "The expression for ΔV [i.e. the change in potential energy when an ion is displaced] is clearly not entirely correct, since the ions must be deformed to a certain extent ... it is possible that the crudeness of the approximations made in dealing with the interactions between the conduction electrons and the lattice vibrations is the cause of the failure of the present theory to include superconductivity. While it is possible that some new physical principle is required to explain the phenomenon of superconductivity, it may well be that the difficulties are essentially mathematical rather than physical, and that, just as a searching analysis of the theory of the equation of state of a gas reveals the possibility of the existence of a liquid phase, a more exact mathematical treatment of the interaction problem would lead to an explanation of superconductivity ... a much improved and more general theory of the interaction between the electrons and the lattice is required."

If we accept (17.1) uncritically then we should expect that the "reduced resistance" r, for each metal: ϱ_T/ϱ_Θ: should be a universal function of the reduced

Table 3.

Θ/T	$F(\Theta/T)$	Θ/T	$F(\Theta/T)$	Θ/T	$F(\Theta/T)$	Θ/T	$F(\Theta/T)$
0	1.0000	4,5	0,3867	9,0	0.06740	14.0	0.01289
0.1	0.9994	4.6	0.3729	9.1	0.06490	14.2	0.012185
0.2	0.9978	4.7	0.3595	9.2	0.06250	14.4	0.011528
0.3	0.9950	4.8	0.3466	9.3	0.06021	14.6	0.010915
0.4	0.9912	4.9	0.3340	9.4	0.05800	14.8	0.010344
0.5	0.9862	5.0	0.3217	9.5	0.05589	15.0	0.0_29805
0.6	0.9803	5.1	0.3098	9,6	0.05386	15,2	0.0_29302
0.7	0.9733	5.2	0.2983	9.7	0.05192	15.4	0.0_28831
0.8	0.9653	5.3	0.2871	9.8	0.05005	15.6	0.0_28389
0.9	0.9563	5.4	0.2763	9.9	0.04826	15.8	0.0_27974
1.0	0.9465	5.5	0.2658	10.0	0.04655	16.0	0.0_27584
1.1	0.9357	5.6	0.2557	10.1	0.04490	16.2	0.0_27218
1.2	0.9241	5.7	0.2460	10.2	0.04332	16.4	0.0_26873
1.3	0.9118	5.8	0.2366	10.3	0.04181	16.6	0.0_26549
1.4	0.8986	5.9	0.2275	10.4	0.04035	16.8	0.0_26243
1.5	0.8848	6.0	0.2187	10.5	0.03896	17.0	0.0_25955
1.6	0.8704	6.1	0.2103	10.6	0.03762	17.2	0.0_25683
1.7	0.8554	6.2	0.2021	10.7	0.03633	17.4	0.0_25427
1.8	0.8398	6.3	0.1942_5	10.8	0.03509	17.6	0.0_25185
1.9	0.8238	6.4	0.1867	10.9	0.03390	17.8	0.0_24956
2.0	0.8073	6.5	0.1795	11.0	0.03276	18.0	0.0_24740
2.1	0.7905	6.6	0.1725	11.1	0.03167	19.0	0.0_23819
2.2	0.7733	6.7	0.1658	11.2	0.03061	20.0	0.0_23111
2.3	0.7559	6.8	0.1593	11.3	0.02960	22	0.0_22125
2.4	0.7383	6.9	0.1531	11.4	0.02863	24	0.0_21500
2.5	0.7205	7.0	0.1471_5	11.5	0.02769	26	0.0_21089
2.6	0.7026	7.1	0.1414	11.6	0.02680	28	0.0_38097
2.7	0.6846	7.2	0.1359	11.7	0.02593	30	0.0_36145
2.8	0.6666	7.3	0.1306	11.8	0.02510	32	0.0_34747
2.9	0.6486	7.4	0.1255_5	11.9	0.02430	34	0.0_33724
3.0	0.6307	7.5	0.1206_7	12.0	0.02353	36	0.0_32963
3.1	0.6128	7.6	0.11599	12.1	0.02279	38	0.0_32387
3.2	0.5950	7.7	0.11150	12.2	0.02208	40	0.0_31944
3.3	0.5775	7.8	0.10719	12.3	0.02139	44	0.0_31328
3.4	0.5600	7.9	0.10306	12.4	0.02073	48	0.0_49375
3.5	0.5428	8.0	0.09909	12.5	0.02009	50	0.0_47964
3.6	0.5259	8.1	0.09529	12.6	0.01948	52	0.0_46806
3.7	0.5091	8.2	0.09165	12.7	0.01889	56	0.0_45061
3.8	0.4927	8.3	0.08816	12.8	0.01832	60	0.0_43841
3.9	0.4766	8.4	0.08480	12.9	0.01777	64	0.0_42967
4.0	0.4608	8.5	0.08159	13.0	0.01725	68	0.0_42328
4.1	0.4453	8.6	0.07851	13.2	0.01624	70	0.0_42073
4.2	0.4301	8.7	0.07555	13.4	0.01531	72	0.0_41852
4.3	0.4153	8.8	0.07272	13.6	0.01445	76	0.0_41492
4.4	0.4008	8.9	0.07000	13.8	0.01364	80	0.0_41215

temperature, $t:T/\Theta$: in fact

$$r = t^5 \int_0^{1/t} \frac{z^5\,dz}{(e^z - 1)(1 - e^{-z})} \bigg/ \int_0^1 \frac{z^5\,dz}{(e^z - 1)(1 - e^{-z})}, \qquad (17.5)$$

assuming that we are not significantly concerned with thermal expansion.

GRÜNEISEN has computed this in the form: $r = 1.056\,(T/\Theta)\,F\,(\Theta/T)$, and the function $F\,(\Theta/T)$ is tabulated in Table 3. That this prediction is *broadly* true is seen from Fig. 24 a, b due to MEISSNER, and from such a graph we can readily estimate a

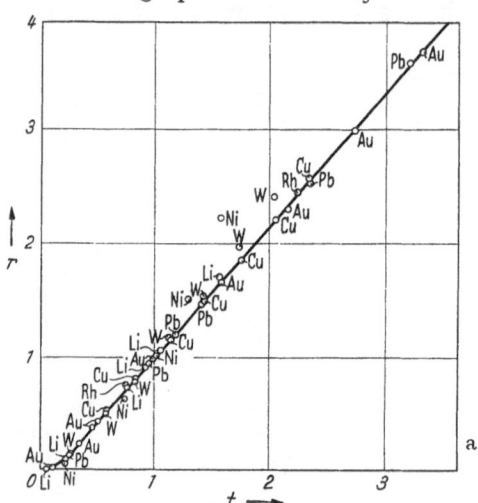
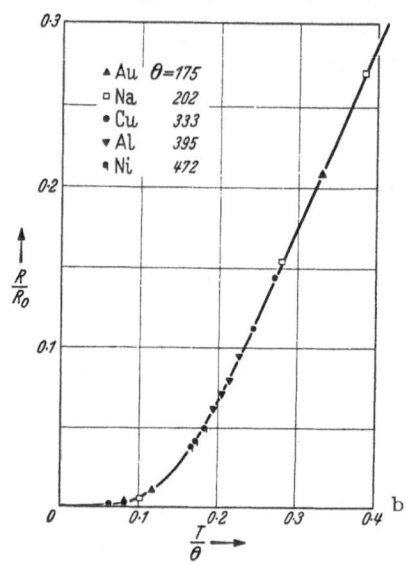

Fig. 24 a and b. (a) Reduced resistance, r, as a function of reduced temperature $t(T/\Theta)$. [After MEISSNER: Handbuch der Experimentalphysik, vol. 11, pt 2, 30, 1935.] (b) Reduced resistance of metals at low temperatures (after MEISSNER). The full curve is the BLOCH-GRÜNEISEN function.

characteristic temperature, Θ, for each metal. In this way GRÜNEISEN[1] deduced Θ values for a considerable number of metals, a selection of which appears in Table 4; these values agree remarkably well with those deduced from specific heat data. However, the great majority of the significant data lie in the range $T/\Theta \gtrsim 0.30$ where the departure from the limiting "classical" law, $\varrho \propto T$, is still fairly small. In the corresponding situation in specific heats it is of no great consequence whether one uses an EINSTEIN or DEBYE model as it is only for $T/\Theta \ll 1$ that the markedly different exponential or cubic temperature dependences become clearly distinguishable. Thus in the electrical case the precise scattering model adopted, or specific assumptions made about the freedom or otherwise of the

Table 4. *Comparison of Θ_D obtained from specific heat data and Θ_R deduced from electrical resistivity.* (After GRÜNEISEN.)

	Metal							
	Na	Cu	Ag	Au	Al	Pb	W	Ta
Θ_D (°K)	159	315 to 330	210 to 225	163 to 186	390	82 to 88	305 to 337	245
Θ_R (°K) . . .	202	333	223	175	395	86	333	228

[1] E. GRÜNEISEN: Ann. Phys., Lpz. **16**, 530 (1933)

conduction electrons, cannot be expected to be at all critical except in the low temperature region; indeed we notice that the relation $\varrho \propto T/M\Theta^2$ for the higher temperature range could already be predicted from WIEN's model. Nevertheless, we must also remember that for temperatures well *above* the characteristic temperature other resistance mechanisms and departures from the scattering model must be expected. Thus, thermal expansion must obviously be considered[1], anharmonicity of lattice vibrations will be expected to lead to a non-linear temperature variation of resistance[2], and a sufficient density of thermally excited lattice defects may arise within say 50 to 100°C of the melting point to influence the resistance quite significantly[3]. In

addition, $s-d$ band scattering may contribute in transition metals[4,5], ([3], page 536). To accommodate departures from linearity at higher temperatures for one reason or another, GRÜNEISEN had to multiply the theoretical formula by an empirical factor $(1 + a_1 T + b_1 T^2)$, which therefore renders less certain the values quoted in Table 4.

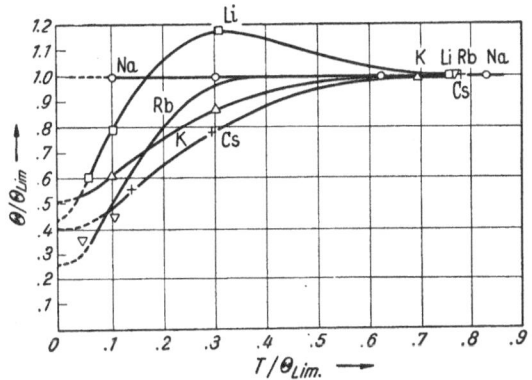

Fig. 25. Relative variation of characteristic temperature Θ deduced from resistance measurements on alkali metals. Θ_{Lim} is apparent value of Θ for $T/\Theta \approx 1$.

18. Experimental comparison with monovalent metals. If therefore we wish to make a satisfactory comparison with experimental data, we ought presumably to restrict ourselves, in the first place at least, to the simple monovalent metals (Li, Na, K, Rb, Cs; Cu, Ag, Au) and to temperatures such that $T/\Theta \lesssim 1$. If the GRÜNEISEN-BLOCH formula were strictly valid the parameter Θ would of course be found to be strictly constant, just as in the corresponding specific heat problem if the DEBYE model were exact. Otherwise we may regard Θ as a temperature-dependent parameter, and KELLY and MACDONALD[6] have considered in some detail the analysis of resistive and calorimetric data in this way. Using experimental results on the alkali metals from the experiments of MACDONALD and MENDELSSOHN[7] together with further data obtained since in our laboratory, Fig. 25 has been drawn showing on normalized scales the behaviour of Θ_R as determined from electrical resistance measurements, by comparison with the BLOCH-GRÜNEISEN formula.

It is immediately evident that sodium appears to behave rather "ideally" since Θ_R is found to be essentially constant from $T/\Theta_R \approx 0.8$ to below $T/\Theta_R \approx 0.1$,

[1] J. MEIXNER: Ann. Phys., Lpz. **38**, 609 (1940). — F. M. KELLY: Canad. J. Phys. **32**, 81 (1954). — It is common in experimental work to normalize the observed resistance, say S_T, of a particular metal specimen to that observed at the ice-point, say $S_{273°K}$. The ratio $S_T/S_{273°K}$ would then equal the resistivity ratio $\varrho_T/\varrho_{273°K}$ if there were no thermal expansion. For many purposes, the volume correction is not of specific importance and is therefore neglected.

[2] J. S. DUGDALE and D. K. C. MACDONALD: Phys. Rev. **96**, 57 (1954).

[3] D. K. C. MACDONALD: J. Chem. Phys. **21**, 177, 2097 (1953). — Report of Bristol Confle. on Defects in Crystalline Solids, p. 383 (Phys. Soc. Lond. 1955).

[4] N. F. MOTT: Proc. Phys. Soc. Lond. **47**, 571 (1935). — Proc. Roy. Soc. Lond., Ser. A **153**, 699 (1936); **156**, 368 (1936).

[5] A. H. WILSON: Proc. Roy. Soc. Lond., Ser. A **167**, 580 (1938).

[6] F. M. KELLY and D. K. C. MACDONALD: Canad. J. Phys. **31**, 147 (1953).

[7] D. K. C. MACDONALD and K. MENDELSSOHN: Proc. Roy. Soc. Lond., Ser. A **202**, 103 (1950).

while Θ_R in the other alkali metals starts to vary quite strongly below $T/\Theta_R \approx 0.3$. An obvious possible source of variation would be deviation of the lattice vibrations from the DEBYE model and there has been considerable theoretical interest recently in the analysis of the proper vibrational spectrum in sodium[1,2]. The influence on

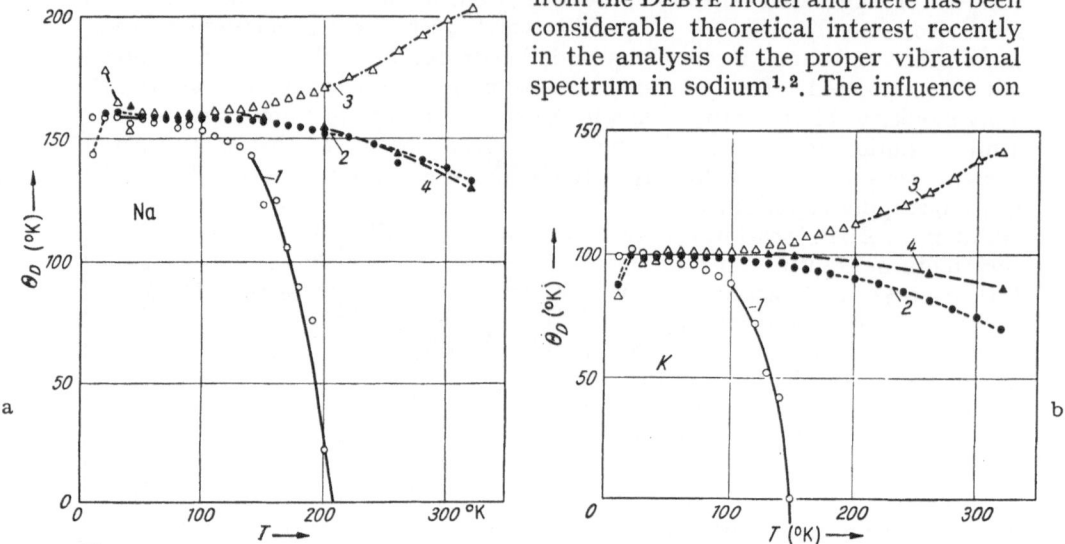

Fig. 26a and b. (a) Variation of Θ in sodium deduced from specific heat data. (See KELLY and MacDONALD: loc. cit., for details of calculation.) (b) Variation of Θ in potassium from specific heat data.

electrical resistance which would result from such a departure has been considered by CORNISH and MacDONALD[3]. More directly, however, curves of Θ_D determined experimentally from specific heat data in sodium and potassium (see Figs. 26a and b) are quite similar to one another suggesting that the difference in Θ_R lies in a variation of *electron* behaviour.

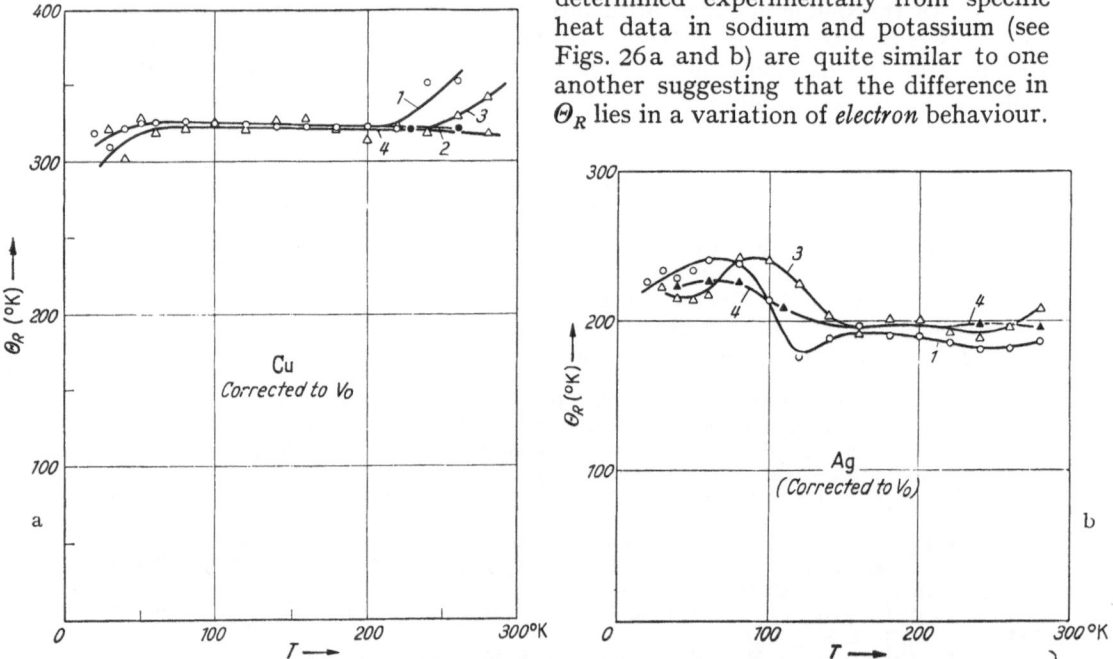

Fig. 27a and b. (a) Θ_R for copper (after KELLY: loc. cit.). (b) Θ_R for silver (after KELLY: loc. cit.)

[1] E. BAUER: Phys. Rev. **92**, 58 (1953).
[2] A. B. BHATIA: Phys. Rev. **97**, 363 (1955).
[3] F. H. J. CORNISH and D. K. C. MacDONALD: Phil. Mag. **42**, 1406 (1951).

The individualistic hump in Θ_R for lithium has so far received no detailed explanation although it may be related to the occurrence of a crystallographic modification reported by BARRETT and TRAUTZ[1]. Curves of Θ_R for copper and silver are also shown in Figs. 27a and b.

Assuming then that we have chosen an appropriate value of Θ_R, we may now turn to the *quantitative* prediction of ϱ by the theory. From WILSON ([1], page 269) we have, for monovalent metals with free electrons,—

$$\sigma = 2.83 \times 10^{-32} \frac{n\,M}{C^2} \frac{\Theta^2}{T} \text{ e.s.u.},$$

for the conductivity when $T \gtrsim 0.5\,\Theta$, where n is the number of electrons (and hence of atoms) per cm³., M is the atomic weight, and C an electron-ion interaction energy in ergs. C is expected to be of the order of ζ, the FERMI energy of the electrons. Setting $C \approx \zeta$ (ζ calculated for a free-electron gas) and rewriting this formula in practical units:

$$\varrho = 8.04 \times 10^{-19} \frac{\zeta^2}{n} \cdot \frac{T}{M\Theta^2} \ \Omega\,\text{cm.} \qquad \text{[cf. (17.2) above]} \qquad (18.1)$$

where ζ is now measured in electron-volts. We then have the data of Table 5. [In the case of the group IB metals (Cu, Ag, Au) we have contented ourselves with data at 0° C since none has a particularly low characteristic temperature and all have high melting points ($\sim$1000° C).]

Table 5. *Comparison of theory and experiment on monovalent1 metals.*

Metal	M	Θ_R	ζ (e.v.)	n	T (°K)	$\varrho\,T$ (Theor. for $C=\zeta$) (Ω cm.)	$\varrho\,T$ (Obs.) (Ω cm.)	$\dfrac{\varrho\,T \text{ (obs.)}}{\varrho\,T \text{ (theor.)}}$
Li....	7	330	4.8	48×10^{21}	250	12.6×10^{-6}	7.7×10^{-6}	0.61
Na ...	23	180	3.2	26	150	6.35	2.17	0.34_4
K	39	114	2.1	14	80	3.99	1.45	0.36_4
Rb ...	85	65	1.8	11	50	3.3	1.92	0.58_2
Cs....	133	45	1.57	9	35	2.77	1.92	0.68_6
Cu ...	64	320	7.0	85×10^{21}	273	1.93×10^{-6}	1.69×10^{-6}	0.88
Ag ...	108	200	5.5	59	273	2.59	1.47	0.57
Au ...	197	200	5.5	59	273	1.42	2.26	1.59

19. Later developments and the influence of electron screening. It is evident that amongst the alkalis, sodium (with potassium a "close second") occupies a privileged position with the lowest ratio of observed to calculated resistivity. We might then interpret this as evidence that the relative interaction energy, C, is lowest for sodium indicating presumably the closest approach to the idealized "free-electron" model. BARDEEN[2] set out to improve the scattering model in a number of ways and found that the data on sodium were then in best (indeed, very close) agreement with the theory he developed, potassium also being satisfactory, while rubidium and caesium show a resistivity experimentally which is considerably too high. BARDEEN allows for the fact that, when the ions are displaced by the lattice-waves setting up a perturbed charge concentration which tends thus to scatter the conduction electrons, the conduction electrons themselves tend to "crowd in" and compensate this charge, giving rise to what we might call dynamic screening. Of course, even in the static condition the electrons tend to screen the ionic charges, and from this point of view we may say that the BLOCH model

---

[1] C. S. BARRETT and O. R. TRAUTZ: Trans. Amer. Inst. Min. Metallurg. Engrs. **175**, 579 (1948). See also W B. PEARSON: Canad. J. Phys. **32**, 708 (1954).

[2] J. BARDEEN: Phys. Rev. **52**, 689 (1937). Cf. also J. BARDEEN and D. PINES: Phys. Rev. **99**, 1140 (1955).

essentially corresponds to almost perfect screening. If an ion, regarded as a point-charge, were entirely unscreened the potential for an electron would be $-\dfrac{e^2}{r}$ and with screening we may write[1] $-\dfrac{e^2}{r}\, e^{-r/b}$; the Bloch model implies $b \approx 0$, and it is on this basis that we obtain (17.1). Houston's[2] approximate theory showed that the low temperature variation of resistivity would depend on the "atomic radius", b, arising in Wentzel's formula (which we may now interpret in terms of conduction electron screening of the ions) the resistance decaying more slowly with larger values of b. Fig. 25 shows that, with the exception of sodium, the resistivity of all the alkali metals falls more slowly at low temperatures than the Bloch formula predicts. More recently Houston[3] has shown that some of MacDonald and Mendelssohn's data on the alkali metals could be fitted quite well by empirical adjustment of the screening constant, b, in his theory. Nordheim[4] also developed a rather detailed extension of the Bloch theory making use of the screened Coulomb potential $-\dfrac{e^2}{r}\, e^{-r/b}$. Similar investigations have been made by Sauter[5], Busse and Sauter[6], and Ziman[7]. The latter has also followed Bardeen in taking into account the so-called "Umklapp" scattering-processes first mentioned by Peierls[8], where an electron in being scattered not only absorbs or emits a quantum of excited vibrational energy ("phonon") but also suffers a Bragg reflexion with the quasi-periodic lattice as a whole. This can be regarded also as modifying the effective scattering "matrix-element" tending to increase large-angle scattering. We have also to remember that (17.1) is *not* an exact mathematical solution for the resistance at all temperatures even on the simplified Bloch model (e.g. cf. Sondheimer[9]), although fortunately (17.1) is a rather good approximation, being mathematically exact on this model at low $(T/\Theta \ll 1)$ and high $(T \gtrsim \Theta)$ temperatures.

All in all, it appears that it may perhaps be difficult to do better, at least in the immediate future, than take as a general theoretical formula for the resistivity of a "simple" metal: (cf. the development by Nordheim[4], although unfortunate misprints appear to occur in the important formulae):

$$\varrho \sim \frac{m^2 e^2}{h^3} \cdot \frac{1}{dc^2} \cdot T^5 \int\limits_0^{\Theta/T} \frac{x^5\, dx}{\{(h\, \Phi)^2 + (x\, kT)^2\}^2\, (e^x - 1)\, (1 - e^{-x})} \tag{19.1}$$

where d is the density, c the velocity of sound in the metal, and Φ an effective "screening temperature" which we define by: $\Phi = \dfrac{A_0}{\pi b}\, \Theta$, where A_0 is the lattice parameter, and b appears in the screened Coulomb potential $-\dfrac{e^2}{r}\, e^{-r/b}$. This may be written alternatively

$$\varrho \propto \left(\frac{\Theta}{\Phi}\right)^4 \cdot \frac{V^{\frac{1}{3}}}{M} \cdot \frac{T^5}{\Theta^6} \int\limits_0^{\Theta/T} \frac{x^5\, dx}{\{1 + x^2\, (T/\Phi)^2\}^2\, (e^x - 1)\, (1 - e^{-x})} . \tag{19.2}$$

[1] Cf. N. F. Mott: Proc. Cambridge Phil. Soc. **32**, 281 (1936) or N. F. Mott and H. Jones ([2], p. 86).
[2] W. V. Houston: Phys. Rev. **34**, 279 (1929).
[3] W. V. Houston: Phys. Rev. **88**, 1321 (1952).
[4] L. Nordheim: Ann. Phys., Lpz. **9**, 607, 641 (1931).
[5] F. Sauter: Ann. Phys., Lpz. **42**, 110 (1942).
[6] C.-A. Busse and F. Sauter: Z. Physik **139**, 440 (1954).
[7] J. M. Ziman: Proc. Roy. Soc. Lond., Ser. A **226**, 436 (1954).
[8] R. Peierls: Ann. Phys., Lpz. **4**, 121 (1930).
[9] E. H. Sondheimer: Proc. Roy. Soc. Lond., Ser. A **203**, 75 (1950).

For very good screening we have $\Theta/\Phi \ll 1$ and we may then write

$$\varrho \propto \left(\frac{\Theta}{\Phi}\right)^4 \cdot \frac{V^{\frac{1}{3}}}{M} \cdot \frac{T^5}{\Theta^6} \int_0^{\Theta/T} \frac{x^5\, dx}{(e^x - 1)\,(1 - e^{-x})} \tag{19.3}$$

identical in form with the BLOCH-GRÜNEISEN equation, noticing, however, that the *absolute value* of the resistance will also tend to zero (through the factor $(\Theta/\Phi)^4$)—as we should expect—for perfect screening. In the general case, we may *always* write

$$\varrho \propto \left(\frac{\Theta}{\Phi}\right)^4 \cdot \frac{V^{\frac{1}{3}}}{M} \cdot \frac{T^5}{\Theta^6} \cdot (124.4), \quad \text{if} \quad \frac{T}{\Phi} \ll 1 \quad \text{as well as} \quad \frac{T}{\Theta} \ll 1$$

and for $T/\Theta \gtrsim 1$:

$$\varrho \propto \frac{V^{\frac{1}{3}}}{M} \cdot \frac{T}{2\,\Theta^2} \left\{ \log\left(1 + \left(\frac{\Theta}{\Phi}\right)^2\right) - \frac{(\Theta/\Phi)^2}{1 + (\Theta/\Phi)^2} \right\}.$$

If also $\Theta/\Phi < 1$ then the latter equation becomes:

$$\varrho \propto \frac{V^{\frac{1}{3}}}{M} \cdot \frac{T}{4\,\Theta^2} \left(\frac{\Theta}{\Phi}\right)^4 \left\{ 1 - \frac{4}{3}\left(\frac{\Theta}{\Phi}\right)^2 + \frac{3}{2}\left(\frac{\Theta}{\Phi}\right)^4 - \cdots \right\}.$$

Thus the resistance-ratio between low and "high" temperatures would be

$$\frac{\varrho_1}{\varrho_2} = \frac{497.6}{\Theta^4} \frac{T_1^5}{T_2} \cdot \left\{ 1 + \frac{4}{3}\left(\frac{\Theta}{\Phi}\right)^2 + \frac{5}{18}\left(\frac{\Theta}{\Phi}\right)^4 \cdots \right\} \tag{19.4}$$

which may be compared with (17.4); evidently as the screening improves $(\Theta/\Phi \to 0)$, the ratio rapidly approaches that predicted by the BLOCH model. If, on the other hand, $\Theta/\Phi > 1$, then $\left\{ \log\left(1 + \left(\frac{\Theta}{\Phi}\right)^2\right) - \frac{(\Theta/\Phi)^2}{1 + (\Theta/\Phi)^2} \right\}$ varies relatively slowly compared with $(\Theta/\Phi)^4$, running from 0.81 to 2.3 as Θ/Φ goes from 2 to 5. Thus we shall make no very great error if we now write for the resistance-ratio between low and high temperatures:

$$\frac{\varrho_1}{\varrho_2} \approx \frac{497.6}{\Phi^4} \frac{T_1^5}{T_2}. \tag{19.5}$$

Consequently, on this basis, we would interpret the relative constancy of Θ (Fig. 25) deduced for sodium on the BLOCH formula as implying that the overall effective screening was good (and therefore consistent with a free electron model). The drop in Θ of the order of 50% in the other metals at low temperatures would imply $\Phi \sim 0.5\,\Theta$ and hence on this model that the screening radius, b, is comparable with the lattice constant (which is roughly the ion diameter). Now MOTT'S calculation, on the THOMAS-FERMI model, assuming one free electron per atom, leads to:

$$\frac{b}{r_0} = \frac{\hbar}{\sqrt{m\,e}} \left(\frac{\pi}{12}\right)^{\frac{1}{6}} \left(\frac{N}{V_0}\right)^{\frac{1}{6}}$$

where N is AVOGADRO's number, (6.06×10^{23}); V_0 is the gramatomic volume (c.c.), and we have set the volume per atom, $\frac{V_0}{N} = \frac{4}{3}\pi r_0^3$. For the alkali metals this would then give:

Element	Li	Na	K	Rb	Cs
b/r_0	0.35	0.31_5	0.28_4	0.27_4	0.26_2

Consequently, to account for the observations we should have to assume that in the heavier alkalis particularly, the electrons were *not* entirely free (so that screening is not fully effective). This hypothesis might be checked rather directly if we could persuade the alkali metals to take into (homogeneous) solid solution atoms of another valency group (such as say Ca or Sr) since on this basis the resistivity due to an atomic concentration, x, of foreign atoms (valency difference Z) should be[1]:

$$\varrho = x \cdot \frac{2\pi Z^2 e^2}{m v^3} \left\{ \log \left(1 + \frac{4 m^2 v^2 b^2}{\hbar^2} \right) - \frac{1}{1 + \frac{\hbar^2}{4 m^2 v^2 b^2}} \right\} \qquad (19.6)$$

where v is the Fermi velocity in the parent metal. Unfortunately, with the exception of lithium, the alkalis will not take metals of other groups into solution.

20. Magneto-resistance. Let us now consider what information we might gain from magnetic measurements. If we apply a magnetic field, say H, to a metal the electrons will experience a force $\frac{e H \cdot u}{c}$, where u is the velocity in the plane perpendicular to H. In the absence of other forces, $\bar{u} = 0$ and hence the magnetic field produces no net force on the electron assembly. If, however, an electric field E has been applied perpendicular to H then $\bar{u} = \frac{e l}{m v} E$ [cf. equation (2.1)] where v is the electron speed (Fermi velocity) and l the mean free path. The electrons will now experience an average force F transverse to E in the plane perpendicular to H given by:

$$F = \frac{e H \cdot \bar{u}}{c} = \frac{e H}{c} \left(\frac{e}{m v} E \right) = \left(\frac{e H l}{c m v} \right) \cdot e E$$

$\left(\text{where } \frac{e H l}{c m v} = \frac{l}{R} \text{ and } R \text{ is the radius of the free electron orbit in the magnetic field} \right)$.

This argument is evidently only valid in this form so long as $\frac{e H l}{c m v} \ll 1$. This Lorentz force, F, will tend to produce a transverse current flow if a completed circuit is provided; otherwise a charge-gradient will result producing an electric field to oppose and balance the force F. This electric field, say E_H, is just the Hall field. Since the electric current, which is directly measurable, is given by $j = N e \bar{u}$ we can on this simple model with measurements of the Hall field, determine the product $N e$. If all electrons in the metal have the *same* "mobility" (i.e. achieve the *same* drift velocity, $\bar{u}$, under the applied electric field, E) then evidently the (average) Lorentz force is the same for each electron and there is "detailed balancing", with E_H giving no net resultant average transverse force on an electron. Consequently, the electrons pursue the same drift paths parallel to E after the application of H as they did before and the conductivity is unaffected. However, if electrons have drift velocities differing by Δu from u, the mean for all the electrons, then we may show that

$$\frac{\varrho_H - \varrho}{\varrho} \sim \left(\frac{e H l}{m c v} \right)^2 \cdot \frac{\overline{\Delta u^2}}{u^2}, \quad \text{i.e.} \quad \sim \left(\frac{l}{R} \right)^2 \cdot \frac{\overline{\Delta u^2}}{u^2} \qquad (20.1)$$

so long as $l/R \ll 1$.

Evidently if $\frac{\Delta u}{u} = 0$, the resistivity is unaltered by a magnetic field in agreement with our previous argument.

---

[1] Cf. Mott and Jones [2], p. 294.

When SOMMERFELD's theory was first developed in full (e.g.[1]) it was thought that the observed magneto-resistive effect in metals might be attributed to the thermal spread of velocities, i.e. $(\overline{\Delta u^2}/u^2) \sim k\,T/\zeta$. However, this predicted an effect very much smaller than was actually observed and also fails to account for a number of other features. BETHE[2] and PEIERLS[3] both suggested that the variation in electron properties might be an intrinsic departure in a given metal from the idealized isotropic free-electron model. On the one hand, the influence of the lattice periodic field may cause electrons having the *same* (FERMI) energy to have different velocities when travelling in different directions; that is, the FERMI surfaces (of constant energy) in momentum space are not spherical. On the other hand, the lattice scattering may also vary with the direction of electron motion relative to the crystal axes—that is to say, the relaxation-time τ is then itself a function of the electron momentum-vector $\boldsymbol{k}$.

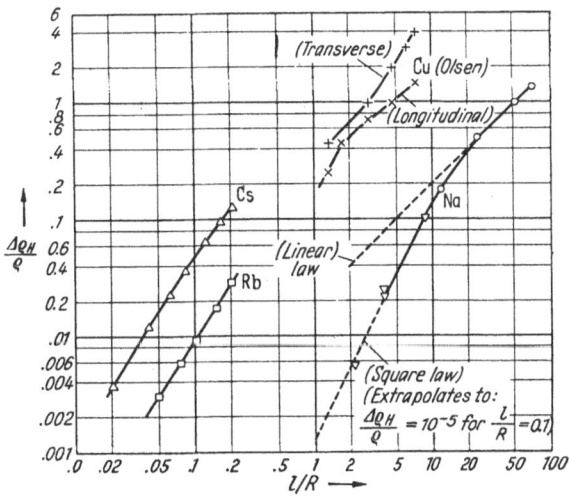

Fig. 28. Magneto-resistance of alkali metals and of copper on normalized scale.

From equation (20.1), we should then expect that by plotting $\Delta\varrho_H/\varrho$ against l/R for different metals a direct comparison of $\overline{\Delta u^2}/u^2$ would result. This has been done in Fig. 28 for the alkali metals Na, Rb and Cs from data taken in liquid helium. l has been computed in each case from the observed conductivity, assuming one free electron per atom. It is immediately clear that on this evidence sodium must be regarded as corresponding exceedingly well to the free electron model since a relatively large value of l/R must be achieved in order to observe a significant magneto-resistance. Extrapolating the data for sodium down to $l/R = 0.1$ where we might then hope to apply (20.1) semi-quantitatively we find $\dfrac{\Delta\varrho_H}{\varrho} \approx 10^{-5}$ and hence $\dfrac{\overline{\Delta u^2}}{u^2} \sim 10^{-3}$, while for rubidium and caesium we should seem forced to recognize $\overline{\Delta u^2}/u^2$ of the order unity. Alternatively, we might transpose the data for Rb and Cs on to that for Na by reducing the number of free electrons per atom to about $\frac{1}{20}$ and $\frac{1}{60}$ respectively in those metals. Either conclusion leads us to recognize again a considerable departure from the ideal behaviour in the heavier alkalis in agreement with our earlier conclusions, although of course these figures must not be taken too literally in magnitude based as they are on free electron formulae. This conclusion is also born out by experiments on thermoelectric power at low temperatures[4].

If we wish to make a wider comparison of the magnetoresistance of metals of all classes it is convenient, as JUSTI[5] and KOHLER[6] have done, to plot directly

[1] A. SOMMERFELD and N. H. FRANK: Rev. Mod. Phys. 3, 1 (1931).

[2] H. BETHE: Nature, Lond. 127, 336 (1931).

[3] R. PEIERLS: Ann. Phys., Lpz. 10, 97 (1931).

[4] D. K. C. MACDONALD and W. B. PEARSON: Proc. Roy. Soc. Lond., Ser. A 219, 373 (1953); 221, 534 (1954).

[5] E. JUSTI and H. SCHEFFERS: Phys. Z. 39, 105 (1938).

[6] M. KOHLER: Phys. Z. 39, 9 (1938).

$(\Delta\varrho_H/\varrho)$ against (H/r) where r is the observed ratio: $\varrho(T)/\varrho(\Theta)$. Fig. 29, taken from Justi's work comprises data from Grüneisen, Grummach, de Haas, Justi and Kapitza, and Justi has drawn certain broad conclusions about the behaviour of metallic groups, suggesting that the magneto-resistance of odd-valent groups tends towards a saturation value while that of even-valent groups continues to increase quadratically with (H/r). At the same time, the magneto-resistance of single crystals can show very remarkable variation with angle of rotation relative to the applied magnetic field (see Fig. 30 for example from the work of Justi and Scheffers on a gold single-crystal and Figs. 31a, b observed

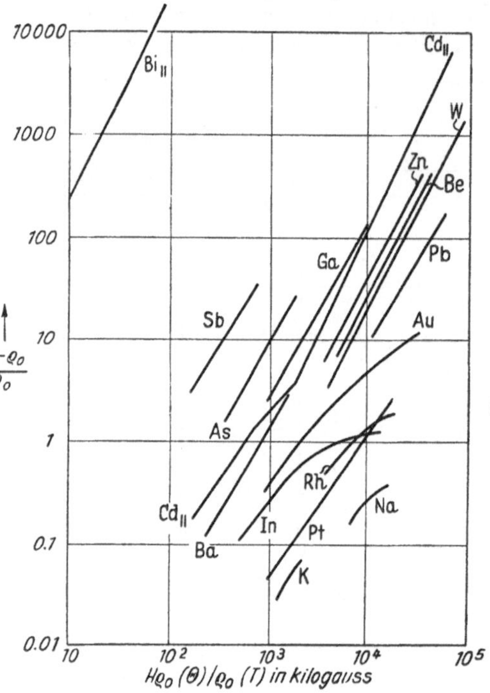

Fig. 29. Magneto-resistance of metals (after Grün-eisen, Grummach, de Haas, Justi and Kapitza).

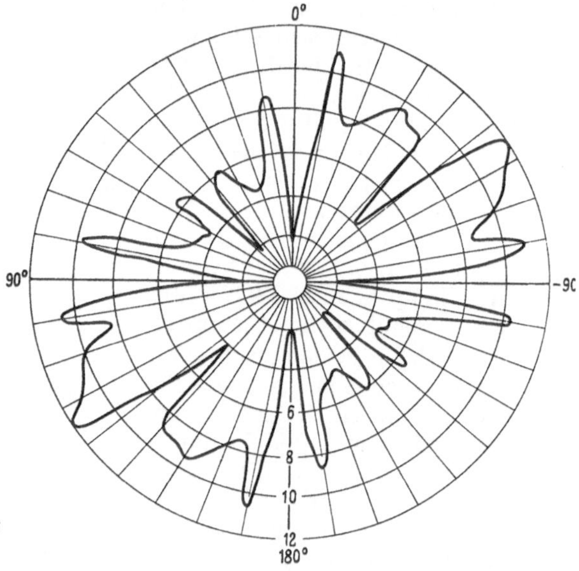

Fig. 30. Variation of magneto-resistance with angular rotation of gold single crystal (after Justi). $T = 4.2°$ K; $H = 18.6$ kgauss. Crystal cut parallel to cubic axis.

by the author on tin). However, it appears that the theory is as yet quite unable to give any detailed explanation of such bizarre behaviour—Wilson says "It does not seem likely that any model simple enough to be tractable theoretically would give a magneto-resistance curve of the complexity of those actually observed".

The dependence of magneto-resistance on the ratio l/R shows immediately that the greatest effects will be attained either through the use of very high magnetic fields to make R small, as Kapitza[1] originally did (attaining fields of 300000 gauss over several cubic centimetres for a short period by momentarily short-circuiting a specially designed generator at the Cavendish Laboratory in Cambridge), or by the use of low temperatures and very pure metals to achieve a large mean free path, l. Recently Olsen[2] at the E.T.H. in Zürich has combined both techniques. Instead of short-circuiting a generator to give a large current for a short period, one may store potential energy over a reasonably long initial

[1] P. Kapitza: Proc. Roy. Soc. Lond., Ser. A **105**, 691 (1924); **115**, 658 (1927); **119**, 358 (1928); **123**, 292 (1929).

[2] J. L. Olsen and L. Rinderer: Nature, Lond. **173**, 682 (1954). — J. L. Olsen: Helv. phys. Acta **26**, 798 (1953).

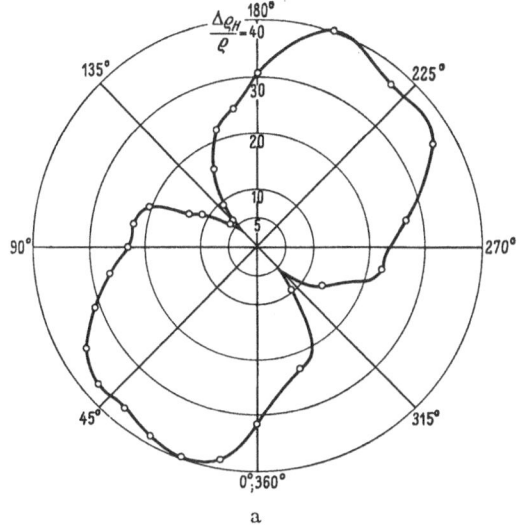

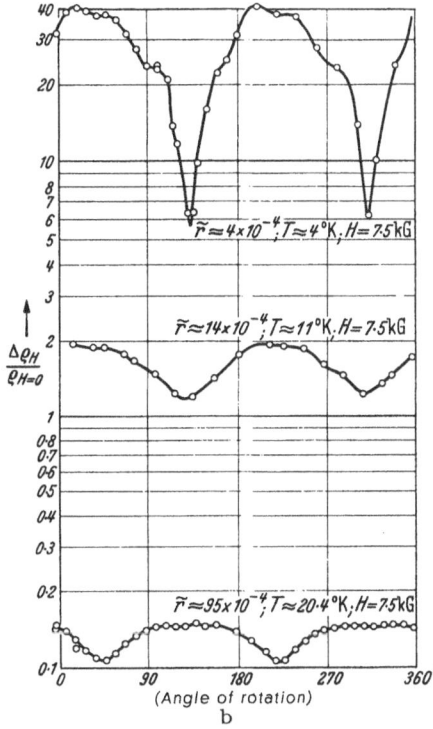

Fig. 31 a and b. (a) Variation of magneto-resistance with angular rotation of tin crystal. $T = 4.2°$ K; $\tilde{r}_T \approx 4 \times 10^{-4}$; $H \approx 7.5$ kgauss ($\tilde{r}_T \equiv \varrho_T/\varrho_{273°K}$). (b) Variation of magneto-resistance in tin crystal.

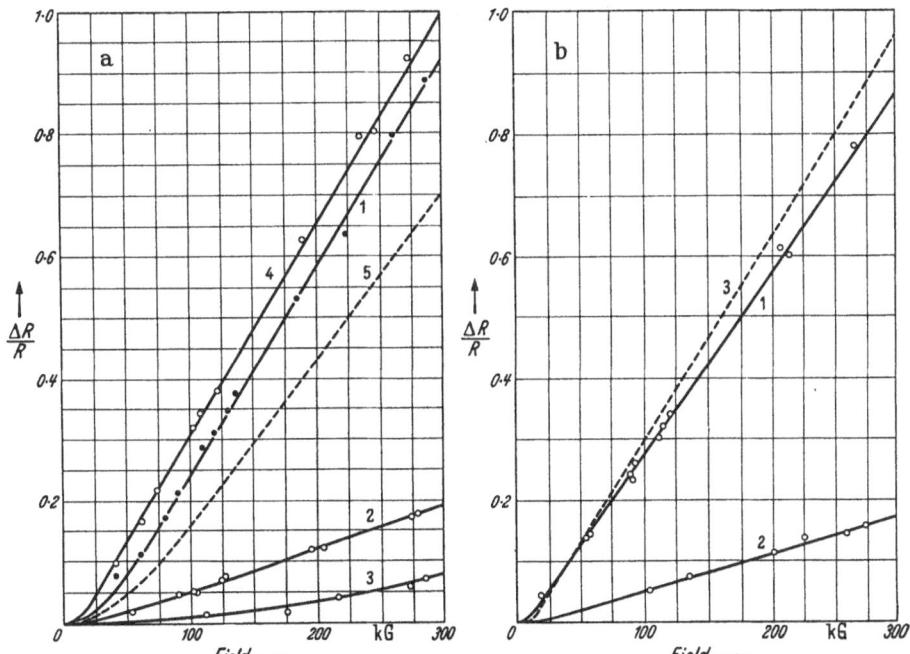

Fig. 32 a and b. Magneto-resistance of cadmium (a) and gallium (b). a: Curve 1, H perpendicular to I, Cd_1, temperature of liquid nitrogen; curve 2, H perpendicular to I, Cd_1, temperature of solid CO_2 and ether; curve 3, H perpendicular to I, Cd_1, room temperature; curve 4, H perpendicular to I, Cd_{II}, temperature of liquid nitrogen; curve 5, H parallel to I, Cd_1, temperature of liquid nitrogen. b: Curve 1, H perpendicular to I, temperature of liquid air; curve 2, H perpendicular to I, temperature of solid CO_2 and ether; curve 3, H parallel to I, temperature of liquid nitrogen (after KAPITZA).

period by charging up a bank of condensers[1], or secondary storage cells as Kapitza did earlier, and then discharge them rapidly through a coil. By placing the coil itself in the refrigerant bath the resistance of the coil may be kept very low, so increasing the current on discharge. These techniques have also been used by Shoenberg[2] in measurements of the de Haas-van Alphen effect (the periodic variation with field strength of magnetic susceptibility observed in certain metals). Olsen used a bank of electrolytic condensers of 3000 μF capacity charged to 360 volts discharged through a coil 1.5 cm. in length with a mean diameter of 0.7 cm., having 750 turns of 0.2 mm. copper wire. No external strengthening was found necessary for the coil for the fields generated of about 150 kilogauss. About 50 cm.³ of liquid helium was evaporated by each discharge through the coil.

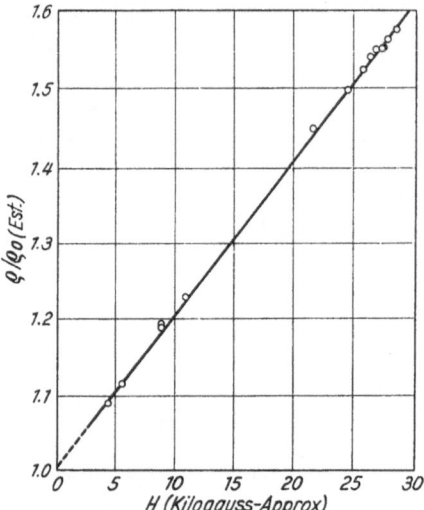

Fig. 33. Transverse magneto-resistance of sodium.
$T = 4.2°$ K; $\bar{r} \approx 2 \times 10^{-4}$.

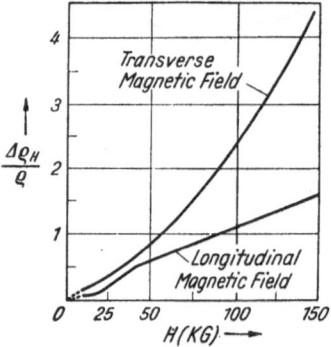

Fig. 34. Magneto-resistance of copper at low temperatures (after Olsen).

Many questions remain as yet unanswered in magneto-resistance of metals[3]. In particular, the linear variation of $\Delta\varrho_H/\varrho$ with H as first found by Kapitza (see Fig. 32; compare also Fig. 33 for sodium by MacDonald, and Olsen's results, Fig. 34, on the longitudinal effect in copper) remains unexplained. Theory generally predicts $\dfrac{\Delta\varrho_H}{\varrho} \approx \dfrac{AH^2}{B + CH^2}$ (e.g. Sondheimer and Wilson[4]; Chambers[5]) and a quasi-linear section will exist for $CH^2 \sim B/3$, but this region appears inadequate to account for the observations. It is perhaps worth remarking that in sodium, which we have concluded approximates rather closely to the ideal free-electron model, linearity of $\Delta\varrho_H/\varrho$ with H (transverse field) does not make its appearance until $l/R \gtrsim 10$ (cf. Fig. 28), while in cadmium, (hex. c.p. structure), on the other hand, $l/R \approx 0.1$ seems to be adequate. We notice also in Olsen's work on copper that with a *longitudinal* field, linear behaviour starts (apparently rather abruptly) at about $l/R \approx 2$, while there appears no sign of this in the *transverse* magneto-resistance up to $l/R \sim 7$. It would evidently be very desirable to extend the work on transverse magneto-resistance in copper to higher values of l/R (which might

[1] Cf. W. J. de Haas and J. Westerdijk: Nature, Lond. **158**, 271 (1946).
[2] D. Shoenberg: Nature, Lond. **170**, 569 (1952); see also Nature, Lond. **171**, 458 (1953): Physica, Haag **19**, 791 (1953).
[3] See e.g. D. K. C. MacDonald and K. Sarginson: Rep. Progr. Phys. **15**, 249 (1952). (A Review with Bibliography).
[4] E. H. Sondheimer and A. H. Wilson: Proc. Roy. Soc. Lond., Ser. A **190**, 435 (1947).
[5] R. G. Chambers: Proc. Phys. Soc. Lond. A **65**, 903 (1952).

be done with existing magnetic fields by using purer metal of greater mean free path) and to make longitudinal measurements on sodium over the available range of l/R. Although when $l/R > 1$ it may be necessary to consider specifically the influence of quantisation of electron-orbits in the magnetic field the example of cadmium just quoted in comparison with sodium suggests that some more "elementary" principle may perhaps be involved which we have not yet appreciated.

21. The size-effect in metals. A further consequence of the almost "ideal" behaviour of sodium is found in the influence of size on the magneto-resistance effect in that metal (MacDonald[1], MacDonald and Sarginson[2], Sondheimer[3], Chambers[4]); we shall first discuss the size-effect itself, and the information we can gain from experiments in this field. If a physical dimension, a, of a metallic specimen is made comparable with the electron mean free path, l, then the observed properties will differ from those of a conventional "bulk" sample. The scattering of the electrons at the surface, which we normally neglect in comparison with that in the volume of the metal, can no longer be ignored. At ambient temperatures, with $l \sim 10^{-6}$ cm., we require extremely thin specimens in order to observe such effects and indeed it may be questioned whether we are entitled to consider the structure then as identical with "bulk" metal. At low temperatures in pure metals, however, l is sufficiently great that no such doubts can arise when we are dealing with samples of perhaps $\frac{1}{10}$ mm. in thickness.

Sondheimer[5] in a valuable review of size-effects in metallic conduction has pointed out that "since the calculation of the mean free path (for electron-scattering) from fundamental principles is highly complicated and involves many drastic approximations, it is desirable to have methods by which l may be estimated directly from observational data". It appears indeed desirable in general that in the theory of metals as many parameters as possible should be determined at least semi-"operationally", in Bridgman's sense, and that every opportunity should be taken to reduce the number of "paper-and-pencil concepts" which occur basically in the theory. Sondheimer observes that there are essentially three broad classes of size-effect conduction phenomena. First, we have the simple case where we observe the increase in resistivity over the value in "bulk" metal of a thin wire or plate ("film") arising from the limitation of normal mean free path, l, by the boundaries of the specimen; such measurements can then be used to deduce the ratio of l to the physical size, say a, of the specimen. Secondly we have the new effects referred to above which arise when a magnetic field is applied to such a specimen. In this case the radius, R, of the electron-orbit in the magnetic field is now involved and we are concerned essentially with the geometrical (and hence classical) relationship of R, l, and a. Not only information about l should now be available but also about R and hence the free-electron momentum. If we are not to be forced to extremely thin specimens of metal and very high magnetic fields, then low temperatures must be used to provide as large a mean free path as possible.

Finally there is the so-called "anomalous skin-effect"[6] in metals, which might more appropriately be called the high frequency size-effect. In this case a

[1] D. K. C. MacDonald: Nature, Lond. **163**, 637 (1949).

[2] D. K. C. MacDonald and K. Sarginson: Nature, Lond. **164**, 920 (1949). — Proc. Roy. Soc. Lond., Ser. A **203**, 223 (1950).

[3] E. H. Sondheimer: Nature, Lond. **164**, 920 (1949). — Phys. Rev. **80**, 401 (1950).

[4] R. G. Chambers: Proc. Roy. Soc. Lond., Ser. A **202**, 378 (1950).

[5] E. H. Sondheimer: Adv. Physics **1**, 1 (1952).

[6] Dr. R. G. Chambers has kindly advised me that the word anomalous should here be taken to mean simply "a new kind of".

dimension, δ, is introduced which corresponds to the depth to which a high-frequency electromagnetic field can penetrate into the metal. So long as $l/\delta \ll 1$, classical theory applies and the high-frequency resistance due to the "skin-effect" is calculated in the normal way. When, however, $l/\delta \sim 1$ the mean free path intrudes directly so that a situation broadly similar to the "normal" size-effect develops, and again low temperatures are essential to a study of this phenomenon.

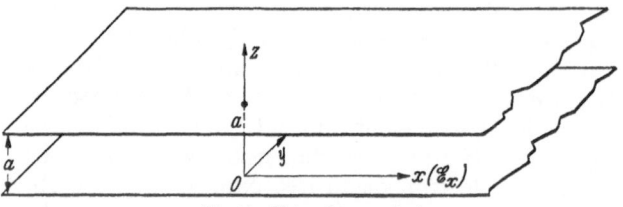

Fig. 35. Thin plate analysis.

The theoretical study of these problems is now always based on the BOLTZMANN equation for the statistical distribution reached by the electrons under the influence of the applied fields and the collisions[1]; the size-limitation enters directly through the appropriate boundary conditions imposed on the solution[2]. If we consider the case of a thin metal plate, thickness a in the x, y plane (see Fig. 35), the BOLTZMANN equation reads:

$$\left(\frac{\partial f}{\partial v_x}\right)\frac{dv_x}{dt} + \left(\frac{\partial f}{\partial z}\right)\frac{dz}{dt} = \left(\frac{df}{dt}\right)_{\text{collisions}} \tag{21.1}$$

where $f(v_x, v_y, v_z; x, y, z)$ is the electron distribution function. Terms such as $\left(\frac{\partial f}{\partial v_y}\right) \cdot \frac{dv_y}{dt}$ do not occur since we are only considering an electric field in the x-direction; similarly only $\partial f/\partial z$ enters since in the other directions, where the film is supposed "infinite", the distribution function must be uniform. If now we assume we may set $\left(\frac{df}{dt}\right)_{\text{coll.}} = -\left(\frac{f-f_0}{\tau}\right)$, where the "relaxation-time" $\tau = l/v$, and write $f = f_0 + f_1(\boldsymbol{v}, z)$ then:

$$\tau v_z \cdot \frac{\partial f_1}{\partial z} + f_1 = -\frac{e \, \mathcal{E}_x \tau}{m}\left(\frac{\partial f_0}{\partial v_x}\right) \tag{21.2}$$

which has the general solution:

$$f_1(\boldsymbol{v}, z) = -\frac{e \, \mathcal{E}_x \tau}{m} \cdot \frac{\partial f_0}{\partial v_x}\{1 + \varphi(\boldsymbol{v})\, e^{-z/\tau v_z}\}$$

where φ is an arbitrary function. The arbitrary function is then to be determined by our choice of boundary conditions at the surfaces ($z = 0$; a) of the film. If one assumes purely *diffuse* scattering at the surfaces then the appropriate solution is:

$$f_1(\boldsymbol{v}, z) = -\frac{e \, \mathcal{E}_x \tau}{m}\frac{\partial f_0}{\partial v_x}\{1 - e^{-z/\tau v_z}\}; \quad (v_z > 0)$$

$$= -\frac{e \, \mathcal{E}_x \tau}{m}\frac{\partial f_0}{\partial v_x}\left\{1 - e^{\frac{a-z}{\tau v_z}}\right\}; \quad (v_z < 0),$$

i.e.

$$\varphi(\boldsymbol{v}) \begin{cases} = -1; & v_z > 0 \\ = -e^{a/\tau v_z}v_z <;0. \end{cases}$$

[1] In Sect. D we shall consider the establishment of the BOLTZMANN equation.
[2] First used by: K. FUCHS: Proc. Cambridge Phil. Soc. **34**, 100 (1938), after a suggestion by R. PEIERLS. Previous theories [e.g. J. J. THOMSON: Proc. Cambridge Phil. Soc. **11**, 120 (1901); L. NORDHEIM: Act. Sci. et Ind. No. 131 Paris: Hermann et Cie.], were based on more approximate free path theories.

With the electron distribution-function thus determined we can now calculate the current density, J_x, and hence obtain the resistivity, ϱ, of the plate. If ϱ_0 be the bulk resistivity, we have:

$$\frac{\varrho}{\varrho_0} = \left\{ 1 - \frac{3}{8}\left(\frac{l}{a}\right) + \frac{3}{2}\left(\frac{l}{a}\right) \int\limits_{1}^{\infty} \left(\frac{1}{x^3} - \frac{1}{x^5}\right) e^{-\frac{a\,x}{l}}\, dx \right\}^{-1} \qquad (21.3)$$

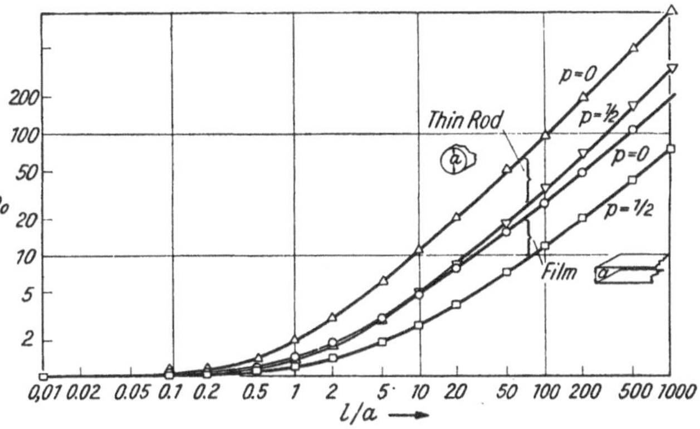

Fig. 36. Size-effect variation of resistivity in thin films and rods.

If $l/a \ll 1$, ("thick" films):

$$\frac{\varrho}{\varrho_0} \approx 1 + \frac{3}{8}\frac{l}{a} \qquad (21.4)$$

and if $l/a \gg 1$, (very thin films):

$$\frac{\varrho}{\varrho_0} \approx \frac{4}{3}\frac{l}{a}\frac{1}{\log(l/a)}. \qquad (21.5)$$

The possibility of some degree of "elastic" scattering may also be included by assuming that a fraction, p, of the electrons undergoes specular reflexion at the surface, the remaining fraction, $(1-p)$, being diffusely reflected. This is of course only an interpolation device and cannot be considered as necessarily corresponding to any physical reality.

Equations (21.4) and (21.5) would now read:

$$\frac{\varrho}{\varrho_0} = 1 + \frac{3}{8}\frac{l}{a}(1-p)$$

and

$$\frac{\varrho}{\varrho_0} = \frac{4}{3}\frac{l}{a}\frac{1-p}{1+p}\frac{1}{\log(l/a)}.$$

In Fig. 36, ϱ/ϱ_0 is shown plotted against l/a for $p = 0$ and $\tfrac{1}{2}$.

In thin rods[1] of diameter a ,the expression corresponding to (21.3) is now:

$$\frac{\varrho}{\varrho_0} = \left\{ 1 - \frac{12}{\pi} \int\limits_{0}^{1} (1-t^2)^{\frac{1}{2}}\, dt \int\limits_{1}^{\infty} e^{-\frac{a\,t\,x}{l}} \left\{ \frac{1}{x^6} - \frac{1}{x^8} \right\}^{\frac{1}{2}} dx \right\}^{-1} \qquad (21.6)$$

---

[1] R. B. DINGLE Proc. Roy. Soc. Lond, Ser., A **201**. 545 (1950). — The theory for a wire of rectangular cross-section has also been given by MacDONALD and SARGINSON: Proc. Roy. Soc. Lond., Ser. A **203**, 223 (1950).

and in the limiting cases this gives:

$$\frac{\varrho}{\varrho_0} = 1 + \frac{3}{4}\frac{l}{a}(1-p); \qquad \left(\frac{l}{a} \ll 1\right) \tag{21.7}$$

and

$$\frac{\varrho}{\varrho_0} = \frac{l}{a}\left(\frac{1-p}{1+p}\right); \qquad \left(\frac{l}{a} \gg 1\right). \tag{21.8}$$

Data for $p = 0, \frac{1}{2}$ are also plotted in Fig. 36, while in Fig. 37 a comparison with some experimental data on sodium is shown.

R. G. CHAMBERS[1] has made a detailed survey of thin film and thin wire experiments and has pointed out that although there was a great volume of previous work on very thin films (e. g. particularly that of LOVELL[2] on thin deposited films of the alkali metals) the experiments of ANDREW[3] on thin rolled films of tin and on wires of mercury were perhaps the first which could be compared properly with the theory (cf. also EUCKEN and FÖRSTER[4]). ANDREW's results-could on the whole be satisfactorily interpreted with the diffuse scattering assumption ($p = 0$), and the values of l were found to be in reasonably good agreement with those determined by PIPPARD[5] and CHAMBERS[6] from high-frequency experiments ("anomalous skin-effect") at frequencies $\sim 10^9$ and 10^{10} c/s down to liquid helium temperatures.

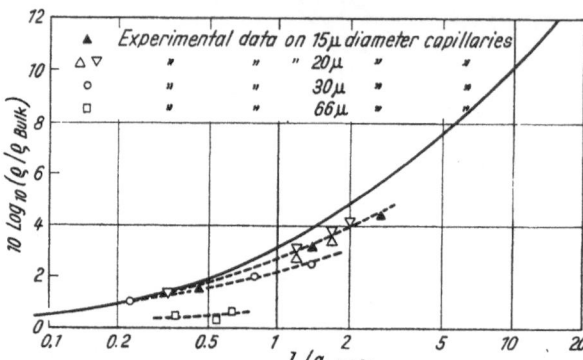

Fig. 37. Size-effect on electrical resistance in sodium "wires" of varying diameter, a. l: mean free path (altered in given specimen by variation of temperature). ———: Theoretical curve assuming diffuse boundary scatttering (after MacDonald and Sarginson).

22. Magneto-resistance size-effect. If now a magnetic field is applied to a thin conductor of size a and the field is such that $2R/a \lesssim 1$, where $2R$ is the diameter of a free electron orbit, then an electron between collisions will tend to spiral around the lines of magnetic force while slowly drifting in the direction of the applied electric field. Thus an electron whose last collision took place further than $2R$ from the walls of the conductor will be unlikely to collide on the next occasion with the walls and in this case the magnetic field will cause a *decrease* in resistance towards the "bulk" resistance-value. On this crude model we can estimate *very* roughly that the conductivity σ of a thin wire and thin plate (see Figs. 38a and b) will be given by:

$$\frac{\sigma_{\text{wire}}}{\sigma_0} \sim \left(1 - \frac{4R}{a}\right)^2 + \left(\frac{8R}{a} - \frac{16R^2}{a^2}\right)\cdot\left(\frac{R}{l}\right)$$

[1] R. G. CHAMBERS: Ph. D. Diss. Cambridge 1951.
[2] A. C. B. LOVELL: Proc. Roy. Soc. Lond., Ser. A **157**, 311 (1936).
[3] E. R. ANDREW: Proc. Phys. Soc. Lond. A **62**, 77 (1949).
[4] A. EUCKEN and F. FÖRSTER: Göttinger Nachr. **1**, 43, 129 (1934). See also Ann. Phys., Lpz. **28**, 603 (1937); **30**, 494 (1937); **33**, 733 (1938); **40**, 121 (1941).
[5] A. B. PIPPARD: Proc. Roy. Soc. Lond., Ser. A **191**, 385 (1947); **203**, 98 (1950).
[6] R. G. CHAMBERS: Nature, Lond. **165**, 239 (1950).

where l is the bulk mean free path and σ_0 the bulk conductivity, i.e.

$$\frac{\sigma_{\text{wire}}}{\sigma_0} \sim 1 - \frac{8R}{a} + \frac{16R^2}{a^2}\left\{1 + \frac{a}{2l}\right\} \tag{22.1}$$

$$\left(\frac{2R}{a} \lesssim 1; \quad \frac{a}{l} \lesssim 1\right)$$

$$\left(\text{i.e. } \frac{\varrho}{\varrho_0} \approx 1 + \frac{8mvc}{e\mathcal{H}a}, \quad \text{for "large" } \mathcal{H}\right)$$

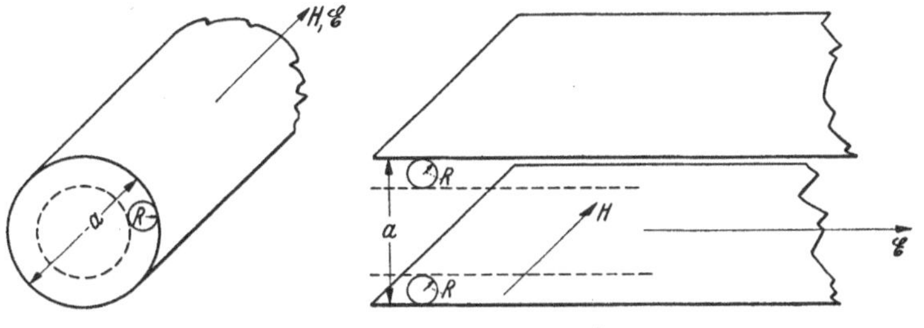

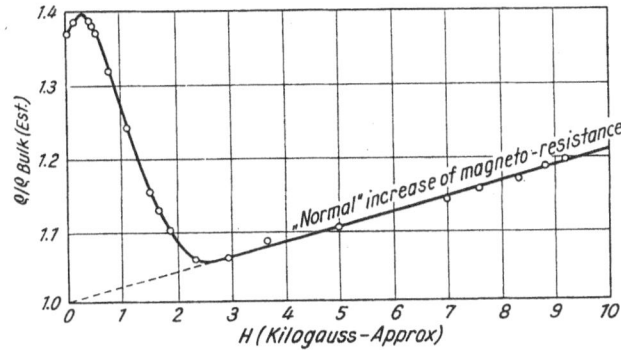

Fig. 38 a and b. (a) To illustrate influence of size-effect in magneto-resistance in thin rods. (b) To illustrate influence of size-effect in magneto-resistance in thin plates.

and correspondingly

$$\frac{\sigma_{\text{film}}}{\sigma_0} \sim 1 - \frac{4R}{a} + \frac{4R^2}{a^2}\left(\frac{a}{l}\right) \tag{22.2}$$

$$\left(\text{i.e. } \frac{\varrho}{\varrho_0} \approx 1 + \frac{4mvc}{e\mathcal{H}a}, \quad \text{for "large" } \mathcal{H}\right).$$

In both cases we have assumed diffuse scattering at the walls. Thus comparison of more exact theory with experiment (e.g. CHAMBERS[1], MacDONALD and SAR-GINSON[2]) would enable one to determine R/a (i.e. effectively mv, the electron

Fig. 39. Influence of size on magneto-resistance in sodium. $T = 4.2°$ K; $\tilde{r}_T \approx 2 \times 10^{-4}$.

momentum) and l/a. If, however, this magnetic effect is to be readily observable it must not be swamped by the bulk magneto-resistance due to deviation from the homogeneous free-electron model as discussed earlier. Thus on the one hand the fact that the effect has only shown itself significantly in sodium is further confirmation of the validity of the free-electron model for that metal while on the other it essentially deprives us, as yet at any rate, of a useful tool. The phenomenon in sodium is shown in Fig. 39.

[1] R. G. CHAMBERS: Proc. Roy. Soc. Lond., Ser. A **202**, 378 (1950).
[2] D. K. C. MacDONALD and K. SARGINSON: Proc. Roy. Soc. Lond., Ser. A **203**, 223 (1950).

23. The "anomalous skin-effect". If we consider equation (21.8) for the resistivity of a very thin wire, $(a/l \ll 1)$, we may write:

$$\varrho \approx \frac{l \, \varrho_0}{a} \left(\frac{1-p}{1+p} \right)$$

but since $\varrho_0 = \dfrac{m \, v}{n \, e^2 \, l}$, this gives $\varrho = \dfrac{m \, v}{n \, e^2 \, a} \left(\dfrac{1-p}{1+p} \right)$, so that the resistance becomes independent of the bulk mean free path. Somewhat similarly[1] when $\delta/l \ll 1$, where δ is the depth of penetration of a high-frequency field (of angular frequency ω), the observed resistivity of the metal becomes *independent* of the bulk d.c. resistivity, while classically at high frequencies the observed resistivity is proportional to $\sqrt{\omega \, \varrho_0}$.

From Pippard's and Chambers' experiments one derives values of the mean free path, l, at a given temperature or alternatively one may take the ratio σ_0/l ($\equiv 1/l\varrho_0$) as characteristic of a given metal. This can otherwise be expressed as an apparent number of free electrons per atom, n/n_a, using the conductivity formula $n = \dfrac{m \, v \, \sigma_0}{e^2 \, l}$. Mean values as given by Chambers are shown in the adjoining table.

Metal	σ_0/l $(\Omega^{-1}\,\mathrm{cm}^{-2})$	n/n_a
Sn	9.5×10^{10}	1.12
Cu	15.4×10^{10}	1.01
Ag	8.5×10^{10}	0.60
Au	8.3×10^{10}	0.58

The chief difficulty in this work, apart from the intrinsic feature of having to resort to the specialized high-frequency techniques, is that the results are highly sensitive to the surface treatment of the metal, depending critically on the method of polishing since of course the resistance being measured is essentially confined to the effective penetration-depth of the high frequency field. Pippard and Chambers[2] have sought the best methods of surface treatment to provide reliable results but it is certainly rather unfortunate that so promising a tool is limited in this way. Pippard[3] has also used these techniques to gain insight into the superconducting state.

24. The resistance-minimum at low temperatures. In the field of electron transport in metals, super-conductivity still presents a major problem of theoretical interpretation although many people believe that Fröhlich and Bardeen have now indicated the essential "mechanism" involved. However, the resistance minimum (e.g. Fig. 40a and b) found in certain metals at low temperatures, first observed in Leiden some 20 years ago, still resists a satisfactory theoretical explanation. This smooth rise of resistance with decreasing temperature is at first sight certainly much less dramatic than the abrupt disappearance of resistance in a superconductor at the transition temperature, but it may perhaps be that an understanding of the minimum will require an advance in fundamental theory similar (and perhaps closely related) to that necessary for superconductivity.

Various patterns of behaviour appear possible in this phenomenon. Thus Gerritsen and Linde find in their extensive investigation of silver alloys

---

[1] Had we chosen the thin plate formula instead of that for a thin wire then a slow logarithmic dependence would still be present. The full theory of the anomalous skin effect, which is quite complex, due to Pippard [Proc. Roy. Soc. Lond., Ser. A **191**, 385 (1947)], Reuter and Sondheimer [Proc. Roy. Soc. Lond. Ser. A **195**, 336 (1948)] shows that the resistivity does in fact become *independent* of the bulk resistivity.

[2] R. G. Chambers and A. B. Pippard: Inst. of Metals Monograph No. 13, p. 281. — R. G. Chambers: Proc. Roy. Soc. Lond., Ser. A **215**, 481 (1952).

[3] Pippard, A. B.: Proc. Roy. Soc. Lond., Ser. A **203**, 195 (1950); **216**, 547 (1953).

(more particularly of maganese in silver) that the resistance, after passing through a minimum passes through a subsequent maximum at some lower temperature and then falls once more with further reduction of temperature (see Figs 41 a and b). On the other hand in our work on various dilute copper alloys we have only observed a resistance-*minimum* and at the lowest temperatures preliminary experiments[1] indicate that the resistance of such alloys ultimately flattens off to become quite constant. In the work at Bristol on the noble metals, MENDOZA and THOMAS also observed only a minimum but in some of their specimens the resistance at very low temperatures (below 1° K) appeared to rise with increasing rapidity.

Naturally, theoretical proposals may be influenced in one direction or another by a particular pattern of experimental behaviour. Thus GORTER in 1938[2] suggested that perhaps a new thermodynamic principle should be invoked — that the velocity of all irreversible processes should vanish at the absolute zero of temperature — with the consequence that all bodies must either become superconducting or perfect insulators at $T = 0$. Thus the resistance of all non-superconductors would ultimately have to rise to infinity at the absolute zero. Although this hypothesis appears today rather unlikely, the results of MENDOZA and THOMAS might yet seem to lend it some support. Again, it may be noted that no resistance minimum in any alkali metal has yet been confirmed; this suggests quite strongly that the phenomenon is in fact confined to certain groups of metals, a behaviour reminiscent

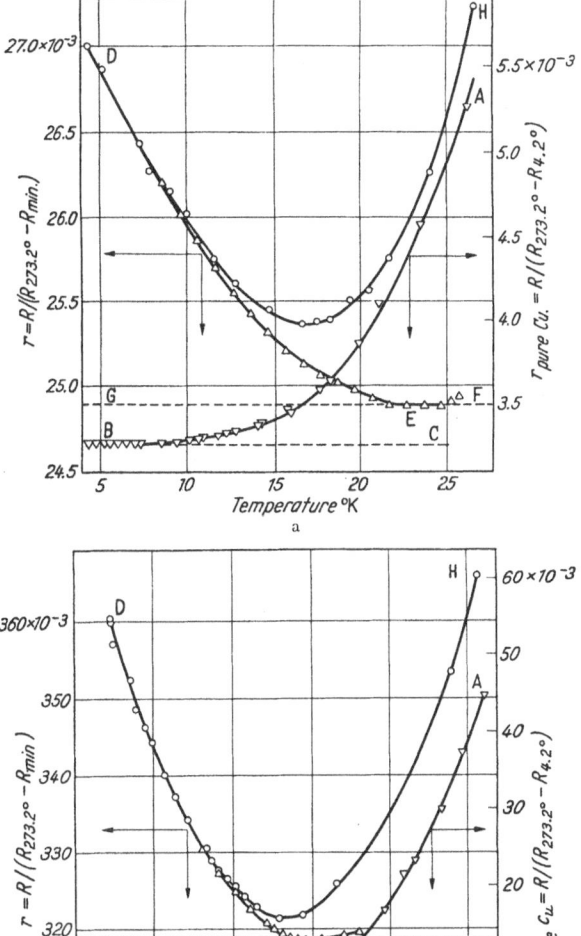

Fig. 40 a and b. (a) Resistance-minimum observed in very dilute alloy of copper with ∼ 0.005 at % Fe (after PEARSON). *H* Resistance of alloy, *A* Resistance of pure copper; *E* "Anomalous" resistance-component of alloy. (b) Resistance-minimum observed in dilute alloy of copper with ∼0.044 at % Fe (after PEARSON). *H* Resistance of alloy; *A* Resistance of pure copper; *E* "Anomalous" resistance-component of alloy.

[1] G. K. WHITE: Canad. J. Phys. **33**, 119 (1955).
[2] G. J. GORTER: Physica, Haag **5**, 483 (1938).

of superconductivity. On the other hand it may be that the appropriate *solute* to evoke the minimum has not been present and we remember that the alkalis with the exception of lithium will only take into stable solid solution members of their own group. In saying, too, that no alkali metal becomes superconductive we must add the qualification: "at least down to temperatures in the order of 0.1° K"; and HEISENBERG's theory originally demanded that at

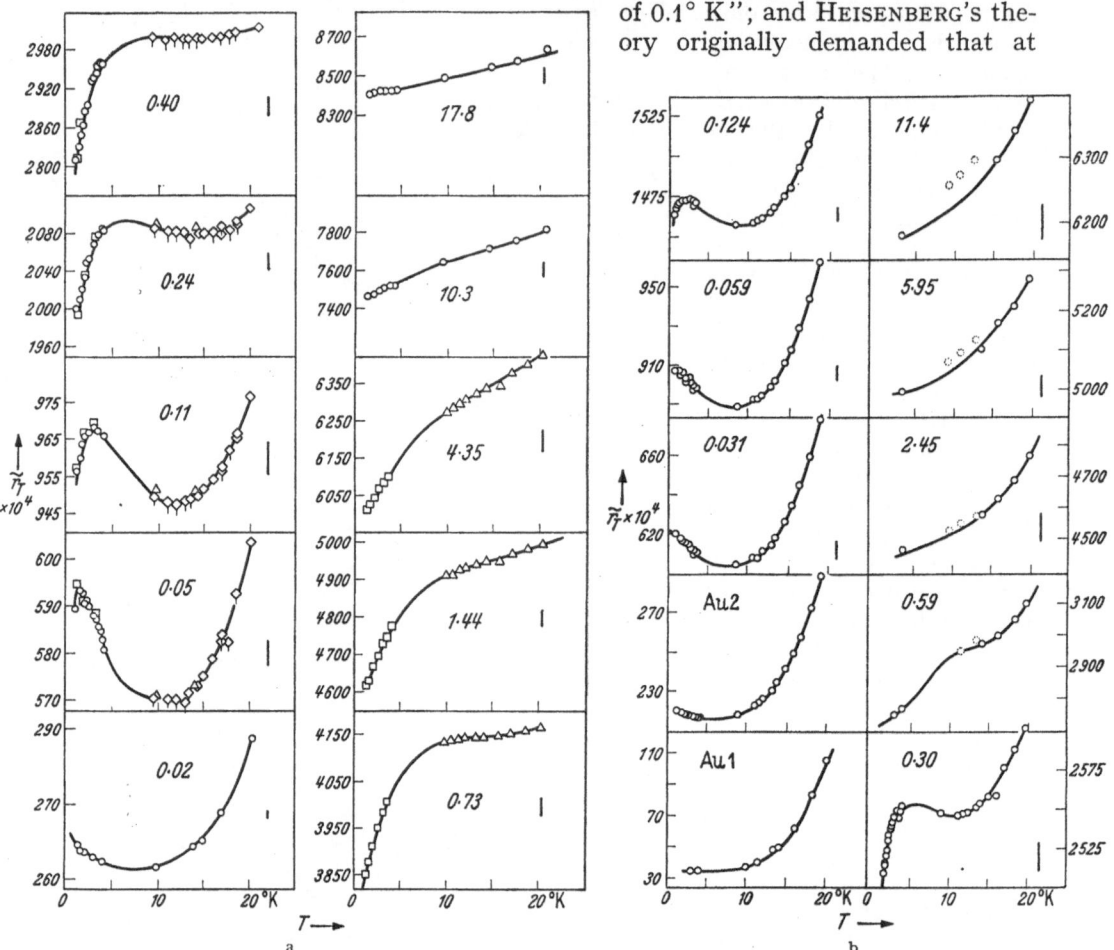

Fig. 41 a and b. (a) Resistance ($\tilde{r}_T = \varrho_T/\varrho_{273}$) of silver-manganese alloys (after GERRITSEN and LINDE). The numbers on each graph give the atomic percentage of manganese. The small vertical lines represent 1% of $\tilde{r}_T$. (b) Resistance ($\tilde{r}_T = \varrho_T/\varrho_{273}$) of gold (Au 1 and Au 2) and of gold-manganese alloys (after GERRITSEN and LINDE). The numbers on each graph give atomic percentage of manganese solute. The small vertical lines represent 1% of $\tilde{r}_T$.

sufficiently low temperatures every metal either become ferromagnetic (? or antiferromagnetic) or superconductive.

We shall then for convenience limit our further discussion of the resistance-minimum to the experiments of MacDONALD and PEARSON (which also include collateral thermoelectric observations), without suggesting that these are comprehensive or necessarily representative of the whole phenomenon. In Fig. 42 are shown experimental results on a range of alloys of tin in copper. It will be seen (Fig. 42b) that the anomalous resistance causing the minimum increases in magnitude with the concentration of tin, reaching a limit, however at about

0.005 at.% tin. This already presents a puzzling problem—to explain why a phenomenon caused by "impurity" atoms in this way should "saturate" at such a low solute concentration. The corresponding curves for the absolute thermo-electric force, of these specimens are shown in Fig. 42a, and it will be seen that, in this same range of concentration, the thermo-e.m.f. appears highly anomalous in relation to the pure metal becoming more normal again, however, at higher concentrations.

The "modern" theory of electron transport in conductors provides an expression for the absolute thermoelectric power S, assuming that the lattice vibrations are negligibly disturbed by either an electric field or tem-

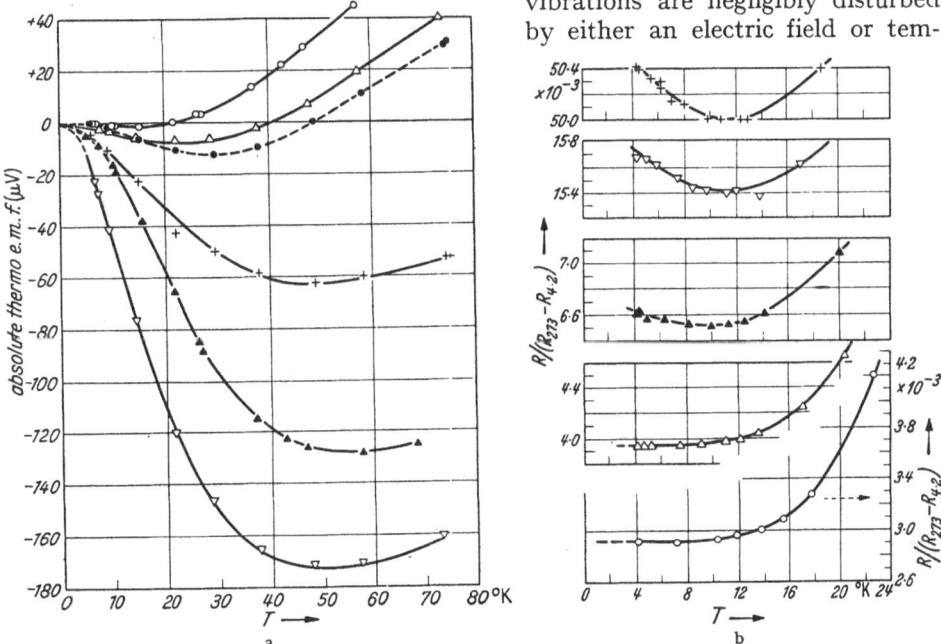

Fig. 42a and b. (a)Absolute thermo-electric force of copper and alloys ○ pure copper; △ pure Cu + 0.0009 atomic % Sn; ▲ pure Cu + 0.0026 atomic % Sn; ▽ pure Cu + 0.0054 atomic % Sn; + pure Cu + 0.026 atomic % Sn; ● pure Cu + 0.01 atomic % Ni. (b) Relative electrical resistance of these copper and copper-tin alloys.

perature-gradient of the magnitude used in experiments, as follows[1]:

$$S = -\frac{\pi^2 k^2 T}{3e} \left(\frac{d \log \varrho(E)}{dE}\right)_{E=\zeta} \tag{24.1}$$

where $\varrho(E)$ is the resistivity experienced by electrons of energy E. Thus on this basis thermoelectric power should in general provide information on the relative variation with electron energy of the scattering cross-section. If we assume that we may write $\varrho_{tot} = \varrho_{norm} + \varrho_{anom}$, where the second term is that directly responsible for the appearance of the anomalous minimum then:

$$S = -\frac{\pi^2 k^2 T}{3e} \left[\frac{\varrho_{norm}}{\varrho_{tot}} \left(\frac{d \log \varrho_{norm}^{(E)}}{dE}\right) + \frac{\varrho_{anom}}{\varrho_{tot}} \left(\frac{d \log \varrho_{anom}^{(E)}}{dE}\right)\right]_{E=\zeta}. \tag{24.2}$$

We can thus see immediately that the anomalous component of S will be greatest when $\varrho_{anom}/\varrho_{tot}$ is maximal and will diminish again if ϱ_{anom} "saturates", while the

---

[1] This expression (cf. e.g. MOTT and JONES [2], p. 310. —WILSON [1], p. 204) itself involves approximations but should be adequate for this discussion.

normal scattering ϱ_{norm} (and hence ϱ_{tot}) continues to increase with added impurity—as indeed observed. A further quantitative comparison is a little more difficult since the precise validity of the factor $\dfrac{\pi^2 k^2 T}{3e}$ is in some doubt[1]. However, we can compare rather directly the values of $d\log \varrho_{anom}(E)/dE$ and $d\log \varrho_{norm}(E)/dE$ within the square brackets in (24.2). On current theories of scattering by lattice vibrations and impurity atoms[2] $d\log\varrho_{norm}(E)/d\log E$ will generally be of order unity. In order then to account for the experimental results on the dilute alloys of copper, it appears[3] that $d\log \varrho_{ancm}(E)/d\log E$ must be $\gtrsim 100$! On this basis we are thus forced to assume that an *extremely* energy-dependent scattering can be produced by certain solutes in dilute concentration. Gerritsen and Korringa assumed that manganese in silver could give rise to a "resonant" scattering level very close to the top of the Fermi surface, but since the phenomenon in copper can be produced by a wide variety of solutes (e.g. Fe, Sn, Pb, Bi, Ge ...) it appears to us rather difficult to carry over such an apparently fortuitous hypothesis to this case.

The assumption that the lattice vibrations are negligibly perturbed in electron transport problems was first questioned by Peierls[4] in relation to electrical conductivity. The electrons acquire momentum continuously from the applied electric field and in a steady state transfer this by collisions to the lattice system. When the transfer is made to excited lattice vibrations ("phonons"), the difficulty arises at low temperatures that the mutual relaxation time of the lattice waves becomes long and it is not clear how effective thermal equilibrium is to be maintained. The problem was raised in a fresh guise in 1945 by Gurevich[5] who pointed out that in thermoelectric power measurements the applied temperature gradient will give rise to a flow of heat by the lattice waves; if there is significant "scattering" of the excited lattice vibrations by the free conduction electrons then the latter will acquire momentum in this way and set up a charge gradient in the conductor which is simply another component of thermoelectric power. The problem has been discussed again recently by a number of workers[6], and it appears that this component of thermoelectric power plays an important role in semiconductors. It would indeed be attractive if one could show that this presumed failure to maintain adequate thermal equilibrium in the lattice vibrations were ultimately responsible for the appearance of the anomalous resistance minimum and corresponding thermoelectric power in these dilute alloys[7]. An obvious difficulty, however, is that of explaining why the situation only appears to arise with the presence of a small concentration of impurity atoms which in fact would rather be expected to *aid* the maintenance of thermal equilibrium by providing independent scattering centres.

[1] Cf. e.g. D. K. C. MacDonald and W. B. Pearson: Proc. Roy. Soc. Lond., Ser. A **219**, 373 (1953); **221**, 534 (1954). — G. T. Pullan: Proc. Roy. Soc. Lond., Ser. A **217**, 280 (1953).

[2] See e.g. N. F. Mott and H. Jones [*2*], p. 311, 312. — A. H. Wilson [*1*], p. 266 et seq. D. K. C. MacDonald and S. K. Roy: Phil. Mag. **44**, 1364 (1953).

[3] D. K. C. MacDonald and W. B. Pearson: Phil. Mag. **45**, 491 (1954). Cf. also D. K. C. MacDonald and W. B. Pearson: Acta met. **3**, 392, 403 (1955).

[4] R. E. Peierls: Ann. Phys., Lpz. **4**, 120 (1930); **10**, 97 (1931).

[5] L. Gurevich: J. Phys. USSR. **9**, 477 (1945); **10**, 67 (1946).

[6] H. P. R. Frederikse: Phys. Rev. **92**, 248 (1953). — D. ter Haar and A. Neaves: Proc. Roy. Soc. Lond., Ser. A **228**, 568 (1955). — C. Herring: Phys. Rev. **96**, 1163 (1954).— P. G. Klemens: Austral. J. Phys. **7**, 520—522 (1954). — D. K. C. MacDonald: Physica, Haag **20**, 996 (1954). I am also indebted to Dr. F. J. Blatt for unpublished discussion.

[7] Cf. also D. K. C. MacDonald, W. B. Pearson and G. K. White: Bull. Int. Inst. Froid (Report on Grenoble Conference, 1954). — Annexe 1955–2, p. 107. — D. K. C. MacDonald: Rapp. 10^e Congr. Solvay.

D. Concluding remarks.

25. The conduction electron assembly and electron-electron interaction. Two outstanding issues in this field today are the need for a more adequate model and analysis of electron-scattering in metals and the resolution of the problems relating to thermal equilibrium. These are not entirely separate questions since, for example, both call for an adequate understanding of the electron assembly behaviour. When LORENTZ first applied statistical methods (the "BOLTZMANN equation") to the theory of electron transport in metals he specifically assumed that electron-electron "collisions" were negligible in comparison with the interaction of electrons and atoms: "..., we shall suppose the collisions with the metallic atoms to preponderate; the number of these encounters will be taken so far to exceed that of the collisions between electrons mutually, that these may be altogether neglected".

Since a collision was then essentially regarded as occurring between billiardball-like particles and since the "classical" radius of the electron ($\sim 10^{-12}-10^{-13}$ cm.) was so much smaller than that of an atom, this view-point was not unreasonable, although the long-range COULOMB interaction must always tend to arouse misgivings. The so-called "one electron" model of later quantum mechanical discussions continued this tradition inasmuch as the motion of any given electron is assumed in first approximation to be independent of the motion of the other electrons. That is to say the "other" electrons are assumed only to give rise to the "self-consistent" average-, or "smeared-out-", potential in which our chosen electrons moves, and correlations in position and energy due to the COULOMB interactions are regarded as small perturbations at most. Using this model to discuss the equilibrium properties of a metal, such as the electronic specific heat, provides a rather satisfactory account of the experimental observations. When, however, we try to improve the model by including the exchange energy[1] we find that the theoretical linear dependence on temperature of the electronic specific heat would now be replaced at low temperatures by variation with $-T/\log a\,T$; yet recent calorimetric experiments by RAYNE[2] confirm the *linear* variation with temperature of specific heat below 1° K in a number of metals.

These difficulties would be essentially removed if we were to assume, as did LANDSBERG[3], that the long-range COULOMB interaction between the electrons could be replaced by a short-range screened COULOMB potential of the form $\frac{e^2}{r}\,e^{-r/b}$ (with $b\sim 10^{-8}$ cm.) similar to that discussed earlier for the electron-ion interaction. BOHM and PINES[4] have recently examined in considerable detail the approximate description of the electron assembly and conclude that the COULOMB interaction is responsible for collective oscillations of the electron fluid as a whole (the "plasma" oscillations) which, however, are *not* excited in metals at ordinary temperatures. These potential collective oscillations (which are of great importance in *other* spheres of interest, such as electronic amplifiers) exhaust a certain number of degrees of freedom and the remainder are found to correspond to a collection of individual electrons interacting through a screened potential

---

[1] J. BARDEEN: Phys. Rev. **50**, 1098 (1936). — E. P. WOHLFARTH: Phil. Mag. **41**, 534 (1950).

[2] J. RAYNE: Phys. Rev. **95**, 1428 (1954).

[3] P. T. LANDSBERG: Proc. Phys. Soc. Lond. A **162**, 49 (1949).

[4] D. BOHM and D. PINES: Phys. Rev. **82**, 625 (1951); **85**, 338 (1952); **92**, 609, 626 (1953). See also D. PINES: Rapp. 10ᵉ Congr. Solvay. — J. BARDEEN and D. PINES: Phys. Rev. **99**, 1140 (1955).

which is indeed closely representable by $\frac{e^2}{r}\, e^{-r/r_c}$, where r_c is about the interatomic distance.

With these conclusions Pines and Bohm essentially restore the *status quo* so that we may with justice discuss the electronic behaviour on a one electron model with a short range interaction. A collision cross-section can now be derived[1] (we recall that this is not analytically possible with an unscreened Coulomb potential) and this is just of the order πr_c^2, i.e. of the same magnitude as that of scattering by an individual ion. However, we have now to bear in mind that while the collision of an electron with an ion can only involve a very small energy interchange, this is not the case for a collision of two similar particles; the Pauli principle then restricts electron-electron collisions effectively to those electrons within the thermal energy "spread" of the Fermi surface. The mean free path for these collisions is thus of order $\left(\frac{\zeta}{kT}\right)^2 \cdot \frac{1}{\pi r_c^2}$; the large factor $\left(\frac{\zeta}{kT}\right)^2$ suggests

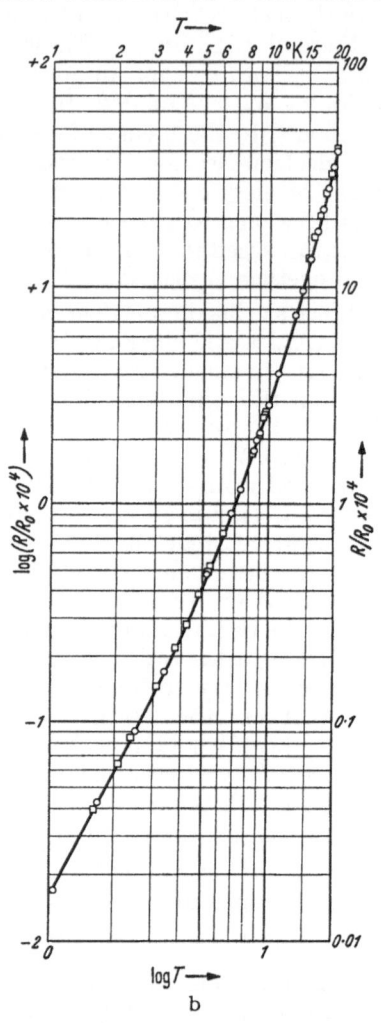

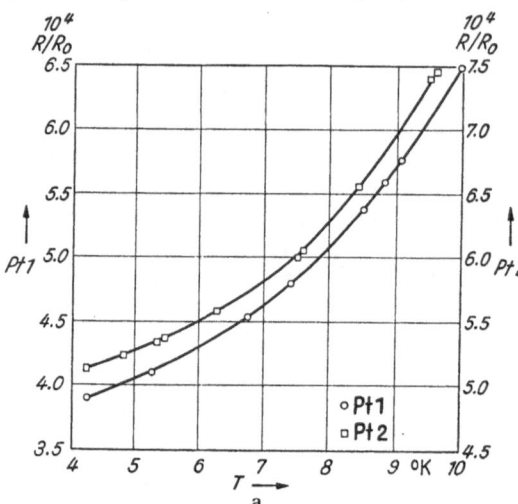

Fig. 43 a and b. (a) Electrical resistance of specimens of platinum at low temperatures (after de Haas and de Boer). (b) Electrical resistance (with residual resistance subtracted) of platinum at low temperatures (after de Haas and de Boer). The logarithmic plot indicates that the resistance at low temperatures tends to a quadratic law.

that electron-electron scattering would generally be negligible in comparison with that due to other mechanisms. If electron-electron interaction were significant in electrical resistance it should give rise to a component varying as T^2; de Haas and de Boer[2] in Leiden observed in fact such a resistive component at low temperatures in certain metals—e.g. platinum, see Figs. 43 a and b and Baber[3] at that time interpreted the resistance component as due to electron-electron interaction between different bands, assuming again an interaction

[1] E. Abrahams: Phys. Rev. **95**, 839 (1954).
[2] W. J. de Haas and J. H. de Boer: Physica, Haag **1**. 609 (1933).
[3] W. G. Baber: Proc. Roy. Soc. Lond., Ser. A **158**, 383 (1937).

potential of the form $-\dfrac{e^2}{r}\,e^{-r/b}$ between a conduction electron and a "hole" in the d-band. WILSON, however, comments: "... the proposed explanation is not very convincing" (see also KEMP et al.[1]).

26. The BOLTZMANN electron-transport equation. Let us now consider in a little detail the assumptions involved in setting up the statistical theory of electron motion in a metal. We assume that the electron-"gas" is sufficiently described by a distribution function $f(x, p_x)\, dx\, dp_x$ for the density of electrons (for convenience we consider only the x coordinate in space and the associated momentum). Thus it should be noticed that we are essentially ignoring individual fluctuations.

Then, quite generally[2], the rate of change of the density, f, with time is given by

$$\frac{\partial f}{\partial t} = -\left[\left(\frac{\partial f}{\partial x}\right)_{p_x}\cdot \dot{x} + \left(\frac{\partial f}{\partial p_x}\right)_x \dot{p}_x + f\left\{\left(\frac{\partial \dot{x}}{\partial x}\right)_{p_x} + \left(\frac{\partial \dot{p}_x}{\partial p_x}\right)_x\right\}\right]. \qquad (26.1)$$

If now the motion of each element (i.e. an electron in this case) in our assembly be describable by a HAMILTONian function of x and p_x then

$$\dot{x} = \frac{\partial H}{\partial p_x}; \qquad \dot{p}_x = -\frac{\partial H}{\partial x}$$

and

$$\frac{\partial \dot{x}}{\partial x} = \frac{\partial^2 H}{\partial x\, \partial p_x} = -\frac{\partial \dot{p}_x}{\partial p_x}$$

and (26.1) reduces to:

$$\frac{\partial f}{\partial t} = -\left[\left(\frac{\partial f}{\partial x}\right)_{p_x}\cdot \dot{x} + \left(\frac{\partial f}{\partial p_x}\right)_x \cdot \dot{p}_x\right] \equiv -\left[\left(\frac{\partial f}{\partial x}\right)_{p_x}\cdot \dot{x} - \left(\frac{\partial f}{\partial p_x}\right)_x \cdot \frac{\partial H}{\partial x}\right]. \qquad (26.2)$$

External electric or magnetic fields applied to the metal can be specified by a HAMILTONian function in this way but not interactions between the electrons themselves or between electrons and the ions forming the lattice. If we then assume that we shall (and can) calculate the rate of change of f produced by these interactions independently, say $\{\partial f/\partial t\}_{\text{collisions}}$, then we may write:

$$\frac{\partial f}{\partial t} = -\left[\left(\frac{\partial f}{\partial x}\right)\dot{x} - \left(\frac{\partial f}{\partial p_x}\right)\frac{\partial H}{\partial x}\right]_{\text{no interaction}} + \left\{\frac{\partial f}{\partial t}\right\}_{\text{collisions}}. \qquad (26.3)$$

Finally if a steady ("statistically stationary") state ensues then f does not depend explicitly on t and we have:

$$\left[\left(\frac{\partial f}{\partial x}\right)\dot{x} - \left(\frac{\partial f}{\partial p_x}\right)\frac{\partial H}{\partial x}\right]_{\text{no interaction}} = \left\{\frac{\partial f}{\partial t}\right\}_{\text{collisions}}. \qquad (26.4)$$

The "solution" of this equation is then carried through on the following assumptions:

1. The influence of an electric and/or magnetic field is specified by the term $\dfrac{\partial H}{\partial x}$ in (26.4) $\left(\text{e.g. with an electric field } \mathcal{E}_x,\ -\dfrac{\partial H}{\partial x} = e\mathcal{E}_x\right)$, and it is assumed that we may write with sufficient accuracy $\partial f/\partial p_x \approx \partial f_0/\partial p_x$.

---

[1] W. R. G. KEMP, P. G. KLEMENS, A. K. SREEDHAR and G. K. WHITE: Phil. Mag. **46**, 811 (1955) and W. R. G. KEMP, P. G. KLEMENS and G. K. WHITE: Austral. J. Phys. **9** (1956).

[2] Cf. e.g. R. C. TOLMAN: Principles of Statistical Mechanics, chap. III, p. 48 et seq. particularly. Oxford University Press 1938.

2. A temperature-gradient $\partial T/\partial x$, applied to the metal is assumed to affect the electron-distribution function directly so that we may write:

$$\frac{\partial f}{\partial x} \approx \frac{\partial f_0}{\partial T} \cdot \frac{\partial T}{\partial x}.$$

3. The collision term is calculated on the basis of a transition probability per unit time, $W(\boldsymbol{k}, \boldsymbol{k}')$, that an electron will be scattered from a state $\boldsymbol{k}$ to $\boldsymbol{k}'$ by the ionic lattice. Then[1], taking into account the exclusion principle:

$$\left\{\frac{\partial f}{\partial t}\right\}_{\text{coll}} = \int \left[W(\boldsymbol{k}', \boldsymbol{k}) f(\boldsymbol{k}') \{1 - f(\boldsymbol{k})\} - W(\boldsymbol{k}, \boldsymbol{k}') f(\boldsymbol{k}) \{1 - f(\boldsymbol{k}')\} \right] d\boldsymbol{k}'. \quad (26.5)$$

For thermal scattering the transition probabilities are calculated (cf. Wilson [2], p. 258 et seq.) assuming equilibrium conditions in the lattice.

4. Any contribution to $\{\partial f/\partial t\}_{\text{coll}}$ from electron-electron collisions is generally ignored (cf. foregoing discussion).

At "high" temperatures the collision calculation may be simplified since a unique relaxation time, τ, exists so that we may write $\left\{\dfrac{\partial f}{\partial t}\right\}_{\text{coll}} = -\dfrac{f - f_0}{\tau}$, and Chambers[2] has shown that a kinetic treatment of the electron-scattering then leads to the same results as the Boltzmann equation. The high temperature solutions for the electrical and thermal conductivities (σ; K) are on the whole in good general agreement with experiment ($\sigma \propto 1/T$; K constant) and the Lorenz number $L \equiv K/\sigma T$ is very well predicted by the theory as discussed earlier in this article. At "low" temperatures, however, objections might be raised against assumptions (2) and (3) above—specifically on the grounds that the electrons do not experience a temperature-gradient, dT/dx, *directly* (as they do an electric field, $\mathcal{E}_r$) but rather are affected by the thermal gradient of the lattice *through the medium of the collisions*[3].

A temperature-gradient also raises the question of maintenance of effective thermal equilibrium mentioned at the close of the previous section. It appears therefore noteworthy that while on the whole the experimental results on electrical conductivity have shown fair agreement with the theory, there remain major quantitative discrepancies in the case of thermal conductivity and thermoelectricity. The theory predicts a pronounced minimum in the thermal conductivity of a pure metal near $T/\Theta \approx 0.25$ of which there is no experimental evidence, and it is difficult to reconcile with theory the relative values of electronic thermal conductivity at high and low temperatures. We have already indicated earlier that there are considerable difficulties in the interpretation of thermoelectric power measurements at low temperatures. On the other hand, it must be noted

---

[1] Cf. Wilson [1], p. 193, following L. Nordheim: Ann. Phys., Lpz. **9**, 607 (1931) (or see Mott and Jones [2], p. 258 et seq.).

[2] R. G. Chambers: Proc. Phys. Soc. A, **65**, 458 (1952).

[3] Indeed the "temperature" of the electrons may be in some question—cf. e.g. Mott and Jones [2], p. 263. "We note that, in the derivation of the above formulae, the transfer of energy from the electrons to the lattice vibrations has not been mentioned. The production of Joule heat is, of course, due to this energy transfer. When an electron collides with a vibrating atom, it can, as we have seen, give up energy $\pm h\nu$; when it collides with a foreign atom in solid solution a very small energy transfer can also take place, since the mass of the foreign atom is not infinitely great compared with that of the electrons. It is not, however, necessary to consider this energy transfer explicitly. If it did not take place, the Joule heat produced by the current would be transferred to the electrons alone, whose temperature would therefore rise rapidly, since their heat capacity is small. The rate of transfer of energy between the electrons and the lattice will thus determine only the difference in the temperatures of the electrons and the lattice vibrations."

that ZIMAN[1] has recently argued that the quantitative inclusion of "Umklapp-prozesse" in the theory can improve the agreement with experiment for thermal conductivity at quite low temperatures without seriously affecting the situation in electrical conductivity. Since "Umklappprozesse" involve electron-interaction with relatively high energy excited lattice vibrations ("phonons") they must ultimately be "frozen out" at sufficiently low temperatures. However, ZIMAN suggests that the temperature required may be very low ($T/\Theta \simeq 1/30$) and consequently a satisfying experimental verification of ZIMAN's thesis may prove very difficult[2] since at such low temperatures the "residual" resistance due to chemical and physical defects is dominant even in extremely pure metals.

In conclusion let us quote from BARDEEN[3] in contrast to our opening remarks taken from WILSON: "... with a deeper understanding of the fundamental causes of resistance in metals and alloys, resistivity measurements promise to play an increasing role in the study of other physical problems".

I should like to thank warmly my colleagues at the National Research Council, Ottawa for many helpful discussions and comments on this article, and also Miss L. Soublière for her able assistance in preparing the manuscript.

Bibliography.

[1] WILSON, A. H.: The Theory of Metals Cambridge University Press, 1st edn 1936; 2nd edn 1953. (Page references are to 2nd edn.)

[2] MOTT, N. F., and H. JONES: The Theory of the Properties of Metals and Alloys. Oxford: Clarendon Press 1936.

[3] SEITZ, F.: The Modern Theory of Solids. New York and London: McGraw-Hill Book Co. 1940.

Some other texts of value and interest in the field are:

BRILLOUIN, L.: Die Quantenstatistik. Berlin: Springer 1931.

FRÖHLICH, H.: Elektronentheorie der Metalle. Berlin: Springer 1936.

HUME-ROTHERY, W.: The Metallic State. Oxford University Press 1931.

JUSTI, E.: Leitfähigkeit und Leitungsmechanismus fester Stoffe. Göttingen: Vanden-hoeck & Ruprecht 1948.

SLATER, J. C.: Quantum Theory of Matter. New York and London: McGraw-Hill Book Co. 1951.

---

[1] J. ZIMAN: Proc. Roy. Soc. Lond., Ser. A **226**, 436 (1954). See Chap. C of this article, p. 176.

[2] S. B. WOODS: Canad. J. Phys. **34** (1956). Cf. also G. K. WHITE, S. B. WOODS and D. K. C. MACDONALD: Proc. Roy. Soc. Lond., Ser. A **1956**.

[3] J. BARDEEN: J. Appl. Phys. **11**, 88 (1940).

Thermal Conductivity of Solids at Low Temperatures.

By

P. G. Klemens.

With 18 Figures.

I. Introduction.

1. Introduction. The most important mechanisms of heat transfer in solids are transfer by the lattice waves and transfer by the conduction electrons, and consequently the materials to be studied are broadly divided into three groups: (i) non-metals, where transfer is only by lattice waves, (ii) metals, where transfer is mainly by the conduction electrons and lattice conduction, though present, is unimportant, and (iii) alloys and other badly conducting metallic solids, where the electronic conductivity is so small that both processes are important.

The theory of the electronic component of thermal conduction forms part of the free electron theory of metals, and one of the earliest successes of that theory was the explanation of a relationship between the electrical and thermal conductivities, observed by Wiedemann and Franz[1] and by Lorenz[2], on the basis of the rough theory of Drude[3], the more refined theory of Lorentz[4] and finally by means of Sommerfeld's[5] theory of a free electron gas obeying Fermi-Dirac statistics. As will be seen in Sect. 13, this relationship can be derived from very general considerations, and requires only a common relaxation time for electrical and thermal conduction.

Measurements of the thermal conductivity of pure metals at low temperatures brought to light cases for which this condition was not satisfied. Measurements were made down to liquid hydrogen temperatures in the late twenties and early thirties in a number of laboratories, most notably by the group under Grüneisen and the Leiden group. Since the thermal conductivity has a maximum, usually not below 10 to 20° K, and since the conductivity below that maximum is determined by crystal imperfections, these measurements gave almost as much information about the intrinsic thermal resistivity as later measurements extending to lower temperatures.

Comparison of the observed intrinsic resistivities with theory was delayed by the absence of a reliable solution of the transport equation for low temperature thermal conduction. The theory predicts the intrinsic thermal resistivity to vary as T^2 at lowest temperatures, in approximate agreement with observations, but the multiplicative constant in that theoretical relation was uncertain. Wilson [60] gave an approximate solution, further discussed by Makinson [61], Sondheimer [64] solved the equation to a higher approximation and confirmed that Wilson's result was not very different from the correct one, and Klemens [69] found the conductivity by means of a numerical solution of the transport equation to differ from Sondheimer's result by only 11%.

[1] G. Wiedemann and R. Franz: Ann. Phys., Lpz. (2) **89**, 497 (1853).
[2] L. Lorenz: Ann. Phys., Lpz. (3) **13**, 422 (1881).
[3] P. Drude: Ann. Phys., Lpz. (4) **1**, 566 (1900).
[4] H. A. Lorentz: Proc. Acad. Sci. Amsterdam **7**, 438, 585, 684 (1904/05)
[5] A. Sommerfeld: Z. Physik **47**, 1 (1928); see also [1].

HULM [92], [93] pointed out substantial discrepancies from theory in the temperature dependence of the intrinsic thermal resistance of all metals considered, and since then further cases of discrepancy have been found. It is now certain that this discrepancy arises from the inadequacy of the simplified theoretical model used, and the necessary modifications of the model have been discussed by KLEMENS [70].

EUCKEN [25] measured the thermal conductivity of non-metals from liquid oxygen to room temperatures and found the thermal conductivity to vary as $1/T$. DEBYE [8] showed that this was expected theoretically; this was confirmed by the quantum-mechanical treatment of PEIERLS [9], [10]. PEIERLS also predicted that the intrinsic thermal resistivity should decrease exponentially with decreasing temperature, because the resistance is due to Umklapp-processes, which become increasingly improbable at low temperatures. PEIERLS' theory was extended by POMERANCHUK [13], [14] and by KLEMENS [20].

Measurements at Leiden [28], [29], [30] at liquid hydrogen temperatures failed to find this exponential variation, for in the substances studied the resistance due to Umklapp-processes was overshadowed by the resistance due to lattice imperfections. However, at liquid helium temperatures the conductivity was found to decrease with decreasing temperature and to depend upon the size of the specimen, due to the scattering of the phonons by the external boundaries.

Experimental work on low temperature thermal conduction was renewed on a larger scale after 1945, particularly at Oxford. Techniques were perfected to cover the temperature range between helium and hydrogen temperatures. Thus MENDELSSOHN and ROSENBERG [85], [86], [87] measured a large number of metallic elements; BERMAN, WILKS and others [5], [39], [41], [42], [43], [46] measured a number of non-metals in the form of large crystals, polycrystalline aggregates and glasses and verified the broad basis of the theory of lattice conduction, including the exponential variation of the conductivity at low temperatures predicted by PEIERLS. Since the reality of Umklapp-processes, both in electron-phonon and phonon-phonon interactions, had been doubted at various times, it was gratifying to find experimental evidence for their existence.

The thermal conductivities of alloys have been measured in various laboratories, and when it was possible to separate the electronic from the lattice component of thermal conductivity, the latter, which is limited by interaction of the lattice waves with the free electrons, and is thus less than the conductivity of non-metals, was found to agree roughly with MAKINSON's theory [61], provided the intrinsic thermal resistivity at low, rather than at high, temperatures is taken as standard of comparison.

Measurements on superconductors in the normal and the superconducting state have made at Leiden, Cambridge, Oxford and in the USA. These measurements can be qualitatively interpreted in terms of a two-fluid theory of electrons, such that the superfluid electrons do not carry entropy and do not interact with lattice waves. Thus in the superconducting state the electronic component of the thermal conductivity is reduced and the lattice component enhanced. In the intermediate state there is additional scattering, both of electrons and of lattice waves, by the phase boundaries. In the absence of a theory of superconductivity no quantitative theoretical predictions are possible, nor is it possible to resolve certain observed inconsistencies.

In addition to testing the current theories, present work on thermal conductivities can give information on the following points: (a) From the magnitudes of the intrinsic electrical and thermal resistivities at low temperatures, from the lattice thermal conductivity and from other conduction properties it is possible to draw conclusions about the electronic band structure and the electron-phonon coupling. (b) The temperature dependence of the lattice thermal conductivity depends upon the frequency variation of the phonon relaxation time; if the thermal resistivity due to imperfections is appreciable, it is possible to draw conclusions about the nature of these imperfections from thermal conductivity data of non-metals and of poorly conducting metals or alloys. (c) Thermal conductivity measurements may throw light on the phenomenon of superconductivity and on the structure of the intermediate state.

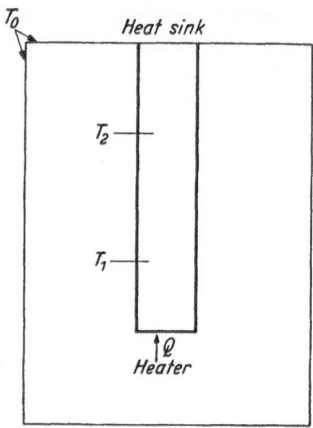

Fig. 1. The "Thermal Potentiometer".

2. Methods of measurement[1]. Thermal conductivity can be measured by stationary or by non-stationary methods. The stationary method, schematically illustrated in Fig. 1, is essentially a thermal potentiometer. Heat is produced at a rate Q in a heater and made to flow through the specimen; when steady conditions have been reached, the temperature difference between two points on the specimen is observed by attached thermometers.

Heat leaks from the heat source and the thermometers to the surrounding shield have to be kept as low as possible, because they falsify the apparent heat current through the specimen; also if there is an appreciable heat leak through the thermometers, contact resistances at the point of attachment may falsify the values of T_1 and T_2.

It is possible to correct for the first, but not the second, of these effects by replacing the specimen by a very bad conductor of known conductivity [39].

While leaks due to gas and solid conduction can be reduced, heat leaks by radiation, increasing rapidly with temperature, make the stationary method somewhat uncertain at high temperatures, but below liquid oxygen temperatures they are not serious and can be corrected for.

In the two-thermometer potentiometric methods the contact resistance between the cold end of the specimen and the shield does not effect the measurements. Nevertheless it is desirable to keep this resistance as low as possible, otherwise it may be impossible to achieve sufficiently low specimen temperatures in the presence of the required heat currents. In some of the earlier work only one thermometer was attached to the specimen, and the conductivity deduced from $T_1 - T_0$; in this case contact resistance can falsify the measurements.

Gas thermometers are frequently used, because they do not require extensive calibration and are not affected by magnetic fields. The two thermometers can be connected to a differential manometer; care must be taken to keep the volume of the two thermometers as equal as possible, and the "dead" volume (capillaries and manometer) as small as possible, so that the appropriate corrections need be treated to first order only. Systems of this type have been described by Hulm [92], Berman [39], Andrews, Webber and Spohr [95], White [88] and Rosenberg [87]. Gas thermometers become increasingly difficult to use below about 2° K, since the thermometer pressure must not exceed the vapour pressure of helium.

---

[1] See also Olsen and Rosenberg [6].

Resistance thermometers have been used by DE NOBEL [109] in the liquid hydrogen range; at lower temperature they are usually not sufficiently sensitive, although ALLEN and MENDOZA [114] have, with some difficulty, used phosphor-bronze resistors in the liquid helium range. Some carbon resistors can also be used at these temperatures and OLSEN and RENTON [134] by using carbon composition thermometers, which they painted directly onto the specimens, have extended measurements using the stationary method down to 0.4° K.

Since the attainment of the steady state takes some time, this method requires good control of the temperature T_0 of the shield. In the temperature ranges attained by pumping a liquid, this is readily achieved by means of a manostat[1], but the intermediate regions present some difficulty. Early measurements therefore fail to cover the region from 5 to 15° K and from 25 to 60° K. A method of continuous temperature control, using liquid helium, hydrogen and nitrogen (or oxygen) has been described by BERMAN [39]; similarly WHITE [88] described a system which avoids the use of liquid hydrogen.

Apart from the steady-state technique of OLSEN and RENTON, measurements below liquid helium temperatures have been done by non-stationary techniques. Thus HEER and DAUNT [140], GOODMAN [138] and NICOL and TSENG [101] have established temperature differences by uneven demagnetisation of two pills of salt, and observed the equalisation of temperature with time, using the pills as thermometers. Some difficulty arises from the internal thermal resistance of the salt, and from contact resistance[2].

Thermal diffusivity can be measured by means of a periodic (preferably sinusoidal) heat input and thermometers of quick response. Such system have been described by HOWLING, MENDOZA and ZIMMERMAN[3] and by WALDRON and HURLIN[4].

In some cases it is desirable to measure the thermal and electrical conductivity of the same specimen, so that the electrical conductivity of a rod of very low resistance must be measured. ROSENBERG [97], [87] and WHITE and WOODS [121] have used a galvanometer amplifier described by MACDONALD[5]; this is satisfactory for potential differences down to about 10^{-6} volts. In the case of specimens whose electrical resistance is too low, it would be possible to use an amplifier incorporating a superconducting modulator and transformer immersed in liquid helium, as used by TEMPLETON[6] to measure the electrical resistance of single crystals of high-purity copper. Alternatively it may be possible to make the specimen the core of a high-frequency coil and to deduce its conductivity from the loss-factor[7].

II. Thermal conductivity of dielectric solids[8].

3. Lattice waves[9]. Thermal conduction in non-metallic solids takes place through the motion of the atoms vibrating about their equilibrium positions in

---

[1] e.g. GILMONT: Analyt. Chem. 23, 157 (1951).

[2] These factors are discussed by E. MENDOZA, Ceremonies LANGEVIN-PERRIN. Paris 1948.

[3] D. H. HOWLING, E. MENDOZA and J. E. ZIMMERMAN: Proc. Roy. Soc. Lond., Ser. A 229, 86 (1955).

[4] S. WALDRON and M. A. HURLIN: Phys. Rev. 91, 447 (1953).

[5] D. K. C. MACDONALD: J. Sci. Instrum. 24, 232 (1947). Cf. also chapter B of MAC-DONALD's contribution to this volume.

[6] I. M. TEMPLETON: J. Sci. Instrum. 32, 172 (1955).

[7] H. E. RORSCHACH: M. I. T. Electronics Research Laboratory, Quarterly Progress Report July 1952.

[8] Previously reviewed by BERMAN [5].

[9] A full discussion will be found in [2], [9] and [10], see also R. E. PEIERLS: Quantum Theory of Solids. Oxford: University Press 1955.

the lattice. This thermal motion can be resolved into travelling plane elastic waves, and for a perfect lattice and harmonic interatomic forces these are the normal modes. In real crystals there is energy transfer between these waves as a result of deviations of the lattice from perfect periodicity and harmonicity. These interactions are responsible for thermal resistivity. For the discussion of thermodynamic properties it is sufficient to assume that such interactions occur, but a detailed knowledge of these interactions is required for the study of thermal conductivity.

For the sake of simplicity consider a crystal lattice of one atom per unit cell. Since the atoms of the crystal lattice are bound not to their equilibrium positions but to their neighbours, which in turn are also free to vibrate, the equations of motion are complicated when expressed in terms of the displacement u_m of the m'th atom. But these equations can be transformed into a set of independent three-dimensional harmonic oscillator equations by the FOURIER transformation

$$u_m = \frac{1}{\sqrt{G}} \sum_k \xi_k \, e^{i k \cdot x} \tag{3.1}$$

provided the interatomic forces are proportional to the relative displacements. Here x is the equilibrium position of the m'th atom (and has thus only a discrete set of values), $G = G_1 G_2 G_3$ is the number of unit cells in the crystal of linear dimensions $G_1 a_1$, $G_2 a_2$, $G_3 a_3$, the a's being the three periodicity vectors, and the wave-vectors k are integral multiples of three vectors b_i / G_i, where the three inverse lattice vectors b_i are defined by

$$(a_i \cdot b_j) = 2\pi \, \delta_{ij}. \tag{3.2}$$

Each of the three-dimensional harmonic oscillators can be transformed linearly to principal axes, so that

$$\dot{u}_m = \frac{1}{\sqrt{G}} \sum_k \sum_{j=1}^{3} \varepsilon_{kj} \, \xi_{kj} \, e^{i k \cdot x} \tag{3.3}$$

where ε_{kj} are three mutually perpendicular vectors defining the three directions of polarization of the waves. The ξ's each satisfy an harmonic oscillator equation, and the frequency ω_{kj} is determined by the three roots of a secular equation involving the interatomic force constants. Thus

$$\xi_{kj} = b_{kj} \, e^{i \omega_{kj} t} + b_{k-j} \, e^{-i \omega_{kj} t} \tag{3.4}$$

where b_{kj} and b_{k-j} are the two arbitrary constants in the solution of the harmonic oscillator equation, so that there are two amplitudes for each wave (k, j). But since all u's must be real, the b's are not all independent, but satisfy.

$$b_{kj} = b^*_{-kj}. \tag{3.5}$$

It is convenient to introduce the notation

$$\omega_{kj} = -\omega_{k-j} = -\omega_{-k-j} = +\omega_{-kj} \tag{3.6}$$

and

$$u_m = \frac{1}{\sqrt{G}} \sum_k \sum_{j=\pm 1}^{\pm 3} \varepsilon_{kj} \, b_{kj} \, e^{i(k \cdot x + \omega_{kj} t)}. \tag{3.7}$$

Since the displacements are defined only for those values of x which correspond to the lattice points, (3.7) is invariant for transformations which add or subtract to a wave-vector k an integral combination of the inverse lattice vectors b_i. Thus there exists in k-space a fundamental region, the first BRILLOUIN zone, which contains G different $\boldsymbol{k}$-values, such that any $\boldsymbol{k}$-value can be transformed into a $\boldsymbol{k}$-value lying in the first zone by adding an integral combination of inverse lattice vectors. This first zone contains all physically different waves, and to each value of $\boldsymbol{k}$ there correspond three normal modes. It follows from (3.6) and from the continuity of ω as a function of $\boldsymbol{k}$ that the derivative $\partial\omega/\partial k_n$ normal to the zone boundary should vanish.

Fig. 2 is a schematic plot of ω against $|k|$ for a fixed direction of $\boldsymbol{k}$. The phase-velocity $\omega/\boldsymbol{k}$ and the group-velocity $\partial\omega/\partial\boldsymbol{k}$ generally differ, but, for low values of k, ω is a linear function of k, and the distinction disappears. The low frequency velocities are the same as the velocities of sound waves derived from the theory of elasticity: as the frequency is decreased the atomicity of the solid becomes progressively less important. There are three branches, corresponding to three mutually perpendicular direc-

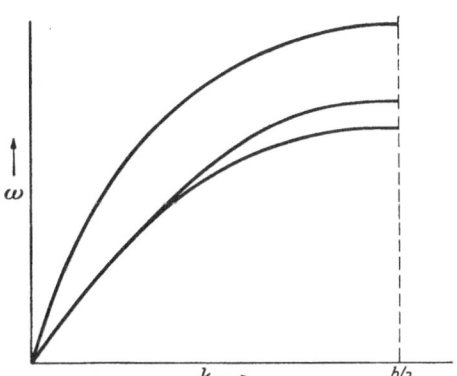

Fig. 2. Frequency versus k for a given direction of $\boldsymbol{k}$; monatomic lattice (schematic).

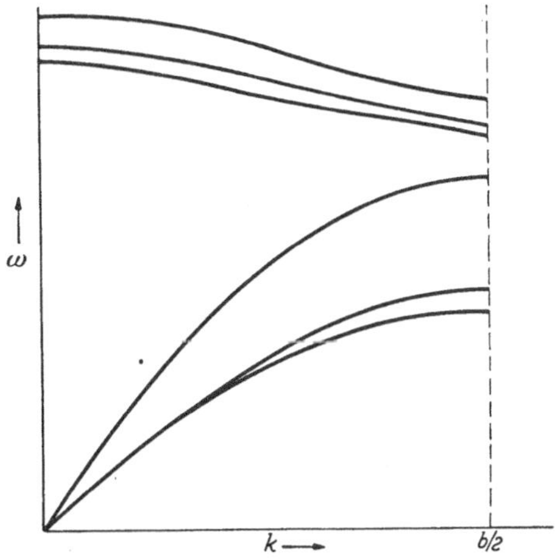

Fig. 3. Frequency versus k for a given direction of $\boldsymbol{k}$; diatomic lattice (schematic).

tions of polarization. At high frequencies the three axes of polarization do not bear any simple relation to the direction of propagation, but at low frequencies their directions are as for the corresponding elastic continuum. The branch of highest frequency is customarily called the longitudinal branch, even if the polarization is not longitudinal.

If there are g atoms per unit cell, there are $3g$ modes to every value of $\boldsymbol{k}$. Fig. 3 is a plot of ω against $|k|$ for a fixed direction of $\boldsymbol{k}$ for the case $g=2$. In addition to the three accoustical branches, for which $\omega(0)=0$, there are branches for which $\omega\neq0$ as $k\to0$, representing relative vibrations of atoms in the same unit cell. These branches are called optical, because in ionic solids such waves interact strongly with electromagnetic radiation. Optical modes can be disregarded for purposes of energy transport at low temperatures, because of their high frequency.

The vibrational energy of the lattice is the sum of the energies of the normal modes

$$E = \sum_{\boldsymbol{k}j} E_{\boldsymbol{k}j} = \frac{M}{2}\sum_{\boldsymbol{k}j}\{\dot{\xi}_{\boldsymbol{k}j}\dot{\xi}_{-\boldsymbol{k}j} + \omega_{\boldsymbol{k}j}^2 \xi_{\boldsymbol{k}j}\xi_{-\boldsymbol{k}j}\} = M\sum_{\boldsymbol{k}j}\omega_{\boldsymbol{k}j}^2 b_{\boldsymbol{k}j} b_{-\boldsymbol{k}j}. \qquad (3.8)$$

The heat flow due to each normal mode, averaged over a cycle, is given by

$$\frac{1}{V} E_{\boldsymbol{k}j}\frac{\partial\omega}{\partial\boldsymbol{k}}, \qquad (3.9)$$

where V is the volume of the crystal; it is thus simply the energy per unit volume multiplied by the group velocity.

Each normal coordinate ξ satisfies a simple harmonic oscillator equation, which can be quantized following standard procedure. Defining

$$a_{kj} = b_{kj} e^{i\omega_{kj}t} \tag{3.10}$$

the symmetrized expression for the energy is

$$E_{kj} = \tfrac{1}{2} M \omega_{kj}^2 (a_{kj} a_{-k-j} + a_{-k-j} a_{kj}). \tag{3.11}$$

The dynamical variable a_{kj} is now a matrix, and a_{-k-j} is its Hermitian conjugate. The momentum conjugate to a_{kj} becomes

$$- i M \omega_{kj} a_{-k-j} \tag{3.12}$$

and the matrices satisfy the commutation relations

$$a_{kj} a_{-k'-j'} - a_{-k'-j'} a_{kj} = \frac{\hbar}{M\omega} \delta_{k,k'} \delta_{j,j'}. \tag{3.13}$$

Since E_{kj} must be diagonal, it can be shown that the only non-vanishing elements of a_k and a_{-k} are[1]

$$\left.\begin{aligned} (a_k)_{N, N-1} &= \sqrt{\frac{\hbar}{M\omega} N}\,, \\ (a_{-k})_{N, N+1} &= \sqrt{\frac{\hbar}{M\omega} (N+1)} \end{aligned}\right\} \tag{3.14}$$

where N is now a quantum number. Thus

$$(E_k)_{N, N} = \hbar \omega (N + \tfrac{1}{2}) \tag{3.15}$$

so that the energy-content of each normal mode is quantized, there being a zero-point energy $\hbar\omega/2$ and N quanta of energy $\hbar\omega$, called phonons. There is a close analogy between the phonon and the photon of radiation theory.

In the present notation (k, j) and $(-k, -j)$, or briefly k and $-k$, denote the same normal mode. To be consistent with (3.6), the following notation is adopted

$$N_{-k} = - (N_k + 1). \tag{3.16}$$

Note that a_k is an annihilation operator and a_{-k} a creation operator. However, if adopting the notation (3.6) and (3.16), it is not necessary to distinguish between these two operators in the formal theory.

There is no limit to the number of phonons of a normal mode. The probability of a mode k being in an energy state E is proportional to $\exp[- E/K T]$, K being the Boltzmann constant. Since the energy states are evenly spaced with intervals $\hbar\omega$, the average value of N_k in thermal equilibrium is

$$\mathscr{N}_k = [e^{\hbar\omega/KT} - 1]^{-1} \tag{3.17}$$

and an assembly of phonons obeys Bose-Einstein statistics.

4. Thermal conductivity of a phonon gas. The quantized lattice waves carry a heat current, which is the sum of the current carried by all normal modes

$$Q = \sum_k N_k \hbar \omega \frac{\partial \omega}{\partial k}. \tag{4.1}$$

---

[1] The mode (k, j) will henceforth be simply designated by k.

Since the modes are distributed symmetrically in k-space and since $\omega(k)$ is also symmetrical, both the zero point energy and an isotropic distribution of phonons do not contribute to the current, in particular $Q=0$ if $N=\mathcal{N}$, that is if the distribution is the equilibrium distribution. Net heat flow arises from an anisotropy in the distribution.

In the absence of processes tending to restore equilibrium a small asymmetry in the distribution and its associated thermal current will persist indefinitely, even in the absence of a temperature gradient, leading to infinite thermal conductivity. In real crystals, without a temperature gradient, a deviation of N from the equilibrium value $\mathcal{N}$ will disappear. Assuming that the return to equilibrium follows an exponential law,

$$\left.\frac{\partial N}{\partial t}\right] = \frac{\mathcal{N}-N}{\tau} = -\frac{n}{\tau} \tag{4.2}$$

where the left-hand side denotes the average rate of change due to interactions, n is the deviation from equilibrium, and (4.2) defines a relaxation time τ.

We have so far considered the phonons to be unlocalized. By a superposition of states it is possible to define wave-packets of momentum range Δk, localized to within a region $\Delta x \sim 1/\Delta k$. We shall disregard the wave-character of the phonons[1] and treat the wave-packets as classical particles moving with the group velocity v_g of the waves; a BOLTZMANN equation can now be set up for the steady state in the presence of a temperature gradient

$$\left.\frac{\partial N}{\partial t}\right] + v_g \cdot \operatorname{grad} T \frac{dN}{dT} = 0. \tag{4.3}$$

The second term denotes the rate of change of N due to the drift motion of phonons in the temperature gradient. To first approximation N is replaced by $\mathcal{N}$ in this term, so that from (4.1)

$$n = -v_g \cdot \operatorname{grad} T \, \tau \frac{\hbar\omega}{KT^2} \frac{e^{\hbar\omega/KT}}{\left(e^{\hbar\omega/KT}-1\right)^2}. \tag{4.4}$$

Substituting into (4.1)

$$Q = \sum_{kj} |\operatorname{grad} T| \cos^2\vartheta \, \tau(k) \, v_g^2(k) \frac{\hbar^2\omega^2}{KT^2} \frac{e^{\hbar\omega/KT}}{\left(e^{\hbar\omega/KT}-1\right)^2}. \tag{4.5}$$

where ϑ is the angle between k and grad T. Since the lattice specific heat per unit volume is

$$S = \sum_{k,j} S_{k,j} = \sum_{k,j} \frac{\hbar^2\omega^2}{KT^2} \frac{e^{\hbar\omega/KT}}{\left(e^{\hbar\omega/KT}-1\right)^2} \tag{4.6}$$

the thermal conductivity becomes

$$\varkappa = \sum_{k,j} S_{k,j} \, v_g^2(k,j) \, \tau(k,j) \cos^2\vartheta. \tag{4.7}$$

If the distribution of states, as well as v_g and τ, are isotropic

$$\varkappa = \tfrac{1}{3} \sum_j S_j(k) \, v_g^2(k) \, \tau(k) \, dk \tag{4.8}$$

where $S_j(k)\,dk$ is the contribution to the specific heat from waves of wavenumber k within dk and polarization j.

---

[1] This is justified provided the temperature gradient is small enough, so that the fractional change of temperature over a distance of the order of a wave-length is small.

The relaxation time τ is obtained by means of a perturbation treatment. Peierls[1] has shown that uncertainty effects can be disregarded, provided

$$\frac{h}{\tau} < KT \tag{4.9}$$

for all the important states; this is satisfied in all cases of practical importance.

5. Interactions due to anharmonicities[2]. It has been assumed in Sect. 3 that the potential energy due to a displacement u is a quadratic function of the relative displacements $u_m - u_{m-l}$, to be summed over all lattice points m and linkages l. The normal modes are then plane waves (3.7). If the potential energy contains terms higher than the quadratic (it cannot contain linear terms for an equilibrium configuration), the plane waves are not normal modes, and energy will be interchanged between them. We shall in particular consider here cubic terms in the potential energy, for these cubic anharmonicities give rise also to thermal expansion [8]. The treatment is easily extended to higher anharmonicities.

The cubic terms in the potential energy are of the form

$$E' = \sum_{m,l,l'} F(l, l')\, u_m\, u_{m-l}\, u_{m-l'} \tag{5.1}$$

which by (3.7) and (3.10) is of the form

$$E' = \sum_{m} \sum_{k, k', k''} G(k, k', k'')\, a_k\, a_{k'}\, a_{k''}\, e^{ix \cdot (k+k'+k'')} \tag{5.2}$$

where F and G are functions depending on the detailed nature of the anharmonicities. Now

$$\sum_{m} e^{ix \cdot (k+k'+k'')} = 0 \tag{5.3a}$$

unless

$$k + k' + k'' = 0 \tag{5.3b}$$

or

$$k + k' + k'' = 2\pi\, b \tag{5.3c}$$

where b is one of the inverse lattice vectors (3.2). One can thus rewrite (5.2) in the form

$$E' = \sum_{k, k', k''} c(k, k', k'')\, q_k\, a_{k'}\, a_{k''} \tag{5.4}$$

where $c(k, k', k'')$ vanishes unless (5.3b) or (5.3c) is fulfilled.

In the quantum-mechanical treatment the a's are matrices (3.14), and the matrix E' consists of a sum of terms, each of which links two states of the system which differ by a change of one in the number of phonons of three modes k, k' and k''. While in (5.4) all operators are formally written as annihilation operators, some of them can be creation operators in virtue of the notation (3.5), (3.6) and (3.16).

If a system is initially in a state i, then the probability of finding it in a state j after time t is given by second order perturbation theory[3] as

$$W_j(t) = 2 E_{ij}^2\, \frac{1 - \cos (E_i - E_j)\, t/\hbar}{(E_i - E_j)^2} \tag{5.5}$$

[1] R. E. Peierls: Helv. phys. Acta 7, suppl. 24 (1934).
[2] See [9], [13] and [14].
[3] P. A. M. Dirac: Proc. Roy. Soc. Lond., Ser. A 112, 661 (1926).

where E_i is the energy of the state i, and E_{ij} the perturbation energy matrix. The state of the crystal is specified by the occupation numbers $N, N', N'' \ldots$ etc. of the modes $\boldsymbol{k}, \boldsymbol{k}', \boldsymbol{k}'' \ldots$ etc. Let i, j and k be three states for which this group of three modes has occupation numbers (N, N', N''), $(N+1, N'+1, N''+1)$ and $(N-1, N'-1, N''-1)$ respectively, all other modes having an unaltered occupation. Then the average rate of change of N is

$$t\,\frac{dN}{dt}\Big] = \sum 2\,(E_{ij}^2 - E_{ik}^2)\,\frac{1 - \cos(\omega + \omega' + \omega'')\,t}{\hbar^2\,(\omega + \omega' + \omega'')^2} \tag{5.6}$$

where the summation is over all such triplets of modes involving $\boldsymbol{k}$, E_{ik} is given by (5.4) and E_{ij} by its Hermitian conjugate, which involves creation operators.

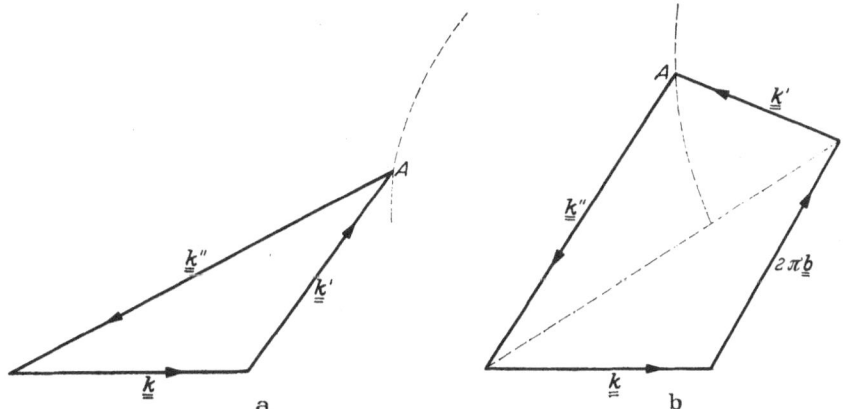

Fig. 4 a and b. The wave-vectors of the interacting triplet of states (a) for ordinary (b) for Umklapp-processes. Dashed line is the locus of A which in three dimensions is a surface of revolution about (a) the vector $\boldsymbol{k}$ or (b) the vector $\boldsymbol{k} + 2\pi\boldsymbol{b}$.

Using the explicit form (3.14) for the a's, this becomes

$$t\,\frac{dN}{dt}\Big] = \frac{2\hbar}{M^3} \sum_{\boldsymbol{k}',\boldsymbol{k}''} \frac{c^2(\boldsymbol{k}, \boldsymbol{k}', \boldsymbol{k}'')}{\omega\,\omega'\,\omega''}\,\frac{1 - \cos(\omega + \omega' + \omega'')\,t}{(\omega + \omega' + \omega'')^2} \times \left.\begin{array}{c} \\ \\ \end{array}\right\}$$
$$\times\,[(N+1)\,(N'+1)\,(N''+1) - N\,N'\,N'']. \tag{5.7}$$

To enumerate the triplets of states which contribute to (5.7) for a fixed state $\boldsymbol{k}$ one can choose a state $\boldsymbol{k}'$, and the state $\boldsymbol{k}''$ is then determined by (5.3). This is, of course, equivalent to choosing a point A in the vector diagram of Fig. 4. However, the resonance factor $[1 - \cos(\omega + \omega' + \omega'')\,t]/(\omega + \omega' + \omega'')^2$ in (5.7) ensures that the only significant contribution comes from those triplets such that A lies an the surface of revolution about the axis $\boldsymbol{k}$ (or $\boldsymbol{k} + 2\pi\boldsymbol{b}$) given by

$$\omega + \omega' + \omega'' = 0 \tag{5.8}$$

which implies conservation of energy in each individual process[1]. Since $\boldsymbol{k}'$ and $\boldsymbol{k}''$ can have three polarizations each, (5.7) must be summed over nine such surfaces.

Replacing summation over states by integration over $\boldsymbol{k}$-space

$$\frac{dN}{dt}\Big] = \frac{2\hbar V}{M^3} \sum_{j',j''} \int dS'\,\frac{c^2(\boldsymbol{k}, \boldsymbol{k}', \boldsymbol{k}'')}{\omega\,\omega'\,\omega''}\,\frac{\pi}{(2\pi)^3}\,[\,|\mathrm{grad}_{\boldsymbol{k}}\,\Delta\omega|]^{-1} \times \left.\begin{array}{c} \\ \\ \end{array}\right\}$$
$$\times\,[(N+1)\,(N'+1)\,(N''+1) - N\,N'\,N''] \tag{5.9}$$

---

[1] This argument is valid only if (4.9) is satisfied.

where $\Delta\omega = \omega + \omega' + \omega''$, dS' is an element of the surface $\Delta\omega = 0$, V is the volume of the crystal and the summation is over all polarizations.

Processes (5.3 c), which do not conserve the wave-vector, have been termed "Umklapp"-processes (flip-over processes) by Peierls [9].

Note that at least one of the frequencies in (5.8) must be negative; this means that the corresponding creation operator is really an annihilation operator, and that the corresponding factor $N+1$ should be replaced by N, or vice versa.

In order that (5.8) and (5.3 b) can be simultaneously satisfied, the ratio ω/k cannot be the same for the three interacting waves; thus if dispersion and anisotropy are weak, the three waves cannot all belong to the same polarization branch, as pointed out by Peierls [9]. Furthermore it was shown by Pomeranchuk [13] that the two conditions could not be satisfied if $|\omega| \ll |\omega'| \sim |\omega''|$ and ω/k exceeds both ω'/k' and ω''/k''; hence a low-frequency longitudinal wave cannot interact with a high-frequency wave. This will be of importance in Sect. 7. Herpin [22] has also discussed this, and another less important, restriction on the types of possible processes. It can be shown similarly that a low-frequency longitudinal wave cannot participate in an Umklapp-process[1].

To determine $c(k, k', k'')$ is generally a complicated undertaking, and can be done only approximately. A rough estimate of the anharmonicity coefficient, and thus of c, can be obtained as follows:

According to Grüneisen's[2] theory of thermal expansion, if the frequencies of all lattice waves is a function of the dilatation Δ, that is if

$$\omega = \omega_0 (1 + \gamma \Delta) \tag{5.10}$$

then the coefficient of thermal expansion becomes

$$\frac{d\Delta}{dT} = \frac{\gamma}{\lambda} S \tag{5.11}$$

where S is the specific heat per unit volume, and λ the compressibility. Thus γ can be derived from thermal expansion data, as well as directly from the variation of the elastic constants with heavy strain, as implied by (5.10). The change in energy of the crystal arising from a dilatation Δ is, from (3.11) and (5.10) and neglecting the order of factors,

$$E' = M \sum_{k} 2\gamma \Delta \omega_0^2 a_k a_{-k}. \tag{5.12}$$

Now a single wave k leads to a dilatation which varies with position, so that

$$\Delta(x) = \sum_{k} \frac{1}{\sqrt{G}} (\varepsilon \cdot k) a_k e^{i k \cdot x}. \tag{5.13}$$

If Δ is not constant, but a function of position, (5.12) can be generalized to

$$E' = M \sum_{k', k''} 2\gamma \omega' \omega'' \sum_{m} \frac{1}{G} \Delta(x) e^{i(k'+k'') \cdot x} a_{k'} a_{k''} \tag{5.14}$$

so that,

$$\left.\begin{aligned} E' &= \frac{2M}{3G^{\frac{3}{2}}} \gamma \sum_{k, k', k''} \omega \omega' \omega'' \left[\frac{\varepsilon \cdot k}{\omega} + \frac{\varepsilon' \cdot k'}{\omega'} + \frac{\varepsilon'' \cdot k''}{\omega''} \right] \times \\ &\times \sum_{m} e^{i(k+k'+k'') \cdot x} a_k a_{k'} a_{k''} \end{aligned}\right\} \tag{5.15}$$

---

[1] S. Simon: Private communication.

[2] E. Grüneisen: Handbuch der Physik, vol. X, p. 1—59. 1926.

and, as an estimation of orders of magnitude, provided (5.3 b) or (5.3 c) are fulfilled

$$c(\boldsymbol{k}, \boldsymbol{k}', \boldsymbol{k}'') \approx \frac{2}{\sqrt{3}} \frac{M\gamma}{G^{\frac{1}{2}}} \frac{\omega\omega'\omega''}{v}. \qquad (5.16)$$

The present treatment disregards any effects arising from special crystal structure, as well as polarization effects.

LEIBFRIED and SCHLOEMANN [24], generalizing the result of a linear chain, have obtained by a different approximation

$$c(\boldsymbol{k}, \boldsymbol{k}', \boldsymbol{k}'') \approx 0.58 \frac{M\gamma}{G^{\frac{1}{2}}} \frac{\omega\omega'\omega''}{v}. \qquad (5.17)$$

6. Scattering by static imperfections. In addition to the mutual interaction of lattice waves arising from the anharmonicity of the interatomic forces there is interaction due to the static imperfections of the crystals, such as faults in the periodicity and static strains. The scattering probability can be calculated by a similar method to that used in Sect. 5: the perturbation energy is expressed in terms of the displacements $\boldsymbol{u}$, which in turn are expressed in terms of the lattice waves (3.7). The terms quadratic in the amplitudes of the lattice waves describe elastic scattering, for if E' can be expressed as

$$E' = \sum_{\boldsymbol{k}, \boldsymbol{k}'} c(\boldsymbol{k}, \boldsymbol{k}') a_{-\boldsymbol{k}'} a_{\boldsymbol{k}} \qquad (6.1)$$

where, since E' should be real and in view of (3.5),

$$c(\boldsymbol{k}, \boldsymbol{k}') = c^*(\boldsymbol{k}', \boldsymbol{k}), \qquad (6.2)$$

it is easily shown, in analogy to (5.7) and (5.9), that

$$\frac{dN}{dt}\bigg] = \frac{2V}{M^2} \frac{\pi}{(2\pi)^3} \int dS' \frac{c^2(\boldsymbol{k}, \boldsymbol{k}')}{\omega\omega'} \frac{dk'}{d\omega'} [(N+1)N' - N(N'+1)] \qquad (6.3)$$

where the integration is over the three surfaces $\omega' = \omega$ in $\boldsymbol{k}'$-space, corresponding to the three possible modes of polarization of $\boldsymbol{k}'$.

A static imperfection can contribute in the following ways to the perturbation HAMILTONian (6.1): a) A small region (or a single atom) has a mass which differs by δM from the normal mass of this region; the change in the kinetic energy is $\delta M \dot{u}^2/2$, which leads to a term in (6.1) such that $c(\boldsymbol{k}, \boldsymbol{k}')$ contains a factor $\delta M \omega\omega'$. b) The elastic constants of some interatomic linkages differ from the normal value, the corresponding coefficient $c(\boldsymbol{k}, \boldsymbol{k}')$ for a single linkage being proportional to kk'. c) As a special case of b), a strain field causes a change of the elastic constants due to the anharmonicity of the interatomic forces; this effect is related to the three-phonon interactions treated in Sect. 5, but in place of one of the lattice waves there is the static strain field.

Using these concepts, the scattering probability can be calculated, though some simplifying assumptions are required to make the calculations tractable. Thus KLEMENS [21] systematically replaces all trigonometric ratios occurring in $c(\boldsymbol{k}, \boldsymbol{k}')$ by their root mean square values, since only c^2 is required in the final result, and treats k as small compared with the reciprocal of the interatomic distance.

This approximation yields a scattering probability which, in virtue of the approximations made, is independent of the polarization of the interacting waves. The following results are obtained in this way:

α) *Point imperfections.* If the disordered region extends over a region small compared with the wavelength of the interacting phonons, there is no interference from different regions of the imperfections or its strain field, and

$$c^2(\boldsymbol{k}, \boldsymbol{k}') = \frac{M^2}{G^2}\,\omega^2\omega'^2\,S^2 \tag{6.4}$$

so that

$$\frac{1}{\tau} = \frac{3\pi\,a^3}{v^3}\,\frac{S^2}{G}\,\omega^4 \tag{6.5}$$

where a is the lattice constant and G the number of unit cells in a volume containing one such imperfection (that is $1/G$ is the concentration of imperfections). The coefficient S^2 is estimated for a number of model imperfections [21], and is usually of the order of unity. The scattering probability varies as the fourth power of frequency in the limit of long wave-lengths, as would be expected from the scattering of sound by a sphere; for higher frequencies the power of the scattering law gradually decreases due to interference.

β) *Single dislocations.* A dislocation consists of a highly disordered cylindrical core, of radius of order a, and an elastic strain field which falls off slowly as $1/r$. The scattering of the core can be approximated by a line of impurity atoms, and interference requires that $k_z = k_z'$, where k_z is the component of $\boldsymbol{k}$ along the cylinder axis. It can be shown that due to the core

$$\frac{1}{\tau} \sim \frac{a^2}{G_0}\,\omega\,k^2 \tag{6.6}$$

where $a^2 G_0$ is an area, perpendicular to the z-axis, which contains only one dislocation (or $a^{-2}G_0^{-1}$ is the number of dislocations per unit area). For waves which are long compared with those of limiting frequency, this scattering is much weaker than the scattering due to the strain field, which yields

$$c^2(\boldsymbol{k}, \boldsymbol{k}') \sim \frac{1}{10}\left(\frac{M\gamma}{a^2 G_0}\right)^2 b^2\,v^4\left(\frac{k\,k'}{|\boldsymbol{k}-\boldsymbol{k}'|}\right)^2 \tag{6.7}$$

where b is the magnitude of the BURGERS vector[1], and where $c^2(\boldsymbol{k}, \boldsymbol{k}')$ vanishes unless $k_z = k_z'$.

Now (6.3) contains the factor $(N+1)N' - N(N'+1) = n' - n$, where n and n' are the deviations from equilibrium. Thus $1/\tau$ contains the factor $1 - n'/n$, which, from (4.4), is equal to $1 - \mu'/\mu$, where μ is the direction cosine of $\boldsymbol{k}$ with respect to grad T. For randomly oriented dislocations one finally obtains

$$\frac{1}{\tau} \sim \frac{1}{20}\,\frac{b^2}{G_0\,a^2}\,\gamma^2\omega. \tag{6.8}$$

The frequency dependence and order of magnitude of this scattering probability had been originally obtained by NABARRO[2] from an optical argument.

It should be noted that the scattering probability varies as the square of the BURGERS vector; thus if dislocations occur in groups of the same sign, as they would if they had been created at a source and held up by some obstacle, their combined scattering is correspondingly stronger.

γ) *Grain boundaries.* If the angle of tilt between the crystals on either side of a given boundary is α, then

$$c^2(\boldsymbol{k}, \boldsymbol{k}') \sim \frac{1}{40}\,\frac{M^2\gamma^2}{a^2 G_1^2}\,\alpha^2\,\frac{k^2\,k'^2}{|\boldsymbol{k}-\boldsymbol{k}'|^2} \tag{6.9}$$

---

[1] For definition of the BURGERS vector, cf. vol. VII, part. I.
[2] F. R. N. NABARRO: Proc. Roy. Soc. Lond., Ser. A **209**, 278 (1951).

and vanishes unless k and k' have the same value for their component tangential to the grain boundary. One obtains after suitable averaging over directions

$$\frac{1}{\tau} \sim \frac{1}{40} \frac{v \alpha^2 \gamma^2}{a G_1} \tag{6.10}$$

where $(a G_1)^{-1}$ is the average number of grain boundaries intersecting a random line per unit distance. The scattering described by (6.10) is specular, and it arises from the mismatch of the regions on either side; there is in addition scattering arising from the disordered region of the grain boundary itself, and this yields

$$\frac{1}{\tau} \sim \frac{a^2 k^2 v \alpha^2 \gamma^2}{a G_1} \tag{6.11}$$

and is thus negligible for waves of frequency less than about $\omega_D/6$, where ω_D is the limiting frequency.

δ) *External boundaries.* In obtaining the normal modes in Sect. 3, it was assumed that the displacements at the boundaries of the crystal are periodic. The actual boundary conditions are far more complicated, so that the external boundaries act as a perturbation and cause scattering. It is generally presumed that the corresponding relaxation time is of the order of

$$\frac{1}{\tau} \sim \frac{v}{L}, \tag{6.12}$$

where L is the shortest linear dimension of the crystal, though no satisfactory derivation has yet been given. However, CASIMIR [11] has calculated the thermal conductivity of a cylindrical body along its axis on the assumption that there are no phonon interactions except at the boundary, where the phonons are scattered isotropically. Comparing the conductivity thus calculated with (4.8), τ is indeed of the magnitude given by (6.12).

7. Equilibrium of phonons in the presence of a temperature gradient[1]. If it is possible to define a relaxation time τ, as in (4.2), which is independent of the occupation numbers of the state k and all the other states, then equilibrium is maintained as described in Sect. 4.

This condition is fulfilled in the case of isotropic elastic scattering, for then the term in N' of (6.3) disappears after summing over all states k'. If scattering is not isotropic, it is still possible to define a relaxation time in an isotropic material, for because of (4.4) one can write (6.3) as

$$-\frac{1}{n} \frac{dN}{dt}\bigg] = \frac{1}{\tau} = \frac{1}{\tau_0} (1 - \bar{\mu}) \tag{7.1}$$

where $\bar{\mu}$ is the average of the cosine of the scattering angle, and

$$\frac{1}{\tau_0} = \frac{2V}{M^2} \frac{\pi}{(2\pi)^3} \int dS' \frac{c^2 (k, k')}{\omega \omega'} \frac{dk'}{d\omega'}. \tag{7.2}$$

A similar averaging procedure is possible in more complicated cases.

The establishment of equilibrium can be regarded as follows: the temperature gradient continuously increases the total momentum of the phonon gas in the direction of heat flow; the scattering processes obliterate the excess momentum; $1/\tau_0$ is proportional to the scattering probability and $(1 - \bar{\mu})$ is a measure of the average change of momentum per collision.

---

[1] The treatment of this section follows closely that given in [20].

Three-phonon processes are more difficult to treat, because the scattering probability (5.9) contains the complicated factor

$$(N+1)(N'+1)(N''+1)-NN'N''=NN'+N'N''+N''N+N+N'+N''+1 \quad (7.3)$$

and because they do not tend to obliterate excess momentum.

It is clear that it is the ability to reduce excess momentum $(\Sigma \boldsymbol{k} N_k)$ which determines the effectiveness of processes to produce thermal resistance. Interaction processes can be thus divided into two groups: those that conserve momentum, namely "ordinary" three-phonon processes obeying (5.3b), and processes which do not conserve momentum. Umklapp-processes, obeying (5.3c), are to be grouped in the latter class, for while there is some degree of correlation between the initial and final momentum, this correlation is sufficiently weak to be disregarded, since $\boldsymbol{b}$ in (5.3c) can have six different values.

It was pointed out by Peierls [9], [10] that momentum-conserving processes themselves cannot maintain equilibrium. It is easily verified that, because of the conservation of energy (5.8), the factor (7.3) vanishes if the N's are given by the equilibrium distribution (3.17). Now if $\boldsymbol{k}$ is conserved as well, so is $\hbar\omega + \boldsymbol{\lambda} \cdot \boldsymbol{k}$, where $\boldsymbol{\lambda}$ is some constant vector, so that the following distribution, of which (3.17) is a special case, is also stationary for processes (5.3b):

$$\mathcal{N}(\lambda) = \left[\exp \frac{\hbar\omega - \boldsymbol{\lambda} \cdot \boldsymbol{k}}{KT} - 1\right]^{-1} \approx \mathcal{N} + \frac{\boldsymbol{\lambda} \cdot \boldsymbol{k}}{KT} \frac{e^{\hbar\omega/KT}}{(e^{\hbar\omega/KT} - 1)^2}. \quad (7.4)$$

The parameter λ determines the anisotropy of the phonon distribution and the total momentum of the phonon gas.

If the phonon distribution deviates from equilibrium, these processes will tend to restore it not to true equilibrium $\mathcal{N}(0)$, but to a quasi-equilibrium distribution $\mathcal{N}(\lambda)$, where λ is determined by the total momentum, which is conserved. Thus processes conserving momentum cannot restore to equilibrium an anisotropic distribution (4.4), which is formally equivalent to (7.4) if

$$\lambda = - v^2 \tau \operatorname{grad} T \frac{\hbar}{T}. \quad (7.5)$$

While processes conserving momentum do not themselves produce thermal resistance, they play a role in the establishment of thermal equilibrium[1] and must be considered. If, in the presence of a temperature gradient, there are ordinary three-phonon processes, and also elastic scattering processes with a relaxation time $\tau'(\omega)$, and if τ' is independent of frequency, then λ is constant, and the stationary distribution (4.4) is of the form (7.4), so that three-phonon processes have no effect. But if, as happens frequently, $\tau'(\omega)$ increases with decreasing frequency, the three-phonon processes will take excess momentum from the low to the high frequencies, and in the steady state $\tau(\omega)$ is no longer the same as $\tau'(\omega)$.

While the form of (7.3) makes it impossible to define a relaxation time for three-phonon processes independent of the n's, it is possible to define a "relaxation time of a single mode" by assuming that all modes, except the mode $\boldsymbol{k}$, have N-values given by (7.4), so that

$$\frac{1}{\sigma(\boldsymbol{k})} = -\frac{1}{n} \frac{dN}{dt} \Bigg\} = \sum_{\boldsymbol{k}', \boldsymbol{k}''} A \left(\mathcal{N}'(\lambda) + \mathcal{N}''(\lambda) + 1\right) \quad (7.6)$$

where A, given by (5.9), is proportional to $\omega \omega' \omega''$.

---

[1] Thus elastic scattering processes cannot produce equilibrium if there is an excess of phonons of a particular frequency; such deviations can be removed by three-phonon processes.

It is also possible to define an "*effective relaxation time*" τ_p in terms of $dN/dt]$ with the n's having those values which they actually take up in the presence of a temperature gradient.

The overall relaxation time τ is given by

$$\frac{1}{\tau} = \frac{1}{\tau_p} + \frac{1}{\tau'} \tag{7.7}$$

where τ' describes the processes which do not conserve $\boldsymbol{k}$; also τ and n are related by (4.4), and τ_p and the n's are related by (5.9). In this way an integral equation can be set up for τ_p or for τ. This has been done by Leibfried and Schlömann [24], who solved their equation by a variational method.

It should be noted that τ thus defined does not describe an exponential decay of the deviation from equilibrium, but only the initial decay of the deviation if the temperature gradient is suddenly removed.

The present treatment follows that of Klemens [20], who effectively replaces the integral equation by a small set of linear algebraic equations, as this approximation yields useful physical insight.

Three-phonon processes involving $\boldsymbol{k}$ can be divided into three groups:

(a) $|\omega| \sim |\omega''| \gg |\omega'|$,

(b) $|\omega| \sim |\omega'| \sim |\omega''|$,

(c) $|\omega| \ll |\omega'| \sim |\omega''|$.

The rate of change of N, to first order in the n's ,can be expressed as

$$\frac{dN}{dt}\bigg] = \sum_{\boldsymbol{k}, \boldsymbol{k}'} A \left[n(\mathcal{N}' + \mathcal{N}'' + 1) + n'(\mathcal{N} + \mathcal{N}'' + 1) + n''(\mathcal{N} + \mathcal{N}' + 1) \right]. \tag{7.8}$$

To enumerate the number of processes contributing to (7.9), one can draw a closed vector diagram as in Fig. 4; given $\boldsymbol{k}$, any vertex A represents a solution of (5.3 b), but only those vertices lying on the surface of revolution (5.8) are admissible. Thus the number of solutions having k' in the range k', dk' is approximately proportional to $k' dk'$.

In the case of processes (a), $\mathcal{N} \approx -\mathcal{N}''$ and since $\boldsymbol{k}$ and $\boldsymbol{k}''$ are nearly in the same direction, the angle between them being of order k'/k, and since n and n'' are of opposite sign, $|n + n''| \sim (k'/k)|n|$. Also to every process having a positive value of ω' corresponds one with a negative value of ω', and since n' is an odd function of ω', the term in n' vanishes on summation. Thus, if $x = \hbar\omega/KT$, etc.,

$$\frac{dN}{dt}\bigg]_{(a)} \propto n \int x' \left[2\mathcal{N} \pm 0\left(\frac{k'}{k}\right) \mathcal{N}' \right] x' \, dx' \tag{7.9}$$

and the greatest contribution to this comes from the large values of ω', that is from processes (b) or (c).

Similarly, in the case of processes (b) and (c), if we consider terms in n only

$$\frac{dN}{dt}\bigg\}_{(c)} \propto n \int x' x'' \left[\frac{\mathcal{N}' \mathcal{N}''}{\mathcal{N}} e^{-x} \right] x' \, dx'. \tag{7.10}$$

Since $|\omega'| \sim |\omega''|$, they have nearly the same value of λ, so that one can write

$$\frac{dN}{dt}\bigg]_{(c)} \propto n \int x' x'' \left[\frac{\mathcal{N}' \mathcal{N}''}{\mathcal{N}} e^{-x} \right] \left(1 - \frac{\lambda(\omega')}{\lambda(\omega)} \right) x' \, dx'. \tag{7.11}$$

The greatest contribution to (7.10) comes from processes $\hbar\omega' \sim KT$ if $\hbar\omega \ll KT$ but comes from processes $\omega' \sim \omega$ if $\hbar\omega \gtrsim KT$. Thus in the latter case the greatest contribution to (7.10) comes from (b)-processes, but since for such processes $\lambda(\omega') \sim \lambda(\omega)$, these processes do not contribute appreciably to (7.11).

To summarize: processes of type (a) are never important; processes of type (b) are never important to $1/\tau$, for if $\hbar\omega \ll KT$, processes (c) are more important and otherwise the interacting states have a common λ-value; processes of type (c) are important if $\hbar\omega \ll KT$, but not otherwise.

Thus

$$\frac{1}{\tau(\omega)} = \frac{1}{\tau'(\omega)} \qquad \text{if} \quad \omega \gtrsim \frac{KT}{\hbar} \qquad (7.12\text{a})$$

and

$$\frac{1}{\tau(\omega)} = \frac{1}{\tau'(\omega)} + \frac{1}{\sigma(\omega)}\left(1 - \frac{\tau(\omega_1)}{\tau(\omega)}\right) \quad \text{if} \quad \omega \ll \frac{KT}{\hbar} \qquad (7.12\text{b})$$

where $\omega_1 = KT/\hbar$, and $\sigma(\omega)$ is defined by (7.6).

In the derivation of eq. (7.12) it was implied that a mode k can take part in processes (c). However, it has been shown by Pomeranchuk [13] that if ω/k is greater than $\omega'/k' \sim \omega''/k''$, (5.3b) and (5.8) cannot be simultaneously satisfied. Therefore a longitudinal mode k cannot participate in (c)-processes, and the establishment of equilibrium of longitudinal modes is fundamentally different from the case of transverse modes.

If k is longitudinal, then the major contribution to $dN/dt]$ comes from processes (b), but now it cannot be assumed that $\lambda(\omega) = \lambda(\omega')$, because the behaviour of λ is different for the different polarization branches. If $\sigma'(\omega)$ is the relaxation time for a single longitudinal mode [again defined by (7.6)], then

$$\frac{1}{\tau_I(\omega)} = \frac{1}{\tau_I'(\omega)} + \frac{1}{\sigma'(\omega)}\left(1 - \frac{\lambda_{II}(\omega)}{\lambda_I(\omega)}\right) \qquad (7.13)$$

where the suffices I and II refer to longitudinal and transverse modes respectively. Eq. (7.13) is based on the assumption that $\tau_I > \tau_{II}$; this will be found to be so in almost all cases, because in general $\sigma' > \sigma$ and if $\omega < \omega_1$, $\tau'(\omega) > \tau'(\omega_1) = \tau(\omega_1)$.

Eq. (7.12) and (7.13) together determine τ for any mode. Thus for longitudinal modes

$$\tau_I(\omega) = \frac{\tau_I'(\omega)\,[\sigma'(\omega) + \tau_{II}(\omega)/C^2]}{\tau_I'(\omega) + \sigma'(\omega)} \qquad (7.14\text{a})$$

where $C = v_I/v_{II}$, τ_I' is the relaxation time of a longitudinal mode for processes not conserving momentum, σ' is the three-phonon relaxation time (7.6) of a single longitudinal mode, and τ_{II} is given below. For transverse modes

$$\tau_{II}(\omega) = \tau_{II}'(\omega) \qquad \text{if} \quad \omega \gtrsim \omega_1, \qquad (7.14\text{b})$$

$$\tau_{II}(\omega) = \frac{\tau_{II}'(\omega)\,[\sigma(\omega) + \tau_{II}'(\omega_1)]}{\tau_{II}'(\omega) + \sigma(\omega)} \quad \text{if} \quad \omega \lesssim \omega_1 \qquad (7.14\text{c})$$

where τ_{II}' is the relaxation time of a transverse mode for processes not conserving momentum, σ is the three-phonon relaxation time (7.6) of a single transverse mode, and $\omega_1 = KT/\hbar$.

The thermal conductivity is obtained by substituting (7.14) into (4.8), so that

$$\varkappa = \varkappa_I + \varkappa_{II}, \qquad (7.15\text{a})$$

$$\varkappa_I = \frac{1}{3}\int\limits_0^\infty S_I(k)\, v_I\, \frac{l_I'\, l_1}{l_I' + l_1}\, dk \qquad (7.15\text{b})$$

where $l'_1 = v_I \tau'_I$, $l_1 = v_I \sigma'$ and S_I is the specific heat of longitudinal waves per wave-number interval; and

$$\varkappa_{II} = \tfrac{1}{3} \int_0^\infty S_{\text{eff}} \, l_{II} \, v_{II} \, dk \qquad (7.15c)$$

where

$$l_{II}(\omega) = l'_{II}(\omega) \qquad \text{if} \quad \omega \gtrless \frac{KT}{\hbar} \qquad (7.16a)$$

$$l_{II}(\omega) = \frac{l'_{II}(\omega) \, [l_2(\omega) + l'_{II}(\omega_1)]}{l'_{II}(\omega) + l_2(\omega)} \qquad \text{if} \quad \omega < \frac{KT}{\hbar} \qquad (7.16b)$$

and

$$S_{\text{eff}} = S_{II} + S_I \frac{l'_I}{l'_I + l_1} \qquad (7.16c)$$

where $l_2 = v_{II} \sigma$, $l'_{II} = v_{II} \tau'_{II}$ and S_{II} is the differential specific heat of the transverse modes.

The relaxation lengths l'_I and l'_{II} have been estimated in Sect. 6; l_1 and l_2 can be obtained from (7.6) and (5.9) by integrating over all admissible k'. An estimation of l_2 was made by LANDAU and RUMER [12] and of l_1 by POMERANCHUK [13]; they can be estimated in the following way: substituting (5.9) into (7.6)

$$-\frac{1}{n} \frac{dN}{dt} \bigg\} = \sum_{j', j''} \frac{4}{3} \frac{V}{G} \frac{\hbar}{(2\pi)^2} \frac{\gamma^2}{Mv^2} \int \omega \, \omega' \, \omega'' \, [\text{grad}_k \, \varDelta \omega]^{-1} \times \\ \times \frac{e^x - 1}{(e^{x'} - 1)\,(e^x - e^{-x'})} \, dS'. \bigg\} \qquad (7.17)$$

Consider now the case when k is a low-frequency longitudinal wave, interacting with transverse waves of about the same frequency. One can thus approximate

$$\sum_{j', j''} = 4, \qquad \int dS' = 4\pi k^2, \qquad \text{grad}_k \, \varDelta \omega = v, \qquad \omega' = \omega'' = \omega,$$

and obtain an order of magnitude result

$$l_1 \sim \frac{1}{4\gamma^2} \, a \, (a\,k)^{-4} \frac{Mv^2}{KT}, \qquad x \ll 1 \qquad (7.18)$$

which roughly agrees with the result of POMERANCHUK.

If one considers the case when k is a low-frequency transverse wave, so that it mainly interacts with waves such that $|x'| = |x''| \sim 1$, $|x| \ll 1$, one can approximate

$$\sum_{j, j'} = 9, \qquad dS' = k_1 dk_1, \qquad \text{grad}_k \, \varDelta \omega = \frac{vk}{k_1}$$

and one obtains

$$l_2 \sim \frac{1}{100\gamma^2} \, a \, (a\,k)^{-1} \frac{Mv^2}{KT} \left(\frac{\Theta}{T}\right)^3, \qquad x \ll 1 \qquad (7.19)$$

in approximate agreement with the result of LANDAU and RUMER.

In the derivation of (7.18) and (7.19) the solid was treated as an isotropic elastic continuum. Different frequency dependences can be expected when one considers different types of crystal symmetry, as was done by HERRING [23], who also showed from similarity considerations that at low frequencies

$$l_1 \propto k^{-n} \, T^{-m}$$

where

$$n + m = 5. \qquad (7.20)$$

8. Thermal conductivity of amorphous solids. The lattice structure of glasses lacks symmetry and periodicity. While a glass cannot be regarded as a solid in respect to its thermodynamic properties, it behaves like a solid in respect to its dynamical properties at low temperatures, for every atom vibrates about a fixed equilibrium position, and the thermal motion can be resolved into normal modes which differ, of course, from plane waves. Any instantaneous displacement can still be resolved into plane waves, but there will be very strong interchange of energy between them. This means that in glasses the mean free path l' of phonons for processes not conserving the wave-vector is very much shorter than in other solids; in particular the mean free paths l_1 and l_2 are now relatively longer, and the thermal conductivity of glass provides a more sensitive test of the general theory of Sect. 7 than does the thermal conductivity of a pure crystal, for which $\varkappa_I$ and terms limited by l_2 can be neglected.

The forces between the atoms will be the same as in the case of the crystalline modification of the substance, if the latter exists. Thus the specific heat of crystalline and vitreous quartz are the same below the softening point[1], and the same can be expected for the elastic constants and the anharmonicities. It is thus possible to treat glass as a crystalline solid except that one must ascribe to it a short mean free path l'.

Kittel [19] suggested that since in a glass there was only short-range order, extending over a few unit cells, l' could be taken as independent of frequency and of the order of a unit cell of the corresponding crystal. Thus (4.8) becomes

$$\varkappa = \tfrac{1}{3} S v l' \tag{8.1}$$

where S is the specific heat, v the velocity (averaged over polarization) of the elastic waves, and $l' = l_0'$, a constant to be determined. He found that $\varkappa$ was indeed proportional to the specific heat from below the softening temperature to about $100°$ K for a variety of silicate glasses, but that below $100°$ K the observed l' increased with decreasing temperature. This increase can be understood qualitatively because the atomic disorder of the vitreous structure becomes progressively less important the longer the wavelength, and the glass responds as a homogeneous elastic solid to very long waves.

Berman [38], [39] concluded from his own measurements, as well as those of Bijl [34] that, if l' is deduced from the observed conductivity at liquid hydrogen and helium temperatures, it varies approximately as T^{-2}. Between 10 and $20°$ K he found $\varkappa$ almost independent of temperature, confirming an earlier result of Wilkinson and Wilks [36].

Klemens [20] has shown that if l' is a constant l_0' for high frequency waves $(ak>1)$, then

$$l' = l_0' (ak)^{-2} \tag{8.2}$$

for low frequency waves $(ak \ll 1)$, where a is of the order of the average distance between the molecular groups forming the glass (e.g. Si—O tetrahedra in the case of quartz glass). This explains why $l' \propto T^{-2}$ at low temperature, but if the observed value of l' is compared with the mean free path at high temperatures, which is identified with l_0', it is found that the observed conductivity at low temperatures is too large by a factor of the order of 50 to 100, depending upon the choice of a.

---

[1] F. Simon: Ann. Phys., Lpz. **63**, 278 (1922). — F. Simon and F. Lange: Z. Physik **38**, 227 (1926).

This discrepancy can be resolved by assuming that l_I' is considerably greater than l_{II}', though both are given by an expression of the form (8.2); that is

$$l_I' = A a (a k)^{-2} \quad \text{if} \quad a k < 1 \atop = A a \qquad\quad \text{if} \quad a k \geq 1 \Bigg\} \tag{8.3}$$

and l_{II}' is given similarly, with a different constant B in place of A. Also in agreement with (7.18)

$$l_1 = C a\, T^{-1} (a k)^{-4} \tag{8.4}$$

where C is a constant to be determined. Substituting into (7.15) and (7.16), one finds that at lowest temperatures

$$\varkappa_I = 3.29\, A\, T\, \frac{K^2}{3 \pi a h}, \tag{8.5 a}$$

then increases more slowly, passes through a maximum at a temperature of about $0.8\,(C/A)^{\frac{1}{4}}(h v/2\pi K a)^{\frac{3}{4}}$, and at higher temperatures varies as $(1/T)^{\frac{1}{2}}$. On the other hand

$$\varkappa_{II} = \tfrac{1}{3} B a v\, S(T) \tag{8.5 b}$$

at high temperatures, and then gradually changes to a linear dependence similar to (8.5 a), with $2B$ replacing A at lowest temperatures.

Fitting this theory to BERMAN's measurements of quartz glass [39], taking $a = 5$ Å as an arbitrary unit of length and taking $v = 2.10^5$ cm./sec.,

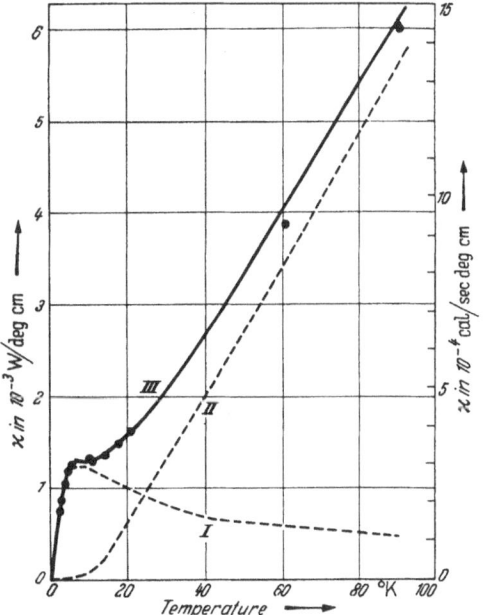

Fig. 5. Thermal conductivity of quartz glass according to KLEMENS [20]. Curve I denotes $\varkappa_I$, curve II $\varkappa_{II}$, curve III $\varkappa = \varkappa_I + \varkappa_{II}$. Full circles denote experimental points of BERMAN [39].

one finds from the high temperature conductivity that $B = 2.4$; the corresponding $\varkappa_{II}$ is plotted in Fig. 5; the discrepancy between $\varkappa$ and $\varkappa_{II}$ can be reconciled by taking $A = 180$, $C = 220$, giving $\varkappa_I$ and $\varkappa_I + \varkappa_{II}$ as shown. With these values for the three constants, good agreement can be obtained with BERMAN's measurements.

It should be noted that while A and B are two empirical constants, the value of C which had to be chosen to fit the observations is about one tenth the magnitude given by the expression (7.18) for l_1. The conductivity is insensitive to the choice of l_2.

According to this theory, $\varkappa$ should be proportional to T at lowest temperatures, or possibly vary less strongly. BIJL [34], however, found that for a number of glasses $\varkappa \propto T^{1.3}$ at liquid helium temperatures. It is not certain how significant this discrepancy is.

The absolute value of the conductivity varies slightly between silicate glasses of different composition, and there is some evidence that the conductivity differs slightly between different specimens of glass of the same composition.

BERMAN also measured a sample of perspex between 2 and 20° K, which showed a similar variation of $\varkappa$ with T.

Rubber has also a disordered structure, so that one may expect l' to be constant at the higher frequencies. The thermal conductivity of a specimen of rubber was measured by Schallamach[1] from 100 to 290° K. In this range it is approximately proportional to the specific heat, and $l'v \approx 6.10^{-4}$ cm.² sec.⁻¹. In the curve of S against T there is a step at 210° K, indicating a gradual transition process involving latent heat. This is reproduced in the curve of $\varkappa$ against T, but on cooling the transition can be delayed for about 50°, and there is a corresponding hysteresis in the thermal conductivity. The measurements do not extend to sufficiently low temperatures to observe a possible lengthening of l' with decreasing frequency.

9. Thermal conductivity of crystals (theory). In the case of good crystals l'_I and l'_II are large compared with l_1 and l_2, so that (7.15) and (7.16) can be considerably simplified by neglecting the terms limited by the magnitude of l_1 and l_2. Thus

$$\varkappa = \tfrac{1}{3} \int_0^\infty S\, v_\text{II}\, l_\text{II}\, dk \tag{9.1a}$$

where

$$S = S_\text{I} + S_\text{II} \tag{9.1b}$$

and

$$\begin{aligned} l_\text{II}(\omega) &= l'_\text{II}(\omega) \quad \text{if} \quad \omega \geq \omega_1 \\ &= l'_\text{II}(\omega_1) \quad \text{if} \quad \omega \leq \omega_1 \end{aligned} \tag{9.1c}$$

where $\omega_1 = K T/\hbar$. This form of l_II avoids divergence difficulties at low frequencies.

The neglect of $\varkappa_\text{I}$ needs some justification, since $l_1 \propto \omega^{-4}$, so that

$$\int S_\text{I}\, v_\text{I}\, l_1\, dk \tag{9.2}$$

diverges at low frequencies. Some cut-off mechanism is provided by the scattering of longitudinal waves by static imperfections, provided l'_I varies at low frequencies more slowly than ω^{-3}. In the absence of a sufficient number of dislocations or grain boundaries[2] $\varkappa_\text{I}$ will be limited by the external dimensions of the crystal, and the greatest contribution to $\varkappa_\text{I}$ will come from phonons of frequency such that $l_1(\omega)$ is of the order of the linear external dimensions of the crystal.

The variation of $\varkappa_\text{I}$ with temperature, specimen size and concentration of point imperfections has been discussed in some detail by Pomeranchuk [14], who believed $\varkappa_\text{I}$ to be the dominant contribution to $\varkappa$. Herring [23] showed that the magnitude of l_1 suggests that $\varkappa_\text{I}$ is not important compared to $\varkappa_\text{II}$, especially in anisotropic crystals where l_1 is less than the values given by (7.18), and the divergence is either weakened or removed. Thus cases for which $\varkappa_\text{I}$ is dominant are unlikely; on the other hand one cannot rule out the possibility that in some cases the contribution from $\varkappa_\text{I}$ may not be negligible.

Thus ordinary three-phonon interactions contribute to the thermal resistance in crystals, but do not explicitly enter into the expression for $\varkappa$, which depends upon the mean free paths of processes not conserving the wave vector. These processes can be classified as a) Umklapp-processes, b) elastic scattering and inelastic scattering by static imperfections, and c) boundary scattering. Elastic scattering has been treated in Sect. 6. Inelastic scattering has been considered by Pomeranchuk [14], who showed that it was not important.

---

[1] A. Schallamach: Proc. Phys. Soc. **53**, 214 (1941).

[2] Point imperfections cannot remove the divergence, since their mean free path varies as ω^{-4} in the limit of low frequencies; see Herpin [22] and Herring [23].

The rate of change due to three-phonon processes is given by (5.9), that is, it is of the form

$$\frac{dN}{dt}\Big] = \sum_{\mathbf{k},\mathbf{k}'} A \{n(\mathcal{N}' + \mathcal{N}'' + 1) + n'(\mathcal{N}'' + \mathcal{N} + 1) + n''(\mathcal{N} + \mathcal{N}' + 1)\}. \qquad (9.3)$$

Consider $dN/dt]$ for Umklapp-processes. The wave-vectors are related by (5.3c), that is

$$\mathbf{k} + \mathbf{k}' + \mathbf{k}'' = 2\pi\mathbf{b}. \qquad (9.4)$$

Since energy must also be conserved, at least two of the interacting states must have values of $\mathbf{k}$ not much less than $\pi\mathbf{b}$. At low temperatures the important processes are those for which $\omega \sim KT/\hbar$ and $\omega' \sim \omega'' \sim K\Theta/\hbar$. The terms in n' and n'' in (9.3) vanish on summing over all interactions, because $\mathbf{b}$ can have six different directions, so that any systematic relation between the signs of n and n' for a given value of $\mathbf{b}$ will be lost on summation; also it is easily seen that each term in n' and n'' is negligible if τ decreases faster than ω^{-1} with increasing frequency. Thus (9.3) can be described by a relaxation time τ_u, and since

$$\mathcal{N}' + \mathcal{N}'' + 1 \approx x\frac{d\mathcal{N}'}{dx'} \quad \text{and} \quad A \propto \frac{\omega\omega'\omega''}{v^2},$$

$$\frac{1}{\tau_u} \propto \frac{1}{v^2}\int \omega\omega'^2 \frac{\omega}{T}\frac{e^{x'}}{(e^{x'}-1)^2}k'\,dk' \qquad (9.5)$$

the integration extending from the smallest possible value of ω' (say $K\Theta/\hbar\alpha$, where α is a number of order unity) to the upper limit of frequencies, of order $K\Theta/\hbar$. Thus

$$\tau_u \propto \Theta^2\omega^{-2}T^{-3}e^{\Theta/aT}. \qquad (9.6)$$

The multiplying factor in (9.6) and the value of the constant α are sensitive to the shape of the BRILLOUIN zone and the law of dispersion of the lattice waves. For a simple cubic crystal without dispersion $\alpha \approx 1.2$, but one would expect α to be larger in real crystals, where dispersion is appreciable.

It was shown by SIMONS (private communication) that conservation of energy and simultaneous satisfaction of (5.3c) prevents the participation of low-frequency longitudinal waves in Umklapp-processes. Thus (9.6) applies to transverse waves only. Since the longitudinal and transverse waves interact strongly by means of ordinary three-phonon processes, this does not influence the overall thermal conductivity, provided $\sigma' \ll v\tau_u$, which is satisfied for all important frequencies.

At high temperatures $(T > \Theta)$, all $\mathcal{N}$'s are proportional to T, so that $\tau_u \propto T^{-1}$, a result first obtained by PEIERLS [9]. LEIBFRIED and SCHLÖMANN [24] have derived the absolute value of τ_u at high temperatures, and thus the thermal conductivity, in terms of fundamental constants. They obtained

$$\varkappa = \frac{3}{10\pi^3}\frac{K^3Ma}{\hbar^3\gamma^2}\frac{\Theta^3}{T}, \qquad T > \Theta \qquad (9.7)$$

but did not derive an absolute value for τ_u at low temperatures.

The scattering by static imperfections has been treated in Sect. 6. It is seen from (9.1) that if $\tau \propto \omega^{-n}$, $\varkappa \propto T^{3-n}$. For point imperfections $\tau \propto \omega^{-4}$; for cylindrical imperfections without a long-range strain field $\tau \propto \omega^{-3}$, but for dislocations $\tau \propto \omega^{-1}$ because of their strain field; for grain boundaries τ is independent of frequency. It is thus possible, in principle, to distinguish between different types of imperfections by the temperature dependence of the thermal conductivity.

The external boundary of a crystal also causes scattering, because the real boundary is different in nature from the idealised boundary assumed in deriving the normal modes of the crystal. In general the mean free path due to boundary scattering is of the order of the smallest linear dimensions of the crystal.

Since boundary scattering is the only process for which the absolute value of the mean free path can be estimated with reasonable certainty, some attention has been paid to the calculation of the effective mean free path. Casimir [11] has calculated the conductivity of an infinitely long cylinder on the assumptions that there are no interaction processes in the interior of the crystal, and thermal equilibrium is attained only at the boundaries, where phonons are absorbed, and re-emitted isotropically. The number of phonons at an interior point and of a given direction is governed by the temperature of their point of emission. This distribution, integrated over all directions, gives the local heat current, and integrating over the cross-section the total heat current is obtained. Thus

$$\varkappa = \tfrac{1}{3} S v L \tag{9.8}$$

where for an infinitely long cylinder of radius R

$$L = 2R \sum_j (v_j)^{-2} \Big/ \Big(\sum_j (v_j)^{-3} \Big)^{\frac{2}{3}}. \tag{9.9}$$

If the phonons are in part reflected specularly at the surface, the effective value of L is increased; it is independent of frequency as long as the coefficient of specular reflection is frequency independent. Berman, Simon and Ziman [46] have made this generalisation, and have also calculated the reduction in the effective value of L when the rod is of finite length only.

It is doubtful, however, whether the quantitative results of any of these calculations should be applied literally except possibly at very low temperatures, because these calculations assume the absence of three-phonon interactions conserving k, while over an appreciable range of the region where the imperfection mean free path l' is larger than L, l_1 and l_2 are still smaller than L, so that the calculations are not necessarily valid. There is as yet no treatment of the effect of the ordinary three-phonon interactions on the thermal conductivity in the size-dependent region.

If there are more than one of the above interaction processes, their combined relaxation time is given by

$$\frac{1}{\tau(\omega)} = \sum_\alpha \frac{1}{\tau_\alpha(\omega)} \tag{9.10}$$

where $\tau_\alpha(\omega)$ is the relaxation time due to interaction processes (α). Since this additivity applies only to each frequency separately, the overall thermal resistance W is not the sum of the thermal resistances due to each process, but in general, if $W_\alpha = 1/\varkappa_\alpha$ is the thermal resistance for process α acting alone, then

$$W \gtrsim \sum_{(\alpha)} W_\alpha \tag{9.11}$$

and the deviation from the additive resistance law is the stronger, the greater the difference between the frequency dependences of the τ_α's, and is greatest when the two resistances are of comparable magnitude. The additive resistance rule is thus a useful qualitative rule, but when the two resistances are comparable, the exact integrals must be evaluated, or a cut-off approximation [20] used.

The thermal conductivities at very low temperatures $(T \ll \Theta)$ in the presence of the following scattering mechanisms, each acting alone, are as follows:

a) External boundary or grain boundaries:

$$\varkappa = \frac{1}{3} S v L = \frac{4 \pi K^4 T^3}{h^3 v^2} L \int_0^\infty \frac{x^4 e^x}{(e^x - 1)^2} \, dx \qquad (9.12)$$

where L is of the order of the shortest linear dimensions of the specimen, or $L \sim l \alpha^2$, l being the average distance between grain boundaries and α the average angle of tilt.

b) Umklapp-processes:

$$\varkappa \propto \left(\frac{\Theta}{T} \right)^2 e^{\Theta/\alpha T}, \qquad T \ll \Theta \qquad (9.13)$$

as follows from (9.6), and $\varkappa$ is given by (9.7) if $T > \Theta$. The absolute value of $\varkappa$ at low temperatures has not yet been estimated.

c) Point imperfections:

$$\varkappa = \frac{1}{T} \frac{h v^2}{(2\pi)^2 3 a^3} \frac{G}{S^2} \left\{ \int_0^1 \frac{x^4 e^x \, dx}{(e^x - 1)^2} + \int_1^\infty \frac{e^x \, dx}{(e^x - 1)^2} \right\} \qquad (9.14)$$

where a^3 is the volume of a unit cell, G^{-1} is the concentration of imperfections per unit cell, and S is the scattering parameter defined in (6.4); S^2 has been estimated for a number of model imperfections [21].

d) Dislocations:

Substituting (6.8) into (9.1)

$$\varkappa \sim \frac{40 K^3}{h^2 v \gamma^2} \left\{ \int_0^1 \frac{x^4 e^x \, dx}{(e^x - 1)^2} + \int_1^\infty \frac{x^3 e^x \, dx}{(e^x - 1)^2} \right\} T^2 \frac{A}{b^2} \qquad (9.15)$$

where b is the magnitude of the BURGERS vector and A^{-1} is the number of dislocations per unit area.

Eq. (9.12) to (9.15) apply only to isotropic materials, and v is the average value over all polarizations.

10. Thermal conductivity of crystals (Observations)[1]. EUCKEN [25] measured the thermal conductivity of a number of dielectric solids down to oxygen temperatures and in a few cases down to hydrogen temperatures; he found generally for crystals that $\varkappa$ varied roughly as T^{-1}, in agreement with (9.7), and that the conductivity was larger for crystals with large Θ-values.

When measurements were extended to temperatures well below Θ the following classes of behaviour were observed: (a) $\varkappa$ increases faster than T^{-1} with decreasing T, until a maximum is reached; at lower temperatures $\varkappa$ is roughly proportional to the specific heat; this is interpreted in terms of Umklapp-resistance and, at lowest temperatures, boundary-resistance. (b) $\varkappa$ varies as T^{-1} or more slowly, with decreasing T a maximum is reached, at lower temperatures boundary resistance predominates; the resistance above the temperature of the maximum is thought to be due to imperfections. (c) In polycrystalline solids the boundary resistance is enhanced, and the maximum is shifted to higher temperatures.

---

[1] See also BERMAN's review [5].

Intrinsic or Umklapp-resistance. A variation of $\varkappa$ faster than T^{-1}, indicating Umklapp-resistance, was observed by Berman in quartz[1] and sapphire [39], in the case of a very pure alkali halide [51] and in rutile (private communication), by Wilks, Webb and Wilkinson in solid helium [42] to [45], and by White and Woods [121] in bismuth—see Sect. 23. In the case of diamond [43], [46] and germanium [50], [121] there is only slight evidence.

Solid helium is particularly interesting, because by varying the density it is possible to vary Θ, so that the dependence of $\varkappa$ on Θ can be compared with the theoretically predicted variation (9.13). Such a comparison can only be very rough, because the variation due to the factor $(\Theta/T)^2$ is swamped by the factor $e^{\Theta/aT}$, and the theory in its present form makes no certain predictions about the value of α. For the various helium specimens, $\varkappa$ can be expressed as a unique function of Θ/T, but the conductivities of other materials do not fit the same curve [44], [24]. But since these differences are not very large, it appears that the present theory is qualitatively correct.

In the vicinity of the maximum it has been found in all these cases that the conductivity is very much less than would have been expected from a combination of boundary and Umklapp-resistances alone—for example see Fig. 6. This has been interpreted as additional resistance due to static imperfections. At first sight it seems a suspicious coincidence that this should be so in all the cases of class (a) so far observed. It should be remembered, however, that crystals will form a continuous range with various amounts of imperfections; if the imperfection resistance is high, they belong to class (b); if the resistance is lower, they belong to class (a) with observable resistance at the maximum—say class (a′), and only for very low imperfection resistances would the belong to class (a) proper. But since Umklapp-resistance decreases very sharply with decreasing temperature, so that the maximum in the curve of $\varkappa$ against T would be very sharp in the case of class (a) proper, class (a′) corresponds to a very wide range of imperfection concentrations, so that it is understandable that with present techniques of crystal growths no case of class (a) proper has yet been observed.

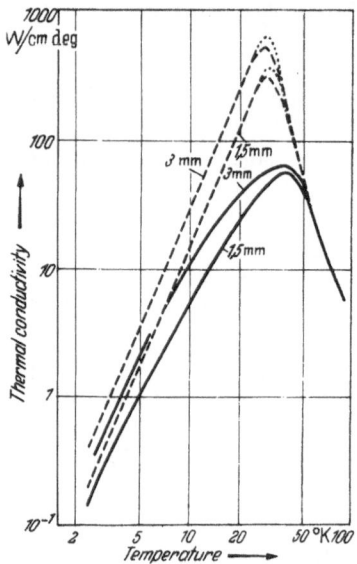

Fig. 6. The thermal conductivity of sapphire single crystals of different diameters according to Berman [39], [5]. Full curve: measured values; dashed curve: resistances combined according to the cut-off approximation [20]; dotted curve: resistances combined by simple addition.

Boundary resistance. Boundary resistance in the case of large crystals was first observed in the liquid helium region by de Haas and Biermasz [30] in the case of quartz, diamond and KCl, and has since been observed for all dielectric solids measured at these temperatures.

If this is the only resistive process acting, the following requirements should be satisfied: (α) the conductivity should be proportional to the lattice specific heat, which means in most cases that is should vary as T^3, (β) the mean free path deduced from the conductivity by substituting into (9.8) should be of the magnitude of the shortest linear dimension of the crystal, and (γ) the conductivity

---

[1] de Haas and Biermasz [29] also observed this variation, though no attention was paid to it.

should vary linearly with the size of the specimen. In the presence of additional resistive processes some or all of these requirements are not fulfilled.

In the case of additional scattering by internal or grain boundaries (α) is satisfied, but the observed mean free path is less than expected from the external dimensions and of course less sensitive to changes in them. Improvement of the agreement with (β) and (γ) is not achieved by lowering the temperature, but by decreasing the external dimensions.

In the case of additional scattering processes whose mean free path increases with decreasing frequency, neither (α), (β) or (γ) is fulfilled, but agreement with (9.8) is improved both by lowering the themperature as well as by decreasing the size of the specimen.

The measurements of DE HAAS and BIERMASZ [30] indicate an additional scattering mechanism with a frequency dependent mean free path. Even at their lowest temperatures ($\sim 2^\circ$ K) $\varkappa$ varies more slowly than T^3, and the discrepancy is greater the larger the crystal, also $\varkappa$ varies more slowly than linearly with the diameter of the specimen. In the case of KCl the deviations from (9.8) have been shown [20] to be consistent with the scattering by point imperfections which must be assumed (see below) to account for the thermal resistance at liquid hydrogen temperatures; since the frequency dependence of boundary and point-imperfection scattering is so different, the effect of the latter processes is appreciable at temperatures surprisingly far below the temperature of the maximum. The deviations from (α), (β) and (γ) in the case of quartz [30], [20], synthetic sapphire [39] and solid helium [44] are probably caused by the same mechanism which prevents the attainment of the theoretical value of the maximum conductivity, discussed above.

BIJL [34] and GARRETT [37] have measured potassium chrome alum from 1.4 to 3.9° K and from 0.16 to 0.29° K respectively. In the former region $\varkappa$ varies as $T^{2.3}$, in the latter as T^3. However, the phonon mean free path calculated from GARRETT's values was only about 0.05 cm, while the diameter of his crystal was 1.5 cm, suggesting that in this case scattering by internal boundaries took place. There has been no experiment on size dependence in this case; presumably $\varkappa$ should be independent of the external dimensions. Similar results were obtained previously by KURTI, ROLLIN and SIMON [31] for potassium chrome alum and iron ammonium alum.

Exhaustive tests of the CASIMIR theory were recently made by BERMAN, SIMON and ZIMAN [46] on diamond. Because of its high DEBYE temperature and the high temperature of its conductivity maximum, it was possible to eliminate other scattering processes in the liquid helium temperature range. On the other hand, the specimens were not long, and the finite length had to be corrected for. It was found that $\varkappa \propto T^{2.8}$ below 6° K; yet the observed mean free path exceeded the value calculated from the size of the specimens and the conductivity varied with specimen size. This was taken to indicate that the scattering at the external boundaries was not completely diffuse, but partly specular. The deviation from the T^3 dependence could thus be explained in terms of an increase with decreasing frequency of the coefficient of specular reflection, which seems reasonable.

BERMAN, FOSTER and ZIMAN [52] have also investigated the size-dependent conductivity of synthetic sapphire crystals—see also [46]. They found a mean free path, slightly smaller than expected from the theory, for rough crystals, but the conductivity was proportional to the diameter, and varied as T^3. For crystals of smooth surface, however, they found a longer effective mean free path and a slower temperature variation, interpreted as in the above case of diamond.

Imperfection resistance. De Haas and Biermasz [29] measured the thermal conductivity of KCl and KBr crystals. They found at liquid hydrogen temperatures $W \propto T$, instead of the exponential variation expected for intrinsic resistance. Klemens [20] suggested that the resistance arises mainly from scattering by point imperfections, in accordance with (9.14). At higher temperatures it was found that W increases more slowly with T; this is consistent with the expected departure from Rayleigh scattering ($l \propto \omega^{-4}$) in the case of higher frequencies. At higher temperatures an appreciable fraction of the total resistance is probably intrinsic. The KCl specimen used by de Haas and Biermasz had impurities of Na^+ and Mg^{++} in concentration of somewhat less than 10^{-4} per atom; also there should be a vacancy in the crystal for every divalent impurity to ensure electric neutrality. Substituting the observed resistance into (9.14), it is found that $\Sigma(S^2/G)$, summed over all types of imperfections, is about 1.2×10^{-4}, in rough agreement with what is known about the impurity content.

The interpretation of the thermal resistance in terms of point imperfection is supported by the fact that Berman [51] has measured the thermal conductivity of LiF crystal of such purity that its resistance was intrinsic, as indicated by an exponential temperature dependence; other alkali halide crystals showed a $W \propto T$ behaviour.

There has been evidence of thermal resistance by static imperfections in the case of a number of crystals of class (a); however, the temperature dependence of the additional resistance is usually not determined sufficiently well to draw conclusions about these imperfections. However, in the case of diamond [43], [46] the extra resistance seems to be temperature independent; hence Klemens[1] has tentatively suggested that this resistance is due to disordered regions of diameter of the order of 50 Å.

Eucken and Kuhn [26] measured the thermal resistance of mixed crystals of KCl and KBr at comparatively high (oxygen) temperatures. The additional resistance due to admixture was approximately independent of temperature. This was also observed in the same temperature region by Devyatkova and Stilbans [47] for KCl crystals with known F-center concentration; the latter experiment would have yielded extremely interesting results if the measurements had been made at liquid hydrogen temperatures.

Berman, Simon, Klemens and Fry [40], [39], [20] have studied the thermal conductivity of a quartz crystal after successive dosages of neutron irradiation, as well as the effect of annealing. Neutron irradiation produced an additional resistance which seems composed of two parts: one part increases with temperature, and this has been ascribed to isolated displacement defects; the other part varies as T^{-n}, when n lies between 1 and 3; this has been ascribed to large disordered regions which arise when an energetic displaced atom produces an avalanche of displacements, in accordance with the theory of radiation damage[2]. Since quartz is a glass-forming substance, the material in the region of such an avalanche has probably become vitreous[3].

Polycrystalline Solids. The thermal conductivity of a solid consisting of many closely packed small crystals is, apart from a temperature independent factor which depends upon the density of packing, the same as the conductivity

---

[1] P. G. Klemens: Phys. Rev. **86**, 1055 (1952).

[2] F. Seitz: Disc. Faraday Soc. No. 5, p. 271 (1949). For a review of radiation damage in solids see, for example, G. H. Kinchin and R. S. Pease: Rep. Progr. Phys. **18**, 1 (1955).

[3] Thermal conductivity measurements of neutron irradiated diamond and synthetic sapphire have now been reported by R. Berman, E. L. Foster and H. M. Rosenberg: Report of Conference on Defects in Crystalline Solids, p. 321. London: Physical Society 1955. In the case of sapphire the interpretation is similar to the case of quartz.

of each small crystal. Since the dimensions of each crystal are small, boundary resistance is correspondingly enhanced, the conductivity is proportional to the specific heat, and the temperature of the maximum is shifted to high temperatures. POMERANCHUK [14] had suggested that the scattering from a small-angle grain boundary varied as ω^2; this introduced the difficulty that it was not certain where the difference lay in principle between such a grain boundary and the boundaries of a polycrystalline solid, which scatter independently of frequency — see for example [5]. It is now known [21] that a low frequencies the disordered region in the immediate vicinity of the grain boundary, which contributes to scattering as discussed by POMERANCHUK, gives rise only to a small part of the total scattering probability; the major part comes from the strain field at large distances, so that a grain boundary scatters independently of frequency. There is thus no difference, in principle, between a small angle grain or mosaic boundary in a coherent crystal and a boundary of a crystallite, except that the former has a smaller scattering probability. This explains the results of GARRETT [37] and of KURTI, ROLLIN and SIMON [31] referred to above.

In the case of compressed powders the size of the crystallites can be estimated and, assuming unit scattering probability at each boundary, the boundary resistance can then be estimated.

KURTI, ROLLIN and SIMON [31] and VAN DIJK and KEESOM [32] found that the conductivity of a compressed powder of iron ammonium alum was about 1/10 of the conductivity of the large crystal [31] which itself apparently had a phonon mean free path of only $\sim$0.05 cm.; the grain size was not given.

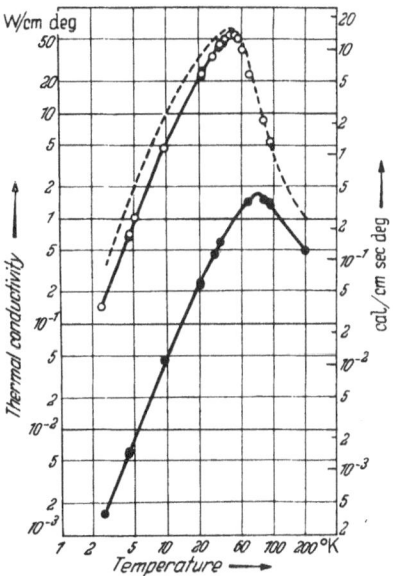

Fig. 7. Thermal conductivity of aluminium oxide according to BERMAN [41]. Dashed curve: 3 mm. diameter single crystal synthetic sapphire. Open circles: same crystal ground to 1.5 mm. diameter. Full circles: sintered alumina.

HUDSON [35] also obtained values for compressed powder of the same salt with crystallites estimated between 10^{-3} and 10^{-2} cm. in size. As pointed out by BERMAN [5] his average phonon mean free path was about 10^{-3} cm. and agreed with the size of the crystallites.

Extensive measurements on polycrystalline solids were made by BERMAN [41]. In Fig. 7 his results for sintered alumina are compared with the results for synthetic sapphire. The size of the crystallites was measured and found to range from 5 to 30×10^{-4} cm. At high temperatures the conductivity is proportional to, and about half of, the conductivity of the large crystals; this factor $\frac{1}{2}$ is thought to arise from the packing geometry. At lowest temperatures the conductivity is about 10^{-2} of the conductivity of the 1.5 mm. specimen, and the phonon mean free path is thus of the order of 20 to 30×10^{-4} cm., in rough agreement with the size of the crystallites; the conductivity varies as $T^{2.7}$, so that the phonon mean free path increases slowly with decreasing frequency, exceeding the geometrical mean free path. Similar results were obtained for sintered beryllia.

BERMAN [41] also measured the thermal conductivity of various graphite samples. The interpretation is not so straight-forward as in the other cases, as the lattice is highly anisotropic. A remarkable feature is that in some cases the conductivity increases faster with temperature than the specific heat. This can be interpreted in terms of a scattering probability which increases with decreasing frequency. This variation is in the opposite sense of what is found usually, and is also theoretically unacceptable. Klemens[1] has discussed the specific heat and thermal conductivity of graphite on the basis of a Debye theory, modified for high anisotropy. He concludes that a variation of the conductivity faster than of the specific heat is possible in the case of flake-shaped crystallites. Smith [48] measured the thermal conductivity of small flakes of natural graphite. He found the conductivity roughly proportional to the specific heat. Since each flake was crossed by slip-lines, so that each perfect region was of the same dimension in each direction, this result is consistent with the above interpretation; however, further work must be done before one can decide which is the correct explanation.

11. Other factors influencing heat transfer. It has been shown that the thermal conductivity can be written in the form

$$\varkappa = \tfrac{1}{3} \sum_i S_i v_i l_i \tag{11.1}$$

where the summation is over all normal lattice modes, S_i being the contribution of each mode to the specific heat, v_i the velocity with which energy is transferred, and l_i a mean free path. Eq. (11.1) should apply not only to conduction by lattice waves, but also to all other unlocalised excitations in the solid. If in general an atomic system is capable of excitation, and if this atomic system is regularly repeated through the volume of the crystal, any property of the atomic system (now a function of position) can be described by plane waves of wave-vector k as in Sect. 3—see for example [10]. To every discrete energy level of the isolated system now corresponds a band of energy values $E(k)$, the broadening depending upon the degree of interaction between neighbouring systems. The velocity of energy transport is the group velocity $(\hbar)^{-1} \partial E / \partial k$. The system of waves will not be completely stationary, but there will be interactions among the waves and between the waves and excitations of a different character. Thus a relaxation time and a mean free path are defined, though sometimes this concept must be used with caution, as in Sect. 7.

Particular cases of such excitations are the lattice waves, discussed previously, and the outer electrons of the atoms in a metal, to be treated in III. and IV. In addition the following excitations can contribute significantly to the specific heat, and may thus contribute appreciably to thermal conduction: spin, magnetic moment, rotation and orientation of molecules and other order-disorder effects, and movement of atoms from site to site. In all these cases the effect on the thermal conductivity can be twofold: there may be an additional component of thermal conductivity, but in so far as the additional excitations interact with other excitations (e.g. lattice waves) they will act on these excitations as an additional mechanism of scattering. This is exemplified in the case of conduction electrons in a lattice. In III. the additional conductivity due to the conduction electrons will be discussed; in IV. it will be shown that the conductivity by the lattice waves is reduced due to interactions with the conduction electrons.

There have been some theoretical discussions on the role of spin interactions and interaction between magnetic moments in the case of paramagnetic solids. Fröhlich and Heitler [16] have resolved the spins into progressive waves as above and have calculated the resulting conductivity, considering the mutual

---

[1] P. G. Klemens: Austral. J. Phys. **6**, 405 (1953).

interactions of the spin waves as the only resistive processes. The spin waves are assumed to obey BOLTZMANN statistics. They found that $\varkappa \propto T^n$, where $-2 < n < -1.5$, if $T \sim 0.01°$ K, and at $0.06°$ K for potassium chrome alum $\varkappa \approx 3 \times 10^{-7}$ watt.cm.$^{-1} \cdot$ deg^{-1}. This is considerably less than the observed conductivity at that temperature.

The same problem has been treated in greater detail by POMERANCHUK [17], while POMERANCHUK and AKHIESER [18] treated conduction by the interaction of magnetic dipoles. The formal theory is the same in both cases; the excitations are quantized waves, and for reasons which are not clear to the present writer, they are assumed to be fermions. The formal theory of thermal conductions is thus analogous to conduction by electrons (see Sect. 13 and 14), and their effect on the lattice conduction is treated in a manner analogous to Sect. 19. They find $\varkappa_m$, the conductivity of the magnetic excitations, to vary as T at lowest temperatures (scattering by imperfection), pass through a maximum and vary as T^{-1} at higher temperatures (scattering by phonons). The mean free path of phonons due to interactions with the magnetic excitations varies as T^{-1}, so that at lowest temperatures the lattice conductivity is unaffected and varies as T^3. They estimate roughly[1] the intrinsic value of $\varkappa_m$ and find that $\varkappa_m$ should exceed the lattice conductivity below about $0.02°$ K, provided, of course, that $\varkappa_m$ is not seriously reduced by static imperfections.

Neither the experiments of KURTI, ROLLIN and SIMON [31] nor those of GARRETT [37] show any signs of a component of conductivity other than the lattice conduction, but this is not at variance with these theories.

On the other hand REZANOV and CHEREPANOV [73] calculated the thermal conductivity of ferromagnetic metals by treating the spin waves as bosons. The role of the spin waves is mainly to scatter electrons, reducing the electronic thermal conductivity. Their theory is formally similar to the theory of Sect. 14.

While the existing theory of conduction by magnetic interactions lacks experimental material to which it can be compared, there exists experimental material of thermal conduction in cases of anomalies in the specific heat, which has not yet been treated theoretically. Thus EUCKEN and SCHRÖDER [27], GERRITSEN and VAN DER STAR [33] and VON SIMSON [53] have measured the thermal conductivity of hydrogen bromide, methane and ammonium chloride respectively. These solids have anomalies in their specific heat due to rotational or orientational states, and corresponding anomalies have been found in their thermal conductivities. These can be understood in terms of (11.1); thus in the case of methane reasonable agreement is found if v is taken to be $E d/\hbar \sim E/k_D \hbar$, and $l \sim d$, where d is the intermolecular distance and E is the energy of the excitations, which is determined by the temperature of the peak in the specific heat.

While pure germanium, as measured by ROSENBERG [50] and by WHITE and WOODS [121], shows the same behaviour as dielectric solids, a strongly impure specimen, measured by ESTERMANN and ZIMMERMANN [49], showed additional resistance which may be due to the scattering of lattice waves by electrons in an impurity band.

12. Thermal conductivity of liquids. The transport properties of liquids can be treated either on the basis of the kinetic theory of gases, extrapolated to the case of high densities and short molecular mean free paths[2] or on the basis of

---

[1] The crudeness of their numerical estimate was commented upon by BERMAN [5].

[2] See, for example, J. O. HIRSCHFELDER, C. F. CURTIS and B. R. BIRD: Molecular Theory of Gases and Liquids. New York: J. Wiley and Sons 1954.

the theory of conduction in solids, extrapolated to the case of high disorder, with a possible additional contribution due to the ability of the molecules to migrate. The second approach would, of course, be similar to the treatment of amorphous solids in Sect. 8.

The thermal conductivities of the following liquefied gases have been measured at low temperatures: liquid argon and nitrogen by UHLIR [54], liquid oxygen in a narrow temperature range by PROSAD [55] and liquid helium I by GRENIER [56] and by BOWERS [57]. The thermal conductivity of liquid helium II between about 0.6° K and the λ-point is determined by two-fluid circulation and presents special problems[1].

Between the λ-point and the boiling point the thermal conductivity of liquid helium was found to be proportional to temperature and to be roughly in agreement with the gas kinetic theory equation

$$\varkappa = \tfrac{5}{2} \eta \, C_v \tag{12.1}$$

η being the viscosity and C_v the specific heat per unit volume.

FAIRBANK and WILKS [58] have measured the thermal conductivity of liquid helium II below 1° K. It is known that at very low temperatures the thermal energy of liquid helium resides almost entirely in the phonon gas; below 0.6° K the thermal conductivity could be expressed as

$$\varkappa = \tfrac{1}{3} S \, v \, L \tag{12.2}$$

where L is fairly constant (increasing slightly with decreasing temperature) and of the order of the diameter of the tube which contained the liquid (0.03 cm.). This suggests thermal conduction by the phonons, their mean free path being limited by the external dimensions of the specimen. This interpretation is also supported by ZIMAN's discussion[2] of the propagation of heat pulses below 0.5° K.

It is, however, remarkable that the mean free path does not appear to be limited by scattering arising from the disordered structure of the liquid, analogous to the scattering of phonons in amorphous solids; the phonon mean free path is given (8.3) by

$$l = A \, a \, (a \, k)^{-2}. \tag{12.3}$$

In quartz glass $A \sim 200$ for longitudinal waves. With the same value of A, l should range from 10^{-2} to 10^{-3} cm. for the important frequencies in the temperature range covered. If the value of A in liquid helium is larger than in quartz glass, this would indicate a higher degree of local ordering in the former. It is not clear why this should be so.

III. Thermal conductivity of metals and alloys: Electronic component.

13. The free electron theory[3]. The present theory of electronic conduction in solids is based on the treatment of BLOCH [59], who in the first instance regards each electron as moving in a periodic potential produced by the metal ions and the other electrons, and then considers the deviation from periodicity due to the vibrations of the lattice as a perturbation.

---

[1] See the article on liquid helium by K. A. G. MENDELSSOHN in vol. XV.
[2] J. M. ZIMAN: Phil. Mag. 45, 100 (1954).
[3] For a detailed treatment see SOMMERFELD and BETHE [1], WILSON [4], MOTT and JONES [3], and also the companion article by H. JONES, vol. XIX of this Encyclopedia.

The wave-function of an electron in a periodic potential is of the form $\psi = u_k(x) e^{ik \cdot x}$, where $u_k(x)$ has the same periodicity as the potential, that is, the periodicity of the crystal lattice. To each k-value correspond two possible electron states (of different spin) extending through the crystal; their energy $E(k)$ is an eigenvalue of the SCHRÖDINGER equation

$$\nabla^2 u + 2i(\boldsymbol{k} \cdot \nabla) u + \left[\frac{2m}{\hbar^2}(E - V) - k^2\right] u = 0. \tag{13.1}$$

The possible values of the wave-vector $\boldsymbol{k}$ are the same as for the lattice waves—see Sect. 3—and depend only on the periodicity and size of the crystal. Additive changes of $\boldsymbol{k}$ by multiples of the inverse lattice vectors $\boldsymbol{b}$, defined by (3.2), leave ψ invariant, and the $\boldsymbol{k}$-space is separated into BRILLOUIN zones by planes which satisfy the condition for BRAGG reflection

$$(\boldsymbol{k} + \pi \boldsymbol{b}) \cdot \boldsymbol{b} = 0. \tag{13.2}$$

The energy $E(\boldsymbol{k})$ depends upon the form of the potential. It is a continuous function of $\boldsymbol{k}$ within each zone, but is discontinuous across a zone boundary (13.2). Also, since $E(\boldsymbol{k}) = E(-\boldsymbol{k})$, the normal derivative of E with respect to $\boldsymbol{k}$ at a zone boundary vanishes. If the potential is constant, $E = \hbar^2 k^2/2m$, and this is continuous across the zone boundaries, which in this case have no physical meaning. If the periodic potential departs only slightly from constancy, $E(\boldsymbol{k})$ will depart only slightly from the free electron value except at the zone boundaries, where the normal derivative vanishes, and a discontinuity is formed. When allocating a $\boldsymbol{k}$-value to a particular electron state of energy E, there is an arbitrariness, but one can choose a particular zone so that the free electron value $\hbar^2 k^2/2m$ matches E as closely as possible. It may be necessary to then adopt an effective electron mass value which differs from the usual mass, and may even be negative.

The values of $E(\boldsymbol{k})$ in a zone trace out a "band" of energy values. At any point at the zone boundary there is discontinuity in energy. Thus two bands may be separated by a "forbidden" region of energy values, but on the other hand the existence of an energy gap at each point of the zone boundary does not exclude the possibility that the bands may overlap.

According to the PAULI principle, the maximum occupation of each state is one. The number of $\boldsymbol{k}$-values in each zone is G, the number of unit cells in the crystal, so that there are $2G$ states. If the number of electrons per unit cell is odd, then at absolute zero at least one zone is only partly filled, and the substance is a metal. If the number of electrons per unit cell is even, it may be an insulator or a metal, depending on whether there is a gap between the highest full band and the next band, or whether there is overlap between these bands.

The equilibrium occupation probability of a state of energy E is

$$f^0 = [e^{(E-\zeta)/KT} + 1]^{-1} = (e^z + 1)^{-1} \tag{13.3}$$

where ζ is a parameter, the FERMI energy, given by the condition

$$\int f^0(E, \zeta) n(E) dE = N \tag{13.3a}$$

where $n(E)dE$ is the number of states in the energy interval E, dE, and N the number of electrons. This makes ζ temperature dependent; in particular, if $KT \ll \zeta$ and N and n are kept constant

$$\zeta = \zeta_0 - \frac{\pi^2}{6} K^2 T^2 \left(\frac{1}{n} \frac{dn}{dE}\right)_\zeta. \tag{13.4}$$

The density of states per unit volume can be expressed as

$$n(E) = \frac{2}{(2\pi)^3} \int \frac{dS}{|\mathrm{grad}_k\,E|} \tag{13.5}$$

where the integration is over the contour $E(\mathbf{k}) = E$ in $\mathbf{k}$-space. If $E(\mathbf{k})$ is a function of $|k|$ only, the contour is spherical and

$$n(E) = \frac{8\pi}{(2\pi)^3}\, k^2 \frac{dk}{dE} = n(k)\frac{dk}{dE}. \tag{13.6}$$

The modulated plane waves are eigenstates only if the potential is purely periodic; in real crystals there are transitions of particles between the eigenstates due to deviations from the perfectly periodic potential. These processes maintain equilibrium. On the other hand, an electric field $\mathbf{F}$ or a temperature gradient ∇T tend to disturb equilibrium. The Boltzmann equation, which is the condition that the actual occupation probability f of states is stationary, becomes

$$\frac{e}{\hbar}\,\mathbf{F}\cdot\frac{\partial f}{\partial \mathbf{k}} + \frac{1}{\hbar}\frac{\partial E}{\partial \mathbf{k}}\cdot\nabla T\frac{df}{dT} = \left.\frac{\partial f}{\partial t}\right] \tag{13.7}$$

where the right-hand side denotes the rate of change due to interaction processes. In the usual approximation f is taken to be f^0 on the left-hand side, and the rate of change due to interactions is described by a relaxation time

$$\left.\frac{\partial f(\mathbf{k})}{\partial t}\right] = \frac{f^0 - f}{\tau(\mathbf{k})} = -\frac{g(\mathbf{k})}{\tau(\mathbf{k})} \tag{13.8}$$

where it is assumed that τ is independent of the deviations g of the state $\mathbf{k}$ and of all the other states.

For an isotropic material and spherical energy contours the Boltzmann eq. (13.7) has the simple solution

$$g(\mathbf{k}) = -\mu\,\tau\,\frac{df}{dE}\frac{1}{\hbar}\frac{dE}{dk}\left\{eF - \nabla T\left(\frac{E}{T} + \frac{d\zeta}{dT}\right)\right\} \tag{13.9}$$

where μ is the direction cosine of $\mathbf{k}$ relative to $\mathbf{F}$, and ∇T is positive if in the same sense as $\mathbf{F}$.

The electric current density is

$$\mathbf{j} = \int e\frac{1}{\hbar}\frac{\partial E}{\partial \mathbf{k}}\,g(\mathbf{k})\,\frac{2}{(2\pi)^3}\,d\mathbf{k} \tag{13.10}$$

and the heat current is

$$Q = \int E\frac{1}{\hbar}\frac{\partial E}{\partial \mathbf{k}}\,g(\mathbf{k})\,\frac{2}{(2\pi)^3}\,d\mathbf{k}. \tag{13.11}$$

The electrical conductivity is defined as $\sigma = j/F$ if $\nabla T = 0$, and is obtained by substituting (13.9) into (13.10). The thermal conductivity is $\varkappa = -Q/\nabla T$ if $\mathbf{j} = 0$. Putting $\mathbf{j} = 0$ into (13.9) and (13.10), a relation is derived between eF and $\nabla T(E/T + d\zeta/dT)$. Substituting this into (3.11) and (3.9), an expression for the thermal conductivity is obtained.

It is noteworthy that (13.10) and (13.11) involve integrals of the following type, which can be expanded if $KT \ll \zeta$:

$$\int F(E)\frac{df^0}{dE}\,dE = -F(\zeta) - \frac{\pi^2}{6}(KT)^2\left(\frac{d^2F}{dE^2}\right)_\zeta. \tag{13.12}$$

While the only important term in σ is the first term in the expansion, in the case of $\varkappa$ thus defined this term vanishes identically and it is necessary to use terms of second order. In this way the electrical and the thermal conductivities become

$$\sigma = \frac{e^2}{12\pi^3} \int \frac{\tau v^2 dS}{|\text{grad}_{\boldsymbol{k}} E|} \tag{13.13}$$

and

$$\varkappa = \frac{K^2 T}{36\pi} \int \frac{\tau v^2 dS}{|\text{grad}_{\boldsymbol{k}} E|} \tag{13.14}$$

where the integration is over the surface $E = \zeta$ in $\boldsymbol{k}$-space, dS is an element of that surface, and $v = (\hbar)^{-1} dE/dk$ is the electron velocity.

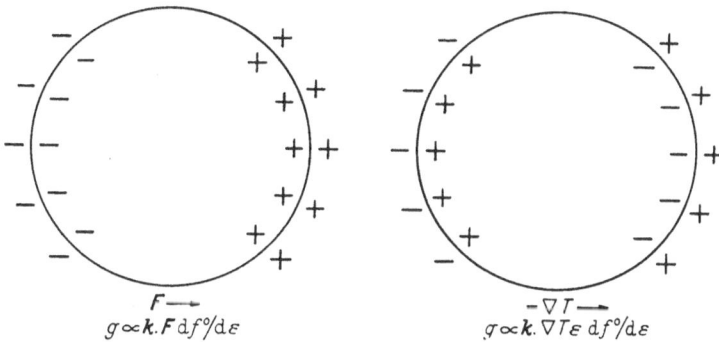

Fig. 8. The FERMI surface in the cases of electrical and thermal conduction; + and − denote regions of positive and negative deviation g of the distribution from equilibrium.

If τ is independent of the form of $g(\boldsymbol{k})$, so that the τ's in (13.13) and (13.14) are the same, the LORENZ number

$$L = \frac{\varkappa}{\sigma T} = \frac{\pi^2}{3}\left(\frac{K}{e}\right)^2. \tag{13.15}$$

becomes a constant independent of the relaxation time and of the band structure. However, this value of L is a consequence of the strong degeneracy of the electron gas; it generally changes as KT becomes comparable to ζ, and in the extreme case of BOLTZMANN statistics, if $\tau \propto v^p$

$$L = \frac{p+5}{2}\left(\frac{K}{e}\right)^2. \tag{13.16}$$

The following argument, apparently due to FRÖHLICH[1], indicates under what conditions one may expect the WIEDEMANN-FRANZ-LORENZ law (L constant) to hold:

In the case of electrical conduction, $g(\boldsymbol{k}) \propto \boldsymbol{k} \cdot \boldsymbol{F} df^0/d\varepsilon$, which is equivalent to a displacement of the FERMI surface in $\boldsymbol{k}$-space—see Fig. 8. Processes restoring equilibrium must take an electron from a point on the FERMI surface to the opposite point ("horizontal" movement). The effective relaxation time is thus given by

$$\frac{1}{\tau} \approx \frac{1}{\tau_0}(1 - \bar{\mu}) \tag{13.17}$$

where τ_0 is the mean free time of an electron and $\bar{\mu}$ is the average of the cosine of the scattering angle.

[1] H. FRÖHLICH: Elektronentheorie der Metalle, p. 205. Berlin: Springer 1936.

Now, in the case of thermal conduction, the dominant term in the deviation from equilibrium is of the form $g(\mathbf{k}) \propto \mathbf{k} \cdot \nabla T \varepsilon df^0/d\varepsilon$; in other words the electrons going in one direction are too hot, those in the other direction too cold. Resistance can be produced in two ways: either by processes changing the direction of an electron, but keeping its energy constant, or by processes changing the energy, but not the direction, that is, either by "horizontal" or "vertical" movement of the electrons on the FERMI surface.

Thus it follows that if the only important processes change the direction, but not the energy, τ will be the same for either form of g, and (13.15) will be satisfied. This will be so, at any temperature, if the electrons are mainly scattered elastically by lattice imperfections, and also at temperatures well above the DEBYE temperature, if the electrons interact with lattice waves, for in the latter case the change in energy on absorbing or emitting a phonon is $K\Theta$, which is small compared with KT, so that the change in ε for each interaction is small.

On the other hand, if $T \ll \Theta$, the change of energy on absorbing or emitting a phonon is of order KT, while the change of direction is only through an angle of order T/Θ. Thus for "vertical" movement $1/\tau$ is of order $1/\tau_0$, while for "horizontal" movement $1/\tau$ is of order $(1-\bar{\mu})/\tau_0$; $\bar{\mu}$ is the average cosine of the scattering angle $(\sim T/\Theta)$, hence $1-\bar{\mu} \sim (T/\Theta)^2$. Consequently, horizontal movement is unimportant compared to vertical movement, and τ for electrical conduction (being determined by horizontal movement) is much larger than τ for thermal conduction (determined by vertical movement).

In the case of scattering by phonons $1/\tau_0$ is roughly proportional to the total number of phonons; thus $1/\tau_0$ varies as T at high temperatures and as T^3 at low temperatures. At high temperatures $1-\bar{\mu} \sim 1$, so that the intrinsic electrical and thermal conductivities, σ_i and $\varkappa_i$, are respectively proportional to $1/T$ and independent of T. At low temperatures $(1-\bar{\mu})/\tau_0$ varies as T^5, so that $\sigma_i \propto T^{-5}$, while $\varkappa_i \propto T^{-2}$, since $1/\tau \propto T^3$.

In addition there is resistance (electrical or thermal) due to lattice imperfections, which has to be added to the intrinsic resistance, thus

$$\frac{1}{\sigma} = \varrho = \varrho_0 + \varrho_i, \tag{13.18a}$$

$$\frac{1}{\varkappa} = W = W_0 + W_i \tag{13.18b}$$

where $\varrho_i = 1/\sigma_i$, $W_i = 1/\varkappa_i$, ϱ_0 is independent of temperature and $W_0 = \varrho_0/LT$.

14. Interaction with lattice waves: The ideal resistance[1]. In a perfectly periodic potential the electrons would not be scattered, thermal equilibrium would not be established and τ would be infinite. Departures from periodicity in real crystals are of two types: static lattice imperfections and lattice vibrations. The scattering due to lattice imperfections can be described, as in (13.8), by a relaxation time τ which is independent of temperature and varies only slowly with E, and the residual resistance can thus be treated by the formal theory of Sect. 13.

The lattice vibrations can be resolved into quantized waves or phonons, as in chapter II, and the interaction between electrons and lattice waves can be regarded as individual processes in which an electron of wave-vector $\mathbf{k}$ and a phonon of wave-vector $\mathbf{q}$ combine to form an electron $\mathbf{k}'$, or vice versa, con-

---

[1] A detailed treatment is found in [1] and [4].

serving energy

$$\Delta E = E + \hbar \omega - E' = 0 \tag{14.1}$$

and satisfying an interference condition

$$\boldsymbol{k} + \boldsymbol{q} = \boldsymbol{k}' \tag{14.2a}$$

or

$$\boldsymbol{k} + \boldsymbol{q} = \boldsymbol{k}' \pm 2\pi \boldsymbol{b} \tag{14.2b}$$

where $\boldsymbol{b}$ is one of the inverse lattice vectors (3.2).

If $N(\boldsymbol{q})$ is the number of phonons of mode $\boldsymbol{q}$ and if $f(\boldsymbol{k})$ is the number of electrons in state $\boldsymbol{k}$, averaged over a group of states about $\boldsymbol{k}$, then the rate of change of $f(\boldsymbol{k})$ due to these interactions can be expressed as

$$t\frac{df}{dt}\bigg] = \frac{4}{9GM\hbar} \sum_{\boldsymbol{q}} C^2 \frac{q^2}{\omega} \left\{ f(\boldsymbol{k}') [1 - f(\boldsymbol{k})] [N(\boldsymbol{q}) + 1] - \right. \\ \left. - f(\boldsymbol{k}) [1 - f(\boldsymbol{k}')] N(\boldsymbol{q}) \right\} \frac{1 - \cos(\Delta E t/\hbar)}{\Delta E^2} \tag{14.3}$$

where C is a parameter, of the dimensions of energy, describing the strength of the electron-phonon interaction.

In the BLOCH theory [59] Umklapp-processes (14.2b) are disregarded and it is assumed that $N(\boldsymbol{q}) = \mathfrak{N}(\boldsymbol{q})$, that is the phonon distribution is assumed to be at equilibrium and to be uninfluenced by deviations of the electron distribution from equilibrium. All energy contours are assumed to be spherical and if in addition $u(x)$ of (13.1) is taken to be spherically symmetrical, then it can be shown that C vanishes for transverse phonons, while for longitudinal phonons it is independent of $\boldsymbol{q}$ and of order ζ.

Substituting (14.3) into (13.7) a transport equation is obtained. One can replace the summation over $\boldsymbol{q}$ by an appropriate integration and the resonance factor by a delta-function; in the case of high degeneracy $E \sim E' \gg \hbar \omega$, so that $\boldsymbol{q}$ is nearly tangential to the energy contour and $E' - E$ can be expanded in powers of q/k. Writing

$$f - f^0 = g = -\mu k \frac{df^0}{dE} c(E) \tag{14.4}$$

where μ is the direction cosine of $\boldsymbol{k}$ relative to $\boldsymbol{F}$ or ∇T, the transport equation is put into the form of the following integral (BLOCH) equation:

$$\mathfrak{S} \cdot c(E) = A(E) \left[K \varepsilon \nabla T + \frac{d\zeta}{dT} \nabla T - e\boldsymbol{F} \right] \tag{14.5}$$

where

$$A(E) = \frac{9\pi Mv}{C^2 \hbar a^3} \frac{1}{k} \left(\frac{dE}{dk} \right)^2 \tag{14.6}$$

and where $\mathfrak{S}$ is the integral operator

$$\mathfrak{S} \cdot c(E) = \int_0^{q_0} q^2 \, dq \, \mathfrak{N}(q) \left\{ \left[\left(1 + \frac{\hbar \omega}{k(dE/dk)} - \frac{q^2}{2k^2} \right) c(E + \hbar \omega) - c(E) \right] \times \right. \\ \left. \times \frac{e^{\varepsilon} + 1}{e^{\varepsilon} + e^{-x}} + \left[\left(1 - \frac{\hbar \omega}{k(dE/dk)} - \frac{q^2}{2k^2} \right) c(E - \hbar \omega) - c(E) \right] \frac{e^{-\varepsilon} + 1}{e^{-\varepsilon} + e^{-x}} \right\}. \tag{14.7}$$

In (14.6) and (14.7) v is the velocity of the phonons, $x = \hbar\omega/KT$ and q_0 is the upper limit of the wave-number of the phonons which can participate in the interactions. If N_a, the number of free electrons per atom, exceeds $\frac{1}{4}$, then $q_0 = K\Theta/\hbar v$, the Debye upper limit, but if $N_a < \frac{1}{4}$, then $q_0 = 2k_\zeta$, as was pointed out by Sondheimer [65]—however, see also [78].

Since the operator $\mathfrak{S}$ is linear, it is possible to separate (14.5) into

$$\mathfrak{S} \cdot c_0 = 1, \tag{14.8a}$$

$$\mathfrak{S} \cdot c_1 = \varepsilon \tag{14.8b}$$

where

$$c(\varepsilon) = \left[\frac{d\zeta}{dT} \nabla T - eF\right] \left(A(\zeta) c_0 + K TA'(\zeta) c_1\right) + K \nabla TA(\zeta) c_1. \tag{14.9}$$

The electrical and thermal currents are respectively given by

$$j = -\frac{8\pi}{3} \frac{e}{(2\pi)^3} \frac{1}{\hbar} \int k^3 \frac{df^0}{dE} c(E) dE, \tag{14.10a}$$

$$Q = -\frac{8\pi}{3} \frac{1}{(2\pi)^3 \hbar} \int k^3 E \frac{df^0}{dE} c(E) dE. \tag{14.10b}$$

The thermal conductivity is defined as $\varkappa = -Q/\nabla T$ on condition that $j = 0$. Defining

$$\left. \begin{aligned}
L_{11} &= \int c_0 \frac{df^0}{dE} k^3 dE = k_\zeta^3 \int c_0 \frac{df^0}{d\varepsilon} d\varepsilon + 0\left(\frac{KT}{\zeta}\right), \\
L_{12} &= \int c_1 \frac{df^0}{dE} k^3 dE = k_\zeta^3 \int c_1 \frac{df^0}{d\varepsilon} d\varepsilon + 0\left(\frac{KT}{\zeta}\right), \\
L_{21} &= \int c_0 \varepsilon \frac{df^0}{dE} k^3 dE = k_\zeta^3 \int c_0 \varepsilon \frac{df^0}{d\varepsilon} d\varepsilon + 0\left(\frac{KT}{\zeta}\right), \\
L_{22} &= \int c_1 \varepsilon \frac{df^0}{dE} k^3 dE = k_\zeta^3 \int c_1 \varepsilon \frac{df^0}{d\varepsilon} d\varepsilon + 0\left(\frac{KT}{\zeta}\right)
\end{aligned} \right\} \tag{14.11}$$

and substituting (14.9) into (14.10), putting $j = 0$ and eliminating $\left[\frac{d\zeta}{dT} \nabla T - eF\right]$ between (14.10a) and (14.10b), it is seen that

$$Q = -\frac{8\pi}{3\hbar(2\pi)^3} A(\zeta) K^2 T \nabla T \left\{ L_{22} - \frac{L_{12}(L_{21} - L_{22} K TA'/A)}{L_{11} - L_{12} K TA'/A} \right\}. \tag{14.12}$$

Since the operator $\mathfrak{S}$ is even in ε except for terms of order (KT/ζ), c_0 is even and c_1 is odd in ε except for terms of that order. Therefore the lowest order terms of L_{11} and L_{22} are both of zero order, while those of L_{12} and L_{21} are of first order in KT/ζ; also $A'KT/A$ is of order KT/ζ. Hence to the lowest (zero) order in KT/ζ the second term in (14.12) can be neglected, and

$$\varkappa = \frac{8\pi}{3(2\pi)^3 \hbar} A(\zeta) K^2 T L_{22}. \tag{14.13}$$

Thus only one integral equation (14.8b) has to be solved to find the thermal conductivity.

Alternatively, if $\bar{c}_0$ and $\bar{c}_1$ are the solutions of

$$\mathfrak{S} \cdot \bar{c}_0 = 1, \tag{14.14a}$$

$$\mathfrak{S} \cdot \bar{c}_1 = E \tag{14.14b}$$

and if coefficients are defined as

$$K_{11} = \int \bar{c}_0 \, A(E) \, \frac{df^0}{dE} \, k^3 \, dE \,,$$

$$K_{12} = \int \bar{c}_1 \, A(E) \, \frac{df^0}{dE} \, k^3 \, dE \,,$$

$$K_{21} = \int \bar{c}_0 \, E \, A(E) \, \frac{df^0}{dE} \, k^3 \, dE \,,$$

$$K_{22} = \int \bar{c}_1 \, E \, A(E) \, \frac{df^0}{dE} \, k^3 \, dE \,,$$

(14.15)

then the thermal conductivity is given by

$$\varkappa = \frac{8\pi}{3(2\pi)^3 \hbar} \, \frac{1}{T} \left\{ K_{22} - \frac{K_{12} K_{21}}{K_{11}} \right\}. \tag{14.16}$$

It is easily seen, by substituting $\bar{c}_1 = \zeta c_0 + K T c_1$ and expanding the coefficients (14.15), that while each coefficient has terms of order zero in (KT/ζ), these terms cancel exactly in $(K_{22} - K_{12} K_{21}/K_{11})$, so that all these coefficients must be evaluated to second order. As long as the integral eqs. (14.8) or (14.14) can be solved exactly, both (14.13) and (14.16) yield the same results; otherwise (14.13) must be preferred, because only one integral equation has to be solved, and because the solution need only be obtained to lowest order.

In the case of electrical conduction there is no difference between the two formulations: $\bar{c}_0 = c_0$ and $K_{11} = A(\zeta) L_{11}$; the electrical conductivity is given by

$$\sigma = \frac{8\pi e^2}{3(2\pi)^3 \hbar} \, A(\zeta) \, L_{11}. \tag{14.17}$$

For the case of *high temperatures* the integral equations are readily solved: $c_0 = \bar{c}_0$ is independent of ε or E, $c_1 = \varepsilon c_0$ and $\bar{c}_1 = E \bar{c}_0$. Only the terms in $q^2/2k^2$ contribute to $\mathfrak{S}$; these terms describe the "horizontal" movement on the FERMI surface. It is easily seen that

$$c_0 = \frac{4 k_\zeta^2 \hbar v}{q_0^4 K T} \tag{14.18}$$

whence σ and $\varkappa$ can easily be calculated. The fact that $c_1 = \varepsilon c_0$, etc., indicates that a relaxation time can be defined which is independent of $c(E)$, so that the WIEDEMANN-FRANZ law (13.15) is obeyed. From (14.18) $\sigma \propto T^{-1}$, so that $\varkappa$ is independent of T (except for higher order terms in KT/ζ) at high temperatures, and is given by

$$\varkappa_\infty = \frac{2\Theta v M K^2}{3\pi \hbar^2} \left(\frac{k}{C} \frac{dE}{dk} \right)^2 \left(\frac{k}{q_0} \right)^2 \tag{14.19}$$

where k and E denote the values at the FERMI level. Note that $(k/q_0)^3 = N_a/2$, except if $N_a < \frac{1}{4}$, when $k/q_0 = \frac{1}{2}$.

However, at *low temperatures* the solution of the integral equations is not so easy, c_0 is no longer constant, and c_1/ε is no longer constant, nor does it equal c_0. Solutions of (14.8) or (14.14) can be found with the aid of the following variational principle due to KOHLER [62], or by methods which can be shown to be equivalent.

One can define a "scalar product" of two functions $\varphi_1(\varepsilon)$ and $\varphi_2(\varepsilon)$ as

$$(\varphi_1, \varphi_2) = (\varphi_2, \varphi_1) = - \int \varphi_1 \frac{df^0}{d\varepsilon} \varphi_2 k^3 d\varepsilon \tag{14.20}$$

and the φ's can be regarded as vectors in the real projection of a Hilbert space. The operator $\mathfrak{S}$ is a linear operator in this space, and can be shown to have the property

$$(\mathfrak{S} \cdot \varphi_1, \varphi_2) = (\varphi_1, \mathfrak{S} \cdot \varphi_2) \tag{14.21 a}$$

$$(\mathfrak{S} \cdot \varphi_1, \varphi_1) \gtreqless 0 \tag{14.21 b}$$

Let φ be any given function, and let c be the solution of

$$\mathfrak{S} \cdot c = \varphi \tag{14.22}$$

then $(\mathfrak{S} \cdot c, c)$ is a maximum for variations of c subject to the normalizing condition

$$(\mathfrak{S} \cdot c, c) = (\varphi, c) . \tag{14.23}$$

For consider the variation

$$\delta(\mathfrak{S} \cdot c, c) = (\mathfrak{S} \cdot \delta c, c) + (\mathfrak{S} \cdot c, \delta c) + (\mathfrak{S} \cdot \delta c, \delta c) \tag{14.24}$$

which, because of (14.23), satisfies

$$\delta(\mathfrak{S} \cdot c, c) = (\varphi, \delta c) . \tag{14.25}$$

From (14.21 a) the first two terms on the right-hand side are equal, and are each equal to $(\varphi, \delta c)$, thus

$$\delta(\mathfrak{S} \cdot c, c) = - (\mathfrak{S} \cdot \delta c, \delta c) \tag{14.26}$$

and this is negative (14.21 b) and of second order in δc. Hence $(\mathfrak{S} \cdot c, c)$, subject to the normalizing condition (14.23), is a maximum if c is the solution of (14.22).

The proof that the integral operator (14.7) satisfies (14.21) is given in Kohler's original paper [62], and is discussed more generally by Wilson [4]. One consequence of (14.21 a) is that $L_{12} = L_{21}$ and $K_{12} = K_{21}$, as required by the thermodynamics of irreversible processes and the principle of microscopic irreversibility; alternatively the variational principle is an expression of the principle of stationary entropy production.

The main advantage of the variational method is that the coefficients L_{11} and L_{22} are stationary, being of the form (14.23). It is, of course, these coefficients, rather than the functions c, which are required for the calculation of the conductivities. Thus if a trial function c_t is chosen which has some adjustable parameters, and if these parameters are adjusted so that $(\mathfrak{S} \cdot c_t, c_t)$ is a maximum subject to $(\mathfrak{S} \cdot c_t, c_t) = (\varphi, c_t)$, then this value of (φ, c_t) differs from the required value (φ, c) only by a term of order $(\delta c, \delta c)$, where $\delta c = c - c_t$. Also if a different class of trial function is chosen, and if the same procedure yields a larger value of (φ, c_t), it is known that this new value is a better approximation. On the other hand the success of this method still depends upon the choice of the trial function: a family of trial functions map out a subspace in the Hilbert space of c; the required solution has a component δc orthogonal to that subspace; the error in (φ, c) is of second order in δc, but if the family of trial functions is badly chosen, δc may still be so large that the second order terms are not small.

If the thermal conductivity is expressed in the form (14.16), then K_{11} and K_{22} are stationary, but K_{12} and K_{21} are not. However, Makinson [68] pointed out that the expression

$$(\bar{c}_0, E) + (\bar{c}_1, 1) - (\bar{c}_0, \mathfrak{S} \cdot \bar{c}_1) = K'_{12}, \tag{14.27}$$

which equals K_{12} if the correct solutions are substituted into it, is stationary for arbitrary small variations of $\bar{c}_0$ and $\bar{c}_1$, and if K'_{12} is used instead of K_{12} in (14.16), the thermal conductivity is stationary. He also pointed out that if the trial function is a polynomial of positive powers of ε, then the expressions for K_{12} and K'_{12} are identical (for restricting the trial function to a polynomial restricts the variation δc), so that the methods of solutions discussed below do in fact give stationary results.

BLOCH [59]—see also [1]— and WILSON [60] have used arguments equivalent to the variational method with trial functions of the form $a_0 + a_1 \varepsilon$ to solve (14.14), though BLOCH only solved (14.14a); KOHLER [62] used the same trial function to solve (14.8); SONDHEIMER [64] solved (14.14) extending the trial

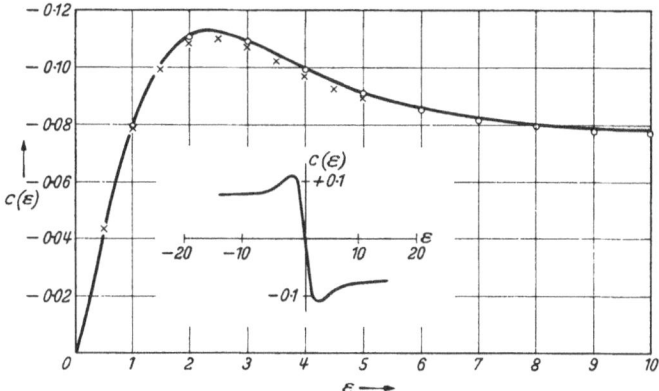

Fig. 9. Solution of the integral equation $\mathfrak{S} \cdot c = \varepsilon \, (K\,T/\hbar v)^3$ according to [69]. Circles and crosses are the discrete values obtained on replacing the integral equation by sets of 10 simultaneous linear equations. The full curve is the result of an iterative procedure, judged to be accurate to better than 0.5%.

function to a cubic expression in ε and KROLL [63] used an apparently arbitrary procedure which, however, is identical to using the variational method with a polynomial trial function. These equations have also been solved with $\mathfrak{S}$ generalized to include a term $c(E)/\tau_0$ in addition to (14.7) to take account of the scattering by static imperfections.

KLEMENS [69] pointed out that in the absence of scattering by static imperfections and in the limit $T \ll \Theta$ the solution of (14.8a) could be well expressed by a polynomial of a few terms in ε, while (14.8b) could not, thus throwing doubt on the results obtained by polynomial trial functions. He therefore solved (14.8b) numerically for lowest temperatures, where the terms in $q^2/2k^2$ in (14.7) can be disregarded (only "vertical" movement important). The numerical solution, shown in Fig. 9, was renormalized by means of (14.23), so that the resulting value of L_{22} is probably quite accurate. It is interesting to note that while $c_1(\varepsilon)$ thus obtained differs radically from SONDHEIMER's trial function, the resultant thermal conductivity exceeds the value derived by SONDHEIMER's third approximation by only 11%.

It should be emphasised that this case is the one for which the error in SONDHEIMER's method is largest; as $1/\tau_0$ and T are increased, $df/dt]$ can be more nearly expressed by a relaxation time and $c_1(\varepsilon)$ can be better represented by a polynomial in ε.

In the absence of imperfection scattering it is easily seen from (14.7) that $c_1 \propto T^{-3}$, so that $W_i \propto T^2$. In common with $1/\varkappa_\infty$ of (14.19), W_i contains the factor $\left(\dfrac{k}{C}\dfrac{dE}{dk}\right)^{-2}$, and since $\left(\dfrac{k_{\zeta}}{q_0}\right)^2 = \left(\dfrac{N_a}{2}\right)^{\frac{2}{3}}$, W_i can be expressed in the form

$$W_i(T) = A\,N_a^{\frac{2}{3}}W_\infty\left(\frac{T}{\Theta}\right)^2 \tag{14.28}$$

where A is a numerical constant derived from the solution of (14.8b).

According to the first approximation of the variational method [60], [62]

$$A = 95.3, \tag{14.29a}$$

according to Sondheimer's third approximation [64]

$$A = 71.6, \tag{14.29b}$$

and according to the numerical solution [69]

$$A = 64.0. \tag{14.29c}$$

Makinson [61] has discussed the thermal conductivity at intermediate temperatures and in the presence of residual resistance. In the first approximation of the variational method, W_0 and W_i are strictly additive; this is because it is assumed that $c_1(\varepsilon) \propto \varepsilon$, while the asumption of a relaxation time also leads to $c_1(\varepsilon) \propto \varepsilon$. The ideal resistance increases with temperature; however, it does not increase uniformly towards W_∞, but, if $N_a = 1$, it goes through a strong maximum ($\sim 1.5\,W_\infty$) at $\Theta/4$. According to Sondheimer's third approximation W_i differs appreciably from the first approximation at lowest temperatures, as illustrated by (14.29b), but as the temperature is increased, it gradually approaches Makinson's curve. In this third approximation the resistances are no longer additive, since for a pure metal $c_1(\varepsilon)$ does not vary as ε. The greater the residual resistance, the more does W approach Makinson's value, because now $c_1(\varepsilon)$ approaches more closely to $c_1(\varepsilon) \propto \varepsilon$. An upper limit to deviations from the additivity of the resistance is thus simply the difference between Makinson's and Sondheimer's values of W_i, and the deviations are always positive. Since Klemens' numerical solution gives even smaller values for W_i, the deviations from additivity can be expected to be slightly larger.

The variational method converges much more rapidly in the case of electrical resistance, since $c_0(\varepsilon)$ is almost constant. In the limit $\varrho_0 = 0$ and $T \to 0$, the first approximation gives the correct value of ϱ_i except for terms of higher order in T/Θ. The third approximation gives slightly lower values for ϱ_i at intermediate temperatures, and there is correspondingly a small positive deviation from the additive resistance (Matthiessen's) rule. The first approximation yields

$$\varrho_i \propto T^5 \int\limits_0^{\Theta/T} \frac{x^5 e^x\,dx}{(e^x - 1)^2} \tag{14.30}$$

and ϱ_0 and ϱ_i are of course additive in this approximation.

Toda [66] has treated electrical and thermal conductivity semi-quantitatively, using the concepts of horizontal and vertical diffusion. Blatt [67] showed that this treatment leads to the same results as the usual treatment in terms of the Bloch equation; it should do so, since the diffusion concepts are implicit in the latter.

15. Comparison with experiments: Monovalent metals. The theory of Sect. 14 applies only to a single band with spherical energy contours in k-space; and thus can only apply to monovalent metals. The magnitude of the conductivities cannot be compared with theory, because it has not been possible to arrive at a reliable theoretical estimate of the electron-phonon interaction constant C, but one can compare the observed temperature dependence of the ideal electrical or thermal resistivities with theory, and one can intercompare the electrical and thermal resistivities.

It is found that monovalent metals satisfy approximately the theoretical relations $W_i(T) \propto T^2$ if $T \ll \Theta$, and $W_i(T) = W_\infty$, independent of T, if $T > \Theta$; similarly $\varrho_i(T) \propto T^5$ if $T \ll \Theta$; and $\varrho_i(T) \propto T$ if $T > \Theta$; thus one can intercompare the four magnitudes $W_i(T_1)/T_1^2$ and $\varrho_i(T_1)/T_1^5$, where $T_1 \ll \Theta$, and $W_i(T_2)$ and $\varrho_i(T_2)/T_2$, where $T_2 > \Theta$. It will be found that there are discrepancies in the relationships between these quantities. There are also descrepancies in respect of the temperature dependence of W_i and ϱ_i at the intermediate temperatures, but these are less important effects; the reasons for the discrepancies between the conductivities at the extreme temperatures will presumably also provide an explanation at intermediate temperatures.

It has been shown that the WIEDEMANN-FRANZ law (13.15) should hold if a relaxation time can be defined independent of the form of $g(\boldsymbol{k})$, and in Sect. 14 it was shown that this is the case at high temperatures; hence at high temperatures irrespective of the band structure

$$T_2 W_i(T_2) = \frac{\varrho_i(T_2)}{L}, \qquad (T_2 > \Theta). \tag{15.1}$$

This is indeed fulfilled for all metals except for those which have an appreciable lattice component of thermal conductivity, and can be regarded as a verification of the general principles of the free electron theory. Thus for purposes of testing the band model only three quantities are independent out of $\varrho(T_1)$, $\varrho(T_2)$, $W(T_2)$ and $W(T_1)$, and there are only two independent relations between them. Any two of the following three are independent

$$\varrho_i(T_1) = 497.6 \, \frac{\varrho(T_2)}{T_2} \, \frac{T_1^5}{\Theta^4}, \tag{15.2}$$

$$W_i(T_1) = 64.0 \, N_a^{\frac{2}{3}} W_i(T_2) \left(\frac{T_1}{\Theta}\right)^2, \tag{15.3}$$

$$\frac{W_i(T_1)}{T_1^2} = \frac{64.0}{497.6} \, N_a^{\frac{2}{3}} \frac{\Theta^2}{L} \, \frac{\varrho_i(T_1)}{T_1^5}. \tag{15.4}$$

Eq. (15.2) is obtained from (14.30), and (15.3) from (14.28) and (14.29c). Eq. (15.4) is best derived by eliminating $\varrho_i(T_2)$ and $W_i(T_2)$ from (15.1), (15.2), (15.3); it could, however, have been derived directly from the theory of Sect. 14 without reference to the conductivities at high temperatures.

The thermal conductivities of a number of monovalent metals have now been measured on specimens of sufficiently low residual resistivity W_0 and to sufficiently low temperatures that the ideal thermal resistivity W_i can be deduced with confidence for the limit of low temperatures. Thus BERMAN and MACDONALD [83], [84] measured sodium and copper; MENDELSSOHN and ROSENBERG [85], [87] measured copper, silver and gold, in addition to various other metals discussed in Sect. 16; WHITE [88], [89], [90] carried out very extensive measurements on a number of specimens each of gold, silver—see Fig. 10—and copper, and MAC-DONALD, WHITE and WOODS [91] measured potassium, rubidium and caesium. In all these cases it was observed that, if N_a is taken to be of order unity and Θ of the order of the DEBYE temperature, then the observed values of $W_i(T_1)$ are much lower, relative to $W_i(T_2)$, than predicted by (15.3). A similar effect has been noted by Hulm [92], [93] and by Andrews, Webber and Spohr [95] for non-monovalent metals, where, however, the quasi-free electron theory does not apply.

However, to test the temperature dependences of ϱ_i and W_i directly by means of (15.2) and (15.3) introduces the difficulty, pointed out by Klemens [70] and by Ziman [71], that the magnitude of $\varrho_i(T_2)$ — hence of $W_i(T_2)$ — may not be given correctly by the Bloch theory of Sect. 14, because that theory disregards Umklapp-processes (14.2b), dispersion of the phonon velocity at high frequencies, and a possible variation of the electron-phonon interaction parameter $C(q)$ with phonon wave-number q. Eq. (15.4), however, is independent of these effects,

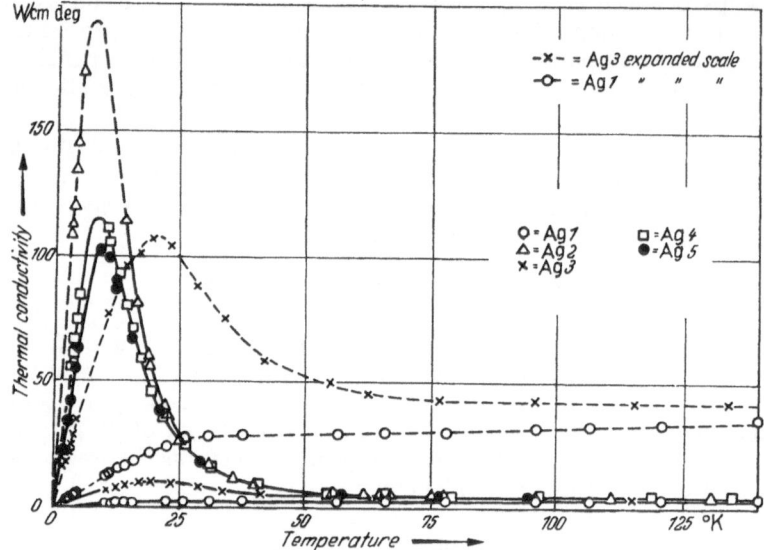

Fig. 10. Thermal conductivity of a pure silver rod in various states of different residual resistivity, according to White [89]. Ag 1—cold drawn, Ag 2—annealed at 650° C, Ag 3—redrawn, Ag 4—reannealed at 650° C, Ag 5—removed from cryostat and then resoldered.

and depends only on the assumption of a spherical Fermi surface, since Umklapp-processes are then negligible at low temperatures, and since phonons of the same order of frequency are involved both in the electrical and thermal resistance processes[1]. Klemens [70] pointed out that there are deviations not only from (15.2) and (15.3), but also from (15.4). Ziman [71], apparently unaware of Klemens' earlier work, attempted to fit both (15.2) and (15.3) by means of a suitable $C(q)$; he could not satisfy both relations, for a reasonable Θ-value, with a common function $C(q)$, which is, of course, an indication that (15.4) is not satisfied.

A point requiring careful consideration is the choice of the values of Θ and N_a in (15.2), (15.3), (15.4). It has been usual to take for Θ the value Θ_D, the Debye temperature deduced from the low-temperature specific heat. Another Θ-value, Θ_R, can be defined as the value which must be substituted into (15.2) to give agreement with the observations. Usually Θ_R is close to Θ_D, but since Θ occurs to the fourth power, this is not a sensitive criterion of fit.

---

[1] The average effective phonon frequency for electrical resistance is twice that for thermal resistance. This could, however, explain the existing discrepancies only if $C(q)$ varies very rapidly with q at lowest frequencies, which is not only unlikely, but would lead to temperature variations different from those observed at lowest temperatures.

BLACKMAN[1] pointed out that since the BLOCH theory assumes interaction between electrons and longitudinal waves only, the appropriate Θ-value in (15.2), and of course also in (15.3) and (15.4), is Θ_L, related in the usual way to the velocity of longitudinal low-frequency phonons, rather than Θ_D, which is related to a mean of $1/v^3$ over all polarizations; for most substances $\Theta_L \sim 1.5\,\Theta_D$. However, if Θ_L is used in (15.2), (15.3), (15.4), there is some improvement in the fit of (15.3), but only at the expense of serious discrepancies in (15.2) and (15.4). The discrepancy in (15.4) is minimized if Θ is taken as small as possible, that is if $\Theta = \Theta_D$. This would imply that conduction electrons interact both with transverse and longitudinal waves. Strong evidence in favour of this coupling scheme is obtained from the magnitude of the lattice component of the thermal conductivity (see Sect. 24).

There has been some confusion in the literature about possible values of N_a, and some authors have suggested an explanation in terms of an "effective" value of N_a less than the actual number of electrons per atom in the band, in analogy to the concept of the "effective number of free electrons", proportional to $k^2\,dE/dk$, which appears in the expressions for the conductivities. However, the factor $N_a^{\frac{2}{3}}$ arises from a factor $(q_0/k_\zeta)^2$ in the expression (14.19) for W_∞, which in turn arises because the effectiveness of horizontal diffusion as a resistance mechanism depends upon the radius of the FERMI surface k_ζ. Thus N_a depends, not on $(dE/dk)_\zeta$, but only on the volume of $\mathbf{k}$-space enclosed by the FERMI surface, that is on the actual number of electrons in the band.

In Table 1 are presented values for the extreme conductivities, as well as the discrepancies from (15.3) and (15.4) for monovalent metals, for which $N_a = 1$, namely

$$D_3 = 64.0\,\frac{W_\infty}{W_i(T_1)}\left(\frac{T_1}{\Theta_D}\right)^2,\tag{15.5}$$

$$D_4 = \frac{64.0}{497.6}\,\frac{\Theta_D^2}{L}\,\frac{\varrho_i(T_1)}{W_i(T_1)\,T_1^3}\tag{15.6}$$

taking Θ_D as the Θ-value. It is seen that even with this Θ-value, which is the lowest which can be justified, $D_4 > 1$ in all cases, which implies that for some reason horizontal diffusion is relatively more effective than vertical diffusion in producing resistance. This can only be explained in terms of non-spherical FERMI surfaces [70]; once this is accepted, it will also explain the interaction between electrons and low-frequency transverse waves.

The effect of deviations of the FERMI surface from spherical shape on the conduction properties will be two-fold: the integrals (13.13) and (13.14) will be affected even if τ is kept fixed, because v and $\mathrm{grad}_{\mathbf{k}} E$ depend on the function $E(\mathbf{k})$, and secondly the relaxation times themselves will be affected. We are not concerned with the former effect, since this is common to all conductivities and disappears in (15.2), (15.3), (15.4), but with the effect on τ. If determined by vertical movement, as in the case of W_i at low temperatures, τ depends only on the local properties of the FERMI surface, and will be relatively unaffected by changes in the shape of the FERMI surface. If, however, τ is determined by horizontal many-step diffusion, as in the case of ϱ_i at low temperatures, it will depend markedly on the shape of the FERMI surface.

The deformations of that surface will be such that the volume enclosed is unchanged, they will have the symmetry of the BRILLOUIN zone and they will be outward along those directions where the zone boundary is closest to the

---

[1] M. BLACKMAN: Proc. Phys. Soc. Lond. A **64**, 681 (1951). Cf. also BLACKMAN s companion article in vol. VII/1.

centre of the zone. It was suggested [70] that such deformations reduce the average path length of an electron diffusing to an opposite point on the surface, hence they enhance horizontal diffusion relative to vertical movement and cause a deviation $D_4 > 1$.

This effect should be particularly marked if the outward deviations are so large that the FERMI surface touches the zone boundary. This is more likely to happen for face-centred than for body-centred cubic metals, since the zone-boundary approaches the centre of the zone more closely. In face-centred structures there would be eight points of contact, of octahedral symmetry. An electron can reach a "neutral" region (i.e. one for which $g = 0$) not only by diffusing to a direction perpendicular to the direction of the field, but also by diffusing to a point of contact. In the latter case the average distance of travel is halved, so that the resistance due to such paths is four times the ordinary resistance, and since the ordinary paths also contribute to the resistance, $D_4 \sim 5$. While for body-centred structures it requires a stronger deformation for the FERMI surface to touch the zone boundary, the change in resistance will be larger if this does occur, for then there are 12 points of contact, the average path length to a point of contact is only one third of the average path lengths to the "neutral" region of a spherical surface, so that $D_4 \sim 10$.

Table 1. *Conduction properties of monovalent metals.*

Element	$\dfrac{W_i\,(T_1)}{T_1^2}$	W_∞	$\dfrac{\varrho_i\,(T_1)}{T_1^5}$	Θ_D	Θ_R	Θ_L	D_3	D_4
Li [80] . . .	5×10^{-4}	0.7	—	~ 350	~ 260	—	~ 0.7	—
Na [83] . .	3.8×10^{-4}	0.73	5.37×10^{-15}	150	200	260	5.3	1.8
K [91] . . .	1.2×10^{-3}	0.7	3.5×10^{-13}	100	70	170	3.7	16
Rb [91] . .	9.2×10^{-3}	1.7	4.5×10^{-12}	60	~ 50	100	3.3	9
Cs [91] . .	2.2×10^{-2}	1.7	6.5×10^{-11}	25(?)	~ 25	—	~ 8	~ 10
Cu [84], [85], [87], [90]	2.55×10^{-5}	0.26	2.64×10^{-16}	315	330	505	6.2	5.4
Ag [85], [87], [89] . . .	6.4×10^{-5}	0.24	1.11×10^{-15}	215	220	340	5.2	4.2
Au [85], [87], [88] . . .	1.3×10^{-5}	0.64	3.9×10^{-15}	170	170	270	5.8	4.5

Remarks: $L = 2.45 \times 10^{-8}$ watt-ohm-deg^2, thermal resistivities are given in watt-units, electrical resistivities in ohm-units, $T_1 \ll \Theta$, Θ_D is Θ-value derived from low-temperature specific heat, Θ_L is the DEBYE temperature of the longitudinal polarization branch, Θ_R is the Θ-value required to fit (15.2), D_3 and D_4 are defined in (15.5) and (15.6).

It appears from inspection of the values D_4 in Table 1 that the FERMI surface touches the zone boundary in the case of the noble (face-centred) metals, and in the case of the two heavy alkali (body-centred) metals rubidium and caesium, while it does not touch in the case of sodium. This agrees with the above conclusion that the FERMI surface is least likely to touch the zone boundary in the case of the light alkali metals.

The case of potassium is anomalous, as it shows the highest value of D_4. A close inspection of the electrical resistance curve[1] reveals that $\varrho_i \propto T^3$ above 6° K, and falls off sharply below that temperature. It is possible that the FERMI surface comes close to, but does not touch, the zone boundary. The low DEBYE temperature and the close approach would result in Umklapp-processes being frozen out only at very low temperatures, apparently below 6° K. What has been taken to be a T^5 variation of ϱ_i below 6° K may really be the exponential

---

[1] D. K. C. MacDONALD and K. MENDELSSOHN: Proc. Roy. Soc. Lond., Ser. A **202**, 103 (1950).

variation due to the freezing out of Umklapp-processes, and the variation $\varrho_i \propto T^5$ may only be attained at lower temperatures and with a value of ϱ/T^5 much lower than the value assumed in the table. The residual resistance, of course, prevents observations of the low values of ϱ_i.

The thermal conductivity of lithium has been measured by BIDWELL [80] down to liquid hydrogen temperatures; W_i varies as T^2 at those temperatures, and it is found that $D_3 \sim 0.7$. The electrical resistance at very low temperatures varies as $T^{4.5}$, instead of T^5 as expected theoretically, so that a comparison between ϱ_i and W_i at low temperatures cannot be made by (15.4). Nevertheless it appears that ϱ_i/W_i is larger than expected from theory. The anomalous behaviour of lithium is quite different from the more regular behaviour of sodium, and the reason for this difference is at present obscure.

There is little independent evidence in support of the hypothesis that the FERMI surface touches the zone boundary, because it seems that electrical resistivity at low temperatures is more sensitive to this eventuality than other observed properties. There is qualitative evidence from the thermoelectric properties of monovalent metals[1] that the band structure deviates violently from the simple model in the case of the noble metals, and in decreasing degree, in the cases of caesium, rubidium and potassium. The change of the electrical resistance in a magnetic field is also sensitive to the geometry of the FERMI surface. According to KOHLER[2] the change of electrical resistivity of a monovalent cubic metal in a strong transverse magnetic field should be isotropic (constant as the transverse field is rotated about the direction of the electric current), provided the FERMI surface does not touch the zone boundary. But the magneto-resistance of gold in a strong field is markedly anisotropic[3] so that in this case, and probably also in the case of other monovalent metals, it appears that the FERMI surface does touch the zone boundary.

The discrepancies D_3 can be understood in terms of Umklapp-processes, which increase the high-temperature resistance relative to $W_i(T_1)$ by a factor of up to 2, and in terms of the dispersion of the lattice waves. The latter effect, decreasing the frequency of the high-energy phonons by a factor of up to 1.5, increases the high-temperature resistance by a factor $\sim(1.5)^2$ because of the factor $v\Theta$ in (14.19). Thus D_3 can range up to about 4 to 5, apart from any effects due to a variation of the interaction parameter $C(q)$.

The absence of a maximum in the observed W_i can be understood in terms of the relative increase in ϱ_i: to a rough approximation $W_i = W_V + W_H$, where W_V and W_H are the resistances arising from vertical and horizontal movement respectively. At low temperatures $W_V \gg W_H$, at high temperatures $W_H \gg W_V$; also $W_H T = \varrho_i/L$. At intermediate temperatures W_V and W_H are comparable, and in that temperature range W_V passes through a maximum, while W_H increases monotonically with temperature. With the ratio of W_H/W_V as determined by the theory of Sect. 14 it so happens that $W_V + W_H$ has a maximum value: it is easily seen that if W_H is increased by a factor of order 4 to 5, this maximum will be greatly reduced and possibly eliminated, so that the absence of the resistance maximum and the high value of D_3 and D_4 are related.

WHITE has observed for silver [89] and copper [90] that $W_i \propto T^n$, where n is slightly larger than 2. This is explained as follows: for $T \ll \Theta$ $W_H \propto T^4$, and

[1] See the companion article by D. K. C. MACDONALD; also D. K. C. MACDONALD and S. K. ROY: Phil. Mag. 44, 1364 (1953). — P. G. KLEMENS: Phil. Mag. 45, 881 (1954) — D. K. C. MACDONALD and W. B. PEARSON: Proc. Roy. Soc. Lond., Ser. A 221, 534 (1954).
[2] M. KOHLER: Ann. Phys., Lpz. 5, 99 (1949).
[3] E. JUSTI and H. SCHEFFERS: Phys. Z. 37, 383, 475 (1936).

$W_V \propto T^2$, so that $W_i \propto T^2(1 + \beta T^2)$. Now $W_H = \varrho_i/TL$. If ϱ_i, relative to W_i, is given by (15.4), the term in βT^2 is negligible except at temperatures where W_V already varies more slowly than T^2, but if ϱ_i is increased by a factor 5, then the term in β is enhanced. Thus one finds for $T = \Theta/20$ that $dW_i/dT = 2.4\,W_i/T$, as against $dW_i/dT = 2.1\,W_i/T$ if β is given by the Bloch theory. The former value is in approximate agreement with White's observations.

White [88], [89], [90] has also carefully measured deviations from the additivity of W_0 and W_i (13.18b), expressing this as an apparent variation of W_i with W_0. He found that at low temperatures W_i apparently increased with increasing W_0. This deviation from (13.18b) is in the same sense as predicted by Sondheimer [64] from his computations, and they can be qualitatively understood as follows: if $W_0 \gg W_i$, Wilson's solution (14.29a) is appropriate, but if $W_i \ll W_0$, the solution is (14.29c), that is W_i is reduced by over 30%. Since in practice W_i is unobservable if $W_i \ll W_0$, the observed range of values of W_i will be less, and White observed a reduction of only about 20% for a reduction in W_0. Furtherfore the index n in $W_i \propto T^n$ decreases with increasing W_0; this can be understood in terms of the above theory, since W_V is increased, but W_H is hardly altered.

16. Comparison with experiments: Other metals. There is at present no detailed theory of the conduction properties of non-monovalent metals which takes account of their band structure, in particular the temperature dependence of ϱ_i at low temperatures is not understood. Therefore it is not possible to quantitatively test the theory for these metals, as was done in Sect. 15. However, some conclusions of the Bloch theory are independent of the detailed band structure, and can be applied to all metals; these are: $W_i \propto T^2$ if $T \ll \Theta$, ϱ_0 should be independent of temperature and $W_0 T = \varrho_0/L$; also if $T > \Theta$, $\varrho_i \propto T$ and $W_i T = \varrho_i/L$. The residual thermal resistivity will be considered in Sect. 17.

Mendelssohn and Rosenberg [85], [86], [87] have made extensive measurements of elemental metals, and in addition to Cu, Ag and Au they measured the conductivity of the following metals below 90° K.

Group II: Be, Mg, Zn, Cd.
Group III: Al, Ga, In, Tl.
Rare Earth: La, Ce.
Group IV: Ti, Zr, Sn, Pb.
Group V: V, Cb, Sb, Ta.
Group VI: Mo, W, U.
Group VII: Mn.
Group VIII: Fe, Co, Ni, Rh, Pd, Ir, Pt.

Hulm [92] measured Hg, In and Ta; Bi was measured by a number of authors and will be discussed in Sect. 23. In addition to the survey measurements of Mendelssohn and Rosenberg, the following metals were studied in some detail: Al by de Nobel [94] and by Andrews, Webber and Spohr [95], Mg and Pd by Kemp, Sreedhar and White [96], [100].

The measurements mentioned here are only the most recent ones extending to liquid helium temperatures; a comprehensive list of all thermal conduction measurements below room temperatures has been compiled by Powell and Blanpied [7], covering work until early 1954. The most notable of these earlier measurements are those of Grüneisen and collaborators (for example [81], [82], [103]), which extended only to liquid hydrogen temperatures, but on specimens which in many cases had a lower residual resistivity than the specimens of Mendelssohn and Rosenberg.

In the case of Be, Ti, Sb, La, Ce, U and Mn the residual resistivity and/or the lattice component of thermal conductivity were so large that W_i could not be reliably determined at low temperatures. In those cases for which W_i could be determined at sufficiently low temperatures, it was verified that W_i varied roughly as T^2, as required theoretically. It is possible to express W_i in the form

$$\frac{W_i(T)}{T^2} = \frac{A W_\infty}{\Theta^2}, \qquad (T \ll \Theta) \tag{16.1}$$

and values of W_i/T^2 and A for various metals are given in Table 2. It is remarkable that for the various elements from different groups the range of variation of A should be so small.

Table 2. *Conduction properties of polyvalent metals.*

Element	W_∞	$\dfrac{W_i(T_1)}{T_1^2}$	Θ	A
Mg	0.62	8.7×10^{-5}	330	16
Zn	0.92	3×10^{-4}	250	20
Cd	1.08	4×10^{-4}	160	10
Hg	3.5	2×10^{-2}	60 (?)	20
Al	0.40	2.7×10^{-5}	400	11
Sn	1.7	3.9×10^{-4}	200	9
Pb	2.8	2.5×10^{-3}	100	9
W	0.6	9×10^{-5}	300	13
Fe	1.2	1×10^{-4}	420	14
Pd	1.3	3.5×10^{-4}	275	20
Pt	1.4	4.3×10^{-4}	230	16

Remarks: W_∞ and W_i in watt$^{-1} \cdot$ cm.deg., Θ in ° K. The values of A given here are only approximate, in view of the unreliability of the Θ-values. Values of $W_i(T_1)$ are taken from the sources quoted in the text, values of W_∞ from [7].

ROSENBERG [99], [87] has studied the anisotropy of the thermal resistivity of Ga; this is similar, though not identical, to the anisotropy of the electrical resistivity.

The only quantitative discussion for a non-monovalent metal given so far is that for Pd by KEMP, KLEMENS, SREEDHAR and WHITE [100]. The electrical and thermal resistivities are approximately additively composed of contributions from $(s-s)$ and $(s-d)$ transitions; at low temperatures both $W_i(s, s)$ and $W_i(s, d)$ vary as T^2, while $\varrho_i(s, s) \propto T^5$ and $\varrho_i(s, d) \propto T^3$; also

$$W_i(s, d)\, T = \frac{\varrho_i(s, d)}{L}, \tag{16.2}$$

that is $(s-d)$ transitions can be described by a relaxation time. In the case of Pd, the s-band has 0.6 electrons per atom, and the FERMI surface in the s-band is thus well removed from the zone boundaries and very nearly spherical; the BLOCH theory of Sect. 14 should therefore hold for $(s-s)$ transitions, but not for $(s-d)$ transitions. Hence from (14.28) and (14.29c)

$$\frac{W_i(s, s)}{T^2} = 64\,(0.6)^{\frac{2}{3}}\,\frac{W_\infty(s, s)}{\Theta_L^2}. \tag{16.3}$$

At lowest temperatures $W_i(s, s)$ and $W_i(s, d)$ are comparable. If it is assumed that the ratio $W_i(s, s)/W_i(s, d)$ is independent of temperature, as it should be approximately if the relative transition probabilities are independent of phonon wave-number, then in (16.1) the factor A should be $64 \times (0.6)^{\frac{3}{2}} = 45$, with Θ taken as $\Theta_L \approx 1.5 \Theta_D$. This is indeed found to be the case, so that it appears that the BLOCH theory is better obeyed for $(s-s)$ transitions in palladium than in monovalent metals.

17. Residual resistance. All static imperfections, both chemical impurities and structural irregularities, scatter electrons in a way describable by a relaxation time, as discussed in Sect. 13, which varies only slowly with electron energy, leading to a residual electrical resistivity ϱ_0, independent of temperature, and a residual thermal resistivity $W_0 = \varrho_0/L T \propto T^{-1}$.

The magnitude of W_0 is not an intrinsic property of the material, but depends, often very markedly, on impurities and the way the specimen has been prepared. The study of the effects of impurities and treatment on the residual resistivity is usually performed more conveniently by electrical resistance measurements than by thermal resistance measurements. Only in exceptional circumstances would thermal resistance measurements offer a special advantage (possibly in experiments on the deformation of rods, or in cases where it is also desired to study the lattice component of the thermal conductivity); in actual fact thermal conductivity measurements have not yet been used for such purposes.

Interest in the residual thermal resistivity itself has up to now centred on two questions: whether $W_0 T$ does indeed equal ϱ_0/L, as the electron theory predicts, in cases when ϱ_0 is independent of temperature, and what is the behaviour of $W_0 T$ in cases where ϱ_0 has an anomalous dependence on temperature.

Since values of ϱ_0 are not completely reproducible, especially in the case of pure metals, a proper comparison of W_0 and ϱ_0 involves measurements of the electrical and thermal resistivities on the same specimen. There have not been very many measurements of this type. KEMP, KLEMENS, SREEDHAR and WHITE have measured the residual electrical and thermal resistivities of pure palladium [*100*] and found agreement between ϱ_0 and $L W_0 T$; the same authors also measured the thermal conductivities of a series of Ag—Pd and Ag—Cd alloys [*119*], and although the presence of an appreciable lattice component made the evaluation of W_0 difficult in some cases, there appeared to be agreement between ϱ_0 and $L W_0 T$ in all those cases where the uncertainties in these values were small. ROSENBERG [*87*] measured the electrical resistance of some of his specimens; in most cases his results were in agreement, or at least not inconsistent, with the theoretical LORENZ ratio. However, in the case of Ti, Zr, V, Mn and Ce he found that at lowest temperatures $\varkappa \propto T$, yet $\varkappa$ was larger than $L T/\varrho$. If this difference is to be ascribed to an appreciable lattice component of thermal conductivity $\varkappa_g$, then $\varkappa_g$ must vary as T. It is not clear whether this explanation should be rejected; if it were, one would be forced to conclude that the LORENZ number exceeded its theoretical value. In view of the wide applicability of (13.15), such a conclusion would be very disturbing, and before reaching it, it may be desirable to investigate these materials more thoroughly.

Anomalies in the behaviour of ϱ_0, in particular an anomalous increase of ϱ with decreasing temperature, have been observed for various metals and seem to be associated with specific impurities[1]. These anomalies have not yet received a full theoretical explanation. There have been some attempts to find corresponding anomalies in the behaviour of $W T$ at lowest temperatures.

A slight increase with decreasing temperatures of $W T$ has been found by WHITE for gold [*88*], silver [*89*] and copper [*90*] for those specimens which had

---

[1] See the companion article by D. K. C. MACDONALD.

high values of ϱ_0, and for which ϱ_0 would thus presumably behave anomalously. Only in the case of copper was the electrical resistance measured, and did show the anomalous behaviour. KEMP, SREEDHAR and WHITE [96] also observed a minimum in WT in the case of magnesium, again in circumstances where a minimum in ϱ was expected. ROSENBERG [97] measured both the electrical and thermal resistivity of the same specimen of magnesium and found both a minimum in ϱ and in WT. Similar measurements on magnesium were carried out by WEBBER and SPOHR [98].

A minimum in WT need not necessarily indicate a minimum in $W_0 T$, because it may arise from the presence of a lattice component of thermal conductivity. If it is assumed that $W_i \propto T^2$ at lowest temperatures, then if $W_0 T$ is constant and in the absence of lattice conductivity, a plot of WT against T^3 should yield a straight line. The lattice component would cause the curve to dip below this line, and to come up to its extrapolated value at $T=0$; a genuine rise of $W_0 T$ with decreasing temperature would cause the curve to rise above the extrapolated line.

ROSENBERG [97] plotted his results for magnesium in this way; his curve is reproduced in Fig. 11, together with a similar curve of WEBBER and SPOHR [98], and a curve obtained from the results which were published, in a different form, by KEMP, SREEDHAR and WHITE [96]. ROSENBERG's curve indicates lattice conductivity, though this does not exclude the possibility of an anomaly in $W_0 T$, WEBBER and SPOHR's curve indicates an anomaly in $W_0 T$, and KEMP, SREEDHAR and WHITE's curve indicates both mechanisms.

NICOL and TSENG [101] measured the thermal conductivity of a specimen of pure copper from 0.25° K upwards. No electrical resistance anomalies are expected from a specimen of such low residual resistance, and the thermal conductivity was found to be proportional to temperature within the limits of experimental uncertainty.

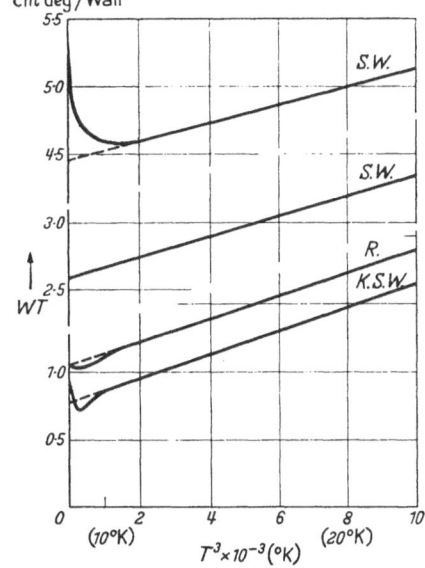

Fig. 11. Anomalies in the thermal conductivity of magnesium; WT plotted against T^3. S.W.= SPOHR and WEBBER [98], R.=ROSENBERG [97], K.S.W.= KEMP, SREEDHAR and WHITE [96].

18. Effect of magnetic fields[1]. The application of a magnetic field will, in general, increase both the electrical and the thermal resistivity, the increase depending on the direction of the field relative to the (electrical or thermal) current. The relative increase is the larger the lower the temperature (or the lower the zero-field resistivity), it is larger for transverse fields than for longitudinal fields and larger for multivalent than for monovalent metals. These general features can be qualitatively understood in terms of the theory, but a quantitative understanding of magneto-resistive effects is largely lacking, because the free electron model, which is so useful as a basis of discussion and extrapolation to more complicated band structures, does not yield any magneto-resistive effects, so that a more complicated model must be adopted.

---

[1] The theory is discussed by WILSON [4].

If a relaxation time τ (13.8) is assumed to exist, the Boltzmann equation (13.7) becomes

$$-\frac{e}{\hbar}\left[\boldsymbol{F}+\frac{1}{c}\,(\boldsymbol{v}\times\boldsymbol{H})\right]\cdot\frac{df^0}{d\boldsymbol{k}}+\boldsymbol{v}\cdot\operatorname{grad}f^0=-\frac{g}{\tau}\tag{18.1}$$

where $\boldsymbol{v}=(\hbar)^{-1}\partial E/\partial\boldsymbol{k}$, c is the velocity of light, and $\boldsymbol{H}$ and $\boldsymbol{F}$ the magnetic and electric fields. The deviation from equilibrium can be written as

$$g(\boldsymbol{k})=-\,\boldsymbol{k}\cdot\boldsymbol{c}(\boldsymbol{k})\,\frac{df^0}{dE}\tag{18.2}$$

and the Boltzmann equation becomes

$$\frac{m}{\hbar\tau}\,\boldsymbol{c}(\boldsymbol{k})-\frac{e}{\hbar c}\left[\boldsymbol{H}\times\boldsymbol{c}(\boldsymbol{k})\right]=\boldsymbol{P}\equiv-e\boldsymbol{F}+T\operatorname{grad}\left(\frac{E-\zeta}{T}\right)\tag{18.3}$$

where $m=\hbar^2 k\,(dE/dk)^{-1}$. The formal solution is obtained by multiplying (18.3) scalarly and vectorially respectively by $\boldsymbol{H}$, and by combining these two equations with (18.3). While the solution in zero field is

$$\boldsymbol{c}_0(\boldsymbol{k})=\frac{\hbar\tau}{m}\,\boldsymbol{P},\tag{18.4}$$

it is more generally

$$\left[\left(\frac{m}{\hbar\tau}\right)^2+\left(\frac{eH}{\hbar c}\right)^2\right]\boldsymbol{c}(\boldsymbol{k})=\frac{m}{\hbar\tau}\,\boldsymbol{P}+\frac{e}{\hbar c}\,[\boldsymbol{H}\times\boldsymbol{P}]+\frac{e^2\tau}{\hbar m c^2}\,\boldsymbol{H}(\boldsymbol{H}\cdot\boldsymbol{P}).\tag{18.5}$$

Keeping fixed the state $\boldsymbol{k}$, and for given directions of $\boldsymbol{H}$ and $\boldsymbol{P}$

$$\frac{(\boldsymbol{c}-\boldsymbol{c}_0)}{\boldsymbol{c}_0}\quad\text{is a function of}\quad\left(\frac{e\tau H}{mc}\right).\tag{18.6}$$

This enables one to set up the following similarity relations, proposed by Kohler [75]. Consider changes in τ due to variations of temperature or residual resistance and the corresponding variations of the magneto-resistive effects. If either τ/m is the same for all states or if the relative change of τ/m is the same for all states, then the similarity relation (18.6) implies that

$$\frac{\Delta R}{R(0)}\quad\text{is a function of}\quad\frac{H}{R(0)}\tag{18.7}$$

and that

$$\frac{\Delta W}{W(0)}\quad\text{is a function of}\quad\frac{H}{W(0)\,T}\tag{18.8}$$

where $R(0)$ and $W(0)$ here denote the electrical and thermal resistivities in zero magnetic field, and ΔR and ΔW the corresponding increases on applying a magnetic field.

Although these similarity rules only apply to changes in τ which increase τ/m for all states in the same proportion, one would expect departures from (18.7) and (18.8) never to be very serious, because even if τ/m varies strongly from state to state, relative changes of τ/m which alter $R(0)$ or $W(0)\,T$ will usually be the same for all states; the most likely exceptions are changes in temperature such that the dominant resistance mechanism changes from scattering by lattice waves (small angle scattering) to scattering by imperfections (large angle scattering) in complicated band structures.

It follows from (18.7) and (18.8) that the relative change of thermal resistance is the greater, the lower the temperature and, in the residual resistance range, the purer the specimen. It is thus much easier to observe magneto-resistive effects at low temperatures.

In principle one can calculate the thermal conductivity in a magnetic field from (18.5) in the same way as it was obtained from (18.4) in Sect. 13. In practice conductivities have been evaluated from (18.4) only in cases of spherical symmetry, but a single band of spherical symmetry gives no change in electrical or thermal resistivities, the only effect being a HALL field. It can be shown quite generally that magneto-resistance vanishes if all states on the FERMI surface have the same relaxation time. Hence a more complex band model must be assumed, and the only one for which magneto-resistive effects have been worked out is the case of two overlapping bands, each having spherical symmetry.

Thus WILSON and SONDHEIMER [74], assuming two such bands with N_1 and N_2 carriers per atom respectively (one band of electrons and one of holes) and assuming that the electrons or holes in each band have a relaxation time (but that τ_1 does not equal τ_2), have obtained the following result for the thermal conductivity in a transverse magnetic field

$$\left. \frac{\varkappa(0) - \varkappa(H)}{\varkappa(H)} = \frac{\varDelta W}{W(0)} = \left(\frac{H}{e\,c\,L_0\,T} \right)^2 \left[\frac{\varkappa_1(0)}{N_1} + \frac{\varkappa_2(0)}{N_2} \right] \frac{\varkappa_1(0)\,\varkappa_2(0)}{[\varkappa_1(0) + \varkappa_2(0)]^2} \times \right.$$
$$\left. \times \left\{ 1 + \left(\frac{H}{e\,c\,L_0\,T} \right)^2 \frac{(N_1 - N_2)^2}{N_1^2 N_2^2} \frac{\varkappa_1^2(0)\,\varkappa_2^2(0)}{[\varkappa_1(0) + \varkappa_2(0)]^2} \right\}^{-1} \right\} \quad (18.9)$$

where $\varkappa_1, \varkappa_2$ are the partial conductivities of each band and where L_0 is the SOMMERFELD value (13.15) of the LORENZ ratio. The derivation of (18.9) is lengthy and the reader is referred to the original paper [74] or to [4]. Thus

$$\frac{\varDelta W}{W(0)} \quad \text{is of the form} \quad \frac{A H^2}{1 + B H^2} \qquad (18.10\,\text{a})$$

except if $N_1 = N_2$, when

$$\frac{\varDelta W}{W(0)} \quad \text{is of the form} \quad A\,H^2. \qquad (18.10\,\text{b})$$

The form (18.10a) leads to a saturation of the magneto-resistive effects. If $\varkappa_1(0)/\varkappa_2(0)$ is constant, (18.9) satisfies the KOHLER rule (18.8), in agreement with the previous general discussion.

Eq. (18.9) was derived on the assumption of a relaxation time which is independent of $g(\boldsymbol{k})$. This assumption leads not only to the WIEDEMANN-FRANZ law in the absence of a magnetic field (i.e. $\varkappa(0)/(\sigma(0) \cdot T) = L_0$), but also leads to the WIEDEMANN-FRANZ law at all field-strength, that is

$$\frac{\varkappa(H)}{\sigma(H)\,T} = L_0 \quad \text{if} \quad \frac{\varkappa(0)}{\sigma(0)\,T} = L_0. \qquad (18.11)$$

The theory can be generalised in the absence of a unique relaxation time in the following way: one can still assume an effective relaxation time, but this relaxation time is now different in the cases of electrical and thermal conduction. Using WILSON's procedure [60], which is equivalent to this modification, SONDHEIMER and WILSON showed that (18.9) was generally valid. The LORENZ ratio is, however, no longer constant in that case, but varies with $\boldsymbol{H}$. Thus if

$$\frac{\varkappa(0)}{\sigma(0) \cdot T} = L(0) < L_0 \qquad (18.12\,\text{a})$$

then assuming for simplicity that $\sigma_1 = \sigma_2 = \sigma/2$ and $\varkappa_1 = \varkappa_2 = \varkappa/2$

$$L(H) = \frac{\varkappa(H)}{\sigma(H) \cdot T} = L(0) \frac{1 + a\,H^2\,\sigma^2(0)}{1 + a\,H^2\,\sigma^2(0)\,(L(0)/L_0)^2} \frac{1 + b\,H^2\,\sigma^2(0)\,(L(0)/L_0)^2}{1 + b\,H^2\,\sigma^2(0)}. \qquad (18.12\,\text{b})$$

Explicit expressions for a and b are given by WILSON and SONDHEIMER; for the present purposes it is sufficient to know that $a > b$, and that $b = 0$ if $N_1 = N_2$.

Thus if $L(0) < L_0$, two cases of behaviour are possible: (i) if $N_1 = N_2$, $L(H)$ increases steadily from $L(0)$ to $L_0^2/L(0)$, the latter value being larger than L_0, (ii) if $N_1 \neq N_2$, $L(H)$ increases from $L(0)$, reaches a maximum, and then drops again to $L(0)$.

The model of SONDHEIMER and WILSON has also been treated by KOHLER [77], who did not assume a relaxation time, but used a first-order variational method. As seen in Sect. 14, this is equivalent to assuming an effective relaxation time (but not necessarily the same for electrical and thermal conduction) so that this method must lead to the same results as those obtained by SONDHEIMER and WILSON, as was indeed the case. Presumably, if the variational calculations are carried out to a higher order, deviations from (18.9) would be found. However, in view of the artificiality of the model, such higher order calculations do not seem worth while at present.

A qualitative discussion of the thermal conductivity in transverse fields was given by KOHLER [76]; his conclusions are consistent with the results of the two-band model.

Increases of thermal resistance of bismuth with magnetic field were observed at high temperatures many years ago. At low temperatures the effect is small [107], in contrast to the rule (18.8); this is readily explicable in terms of the large lattice component (see also Sect. 23).

GRÜNEISEN and ADENSTEDT [103] observed the thermomagnetic effect at liquid hydrogen temperatures on a tungsten single crystal, and on copper, silver, beryllium (single crystal) and platinum. They found the relative increases $\Delta W/W_0$ generally somewhat smaller than the corresponding increases in electrical resistance. Correspondingly $\varkappa/\sigma T$ increased with increasing field; this effect was first ascribed to an appreciable lattice component, but extension to higher fields showed that this increase was an electronic effect; in this way we have a qualitative verification of (18.12b). For low fields $\Delta W \propto H^2$, but increased more slowly (linearly) for higher fields. No sign of saturation was found.

In the case of tungsten and beryllium ΔW was found to be anisotropic, changing as the direction of H was rotated about the (constant) direction of flow. Also $\Delta W/W(0)$ was greatest for these two highly anisotropic metals. Further measurements on beryllium were made by GRÜNEISEN and ERFLING [104].

DE HAAS and DE NOBEL [108] and later DE NOBEL [109] measured the thermal conductivity of tungsten (single crystal) in the range 14 to 20° K in fields up to 36 kilogauss.

Measurements at liquid helium temperatures were made by HULM [92] for tin, and by MENDELSSOHN and ROSENBERG [111], [112] for cadmium, zinc, tin, lead and gallium. While the measurements at liquid helium temperatures involve a greater relative change of T than the earlier measurements of GRÜNEISEN et al. and of DE HAAS et al., the range of variation of $W(0)T$ was less, being in the residual resistance region. Thus measurements at helium temperatures do not necessarily constitute a very sensitive test of the KOHLER rule (18.8). MENDELSSOHN and ROSENBERG did, in fact, find deviations from the KOHLER rule at high fields, but these are not very large and could be explained in terms of a lattice component which, in the case of zero field, would have to be only a few per cent of the total conductivity. At liquid hydrogen temperatures, however, KOHLER [75], [77] showed that there is good agreement with (18.8) for tungsten [108] and for beryllium [104].

One can summarize the experimental results as follows:

(i) KOHLER's rule (18.8) is satisfied within the limits of uncertainty of the lattice component; however it has not yet been tested over a temperature range such that the dominant resistance process changes from imperfection scattering to phonon scattering.

(ii) There is strong anisotropy in the magneto-resistive effects of metals for those cases where measurements on single crystals have been made, irrespective of whether the crystals show cubic symmetry or not. The anisotropy is similar to the anisotropy of the electrical resistance change, but does not parallel it completely [i.e. $L(H)$ is also anisotropic].

(iii) The LORENZ number $L(H)$ is not independent of H at liquid hydrogen temperatures [where $L(0) \neq L_0$], but at first increases with increasing field. There has not yet been a detailed study of $L(H)$ in the liquid helium region, where one may expect $L(H) = L(0) = L_0$.

(iv) The dependence of $W(H)$ cannot be represented by an equation of the form (18.9) or (18.10).

Up to now the discussion has assumed fields not so strong that the quantization of the electron states is important, in other words the magnetic field was assumed to be weak enough for the diamagnetism not to be affected by the DE HAAS—VAN ALPHEN effect. The existence of oscillatory variations of the electrical resistivity of some metals for field-strengths for which the DE HAAS—VAN ALPHEN effect occurs has been demonstrated, and STEEL and BABISKIN[1] have correspondingly found an oscillatory thermal conductivity for bismuth. A theoretical treatment had been given by DAVYDOV and POMERANCHUK[2].

IV. Thermal conductivity of metals and alloys: Lattice component.

19. Scattering of phonons by electrons[3]. The interaction between electrons and phonons, treated in Sect. 14, not only changes the electron distribution function f, but also the phonon distribution function N. The change of f can be expressed (14.3) as the sum of the change due to processes in which an electron and a phonon combine to form another electron, or the reverse processes. A similar expression will also give the rate of change of N, but while in the former case k is kept constant and the summation extends over all q, now q is kept fixed and the summation is over all k. Thus the rate of change of N due to the interaction with the conduction electrons is given by

$$t \frac{dN}{dt}\bigg]_e = \frac{4}{9GM\hbar} C^2(q) \frac{q^2}{\omega} \sum_k \{f(k')[1-f(k)][N(q)+1] - \\ -f(k)[1-f(k')]N(q)\} \frac{1-\cos(\Delta E t/\hbar)}{\Delta E^2}\Bigg\} \qquad (19.1)$$

where the interference conditions (14.2) are satisfied and where ΔE is defined as in (14.1). This is readily derived from (14.3) if it is considered that a process which increases $f(k)$ by one also increases $N(q)$ by one. In the summation over all electron states k a factor 2 must be included for the spin degeneracy.

It will be assumed here, and justified in Sect. 21, that for purposes of calculating the lattice component of the thermal conductivity, the electron distribution $f(k)$ can be taken to have the equilibrium value $f^0(k)$, so that the

[1] M. C. STEEL and J. BABISKIN: Phys. Rev. **98**, 359 (1955).
[2] B. DAVYDOV and I. POMERANCHUK: J. Phys. USSR. **2**, 147 (1940).
[3] See BETHE [1].

expression in the curly brackets of (19.1) takes the simple form

$$n(q)\left[f^0(k') - f^0(k)\right] = n(q)\left[\frac{1}{e^{\varepsilon+x}+1} - \frac{1}{e^\varepsilon+1}\right] \tag{19.2}$$

where n is the deviation of N from equilibrium. Replacing the resonance factor by a delta function and integrating over k, BETHE [1] has shown that

$$-\frac{n}{\tau_{pe}} = -\frac{1}{n}\frac{dN(q)}{dt}\bigg]_e = \frac{2C^2\hbar}{9\pi GM}\, q\, k_\zeta^2 \left(\frac{dE}{dk}\right)_\zeta. \tag{19.3}$$

In the derivation of (19.3) it was assumed that the contour $E=\zeta$ in k-space is spherical, that the phonon q can interact directly with the conduction electrons, and that the electron-phonon interactions conserve the wave-vector, that is Umklapp-processes (14.2b) are excluded.

The neglect of Umklapp-processes at low temperatures can be justified as follows: in the derivation of (19.3) from (19.1) every element of the FERMI surface contributes in an additive way. At low temperatures Umklapp-processes can only involve those electron states k which are near the zone boundary. Since only a small fraction of the FERMI surface is near the zone boundary, the contribution of Umklapp-processes to the rate of change of $N(q)$ is negligible.

Since every element of the FERMI surface contributes additively to $dN/dt]_e$, the generalisation of (19.3) for a non-spherical FERMI surface or for a FERMI surface extending over more than one zone is obviously

$$-\frac{n}{\tau_{pe}} = -\frac{1}{n}\frac{dN}{dt}\bigg]_e = \frac{2\hbar}{9\pi GM}\, q\, \frac{1}{4\pi}\int\left(\frac{dE}{dk}\right)_\zeta^2 C^2(q)\, dS \tag{19.4}$$

where dS is an element of surface area of the FERMI surface in k-space, $(dE/dk)_\zeta$ is the normal derivative, and the integration is over the FERMI surface. It should be noted that the electron-phonon interaction parameter $C(q)$ is in general not only a function of q, but would also depend upon the position of k on the FERMI surface, and possibly even on the orientation of q. It will be noted that in such complicated cases the calculation of the electronic conduction properties would present a formidable task; the calculation of $dN/dt]_e$ is nevertheless relatively straightforward.

We must now consider the polarization of the phonons. The BLOCH theory assumes that transverse phonons cannot directly interact with the conduction electrons. This has been sometimes thought to imply that the contribution of the transverse waves to the lattice thermal conductivity would not be influenced by the electron interaction, so that the lattice conductivity would be almost as large as the thermal conductivity of an equivalent dielectric solid. If it is considered, however, that the transverse and longitudinal phonons interact by means of three-phonon processes conserving wave-vector, which tend to equalise the parameter τ in (7.5), it is clear that the effective relaxation time of the longitudinal and transverse phonons respectively is

$$\left.\begin{aligned}
\tau_{\mathrm{I}} &= \frac{\tau_{pe}\left[\sigma' + \tau_{\mathrm{II}}\right]}{\tau_{pe} + \sigma'} && \text{if} \quad \omega < \frac{KT}{\hbar}, \\[4pt]
&= \tau_{pe} && \text{if} \quad \omega \geq \frac{KT}{\hbar}, \\[4pt]
\tau_{\mathrm{II}} &= \tau_{pe}\left(\frac{KT}{\hbar}\right) + \sigma && \text{if} \quad \omega < \frac{KT}{\hbar}, \\[4pt]
&= \tau_{pe} + \sigma && \text{if} \quad \omega \geq \frac{KT}{\hbar}
\end{aligned}\right\} \tag{19.5}$$

where σ and σ' are defined as in (7.6) for transverse and longitudinal phonons respectively, and the derivation of (19.5) is analogous to the derivation of (7.16). In particular if τ_{pe} is much bigger than either σ or σ', so that the coupling between longitudinal and transverse waves is very strong, $\tau_I \approx \tau_{II} \approx \tau_{pe}$, the effective relaxation times of waves of all polarizations being the same and governed by the scattering of longitudinal waves.

20. The lattice thermal conductivity. The lattice component of the thermal conductivity of metals and alloys can be described in terms of the theory of thermal conduction of non-metals, but with the scattering of phonons by electrons (19.3) acting as an additional resistive process. Since $\tau_{pe} \propto q^{-1}$, the resistance W_E due to scattering by electrons varies as T^{-2}—see Sect. 9.

The details of the theory depend upon the coupling between the conduction electrons and the longitudinal and transverse phonons. There are three possibilities:

(a) $C_L^2 = C_T^2$: longitudinal and transverse phonons interact equally strongly with the conduction electrons. This case was treated by MAKINSON [61].

(b) $C_L^2 > C_T^2$ (in particular $C_T^2 = 0$) and $\tau_{pe} \gg \sigma'$: the conduction electrons interact mainly (or only) with longitudinal waves, but the transverse waves are closely coupled to the longitudinal waves by ordinary three-phonon processes (5.3b).

(c) $C_L^2 > C_T^2$, and σ' is greater than, or of the order of, τ_{pe}: the transverse waves do not interact as strongly (or not at all) with the conduction electrons and are only loosely coupled to the longitudinal waves.

It should be noted that cases (a) and (b) lead to the same value of the thermal conductivity, if the latter is expressed in terms of C_L^2. In the case (c) there arises an additional component of conductivity, analogous to the component $\varkappa_I$ in the case of non-metals, because of the coupling relaxation time σ'. In this case, however, there are no divergence difficulties at low frequencies. The additional component is thus easily derived, though we shall not do so here, because all the metals and alloys studied so far appear to belong to case (a).

If it is assumed that scattering by electrons is the only important resistive process, then it is easily seen, by substituting τ of (19.5) and (19.3) into (4.8) and neglecting σ' and σ, that the lattice component of the thermal conductivity is

$$\varkappa_g = (W_E)^{-1} = 7.18 \frac{8 \pi^2 K^3 MG}{h^3 C_L^2} \left(\frac{1}{k} \frac{dE}{dk} \right)_\zeta^2 T^2 \quad \text{if} \quad T \ll \Theta. \tag{20.1}$$

Note that this expression is independent of $\dot{\Theta}$.

Since the electron-phonon interaction constant is not known, MAKINSON [61] eliminated it by expressing W_E in terms of $\varkappa_\infty$, the electronic ideal thermal conductivity at high temperatures:

$$(W_E)^{-1} = \frac{27}{4 \pi^2 N_a^2} 7.18 \left(\frac{T}{\Theta} \right)^2 \varkappa_\infty \tag{20.2}$$

where N_a is the number of electrons per atom.

It has been shown in Sect. 15 that the BLOCH theory does not reproduce correctly the temperature dependence of $\varkappa_i$, the ideal electronic thermal conductivity, and that this discrepancy arises mainly because it fails to take account of Umklapp-processes and the dispersion of the lattice waves, which are important in determining $\varkappa_\infty$, while the expression for $\varkappa_i$ at low temperatures is not influenced by these effects It thus seems better to compare W_E with the low-temperature limit of $\varkappa_i$, as was done by KLEMENS [72]. In this way one compares two magnitudes which are determined by the same processes, and one also eliminates the effect of any slow variation of C with q. From (15.2) and (20.2) one obtains in the case of a spherical Fermi surface

$$(W_E)^{-1} = 313 \, \varkappa_i(T) \left(\frac{T}{\Theta}\right)^4 N_a^{-\frac{1}{3}} \quad (T \ll \Theta). \tag{20.3}$$

In the derivation of (20.2) and (20.3) it was assumed that coupling is of case (a), the conduction electrons interacting equally with waves of all polarizations. This has two consequences. Firstly, since all modes interact with the electrons, the Θ-value in the expression for $\varkappa_i$ is an average over all polarizations, and is thus approximately equal to Θ_D, the Debye temperature deduced from the low-temperature specific heat. Now W_E of (20.1) is independent of Θ, so that the appropriate Θ-value in (20.3) is Θ_D. Secondly, $C_L^2 = C_T^2 = C^2/3$, where C^2 is the value of the coupling constant appearing in the expressions for $\varkappa_i$ in Sect. 14. This has been taken account of in the derivation of (20.2) and (20.3).

Consider now coupling according to case (b). Now $C_T^2 = 0$ and $C_L^2 = C^2$, where C^2 is the value of the coupling constant in Sect. 14. This introduces a further factor 3 into (20.2) and (20.3), the lattice conductivity being reduced relative to the electronic conductivity. Furthermore the appropriate Θ-value in (20.3) is now Θ_L, the Debye temperature of the longitudinal branch. Thus for case (b)

$$(W_E)^{-1} = 105 \, \varkappa_i(T) \left(\frac{T}{\Theta_L}\right)^4 N_a^{-\frac{1}{3}} \quad (T \ll \Theta). \tag{20.4}$$

Since the value of W_E relative to $\varkappa_i$ differs by a factor of about 15 to 20, depending on whether (20.3) or (20.4) is the appropriate comparison formula, it is possible to discriminate experimentally between cases (a) and (b) by measuring $\varkappa_g$ relative to $\varkappa_i$.

It now remains to discuss what are the appropriate values of N_a. Since $\varkappa_g$ of (20.1) varies as $k_\zeta^{-2}(dE/dk)_\zeta^2$, while $\varkappa_\infty \propto k_\zeta^4(dE/dk)_\zeta^2$ and if $T \ll \Theta$ $\varkappa_i \propto k_\zeta^2 (dE/dk)_\zeta^2$, it follows that $N_a \propto k_\zeta^3$, so that N_a is the actual number of electrons per atom enclosed within the Fermi surface; as in Sect. 15, N_a is independent of (dE/dk).

It should also be emphasised, since some confusion has arisen in the literature on that point, that (20.2), (20.3) and (20.4) compare W_E and W_i of the *same* substance. If the electron concentration is altered, say by alloying, W_i alters as well as W_E, and the variation of W_E with N_a is not obtained by varying N_a in these equations and keeping $\varkappa_i$ or $\varkappa_\infty$ constant. The variation of W_E with electron concentration, or k_ζ, is directly given by (20.1) for spherical Fermi surfaces, and by (19.4) in more general cases.

Klemens [72] has discussed the variation of W_E with electron concentration for a single band with spherical energy contours, treating C^2 as constant. At low electron concentrations $E \propto k^2$, so that W_E is constant as N_a tends to zero. Near the zone boundary dE/dk decreases below the free electron value, so that W_E increases, but it decreases again as the zone is filled, because the area of Fermi surface decreases.

However, since C^2 may be expected to increase with increasing electron concentration, because $E(\boldsymbol{k})$ progressively deviates from the free-electron value, W_E is seen to tend to zero as N_a tends to zero, to increase with increasing N_a, to reach a maximum and finally to tend to zero when the zone is filled. At the present state of development of the theory, it is unfortunately not possible to make more definite predictions about the variation of W_E with electron concentration.

The resistance W_E thus obtained is only one of several resistive processes; in addition there are the resistive processes discussed in Sect. 9. To a degree of approximation sufficient for most purposes W_g is obtained by simply adding W_E to (9.11). At very low temperatures the boundary resistance $W_B \propto T^{-3}$ must become important. The resistance due to single dislocations has the same temperature dependence as W_E and can be comparable with it over a wide range of temperatures. Resistance due to point imperfections and due to three-phonon Umklapp-processes will become important at higher temperatures.

21. Equilibrium between electrons and phonons. The rate of change of the electron and the phonon distribution functions due to their mutual interactions is given by (14.3) and (19.1) respectively. Both these expressions are additively composed of terms describing processes in which an electron k and a phonon q combine to form an electron k', or vice versa, and each term contains the factor

$$f'(1-f)(N+1) - f(1-f')N. \tag{21.1}$$

It is easily seen that this expressions vanishes if

$$f = (e^\varepsilon + 1)^{-1} \tag{21.2a}$$

$$N = (e^x - 1)^{-1} \tag{21.2b}$$

and if

$$\varepsilon + x = \varepsilon' \tag{21.2c}$$

for each interaction. In particular, because of the conservation of energy (14.1), the expression (21.1) vanishes if $\varepsilon = (E-\zeta)/KT$, $x = \hbar\omega/KT$, so that (21.2a), (21.2b) are the equilibrium distributions; this was to be expected.

It was noted by PEIERLS[1] that if Umklapp-processes (14.2b) are excluded from consideration, so that $k + q = k'$, the quantities $(E - \lambda \cdot k)/KT$ and $(\hbar\omega - \lambda \cdot q)/KT$ can be identified with ε and x respectively, and (21.2c) would still be satisfied for each interaction. Consequently in the case of the following anisotropic distributions

$$\overline{f^0}(k, \lambda) = \left[\exp\frac{E - \lambda \cdot k - \zeta}{KT} + 1\right]^{-1}, \tag{21.3a}$$

$$\overline{N}(q, \lambda) = \left[\exp\frac{\hbar\omega - \lambda \cdot q}{KT} - 1\right]^{-1} \tag{21.3b}$$

where λ is a constant and common to both distributions, the rate of change of both distributions due to the electron-phonon interactions vanishes. Hence (21.3a), (21.3b) describe quasi-equilibrium distributions stationary for inter-action processes conserving the total wave vector $J = \sum k + \sum q$. It was pointed out in Sect. 7 that (21.3b) is also stationary for "ordinary" three-phonon inter-actions which conserve $\sum q$ and thus J. The same considerations clearly apply to all processes which conserve J.

In reality, of course, there will also be processes which do not conserve J; these would disturb the quasi-equilibrium (21.3) and tend to restore true equi-librium ($\lambda = 0$).

The quasi-equilibrium value of J is a linear function of λ. If the value of J for the electron-phonon system is kept fixed, then in the absence of other pro-cesses the mutual interactions will tend to bring the electron and phonon distri-butions to the quasi-equilibrium (21.3) with the appropriate value of λ.

---

[1] R. E. PEIERLS: Ann. Phys. **4**, 121; **5**, 244 (1930); **12**, 154 (1932).

If the electron distribution is given by (21.3a) with $\lambda = c$ and the phonon distribution by (21.3b) with $\lambda = b$, where c and b are two vectors in a common direction, then the rate of change of f due to phonon interactions is

$$\frac{df}{dt}\bigg]_p = \frac{c-b}{c}\frac{df}{dt}\bigg\}_p \qquad (21.4)$$

where $df/dt\}_p$ is the rate of change if the phonon distribution is at equilibrium, that is the rate of change as calculated in Sect. 14. Similarly the rate of change of N due to electron interactions is

$$\frac{dN}{dt}\bigg]_e = \frac{b-c}{b}\frac{dN}{dt}\bigg\}_e \qquad (21.5)$$

where $dN/dt\}_e$ is the rate of change if the electrons are at equilibrium, that is the rate of change as calculated in Sect. 19. If c and b are not in the same direction, the corresponding expressions are more complicated.

This principle was applied by Peierls to the intrinsic electrical conduction at low temperatures. The electric field tends to increase J at a uniform rate, the electron-phonon interactions conserve J, so that balance can be obtained only by phonon-phonon interactions which do not conserve q, that is by the same interactions as produce the thermal resistance in Sect. 7. Thus in the steady state $b \neq 0$, and τ_{ep}, the relaxation time for electrons due to interactions with phonons, is according to (21.4) increased over the value calculated by the Bloch theory. If σ_B is the conductivity calculated by the Bloch theory, assuming $b = 0$, then σ is given by (21.4) as

$$\sigma = \sigma_B \frac{c}{c-b} \qquad (21.6)$$

while the condition that the phonon distribution should be stationary gives

$$\frac{b}{c} = \frac{\tau_{bb}}{\tau_{pe} + \tau_{pp}} \qquad (21.7)$$

where τ_{pp} is the effective relaxation time for phonons due to phonon-phonon interactions (7.14) and τ_{pe} is given by (19.3) or (19.4). This result has been derived on the assumption that b is independent of the phonon frequency. A more general treatment[1], apart from a very minor difference between τ_{pp} and the expression (7.14), leads to the following result

$$\frac{\sigma}{\sigma_B} = \int_0^\infty \frac{x^4\,dx}{e^x-1} \bigg/ \int_0^\infty \frac{x^4}{e^x-1}\left[1 + \frac{\tau_{pp}(x)}{\tau_{pe}(x)}\right]^{-1} dx. \qquad (21.8)$$

To investigate deviations of σ from σ_B, both authors used for τ_{pp} values derived from the thermal conductivity of NaCl; Peierls concluded that deviations should be marked below $50°$ K, while Klemens, considering that $\tau_{pp} \propto x^{-4}$, suggested that deviations would not be marked above $10°$ K.

It was pointed out by Peierls that if the Fermi surface touches the zone boundary, so that Umklapp-processes are responsible even at lowest temperatures for the greater portion of the ideal electrical resistance, the above considerations do not apply, and no deviations from the T^5 variation of ϱ_i should occur. Peierls concluded from the absence of such a deviation that the Fermi surface of monovalent metals does touch the zone boundary, while Klemens claimed

---

[1] P. G. Klemens: Proc. Phys. Soc. Lond. A **64**, 1030 (1951).

that since taking account of the frequency variation of τ_{pp} would lower the critical temperature, this conclusion is not compelling. Since then, two further developments have occurred which have a bearing on this question. As seen in Sect. 15, there is evidence from the low-temperature LORENZ ratio $\varrho_i/W_i T$ that in the case of the noble metals and the heavy alkali metals the FERMI surface does touch the zone boundary: this would greatly reduce any deviations from $\sigma \propto T^{-5}$ due to quasi-equilibrium effects for these metals. Also it is known from a study of the lattice component of the thermal conductivity (see Sect. 24) that τ_{pp} is usually smaller for metals than for ionic solids, that it is sensitive to deformations and annealing, and that dislocations play an important part in the scattering of phonons.

The coupling between electrons and phonons also results in a contribution to the thermoelectric effect. An electric current flowing in a metal corresponds to an electron distribution (21.3a) with non-vanishing λ. The interaction between electrons and phonons will tend to bring the phonon distribution into quasi-equilibrium (21.3b) with the same value of λ; if $\tau_{pp} \gg \tau_{pe}$ both distributions will have a common value of λ. In this way a phonon energy current, proportional to the electric current, is obtained, and hence a lattice component to the PELTIER and THOMSON coefficients[1]. For a band of quasi-free electrons, if the phonons are strongly coupled to the electrons, the phonon component of the THOMSON coefficient is

$$\mu_P = \frac{1}{3Ne} T \frac{dS}{dT} \qquad (\tau_{pp} \gg \tau_{pe}) \tag{21.9}$$

where e is the charge of the carrier, S the lattice specific heat and N the density of free electrons (effective number).

If τ_{pp} is comparable to τ_{pe}, and if the frequency dependence of τ_{pp} and τ_{pe} is considered, (21.9) is generalized to

$$\mu_P = \frac{1}{3Ne} T \frac{d}{dT} \left\{ \int S(\omega) \frac{\tau_{pp}(\omega)}{\tau_{pp}(\omega) + \tau_{pe}(\omega)} d\omega \right\} \tag{21.10}$$

where $S(\omega) d\omega$ is the contribution to the lattice specific heat from modes of frequency $\omega, d\omega$. The same result was obtained, in a different manner, by BLATT[2] and by MACDONALD[3]. The phonon contribution would be expected to become important above say 10° K, but would be reduced again at higher temperatures by the factor $\tau_{pp}/(\tau_{pp} + \tau_{pe})$. There exists, as yet, not systematic experimental study of this effect, nor of the deviations of the electrical conductivity from σ_B, though such a study could provide valuable information on the coupling ratio τ_{pp}/τ_{pe}.

In the case of semi-conductors there is also a lattice component of the thermoelectric effects. The theoretical treatment is complicated in this case by the fact that the electrons interact only with phonons of very low frequency ($\hbar\omega \ll KT$), so that it is necessary to consider explicitly not only the coupling between electrons and phonons, but also the coupling of low frequency phonons with the phonons of thermal frequencies. This has been discussed fully by HERRING[4].

In the case of thermal conduction one would, at first sight, expect a similar interrelation between the deviations from equilibrium of the electrons and the phonons. Thus in Sect. 14 we have calculated $df/dt]_p$ for heat conduction on

[1] See P. G. KLEMENS: Austral. J. Phys. 7, 520 (1954).
[2] F. J. BLATT: Private communication.
[3] D. K. C. MACDONALD: Physica, Haag 20, 996 (1954).
[4] C. HERRING: Phys. Rev. 96, 1163 (1954), see also [23].

the assumption that the phonons are at equilibrium, which is not so, as seen in Sect. 19. Conversely, in Sect. 19, $dN/dt]_e$ was obtained, assuming the electrons to be at equilibrium. It must now be shown that each system, acting on the other, tends to restore it to true equilibrium, irrespective of its own deviation from equilibrium.

While in the case of electrical conduction in metals or in the case of thermal conduction in non-metals the field (or temperature gradient) produced a uniform increase of $\boldsymbol{J}$ which has to be balanced by processes not conserving $\boldsymbol{J}$, in the case of thermal conduction the increase in $\boldsymbol{J}$ is balanced by the thermoelectric field which is built up in consequence of the condition that the electric current should vanish.

The vanishing of the electric current implies that $c(\varepsilon)$ is an almost odd function of ε. On the other hand $b(x)$, though in general a function of x, has the same sign for all values of x. Quasi-equilibrium implies that $b=c$ for all values of x and ε. Consider a particular mode $\boldsymbol{q}$, interacting with electron states $\boldsymbol{k}$: in the summation of (19.1) over all states $\boldsymbol{k}$, states with positive and negative ε will be equally represented, so that the average value of $c(\varepsilon)$ is zero; this has been pointed out by Bethe [1]. Thus if the electric current vanishes, the deviation of the electron distribution from equilibrium has no effect on $dN/dt]_e$, except for higher order terms in KT/ζ. This justifies the omission of terms in $g(\boldsymbol{k})$ in (19.2) for purposes of calculating the lattice component of the thermal conductivity, and the result of Sect. 20.

The matter is more complicated in the reverse case, the effect of $b(x)$ on $df/dt]_p$. Bethe [1] has shown that the Boltzmann equation (14.5) is modified:

$$\mathfrak{S} \cdot c = A\left[K\,\varepsilon \nabla T + \frac{d\zeta}{dT}\nabla T - eF\right] + B\nabla T \tag{21.11}$$

where

$$B\nabla T = -\frac{1}{2k^2}\int\limits_{0}^{q_0}\frac{q^4\,dq}{e^x-1}\,b(x)\left[\frac{e^\varepsilon+1}{e^\varepsilon+e^{-x}}+\frac{e^{-\varepsilon}+1}{e^{-\varepsilon}+e^{-x}}\right]dx. \tag{21.12}$$

The term $B\nabla T$ is an almost even function of ε, and Bethe stated that it would thus influence the thermal conductivity only to higher order. Makinson [61] pointed out that $B\nabla T$ may be sufficiently large so that even the higher-order effects would be appreciable ,and he investigated the effect of this term on the thermal conductivity. The following considerations lead to a quick estimate of the magnitude of the effect:

Since $B\nabla T$ is almost even in ε, the expression $(d\zeta/dT)\nabla T - eF + B\nabla T$ takes the place of $(d\zeta/dT)\nabla T - eF$ in the treatment of Sect. 14. To first order this does not change the thermal conductivity, but changes the electric field $\boldsymbol{F}$. This is, of course, only another way of treating the effect of the phonons on the thermoelectric power, and this is the method of calculating μ_P adopted by Blatt.

There is, however, a change in the higher order terms in the expression (14.12) of $\varkappa_e$, for the expression A'/A should now be replaced by

$$\frac{A'\left(\dfrac{d\zeta}{dT}\nabla T - eF\right) + B'\nabla T}{A\left(\dfrac{d\zeta}{dT}\nabla T - eF\right) + B\nabla T}. \tag{21.13}$$

Since $\boldsymbol{j}=0$,

$$A\left(\frac{d\zeta}{dT}\nabla T - eF\right) + B\nabla T = -A\,K\nabla T\,\frac{L_{21}}{L_{11}} \tag{21.14}$$

so that, after some reduction, one can show that the expression (14.12) for $\varkappa_e$ should be corrected, because of the replacement of A'/A in (21.13), by

$$\delta\varkappa_e = \varkappa_e\left[\frac{BT}{A}\left(\frac{A'}{A} - \frac{B'}{B}\right)\right].$$

(21.15)

It should be noted that (21.15) is independent of the form of $\mathfrak{S}$, that is independent of the temperature (except for its explicit dependence on T and on B), and independent of the relative contributions of W_0 and W_i to W_e. For a gas of quasi-free electrons $A'/A = 1/2\zeta$, $B'/B = -1/\zeta$, so that

$$\frac{\delta\varkappa_e}{\varkappa_e} = \frac{3KT}{2\zeta}\frac{B}{AK}.$$

(21.16)

Here B is given by (21.12) with $\varepsilon = 0$.

One can evaluate (21.16), relating A to $\varkappa_i$ by (14.16) and B to $\varkappa_g$, since (4.8) can be expressed as

$$Q_g = -\varkappa_g\nabla T = \frac{1}{(2\pi)^3}\frac{4\pi}{3}\frac{\hbar v^2}{KT}\left(\frac{KT}{\hbar v}\right)^5\int\frac{x^4 b(x)\, dx}{(e^x - 1)(1 - e^{-x})}.$$

(21.17)

Substituting into (21.16) and noting that in the limit of small T and in the absence of residual resistivity [from (14.13), (14.19), (14.28) and (14.29c)]

$$L_{22} = 0.13\left(\frac{k}{q_0}\right)^3\left(\frac{\Theta}{T}\right)^3,$$

(21.18)

one finds after some reduction that

$$\frac{\delta\varkappa_e}{\varkappa_e} \approx -\frac{K\Theta}{\zeta}\frac{\varkappa_g}{\varkappa_i}\qquad (T\ll\Theta)$$

(21.19)

and

$$\frac{\delta\varkappa_e}{\varkappa_e} \approx -\frac{KT}{\zeta}\frac{\varkappa_g}{\varkappa_i}\qquad (T>\Theta).$$

(21.20)

Since $\varkappa_g\ll\varkappa_i$ and $K\Theta\ll\zeta$, $\delta\varkappa_e$ is always an unobservably small fraction of $\varkappa_e$.

MAKINSON [61] evaluated $\delta\varkappa_e$ for $T>\Theta$ and also for $T\ll\Theta$ if $W_0>W_i$, and his results agree in order of magnitude with (21.19)[1].

Thus, when calculating the thermal conductivity, one can disregard the deviation of the electron distribution from equilibrium for purposes of calculating the lattice component, and no appreciable error is incurred in the calculation of the electronic component by disregarding the deviation from equilibrium of the phonon distribution.

22. Separation of electronic and lattice components. The total thermal conductivity is

$$\varkappa = \varkappa_e + \varkappa_g$$

(22.1)

where $\varkappa_e$ and $\varkappa_g$ are the electronic and lattice components; also

$$\frac{1}{\varkappa_e} = W_e = W_0 + W_i$$

(22.2)

where W_0 and W_i are the residual and ideal thermal resistivities respectively; similarly, in view of (9.10) and (9.11),

$$\frac{1}{\varkappa_g} = W_g \approx W_E + W_B + \sum_{(\alpha)} W_\alpha$$

(22.3)

---

[1] Except that he omitted a factor $\int dx\, x^5 e^x/(e^x - 1)^2$ in the denominator of his final result for low temperatures, thus overestimating $\delta\varkappa_e/\varkappa_e$ in that case by a factor of order 100.

17*

where W_E is the resistivity due to scattering by conduction electrons, W_B is the boundary resistance and W_α are resistivities due to various static imperfections. The electrical resistance is

$$\varrho = \varrho_0 + \varrho_i \qquad (22.4)$$

ϱ_0 and ϱ_i being the residual and ideal resistivities respectively. The electrical and thermal residual resistivities are related by the Wiedemann-Franz law

$$\varrho_0 = L W_0 T, \quad \text{hence} \quad W_0 \propto T^{-1} \qquad (22.5)$$

where L is the theoretical Lorenz number (13.15); but generally, in zero magnetic field

$$\varrho \leq L W_e T. \qquad (22.6)$$

The problem is to deduce $\varkappa_e$ and $\varkappa_g$ separately from a knowledge of the overall thermal conductivity, and from ancillary measurements. The lattice component $\varkappa_g$ can, of course, only be deduced with reasonable accuracy if it forms 'an appreciable fraction of $\varkappa$. Even then there is no generally applicable method: there are available a number of different methods, each based on some assumption which is not always satisfied, and which in many cases cannot be tested, so that, depending on circumstances, a judicious choice of the best method must be made for each case, and the result must be judged by plausibility and consistency criteria.

If the Wiedemann-Franz law were obeyed for all temperatures, $\varkappa_e$ could be easily obtained from electrical resistance measurements. This is, however, not so, but the inequality (22.6) enables one to obtain a lower limit for $\varkappa_g$. Since this lower limit can turn out to be negative, this method has limited applicability.

There are cases where $\varkappa/\sigma T$ exceeds L by a large factor, so that $\varkappa_g$ is much greater than $\varkappa_e$ and can be readily determined, with only a small relative uncertainty due to the uncertainty in $\varkappa_e$. This is so in the case of bismuth at low temperatures, and to a lesser degree in the case of other semi-metals and, of course, in the case of semi-conductors.

If $W_0 \gg W_i$, so that from (22.5) $\varrho = L W_e T$, it is possible to arrive at a reliable estimate of $\varkappa_e$, provided the electrical and thermal resistivities are measured on an identical specimen. If $\varkappa_g$ is then comparable with $\varkappa_e$, it can be determined fairly reliably. In most metals, including all the monovalent metals, $\varkappa_i \gg \varkappa_g$ over most temperatures, so that it is not sufficient to merely have $W_0 > W_i$, but W_0 must be sufficiently large so that W_0 and W_g are comparable. It is possible to increase W_0 by suitable alloying, and this method has been used by many authors to evaluate $\varkappa_g$ [82], [92], [115], [116], [117], [119], [145].

If it is desired to extend the separation to higher temperatures, where W_i is no longer negligible, it is necessary to know W_i, so that W_e can be obtained from (22.2) if ϱ_0 or W_0 is known. Clearly W_i cannot be obtained from thermal conductivity measurements on the same specimen, since the effects of $\varkappa_g$ and W_i cannot be separated, and some assumption about W_i must be made. It may be possible to assume that W_i is the same as for the parent metal in the pure state (low W_0), where a direct measurement is possible. There are, however, two difficulties: firstly, W_i is not independent of W_0, but, as seen in Sect. 14, it changes by a factor of up to 1.5 as W_0 changes from $W_0 < W_i$ to $W_0 > W_i$; secondly, if W_0 is increased by alloying, the electronic band structure and possibly the lattice properties are changed, so that all electronic conduction properties are altered, including W_i of course. This difficulty is discussed elsewhere [119].

It is possible to deduce $\varkappa_0 = 1/W_0$ without electrical resistance measurements if it is assumed that $\varkappa_g \propto T^2$, as it would be if W_E were the principal resistance. This will often be the case at liquid helium temperatures for reasonable grain size, so that $W_B \ll W_E$; since $W_E \propto T^{-2}$ other resistive processes will become important only at higher temperatures. It is then possible to express $\varkappa$ in the form $aT + bT^2$, where the first term describes the electronic and the second term the lattice component. This method was used by KEMP, KLEMENS, SREEDHAR and WHITE [118] and also by SLADEK [145]. It does, however, require accurate measurements of $\varkappa$ and T, and for specimens of low overall thermal conductivities discrepancies were found between $\varkappa_0$ thus obtained and ϱ_0 measured electrically; these discrepancies may be the result of small heat leaks which become important for specimens of low overall conductivity [119].

There are other methods of deducing $\varkappa_g$ which depend upon the fact that $\varkappa_e$ is reduced in a magnetic field, but that $\varkappa_g$ is unaltered[1]. This method is only useful when $\varkappa_e$ can be appreciably reduced by moderate magnetic fields, that is mainly in the case of anisotropic metals. Even then there is the difficulty that the variation of W_e with H is not known. The following methods have been proposed and applied in some cases:

(a) Since W_e increases with H, so that $\varkappa_e$ should become very small for large H, the curve of W against H shows saturation, and the saturation value of W would be W_g. However, since for some models $W_e(H)$ shows saturation (18.10b) one cannot be sure that the observed saturation value of $W(H)$ should be identified with W_g. According to KOHLER [76] $\varkappa_e \propto H^{-2}$ for divalent metals in strong fields, while other metals would show saturation. Thus for divalent metals a plot of $\varkappa(H)$ against $1/H^2$ should be a straight line, and its extrapolation to $H^{-2} = 0$ should give $\varkappa_g$. One can check this extrapolation for anisotropic single crystals by varying the orientation of the magnetic field, keeping the direction of heat flow fixed. Using the results for Be of GRÜNEISEN and ADENSTEDT [103] and of GRÜNEISEN and ERFLING [104] he found discrepancies in the extrapolated values of $\varkappa_g$ which could be explained in terms of inaccuracies in the extrapolation procedure.

(b) If the electrical and thermal conductivities are measured simultaneously in magnetic fields, the deduction of $\varkappa_g$ is more certain. If $L_e = \varkappa_e/\sigma T$ were independent of H, the separation of $\varkappa_g$ would then be easy. This method was proposed by GRÜNEISEN and ADENSTEDT, but it was soon seen that it was not reliable, since L_e is not independent of H, as was later deduced theoretically [74]. Only in the limiting case $W_0 > W_i$, where $L_e = L$, would $L_e(H)$ be independent of H. This method could be useful in cases where $W_0 > W_i$ but $W_g > W_0$. In divalent metals, however, where $\varkappa_e$ and σ both vary as H^{-2}, L_e should tend to a constant value for strong fields, and a plot of $\varkappa$ against σ for various values of H should, when linearly extrapolated to $\sigma = 0$, give $\varkappa_g$. KOHLER [76], using measurements on Be, obtained approximately the same value of $\varkappa_g$ by this method as by plotting $\varkappa_e$ against $1/H^2$.

(c) In anisotropic metals one can plot $\varkappa$ against σ for various values of H, and by rotating H about the direction of flow, the value of $\varkappa_g$ thus derived can be checked for consistency. This method was used by GRÜNEISEN, RAUSCH and WEISS [106] for Bi at liquid oxygen temperatures.

It is seen that there is no unique solution to the problem of evaluating $\varkappa_g$. Clearly the more ancillary measurements are made, the more reliable is the

---

[1] The improbability of $\varkappa_g$ being changed by a magnetic field has been discussed by WILSON and SONDHEIMER [74].

deduced value of $\varkappa_g$; but for different cases there are different degrees of difficulty: while in some cases it is sufficient to measure $\varkappa$ as a function of T, there are other cases where measurements of thermal and electrical conductivities in various magnetic fields for various orientations could still leave appreciable uncertainties in $\varkappa_g$.

23. Comparison with experiments: Pure metals. Knowledge of the lattice component of pure metals is obtained in three ways: (a) in some cases $\varkappa_e$ is clearly negligible and the lattice component is the measured thermal conductivity, (b) the lattice component is derived by extrapolation from the lattice component of dilute alloys, and (c) it is deduced from measurements in magnetic fields.

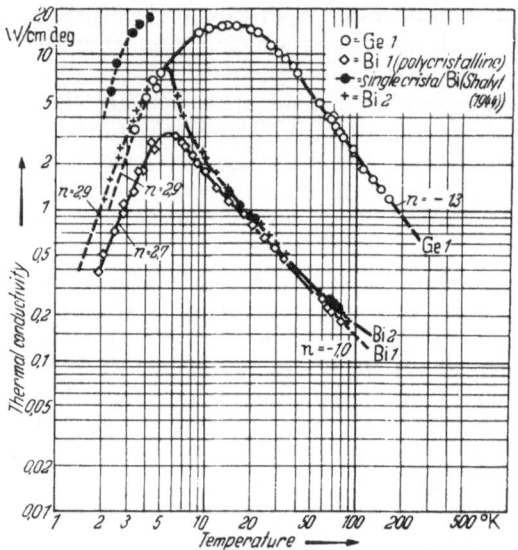

In the case of bismuth, $\varkappa_e$ can be neglected below about 80° K. This conclusion was reached theoretically by AKIEZER and POMERANCHUK [78]; it is experimentally derived from the large values of $\varkappa/\sigma T$, as well as from the insensitivity of the thermal conductivity to magnetic fields [102], [107], [110].

The thermal conductivity of bismuth was measured down to liquid hydrogen temperatures by DE HAAS, GERRITSEN and CAPEL [107] and to liquid helium temperatures by SHALYT [110]; the latter measurements are shown in Fig. 12. At liquid hydrogen temperatures $\varkappa_g$ was only a few percent of $\varkappa$, at liquid helium temperatures $\varkappa_e$ could only be deduced from electrical measurements. WHITE and WOODS [121] made a similar measurements covering a wider range of temperatures—see Fig. 12.

Fig. 12. Thermal conductivity of a germanium specimen and two bismuth specimens, according to WHITE and WOODS [121]; SHALYT's [110] measurements on a single bismuth crystal are included for comparison; n denotes the power of the observed variation $\varkappa \propto T^n$.

The temperature dependence of $\varkappa_g$ thus found is similar in character to that of non-metallic solids. The rapid increase of $\varkappa$ with decreasing temperature below 20° K is an indication of the importance of Umklapp-processes in producing thermal resistance; only in the case of WHITE and WOODS' specimen Bi 1, which was less pure and was cooled quickly after casting, does it appear that the Umklapp-resistance is overshadowed below about 10° K by resistance due to some imperfections (apparently point imperfections).

The case of antimony is similar to bismuth, except that, even at low temperatures, the electronic component is not negligible. RAUSCH [105], using the anisotropy of the magneto-resistive effect, deduced the magnitude of the lattice component at liquid oxygen temperatures for a single crystal. The lattice component was of similar magnitude to the electronic component, the ratio varying with direction.

ROSENBERG [87] measured the thermal conductivity of an antimony specimen below 40° K. This case resembles a dilute alloy (see Sect. 24): $\varkappa_e$ varies as T, being determined by imperfection scattering, while $\varkappa_g$ varies approximately as T^2 at liquid helium temperatures, changing to a slower variation at higher temperatures, as scattering processes other than scattering by electrons become important. From this data it appears that $W_E \approx 20 \ T^{-2}$ watt^{-1} cm. deg.

In the case of semiconductors the electronic component is negligible. The thermal conductivity of germanium has been measured by ESTERMANN and ZIMMERMAN [49], who noted large changes due to impurities, referred to in Sect. 11, by ROSENBERG [50], who also measured silicon, and by WHITE and WOODS [121]. The results of the last-mentioned authors are shown in Fig. 12; the curve is similar to the curve for germanium obtained by ROSENBERG. There is no clear indication of Umklapp-resistance, though $\varkappa$ varies faster than as T^{-1} between 20 and 100° K. If the Umklapp-resistance is hidden by resistance due to point imperfections, the concentration of imperfections would have to be of the order of 10^{-5} per atom; this large concentration cannot be explained in terms of donor or acceptor impurities; the point imperfections may be vacancies, but it is not clear why they should be so numerous[1].

In the case of bismuth, germanium and silicon one can attribute the variation of thermal conductivity with temperature below the maximum to the effect of boundary scattering.

An attempt by DE HAAS and DE NOBEL [108], [109] to separate $\varkappa_g$ of a tungsten single crystal at liquid hydrogen temperatures by measuring the electrical and thermal conductivities in magnetic fields and extrapolating to high fields proved unsuccessful, because of the variations of $\varkappa_e/\sigma T$ at high field strength; the lattice conductivity at these temperatures appears to be considerably smaller than the lowest overall conductivity measured (0.4 watt units), which is surprising.

The measurements of GRÜNEISEN and ADENSTEDT [103] and GRÜNEISEN and ERFLING [104] of the thermal and electrical conductivities of beryllium single crystals in magnetic fields at liquid hydrogen temperatures do permit a fairly reliable evaluation of $\varkappa_g$. This has been discussed by KOHLER [76], both by extrapolating a curve of $\varkappa(H)$ against $\sigma(H)$, and by extrapolating a curve of $\varkappa(H)$ against $1/H^2$ for H tending to infinity. Typical values at liquid hydrogen temperatures are in the range of 0.12 to 0.17 watt units; a similar procedure for liquid oxygen temperatures proved inconclusive.

WHITE and WOODS [121] have measured the thermal conductivity of sintered beryllium rods of high residual electronic resistivity, and obtained $\varkappa_g$ as for an alloy; they found $\varkappa_g \approx 2 \times 10^{-4} T^2$, which is lower than the values obtained on single crystals by the magnetic method. It is not surprising that the lattice conductivity of a sintered specimen should be reduced to about half the value of a single crystal; it does serve to emphasise, however, that the resistance W_g obtained for impure materials cannot be identified confidently with W_E, even if it varies as T^{-2}.

The evaluation of W_g of other dilute alloys, and the way this can give information about W_g for pure metals will be discussed in Sect. 24. In view of the uncertainty of the electronic band structure of all but the monovalent metals it does not appear profitable to compare the various estimates of W_E with the theoretical expressions of Sect. 19 and 20, except for monovalent metals, which is done in Sect. 24 below.

24. Comparison with experiments: Alloys. While the thermal conductivity of a large number of different alloys has been measured[2], most of which have

---

[1] E. V. MIELCZAREK (private communication) has measured the thermal conductivity of indium antimonide; her results are similar to those on germanium.

[2] POWELL and BLANPIED [7] give a comprehensive bibliography; see also OLSEN and ROSENBERG [6].

presumably an appreciable lattice component, these measurements are often not sufficiently extensive to permit a clear evaluation of $\varkappa_g$. It is only in recent years that experiments have been designed for the study of the lattice component, and it is more recently still that the great sensitivity of $\varkappa_g$ to the physical state of the specimen has been appreciated. It is for this reason that many early measurements do not yield full information about $\varkappa_g$ and could be repeated and extended with profit.

Grüneisen and Reddemann [82] measured $\varkappa$ and σ of a number of alloys of Cu, Ag and Au at oxygen and hydrogen temperatures. They demonstrated that $\varkappa_g$ is indeed appreciable. In view of the uncertainties of the theoretical picture at that time, they refrained from deducing the magnitude of $\varkappa_g$, Borelius[1] deduced $\varkappa_g$ from these measurements for a number of gold alloys, and by extrapolation, for gold itself, but emphasised that his conclusions were provisional. This value of W_g cannot be identified with W_E, but in the light of present knowledge it is probably due to static imperfections.

Hulm [92] deduced the lattice component of an alloy Sn 96—Hg 4 by the same method at liquid helium temperatures and found $\varkappa_g \approx 3 \times 10^{-3} T^3$ watt cm.$^{-1}$ deg.$^{-1}$. He also studied the alloy Cu 80—Ni 20 [115], obtaining $\varkappa_g \approx 2.2 \times 10^{-4} T^2$, and discussed the lattice component of some alloys from earlier measurements of Karweil and Schaefer [113], of which the most interesting is Cu 64—Ni 16—Zn 20, which showed $\varkappa_g \approx 3.8 \times 10^{-4} T^2$.

Berman [116] measured $\varkappa_g$ for an alloy Cu 60—Ni 40 and found $\varkappa_g \approx 2 \times 10^{-4} T^2$ below 10° K. He studied the departure of $\varkappa_g$ from a T^2 variation, which he provisionally ascribed to point imperfections and obtained similar results for German silver and stainless steel.

Estermann and Zimmerman [117] studied $\varkappa_g$ for Cu 90—Ni 10 both in the cold-worked and the annealed state. For the annealed specimen they found $\varkappa_g \approx 3.9 \times 10^{-4} T^2$, while for the cold-worked specimen $\varkappa_g \approx 1.5 \times 10^{-4} T^2$. Thus the additional imperfections introduced by cold work have a phonon scattering probability proportional to the wave-number q, and the additional resistance has the same temperature dependence as W_E. Since one cannot be sure that annealing completely removes these imperfections, one must regard all measured values of W_E as upper limits only.

Estermann and Zimmerman also deduced from the measurements of Wilkinson and Wilks [36] that for Cu 70—Ni 30 $\varkappa_g \sim 1 \times 10^{-4} T^2$, though this value cannot be regarded as well established.

Klemens [72] concluded from these measurements by extrapolation that for pure copper $W_E \lesssim 2 \times 10^3 T^{-2}$ watt^{-1} cm. deg., and comparing this with (20.3) and (20.4) respectively, he concluded that in the case of copper the conduction electrons interact not only with the longitudinal, but also with the transverse phonons, though possibly less strongly with the latter. This is the same conclusion as had been reached from a consideration of the electronic conduction properties alone— see Sect. 15.

Kemp, Klemens, Sreedhar and White [118] deduced $\varkappa_g$ for a series of Ag—Pd alloys, and by extrapolation reached the same conclusion for the case of silver. White and Woods [120], [121], by measuring $\varkappa_g$ for a very dilute copper alloy (0.056% Fe), reduced the upper limit of W_E for pure copper to the value of $6 \times 10^2 T^{-2}$ watt^{-1} cm. deg., which is even slightly less than the value given by (20.3). Kemp, Klemens, Sreedhar and White [119] extended their measurements to alloys of high palladium content, as well as to Ag—Cd alloys.

---

[1] G. Borelius: Ann. Phys., Lpz. **29**, 251 (1937).

Sladek [145] has measured $\varkappa_g$ for a number of In—Tl alloys (up to 50% Tl) in the normal as well as in the superconducting[1] state.

The measurements of Kemp et al. confirmed that W_D, the resistance due to cold work, varied as T^{-2}. The temperature dependence of this resistance indicates scattering of phonons by isolated dislocations. However, if the magnitude of W_D is compared with (9.15), the number of single dislocations per unit area required to produce this resistance would be of the order of 10^{13} cm.$^{-2}$. Not only is such a density of dislocations in conflict with other evidence[2], but it is so high that the picture of isolated dislocations would no longer hold, leading

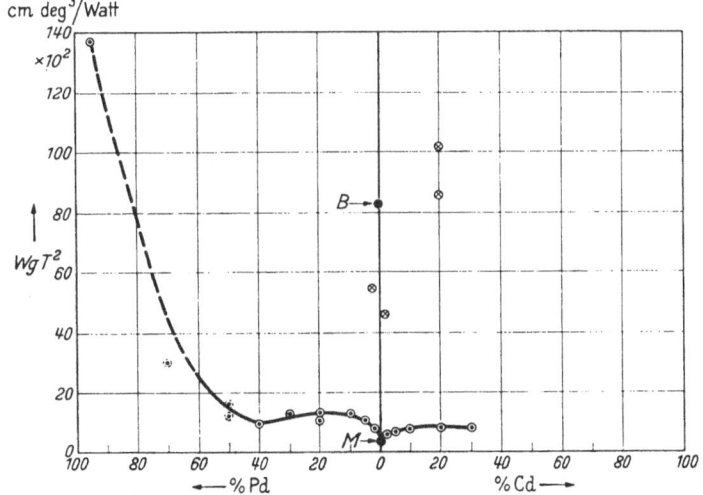

Fig. 13. The variation of lattice resistance $W_g T^2$ of silver alloys with palladium or cadmium concentration, according to Kemp, Klemens, Sreedhar and White [119]; ⊙ annealed specimens, ⊗ strained specimens. Points B and M denote the values of $W_g T^2$, derived from W_i for pure silver using (20.4) (Bloch coupling scheme) and (20.3) (Makinson coupling scheme) respectively.

to a different temperature dependence of W_D. It must thus be concluded [21], [119] that the scattering arises mainly from groups of dislocations, each group containing N dislocations of the same sign, for the scattering cross-section of such a group is proportional to N^2 (hence proportional to N per single dislocation). To account for the magnitude of W_D in conformity with other estimates of dislocation density (which are mainly grouped about 10^{11}) N must be of order 50 to 100, as discussed in [119].

Let us consider now the variation of $W_g T^2$ with composition for annealed alloys. The values of $W_g T^2$ in the liquid helium range are plotted as a function of electron concentration in Fig. 13 for the silver-palladium and silver-cadmium series. There are two major features in this curve: the value of $W_g T^2$ increases strongly with increasing palladium concentration above about 50% palladium, and the curve shows a cusp in the vicinity of pure silver. The first effect is readily understood in terms of (19.4): at about 50% Pd holes appear in the d-band, and their number increases with increasing concentration of palladium; the lattice waves are thus scattered not only by the portion of the Fermi surface in the s-band, but also by the section in the d-band. However, the cusp in the curve can hardly be explained in terms of variations of the electronic band structure.

[1] These measurements will be discussed in V.
[2] See, for example, A. H. Cottrell: Dislocations and Plastic Flow in Crystals. Oxford: University Press 1953.

The increase of W_g with increasing impurity concentration (irrespective of the valency of the impurities) for dilute alloys must be ascribed to scattering of lattice waves by imperfections associated with the impurities. Since the resistance varies as T^{-2}, this cannot be scattering by the impurities themselves, but more probably it is scattering by dislocation. That impurities should have associated with them dislocations which are firmly anchored and cannot be removed by annealing, the dislocation density increasing with increasing impurity concentration, is not improbable in view of the known occurrence of the inverse effect, the segregation of impurities along dislocation lines.

Sladek [145] also observed an increase of $W_g T^2$ with increasing thallium content in indium. Since both elements have the same valency, this cannot be explained in terms of changing the electron concentration, and the above effect may at least contribute to the increase in $W_g T^2$. The same effect may also be responsible for the difference of $W_g T^2$ between White and Woods' dilute copper-iron alloy and Estermann and Zimmerman's Cu 90—Ni 10. More work must be done to elucidate this question. One possible approach would be to measure the thermoelectric power of these alloys. If it is possible to separate the phonon contribution of the Thomson coefficient (21.10) from the electronic component, then the quantity τ_{pe}/τ_{pp} could be obtained, which is of course proportional to W_D/W_E.

Another aspect of W_g which deserves further study is its temperature variation above $10°$ K, where $\varkappa_g$ varies more slowly then T^2. This is no doubt due to some static imperfections, but further experimental and theoretical studies are necessary to elucidate their nature.

V. Thermal conductivity of superconductors[1].

25. The two-fluid model[2]. While a full understanding of the thermal conduction properties of superconductors can only follow from a detailed microscopic theory of superconductivity, it is possible to understand them qualitatively in terms of the two-fluid model[3], which, while not explaining the phenomena, is a convenient scheme for their description and one to which a successful microscopic theory will presumably have to conform.

The two-fluid model assumes that a fraction $(1-x)$ of the Fermi surface is modified, the electrons on the surface condensing into a lower state, and that there cannot be any thermal excitation of the electrons in those modified regions, though the fraction of the Fermi surface thus affected is a function of temperature and increases as the temperature is lowered. The superconductive regions can be orientated in such a way as to yield a supercurrent. Since the superconductive regions give zero contribution to the entropy, there being no thermal excitation, the Thomson coefficient for supercurrents is zero, and the entropy is less than it would be in the normal state. Similarly the electronic thermal conductivity is reduced, since only the normal fraction of the Fermi surface contributes towards it. On the other hand the lattice component of the thermal conductivity is enhanced, since the lattice waves can only be scattered by the electrons on the normal fraction of the Fermi surface.

---

[1] See also the companion articles by Serin and Bardeen in vol. XV of this Encyclopedia.

[2] See also D. Shoenberg: Superconductivity, 2nd ed. Cambridge: University Press 1952.

[3] Gorter and Casimir [122] discussed the thermodynamic properties of superconductors in terms of this model; Heisenberg [123] attempted to justify the model and extended its application to thermal conduction.

Following GORTER and CASIMIR [122] one assumes that the modification of the superconducting fraction of the FERMI surface has associated with it a latent heat; thus the HELMHOLTZ free energy, usually of the form $F = -\frac{1}{2}\gamma T^2$, is now modified to

$$F = -\tfrac{1}{2} x^r \gamma T^2 - (1 - x)\beta. \tag{25.1}$$

The first term is the contribution from the normal, the second that from the superconducting region; the latter is due to the latent heat, there being no continuum of states available for thermal excitation. The condition $(\partial F/\partial x)_T = 0$ gives x as a function of T. It is not possible to assume simply $r = 1$, as it would be in the absence of interaction between the n- and the s-regions, as this would not generally satisfy $(\partial F/\partial x)_T = 0$. If one takes $r = \frac{1}{2}$, the observed thermodynamic properties are approximately reproduced, that is

$$x = \left(\frac{\gamma T^2}{4\beta}\right)^2 = \left(\frac{T}{T_c}\right)^4 \tag{25.2}$$

where T_c is the transition temperature, for which $x = 1$. The specific heat per unit volume is

$$C = -T\frac{d^2 F}{dT^2} = 3\gamma\frac{T^3}{T_c^3}, \tag{25.3}$$

but the specific heat due to the thermal excitation of the electrons in the normal region is

$$C_n = -T\left(\frac{d^2 F}{dT^2}\right)_x = \gamma\frac{T^3}{T_c^3}. \tag{25.4}$$

The difference $C - C_n$ is ascribed to the change of energy as electrons change their phase (from being in an s-region to being in an n-region).

The thermal conductivity of normal metals can be written in the form [compare with (13.14)]

$$\varkappa = \tfrac{1}{3} C v l \tag{25.5}$$

where $C = \gamma T$ is the electronic specific heat, v is the velocity of electrons of FERMI energy (here assumed isotropic) and $l = v\tau$ is the effective electron mean free path. HEISENBERG [123] has applied the two-fluid model to thermal conduction: the electronic thermal conductivity in the superconducting state differs from that in the normal state because C, and possibly l, are altered. For C one should now use C_n of (25.4), because one is concerned with the transport of energy by electrons which remain normal when passing along a temperature gradient, rather than with a change of energy due to a change of phase; this point will be amplified below. Thus

$$\frac{\varkappa_{es}}{\varkappa_{en}} = \left(\frac{T}{T_c}\right)^2 \left(\frac{l_s}{l_n}\right) \tag{25.6}$$

where $\varkappa_e$ is the electronic component of the thermal conductivity and the suffixes s and n refer to the normal and superconducting states respectively. It should be noted that had C of (25.3) been used instead of C_n to calculate $\varkappa_{es}$, there would have been a discontinuity at the transition temperature in the curve $\varkappa_e$ against T, similar to that observed for the specific heat, for $l_s(T_c)$ must equal $l_n(T_c)$.

Some assumption has to be made about the behaviour of l_s; l_s could differ from l_n because, when an electron is scattered from a state in the n-region, there are fewer final states available to it, since the electron states in the s-region are modified. The ratio l_s/l_n should depend upon the mechanism of scattering.

Heisenberg suggests that for scattering by static imperfections

$$l_s = \frac{2l_n}{(1+x)} \tag{25.7}$$

so that $l_s > l_n$. This form would be appropriate if scattering were isotropic and if a fraction $x/2$ of the possible final states were blocked by electrons in the s-states. Thus, if the thermal resistance is mainly due to static imperfections, the fractional change of the conductivity, on going from the normal to the superconducting state, is

$$f = \frac{W_{on}}{W_{os}} = \frac{\varkappa_{os}}{\varkappa_{on}} = \frac{2(T/T_c)^2}{1+(T/T_c)^4}. \tag{25.8}$$

Consider now the case when the thermal resistance is mainly due to lattice waves: at low temperatures an electron, interacting with a phonon, does not change its "horizontal" position on the Fermi surface by a large amount—see Sect. 13. It thus appears that an electron in the n-region will, in the majority of cases, remain in the n-region after an interaction, so that $l_s = l_n$. Hence the change in the ideal thermal resistance should be

$$g = \frac{W_{in}}{W_{is}} = \frac{\varkappa_{is}}{\varkappa_{in}} = \left(\frac{T}{T_c}\right)^2. \tag{25.9}$$

One would hardly expect (25.8) and (25.9) to give the temperature dependence of the ratios f and g exactly, though one would expect these equations to give at least a qualitative description of their variation. It will be seen in Sect. 26 that this is so for (25.8), but not for (25.9).

The two-fluid model in the explicit form given above reproduces the thermodynamic properties only at temperatures above about $T_c/2$; at lower temperatures the specific heat decreases exponentially with decreasing temperature, and in view of (25.5) one would expect $\varkappa_{es}$ to behave similarly.

It now remains to be shown that (25.5), with the value C_n of (25.4) and appropriate values of l_s, does indeed describe the electronic thermal conductivity. Let us consider the derivation of (25.5) for normal metals. The thermal conductivity is defined as $-Q/\nabla T$, Q being the heat current, under the supplementary condition that the electric current $j = 0$. In superconducting metals $j = j_n + j_s$, where j_n and j_s are the contributions to j from the n-regions and s-regions respectively of the Fermi surface. If Q_n is the heat current due to the normal electrons under the condition $j_n = 0$, there now arises an additional heat flow Q_s due to the fact that $j = 0$, but $j_n \neq 0$. The existence of this heat flow due to circulation was suggested by Ginsburg[1] and the idea was later revived by Mendelssohn and Olsen [132].

The additional heat flow Q_s has been estimated by Klemens [124] as follows: Q_s consists of two parts, the change in Q_n because $j_n \neq 0$, and the heat transported by virtue of the fact that energy is required to raise an electron from an s-region to an n-region and is given up in the reverse process. It is the latter effect which has been disregarded in (25.6) by using C_n in (25.5), and the following justifies this procedure.

Since $j = 0$, $|j_n| = |j_s|$, and since in a superconductor in the steady state $F = 0$ or is at least small[2], $|j_n|$ is of order $L_{11}(KT/\zeta)K\nabla T$, where L_{11} is the transport coefficient (14.11) extended to include scattering by static imperfections, but modified to consider only contributions to the current from the n-region. The

[1] N. L. Ginsburg: J. Phys. USSR. 8, 148 (1944).
[2] Ginsburg assumed $F - V\zeta = 0$; this would change Q_s, but not its order of magnitude.

heat transported by j_n is of order $L_{11}K^2T(KT/\zeta)^2\nabla T$, which is smaller than Q_n by a factor of order $(KT/\zeta)^2$. The second contribution to Q_s is of order j_sKT_c, since the latent heat per electron for the transition from normal to super-conducting is of order KT_c. The second contribution to Q_s is thus of order $L_{11}(KT/\zeta)K^2T_c\nabla T$, and is greater than the first. Thus Q_s is smaller than Q_n by a factor of order KT_c/ζ, the circulation mechanism does not contribute appreciably to the total electronic thermal conductivity[1], and in (25.5) only the component C_n of the specific heat contributes to the heat transport.

Since the electrons in the s-region of the FERMI surface cannot be thermally excited into a continuum of states, it follows that lattice waves can be scattered only by the electrons of the n-region, and not by the s-regions of the FERMI surface. Thus

$$\frac{W_{Es}}{W_{En}} = x = \left(\frac{T}{T_c}\right)^4 \qquad (25.10)$$

and if $\varkappa_g$ is limited by the interaction with conduction electrons $\varkappa_{gs}/\varkappa_{gn} = 1/x$. However, the lattice resistance due to phonon-phonon interactions should be unchanged by the transition from the normal to the superconducting state.

It is thus seen that $\varkappa_{es} < \varkappa_{en}$, but $\varkappa_{gs} > \varkappa_{gn}$, so that, depending on the circumstances, $\varkappa_s$ may be smaller or larger than $\varkappa_n$.

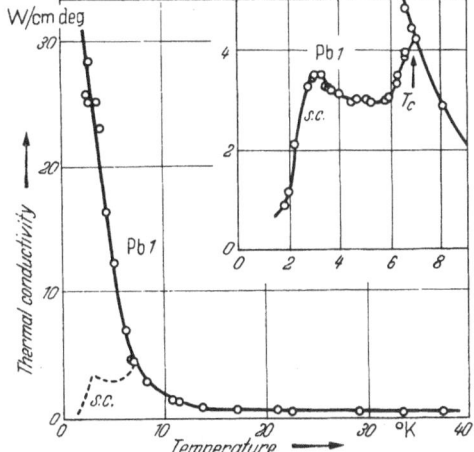

Fig. 14. The thermal conductivity of a lead single crystal, according to ROSENBERG [87]. The inset shows the curve for the superconducting state on a larger scale.

26. Thermal conductivity in the superconducting state. The thermal conductivity of superconductors has been measured both in the superconducting and in the normal states. The latter measurements are made in a magnetic field above the threshhold value and, if necessary, are reduced to zero field strength by extrapolation (e.g. HULM [92]). Phenomena associated with the transition from the normal to the superconducting states will be discussed in Sect. 27.

Observations on the thermal conductivity in the superconducting state can be classified into (a) cases where $\varkappa_g < \varkappa_e$ and $W_i > W_0$, (b) cases where $\varkappa_g < \varkappa_e$ and $W_0 > W_i$, (c) cases where $\varkappa_g$ is negligible in the normal state but appreciable in the superconducting state, and (d) cases where $\varkappa_g$ is appreciable both in the normal and in the superconducting states. There are, of course, cases inter-mediate between any of the above classes, and their interpretation is correspond-ingly uncertain.

$\alpha)$ *The ideal resistance in the superconducting state.* In order that $W_i > W_0$ below T_c, the specimen must be pure and the transition temperature must be reasonably high. This condition has so far been fulfilled only in the case of lead and mercury. Lead has been measured by BREMMER, DE HAAS and RADE-MAKERS [126], [129], [130], by MENDELSSOHN and PONTIUS [131] and by MENDELS-SOHN, OLSEN and ROSENBERG [132], [133], [86], [87]. Mercury has been measured by DE HAAS and BREMMER [127] and by HULM [92].

Fig. 14 shows the thermal conductivity of lead according to ROSENBERG [87], and is similar to the curve of DE HAAS and RADEMAKERS [129], except that

---

[1] However, this argument has been criticized by MENDELSSOHN [136].

the maximum of $\varkappa_s$ at about $3°$ K is somewhat higher in the latter case. Fig. 15 shows a plot of $\varkappa_s/\varkappa_n$ against T/T_c for a number of mercury specimens of HULM [92]; his curve of $\varkappa_s$ against T for pure mercury is similar in form to the curve of Fig. 14 for lead.

According to the considerations leading to (25.9), $\varkappa_s$ should be independent of T just below T_c, until scattering by imperfections becomes important, and in general $\varkappa_s$ should be of the form

$$\frac{1}{\varkappa_s} = W_s = W_{is} + W_{os} \tag{26.1}$$

where $W_{is} = W_{in}(T_c)$ is independent of temperature and W_{os} is related to W_{on} by (15.8), while $W_{on} \propto T^{-1}$. For the present purposes it is important to note thas W_{os} should increase monotonically with decreasing temperature, and so should W_s.

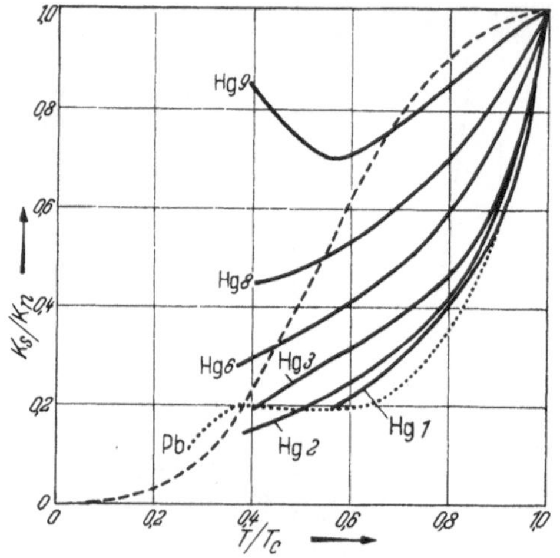

Fig. 15. Ratio $\varkappa_s/\varkappa_n$ plotted against reduced temperature T/T_c of a number of mercury specimens, numbered in order of increasing residual resistivity, according to HULM [92], as well as the lead specimen (dotted curve) of DE HAAS and RADEMAKERS [129]. The dashed curve is HEISENBERG's [123] f-function (25.8).

The observed behaviour of $\varkappa_s$ does not conform to these predictions. Immediately below T_c $\varkappa_s \propto T^3$, so that $g = (T/T_c)^5$, in contrast to (25.9). At lower temperatures $\varkappa_s$ does not decrease steadily with decreasing temperature, but increases again and then decreases at a temperature such that W_{on} is comparable to W_{in}.

In interpreting the observed behaviour of $\varkappa_s$, various points of view are possible. The one favoured by the present writer is the following: W_{is} is approximately described by (25.9), but for reasons which are not known, and presumably outside the scope of the two-fluid treatment, (25.9) does not describe g immediately below T_c: just below T_c the actual g-function is smaller than (25.9), but not by a very large amount [the observed values of $\varkappa_s$ are not less than $\frac{3}{4}\varkappa(T_c)$ at the minimum]; at lower temperatures $\varkappa_s$ increases again to its theoretical value $\varkappa(T_c)$, and decreases monotonically at still lower temperatures due to W_{os}.

Another point of view is that $g = (T/T_c)^5$ over a wide range of temperatures, but that the ideal and imperfection resistances do not combine additively, as in (26.1), so that W_s is not a monotonic function, even though W_{os} and W_{is} are. This point of view raises two difficulties: why should l_{is} be so much smaller than l_{in}, and why should (26.1) break down so violently at intermediate temperatures that W_s decreases with decreasing temperatures, even though both W_{is} and W_{os} increase?

Yet another possibility would be to ascribe the maximum of $\varkappa_s$ to an enhanced lattice conductivity in the superconducting state; however, this is ruled out for at least one specimen of lead [86] by the measurements of OLSEN and RENTON [134] on the same specimen, for they observed $\varkappa_{gs}$ at lowest temperatures to be limited by boundary scattering; this gives an upper limit to $\varkappa_{gs}$ at higher temperatures which is too low to account for the peak in the $\varkappa_s$ curve at about $3°$ K.

β) *The residual resistance in the superconducting state.* In the case of Sn, In, Ta, Tl, V and Cb the normal state electronic thermal conductivity, for those specimens which have been studied, is determined at T_c and below by imperfection scattering, and the same applies, of course, to the various alloys and to impure specimens of Pb and Hg. These specimens can thus be used to test eq. (25.8) for $\varkappa_{es}$, except where $\varkappa_{gs}$ is appreciable and complicates the picture. As long as lattice conduction is unimportant, the ratio $\varkappa_s/\varkappa_n$ should agree with (25.8); if $\varkappa_s/\varkappa_n$ is larger than expected, the difference is ascribed to $\varkappa_{gs}$, though it is usually not possible to prove that this is so.

RADEMAKERS [130] and HULM [92] have measured $\varkappa_s$ for tin down to about $T_c/3$. In Fig. 16 are plotted HULM's values of $\varkappa_s/\varkappa_n$; the specimens are numbered in increasing order of W_0. The curves for Sn 2 and Sn 3 are practically coincident, even though their values of W_{0n} differ by a factor of about 2. The high values of $\varkappa_s/\varkappa_n$ are ascribed to $\varkappa_{gs}$; since these specimens have appreciable lattice conduction even in the normal state, this seems a plausible interpretation.

Similar results were obtained by HULM [92] and by SLADEK [145] for indium. However, in all cases it is found that the observed f-function decreases more rapidly with temperature just below T_c than the function (25.8). It appears not impossible that this departure from the conclusions of the two-fluid theory just below T_c is related to the similar departure already noted for the g-function. On the other hand, SLADEK suggested the following form for $\varkappa_{es}/\varkappa_{en}$

$$f = \frac{3\,(T/T_c)^2}{2 + (T/T_c)^4} \qquad (26.2)$$

and based this on an assumption about l_s which seems no more artifical than (25.7).

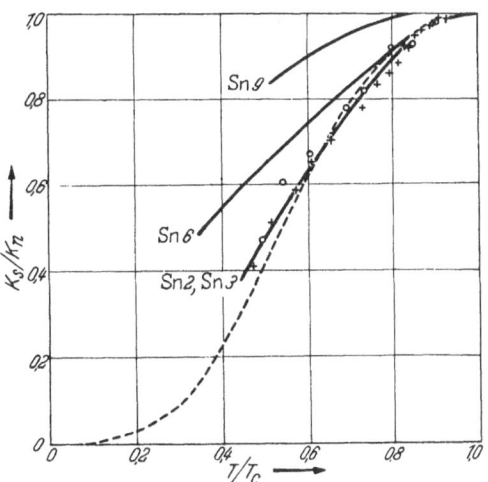

Fig. 16. Ratio $\varkappa_s/\varkappa_n$ plotted against reduced temperature T/T_c for tin specimens, according to HULM [92]; points 0, Sn 2, and +, Sn 3, show typical scatter; dashed line HEISENBERG's [123] f-function (25.8).

Substantially similar results for $\varkappa_{os}$ have been obtained by HULM [92] for tantalum, and by ROSENBERG [87] for tin, indium, thallium, tantalum and vanadium; the case of columbium was complicated by frozen-in magnetic flux.

There is not much data on $\varkappa_{es}/\varkappa_{en}$ below about $T_c/2$, partly because not many measurements extend to sufficiently low temperatures, and partly because of the increasing importance of $\varkappa_{gs}$ as the temperature is lowered.

Measurements of $\varkappa_s$ below about 1° K, that is in the demagnetisation region, have been made by HEER and DAUNT [140], OLSEN and RENTON [134], GOODMAN [138] and MENDELSSOHN and RENTON [135]. For tin and tantalum, HEER and DAUNT found indications that $\varkappa_s/\varkappa_n$ decreased faster with temperature than it should according to (25.6) and (25.8). This was confirmed by GOODMAN's extensive work on tin [138], who found an exponential decrease with temperature of $\varkappa_s$ for two pure specimens, changing at lower temperatures to a T^3 variation. The latter variation is ascribed to lattice conduction, while at higher temperatures $\varkappa_{es}$ is important; it is $\varkappa_{es}$ which varies exponentially[1]. OLSEN and RENTON [134] measured a specimen of lead, which was previously measured at helium temperatures [86], down to 0.4° K: below 1° K they found $\varkappa_s \propto T^3$ (presumably lattice

---

[1] Further measurements on tin have been reported by S. J. LAREDO: Proc. Roy. Soc. Lond., Ser. A **229**, 473 (1955).

conduction), but just at the upper limit of their temperature range they found indications of a faster variation which is confirmed if one joins up their measurements with those at higher temperatures. This is therefore a probable case of $\varkappa_{es}$ decreasing exponentially with temperature. Mendelssohn and Renton [135], who measured Sn, In, Tl, Al, Ta and Cb from 0.4 to 1° K, found an exponential variation of $\varkappa_s$ in the case of thallium. In the other cases the effect appears to be masked by lattice conduction.

The exponential variation of $\varkappa_{es}$ at low temperatures is presumably related to a similar variation in the specific heat. Indications of such a variation of the specific heat of superconductors well below the transition temperature have been observed in some cases[1]. Thus it seems likely that the Gorter-Casimir two-fluid theory breaks down at very low temperatures.

γ) *The lattice component of thermal conductivity.* Even if $\varkappa_g$ is too small to be observed in the normal state, in the superconducting state W_E decreases very rapidly with temperature, so that at very low temperatures $\varkappa_{gs}$ is only limited by the scattering of phonons by static imperfections or boundaries. Thus experiments below 1° K usually give clear indications of lattice conduction.

A T^3-variation of $\varkappa_s$ at lowest temperatures has been found by Olsen and Renton [134] for lead, by Mendelssohn and Renton [135] for tin and indium, by Goodman [138] for tin and by Laredo and Shoenberg [139] for tin. In general the Oxford group [134], [135], [136] deduced values for W_B which were five to ten times higher than values expected from the external dimensions of their specimens (9.8)[2], while the Cambridge group [138], [139] obtained values of W_B in rough agreement with the external dimensions. It is possible that the Oxford specimens had considerably more grain boundaries; another possibility is a reduction in $\varkappa_g$ due to frozen-in magnetic flux—see Sect. 27.

In the case of Goodman's impure tin specimens $\varkappa$ varies more slowly than T^3 except at the lower end of his temperature range; at higher temperatures $\varkappa_g \propto T^2$. Goodman presumes the dominant scattering mechanism in the T^2 region to be scattering by the free electrons. This is unlikely since from (25.10) W_{Es} should vary as T^2, not as T^{-2}. However, it has been shown in Sect. 24 that dislocations can play an important part in determining $\varkappa_g$, even in normal metals where W_E is not quenched. It thus seems quite possible that in the T^2 region $\varkappa_{gs}$ is limited mainly by dislocations. A similar interpretation would apply for a tantalum specimen of Renton [137], who also found $\varkappa_{gs} \propto T^2$ below 1° K.

The observations at very low temperatures do not give a value of W_{Es} (except an upper limit which is probably very much larger than W_{Es}), because of the importance of phonon-phonon interactions, which are not influenced by the superconducting behaviour. It is only possible to observe W_{Es} over a limited temperature range below T_c, and it has been determined in this way by Hulm [92] for Sn 96—Hg 4, by Olsen [133] for lead-bismuth alloys previously measured by Mendelssohn and Olsen [132] and by Sladek [145] for indium-thallium alloys.

These observations yield a conflicting picture of the behaviour of the ratio $h = W_{En}/W_{Es}$: thus Hulm suggested $h \approx (T_c/T)^2$, Olsen $h \approx (T_c/T)^6$ and Sladek, whose measurements seem best suited for the evaluation of h, does not obtain a simple power law, nor the same curve of h against T/T_c for all his specimens, but gets a series of curves for h, all in the vicinity of $h = (T_c/T)^4$, but too high just below T_c and tending to become too low at lower temperatures.

---

[1] See the companion articles by Keesom and Pearlman in this volume and by Serin in vol. XV.

[2] Further results below 1° K for a number of superconducting elements have been reported by K. Mendelssohn and C. A. Renton: Proc. Roy. Soc. Lond., Ser. A **230**, 157 (1955).

There are, however, uncertainties in the interpretation: thus W_g could easily have components other than W_E (this is certainly so at lowest temperatures and may be so even if $W_g \propto T^{-2}$), which would tend to lower h, and the separation of $\varkappa_s$ into $\varkappa_{gs}$ and $\varkappa_{es}$ involves the assumption that the ratio $f = \varkappa_{os}/\varkappa_{on}$ is independent of alloy composition. Nevertheless there are probably real discrepancies from $h = (T_c/T)^4$, particularly just below T_c, just as there are discrepancies just below T_c in the case of $\varkappa_{is}$.

It should be mentioned that OLSEN's [133] interpretation of the thermal conductivity of lead-bismuth alloys is not without difficulties. Fig. 17 shows the thermal conductivity of these alloys as a function of temperature in both the normal and the superconducting states. It is easily seen that if the increase of $\varkappa_s$ for the alloys 0.2% Bi and 0.5% Bi over the values of $\varkappa_s$ for alloys of low Bi-content is to be explained in terms of enhanced lattice conduction, then $\varkappa_{gs}$ for these two alloys is higher than $\varkappa_{gs}$ for the alloys 0.1% Bi and 0.02% Bi, and possibly even higher than $\varkappa_{gs}$ for pure lead. It is, of course, possible to explain this by assuming some imperfections to be present in the more dilute alloys and not in the more concentrated alloys, though this is at variance with what is usually experienced. A re-examination of this alloy series, paying attention to the possible causes of this anomaly, seems therefore very desirable.

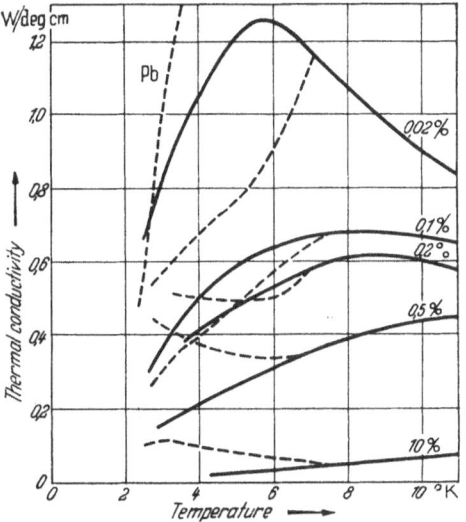

Fig. 17. Thermal conductivity of some lead-bismuth alloys (bismuth content in per cent as marked on the curves) in the normal state (full curve) and the superconducting state (broken curve), according to MENDELSSOHN and OLSEN [132].

27. Thermal conductivity in the intermediate state. The transition from the superconducting to the normal state of a superconductor by the application of a magnetic field is sharp only in the case of pure elements and provided the specimen is in the shape of a long cylinder and the field is longitudinal. In other cases the transition is gradual; an increase in the magnetic field causes a gradual increase of flux in the specimen, until all the material is normal. When the field is then removed, the material does not return to the original superconducting state, but some magnetic flux remains frozen in.

The intermediate state of the material is not a homogeneous state, but the material consists of a mixture of normal and superconducting regions, the former of high flux density (above the critical field) and the latter of zero flux. Since the lines of flux are continuous, the structure of the intermediate state is dominantly one of filaments or layers, alternately normal and superconducting, lying in the direction of the field[1].

A number of measurements have been made of the thermal conductivity of superconductors in the intermediate state, with the specimens in the shape of long cylinders. In the case of longitudinal fields there will usually not be a marked mixing of the two phases expect in the case of alloys, when the normal state

---

[1] A description of intermediate state phenomena is given, for example, by D. SHOENBERG, Superconductivity, 2nd. ed. Cambridge: University Press 1952.

inclusions will be mainly filaments running the length of the specimen. In the case of transverse fields, however, the specimen will readily break up into a mixture of two phases and the normal inclusions would then be dominantly layers perpendicular to the cylinder axis and thus to the direction of heat flow; the thickness of the individual regions may be of the order of 10^{-2} cm.

In the case of longitudinal fields, with normal and superconducting filaments along the direction of heat flow, one would expect the overall thermal conductivity to be given by the average

$$\varkappa = x_n \varkappa_n + (1 - x_n) \varkappa_s \tag{27.1}$$

where x_n is the fraction of normal material and can be deduced from flux measurements. Similarly, in the case of transverse fields, the resistances are averaged, so that

$$W = x_n W_n + (1 - x_n) W_s. \tag{27.2}$$

In consequence of either (27.1) or (27.2), the conductivity in the intermediate state (either with subcritical field strength or with frozen-in flux) should be

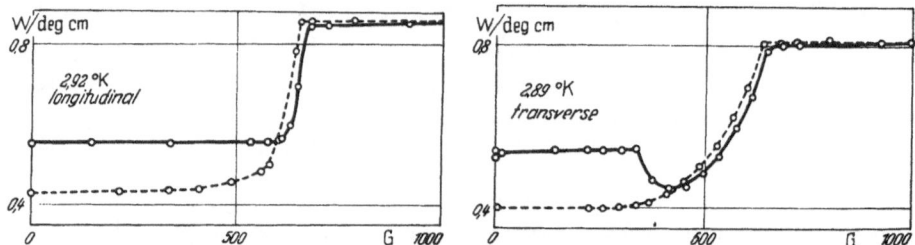

Fig. 18. Variation of the thermal conductivity of Pb-Bi 0.02% with longitudinal and transverse magnetic fields at about 2.9° K, according to Mendelssohn and Olsen [132]. Full line: field increasing; broken line: field decreasing.

intermediate between the normal conductivity $\varkappa_n$ and $\varkappa_s$ as measured in the virgin superconducting state. Within the bounds of this restriction hysteresis effects are possible.

Such behaviour was indeed observed in many early measurements, such as those of Mendelssohn and Pontius [131] and of de Haas and Rademakers [129] in transverse fields, and those of Hulm [92] in longitudinal fields. Later measurements, however, showed variations of $\varkappa$ with magnetic field strength H which could not be reconciled with either (27.1) or (27.2). Mendelssohn and Olsen [132] found, in the case of some lead-bismuth alloys and of columbium, that the thermal resistance passed through a maximum value, on applying a field, which was higher than the resistance in either the normal or the superconducting state; a typical case is illustrated in Fig. 18. Also, on removal of the field, $\varkappa$ did not return to the original value $\varkappa_s$ (this itself is not remarkable), but to a value again lower than either $\varkappa_s$ or $\varkappa_n$. These anomalies seem more likely to occur the lower the temperature and the higher the impurity content, but this is not a general rule, as will be seen below. The explanation suggested by these authors is in terms of a heat flow mechanism which occurs in the superconducting state and is additional to electronic conduction; in the intermediate state this mechanism would be inhibited. They suggested a two- fluid circulation for this additional mechanism; it seems more likely that it is lattice conduction, as will be discussed below.

Another explanation, remarkable for its ingenuity, though not now taken seriously, was suggested by Cornish and Olsen [125]: in place of averaging the overall thermal resistances, as in (27.2), W_g and W_e should be separately averaged; in this way it is possible to obtain an intermediate state resistance higher than either W_s or W_n. This implies that the electrons and the lattice are loosely coupled, so that they can have different temperatures with slow interchange of energy between them. This is a very unlikely state of affairs; furthermore, it is not possible to explain all intermediate state anomalies in this way.

The anomalies found by MENDELSSOHN and OLSEN were mainly, but not exclusively, in transverse fields. Further cases of anomalies in transverse fields were found later, in particular by WEBBER and SPOHR [141] and by OLSEN and RENTON [134] for pure lead, by DETWILER and FAIRBANK [143] for pure tin and indium (but not for tin containing 0.134% Bi), and by WEBBER and SPOHR [142] and by HULM [144] for mercury.

It was suggested by a number of authors that the anomalously high thermal resistance in the intermediate state could be due to the scattering of either electrons or phonons by the internal boundaries between normal and superconducting regions. In the case of the observations on pure metals at liquid helium temperatures, where apparently the electronic conductivity is important in both the superconducting and the normal state, the only possible explanation seems to be in terms of the scattering of electrons by the boundaries.

This additional resistive process was discussed by HULM [144]. In the normal region $\varkappa_e$ is higher than in the superconducting region. In the latter only a fraction x of the electrons takes part in heat transport: these can pass freely from the superconducting to the normal metal; but a fraction $1 - x$ electrons crossing the boundary into the normal metal are converted into normal electrons at the boundary. They take up the temperature of the boundary. Consider now a temperature gradient perpendicular to the boundary, and a point in the normal material a short distance $r < l_n$ from the boundary, where l_n is the electron mean free path. A fraction x of the electrons have a temperature $T - l_n \operatorname{grad} T$, but a fraction $1 - x$ have a temperature $T - r \operatorname{grad} T$. Thus in a region of depth of order l_n the overall heat flow is reduced, and this is equivalent to having a layer of thickness l_n for which the thermal resistance is of order $W_{en}/x = W_{en}(T_c/T)^4$. For the specimen as a whole this yields an additional thermal resistance (if $\varkappa_{gn}$ is negligible)

$$\Delta W \approx \frac{l_n}{Z} W_{en} \left(\frac{T_c}{T}\right)^4 = \frac{l_{nc} W_{enc}}{Z} \left(\frac{T_c}{T}\right)^5 \qquad (27.3)$$

where l_{nc} and W_{enc} refer to the values at temperature T_c, and Z is the average thickness of the normal layers.

HULM found that he could describe his own observations [144], as well as those by DETWILER and FAIRBANK [143], in terms of the additional resistance (27.3); it appears that WEBBER and SPOHR's [142] result can be described similarly. It should be noted that this resistance mechanism would be important only for transverse fields.

There are, however, also cases of anomalous resistances which cannot be interpreted in this way, mainly because electronic conduction is unimportant in the superconducting state. These are the anomalies observed below 1° K by OLSEN and RENTON [134] on a lead single crystal in transverse fields, anomalies in the thermal conductivity of tantalum and columbium due to frozen-in flux and anomalous resistance maxima in tin and thallium observed by MENDELSSOHN and RENTON [135], a similar effect found by LAREDO and PIPPARD [139], [146] in tin, as well as the observations by RENTON [137] on a number of materials.

Another set of observations of anomalous resistance increases which cannot be explained by means of electron scattering are the observations on indium-thallium alloys in longitudinal fields by SLADEK [145], as well as some earlier observations of lead-bismuth alloys in longitudinal fields by Mendelssohn and Olsen [132].

Sladek has interpreted the additional resistance as an increase in W_{gs} which arises because the phonon mean free path is much shorter in the normal regions than in the superconducting region. The boundaries thus emit phonons isotropically into the superconducting region, so that, for purposes of lattice conduction in the superconducting regions, the boundaries play the same role as the external boundaries in the Casimir theory—see Sect. 9. This mechanism would be about equally effective whichever the direction of the boundaries relative to the flow, that is the additional resistance should be independent of whether the field is longitudinal or transverse, which is in accord with Sladek's observations.

The same explanation was given independently by Renton [137] and by Laredo and Pippard [146].

With two mechanisms of additional resistance, which can of course in some cases be of comparable importance, acting in the intermediate state, the effects can be quite complicated and their interpretation correspondingly uncertain. On the other hand it is possible, by selecting conditions which allow an interpretation, to throw some light on the fine structure of the intermediate state: see for example [146].

Acknowledgments. The author wishes to thank Dr. R. Berman, Mr. A. F. A. Harper, Dr. P. G. Harper, Mr. W. R. G. Kemp, Dr. R. E. B. Makinson and Dr. G. K. White for their advice and assistance in the preparation of this review; he is also indebted to the many authors who informed him of their work before publication.

References.

Review Papers and Books.

[1] Sommerfeld, A., u. H. Bethe: Electron Theory of Metals. In Handbuch der Physik, Vol. 24/2. 1933.
[2] Born, M., u. M. Goeppert-Mayer: Dynamical Theory of Crystal Lattices. In Handbuch der Physik, Vol. 24/2. 1933.
[3] Mott, N. F., and H. Jones: Theory of the Properties of Metals and Alloys. Oxford: University Press 1936.
[4] Wilson, A. H.: Theory of Metals, 2nd ed. Cambridge: University Press 1953.
[5] Berman, R.: Adv. Physics 2, 103 (1953); thermal conductivity of dielectric solids at low temperatures.
[6] Olsen, J. L., and H. M. Rosenberg: Adv. Physics 2, 28 (1953); thermal conductivity of metals at low temperatures.
[7] Powell, R. L., and W. A. Blanpied: Thermal Conductivity of Metals and Alloys at Low Temperatures—A Review of the Literature. United States National Bureau of Standards, Circular No. 556 (1954).

Thermal Conductivity of Non-metals, Theory.

[8] Debye, P.: Equation of State and the Quantum Hypothesis with an Appendix on Thermal Conduction, in: Vorträge über die kinetische Theorie der Materie und Elektrizität. Berlin: Teubner 1914.
[9] Peierls, R.: Ann. Phys., Lpz. 3, 1055 (1929); kinetic theory of thermal conduction in dielectric crystals.
[10] Peierls, R.: Ann. Inst. Poincaré 5, 177 (1935); lattice periodicity and properties of solids
[11] Casimir, H. B. G.: Physica, Haag 5, 495 (1938); thermal resistance at low temperatures due to external boundaries.
[12] Landau, L., and G. Rumer: Phys. Z. Sowjet. 11, 18 (1937); scattering of transverse waves due to cubic anharmonicities.
[13] Pomeranchuk, I.: J. Phys. USSR. 4, 259 (1941); insufficiency of three-phonon interactions, acting alone, to maintain equilibrium of low-frequency longitudinal waves at high temperatures.

[14] POMERANCHUK, I.: J. Phys. USSR. 6, 237 (1942); thermal conduction at low temperatures, role of longitudinal waves, effects of boundary and imperfections.

[15] POMERANCHUK, I.: J. Phys. USSR. 7, 197 (1943); effect of four-phonon interactions.

[16] FROEHLICH, H., and W. HEITLER: Proc. Roy. Soc. Lond., Ser. A 155, 640 (1936); thermal conductivity of paramagnetic crystals.

[17] POMERANCHUK, I.: J. Phys. USSR. 4, 357 (1941); thermal conductivity of paramagnetic solids at low temperatures.

[18] AKHIESER, A, and I. POMERANCHUK: J. Phys. USSR. 8, 216 (1944); thermal conductivity of salts used for magnetic cooling.

[19] KITTEL, C.: Phys. Rev. 75, 972 (1949); thermal conductivity of glasses.

[20] KLEMENS, P. G.: Proc. Roy. Soc. Lond., Ser. A 208, 108 (1951); thermal conductivity of dielectric solids at low temperatures.

[21] KLEMENS, P. G.: Proc. Phys. Soc. Lond. A 68, 1113 (1955); scattering of low-frequency phonons by static imperfections.

[22] HERPIN, A.: Ann. Physique 7, 91 (1952); three-phonon interactions and thermal conduction.

[23] HERRING, C.: Phys. Rev. 95, 954 (1954); role of low-energy phonons in thermal conduction.

[24] LEIBFRIED, G., u. E. SCHLOEMANN: Nachr. Akad. Wiss. Göttingen IIa, No. 4, 71 (1954); thermal conductivity of dielectric solids by a variational technique.

Thermal Conductivity of Non-metals, Experimental.

[25] EUCKEN, A.: Ann. Phys., Lpz. 34, 185 (1911). — Verh. dtsch. phys. Ges. 13, 829 (1911). — Z. Physik 12, 1005 (1911); thermal conductivity of various crystals below room temperature.

[26] EUCKEN, A., u. G. KUHN: Z. phys. Chem. 134, 193 (1928); thermal conductivity of mixed crystals.

[27] EUCKEN, A., u. E. SCHROEDER: Ann. Phys., Lpz. 36, 609 (1939); thermal conductivity of solidified benzene, hydrogen bromide and nitrous oxide.

[28] HAAS, W. J. DE, and T. BIERMASZ: Physica, Haag 2, 673 (1935); low temperature thermal conductivity of quartz.

[29] HAAS, W. J. DE, and T. BIERMASZ: Physica, Haag 4, 752 (1937); low temperature thermal conductivity of quartz, potassium chloride and potassium bromide.

[30] HAAS, W. J. DE, and T. BIERMASZ: Physica, Haag 5, 47, 320, 619 (1938); boundary resistance observed for quartz, diamond, potassium chloride.

[31] KURTI, N., B. V. ROLLIN and F. SIMON: Physica, Haag 3, 266 (1936); thermal conductivity of single crystals of potassium chrome alum and iron ammonium alum below 0.2° K.

[32] DIJK, H. VAN, and W. H. KEESOM: Physica, Haag 7, 970 (1940); thermal conductivity or iron ammonium alum (compressed).

[33] GERRITSEN, A. N., and P. VAN DER STAR: Physica, Haag 9, 503 (1942); thermal conductivity of solid methane.

[34] BIJL, D.: Physica, Haag 14, 684 (1949); thermal conductivity of potassium chrome alum and some glasses.

[35] HUDSON, R. P.: Thesis, Oxford (1949); thermal conductivity of iron ammonium alum (compressed) below 0.2° K.

[36] WILKINSON, K. R., and J. WILKS: J. Sci. Instrum. 26, 19 (1949); thermal conductivity of glass and some technical alloys.

[37] GARRETT, C. B. G.: Phil. Mag. 41, 621 (1950); thermal conductivity of potassium chrome alum below 0.3° K.

[38] BERMAN, R.: Phys. Rev. 76, 315 (1949); thermal conductivity of quartz glass at low temperature.

[39] BERMAN, R.: Proc. Roy. Soc. Lond., Ser. A 208, 90 (1951); thermal conductivity of glass, quartz, neutron-irradiated quartz and synthetic sapphire.

[40] BERMAN, R., F. E. SIMON, P. G. KLEMENS and T. M. FRY: Nature, Lond. 166, 277 (1950); thermal conductivity of neutron-irradiated quartz.

[41] BERMAN, R.: Proc. Phys. Soc. Lond. A 65, 1029 (1952); thermal conductivity of polycrystalline alumina, beryllia and graphite.

[42] WILKINSON, K. R., and J. WILKS: Proc. Phys. Soc. Lond., A 64, 89 (1951); thermal conductivity of solid helium.

[43] BERMAN, R., F. E. SIMON and J. WILKS: Nature, Lond. 168, 277 (1951); Umklapp-resistance in various dielectric solids.

[44] WEBB, F. J., K. R. WILKINSON and J. WILKS: Proc. Roy. Soc. Lond., Ser. A **214**, 546 (1953); thermal conductivity of solid helium.

[45] WEBB, F. J., and J. WILKS: Phil. Mag. **44**, 664 (1953); thermal conductivity of solid helium.

[46] BERMAN, R., F. E. SIMON and J. M. ZIMAN: Proc. Roy. Soc. Lond., Ser. A **220**, 171 (1953); thermal conductivity and size effect of diamond.

[47] DEVYATKOVA, D., and L. S. STILBANS: Zh. Tekh. Fiz. **22**, 968 (1952); influence of F-centres on thermal conductivity of potassium chloride.

[48] SMITH, A. W.: Phys. Rev. **95**, 1095 (1954); thermal conductivity of graphite.

[49] ESTERMANN, T., and J. E. ZIMMERMAN: Technical Report 6, O.N.R. Carnegie Institute of Technology, USA. (1951).

[50] ROSENBERG, H. M.: Proc. Phys. Soc. Lond. A **67**, 837 (1954); thermal conductivity of silicon and germanium.

[51] BERMAN, R.: To be published; thermal conductivity of some alkali halides.

[52] BERMAN, R., E. L. FOSTER and J. M. ZIMAN: Proc. Roy. Soc. Lond.. Ser. A **231**, 130 (1955); thermal conductivity and size effect of artificial sapphires.

[53] SIMSON, C. v.: Naturwiss. **38**, 559 (1951); thermal conductivity of ammonium chloride.

[54] UHLIR, A.: J. Chem. Phys. **20**, 463 (1953); thermal conductivity of fluid argon and nitrogen.

[55] PROSAD, S.: Brit. J. Appl. Phys. **3**, 58 (1952); thermal conductivity of liquid oxygen.

[56] GRENIER, C.: Phys. Rev. **83**, 598 (1951); thermal conductivity of liquid helium I.

[57] BOWERS, R.: Proc. Phys. Soc. Lond., A **65**, 511 (1952); thermal conductivity of liquid helium I.

[58] FAIRBANK, H. A., and J. WILKS: Phys. Rev. **95**, 277 (1954); Proc. Roy. Soc. Lond., Ser. A **231**, 545 (1955); thermal conductivity of liquid helium below 1° K.

Thermal Conductivity of Metals, Theory.

[59] BLOCH, F.: Z. Physik **52**, 555 (1928); interaction between electrons and lattice, electronic conduction in metals.

[60] WILSON, A. H.: Proc. Camb. Phil. Soc. **33**, 371 (1937); first order approximation to solution of transport equation for electrical and thermal conduction.

[61] MAKINSON, R. E. B.: Proc. Camb. Phil. Soc. **34**, 474 (1938); numerical evaluation of electronic thermal conductivity, treatment of lattice component, influence of anisotropy of phonon distribution on the electronic component.

[62] KOHLER, M.: Z. Physik **124**, 772 (1948); **125**, 679 (1949); solution of transport equation by variational method and application to electronic conduction properties.

[63] KROLL, W.: Sci. Papers Inst. Phys. Chem. Res. Tokyo **34**, 194 (1938); solution of transport equation by successive approximations.

[64] SONDHEIMER, E. H.: Proc. Roy. Soc. Lond., Ser. A **203**, 75 (1950); solution of transport equation to third order by the variational method.

[65] SONDHEIMER, E. H.: Proc. Phys. Soc. Lond., Ser. A **65**, 561, 562 (1952); modification of theory for small number of conduction electrons.

[66] TODA, M.: J. Phys. Soc. Japan **8**, 339 (1953); diffusion on the Fermi surface and conductivity.

[67] BLATT, F. J.: J. Phys. Soc. Japan **9**, 444 (1954); diffusion on the Fermi surface, evaluation of thermal conductivity by numerical solution.

[68] MAKINSON, R. E. B.: Proc. Phys. Soc. Lond., Ser. A **67**, 290 (1954); comment on stationarity in SONDHEIMER's variational treatment.

[69] KLEMENS, P. G.: Proc. Phys. Soc. Lond., Ser. A **67**, 194 (1954); Austral. J. Phys. **7**, 64 (1954); numerical solution of transport equation for pure metals at low temperatures.

[70] KLEMENS, P. G.: Proc. Phys. Soc. Lond., Ser. A **67**, 194 (1954); Austral. J. Phys. **7**, 70 (1954); modification of electronic band structure of monovalent metals to account for low temperature conduction properties.

[71] ZIMAN, J. M.: Proc. Roy. Soc. Lond., Ser. A **226**, 436 (1954); modification of electron-phonon interaction parameter to account for conduction properties.

[72] KLEMENS, P. G.: Austral. J. Phys. **7**, 57 (1954); lattice component of thermal conductivity and electron-phonon interaction.

[73] REZANOV, A. J., and V. J. CHEREPANOV: Proc. Acad. Sci. USSR. **93**, 641 (1953); low temperature thermal conductivity of ferromagnetic metals.

[74] SONDHEIMER, E. H., and A. H. WILSON: Proc. Roy. Soc. Lond., Ser. A **190**, 435 (1947); magneto-resistance effects in metals assuming relaxation times.

[75] KOHLER, M.: Naturwiss. **36**, 186 (1949); similarity rule for thermal resistance in magnetic fields.

[76] KOHLER, M.: Ann. Phys., Lpz. **5**, 181 (1949); thermal conduction in strong magnetic fields, extrapolation for the lattice component.

[77] KOHLER, M.: Ann. Phys., Lpz. **6**, 18 (1949); magneto-resistance effects, first order variational treatment.

[78] AKHIESER, A, and I. POMERANCHUK: J. Phys. USSR. **9**, 93 (1945); thermal conductivity of bismuth.

Thermal Conductivity of Metals, Experimental.

[79] MEISSNER, W.: Ann. Phys., Lpz. **47**, 1001 (1915). — Z. Physik **2**, 373 (1915); thermal conductivity of Li, Au and Pt down to liquid hydrogen temperatures.

[80] BIDWELL, C. C.: Phys. Rev. **27**, 819 (1926); **28**, 584 (1926); thermal conductivity of lithium and sodium.

[81] GRÜNEISEN, E., u. E. GOENS: Z. Physik **44**, 615 (1927); thermal conductivity down to liquid hydrogen temperatures of a wide range of metals of high purity.

[82] GRÜNEISEN, E., u. H. REDDEMANN: Ann. Phys., Lpz. **20**, 843 (1934); thermal conductivity of alloys, role of lattice conduction.

[83] BERMAN, R., and D. K. C. MacDONALD: Proc. Roy. Soc. Lond., Ser. A **209**, 368 (1951); low temperature thermal conductivity of sodium.

[84] BERMAN, R., and D. K. C. MacDONALD: Proc. Roy. Soc. Lond., Ser. A **211**, 122 (1952); low temperature thermal conductivity of copper.

[85] MENDELSSOHN, K., and H. M. ROSENBERG: Proc. Phys. Soc. Lond., Ser. A **65**, 385 (1952); low temperature thermal conductivity of Cu, Ag, Au, Mg, Zn, Cd, Al and In.

[86] MENDELSSOHN, K., and H. M. ROSENBERG: Proc. Phys. Soc. Lond., Ser. A **65**, 388 (1952); low temperature thermal conductivity of Ti, Mn, Fe, Ni, Zr, Cb, Mo, Rh, Pd, Ta, W, Ir, Pt, U and Pb.

[87] ROSENBERG, H. M.: Phil. Trans. Roy. Soc. Lond., Ser. A **247**, 441 (1955); low temperature thermal conductivity of Cu, Ag, Au, Be, Mg, Zn, Cd, Al, Ga, In ,Tl, Ti, Zr, Sn, Pb, V, Cb, Ta, Sb, Mo, W, Mn, Fe, Co, Ni, Rh, Pd, Ir, Pt, Ce, La and U.

[88] WHITE, G. K.: Proc. Phys. Soc. Lond. Ser. A **66**, 559 (1953); low temperature thermal conductivity of gold.

[89] WHITE, G. K.: Proc. Phys. Soc. Lond., Ser. A **66**, 844 (1953); low temperature thermal conductivity of silver.

[90] WHITE, G. K.: Austral. J. Phys. **6**, 397 (1953); low temperature thermal conductivity of copper.

[91] MacDONALD, D. K. C., G. K. WHITE and S. B. WOODS: Proc. Roy. Soc. Lond., to be published; thermal conductivity of K, Rb and Cs at low temperatures.

[92] HULM, J. K.: Proc. Roy. Soc. Lond., Ser. A **204**, 98 (1950); thermal conductivity of superconductors (Sn, Hg, In and Ta and dilute alloys) at liquid helium temperatures in normal and superconducting state.

[93] HULM, J. K.: Proc. Phys. Soc. Lond., Ser. A **65**, 227 (1952); discussion of discrepancy between theory and experiments of ideal thermal resistivity.

[94] NOBEL, J. DE: Physica, Haag **17**, 551 (1951); low temperature thermal conductivity of Al, Fe, Ni, various steels and monel.

[95] ANDREWS, F. A., R. T. WEBBER and D. A. SPOHR: Phys. Rev. **84**, 994 (1951); low temperature thermal conductivity of aluminium.

[96] KEMP, W. R. G., A. K. SREEDHAR and G. K. WHITE: Proc. Phys. Soc. Lond., Ser. A **66**, 1077 (1953); low temperature thermal conductivity of magnesium.

[97] ROSENBERG, H. M.: Phil. Mag. **45**, 73 (1954); thermal and electrical conductivity of magnesium.

[98] SPOHR, D. A., and R. T. WEBBER: Phys. Rev. **95**, 602 (1954) and private communication; thermal and electrical conductivity of magnesium.

[99] ROSENBERG, H. M.: Phil. Mag. **45**, 767 (1954); anisotropy of thermal resistivity of gallium single crystals.

[100] KEMP, W. R. G., P. G. KLEMENS, A. K. SREEDHAR and G. K. WHITE: Phil. Mag. **46**, 811 (1955); low temperature thermal conductivity of palladium.

[101] NICOL, J., and T. P. TSENG: Phys. Rev. **92**, 1062 (1953); thermal conductivity of copper between 0.25 and 4.2° K.

[102] REDDEMANN, H.: Ann. Phys., Lpz. **20**, 441 (1934); thermal conductivity of a Bi single crystal in magnetic fields.

[103] GRÜNEISEN, E., and H. ADENSTEDT: Ann. Phys., Lpz. **29**, 597 (1937); **31**, 714 (1938); effect of transverse magnetic fields on thermal conductivity of some pure metals at liquid hydrogen temperatures.

[104] GRÜNEISEN, E., and H. D. ERFLING: Ann. Phys., Lpz. **38**, 399 (1940); electrical and thermal resistance of Be single crystals in transverse magnetic fields.

[105] RAUSCH, K.: Ann. Phys., Lpz. **1**, 190 (1947); thermal conductivity of Sb single crystal in magnetic fields.

[106] GRÜNEISEN, E., K. RAUSCH and K. WEISS: Ann. Phys., Lpz. **7**, 1 (1950); electrical and thermal resistance of Bi single crystals in magnetic fields.

[107] HAAS, W. J. DE, A. N. GERRITSEN and W. H. CAPEL: Physica, Haag **3**, 1143 (1936); thermal resistance of Bi single crystals in magnetic fields.

[108] HAAS, W. J. DE, and J. DE NOBEL: Physica, Haag **5**, 449 (1938); electrical and thermal resistance of W single crystal in magnetic fields.

[109] NOBEL, J. DE: Physica, Haag **15**, 532 (1949); thermal and electrical resistance of W single crystal in high magnetic fields.

[110] SHALYT, S.: J. Phys. USSR. **8**, 315 (1944); thermal conductivity of bismuth and effect of magnetic fields down to liquid helium temperatures.

[111] MENDELSSOHN, K., and H. M. ROSENBERG: Proc. Phys. Soc. Lond., Ser. A **64**, 1057 (1951); thermal conductivity of cadmium in magnetic fields at liquid helium temperatures.

[112] MENDELSSOHN, K., and H. M. ROSENBERG: Proc. Roy. Soc. Lond., Ser. A **218**, 190 (1953); thermal conductivity of various metals in magnetic fields at liquid helium temperatures, in particular Zn, Cd, Sn, Pb and Ga.

[113] KARWEIL, J., u. K. SCHÄFER: Ann. Phys., Lpz. **36**, 567 (1939); thermal conductivity of various alloys between 3 and 30° K.

[114] ALLEN, J. F., and E. MENDOZA: Proc. Camb. Phil. Soc. **44**, 280 (1948); thermal conductivity of copper and German silver at liquid helium temperatures.

[115] HULM, J. K.: Proc. Phys. Soc. Lond., Ser. A **64**, 207 (1951); thermal conductivity and lattice component of alloy Cu 90 — Ni 10.

[116] BERMAN, R.: Phil. Mag. **42**, 642 (1951); low temperature thermal conductivity and lattice component of German silver, stainless steel and Cu 60 — Ni 40.

[117] ESTERMANN, I., and J. E. ZIMMERMAN: J. Appl. Phys. **23**, 578 (1952); low temperature thermal conductivity and lattice component of monel, inconel, stainless steel and Cu 90 — Ni 10 and effect of cold work.

[118] KEMP, W. R. G., P. G. KLEMENS, A. K. SREEDHAR and G. K. WHITE: Proc. Phys. Soc. Lond., Ser. A **67**, 728 (1954); lattice component of thermal conductivity of Ag-Pd alloys.

[119] KEMP, W. R. G., P. G. KLEMENS, A. K. SREEDHAR and G. K. WHITE: Proc. Roy. Soc. Lond., Ser. A **233**, 41 (1956); thermal and electrical conductivities of Ag-Pd and Ag-Cd alloys.

[120] WHITE, G. K., and S. B. WOODS: Phil. Mag. **45**, 1343 (1954); lattice component of thermal conductivity of a dilute copper alloy.

[121] WHITE, G. K., and S. B. WOODS: Canad. J. Phys. **33**, 58 (1955); low temperature thermal conductivity and lattice component of copper alloys, Bi, Ge and Be.

Thermal Conductivity of Superconductors.

[122] GORTER, C. J., u. H. B. G. CASIMIR: Phys. Z. **35**, 963 (1934). — Z. techn. Phys. **15**, 539 (1934); two-fluid model of superconductivity.

[123] HEISENBERG, W.: Z. Naturforsch. **3a**, 65 (1948); two-fluid model, including application to thermal conductivity of superconductors.

[124] KLEMENS, P. G.: Proc. Phys. Soc. Lond., Ser. A **66**, 576 (1953); electronic thermal conduction in superconductors and two-fluid circulation.

[125] CORNISH, F. H. J., and J. L. OLSEN: Helv. Phys. Acta **26**, 369 (1953); calculation of thermal conductivity of intermediate state, assuming loose coupling between electron and lattice temperatures.

[126] BREMMER, H., and W. J. DE HAAS: Physica, Haag **3**, 672 (1936); thermal conductivity of Pb and Cu.

[127] HAAS, W. J. DE, and H. BREMMER: Physica, Haag **3**, 687 (1936); thermal conductivity of mercury.

[128] BREMMER, H., and W. J. DE HAAS: Physica, Haag **3**, 692 (1936); thermal conductivity of some superconducting alloys.

[129] HAAS, W. J. DE, and A. RADEMAKERS: Physica, Haag **7**, 992 (1940); thermal conductivity of lead in superconducting and normal states.

[130] RADEMAKERS, A.: Physica, Haag **15**, 849 (1949); thermal conductivity of Pb and Sn in superconducting and normal states.

[131] MENDELSSOHN, K., and R. B. PONTIUS: Phil. Mag. **24**, 777 (1937); change of thermal conductivity of lead on going from normal to superconducting state.

[132] MENDELSSOHN, K., and J. L. OLSEN: Proc. Phys. Soc. Lond., Ser. A 63, 2 and 1182 (1950); Phys. Rev. 80, 859 (1950); thermal conductivity of lead and lead-bismuth alloys and intermediate state anomalies.

[133] OLSEN, J. L.: Proc. Phys. Soc. Lond., Ser. A 65, 518 (1952); thermal conductivity of lead-bismuth alloys and discussion in terms of enhanced lattice conduction.

[134] OLSEN, J. L., and C. A. RENTON: Phil. Mag. 43, 946 (1952); thermal conductivity of lead below 1° K.

[135] MENDELSSOHN, K., and C. A. RENTON: Phil. Mag. 44, 776 (1953); thermal conductivity of Sn, In, Tl, Ta, Cb and Al below 1° K.

[136] MENDELSSOHN. K.: Physica, Haag 19, 775 (1953); review of Oxford work on thermal conductivity of superconductors; discussion.

[137] RENTON, C. A.: Phil. Mag. 46, 47 (1955); effect of magnetic fields and frozen-in flux on the thermal conductivity of superconductors.

[138] GOODMAN, B. B.: Proc. Phys. Soc. Lond., Ser. A 66, 217 (1953); thermal conductivity of tin below 1° K.

[139] SHOENBERG, D.: Physica, Haag 19, 788 (1953); review of Cambridge work on the thermal conductivity of superconductors and size effect below 1° K.

[140] HEER, C. V., and J. G. DAUNT: Phys. Rev. 76, 854 (1949); thermal conductivity of Sn and Ta below 1° K.

[141] WEBBER, R. T., and D. A. SPOHR: Phys. Rev. 84, 384 (1951); thermal conductivity of lead in the intermediate state.

[142] WEBBER, R. T., and D. A. SPOHR: Phys. Rev. 91, 414 (1953); thermal conductivity of mercury in the intermediate state.

[143] DETWILER, D. P., and H. A. FAIRBANK: Phys. Rev. 86, 574 (1952); 88, 1049 (1952); thermal conductivity of tin and indium in the intermediate state.

[144] HULM, J. K.: Phys. Rev. 90, 1116 (1953); thermal conductivity of mercury in the intermediate state.

[145] SLADEK, W. J.: Phys. Rev. 91, 1280 (1953); 97, 902 (1955); thermal conductivity of indium-thallium alloys, study of lattice conduction and of intermediate state.

[146] LAREDO, S. J., and A. B. PIPPARD: Proc. Camb. Phil. Soc. 51, 368 (1955); extra resistivity in the intermediate state due to scattering of phonons.

Low Temperature Heat Capacity of Solids.

By

P. H. KEESOM and N. PEARLMAN.

With 29 Figures.

Introduction.

Since the publication of EUCKEN's ,,Energie und Wärmeinhalt" [1] in 1929 our knowledge of the heat capacity of solids has increased tremendously, especially in the very low temperature region.

Precision calorimetry in the liquid helium range started in 1930 with the development of the very sensitive phosphor-bronze thermometer. This led shortly to the discovery of the change in heat capacity connected with the superconductive phase transition. Soon thereafter the electronic heat capacity in metals was discovered and found to be in reasonable agreement with theoretical predictions. On the theoretical side the theory of lattice vibrations was worked out in detail for several elements and the electronic heat capacity was treated in greater detail, using the newly developed theory of band structure in solids.

Other interesting phenomena influencing the heat capacity have been investigated at low temperatures, such as the spin system of paramagnetic salts and the antiferromagnetic cooperative phenomena. The field of investigations of heat capacity is now very wide and a large variety of related properties is of interest. We have limited ourselves in this survey mostly to calorimetric results obtained in the liquid helium range. For a comparison with the theory of lattice vibrations it appeared necessary to include some of the data obtained in the liquid hydrogen range, particularly because deviations from the " T^3 Law " (this law appears to be valid for most solids in the region below 4° K), are most noticeable up to around 20° K.

In the next volume of this Encyclopedia several chapters are devoted to adiabatic demagnetisation and to superconductivity, so only a very short discussion of these phenomena, as they are related to heat capacity problems, appears necessary here.

A. Theory.

The theory of heat capacity is discussed in detail elsewhere in the Encyclopedia (see BLACKMAN's article in vol. VII, part 1) and is also to be found in many books on statistical mechanics, the theory of solids, etc. We therefore give here only the main results that will be needed in discussing the experimental results at low temperatures and a brief account of the physical principles underlying them.

1. Modes of thermal excitation. The heat capacity of a solid at constant volume, c_v, is the temperature derivative of its internal energy. Its calculation thus involves information about the manner in which the internal energy is distributed among the different modes of thermal excitation. Conversely, values of the heat capacity, particularly at low temperatures, can often provide this information.

The quantity usually measured in solids is c_p, the heat capacity at constant pressure, rather than c_v. The difference between the two is given by the thermodynamic relation

$$c_p - c_v = T V \alpha_v^2 / K,\qquad(1.1)$$

where

$$\alpha_v = \frac{1}{V}\left(\frac{\partial V}{\partial T}\right)_p\qquad(1.2)$$

is the volume coefficient of thermal expansion,

$$K = -\frac{1}{V}\left(\frac{\partial V}{\partial p}\right)_T\qquad(1.3)$$

is the isothermal compressibility and T is absolute temperature. Since neither α_v nor K is known over a wide range of temperature for many substances, the determination of c_v is subject to uncertainty at temperatures for which $c_p - c_v$ is not small compared to c_v. However, α_v goes to zero at absolute zero and is very small at low temperatures, so $c_p - c_v$ is negligibly small in the temperature range in which we are interested. Since also $(c_p - c_v)/c_v$ is much smaller than the usual experimental error, we may assume that c_p and c_v are equal in this temperature range. We therefore omit the subscripts $_v$ and $_p$ in what follows.

To a first approximation, each of the types of modes of thermal excitation can be discussed independently, with interactions among them taken into account in a higher order of approximation. Among the types which must be considered we shall distinguish as the *lattice modes*, those dynamical degrees of freedom associated with the vibrations about their equilibrium positions of the lattice constituents (atoms, ions, molecules). If the lattice constituents are molecules, the lattice modes refer to vibrations of the molecules as units, while there may also be *molecular modes* associated with the vibrations of atoms or ions within the molecules. Molecular modes will occur in a crystal if the interatomic forces in a group of atoms are stronger than the forces between atoms in adjoining groups.

In metals, it would be expected that degrees of freedom are associated with the motion of the "free" electrons, and these can be called *electronic modes*. In some solids, a degenerate electronic energy level is split by local electric or magnetic fields into discrete levels, and the transitions between them can be described as *excitation modes*. They are often referred to as "SCHOTTKY transitions". This term can also be applied to those cases where transitions can occur between the ground state and low-lying excited states, as apparently occurs in the rare-earth elements.

Ordinarily, modes of a certain type associated with a particular atom[1] are excited independently of the excitation of similar modes in any other atom. In certain cases, however, the probability of excitation depends strongly on the number of similar modes which have already been excited. Excitation of all equivalent modes therefore proceeds extremely rapidly once the first such modes have been excited. Such excitations are called *cooperative phenomena*. The energy of the solid increases sharply in a small interval below a critical temperature T_c, where $k T_c$ is of the order of the energy of one of the modes involved

---

[1] Where it is possible to do so without confusion, we henceforth replace the somewhat clumsy term "lattice constituent" by "atom", when the possible structure of the constituent is irrelevant.

(k is Boltzmann's constant). The heat capacity therefore exhibits a peak in the neighborhood of T_c superimposed on the variation due to the excitation of the other modes. One of the most striking of these transitions is that occurring at the λ point of liquid helium, which has given the generic name "λ transition" to these phenomena. Fig. 28 illustrates a heat capacity peak of this type in an antiferromagnetic substance.

In I and II below we discuss the theoretical treatment of the lattice and electronic modes respectively. The other types which are of interest will be treated at the appropriate places in Part C, where the experimental data are summarized.

I. Lattice modes.

2. Normal modes. We first consider the lattice modes, which represent the only one of the types described above which occurs in all solids. The thermal modes of the atoms are of course extremely complex, and in general, the potential energy contains cross-products of coordinates of pairs of atoms. For small amplitudes of vibration about the equilibrium positions it should be a reasonable approximation to consider the binding forces between atoms as harmonic. In this case their coordinates can be replaced by linear combinations ("normal coordinates") so chosen that in terms of them both kinetic and potential energies are sums of squares. Since there are now no cross-product terms, the resulting expression for the energy is that of an assemblage of independent harmonic oscillators. For N atoms the number of equivalent linear harmonic oscillators is $3N$, corresponding to their motion in three independent directions in space.

3. Classical treatment. If the classical equipartition law is applied[1] the energy of each linear oscillator is kT. Hence for N lattice constituents the contribution of the lattice modes to the heat capacity should be

$$c_L = 3Nk \qquad (3.1)$$

and in the particular case of the elements, where the lattice constituents are atoms, the heat capacity of a gram-atom, or the atomic heat, will be

$$C_L = 3R, \qquad (3.2)$$

where R is the gas constant per mole, since $R = N_0 k$, where N_0 is Avogadro's number. This corresponds to the experimental observations summed up in the Law of Dulong and Petit[2] that the atomic heat of many elements at room temperature is of about this magnitude.

If anharmonic terms are present in the interatomic potential (forces proportional to the square, cube, ... of the interatomic distance) the normal mode analysis of the atomic motions is no longer exact. It can be used as a first order approximation however, if the anharmonic terms are small compared to the harmonic. The anharmonic terms can then be treated by perturbation methods and they are found to contribute an additional amount to the heat capacity[3] so that C_L should be larger than $3R$. It is not certain whether this anharmonic contribution to the heat capacity has actually been observed, since it will be appreciable only at temperatures so high that $C_p - C_v$ is difficult to determine accurately. Furthermore, as will be shown later, there is another possible source for additional heat capacity, at least in metals, namely the conduction electrons.

---

[1] L. Boltzmann: Wien. Sitzgsber. (2) **63**, 712 (1871). — Wiss. Abh. vol. 1, pp. 288—308.
[2] P. L. Dulong and A. T. Petit: Ann. chim. phys. (2) **10**, 395 (1819).
[3] M. Born and E. Brody: Z. Physik **6**, 132 (1921).

The magnitude of the anharmonic contribution to the interatomic potential should itself depend on temperature through the dependence of the vibration amplitudes on temperature. Hence at very low temperatures, where the total thermal expansion is negligible and therefore the vibration amplitudes are small, it should be possible to neglect the anharmonic terms in the interatomic potential and assume that the synthesis of atomic vibrations from superimposed normal modes holds exactly.

4. EINSTEIN theory. It was known by about 1900 that some elements (for instance, carbon in the form diamond) have smaller atomic heat than $3R$ at room temperature and that the atomic heat for other elements decreases below this value at lower temperatures[1]. In order to account for these facts, EINSTEIN [2] replaced the classical value, kT, of the mean oscillator energy by PLANCK's expression

$$\bar{\varepsilon} = \frac{h\nu}{e^{h\nu/kT} - 1}. \tag{4.1}$$

The classical value is derived on the basis of a continuous range of oscillator energy values, while (4.1) rests on the assumption that only discrete levels $nh\nu$, $n = 1, 2, \ldots$ are permitted, where h is PLANCK's constant and ν is the oscillator frequency. With the simplifying assumption that all the oscillators have the same frequency ν_0, EINSTEIN found

$$C_L = 3R\left(\frac{h\nu_0}{kT}\right)^2 \frac{e^{h\nu_0/kT}}{(e^{h\nu_0/kT} - 1)^2} = 3R\,E\,(h\nu_0/kT) \tag{4.2}$$

which serves to define the "EINSTEIN function" $E(h\nu_0/kT)$. As the argument approaches zero, this function approaches unity, so in the high temperature limit, $kT \gg h\nu_0$, (4.2) approaches the classical value. The equation also accounts for the decrease in C_L below room temperature, but it requires that C_L should approach zero exponentially as T approaches zero. Measurements by NERNST[2] and others at liquid hydrogen temperatures showed, however, that the actual approach to zero was much more gradual. NERNST and LINDEMANN[3] found that agreement with low temperature results was considerably improved if (4.2) was replaced by

$$C_L = 3R \cdot \tfrac{1}{2}\left[E\,(h\nu_0/kT) + E\,(h\nu_0/2kT)\right] \tag{4.3}$$

but since this was justified by the introduction of "half-quanta" it must be considered an essentially empirical expression. The success of this approximation is probably connected with the fact that it happens to be close to the best fit obtainable with two terms of the EINSTEIN type to the continuous spectrum of vibration frequencies discussed below [3][4].

EINSTEIN himself pointed out[5] that a better approximation to the heat capacity of the lattice modes would require replacing the assumed single frequency by a distribution of the frequencies which actually occur. Because of the tremendous number of modes it should be possible to write the sum of the EINSTEIN functions of these frequencies as an integral, so that (4.2) would become

$$C_L = k \int_0^{\nu_M} g\,(\nu)\,E\,(h\nu/kT)\,d\nu \tag{4.4}$$

[1] U. BEHN: Wied. Ann. **66**, 237 (1898).

[2] W. NERNST: Bibliography in Appendix 3 of "The New Heat Theorem". New York: E. P. Dutton & Company 1926.

[3] W. NERNST and F. A. LINDEMANN: Z. Elektrochem. **17**, 817 (1911).

[4] E. KATZ: J. Chem. Phys. **19**, 488 (1951).

[5] A. EINSTEIN: Ann. Phys. **22**, 180 (1907).

where the upper limit of the integral is determined by

$$3 N_0 = \int_0^{\nu_M} g(\nu)\, d\nu. \tag{4.5}$$

The problem of finding the distribution function $g(\nu)$ (also called the vibration spectrum or frequency distribution) was attacked by two different methods, the results of which were published almost simultaneously by DEBYE [3] and by BORN and VON KÁRMÁN [4].

5. DEBYE theory. DEBYE's solution was based on a discussion of wave propagation in an infinite homogeneous isotropic elastic continuum, without examination of the normal modes in detail. His results can be more easily derived by a method due to BORN and VON KÁRMÁN[1] in which the momenta of the quanta corresponding to the waves are assumed to be uniformly distributed in momentum space. Hence we can write

$$g'(p)\, dp = (4\pi V/h^3)\, p^2\, dp \tag{5.1}$$

which can be transformed to the distribution in frequency

$$g(\nu)\, d\nu = (4\pi V/\bar{v}^3)\, \nu^2\, d\nu \tag{5.2}$$

where $\bar{v}$ is a mean velocity given by

$$\frac{1}{\bar{v}^3} = \frac{1}{v_L^3} + \frac{2}{v_T^3} \tag{5.3}$$

with v_L and v_T the velocities of longitudinal and transverse waves, respectively. The maximum frequency is then given by

$$\nu_M = \left(\frac{9N}{4\pi V}\right)^{\frac{1}{3}} \bar{v}. \tag{5.4}$$

In terms of the dimensionless variable $x = h\nu/kT$ and the parameter Θ_0 defined by

$$h\nu_M = k\Theta_0 \tag{5.5}$$

the molar heat capacity becomes

$$C_L = 3 n R \left\{ 3 \left(\frac{T}{\Theta_0}\right)^3 \int_0^{x_M} \frac{x^4 e^x}{(e^x - 1)^2}\, dx \right\} = 3 n R\, D\left(\frac{\Theta_0}{T}\right) \tag{5.6}$$

where n is the number of atoms per molecule and $D(\Theta_0/T)$ is the DEBYE function. The function D and also the EINSTEIN function E [see (4.2)] has been tabulated[2] and a nomogram for determining C_L for given T, Θ_0, has been given by EUCKEN[3].

At high temperatures x_M approaches zero and D approaches unity, so C_L approaches the classical value. At low temperatures the integral approaches the value $4\pi^4/15$ so that for T less than about $\Theta_0/12$, C_L is given by

$$C_L = n\,(12\pi^4 R/5)\,(T/\Theta_0)^3 = 1944\, n\,(T/\Theta_0)^3 \quad \text{joules/mole degree.} \tag{5.7}$$

According to the DEBYE theory, C_L is thus given by the universal function $D(\Theta_0/T)$ of the single parameter Θ_0. Moreover, Θ_0 can be calculated from the elastic constants of the material, using (5.4) and (5.5).

---

[1] M. BORN and T. VON KÁRMÁN: Phys. Z. **14**, 15 (1913).
[2] LANDOLT-BÖRNSTEIN: Physikalisch-Chemische Tabellen, 1. Erg.-Bd., S. 702—707. 1927.
[3] A. EUCKEN [1], facing p. 230.

In the case of a crystalline solid, however, this calculation is complicated by the fact that the velocities of the elastic waves depend on the direction of propagation, unlike the situation in an isotropic medium. Furthermore, the waves are purely longitudinal or transverse only in certain special directions in the crystal. The definition of $\bar{v}$ in (5.3) must accordingly be generalized to

$$\frac{1}{\bar{v}^3} = \frac{1}{4\pi} \int \sum_{i=1}^{3} \frac{1}{v_i^3(\vartheta, \varphi)} \, d\Omega \tag{5.8}$$

where $d\Omega$ is the element of solid angle and v_i are the velocities of the three different types of wave which can be propagated in each direction. Various approximation methods have been proposed for carrying out this calculation in cubic crystals in terms of the elastic constants[1]. The value of the DEBYE parameter corresponding to the elastic constants, $\Theta_0(E)$ can then be found from

$$\Theta_0(E) = \frac{2.514 \times 10^{-3} \, \bar{v}}{V_m^{\frac{1}{3}}} \, {}^\circ\mathrm{K} \tag{5.9}$$

if $\bar{v}$ is in cm/sec. Comparison with Θ_0 values determined from thermal measurements using (5.7) is not entirely unambiguous, however, since values of the elastic constants at 0° K should be used in calculating $\bar{v}$, and low temperature elastic constants have been measured for very few substances.

A more fundamental difficulty in the application of the DEBYE theory to crystalline materials is connected with the form of $g(\nu)$ given by (5.2). At high temperature $(T > \Theta_0)$ C_L should not depend strongly on the form of $g(\nu)$ since all the modes are excited. At very low temperatures $(T \ll \Theta_0)$, on the other hand, only the extreme low-frequency end of the spectrum, which corresponds to waves of very long wavelength, will be excited. The propagation of waves whose wavelength is long compared to interatomic distances should not be affected by details of the atomic arrangement and interatomic forces. Hence the DEBYE calculation should be rigorously correct as applied to these waves, and at the low frequency limit $g(\nu)$ should be proportional to ν^2. The theory can give no information, however, about the degree to which the actual lattice structure will result in departures from the parabolic form at slightly higher frequencies. Thus it is possible that C_L may depart from the values given by (5.6) at temperatures which are neither high nor very low, compared to Θ_0.

6. Lattice theory. The method introduced by BORN and VON KÁRMÁN [4] is designed to calculate the actual form of $g(\nu)$ by explicitly taking into account the geometry of the lattice and the interatomic forces. It is based on the fact that the frequencies of the normal coordinate oscillators (normal modes) occur as the roots of the secular determinant of the normal coordinate transformation. These determinants are of order $3s$ where s is the number of atoms per unit cell of the lattice and there is one determinant per unit cell, so approximation methods must still be used in evaluating $g(\nu)$. BORN and VON KÁRMÁN[2] used a method essentially similar to that of (5.1) and (5.2) to show that their analysis agreed with DEBYE's in predicting that C_L is proportional to T^3 at very low temperatures. The labor involved in applying their methods to the more general case of higher temperatures in three-dimensional lattices, together with the observation that the parabolic form of the frequency distribution seemed to lead

---

[1] L. HOPF and G. LECHNER: Verhandl. dtsch. phys. Ges. **16**, 643 (1914). — A. B. BHATIA and G. E. TAUBER: Phil. Mag. **45**, 1211 (1954). — S. L. QUIMBY and P. M. SUTTON: Phys. Rev. **91**, 1122 (1953).

[2] M. BORN and T. VON KÁRMÁN: Phys. Z. **14**, 65 (1913).

to excellent agreement with experimental heat capacity results, led to the neglect of their method for many years. Within the last twenty years, however, it has become apparent that it is impossible to describe the temperature variation of C_L in detail by the universal function $D(\Theta_0/T)$, especially at temperatures considerably below Θ_0. If on the other hand (5.6) is used instead to define a parameter Θ as a function of T in terms of measured values of C_L, it is found that $\Theta(T)$ exhibits significant variations from the constant value predicted by DEBYE's theory. Furthermore, the variation of Θ with T is not the same for all substances. It has become usual to express the results of calculations of C_L in terms of the corresponding $\Theta(T)$ curve, which can then be compared with the curve calculated from the measured values of C_L.

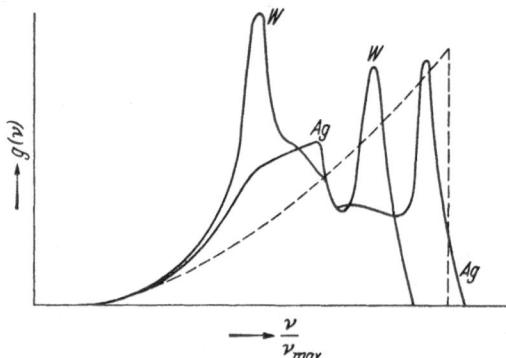

Fig. 1. Typical vibration spectra. The solid curves refer to Ag (see reference 17.13) and to W (see reference 27.1). The dashed curve is the DEBYE parabolic spectrum having the same area as the two solid curves.

The first calculations of this type were carried out by BLACKMAN starting in 1935 [5]. He obtained the roots of a representative sample of secular determinants of the simple cubic lattice and smoothed the resulting histogram to get an approximate distribution function for the normal mode frequencies. His method, together with others that have been used in applying the approach of BORN and VON KÁRMÁN, have recently been described by BORN and HUANG [6].

The detailed forms of the distribution function found by these methods, and the corresponding $\Theta(T)$ curves, depend of course on the particular lattice investigated. Certain general features appear, however, in all cases. In the first place, as would be expected, the spectrum is proportional to ν^2 in the low frequency limit. Corresponding to this, the $\Theta(T)$ curve approaches the absolute zero with zero slope. Therefore Θ is constant (and equal to Θ_0) in a small region between $0°$ K and a temperature we can designate T_B. According to BLACKMAN[1] T_B/Θ_0 is of the order of $1/100$ to $1/50$, and he refers to temperatures up to T_B as the "true T^3 region". At frequencies immediately above those for which the spectra are parabolic they rise faster than ν^2 so that a dip occurs in the $\Theta(T)$ curve above T_B. At somewhat higher frequencies, a peak occurs in the spectrum, and a corresponding minimum in the $\Theta(T)$ curve. In Fig. 1 are shown two typical calculated vibration spectra, together with the parabolic DEBYE spectrum for comparison. We shall be interested in this article mainly in the region below the first peak in the spectra, which generally occurs in the vicinity of half the maximum frequency[2].

II. Electronic modes.

7. Classical free electron theory. The early theories of the conduction of electricity in metals introduced by DRUDE[3] and others at the end of the nineteenth century used the notion of "free" electrons moving within the metal as a perfect

---

[1] M. BLACKMAN: Rep. Progr. Phys. **8**, 10 (1941).

[2] Problems reviewed in this section are treated in more detail in both G. LEIBFRIED's and M. BLACKMAN's contributions to vol. VII, part 1, of this Encyclopedia.

[3] P. DRUDE: Ann. Phys. **1**, 566 (1900).

gas. The application of methods which had been developed in connection with the kinetic theory of gases led to results which bore at least a qualitative resemblance to experiment. However, a serious discrepancy occurred in the heat capacity, for by the same argument as that leading to (3.2) the classical contribution of the electrons to the atomic heat should be

$$C_E = \tfrac{3}{2} n_a R \qquad (7.1)$$

where n_a is the number of free electrons per atom. Accordingly, the total C for metals at room temperature should be about $\tfrac{9}{2} R$, because n_a must be of the order of unity to account for the electrical properties. However, there is no tendency for the atomic heat of metals to be appreciably higher than that of insulators at room temperature, C for both being about $3R$. Since this is also the expected value of C_L, the lattice mode contribution, the observed values of C_E must be very small compared to the classical value given by (7.1).

8. SOMMERFELD theory. The solution of this difficulty is due to SOMMERFELD [7]. We saw in Sect. 4 above how EINSTEIN accounted for the decrease in C_L with temperature by replacing the classical equipartition value of the oscillator mean energy by the value calculated by PLANCK on the basis of his quantum hypothesis. This corresponds to replacing the MAXWELL-BOLTZMANN distribution function

$$f_{\mathrm{MB}}(\varepsilon) = \mathrm{const}\ e^{-\varepsilon/kT} \qquad (8.1)$$

from which the classical equipartition value, kT, of the mean oscillator energy is derived, by the BOSE-EINSTEIN distribution function

$$f_{\mathrm{BE}}(\varepsilon) = \frac{1}{e^{\varepsilon/kT} - 1} \qquad (8.2)$$

which is appropriate for particles with integral spin (in this case, zero spin). SOMMERFELD discussed the properties of electrons in metals using for them the FERMI-DIRAC distribution function

$$f_{\mathrm{FD}}(\varepsilon) = \frac{1}{e^{\frac{\varepsilon-\zeta}{kT}} + 1} \qquad (8.3)$$

which is appropriate for particles with half-integral spin (in this case, spin equal to $\hbar/2$). The normalizing constant which appears in the M-B and F-D distribution functions[1] occurs in the latter as the energy ζ, which is the chemical potential of the electrons. The energy ζ is often referred to as the FERMI level of the electron distribution. It has the formal significance, evident from (8.3) that

$$f_{\mathrm{FD}}(\zeta) = \tfrac{1}{2}. \qquad (8.4)$$

It can also be seen from (8.3) that f_{FD} decreases from nearly unity at $\varepsilon = \zeta - \Delta\varepsilon$ to nearly zero at $\varepsilon = \zeta + \Delta\varepsilon$ where $\Delta\varepsilon \approx kT$. Correspondingly, $|d f_{\mathrm{FD}}/d\varepsilon|$ is large only in this energy range.

By the same argument as was used in connection with (5.1), the density per cm.^{3} of electron levels in momentum space can be written

$$g'(p)\,dp = 2\,(4\pi/h^3)\,p^2\,dp \qquad (8.5)$$

where the factor two occurs since there are two levels for each cell in momentum space, corresponding to the two directions for electron spin. For perfectly free

---

[1] No normalizing constant appears in the B-E distribution because in this case the number of particles (quanta) is not fixed.

electrons (electrons moving in a region of constant potential) the energy levels are given by

$$\varepsilon = \frac{h^2 \sigma^2}{2m} = \frac{p^2}{2m} \tag{8.6}$$

where σ is the wave-number, equal to the reciprocal of wavelength. This is the same as the classical expression, and (8.5) can be transformed to

$$g(\varepsilon) \, d\varepsilon = 4\pi \, (2m/h^2)^{\frac{3}{2}} \, \sqrt{\varepsilon} \, d\varepsilon. \tag{8.7}$$

The number of electrons per cm.^{3} is then

$$n = 4\pi \, (2m/h^2)^{\frac{3}{2}} \int\limits_0^\infty f_{\mathrm{FD}}(\varepsilon) \, \sqrt{\varepsilon} \, d\varepsilon. \tag{8.8}$$

At absolute zero, $f_{\mathrm{FD}} = 1$ for $\varepsilon < \zeta_0$, the FERMI level at the absolute zero, and $f_{\mathrm{FD}} = 0$ for $\varepsilon > \zeta_0$, so

$$n = 4\pi \, (2m/h^2)^{\frac{3}{2}} \int\limits_0^{\zeta_0} \sqrt{\varepsilon} \, d\varepsilon \tag{8.9}$$

which can be solved for ζ_0:

$$\zeta_0 = \frac{h^2}{2m} \left(\frac{3n}{8\pi}\right)^{\frac{2}{3}} = \frac{h^2}{2m} \left(\frac{3N_0}{8\pi}\right)^{\frac{2}{3}} \left(\frac{n_a}{V_m}\right)^{\frac{2}{3}} = 25.8 \left(\frac{n_a}{V_m}\right)^{\frac{2}{3}} \mathrm{eV} \tag{8.10}$$

where V_m is the molar volume in cm^3. Since k is 0.863×10^{-4} eV/degree and ζ_0 from (8.10) will be of the order of eV for metals, the transition energy range in which f_{FD} is appreciably different from zero or unity is small even at high temperatures. Since according to the PAULI exclusion principle an electron can be excited only to an unoccupied level, and since the thermal energy available for excitation at temperature T is about kT, only those electrons within about kT of the FERMI level can be thermally excited at temperature T. The number of electrons in this energy range will be roughly the fraction kT/ζ_0 of the total number, and each will have heat capacity equal to $3k/2$, so the total electronic heat capacity per mole will be of the order

$$C_E \approx (kT/\zeta_0) \, n_a R. \tag{8.11}$$

SOMMERFELD's more accurate calculation, taking into account the small variation of the FERMI level with temperature gives $\pi^2/2$ for the proportionality constant in (8.11) so that

$$C_E = 0.136 \, V_m^{\frac{2}{3}} \, n_a^{\frac{1}{3}} \, T \quad \text{millijoule/mole degree} = \gamma T. \tag{8.12}$$

This accounts for the observation that $C_L \approx C$ at room temperature. However, since C_L is proportional to T^3 as T approaches zero, while C_E is proportional to T, at sufficiently low temperature C_E will be the dominant contribution in metals.

9. Band theory [8], [9]. It is hardly to be expected, however, that treating the electrons as perfectly free will be an adequate approximation for all metals. The energy levels of (8.6) are appropriate only for a particle with constant potential energy. The actual potential energy of an electron will in general not be constant, but will depend both on the arrangement of the lattice ions and also on the distribution of the other electrons. Its determination is thus a "self-consistent field" problem of the sort treated by HARTREE. The SOMMERFELD treatment assuming constant potential is a first approximation to the problem. The next approximation can be based on assuming that the potential due to the electrons is constant, and using that of the periodic lattice of positive ions in the potential energy term of the SCHRÖDINGER equation. Various methods have been used for finding approximate solutions which make possible at least qualitative discussion of the behaviour of electrons in real metals.

The first of these, due to BLOCH, has been called the "tight-binding" approximation since it builds up wave functions for an electron in a crystal from atomic wave functions surrounding the lattice ions. The overlapping of wave functions of adjacent ions has the consequence that the discrete atomic levels spread out into bands of levels in the crystal, the width of the bands depending on the amount of overlap. Hence the bands corresponding to the levels of the innermost electrons will be narrow, while those arising from the levels of the valence electrons, and from excited levels, may be so broad that successive bands overlap. If two bands which are adjacent in energy do not overlap, there will be an energy region between the highest energy of the lower band and the lowest energy of the higher band in which there are no allowed energy levels for an electron in the crystal.

The method due to BRILLOUIN treats the periodic lattice ion potential as a perturbation so that the crystal wave functions are derived from the plane waves which are the solutions of the free electron problem ("weak-binding" approximation). The energy is then a function of the wave-vector, not merely of its magnitude as in (8.6). In this treatment also, gaps may appear in the energy (so-called "forbidden regions") for which no eigenvalues exist. These gaps arise as a consequence of discontinuities in the dependence of energy on momentum which correspond to the circumstance that propagation through the lattice is impossible for electron waves with wave-vectors which satisfy the BRAGG reflection condition.

Cases may thus arise in which the valence electrons just fill a permitted band of energy levels in a crystal in which a forbidden energy region occurs between the highest occupied energy band and the next higher permitted band, so that the FERMI level is at the top of the filled permitted band or in the forbidden region. Such a crystal will be an insulator, and the electrons will not contribute significantly to the heat capacity when kT is less than the width of the forbidden region.

In metals, on the other hand, the FERMI level is within the permitted band, so that there are unoccupied permitted states immediately above the highest occupied levels. The electronic heat capacity will no longer be given by (8.12) in general, since $g(\varepsilon)$ need not have the form of (8.7). It can be shown that for any $g(\varepsilon)$ the heat capacity per electron is

$$c_E = \frac{\pi^2}{3} k^2 T \frac{g(\zeta)}{n} \tag{9.1}$$

where n is the number of electrons per cm³. The electronic heat capacity per mole is thus

$$C_E = \frac{\pi^2}{3} kT \frac{g(\zeta)}{L} R = \gamma T \tag{9.2}$$

where L is the number of atoms per cm³, $L = N_0/V_m$. Thus the density of electronic states at the FERMI level is related to the coefficient of the electronic atomic heat by

$$g(\zeta)/L = g_a(\zeta) = 0.424\,\gamma \text{ levels/eV atom} \tag{9.3}$$

if γ is in millijoules/mole degree².

The ratio

$$\varrho_E = \gamma_{\text{band}}/\gamma_{\text{free}} \tag{9.4}$$

where the numerator is given by (9.2) and the denominator by (8.12) is thus a measure of the departure of the actual electron level distribution from that of free electrons. An empirical estimate of ϱ_E can be made by using the calorimetrically observed value of γ in the numerator of (9.4).

The density of levels depends on the electron-lattice interaction. The rate of change of electron momentum in the crystal due to the application of a force will in general be different from that of free electrons as a result of this interaction. This difference can be accounted for by assigning an "effective mass", m^*, to the crystal electrons which in general will be different from m, the free electron mass. For our purposes we may treat m^* as a scalar, although the effective mass is in fact a tensor because of the spatial dependence of the lattice-electron interaction.

Near the edge of a band, $g_a(\varepsilon)$ may still be proportional to the square root of the energy difference from the band edge ("normal" band) as in (8.7). The numerical value of the level density will depend on the value of m^* since this, rather than m is the appropriate mass value in the numerator of (8.7). If we introduce the "effective mass ratio"

$$\mu = m^*/m \tag{9.5}$$

then for such a normal band the distance of the FERMI level from the band edge will be given by

$$\zeta_0 = (25.8/\mu)\,(n_a/V_m)^{\frac{2}{3}}\,\text{eV} \tag{9.6}$$

and the electronic atomic heat will be

$$\gamma T = C_E = 0.136\,\mu\,V_m^{\frac{2}{3}}\,n_a^{\frac{1}{3}}\,T \text{ millijoules/mole degree}^2. \tag{9.7}$$

If the FERMI level is near the bottom of a normal band n_a is the number of electrons per atom in the band. If it is near the top of a normal band, n_a is the number of "holes" per atom in the band, that is, the number of electrons per atom lacking from the number present if the band were full. In either case, if an estimate of n_a is available μ can be estimated from the calorimetrically determined value of γ using (9.7):

$$\mu = 7.3\,(n_a\,V_m^2)^{-\frac{1}{3}}\,\gamma \tag{9.8}$$

if γ is in millijoules/mole degree^2. Equivalently, the observed value of ϱ_E will also be equal to μ for a normal band.

The lattice-electron interaction affects not only C_E through the introduction of μ but also C_L by virtue of the contribution of the electrons to the binding of the lattice and hence to the elastic constants. This effect has recently been discussed by DE LAUNAY[1] for the two extreme cases in which the electrons either follow the thermal motion of the lattice ions exactly, or do not partake of this motion appreciably. For both cases he gives expressions for Θ_0 which make explicit its dependence on the elastic constants which is indicated in (5.9). A different aspect of the interaction has recently been investigated by BUCKINGHAM and SCHAFROTH[2]. The band structure discussed above is derived from the potential of a static lattice. When the lattice vibrations are taken into account, BUCKINGHAM and SCHAFROTH find that $g(\zeta)$ is increased, thus increasing the electronic atomic heat.

[1] J. DE LAUNAY: J. Chem. Phys. **21**, 1975 (1953).
[2] M. J. BUCKINGHAM and M. R. SCHAFROTH: Proc. Phys. Soc. Lond., Ser. A **67**, 828 (1954).

As was mentioned above both the "tight-binding" and "weak-binding" approximations treat the electron-electron interactions as a relatively small perturbation compared to the interactions between electrons and lattice. Since the former has in fact a long range because of the rather weak dependence on distance of COULOMB force it is surprising that its neglect leads to results which are qualitatively correct and even in some cases quantitatively correct.

What would appear to be the natural first step towards improving the approximation, however, turns out to give results which are not at all borne out by experiment. This step is to use determinantal products of one-electron wave functions in place of simple products, since the former satisfy the symmetry requirements of the exclusion principle while the latter do not. This procedure introduces an exchange energy term which has the effect of reducing $g_a(\zeta)$ with the result that the electronic atomic heat has the form

$$C_{E(e)} = C_E/(a - b \log T) \tag{9.9}$$

where $C_{E(e)}$ is the electronic atomic heat including the effects of exchange energy and C_E is the free electron value (8.12). This equation was derived by WOHLFARTH[1], who also discussed earlier calculations of $C_{E(e)}$. He found that $C_{E(e)}/C_E$ was about 1/9 for sodium at 1° K. Since the observed ratio ϱ_E is actually of the order of unity and the linear dependence of C_E on temperature is well established for many metals, WOHLFARTH concludes that this treatment is inadequate for representing the actual electron-electron interaction in metals.

Several attempts have been made to account for this discrepancy. WIGNER[2] pointed out that the use of one-electron wave functions in calculating exchange energy might result in an overestimate, since the proper many-electron wave functions should depend on interelectronic distance, while the one-electron wave functions do not. WIGNER calculated the difference, termed "correlation energy", in exchange energy using the two types of wave function, and suggested that the inclusion of the correlation term might lead to linear dependence of $C_{E(e)}$ on temperature.

In order to avoid the difficulties associated with the systematic inclusion of the correlation term WOHLFARTH[1] used the suggestion made by LANDSBERG in another connection that the ordinary COULOMB potential be replaced by the screened potential

$$V(r) = (e^2/r) \, e^{-r/\lambda}. \tag{9.10}$$

If λ is of the order of interatomic distances, which was found to be the case by LANDSBERG in his discussion of soft X-ray emission of sodium, WOHLFARTH showed that $C_{E(e)}$ is proportional to temperature rather than having the form (9.9).

A more rigorous justification of the screened potential (9.10) follows from the "collective electron" treatment of electron systems of BOHM and PINES[3], as applied by PINES[4] to electrons in metals. In this treatment electron-electron interactions are considered in two parts: long-range, which correspond to organized modes analogous to plasma oscillations, and short-range, which turn out to be similar to those of free particles interacting with a screened potential like (9.10). The first part can be neglected compared to the second, which accounts for the success of the SOMMERFELD treatment and of band calculations neglecting exchange. PINES finds that taking the electron-electron interaction into account in this way for sodium leads to the value 0.82 for $C_{E(e)}/C_E$.

[1] E. P. WOHLFARTH: Phil. Mag. 41, 534 (1950).
[2] E. WIGNER: Trans. Faraday Soc. 34, 678 (1938).
[3] D. BOHM and D. PINES: Phys. Rev. 80, 903 (1950); 85, 338 (1952); 92, 609 (1953).
[4] D. PINES: Phys. Rev. 92, 626 (1953).

B. Experimental techniques.

10. The Nernst-Eucken method. The heat capacity of a specimen is defined by

$$c(T) = \lim_{\Delta T \to 0} (\Delta Q / \Delta T) \qquad (10.1)$$

and so can be obtained directly by measuring the temperature increase resulting from the addition of a known quantity of heat. Indirect methods for determining the heat capacity are based on theoretical connections between this and other properties. Among these relations can be mentioned that between the elastic constants and the lattice heat capacity (see Sect. 5 and 6) and that between the electronic heat capacity and the magnetic threshold curve in superconductive materials (see Sect. 33). In adiabatic demagnetization experiments[1] the relation between the entropy and the temperature can be established and thence the heat capacity as a function of temperature. If the sample temperature is varied periodically or short pulses of heat are introduced the velocity with which temperature variations are propagated can be used to find the heat capacity if the thermal conductivity is known[2]. In this survey we shall not discuss further these or other related phenomena but will restrict ourselves almost exclusively to the direct method.

The method introduced by Nernst and Eucken[3] has been used in nearly all low temperature calorimetric work. In this method a calorimeter containing the sample under investigation is isolated from its surroundings by means of a high vacuum and the temperature rise ΔT resulting from the introduction of a known quantity of heat ΔQ is measured. The calorimeter thus consists of the sample and certain addenda, namely everything that is in good thermal contact with the sample during the calorimetric measurements. The primary measurements yield the heat capacity of the calorimeter, c_{cal} so that it is necessary to determine that of the addenda, c_{add} separately in order to obtain c_{sam}, the heat capacity of the sample.

In part I below we discuss some of the elements involved in the design of calorimeters and their component parts; in II some of the procedures involved in making measurements are described.

I. Calorimeter.

11. Types of calorimeter. One of the chief considerations in calorimeter design is to keep c_{add} small. If the sample is available as a single solid ingot this can be accomplished by applying the heater and thermometer directly to it. The heater and thermometer together with the glue used in attaching them so as to provide good thermal contact with the sample will in this case comprise the addenda. With material of high Θ_0 the relatively large specific heat c_g (heat capacity per gram) of commonly used glues (see Sect. 39) may make c_{add} comparable to or larger than c_{sam} at low temperatures so that accurate determination of c_{add} may be necessary even with this arrangement.

Attaching heater and thermometer to the outside of the ingot may be the only possible technique if the material is difficult to machine. If a hole can

---

[1] See article by D. de Klerk in vol. XV of this Encyclopedia.
[2] D. H. Howling, E. Mendoza and J. E. Zimmerman: Proc. Roy. Soc. Lond., Ser. A **229**, 86 (1955).
[3] W. Nernst: Berl. Ber. **262**, 306 (1910). — Ann. Phys. **36**, 395 (1911). — A. Eucken: Phys. Z. **10**, 586 (1910).

be drilled in the ingot the alternative is available of preparing a "core" containing the heater and thermometer, which can be glued into the hole or screwed into a tapped hole. The latter procedure, if it provides sufficient thermal contact, has the advantage of obviating the use of glue. The use of a core is especially advantageous for a series of measurements on different samples since then one determination of c_{add} will serve for the entire series if the amount of glue used can be kept the same for each sample.

In using solid ingots, another problem arises besides that of thermal contact between sample and heater and thermometer. This is the rate at which the sample approaches a homogeneous temperature distribution after heat has been supplied locally, externally or internally when a core is used. This rate is determined by τ, the thermal relaxation time, the order of magnitude of which is given by

$$\tau \approx \frac{\varrho\, c_g L^2}{\varkappa} \approx \frac{c_{sam}}{\varkappa L} \qquad (11.1)$$

where ϱ is the density and $\varkappa$ the thermal conductivity of the sample, and L is a dimension of the sample. It is in general advantageous to keep τ small. The reduction of data may be more complicated than otherwise if τ is longer than the response time of the instrument used to record temperature variations of the sample. A more important limitation on τ is set by the inevitable heat leak, $\dot{Q}_L$, from the sample arising from imperfect insulation between the calorimeter and the low temperature bath. In order to reduce the error arising from the heat leak, $\tau \dot{Q}_L$ should be less than ΔQ, but ΔQ cannot be made arbitrarily large since then ΔT increases and the measurement may not conform to the fundamental definition (10.1).

On the other hand, the thermal relaxation time can be reduced by using a smaller sample only at the expense of reducing c_{sam}, the limit of this reduction being set by c_{add}. For metals, $\varkappa$ is sufficiently large at low temperatures to make τ small enough for samples of reasonable size. Although $\varkappa$ decreases with T, c_g ordinarily will decrease at least as fast, so τ should not increase as T decreases. In some materials, however, modes of excitation other than lattice and electronic may keep c_g large while $\varkappa$ is decreasing, thus making τ too large for satisfactory measurements.

In this case, τ can be reduced by using the material in the form of small pieces or powder in a vessel which contains helium gas at low pressure as well as the thermometer and heater. This technique must also be used with material which cannot be obtained in sufficiently large ingots, or which will deteriorate under evacuation (for instance, hydrated salts). Since the vessel is now part of the calorimeter, together with the exchange gas, c_{add} is usually larger than with the previous arrangements. Another difficulty which has been observed is that some of the gas may be desorbed from the sample while it is being heated so that ΔT will be smaller than should correspond to the amount of heat supplied. The method has the advantage, however, that the vessel itself or a small container soldered to it can be used as a gas thermometer for calibrating the sample thermometer in the temperature region between the normal boiling point of helium and the lowest temperatures obtainable with hydrogen (4 to 10° K).

12. Thermometers. Every physical property which is temperature dependent in the temperature region of interest, $\pi(T)$, can be used as a thermometer but it must fulfil certain conditions in order to be useful in calorimetry. The property should be easy to measure with sufficient accuracy, reproducible at least during the time of measurements, and the temperature sensitivity, $(1/\pi)\,(d\pi/dT)$ should be high. The thermometer heat capacity should be small relative to c_{sam}, it should be easy to mount with good thermal contact with the sample, and it should not cause large heat leak between calorimeter and surroundings. Temperature measurements should not generate excessive amounts of heat. It is also desirable, but not essential in all cases, that $\pi(T)$ be independent of magnetic field and that it be reproducible after cycling between high and low temperatures.

In terms of the thermometric property $\pi(T)$, the heat capacity is

$$c = (dQ/d\pi)\,(d\pi/dT).   \tag{12.1}$$

Hence $d\pi/dT$ determines directly the accuracy with which c can be measured. For this reason it is convenient if a theoretical or experimental formula can be established for $\pi(T)$. If such a formula is available, fewer calibration points will be required, and it may be unnecessary to carry out the cumbersome calibration between 4 and 10° K.

Between 1 and 20° K resistance thermometers have been used almost exclusively, mainly because measurement of resistance is fast and accurate and the heat capacity of the resistor can be kept small compared to c_{sam}. Most resistance thermometers are wires which can be used with any of the types of calorimeter discussed in the previous section. Another type which has recently been used with success is the commercial carbon compound radio resistor. We discuss briefly below some of the more common resistance thermometers.

For metals the temperature sensitivity of resistance, $(1/R)\,(dR/dT)$, approaches zero between 10 and 15° K. Gold wire has been used and platinum wire is still used in the hydrogen temperature region. Lead is the only pure metal with sufficient temperature sensitivity down to 9° K [1]. Lead wire can be made easily by forcing molten lead through a small hole. It has the serious disadvantage of very low breaking strength, so mounting and cooling to low temperatures must be done with care.

Alloys such as constantan and manganin, and in England "Eureka wire" have been used from 1° K upwards. The temperature sensitivity is small however ($\sim 10^{-3}\,\mathrm{deg^{-1}}$) so that the temperature increase upon heating, ΔT, must be large. When these alloys contain superconductive elements as components or impurities, anomalies may occur in $R(T)$ at the transition temperatures, so careful calibrations must be made there.

Phosphor-bronze[2] and leaded-copper thermometer wires have been used with great success in the helium temperature range. These wires are essentially copper through which is distributed 0.1% lead. The lead is presumably not in solid solution but occurs as small particles or veins. Below the normal superconducting transition point of lead, 7.3° K, the temperature sensitivity of the resistance may be of the order of unity, so that temperature differences of 10^{-5} ° K can be measured. These thermometers have some serious disadvantages, however. In the first place, the fabrication of wire with a high temperature sensitivity seems to be a difficult art. Further, the resistance is current and magnetic field dependent, so both must be kept constant during measurement. As a

---

[1] EUCKEN [1], p. 64 gives a table for R and dR/dT of pure lead as functions of T from 9° K to room temperature. We have, however, found that careful calibration of lead thermometer wire is necessary in the hydrogen region as the table cannot be relied upon in this temperature range.

[2] W. H. KEESOM and J. N. VAN DEN ENDE: Commun. Kamerlingh Onnes Lab. Univ. Leiden no. 203c, also in Proc. Kon. Akad. Wetensch. **32**, 1171 (1929).

consequence of the random distribution of the lead through the wire it is impossible to predict the nature of $R(T)$, and even two pieces of the same wire may differ appreciably in this respect. Finally, temperature cycling, handling, and heat treatment at elevated temperatures will all affect $R(T)$. Careful recalibration at helium temperatures for all currents and magnetic fields employed is therefore necessary each time such a wire is used. Despite all these limitations, these wires have been among the most useful low temperature thermometers.

Carbon resistors of various types have also been widely used in low temperature calorimetry. The temperature sensitivity is negative and may be of the order of unity or larger. They can be "home made" by painting a suspension of carbon particles such as aquadag or india ink on a base, for instance tissue paper[1]. The resistance of such a carbon strip is temperature dependent but the thermometers are not very stable; cycling between room temperature and. helium temperatures may change the resistance by orders of magnitude. This type of instability may sometimes be avoided by ageing or heat treating the thermometer before use.

Commercial carbon compound radio resistors, especially made to be temperature insensitive at room temperature, were found some years ago to be excellent thermometers at low temperatures[2]. They are inexpensive, rugged and one-half and one-tenth watt sizes are small enough, and have low enough heat capacity to make them useful in calorimetry. They respond well to temperature cycling and an approximate formula for $R(T)$ is available:

$$\left(\frac{\text{Log } R}{T}\right)^{\frac{1}{2}} = a + b \text{ Log } R. \tag{12.2}$$

This formula is not accurate enough to be used with values of a and b taken from two calibration points, but a convenient deviation curve may be derived from it. A plot of $\sqrt{\dfrac{\text{Log } R}{T}}$ against $\text{Log } R$ is used to find the best estimate of the constant b and the constant a calculated from (12.2) with this value of b is then plotted as a function of T. The smoothed $a(T)$ curve is used in interpolation. Eq. (12.2) can also be used with measurements on a particular resistor at room and liquid nitrogen temperatures to get a rough prediction of resistance values at helium and hydrogen temperatures by extrapolation. Such a prediction can be made from room temperature values alone from a three-constant formula given by CLEMENT and QUINNELL[3]

$$\text{Log } R + K/\text{Log } R = A + B/T \tag{12.3}$$

with

$$A \pm 3\% = 1.62 \text{ Log } R_{290} + 0.27, \tag{12.4}$$

$$B \pm 9\% = 1.60 \text{ Log } R_{290} + 0.48, \tag{12.5}$$

$$K \pm 6\% = 0.594 (\text{Log } R_{290})^2 + 0.377 \text{ Log } R_{290} - 0.121. \tag{12.6}$$

Because of their size and shape such resistors are most useful with calorimeters containing a core, or with exchange gas in a vessel. They are manufactured with an insulating layer which may be ground away before use to improve thermal contact with the temperature sensitive part of the resistor.

---

[1] W. F. GIAUQUE: Industr. Engng. Chem. **28**, 743 (1936).

[2] J. R. CLEMENT and E. H. QUINNELL: Phys. Rev. **79**, 1028 (1950). — M. W. ZEMANSKY and H. A. BOORSE: Proceedings of the Oxford Conference on Low Temperatures 1951, unpublished.

[3] J. R. CLEMENT and E. H. QUINNELL: Rev. Sci. Instrum. **23**, 213 (1952).

Because of the low thermal conductivity of carbon it is necessary to keep the heat dissipated in the resistor small so that its temperature will not be appreciably higher than that of the specimen. BERMAN[1] has given a relation between the resistor temperature rise and the power P dissipated in it:

$$dP/dT = 39.6\ T^{1.6}\ \text{microwatts/degree.} \tag{12.7}$$

This applies to a nominal 100 ohm $\frac{1}{2}$ watt resistor; BERMAN found that dP/dT was smaller for a nominal 20 ohm $\frac{1}{2}$ watt resistor. It would presumably be larger for smaller resistors ($\frac{1}{4}$ or $\frac{1}{10}$ watt).

The resistance of carbon thermometers, like that of phosphor-bronze wire, is magnetic field dependent. While this dependence is erratic in the latter case since it is determined by the distribution of superconducting veins in the wire, it seems to be a well-defined property of carbon resistors. CLEMENT and QUINNELL[2] expressed the dependence in the form

$$d\,(R/R_0)/d\,(H^2) = (a + b\ \text{Log}\ R_{290})\ T^{-1.5} \tag{12.8}$$

for carbon-composition resistors, where H is the applied field and R_0 is the resistance in zero field at temperature T. GIAUQUE et al.[3] also found that the resistance of their amorphous carbon thermometers varied linearly with H^2 at constant temperature, although the dependence on resistance and temperature is not that of (12.8).

Another type of resistance thermometer is the semiconductor germanium employed by ESTERMANN et al.[4]. Like carbon resistors, semiconductors have a negative value of $(1/R)\,(dR/dT)$, but the resistivity of pure Ge is much too large at helium temperatures to enable it to be used as a resistance thermometer. When alloyed with a small concentration of a suitable impurity, however, the resistivity decreases while the temperature sensitivity remains large. The thermometers used by ESTERMANN et al., for instance, had resistances of several hundred ohms at 4.2° K and of several thousand ohms at about 2° K. While such thermometers have no marked superiority over carbon compound resistors, they have the disadvantage that they are not nearly so rugged; further, rather critical metalurgical procedures are involved in their preparation.

The most convenient way to apply heat to a calorimeter in the liquid helium temperature range is to use JOULE heat developed in a resistance wire. For this application the low temperature sensitivity of constantan, manganin, etc. is an advantage. It is possible to use a resistance thermometer itself as heater, but this has the disadvantage that during the heating period and shortly thereafter the thermometer temperature will be higher than that of the sample.

At temperatures below 1° K this manner of heating may no longer be suitable, especially with samples such as paramagnetic salts which have long thermal relaxation times. Heating by gamma rays and by induction has been used successfully for such measurements.

II. Measuring procedures.

13. Liquid helium temperatures. In this section we describe as examples of the design principles discussed in part I, two types of calorimeter and associated measuring apparatus used in the low temperature laboratories at Leiden and Oxford.

[1] R. BERMAN: Rev. Sci. Instrum. **25**, 94 (1954).

[2] J. R. CLEMENT and E. H. QUINNELL: Rev. Sci. Instrum. **23**, 213 (1952).

[3] W. F. GIAUQUE, J. W. STOUT and C. W. CLARK: J. Amer. Chem. Soc. **60**, 1053 (1938).

[4] I. ESTERMANN, S. A. FRIEDBERG and J. E. GOLDMAN: Phys. Rev. **87**, 582 (1952).

Fig. 2[1] shows a typical arrangement used at Leiden with material which permits the insertion of a core. The calorimeter (sample plus core) is contained in a vacuum vessel surrounded by liquid helium in a dewar. The liquid helium dewar is in turn surrounded by a dewar (not shown) containing liquid hydrogen. In order to minimize thermal contact between the calorimeter and the bath the calorimeter rests on sharp points in the vacuum vessel. Alternatively, it may be hung from thin threads attached to a hook in the top of the vacuum vessel.

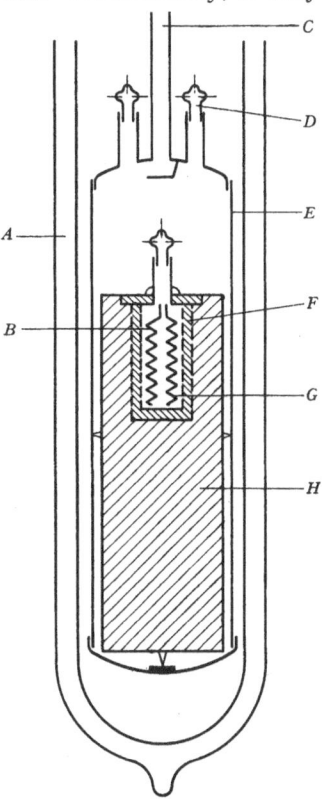

In order to calibrate the thermometer against the vapor pressure of liquid helium a small amount of helium gas is admitted into the vacuum vessel through the pumping tube, thus assuring thermal equilibrium between the calorimeter and the bath. After calibration the vessel is pumped to a pressure of the order of 10^{-6} mm Hg[2] or less and when the calorimeter is sufficiently isolated thermally from the bath calorimetric measurements are begun. The resistance of the thermometer (and hence its temperature) is measured at regular intervals, perhaps 10 or 20 seconds, to obtain the rate of temperature drift (fore period). The calorimeter is then heated for a measured length of time (heating period) and the amount of heat supplied determined by measuring the current through and the potential drop over the heater. After heating is stopped the temperature is followed again to establish the new temperature drift rate (after period). If the drifts during the fore and after periods are small compared to the increase in temperature resulting from heating, the fore and after periods can be extrapolated to the middle of the heating period. The difference of these extrapolated values then gives the temperature increase. If the drifts are not small the insulation of the calorimeter has not been satisfactory and one of the schemes[3] which have been devised to be applied in this situation in order to determine the "ideal" temperature increase must be employed.

Fig. 2. Typical calorimetric apparatus used at Leiden (after KEESOM and VAN DEN ENDE). *A* cryostat, *B* thermometer, *C* to diffusion pump, *D* wire inlets, *E* vacuum vessel, *F* core, *G* heater, *H* calorimeter.

A calorimeter used at Oxford[4], which includes a vessel containing exchange gas, is illustrated in Fig. 3. In this apparatus the liquid helium is produced by the SIMON expansion method in the upper chamber. Helium is introduced into the calorimeter vessel through the capillary and condensed there so that the vessel serves as a vapor pressure thermometer below 4° K and as the bulb

[1] W. H. KEESOM and J. N. VAN DEN ENDE: Proc. Kon. Akad. Wetensch. **35**, 143 (1932), also in Commun. Kamerlingh Onnes Lab. Univ. Leiden, no. 219b.

[2] This pressure is measured in the pumping system at room temperature which is connected to the vacuum can. M. P. GARFUNKEL and A. WEXLER [Rev. Sci. Instrum. **25**, 170 (1954)] report measurements with a helium mass spectrometer in which they determined the actual pressure in their vacuum can at liquid helium temperatures to be less than 10^{-8} mm Hg.

[3] See for instance W. H. KEESOM and J. A. KOK: Proc. Kon. Akad. Wetensch. **35**, 294 (1932), also in Commun. Kamerlingh Onnes Lab. Univ. Leiden, no. 219c.

[4] D. H. PARKINSON, F. E. SIMON and F. H. SPEDDING: Proc. Roy. Soc. Lond., Ser. A **207**, 137 (1951).

of a constant volume gas thermometer between 4 and 10° K. After calibration of the thermometer the liquid is removed by pumping and then a small amount of helium gas readmitted to ensure thermal equilibrium in the calorimeter vessel. The measuring procedure is then similar to that described above.

An interesting design has recently been reported by RAMANATHAN and SRINIVASAN[1], in which it is possible to cool the calorimeter without introducing exchange gas into the vacuum vessel. This vessel is sealed while it contains a small amount of air so that the sample is easily cooled to liquid nitrogen temperatures. Cooling to liquid helium temperatures thereafter freezes the air and results in an excellent vacuum. The under surface of the sample and the inner surface of the vacuum vessel are highly polished. In order to establish equilibrium between the sample and the bath for calibration, and for changing sample temperature between heat capacity measurements, the sample is pressed into contact with the bottom of the vacuum vessel. Thermal contact can then be broken for heat capacity measurements by lifting the sample. Difficulties connected with adsorption of exchange gas on the surface of the sample can thus be avoided with this arrangement.

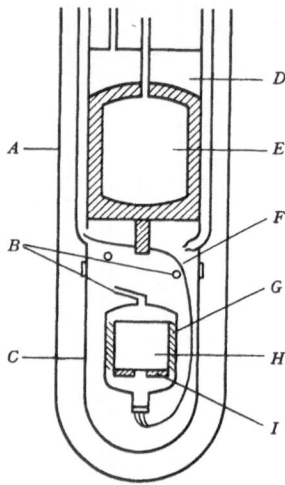

Fig. 3. Expansion liquefier adapted for calorimetry at Oxford (after PARKINSON *et al.*). *A* vacuum case, *B* capillary, *C* shield, *D* hydrogen chamber, *E* helium expansion chamber, *F* leads, *G* thermometer, *H* specimen, *I* heater.

14. Temperatures below 1° K. Several measurements of heat capacity below 1° K of paramagnetic salts have been reported. These salts have very large molar heats at low temperatures (see Sect. 34 and 35) so that they present problems of measurement not found with metals, for instance. Because of the large heat capacity of a typical sample, the value of c_{add} or of the heat leak is of less concern than the fact that the thermal relaxation time will in general be rather large. Hence, as was mentioned in Sect. 12, JOULE heating cannot be used and heat must be supplied by γ-rays or induction instead.

Two measurements on metals have also been reported. Here the heat leak problem is paramount because of the small value of c_{sam}. RAYNE[2] incorporated some paramagnetic salt in his calorimeter, providing good thermal contact by soldering one end of a thick copper wire to his metal sample and the other end to a vane system imbedded in a salt pill. Below 1° K the heat capacity of the salt is proportional to $1/T^2$ (see Sect. 34) while that of the metal is proportional to T as the electronic heat capacity contribution is preponderant in this temperature range. The total heat capacity will thus have the form

$$c_{cal} = \frac{a}{T^2} + \gamma \frac{m}{M} T \qquad (14.1)$$

where m is the mass of the metal sample, M is its atomic weight and γ the coefficient of the electronic term in its atomic heat. RAYNE therefore plotted $c_{cal} T^2$ versus T^3 and found that with the exception of sodium his results asymptotically approached straight lines at his lowest temperatures. The slope of these lines then give the value of γ.

[1] K. G. RAMANATHAN and T. M. SRINIVASAN: Phil. Mag. **46**, 338 (1955).
[2] J. RAYNE: Phys. Rev. **95**, 1428 (1954).

SAMOILOV[1] used instead two thin wires to link the sample thermally to the paramagnetic salt in his measurements on cadmium in the normal and super-conducting state. The resulting thermal coupling was so weak that the heat supplied by the heater during the heating period raised the temperature only of the sample. The heat leak during the heating period was small enough to be accounted for in the usual way by observing temperature drift rates during the fore and after periods. This method has the advantage over that of RAYNE that it gives the heat capacity of the sample directly. A slight disadvantage of the weak coupling between salt and sample is the long time, about an hour, before the sample reached the lowest temperatures after demagnetization of the salt. This method might be improved further by the use of a thermal switch based on the fact that the pure superconductive metals, e.g. lead, have much lower thermal conductivity in the superconducting than in the normal state.

C. Experimental results.

15. Reduction of the data. In the following sections we present a survey of the published data on heat capacity of solids in the helium temperature range. Part I includes data relating to the lattice and electronic atomic heat of the elements (excluding the solidified gases) arranged according to the Groups of the Periodic Table. In Part II we discuss other sources of heat capacity at low temperatures, as well as measurements on various compounds.

In Table 1 we present what we believe to be the best available estimates of Θ_0 and γ as defined by (5.7) and (9.2) respectively. For each Group we also give in tabular form the results of individual investigations together with a discussion of their particular features of interest. The tabulated values of Θ_0 and γ were recalculated from the original data wherever possible. They were obtained from fitting by least squares equations of the form

$$(C/T) = \alpha\, T^2 + \gamma \tag{15.1}$$

where according to (5.7)

$$\Theta_0 = (1.944/\alpha)^{\frac{1}{3}} \times 100^\circ \text{ K} \tag{15.2}$$

if α is given in millijoules/mole degree[4]. T_B is the estimated upper limit of the true T^3 region. For the superconductive elements, values of γ obtained from threshold field measurements (see IIa below) which are taken from the recent compilation by EISENSTEIN [10], are included for comparison with the calori-metric values. In addition, values of γ derived from calculations of the electronic band structure, where available, are discussed in the text below. References to these calculations will be found in the recent review articles of RAYNOR [11], DAUNT [12], and MOTT [13].

The values of Θ_0 and γ given for the superconductive elements are derived from measurements made on the normal state. The lattice atomic heat in the superconducting state should correspond to the same value of Θ_0, but it is not certain at present that $C_{E(s)}$, the electronic contribution in the superconducting state, can be expressed for all elements in a unique form with well-defined para-meters for each element. Expressions which have been proposed for $C_{E(s)}$ and difficulties connected with their application are discussed in Sect. 33 below. We therefore refrain from giving analytic expressions for $C_{E(s)}$ and refer instead to the original papers where the data may be found.

The data in the following sections include those reported up to July 1, 1955. Later publications are noted in the Appendix, Sect. 40.

---

[1] B. N. SAMOILOV: C. R. Acad. Sci. URSS. **86**, 281 (1952).

Table 1. *Atomic heat of the elements at low temperatures.*

The Atomic Number stands on the left, the Symbol on the right of the field. Below these are given Θ_0 (°K) and (boldface) γ (millijoules/mole deg²). Parenthesis indicates uncertain value. For superconductors, Θ_0 refers to the normal state and γ is the best estimate from calorimetric data.

Period	Ia	IIa	IIIb	IVb	Vb	VIb	VIIb	VIII	VIII	VIII	Ib	IIb	IIIa	IVa	Va	VIa	VIIa	0
1	1 H																	2 He
2	3 Li	4 Be 1160 **0.226**											5 B	6 C Diamond (2000) Graphite 391	7 N	8 O	9 F	10 Ne
3	11 Na 158 **1.8**	12 Mg 406 **(1.35)**											13 Al 418 **1.46**	14 Si 658	15 P	16 S	17 Cl	18 A
4	19 K	20 Ca (219) **(0.38)**	21 Sc	22 Ti 278 **3.34**	23 V 273 **8.83**	24 Cr 402 **1.54**	25 Mn **13.8**	26 Fe 467 **5.0**	27 Co 445 **5.0**	28 Ni 456 **7.4**	29 Cu 339 **0.72**	30 Zn 308 **0.66**	31 Ga	32 Ge 366	33 As	34 Se	35 Br	36 Kr
5	37 Rb	38 Sr	39 Y	40 Zr 270 **2.95**	41 Nb 252 **8.5**	42 Mo 425 **2.14**	43 Tc	44 Ru	45 Rh	46 Pd 275 **10.7**	47 Ag 225 **0.66**	48 Cd 300 **0.71**	49 In 109 **1.81**	50 Sn White 189 / Gray 212 **1.82**	51 Sb	52 Te	53 I	54 Xe
6	55 Cs	56 Ba	57* La 132 **6.7**	72 Hf	73 Ta 231 **5.44**	74 W (379) **1.48**	75 Re	76 Os	77 Ir	78 Pt 229 **6.8**	79 Au 165 **0.74**	80 Hg (60-90)	81 Tl 89 **3.1**	82 Pb 94.5 **3.0**	83 Bi 117 **(<0.08)**	84 Po	85 At	86 Rn
7	87 Fr	88 Ra	89** Ac															

* 58—71 Lanthanide Rare Earths
** 90—103 Actinide Rare Earths

I. Lattice and electronic atomic heats of the elements[1].

16. Group I a. The only results in the liquid helium temperature region which have been reported for elements in this Group are measurements on sodium.

Sodium. PICKARD and SIMON measured the atomic heat between 2 and 25° K and found a peak around 7° K. Fig. 4 gives a plot of the DEBYE Θ versus the absolute temperature T; the minimum in Θ at 7° K corresponds to the peak in the atomic heat. Unpublished results of HILL and SMITH and of PARKINSON (see reference 16.2), however, do not confirm this peak.

RAYNE measured the atomic heat below 1° K and found a peak between 0.8 and 0.9° K. He suggested that this might be connected with a transformation from the body-centered cubic lattice, stable above this transition temperature, to the face-centered cubic lat-

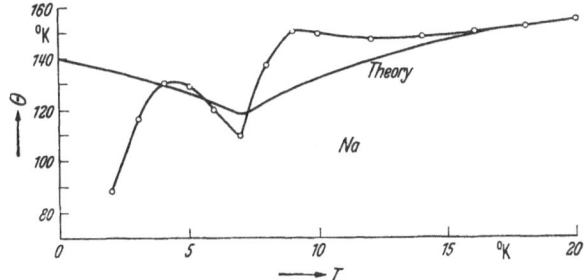

tice. This, or another common mechanism, might conceivably account as well for the peak observed around 7° K. That the two peaks occur at different temperatures in different samples might be a consequence of different external parameters such as impurities, strains or cooling.

Fig. 4. Θ versus T for sodium. O experimental points (reference 16.3). The theoretical curve is due to BHATIA (reference 16.1).

The result of a calculation by BHATIA of the lattice atomic heat alone is also plotted in Fig. 4. He concludes that there is excellent agreement except for the peak. The difference between the two curves below 4° K is very likely due to the electronic atomic heat contribution.

Because of its relatively simple electronic configuration, the electronic band structure of sodium has been the subject of extensive theoretical investigation. BARDEEN found that the free-electron treatment with effective mass ratio of 0.95 was a reasonable approximation to his calculated band structure. PINES calculated the effect of taking the electron exchange energy into account by means of the "collective electron" approach and found the corresponding C_E to be 0.82 of the free-electron value. BUCKINGHAM and SCHAFROTH estimated the value of the interaction parameter occurring in their theory of lattice-electron interaction from the unpublished results of PARKINSON referred to above, which they quote as indicating C_E to be about twice the free-electron value. Unambiguous data in the liquid helium temperature range and below would thus be very useful in assessing the applicability of these various theoretical treatments.

16.1 A. B. BHATIA, Phys. Rev. **97**, 363 (1955): Na; vibration spectrum.

16.2 M. J. BUCKINGHAM and M. R. SCHAFROTH, Proc. Phys. Soc. Lond. A **67**, 828 (1954): electron-lattice interaction.

16.3 G. L. PICKARD and F. SIMON, Proc. Phys. Soc. Lond. **61**, 1 (1948): Na, Hg; 3 to 90° K, Pd; 2—22° K.

---

[1] References to data listed in the Tables and to data which are commented on in the text are collected at the ends of the respective sections except for theoretical calculations of γ, references to which will be found in [*11*], [*12*], [*13*]. For convenience in cross-reference, each item is identified by a number giving the section in which it appears as well as the order of listing in that section, e.g. 16.3. The lists are not exhaustive, especially for work above 10° K. Data in this temperature range are referred to, in general, only when they can be compared with data below 10° K. Extensive lists of references covering all temperatures can be found in the compilations of SHIFFMAN [*14*] and of SHULL and SINKE [*15*].

16.4 D. Pines, Phys. Rev. **92**, 626 (1953): "Collective electron" treatment, including application to electronic atomic heat.

16.5 J. Rayne, Phys. Rev. **95**, 1428 (1954): Na, Cu, Ag, Mo, W, Pd, Pt; below 1° K.

16.6 F. Simon and W. Zeidler, Z. phys. Chem. **123**, 383 (1926): Na, K; 14—280° K, Mo, Pt; 16—300° K.

17. Group Ib. α) Copper. Agreement among the different investigators appears to be satisfactory. The results of Estermann *et al.* are higher than the average, but their sample was not very pure and their results have a rather large scatter. The results of Keesom and Kok are about 5% higher than those of Corak *et al.*, but the scatter in the former is also of about this order. Experimental values of $\Theta(T)$ are given in Fig. 5 together with the curve calculated by Leighton from his theoretically determined vibration spectrum and the agreement is excellent.

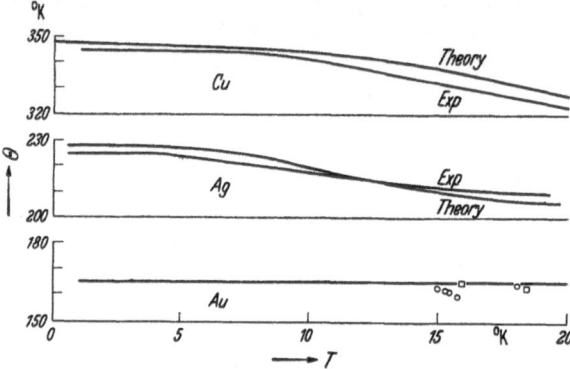

Fig. 5. Θ versus T for Cu, Ag and Au. The theoretical curves are due to Leighton (reference 17.13). ○: Clusius and Harteck (reference 17.2); □: Giauque and Meads (reference 17.7).

The elastic constants of copper have recently been measured by Overton and Gaffney from the normal boiling point of helium to room temperature. The values of the elastic constants extrapolated to 0° K together with de Launay's theoretical calculations (mentioned above in Sect. 9) give Θ_0 equal to 340° K which is in excellent agreement with the calorimetric values.

The value of γ for free electrons in a univalent metal with the atomic volume of copper as calculated from (8.12) is only 0.502 millijoules/mole degree², so that the ϱ_E is 1.4. The calculated energy band for copper valence electrons has approximately the "normal" form given by (8.7) at the Fermi level, so that this number may also be taken as roughly the effective mass. The shape of the energy band has been calculated approximately by Jones, from the form of the Brillouin zone. A more detailed calculation has been carried out by Krutter and others on the basis of the $3d$ and $3s$ wave functions. The former gives $g_a(\zeta) = 0.308$ and the latter gives $g_a(\zeta) = 0.316$; by (9.3) these correspond to $\gamma = 0.73$ and 0.75 millijoules/mole degree² respectively, so that there is excellent agreement with the calorimetric values of Keesom and Kok and of Rayne. Those of Estermann *et al.* and of Corak *et al.* are higher and somewhat lower, respectively.

β) Silver. The electronic contribution to the atomic heat was first discovered by Keesom and Kok in their measurements on silver. They found that $\Theta(T)$ computed from C, the total measured atomic heat, decreased sharply in the helium temperature range, but that C could be represented well by the sum of terms proportional to T^3 and to T. Their measurements extended into the intermediate temperature region between the helium and hydrogen ranges as well, but they reported that difficulties with thermometer calibration in this temperature region made their results there largely qualitative. Nevertheless, attempts have been made by others to find a theoretical explanation for an apparent peak in $\Theta(T)$ from their data at about 5° K. Katz discussed the effect on $\Theta(T)$ of superimposing Einstein peaks on a Debye vibration spectrum, while BHATIA and HORTON recently investigated the possibility of getting such a peak in $\Theta(T)$ by suitable choice of elastic constants. Neither these latter authors nor LEIGHTON found, however, that such a peak occurred in their calculated $\Theta(T)$ curves for the actual elastic constants of silver. Later measurements by KEESOM and PEARLMAN and by CORAK et al. make it appear probable that such a peak does not exist. Fig. 5 also gives $\Theta(T)$ for silver from 1 to 20° K. The experimental curve and that computed by LEIGHTON from his theoretical vibration spectrum are in excellent agreement.

Another apparent anomaly, connected with the electronic heat capacity, was reported by KEESOM and PEARLMAN in 1952. In a more recent publication (1955) this was traced to deviations of the accepted 1948 temperature scale from the absolute scale. A careful discussion of the effects of such deviations is given by CORAK et al. Their value of γ is somewhat lower than the free electron value, 0.644 millijoules/mole degree², while the other γ values reported are somewhat higher. The ratio ϱ_E is much closer to unity for silver than for copper.

γ) Gold. Results are now available from 1° K upwards. The ratio ϱ_E is 1.16, which is between those for silver and copper.

Table 2. *Low temperature atomic heat of the Elements-Group I b.*

Element	Θ_0(°K)	range (°K)	T_B(°K)	γ (millijoules/mole degree²)	reference
Cu	334	1 — 20	7	0.73	17.12
	352	1 — 4		0.92	17.4
		< 1		0.72	16.5
	344	1 — 5		0.69	17.3
		> 10			17.7
Ag	230	1 — 10	4	0.67	17.10
	225	1 — 4		0.66	17.11
		< 1		0.68	16.5
	225	1 — 5		0.61	17.3
	229	> 1		0.65	17.16
		> 10			17.14, 17.2, 17.5
Au	163	> 12	20		17.2
	165	> 16			17.6
	165	1 — 5		0.74	17.3

17.1 A. B. BHATIA and G. K. HORTON, Phys. Rev. **98**, 1715 (1955): theoretical vibration spectrum for face-centered cubic lattice, with application to silver.

17.2 K. CLUSIUS and P. HARTECK, Z. phys. Chem. **134**, 243 (1928): Ag, Zn; 12—200° K, Au, Ga; 15—200° K.

17.3 W. S. CORAK, M. P. GARFUNKEL, C. B. SATTERTHWAITE and A. WEXLER, Phys. Rev. **98**, 1699 (1955): Cu, Ag, Au; 1—5° K.

17.4 I. ESTERMANN, S. A. FRIEDBERG and J. E. GOLDMAN, Phys. Rev. **87**, 582 (1952): Cu, Mg, Ti, Zr, Cr; 2—4° K.

17.5 A. EUCKEN, K. CLUSIUS and H. WOITINCK, Z. anorg. allg. Chem. **203**, 39 (1931): Ag; 12—200° K.

17.6 T. H. GEBALLE and W. F. GIAUQUE, J. Amer. Chem. Soc. **74**, 2368 (1952): Au; 16—300° K.

17.7 W. F. GIAUQUE and P. F. MEADS, J. Amer. Chem. Soc. **63**, 1897 (1941): Cu, Al; 15—300° K.

17.8 H. JONES, Proc. Phys. Soc. Lond. **49**, 250 (1937): Electronic energy levels for face-centered and body-centered cubic lattices; application to Cu.

17.9 E. KATZ, J. Chem. Phys. **19**, 488 (1951): superposition of EINSTEIN peaks on DEBYE vibration spectrum, derived from empirical $\Theta(T)$ curves; application to Ag.

17.10 W. H. Keesom and J. A. Kok, Proc. Kon. Ned. Akad. Wetensch. **35**, 301 (1932), also in Commun. Kamerlingh Onnes Lab. Univ. Leiden no. 219d: Ag; 1—20° K. Physica, Haag **1**, 770 (1933), also in Commun. Kamerlingh Onnes Lab. Univ. Leiden no. 232d: Ag, Zn; 1—5° K.

17.11 P. H. Keesom and N. Pearlman, Phys. Rev. **88**, 140 (1952); **98**, 548 (1955): Ag; 1—4° K.

17.12 J. A. Kok and W. H. Keesom, Physica, Haag **3**, 1035 (1936), also in Commun. Kamerlingh Onnes Lab. Univ. Leiden no. 245a: Cu, Pt; 1—20° K.

17.13 R. B. Leighton, Revs. Mod. Phys. **20**, 165 (1948): Theory of lattice atomic heat for face-centered cubic lattice; application to Ag.

17.14 P. F. Meads, W. R. Forsythe and W. F. Giauque, J. Amer. Chem. Soc. **63**, 1902 (1941): Ag, Pb; 15—300° K.

17.15 W. C. Overton Jr. and J. Gaffney, Phys. Rev. **98**, 969 (1955): Measurement of elastic constants of Cu between 4 and 300° K; calculation of Θ_0.

17.16 B. Yates and F. E. Hoare, private communication: Ag, Pd; above 1° K.

18. Group IIa. α) *Beryllium*. The older work of Cristescu and Simon indicated a peak in the atomic heat around 11° K, but the newer results of Hill and Smith do not confirm this.

Since the $2s$ energy band can accomodate just two electrons, the fact that divalent beryllium is a metal rather than an insulator implies that the $2s$ and $2p$ bands must overlap. Herring and Hill have calculated the electronic band structure, which is shown in Fig. 6. The minimum following the peak is an indication of the overlap and since ζ_0 falls near the minimum it would be expected that ϱ_E is less than unity. This expectation is borne out; the observed value is 0.47 while that calculated from the curve is 0.39. The two parabolas in Fig. 6 are respectively the extrapolation of the normal portion of the band near the lower edge (upper parabola, corresponding to $\mu = 1.6$) and the free electron level density (lower parabola). Since the band is far from normal at the Fermi level, it would be invalid in this case to interpret ϱ_E as an estimate of μ.

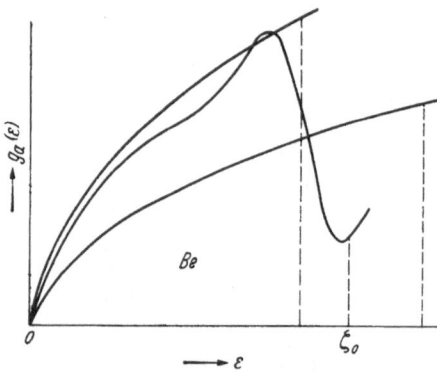

Fig. 6. Electronic band structure of Be, after Herring and Hill, Phys. Rev. 58, 132 (1940). The two parabolas correspond to normal bands with $\mu = 1.62$ for the upper, $\mu = 1.00$ for the lower.

β) *Magnesium*. The spread in the data in the whole temperature range is large. It would appear to be desirable to have more experimental data, especially measurements on one sample covering the whole temperature range.

γ) *Calcium*. The values of Θ and γ fit the data between 10 and 20° K. Without results at lower temperatures, however, it is impossible to judge if the true T^3 region has been reached.

Table 3. *Low temperature atomic heat of the Elements-Group IIa.*

Element	Θ_0(°K)	range (°K)	T_B(°K)	γ (millijoules/mole degree²)	reference
Be	1160	4—300 >10	20	0.226	18.4 18.3
Mg	(330—390)	2—4 >10		(1.35)	17.4 18.1, 18.2
Ca	(219)	>10		0.38	18.1

18.1 K. Clusius and J. V. Vaughen, J. Amer. Chem. Soc. **52**, 4686 (1930): Ca; 10—200° K, Mg; 10—300° K, Tl; 10—250° K.

18.2 R. S. Craig, C. A. Krier, L. W. Coffer, E. A. Bates and W. E. Wallace, J. Amer. Chem. Soc. **76**, 238 (1954): Mg, Cd; 12—320° K.

18.3 S. Cristescu and F. Simon, Z. phys. Chem. Abt. B **25**, 273 (1934): Be; 10—300° K, Ge; 10—200° K, Hf; 13—210° K.

18.4 R. W. Hill and P. Smith, Phil. Mag. **44**, 636 (1953): Be; 4—300° K.

19. Group IIb. *α) Zinc.* Around 4° K the results of Silvidi and Daunt are slightly higher than those of Keesom and van den Ende. The corresponding

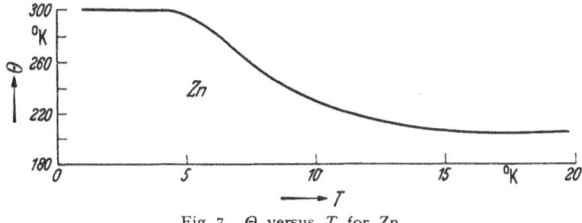

Fig. 7. Θ versus T for Zn.

Θ_0 values are 296° K and 321° K respectively. There is good agreement at higher temperatures among the results of different investigations. It is striking that Θ is nearly constant between 12 and 20° K (see Fig. 7), but with a very different

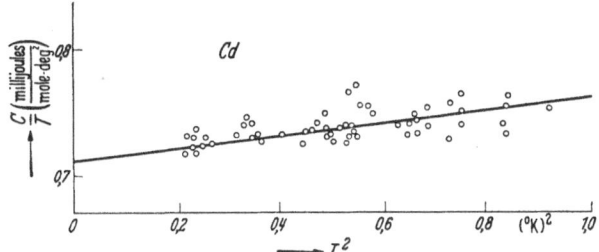

Fig. 8. C/T versus T^2 for Cd below 1° K, after Samoilov (reference 19.5).

value from that in the helium region. This is thus an excellent example of a "pseudo-T^3 region" in constrast to the true T^3 region below 5° K.

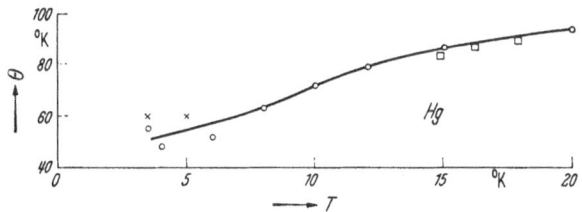

Fig. 9. Θ versus T for Hg. Smoothed curve and ◯: Pickard and Simon (reference 16.3); ×: Kamerlingh Onnes and Holst (reference 19.2); ☐: Busey and Giauque (reference 19.1).

β) Cadmium. The only measurements below hydrogen temperatures appear to be those of Samoilov below 1° K. His results are given in Fig. 8 as a plot of C/T versus T^2. The Θ values in the hydrogen region are less than half the value of Θ_0 determined by Samoilov.

γ) Mercury. Pickard and Simon's smoothed results are given in Fig. 9. We also include two points measured by Kamerlingh Onnes and Holst as these are the first calorimetric data in the liquid helium range. In order to separate the electronic and lattice terms data at lower temperatures are necessary.

Table 4. *Low temperature atomic heat of the Elements-Group IIb.*

Element	Θ_0(°K)	range (°K)	T_B(°K)	γ (millijoules/mole degree²)		reference
				calorimetric	magnetic	
Zn	321	1—20	5	0.654		19.3
		1—5				17.10
	296	1—4		0.66		19.6
		>12				17.2
					0.48—0.59	[*10*]
Cd	300	<1		0.71		19.5
		>12				18.2
					0.71	[*10*]
Hg	(60—90)	3—90				16.3
		>15				19.1
					1.6—2.2	[*10*]

19.1 R. H. BUSEY and W. F. GIAUQUE, J. Amer. Chem. Soc. **75**, 806 (1953): Hg; 15—300° K.

19.2 H. KAMERLINGH ONNES and G. HOLST, Commun. Phys. Lab. Univ. Leiden (1914), No. 142c: Hg; 4° K.

19.3 W. H. KEESOM and J. N. VAN DEN ENDE, Proc. Kon. Ned. Akad. Wetensch. **35**, 143 (1932), also in Commun. Kamerlingh Onnes Lab. Univ. Leiden, no. 219b: Sn, Zn; 1—20° K.

19.4 F. LANGE and F. SIMON, Z. phys. Chem. **134**, 372 (1928): Cd; 10—300° K.

19.5 B. N. SAMOILOV, Dokl. Akad. Nauk SSSR. **86**, 281 (1952): Cd; below 1° K.

19.6 A. A. SILVIDI and J. G. DAUNT, Phys. Rev. **77**, 125 (1950): Zn, W; 1—4° K.

20. Group IIIa. Lanthanum and lanthanide rare earths.

The only elements of this Group which have been measured in the liquid helium region are lanthanum and the associated rare earths cerium, praseodymium and neodymium. The results of PARKINSON et al. below 35° K are shown in Fig. 10. The atomic heat of lanthanum is the lowest of the four elements and its variation with temperature is smooth except for the jump at the superconducting transition at 4.37° K (just visible on the Figure). The true T^3 region appears to extend to about 6° K. The jump in the atomic heat at T_0, the normal transition point should therefore be related to the value of γ by

$$\gamma = (\Delta C)_{T_0}/2\,T_0 \qquad (20.1)$$

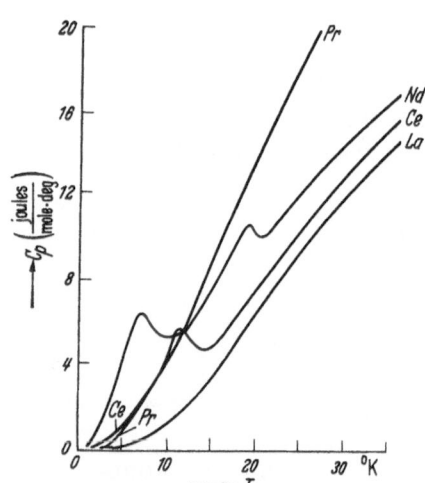

Fig. 10. *C* versus *T* for La, Ce, Pr and Nd, after PARKINSON et al. (reference 20.1).

(see Sect. 33) if the threshold magnetic field curve is parabolic. The value of $(\Delta C)_{T_0}$ obtained by extrapolation from the atomic heat curves below and above T_0 is 58 millijoules/mole degree, which by (20.1) corresponds to the same γ value as that observed in C_n, the atomic heat in the normal state (see Table 5).

The atomic heats of the three rare earths, however, are anomalous. There are peaks in the atomic heats of cerium and neodymium, and while praseody-

Table 5. *Low temperature atomic heat of the Elements-Group III a.*

Element	Θ_0(°K)	range (°K)	T_B(°K)	γ (millijoules/mole degree²)		reference
				calorimetric	magnetic	
La	132	2—180	6	6.7 6.6[1]		20.1 20.1

mium has no peak its atomic heat is the highest of the four elements above about 11° K. The interpretation of these results is complicated by the fact that the atomic heat may depend on the crystal structure present (cubic or hexagonal close-packed) and in the case of cerium at least, the magnitude of the low temperature peak depends on the rate of cooling of the specimen. Cerium also exhibits anomalous behavior between about 90 and 170° K. In two samples irregular results with hysteresis effects were observed in this temperature range; in one of the samples there was a large peak which depended strongly on the rate of cooling and on the thermal treatment.

In view of these complications it is clearly difficult to account for these anomalies in a quantitative manner. PARKINSON *et al.* assume that because of the very similar crystal structures and electronic configurations the four elements have approximately the same lattice atomic heats. Lanthanum has no $4f$ electrons and cerium, praseodymium and neodymium have 1, 2 and 3, $4f$ electrons respectively, so they ascribe the differences between the atomic heats of the latter elements and that of lanthanum to heat capacity contributions from the $4f$ electrons. Since these electrons are rather deep in the atom cores (the valence electrons for all four elements are $6s$) there is presumably not enough overlapping between $4f$ wave functions on adjacent atoms to form $4f$ bands, but some of the degeneracy of these electronic states may be removed by the crystalline field. Transitions between the resulting levels would then be responsible for the excess atomic heat compared to lanthanum.

This mechanism accounts for the heat capacity anomalies in magnetically dilute hydrated paramagnetic salts (see Sect. 35 below). In the rare earth elements the magnetic interaction would be expected to complicate the analysis. PARKINSON *et al.* compared their observed excess atomic heat with contributions calculated from level splittings derived from measurements on the corresponding hydrated sulphates. They conclude that their explanation is correct in principle since the agreement is as good as could be expected in view of the complicating factors involved.

20.1 D. H. PARKINSON, F. E. SIMON and F. H. SPEDDING, Proc. Roy. Soc. Lond., Ser. A **207**, 137 (1951): La, Ce, Pr, Nd; 2—180° K.

21. Group IIIb. α) *Aluminum.* The results of KEESOM and KOK between 1 and 20° K can be represented very well by the sum of a cubic and a linear term with the corresponding Θ_0 and γ values listed in Table 6. If, however, the data between 1 and 4° K are considered by themselves a slightly different set is found: $\Theta_0 = 511°$ K and $\gamma = 1.56$ millijoules/mole degree². The electronic term is so large, however, that this latter value of Θ_0 is uncertain. Although in general it is dangerous to combine the hydrogen with the helium data to obtain a constant Θ over the whole range, this appears to be the preferable choice in this case.

The results of GIAUQUE and MEADS in the hydrogen range are about 10% lower than those discussed above.

---

[1] Calculated from (20.1), using observed values of $(\Delta C)_{T_0}$.

The electronic band structure of aluminum has been investigated by MATYAS and later by LEIGH. RAYNOR [11] gives a curve of $g_a(\varepsilon)$ versus energy based on MATYAS' calculations which shows that there is overlap between the first and second zones and that the FERMI level occurs at an energy where the contribution from both zones is appreciable, so that ϱ_E should be greater than unity. The value of $g_a(\zeta)$ obtained from a numerical integration of this curve is 0.87 which corresponds to a ϱ_E of 2.2, while that derived from the measurements of KOK and KEESOM is 1.6 (assuming $n_a = 3$). LEIGH has pointed out in connection

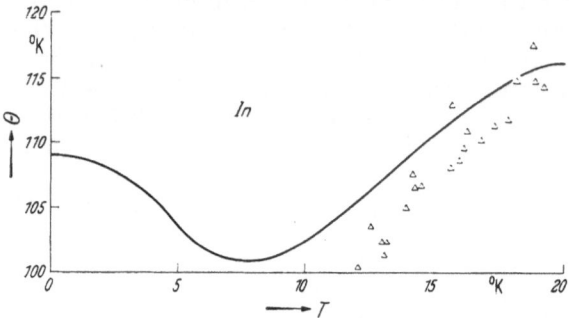

with his discussion of the elastic constants of aluminum that the band structure is probably different from that calculated by MATYAS. LEIGH did not calculate $g_a(\zeta)$, however, but used the value 0.619, which corresponds to the measurements of KOK and KEESOM, as a known parameter in his calculations.

Fig. 11. Θ versus T for In. Solid line: CLEMENT and QUINNELL (reference 21.2); $\triangle$: CLUSIUS and SCHACHINGER (reference 21.3).

To the value of $(\Delta C)_{T_0}$ observed by KOK and KEE-SOM, 1.9 millijoules/mole degree there corresponds by (20.1) the γ value 0.85 millijoules/mole degree². This is smaller than both the value obtained from C_n and that deduced from the magnetic threshold curve.

β) Gallium. Only a few points are available in the hydrogen range. DE SORBO remarks that the data between 15 and 34° K fit a T^2 law.

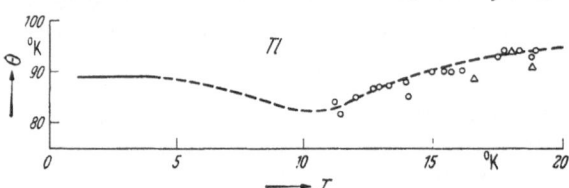

γ) Indium. Instead of using the method described in Part B above, CLEMENT and QUINNELL heated their indium sample continuously and derived the heat capacity from the rate of change of temperature. They made a careful analysis of their results using for the lattice con-

Fig. 12. Θ versus T for Tl. Solid line and $\triangle$: KEESOM and KOK (reference 21.6); $\bigcirc$: CLUSIUS and VAUGHEN (reference 18.1).

tribution the sum of a term proportional to T^3 and a term proportional to T^5. They found the coefficient of the T^5 term to be small, so that Θ does not vary appreciably in the helium range. In Fig. 11 the solid line gives their values of Θ while the results of CLUSIUS and SCHACHINGER are plotted separately as points. There appears to be some disagreement, except at 20° K.

CLEMENT and QUINNELL found $(\Delta C)_{T_0}$ to be 9.75 millijoules/mole degree. By (20.1) this corresponds to $\gamma = 1.44$ millijoules/mole degree² which again is lower than the calorimetric and magnetic value. For indium $T_B > T_0$ but the magnetic threshold curve is not parabolic in the neighborhood of T_0.

δ) Thallium. The variation of Θ with T is given in Fig. 12.

KEESOM and KOK found the value 6.19 millijoules/mole degree for $(\Delta C)_{T_0}$, the jump in the atomic heat at the normal transition point, $T_0 = 2.36°$ K. By (20.1) this corresponds to $\gamma = 1.3$ millijoules/mole degree² which agrees excellently with the value obtained from magnetic threshold field measurements. The calorimetric value which we obtained from a least squares fit to the points of KEESOM and Kok above T_0 is much larger, however (see Table 6). There appears to be no obvious reason for this discrepancy since $T_B > T_0$ and the magnetic threshold curve is approximately parabolic so that both of the assumptions on which (20.1) is based are essentially valid.

Table 6. *Low temperature atomic heat of the Elements-Group IIIb.*

Element	Θ_0 (°K)	range (°K)	T_B (°K)	γ (millijoules/mole degree²)		reference
				calorimetric	magnetic	
B		>13				21.5
Al	418	$1-20$	20	1.46		21.7
				0.85 [1]		21.7
		>15				17.7
					1.2	[10]
Ga		>15				17.2, 21.1
					0.38	[10]
In	109	$1-20$	4	1.81		21.2
				1.44 [1]		21.2
					1.8	[10]
		>12				21.3
Tl	89	$1-20$	4	3.1		21.6
				1.31 [1]		21.6
		>10				18.1, 21.4
					1.3	[10]

21.1 G. B. Adams Jr., H. L. Johnston and E. C. Kerr, J. Amer. Chem. Soc. 74, 4784 (1952): Ga; 15—320° K.

21.2 J. R. Clement and E. H. Quinnell, Phys. Rev. 92, 258 (1953): In; 1—20° K.

21.3 K. Clusius and L. Schachinger, Z. angew. Phys. 4, 442 (1952): In; 12—273° K.

21.4 J. F. G. Hicks Jr., J. G. Hooley and C. C. Stephenson, J. Amer. Chem. Soc. 66, 1064 (1944): Tl; 14—300° K.

21.5 H. L. Johnston, H. N. Hersh and E. C. Kerr, J. Amer. Chem. Soc. 73, 1112 (1951): B; 13—305° K.

21.6 W. H. Keesom and J. A. Kok, Physica, Haag 1, 175, 503, 595 (1934), also in Commun. Kamerlingh Onnes Lab. Univ. Leiden, nos. 230c, 230e, 232a: Tl; 1—4° K.

21.7 J. A. Kok and W. H. Keesom, Physica, Haag 4, 835 (1937), also in Commun. Kamerlingh Onnes Lab. Univ. Leiden no. 248e: Al; 1—20° K.

21.8 W. de Sorbo, J. Chem. Phys. 21, 168 (1953): Ga; discussion of T^2 dependence.

22. Group IVa. This Group has been the subject of extensive study for a variety of reasons. Two of the elements each occur in two different crystal lattices at low temperatures. Two become superconducting and especially for tin, the thermodynamic consequences of the transition between the superconducting and the normal states have been studied in detail. Finally the lattice atomic heat of the diamond lattice has been investigated theoretically, and also that of the graphite modification of carbon.

α) Carbon-Graphite. Graphite crystallizes in a typical layer lattice, in which the atoms in certain planes have smaller interatomic distance, and are more strongly bound by interatomic forces, than is the case for atoms in adjoining planes. Much theoretical work has been done towards elucidating the effect of this anisotropy on the low temperature heat capacity. In addition, the electronic band structure has been investigated.

---

[1] Calculated from (20.1), using observed values of $(\Delta C)_{T_0}$.

The assumption of isotropic distribution of lattice vibration momentum vectors used in the discussion of Sect. 5 is no longer tenable except at very low frequencies. The anisotropic distribution appropriate to the graphite lattice seems to give rise to the expectation of a quadratic dependence of the lattice atomic heat on temperature, over a certain temperature range. Various authors, however, are in disagreement as to the form of the resulting vibration spectrum and thus of the temperature range covered by the T^2 dependence. On the other hand, there seems to be agreement that at sufficiently low temperatures a transition to cubic dependence should be observed, but again, the location of this transition seems to depend on the manner in which the calculation is performed. There

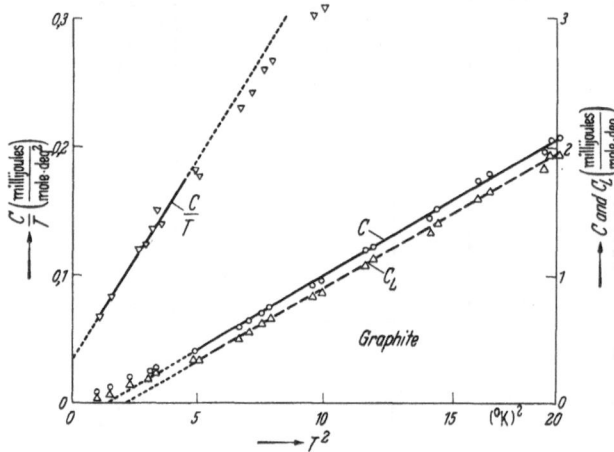

Fig. 13. C/T, C and C_L versus T^2 for graphite. After KEESOM and PEARLMAN (reference 22.14).

is the further difficulty that quantitative prediction from theoretical results demands knowledge of elastic properties (in order to evaluate the interatomic forces) which have not been measured for graphite.

Data obtained by DE SORBO and TYLER between 13 and 300° K when plotted as Log C versus Log T show a region of T^2 dependence between 13 and 54° K. In a similar plot, the data of BERGENLID et al. between 1.5 and 100° K give a smooth variation of the exponent of T between 1.8 at 90° K and 2.4 in the helium temperature region. Measurements of KEESOM and PEARLMAN between 1.0 and 20° K agree very well with those of BERGENLID et al. These results are given in Fig. 13, in the form of C/T versus T^2 (left-hand scale) and C and C_L versus T^2 (right-hand scale). It appears that below 2° K the data can be represented by the sum of a cubic and a linear term, the latter presumably the electronic contribution. This term agrees within an order of magnitude with that calculated by KOMATSU and NAGAMIYA. The transition temperature is very close to that recently calculated by KOMATSU, though this may be fortuitous since he has adjusted his results to agree in magnitude with the experimental data of BERGENLID et al. The departure of the data above 2° K from a simple quadratic dependence on temperature may be the analog to the departure from cubic dependence above the true T^3 region in isotropic lattices. The effect of the anisotropy can be seen in the deficit in heat capacity compared to extrapolated cubic dependence, whereas in isotropic lattices an excess is almost always observed above the true T^3 region.

β) Carbon-Diamond. The atomic heat of diamond has not been measured below 20° K. Because large single crystals are not readily available, small pieces in a calorimeter containing exchange gas must be used. Diamond has a very high DEBYE Θ (1800 to 2000° K) so that the heat capacity of the addenda will be a large fraction of the total heat capacity at low temperatures, thus reducing the accuracy with which c_{sam} can be determined. The most recent extensive measurements are those of DE SORBO, whose sample consisted of almost 7 moles of "fragmented boart", particle size about 2 mm. Because of the difficulties discussed above his experimental error was rather large below 70° K and this may account for his observation of a peak in Θ at about 60° K. The occurrence of such a peak would be surprising, since as is discussed below, the Θ versus T curves which have been observed down to the true T^3 region for several other substances which crystallize in the diamond lattice show no signs of a peak. DE SORBO's data are consistently lower than the earlier results of PITZER, but this difference may arise from small concentrations of impurities in each of the samples.

γ) Silicon. The atomic heat of pure silicon has not been measured at helium temperatures. Since silicon is a semiconductor, very small impurity concentrations will result in relatively large changes in the electronic distribution with corresponding effects in the heat capacity. PEARLMAN and KEESOM measured the atomic heat between 1 and 100° K of silicon containing about 2 parts in 10^5 boron, which provided about 10^{18} electrical carriers per cm^3. These carriers apparently are a degenerate assemblage at low temperatures and so give a contribution to the heat capacity which varies linearly with temperature. Because Θ_0 is high, the electronic contribution is 50% of the total at about 1° K. The value of Θ_0 is very close to that calculated from the elastic constants by a modification of the method of HOPF and LECHNER (see Sect. 5).

The effect on the heat capacity of irradiation by neutrons in a nuclear reactor has been investigated by KEESOM et al. Exposure to a total flux of 5×10^{18} neutrons per cm^2 was found to produce two effects: a) the value of Θ_0 decreased by about 3%; b) the linear electronic term vanished, within the accuracy of measurement. Annealing the sample after irradiation at temperatures up to about 500° C produced no change in the low temperature atomic heat, but annealing at 780° C restored the electronic term without changing the post-irradiation value of Θ_0 appreciably. These changes can be interpreted in terms of the effects on the electrical properties of silicon of neutron irradiation (see reference 22.12 for references to this work). The lattice disorder (vacant sites and interstitial atoms) resulting from the irradiation apparently results in new levels in the forbidden energy gap between the valence electron band and the higher conduction electron band. These levels serve to trap the conduction electrons and holes at low temperatures, thus accounting for the disappearance of the linear term in the atomic heat due to the carriers (holes in this case since the sample was a p-type semiconductor before irradiation). Annealing at sufficiently high temperatures reduces the irradiation-produced lattice disorder and so releases the carriers from the trapping levels, thereby restoring their contribution to the atomic heat.

The interpretation of the change in Θ_0 produced by irradiation, which is not restored by annealing, is not as clear. It is possible that annealing does not restore the lattice to its pre-irradiation state, but instead causes clustering of the vacant sites and interstitial atoms. This could reduce the number of trapping levels without materially affecting the interatomic forces on which Θ_0 depends.

δ) Germanium. The data of HILL and PARKINSON and of KEESOM and PEARLMAN are in good agreement at temperatures where they overlap. The low value for Θ_0 reported by ESTERMANN and FRIEDBERG (268° K) was apparently due to the large amount of scatter in their data.

An electronic component in the low temperature atomic heat of germanium has not been observed even with material more impure than the silicon samples discussed above. This is not surprising because the effective masses of the carriers are of the same order of magnitude in silicon and germanium. The electronic

Table 7. *Low temperature atomic heat of the Elements-Group IV a.*

Element	Θ_0 (°K)	range (°K)	T_B (°K)	γ (millijoules/mole degree²) calorimetric	magnetic	reference
C (graphite)	391	1—20 1.5—200 >13	2	0.031		22.14 22.1 22.23
C (diamond)	(2000)	>70 >20 drop method				22.19 22.22 22.2
Si	658	1—100	5			22.18
Ge	366	1—4 2—4 4—170 >20	4			22.13 22.4 22.6 22.5, 18.3
Sn (gray)	(200)	7—110 >9				(see text) 22.6 22.17
Sn (white)	189	1—4 3.5—3.9 1—20 1—4 >9	3	1.82 1.4 [1] 1.4 [1]	1.86	22.11 22.10 19.3 22.20 22.17 [10]
Pb	94.5	1—70 1—20 6.7—7.7 >15	4	3.0 3.64 [1]	2.9	22.7 22.9 22.3 17.14 [10]

contribution is directly proportional to the effective mass ratio and proportional to the cube root of impurity concentration [see (9.7)] so that it will be about the same in both substances. The lattice component, however, is almost six times larger in germanium than in silicon so that measurements would have to be made at temperatures below 1° K for the electronic contribution to be detected.

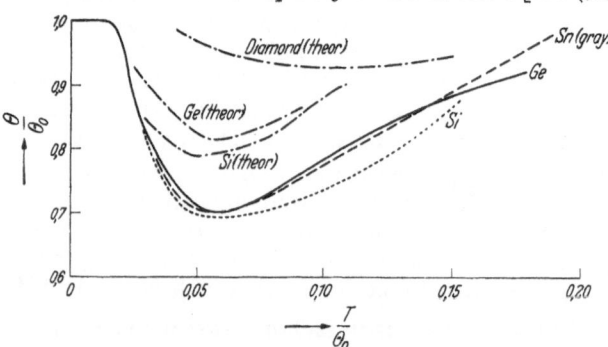

Fig. 14. Θ/Θ_0 versus T/Θ_0 for elements having the diamond lattice. The theoretical curve for diamond is due to SMITH (reference 22.21) and those for Ge and Si are due to HSIEH (reference 22.8).

ε) *Gray tin.* Tin exists in several allotropic forms. The α-modification, gray tin, has the diamond crystal structure and is stable below 18° C. Since the density of gray tin is less than 80% of that of metallic tin (β-tin, white tin) which is stable between 18 and 170° C, massive crystals of metallic tin decompose into

---

[1] Calculated from (20.1), using observed values of $(\Delta C)_{T_0}$.

a fine powder upon transforming into gray tin. The usual problems connected with calorimetric measurements on powder in a calorimeter vessel containing exchange gas prevented HILL and PARKINSON from obtaining accurate heat capacity data below $7°$ K.

ζ) *The diamond lattice.* Except for diamond itself, the substances which have the diamond crystal structure show very similar temperature variations in their low temperature molar heats. The dependence of Θ/Θ_0 on T/Θ_0 is plotted in Fig. 14 for germanium, silicon and gray tin. The molar heat of indium anti-monide has been measured between 1 and $20°$ K (unpublished results of KEESOM and PEARLMAN), and Θ_0 found to be $200°$ K. This substance also has the diamond crystal structure and its lattice constant is almost identical with that of gray tin (6.45 Å compared to 6.46 Å). The atomic masses of In and Sb are close to that of Sn. If Θ_0 is assumed to be

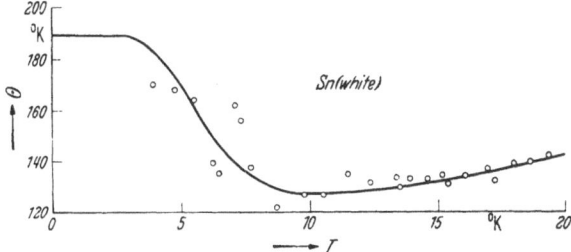

Fig. 15. Θ versus T for white Sn, after KEESOM and VAN DEN ENDE (reference 19.3).

$200°$ K for gray tin as well, then Θ/Θ_0 versus T/Θ_0 for InSb and gray tin are identical up to T/Θ_0 equal to 0.10. The curves for silicon and germanium are also similar up to the common minimum at about $T/\Theta_0 = 0.06$. Corresponding curves have been calculated on the basis of the BORN-VON KÁRMÁN theory, by SMITH for diamond and by HSIEH for germanium and silicon, and are shown in Fig. 14 for comparison (dot-dash lines). The experimental points of PITZER for the atomic heat of diamond lie fairly close to the theoretical curce.

η) *Metallic tin.* Since the normal superconductive transition point of metallic tin is conveniently located in the helium temperature region ($T_0 = 3.73°$ K) its thermo-dynamic consequences have been studied extensively at

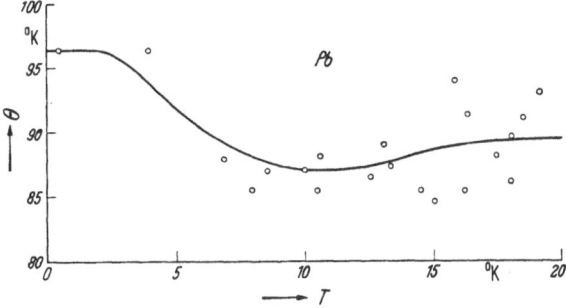

Fig. 16. Θ versus T for Pb after HOROWITZ *et al.* (reference 23.7)

Leiden. The value of γ which follows from $(\Delta C)_{T_0}$ as observed by KEESOM and KOK (10 millijoules/mole degree) and (20.1), is 1.4 millijoules/mole degree^2. This is somewhat lower than the calorimetric value, which is to be expected in this case since $T_B < T_0$ so that $C_n(T_0)$ is larger than is assumed in the derivation of (20.1). Hence $(\Delta C)_{T_0}$ will be too small and the same should be true of γ calculated from (20.1). For further details concerning calorimetric investigations of the super-conductive transition of tin we refer to Sect. 31 to 33 below and to the article on superconductivity in the next volume of this Encyclopedia.

The relation between Θ and T in the normal state up to $20°$ K is shown in Fig. 15.

ϑ) *Lead.* Following HOROWITZ *et al.* we give in Fig. 16 the variation of Θ with T, together with a theoretical curve deduced from LEIGHTON's work (see reference 17.13). The results of the different investigators have a considerable spread but fall around the curve.

It is clear from Fig. 16 that T_B is less than T_0 for lead $(T_0 = 7.23°$ K). The argument given above for tin concerning the relation between γ determined from $(\varDelta C)_{T_0}$ and (20.1) and that directly observed from C_n should therefore apply in this case as well. The measurements of Clement and Quinnell in the neighborhood of T_0 give the value 52.7 millijoules/mole degree for $(\varDelta C)_{T_0}$ which corresponds to $\gamma = 3.64$ millijoules/mole degree². However, this is higher than the value calculated from the C_n data of Horowitz et al. This discrepancy may be attributable to the departure from parabolic form of the threshold magnetic field curve near T_0.

22.1 U. Bergenlid, R. W. Hill, F. J. Webb and J. Wilks, Phil. Mag. **45**, 851 (1954): graphite; 1.5—100° K.

22.2 R. Berman and J. Poulter, J. Chem. Phys. **21**, 1906 (1953): diamond (drop method); difference in heat content between 90 and 290° K and between 4 and 90° K.

22.3 J. R. Clement and E. H. Quinnell, Phys. Rev. **85**, 502 (1952): Pb; 6.7—7.7° K.

22.4 I. Estermann and S. A. Friedberg, Phys. Rev. **85**, 715 (1952): Ge; 2—4° K.

22.5 I. Estermann and J. R. Weertman, J. Chem. Phys. **20**, 972 (1952): Ge; 20—200° K.

22.6 R. W. Hill and D. H. Parkinson, Phil. Mag. **43**, 309 (1952): Ge; 4—170° K, gray tin; 7—110° K.

22.7 M. Horowitz, A. A. Silvidi, S. F. Malakker and J. G. Daunt, Phys. Rev. **88**, 1182 (1952): Pb; 1—17° K.

22.8 Y. C. Hsieh, J. Chem. Phys. **22**, 306 (1954): Ge and Si; theoretical vibration spectra.

22.9 W. H. Keesom and J. N. van den Ende, Proc. Kon. Ned. Akad. Wetensch. **33**, 243 (1930) and **34**, 210 (1931), also in Commun. Kamerlingh Onnes Lab. Univ. Leiden nos. 203d and 213c: Pb, Bi; 1—20° K.

22.10 W. H. Keesom and J. A. Kok, Proc. Kon. Ned. Akad. Wetensch. **35**, 743 (1932), also in Commun. Kamerlingh Onnes Lab. Univ. Leiden, no. 221e: Metallic tin; 3.5—3.9° K.

22.11 W. H. Keesom and P. H. van Laer, Physica, Haag **5**, 193 (1938), also in Commun. Kamerlingh Onnes Lab. Univ. Leiden no. 252b: metallic tin; 1—4° K.

22.12 P. H. Keesom, K. Lark-Horovitz and N. Pearlman, Science, Lancaster, Pa. **116**, 630 (1952): neutron irradiated and annealed Si; 1—4° K.

22.13 P. H. Keesom and N. Pearlman, Phys. Rev. **91**, 1347 (1953): Ge; 1—4° K.

22.14 P. H. Keesom and N. Pearlman, Phys. Rev. **99**, 1119 (1955): graphite; 1—20° K.

22.15 K. Komatsu and T. Nagamiya, J. Phys. Soc. Japan **6**, 438 (1951): graphite; theoretical calculation of lattice and electronic atomic heat.

22.16 K. Komatsu, J. Phys. Soc. Japan **10**, 346 (1955): graphite; lattice atomic heat theory, including references to earlier work.

22.17 F. Lange, Z. phys. Chem. **110**, 343 (1924): gray and metallic tin; 9—280° K, W; 20—91° K.

22.18 N. Pearlman and P. H. Keesom, Phys. Rev. **88**, 398 (1952): Si; 1—100° K.

22.19 K. S. Pitzer, J. Chem. Phys. **6**, 68 (1938): diamond; 70—300° K.

22.20 K. G. Ramanathan and T. M. Srinivasan, Phil. Mag. **46**, 338 (1955): metallic tin; 1—4° K.

22.21 H. M. J. Smith, Phil. Trans. Roy. Soc. Lond., Ser. A **241**, 105 (1948): diamond; theoretical vibration spectrum.

22.22 W. de Sorbo, J. Chem. Phys. **21**, 876 (1953): diamond; 20—300° K.

22.23 W. de Sorbo and W. W. Tyler, J. Chem. Phys. **21**, 1660 (1953): graphite; 13—300° K.

23. Group IVb. α) Titanium.
Measurements have been reported between 2 and 4° K and above 14° K, but the two sets of data do not appear consistent. Table 8 includes only the helium range data. It would be desirable to have data on one sample covering the entire temperature range.

β) Zirconium. Data are also available between 2 and 4° K and above 14° K, but in this case they are in reasonable agreement. The Debye Θ decreases from 270° K in the helium region to 254° K in the hydrogen region.

Table 8. *Low temperature atomic of the elements-Group IVb.*

Element	Θ_0(°K)	range (°K)	T_B(°K)	γ (millijoules/mole degree²)		reference
				calorimetric	magnetic	
Ti	278	2—4 >15	4	3.34	0.46	17.4 23.1 [10]
Zr	270	2—4 >14	4	2.95	1.64	17.4 23.2 [10]

23.1 C. W. KOTHEN and H. L. JOHNSTON, J. Amer. Chem. Soc. **75**, 3101 (1953): Ti; 15—305° K.

23.2 G. B. SKINNER and H. L. JOHNSTON, J. Amer. Chem. Soc. **73**, 4549 (1951): Zr; 14—300° K.

24. Group Va. Of the elements in this Group only bismuth has been measured in the helium temperature region; antimony has been measured down to 13° K.

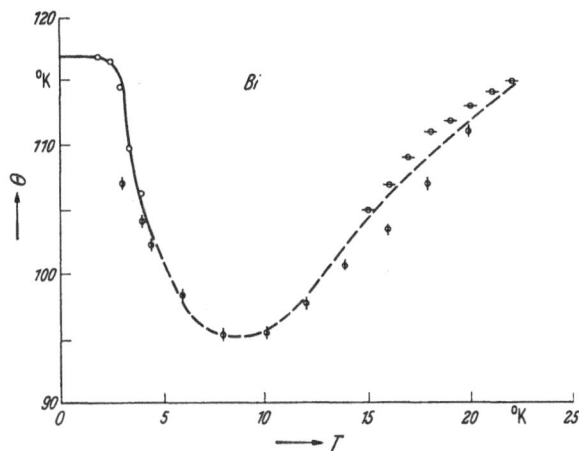

Fig. 17. Θ versus T for Bi after KEESOM and PEARLMAN (reference 24.2). —O—: ARMSTRONG and GRAYSON-SMITH (reference 24.1); ⦶ : KEESOM and VAN DEN ENDE (reference 22.9).

α) *Bismuth.* The variation of Θ with T, which is shown in Fig. 17, appears to be normal. The results of KEESOM and VAN DEN ENDE are about 10% higher than those of ARMSTRONG and GRAYSON-SMITH in the hydrogen region.

Table 9. *Low temperature atomic heat of the Elements-Group Va.*

Element	Θ_0(°K)	range (°K)	T_B(°K)	γ (millijoules/mole degree²)	reference
Bi	117	1—4 1—20 >14	2.3	0.078	24.2 22.9 24.1

24.1 L. D. ARMSTRONG and H. GRAYSON-SMITH, Canad. J. Res. A **27**, 9 (1949): Bi; 14—22° K.

24.2 P. H. KEESOM and N. PEARLMAN, Phys. Rev. **96**, 897 (1954): Bi; 1—4° K.

24.3 W. DE SORBO, Acta metallurgica **1**, 503 (1954): Sb; 13—70° K.

25. Group Vb. These elements are of interest for investigation of the thermodynamics of the phase transition from the normal to the superconducting state. They have the advantage that their Θ_0 values are high, as well as their normal transition temperatures. The former correspond to low values of C_L, so that the electronic atomic heat, which is of major interest in the transition, is a large fraction of the total. High transition temperatures permit measurements down to low values of the reduced temperature T/T_0, where T_0 is the normal transition temperature. A slight disadvantage, however, is that T_0 for vanadium and niobium falls in the "intermediate" temperature region between 4 and 10° K where no cooling bath is available.

One difficulty that is apparent with measurements on metals in this Group is that different samples give different results without apparent reason, even if measured by the same group of investigators.

α) *Tantalum.* The data of WORLEY et al. on tantalum in the normal state agree with those of KEESOM and DESIRANT above $T^2 = 6$, but the former find higher values of C/T below this (see Fig. 26). In a discussion of these results at the Low Temperature Conference held in Houston in December 1953, it was suggested that the magnetic field used by WORLEY et al. might not have been high enough, so that the sample perhaps was partially in the intermediate state at the lowest temperatures.

The results of KEESOM and DESIRANT above the boiling point of helium appear to be too high, probably due to difficulties in extrapolating their resistance thermometer calibration.

These measurements and those of DESIRANT and MENDELSSOHN (see references 25.3 and 25.5 for the latter) give values of $(\Delta C)_{T_0}$ of 34 and 40 millijoules/mole degree at T_0 equal to 4.0 and 4.4° K, respectively. By (20.1) these correspond to γ values of 4.2 and 4 millijoules/mole degree^2, respectively, which are lower than both the calorimetric and magnetic values.

β) *Vanadium.* WORLEY et al. observed positive deviations in C/T at their lowest temperatures similar to those mentioned for tantalum. The spread in their value of Θ_0 refers to measurements on two different samples, the lower value corresponding to the purer sample. SATTERTHWAITE et al. do not report deviations in C/T versus T^2. They do not give an analysis of the impurities in their sample, which had a T_0 value about 0.2° K higher than that of the purer sample of WORLEY et al.

An interesting indirect method was used by CORAK et al. to obtain their earlier estimate of γ, since they had not measured C_n and therefore could not evaluate $(\Delta C)_{T_0}$ directly. They calculated $S_n(T_0)$, the entropy of the normal state at T_0 on the assumption that C_n had the normal form below T_0. By (32.2) this can be set equal to $S_s(T_0)$, the entropy of the superconducting state at T_0, since H_c, the threshold magnetic field vanishes at T_0. In the resulting equation which also assumes $S_n(0° \text{K}) = S_s(0° \text{K})$ $(=0$ by NERNST's Theorem),

$$S_n(T_0) = \frac{\alpha}{3} T_0^3 + \gamma T_0 = S_s(T_0), \tag{25.1}$$

they estimated α from (15.2) and the average of the Θ_0 values reported by WORLEY et al. Since they were able to calculate $S_s(T_0)$ from their measurements of C_s, the atomic heat of the superconducting state below T_0, they were able to solve for γ.

γ) *Niobium.* Above 12° K the results deviate from the usual behavior of C_L, as Θ seems to increase above Θ_0 without any intervening decrease[1].

---

[1] ZEMANSKY indicated in private discussion that these data may need revision.

Table 10. *Low Temperature Atomic Heat of the Elements-Group V b.*

Element	$\Theta_0(°K)$	range (°K)	$T_B(°K)$	γ (millijoules/mole degree²) calorimetric	magnetic	reference
V	(273—321) 338	1—5 1—5	5	(8.6—8.9) 9.26 8.8 [1]	6.28	25.8 25.6 25.2 [10]
Nb	252	2—20	10	8.5		25.1
Ta	246 (213—225)	1—20 2—5 3.5—5.5	5	5.94 (5.0—5.4) 4 [2] 4.2 [2]	8.0	25.4 25.7 25.3, 25.5 25.4 [10]

25.1 A. BROWN, M. W. ZEMANSKY and H. A. BOORSE, Phys. Rev. **86**, 134 (1952): Nb; 2—20° K.

25.2 W. S. CORAK, B. B. GOODMAN, C. B. SATTERTHWAITE and A. WEXLER, Phys. Rev. **96**, 1442 (1954): V (superconducting); 1—5° K.

25.3 M. DESIRANT, Report of an International Conference on Fundamental Particles and Low Temperatures (The Physical Society, London, 1947), Vol. II, p. 124: Ta; 3.5 to 5.5° K.

25.4 W. H. KEESOM and M. DESIRANT, Physica, Haag **8**, 273 (1941), also in Commun. Kamerlingh Onnes Lab. Univ. Leiden no. 257b: Ta; 1—5° K.

25.5 K. MENDELSSOHN, Nature, Lond. **148**, 316 (1941): Ta; 3.5—5.5° K.

25.6 C. B. SATTERTHWAITE, W. S. CORAK, B. B. GOODMAN and A. WEXLER, Bull. Amer. Phys. Soc. **30**, No. 4, 34 (1955): V (normal); 1—5° K.

25.7 R. D. WORLEY, M. W. ZEMANSKY and H. A. BOORSE, Phys. Rev. **91**, 1567 (1953): Ta; 2—5° K.

25.8 R. D. WORLEY, M. W. ZEMANSKY and H. A. BOORSE, Abstracts of the Third International Conference on Low Temperature Physics and Chemistry (The Rice Institute, Houston, Texas, 1953, unpublished), p. 57: Ta and V; 2—5° K. Further details concerning these measurements are given in J. EISENSTEIN, Rev. Mod. Phys. **26**, 277 (1954).

26. Group VI a. No measurements in the helium temperature range are available for elements in this Group.

26.1 C. M. SLANSKY and L. V. COULTER, J. Amer. Chem. Soc. **61**, 564 (1939): Te; 14—300° K.

26.2 W. DE SORBO, J. Chem. Phys. **21**, 1144 (1953): Se; 15—300° K.

27. Group VI b. α) *Chromium.* If the method of least squares is applied to the data of ESTERMANN et al. in the liquid helium range a slightly lower value of Θ_0 is obtained than that given by these authors. Neither value is very accurate, however, as the electronic contribution to the atomic heat is large and there is considerable spread in the experimental points. The hydrogen range data are reported only in an abstract from which quantitative conclusions cannot be drawn.

β) *Molybdenum.* HOROWITZ and DAUNT obtained a curve of Θ versus T (see Fig. 18, curve I) by using a plot of C/T versus T^2 to determine C_E, finding $C_L = C - C_E$ for each measured point and then calculating Θ from each value of C_L. Above 18° K they used the data of SIMON and ZEIDLER. The one point shown in Fig. 18 at about 1.5° K is an average of all their Θ values below 4° K, which show a considerable spread. If, however, the slope of the least squares line

[1] Indirect method—see text.
[2] Calculated from (20.1), using measured values of $(\Delta C)_{T_0}$.

fitted to their values of C/T versus T^2 is used to find Θ_0, we get the value given in Table 11, which corresponds to curve II in Fig. 18. This curve resembles the behavior of Θ versus T of most other elements more closely than does curve I.

$\gamma)$ *Tungsten (Wolfram)*. The data of SILVIDI and DAUNT (see reference 19.6) have been superseded by the newer results of HOROWITZ and DAUNT. A variety of alternatives is available in interpreting these data, which are plotted in Fig. 19 up to 20° K, with the liquid helium temperature range points plotted separately

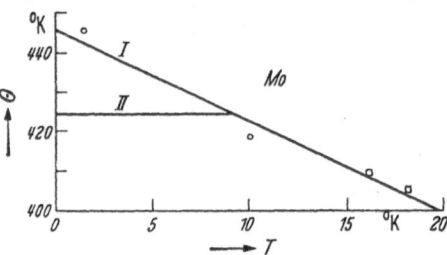

Fig. 18. Θ versus T for Mo. O: HOROWITZ and DAUNT (reference 27.2); □: SIMON and ZEIDLER (reference 16.6). For solid lines see text.

on a larger scale on the inset. The least squares fit for all the points is shown as a solid line on both the main graph and the inset. This corresponds to the second entry under W in Table 11 ($\Theta_0 = 379°$ K, $\gamma = 1.2$ millijoules/mole degree[2]). The dashed line in the inset has the same slope as the solid line with an intercept corresponding to RAYNE's value of γ (1.48 millijoules/mole degree[2]). Neither line fits the points in the inset very well, but the least squares fit to these points alone, shown as the dash-dot line in the inset, corresponds to much lower Θ_0 and γ values ($\Theta_0 = 279°$ K, $\gamma = 0.8$ millijoules/mole degree[2]) which are given as the third entry under W in Table 11. HOROWITZ and DAUNT estimated Θ_0 as 250° K by the method described above in the paragraph on molybdenum; their estimate of γ is 0.75 millijoules/mole degree[2] (first entry in Table 11).

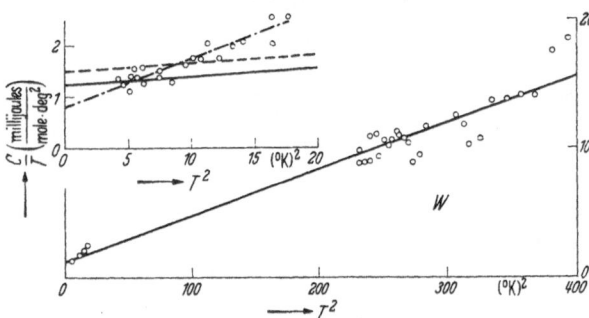

Fig. 19. C/T versus T^2 for W. O: HOROWITZ and DAUNT (reference 27.2); for lines see text.

These low values of Θ_0 would result in a $\Theta(T)$ curve which rises immediately after the true T^3 region rather than falling as is the case with other elements. As mentioned below the calculated vibration spectrum for tungsten also corresponds to a much higher value of Θ_0. These observations, and the scatter of the points below 4° K lead us to the conclusion that the best available estimates of Θ_0 and γ are 379° K, corresponding to the slope of the solid line in Fig. 19, and RAYNE's value, 1.48 millijoules/mole degree[2], respectively.

FINE has calculated the vibration spectrum for the body-centered lattice and applied the results to tungsten. The parabolic portion of the spectrum at the low frequency end corresponds to a Θ_0 of 367° K, which is in reasonable agreement with the value given in Table 1 (p. 302). FINE was able to compare his calculation only with experimental data above 26° K, from which it appeared that the calculated heat capacity was too low. The experimental data above 26° K are higher than would follow from a simple DEBYE spectrum with Θ equal to 367° K, so that Θ apparently decreases below Θ_0 at temperatures above 20° K.

MANNING and CHODOROW calculated the electron distribution and found $g_a(\zeta)$ equal to 0.858 levels/eV atom. The value calculated with RAYNE's value of γ in (9.3) is 0.627. By using the same distribution with one electron less, they estimated $g_a(\zeta)$ for tantalum to be 1.09, whereas the value calculated from the tabulated γ (see Table 10 in Sect. 25) is 2.3.

Table 11. *Low temperature atomic heat of the Elements-Group VIb.*

Element	Θ_0(°K)	range (°K)	T_B(°K)	γ (millijoules/mole degree²)	reference
Cr	402	1—4 >10	4	1.54	17.4 27.4
Mo	425	1—10 <1 >16	10	2.09 2.10	27.2 16.5 16.6
W	250 379 279	2—20 <1 >20	?	0.75 1.2 0.8 1.48	27.2 (see text) (see text) 16.5 22.17

27.1 P. C. Fine, Phys. Rev. **56**, 355 (1939): W; calculation of vibration spectrum.
27.2 M. Horowitz and J. G. Daunt, Phys. Rev. **91**, 1099 (1953): Mo; 1—10° K, W; 2—20° K.
27.3 M. F. Manning and M. I. Chodorow, Phys. Rev. **56**, 787 (1939): W and Ta; calculation of distribution of electrons.
27.4 J. Weertman, D. Burk and J. E. Goldman, Phys. Rev. **86**, 628 (1952): Cr; 10—300° K.

28. Group VIIa and b. *Manganese.* Of the elements in these Groups only manganese has been measured in the liquid helium range. Guthrie *et al.* have recently reported the value 13.8 millijoules/mole degree² for γ, with a probable error of about 5%, but they gave no Θ_0 value. The earlier data above 10° K indicate a Θ value of 417° K, but it is doubtful if this can be extrapolated to lower temperatures.

Table 12. *Low temperature atomic heat of the Elements-Groups VIIa and b.*

Element	Θ_0(°K)	range (°K)	T_B(°K)	γ (millijoules/mole degree²)	reference
Mn		2—4 >14		13.8	28.3 28.1, 28.2

28.1 L. D. Armstrong and H. Grayson-Smith, Canad. J. Res. A **27**, 9 (1949): Mn; 14—22° K.
28.2 R. G. Elson, H. Grayson-Smith and J. O. Wilhelm, Canad. J. Res. A **18**, 83 (1940): Mn; 16—40° K.
28.3 G. Guthrie, S. A. Friedberg and J. E. Goldman, Phys. Rev. **98**, 1181 (1955): Mn (electronic contribution only); 2—4° K.

29. Group VIII. These elements occur in sets of three at the ends of the three transition series, which correspond to the filling of the $3d$, $4d$ and $5d$ electron shells, respectively. Many of the properties of the transition elements, such as the ferromagnetism of iron, cobalt and nickel, have been related to the structure of the d-levels. The information which can be gained about this structure form the γ values of all the transition elements will be discussed in the next section.

α) Iron. The results of Duyckaerts and of Keesom and Kurrelmeyer are in excellent agreement, but those of Eucken and Werth appear to be about 10% too low. Keesom and Kurrelmeyer noticed a small irregularity around

13.5° K (just below $T^2 = 200$ in Fig. 20) and attributed this to a minute quantity of hydrogen in the cavities of their sample.

MANNING found from his calculation of the band structure the value 1.9 millijoules/mole degree² for γ, compared with 5.0 from the atomic heat.

β) Cobalt. In the overlapping temperature region the results of DUYCKAERTS and of CLUSIUS and SCHACHINGER agree excellently. The $\Theta(T)$ curve is normal; after a constant Θ_0 value in the true T^3 region it decreases. In their analysis of the higher temperature data, CLUSIUS and SCHACHINGER conclude that the electronic contribution is noticeable up to room temperature.

γ) Nickel. Agreement between the different sets of data is reasonable; the results of CLUSIUS and GOLDMAN in the hydrogen range are about 2% higher than those of KEESOM and CLARK. The true T^3 region seems to extend up to 20° K and above this temperature Θ begins to decrease.

FLETCHER's value of 7.1 millijoules/mole degree² for γ from his band structure calculation agrees very well with the value 7.4 obtained calorimetrically.

δ) Palladium. There is reasonable agreement among the different sets of data in the hydrogen region, with the true T^3 region extending to about 17° K.

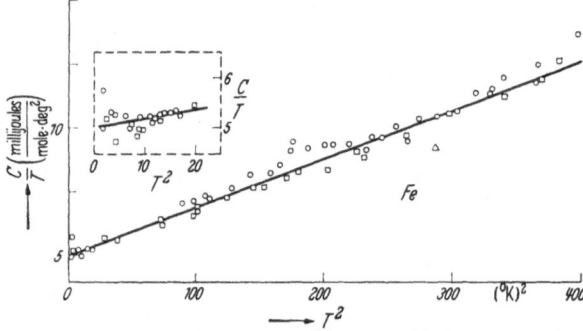

Fig. 20. C/T versus T^2 for Fe: O: KEESOM and KURRELMEYER (reference 29.9); □: DUYCKAERTS (reference 29.5); △: EUCKEN and WERTH (reference 29.7).

There is some spread in the reported γ values, with that of PICKARD and SIMON being rather higher than those of RAYNE and of YATES and HOARE.

ε) Platinum. All the data for the whole temperature range are in excellent agreement. The γ values of KEESOM and KOK above 1° K and of RAYNE below 1° K agree within the experimental error.

Table 13. *Low temperature atomic heat of the Elements-Group VIII.*

Element	Θ_0(°K)	range (°K)	T_B(°K)	γ (millijoules/mole degree²)	reference
Fe	469	1—20	20	5.0	29.5
	466	1—20	20	5.1	29.9
		>15			29.7
Co	445	2—18	20	5.0	29.6
		>15			29.4
Ni	459	10—26	19	7.6	29.2
	456	1—19	19	7.3	29.8
		>10			29.1, 29.7
Pd	275	2—22	17	13	16.3, 29.10
	273	>1		9.30	17.16
		<1		10.7	16.5
		>14			29.3
Pt	229	1—20	20	6.6	17.12
		<1		6.9	16.5
		>17			16.6

29.1 R. H. BUSEY and W. F. GIAUQUE, J. Amer. Chem. Soc. **74**, 3157 (1952): Ni; 13—300° K.

29.2 K. CLUSIUS and J. GOLDMAN, Z. phys. Chem. Abt. B **31**, 256 (1936): Ni; 10—26° K.

29.3 K. CLUSIUS and L. SCHACHINGER, Z. Naturforsch. 2a, 90 (1947): Pd; 14—270° K.

29.4 K. CLUSIUS and L. SCHACHINGER, Z. Naturforsch. 7a, 185 (1952): Co; 15—270° K.

29.5 G. DUYCKAERTS, Physica, Haag **6**, 401 (1939), also in C. R. Acad. Sci. Paris **208**, 979 (1939): Fe; 1—20° K.

29.6 G. DUYCKAERTS, Physica, Haag **6**, 817 (1939): Co; 2—18° K.

29.7 A. EUCKEN and H. WERTH, Z. anorg. u. angew. Chem. **188**, 152 (1930): Fe, Ni; 15—205° K.

29.8 W. H. KEESOM and C. W. CLARK, Physica, Haag **2**, 513 (1935), also in Commun. Kamerlingh Onnes Lab. Univ. Leiden no. 235e: Ni; 1—19° K.

29.9 W. H. KEESOM and B. KURRELMEYER, Physica, Haag **6**, 633 (1939), also in Commun. Kamerlingh Onnes Lab. Univ. Leiden no. 257a: Fe; 1—20° K.

29.10 G. L. PICKARD, Nature, Lond. **138**, 123 (1936): Pd; 2—22° K.

30. Electronic heat capacity in the transition metals and their alloys. A striking disparity is evident in Table 1 between the γ values of the transition metals and those of the non-transition metals. The average of 15 entries in the former group is 5.8 millijoules/mole degree^2 while that of 14 in the latter group is only 1.2. If, as is reasonable in terms of the interpretation to be discussed later, the values of the three ferromagnetic elements Fe, Co and Ni are doubled, the disparity becomes greater; the average for the transition metals so weighted is 7.2.

The relationship between the high electronic atomic heat of the transition metals and the d-bands was first pointed out by MOTT[1]. The value of $g_a(\zeta)$ and therefore of the electronic atomic heat would be expected to be high in d-bands for two reasons. In the first place, the spatial extension of the d-electron wave functions is not as large as that of the valence s-electron wave functions so that there is less overlap between the d-electron wave functions on adjacent atoms. Hence the d-bands should be narrower than the valence s-bands. Furthermore, the former must accomodate ten electrons while the latter accomodate only two. Thus if it is assumed that in the transition metals the d-band and the valence s-band overlap and that the FERMI level falls within the d-band these two factors could account qualitatively for the observed difference in γ values between the transition and non-transition metals.

For a quantitative discussion information is required on the actual structure of the d-band. This structure has been calculated for the $3d$-band by KRUTTER[2] and later by SLATER[3] and the latter's results are given in Fig. 21. The vertical dotted lines indicate the points at which the $3d$-band accomodates the indicated number of electrons. The overlapping $4s$-band is also shown, and it is clear that this band makes only a small contribution to $g_a(\zeta)$ when the FERMI level falls in the $3d$-band. This curve has been the basis of an extensive recent discussion

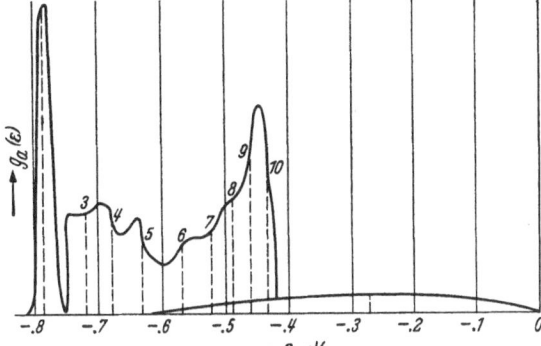

Fig. 21. $g_a(\varepsilon)$ versus ε for the $3d$ and $4s$ bands, after SLATER, Phys. Rev. 49, 537 (1936).

---

[1] N. F. MOTT: Proc. Phys. Soc. Lond. **47**, 571 (1935).
[2] H. M. KRUTTER: Phys. Rev. **48**, 664 (1935).
[3] J. C. SLATER: Phys. Rev. **49**, 537 (1936).

by DAUNT [12] who plots it with n_v, the number of valence electrons per atom, rather than energy as abcissa (see Fig. 22). The ordinate is γ which by (9.3) is proportional to $g_a(\zeta)$. This equation must be modified for application to the ferromagnetic metals Fe, Co and Ni. The d-band is presumed to split into two sub-bands, one for each direction of spin. In the ferromagnetic metals one of these sub-bands is completely filled with five electrons and shifted to lower energy than the other. The valence electrons in excess of five are distributed between the higher of the d-sub-bands and the overlapping s-band. Since the filled sub-band makes no contribution to the heat capacity we have instead of (9.3)

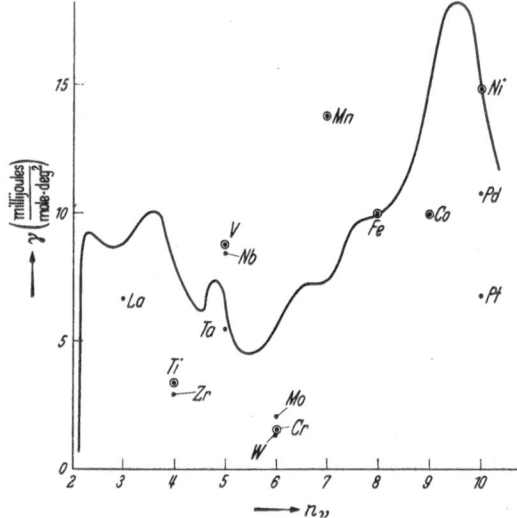

$$g_a(\zeta)_{\frac{1}{2}\cdot 3d} + g_a(\zeta)_{4s} = 0.424\,\gamma. \quad (30.1)$$

As was mentioned above $g_a(\zeta)_{4s}$ is small compared to $g_a(\zeta)_{3d}$; and since

$$g_a(\zeta)_{\frac{1}{2}\cdot 3d} = \tfrac{1}{2} g_a(\zeta)_{3d} \quad (30.2)$$

we have approximately for the ferromagnetic metals

$$g_a(\zeta)_{3d} = 2 \times 0.424\,\gamma. \quad (30.3)$$

Hence the values plotted in Fig. 22 for these metals are twice the calorimetric γ values.

The comparison in Fig. 22 between the calorimetric values and the $3d$-band level density curve based on SLATER's calculation is not entirely quantitative since, following DAUNT, the curve has been normalized so as to pass through

Fig. 22. γ versus n_v for the transition metals, after DAUNT [12]. Those in the fourth period are circled.

the observed γ for iron. The agreement would be expected to be best for the $3d$ transition metals, whose γ values are circled in Fig. 22, but except for nickel they do not seem to be closer to the curve than those of the $4d$ and $5d$ series. The recently reported value for manganese (see reference 28.3) is especially high; DAUNT has discussed the possibility of a large maximum in the level density curve at $n_v = 7$. The uniformly low values for the VIb elements are discussed by DAUNT in terms of a deep minimum in the d-bands at $n_v = 6$, for which there appears to be other evidence as well. This feature would seem to be common to the $3d$, $4d$ and $5d$ bands while at other values of n_v there are probably significant differences among them, especially at $n_v = 10$.

The electronic heat capacity of alloys of transition with non-transition metals can also be discussed in terms of the band structure of Fig. 21. In the case of the ferromagnetic metals the difference in occupancy between the filled and unfilled $3d$-sub-bands can be estimated from the saturation magnetic moment which follows from it. In nickel, for instance, this leads to the value 0.6 for the number n_a of holes per atom in the unfilled $3d$ band. There must therefore also be 0.6 electron per atom in the $4s$-band. Copper, with one valence electron more than nickel has the same crystal structure with a lattice constant only 3% larger, and forms solid solutions with nickel in all concentrations. If it is assumed that the electron distribution changes smoothly from that of nickel to that of copper as the concentration χ of copper increases from zero to one, the extra electrons contributed by the added copper atoms will fill the empty levels in the $3d$ and $4s$-bands. Since as we have seen $g_a(\varepsilon)_{3d}$ is much greater than $g_a(\varepsilon)_{4s}$ below the upper edge of the $3d$-band, most of the added electrons will go into the $3d$-levels. If the filling of the $4s$-levels is entirely neglected, the $3d$-band will be full at $\chi = 0.6$ which thus appears as a critical concentration below which the alloys should be nickel-like and above which they should be copper-like (full $3d$-band, electrons in the $4s$-band).

Some of the magnetic properties of copper-nickel alloys behave in this manner, but the molar heat of a series of alloys measured by KEESOM and KURRELMEYER[1] does not. Their γ values for alloys with 20, 40, 60 and 80% copper as well as those for pure nickel and copper are shown as the points in Fig. 23 and even that for the 60% copper alloy is about as large as that for pure nickel. The curve in Fig. 23 was calculated by WOHLFARTH[2] who took into account the filling of the

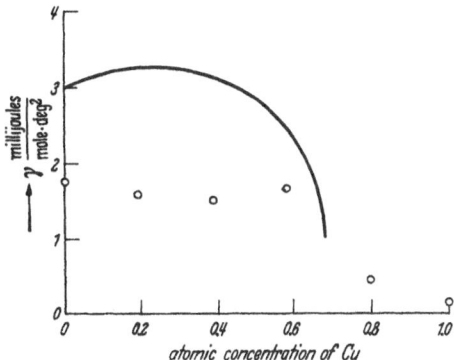

4s-band, assuming it and the $3d$-band to have normal form near the band edge, and that $g_a(\zeta)_{\frac{1}{2}\cdot 3d} = 10 \times g_a(\zeta)_{4s}$. This corresponds to a critical copper concentration of 0.69 instead of 0.6 occuring when the $4s$-band is neglected so that the theoretical curve drops sharply at about this value of χ. The relative insensitivity of the curve to χ below the critical value is a consequence of WOHLFARTH's application of STONER's treatment of ferromagnetism, in which the dependence of the exchange interaction on magnetization is taken into account. Thus the 20, 40 and 60% points agree roughly in relative position, if not in magnitude, with the curve. The point

Fig. 23. γ versus x, the atomic concentration of Cu in Cu-Ni alloys. ⊙: KEESOM and KURRELMEYER, Physica, Haag 7, 1003 (1940); curve: WOHLFARTH, Proc. Roy. Soc. Lond., Ser. A 195, 434 (1949).

at $\chi = 0.8$ is still about 2.5 times as large as the γ value for pure copper, but WOHLFARTH's treatment cannot be applied to such high values of χ so that this discrepancy remains unaccounted for.

KEESOM and KURRELMEYER[1] also observed a peculiarity in the molar heats of their alloys below about 2.5° K. Their data from 2.5 to 20° K on all the copper-nickel alloys and also on 15, 20 and 50% alloys of iron with nickel could be represented by straight lines on plots of C/T versus T^2. Below $T^2 \approx 6$ there were systematic positive deviations from the respective lines. At about 1.5° K these deviations were roughly 10% for the three iron-nickel alloys and increased from 8% for the 20% copper-nickel alloy to 36% in the 80% alloy. These deviations are outside experimental error but no explanation appears ever to have been attempted.

II. Other sources of heat capacity.

a) Superconductivity[3].

31. The superconductive transition. In 1911 KAMERLINGH ONNES discovered superconductivity in mercury. A short while later, he and HOLST measured the specific heat of mercury above and below the transition temperature in an attempt

---

[1] W. H. KEESOM and B. KURRELMEYER: Physica, Haag 7, 1003 (1940), also in Commun. Kamerlingh Onnes Lab. Univ. Leiden, no. 260d.

[2] E. P. WOHLFARTH: Proc. Roy. Soc. Lond., Ser. A 195, 434 (1949).

[3] For a fuller discussion of superconductivity, see the articles by BARDEEN and by SERIN in vol. XV of this Encyclopedia, also [16] and [17].

to correlate changes in other properties with the abrupt loss in resistivity, but they observed no striking change. In 1931 Keesom and van den Ende took up this problem again, using the then recently developed sensitive phosphor-bronze thermometer. The increased resolution enabled them to detect a discontinuity in the atomic heat of tin at the transition temperature.

Since that time a considerable amount of work has been done on this problem. A thermodynamical theory has been completed, connecting latent heat, changes in entropy and heat capacity with the magnetic threshold curve. Heat capacity measurements of several elements and alloys have been performed and found to be in excellent agreement with threshold curves in many cases.

The subject is not closed, however, mainly due to the lack of a theory on the molecular level which is able to describe the transition phenomena in a satisfactory way.

We use here a very simplified approach in which the transition is assumed to be perfectly sharp, so that the intermediate state is completely neglected. That is, the material is assumed to be completely superconducting when the magnetic field is less than a well defined value H_c (the critical field), where H_c is a function of temperature. The critical field has its maximum value, H_0, at the absolute zero and decreases to zero at the normal transition temperature T_0. We neglect demagnetization effects in the normal state, and for the superconductive state we assume a complete Meissner-Ochsenfeld effect:

$$B = 0; \quad M_s = - V_m H_c/4\pi \tag{31.1}$$

where M_s is the magnetic moment per mole.

32. Thermodynamic treatment. The transition from the superconducting to the normal state ($s \rightleftharpoons n$) is assumed to be reversible in the thermodynamical sense. Hence using the Clausius-Clapeyron formula we obtain the latent heat of the isothermal transition

$$L_{s \to n} = T(S_n - S_s) = T \frac{dH_c}{dT} (M_s - M_n) = \frac{-T V_m H_c}{4\pi} \frac{dH_c}{dT}. \tag{32.1}$$

Hence the difference in entropy is

$$\Delta S = S_s - S_n = \frac{V_m H_c}{4\pi} \frac{dH_c}{dT} \tag{32.2}$$

and the difference in atomic heat is

$$\Delta C = C_s - C_n = \frac{V_m T}{4\pi} \left[H_c \frac{d^2 H_c}{dT^2} + \left(\frac{dH_c}{dT} \right)^2 \right]. \tag{32.3}$$

At $T = T_0$, $H_c = 0$, this becomes a transition of the second order since then $L = \Delta S = 0$. In this case (32.3) becomes

$$(\Delta T)_{T_0} = \frac{V_m T_0}{4\pi} \left(\frac{dH_c}{dT} \right)^2_{T_0} \tag{32.4}$$

which is Rutger's formula.

These formulas have been subjected to careful study and found to be valid for several pure metals within experimental error. In addition, reversibility has been verified by a direct method, for tin[1] and thallium[2]. In Fig. 24 is a drawing

[1] P. H. van Laer and W. H. Keesom: Physica, Haag **5**, 993 (1938), also in Commun. Kamerlingh Onnes Lab., Univ. Leiden, no. 254e.

[2] W. H. Keesom and J. A. Kok: Physica, Haag **1**, 503, 595 (1934), also in Commun. Kamerlingh Onnes Lab. Univ. Leiden, nos. 230e and 232a.

of results for tin of KEESOM and VAN LAER[1]. The solid line represents the difference $C_s - C_n$ obtained calorimetrically and the dotted line, the same difference calculated from the threshold curve measured by DE HAAS and ENGELKES[2],

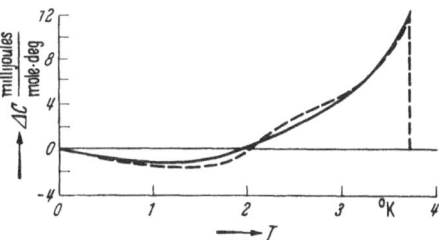

using (32.3). KEESOM and VAN LAER remark: "Taking into consideration that it is very difficult to determine the second derivative accurately and that especially at lower temperatures the second term in [our eq. (32.3)] becomes more important, the agreement between the observed and calculated values is satisfactory." The drawing shows that ΔC reverses sign, which is necessary because ΔS is zero at $0°$ K as well as at T_0.

Fig. 24. ΔC versus T for Sn in normal and superconducting states, after KEESOM and VAN LAER, Physica, Haag 5, 193 (1938). Solid line: calorimetric; dashed line: magnetic.

33. Electronic heat capacity in the normal and superconducting states. In Fig. 25 are plotted C_n/T and C_s/T versus T^2, calculated from the data of KEESOM and VAN LAER for tin, where C_n and C_s are the atomic heats in the normal and superconducting states, respectively. KOK[3] approximated the relation between atomic heat and temperature by the solid lines in Fig. 25. Thus he wrote for C_n the usual expression

$$C_n = \alpha T^3 + \gamma T \quad (33.1)$$

and since the straight line representing C_s goes through the origin,

$$C_s = A T^3. \quad (33.2)$$

Thus

$$\left.\begin{aligned}\Delta C &= (A - \alpha) T^3 - \gamma T \\ &= \frac{V_m T}{4\pi} \frac{d}{dT}\left(H_c \frac{dH_c}{dT}\right).\end{aligned}\right\} \quad (33.3)$$

Integrating this equation twice leads to a parabolic threshold curve

$$H_c = H_0 [1 - (T/T_0)^2]. \quad (33.4)$$

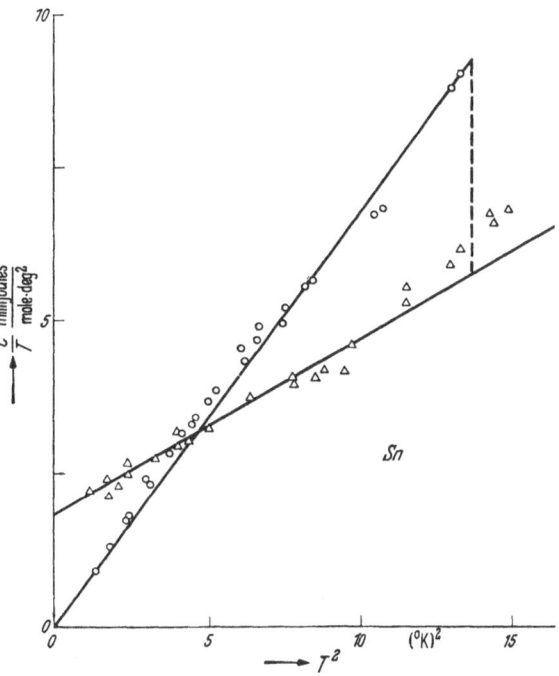

Fig. 25. C/T versus T^2 for Sn. O: C_s/T; $\triangle$: C_n/T.

Such an equation had earlier been found by TUYN and KAMERLINGH ONNES[4] to represent their results for several elements. Resubstituting (33.4) in (33.3)

---

[1] W. H. KEESOM and P. H. VAN LAER: Physica, Haag 5, 193 (1938), also in Commun. Kamerlingh Onnes Lab. Univ. Leiden, no. 252b.

[2] W. J. DE HAAS and A. D. ENGELKES: Physica, Haag 4, 325 (1937), also in Commun. Kamerlingh Onnes Lab. Univ. Leiden, no. 247d.

[3] J. A. KOK: Physica, Haag 1, 1103 (1934), also in Commun. Kamerlingh Onnes Lab. Univ. Leiden, Suppl. no. 77a.

[4] W. TUYN and H. KAMERLINGH ONNES: J. Franklin Inst. 201, 379 (1926), also in Commun. Phys. Lab. Univ. Leiden, no. 174a.

leads to

$$\Delta C = \frac{V_m H_0^2 T}{2\pi T_0^2}\left(3\,\frac{T^2}{T_0^2} - 1\right) = \gamma\, T\left[3\left(\frac{T}{T_0}\right)^2 - 1\right] \tag{33.5}$$

with

$$\gamma = \frac{V_m H_0^2}{2\pi T_0^2}. \tag{33.6}$$

For $T = T_0$,

$$(\Delta C)_{T_0} = 2\gamma\, T_0 = \frac{V_m T_0}{4\pi}\left(\frac{dH_c}{dT}\right)^2_{T_0}. \tag{33.7}$$

From these equations we see several ways in which γ, the coefficient of the electronic atomic heat in the normal state, C_{nE}, can be obtained. It is connected with V_m, H_0, and T_0 by (33.6); with the jump in the atomic heat at the normal transition point, and also with the initial slope of the threshold curve at T_0 by (33.7).

It is also possible to evaluate γ without using the special parabolic form of the threshold curve, since from (33.3) it follows that

$$\lim_{T\to 0}(\Delta C/T) = \gamma \tag{33.8}$$

provided that C_s approaches zero faster than linearly. Thus γ can be obtained from the limiting slope at $0°$ K of ΔS versus T, using (32.2). The advantage of this procedure, that the special assumption of (33.2) concerning C_s need not be made, is to some extent countered by the necessity of making threshold field measurements to very low temperatures in order to get accurate values of the limiting slope.

We refer to EISENSTEIN's recent review [10] for a comparison between calorimetric and magnetic determinations of γ. A column of γ_{magn} values taken from his article is included in our Tables of the atomic heat of the elements (part I above). EISENSTEIN concluded that the electronic atomic heats obtained from the threshold curve data are frequently in good agreement with calorimetric determinations. However, in some cases discrepancies up to an order of magnitude exist, with the magnetic values usually lower than the calorimetric. From the entries in Tables 5, 6, 7 and 10 above, the same appears to be true of γ values calculated from measurements of $(\Delta C)_{T_0}$ using (33.7) [given earlier as (20.1)].

There are several reasons why discrepancies might be expected. To begin with we assumed thermodynamic reversibility, which appears to have been demonstrated in pure metals but not in alloys or impure metals (chemical or physical impurities). If the MEISSNER-OCHSENFELD effect is not complete, reversibility is doubtful. Furthermore, threshold curves are obtained either by observing the expulsion of the magnetic field from the specimen or the vanishing of its resistance. Those of the latter type can lead to threshold field values higher than are appropriate to the bulk of the material.

Also, the assumptions of KOK as to the temperature dependence of the heat capacity, which were discussed above, may be over-simplified. Already in the case of tin (see Fig. 25 taken from reference 22.11) it is clear that these simple relations do not hold, as the experimental points deviate from the straight lines. Apparently the BLACKMAN temperature T_B is lower than T_0 so that above $3°$ K (33.1) does not hold. This objection should not be too serious, however, since it would be expected that the lattice contribution in the superconducting state equals that in the normal state. From this it would follow that the positive deviations of C_n/T from the straight line above $T^2 = 10$ should be paralleled by similar deviations in C_s/T, but these are not observed if the line for C_s/T is drawn as in Fig. 25.

As far as KOK's second assumption is concerned, (33.2) implies a cubic dependence for C_{sE}, the electronic atomic heat in the superconducting state. Using (33.2) and (33.5) we can write

$$C_{sE} = 3\gamma\, T_0\, (T/T_0)^3. \tag{33.9}$$

But even in the case of tin, from which data KOK originally derived these relations, a devivation from the straight line on the plot of C_s/T versus T^2 (see Fig. 25) can be seen around $T^2 = 5$. In Fig. 26 the results for tantalum of WORLEY et al. (see reference 25.7) and of KEESOM and DESIRANT (see reference 25.4)

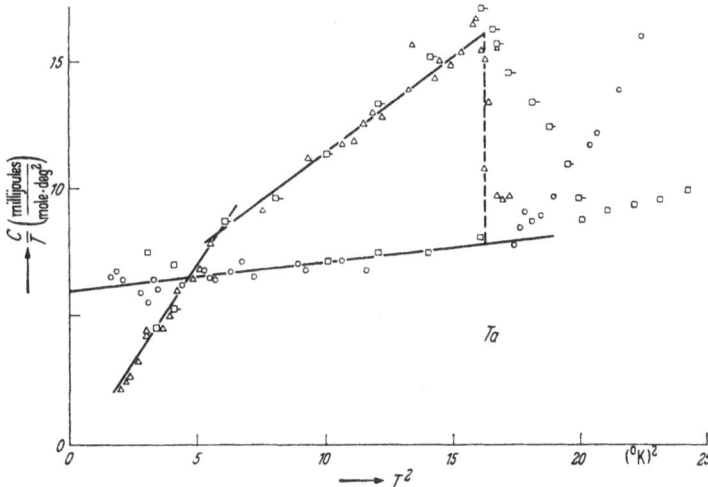

Fig. 26. C/T versus T^2 for Ta. $\bigcirc$: C_n/T and $\triangle$: C_s/T, KEESOM and DESIRANT (reference 25.4); $\square$: C_n/T and $\square$–: $C_s T$, WORLEY et al. (reference 25.7).

are combined, and there a definite break in the line is observed. A similar break can also be seen in the data for vanadium. To fit the latter, CORAK et al.[1] proposed an equation of the form

$$C_{sE} = \gamma\, T_0\, a\, e^{-b/T}. \tag{33.10}$$

The constants a and b are not independent; a relation between them can be obtained from (32.2).

In addition to being empirically applicable, CORAK et al. point out that (33.10) is of the form that would be expected to hold for a theory in which the levels occupied by the superconducting electrons are separated by a gap from those occupied by the normal electrons. These authors remark that the energy gap as deduced from (33.10) is of the order of kT_0.

It should be remarked that the failure of KOK's assumptions embodied in (33.1), (33.2) and (33.4) should not result in discrepancies between magnetically and calorimetrically determined values of γ if (33.8) and (33.2) are used, since as was mentioned above those assumptions are not involved in the latter equations. As was remarked earlier, however, the difficulties involved in measuring

---

[1] W. S. CORAK, B. B. GOODMAN, C. B. SATTERTHWAITE and A. WEXLER: Phys. Rev. **96**, 1442 (1954).

magnetic threshold curves at very low temperatures may introduce some uncertainty in this method. It would thus appear that the least ambiguous way to compare the magnetic and calorimetric aspects of the thermodynamic treatment is to apply (32.4) directly, since no special assumptions are involved and measurements of both $(\Delta C)_{T_0}$ and $(dH_c/dT)_{T_0}$ are straightforward. SHOENBERG has tabulated the data necessary for such a comparison (Table 1, p. 62 of reference [17]) and in all cases where both sets of data are reliable the agreement is excellent.

b) Excitation modes.

34. Heat capacity "anomaly". We consider a system which has available a group of levels separated from the ground state by energies $\delta_1, \delta_2, \ldots \delta_m$, with degeneracies $g_0 g_1, g_0 g_2, \ldots g_0 g_m$, where g_0 is the degeneracy of the ground state. The average energy of the system is then

$$\bar{\delta} = \frac{\sum\limits_{i=1}^{m} g_i \delta_i e^{-\beta \delta_i}}{1 + \sum\limits_{i=1}^{m} g_i e^{-\beta \delta_i}} \tag{34.1}$$

where

$$\beta = hc/kT \tag{34.2}$$

with c the velocity of light, if the unit of δ_i is cm^{-1}. A mole of such systems which do not interact with each other will have a mean energy which is just N_0 times the expression (34.1). Introducing the dimensionless parameter

$$\tau_i = \beta \delta_i \tag{34.3}$$

the molar heat arising from transitions among the levels is found to be

$$C = R \left\{ \frac{\sum\limits_{i=1}^{m} g_i \tau_i^2 e^{-\tau_i} + \sum\sum\limits_{i \neq j} g_i g_j (\tau_i - \tau_j)^2 e^{-(\tau_i + \tau_j)}}{\left(1 + \sum\limits_{i=1}^{m} g_i e^{-\tau_i}\right)^2} \right\}. \tag{34.4}$$

This can also be written in terms of the commonly used "splitting parameter" a,

$$a_i = (hc/k) \delta_i = 1.44 \delta_i \, °K. \tag{34.5}$$

In the simplest case of two levels separated by δ cm^{-1} and splitting parameter a °K, and with equal degeneracy, (34.4) reduces to

$$C = R \left\{ \frac{(a/T)^2 e^{a/T}}{(1 + e^{a/T})^2} \right\}. \tag{34.6}$$

This function has a maximum, equal to $0.43 R$, at $T = 0.416 a$. For $T \gg a$ it reduces to

$$C = \frac{a^2 R}{4} \cdot \frac{1}{T^2} \tag{34.7}$$

while for $T \ll a$, it becomes

$$C = a^2 R \cdot \frac{e^{-a/T}}{T^2}. \tag{34.8}$$

This treatment, which was developed by SCHOTTKY[1], was used by SIMON[2] to account for deviations from the DEBYE formula (5.6) in the atomic heat of Li, K, Na, Si, gray tin and diamond. The observed atomic heats in all these cases, however, are monotonic functions of T. BLACKMAN has pointed out[3] that any monotonic variation of C with T can be accounted for by a suitable non-parabolic vibration spectrum. It is thus necessary to consider the possibility of an energy level scheme such as that discussed above only when there are peaks in the heat capacity. The case of some of the rare earth elements, in which such peaks have been associated (see reference 20.1) with transitions among levels arising from the atomic $4f$ levels split by the crystalline field, has been mentioned in Sect. 20 above.

35. SCHOTTKY transitions in paramagnetic ions. The most extensive application of the SCHOTTKY analysis at low temperatures has been to the levels occurring in salts containing non-interacting paramagnetic ions. Many such salts, principally alums and TUTTON salts, in which the water of crystallization provides the necessary dilution of paramagnetic ions, have been used to reach extremely low temperatures (down to about 10^{-3} °K) by adiabatic demagnetization. Since some of these heat capacity results are described in the article on adiabatic demagnetization in vol. XV of this Encyclopedia, we discuss here instead some measurements on similar salts which have not been used for magnetic cooling.

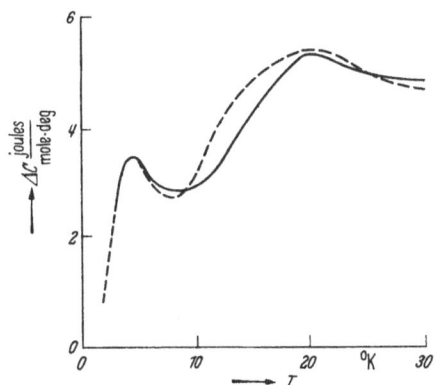

Fig. 27. Anomalous specific heat of ferrous ammonium sulphate, after HILL and SMITH, Proc. Phys. Soc., Lond. A 66, 228 (1953). Solid line: specific heat difference between zinc and ferrous ammonium sulphates; dashed line: theoretical anomalous specific heat calculated from postulated level scheme.

HILL and SMITH[4] investigated the levels of the ferrous ion in the TUTTON salt $Fe(NH_4)_2(SO_4)_2 \cdot 6H_2O$ by comparing its molar heat with that of $Zn(NH_4)_2(SO_4)_2 \cdot 6H_2O$. Since the two salts differ only in the presence of the paramagnetic Fe^{++} ion in the former and the diamagnetic Zn^{++} ion in the latter, they assumed that the difference in molar heat, which is plotted as the solid line in Fig. 27, could be attributed to the excitation modes of the ferrous ion. The dotted line in Fig. 27 gives the heat capacity calculated from (34.4) with $m = 2$, $\delta_1 = 6.5$ cm^{-1}, $\delta_2 = 38$ cm^{-1}, $g_0 = g_1 = 1$, $g_2 = 3$. Although the agreement with the measured molar heat difference is good, the authors point out that it does not unambiguously confirm their level scheme.

LYON and GIAUQUE[5] measured the heat capacity of $FeSO_4 \cdot 7H_2O$ between 1 and 20° K and also found two peaks, one at 2° K and the other at 15° K.

An even more complicated system $(Cr_3(CH_3COO)_6(OH)_2)Cl \cdot 8H_2O$, which contains three paramagnetic chromium ions per molecule, was investigated by WUCHER and WASSCHER[6]. The spins of these three ions interact and give rise to a very complicated level diagram.

[1] W. SCHOTTKY: Phys. Z. **22**, 1 (1921); **23**, 9, 448 (1922).
[2] F. E. SIMON: Ergebn. exakt. Naturw. **9**, 222 (1930).
[3] M. BLACKMAN: Rep. Progr. Phys. **8**, 11 (1941).
[4] R. W. HILL and P. L. SMITH: Proc. Phys. Soc. Lond. A **66**, 228 (1953).
[5] D. N. LYON and W. F. GIAUQUE: J. Amer. Chem. Soc. **71**, 1647 (1949).
[6] J. WUCHER and J. D. WASSCHER: Physica, Haag **20**, 721 (1954), also in Commun. Kamerlingh Onnes Lab. Univ. Leiden, no. 296a.

Two rare earth salts, $Sm_2(SO_4)_3 \cdot 8H_2O$ and $Nd_2(SO_4)_3 \cdot 8H_2O$ have been studied by AHLBERG et al.[1]. They found a level splitting of less than 1 cm^{-1} in both, and an additional splitting of 77 cm^{-1} in the latter salt, from measurements between 3 and 40° K.

c) Cooperative phenomena.

36. Cooperative transitions and the associated heat capacity peak. A general discussion of cooperative phenomena would lead us too far afield, so we refer instead to specialized articles on this subject [18]. As already described in Sect. 1 the modes of excitation involved are such that the probability of excitation depends on the existing degree of excitation of these modes in the solid. Associated with such a transition is a peak in the heat capacity which occurs at $T = T_c$, where kT_c is of the order of the excitation energy involved (k is BOLTZMANN's constant). According to theory the peak should be asymmetrical with a long tail below T_c and a sharp drop from the maximum at T_c.

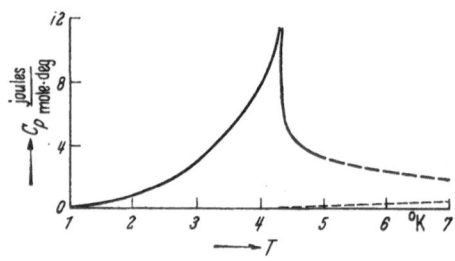

Fig. 28. Molar heat of $CuCl_2 \cdot 2H_2O$ versus T, after FRIEDBERG, Physica, Haag **18**, 714 (1952). Dotted line: lattice molar heat, proportional to T^3.

Among the transitions for which such peaks have been found are: order-disorder transitions in alloys; paramagnetic-ferromagnetic transitions at the CURIE point in ferromagnetic materials; free rotation-hindered rotation transitions in methane, hydrogen and other molecular solids; molecular orientation transition in ammonium chloride; and spin alignment corresponding to the antiferromagnetic state in varous salts.

As an example of the last-named transition, many cases of which have been investigated at low temperatures, we give in Fig. 28 the specific heat of $CuCl_2 \cdot 2H_2O$ as reported by FRIEDBERG[2]. By correcting for the lattice heat capacity he found the total entropy change connected with the peak to be 5.45 joules/mole degree, compared with the theoretical value $R \log 2 = 5.78$. About 30% of the entropy change occurs above the transition point, $T_c = 4.3°$ K, whereas according to the simpler theories of antiferromagnetism it should be confined to $T < T_c$.

d) Size effect.

37. Surface contribution to the lattice heat capacity. In the derivation of the lattice heat capacity the volume of the specimen is assumed to be so large that surface effects can be neglected. If they are included, the vibration spectrum is modified by the addition of a term proportional to the frequency ν and to the surface S:

$$g(\nu) = \frac{4\pi V}{\bar{v}^3} \nu^2 + \frac{\pi S}{2\bar{v}^2} \nu + \cdots. \tag{37.1}$$

The first term, as we have seen, leads to the T^3 law at low temperatures; the second term introduces a T^2 dependence so that

$$C_L = \alpha T^3 + \beta T^2. \tag{37.2}$$

[1] J. E. AHLBERG, E. R. BLANCHARD and W. O. LUNDBERG: J. Chem. Phys. **5**, 554 (1937).
[2] S. A. FRIEDBERG: Physica, Haag **18**, 714 (1952), also in Commun. Kamerlingh Onnes Lab. Univ. Leiden, no. 289d.

Montroll[1] defines $r = \alpha\, T^3/\beta\, T^2$ and derives the expression

$$r \approx 26\, (T/\Theta_0)\, (V_m/S)\, (\varrho/M)^{\frac{1}{3}}\, N_0^{\frac{1}{3}} \qquad (37.3)$$

where Θ_0 is the Debye parameter for the bulk material, S the surface area, ϱ the density, and M the molecular weight. Koppe[2] derived an equivalent formula which he applied also to the contribution to the heat capacity of interfaces between two phases. He found that this contribution depends on the ratio of their elastic moduli.

Patterson et al.[3] have recently investigated the size effect in rocksalt (NaCl). They compared the specific heat of the bulk material with that of three samples of different particle size (specific surfaces between 38 and 59 m²/g). There were differences in the specific heat of between 5 and 10%, but the measured size effect was 3 to 4 times larger than would follow from (37.3). The authors attribute this discrepancy to surface roughness and non-stoichiometry, especially because two of the samples, with the same surface area but prepared in different ways, gave differences comparable with the total surface effect.

III. Miscellaneous measurements.

38. Compounds. We have already discussed several compounds, such as the alloys of nickel with copper and iron (Sect. 30), and several salts in the sections on cooperative phenomena and excitation modes. Only a few other compounds have been investigated in the helium temperature range.

α) *KCl*. The molar heat of KCl[4,5] is due entirely to the lattice, as it is proportional to T^3 below 4° K, with $\Theta_0 = 233°$ K. This agrees fairly well with estimates calculated from the vibration spectrum and from elastic constants. The molar heat increases above the extrapolated T^3 value in the hydrogen region, $\Theta(T)$ decreasing to 210° K at 20° K.

β) *LiF₂*. Recent measurements of the molar heat of LiF₂ show that its heat capacity is likewise due entirely to the lattice at low temperatures[6]. The molar heat is proportional to T^3 up to 20° K, with $\Theta_0 = 736°$ K.

γ) *TiO₂ (rutile)*. The molar heat of pure rutile[7] is also proportional to T^3 in the helium temperature range. The value of Θ_0 is 758° K and $\Theta(T)$ decreases from 681° K at 10° K to 450° K at 20° K. This appears to agree with the results of Dugdale et al.[8] where the two measurements overlap. When the material is slightly reduced (oxygen loss less than $\frac{1}{2}$%) it exhibits a large increase in molar heat which is constant in the helium range. For about $\frac{1}{2}$% reduction, the excess is about 1000 times C_L of the original material at 1° K. Reduced rutile is a semiconductor and it is possible that the conduction electrons give the classical

---

[1] E. W. Montroll: J. Chem. Phys. **18**, 183 (1950).

[2] H. Koppe: J. Chem. Phys. **18**, 638 (1950).

[3] D. Patterson, J. A. Morrison and F. W. Thompson: Canad. J. Chem. **33**, 240 (1955).

[4] W. H. Keesom and C. W. Clark: Physica, Haag **2**, 698 (1935), also in Commun. Kamerlingh Onnes Lab. Univ. Leiden, no. 238c.

[5] P. H. Keesom and N. Pearlman: Phys. Rev. **91**, 1354 (1953).

[6] G. O. Jones and D. L. Martin: Phil. Mag. **45**, 649 (1954).

[7] P. H. Keesom and N. Pearlman: Abstracts of the Third International Conference on Low Temperature Physics and Chemistry (Houston, 1953), unpublished. — A small constant term observed below 4° K is believed to be due to a very small concentration of electrons arising from a slight departure from stoichiometry in the "pure" crystal.

[8] J. S. Dugdale, J. A. Morrison and D. Patterson: Proc. Roy. Soc. Lond., Ser. A **224**, 228 (1954).

contribution to the heat capacity [see (7.1)] because of their high effective mass and low concentration. Further experiments are needed to substantiate this assumption.

δ) *Organic compounds.* AHLBERG et al.[1] have investigated several organic compounds between 2 and 90° K. Their results up to 6° K can be approximately represented by cubic expressions:

methyl alcohol: CH_3OH $C = 0.65\ T^3$ millijoules/mole degree

Benzene: C_6H_6 $C = 1.2\ \ T^3$ millijoules/mole degree

Glycerol: $C_3H_5(OH)_3$ crystalline $C = 0.69\ T^3$ millijoules/mole degree

glassy $C = 1.3\ \ T^3$ millijoules/mole degree.

The heat capacity of glassy glycerol had earlier been found to be larger than that of the crystalline substance above 10° K by SIMON and LANGE[2], who dis-cussed the implications of this with respect to the third law of thermo-dynamics.

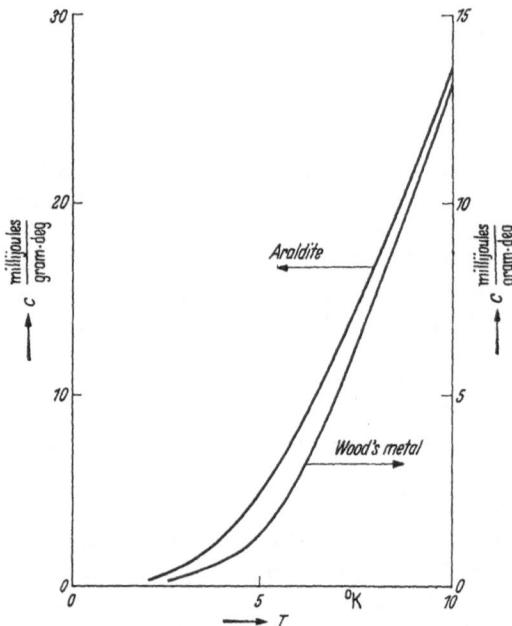

Fig. 29. Specific heats of Araldite and WOOD's metal versus temperature after PARKINSON and QUARRINGTON.

39. Calorimeter materials. Various glues and low melting point solders are used in the construction of calorimeters. Some of these materials have specific heats (heat capacity per gram) much larger than that of the specimens being investigated, so that even the small quantities used can make a significant contribution to the total calorimeter heat capacity. Fairly accurate values of their specific heats are thus required in order to make reliable corrections to the measured heat capacity of the calorimeter.

The specific heats of WOOD's metal and of araldite, a thermo-setting polymer, have recently been measured by PARKINSON and QUARRINGTON[3]. Their results below 10° K are reproduced in Fig. 29. Between 10 and 20° K the respective specific heats can be represented by the formulas

Araldite: $c_g = 5.39\ T - 26.7$ millijoules/gram degree

WOOD's metal: $c_g = 3.26\ T - 19.2$ millijoules/gram degree.

An approximate formula for the specific heat of red glyptal varnish, which represents three values at 4, 10 and 15° K (see reference 22.18) is

$$c_g = 0.22\ T^2 \text{ millijoules/gram degree.}$$

[1] J. E. AHLBERG, E. R. BLANCHARD and W. O. LUNDBERG: J. Chem. Phys. **5**, 539 (1937).
[2] F. SIMON and F. LANGE: Z. Physik **38**, 227 (1926).
[3] D. H. PARKINSON and J. E. QUARRINGTON: Brit. J. Appl. Phys. **5**, 219 (1954).

HILL and SMITH[1] have reported the values for the specific heat of Formite Bakelite varnish which are given in Table 14.

Table 14. *Specific heat of formite bakelite varnish.*

$T(°K)$	c_g (millijoules/gram degree)	$T(°K)$	c_g (millijoules/gram degree)	$T(°K)$	c_g (millijoules/gram degree)
4	4.6	30	121	70	343
10	19	40	180	80	414
15	42	50	238	90	439
20	71	60	285		

The authors would like to express their appreciation for the support rendered by the US Army Signal Corps and the US Atomic Energy Commission, respectively, during the preparation of the manuscript.

Appendix.

40. Notes on recent publications. SMITH[2] has measured the atomic heat of magnesium from 1 to 20° K, and that of zinc from 4 to 20° K. He finds for magnesium $\Theta_0 = 406°$ K, $\gamma = 1.32$ millijoules/mole degree[2]. His results on zinc are in good agreement with those of KEESOM and VAN DEN ENDE (reference 19.3) where the temperature ranges overlap and join on smoothly to those of SILVIDI and DAUNT (reference 19.6) below 4° K.

MARTIN[3] has reported measurements between 2 and 30° K on lithium fluoride, sodium chloride and zinc sulphide. His results on LiF were reported earlier (see Sect. 38). For NaCl, Θ rises from about 295° K at 20° K to about 330° K at 4° K but $|d\Theta/dT|$ is still much larger than zero at 4° K so the true T^3 region has not yet been reached. In ZnS, Θ_0 is about 315° K, with $T_B \approx 8°$ K; Θ decreases to about 260° K at 20° K.

The measurements on tantalum and vanadium reported by WORLEY et al. in short communications (references 25.7 and 25.8) have now been published in detail[4]. For normal vanadium they find $\Theta_0 = 273°$ K, $\gamma = 8.83$ millijoules/mole degree[2]; for normal tantalum, $\Theta_0 = 231°$ K, $\gamma = 5.44$ millijoules/mole degree[2]. In the superconducting states, the atomic heats could be represented by:

$$V : C_s = 7.07\ T^2 - 8.24\ T \text{ millijoules/mole degree}$$
$$T_a : C_s = 5.57\ T^2 - 6.07\ T \text{ millijoules/mole degree.}$$

RAMANATHAN and SRINIVASAN[5] have reported new measurements on bismuth which agree well with those of KEESOM and PEARLMAN (reference 24.2). They give the values $\Theta_0 = 120°$ K, $\gamma = (0.0477 \pm 0.0611)$ millijoules/mole degree[2], with $T_B \approx 2°$ K.

New measurements on gray tin and potassium chloride have been published by WEBB and WILKS[6] who also describe in greater detail the graphite measurements earlier reported by BERGENLID, HILL, WEBB and WILKS (reference 22.1). The results on KCl, for which they report $\Theta_0 = 236°$ K, are in excellent agreement with those reported earlier by KEESOM and PEARLMAN[7]. For gray tin they give $\Theta_0 = 212°$ K, with $T_B \approx 3°$ K.

[1] R. W. HILL and P. L. SMITH: Phil. Mag. **44**, 636 (1953).
[2] P. L. SMITH: Phil. Mag. **46**, 744 (1955).
[3] D. L. MARTIN: Phil. Mag. **46**, 751 (1955).
[4] R. D. WORLEY, M. W. ZEMANSKY and H. A. BOORSE: Phys. Rev. **99**, 447 (1955).
[5] K. G. RAMANATHAN and T. M. SRINIVASAN: Phys. Rev. **99**, 442 (1955).
[6] F. J. WEBB and J. WILKS: Proc. Roy. Soc. Lond., Ser. A **230**, 549 (1955).
[7] P. H. KEESOM and N. PEARLMAN: Phys. Rev. **91**, 1354 (1953).

The measurements on Na mentioned in reference 16.2 have been published by PARKINSON and QUARRINGTON[1]. They find $\Theta_0 = 158°$ K, $\gamma = 1.8$ milli-joules/mole degree[2] (which corresponds to $\varrho_E = 1.6$) and $T_B = 3°$ K. Above T_B, $\Theta(T)$ decreases to a minimum of $147°$ K at $6°$ K, then rises to $155°$ K at $20°$ K.

General References.

[1] EUCKEN, A.: Energie und Wärmeinhalt, Bd. 8, Teil 1 im Handbuch der Experimental-physik, Herausg. W. WIEN u. F. HARMS. Leipzig: Akademische Verlagsgesellschaft 1929. — An encyclopedic work on all aspects of heat capacity problems, including experimental techniques, results of experiment and theory, together with an extensive bibliography.

[2] EINSTEIN, A.: Die Plancksche Theorie der Strahlung und die Theorie der spezifischen Wärme. Ann. Phys. **22**, 180—190 (1907). — Berichtigung zu: die Plancksche Theorie usw. Ann. Phys. **22**, 800 (1907). — Eine Beziehung zwischen dem elastischen Verhalten und der spezifischen Wärme bei festen Körpern mit einatomigen Molekül. Ann. Phys. **34**, 170—174 (1911). — The first of these papers is historically important as marking two great theoretical advances: the application of PLANCK's quantum theory outside its original domain; and the explanation of the failure of the equipartition law as evidenced by atomic heat values for certain elements which are lower than that predicted by the law of DULONG and PETIT.

[3] DEBYE, P.: Zur Theorie der spezifischen Wärme. Ann. Physik **39**, 789—839 (1912). — This paper contains DEBYE's derivation of his T^3 law for heat capacity at low tempera-tures. Besides being of historical interest, his conclusions remain valid, in a restricted temperature range, even in more detailed theories.

[4] BORN, M., u. T. v. KÁRMÁN: Über Schwingungen in Raumgittern. Phys. Z. **13**, 297—309 (1912). — Zur Theorie der spezifischen Wärme. Phys. Z. **14**, 15—21 (1913). — Über die Theorie der Verteilung der Eigenschwingungen von Punktgittern. Phys. Z. **14**, 65—71 (1913). — The first and third of these papers laid the foundation of the lattice theory which has since been developed by many workers, especially BLACKMAN (see reference [5]). In the second the T^3 law is shown to follow from the lattice theory at very low temperatures.

[5] BLACKMAN, M.: The Theory of the Specific Heat of Solids. Repts. Progr. Phys. **8**, 11—30 (1941). — A review of the experimental values of atomic heat at low tempera-tures, as compared with calculations based on DEBYE's theory using elastic constants, and with the lattice theory as developed by BLACKMAN and others. The apparent agreement with the DEBYE theory is shown in many cases to be largely fortuitous. References are included to BLACKMAN's original calculations of vibration spectra and to other calculations published up to 1940.

[6] BORN, M., and K. HUANG: Dynamical Theory of Crystal Lattices. London: Oxford University Press 1954. — Discussion of the application of lattice theory to a variety of problems. Sect. 4 through 6 treat thermodynamical applications, including heat capacity. A review of the methods which have been devised to evaluate vibration spectra is included, with references.

[7] SOMMERFELD, A.: Zur Elektronentheorie der Metalle. Naturwiss. **15**, 825—832 (1927). — Zur Elektronentheorie der Metalle auf Grund der FERMIschen Statistik I. Allgemeines, Strömungs- und Austrittsvorgänge. Z. Physik **47**, 1—32 (1928). — The first of these papers is a shorter summary of the results presented in detail in the second. SOMMER-FELD's formulas for the electronic heat capacity appears for the first time in the latter, although a qualitative discussion on the basis of the temperature variation of a de-generate electron gas is given in the former.

[8] SOMMERFELD, A., u. H. BETHE: Elektronentheorie der Metalle. Handbuch der Phy-sik, Bd. 24, Teil 2, S. 333—622. Berlin: Springer 1933. — Classical account of the electron theory of metals, including SOMMERFELD's free electron treatment and the applications of band theory as well.

[9] SEITZ, F.: The Modern Theory of Solids. New York and London: The McGraw-Hill Book Company, Inc. 1940. — A valuable text on many aspects of solid state theory. Chapter III is devoted to specific heat of simple solids, including a discussion of vibra-tion spectra; Chapter IV sect. 27 treats the electronic heat capacity in non-transition metals and sect. 28, that in transition metals. The theory of the band approximation and approximation methods involved in band calculations are discussed in Chapters VIII and IX respectively and typical band structures in metals and other solids are described in Chapter XIII.

---

[1] D. H. PARKINSON and J. E. QUARRINGTON: Proc. Phys. Soc. Lond. A **68**, 762 (1955).

[10] Eisenstein, J.: Superconducting elements. Revs. Mod. Phys. 26, 277—291 (1954). — A review of magnetic and calorimetric data on superconducting elements, with the data for each element discussed in some detail. A summarizing table includes the crystal structure, normal transition temperature, critical field at absolute zero and magnetic and calorimetric values of γ.

[11] Raynor, G. V.: The band structure of metals. Repts. Progr. Phys. 15, 173—248 (1952). — A comprehensive survey of theoretical calculations and experimental evidence on metallic band structures. Other types of data receive more attention than the electronic heat capacity.

[12] Daunt, J. G.: The electronic specific heat of metals. In Progress in Low Temperature Physics, ed. C. J. Gorter, vol. 1, pp. 202—223. Amsterdam and New York: North-Holland Publishing Company and Interscience Publishers, Inc. 1955. — A detailed discussion of the relation of the band structure to the electronic heat capacity, including magnetic evidence from superconductive elements. Extensive references are appended.

[13] Mott, N. F.: Recent advances in the electron theory of metals. In Progress in Metal Physics, vol. 3, pp. 76—114. London and New York: Pergamon Press Ltd. and Interscience Publishers, Inc. 1952. — A comprehensive survey of many aspects of the theory, including heat capacity.

[14] Shiffman, C. A.: The Heat Capacities of the Elements below Room Temperature. Schenectady: General Electric Research Laboratory 1952.

[15] Shull, D. R., and G. C. Sinke: The Thermodynamic Properties of the Elements in their Standard States. Midland, Michigan: The Dow Chemical Company 1955. — This and the previous reference tabulate values of the thermodynamic properties and include extensive lists of references. Those in reference [14] are especially conveniently arranged.

[16] Serin, B.: The Magnetic threshold curve of superconductors. In Progress in Low Temperature Physics, ed. C. J. Gorter, vol. 1, pp. 138—150. Amsterdam and New York: North-Holland Publishing Company and Interscience Publishers, Inc. 1955. — The relation between the threshold curve and thermodynamic properties is discussed, with data on tin treated in detail.

[17] Shoenberg, D.: Superconductivity. Cambridge: Cambridge University Press 1952. — In this text both experimental and theoretical aspects are discussed. Tables and graphs of data are included, with a detailed bibliography of recent work.

[18] Nix, F. C., and W. Shockley: Order-disorder transformations in alloys. Rev. Mod. Phys. 10, 1—17 (1938).
Nakamura, T.: Statistical Mechanics of cooperative phenomena. Progr. Theor. Phys. 7, 241—254 (1952).
Kikuchi, R.: A theory of cooperative phenomena. Phys. Rev. 81, 988—1003 (1951).
Ter Haar, D.: Elements of Statistical Mechanics. New York: Rinehart & Company, Inc. 1954. Chapt. XIII, pp. 251—295. Cooperative phenomena.
Smoluchowski, R., editor: Phase Transformations in Solids. New York and London: John Wiley & Sons, Inc. and Chapman & Hall, Ltd. 1951. — Ter Haar gives a good account of the basic theory together with some applications. Nakamura and Kikuchi present detailed discussions of particular forms of the theory. The problem of order-disorder, both theoretical and experimental, is considered by Nix and Shockley. The last entry is a symposium in which a variety of cooperative phenomena are treated.

Sachverzeichnis.
(Deutsch-Englisch.)

Bei gleicher Schreibweise in beiden Sprachen sind die Stichwörter einmal aufgeführt.

22b

Subject Index.

(English-German.)

Where English and German spelling of a word is identical the German version is omitted.